The Palgrave Handbook of Chinese Language Studies

Zhengdao Ye
Editor

The Palgrave Handbook of Chinese Language Studies

With 50 Figures and 73 Tables

palgrave
macmillan

Editor
Zhengdao Ye
School of Literature, Languages and Linguistics
The Australian National University
Canberra, Australia

ISBN 978-981-16-0923-7   ISBN 978-981-16-0924-4 (eBook)
ISBN 978-981-16-0924-4 (print and electronic bundle)
https://doi.org/10.1007/978-981-16-0924-4

© Springer Nature Singapore Pte Ltd. 2022
This work is subject to copyright. All rights are reserved by the Publisher, whether the whole or part of the material is concerned, specifically the rights of translation, reprinting, reuse of illustrations, recitation, broadcasting, reproduction on microfilms or in any other physical way, and transmission or information storage and retrieval, electronic adaptation, computer software, or by similar or dissimilar methodology now known or hereafter developed.
The use of general descriptive names, registered names, trademarks, service marks, etc. in this publication does not imply, even in the absence of a specific statement, that such names are exempt from the relevant protective laws and regulations and therefore free for general use.
The publisher, the authors and the editors are safe to assume that the advice and information in this book are believed to be true and accurate at the date of publication. Neither the publisher nor the authors or the editors give a warranty, expressed or implied, with respect to the material contained herein or for any errors or omissions that may have been made. The publisher remains neutral with regard to jurisdictional claims in published maps and institutional affiliations.

This Palgrave Macmillan imprint is published by the registered company Springer Nature Singapore Pte Ltd.
The registered company address is: 152 Beach Road, #21-01/04 Gateway East, Singapore 189721, Singapore

Preface

The idea for this handbook arose during a conversation I had with Sara Crowley Vigneau when I visited the Shanghai Office of Springer Nature in January 2017. Sara was the Senior Acquisitions Editor for the Asia and Pacific region at Palgrave Macmillan at the time. As Sara and I were looking out of the window at the ever-changing skyline, marveling at the unprecedented pace and scale at which Chinese society had changed, and reflecting on China's place in the world, we came to the conclusion that there was a need for a handbook that could offer a window into what was happening in the linguistic sphere and showcase the latest research into the diverse and evolving linguistic phenomena resulting from the intensified interactions between the Sinophone world and other lingua-spheres. These dynamic interactions could be those between languages, concepts, and ideas; they could also be those of theories and models for studying them. So, the initial idea for the handbook was formed.

As the idea matured, it was evident to both Sara and me that situating Chinese language studies in a global context would require an international editorial team who could bring together established, mid-career, and emerging scholars across the Asia-Pacific, Europe and the USA.

I am grateful to Quan Wan from the Chinese Academy of Social Sciences, Li Yu from Williams College (Massachusetts), and Jock Wong from the National University of Singapore for joining the team and shaping the direction of each part and the whole volume.

The handbook is divided into five parts. The chapters in Part I ("New Research Trends in Chinese Linguistic Research") present fast-growing research areas in Chinese linguistics, particularly those undertaken by scholars based in China. Part II ("Interactions of Sinitic Languages") focuses on language-contact situations inside and outside China. The chapters in Part III ("Meaning, Culture, Translation") explore the meanings of key cultural concepts, and how ideas move between Chinese and English through translation across various genres. Part IV ("New Trends in Teaching Chinese as a Foreign Language") covers new ideas and practices relating to teaching the Chinese language and culture. The final part, Part V ("Transference from Chinese to English"), explores dynamic interactions between varieties of Chinese and varieties of English, as they play out in multilingual sites and settings.

In its scope, focus, and perspectives, the handbook complements more structurally and theoretically oriented classic handbooks on Chinese linguistics, such as *The Blackwell Handbook of Chinese Linguistics* edited by C.-T. James Huang, Y. H. Audrey Li, and Andrew Simpson (John Wiley & Sons, 2014) and *The Oxford Handbook of Chinese Linguistics* edited by William S.Y. Wang and Chaofen Sun (Oxford University Press, 2015). In its effort to reflect the dynamic interactions between Chinese languages, and how they interact with varieties of English across different settings and through various media, the present collection complements the more socially and culturally oriented *Routledge Handbook of Chinese Applied Linguistics* edited by Chu-Ren Huang, Jing-Schmidt Zhuo, and Barbara Meisterernst (Routledge, 2019).

Of course, this handbook in its final form departs in some ways from the initial table of contents, which originally included topics such as Chinese-English machine translation, Chinese-English bilingual dictionaries, Mandarin-speaking children in Britain, the subtitling of Chinese films in English as cross-cultural mediation, and Sinitic languages along the Belt and Road. Due to various factors, especially the upheavals of COVID, several authors had to withdraw under difficult circumstances. But the distinctive features of the handbook, I believe, have been preserved. These features include having students of Chinese and Chinese language studies as the primary reader; attending to those aspects of Sinitic languages less discussed in the available English scholarship; showcasing a meaning-oriented approach to Chinese language studies; and involving more contributors from China and more emerging scholars who are branching out into new areas of linguistic research. All in all, I believe the handbook represents a transnational effort to make research on Sinitic languages more accessible to the international scientific community.

I should point out that not all reviewers of the initial proposal for the handbook favored the idea of involving more scholars from China, citing a scholarly tradition too different to accommodate in an English publication. My own position is that a book on Chinese language studies needs to include perspectives and approaches from scholars working in a Chinese linguistic and cultural environment, and allow readers to form their own opinions. Inevitably, there will be language barriers and differences in approaches to Chinese language studies. However, those differences themselves are quite revealing of the broader institutional context and social environment in which researchers are situated, and, therefore, help the reader to understand what drives research directions. For example, understanding the institutional tradition initiated by Liu Fu (Liu Bannong), Chao Yuen-ren, Wang Li, Luo Changpei, Wu Zongji, and others is crucial to understanding the kind of phonetic research undertaken in China (see the chapter on "Studies in Chinese Phonetics" by Aijun Li and Chunyu Ge). If there is such a thing as Chinese linguistics with Chinese characteristics, we need to understand what those characteristics are and how they came about.

Language barriers presented a real challenge. The editorial team is especially grateful to Tony Edwards, Janet Davey, Wendi Xue, Xinjia Peng, Luyuan Zhao, Huade Huang, and Liska Fell for their translation efforts. Particular thanks must also go to Wendi Xue and Tony Edwards for their extra assistance and support throughout the process of bringing this work to press.

This handbook would not have been possible without the assistance of many people. I thank Sara for suggesting the handbook as one of the Major Reference Works (MRW), a signature publication of Springer Nature. I appreciate the conversations Mokshika Gaur, Acquisition Editor of MRW, Sara, and I had about the production process. I am grateful to the MRW team. In particular, I want to thank Audrey Wong-Hillman, Editor of MRW, for providing editorial guidance. I want to thank Project Coordinators Sindhu Ramachandran Salmanul Faris Nedum Palli, and Niraja Deshmukh for their dedication, quick communication, and patience.

I also want to thank the reviewers for their time and valuable comments and suggestions.

What is presented in the volume is the state of research at a particular point in time. Language evolves, and so does the way we study them. This handbook is a living book in the sense that the authors can regularly update their chapters and incorporate new advances in a particular research area. I encourage readers to check updates of the chapters regularly.

Canberra, Australia   Zhengdao Ye

Contents

Part I New Research Trends in Chinese Linguistics 1

1. **New Research Trends in Chinese Linguistics: Introduction** 3
   Quan Wan

2. **Studies in Chinese Phonetics** 9
   Aijun Li and Chunyu Ge

3. **Cognitive Linguistics in China** 41
   Quan Wan and Zhong Lin

4. **Neurolinguistics in China** 67
   Yiming Yang 杨亦鸣, Zude Zhu 朱祖德, and Qingrong Chen 陈庆荣

5. **Linguistic Typology in China** 115
   Zhengda Tang and Yue Wu

6. **Development of the Semantic Map Model in China** 133
   Ding Zhang

7. **Advances in the Study of the Grammaticalization of Sinitic Dialects** 155
   Huayong Lin and Niting Yan

8. **Recent Advances in the Study of Chinese Historical Syntax** 187
   Changcai Zhao

9. **The Sinicization of Buddhist Scriptural Language** 213
   Guanming Zhu

10. **Language Strategies in China** 253
    Xiaobing Fang and Yanhong Ge

11. **Internet Language Study in China** 275
    Minzhe Chen, Shuqing Liu, and Shuyu Zhang

Part II Interactions of Sinitic Languages **297**

12 Interactions of Sinitic Languages: Introduction 299
    Zhengdao Ye

13 Contact-Induced Change in the Languages of Southern China ... 303
    Fuxiang Wu and Yang Huang

14 The Evolution of Chinese Grammar from the Perspective of
    Language Contact 333
    Yonglong Yang and Jingting Zhang

15 Interactions of Sinitic Languages in the Philippines: Sinicization,
    Filipinization, and Sino-Philippine Language Creation 369
    Wilkinson Daniel Wong Gonzales

16 The Expansion of Cantonese Over the Last Two Centuries 409
    Hilário de Sousa

17 Interactions Between Min and Other Sinitic Languages: Genetic
    Inheritance and Areal Patterns 441
    Ruiqing Shen

Part III Meaning, Culture, and Translation **457**

18 Meaning, Culture, and Translation: Introduction 459
    Zhengdao Ye

19 Migrating Concepts in Chinese 465
    James Underhill and Mariarosaria Gianninoto

20 Chinese Cultural Keywords 491
    Paweł Kornacki

21 The Semantics of Kinship in Sinitic Languages 513
    Wendi Xue and Zhengdao Ye

22 Translating Legal Language Between Chinese and English 539
    Deborah Cao

23 Chinese Children's Literature in English Translation 551
    Minjie Chen and Helen Wang

24 Translating Films About Ethnic Minorities in China into English ... 603
    Haina Jin

25 Subtitling in Chinese: Reflections on Emergent Subtitling
    Cultures in Contemporary China 617
    Dingkun Wang

Contents xi

Part IV New Trends in Teaching Chinese as a Foreign Language ... **637**

26 New Trends in Teaching Chinese as a Foreign Language: Introduction 639
Li Yu

27 Teacher Training in the Field of Teaching Chinese as a Foreign Language in the United States 647
Jianhua Bai

28 The History and Development of Assessment of Chinese as a Second Language 675
Dan Nie 聂丹 and Qifeng Zhao 赵琪凤

29 Technology in Chinese Language Teaching 701
Shijuan Liu and Jun Da

30 Innovating the Design of Lower-Level Reading and Writing Curriculum for the Digital Age 743
Junqing Jia and Zhini Zeng

31 Developing Advanced CFL Learners' Academic Writing Skills: Theory and Practice 775
Yang Wang

32 Calligraphy Education in Teaching Chinese as a Second Language 813
Yu Li

33 The Performed Culture Approach 843
Li Yu

Part V Transference from Chinese to English **877**

34 Transference of Chinese to English: Introduction 879
Jock Onn Wong and Zhengdao Ye

35 The "Chineseness" of Singlish 883
Jock Onn Wong

36 Chinese Languages and Malaysian English: Contact and Competition 909
Siew Imm Tan

37 Chinese as a Mother Tongue in the Context of Global English Business Communication 935
Per Durst-Andersen and Xia Zhang

38 **Chinese Ethnolinguistic Influences on Academic English as a Lingua Franca** 975
Feng Cao

39 **A Sketch of Mandarin Loanword Use from *kowtow* to *gaokao*** ... 1003
Will Peyton

Index 1023

About the Editor

Zhengdao Ye is a senior lecturer in the School of Literature, Languages and Linguistics at the Australian National University (ANU), Canberra. She obtained her BA from the Department of Chinese Linguistics and Literature, East China Normal University in Shanghai, and her MA and PhD in linguistics from ANU. Her teaching and research interests encompass Chinese linguistics, semantics and typology, cognitive linguistics, the language of emotions, cross-cultural communication, meaning and culture, translation across languages, and second language usage. She has lectured extensively on these topics in Australia and overseas. She has been a guest speaker and lecturer at Aarhus University, Copenhagen Business School, Griffith University, La Sapienza University, La Trobe University, Le Centre de recherches linguistiques sur l'Asie orientale, Monash University, Roskilde University, Rouen Normandy University, Universidad Nacional Autónoma de México, the University of New England, and the University of Strasbourg. She is the editor of *The Semantics of Nouns* (Oxford University Press, 2017), co-editor (with Cliff Goddard) of *'Happiness' and 'Pain' across Languages and Cultures* (John Benjamins, 2016), and co-editor (with Helen Bromhead) of *Meaning, Life and Culture: In Conversation with Anna Wierzbicka* (ANU Press, 2020).

About the Section Editors

Quan Wan obtained his PhD from the Graduate School of the Chinese Academy of Social Sciences in 2010 and is now a research fellow at the Institute of Linguistics, Chinese Academy of Social Sciences. He is the deputy editor-in-chief of *Contemporary Linguistics* and the director of the editorial office. His monograph, *The Properties and Functions of DE*, won second prize in the 20th Lü Shuxiang Linguistics Prize. His current research focuses on interactional linguistics, usage-based grammar, and Chinese sentence structure.

Jock Onn Wong obtained his PhD in linguistics from the Australian National University under the supervision of Anna Wierzbicka. His PhD research led to a semantic and cultural interpretation of Singapore English using natural semantic metalanguage and formed the basis for his monograph *The Culture of Singapore English*, published by Cambridge University Press in 2014 and shortlisted for the Edward Sapir Book Prize in 2015. A linguist by training and an educator by vocation, Jock Wong is a senior lecturer at the Centre for English Language Communication, where he teaches semantics and academic writing. He has published widely on semantics, pragmatics, interactions between Sinitic languages and Singaporean English, and ELT (English language teaching).

Zhengdao Ye is a senior lecturer in the School of Literature, Languages and Linguistics at the Australian National University (ANU), Canberra. She obtained her BA from the Department of Chinese Linguistics and Literature, East China Normal University in Shanghai, and her MA and PhD in linguistics from ANU. Her teaching and research interests encompass Chinese linguistics, semantics and typology, cognitive linguistics, the language of emotions, cross-cultural communication, meaning and culture, translation across languages, and second language and usage. She is the editor of *The Semantics of Nouns* (Oxford University Press, 2017), co-editor (with Cliff Goddard) of *'Happiness' and 'Pain' across Languages and Cultures* (John Benjamins, 2016), and co-editor (with Helen Bromhead) of *Meaning, Life and Culture: In Conversation with Anna Wierzbicka* (ANU Press, 2020).

Li Yu is Professor of Chinese and Chair of the Department of Asian Languages, Literatures, and Cultures at Williams College. Specializing in Chinese language pedagogy and cultural history, she is committed to helping learners achieve proficiency in Chinese and function successfully in Chinese cultural spheres. She is also an experienced teacher trainer and conducts research on the history of reading and reading pedagogy in late imperial China. Li Yu served as Director of Chinese at the ALLEX Summer Teacher Training Institute at Washington University in St. Louis and as visiting faculty and teacher trainer in the Training Program for Teachers of Chinese of the Ohio State University's SPEAC. Her published work includes over a dozen book chapters and peer-reviewed journal articles. She holds a BA in teaching Chinese as a foreign language from East China Normal University (Shanghai, China), an MA in Chinese language pedagogy, and a PhD in Chinese language pedagogy and cultural history from the Ohio State University.

Contributors

Jianhua Bai Kenyon College, Gambier, OH, USA

Deborah Cao Griffith University, Brisbane, Australia

Feng Cao Centre for English Language Communication, National University of Singapore, Singapore, Singapore

Minjie Chen Princeton University Library, Princeton, NJ, USA

Minzhe Chen Hunan Normal University, Changsha, China

Qingrong Chen Collaborative Innovation Center for Language Ability, School of Linguistic Sciences and Arts, Jiangsu Normal University, Xuzhou, China

Jun Da Middle Tennessee State University, Murfreesboro, TN, USA

Hilário de Sousa EHESS – CRLAO, Paris, France

Per Durst-Andersen Copenhagen Business School, Copenhagen, Denmark

Xiaobing Fang China Center for Linguistic and Strategic Studies, Nanjing University, Nanjing, China

Chunyu Ge Department of Linguistics and Modern Languages, The Chinese University of Hong Kong, Hong Kong, China

Yanhong Ge China Center for Linguistic and Strategic Studies, Nanjing University, Nanjing, China

Mariarosaria Gianninoto Université Paul Valéry, Montpellier, France

Wilkinson Daniel Wong Gonzales University of Michigan, Ann Arbor, MI, USA

Yang Huang Southwest Jiaotong University, Chengdu, China

Junqing Jia Hamilton College, Clinton, NY, USA

Haina Jin School of Foreign Languages and Cultures, Communication University of China, Beijing, China

Paweł Kornacki Institute of English Studies, University of Warsaw, Warsaw, Poland

Aijun Li Institute of Linguistics, Chinese Academy of Social Sciences, Beijing, China

Yu Li Department of Modern Languages and Literatures, Loyola Marymount University, Los Angeles, CA, USA

Huayong Lin Department of Chinese, Sun Yat-sen University, Guangzhou, China

Zhong Lin School of Foreign Languages, Chang'an University, Xi'an, China

Shijuan Liu Indiana University of Pennsylvania, Indiana, PA, USA

Shuqing Liu Guangdong Polytechnic, Foshan, China

Dan Nie Beijing Language and Culture University, Beijing, China

Will Peyton The Australian National University, Canberra, ACT, Australia

Ruiqing Shen Department of Chinese Studies, National University of Singapore, Singapore, Singapore

Siew Imm Tan Faculty of Education, University of Canberra, Canberra, ACT, Australia

Zhengda Tang Institute of Linguistics, Chinese Academy of Social Sciences, Beijing, China

James Underhill Rouen University, Rouen, France

Quan Wan Institute of Linguistics, Chinese Academy of Social Sciences, Beijing, China

Dingkun Wang University of Hong Kong, HKSAR, Hong Kong, China

Helen Wang The British Museum, London, UK

Yang Wang East Asian Studies, Brown University, Providence, RI, USA

Jock Onn Wong Centre for English Language Communication, National University of Singapore, Singapore, Singapore

Fuxiang Wu Beijing Language and Culture University, Beijing, China

Yue Wu School of Humanities, Hangzhou Normal University, Hangzhou, China

Wendi Xue Australian National University, Canberra, Australia

Niting Yan Department of Chinese, Sun Yat-sen University, Guangzhou, China

Yiming Yang Collaborative Innovation Center for Language Ability, School of Linguistic Sciences and Arts, Jiangsu Normal University, Xuzhou, China

Yonglong Yang Institute of Linguistics, Chinese Academy of Social Sciences, Beijing, China

Zhengdao Ye School of Literature, Languages and Linguistics, The Australian National University, Canberra, Australia

Li Yu Department of Asian Studies, Williams College, Williamstown, MA, USA

Zhini Zeng University of Mississippi, Oxford, MS, USA

Ding Zhang Institute of Linguistics, Chinese Academy of Social Sciences, Beijing, China

Jingting Zhang Institute of Linguistics, Chinese Academy of Social Sciences, Beijing, China

Shuyu Zhang The Australian National University, Canberra, Australia

Xia Zhang Copenhagen Business School, Copenhagen, Denmark

Changcai Zhao Institute of Linguistics, Chinese Academy of Social Sciences, Beijing, China

Qifeng Zhao Beijing Language and Culture University, Beijing, China

Guanming Zhu Renmin University of China, Beijing, China

Zude Zhu Collaborative Innovation Center for Language Ability, School of Linguistic Sciences and Arts, Jiangsu Normal University, Xuzhou, China

Part I

New Research Trends in Chinese Linguistics

New Research Trends in Chinese Linguistics: Introduction

Quan Wan

Contents

Introduction 3
Chapter Summaries 4
Conclusion 7

Abstract

This chapter summarizes the chapters in Section 1 of *The Palgrave Handbook of Chinese Language Studies* which discuss recent developments in various fields of Chinese linguistics in China.

Keywords

Chinese linguistics · Chinese languages

Introduction

This section consists of ten chapters. They aim to introduce recent developments in various fields of Chinese linguistics in China. These chapters cover five areas: phonology and prosody, cognition and neuropsychology, dialectology and typology, history and culture, and society and communication.

Q. Wan (✉)
Institute of Linguistics, Chinese Academy of Social Sciences, Beijing, China

© The Author(s), under exclusive licence to Springer Nature Singapore Pte Ltd. 2022
Z. Ye (ed.), *The Palgrave Handbook of Chinese Language Studies*,
https://doi.org/10.1007/978-981-16-0924-4_9

Chapter Summaries

The first two contributions are about phonology and prosody studies. Chapter 2, "Studies in Chinese Phonetics" by Li Aijun 李爱军 and Ge Chunyu 葛淳宇 outline the historical development of Chinese phonology. They point out that in the early nineteenth century, Liu Fu 刘复 and Chao Yuen Ren were among the first linguists in the world to use a kymograph to measure the frequency of tones. Their efforts initiated a new phase in the study of phonetics in China. This chapter provides a synopsis of experimental phonetic studies in China over the last century. It starts with a review of the development of studies in Chinese phonetics, followed by an overview of research in speech production, acoustic phonetics, and speech perception. Application-oriented research is described in the third section which looks at speech pathology, forensic phonetics, speech acquisition and cognitive development in children, and speech learning of Chinese as a second language.

The next two chapters deal with cognitive and psychological research on the Chinese language. In ▶ Chap. 3, "Cognitive Linguistics in China", Wan Quan 完权 and Lin Zhong 林忠 review the history of Chinese cognitive linguistics. Ten years after its establishment in the 1970s, cognitive linguistics was introduced to mainland China. Since 2000, cognitive linguistics in China has continued, with the publication of hundreds of academic works and thousands of articles that cover many topics. This chapter first provides a brief introduction of its history, followed by an analysis of the characteristic features of the Chinese language and illustrations of Chinese as an ideal language to be explored from the perspective of cognition. It then discusses the most important findings by Chinese linguists, including (un)boundedness, part of speech, metaphor and metonymy, blending and haplology, and construction grammar and particles. It concludes that a distinctive paradigm has been formed in terms of a combination of cognition, pragmatics, and functionalism in the cognitive linguistic research in China.

"Neurolinguistics and Psycholinguistics in China" by Yang Yiming 杨亦鸣, Zhu Zude 朱祖德, and Chen Qingrong 陈庆荣 investigates neurolinguistics, a new interdisciplinary area that explores the relationship between language and the brain. Neurolinguistics in China was established in the middle 1990s. Based on research into neural mechanisms during the early stages of its introduction to China, neurolinguistics has developed rapidly and remarkable progress has been made in terms of mental lexicon, grammar, semantics, phonetics, language acquisition and development, and the neural mechanism of language ability of people with cognitive impairment. Future research directions include artificial intelligence, refinement of research methodologies, and the need to establish a research training program to support the development of neurolinguistics in China.

The next three chapters concern linguistic typology. In ▶ Chap. 5, "Linguistic Typology in China," Tang Zhengda 唐正大 and Wu Yue 吴越 discuss how seminal research by Greenberg and Comrie was first introduced to China through translation and how linguistic typology has drawn increasing attention among linguists and inspired studies that have discovered and rediscovered language facts. This introduction of methods of cross-linguistic comparisons, typological classifications, and implicational universals has also led Chinese linguists to uncover cross-categorical correlations within languages. A vast amount of research has been published under the rubric of linguistic typology is associated in one way or another with corresponding features of genetically or geographically related languages. In Chinese academic circles, however, most typological studies still concern descriptions of Sinitic languages from a typological perspective. Unbiased studies of cross-linguistically comparable categories based on data of balanced samples are still underdeveloped; the same is true for fine-grained studies of non-native languages, especially those from outside China. The authors suggest that recently published reference grammar can contribute to theoretical and empirical aspects of linguistic typology. Additionally, Chinese typologists are also contributing their own innovative approaches in the development of typological theory and methodology.

In "The Semantic Map Model in China," Zhang Ding 张定 summarizes the spread and application of the Semantic Map Model (SMM) in China, an emerging research area among Chinese scholars. It retraces the introduction of SMM into China and demonstrates the various domains to which Chinese researchers have applied the semantic map. Moreover, it also analyses the research methods employed by Chinese scholars doing SMM-based studies, with an account of how the Chinese academia responds to such state-of-art issues as diachronic semantic maps and "second-generation" semantic maps. This is followed by a discussion of current debates and future directions of SMM-based research in China.

Lin Huayong 林华勇 and Yan Niting 颜铌婷 discuss advances in the study of the grammaticalization of Sinitic languages. China has a long history of the study on lexeme abstraction with its original theories. From the background of such an academic tradition, grammaticalization theory was quickly accepted by Chinese linguists. They have applied grammaticalization theory extensively in the study of dialect grammar, which has become an important bridge connecting diachrony and synchrony. Since its introduction to China 30 years ago, grammaticalization theory has ushered in a fruitful period in the study of dialect grammar. Based on the works of Chinese scholars, this chapter reviews the application of the theory in the studies of Chinese dialects from three aspects: materials, contents, and methods, and make suggestions on future research directions.

The next two chapters are about the historical and cultural study of the Chinese language. In ▶ Chap. 8, "Recent Advances in the Study of Chinese Historical Syntax," Changcai Zhao discusses achievements in the study of Chinese historical grammar, which has become one of the most vibrant research fields in Chinese linguistics. The introduction of grammaticalization and lexicalization in the study of Chinese historical grammar has made many important findings. Some scholars have also introduced the theory of construction grammar and attempted to explain related phenomena in Chinese historical grammar from this perspective. The attention paid to the historical and cultural background of grammatical changes has been unprecedentedly, which has led to some special grammatical phenomena in Chinese being studied from the perspective of language contact. In the past 30 years, many important achievements have been made, such as the classification of verbs in ancient Chinese, the historical evolution of Chinese basic word order, and the generation mechanism of Chinese verbal aspects.

In "The Sinicization of Buddist Scriptural Langauge," Zhu Guanming 朱冠明 explores the influence of Buddhism on the Chinese language. For thousands of years, Chinese translations of Buddhist scriptures have had an enormous influence in China. Furthermore, the language used in the translation of Buddhist scriptures has unique characteristics that are quite different from the domestic written language. The Buddhist scripture language became mixed with vernacular elements and many foreign language elements. These foreign elements in the Buddhist scriptures have gradually become familiar to Chinese people, and have been accepted and integrated into spoken Chinese. This chapter discusses three completely different modes of the Sinicization of Buddhist Scripture language by taking three Buddhist source terms *ziji* 自己, *xianzai* 现在, and *shengse* 生色 as examples. This chapter also outlines several aspects of Sinicization including the processes and rules by which words derived from Buddhist scriptures have become integrated into Chinese and, the factors that determine the Sinicization of each example along with their differences.

The last two chapters relate to social and communication issues. Fang Xiaobin 方小兵 and Ge Yanhong 葛燕红 give a comprehensive review and summary of the study of language strategy emerged as a new area of research in China in the twenty-first century. At the beginning, the studies were generally driven by the problem-solving needs of the country that was going through a fast process of modernization. Under such conditions, the application of the established theories in language policy and planning (LPP) was frequently implicit and unreflective while the pragmatic approach integrating knowledge from various sources seemed to be effective. The chapter summarizes the study of language strategies with their social effects from the theoretic perspective of LPP. The studies are seen to be an outgrowth of the twentieth century scholarship in LPP but seem to have elevated to a new level of multi-disciplinary and cross-disciplinary development. The major issues that the studies attempt to deal with are seen to be of both theoretically important and broadly relevant beyond the Chinese situation. The economic approach by some of language strategy studies is evaluated in terms of their innovation in theoretic development and practical application.

In "Internet Language Study in China," Chen Minzhe 陈敏哲, Liu Shuqing 刘书晴, and Shuyu Zhang 张舒妤 investigate how Chinese netspeak emerged in the wake of the Internet's arrival in China. Internet language has continued to develop over almost three decades and continues to prosper thanks to a high Internet penetration rate and the prevalence of social media in contemporary China. Research on Chinese netspeak took off just before the turn of the century. After almost 30 years, it is now a robust field of study that engages the next generation of Chinese linguistic researchers and draws attention from multiple disciplines. With an objective to identify research trends and relevant research interests for future Chinese netspeak studies, this chapter revisits previous research published in Chinese academic periodicals and sourced for the *Chinese Social Science Citation Index* (CSSCI) from 1994 – the first year China had access to the global Internet – to 2019. This chapter concludes by calling for

Chinese netspeak research to focus on strengthening its theoretical foundation, broadening its research scope, and enhancing its research focus to promote the future development of the field.

Conclusion

Although this volume has reviewed many new research trends in Chinese Linguistics, there are still some active fields that have not been included, such as computational linguistics, interactive linguistics, and language research in unearthed documents. We look forward to the opportunity to report on new developments in these fields in the future.

Studies in Chinese Phonetics

Aijun Li and Chunyu Ge

Contents

History of Phonetic Research in China	11
The Beginning of Experimental Phonetics	12
Formation and Development of Modern Phonetics	15
Overview of Phonetic Research in China	19
Speech Production Mechanism	20
Acoustic Phonetic Study	22
Speech Perception	26
Interdisciplinary Research in Phonetics	28
Speech Pathology	28
Forensic Phonetics	30
Speech Acquisition and Cognitive Development in Infants and Young Children	31
Speech Learning of Chinese as a Second Language	33
Concluding Remarks	34
References	35

Abstract

The experimental study of Chinese phonetics started in the early twentieth century in the context of a long philological tradition in China. Its development was interrupted by the war and political events during the first half of the twentieth century, but resumed and eventually flourished after the late 1970s. The works of Chao Yuenren, Wu Zongji, and other phoneticians have laid the foundations for in-depth studies of Chinese phonetics. Over several decades, studies of Chinese phonetics have covered all three fundamental aspects of

A. Li (✉)
Institute of Linguistics, Chinese Academy of Social Sciences, Beijing, China
e-mail: liaj@cass.org.cn

C. Ge
Department of Linguistics and Modern Languages, The Chinese University of Hong Kong, Hong Kong, China
e-mail: chunyuge@cuhk.edu.hk

speech, i.e., speech production, acoustic phonetics, and speech perception. Particular attention has been paid to significant phonetic features of Chinese, such as tone and intonation. Phonetic studies of Chinese have also stimulated interests in interdisciplinary research. Speech pathology, forensic phonetics, speech acquisition, and phonetic learning of Chinese as a second language have been explored vigorously. Further studies in Chinese phonetics along these lines with integration with other disciplines and state-of-the-art technologies will not only deepen our understanding of Chinese, but of human language in general.

Keywords

Chinese · Phonetics · Production · Acoustics · Perception · Acquisition

With the study of sounds in ancient China emerging in the fourth century, the philological tradition has continued since then in what is known as historical Chinese phonology. The earliest explication of the sound system in Chinese could be traced back to *Qieyun*, a Chinese rhyme dictionary created by LU Fayan 陆法言 in 601 during the Sui dynasty (581–618 CE). The book contains over 12,000 character entries, divided into five volumes, corresponding to the four tonal categories. These entries are divided into about 200 final rhyme groups, with their pronunciations given in the *fanqie* formula. No clear descriptions of tonal values were offered in the book, except highly impressionistic terms such as "light and shallow" and "heavy and hollow," found in the only extant preface to *Qieyun*. It is worth mentioning that SHEN Yue 沈约 (441–513 CE) of the earlier Six Dynasties period (220–589 CE) was the first one who recognized the existence of the four tonal categories in the *Treaties on Four Tones* that he compiled. This book is probably the earliest work on tones, but is no longer available. Ancient Chinese scholars made enormous effort to classify the Chinese language according to the syllable structure. The Chinese characters, which are entirely monosyllabic, are analyzed by the scholars and arranged in diagrams called *Dengyun Tu* (等韵图). In this kind of diagram, a syllable is represented by its initial in the column, and its final in the row. Initials are grouped according to their places of articulation, and the finals according to the height of the nucleus. This method represents the ancient scholars' attempt to understand the phonetics of Chinese by introspection. Another point to note is a story recorded by the scientist, SHEN Kuo 沈括 (1031–1095 CE), during the Song dynasty. The story said the local officials employed an instrument, which could be put into a mute person's throat to allow him to speak in the court. This is an early and interesting application of phonetics in the ancient Chinese court.

For about a thousand years, Chinese historical phonologists mainly focused on the classification of tones and did not make much significant progress in identifying tonal values. The situation started to change with the arrival of missionaries during the Ming and Qing periods (1368–1911 CE). They worked with a small number of Chinese scholars to transcribe Chinese sounds with the Roman alphabet, including for Chinese dialects, and produced Chinese textbooks with phonetic transcriptions. Most philologists, however, immersed themselves in work of tonogenesis or evolution of tones, disregarding the actual phonetic values of tones. They made no attempt to discover principles of speech production, but it is fair to say that the system of sound changes and evolutionary paths they developed from the Middle Chinese period onward is a remarkable accomplishment, which, in terms of its theoretical significance, is on par with distinctive feature theory as espoused in the SPE model of generative phonology.

The development of phonetics was driven by the intention to romanize Chinese characters. The intellectuals in the young Republic of China believed that the complex Chinese characters were hard for ordinary people, and caused the poverty of the nation. Romanization of Chinese characters was their first step in public education. A thorough and detailed understanding of the Chinese languages, especially the dialects spoken by most of the population, was extremely in need. While phonetic studies before the 1920s relied heavily on skills of hearing and imitating speech sounds within the purview of articulatory phonetics, LIU Fu 刘复 (LIU Bannong 刘半农) and CHAO Yuen-Ren 赵元任, who are the pioneers of the New Culture Movement in China, were among the first to use a kymograph in the measurement of the frequency of tones in the early nineteenth century. Their pioneering efforts started a new phase in the experimental study of speech in China.

This chapter seeks, in four sections, to provide a synopsis of the centennial history of experimental phonetic studies in China, surveying important findings, scholars and milestones. It starts with a review of the overall development of studies in Chinese phonetics, followed by an overview of phonetic research in speech production, acoustic phonetics and speech perception. Application-oriented research is described in the third section on speech pathology, forensic phonetics, speech acquisition and cognitive development in children, and speech learning of Chinese as a second language. The fourth section offers concluding remarks. Due to space limitations, this introduction of phonetic research itself and application-oriented research is succinct and highly selective. Readers are referred to the references for more details about the research discussed in this chapter.

History of Phonetic Research in China

This section gives a comprehensive summary of the history of phonetic research in China, which began almost at the same time as that in Europe. The adoption of the experimental approach in studying Chinese concentrated mainly on lexical tones in the early days. Later research was largely conducted in a few prestigious research institutions, most notably the Institute of Linguistics and Beijing University. The research areas and topics have been greatly expanded in the past few decades, and a wide array of equipment and experimental paradigms have been exploited.

The Beginning of Experimental Phonetics

Experimental study of phonetics in China began in the 1920s and grew rapidly for two decades from the May Fourth Movement in 1919 till the outbreak of China's war against Japan in 1937. In 1924, Liu Fu did research on Chinese tones with a Tone Inference Scale he designed (Fig. 1 and 2), which could measure tone frequency on waves drawn with a kymograph. He completed his doctoral thesis "An Experimental Study of Chinese Tones" (*Étude expérimentale sur les tons du chinois*) in Paris, which was later revised and published in China under the title *Experimental Study of the Four Tones*. Liu's thesis is the first one in experimental phonetics ever written by a Chinese. In 1927, WANG Li 王力 also finished his doctoral thesis in experimental phonetics entitled "An Experimental Study of the Sound System in Bobai Dialect" in France, which was published in 1932 (Wang 1932). Liu Fu returned to China in fall 1925 and brought with him equipment to conduct phonetic experiments before setting up the Phonetics and Music Lab in the Institute of Liberal Arts at Peking University. He died of an acute illness at the age of 44 in Beijing on July 14, 1935, after a linguistic field trip in Inner Mongolia.

LUO Changpei 罗常培 took over as director of the lab shortly after. He advocated using "experiments to compensate for the inadequacy of hearing" and was instrumental in prompting a shift in the study of Mandarin from the articulation-oriented tradition to the experimentation-based approach of modern phonetics (Fig. 3). In his research on Chinese historical phonology, Luo adopted the comparative method, commonly used in the study of Indo-European languages, and worked on materials from loanwords and modern dialects. Among his works, *An Outline of General Phonetics*, coedited by Luo Changpei and WANG Jun 王均 (1957), laid out principles of phonetics as part of its description of speech data in Chinese, and is still regarded one of the best readers in phonetics.

The founding of Academia Sinica in Nanking in 1928 was an important milestone in promoting scholarly research in sciences and humanities. Chao Yuen-Ren was appointed the Head of the Language Section at the Institute of History and Philology and put investigating Chinese dialects and establishing the phonetics lab high on the

Fig. 1 Prof. Liu Fu recording Chinese dialectal sounds

大型声调推断尺
Large scale Tone inference scale

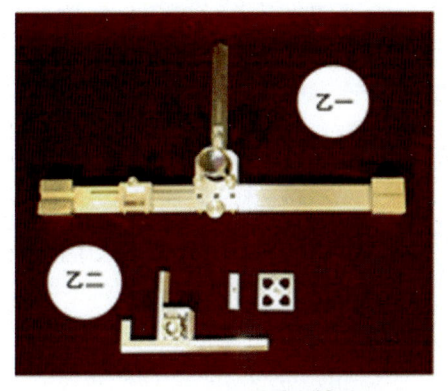

"乙一"、"乙二"声调推断尺
"Yiyi" and "Yier"
Tone inference scale

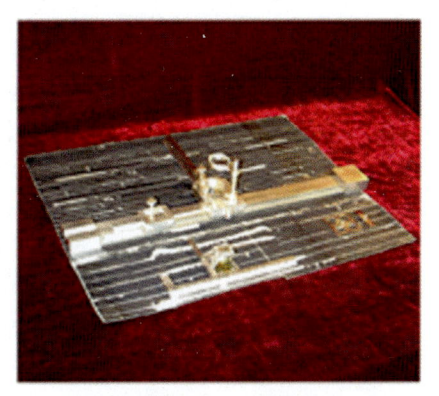

用声调推断尺测量浪线图
Measure Kymograph using
Tone inference scale

Fig. 2 Tone inference scales Yiyi (type I) and Yier (portable type II) invented by Liu Fu

agenda, with priority given to the development of phonetics as a field. As illustrated in Fig. 4, Chao also enthusiastically conducted fieldwork on various Chinese dialects. The field work was summarized in a seminal work entitled *Studies in Modern Wu Dialects* (Chao 1956), which was initially published in 1928. In this monograph, he transcribed Wu Chinese dialects in an adopted version of IPA. Tones were transcribed in a time-pitch graph using a sliding pitch pipe. The book was considered the epitome of fieldwork on Chinese dialects. In 1930, Chao invented the "tone letters" used in the IPA for the transcription of tones (Chao 1930/2002: 713–717). Chao's phonetic work was also manifested in his contribution to phonological theory. In a classic paper on structural linguistics entitled "The non-uniqueness of phonemic solutions of phonetic systems" (Chao 1934/2002: 750–795), Chao argues that a phonemic analysis serves multiple functions and there is no unique solution that serves all functions. Chao established the phonetics lab in 1935 and invited WU Zongji 吴宗济 as his assistant to carry out phonetic research. Their work was cut short due to the outbreak of China's war against Japan.

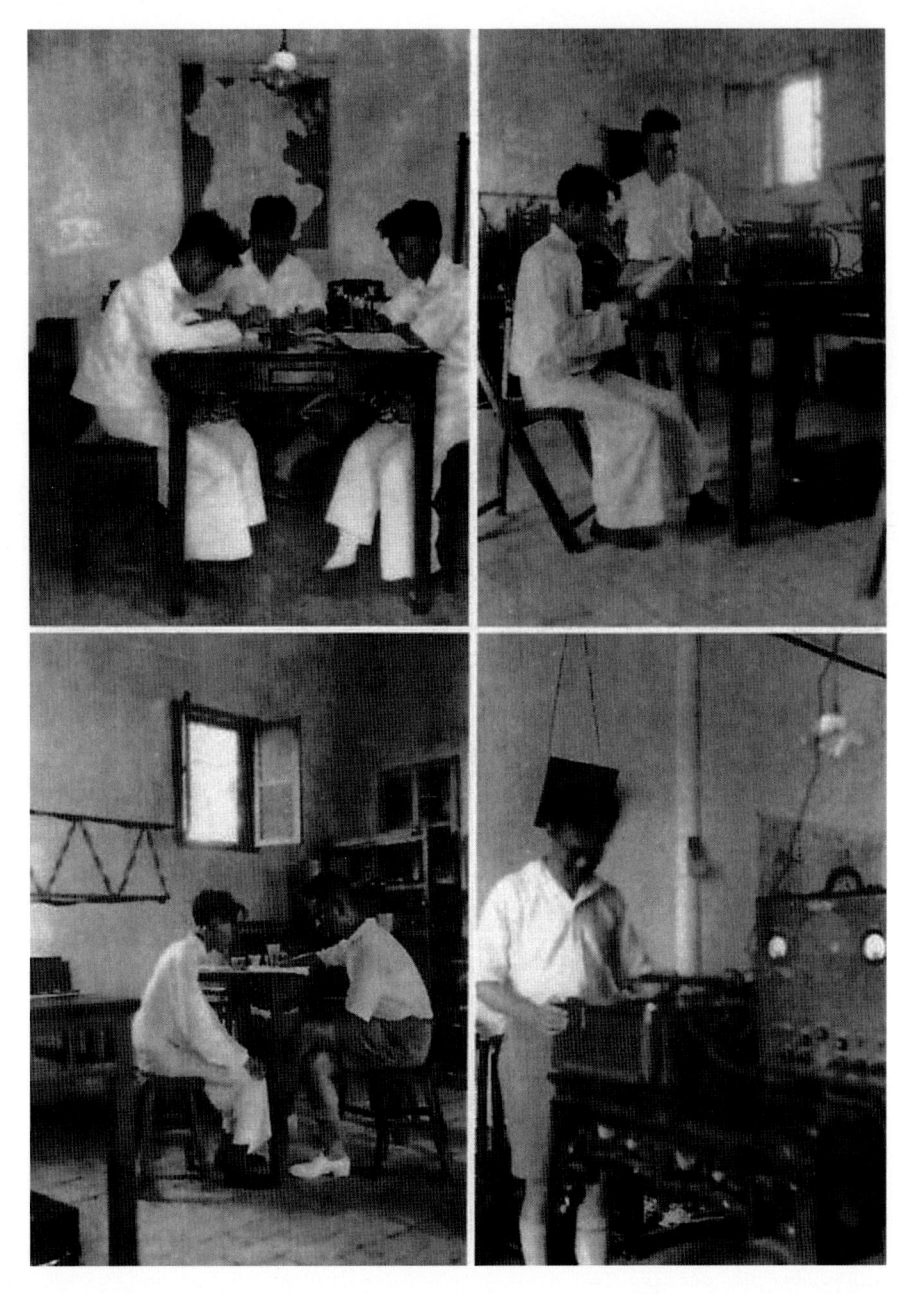

Fig. 3 Prof. Luo Changpei (wearing glasses) investigating the Huizhou dialect in Anhui Province (1932)

Fig. 4 Chao Yuen-ren at fieldwork in Southern China (1936)

Formation and Development of Modern Phonetics

After a period of stagnation due to the war, phonetic research began to recover. The formation and development of modern phonetics as a field of scholarly inquiry can be traced back to the founding of Institute of Linguistics in 1950, which was then part of the Division of Philosophy and Social Sciences at the Chinese Academy of Sciences. The development can be divided into four periods: 1949–late 1960s, 1970s–1980s, 1990s–2010, and 2011–present.

Period 1: 1949–Late 1960s

Shortly after the founding of the Chinese Academy of Sciences, at the request of then premier ZHOU Enlai 周恩来, Luo Changpei – then director of Institute of Liberal Arts at Peking University – was charged with the founding of the Institute of Linguistics in June 1950 and served as its first director. Within the institute, a phonetic research group, the predecessor of the phonetics lab, was assembled on the basis of the Phonetics and Music Lab at Peking University. Wu Zongji joined the

Institute at the invitation of Director Luo Changpei in 1956, whose arrival set in motion the upcoming experimental phonetic research in the Institute. In its early days, the phonetics lab was far from well-equipped for the development of the field in that the equipment mainly comprised of a kymograph, tone inference scales Yiyi and Yier (Fig. 2), tuning forks, a sliding pitch pipe, and a wire recorder. The Institute made a decision in 1958 to purchase speech analysis equipment such as a frequency meter, oscillograph, and sound spectrograph in order to boost research in acoustic phonetics. Wu Zongji also designed a palatography device to identify which parts of the mouth are used when pronouncing different sounds. The focus of research in this period was on the articulatory and acoustic properties of consonants and vowels in *Putonghua* (Standard Chinese), which was made possible by research tools in acoustics and physiology, and in particular, the availability of X-ray photography through collaboration with a local hospital. Major findings include a simple method to calculate vowel formants proposed by Wu Zongji (1964), and static X-ray pronunciation data, palatograms, and tongue position diagrams of phones in Putonghua, all produced by BAO Huaiqiao 鲍怀翘 and Wu Zongji. Figure 5 illustrates the X-ray photos and palatographs acquired by Wu. The five-volume *Experimental Study of Putonghua* and *Putonghua Sound Diagrams* published in 1963, compiled by ZHOU Dianfu 周殿福 and Wu Zongji (1963), were also important accomplishments of this period.

Zhou Dianfu undertook studies in diction. He worked with the famous Chinese crosstalk performer HOU Baolin 侯宝林 and the Peking opera master HAO Shouchen 郝寿臣, and lectured to young performers.

LIN Maocan 林茂灿 specialized in the study of tones in Putonghua. He carried out extensive acoustic experiments on tones in monosyllables and polysyllables using the visible pitch display he designed. His paper "Visible pitch display and the acoustic properties of tones in Putonghua" (Lin 1965) represented an important step forward in uncovering the acoustic properties of tones via visualization of tonal contours.

LIN Tao 林焘 from Peking University paid close attention to the relationship between phonetics and syntax/semantics. The issue that he was concerned with was the relationship between neutral tone and syntactic structures in modern Chinese (Lin 1962).

Period 2: 1970s–1980s

The second period was characterized by a full resumption of phonetic research, followed by remarkable progress made after a hiatus due to the Cultural Revolution (1966–1976). Wu Zongji and Lin Tao provided effective leadership respectively in their phonetics labs. In 1978, Lin Tao reopened the phonetics lab in the Chinese department at Peking University while Wu Zongji assumed the directorship of the phonetics lab in the Institute of Linguistics, now affiliated with the newly created Chinese Academy of Social Sciences in 1977.

For the first time, large-scale studies of the phonetic system in Putonghua were underway in the phonetics lab in the Institute of Linguistics. Under the leadership of Wu, researchers in his lab used a sound spectrograph to conduct experiments on

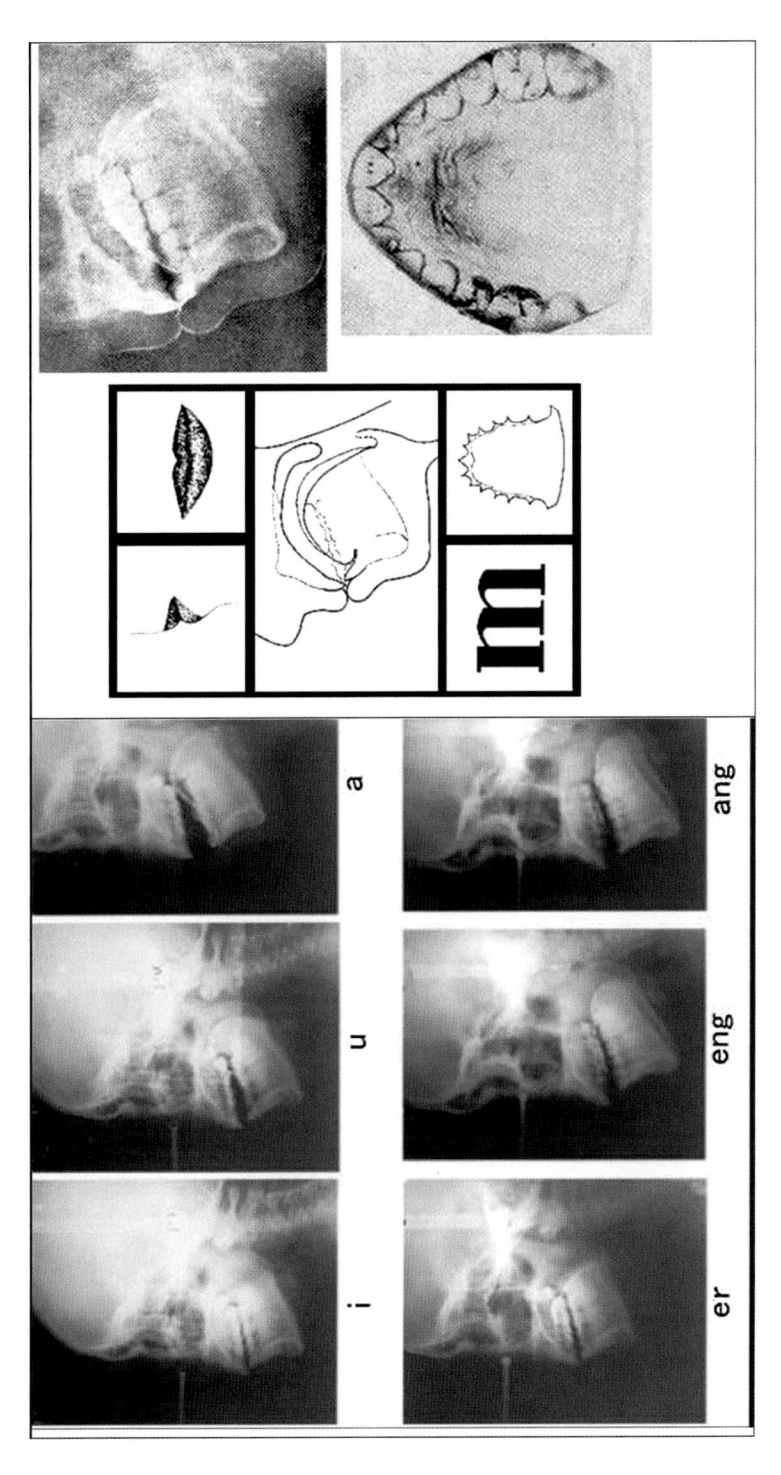

Fig. 5 Examples of X-ray photos of Chinese finals (left) and comprehensive diagram of articulation of [m] (right). (From Zhou and Wu 1963)

monosyllables and disyllables in Putonghua for analyses of the acoustic properties of subsyllabic components of initials, finals, and tones. Research findings were produced in abundance, which were summarized in a series of *"Notes on experimental phonetics"* published in the core journal *Zhongguo Yuwen* (Studies of the Chinese Language) over a period of 5 years (Wu et al. 1979). Another major work was *Monosyllable Spectrograms of Chinese Putonghua*, edited by Wu (Wu 1989), which became a valuable reference for the study of syllables in Putonghua.

Lin Tao and William S-Y. Wang 王士元 (1984) presented an early study on tone perception, in which by varying frequency and duration values of the second syllable in disyllabic combinations, they examined how listeners perceived the tone in the first syllable.

Rule-based speech synthesis was a highlight in this period. A serious interest in synthesis was developed, thanks to the detailed descriptions of acoustic features of Putonghua. YANG Shun'an 杨顺安 was responsible for development of the first ever rule-based formant synthesizer for Putonghua.

Research began to be reported at international conferences too. In 1979, WU Zongji attended the 9th International Congress of Phonetic Sciences (ICPhS) held in Demark and presented a simple method to calculate vowel formants. He was elected to the ICPhS permanent council at the conference, joining the most esteemed phoneticians of his time.

During this period, extensive work on both the acoustics and physiology of speech was carried out. While a linguistic approach was highlighted in the analysis of speech in order to unveil the articulatory and acoustic features of speech in Putonghua, attention was also given to the needs of speech engineering relating to data and knowledge of language.

In addition to Putonghua, experimental methods were also utilized in the analysis of Chinese dialects in a very limited way. For example, in the early 1980s, Lin Tao and students from the Chinese department at Peking University conducted a survey on the Beijing dialect spoken in the city of Beijing, its surrounding suburbs, and a few nearby regions. They transcribed and recorded the speech data.

Several important monographs and edited volumes emerged in this period, mostly by researchers in the Institute of Linguistics. *An Outline of Experimental Phonetics*, coedited by Wu Zongji and Lin Maocan, provided a comprehensive introduction to the main areas of phonetics and an overview of experimental phonetic studies of Putonghua (Wu and Lin 1989). A revision was published with Bao and Lin as coeditors (Bao and Lin 2016). *An Acoustic-Phonetics Oriented Speech Synthesis Technology in Putonghua* by Yang Shun'an (1994), despite its short length, detailed rules and parameters used in the highly acclaimed formant-based speech synthesizer of Putonghua designed by the author. Unfortunately, the monograph was published posthumously.

Across the city at Beijing University, a volume of experimental work on the Beijing dialect was compiled by Lin Tao and WANG Lijia 王理嘉 (1985). It contained acoustic studies of several important topics in the Beijing dialect, namely, the neutral tone, the r-suffix, tone and intonation, and duration distribution in diphthongs, and is still an important reference in experimental phonetics.

Period 3: 1990s–2010
Phonetic research continued to thrive in the 1990s. One significant shift was evident in its increasingly closer connection with speech engineering. The object of inquiry was moving from isolated speech units to naturally occurring continuous utterances, and from segments alone to the inclusion of suprasegmentals. As a result, much effort was put into the study of coarticulation, sound change, stress, rhyme, intonation, and prosodic structure. As computers became more powerful and advanced tools of speech analysis more available, phonetic sciences were developing at an accelerated pace into the twenty-first century.

Period 4: 2011–Present
The recurring theme in this period was innovation and solving previously unsolvable problems. More phonetic research projects were funded by the national program on key basic research, with new developments in interdisciplinary work. At the same time, the scope of phonetic research continued to expand as interdisciplinary research in phonetics began to gain momentum. Thanks to strong support from the national research and talent programs, new phonetics labs were founded in research institutes and universities, with the quantity and quality of research immensely enhanced. All the above laid a solid foundation for more interdisciplinary and cross-disciplinary work in the future.

Specifically, different stages in speech communication as schematically represented in the speech chain were being intensively investigated, from the speech production and articulation mechanism to speech perception and speech and cognitive development. Other areas of research included speech pathology, phonetic study of Chinese dialects, and developments for speech and language research platforms. Equipment and tools used in research were more diversified. In addition to tools for acoustic analysis, equipment was also available for psychological and physiological analyses, such as electroencephalography (ERP), eye-tracking, electromagnetic articulography (EMA), the NDI wave speech research system, high-speed glottal photography, and ultrasound system.

In sum, the past seven decades from 1950 to the present have witnessed the development and thriving of modern phonetic research in China. Despite different priorities in the four periods, phonetic research in China has been moving at an accelerating pace largely thanks to the establishment of key research institutes, strong leadership of senior scholars, funding support from the government, and increasing collaborations with other disciplines.

Overview of Phonetic Research in China

Phonetic research in China has covered all three aspects of phonetics, namely, articulatory, acoustic and auditory. Both segmental and suprasegmental properties have been investigated from these three perspectives. Tones, in particular, have been extensively studied. Earlier research is mostly about Putonghua, the standard variety of Mandarin, while studies of other Chinese dialects and ethnic minority languages in China have been gaining momentum. This section offers an overview of phonetic research in China in three parts, concentrating on speech production, acoustics of speech, and speech perception, respectively.

Speech Production Mechanism

Research on the kinematics of speech production has focused on two processes, phonation and articulation. Phonation is the process by which airflow from the lung causes vibration of the vocal folds to generate the quasi-periodic glottal waveform. Articulation is the process by which the glottal waveform is modulated by the response characteristics of the vocal tract, resulting in different speech sounds.

Glottal vibration can be described by characteristics in the time and frequency domains. Characteristics in the time domain reflect the speed at which the vocal folds vibrate, corresponding acoustically to the fundamental frequency (F0). Characteristics in the frequency domain reflects the manner in which the vocal folds vibrate and can be described by two parameters, open quotient (OQ) and speed quotient (SQ). BAO Huaiqiao 鲍怀翘 was the first researcher to show a serious interest in the phonation types. He and ZHOU Zhizhi 周植志 conducted the first study of phonation types of consonants in the Wa language in 1990 (Bao and Zhou 1990). Following Bao, KONG Jiangping 孔江平 carried out extensive research on phonation (Kong 2003). He identified the most common phonation types and offered detailed acoustic analyses of seven different types. Like Bao, he also analyzed the phonation types of consonants and vowels (and their acoustic properties) in different ethnic minority languages, including Hani, Miao, Liangshan Yi, Axi Yi, Zaiwa, and Jingpo (Kong 2001). ZHU Xiaonong 朱晓农 focused more on the linguistic functions that phonation types could be recruited to perform by pointing out that phonation types can be contrastive in defining "pitch register" in phonology (Zhu 2009).

The process of articulation is concerned with the vocal tract resonance. There are three main areas of research.

The first area is the study in the kinematics of articulation. Zhou Dianfu and Wu Zongji worked with X-ray photography and palatographical devices to record the shape of the vocal tract, contact of the tongue with the palate, and the shape of the lips for each of the 22 consonants, 10 simple vowels, and 2 nasals (used as syllable endings) in Putonghua (Fig. 6). The results were presented in *The Atlas of Sound Spectrograms in Putonghua* (Zhou and Wu 1963). Two decades later, the publication of *Characteristics of Articulatory Gestures in Putonghua* (X-ray videotapes) by Bao Huaiqiao and YANG Lili 杨力立 (1985) represented an important step forward, in its comparison with Zhou and Wu (1963), in that it contains the kinematic trajectories of more complex articulatory configurations in monosyllables, disyllables, and syllables with the r-suffix. Compilation of the articulatory data received a major boost when the first dynamic electropalatography (EPG) database – containing dynamic EPG data collected in the production of monosyllables, disyllables, syllables in neutral tone, syllables with the r-suffix, ancient Chinese poems, sentences,

Fig. 6 Professor WU Zongji operating his palatographical device designed in the 1950s

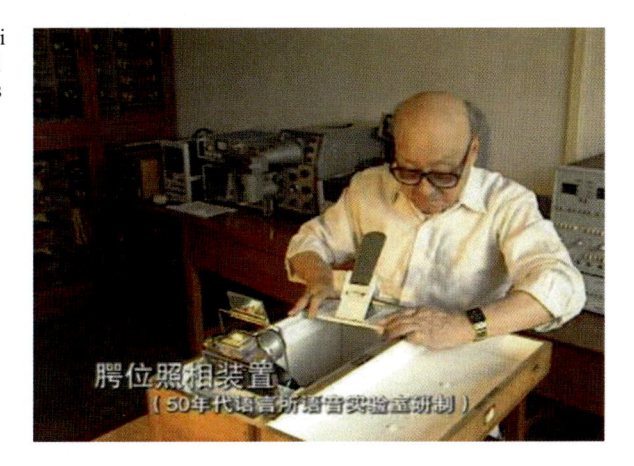

and dialogues – was created at the Institute of Ethnology and Anthropology (Bao and Zheng 2011). Soon after, similar databases were set up at the Institute of Linguistics, Peking University and Fudan University.

In recent years, EMA has been used more often than before. For example, HU Fang 胡方 from the Institute of Linguistics at the Chinese Academy of Social Sciences was able to study with EMA the articulatory characteristics and tongue movement patterns of vowels in the Ningbo dialect (Hu 2014). LI Aijun 李爱军 used EMA to investigate the vowel articulatory spaces of Chinese and Japanese speakers in their speech under different emotions by examining the patterns of tongue movement (Li 2015).

The second area is the study of coarticulation, which refers to the effect of articulatory overlapping of neighboring segments. XU Yi 许毅 distinguished four levels of segmental junctures in Putonghua in his description of how closely neighboring segments are related: close juncture between segments within a syllable, syllabic juncture between syllables, rhythmic juncture between rhythmic units and pausal juncture in and between utterances (Xu 1986). Research in coarticulation started at the segmental level first. Research by Wu Zongji and Sun Guohua 孙国华 (1990) was the earliest study of segmental coordination and they examined the coarticulatory effects of unaspirated stops and fricatives within and between syllables in disyllabic words (C1V1#C2V2). Lin Maocan and YAN Jingzhu 颜景助 (1994) followed up with an experiment on coarticulation between vowels and nasals in zero-initial syllables with nasal endings (i.e., with the syllable structure VN). Development of the EPG technique and creation of the dynamic EPG database have made available a wealth of EPG data for the study of coordination. For example, consonant production in different segmental contexts was examined on the basis of EPG data by Bao Huaiqiao and ZHENG Yuling 郑玉玲 (2011).

Most studies on coarticulation zeroed in on the articulatory overlapping and acoustic consequences of neighboring segments in the production of consonants and vowels in many languages. Chinese researchers turned their attention to the coarticulation of suprasegmental features such as tones. Chao discussed interaction of tones in his most influential book *A Grammar of Spoken Chinese* (Chao 1968/ 1979). He noted that in the combination of two high falling tones in Mandarin, the preceding falling tone did not fall to a lower pitch level whereas the following falling tone did not start at a higher pitch level, due to the influence of the high pitch onset in the second tone on the low pitch offset in the first tone, and vice versa. Wu Zongji described the basic pitch contour patterns for the 16 disyllabic tonal combinations out of the four lexical tones in Putonghua, and based on the disyllabic patterns, he further proposed tone sandhi rules for the trisyllabic and quadrisyllabic units (Wu 1982). More studies on tonal coordination followed. In this line of research, it has been found that anticipatory and carryover effects are both present but differ in magnitude and in nature in that the former are relatively small and mostly dissimilatory while the latter are large and mostly assimilatory (Xu 1997).

The third area is modeling of speech production. In terms of articulatory models, the finite element method was employed to simulate coordinated activities of articulators and muscles based on MRI data and model the human speech production process (Dang and Honda 2004).

Acoustic Phonetic Study

Unlike Indo-European languages, Chinese language in general uses pitch variations to distinguish lexical meanings, resulting in complex interactions of tones. In addition to extensive research on tones, acoustic phonetic studies were also carried out on consonants and vowels in a wide variety of Chinese dialects, which revealed some very interesting facts. An overview of the acoustic phonetic studies of vowels, consonants, tones (including neutral tone), and intonation is offered in this section.

The study of vowels mainly focused on two areas. The first area was concerned with the analysis of acoustic parameters of vowels, including formant frequencies, formant movement trajectories, and duration. As early as the 1960s, Wu Zongji proposed formulas to calculate vowel formant frequencies, bandwidths, and amplitudes and used them in his calculations of formants for the 10 simple vowels in three groups of speakers – male, female adults, and children – in Putonghua (Wu 1964). He plotted vowel diagrams based on the calculated vowel formants. In addition to monophthongs, complex vowels (diphthongs and triphthongs) were investigated at length in relation to their dynamic formant movement, properties in time and frequency domains, and spectral features (Yang and Cao 1984). While vigorous studies on Putonghua continue, interest in the acoustic analysis of vowels in Chinese dialects has recently grown significantly. Examples include devoicing of vowels in the Shanghai dialect (Xu 1990), and the acoustic analysis of diphthongs in the Ningbo dialects (Hu 2014).

The second area involved normalization of formants. Normalization was necessary in acoustic-phonetic analysis of speech to account for between-speaker and cross-dialectal variations in order to derive meaningful linguistic information from the acoustic signals. For example, LING Feng 凌锋 suggested that the normalization method used by Gunnar Fant worked better for gender-induced variations and also proposed a modified normalization method based on a standard score to smooth out cross-dialectal variations (Ling 2008).

Consonants in Chinese have also been studied from different perspectives. In a series of early studies of Putonghua, Wu Zongji analyzed formants of voiced consonants, concentration areas of acoustic energy in voiceless consonants, and acoustic properties of zero-initials in onsetless syllables; he also conducted studies on aspirated and unaspirated obstruents in terms of their differences in airflow, air pressure, and acoustic properties such as voice onset time (VOT), sound intensity, and spectrums (Wu 1992). Marilyn Chen investigated acoustic properties of nasal endings ([n]and [ŋ]) in Putonghua in search of acoustic cues for nasalized vowels and nasals (V-N) in the acoustic signals (Chen 2000). Chen proposed the A1-P1 and A1-P0 parameters as a measurement of nasalization, while Fang (2004) exploited the zero-pole model. Approaches other than acoustic analysis have also emerged in the study of consonants. Hu Fang analyzed EPG, EMA, and acoustic data in his detailed study of the three sibilants (both aspirated and unaspirated) in Putonghua (Hu 2008).

Consonants in the Wu dialects have received special attention, due to the three-way contrast in stops shared by many Wu dialects. Cao and Maddieson (1992) conducted acoustic and aerodynamic analyses of voicing in four Wu dialects and observed that voicing in Wu dialects is not cued by the vibration of vocal folds during the closure period of the voiced stops; instead, it was manifested in the breathy voice of the following vowel.

Chinese is well-known as a tone language. The four tonal categories were referred to as "Ping," "Shang," "Qu," and "Ru" in Middle Chinese. The nature of the four tones remained murky and oftentimes mysterious – their descriptions were solely based on perceptual impressions of high, low, long, and short – until the 1920s when experimental methods were introduced in the measurement of tones in Chinese by Liu Fu, Wang Li, and Yuen-Ren Chao. In *"An Experimental Study of Chinese Tones,"* Liu Fu measured the tones in 12 Chinese dialects using a kymograph and Tone Reference Scale. In 1931, he created *"A Chart of Phonetic Symbols for Investigating Chinese Dialects."* BAI Dizhou 白涤洲, the student of Liu Fu, undertook investigations of dialects in the Guanzhong regions and offered a systematic analysis of the acoustic properties of tones and the evolution of the "Ru" tone in a series of articles, including "An Experimental Study of Tone in Guangzhong." Yuen-Ren Chao proposed the "tone letters" used in the IPA for the transcription of tones, in which the normal pitch range is divided into four equal parts, marked by five pitch levels, numbered 1, 2, 3, 4, 5, with 1 being the lowest and 5 the highest (Chao 1930/2002).

There are abundant studies of tones both in Mandarin Chinese and other Chinese dialects. Some researchers focused on the analysis of acoustic features such as F0, duration and amplitude of tones in isolation and in a sequence, as in Lin and Yan (1992). An interesting development in recent years is the study of phonation types as a cue for tonal contrast. Evidence was presented to show that tones can also be distinguished using phonation types, against the previous assumption that pitch is the sole cue of tonal contrast for tone languages like Mandarin Chinese (Kong 2001).

Another area was more methodology-oriented with a focus on normalization of F0. Various normalization methods were proposed, as reviewed in detail by Zhu Xiaonong. He made a comprehensive comparison of six methods using data from speakers of the Shanghai dialect, and the logarithmic z-score transform had the best performance in terms of normalization index and dispersion coefficient (Zhu 2004).

In Putonghua, a special phenomenon called the neutral tone received ample attention in two aspects. The first aspect involves the status of the neutral tone in the Mandarin phonological system, to wit: whether the neutral tone is a tonal category or part of the metrical system. A syllable is said to be in the neutral tone when it does not carry one of the four lexical tones in Mandarin. According to Chao (1979), the neutral tone is related to weak stress. When a syllable is in weak stress, its tonal range is almost reduced to zero and its duration significantly shortened. Lu and Wang (2005) distinguished "neutral tone" from "weak stress," reserving the former to refer to a neutralized tone in the tonal system and the latter to refer to an unstressed syllable in the metrical system. The second aspect involves studies on the acoustic properties of syllables with the neutral tone. In particular, Li considered the effects of prosodic boundary and information structure on the phonetic realization of the neutral tone. Pitch and durational properties of the neutral tone in different prosodic contexts were examined (Li 2017).

Since both tone and intonation exploit pitch as its primary acoustic cue, the interaction of tone and intonation in Chinese is an intriguing issue in intonational phonology. Since the seminal work of Chao (1932/2002, 1933/2002), a rich body of research, both descriptive and experimental, has been produced to advance the understanding of the linguistic functions and physical properties of tone and intonation in Mandarin Chinese, especially regarding the interaction of tone and intonation. Chao came up with two well-known metaphors for characterizing the ways tone and intonation interact: the "rubber band effect" and the "small ripple and big wave" theory. According to the latter, tone and intonation are related in the form of superimposition – either successive or consecutive – just like small waves sitting on top of big waves (Chao 1933/2002: 198–220).

Wu Zongji, who inherited and expanded Chao's theory, proposed the "transposition model" of intonation, which accounts for obligatory and optional tone sandhi patterns in Chinese (Wu 1982). In his model, an intonational phrase is composed of one or more tone groups at the phrase level and an additional boundary tone. Between tone groups, the pitch range (in semitones) does not fluctuate much, and within each tone group, the interaction of tones follows relatively fixed patterns. This is the obligatory tone sandhi of the model. Paralinguistic factors such as focus and emotions could cause the pitch range of a tone group to shift to a new pitch key, which is the optional tone sandhi of the model. SHEN Jiong 沈炯 characterized Mandarin intonation in terms of the upper line and lower line of F0 that define a pitch register, and argued that the two lines can be manipulated independently of each other in different intonational patterns (Shen 1992). Xu (2005) proposes the Parallel Encoding and Target Approximation (PENTA) model of speech prosody, which is a framework for conceptually and computationally linking communicative meanings to fine-grained prosodic details, based on an articulatory-functional view of speech. In a monograph, *The experimental study of tone and intonation in Chinese* (Lin

2012), Lin took Chao's insights as a point of departure and explicitly adopted the autosegmental-metrical (AM) model of intonation. In this model, focal prominence and boundary tones are the two key elements in describing intonation in Chinese. For example, the difference between declarative intonation in statements and interrogative intonation in questions without sentence-final question particles resides in the boundary tone, which is realized acoustically as variations in pitch register and slope of the contour. SHI Feng 石锋 looked at intonation from a broader perspective and proposed a systematic method to define an "intonation pattern" with three parameters – F0 contours, pause-lengthening ratio, and sound intensity (Shi 2013). In his study of declarative and interrogative intonation patterns in Putonghua, Cantonese, and Korean, Shi tried to figure out cross-linguistic patterns in intonation in terms of the quantitative measurements of F0, duration, and intensity.

Focus is probably the topic in the study of intonation that has received the most attention. Xu (1999) looks into effects of tone and focus on the alignment of F0 contours and found that focus exerts influence on the pitch range of different components in different ways. The pitch range of syllables before the focal position remains unmodulated, and the pitch range is dramatically expanded for syllables in the focal position and compressed for syllables after focus. He termed the last phenomenon "post-focus compression" (PFC). Other studies by Jia, Li, and Chen (2009) analyzed phonetic realizations of different types of focus and situations in which there were one, two, or multiple foci in the utterance. In CAO Wen 曹文's production and perception study, he approached the prosodic realization of focal accent in short utterances in Chinese by adopting a perception test using synthetic materials (Cao 2010). He concluded that the most crucial cue in conveying focus is the extent of changes in F0 maxima, with duration coming second.

In addition to focus, pauses (or breaks before a prosodic boundary) were another closely examined area in the study in intonation. Main areas of research included acoustic realizations of pause, such as changes in F0 before and after a pause (Shen 1992), the duration of syllable rhyme before a pause and the duration of the syllable initial after a pause (Wang, Yang and Lü 2004) and also prosodic transcription of pauses (i.e., break indices) (Li 2002).

Experimental studies on the influence of emotions on intonation patters have flourished in recent years. Li undertook an extensive study on the role of intonation in conveying emotion in a tone language like Chinese, with focus on F0 levels and pitch contours in what she termed "successive addition boundary tone" (Li 2015). She proposed that the boundary tone is composed of two components – the base tone of the syllable and an additional contour – in expressing basic emotions such as happiness and anger. She also offered a phonological representation with phonetic descriptions.

Stepping further into higher prosodic levels, the need to understand intention understanding and generation in human-machine interactions called for greater integration of discourse-level prosodic information in spoken dialogue understanding systems. In a series of studies on the interface of prosody and discourse, Li, Jia, and their collaborators conducted detailed analyses on prosodic features in connection with discourse structure, information structure, and speech acts, which is still an ongoing enterprise.

Speech Perception

One important area of research in speech perception is the mode of speech perception, whether it is categorical or continuous. Categorical perception refers to the phenomenon that speech stimuli along a continuum are perceived as being discrete linguistic categories. Continuous perception is the opposite. Early studies focused on consonants and vowels at the segmental level. The perception of vowels was generally considered to be continuous, while consonants in Chinese are categorical in perception, as indicated in a study by XI Jie 席洁, JIANG Wei 姜薇, ZHANG Linjun 张林军, and SHU Hua 舒华 (2009).

Perception of Chinese tones was approached from two perspectives: modes of perception and perceptual cues. William S-Y. Wang was the first to demonstrate that tone perception in Putonghua was categorical. In Wang (1976), he investigated the perception of tones in Putonghua by manipulating stimuli along a continuum of Yinping (Tone 1, high-level tone) and Yangping (Tone 2, rising tone) on the syllable /i/ in identification and discrimination tasks. Results showed crossovers on the identification curve, which corresponds to the perceptual boundary of the two tones, and prominent peaks on the discrimination curve. These are general characteristics of categorical perception. WANG Yunjia 王韫佳 and QIN Xihang 覃夕航 hypothesized that tones having distinct pitch contours exhibit a strong tendency for categorical perception, while the clear perceptual boundary does not exist for tones with similar pitch contours (Wang and Qin 2015). Generally speaking, research so far has established that the perception of Yinping and Yangping tones and that of Yinping and Qusheng (Tone 4, falling tone) tones are categorical, but the perception of Yangping and Shangsheng tones is complicated due to the similarity of surface pitch contours. Categorical perception only occurs with availability of specific acoustic cues.

Research on perceptual cues in tonal perception has discovered that fundamental frequency (F0) is the key to the perception of tones. In addition to F0, other acoustic properties such as syllable duration and phonation types (Kong 2001, 2003) also play a role. Contextual factors also contribute to tonal perception. In "*Issues in Tonal Perception*," Lin and Wang's (1984) study addresses these factors. For example, in a disyllablic word the F0 onset of the second tone can influence the listener's judgment of the first tone. The higher the F0 onset of the second tone, the more likely the first tone is to be heard as Yangping (Tone 2, rising tone), or Shangsheng (Tone 3, low dipping tone) if the first syllable itself has a longer duration. Linguistic background of the subject participating in the listening tasks has also been proved to be an important factor in tonal perception in Putonghua (Peng et al. 2010).

The third tone (Shangsheng, low dipping tone) in Putonghua is considered a special category largely because of its contour shape of a low falling followed by a rise, traditionally described as "214" in Chao's system. As a result, plentiful studies have been undertaken to evaluate effects of the onset, turning point, and offset of the pitch contour on the perception of the third tone. The falling and rising components of the pitch contour have also been investigated. For example, Lin started out with a claim that the onset and offset of the pitch contours did not provide crucial information on the perception of the third tones (Lin 1965). Shen and Lin (1991) conducted a perceptual study of Tone 2 and Tone 3 in Putonghua, the two tones said to be most confusable among the four lexical tones due to their similar concave shapes. Their results showed that the distinction between Tone 2 and Tone 3 was cued by the timing of the turning point, which is correlated with the degree of the initial falling F0. In addition to perceptual cues, the third tone sandhi and its underlying phonological form were also actively studied. For example, Cao (1995) discusses various surface patterns after application of the third tone sandhi and its phonological, semantic, and syntactic constraints. In terms of its underlying phonological form, there is strong consensus that the third tone, despite its concave shape in pitch contour, is a low tone. Disagreement centers on whether it is a level tone. Cao (2010) points out, based on results of a perception test of level tones, that one should be cautious of describing the third tone as a "low level" tone.

With the developments in cognitive neuroscience, new experimental methods were adopted to explore the cognitive and neural mechanisms underlying speech perception. In terms of language processing, studies of event-related potential (ERP) revealed that in speech perceptions, segments (i.e., consonants and vowels) and suprasegments (e.g., tones) are handled by different processing mechanisms (Li et al. 2010), activating different regions in the brain. Most of the processing of the segmental information happens in the left hemisphere, and no such strong left-lateralization has been found for the processing of the tonal information (Liu et al. 2006). Adoption of ERP and fMRI in studies of tonal perception made it possible to examine neural responses to within-category and across-category stimuli. Experimental results showed that within-category deviants and across-category deviants both triggered mismatch negativity (MMN), but the former induced larger electrophysiological responses (Xi et al. 2010). In a recent study on categorical processing of Chinese lexical tone, Si et al. (2017) conducted cortical surface recordings in surgical patients and revealed a cooperative cortical network along with its dynamics responsible for this categorical perception. Based on an oddball paradigm, they found amplified neural dissimilarity for cross-category tone pairs, over cortical sites covering both the ventral and dorsal streams of speech processing, but no such effect has been found for within category tone pairs.

Phonetic research of the last half century has made significant progress in almost all aspects of speech events that constitute the Speech Chain, from speech production to the acoustics of speech and speech perception. Topics such as tones and intonation in Chinese, perception of Chinese lexical tones, and phonation types have been vigorously pursued. While extensive research has been done on Putonghua or the Mandarin variety of Chinese, other Chinese dialects and minority languages spoken in China were also investigated using experimental methods of phonetics. While experimentally designed speech data are still dominant in current studies, increasing effort has been made to exploit naturally occurring spoken dialogues to explore topics such as utterance- and discourse-level prosody.

Interdisciplinary Research in Phonetics

Apart from research of phonetics in itself, phonetics has attracted attention from various disciplines. The interdisciplinary research of phonetics and the application of phonetics have been onverarching recently. The wide-ranging applications of phonetic research have been well attested in a variety of interdisciplinary fields such as speech pathology, forensics, language acquisition, language pedagogy, and artificial intelligence. The natural maturation of linguistic ability and normal speech mechanisms are fundamental to the well-being of human beings. Curiosity regarding how children acquire their mother tongues leads to the study of language acquisition, while the diseases and physical disorders faced by many patients urge the researchers to study speech pathology. Phonetics can also be employed to assist with forensic evidence. This section covers developments in these areas in China.

Speech Pathology

Pathological speech usually refers to distorted speech resulting from abnormalcies in voice or in the articulatory mechanisms due to diseases or other physical or biological disorder of the speech production system. Speech pathology originated in the medical field. Content wise, early studies mainly focused on three areas. The first area was concerned with the characteristics of pathological speech. For example, WANG Jianhua 王建华 et al. analyzed characteristics of Putonghua speech sounds produced by cleft palate patients based on transcriptions of their pronunciations of syllable initials and finals in Putonghua (Wang et al. 2003). These studies provided theoretical and clinical support for speech assessment and therapy for patients with cleft palates. The second area was the assessment and detection of pathological speech. *The Word List for Cleft Palate Speech Intelligibility*, developed for assessing speech disorder in patients with cleft palates, contains minimal pairs of speech sounds and serves as an effective tool for qualitative and quantitative evaluations of speech intelligibility (Jiang et al. 2009). The third area was related to treatment and training, for example, treatment of patients with cleft palates (Wang 2013).

In terms of research methods and techniques, the spectrograph was used in the early analysis of pathological speech, but not on a large scale. More frequently used methods of diagnosis included stroboscopic measurement, laryngoscopy, endoscopy, and laryngeal electromyogram to locate abnormalcies for treatment. These procedures have some shortcomings. Most of them are invasive and can cause discomfort. Their execution is dependent upon the patients' health conditions and their willingness to cooperate. Diagnosis results are based on a doctor's subjective judgments and experience, which can sometimes be unreliable. Early detection is not likely with these methods, which can delay treatment. As speech pathology became more interdisciplinary, methods and tools phonetic research were gradually adopted. One approach was to track changes in perceptual judgments by modifying acoustic parameters of pathological speech using a voice synthesizer. Parametric modifications such as slight reduction of F0 and amplitude, and standardized glottal noise energy turned out to be effective in evaluating voice quality, due to its multidimensional properties. An attempt in this direction was made by Huang and Wan (2008), which showed that perceptual judgments of voice quality and the acoustic analysis of speech worked together to yield the most reliable evaluations of voice quality. Based on theories of acoustics and anatomy, early diagnosis can be made by adopting computer technology and modern digital signal processing methods to conduct time-frequency analysis and multi-parameter pattern classification on voice data (Peng 2008). Analysis of the acoustic properties of some vowels and consonants in pathological speech (atypical speech sounds) allowed patients with congenital velopharyngeal Insufficiency (VPI) to be diagnosed and treated (Zhou et al. 2012).

As researchers learned more about phonetic features, studies of the feature-based identification of pathological speech started to emerge, as exemplified in the identification of vocal cord nodules and polyps and normal speech using support-vector machines (Yuan et al. 2015). Developments in pattern recognition technologies made it possible for nondestructive testing methods to be used in the automatic evaluation of pathological speech disorders so that testing results can be more objective. Pathological speech recognition is also replacing traditionally used subjective evaluation methods and it has become a hot research topic.

Along the same line, studies of pathological speech are also gradually expanding their subject populations from adults to children. Hearing impairment is the focus of pathological speech research in children. In terms of research methods, most studies are comparing the speech output of normal hearing (NH) children speaking Putonghua with the speech development of hearing impaired children who are wearing hearing aids or have received cochlear implants (CI). Content wise, the majority of research was concerned with speech output. At the segmental level, Yang and Xu (2017) conducted an analysis of formant movement patterns for vowels produced by cochlear implant children. Other studies looked into the speech forming ability of children with hearing loss in different age groups, their speech intelligibility and the development of mouth and tongue movements. At the suprasegmental level, some research showed that currently artificial cochleae are not efficient in decoding pitch information and identifying Chinese tones. Most studies found that children with hearing loss produced the Yingping tone (Tone 1) with the best accuracy, demonstrating good speech forming ability. It is also reported that the most difficult tone for hearing impaired children is Yangping (Tone 2) (Han et al. 2007).

Another actively researched area is the relationship between the age of hearing impaired children and their speech development. Most studies in this area found that speech development is better in hearing impaired children who received artificial cochlear implantation before age 3 or 4 than after. Han et al. (2007) investigated tone production performance of native Mandarin Chinese-speaking children with cochlear implants and evaluated the effects of age at implantation and duration of implant use on tone production in those children. Their results showed that an increased duration of implant use might facilitate tone production, but the age of implantation appears to have a negative effect on tone production in cochlear implant children (the later children receive cochlear plantation, the poorer the performance). Therefore, early implantation might be beneficial to tone production in prelingually deaf children whose native language is a tone language.

Forensic Phonetics

Forensic phonetics is an interdisciplinary research discipline that aims to analyze and compare speech characteristics for the purposes of speaker identification, decoding spoken messages, analysis of emotions in voice, authentication of recordings, and related. In practice, theories and methods in the field of forensic phonetics are often known as voiceprint identification technology. Despite its brief history in China, its development has been fast.

In 1988, Yue Junfa from the Documents Inspection Department at the Criminal Investigation Police University acquired the Kay Elemetrics 7800 Sonagraph and started research and teaching in forensic phonetics. He is regarded as the founder of voiceprint identification in China. A year later, the Center of the Material Evidence Identification affiliated with the Ministry of Public Security, known as the Second Institute of the Ministry of Public Security at the time, also commenced forensic phonetic research after acquiring a newer model of the Kay Elemetrics 5500 Sonagraph. In 1992, the Center completed a key research project of the Ministry of Public Security called *Application of the 5500 Sonagraph in Voiceprint Identification*. The project recruited 60 speakers (30 males and 30 females) who shared a lot of similar characteristics such as age, educational background, birthplace, and long-term residence. Researchers examined their pronunciations of Putonghua in normal speaking conditions and performed a statistical analysis of the acoustic speech data in order to establish the methods and procedures for voiceprint identification in normal speaking condition. The accuracy rate of identification reached 100%. This was the beginning of standardized forensic phonetic research in China. Since then, the voiceprint identification technology has started to be employed in more than 70 court labs throughout the country.

In 1996, the Center of the Material Evidence Identification started to undertake studies of atypical speech and other variations in voiceprint identification as part of the research work carried out for the project *Voiceprint Identification Key Technology and Speaker Recognition System*, funded by the 9th Five Years Key Programs for Science and Technology Development of China. In order to improve the methods and procedures used in voiceprint identification, researchers at the Center analyzed disguised voices and other speech variations caused by colds, different recording devices, dialect accents, whispering, imitating, and impersonating. A proprietary speech workstation was also designed as part of the above-mentioned project. In addition to analyzing acoustic speech data, the workstation can also perform noise reduction for noisy speech and digital amplification for weak speech signals. It outperforms similar products abroad in terms of functions and practicality. The successful deployment of the proprietary speech workstation was seen as one of the most important advancements in forensic phonetics in China.

In 2006, the Center of the Material Evidence Identification with its voiceprint identification lab was accredited by the China National Accreditation Service for Conformity Assessment (CNAS), which is seen as an important milestone in the standardization of methods and procedures in voiceprint identification technology. The publication of one national standard and four industry standards in 2017 proved to be another milestone. Recently, the increasing integration of forensic phonetics and artificial intelligence technology, supported by the Intelligent Speech Technology Key Lab of the Ministry of Public Security, is going to play a key role in the deployment of "Smart Courtrooms."

Speech Acquisition and Cognitive Development in Infants and Young Children

Compared to other areas of phonetic research in China, studies of speech acquisition and cognitive development in Chinese-speaking infants and young children did not appear until recently, with early studies focusing on production research of individuals for the purpose of understanding the development sequence of pronunciation and error types. Work by LI Yuming 李宇明, who conducted research on speech acquisition of infants of 1 to 4 months old, is one of the earliest studies of children language acquisition in China (Li 1991). LI Wei 李嵬 and others analyzed the acquisition sequence of phonemes in Putonghua and error types in 129 children of 1.5–4.5 years old in a comprehensive study of children's speech acquisition (Li et al. 2000). In recent years, research teams in China have adopted more advanced methodologies and produced substantial results on children's speech acquisition and cognitive development based on large-scale corpora.

Research on infant and young children's speech included analysis of error rates and types of vowels, consonants, and tones produced by children in different age groups based on big data and speech corpora. The children's speech research team, led by Li Aijun and Gao Jun GAO Jun 高军, at Institute of Linguistics, the Chinese Academy of Social Sciences, has been collecting large amount of children's speech data since 2008. They analyzed pronunciation data in picture naming tasks of more than 4000 children of 1.5–6 years with normal hearing in the Beijing area and established pronunciation testing standards and norms for Putonghua-speaking children with normal hearing. Gao and Shi (2019) conducted acoustic analysis of the speech output in the children's speech database they created. They found that in terms of error rates, most errors happened in the initials, some in the rhymes and a few in the tones. In terms of the acquisition sequence, acquisition of the tones seemed to occur early and that of the initials late. They also found that acquisition of the single tones and that of the first tone in a disyllabic (i.e., bitonal) sequence followed different acquisition mechanisms and that by the age of 2, children have already acquired the phonological rule of the neutral tone.

Studies of speech perception in infants and young children have been lacking, especially regards their ability to discriminate tones. The children's speech research team at Institute of Linguistics has started to undertake studies of tonal perception in infants and toddlers since 2008 and published a series of research reports (Shi et al. 2017). Among them, Cao, Li, and Fang developed a language acquisition model based on the interconnected network model structure and simulated the acquisition mechanisms in the process of establishing phonemic categories in Putonghua by infants and toddlers (Cao et al. 2017). The published results on speech perception so far have raised doubts about whether "perceptual reorganization" happen in the early stages of infants' speech development. Perceptual reorganization refers to the phenomenon that infants are able to discriminate native speech sounds from non-native ones, but after 6 months their ability to perceive non-native speech contrasts starts to decline or disappear and their ability to perceive native speech contrasts continues to grow and strengthen. While some studies argued for the perceptual reorganization hypothesis, citing that reorganization has been found for lexical tones between 6 and 9 months of age (Mattock and Burnham 2006), others failed to reduplicate similar reorganization effects (Shi et al. 2017). Yet other studies contended that infants do not possess the innate ability to discriminate phonetic categories in their native language, especially when the two categories are acoustically similar and would require a long time to acquire the contrast (Fan et al. 2018).

Currently there have been few studies on speech perception and development in toddlers and preschool children. Some research examined perceived accuracy of monosyllabic Mandarin tones produced by 3- to 5-year-old children growing up in Taiwan and in the United States and found that none of the four tones produced were adultlike. In terms of the accuracy of the tones, more errors are made in Third Tone than other tones, which appeared to follow the order of articulatory complexity for the producing the tones (Wong 2013). In other studies (e.g., Xi et al. 2009), 6- to 7-year-old children were shown to have exhibited adultlike behavior in perceiving Mandarin tones categorically.

One interesting phenomenon in children's speech perception is "perceptual asymmetry," which refers to a situation in which the presentation order of stimuli could impact infants' responses in experiments, to wit, the subject performs better in one order than the opposite order. Perceptual asymmetry can happen in the perception of consonants and vowels and that of suprasegmental features such as tone. In experiments testing discrimination of Tone 1 and Tone 3 in Mandarin, infants demonstrated better discriminatory ability when trained on Tone 1 before Tone 3. The mechanism behind the perceptual asymmetry remains unclear. Speculatively, it might be related to the statistical distribution of the input. For example, when infants are learning a low-frequency tone category, they will establish a phonological feature for that "atypical" token to encode the contrast between low-frequency and high-frequency categories. However, when infants had learned the high-frequency tone category first, they would likely have treated the low-frequency tone category they learned later as an "atypical" exception of the high-frequency tone category, resulting in nondiscrimination of the two tone categories. Therefore, the statistical distribution of the phonological features in the input speech could impact infants' speech perception ability. Some research has demonstrated that infants are very sensitive to the statistical distribution of the input speech, suggesting that they have access to a powerful mechanism for the computation of statistical properties of the language input.

Speech Learning of Chinese as a Second Language

Tone has proved to be the most difficult part of the phonology for learners in the acquisition of Chinese as a second language. Lin remarked that "foreign accent" was not necessarily caused by initials and rhymes, but rather by tones and higher prosodic units (Lin 1996). As a result, tone has been the most intensively examined category in L2 speech learning of Chinese.

Studies of the acquisition of tones have been primarily concerned with errors in production and perception. The former dealt with the acoustic analysis (F0) of L2 tones and the order of tone acquisition. As these studies adeptly pointed out, when foreigners learn to speak Chinese, they have trouble with pitch register, not pitch contour. Errors in pitch contour happen when the F0 contour of a tone produced by L2 learners deviates from the canonical contour shape of the tone, as in the case of a level tone being said as either a rising or falling tone, or a rising tone being said as either a falling or level tone. Errors in pitch register happen when the pitch contour in L2 speech looks fine, but the pitch register ends up either too high or too low, to wit, a high-level tone becomes a low-level tone, or a falling tone becomes a half falling tone (Wang 1995).

Studies of production errors have been plentiful, targeting L2 learners from L1 backgrounds such as American English speakers (Wang 1995). These studies analyzed L2 productions of the four tones in Putonghua in great detail and found that errors in L2 tones could be traced back to both pitch contour and pitch register, with different tones manifesting different error patterns.

Perception errors in L2 speech have been approached through two aspects of tonal perception, largely following studies of speech perception in L1. One aspect was to investigate the perception mode of L2 speakers, using continuous synthetic acoustic stimuli with modified F0 values (Zhang 2010). Zhang (2010) sought to understand how Chinese tones were perceived by L2 learners at different proficiency levels from Thailand, Japan, and Korean. He found that true beginners perceived tones continuously and Thai speakers at the elementary level demonstrated characteristics of categorical perception – not as clear as L1 Chinese speakers, but L2 learners at the intermediate level perceived tones more categorically. Another aspect was to find out the perceptual cues that L2 learners use. Between the two pitch attributes of tones, pitch height, and pitch contour, L2 learners and L1 Chinese speakers exploit different strategies. L2 learners relied more on changes in pitch height whereas L1 speakers on changes in pitch contour.

L2 acquisition of intonation has also been examined vigorously. Some of the topics being explored include declarative intonation produced by L2 speakers, interrogative intonation, acoustic features such as the pitch and duration of declarative intonation and interrogative intonation, L2 acquisition of prosodic boundaries, and L2 acquisition of focus in declarative intonation. Specifically, a series of studies (e.g., Shi et al. 2015) working in the intonation pattern framework developed by Shi Feng and his colleagues investigated error patterns in pitch, duration, and intensity, three important parameters of intonation, in utterances in interrogative intonation produced by L2 learners. Their results showed L2 learners differ greatly from L1 speakers in terms of the movement of the F0 contours and boundary-related lengthening. LI Zhiqiang 李智强 published a monograph on the acquisition and teaching of Mandarin Chinese phonology, in which he advocated the integration of intonation into pronunciation teaching and proposed a model of phonological acquisition in Putonghua based on speech production and perception (Li 2018). In his analysis, L2 speech teaching and learning in Chinese should focus on two key features of Chinese, tone and syllable. In order to cultivate L2 learners' ability to speak Chinese fluently and naturally, L2 speech teaching would benefit from guided practice of prosodic features of the language such as intonation in larger linguistic units than words and phrases.

In addition to suprasegmental features, acquisition of segments has attracted researchers' attention too. Some studies investigated the error patterns and degree of difficulty in the acquisition of vowels in Putonghua by L2 learners from different L1 backgrounds. They also compared perceptual similarities and acoustic similarities of the two vowel systems in the target language L2 and the native language L1. For consonants, most studies dealt with stops, affricates, and nasal endings in Putonghua. Wang, 王韫佳, for example, studied perception and production of nasal endings, aspirated and unaspirated consonants by Japanese-speaking L2 learners (Wang 2002). She found that L2 learners' perception of VN syllables in Putonghua is affected by the nuclear vowel: higher accuracy is obtained when the nuclear vowels are acoustically very different. A similar tendency does not materialize in production data. She also found that the voicing contrast in Japanese influences the Japanese speakers when they process aspirated and unaspirated consonants in Putonghua.

Concluding Remarks

Phonetics in China has undergone enormous advancements in all aspects in the last century, especially the last several decades. It has not only deepened our understanding of Chinese languages themselves but has made remarkable contributions to the understanding of human language. The most noteworthy feature of Chinese is the complex tonal system. Since the initial application of instruments to study the acoustic nature of Chinese tones, the phonetic study of tone has been representative of Chinese phonetics. The tone letter system proposed by Chao has been extremely handy in transcribing tonal values. The interaction of tone and intonation in Chinese is intriguing and has led to a large amount of research. The most recent studies have looked into the representation and processing of tone in the brain.

Phonetic research up until now has become more interdisciplinary and better integrated with other disciplines than ever before. Availability of advanced research instruments has allowed researchers to begin explorations of speech production and perception mechanisms at the neurological level by taking advantage of the most salient features of Chinese. There will be more acoustic studies on Chinese dialects and ethnic minority languages in China. With the advent of new technologies, research on articulatory phonetics will be more fine-grained, and the online processing of speech will be investigated more deeply in the domain of auditory phonetics. Better integration of phonetics into neuroscience holds great promise for shedding new light on the understanding of the neural mechanism of speech production and perception. Last but not least, phonetic research will find more applications in every corner of our life. It is certain that phonetic research of Chinese and languages in China will continue to provide new insights and perspectives to the general understanding of human languages.

References

Bao, Huaiqiao (鲍怀翘), and Maocan Lin (林茂灿). 2016. *Shiyan yuyinxue gaiyao zengdingban* 实验语音学概要《增订版》 [*An outline of experimental phonetics - revised edition*]. Beijing: Beijing daxue chubanshe 北京:北京大学出版社 [Beijing: Peking University Press].

Bao, Huaiqiao (鲍怀翘), and Lili Yang (杨力立). 1985. *Putonghua fayinqiguan dongzuotexing* (X guang luxiangdai) 普通话发音器官动作特性 (X光录像带) [*Characteristics of articulatory gestures in Putonghua" (X-ray videotapes)*]. Beijing: Beijing Yuyanxueyuan Chubanshe 北京: 北京语言学院出版社 [Beijing: Beijing Language and Culture College Press].

Bao, Huaiqiao (鲍怀翘), and Yuling Zheng (郑玉玲). 2011. Putonghua dongtai yuewei yanjiu 普通话动态腭位研究 [Electropalatography studies on articulation of standard Chinese]. *Nanjing Shifandaxue Wenxueyuan Xuebao* 南京师范大学文学院学报 [*Journal of School of Chinese Language and Culture Nanjing Normal University*] 3: 1–11.

Bao, Huaiqiao (鲍怀翘), and Zhizhi Zhou (周植志). 1990. Wayu zhuosongqi shengxuetezheng fenxi 佤语浊送气声学特征分析 [An acoustical analysis of voiced aspiration in Wa language]. *Minzu Yuwen* 民族语文 [*Minority Language of China*] 2: 62–70.

Cao, Jianfen (曹剑芬). 1995. Lianxu biandiao yu qingzhong duili 连续变调与轻重对立 [Tone sandhi and stress]. *Zhongguo yuwen* 中国语文 [*Studies of the Chinese Language*] 4: 312–320.

Cao, Wen (曹文). 2010. *Hanyu jiaodianzhongyin de yunlv shixian* 汉语焦点重音的韵律实现 [*Realization of prosody in focus in Chinese*]. Beijing: Beijing Yuyan Daxue Chubanshe 北京:北京语言大学出版社 [Beijing: Beijing Language and Culture University Press].

Cao, J., and I. Maddieson. 1992. An exploration of phonation types in Wu dialects of Chinese. *Journal of Phonetics* 20: 77–92.

Cao, Mengxue, (曹梦雪), Aijun Li (李爱军), and Fangqiang Fang (方强). 2017. Yingyouer muyu yinweifanchou xidejizhi de jianmo yanjiu 婴幼儿母语音位范畴习得机制的建模研究 [A modeling study of infants' categorical phonemic perception mechanism in first language acquisition]. *Zhongguo yuwen* 中国语文 [*Studies of the Chinese Language*] 3: 116–127+130.

Chao, Yuenren (赵元任). 1956. *Xiandai Wuyu de yanjiu* 现代吴语的研究 [*Studies in the modern Wu dialects*]. Beijing: kexue chubanshe 北京: 科学出版社 [Beijing: Science Press].

——— (赵元任). 1979. *Hanyu kouyu yufa* 汉语口语语法 [*A Grammar of Spoken Chinese*]. Beijing: Shangwu chubanshe 北京:商务出版社 [Beijing: The Commercial Press].

——— (赵元任). 2002. *Zhao yuanren yuyanxue lunwenji*《赵元任语言学论文集》[*Linguistics essay by Yuenren Chao*] Beijing: shangwu yinshu guan 北京: 商务印书馆 [Beijing: The Commercial Press].

Chen, M.Y. 2000. Acoustic analysis of simple vowels preceding a nasal in standard Chinese. *Journal of Phonetics* 28 (1): 43–67.

Dang, J., and K. Honda. 2004. Construction and control of a physiological articulatory model. *Journal of the Acoustical Society of America* 115 (2): 853–870.

Fang, Qiang (方强). 2004. *Putonghua biyin de yanjiu* 普通话鼻音的研究 [*Nasal Sounds of Mandarin*]. Zhongguo shehui kexueyuan shuoshi xuewei lunwen 中国社会科学院士学位论文 [CASS Master's Essay].

Fan, S., A. Li, and A. Chen. 2018. Perception of lexical neutral tone among adults and infants. *Frontiers in Psychology* 9: 322.

Gao, J., and R. Shi. 2019. Perceptual development of phonetic categories in early infancy: Consonants, vowels and lexical tones. In *Prosodic studies: Challenges and prospects*, ed. Hongmin Zhang and Youyong Qian, to appear. London: Routledge.

Han, D., N. Zhou, Y. Li, et al. 2007. Tone production of Mandarin Chinese speaking children with cochlear implants. *International Journal of Pediatric Otorhinolaryngology* 71 (6): 875–880.

Hu, F. 2008. The three sibilants in standard Chinese. *The 8th international seminar on speech production*, 105–108.

Hu, Fang. 2014. *A phonetic study on the vowels in Ningbo Chinese*. Beijing: China Social Science Press.

Huang, Shaoming (黄昭鸣), and Ping Wan (万萍). 2008. Sangyin shengxuecanshu yu sangyinyinzhi de xiangguanyanjiu 嗓音声学参数与嗓音音质的相关研究 [On the Relationship between voice acoustic parameters and vocal quality]. *Linchuang erbiyanhou toujing waike zazhi* 临床耳鼻咽喉头颈外科杂志 [*Journal of Clinical Otorhinolaryngology Head and Neck Surgery*] 6: 251–255.

Jia, Yuan (贾媛), Aijun Li (李爱军), and Yiya Chen (陈轶亚). 2009. Hanyu "shi" he "lian" bianji de jiaodianchengfen yuyin tezheng yanjiu 汉语"是"和"连"标记的焦点成分语音特征研究 [Phonetic properties of "Shi" and "Lian" marked constituents in Chinese]. *Qinghua daxue xuebao (ziran kexue ban)* 清华大学学报 (自然科学版) [*Journal of Tsinghua University (Science and Technology)*] s1: 1294–1301.

Jiang, Chenghui (姜成惠), Xinghui Shi (施星辉), Wanzhong Lin (万林), Hongbing Jiang (江宏兵), and Hua Yuan (袁华). 2009. Elie huanzhe putonghua zuixiao yuyindui tezheng chubuyanjiu 腭裂患者普通话最小语音对特征初步研究 [The characteristics of minimal pairs for cleft palate speech in Chinese Mandarin]. *Tinglixue ji yanyujibing zazhi* 听力学及言语疾病杂志 [*Journal of Audiology and Speech Pathology*] 17(4): 347–350.

Kong, Jiangping (孔江平). 2001. *Lun yuyan fasheng* 论语言发声 [*On language phonation*]. Beijing: Zhongyang minzu daxue chubanshe 北京:中央民族大学出版社 [Beijing: China Minzu University Press].

——— (孔江平). 2003. Sangyin fasheng leixing de shengxue xignzhi ji canshu hecheng 嗓音发声类型的声学性质及参数合成 [A study on the acoustic properties of phonation types and parameter synthesis]. *Diliujie quanguo xiandai yuyinxue xueshu huiyi* 第六届全国现代语音学学术会议 [*The 6th phonetics conference of China*].

Li, Yuming (李宇明). 1991. 1–120 tian yinger fayin yanjiu 1–120 天婴儿发音研究 [A case study of baby's phonological development from 1 to 120 days old]. *Xinli kexue* 心理科学 [*Phychological Science*] 5: 21–25.

Li, A. 2002. Chinese prosody and prosodic labeling of spontaneous speech. *Speech Prosody*.

Li, Aijun. 2015. *Encoding and decoding of emotional speech: A cross-cultural and multimodal study between Chinese and Japanese (prosody, phonology and phonetics)*. 1st ed. Springer.

——— (李爱军). 2017. Putonghua Butong Xinxi Jiegou Zhong qingsheng de yuyin texing 普通话不同信息结构中轻声的语音特性 [Phonetic correlates of neutral tone in different information structures]. *Dangdai yuyanxue* 当代语言学 [*Contemporary Linguistics*] 19(3): 348–378.

Li, Zhiqiang (李智强). 2018. *Yuyin xide yu jiaoxue yanjiu* 语音习得与教学研究 [*Research on Chinese pronunciation: Acquisition and teaching*]. Beijing: Beijing Yuyan Daxue Chubanshe 北京:北京语言大学出版社 [Beijing: Beijing Language and Culture University Press].

Li, Wei (李嵬), Hua Zhu (祝华), Barbara Dodd, Tao Jiang (姜涛), Danlin Peng (彭聃龄), and Hua Shu (舒华). 2000. Shuo Putonghua ertong de yuyin xide 说普通话儿童的语音习得 [Phonological acquisition of Putonghua speaking children]. *Xinli xuebao* 心理学报 [*Acta Psychologica Sinica*] 32(2): 170–176.

Li, X., J.T. Gandour, T. Talavage, et al. 2010. Hemispheric asymmetries in phonological processing of tones versus segmental units. *Neuroreport* 21 (10): 690.

Lin, Tao (林焘). 1962. Xiandai hanyu qingyin he yufa jiegou de guanxi 现代汉语轻音和句法结构的关系 [Neutral tone and syntactic structures in modern Chinese]. *Zhongguo yuwen* 中国语文 [*Studies of the Chinese Language*] 7.

Lin, Maocan (林茂灿). 1965. Yingao xianshiqi yu putonghua shengdiao yingao texing 音高显示器与普通话声调音高特性 [Visible pitch display and the acoustic properties of tones in Putonghua]. *Shengxue xuebao* 声学学报 [*Acta Acustica*] 1: 10–17.

Lin, Tao (林焘). 1996. Yuyin yanjiu he duiwai hanyu jiaoxue 语音研究和对外汉语教学 [Phonetics research and teaching Chinese as foreign languages]. *Shijie hanyujiaoxue* 世界汉语教学 [*Chinese Teaching in the World*] 3: 20–23.

Lin, Maocan (林茂灿). 2012. Hanyu yudiao shiyan yanjiu 汉语语调实验研究 [*The experimental study of tone and intonation in Chinese*]. Beijing: Zhongguo shehui kexue chubanshe [Beijing: Chinese Social Sciences Press].

Lin, Tao (林焘), and Lijia Wang (王理嘉). 1985. *Beijing Yuyin Shiyanlu* 北京语音实验录 [*An Experimental study of the Beijing speech*]. Beijing: Beijing daxue chubanshe 北京:北京大学出版社 [Beijing: Peking University Press].

Lin, Tao (林焘), and S-Y Wang William (王士元). 1984. Shengdiao ganzhi wenti 声调感知问题 [Perception of tone]. *Zhongguo yuyan xuebao* 中国语言学报 [*Journal of Chinese Linguistics*] 2: 59–69.

Lin, Maocan (林茂灿), and Jingzhu Yan (颜景助). 1994. Putonghua daibiwei lingshengmu yinjie zhong de xietongfayin 普通话带鼻尾零声母音节中的协同发音 [An experiment on the coarticulation between vowels and nasals in zero-initial syllables with nasal endings (VN)]. *Yingyong shengxue* 应用声学 [*Journal of Applied Acoustics*] 1: 12–20.

Ling, Feng (凌锋). 2008. Sanzhong yuanyin guizheng fangfa zai suzhouhua yuanyin yanjiu zhong de yingyong yu bijiao 三种元音规整方法在苏州话元音研究中的应用与比较 [Applications and comparisons of three vowel adjustment methods in Suzhou vowel research]. *Zhongguo yuyinxue xueshu huiyi ji qinghe wuzongji xiansheng baisui huadan yuyin kexue qianyan wenti guoji yantaohui* 中国语音学学术会议暨庆贺吴宗济先生百岁华诞语音科学前沿问题国际研讨会 [*China conference on phonetics and the international symposium on the frontiers of speech science in Mr. Wu Zongji's 100th birthday*].

Liu, L., D. Peng, G. Ding, et al. 2006. Dissociation in the neural basis underlying Chinese tone and vowel production. *NeuroImage* 29 (2): 515–523.

Lu, Jilun (路继伦), and Jialing Wang (王嘉龄). 2005. Guanyu qingsheng de jieding 关于轻声的界定 [On defining "Qingsheng"]. *Dangdai yuyanxue* 当代语言学 [*Contemporary Linguistics*] 7(2): 107–112.

Luo, Changpei (罗常培), and Jun Wang (王均). 1957. *Putong yuyinxue gangyao* 普通语音学纲要 [*An outline of general phonetics*]. Beijing: kexue chubanshe 北京:科学出版社 [Beijing: Science Press].

Mattock, K., and D. Burnham. 2006. Chinese and English infants' tone perception: Evidence for perceptual reorganization. *Infancy* 10 (3): 241–265.

Peng, Ce (彭策). 2008. *Jiyu shengxue yu xiaoboshang ji zihuigui moxing de bingtai sangyin zhenduan xinfangfa yanjiu* 基于声学与小波熵及自回归模型的病态嗓音诊断新方法研究 [*Study on novel method of pathological voice diagnosis based on acoustics, wavelet entropy and auto-regressive model*]. Tianjin daxue boshi xuewei lunwen 天津大学博士学位论文 [Tianjin University Dissertation].

Peng, G., H.Y. Zheng, T. Gong, et al. 2010. The influence of language experience on categorical perception of pitch contours. *Journal of Phonetics* 38 (4): 616–624.

Shen, Jiong (沈炯). 1992. Hanyu yudiao moxing chuyi 汉语语调模型刍议 [Model of Chinese intonation]. *Yuwen yanjiu* 语文研究 [*Linguistic Research*] 4: 16–24.

Shen, X.S., and M. Lin. 1991. A perceptual study of Mandarin tones 2 and 3. *Language and Speech* 34 (2): 145–156.

Shi, Feng (石锋). 2013. *Yudiao geju-shiyan yuyanxue de dianjishi* 语调格局-实验语言学的奠基石 [*The Intonation pattern- foundation of experimental linguistics*]. Shanghai: shangwu yinshu guan 上海:商务印书馆 Shanghai: The Commercial Press.

Shi, Feng (石锋), Baoying Wen (温宝莹), and Yajuan Han (韩亚娟). 2015. Riben xuesheng hanyu yiwenju yudiao xide shiyan duibi fenxi 日本学生汉语疑问句语调习得的实验研究 [A Study on the Japanese learner's acquisition of interrogative sentence intonation]. *Dishisanjie Quanguo renji yuyin tongxun xueshu huiyi* 第十三届全国人机语音通讯学术会议 [*The 13th national conference on man- machine speech communication*].

Shi, R., J. Gao, A. Achim, et al. 2017. Perception and representation of lexical tones in native Mandarin-learning infants and toddlers. *Frontiers in Psychology* 8: 1–11.

Si, X., W. Zhou, and B. Hong. 2017. Cooperative cortical network for categorical processing of Chinese lexical tone. *Proceedings of the National Academy of Sciences* 114 (46): 12303–12308.

Wang, Bei (王), Yang, Yufang (杨玉芳) and Lv, Shinan (吕士). 2004. Hanyu yunlv cengji jiegou bianjie de shengxue fenxi 汉语韵律级结构界的声学分析 [An Acoustic Analysis of Prosodic Hierarchical Boundaries of Chinese]. *Shengxue xuebao* 声学学报 [*ACTA ACUSTICA*], 2004.1:29–36.

Wang, Li (王力). 1932. *Bobai fangyan shiyanlu* 博白方言实验录 [*An Experimental Study of the Bobai Dialect*]. Bali daxue boshi xuewei lunwen 巴黎大学博士学位论文 [Université de Paris Dissertation].

Wang, S.Y. 1976. Language change. *Annals of the New York Academy Sciences* 208: 61–729.

Wang, Yunjia (王韫佳). 1995. Yetan meiguoren xuexi hanyu shengdiao 也谈美国人学习汉语声调 [On American Chinese leaners' study of intonation]. *Yuyanjiaoxue yu yanjiu* 语言教学与研究 [*Language Teaching and Linguistic Studies*] 3: 126–140.

——— (王韫佳). 2002. Riben xuexizhe ganzhi he chansheng putonghua biyin yunmu de shiyan yanjiu 日本学习者感知和产生普通话鼻音韵母的实验研究 [An experimental study on the perception and production of nasal codas by Japanese learners of Chinese Putonghua]. *Shijie hanyujiaoxue* 世界汉语教学 [*Chinese Teaching in the World*] 2: 47–60.

Wang, Guoming (王国民). 2013. *Chun'e'lie xiufushu yu yuyin zhiliao* 唇腭裂修复术与语音治疗 [*Surgical repair and speech therapy for clefts*]. Shanghai: shanghai shijie tushu chugan gongsi 上海：上海世界图书出版公司 [Shanghai: Shanghai World Book].

Wang, Yunjia (王韫佳), and Xihang Qin (覃夕航). 2015. Putonghua danzidiao yangping he shangsheng de bianren ji qufen 普通话单字调阳平和上声的辨认及区分 [Grcognization of Tone2 and Tone3 in Mandarin]. *Yuyan kexue* 语言科学 [*Linguistic Sciences*] 4: 337–352.

Wang, Jianhua (王建华), Juan Liu (刘娟), Yong Wang (王勇), Aiqin Lv (吕艾芹), and Renhua Zhang (张仁华). 2003. Elie putonghua yuyin de tedian yu fenxi 腭裂普通话语音的特点与分析 [The feature of cleft palate in Mandarin]. *Linchuang kouqiang yixue zazhi* 临床口腔医学杂志 [*Journal of Clinical Stomatology*] 19(8): 464–466.

Wong, P. 2013. Perceptual evidence for protracted development in monosyllabic Mandarin lexical tone production in preschool children in Taiwan. *Journal of the Acoustical Society of America* 133 (1): 434–443.

Wu, Zongji (吴宗济). 1964. Putonghua yuanyin he fuyin de pinpu fenxi ji gongzhenfeng de cesuan 普通话元音和辅音的频谱分析及共振峰的测算 [The spectrogram analysis of the vowels and consonants in standard colloquial]. *Shengxue xuebao* 声学学报 [*Acta Acustica*] 1(1): 33–40.

——— (吴宗济). 1982. Putonghua yujv zhong de shengdiao bianhua 普通话语句中的声调变化 [Pitch contour in Mandarin]. *Zhongguo yuwen* 中国语文 [*Studies of the Chinese Language*] 6: 439–450.

——— (吴宗济). 1989. Putonghua danyinjie yutuce 普通话单音节语图册 [*Chinese Putonghua monosyllable spectrograms*]. Beijing: zhongguo shehui kexue chubanshe 北京：中国社会科学出版社 [Beijing: China Social Sciences Press].

——— (吴宗济). 1992. Putonghua lingshengmu yinjie qishiduan de shengxue fenxi 普通话零声母音节起始段的声学分析 [An Acoustic analysis of the initial segment of zero initial syllables in Mandarin]. *Disanjie quanguo renji yuyin tongxun huiyi* 第三届全国人机语音通讯会议 [*The 3th national conference on man-machine speech communication*].

Wu, Zongji (吴宗济), and Maocan Lin (林茂灿). 1989. *Shiyan yuyinxue gaiyao* 实验语音学概要 [*An outline of experimental phonetics*]. Beijing: Gaodeng jiaoyu chubanshe 北京：高等教育出版社 [Beijing: Higher Education Press].

Wu, Zongji (吴宗济), and Guohua Sun (孙国华). 1990. Putonghua qingcayin xietongfayin de shengxue moshi 普通话清擦音协同发音的声学模式 [Acoustic coarticulation patterns of voiceless fricatives in CVCV in standard Chinese]. *Yuyin yanjiu baogao* 语音研究报告 [*Annual phonetic research report*].

Wu, Zongji (吴宗济), Jianfen Cao (曹剑芬), and Lili Yang (杨力立). 1979. Shiyan yuyinxue zhishi jianghua 实验语音学知识讲话 [Notes on experimental phonetics]. *Zhongguo yuwen* 中国语文 [*Studies of the Chinese Language*] 1979连载于第1,2,4,5,6 期, 1,2,4,5,6.

Xi, Jie (席洁), Wei Jiang (姜薇), Linjun Zhang (张林军), and Hua Shu (舒华). 2009. Hanyu yuyin fanchouxing zhijue jiqi fazhan 汉语语音范畴性知觉及其发展 [Categorical perception of VOT and lexical tones in Chinese and the developmental course]. *Xinli xuebao* 心理学报 [*Acta Psychological Sinica*] 41(7): 572–579.

Xi, J., L. Zhang, H. Shu, et al. 2010. Categorical perception of lexical tones in Chinese revealed by mismatch negativity (MMN). *Neuroscience* 170 (1): 223–231.

Xu, Yi (许毅). 1986. Putonghua yinlian de shengxue yuyinxue tezheng 普通话音联的声学语音学特性 [Acoustic properties of juncture in Mandarin]. *Zhongguo yuwen* 中国语文 [*Studies of the Chinese Language*] 5: 353–360.

Xu, Yunyang (徐云扬). 1990. Shanghaihua yuanyin qinghua de yanjiu 上海话元音清化的研究 [A study of the unvoicing of vowels in the Shanghai dialect]. *Dangdai yuyanxue* 当代语言学 [*Contemporary Linguistics*] 3: 19–34.

Xu, Y. 1997. Contextual tonal variations in Mandarin. *Journal of Phonetics* 25 (1): 1–83.

———. 1999. Effects of tone and focus on the formation and alignment of F0 contours. *Journal of Phonetics* 27 (1): 55–105.

———. 2005. Speech melody as articulatorily implemented communicative functions. *Speech Communication* 46 (3): 220–251.

Yang, Shun'an (杨顺安). 1994. *Mianxiang shengxue yuyinxue de putonghua yuyin hecheng jishu* 面向声学语音学的普通话语音合成技术 [*An acoustic-phonetics oriented speech synthesis technology in Putonghua*]. Beijing: zhongguo shehui kexue chubanshe 北京: 中国社会科学出版社 [Beijing: China Social Sciences Press].

Yang, Shun'an (杨顺安) and Cao, Jianfen (曹剑芬). 1984. Putonghua erhe yuanyin de dongtai texing 普通话二合元音的动态特性 [Dynamic properties of diphthongs and compound vowels]. *Yuyan yanjiu* 语言研究 [*Studies in Language and Linguistics*], 1984.1:15–22.

Yang, J., and L. Xu. 2017. Mandarin compound vowels produced by prelingually deafened children with cochlear implants. *International Journal of Pediatric Otorhinolaryngology* 97: 143–149.

Yuan, Yue (袁悦), Lingling Gu (顾玲玲), Jingya Chang (常静雅), Xiaojun Zhang (张晓俊), and Zhi Tao (陶智). 2015. Jingyan motai fenjiefa shibie shengdai xirou heshengdai nangzhong de yanjiu 经验模态分解法识别声带息肉和声带囊肿的研究 [Empirical Model decomposition in the research of vocal polyp and vocal cyst]. *Xinxihua yanjiu* 信息化研究 [*Informatization Resaerch*] 2: 27–32.

Zhang, Linjun (张林军). 2010. Muyu jingyan dui liuxuesheng hanyu shengdiao fanchouhua zhiju de yingxiang 母语经验对留学生汉语声调范畴化知觉的影响 [On the influence of native language on the categorical perception of Chinese tones]. *Huawen jiaoxue yu yanjiu* 华文教学与研究 [*TCSOL Studies*] 2: 15–20.

Zhou, Dianfu (周殿福), and Zongji Wu (吴宗济). 1963. *Putonghua fayin tupu* 普通话发音图谱 [*The atlas of sound spectrograms in Putonghua*]. Beijing: shangwu yinshu guan 北京: 商务印书馆 [Beijing: The Commercial Press].

Zhou, Haiyan (周海燕), Xuguang Xu (徐旭光), Guomin Wang (王国民), Zhuang Si (姒壮), and Zhiyong Lv (吕智勇). 2012. E yan bihe buquan huanzhe yichang shengxue tedian de fenxi yanjiu 腭咽闭合不全患者异常声学特点的分析研究 [An analysis of Velopharyneal incompetent' abnormal speech characteristics]. *Kouqiang yixue yanjiu* 口腔医学研究 [*Journal of Oral Science Research*] 28(5): 457–463.

Zhu, Xiaonong (朱晓农). 2004. Jipin guiyihua – ruhe chuli shengdiao de suiji chayi 基频归一化——如何处理声调的随机差异? [F0 normalization: How to deal with between-speaker tonal variations?]. *Yuyan kexue* 语言科学 [*Linguistic Sciences*] 2: 3–19.

——— (朱晓农). 2009. Fashengtai de yuyanxue gongneng 发声态的语言学功能 [Phonation types: Phonological categories and linguistic functions]. *Yuyan yanjiu* 语言研究 [*Studies in Language and Linguistics*] 3: 1–19.

Cognitive Linguistics in China

Quan Wan and Zhong Lin

Contents

Introduction: History of Cognitive Linguistics in China	42
The Chinese Context	44
Features of the Chinese Language	45
Linguistic Research Traditions in China	46
Achievement in Chinese Cognitive Linguistics	48
(Un)boundedness	48
Parts of Speech and Word Categories	49
Metaphor and Metonymy	51
Blending and Haplology	53
Construction Grammar	55
de (的)-Construction	57
Conclusion: Research Features of Chinese Cognitive Linguistics	59
References	60

Abstract

This chapter provides an overview of cognitive linguistics as a field of research in China. It first provides a brief introduction of its history, followed by an analysis of the characteristic features of the Chinese language and illustrations of Chinese as an ideal language to be explored from the perspective of cognition. It then discusses the most important findings by Chinese linguists, including (un)boundedness, parts of speech, metaphor and metonymy, blending and haplology, construction grammar and particles, and so on. It concludes that a distinctive paradigm has been formed in terms of a combination of cognition, pragmatics, and functionalism in the cognitive linguistic study in China.

Q. Wan
Institute of Linguistics, Chinese Academy of Social Sciences, Beijing, China

Z. Lin (✉)
School of Foreign Languages, Chang'an University, Xi'an, China

Keywords

Cognitive linguistics · Chinese language · Chinese grammar

Abbreviations

1	First person
2	Second person
3	Third person
ASSOC	Associative phrase (Associative phrase denotes a type of modification where two noun phrases (NPs) are linked by the particle -de. The first noun phrase together with the particle -de is the associative phrase. The name associative phrase indicates that two noun phrases are "associated" or "connected" in some way; the precise meaning of the association or connection is determined entirely by the meanings of the two noun phrases involved (Li and Thompson 1981: 113).)
CLF	Classifier
COP	Copula
DUR	Durative aspect
EXP	Experiential aspect
GEN	Genitive
NMLZ	Nominalizer (According to Li and Thompson (1981: 113–123, 579–587), Chinese particle de, which is the most commonly used word in Chinese, is mainly used in nominalization (NMLZ), genitive/possessive phrase (GEN), and associative phrase (ASSOC))
PASS	Passive
PFV	Perfective
REDUP	Reduplicative
SG	Singular
S	Clause or sentence
VP	Verb phrase

Introduction: History of Cognitive Linguistics in China

Ten years after its establishment in the 1970s, cognitive linguistics was introduced to and applied in mainland China. In 1982, an interview with George Lakoff and Charles Fillmore by a Chinese scholar was published in *Linguistics Abroad* (renamed *Contemporary Linguistics* in 1998), a prestigious linguistic journal in China, marking the first time Chinese scholars came into close contact with cognitive linguistics. Over the past 30 years, cognitive linguistic research on the Chinese language has undergone rapid development, from initial introduction and application of theory to theoretical retrospection and modification. Today, cognitive linguistics has become a prominent field of research in China, accepted and defended by many grammarians, semanticists, historical linguists, and typologists.

Shen Jiaxuan 沈家煊 (Shen 1985) was the first researcher to apply methods of cognitive linguistics to the study of Chinese in mainland China. He introduced and elucidated such key terms as Figure, Ground, Path, and Point of Orientation (cf. Talmy 1975). Shen (1985) also contrasted the representation of space in Chinese and English from the perspective of subjectivity in cognitive psychology. However, significant as it was, that study failed to receive scholarly attention at the time.

Several years later, *Linguistics Abroad* started to promote cognitive linguistics in China. Editors published a series of translated articles which used the cognitive linguistic approach to analyze the Chinese language. Their authors included Chinese American scholars James H-Y Tai 戴浩一 (Tai 1988, 1990/1991) and Frank F. S. Hsieh 谢信一 (Hsieh 1989, 1991, 1992a, b). Tai's works adopted basic concepts in cognitive linguistics, such as Iconic Motivation, Spatial Metaphor, Non-objectivist Approach, and Symbolic Representation of Conceptualized Reality, and developed the Principle of Temporal Sequence (PTS) and the Principle of Temporal Scope (PTSC). Hsieh systematically argued for the interaction of iconicity and abstractness of temporal expressions in Chinese, based on Ronald Langacker's Cognitive Grammar theory, which includes notions relating to Scene, Image, Schema, Profile, Trajectory, and Landmark. Tai and Hsieh both conducted cognition-based study on Chinese grammar and agreed on different mapping in grammatical structure, because different cultures contribute to different conceptual structures. The research methods used in their papers were new to the linguists in Mainland China and stimulated much interest among Chinese scholars.

Subsequently, *Linguistics Abroad* translated Langacker's (1991) work to introduce cognitive linguistic methodology. It also published several reviews, including Liao Qiuzhong 廖秋忠 (1991), Shen (1993, 1994), and Shi Yuzhi 石毓智 (1995b). They provide detailed explanation of cognitive theories, including John Taylor's Linguistic Categorization, John Haiman's Natural Syntax, Langacker's (1987b, c) Cognitive Grammar, and George Lakoff's (1987) Cognitive Semantics.

These articles inspired many researchers to adopt principles and methods of cognitive linguistics to investigate Chinese syntax and semantics. Several major researches appeared in key Chinese journals, such as *Studies of the Chinese Language* (see Liu Ningshen 刘宁生 1994, 1995; Yuan Yulin 袁毓林 1995; Shi 1995a; Shen 1995a, 1997a, b; and Zhang Min 张敏 1997). Later on, China's cognitive linguistics reached its first peak with the publication of several significant works (Zhang 1998; Shen 1999a; Shi 2000, 2001[1992]), which center on traditional Chinese topics, such as number category, parts of speech, and the construction of *de* (的), showing that these difficult topics can be better analyzed and understood within the framework of cognitive linguistics than that of traditional grammar and structuralism.

Additionally, another group of scholars devoted themselves to theoretical introduction and retrospection: Lin Shuwu 林书武 (Lin 1997) introduced metaphor theory; Wen Xu 文旭 (Wen 1999) introduced several key cognitive research areas; and Wang Yin 王寅 (Wang 2002) discussed the philosophical basis of cognitive linguistics. Many introductory works also started appearing, promoting the learning and research in cognitive linguistics. These included *Cognitive Pragmatics* by Xiong

Xueliang 熊学亮 (Xiong 1999), *An Introduction to Cognitive Linguistics* by Zhao Yanfang 赵艳芳 (Zhao 2000), *Cognitive Linguistics and Metaphor* by Lan Chun 蓝纯 (Lan 2005), *An Introduction to Cognitive Semantics* by Shu Dingfang 束定芳 (Shu 2008), and *Researches on Construction Grammar* by Wang Yin 王寅 (Wang 2011).

Some publishing houses in China also reprinted seminal works on cognitive linguistics, e.g., Langacker (1987c, 1991), Taylor (1989), Sweetser (1990), Ungerer and Schmid (1996), Talmy (2000), and Croft and Cruse (2004). Most of these works are prefaced with a detailed introduction written in Chinese.

Meanwhile, conferences and lectures on cognitive linguistics were frequently organized in China. The First China Cognitive Linguistics Conference was held at the Shanghai International Studies University. George Lakoff, the founding father of conceptual metaphor theory, was invited to deliver lectures at universities in Beijing in 2005 at the invitation of the China International Forum on Cognitive Linguistics (CIFCL). Since then, one internationally renowned scholar has been invited to Beijing each year to deliver ten lectures, their speeches being compiled into books. This initiative has greatly promoted cognitive linguistics in China and led to the founding of the China Cognitive Linguistics Association (CCLA).

Since 2000, cognitive linguistics in China has continued, with the publication of hundreds of academic works and thousands of articles, covering a great many topics, such as prototype, schema, mental space, concept integration, metaphor and metonymy, reference point, boundedness, and construction, among which metaphor is an ever-lasting "hot topic." In particular, research focus began to turn to construction grammar, and researchers tried to combine cognitive theories with other principles from rhetoric, sociolinguistics, computational linguistics, corpus linguistics, discourse analysis, second language acquisition, translation and interpretation, and even cultural studies and literary studies.

In recent years, despite the slightly reduced number of publications, cognitive linguistic research has been exploring more profound issues concerning the Chinese language. New insights are offered to explain some classic topics in Chinese linguistics, such as grammatical metaphor and metonymy (Shen 2006a), agent-patient relation and constructional semantics conducted by Zhang Bojiang 张伯江 (Zhang 2009), co-occurrence of monosyllables and disyllables in Mandarin Chinese by Ke Hang 柯航 (Ke 2012), nouns and verbs (Shen 2016), construction of the particle *de* in Chinese by Wan Quan 完权 (Wan 2016), and run-on sentences (流水句, liúshuǐ jù, or sentences in progress) in Chinese (Shen 2017a). In particular, Shen (2016) proposes an inclusive model of Chinese verbs and nouns, which deserves attention for further exploration.

The Chinese Context

The development of cognitive linguistics as a new branch of linguistic research grew out of its founders' dissatisfaction with generative linguistics. Yet in the Chinese context, the acceptance of cognitive linguistic theories and approaches seems natural, owing to the inherent features of Chinese and the Chinese research traditions. This section briefly discusses these features and traditions.

Features of the Chinese Language

As a typical isolating language, Chinese relies on analytic devices of functional words rather than on those of morphological markers. This feature makes it somewhat difficult for structural linguists to analyze Chinese grammar. Yet, cognitive linguistics provides a key to dealing with this difficulty. For example, traditional Chinese grammarians recognize that Chinese is syntactically parataxis, which depends on inherent meanings of words and logic to construct sentences. This feature is closely associated with the Chinese way of thinking. The following example suffices to show that, without personal pronouns and conjunctions, a sentence can still express meaning.

(1) 需要 帮忙 的 事, 说 一 声
xūyào bāngmáng de shì, shuō yī shēng.
need help NMLZ matter, say one CLF
'(If you) need help, tell (me).'

It is agreed that parataxis in Chinese is related to the holistic Chinese way of thinking. However, how parataxis works remained a puzzle for traditional grammar, until Concept Integration Theory, proposed by Fauconnier and Turner (2003), was used to analyze it. Using the theory, Shen (2006a) presents the analysis of (2) as a process of concept integration, as shown in (3):

(2) 他 被 后面 的 司机 按了 一 喇叭。
tā bèi hòumiàn de sījī àn-le yī lǎbā
3SG PASS behind ASSOC driver punch-PFV one horn.
'He was beeped by the driver behind.'

(3) 他被 后面 的 司机 警告了 一 下 + 后面 的 司机 按了 一 喇叭
tā bei hòumiàn de sījī jǐngào-le yī xià + hòumiàn de sījī àn-le yī lǎbā
3SG PASS behind ASSOC driver warn-PFV one CLF + behind ASSOC driver punch-PFV one horn
'The driver behind him gave him a warning.' + 'The driver behind him pressed the horn.'

According to Jiang Lansheng 江蓝生 (Jiang 2008), such a process of concept integration can be easily understood. It is also applicable in other linguistic phenomena, such as "*chādiǎn'r* (差点儿, almost)+VP=*chādiǎn'r měi* (差点儿没, almost not) + VP" and "VP + *zhīqián* (之前, before)=*měi* (没, not) VP + *zhīqián* (之前, before)." In both constructions, the expressions on both sides, although seemingly contrary in form, actually mean the same.

Some features of the Chinese language are also useful in discovering areas not easily found in Indo-European languages. For example, syntactic iconicity is represented prominently in Chinese in many ways, one of which is the Principle of Temporal Sequence, as explained in (Tai 1988). This principle explains the sequence of both clauses (4) and verb phrases (5).

(4a) 我 吃过 饭(S_1), 你 再 打 电话 给 我(S_2)。
wǒ chī-guò fàn, nǐ zài dǎ diànhuà gěi wǒ.
1SG eat-EXP meal, 2SG again make call BEN 1SG
'Call me (s_2) after I have finished dinner (s_1).'
*你再打电话给我(S_2), 我吃过饭(S_1)。

(5) 我 吃过 饭(VP_1)再 打 电话 给 你(VP_2)。
wǒ chī-guò fàn zài dǎ diànhuà gěi nǐ
1SG eat-EXP meal again make call BEN 1SG
'I will call you (VP_2) after finishing dinner (VP_1).'
*我再打电话给你(VP_2) 吃过饭(VP_1)。

The above two sentences serve as good examples to illustrate the explanatory power of cognitive linguistics.

Linguistic Research Traditions in China

Cognitive linguistics rests upon the philosophical idea of embodiment. That is, thoughts are derived from interactions between humans and the outside world and from experiences gained through physical contact with it. This cognitive perspective is compatible with mainstream philosophical notions rooted in ancient China. The traditional Chinese notion of *géwùzhìzhì* 格物致知 ("the extension of knowledge through investigation of things") addresses epistemology, which had a far-reaching impact on Chinese scholars (Fung 1966: 305). Marxist philosophy, a significant strain in Western thought, which stresses that "Genuine knowledge comes from practice," influences modern Chinese people deeply (Tian 2009). All of these notions paved the way for Chinese researchers to accept cognitive experientialism.

Cognitive linguistics does not consider language as a separate cognitive domain or syntax as a separate language module. This means, from the perspective of cognitive linguistics, there are no clear boundaries between syntax, semantics, and pragmatics. Moreover, cognitive linguistics pays close attention to meaning, and attempts to study linguistic structure from the standpoint of semantics. In fact, similar integrated approaches were adopted by Chinese linguists in ancient China, and the studies of grammar, semantics, and pragmatics were never separated, but rather they were brought together under the name of philology训诂学. In that tradition, a major task was to interpret meanings. Important works include conceptually oriented *Erya*《尔雅》, a semantic classification dictionary, and the usage-oriented *Maoshi guxun zhuan*《毛诗故训传》, an annotation of classical poems, both produced in the Han Dynasty (202 BC–220 AD). Both traditions pay special attention to functional words, particularly their meanings in context, a representative work *Zhuyuci*《助语词》(*Particles*) by Lu Yiwei 卢以纬 in Yuan Dynasty (1271 – 1368 CE). The attention given to functional words is resonant with the usage-based grammar proposed by cognitive linguistics, which continues to this day. For example, *Xiandai Hanyu Babai Ci*《现代汉语八百词》(*Eight Hundred Words in Modern Chinese*), the classic work on modern Chinese grammar, is a dictionary centering on the meanings and usage of function words. Also Chao Yuen-Ren 赵元任 (Chao 1968) classifies words according to meaning and argues that demarcations of grammatical categories are relative. These views have influenced Chinese linguistics, and contemporary Chinese linguists find the "integrative nature" of cognition easy to accept.

Cognitive semantics focuses on conceptual structure and its formation. It places emphasis on subjectivity. When viewed from different angles by two persons, the same object may result in two different mental images, and images in language may vary according to different construals – profiling, perspective-taking, and level of abstraction – and eventually lead to different grammatical forms. The following line from a Chinese poem captures this well, "*héngkàn chénglǐng cè chéngfēng*" 横看成岭侧成峰, which literally means "From the side, a whole range; from the end, a single peak; Far, near, high, low – no two parts alike." This linguistic view that the subject and the object are integrated is inherent in the Chinese language. For this reason, Chinese scholars generally accept the idea of the "subjectivity of language."

Construction grammar, which is an emerging paradigm in cognitive linguistics, focuses on meanings beyond structural ones. Useful in the explanation of Chinese-specific grammatical patterns, it became popular shortly after being applied in the analysis of the Chinese language. As pointed out by Zhang (2018), Chinese grammarians have never neglected the concept of construction. Many scholars, such as Ding Shengshu 丁声树 (Ding 1935) and Lü Shuxiang 吕叔湘 (Lü 1941), have attempted to explore the hidden grammatical elements underlying a particular structure. The history of "disposal construction" studies can illustrate this tradition. Wang Li 王力 (1943) argued that the *ba* (把, disposal)-construction, which previous studies simply regarded as a syntactic device for placing the direct object before the verb, actually conveys a different meaning. A sentence without *ba* is just a normal narration, while the *ba*-construction conveys an additional sense of disposition or manipulation. Such a disposal reading does not come from the word *ba* itself; rather it results from the whole *ba*-construction. From then on, sentences with the preposition *ba* have been called disposal construction or *ba*-construction.

Decades of research in Chinese grammar have fully described many Chinese-specific grammatical patterns, resulting in many traditional topics and perennial ones, such as the *ba*-construction, the *Zai* (在, exist)-construction, and the ditransitive construction. All those grammatical constructions have been analyzed using the construction grammar theory, which has become a popular and useful tool for analyzing Chinese linguists and is gradually occupying a prominent place in cognitive linguistic research.

Achievement in Chinese Cognitive Linguistics

(Un)boundedness

Langacker (1987a) argues that entity and action, which are conceptually represented by NP and VP, respectively, have a distinction between boundedness and unboundedness. The linguistic representation is that bounded individual nouns (e.g., apple) can follow the article a/an, while unbounded mass nouns (e.g., water) cannot. Likewise, bounded action verbs (e.g., jump) are durative, while unbounded state verbs (e.g., like) are not.

(6)

	Unbounded	Bounded
Noun	Water/*a water	Apple/an apple
Verb	Like/*He is liking it.	Jump/He is jumping on it.

Shen (1995a, 2004b) further points out that in Chinese, boundedness also pertains to adjectives in terms of description of properties. For example:

(7)

Unbounded adjectives	Bounded adjectives
红 hóng red	红彤彤 hóng tóng tóng red REDUP
红 脸 hóng liǎn red face	*红彤彤 脸 * hóng tóng tóng liǎn
*红 一 张 脸 *hóng yī zhāng liǎn	红彤彤 一 张 脸 hóng tóng tóng yī zhāng liǎn red　　　　　one CLF　face a red face
干净 gān jìng clean	干干净净 gàn gān jìng jìng clean REDUP
干净 衣服 gān jìng yīfu clean dress	*干干净净 衣服 *gàn gān jìng jìng yīfu
*干净 一 件 衣服 *gān jìng yī jiàn yīfu	干干净净 一 件 衣服 gàn gān jìng jìng yī　jiàn yīfu clean　　　　　one CLF　dress a clean dress

Chinese adjectives possess two morphological forms: a simple form and a reduplicative form. The former is unbounded in nature and the latter is bounded. Furthermore, while the former refers to property, the latter depicts manner. This is rarely seen in English adjectives, except when *red* and *reddish* are considered. Yet in Chinese, this morphological distinction is so evident and systematic that they deserve further attention. In contrast, Chinese nouns and verbs lack morphological changes. This feature has inspired Shen to investigate Chinese word categories and led to this significant finding concerning Chinese adjectives (Shen 2016).

Parts of Speech and Word Categories

In the history of Chinese grammatical research, there have been several major debates about the determination of parts of speech. But researchers have reached little consensus. Technically speaking, this is due to the lack of morphological markers in Chinese. But from a theoretical point, methods in linguistic research need to change, too. In this context, the proposal of prototypical categorization developed in cognitive linguistics has revolutionized the field.

The difficulty of demarcating parts of speech in Chinese lies in ambiguity and the indefinite number of categories, for which structuralism is incapable of offering a satisfactory explanation. For example, Zhu Dexi 朱德熙 used the degree adverb *hěn* (很, very) to distinguish verbs from adjectives (Zhu 1982). In his theory, words that are modified by *hen* but cannot be followed by objects are adjectives; words that are not modified by *hen* but can be followed by objects are verbs. However, exceptions abound, as shown in (8) and (9):

(8) Adjectives that cannot be modified by *hěn*:
shàbái (煞白, pale), *bīngliáng* (冰凉, freezing cold), *tōnghóng* (通红, blushed), *jīngguāng* (精光, empty). . . . . .

(9) Adjectives followed by objects:

红了脸	空着手
hóng-le liǎn	*kōng-zhe shǒu*
red-PFV face	empty-DUR hand
"face blushes"	"empty-handed"

瞎了眼睛	高你一头
xiā-le yǎnjīng	*gāo nǐ yī tóu*
blind-PFV eyes	tall 2SG one head
"eyes become blind"	"one head taller than you"

In Chinese, nearly all kinds of distributional features of parts of speech are challenged. Moreover, Chinese parts of speech do not correspond to syntactic properties, as shown in the following figure (adapted from Zhu 1985):

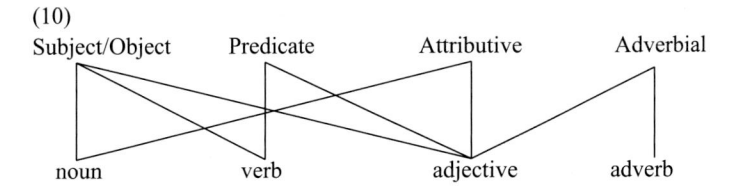

(10)
Subject/Object   Predicate   Attributive   Adverbial

noun   verb   adjective   adverb

As Yuan (1995: 166) points out, this seemingly complex mapping is due to the fact that feature-based categories are not applicable to Chinese. From the prototype category view of cognitive linguistics, the definition of part of speech should be based on the typical members of a part of speech. For example,

(11) a. Nouns are normally used as typical subjects or objects and are generally not modified by adverbs.

    b. Verbs are normally used as predicates, but they are not modified by degree adverbs, or they can still be followed by objects while being modified by degree adverbs.

    c. Adjectives are normally used as predicates and complements and cannot be followed by objects while being modified by adverbs of degree.

Based on the prototype view of parts of speech, Yuan (2009) selects a set of distributional features of the typical members of function words in Chinese, and assigns different weight value to them according to their importance to the related word class, and then checks and corrects them, and finally constructs a set of distribution scales for measuring the degree of membership of these word classes. This approach highlights the fuzziness of category boundaries, but the analysis process is very complicated.

From another point of view, one can see an elegant picture. The fact is that Chinese verbs and adjectives can conventionally be both predicate and subject/object without any morphological changes. As Zhu (1985: 5) writes: "that is an important feature in Chinese, different from European languages." Zhu (1985:7) cites the following examples of English-Chinese comparison to reflect this feature.

12) a. He flies a plane.	他 开 飞机。
	tā kāi fēijī
	3SG fly plane
	"He flies a plane."
b. To fly a plane is easy	开 飞机 容易
	kāi fēijī róngyì
	fly plane easy
	"To fly a plane is easy."

The following examples illustrate contrastively that both nouns and verbs can be used as objects.

| (13) a. 我 想 家$_n$, 还 想 吃$_v$。 |
| wǒ xiǎng jiā, hái xiǎng chī |
| 1SG miss family also want eat |
| 'I miss my family, and I also miss food (at home).' |
| b. 我 怕 爸$_n$, 是 怕 打$_v$。 |
| wǒ pà bà shì pà dǎ |
| 1SG scare father, COP scare beat |
| "I am scared of my father because I'm afraid he'll beat me." |

Following the principle of economy in theory, Shen (2007, 2016, 2017c) has proposed a new method for explaining those linguistic phenomena, namely, the inclusion model of Chinese parts of speech. While noun and verb are in exclusive opposition in English, in Chinese they are in inclusive opposition, as shown in (14):

```
   _____         _____
  /     __    \       /           \
 | noun(  )verb|     | noun ( verb )|
  \___/  \____/       _____/
      English              Chinese
```

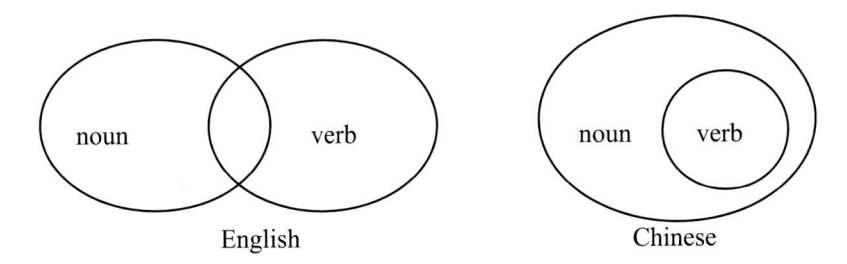

In this model, Chinese verbs are a sub-type of nouns for expressing dynamic meanings. Shen (2017b) argues that while nouns and verbs are two separate syntactic classes in Indo-European languages, in Chinese they are a pair of pragmatic classes (reference and predication), and not in opposition to each other. Chinese nouns constitute a super-noun category, with verbs as its sub-category. In other words, all verbs in Chinese are actually verbal nouns. This characteristic feature of Chinese can be explained from the perspectives of cognition, language typology, grammaticalization, philosophical background, and experimental studies of word classes. Furthermore, the semantic basis for inclusion is privative opposites (Trubetzkoy 1958) instead of parallel relations. This is also a further development of John Lyons's notion of class inclusion (see Lyons 1977: 156).

It is asserted that Westerners and Chinese have different concept of category, for they view category relations differently (Shen 2017b). In the West, people stress that two categories exist only when A and B are separate, whereas Chinese speakers stress that two categories can exist when A contains B. Logically, the separation of A and B may result in their intersection, but if the relation between A and B is inclusive, no intersection is possible, as B becomes a subset of A. Separation is analogous to dispersion, but inclusion differs from continuity; although inclusiveness exists within the category of continuity, the two poles remain separate and are not in an inclusive relationship.

Metaphor and Metonymy

Given that Chinese verbs belong to the category of nouns, nominalization of verbs does not exist. In his work *Shiming* (《释名》, *Interpretation of Names*), Liu Xi from East Han Dynasty (202 BCE–220 CE) considered names as designation of both objects (e.g., "sky," "ground," "mountain," "water") and actions and states (e.g., "run," "walk," "watch," "listen").

According to the concepts of cognitive linguistics, an abstract activity can be understood as a concrete thing through ontological metaphor, namely, actions and states are entities. Lakoff and Johnson (1980: 26, 31) illustrate ontological metaphor as follows.

(15) INFLATION IS AN ENTITY.
RUNNING AS A SUBSTANCE

It is worth noticing that in the above examples, both INFLATION and RUNNING are nominal. However, in Chinese, *péngzhàng* (膨胀, inflation/inflate) and *pǎo* (跑, running/run) are both nouns and verbs because Chinese verbs are a subcategory of nouns. For speakers of Chinese, actions and states are composed by concepts of entity (Zhu 1985), without the process of concept realization. They are originally entities without any morphological change or inflection. The ontological metaphor in Chinese is as follows.

(16) INFLATE IS AN ENTITY.	(膨胀　　是 一 个 实体。)
	péngzhàng shì yī gè shítǐ
	inflate COP one CLF entity
RUN AS A SUBSTANCE	(跑 作为　一 个 实体)
	pǎo zuòwéi yī gè shítǐ
	run as　one CLF substance

It is because the speaker in Chinese-style ontological metaphor directly regards actions and states as entities that the misuse of nouns and verbs/adjectives accounts for 68% of all errors among Chinese learners of English (Shen 2018). According to the statistics provided by Xia Lixin 夏立新 (Xia 2015), among the gerund-related errors made by Chinese learners of English, the most frequent category is the misuse of original form of verb where gerund or present particle should be used. (e.g., * *Do anything is the same as learning English*). To better understand the feature, it seems worthwhile to quote a line from a very popular song, "jiu zhege feel beier shuang" 就这个feel 倍儿爽 ("The feeling is just awesome"). Inferring from the whole lyric, the meaning of the word "feel" is "feeling," not "touch" or "impression." In this line, the original form of the English verb "feel" is used as the subject with the demonstrative adjective zhe (这, this). In this case, the English verb is used as a noun in the Chinese sentence, which is conceptual as well as a metaphor of linguistic form. This finding is another contribution of Chinese language to cognitive metaphor.

There are many other valuable findings on syntactic metaphor in Chinese, for example, the semantic mapping of the conceptual domains of acting, knowing, and uttering in Chinese compound sentences (Shen 2003); mapping of Chinese constructions "*yī* (一, one) ……*bù* (不, no)" and "*zài* (再, again) ……*yě* (也, too)" between syntactic, semantic, and pragmatic domains (Shen 1995b); and mapping between form and meaning of similar semantic structure of *tōu* (偷, steal) or *qiǎng* (抢, rob) with different conceptual structures (Shen 2000). Likewise, Yuan (2004) and Zhou Ren 周韧 (2011) argued that scope adverbs of *mǎn* (满, full) and *quán* (全, complete), despite their similar meaning, are used differently due to differences in CONTAINER and SET metaphors.

There are also fruitful studies on metonymy. For example, metonymy models explain the lack of the head in the "*de* (的, GEN/NMLZ/ASSOC) construction," (Shen

1999b) which is referred to as "transferred designation" (Zhu 1983). In most cases, the explanation works; but in others it does not:

(17) a.经理 的	(= 经理 的 外套)
jīnglǐ de	jīnglǐ de wàitào
manager GEN	manager- GEN coat
"manager's"	"manager's coat"
b.*经理 的	(= 经理 的 身份)
jīnglǐ de	jīnglǐ de shēnfèn
manager GEN	manager GEN identity
"manager's"	"as a manager"

(18) a.灰姑娘 的	(= 灰姑娘 的 水晶鞋)
huīgūniáng de	huīgūniáng de shuǐjīngxié
Cinderella GEN	Cinderella GEN glass slipper
"Cinderella's"	"Cinderella's glass slipper"
b.*灰姑娘 的	(= 灰姑娘 的 故事)
huīgūniáng de	huīgūniáng de gùshi
Cinderella GEN	Cinderella GEN story
"Cinderella's"	"Cinderella story"

As argued in Shen (1996b), in a cognitive framework of mental gestalt structure, salient entities can refer to unsalient ones via metonymy. Possessor-Possessed is a common cognitive framework, and therefore the concepts in (17a) and (18a) can be established. However, "manager" and "identity" cannot be thought of as being in one cognitive framework, and therefore (17b) and (18b) cannot be established.

Metonymy is also used to account for other linguistic phenomena, such as the lexicalization and grammaticalization of *búguò* (不过, just) (Shen 2004a) and the rhetorical phenomenon of Chinese "continuous metonymy" (Wan 2009).

Blending and Haplology

It is generally agreed that Chinese word formation is largely compounding-based and one can never rely on the components to decipher the meaning of compounds. For example, *diànchē* (电车, trolleybus) is not equal to *diàn* (电, electricity) + *chē* (车, vehicle). Structuralism has long held that Chinese sentence formation is basically the same as that of phrases (Zhu 1985), that is, simple Chinese sentences are constructed through compounding, too. However, the construction of compounding sentences had been neglected, until it was explained with the concepts of blending and haplology by Shen (2006b, c), who drew inspiration from the Concept Integration Theory proposed by Fauconnier and Turner (2003).

A classic sentence in Chinese grammar is "*Wáng Miǎn sǐle fùqin*" (王冕死了父亲; lit. "Wang Mian died his father"). In this sentence, the intransitive verb *sǐ* (死, die) is followed by an object. Furthermore, this sentence does not mean simply

"Wang Mian's father has died"; it includes a sense of loss. In spite of various analyses, a consensus had not been reached before Shen (2006c) put forward the analogical blending in grammatical metaphor for the analysis. When a new meaning needs to be expressed, while there is no ready-made expression, it could be compounded by conceptual blending to form a new expression. The compound process conforms to the following matrix.

(19) S_a	S_b
S_x	— ←S_{xb}
S_a. 王冕　的　某物　丢了。	S_b. 王冕　丢了　某物。
Wáng Miǎn de mǒuwù diūle.	Wáng Miǎn diūle mǒuwù.
Wáng Miǎn GEN something lost-PFV	Wáng Miǎn loose-PFV something.
"Wang Mian's possession lost."	"Wang Mian lost his possession."
S_x. 王冕　的　父亲　死了。	S_y. — ←S_{xb} 王冕　死了 父亲。
Wáng Miǎn de fùqin sǐle	Wáng Miǎn sǐle fùqin
Wáng Miǎn GEN father die-PFV	Wáng Miǎn die-PFV father
"Wang Mian's father died."	"Wang Mian lost his father."

In (19), the horizontal and vertical axes share similarities; so there exists syntactic metonymy. Structures of S_b and lexical items of S_x combine to form a new sentence, S_y. This blending process also causes the emergent meaning of "loss/gain," which is the same as that of Chinese phrase formation, as shown below:

(20) a.火	b.火车
huǒ	huǒchē
fire	railway
x.电	y. (electric vehicle) ←xb 电车
diàn	diànchē
electricity	trolleybus

The other method of sentence formation is haplology, which is mainly used in syntactic metonymy, as was shown in Examples (2–3). A brief transcription is as follows:

(20) 他被后面的司机警告了一下 + 后面的司机按了一喇叭
＝ 他被后面的司机．．．．．．．．．．．．．．．．．．．．．．．．．．．．．．按了一喇叭。
"The driver behind him gave him a warning." + "The driver behind him pressed the horn."
"The driver behind him pressed the horn to warn him."

Blending and haplology often work together in grammatical constructions. Ye Jianjun 叶建军 and Yue Yao 乐耀 argued that the compounding of sentence formation can also account for other Chinese structures, such as analogous construction in contemporary Chinese (Ye 2008) or in modern Chinese (Yue 2014).

The above analysis uses concept blending, which is based on grammatical metaphor and metonymy. It adopts key principles of the conceptual integration theory, but does not mechanically apply its mental space framework. This is because of the inherent features of the Chinese language. Liu Tanzhou 刘探宙 (2018) sketched the following diagram strictly following Fauconnier's model.

(21) 王冕　死了　　父亲。
　　 Wáng　Miǎn sǐle　fùqīn.
　　 Wáng　Miǎn die-PFV father
　　 "Wang Mian lost his father."

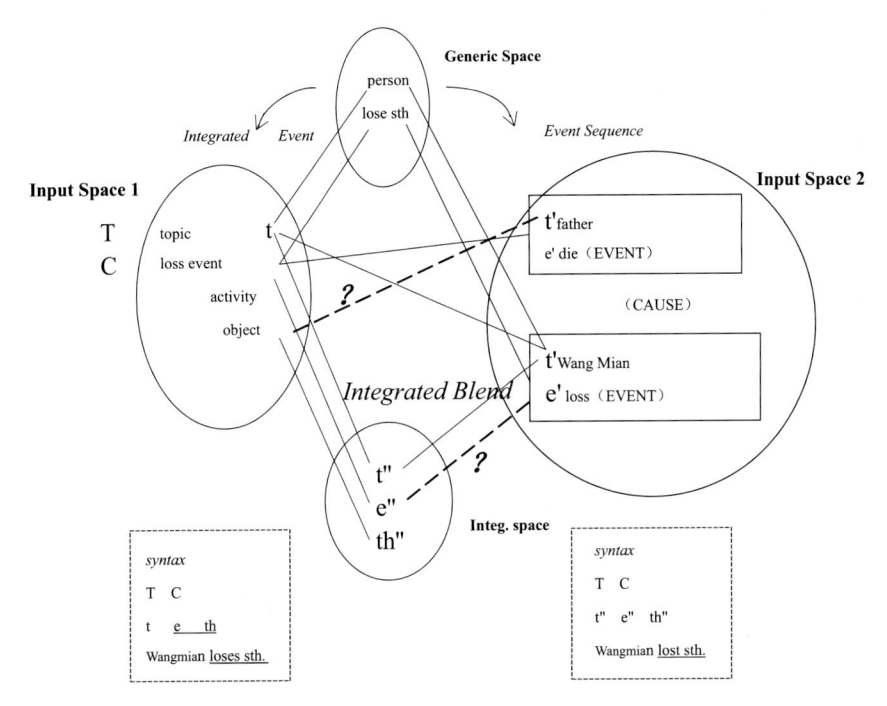

[Diagram: Conceptual blending model with Generic Space (person, lose sth), Input Space 1 (T, C; topic, loss event, activity, object), Input Space 2 (t' father, e' die (EVENT), (CAUSE), t' Wang Mian, e' loss (EVENT)), Integrated Blend (t", e", th"), and two Integ. space boxes showing syntax T C / t e th "Wangmian loses sth." and syntax T C / t" e" th" "Wangmian lost sth."]

Liu (2018) maintains that when Fauconnier's model is adopted, considering the inheritance of emergent structures, "*fùqīn* (父亲, father)" cannot be inherited by the new structure, or the new meaning cannot emerge. She concludes that this dilemma is ultimately due to differences between the topic-comment structure in Chinese and the argument structure in English.

Construction Grammar

The above example of "*Wáng Miǎn sǐle fùqīn*" (王冕死了父亲, "Wang Mian lost his father") can also be regarded as an example of construction grammar, since construction grammar endeavors to interpret the meaning of construction, which is independent of the words in the sentence (Goldberg 1995: 1). Given that construction grammar is advantageous to deciphering meanings in Chinese-specific syntactic structures and that there are many traditional topics related to the meaning of syntactic structure in the study of Chinese grammar, it has been readily accepted and has become popular in Chinese cognitive linguistics.

Zhang (1999) was among the first scholars who introduced construction grammar to China. He cited examples of ditransitive construction and argued that in Chinese, the function of construction is greater than that of valency requirements for verbal semantics. This means that ditransitive constructions are able to express the meaning of "giving" without the verb "give" or even with verbs which do not contain the meaning of "giving":

(22)单位 分了 我 一 套 房子。
dānwèi fēnle wǒ yī tào fángzi
organization arrange-PFV me one CLF apartment
"My organization allocated an apartment to me."

(23)老王 答应 我 两 张 电影 票。
Lǎowáng dāying wǒ liǎng zhāng diànyǐng piào
Laowang promise me two CLF movie ticket
"Laowang promised me two movie tickets."

Even in idiomatic expressions, when a speaker intends to express the idea of "giving" and employs the ditransitive construction, the syntactic requirements have to be met. Therefore, in Example (24), a quantifier is placed before a non-nominal word, and in Example (25), there is a dummy recipient of verbs, i.e., *tā* (它, it).

(24)打 他 一 个 冷不防。
dǎ tā yī gè lěngbufang
beat him one CLF unawares
"Take him unawares."

(25)玩 它 个 痛快。
wán tā gè tòngkuài
enjoy it CLF satisfied
"have a heavenly time."

Zhang's use of construction grammar theory proves that construction meaning is the basis of Chinese syntax, which originates from the research traditions established by Lü (1948), and 10 years later found its resonance with Western construction grammar. A thorough explanation of this can be found in Shen (2000).

Proponents of construction grammar do not believe in universal word categories or grammatical relations. Rather, they argue that construction grammar is language-specific and construction is the minimum unit of grammar, which defines grammar relations in a language (Croft 2001). This view is shared by Chinese linguists, who set out to study constructions specific to Chinese. It also encouraged Chinese researchers to move away from perspectives based on Indo-European languages.

3 Cognitive Linguistics in China

For example, Shen (1999c) finds that the "*zai* (在, at/in)-construction" and the "*gei* (给, give)-construction" are parallel in syntax and semantics. Also Shen (2002), starting from the meaning of "subjective disposal," argues that "*ba* (把, disposal)-construction" and its meaning can only be understood by applying the notions of construction grammar. Liu (2005) finds that in the "*lian* (连, even)-construction," even when the contrastive topic is not present, the influence of topic structure persists. This indicates the dominant role of topic structure in Chinese syntax. Biq Yung-O 毕永娥 (Biq 2007) explores the interactions between construction and meaning of Chinese indefinite quantifiers in aspects of the distribution and evolution of such constructions in spoken Chinese.

de (的)-Construction

Cognitive linguistics is useful for analyzing Chinese function words such as *de* (的, ASSOC/GEN/NMLZ), which is put between the modifier and the head noun. According to Li and Thompson (1981: 113–123, 579–587), the modifier can be a nominal phrase in an associative phrase (ASSOC) or a genitive/possessive phrase (GEN) and a relative clause or an attributive adjective in a modifying phrase (NMLZ), in which *de* can be called a nominalizer.

It is not only the most commonly used function word but also one of the most difficult ones for researchers. It has been analyzed in different grammar theories. Decades of research failed to offer a convincing explanation until cognitive linguistics brought new insights.

Zhu (1956) pointed out the following minimal pairs but did not give any explanation:

(26) 聪明　　孩子	*聪明　　动物	聪明　　的 动物
cōngmíng háizi	cōngmíng dòngwù	cōngmíng de dòngwù
clever children	clever animals	clever NMLZ animals

These facts were explored by Zhang (1998) from the perspective of iconicity of distance. He holds that people normally use the phrase *cōngming* (聪明, clever) to describe children. Consequently, there is a high degree of integration between the attributive and the head. However, people seldom use *cōngming* (聪明, clever) to describe animals. So there is a big conceptual distance between people and animals. If "animal" and "clever" are put together, the structure is not stable if a "*de*" is not inserted to match the form to the meaning. This accounts for the acceptability of "*cōngming de dòngwù*" (聪明的动物, clever animals).

Shen and Wang (2000) further maintained that the inserted *de* makes the noun phrase a reference-point construction, because the attributive requires a higher level of informativity (Taylor 1994: 225), accessibility, and salience than the head word. More examples are given in (27) and (28):

(27)教堂　旁边　的 自行车	*自行车 旁边　的 教堂
jiàotáng pángbiān de zìxíngchē	zìxíngchē pángbiān de jiàotáng
church nearly ASSOC bicycle	bicycle nearby ASSOC church
the bicycle next to the church	* the church next to the bicycle

书桌　　的 抽屉	*抽屉　的　书桌
shūzhuō de chōuti	chōuti de shūzhuō
desk ASSOC drawer	drawer ASSOC desk
drawer of the desk	* desk of the drawer

(28)我们　抵抗　敌人。→	我们　的 抵抗 →	*敌人 的　抵抗
wǒmen dǐkàng dírén	wǒmen de dǐkàng	dírén de dǐkàng
1 PL resist enemies	1PL GEN resistance	enemies GEN resistance
we resist enemies.	our resistance	*enemies' resistance
我们　消灭　敌人。→	*我们　的 消灭 →	敌人 的 消灭
wǒmen xiāomiè dírén	wǒmen de xiāomiè	dírén de xiāomiè
1PL eliminate enemies	1PL GEN elimination	enemies GEN elimination
we eliminate enemies.	* our elimination	enemies' elimination

After the application of the reference-point construction to the *de*-construction, it was further adopted by Wan (2016) to study texts. Also using Langacker's (2008) Grounding theory, Wan (2016) explains the linguistic phenomenon of attributive shifting, which was first proposed by Chao (1968, 286):

(29)
a. 那 位　戴 眼镜儿 的 先生　　是 谁?
nà wèi dài yǎnjìngér de xiānsheng shì shéi
that CLF wear glasses NMLZ gentleman is who
"Who is that gentleman (who incidentally is) with glasses?" ("with glasses" is used descriptively.)

b.
戴 眼镜儿 的 那 位 先生　　是 谁?
dài yǎnjìngér de nà wèi xiānsheng shì shéi
wear glasses NMLZ that CLF gentleman is who
"Who is that gentleman who is wearing glasses (not the one who is not wearing glasses)?" ("with glasses" is used restrictively.)

Changes in the order of the deictic attributive and the *de*-construction lead to changes of reference-point construction (Langacker 1993) in the noun phrase. Evidently, the two different structures need to be used in different contexts, depending on different models of cognitive grounding. One is deictic grounding, as shown in Example (29a) for the type of text with a new reference point and a new target, in which the deictic attributive precedes the *de*-attributive. The other is retrospective grounding, as shown in Example (29b) for the reoccurrence of the deictic word appearing in previous texts, in which the *de*-attributive is ahead of the deictic attributive.

Another classic topic also concerns multiple attributives, as explained by Yuan (1999).

(30) 小 的 瓷 碗儿	*小 瓷 的 碗儿
xiǎo de cí wǎnr	xiǎo cí. de wǎnr
small NMLZ porcelain bowl	small porcelain NMLZ bowl
small porcelain bowl	* small porcelain bowl

瓷 的 小 碗儿	*瓷 小 的 碗儿
cí de xiǎo wǎnr	cí xiǎo de wǎnr
porcelain NMLZ small bowl	porcelain small NMLZ bowl
porcelain small bowl	* porcelain small bowl

To explain this, Yuan (1999) argues that there are few binary oppositions for *xiǎo* (小, small) (size: big, small), whereas there are many more binary oppositions for *cí* (瓷, porcelain) (material: iron, silver, wood, porcelain). Consequently, *xiǎo* (小, small), with less processing difficulty, appears ahead of other elements; and *cí* (瓷, porcelain), which requires more processing effort, is placed after *de*. Coincidentally, in explaining how sentences are made in Chinese, Lu Bingfu 陆丙甫 (1993, 2008) also uses a similar principle, known as the Identifiability Precedence Principle, which shares the same cognitive principle underlying the formation of Chinese phrases. Lu's principle falls under the category of iconicity.

Conclusion: Research Features of Chinese Cognitive Linguistics

It is evident that the theories that have influenced Chinese cognitive linguistics include Langacker's Cognitive Grammar, Talmy's Cognitive Semantics, Lakoff's Cognitive Metaphor, Fauconnier's Concept Integration, and Goldberg's Construction Grammar.

Over the three decades of absorption, internalization, and development, China's cognitive linguistics has gradually developed its own characteristics. It focuses on the explanation of linguistic facts instead of hastily proposing theoretical frameworks and adopts different approaches and tools, as long as they have explanatory power, rather than relying solely on one particular theory. It has developed an approach that integrates cognition, language use, and language function. This is clearly exhibited in the two volumes of *Function, Pragmatics and Cognition of Modern Chinese Grammar* (Shen 2005; Zhang 2016).

However, in order to make further progress, it is necessary to develop general linguistic theories. Particularly, it is important to systematize theories for the Sinitic languages and draw insights from traditional Chinese philosophy so as to make due contributions to general linguistics. For this purpose, Shen (2017b) made an enlightening exploration, in which he examined the different criteria for Chinese native speakers and Westerners in construing the relationship between two conceptual categories (e.g., category A and category B). He proposed that in Western classical theory of categorization, two categories exist only when they are separate, whereas for Chinese native speakers, the case holds when one contains the other. Differences between the two views on the construction of category render a better explanation for Chinese grammatical categories, as well as a better interpretation of differences between Chinese native speakers and their Western counterparts in thought, behavior, and culture. It can be expected that such a view on the construction of category may inspire more cross-linguistic and cognitive linguistic studies.

References

Biq, Yung-O (毕永娥). 2007. Budingliangci ciyi yu goushi de hudong 不定量词词义与构式的互动 [Chinese measure words for unspecified quantities: Interaction between lexical meaning and construction]. *Zhongguo Yuwen* 中国语文 [*Studies of the Chinese Language*] 6: 507–515.

Chao, Yuen Ren (赵元任). 1968. *A grammar of spoken Chinese*. Berkeley/Los Angeles: University of California Press.

Croft, William. 2001. *Radical construction grammar: Syntactic theory in typological perspective*. Oxford/New York: Oxford University Press.

Croft, William, and D. Alan Cruse. 2004. *Cognitive linguistics: Conceptualization and construal operations*. Cambridge: Cambridge University Press.

Ding, Shengshu (丁声树). 1935. Shi foudingci "fu" "bu" 释否定词"弗""不" [The particles of negation *fu* and *bu*]. In *Qingzhu Cai Yuanpei Xiansheng Liushiwu Sui Lunwenji* (xiace) 庆祝蔡元培先生六十五岁论文集 (下册) [*A collection of writings in honour of Yuanpei Cai on his 65th birthday, Part 2*], 967–996. Guoli Zhongyang Yanjiuyuan Lishi Yuyan Yanjiusuo Jikan Waibian Diyi Zhong 国立中央研究院历史语言研究所集刊外编第一种 [Supplementary issue of the bulletin of institute of history and philology, Academia Sinica, vol. 1].

Fauconnier, Gilles, and Mark Turner. 2003. *The way we think: Conceptual blending and the mind's hidden complexities*. New York: Basic Books.

Fung, Yu-Lan (冯友兰). 1966. *A short history of Chinese philosophy*. New York: The Free Press.

Goldberg, Adele E. 1995. *Construction: A construction grammar approach to argument structure*. Chicago: University Chicago Press.

Hsieh, Hsin-I (谢信一). 1989. Time and imagery in Chinese. In *Functionalism and Chinese grammar*, Chinese language teachers association monograph series, no.1, ed. Tai, James H-Y and Frank F. S. Hsueh, 45–94. 1991/1992 Hanyu zhong de shijian he yixiang (shang) (zhong) (xia) 汉语中的时间和意象(上)、(中)、(下) (trans: Ye Feisheng 叶蜚声). *Guowai Yuyanxue* 国外语言学 [*Linguistics Abroad*] 1991.4: 12–20; 1992.1: 12–20; 1992.2: 12–20.

Jiang, Lansheng (江蓝生). 2008. Gainian diejia yu goushi zhenghe 概念叠加与构式整合 [Semantic accumulation and constructional integration: An explanation of the asymmetry between affirmation and negation]. *Zhongguo Yuwen* 中国语文 [*Studies of the Chinese Language*] 6: 483–497.

Ke, Hang (柯航). 2012. *Xiandai Hanyu Danshuang Yinjie Dapei Yanjiu* 现代汉语单双音节搭配研究 [*A study of monosyllabic and disyllabic patterns in contemporary Chinese*]. Beijing: Shangwu Yinshuguan 北京: 商务印书馆 [Beijing: The Commercial Press].

Lakoff, George. 1987. *Women, fire, and dangerous things*. Chicago: The University of Chicago Press.

Lakoff, George, and Mark Johnson. 1980. *Metaphors we live by*. Chicago: The University of Chicago Press.

Lan, Chun (蓝纯). 2005. Renzhi Yuyanxue yu Yuyan Yanjiu 认知语言学与语言研究 Cognitive Linguistics and Metaphor. Beijing: Waiyu Jiaoxue yu Yanjiu Chubanshe 北京:外语教学与研究出版社 [Beijing: Foreign Language Teaching and Research Press].

Langacker, Ronald. 1987a. Nouns and verbs. *Language* 63 (1): 53–94.

———. 1987b. The cognitive perspective. *The Newsletter on the Center for Research in Language* 1 (3): 1–16.
———. 1987c. *Foundations of cognitive grammar*. Vol. 1 & 2. Stanford: Stanford University Press.
———. 1991. Yuyan Yanjiu Zhong de Renzhiguan 语言研究中的认知观 (trans: Shen Jiaxuan 沈家煊). *Guowai Yuyanxue* 国外语言学 [*Linguistics Abroad*] 1987 4: 1–6.
———. 1993. Reference point constructions. *Cognitive Linguistics* 4 (1): 1–38.
———. 2008. *Cognitive grammar: A basic introduction*. New York: Oxford University Press.
Li, Charles N., and Sandra A. Thompson. 1981. *Mandarin Chinese: A functional reference grammar*. Berkeley: University of California Press.
Liao, Qiuzhong (廖秋忠). 1991.《Yuyan de fanchouhua: yuyanxue lilun zhong de dianxing》pingjie 《语言的范畴化:语言学理论中的典型》评介 [A review of *Linguistic Categorization: Prototypes in Linguistic Theory*]. *Guowai Yuyanxue* 国外语言学 [*Linguistic Abroad*] 4: 17–26.
Lin, Shuwu (林书武). 1997. Guowai yinyu yanjiu zongshu 国外隐喻研究综述 [A survey of the study of metaphor abroad]. *Waiyu Jiaoxue yu Yanjiu* 外语教学与研究 [Foreign Language Teaching and Research]. 1: 11–19.
Liu, Ningsheng (刘宁生). 1994. Hanyu zenyang biaoda wuti de kongjian guanxi 汉语怎样表达物体的空间关系 [How Chinese structures space]. *Zhongguo Yuwen* 中国语文 [Studies of the Chinese language]. 3: 169–179.
——— (刘宁生). 1995. Hanyu pianzheng jiegou de renzhi jichu jiqi zai yuyan leixingxue de yiyi 汉语偏正结构的认知基础及其在语言类型学的意义 [On the cognitive basis of Chinese subordinative constructions and its typological significance]. *Zhongguo Yuwen* 中国语文 [*Studies of the Chinese Language*] 2: 81–89.
Liu, Tanzhou (刘探宙). 2018. *Shuo "Wangmian Sile Fuqin" Ju* 说"王冕死了父亲"句 [*Studies of the sentence "Wang Mian Died His Father"*]. Shanghai: Xuelin Chubanshe 上海: 学林出版社 [Shanghai: Academia].
Lü, Hsiang (吕叔湘, Lü Shuxiang). 1941. Lun *wu* yu *wu* 论毋与勿 [The particles of negation *Wu²* and *Wu⁴*]. In *Huaxi Xiehe Daxue Zhongguo Wenhua Yanjiusuo Jikan* 华西协合大学中国文化研究所集刊 [*Studia Serica, Journal of the Chinese Cultural Studies Research Institute, West China Union University*] 1(4):85–117. In *Hanyu Yufa Lunwenji (Zengdingben)* 汉语语法论文集 [*Collected papers on Chinese grammar, revised edition*]. Beijing: Shangwu Yinshuguan 北京: 商务印书馆 [Beijing: The Commercial Press] 1984.
——— (吕叔湘, Lü Shuxiang). 1948. Ba zi yongfa de yanjiu 把字用法的研究 [On the uses of the pre-transitive *PA*]. In *Zhongguo Wenhua Yanjiu Huikan* (di 8) 中国文化研究汇刊(第8卷) [*Bulletin of Chinese studies*]. In *Hanyu Yufa Lunwenji (Zengdingben)* 汉语语法论文集 [*Collected papers on Chinese grammar, revised edition*]. Beijing: Shangwu Yinshuguan 北京: 商务印书馆 [Beijing: The Commercial Press] 1984.
Lu, Bingfu (陆丙甫). 1993. *Hexin Tuidao Yufa* 核心推导语法 [*The head-oriented grammar*]. Shanghai: Shanghai Jiaoyu Chubanshe 上海: 上海教育出版社 [Shanghai: Shanghai Educational Publishing House].
——— (陆丙甫). 2008. Yuxu youshi de renzhi jieshi 语序优势的认知解释 [Word order dominance and its cognitive explanation]. *Dangdai Yuyanxue* 当代语言学 [*Contemporary Linguistics*] 1: 1–15, 2: 132–138.
Lyons, John. 1977. *Semanitics*. Vol. 2. Cambridge: Cambridge University Press.
Shen, Jiaxuan (沈家煊). 1985. Yinghan kongjian gainian de biaoda xingshi 英汉空间概念的表达形式 [Expression of spatial concepts in English and Chinese]. *Waiguo Yuwen Jiaoxue*《外国语文教学》[*Foreign Language and Literature*] 4: 33–40.
——— (沈家煊). 1993. Jufa de xiangsixing wenti 句法的象似性问题 [A survey of studies of iconicity in syntax]. *Waiyu Jiaoxue yu Yanjiu* 外语教学与研究 [Foreign Language Teaching and Research] 1: 2–8.
——— (沈家煊). 1994. R. W. Langacker de renzhi yufa R. W. Langacker 的认知语法 [R. W. Langacker's cognitive grammar]. *Guowai Yuyanxue* 国外语言学 [Linguistics Abroad] 1: 12–20.
——— (沈家煊). 1995a. Youjie yu wujie 有界与无界 [Boundedness and unboundedness]. *Zhongguo Yuwen* 中国语文 [*Studies of the Chinese Language*] 5: 367–380.

——— (沈家煊). 1995b. Zhengfu diandao he yuyong dengji 正负颠倒和语用等级 [Positive-negative reversal and the pragmatic scale]. In *Yufa yanjiu he tansuo* 语法研究和探索(七) [*Grammar research and exploration I*, vol. 7], 237–244. Beijing: Shangwu Yinshuguan 北京: 商务印书馆 [Beijing: The Commercial Press].

——— (沈家煊). 1997a. Xingrongci jufa gongneng de biaoji moshi 形容词句法功能的标记模式 [The syntactic functions of Chinese adjectives]. Zhongguo Yuwen 中国语文 [Studies of the Chinese Language] 4: 242–250.

——— (沈家煊). 1997b. Yuyong ·renzhi ·yanwaiyi 语用·认知·言外义 [Pragmatics, cognition, and implicature]. *Waiyu yuwai yu jiaoxue* 外语与外语教学 [*Foreign Languages and Their Teaching*] 4: 10–12.

——— (沈家煊). 1999a. *Buduicheng he Biaojilun* 不对称和标记论 [*Asymmetry and the markedness theory*]. Nanchang: Jiangxi Jiaoyu Chubanshe 南昌:江西教育出版社 [Nanchang: Jangxi Education Publishing House].

——— (沈家煊). 1999b. Zhuanzhi he zhuanyu 转指和转喻 [A metonymic model of transferred designation of *de*-constructions in Mandarin Chinese]. Dangdai Yuyanxue 当代语言学 [Contemporary Linguistics] 1: 3–15.

——— (沈家煊). 1999c. "Zai" zi ju he "gei" zi ju "在"字句和"给"字句 [Sentences with *Zai* and *Gei* in Mandarin Chinese]. Zhongguo Yuwen 中国语文 [Studies of the Chinese Language] 2: 94–102.

——— (沈家煊). 2000. Jushi he peijia 句式和配价 [Valency and sentences]. Zhongguo Yuwen 中国语文 [Studies of the Chinese Language] 4: 291–297.

——— (沈家煊). 2002. Ruhe chuzhi "chuzhishi "——lun ba zi ju de zhuguanxing 如何处置"处置式"——论把字句的主观性 [Can the disposal construction be disposed of? On the subjectivity of the *Ba* construction in Mandarin Chinese]. *Zhongguo Yuwen* 中国语文 [*Studies of the Chinese Language*] 5: 387–399.

——— (沈家煊). 2003. Fuju San Yu "Xing, Zhi, Yan" 复句三域"行、知、言" [Compound sentences in three conceptual domains: acting, knowing, and utering]. Zhongguo Yuwen 中国语文 [Studies of the Chinese Language] 2003.3: 195–204.

——— (沈家煊). 2004a. Shuo "bu guo " 说"不过" [Remarks on *Buguo*]. *Qinghua daxue xuebao (zhexue shehui kexue ban)* 清华大学学报(哲学社会科学版) [*Journal of Tsinghua University (Philosophy and Social Sciences)*] 5: 30–37.

——— (沈家煊). 2004b. Zai tan "youjie" yu "wujie" 再谈"有界"与"无界" [Further remarks on boundedness and unboundedness]. *Yuyanxue Luncong* 语言学论丛(第三十辑) [*Essays on linguistics*, vol. 30], 40–54. Beijing: Shangwu Yinshuguan 北京: 商务印书馆 [Beijing: The Commercial Press].

——— (沈家煊) (ed.) 2005. *Xiandai Hanyu Yufa de Gongneng Yuyong Renzhi Yanjiu* 现代汉语语法的功能、语用、认知研究 [*A study of the function, pragmatics and cognition of modern Chinese grammar*]. Beijing: Shangwu Yinshuguan 北京: 商务印书馆 [Beijing: The Commercial Press].

——— (沈家煊). 2006a. *Renzhi yu Hanyu Yufa Yanjiu* 认知与汉语语法研究 [*Cognition and Chinese grammar*]. Beijing: Shangwu Yinshuguan 北京: 商务印书馆 [Beijing: The Commercial Press].

——— (沈家煊). 2006b. "Rouhe" he "Jieda" "糅合"和"截搭" [Blending and haplology]. *Shijie Hanyu Jiaoxue* 世界汉语教学 [*Chinese Teaching in the World*] 4: 5–12.

——— (沈家煊). 2006c. "*Wangmian si le fuqin*" de shengcheng fangshi "王冕死了父亲"的生成方式——兼说汉语"糅合"造句 [The generative mechanism of sentences like "*Wangmian died father*": Sentence generation by blending in Chinese]. *Zhongguo Yuwen* 中国语文 [*Studies of the Chinese Language*] 4: 291–300.

——— (沈家煊). 2007. Hanyu li de mingci he dongci 汉语里的名词和动词 [On nouns and verbs in Chinese]. *Hanzangyu xuebao* 汉藏语学报 [*Journal of Sino-Tibetan Linguistics*] 1: 27–47.

——— (沈家煊). 2016. *Mingci he Dongci* 名词和动词 [*Nouns and Verbs*]. Beijing: Shangwu Yinshuguan 北京: 商务印书馆 [Beijing: The Commercial Press].

——— (沈家煊). 2017a.《*Fanhua*》*Yuyan Zhaji*《繁花》语言札记 [Linguistic notes on the novel "Flowers"]. Nanchang: Ershiyi Shiji Chubanshe Jituan 南昌:二十一世纪出版社集团 [Nanchang: 21st Century Publishing Group].

——— (沈家煊). 2017b. Cong yuyan kan zhongxifang de fanchouguan 从语言看中西方的范畴观 [Western and Chinese views of categories seen from a linguistic perspective]. *Zhongguo Shehui Kexue* 中国社会科学 [*Social Sciences in China*] 7: 131–143.

——— (沈家煊). 2017c. Nouns and verbs: Evolution of grammatical forms. In *New horizons in evolutionary linguistics*, Journal of Chinese Linguistics monograph series 27, ed. Peng Gang (彭刚) and Wang Feng (汪锋), 222–253. Hong Kong: Chinese University Press.

——— (沈家煊). 2018. Hanyu "mingdong baohan" geju dui yingyu xuexi de fuqianyi 汉语"名动包含"格局对英语学习的负迁移 [Negative transfer of Chinese nominalism in learning English as a second language]. *Waiguo Yuyan Wenxue* 外国语言文学 [*Foreign Language and Literature Studies*] 1: 4–22.

Shen, Jiaxuan (沈家煊) and Dongmei Wang (王冬梅). 2000. "N *de* V" he "canzhaoti – mubiao" goushi "N的V" 和"参照体—目标"构式 ["N *de* V" as a reference point construction in Mandarin Chinese]. *Shijie Hanyu Jiaoxue* 世界汉语教学 [*Chinese Teaching in the World*] 4: 25–32.

Shi, Yuzhi (石毓智). 1995a. Shijian de yiweixing dui jieci yansheng de yingxiang 时间的一维性对介词衍生的影响 [Effect of the one-dimensionality of time on emergence of prepositions]. *Zhongguo Yuwen* 中国语文 [*Studies of the Chinese Language*] 1: 1–10

——— (石毓智). 1995b.《Nüren、huo he weixian shiwu ——fanchou jieshi le siwei de shenme aomi》pingjia《女人、火和危险事物——范畴揭示了思维的什么奥秘》评价 [A review of *women fire dangerous things: What categories reveal about the mind*]. *Guowai Yuyanxue* 国外语言学 [*Linguistic Abroad*] 2: 17–22.

——— (石毓智). 2000. *Yufa de Renzhi Yuyi Jichu* 语法的认知语义基础 [*Cognitive and semantic foundations of grammar*]. Nanchang: Jiangxi Jiaoyu Chubanshe 南昌:江西教育出版社 [Nanchang: Jangxi Education Publishing House].

——— (石毓智). 2001/1992. *Kending he Fouding de Duicheng yu Buduicheng* (Zengdingben) 肯定和否定的对称与不对称(增订本) [*Symmetry and asymmetry between affirmation and negation*, 2nd ed.]. Beijing: Beijing Yuyan Wenhua Daxue Chubanshe 北京:北京语言文化大学出版社 [Beijing: Beijing Language and Culture University Press]. 1992. 1st ed. Taibei: Xuesheng Shuju 学生书局 [Taipei: Student Book Co., Ltd.].

Shu, Dingfang (束定芳). 2008. Renzhi Yuyixue 认知语义学 [An Introduction to Cognitive Semantics]. Shanghai: Shanghai Waiyu Jiaoyu Chubanshe 上海: 上海外语教育出版社 [Shanghai: Shanghai Foreigh Language Education Press].

Sweetser, Eve. 1990. *From etymology to pragmatics: Metaphorical and cultural aspects of semantic structure*. Cambridge: Cambridge University Press.

Tai, James H-Y (戴浩一). 1985. Temporal sequence and Chinese word order. In *Iconicity in syntax*, ed. John Haiman, 49–72. Amsterdam/Philadelphia: John Benjamins.

——— (戴浩一). 1988. Shijian shunxu he hanyu de yuxu 时间顺序和汉语的语序 (trans: Huang He 黄河). *Guowai Yuyanxue* 国外语言学 [*Linguistics Abroad*] 1: 10–20.

———(戴浩一). 1990/1991. Yi renzhi wei jichu de hanyu gongneng yufa chuyi 以认知为基础的汉语功能语法刍议 (trans: Ye Feisheng叶蜚声). *Guowai Yuyanxue* 国外语言学 [*Linguistics Abroad*] 4: 21–27, 1991 1: 25–33.

———(戴浩一). 1989. Toward a cognition-based functional grammar of Chinese. In *Functionalism and Chinese grammar*, ed. Tai, James H-Y (戴浩一) and Frank F. S. Hsueh (谢信一), 187–226. South Orange: Seton Hall University, Chinese Language Teachers Association.

Talmy, Leonard. 1975. Semantics and syntax of motion. In *Syntax and semantics*, vol. 4, 181–238. New York: Academic.

———. 2000. *Toward a cognitive semantics*. Vol. 1 & 2. Cambridge, MA: The MIT Press.

Taylor, John. 1989. *Linguistic categorization: Prototypes in linguistic theory*. Oxford: Oxford University Press.

———. 1994. "Subjective" and "objective" readings of possessor nominals. *Cognitive Linguistics* 5 (3): 201–242.

Tian, Chenshan. 2009. Development of dialectical materialism in China. In *Routledge history of Chinese philosophy*, ed. Bo Mou, 512–538. London/New York: Routledge.

Trubetzkoy, Nikolai. 1958. *Grundzüge der Phonologie*. Gottingen: Vandenhoeck and Ruprecht.

Ungerer, Friedrich, and Hans-Jorg Schmid. 1996. *An introduction to cognitive linguistics*. London: Longman.

Wan, Quan (完权). 2009. Lianxu jiedai 连续借代 [Contiguous Metonymy]. *Xiuci Xuexi* 修辞学习 [Rhetoric Learning] 3: 24–29.

——— (完权). 2016. "De "de Xingzhi yu Gongneng "的"的性质与功能 [*The properties and functions of De*]. Beijing: Shangwu Yinshuguan 北京: 商务印书馆 [Beijing: The Commercial Press].

Wang, Li (王力). 1943. *Zhongguo Yufa Lilun* 中国语法理论 [*A theory Chinese grammar*]. Beijing: Zhonghua Shuju 北京: 中华书局 [Beijing: Zhonghua Book Company].

Wang, Yin (王寅). 2002. Renzhi yuyanxue de zhexue jichu:tiyan zhexue 认知语言学的哲学基础: 体验哲学 [The philosophical basis for cognitive linguistics: Embodied philosophy]. *Waiyu Jiaoxue yu Yanjiu* 外语教学与研究 [*Foreign Language Teaching and Research*] 2: 82–89.

——— (王寅). 2011. *Goushi Yufa Yanjiu* 构式语法研究 [*Researches on construction grammar, vol. I & II*]. Shanghai: Shanghai Waiyu Jiaoyu Chubanshe 上海: 上海外语教育出版社 [Shanghai: Shanghai Foreign Language Education Press].

Wen, Xu (文旭). 1999. Guowai renzhi yuyanxue yanjiu zongguan 国外认知语言学研究综观 [A survey of overseas studies in linguistics studies]. *Waiguoyu* 外国语 [*Journal of Foreign Languages*] 1: 34–40.

Xia, Lixin (夏立新). 2015. *Neixiangxing Han-Ying Xuexi Cidian de Duowei Yiyi Moshi Yanjiu* 内向型汉英学习词典的多维译义模式研究 [A study of a multi-dimensional definition model of the Chinese-English dictionary for Chinese EFL learners]. Beijing: Shangwu Yinshuguan 北京: 商务印书馆 [Beijing: The Commercial Press].

Xiong, Xueliang (熊学亮). 1999. *Renzhi Yuyongxue Gailun* 认知语用学概论 [*Cognitive pragmatics*]. Shanghai: Shanghai Waiyu Jiaoyu Chubanshe 上海: 上海外语教育出版社 [Shanghai: Shanghai Foreign Language Education Press].

Ye, Jianjun (叶建军). 2008. 《Zutang Ji》zhong sizhong rouhe jushi 《祖堂集》中四种糅合句式 [On the mingled phenomena of sentence pattern from four kind of sentence pattern in *Zutang Ji*]. *Yuyan Yanjiu* 语言研究 [*Studies in Language and Linguistics*] 1: 94–99.

Yuan, Yulin (袁毓林). 1995. Cilei fanchou de jiazu xiangsixing 词类范畴的家族相似性 [On the family resemblance of word-class category]. *Zhongguo Shehui Kexue* 中国社会科学 [*Social Sciences in China*] 1: 154–170.

——— (袁毓林). 1999. Dingyu shunxu de renzhi jieshi jiqi lilun yunhan 定语顺序的认知解释及其理论蕴涵 [A cognitive explanation of order of modifier and its theoretical implication]. *Zhongguo Shehui Kexue* 中国社会科学 [*Social Sciences in China*] 2: 185–201.

——— (袁毓林). 2004. Rongqi yinyu he taojian yinyu ji xiangguan de yufa xianxiang 容器隐喻和套件隐喻及相关的语法现象——词语同现限制的认知解释和计算分析 [On some grammatical phenomena related to the container metaphor and set metaphor: Towards a cognitive explanation and a computational analysis of co-occurrence restriction of words]. *Zhongguo Yuwen* 中国语文 [*Studies of the Chinese Language*] 3: 195–208.

Yuan, Yulin (袁毓林), Hui Ma (马辉), Ren Zhou (周韧), and Hong Cao (曹宏). 2009. *Hanyu Cilei Huafen Shouce* 汉语词类划分手册 [*A handbook of Chinese word classes*]. Beijing: Beijing Yuyan Wenhua Daxue Chubanshe 北京:北京语言大学出版社 [Beijing: Beijing Language and Culture University Press].

Yue, Yao (乐耀). 2014. Lun xiandai hanyu de binixing duidaiju 论现代汉语的比拟型对待句 [Analogical *dui* construction in modern Chinese]. *Zhongguo Yuwen* 中国语文 [*Studies of the Chinese Language*] 1: 35–47.

Zhang, Min (张敏). 1997. Cong leixingxue he renzhi yufa de jiaodu kan hanyu chongdie xianxiang 从类型学和认知语法的角度看汉语重叠现象 [Reduplication in perspectives of typology and cognitive grammar]. *Guowai Yuyanxue* 国外语言学 [*Linguistics Abroad*] 2: 37–45.

——— (张敏). 1998. *Renzhi Yuyanxue yu Hanyu Mingci Duanyu* 认知语言学与汉语名词短语 [*Cognitive linguistics and Chinese noun phrase*]. Beijing: zhongguo Shehui Kexue Chubanshe 北京: 中国社会科学出版社 [Beijing: China Social Sciences Press].

Zhang, Bojiang (张伯江). 1999. Xiandai hanyu de shuangjiwu jiegoushi 现代汉语的双及物结构式 [The ditransitive construction in Mandarin Chinese]. *Zhongguo Yuwen* 中国语文 [*Studies of the Chinese Language*] 3: 115–184.

——— (张伯江). 2009. *Cong Shishou Guanxi Dao Jushi Yuyi* 从施受关系到句式语义 [*Argument structure and grammatical construction*]. Beijing: Shangwu Yinshuguan 北京: 商务印书馆 [Beijing: The Commercial Press].

——— (张伯江) ed. 2016. *Xiandai Hanyu Yufa de Gongneng Yuyong Renzhi Yanjiu* (er) 现代汉语语法的功能、语用、认知研究 (二) [*A study of the functions, pragmatics and cognition of modern Chinese grammar*, vol. 2]. Beijing: Shangwu Yinshuguan 北京: 商务印书馆 [Beijing: The Commercial Press].

——— (张伯江). 2018. Goushi yufa yingyong yu hanyu yanjiu de ruogan sikao 构式语法应用于汉语研究的若干思考 [Some reflections on the application of construction grammar in Chinese studies]. *Yuyan Jiaoxue yu Yanjiu* 语言教学与研究 [*Language Teaching and Linguistic Studies*] 4: 2–11.

Zhao, Yanfang (赵艳芳). 2000. *Renzhi Yuyanxue Gailun* 认知语言学概论 [*An introduction to cognitive linguistics*]. Shanghai: Shanghai Waiyu Jiaoyu Chubanshe 上海: 上海外语教育出版社 [Shanghai: Shanghai Foreign Language Education Press].

Zhou, Ren (周韧). 2011. "Quan" de zhengtixing yuyi tezheng jiqi jufa houguo "全"的整体性语义特征及其句法后果 [The semantic feature "integrity" of *Quan* and its syntactic consequences]. *Zhongguo Yuwen* 中国语文 [*Studies of the Chinese Language*] 2: 133–144.

Zhu, Dexi (朱德熙, Chu, The-hsi). 1956. Xiandai hanyu xingrongci yanjiu 现代汉语形容词研究 [A study of the adjective in Modern Chinese]. *Yuyan Yanjiu* 语言研究 [*Linguistic Researches*] 1: 83–112.

——— (朱德熙). 1982. *Yufa Jiangyi* 语法讲义 [*Lecture notes on grammar*]. Beijing: Shangwu Yinshuguan 北京: 商务印书馆 [Beijing: The Commercial Press].

——— (朱德熙). 1983. Zizhi he zhuanzhi: Hanyu mingcihua biaoji "de, zhe, suo, zhi" de yufa gongneng he yuyi gongneng 自指和转指——汉语名词化标记"的、者、所、之"的语法功能和语义功能 [Self designation and non-self designation: The syntactic and semantic functions of Chinese nominal marker de, zhe, suo and zhi]. *Fangyan* 方言 [*Dialect*]1983.1:16–31.

——— (朱德熙). 1985. *Yufa Dawen* 语法答问 [*Replies to the questions on grammar*]. Beijing: Shangwu Yinshuguan 北京: 商务印书馆 [Beijing: The Commercial Press].

Neurolinguistics in China

Yiming Yang 杨亦鸣, Zude Zhu 朱祖德, and
Qingrong Chen 陈庆荣

Contents

Introduction	68
The Foundational Period of Neurolinguistics in China	69
Aphasia Research: Studying the Relationship Between Brain Injury and Language Disorders	70
Chinese Language Processing Research: Studying Cognitive Mechanisms for Chinese Characters, Words, Sentences, and Text	71
Translating Neurolinguistics Research: Introducing Neurolinguistic Theories, Methods, and Case Studies to China	73
Establishing the Neurolinguistics Discipline in China	75
The Development of Neurolinguistics in China	76
Disciplinary Expansion and the Researcher Development Program	77
Research Areas in Chinese Neurolinguistics	79
The Chinese Mental Lexicon	79
Syntax	83
Semantics	86
Pragmatics	88
Phonology	90
Language Acquisition, Development, and Decline	91
Language Use in Atypical Populations	93
Second Language Learning and Bilingualism	95
Prospects for Future Research	97
References	99

Translated by Janet Davey, The Australian National University

Y. Yang 杨亦鸣 (✉) · Z. Zhu 朱祖德 · Q. Chen 陈庆荣
Collaborative Innovation Center for Language Ability, School of Linguistic Sciences and Arts, Jiangsu Normal University, Xuzhou, China
e-mail: yangym@jsnu.edu.cn

Abstract

Neurolinguistics is a nascent interdisciplinary field that studies language and the human brain at the frontier of modern scientific research. The emergence of neurolinguistics in China is a recent development, only established as a distinct discipline in the mid- to late 1990s. That said, over the past two decades Chinese neurolinguistics has achieved notable progress, which has been driven by advances in scientific research, the development of research teams, and disciplinary integration in China. Chinese neurolinguistics research has yielded new insights into Chinese syntax, semantics, and phonology; the Chinese mental lexicon; language acquisition and decline; second language learning; and neural mechanisms of language processing in atypical populations. In the future, Chinese neurolinguistics should give more focus to incorporating the latest developments in global research, especially in the areas of linguistic theory, artificial intelligence (AI), and human brain projects. The discipline should also adopt new research methods and techniques, and establish a professional development scheme for researchers. This will support the long-term development of neurolinguistics in China.

Keywords

Chinese neurolinguistics · Mental lexicon · Language processing

Introduction

Language is not only a symbolic system that links sound with meaning but also a function of the human brain. The initial state of language, genomic expression, is realized through the neural circuits of the brain. The rise of cognitive science-based approaches to studying language abilities and language behavior has facilitated a rapid proliferation of research on the neural mechanisms of language, ultimately leading to the establishment of neurolinguistics as a transdisciplinary field (Yang 2002). Neurolinguistics studies the relationship between language and the brain. It combines linguistics with brain science, neuroscience, neurobiology, and other disciplines in order to explore the neural and cognitive mechanisms of human language acquisition, comprehension, and production (Ahlsén 2006).

Today, neurolinguistics is already a mature research discipline in developed Western countries. It acts as a convergence point for scientific frontiers across contemporary linguistic theory, brain science, psychology, artificial intelligence (AI), and other fields (Yang 2012). In China, however, the neurolinguistics field emerged only relatively recently, with research on neural mechanisms of language beginning to appear in the latter half of the twentieth century. The first Chinese studies of the neural mechanisms of language were conducted by clinical neuroscientists and psychologists. Clinical neuroscientists mostly studied aphasia, focusing on the correspondence between brain lesions and language impairment. They also explored the pathogenesis of aphasia to assist with the diagnosis, treatment, and rehabilitation of aphasic patients. Psychologists, on the other hand, mainly investigated the relationship between the brain and cognitive processing of characters, words, sentences, and text. While this early research was essential and influential, it was not aimed at revealing neural mechanisms of language. In terms of research methods, the studies typically drew on clinical neuropsychological assessments of aphasic patients as well as classic psychological experiments of human behavior. However, clinical neuropsychological assessments cannot provide a dynamic view of how neural mechanisms of language operate in healthy individuals in real time. Similarly, behavioral experiments draw inferences about the psychological processing of language via measurements that reflect only the end state rather than the process itself, and these measurements cannot directly assess neural responses.

It was only at the start of the 1980s that Chinese linguists began to focus on neurolinguistics research. Initially, this was just a small number of researchers introducing existing work from abroad; the linguistics community in China had yet to develop a truly neurolinguistic approach to its own research. True neurolinguistics research in China only began in the mid- to late 1990s. Over time a group of specialists emerged who had benefited from a rigorous training in linguistics, acquired knowledge of brain science and psychological theories, and mastered experimental techniques. This allowed the linguistics community to establish a Chinese neurolinguistics research team. Moreover, they founded the academic journal *Linguistic Sciences*, which covers basic linguistic theories, neurolinguistics, and other aspects of applied research on language. These developments facilitated the formation and development of the neurolinguistics discipline in China.

The Foundational Period of Neurolinguistics in China

Language is fundamental to human beings; it is what distinguishes us from other animals. The importance of human language was recognized by different academic disciplines in China during the 1980s. The joint work of these disciplines informed the establishment of neurolinguistics as a new branch of study. The neuroscience community focused its research on studying language disorders, especially aphasia, from the perspective of clinical practice. The world of psychology, meanwhile, studied the psychological processes involved in reading characters, words, and sentences, and oriented research toward neural mechanisms of language processing in the brain. Many scholars in the linguistics community were aware that a new discipline was developing in the West. Introductory work by these scholars that translated existing research into Chinese allowed the core research question of neurolinguistics to be identified, laying the groundwork for its establishment as a discipline in China.[1]

---

[1] *Translator's Note*: In this entry, "Chinese language" or "Chinese" refers to Mandarin Chinese unless specified otherwise. The official, standardized form of the Chinese language spoken in mainland China is known as *Putonghua* 普通话 or Modern Standard Chinese (MSC). MSC pronunciation is based on the Beijing dialect of Mandarin, and so "Mandarin" and "MSC/Putonghua" are often used interchangeably.

Aphasia Research: Studying the Relationship Between Brain Injury and Language Disorders

Neurolinguistics developed from the study of language disorders such as aphasia. Numerous insights generated from research into language impairment shaped the establishment and development of neurolinguistics as a distinct discipline in China. In 1938, R. S. Lyman reported a case of alexia with agraphia in a Mandarin Chinese speaker. Later, Wang Xinde 王新德 and colleagues documented language impairments in Chinese people with pure alexia or pure agraphia. In a series of studies, the researchers examined patients' performance on mirror-writing tasks with Chinese characters, reporting their findings in 1959, 1981, and 1985 (Wang 1994). In 1994, Cui Gang 崔刚 analyzed lexical deficits in Chinese-speaking patients with Broca's aphasia. After this, studies of language disorders began to appear in large number in China, most of which focused on the location of brain damage and mechanisms that gave rise to language disorders.

Among the studies of pure alexia, the majority of findings appear to show that it is caused by lesions in the occipital lobe, parietal lobe, and a posterior section of the corpus callosum known as the splenium (Gao 1993). However, Chen Haibo 陈海波 and Wang Xinde (Chen and Wang 1992) found that a lesion in the left posterior inferior temporal gyrus (ITG) only resulted in reading impairments in Chinese. They believed that this may be because the left posterior ITG and its surrounding region are responsible for visual word memory, a form of memory that holds greater significance for logographic Chinese characters than alphabetic writing systems. Taking a different view, Hu Chaoqun 胡超群 (Hu 1989) and Gao Surong 高素荣 (Gao 1993, 150–151) believed that reading impairments in patients with alexia were caused by damage to the links between the representations of a character's form, pronunciation, and meaning.

Researchers have also devoted considerable attention to the question of whether there is a writing center in the brain analogous to its reading center. Sigmund Exner proposed that the left posterior middle frontal gyrus (MFG) is the brain's writing center (Gao 1993, 31–60). Yet the findings of later studies did not support this view. For example, Gao (1996) analyzed the pathogenesis of agraphia in terms of the form, pronunciation, and meaning of Chinese characters. Gao held that the patients' writing errors were mostly deviations from the correct form, pronunciation, or meaning that occurred during character recall. In other words, the patients produced characters that were not related to the specified characters in one or more of form, sound, or meaning.

Another important breakthrough in the study of Chinese aphasia came through the development of assessment scales for Mandarin aphasia. These have proved to be a valuable tool for the diagnosis, evaluation, and rehabilitation of aphasic patients. The most systematic and comprehensive Chinese aphasia assessment scale to date is the Aphasia Battery of Chinese (ABC). Developed by Gao and colleagues, the ABC adapts the Boston Diagnostic Aphasia Evaluation (BDAE) test used in the USA (Gao 1993). Because the ABC "took the linguistic features of Chinese into account and reflected the nature of Chinese aphasia" (Liang 2004, 142), it was "the leading test for a long period of time" (Zhang et al. 2005, 5). Other researchers developed special diagnostic scales for Mandarin dysgraphia (Liu and Liang 1996) and agrammatism (Zhao et al. 2002). Meanwhile, Qiu Weihong 丘卫红, Dou Zulin 窦祖林, and Wan Guifang 万桂芳 (Qiu et al. 2000) devised an aphasia screening test for Cantonese. Assessment scales such as these have made a valuable contribution to the study of Chinese aphasia. However, most are based on assessment scales developed and used in other countries. Although certain modifications have been made in light of the distinctive features of Chinese, the scales cannot provide an in-depth assessment of all aspects of the language-learning system for Chinese. Moreover, a failure to select the most representative characteristics of language learning as assessment criteria has restricted somewhat the scales' ability to evaluate language abilities in a comprehensive and effective manner.

Revealing the neural mechanisms of language has never been the overarching goal of Chinese aphasia research. Instead, Chinese aphasia studies have typically described clinical manifestations of language impairment, aiming to determine the lesion location, and pathogenesis of aphasia to promote better diagnostic, evaluation, and treatment outcomes. Yet some aspects of the relationship between language and neural mechanisms have been indirectly explored in aphasia studies. To date the most significant contribution of this research has been in establishing a tentative link between brain injury and language impairment. This has advanced understanding of processing mechanisms and regions associated with language function in the brain.

Chinese Language Processing Research: Studying Cognitive Mechanisms for Chinese Characters, Words, Sentences, and Text

Research into cognitive mechanisms of Chinese language processing proceeds from the study of character processing mechanisms. Chinese character processing research is centered on two main topics. The first examines interactions between character form, sound, and meaning in the recognition process, that is, whether pronunciation acts as an intermediary between character form and semantic meaning during recognition. The second topic investigates the relationship between Chinese character processing and the two hemispheres of the brain, namely whether recognition occurs through "dual hemispheric character processing," namely the relative contribution of each hemisphere to character recognition.

Identifying the stage in the semantic comprehension process at which the pronunciation of a character is activated is an issue that has received widespread attention from the psychology community. Experimental findings have shown that there are three potential sequences of activation following the initial activation of the glyph: (i) the pronunciation and meaning of a character are activated simultaneously (Peng et al. 1985); (ii) pronunciation is activated prior to meaning (Perfetti and Tan 1998); or (iii) meaning is activated before pronunciation (Zhou 1997). Studies by Chen Baoguo 陈宝国 and Peng Danling 彭聃龄, later also with Wang Lixin 王立新 (Chen and Peng 2001; Chen et al. 2003), have shown that whether pronunciation acts as an intermediary in the semantic comprehension process is further modulated by character frequency. Hu (1989, 1992) noted that in reading, processing of the form, sound, and meaning of characters starts with visual perception of a character's form. Links are then established between a character's form and sound, and between form and meaning, respectively. As such, recognition and reading of Chinese characters and words can occur via direct comprehension of meaning from the written form; it is not necessary to go from form to pronunciation and then to meaning. This shows that the link between character form and meaning is stronger than the one between form and pronunciation. Yin Wengang 尹文刚 (Yin 1990) and Gao (1993) reached similar conclusions. In reality, however, the brain's mechanisms for processing a Chinese character's form, pronunciation, and meaning are likely to function in a far more complex manner. In a study that measured event-related potentials (ERPs) during a semantic association task, Wei Jinghan 魏景汉 and colleagues (Wei et al. 1995) identified a positive slow wave as an indication of character recognition. The positive slow wave component is related to multiple pieces of information about the character, including perception of written form and pronunciation, and the connection between pronunciation and meaning. Hence, the relationship between form, sound, and meaning in character processing cannot simply be summed up as a question of whether meaning is obtained through phonological processing or not. Chinese researchers Yang Yiming 杨亦鸣 and Cao Ming 曹明 (Yang and Cao 1998) highlighted the fact that surface alexia (form–meaning) and phonological alexia (form–sound) both exist as distinct conditions challenges, the idea that a Chinese character's form and meaning are strongly linked while its form and pronunciation are only weakly so. In their article, Yang and Cao regard language as a symbolic system that links sound and meaning, and see writing as a body of linguistic signs with symbolic characteristics. This view informed their discussion of the universal relationship between linguistic form, sound, and meaning on the one hand, and unique aspects of Chinese characters as linguistic signs on the other. By moving away from a focus on character form, sound, and meaning to study processes of character recognition and reading in terms of the mental lexicon of Chinese, their article made a significant contribution to the neurolinguistics literature.

There are also two different views on how cognitive processing of Chinese characters relates to the two hemispheres of the brain. One view holds that the left hemisphere plays the major role in recognizing and reading characters (Zhang and Peng 1984), while the other is that there is balanced involvement of the two hemispheres in the recognition process, that is, dual hemispheric character processing (Yin 1984; Guo and Yang 1995). During their time working in the USA, Ovid J.-L Tzeng 曾志朗 and William S.-Y. Wang 王士元 extensively studied processing mechanisms for written text in the brain (Tzeng and Wang 1983). They found a right hemisphere processing advantage for recognition of a single character, but superior processing in the left hemisphere for words composed of multiple characters. Tzeng and other scholars believed this may be because visual recognition plays a more decisive role in identifying a single character, and hence the right hemisphere is involved to a greater degree. By contrast, the need to process multiple pieces of phonological, morphological, and semantic information contained in a multicharacter word translates into a left hemisphere advantage. Moreover, the processing route from character form to character meaning must be mediated by pronunciation. Additional work on character recognition processing was carried out by Luo Yuejia 罗跃嘉 and colleagues (Luo et al. 2001) in a study that used ERPs to compare recognition of novel and familiar Chinese characters across visual and auditory processing routes. They found a right hemisphere advantage for auditory recognition of characters, whereas there was greater activation in the left parietal lobe, left posterior temporal lobe, and right occipital lobe during visual processing. The result shows that identifying differences between the two hemispheres of the brain cannot be regarded as merely a question of which hemisphere has the dominant role in processing. Brain mechanisms involved in processing Chinese characters should instead be considered from multiple perspectives, and not least in terms of the course and different routes of cognitive processing.

A few studies have examined the reading process for Chinese sentences and larger texts. Research by Peng and Liu Songlin 刘松林 (Peng and Liu 1993) found a weak interaction between syntax and semantics when reading sentences, rather than syntax being completely independent of semantics. Chen Yongming 陈永明 and Cui Yao 崔耀 (Chen and Cui 1994) observed a processing advantage for the first mentioned pronoun in a sentence, which they concluded helps to establish a coherent cognitive representation during reading. Meanwhile, a series of studies led by Wang Suiping 王穗苹 has shown that background information in longer texts is rapidly activated and incorporated into a reader's online semantic representations. This is an indication that in Mandarin Chinese – as in Indo-European languages – information is processed online in linguistic comprehension (Wang and Morey 2001; Wang et al. 2001).

Compared with clinical aphasia research, the study of neural mechanisms for Chinese language processing is more advanced in the fields of psychology and linguistics. Psychology has already begun to purposely examine linguistic processing mechanisms, while many linguists have touched on neural mechanisms of language as part of their research. The results of such studies undoubtedly act as a reference point and help motivate further neurolinguistics research on the mental lexicon of Chinese. However, rarely have they investigated language mechanisms in the brain directly.

Translating Neurolinguistics Research: Introducing Neurolinguistic Theories, Methods, and Case Studies to China

Western neurolinguistics made significant progress in the 1970s and 1980s. In 1975, renowned Russian neuropsychologist Alexander Luria published *Basic Problems of Neurolinguistics*, a seminal work in neurolinguistics. The American behavioral neurologist Norman Geschwind revived the Wernicke-Lichtheim model of aphasia to formulate the Wernicke-Lichtheim-Geschwind (or Wernicke-Geschwind) neurological model of language. Geschwind also proposed a new classification system for aphasias and introduced experimental methods from psychology into aphasia research. American neuropsychologists Harold Goodglass and Edith Kaplan

(1972, 1983) developed and later revised the BDAE for aphasia assessment. These developments led to the establishment of two academic journals: *Brain and Language* in 1974 and the *Journal of Neurolinguistics* in 1985. Over time, the growing influence of neurolinguistics attracted the attention of Chinese linguistic researchers as well as scholars from other fields interested in neural mechanisms of language. These scholars began the preliminary work of translating and interpreting existing neurolinguistics research. In 1980, Chinese researchers Li Jiarong 李家荣 and Li Yunxing 李运兴 collated and published findings that American linguist Professor Thomas Scovel had presented at the third annual meeting of the Chinese Psychological Association in a lecture titled "Recent Trends in Neurolinguistics Research." Linguist Zhao Jisheng 赵吉生 provided a Chinese language introduction to neurolinguistics in 1984 by translating "Prospects for Neurolinguistic Theory," a journal article written by Canadian neurologist David Caplan of the Ottawa Civic Hospital in 1981. In 1987, Zhao and Wei Zhiqiang 卫志强 published a translation of *Basic Problems of Neurolinguistics*. The most notable feature of Luria's work is that its study of neurolinguistics was conducted entirely according to the syntagmatic and paradigmatic framework put forward by Swiss linguist Ferdinand de Saussure. This approach has had a decisive impact on the neurolinguistics discipline. The Chinese translation of Luria's book, titled *Neurolinguistics*, was the first monograph on neurolinguistics available in China. Its publication "promoted the spread of neurolinguistic theories in China" (Liang 2004, 144). Linguist Shen Jiaxuan 沈家煊 (Shen 1989) subsequently introduced aphasia studies to Chinese neurolinguistics, while Wei (1994) reexamined existing findings in an effort to elucidate the essence of neurolinguistics. Wei had already come to the realization that "there remain many unresolved problems in language science," and that finding an "ultimate solution to these problems will depend on developing neurolinguistics in great depth" (Wei 1994, 43). In 1997, Wang Dechun 王德春 published *Neurolinguistics*, which was the first Chinese work intending to provide a systematic introduction to neurolinguistics (Wang 1997).[2] In the early 1990s, Gui Shichun 桂诗春 published *An Outline of Experimental Psycholinguistics*, and followed up with an expanded version, *New Psycholinguistics*, in 2000 (Gui 1991, 2000). Gui's two books outlined current trends and research methods in experimental psycholinguistics outside China. Its attention to the biological and physiological basis of language also helped inspire new neurolinguistics research. In *Neurolinguistics: A Chinese Perspective*, Tzeng and Daisy Lan Hung 洪兰 (Hung and Tzeng 1996) compared the development of neurolinguistics across languages and cultures, reviewing the state of research in China and internationally, and identifying some general problems faced by the discipline. All of the aforementioned works contributed to the establishment of neurolinguistics in China. Over the years, an increasing number of Chinese linguistic researchers have become interested and involved in neurolinguistics through these translated works.

---

[2] *Translator's Note*: Wang's *Neurolinguistics* is not to be confused with the Chinese translation of Luria's *Basic Problems of Neurolinguistics*, also titled *Neurolinguistics*.

Establishing the Neurolinguistics Discipline in China

The advent of neurolinguistics as a discipline in China was the culmination of linguists' efforts over many years. Adopting a research perspective informed by fundamental theories in linguistics, Chinese researchers studied language in the brain, established several academic journals, and expanded research training centers and research teams.

Chomsky's transformational grammar, one of the most influential linguistic theories of the twentieth century, holds that the sentences we read are represented at two different levels. A sentence is a specific linguistic form, known as a surface structure, which is derived, projected, or otherwise generated from the cognitive level of deep structure via transformations. In particular, the transformation process allows the same deep structure to be mapped onto different surface structures. As a result, passive constructions are generally more difficult to process than active constructions, because the surface structure differs from the deep structure. However, in a 1997 journal article, "A Comparative Study of the Production and Comprehension of Chinese Active and Passive Constructions in Patients with Subcortical Aphasia," Yang Yiming and Cao Ming found that patients with subcortical aphasia completing a sentence-picture matching task were faster matching sentences with *ba-* 把 or *bei-* 被 constructions than typical active sentences (Yang and Cao 1997).[3] Comparing *ba-* and *bei*-constructions, the aphasic patients were quicker at sentence-picture matching for *ba*-constructions but made more errors. Yang and Cao's results ran counter to the assumption that response times would increase as the task of transforming deep structure into surface structure becomes more difficult. Instead, their findings suggested that subcortical structures are extensively involved in the storage and retrieval of syntactic structures. The way that active and passive syntactic structures are stored and retrieved in the brain, then, does not reflect a transformational relationship between deep and surface structure. On the contrary, it points to the existence of parallel mechanisms of storage and retrieval which are relatively independent yet interconnected. This conclusion challenged a key basis of support for transformational grammar.

Yang and Cao's article combined the latest linguistic theories with research methods from brain science to investigate neural mechanisms of language using a Chinese language corpus. Its aim to reveal the nature and rules that govern language itself demonstrated that it is "neurolinguistics research with true significance" (Liang 2004, 149). The Chinese academic community considered Yang and Cao (1997) a formative study as "publication of this article played a major role in the establishment of neurolinguistics as a discipline in mainland China" (Lin and Gu 2002, 401). Furthermore, the study prompted the realization among linguists that many research

---

[3] *Ba* 把 and *bei* 被 structures are two important and commonly used constructions in Chinese. *Bei* is a passive marker and hence *bei*-constructions express the passive voice. In *ba*-constructions, the object is placed in a preverbal position and directly follows *ba*. Both structures have a telicity requirement that restricts the type of VP they select.

methods used in theoretical linguistics are merely theoretical deductions. If hypotheses in theoretical linguistics are to be considered genuine scientific theories, then they must be tested and revised through empirical research. This conclusion has served as the point of departure for Chinese neurolinguistics ever since (Yang 2007).

The establishment in 1996 of the Institute of Linguistics at Xuzhou Normal University, now Jiangsu Normal University (JSNU), marked a significant step forward in the effort to increase the number of scholars working in neurolinguistics and develop research talent. The institute's linguistics and applied linguistics departments specified neurolinguistics as their research direction and actively enrolled postgraduate students in their programs. Taking initiative in Chinese academia, neurolinguistics research conducted at the Institute of Linguistics pioneered the use of a Mandarin Chinese corpus and informed its approach with contemporary linguistic theories. Since 2001, the institute has worked in partnership with associated universities to enroll and train doctoral and postdoctoral students in neurolinguistics, and establish a Chinese neurolinguistics research team. In 2002, the Institute of Linguistics launched *Language Science*, an academic journal that covers major theories in linguistics, neurolinguistics research, and applied research in other languages. The journal gave neurolinguistics an entry point into research publication in China.

The Development of Neurolinguistics in China

Unlike neurolinguistics in Western countries – which historically relied on patients with language deficits to study language processing mechanisms – China's nascent neurolinguistics discipline quickly caught up to the wave of research that was using noninvasive neuroimaging technology. It was not long, for example, before Yang Yiming was employing ERP and functional magnetic resonance imagining (fMRI) technology to study one of the major and fundamental questions in linguistics: whether there are distinct neural representations for nouns and verbs in the mental lexicon. Yang's study, "Neurolinguistics and the Chinese Mental Lexicon Research Program" (2001), was the first neurolinguistics research project to receive funding from the National Social Science Foundation. In the study, Yang found that the classification of nouns and verbs occurs on the basis of grammatical properties rather than semantic properties (Yang 2002). Members of the study's neurolinguistics research team soon took up leading positions at renowned Chinese institutions including Zhejiang University, Nanjing Normal University, Central China Normal University, and the Institute of International Relations of the People's Liberation Army (PLA). In turn, this led a number of scholars who had been working in theoretical and applied linguistics to reorient their research toward neurolinguistics. Their work has yielded a series of new findings on aspects of language acquisition and comprehension. Many of these results came about through the use of neurophysiological and neuroimaging technologies, and by studying both healthy individuals and people with aphasia, reading deficits, dysphemia, and other forms of atypical language use.

By 2010, Chinese neurolinguistics had made significant progress as an emerging discipline. Yang Yiming secured funding in the first round of grants for major projects in both basic research and interdisciplinary studies, awarded by the National Social Science Foundation through its "Neurolinguistics Research and Development Program." This helped promote the rapid development of neurolinguistics in China, as it consolidated the foundations of the discipline, provided funding and support for training talented scholars, and drove advances in scientific research. The scope of neurolinguistics research expanded to cover a great many topics, including language acquisition, production and comprehension, language deficits, and language decline. The objects of study in neurolinguistics similarly expanded beyond typical language to include not only different forms of atypical language – such as occurs in dysphemia and autism – but also study participants from deaf, elderly, and youth populations. The availability and potential applications of research technologies also increased considerably over this period. The early days of EEG and fMRI gave way to research that adopted enhanced structural imaging technologies to study the neural basis of language, most commonly diffusion tensor imaging (DTI). There was also a rise in the use of technologies capable of nerve stimulation such as transcranial magnetic stimulation (TMS), along with a shift away from studies relying on a single technology toward multimodal research. These research developments and integrations have allowed us to understand neural mechanisms of language from a range of perspectives.

Disciplinary Expansion and the Researcher Development Program

In 1994, JSNU took the initiative by enrolling postgraduate students into the nascent field of neurolinguistics. Building on this foundation, the university partnered with other institutions in 2001 to recruit doctoral and postdoctoral students into the discipline. Over time, they established the first dedicated neurolinguistics research team in China. This team has had a formative influence on talented researchers working in the area and the discipline as a whole. Placing a high value on domestic and international research exchanges and partnerships, the research team has conducted joint research projects with University College London, Harvard University, the Massachusetts Institute of Technology, University of Southern California, University of Hong Kong, Zhejiang University, Nanjing Normal University, Institute of International Relations of the PLA, and Central China Normal University, among others. The team has also provided funding for its members to complete relevant degrees and specializations at eminent foreign universities including the University of Groningen, University of Potsdam, Hebrew University of Israel, University of Texas at Dallas, University of Southern California, and University College London. The financial support has allowed JSNU's neurolinguistics research team to study the latest theories, technologies, and research methods in neurolinguistics. In addition to its academic development program for full-time students, in 2015 the university obtained approval from China's Ministry of Education (MOE) to set up the first experimental teaching demonstration center for neurolinguistics research. The center holds annual summer camps for exceptional students from across China who are interested in neurolinguistics, giving them the opportunity to exchange ideas and learn from one another.

The JSNU research team has also given particular focus to adopting the latest research methods and techniques, consulting the world's leading neurolinguistics laboratories and steering the development of neurolinguistics research laboratories in China. In 2005, the research team established the Jiangsu Provincial Key Laboratory of Language Science and Neurocognitive Engineering for neurolinguistics research. In terms of research equipment, the laboratory installed China's first Neuroscan SynAmps 2 256-channel ERP workstation in 2004, which was one of only seven in the world at the time. In 2013, the laboratory purchased a GE750 MRI scanner, the first in China to be used for the express purpose of studying neural mechanisms of language. This was followed by the procurement of a TMS machine in 2018. In addition, the JSNU research team was among those selected in the first group of outstanding scientific and technological innovation teams to receive support from Jiangsu Province's "Qinglan Project" in 2006. With the aim of furthering research into the neural mechanisms of language, the JSNU team then founded the first Chinese institute with "language science" in its title in 2008. The following year it established a key research base for the study of philosophy and social science, catering to Jiangsu's regular higher education institutions.

The establishment of the Collaborative Innovation Center for Language Ability at JSNU was particularly noteworthy. In 2014, the university obtained official support to develop the center into the Jiangsu Provincial Collaborative Innovation Center for Language Ability. Further approval from provincial authorities was granted in 2019 for the center to be upgraded into a National Collaborative Innovation Center for Language Ability. Establishment of the center has allowed institutions including the MOE's Institute of Applied Linguistics, Peking University, Tsinghua University, Beijing Foreign Studies University, the Institute of Ethnology and Anthropology of the Chinese Academy of Social Sciences (CASS), the Institute of Psychology of the Chinese Academy of Sciences (CAS), Beijing CiWang Technology, the China Rehabilitation Research Center, and the Chinese University of Hong Kong to collaborate on research projects. Taking the study of human language abilities as a common starting point, the institutional collaboration facilitated by the center has helped address key issues in language research including the identification of universal cognitive mechanisms for language and efforts to improve language abilities among the Chinese population.

Since 1997, the JSNU research team has published over a hundred neurolinguistics research papers in leading international journals including *Brain, Cognition, Alzheimer's & Dementia, Brain and Language*, and the *Journal of Neurolinguistics*. The team has also published in reputable Chinese journals such as *Social Sciences in China, Science China, Studies of the Chinese Language, Foreign Language Teaching and Research*, and *Acta Psychologica Sinica*. The team has undertaken dozens of research projects funded by the National Natural Science Foundation of China (NSFC), the National Social Science Foundation of China (SSFC), and the National Basic Research Program of China (973 Program).

Aside from JSNU, several other Chinese tertiary institutions also launched neurolinguistics research programs during this period including Nanjing Normal

University, Zhejiang University, Hunan University, and the PLA's Institute of International Relations. Some other institutions added neurolinguistics courses to their offerings despite it not being one of their established disciplines. Peking University, for example, began to offer an elective course in neurolinguistics available to all students. Furthermore, neurolinguistics-related research teams were set up at universities including Central China Normal University, Southeast University, Nanjing University of Aeronautics and Astronautics, Dalian University of Technology, and Nankai University. Academics working in psychology also established their own research teams at a number of other research and tertiary institutions across China such as the Institute of Psychology of CAS, Peking University, Tsinghua University, Beijing Normal University, South China Normal University, Capital Normal University, Tianjin Normal University, Southwest University, Hebei Normal University, and Shanghai Normal University. Two research teams at Nanjing Normal University are of particular note. In addition to their main work of academic training and development, the team led by Professor Liang Dandan 梁丹丹 convened numerous study sessions on language impairments associated with autism, while the team of Professor Zhang Hui 张辉 organized training courses on research methods in neurolinguistics. In both cases, the teams provided teachers and students with important opportunities for learning and discussion.

The growth in multidisciplinary research teams has significantly enhanced the academic standards and skills of researchers. This is reflected in not only the diverse range of perspectives adopted in contemporary neurolinguistics research, but also in the way that researchers from different fields collaborate, draw on, and extend each other's work. This demonstrates the truly transdisciplinary nature of neurolinguistics. It is further illustrated by the fact that many scholars working in psychology, medicine, and cognitive neuroscience have acquired a substantial knowledge base in linguistics and its theories. Equally, those in linguistics have been able to use clinical methods from neurology, theoretical knowledge and experimental paradigms in psychology, and neurophysiological and brain imaging technologies such as ERPs and fMRI. Recently, a neurolinguistics research branch of the "Society on Modernization of Chinese Language" was founded at JSNU in 2018. The new research branch was established so that experts from linguistics, psychology, brain science, and related fields could come together to discuss their work and exchange ideas. It also gave those working in neurolinguistics a dedicated academic organization of their very own.

Research Areas in Chinese Neurolinguistics

The Chinese Mental Lexicon

Research on the mental lexicon, which investigates ways in which lexical information is stored and retrieved in the brain, is a major area in the study of human cognition (Yang and Cao 1998). Looking specifically at Chinese, Yang and Cao (1998, 2000) systematically investigated several aspects of the mental lexicon.

Adopting neurolinguistic methods, they not only examined storage and retrieval mechanisms in subsystems of the Chinese mental lexicon, but also studied neural mechanisms associated with recognizing characters, integrating linguistic form, sound and meaning, and storing and retrieving linguistic units more generally. The researchers "thereby incorporated existing findings from psychology and cognition research that had been based on intuition and encapsulated them in a study of the Chinese language grounded in neuroscience and empirical cognition research. This helped promote the formation of neurolinguistics in China and opened up new avenues for research" (JSPOPSS 2004, 240).

In their earlier study, Yang and Cao (1998) conducted a clinical survey of 12 patients with Chinese language aphasia. Most of the different types of aphasia and alexia were represented in the data. Participant aphasias included: Broca's aphasia, Wernicke's aphasia, transcortical motor aphasia, conduction aphasia, transcortical sensory aphasia, anomic aphasia, subcortical aphasia, and global aphasia. The survey results showed that, "the Chinese mental lexicon has coexisting and interrelated—yet independent—subsystems for storing a character's written form, pronunciation, and meaning. The appearance of phonological alexia (form–sound), surface alexia (form–meaning), and deep alexia (form–sound–meaning) is then an indication of damage to the orthographic, phonological or semantic subsystem respectively" (Yang and Cao 1998, 417). The authors' discussion challenges the cursory conclusion that the mental lexicon will necessarily be very different for the Chinese language than it is for Indo-European languages. It also counters the claim that the link between a Chinese character's written form and its meaning is strong while that between its form and pronunciation is weak. Yang and Cao instead take a broader perspective, one which outlines the connection between form, pronunciation, and meaning of Chinese characters in the mental lexicon, and furthermore understands the nature of the difference between Chinese and Indo-European languages. In a second, follow-up study, Yang and Cao (2000) further analyzed the structure of semantic, phonological, and orthographic representations in the Chinese mental lexicon, and aspects of storage and retrieval processes. Based on findings from their earlier clinical survey of aphasic patients, semantic fields in the mental lexicon are thought to play a key role in word storage. In particular, coordinate terms appear to have the strongest semantic connection. In the phonological subsystem, words that share initials, finals, or tones appear to have the strongest relation. Likewise, words with similar orthographic forms seem to be stored in close proximity. Observations of patients with alexia who display deficits for particular word groupings suggests that some words in the Chinese mental lexicon may be stored in a relatively integrated manner but retrieved via a diffuse activation mode.

In addition to the relative importance of form, sound, and meaning in lexical access, another major question concerning the Chinese mental lexicon is whether entries are stored as single characters (*zi* 字), morphemes (*yusu* 语素), or phrases (*duanyu* 短语). Yang (2007) and Zhang Shanshan 张珊珊 and colleagues (Zhang et al. 2006) conducted behavioral and ERP experiments with aphasic patients and healthy individuals to explore this issue. The experimental results showed that it is far more likely that words (*ci* 词) are the primary unit for lexical access, rather than single characters, morphemes, or phrases. In saying that, bound morphemes may also be stored in the mental lexicon as nonfixed lexical items. If it is true that words are the primary unit, then the way that *liheci* 离合词 "separable words" are represented in the mental lexicon is an interesting question that deserves further exploration.[4] The linguistics community has long been divided over whether *liheci*, a distinctive feature of the Chinese language, are best categorized as words or phrases. Those who support the view that *liheci* are words point to the fact that they have a single, unified meaning. Those of the opposing view argue that the very nature of *liheci* being separable means that they violate the principle of lexical indivisibility. There are also some who believe that *liheci* are fundamentally words but become phrases once other elements are inserted between the constituent characters. Zhang and Jiang Huo 江火 (2010) were the first to use ERPs to investigate differences in processing compound words, *liheci*, and phrases. They found that *liheci* and compound words differ in their storage format and semantic integration process, with compound words stored and semantically integrated in a manner akin to that for phrases. This evidence therefore does not support the view that a *liheci* is "a word when its characters are adjacent, a phrase when its characters have been separated." In addition, Yang and Gu Jiexin 顾介鑫 (Gu and Yang 2010) found that compound words in Chinese can form a continuous distribution according to the productivity of different compounding structures. In particular, subordinate compounding structures are relatively more productive (i.e., yield a greater number of compounds) than coordinative compounding structures.[5] The productivity of each compounding structure is indicated by the P600 amplitude, with a higher peak corresponding to greater productivity. In a more recent study, Gu and colleagues (Gu et al. 2018) found that the syntactic ambiguity of *liheci* leads to multiple possibilities for segmentation of syllabic units. This manifests as an increase in P200 amplitude on neuro-electrophysiological indicators.

But if words are the unit of storage and retrieval in the mental lexicon, does that mean that word classes have neural correlates in the brain? In Chinese neurolinguistics, the most important research topic in this area revolves around classification criteria for parts of speech. Current theoretical research essentially regard the grammatical function of words as the basic criterion governing part of speech classifications. However, evidence from the many experimental studies on neural

---

[4] *Translator's Note*: *Liheci* 离合词 "separable words" are a special type of Chinese verb. The unique syntactic feature of *liheci* is that its two characters can be separated by other elements such as modifiers and complements to form grammatically acceptable sentences, hence the name. For example, to express the meaning "we have met three times" using the verb *jianmian* 见面 "to meet," the complement is inserted between the two characters *jian* 见 "see" and *mian* 面 "face": *wo men **jian** guo san ci **mian*** 我们见过三次面 'we see GUO three time face'.

[5] *Translator's Note*: Subordinate compounds, in which one morpheme modifies the other, tend to be similar in Chinese and English. For example, the Chinese word *fa-shua* 发刷 "hair-brush" matches the English word "hairbrush." Conversely, coordinative compounds, in which both morphemes contribute equally to word meaning, are common in Chinese but rare in English (and usually hyphenated). Chinese examples include *fu-mu* 父母 "father-mother" meaning "parents," and *shi-wu* 事物 "matter-thing" meaning "thing" or "matter."

mechanisms for verbs and nouns conducted around the world suggests that while the two word classes can be differentiated within the brain, the dissociation is done primarily on the basis of semantic properties (Bates et al. 1991). If so, this brings into question existing findings on part of speech classifications, and even the organizational principles of the grammatical system as a whole. For the moment though, such a conclusion is premature; experimental studies of nouns and verbs to date have mostly focused on semantic rather than grammatical properties. Among the neurolinguistics research that has examined nouns and verbs from a grammatical perspective, two studies led by Yang (Yang et al. 2002) and Liu Tao 刘涛 (Liu et al. 2008), respectively, have investigated the role of the cranial nerve in noun-verb dissociation in Chinese using fMRI and ERPs. Both sets of results showed that there are distinct neural mechanisms for nouns and verbs, at least in terms of grammar. Similarly, Xia Quansheng 夏全胜, Lü Yong 吕勇, and Shi Feng 石锋 (Xia et al. 2012) found that noun-verb dissociation occurs as a result of differences in syntactic properties, not semantic dimensions such as concreteness. Yu et al. (2012) and Feng et al. (2020) followed up these results with fMRI studies that have shown that the left inferior frontal gyrus (IFG) has a higher degree of activation during verb processing than during noun comprehension and production. Yang and Wang Tao 王涛 (Wang and Yang 2017a) provided further support for noun-verb dissociation in their study of Chinese-English bilinguals. Overall, the aforementioned studies confirm that grammatical criteria for part of speech classifications have a neurological basis in the brain. The findings are of great significance for resolving theoretical debates in this area (Shen and Le 2013).

Researchers have observed other grammatical class dissociations in addition to that for nouns and verbs. In an fMRI study, Liang Dandan and colleagues (Liang et al. 2006) found that nouns, verbs, and adjectives are each supported by distinct neural correlates, which are derived from the different grammatical functions of these three word classes. Chou et al. (2012) observed that syntactic processing of noun phrases involving classifiers recruits different areas of the left IFG than does semantic processing. Yang (2007) explored subcategories of modern Chinese classifiers by investigating classifier use in people with aphasia. Yang found that nominal and verbal classifiers are supported by distinct neurophysiological mechanisms, which are located in the left temporal lobe and/or along frontal and temporal lobe pathways. This result shows that nominal classifiers and verbal classifiers have corresponding neurophysiological mechanisms distinguishable on the basis of grammatical criteria. Conversely, specialized and borrowed classifiers – which are encompassed by the broader classes of nominal and verbal classifiers but categorized according to other (nongrammatical) criteria – do not have corresponding neurophysiological mechanisms. Meanwhile, Chinese-speaking patients with grammatical deficits (Chinese agrammatism) show impairments in, firstly, their ability to generate certain parts of speech such as function words (You et al. 2005) and verbs (Xu et al. 2006), and, secondly, their ability to position prepositions correctly in a sentence (Sun et al. 2007). Researchers believe that patients with the same type of aphasia will display similar grammatical impairments. A study of aphasic patients by Tzeng, Chen, and Hung (1991) found that patients with Chinese-specific aphasia exhibited clear absence and/or substitution errors when generating nominal classifiers, but those with Broca's aphasia were often able to substitute the correct form with a general classifier. On the other hand, patients with Wernicke's aphasia exhibited irregular patterns of substitution errors. Their selection and processing of nominal classifiers was also less systematic than in other types of aphasia.

Confirmation of a neurological basis for parts of speech prompted scholars to begin exploring neural representation of individual word classes in greater depth. This has included the study of semantic categories. For example, Bi et al. (2011) studied the performance of a patient with left anterior temporal lobe damage on a picture-naming task. The results showed that the degree of impairment in producing the name was specific to its semantic category. In particular, the poorest naming performance was for people- and place-related categories. The next most impaired category was inanimate objects. The patient's ability to name animals, on the other hand, remained unaffected. Here it is worth noting that the naming deficit only manifested at the moment of production, an indication that semantic-category-specific deficits arise during lexical retrieval and not earlier in the concept representation stage. Yet the semantic-category-specific effects found in lesion-deficit research – including the animate/inanimate distinction identified by Bi and colleagues – were not seen in a study of healthy individuals (Mo et al. 2005).

In research into abstract concepts, a recent study by Wang et al. (2018) collected BOLD (blood-oxygenation-level-dependent) fMRI responses for 350 abstract words. They found that the linguistic contextual similarity of abstract words, which was reflected in language-corpus-based co-occurrence patterns, correlated with neural activation patterns in the parietal lobe, precentral gyrus, and other areas of high-level linguistic processing. Similarity in semantic features, on the other hand, was associated with distributed activation in regions including the bilateral temporoparietal junction (TPJ), left middle temporal cortex, and left inferior frontal cortex. These results show that both similarities in linguistic contextual information and semantic features influence the representation of abstract concepts in the brain.

Syntax

Yang Yiming's research team used neurolinguistic methods to demonstrate that, contrary to the ideas of transformational grammar, the human brain does not perform syntactic processing using transformation mechanisms on deep and surface structure (Yang and Cao 1997). Following this result, Yang's team established the neurological basis of trace theory, which is now one of the leading theories in generative grammar. According to trace theory, movement of a constituent from one syntactic position to another leaves an empty trace (i.e., one with no phonetic content) in its former position. The moved constituent and its trace then combine to form a syntactic chain. Here it is worth noting that some similar theories, such as relational grammar and lexical-functional grammar, do not include traces. Generative grammar itself discarded the idea of traces during its reductionist period. But, ultimately, the key issue is not whether traces are included in the theory, but whether traces have a neurophysiological basis in the brain. To investigate this question, Yang and Liu (2013) conducted an ERP study in which participants were presented with topicalization constructions in Mandarin Chinese. The researchers found that processing topicalized constructions elicited three electrophysiological responses: a sustained anterior negativity (SAN), a negativity associated with the verb, and a P600 at the sentence-final position. Together these ERP components indicate that a sentence-initial topic leaves a trace in the position out of which it moves – the sentence-final position. In particular, the P600 is triggered following the establishment of syntactic dependency between the initial moved constituent and its final trace. This form of syntactic integration processing is additionally modulated by working memory load, with greater working memory load associated with higher P600 amplitude (Liu and Jiang 2016).

Yang and Liu's findings inspired a series of follow-up studies that sought to determine the psychological reality of linguistic theories using cognitive neuroscience techniques. These studies have found evidence supporting the existence of universal grammar in Chinese. Yet they have also shown that the Chinese language may be distinctive in some respects. The latter finding can be attributed to differences in the specific parameters of Chinese within the universal grammar framework.

Regarding the universality of language, a major issue research topic concerns the neurological basis of Chomsky's hierarchy of grammars. Ding et al. (2015) has directly examined this question. The researchers found that the hierarchical structure that emerges from recursive composition in language has parallel representation in the brain in the form of neural oscillations of different rhythms, which are realized in the bilateral frontotemporal network. Their study showed that retrieval of hierarchical syntactic structures is a necessary step in language processing. It further showcased new techniques that could be adopted in subsequent research into neural representations of syntactic structure, and that could also be used to describe language acquisition in childhood. In a more recent study, Sheng et al. (2019) found that the left superior temporal gyrus (STG) is involved in language processing at word, phrase, and sentence levels. The researchers also recorded activity in the right motor area that demonstrated its sensitivity to the rhythm of monosyllabic words with clear acoustic boundaries. Processing phrases and sentences, on the other hand, was associated with the left temporal lobe and left IFG, respectively.

In an fMRI study, Yang (2007) examined response mechanisms in the brain when Chinese nouns, verbs, and adjectives acted as modifiers for the same noun head. The results showed that the left MFG – which is mainly responsible for syntactic analysis and processing – was only activated when verbs acted as the modifier. This suggests that verbs require complex syntactic operations to act as modifiers. That is, the syntactic properties of verbs need to be altered in order to achieve a complete transformation of grammatical function. While this operation has a basis in neural mechanisms, it does not appear to be expressed at the phonetic level. What this ultimately shows is that Chinese verbs are subject to an abstract morphological finiteness. Another study by Yang, working with Feng Shiwen 封世文 (Feng and

Yang 2011), combined fMRI technology with Chinese case studies to explore the plausibility of the Chinese light verbs hypothesis and its associated theoretical framework. The results confirmed the hypothesis that the left frontal cortex contributes to processing light verbs in Chinese. More recently, Wang et al. (2017) used ERPs to conduct the first investigation of the time course for processing Chinese elliptical constructions. They observed differences in amplitude for the N400 and P600 components generated by "*ye* 也 + modal verb," "*yeshi* 也是," and null object constructions (NOC).[6] In particular, the P600 component had a higher amplitude when generated by *yeshi*-constructions as opposed to NOC, suggesting that NOCs are a form of VP-ellipsis.

Related research has also helped address a number of debates concerning the distinctiveness of Chinese grammar. An aspect of this research effort has involved examining some unique features of Chinese relative clauses. A behavioral study by Hsiao and Gibson (2003) offered the first evidence that in Chinese, the processing of subject relations in relative clauses is comparatively more difficult than processing object relations. This stands in contrast to Indo-European languages, in which the opposite is true. Hsiao and Gibson's conclusion drew the attention of scholars researching the Chinese language. For example, it motivated Zhang Qiang 张强 and Yang (Zhang and Yang 2010) to conduct an ERP study using a corpus of common words, as they sought to improve upon the low ecological validity of previous behavioral studies (e.g., Chen and Ning 2008; Zhou et al. 2010) that used corpora of complex words seldom encountered in everyday life. The results revealed that presentation of subject relative clauses resulted in N400 or sustained negativity components of greater amplitude than those induced by object relative clauses. Liu, Yang, and Zhou Tongquan 周统权 (Liu et al. 2011) then revealed that Chinese shows a processing advantage for subject relative clauses once the temporary ambiguity of the syntactic structure is resolved with the inclusion of a sentence-initial demonstrative classifier phrase.[7] Taken together, these studies show that it is too simplistic to conclude that there is a general processing advantage for object relative clauses in Chinese.

A second research question relating to the distinctiveness of Chinese grammar is the extent of autonomy between syntactic and semantic processing. The lack of morphological markers in Chinese has motivated a theoretical concept of paratactic grammar, that is, that grammatical relations in Chinese depend on semantic relations.

In order to assess the relative autonomy of syntactic and semantic processing, Wang et al. (2008) compared normal sentences with those that contained either semantic violations or both semantic and syntactic violations. They found that syntactic violations led to difficulties with syntactic integration and were associated

---

[6] *Translator's Note*: In Chinese, VP-ellipsis is often introduced with a "*ye* 也 + modal auxiliary verb" structure or "*yeshi* 也是" structure, as the adverb *ye* means "also" or "too" and *yeshi* expresses the meaning that something "is also true."

[7] *Translator's Note*: A demonstrative classifier phrase is a demonstrative with a classifier. For example, *zhe zhong* 这种 "this CF," which means "this kind (of)."

with stronger activation in Brodmann area 44 (BA44). These effects were independent of the semantic integration process, a result consistent with findings for Indo-European languages. Chou et al. (2012) also observed that different regions of the left inferior frontal cortex were involved in syntactic compared to semantic processing of numerative classifier phrases. Other studies using a priming paradigm have found syntactic priming effects that are independent of semantics in sentence comprehension (Chen et al. 2013), and sentence production (Huang et al. 2016). These results indicate that effects of autonomous syntactic processing and syntactic representations are both observable in Chinese.

In addition to the aforementioned processing mechanisms, all of which are consistent with findings for Indo-European languages, neurolinguistics research has discovered a number of phenomena particular to Chinese grammar. For example, Yang, Wu, and Zhou (2015) found that syntactic processing does not take precedence over semantic processing in Chinese. Several research groups have similarly demonstrated that syntactic analysis is not necessarily a prerequisite for semantic processing, as semantic processing is still able to proceed even in cases where syntactic analysis cannot be performed as normal (Ye et al. 2006), or else have shown that semantic processing at higher syntactic levels is still able to continue even when processing at lower syntactic levels fails (Zhang et al. 2010). These distinctive findings might be attributable to the lack of morphological markers in Chinese, as the early processing of parts of speech thought to occur in Indo-European languages is in fact closely related to the presence of morphological markers (Friederici 2002).

Furthermore, in the case of Chinese reading comprehension, the processing of syntactic information and semantic information is not carried out independently in simple parallel or separate stages but is interlinked. Work by Zhang et al. (2011) found that the appearance of semantic violations across multiple syntactic levels in the course of reading sentences has a cumulative effect, in that there are greater increases in N400 and P600 amplitudes compared to when semantic violations occur at only one syntactic level. This result reveals that semantic processing of high-level syntax does not simply override semantic processing at lower levels. More recently, Gu Jiexin, Zhou Xin 周昕, and Weng Jingqi 翁婧琦 (2018) showed that syntactic information and semantic information jointly influence the interpretation of syntactic ambiguity. Moreover, Li and Yang (2010b) found that prosodic boundaries may be used to guide syntactic analysis, and can be integrated with contextual information in real time during comprehension of spoken text. Wang et al. (2012a, b) also showed that information structure regulates the depth of syntactic processing.

Semantics

Clinical surveys of Chinese-speaking patients with Wernicke's aphasia provided the earliest evidence that the left posterior STG is closely associated with semantic processing (Gao 1993). This and similar findings obtained through the study of aphasia have since been confirmed in fMRI studies. Whether examining semantic processing at the word level (Tan et al. 2000), phrase level (Luke et al. 2002), or sentence level (Mo et al. 2005; Zhu et al. 2009), fMRI-based research has found activation in areas of temporal and frontal lobes around the lateral sulcus. Based on documented differences between Chinese and Indo-European languages, some researchers have suggested that the left MFG (BA9) could be a key region for Chinese word reading. However, establishing the specificity or otherwise of Chinese in this regard would require that multiple languages be directly compared within the one study, and additionally, that alternative explanations such as task demands could be ruled out. In fact, a recent meta-analysis of fMRI results by Zhao et al. (2017) suggests that left MFG activity is modulated by task demands for general cognitive processing, not by the specificity of Chinese character stimuli. Hence, further research is needed to establish whether Chinese language processing involves specialized brain regions.

Unlike semantic retrieval of individual characters, reading comprehension requires establishing a cohesive semantic representation that successively integrates word meanings with the previous text. A study of Chinese sentences by Zhu et al. (2009) found that an increase in the level of difficulty of semantic integration was associated with stronger activation in the left IFG. Subsequent research has shown that the involvement of the left IFG in semantic integration is not affected by task strategy (Zhu et al. 2012). Another study conducted a regression analysis of the N400 and BOLD signal using simultaneously recorded EEG and fMRI data. Results of the analysis confirmed that the left IFG plays an important role in semantic integration (Zhu et al. 2019a). An increase in the difficulty of text integration has also been shown to lead to greater working memory load (Yang et al. 2018).

Data analysis in studies of sentence reading is most often based on conditional averaging. This analytical technique makes it difficult to observe how variation in lexical and semantic features of language stimuli, and in semantic integration, affects processing. However, by using trial-by-trial integration of simultaneously recorded EEG and fMRI signals, Zhu, Bastiaansen et al. (2019a) was able to reveal that the bilateral supramarginal gyrus (SMG) is sensitive to trial-level fluctuations in the difficulty of semantic integration, an effect which would otherwise have been concealed by conditional averaging. Bilateral SMG sensitivity may reflect increases in phonological loop/working memory load resulting from semantic integration difficulties.

The study of brain injury and applied research on neuromodulation offer another window into semantic processing in the brain. A lesion study by Han et al. (2013) found that more extensive damage to and reduced white matter (WM) integrity in the left inferior fronto-occipital fasciculus, left anterior thalamic radiation, and left uncinate fasciculus were associated with more noticeable semantic deficits. Mirman et al. (2015) showed that semantic and phonological deficits were related to structural damage in different neural pathways. Zhu et al. (2015) and Zhang et al. (2019), meanwhile, used TMS to create temporary "virtual lesions" in the left IFG and left middle temporal gyrus (MTG). The resultant localized disruption to functioning helped confirm that these two regions play a major role in semantic processing.

Any neural mechanism involved in reading comprehension must be capable of integrating lexical and contextual information. An ERP study by Wang, Zhu, and Bastiaansen (2012b) found that an increase in the difficulty of semantic integration leads to increases in not only N400 amplitude, but also the power of beta band EEG signals (19–25 Hz). In comparison to these results for sentences containing semantic violations, semantically acceptable sentences that could be integrated successfully resulted in power increases in the gamma band. Compared to previous findings that had only observed changes in N400 amplitude, these results demonstrate that time-frequency analysis is a viable new approach for investigating language processing mechanisms. Subsequent research has shown that in cases where sentences are presented under strict constraints, the provision of contextual information allows words to be predicted with greater certainty. More particularly, at the anticipatory stage of target word presentation, a strong-constraint condition induces a larger SAN than a weak-constraint condition. This is accompanied by a decrease in the time frequency of the beta band (Li et al. 2017). Zhou Tongquan and Zhou Siruo 周思若 (Zhou and Zhou 2018) provided further evidence of this expectation mechanism operating in the course of reading modifier-head NPs. Characteristics of information structure have also been shown to affect semantic processing. Wang, Hagoort, and Yang (2009) found that semantic violations induce the typical N400 amplitude increase when the target word is in focus, but that the effect on N400 amplitude is significantly reduced at nonfocused positions. This suggests that information structure enhances semantic processing through the allocation of greater resources to the most important pieces of information.

Pragmatics

Although pragmatics can be distinguished from semantics as far as linguistic theory is concerned, the neurological basis and specific mechanisms of language function in the brain remain unclear. Drawing on the notion that language use will influence verb classifications, Yang (2007) divided adverbs into two categories, "dynamic adverbs" (e.g., *ye* 也 "also," *cai* 才 "just," *zhi* 只 "only") and "static adverbs" (e.g., *gewai* 格外 "especially," *huxiang* 互相 "mutually"). Based on this distinction, Yang applied neuropsychological techniques to investigate neural mechanisms of adverb use. The premise was that if neurolinguistic evidence supported a theoretical classification of adverbs according to use, this would also indirectly establish the existence of neuropsychological mechanisms for language use in general. The results showed that lesions in the left frontal lobe led to difficulties in using dynamic adverbs, whereas lesions in the left temporal and parietal lobes led to issues with static adverbs. Thus, there appears to be a double dissociation between dynamic and static adverbs in the brain, offering support for the psychological reality of adverb use categories.

Another important aspect of pragmatics looks at processing mechanisms for idioms and metaphors. A study of two-part allegorical Chinese sayings (*xiehouyu* 歇后语) by Zhang, Jiang et al. (2013) found that sayings that were highly familiar led to less significant amplitude changes in N170 and N380 components.[8] In subsequent work, Zhang, Yang et al. (2015) observed different neural responses to Chinese four-character set phrases (*chengyu* 成语) and nonidiomatic phrases. Overall, nonidiomatic phrases elicited greater N250 and N400 responses than set phrases. But set phrases with a lower degree of compositionality were associated with greater N250 and N400 amplitude changes. These changes may relate to early access and late integration processes, respectively. ERPs have also been used to examine neural mechanisms for metaphor recognition in Chinese. In one such study, Wang Xiaolu 王小潞 (Wang 2009) found that metaphor recognition has a distinct basis in patterns of neural activity, and that both hemispheres of the brain are involved in metaphor processing. This study is of particular significance; it was the first in China to examine processing mechanisms for metaphors using ERPs. Subsequent work by Li Ying 李莹 and colleagues (Li et al. 2016) found that judgements of sentences containing derived metaphors were associated with higher cortical activation than judgements of ordinary sentences. Cortical activation was mainly distributed across the left MTG/STG and the left IFG. Gu Jiexin, Yu Fangfang 余芳芳, and Wang Jinjun 王进军 (Gu et al. 2019) recently studied neural mechanisms of metaphor production in Chinese. Key regions identified as being involved in metaphor production include the left angular gyrus, left MTG, bilateral cingulate cortex, right ACC, and right STG.

Zhao Ming 赵鸣 and colleagues (Zhao et al. 2012) found that pragmatic reasoning and comprehension of conversational implicature require more complex processing than literal comprehension. This is reflected in increases in EEG amplitude. In subsequent research, Zhao et al. (2015) considered the effect of individual differences on the mechanism of scalar implicature processing. The researchers discovered that ERP amplitude increases elicited by underinformative sentences were more pronounced in the high pragmatic ability group compared to the low pragmatic ability group. Another ERP study by Jiang, Li, and Zhou (2013) found that honorifics and other pragmatic features influence the comprehension of pronouns in real time. Finally, Feng et al. (2017) investigated the effect of contextual relevance on pragmatic inference during conversation. The results showed that, in comparison to direct replies, indirect replies lead to increased activation in classical language areas (such as the bilateral IFG and MTG) and theory-of-mind processing areas (such as the bilateral TPJ, dorsal medial frontal cortex, and precuneus). They also noted greater functional connectivity between these two areas for indirect replies.

---

[8] *Translator's Note*: *Xiehouyu* 歇后语 are a kind of witty or humorous Chinese saying that is expressed in two parts. The first part of the *xiehouyu* is always stated and usually describes a scenario. The second part, often left unstated, contains the allegory or message. For example, a speaker may only state the first part of the expression *rou bao zi da gou— you qu wu hui* 肉包子打狗—有去无回 "to throw a pork bun at a dog—one can never get back what is given away," expecting that the other party is already familiar with the saying and knows the second part. *Xiehouyu* are not to be confused with the more widely known *chengyu* 成语. *Chengyu* are set phrases or idiomatic expressions, almost always four characters in length, which typically follow linguistic conventions of Classical Chinese. *Chengyu* often refer to and/or originate in ancient Chinese history and literature. An example of a *chengyu* is *dui niu tan qin* 对牛弹琴 "play the lute to a cow," which has a similar meaning to the English saying "cast pearls before swine."

Phonology

Language is a system of symbols that link sound with meaning. Even though sound and meaning ultimately cannot be separated, it is essential to try to analyze them in isolation in order to investigate the basis of phonological processing. For example, Yan Pei 晏培 and Gao Surong (Yan and Gao 2000) observed that Chinese-speaking patients with transcortical sensory aphasia or deep aphasia are able to activate the semantic system through visual processing, but are unable to derive a character's correct pronunciation due to issues with grapheme-to-phoneme conversion. Examining Broca's aphasia and conduction aphasia, Cui Gang 崔刚 (Cui 1999) identified phoneme substitution errors as the most common error. Cui believed that this may be because the Mandarin syllable structure, consisting of an initial (*shengmu* 声母, i.e., onset, an initial consonant or zero initial) and a rime (*yunmu* 韵母, the remainder of the syllable excluding the initial), significantly restricts articulation possibilities; even those with language deficits will make an effort to produce language that conforms to this syllable structure. In terms of phonological processing of Chinese characters and compound words, Peng, Deng, and Chen (2003) and Zhang et al. (2004) showed that processing mainly involves the temporal lobe, with coordinate processes performed by the parietal and occipital cortices, such that these three regions together form a temporo-parieto-occipital phonological processing network. Adopting a priming paradigm, Yu et al. (2015) found that the bilateral basal ganglia shows clear neural adaptation in response to phonemic repetition, while the bilateral STG is sensitive to syllabic repetition. The identification of distinct representations of phonemes and syllables in Chinese production is consistent with evidence from Indo-European languages.

As Mandarin is a tonal language, lexical tone is a major research area in Chinese phonology. In a study of Mandarin lexical tone and sentence intonation, Liang Jie 梁洁 (Liang 2002) comparatively analyzed data obtained from an aphasic patient and five healthy participants as they read out declarative sentences. Based on the results, Liang proposed that declarative sentences in Modern Standard Chinese are characterized by a falling intonation pattern. He additionally suggested that Mandarin tones and intonation be classified into two separate tiers: a higher tonal level and lower intonation level, with the four tones influenced by intonation, distribution, speech rate, and vocal range. A research team led by Chen Hsuan-Chih 陈烜之 later showed that tone information specifying word meaning contributes to semantic processing within approximately 300 milliseconds of word onset (Schrimer et al. 2005).

In research on the functional organization of lexical tone processing in the brain, Luo et al. (2006) found that the right hemisphere dominates early processing of acoustic information, while the left hemisphere dominates late processing of semantic information. Zhang et al. (2016) identified the bilateral STG as a more specific area of tone processing. Feng et al. (2018b) then found through representational similarity analysis that the left STG and left inferior parietal lobule are associated with two dimensions of tone perception: pitch height and pitch direction. These two regions thereby support speech representation and categorization processing. Liang and Du (2018) found through activation likelihood estimation meta-analysis that only speakers of a tonal language, such as Mandarin, process lexical tones using the left anterior STG. They also found that processing of phonemes, tones, and prosody in this region is functionally organized along an anterior-posterior axis. While the left anterior STG appears not to be engaged in lexical tone processing during visual word recognition (Kwok et al. 2015), Si, Zhou, and Hong (2017) used intracranial recordings to show that it is involved in auditory tone processing. They also recorded activation in the MTG and motor cortex. Finally, Ge et al. (2015) found by means of a comparative natural language task that the bilateral anterior temporal lobe contributes to comprehension of tonal languages.

Prosody is another important aspect of phonology. Chinese research in this area by Li Weijun 李卫君 and Yang Yufang 杨玉芳 (Li and Yang 2010a) found that the EEG response elicited by hierarchical prosodic boundaries are influenced by prosodic boundary cues. In particular, a sentence-final boundary that indicates closure of the preceding sentence and anticipation of the forthcoming sentence induces a closure positive shift (CPS), whereas a sentence-final boundary that only indicates closure of the previous sentence generates a P3 component. Furthermore, CPS amplitude and onset latency is greater for prosodic boundaries at higher levels of the prosodic hierarchy (Yang 2015). For texts of larger scale, all prosodic boundaries can evoke CPS, but its onset latency will change systematically with the level (Li and Yang 2010b). Furthermore, Li and Yang (2010a) also discovered that prosodic boundaries influence syntactic analysis. Instances when the boundary position is inconsistent with the text structure generate a left anterior negativity (LAN) of greater amplitude. Moreover, the processing difficulty increases in cases where the prosodic pattern is incorrect, regardless of whether it conforms to the given syntactic structure or not (Zhang et al. 2012, 2013). A recent study has also demonstrated that rhythm regularity facilitates Mandarin speech comprehension (Li et al. 2018).

Comprehension of classic Chinese poetry also relies heavily on prosodic processing. Chen et al. (2016) found that top-down rhyme scheme expectations affect the P200 response, influencing early phonological coding before lexical-semantic processing. This shows that top-down prosodic expectations can modulate early phonological processing in visual word recognition. Furthermore, in the N400 to P600 response time window, semantic comprehension and integration processing of Chinese poetry is modulated by expectations based on rhyme scheme. In a later study, Chen Qingrong and Yang Yiming (Chen and Yang 2017) discovered that rhyme effects are sustained during the reading of classic Chinese poems, which shows that prosodic expectation controls generation of the rhyme scheme in early processing, and constrains semantic comprehension in later stages.

Language Acquisition, Development, and Decline

In the study of typical language acquisition in children, the early stages of research – which came mainly in the form of behavior observation – concentrated on acquisition of vocabulary, phonology, semantics, syntax, and sentence structures in normally developing children (Li and Chen 1997). Considering neural mechanisms of typical language development, Zhao et al. (2014) found that Chinese-speaking children already display a relatively specific N170 response for Chinese characters by the age of seven. As children develop their reading skills, greater sensitivity to orthographic regularity facilitates more efficient lexical classification. This change is accompanied by a moderated N170 response (Zhao et al. 2019). Xia et al. (2017) reported a positive correlation between children's oral word reading ability and cortical thickness of the bilateral STG, left ITG, and left superior frontal gyrus (SFG). The researchers also noted that different cortical regions are involved in processing phonological representations, orthographic-to-phonological mapping skills, and the development of phonological awareness. Another study by Su et al. (2018) looking at Chinese children aged four to ten found that reading development depends on WM integrity of the arcuate fasciculus along with the rate of vocabulary acquisition. However, the initial size of a child's vocabulary was not reliably predictive of changes in the left arcuate fasciculus. This suggests that language-learning experience promotes structural development of the brain in children and adolescents to some extent.

In contrast to language development studies, research on language decline receives little attention. In one recent study examining language change in an elderly population, Zhu et al. (2019b) controlled for a number of potential confounding variables including years of education, verbal fluency, verbal intelligence, and working memory. ERP results revealed a delayed peak latency and smaller N400 effect in older participants compared to younger participants. The fact that the N400 had a reduced amplitude and was restricted to posterior regions in older adults also suggests that semantic integration ability gradually declines with age. This age-related decline was also observed by Zhu et al. (2017). Using fMRI, the researchers found that activation in regions linked to semantic integration decreased significantly with age, namely, in the left MFG and right IFG. Moreover, the reduction in neural activity was closely associated with structural atrophy. Other studies have shown a similar relationship between lower WM integrity and semantic deficits (Fang et al. 2015), or else observed that late-maturing fiber tracts in the language network are more affected by aging than early-maturing fibers (Yang et al. 2014). Zhu, Hou, and Yang (2018) found a gradual age-related decline in semantic integration ability. Interestingly, even though the older adult participants showed a delayed P600 peak and reduced processing efficiency during sentence comprehension, syntactic processing itself was maintained. This indicates that aging affects different aspects of language function in different ways. Subsequent work by Xiao Rong 肖蓉, Liang Dandan, and Li Shanpeng 李善鹏 (Xiao et al. 2020) on age-related changes in lexical tone perception found that the mismatch negativity (MMN) response intensity elicited by tones involving category changes and nonspeech intonation was weaker in older adults. Conversely, the MMN for tones that did not involve category changes showed no reduction in intensity. Among studies into language rehabilitation, Zhou et al. (2018) observed that plasticity mechanisms in neural networks can effectively promote language recovery in patients with aphasia. The following year, Tang et al. (2019) published the first evidence that language ability can compensate for cognitive decline, demonstrating that language-related cognitive training enhances language and cognitive function in patients with mild vascular cognitive impairment.

Language Use in Atypical Populations

Dyslexia

The study of language processing in atypical populations has concentrated on individuals with conditions such as dyslexia, autism, stuttering, or deafness. Dyslexia is a characteristic language disorder. Developmental dyslexia is a condition that occurs in children who have normal intelligence, normal comprehension skills, and show no neurological or organic deficit, but nevertheless have reading difficulties (Lyon et al. 2003). For Indo-European languages, the underlying cause of developmental dyslexia is thought to be a phonological deficit and an additional naming-speed deficit. However, research by Wu Sina 吴思娜, Shu Hua 舒华, and Wang Yu 王彧 (Wu et al. 2004) has shown that developmental dyslexia in Chinese may arise due to deficits not only in phonological processing but also in morphological processing. This suggests that the condition takes a unique form in Chinese speakers. Following up on these results, Tong et al. (2017) found that the appearance of spelling difficulties in Chinese-speaking children with dyslexia is associated with morphological awareness, which can be assessed using visual reproduction tasks. Orthographic skills are also significantly poorer in children with dyslexia compared to normally developing children. This result has major implications for Chinese, as visuo-orthographic skills play an important role in reading ability among Chinese children with dyslexia (Li et al. 2012a).

This raises the question of whether developmental dyslexia has a universal neurological basis or one that is language specific. Neural substrates associated with reading impairment in Chinese were explored by Siok et al. (2004) in an fMRI study. They found that developmental dyslexia usually arose following functional disruption of the left MFG. Reading impairments manifested as two deficits, one associated with grapheme-to-phoneme conversion and another in orthography-to-semantics mapping. This is a novel result, as existing literature on reading impairment in Western orthographic systems has generally only identified the grapheme-to-phoneme conversion deficit. On the basis of this, the researchers suggest that the pathophysiology of developmental dyslexia is not universal but dependent on culture. Differences in developmental dyslexia across languages have been further explored by Hu et al. (2010). They compared the performance of Chinese and English speakers with developmental dyslexia and matched controls on a lexical-semantic matching task using English language materials and equivalent Chinese translations. Although clear differences in regions of neural activity were observed for normal Chinese and English readers, the brain activation patterns of the Chinese and English dyslexic participants were very similar. This was the first time that shared neural correlates of dyslexia had been identified in Chinese and English populations, an indication that dyslexia is not affected by specific languages or orthographic systems but influenced by shared cognitive and environmental factors. Among other studies to identify neural correlates of developmental dyslexia, Zhou et al. (2015) found that children with dyslexia have deficits in functional connectivity between areas associated with reading fluency, in particular, between the MFG and dorsal and ventral pathways of the visual cortex. Su et al. (2018) also observed damage to WM fibers in areas including the left arcuate fasciculus in people with dyslexia.

Stuttering

Research into people who stutter has revealed differences to normally developing individuals in some aspects of language processing and its neural correlates. Lu, Peng, et al. (2010b) found that people with a stutter had dual-stream processing deficits for speech motor planning and execution, and functional and structural anomalies in the basal ganglia-thalamocortical circuit. A recent study by Li Weijun and colleagues (Li et al. 2018) that examined processing of ambiguous phrases revealed no significant differences in the CPS (an ERP that reflects prosodic segmentation) for people who stutter compared to fluent speakers of Chinese. Another study found that people who stutter have a higher sensitivity to prosodic priming than other adults, and that this is lateralized to the left hemisphere (Liu et al. 2018).

High-Functioning Autism

Children with high-functioning autism (HFA) are another group that exhibit atypical development of language skills. Wang et al. (2017) noted that the fact that different modes of pitch processing for speech compared to nonspeech stimuli are observed in individuals with autism may be due to speech-specific deficits in the categorical perception of lexical tones. Song et al. (2017) found that children with HFA have difficulty understanding verb argument structure for psychological verbs and produce sentences or clauses of lower argument complexity. Finally, Zheng, Jia, and Liang (2015) observed that children with HFA perform more poorly on metonymy comprehension than metaphor comprehension.

Deafness and Other Hearing Impairments

Sign languages are the main means by which people who are deaf or hard of hearing converse and communicate. Just like spoken languages, sign languages have their own phonology, lexicon, grammar, and other linguistic features. They are independent languages that use a visual modality to convey meaning. Research into sign languages by Liu Junfei 刘俊飞 (Liu 2010) found that comprehension of a Chinese sign language mainly activates the frontal-temporal language network. Sign language comprehension thus shows similarities to spoken language processing in normally hearing people. Sun Zhaocui 孙照翠 (Sun 2011) likewise observed that semantic violations in sign language elicit a similar N400 response in hearing-impaired people as spoken language semantic violations do in people with normal hearing. However, features of sign languages differ from languages of an auditory modality in a number of other respects. First, as noted by Hu et al. (2011), sign languages use visual-spatial symbols, which makes spatial information very important. In an fMRI study, the researchers found greater activation in regions associated with visuospatial processing during word signing, namely in the left IFG and inferior parietal lobule. In another study examining the role of spatial information, Hu Yanchao 胡彦超 (Hu 2011) compared spatial information expressed by hearing-impaired people using either prepositions or classifier constructions. While prepositions are expressed in the same way in sign languages as in spoken languages, classifier constructions in sign languages are quite distinctive. Hu observed more significant activation in the bilateral parietal lobule and areas of the right hemisphere for classifier constructions compared to prepositions. Another potential difference was identified by Li Pei 李培, Liu Junfei, and Yang Yiming (Li et al. 2017), who found that hearing-impaired people only show limited processing of phonological information when reading Chinese. But they may compensate by drawing on other sources of information to a greater extent, such as information about character form and semantic meaning. Additionally, Li et al. (2014) observed greater activation in language processing areas of the right hemisphere during phonological processing tasks for participants who were congenitally deaf compared to those with normal hearing. Pan et al. (2015) also showed that deaf people activate sign language translations of parafoveal words during lexical processing.

Numerous studies have documented plasticity in deaf and other hearing-impaired populations following partial or total hearing loss. For example, Li, Ding, et al. (2012c) recorded a significant reduction in fractional anisotropy (FA) values in the bilateral superior temporal cortex and the splenium of corpus callosum in people who were deaf compared to those with normal hearing. In addition, Li, Li, et al. (2012b) found hemispheric asymmetries in gray matter volume in deaf participants, with right asymmetry in the cerebellum and left asymmetry in the posterior cingulate cortex (PCC). The deaf participants also showed noticeable cortical thinning in other regions including the left precentral gyrus, left superior occipital gyrus, and left fusiform gyrus. Work by Li et al. (2013) and Wang et al. (2015) has shown that connections between the auditory cortex and other areas of the brain are weaker in deaf individuals compared to normally hearing people. Yet, as shown by Li, Emmorey, et al. (2016), the opposite was true for connections between the visual cortex and language areas of the brain, a result most likely related to sign language experience. This kind of plasticity also affects the success of cochlear implant surgery. Feng et al. (2018a) found that auditory association and cognitive brain regions unaffected by auditory deprivation was a valuable source of information for predicting variability in postsurgical speech perception outcomes. Prior to this, Liu Junfei, Xu Mengjie 许梦杰, and Yang Yiming (Liu et al. 2018) argued that not every child with prelingual deafness should be required to receive cochlear implants just so that they can learn spoken language. They proposed that rather than people who are deaf losing their "mother tongue," sign and spoken language could be learned at the same time where possible. This would have significant practical benefits for communication between people with normal hearing and those with hearing impairments.

Second Language Learning and Bilingualism

Scholars in China have conducted a large number of studies of second language (L2) learning. Early research focused on whether the native language (L1) and L2 have shared or separate representations in the brain. Ye Caiyan 叶彩燕 (Ye 2004) proposed that bilingual speakers have two completely different language systems, but there are ample opportunities for the two systems to interact. Cai Houde 蔡厚德 (Cai 2008) noted that for relatively proficient Chinese-English bilinguals, both hemispheres play a role in semantic access for L1, whereas L2 semantic access mainly involves the right hemisphere. Furthermore, semantic access in cross-language conditions is facilitated by interhemispheric cooperation. In an fMRI study, Dong Qi 董奇, Xue Gui 薛贵, and Jin Zhen 金真 (Dong et al. 2004) observed that cortical representation of L2 phonological processing gradually develops in conjunction with learning experience. Subsequent research has since moved beyond the simplistic debate over whether L1 and L2 have shared or separate representations in the brain. Researchers now look to study the characteristics of L2 language ability at multiple levels (e.g., lexical level and sentence level) and investigate ways in which factors such as language proficiency and cognitive control affect particular mechanisms of L2 learning and other aspects of language function (Li and Li 2019; Fu 2017).

Studying L2 vocabulary acquisition mainly involves investigating the development of L2 lexical knowledge and vocabulary learning skills (Sun 2002); mental representation of the L2 lexicon (Zhang 2010); and the relationship between L2 vocabulary acquisition and lexical representations (Sun 2004). Yang Yiming and Geng Libo 耿立波 (Yang and Geng 2008) were among the first to use ERPs to study the neuropsychological basis of L2 vocabulary acquisition. In their results, it could be observed from the N400 component that while vocabulary size – the depth and breadth of semantic word knowledge – was significantly weaker in English (L2) than in Chinese (L1), there was no difference between L1 and L2 in terms of speed of semantic access. The finding that lexical knowledge has clear psychological reality has since been verified and extended in subsequent research. For example, Zhang et al. (2014) showed that resting-state functional connectivity (RSFC) within the reading network correlates with L1 and L2 reading abilities. Li et al. (2019) found that learning L2 vocabulary causes neural activation patterns for L2 lexical processing to become more similar to neural patterns for L1 reading, with greater pattern similarity after a 12-day L2 learning period compared to a 4-day period. Chen et al. (2017), however, found that L2 lexical learning can also be independent of exposure duration. Furthermore, learning rate was expedited by contextual information; greater constraints on sentence context were associated with more rapid learning. This effect also became more pronounced with higher L2 proficiency.

Looking at L2 reading performance at a sentence level, Geng Libo and colleagues (Geng et al. 2008) used ERPs to explore mechanisms for processing L1 and L2 sentences in bilinguals. The researchers observed a trend of complementary implicit and explicit processing for the two types of sentences. In a later study, Geng and Yang Yiming (Geng and Yang 2014) found that L2 learners devote greater cognitive resources to word comprehension when reading L2 sentences compared to L1 sentences. L2 readers are also less effective at semantic integration in sentence processing when reading L2 as opposed to L1. Chang Xin 常欣 and Wang Pei 王沛 (Chang and Wang 2013) further observed that semantic processing precedes syntactic processing for L2 sentences. This is consistent with the results of previous studies, which found that processing of English sentences in Chinese university students is semantically driven (Chang and Gao 2009).

Two factors that will influence L2 learning are proficiency level and structural similarity between L1 and L2. Chen Baoguo and Gao Yiwen 高怡文 (Chen and Gao 2009), for example, found that the degree of similarity between L1 and L2 grammatical structure explained to some extent the variation in scores observed for grammatical judgments of different sentence types. L1-L2 similarity in syntactic structure also appears to interact with the speaker's level of proficiency. As shown by Chen and Guo Jingjing 郭晶晶 (Guo and Chen 2011), the greater the syntactic similarity between the two languages, the more significant the effect on L2 proficiency. Feng et al. (2015) found that L2 proficiency is further associated with global network efficiency and local activation within neural networks.

Interestingly, not only does an L1 affect L2 learning, but L2 learning likewise influences L1 processing. Research by Mei et al. (2014) found that training in a new language affects L1 processing when it involves semantics, but not when the training relates to visual forms or phonology. In a follow-up study, the researchers observed that when adult English speakers read L1 English words, those without Chinese language experience showed left laterality in the posterior fusiform region. Conversely, native English speakers with Chinese experience showed greater recruitment of the right posterior fusiform cortex, a result similar to the right fusiform laterality observed in native Chinese speakers (Mei et al. 2015). These findings suggest that long-term L2 experience affects the fusiform laterality of L1 reading.

Another important topic in bilingualism research is language control. The dorsal anterior cingulate cortex (ACC) and left caudate were identified by Li, Emmorey, et al. (2016) as two areas that play an important role in bilingual control. Specifically, they found that the dorsal ACC monitors and supports processing of the target language, while response selection is controlled by the left caudate when language is more difficult to comprehend. Tan et al. (2011) also found that activation levels in the left caudate and left fusiform gyrus during an L2 lexical decision task predicted L2 reading proficiency one year later. An ERP study by Wu and Thierry (2017) observed that when native Chinese speakers were asked to name pictures in either L1 Chinese or L2 English, the contingent negative variation (CNV) elicited in the response preparation period was of a greater amplitude for L2. This demonstrates that L2 production is subject to proactive interference. Finally, the nature of the causal relationship between language control and brain structure was clarified by Wang Tao and Yang Yiming (Wang and Yang 2017b). In an fMRI study assessing healthy individuals and a nonaphasia patient lacking language control, the researchers found that damage to the basal ganglia led to a loss of language control in the patient. This was the first time that a study conducted in China offered causal evidence identifying neural correlates of bilingual language control.

Prospects for Future Research

Following decades of research work, the neurolinguistics discipline in China has achieved notable progress. Both the quantity and quality of publications have improved significantly, and the global influence of Chinese neurolinguistics has seen a marked rise. However, measured against the ambitious goals pursued by international counterparts, it is clear that Chinese neurolinguistics has a long way to go.

First of all, neurolinguistics in China must be unswervingly targeted toward cutting-edge global research. Although Chinese research has made progress in certain areas, the majority of work still revolves around models proposed by neurolinguists from European countries and the USA. Researchers now understand that language comprehension and production involves certain areas of the brain. They have also moved beyond studying language function at the cortical level to examine ways in which other areas such as the cerebellum and brain nuclei contribute to language processing. Nonetheless, a complete explanatory model of language mechanisms in the brain remains elusive. This presents a significant challenge to global neurolinguistics research. But it is also a major opportunity for the development of neurolinguistics in China.

Second, researchers in China must have the courage to undertake major scientific projects and establish an interdisciplinary field that is tailored to a Chinese context and gives China a voice in the international academic community. AI and human brain projects are two areas at the forefront of global competition in scientific research. Both are closely related to neurolinguistics. Looking around the globe, there is still much to be improved upon in the new generation of AI. AI's ability to solve specific tasks is vastly different from the general intelligence of human beings. The key to more advanced AI is to enable machines to obtain true mastery of language in the manner that humans can. But achieving a real breakthrough in AI requires first designing an AI framework that puts language comprehension at the heart of research. In turn, this requires that neurolinguistics adopts new frameworks. There is already an extensive body of research on language mechanisms at a behavioral level. There are also quite a number of studies conducted at a systems level, examining language-related brain function, the structure of gray matter, and the emerging area of changes in WM fiber structure. However, research of language mechanisms conducted at a cellular level is currently almost nonexistent. Similarly, molecular-level research is limited to studying links between language deficits and a small number of genes. A complete account of language that is able to explain the influences of various factors ranging from individual molecules through to human behavior has yet to be established. But it is only by formulating such a model of human language that AI and human brain projects will be able to advance to a stage where they truly reflect what it means to be human.

Third, neurolinguistics in China must keep abreast of the latest research methods and techniques, and establish a complete, systematic, and innovative academic training and professional development scheme. Neurolinguistics research needs both a solid theoretical foundation and cutting-edge research techniques. The rapid development of brain science has been accompanied by many new innovations and changes in the way that technology is used to study language mechanisms in the brain at multiple levels. If neurolinguistics is to establish a theoretical model that spans the molecular to the behavioral level of linguistics, then it must become proficient in the use of suitable techniques and methodologies, and furthermore advance a comprehensive empirical approach to research.

As for academic training and research team development, the goals and research questions of neurolinguistics underlie its transdisciplinary nature. Accordingly, academic and professional training should also be multidisciplinary, focused on developing talented scholars who are knowledgeable in multiple fields including linguistics, brain science, AI, and psychology. At the same time, it must also attract researchers from various related disciplines to the field of neurolinguistics and work to combine their areas of expertise. Researchers with a background in linguistics, for instance, are advantaged by their grasp of language materials and theoretical analysis. Those coming from neuroscience, psychology, brain science, imaging, and statistics, meanwhile, are skilled in areas such as case study data collection, experimental design, data processing, and operation of scientific instruments. By complementing each other's strengths and learning from one another, these researchers would certainly be capable of generating new insights into the relationship between language and the brain, ones that will astonish the international academic community.

However, we must also take particular care to ensure that this kind of interdisciplinary collaboration is not merely a superficial integration of linguistic concepts and neuroscientific gadgets. As pointed out by Hickok and Poeppel (2004), the deeper requirement for true disciplinary integration calls for each branch of study to free itself from the limitations of its own field, actively engage with others' research, and combine various disciplinary approaches with confidence. In the future, new explanations of the unified nature of the relationship between language and the brain will be proposed. These explanations will come about through organic integration of findings from linguistics, psycholinguistics, and research on neurobiological mechanisms. Accordingly, academic training should focus on providing students of linguistics with knowledge that transcends individual disciplines. As part of this, course offerings in linguistics-related areas including psychology, brain science, neurobiology, medicine, statistics, and AI need to be expanded. Enacting these measures would help promote the establishment of a professional development scheme for linguistics researchers, one that is oriented to the future needs of the discipline and supports the broader goal of advancing Chinese neurolinguistics.

References

Ahlsén, Elisabeth. 2006. *Introduction to neurolinguistics*. Amsterdam: John Benjamins Publishing Company.

Bates, Elizabeth, Sylvia Chen, Ovid J.-L. Tzeng, Ping Li, and Meiti Opie. 1991. The noun-verb problem in Chinese aphasia. *Brain and Language* 41: 203–233. https://doi.org/10.1016/0093-934X(91)90153-R.

Bi, Yanchao, Tao Wei, Chenxing Wu, Zaizhu Han, Tao Jiang, and Alfonso Caramazza. 2011. The role of the left anterior temporal lobe in language processing revisited: Evidence from an individual with ATL resection. *Cortex* 47: 575–587. https://doi.org/10.1016/j.cortex.2009.12.002.

Cai, Houde (蔡厚德). 2008. Zhongying shuangyuzhe cihui yuyi tongda de danao gongneng piancehua yu hezuo xiaoying 中-英双语者词汇语义通达的大脑功能偏侧化与合作效应

[Lateralization and cooperation effect of lexical semantic access in Chinese-English bilinguals]. *Xinli Kexue* 心理科学 [*Psychological Science*] 6: 1394–1397.

Caplan, David, and Jisheng Zhao (赵吉生). 1984. Shenjing yuyanxue lilun de qianjing 神经语言学理论的前景 [Prospects for neurolinguistic theory] *Guowai Yuyanxue* 国外语言学 [*Contemporary Linguistics*] 2: 31–33.

Chang, Xin (常欣), and Pei Wang (王沛). 2013. Jiyu ERP de eryu juzi jiagong guochengderenzhi shenjing jizhi yanjiu jinzhan ji qi qishi 基于 ERP 的二语句子加工过程的认知神经机制研究进展及其启示 [Research progress and enlightenment on cognitive neural mechanism of second language sentence processing based on ERP]. *Xinli Kexue* 心理科学 [*Psychological Science*] 2: 279–283.

Chang, Xin (常欣), and Shenchun Gao (高申春). 2009. Zhongguo daxuesheng yingyu juzi de xinli jizhi: Yi beidongju weili 中国大学生英语句子加工的心理机制—以被动句为例 [The psychological mechanism of Chinese college students' English sentence processing: A case study of passive sentences]. *Xinli Kexue* 心理科学 [*Psychological Science*] 6: 1363–1367.

Chen, Baoguo (陈宝国), and Aihua Ning (宁爱华). 2008. Hanyu zhuyu he binyu guanxicongju jiagong nandu de bijiao 汉语主语和宾语关系从句加工难度的比较 [The comparison of processing difficulty between Chinese subject-relative and object-relative clauses]. *Yingyong Xinlixue* 应用心理学 [*Chinese Journal Of Applied Psychology*] 1: 29–34.

Chen, Baoguo (陈宝国), and Yiwen Gao (高怡文). 2009. Gongzuo jiyi rongliang de xianzhi dui dier yuyan yufa jiagong de yingxiang 工作记忆容量的限制对第二语言语法加工的影响 [Effects of limited working memory capacity on second language grammar processing]. *Waiyu Jiaoxue yu Yanjiu* 外语教学与研究 [*Foreign Language Teaching and Research*] 1: 38–45.

Chen, Baoguo (陈宝国), and Danling Peng (彭聃龄). 2001. Hanzi shibie zhong xingyinyi jihuo shijian jincheng de yanjiu (I) 汉字识别中形音义激活时间进程的研究 (I) [Study on activation time process of shape, sound and meaning in Chinese character recognition (I)]. *Xinli Xuebao* 心理学报 [*Acta Psychologica Sinica*] 1: 1–6.

Chen, Baoguo (陈宝国), Lixin Wang (王立新), and Danling Peng (彭聃龄). 2003. Hanzi shibie zhong xingyinyi jihuo shijian jincheng de yanjiu (II) 汉字识别中形音义激活时间进程的研究 (II) [Study on activation time process of shape, sound and meaning in Chinese character recognition (II)]. *Xinli Xuebao* 心理学报 [*Acta Psychologica Sinica*] 5: 576–581.

Chen, Baoguo, Tengfei Ma, Lijuan Liang, and Huanhuan Liu. 2017. Rapid L2 word learning through high constraint sentence context: An event-related potential study. *Frontiers in Psychology* 8: 2285. https://doi.org/10.3389/fpsyg.2017.02285.

Chen, Haibo (陈海波), and Xinde Wang (王新德). 1992. Hanyu shiduzheng de yanjiu 汉语失读症的研究 [The study of Chinese dyslexia]. *Zhonghua Jingshenke Zazhi* 中华精神科杂志 [*Chinese Journal of Psychiatry*] 3: 139–142.

Chen, Qingrong (陈庆荣), and Yiming Yang (杨亦鸣). 2017. Gushi yuedu de renzhi jizhi: laizi yandong de zhengju 古诗阅读的认知机制:来自眼动的证据 [Cognitive mechanism of ancient poetry reading: evidence from eye movement]. *Zhongguo Shehui Kexue* 中国社会科学 [*Chinese Social Sciences*] 3: 48–76.

Chen, Qingrong, Jingjing Zhang, Xiaodong Xu, Christoph Scheepers, Yiming Yang, and Michael K. Tanenhaus. 2016. Prosodic expectations in silent reading: ERP evidence from rhyme scheme and semantic congruence in classic Chinese poems. *Cognition* 154: 11–21. https://doi.org/10.1016/j.cognition.2016.05.007.

Chen, Qingrong, Xiaodong Xu, Dingliang Tan, Jingjing Zhang, and Yuan Zhong. 2013. Syntactic priming in Chinese sentence comprehension: Evidences from event-related potentials. *Brain and Cognition* 83: 142–152. https://doi.org/10.1016/j.bandc.2013.07.005.

Chen, Yongming (陈永明), and Yao Cui (崔耀). 1994. Juzi xiantishu de canyuzhe zai ketiquxing shangde youshi xianxiang 句子先提述的参与者在可提取性上的优势现象 [Sentence first refers to the dominant phenomenon of participants in extractability]. *Xinli Xuebao* 心理学报 [*Acta Psychologica Sinica*] 2: 113–120.

Chou, Tai-Li, Shu-Hui Lee, Shao-Min Hung, and Hsuan-Chih Chen. 2012. The role of inferior frontal gyrus in processing Chinese classifiers. *Neuropsychologi* 50: 1408–1415. https://doi.org/10.1016/j.neuropsychologia.2012.02.025.

Cui, Gang (崔刚). 1994. Buluokashi shiyuzheng shili yanjiu 布洛卡氏失语症实例研究 [The case study of Broca's aphasia]. *Waiyu Jiaoxue yu Yanjiu* 外语教学与研究 [*Foreign Language Teaching and Research*] 1: 27–33.

———. 1999. Buluokashi yu chuandaoxing Shiyuzheng huanzhe de yuyin zhangai 布洛卡氏与传导性失语症患者的语音障碍 [Phonological impairment in Broca and conductive aphasia]. *Waiyu Jiaoxue yu Yanjiu* 外语教学与研究 [*Foreign Language Teaching and Research*] 3: 24–29.

Ding, Nai, Lucia Melloni, Hang Zhang, Xing Tian, and David Poeppel. 2015. Cortical tracking of hierarchical linguistic structures in connected speech. *Nature Neuroscience* 19 (1): 158–164. https://doi.org/10.1038/nn.4186.

Dong, Qi (董奇), Gui Xue (薛贵), and Zhen Jin (金真). 2004. Yuyan jingyan dui danao jihuo de yingxiang: Laizi dieryuyan chuxuezhe de zhengju (yingwen) 语言经验对大脑激活的影响:来自第二语言初学者的证据(英文) [Effects of language experience on brain activation: Evidence from beginners in second language (English)]. *Xinli Xuebao* 心理学报 [*Acta Psychologica Sinica*] 4: 448–454.

Fang, Yuxing, Zaizhu Han, Suyu Zhong, Gaolang Gong, Luping Song, Fangsong Liu, Ruiwang Huang, Du Xiaoxia, Rong Sun, Qiang Wang, Yong He, and Yanchao Bi. 2015. The semantic anatomical network: Evidence from healthy and brain-damaged patient populations. *Human Brain Mapping* 36: 3498–3515. https://doi.org/10.1002/hbm.22858.

Feng, Gangyi, Erin M. Ingvalson, Tina M. Grieco-Calub, Megan Y. Roberts, Maura E. Ryan, Patrick Birmingham, Delilah Burrowes, Nancy M. Young, and Patrick C.M. Wong. 2018a. Neural preservation underlies speech improvement from auditory deprivation in young cochlear implant recipients. *Proceedings of the National Academy of Sciences of the United States of America* 115 (5): E1022–E1031. https://doi.org/10.1073/pnas.1717603115.

Feng, Gangyi, Hsuan-Chih Chen, Zude Zhu, Yong He, and Suiping Wang. 2015. Dynamic brain architectures in local brain activity and functional network efficiency associate with efficient reading in bilinguals. *Neuroimage* 119: 103–118. https://doi.org/10.1016/j.neuroimage.2015.05.100.

Feng, Gangyi, Zhenzhong Gan, Suiping Wang, Patrick C.M. Wong, and Bharath Chandrasekaran. 2018b. Task-general and acoustic-invariant neural representation of speech categories in the human brain. *Cerebral Cortex* 28: 3241–3254. https://doi.org/10.1093/cercor/bhx195.

Feng, Shiwen, Rounan Qi, Jing Yang, Aanya Yu, and Yiming Yang. 2020. Neural correlates for nouns and verbs in phrases during syntactic and semantic processing: An fMRI study. *Journal of Neurolinguistics* 53: 100860.

Feng, Shiwen (封世文), and Yiming Yang (杨亦鸣). 2011. Jiyu gongnengxing cigongzheng chengxiang de hanyu qingdongci jiqishenjing jizhi yanjiu 基于功能性磁共振成像的汉语轻动词及其神经机制研究 [The light verb in Chinese and its neural mechanisms: An fMRI study]. *Yuyan Wenzi Yingyong* 语言文字应用 [*Applied Linguistics*] 2: 45–55.

Feng, Wangshu, Yue Wu, Catherine Jan, Hongbo Yu, Xiaoming Jiang, and X. Zhou. 2017. Effects of contextual relevance on pragmatic inference during conversation: An fMRI study. *Brain and Language* 171: 52–61. https://doi.org/10.1016/j.bandl.2017.04.005.

Friederici, Angela D. 2002. Towards a neural basis of auditory sentence processing. *Trends in Cognitive Science* 6: 78–84. https://doi.org/10.1016/S1364-6613(00)01839-8.

Fu, Mengling (付梦玲). 2017. *Dui Butong Shuliandu de Zhongguo Yingyu Xuexizhe Jiagong Yingyu Duanyu Dongci de Shijian Xiangguan Dianwei Yanjiu* 对不同熟练度的中国英语学习者加工英语短语动词的事件相关电位研究 [*A study on the event-related potentials of processing English phrasal verbs in Chinese English learners with different proficiency*]. Nanjing Shifan Daxue Shuoshi Xuewei Lunwen 南京师范大学硕士学位 [Master's thesis, Nanjing Normal University].

Gao, Surong (高素荣). 1993. *Shiyuzheng* 失语症 [*Aphasia*]. Beijing: Beijing Yike Daxue and Zhongguo Xiehe Yike Daxue Lianhe Chubanshe 北京: 北京医科大学 and 中国协和医科大学联合出版社 [Beijing: Beijing Medical University and Peking Union Medical College Joint Press].

———. 1996. Hanyu shiyu jianchafa de linchuang yingyong: 199 li cozhonghou shiyu 汉语失语检查法的临床应用: 199例卒中后失语 [Clinical application of Chinese aphasia test: 199 cases of aphasia after stroke]. *Cuzhong yu Shenjing Jibing* 卒中与神经疾病 [*Stroke and Nervous Diseases*] 2: 57–60.

Ge, Jianqiao, Gang Peng, Bingjiang Lyu, Yi Wang, Yan Zhuo, Zhendong Niu, Li Hai Tan, Alexander P. Leff, and Jia-Hong Gao. 2015. Cross-language differences in the brain network subserving intelligible speech. *Proceedings of the National Academy of Sciences of the United States of America* 112 (10): 2972–2977. https://doi.org/10.1073/pnas.1416000112.

Geng, Libo (耿立波), and Yiming Yang (杨亦鸣). 2014. Jiyu ERP shiyan de eryu juzi yuedu nengli de xinli xianshixing yanjiu 基于ERP实验的二语句子阅读能力的心理现实性研究 [A study on the psychological reality of second language sentence reading ability based on ERP experiment]. *Waiyu Jiaoxue Lilun yu Shijain* 外语教学理论与实践 [*Foreign Language Teaching Theory and Practice*] 3: 23–29.

Geng, Libo (耿立波), Liang Yu (于亮), Shanshan Zhang (张珊珊), and Yiming Yang (杨亦鸣). 2008. Yingyong shijian xiangguan dianwei jishu kaocha shuangyu lijiezhong neiyin jiagong he waixian jiagong de hubu 应用事件相关电位技术考察双语理解中内隐加工与外显加工的互补 [An event-related potential study on complementation of implicit processing and explicit processing in reading bi-language]. *Zhongguo Zuzhi Gongcheng yu linchuang kangfu* 中国组织工程研究与临床康复 [*Tissue Engineering Research and Clinical Rehabilitation in China*] 30: 5873–5876.

Gu, Jiexin (顾介鑫), and Yiming Yang (杨亦鸣). 2010. Fuhe goucifa nengchanxing jiqi shenjing dianshenglixue yanjiu 复合构词法能产性及其神经电生理学研究 [A neuro-electrophysiological study on productivity of Chinese compounding]. *Yuyan Wenzi Yingyong* 语言文字应用 [*Applied Linguistics*] 3: 98–107.

Gu, Jiexin (顾介鑫), Fangfang Yu (余芳芳), and Jinjun Wang (王进军). 2019. Yanyu chanchu zhong yinyu jiagong de nao gongneng chengxiang yanjiu 言语产出中隐喻加工的脑功能成像研究 [A functional magnetic resonance imaging study: On metaphorical processes during language production]. *Dangdai Xiucixue* 当代修辞学 [*Modern Rhetoric*] 4: 10–18.

Gu, Jiexin (顾介鑫), Xin Zhou (周昕), and Jingqi Weng (翁婧琦). 2018. 'Guanian xiaofang de yeye' lei jufa qiyi jiagong de nao gongneng chengxiang yanjiu "挂念小芳的爷爷"类句法歧义加工的脑功能成像研究 [Brain functional imaging study on syntactic ambiguity processing in the category of 'grandpa who worries about xiaofang']. *Yuyan Kexue* 语言科学 [*Linguistic Sciences*] 6: 647–662.

Gu, Jiexin (顾介鑫), Yiming Yang (杨亦鸣), Yong Ma (马勇), and Zhaoyong Shen (沈兆勇). 2018. Liheci de yinxi tezheng jiqi shenjing jichu yanjiu 离合词的音系特征及其神经基础研究 [A study on phonological character of split words and its neural basis]. *Yuyan Yanjiu* 语言研究 [*Studies in Language and Linguistics*] 1: 63–73.

Gui, Shichun (桂诗春). 1991. *Shiyan Xinli Yuyanxue Gangyao* 实验心理语言学纲要 [*An outline of experimental psycholinguistics*]. Changsha: Hunan Jiaoyu Chubanshe 长沙: 湖南教育出版社 [Changsha: Hunan Education Press].

———. 2000. *Xinbian Xinli Yuyanxue* 新编心理语言学 [*Psycholinguistics (New Edition)*]. Shanghai: Shanghai Waiyu Jiaoyu Chubanshe 上海: 上海外语教育出版社 [Shanghai: Shanghai Foreign Language Education Press].

Guo, Jingjing (郭晶晶), and Baoguo Chen (陈宝国). 2011. Han-Ying jufa jiegou xaingsixing yu dieryuyan de shuliandu dui dier yuyan jufa jiagong de yingxiang 汉、英句法结构相似性与第二语言的熟练度对第二语言句法加工的影响 [The influence of syntactic similarity between Chinese and English and the proficiency of the second language on syntactic processing of the second language]. *Xinli Kexue* 心理科学 [*Psychological Science*] 3: 571–575.

Guo, Kejiao (郭可教), and Zhiqi Yang (杨奇志). 1995. Hanzi renzhi de funaoxiaoying de shiyan yanjiu 汉字认知的"复脑效应"的实验研究 ["Both-Hemisphere Effect" in the cognition of Chinese characters]. *Xinli Xuebao* 心理学报 [*Acta Psychologica Sinica*] 1: 78–83.

Goodglass, Harold, and Edith Kaplan. 1972. *Boston Diagnostic Aphasia Examination (BDAE)*. Philadelphia: Lea & Febiger.

———. 1983. *The assessment of aphasia and related disorders*. 2nd ed. Philadelphia: Liea & Febiger.

Han, Zaizhu, Yujun Ma, Gaolang Gong, Yong He, Alfonso Caramazza, and Yanchao Bi. 2013. White matter structural connectivity underlying semantic processing: Evidence from brain damaged patients. *Brain* 136: 2952–2965. https://doi.org/10.1093/brain/awt205.

Hickok, Gregory, and David Poeppel. 2004. Dorsal and ventral streams: A framework for understanding aspects of the functional anatomy of language. *Cognition* 92: 67–99. https://doi.org/10.1016/j.cognition.2003.10.011.

Hu, Chaoqun (胡超群). 1989. Shidu huanzhe yuedu guocheng zhong hanyuci de xingyinyi sanwei guanxi de tantao 失读患者阅读过程中汉语词的形、音、义三维关系的探讨 [Discussion on the three-dimensional relationship of shape, sound and meaning of Chinese words in the reading process of dyslexic patients]. *Xinli Xuebao* 心理学报 [*Acta Psychologica Sinica*] 1: 41–46.

———. 1992. Shidu bingren yuju pianzhang yuedu zhong xingyinyi guanxi de tantao 失读病人语句、篇章阅读中形、音、义关系的探讨 [Discussion on the relationship of form, sound and meaning in sentence and text reading for dyslexic patients]. *Zhongguo Yuwen* 中国语文 [*Studies of the Chinese Language*] 3: 191–194.

Hu, Wei, Hwee Ling Lee, Qiang Zhang, Tao Liu, Libo Geng, Mohamed L. Seghier, Clare Shakeshaft, Tae Twomey, David W. Green, Yiming Yang, and Cathy J. Price. 2010. Developmental dyslexia in Chinese and English populations: Dissociating the effect of dyslexia from language differences. *Brain* 133: 1694–1706. https://doi.org/10.1093/brain/awq106.

Hu, Yanchao (胡彦超). 2011. *Zhongguo Shouyu Kongjian Guanxi Jiagong de Shenjing Jichu* 中国手语空间关系加工的神经基础 [*Neural basis of spatial relationship processing in Chinese sign language*]. Jiangsu Shifan Daxue Shuoshi Xuewei Lunwen 江苏师范大学硕士学位论文 [Master's thesis, Jiangsu Normal University].

Hu, Zhiguo, Wenjing Wang, Hongyan Liu, Danling Peng, Yanhui Yang, Kuncheng Li, John X. Zhang, and Guosheng Ding. 2011. Brain activations associated with sign production using word and picture inputs in deaf signers. *Brain and Language* 116: 64–70. https://doi.org/10.1016/j.bandl.2010.11.006.

Huang, Jian, Martin J. Pickering, Juanhua Yang, Suiping Wang, and Holly P. Branigan. 2016. The independence of syntactic processing in Mandarin: Evidence from structural priming. *Journal of Memory and Language* 91: 81–98. https://doi.org/10.1016/j.jml.2016.02.005.

Hung, Daisy L., and Ovid J.-L. Tzeng. 1996. Neurolinguistics: A Chinese perspective. In *New horizons in Chinese linguistics*, ed. C.-T. James Huang and Yen-hui Audrey Li. Springer.

Hsiao, Franny, and Edward Gibson. 2003. Processing relative clauses in Chinese. *Cognition* 90: 3–27. https://doi.org/10.1016/s0010-0277(03)00124-0.

Jiang, Xiaoming, Yi Li, and Xiaolin Zhou. 2013. Is it over-respectful or disrespectful? Differential patterns of brain activity in perceiving pragmatic violation of social status information during utterance comprehension. *Neuropsychologia* 51: 2210–2223. https://doi.org/10.1016/j.neuropsychologia.2013.07.021.

JSPOPSS (Jiangsu Planning Office of Philosophy and Social Science) Jiangsu Sheng Zhexue Shehui Kexue Guihua Bangongshi 江苏省哲学社会科学规划办公室. 2004. Yibu zai woguo shenjing yuyanxue lingyu you kaituo yiyi de yanjiu wenxian 一部在我国神经语言学领域有开拓意义的研究文献 [A Research Document with Pioneering Significance in China's neurolinguistics]. *Jiangsu Shehui Kexue* 江苏社会科学 [*Jiangsu Social Sciences*] 2: 240.

Kwok, Veronica P.Y., Tianfu Wang, Siping Chen, Kofi Yakpo, Linlin Zhu, Peter T. Fox, and Li Hai Tan. 2015. Neural signatures of lexical tone reading. *Human Brain Mapping* 36 (1): 304–312. https://doi.org/10.1002/hbm.22629.

Li, Degao, Kejuan Gao, Xueyun Wu, Xiaojun Chen, Xiaona Zhang, Ling Li, and Weiwei He. 2013. Deaf and dard of hearing adolescents' processing of pictures and written words for taxonomic categories in a priming task of semantic categorization. *American Annals of the Deaf* 158: 426–437. https://doi.org/10.1353/aad.2013.0040.

Li, Hong, Hua Shu, Catherine McBride-Chang, Hongyun Liu, and Hong Peng. 2012a. Chinese children's character recognition: Visuo-orthographic, phonological processing and morphological skills. *Journal of Research in Reading* 35: 287–307. https://doi.org/10.1111/j.1467-9817.2010.01460.x.

Li, Huiling, Jing Qu, Chuansheng Chen, Yanjun Chen, Gui Xue, Lei Zhang, Chengrou Lu, and Leilei Mei. 2019. Lexical learning in a new language leads to neural pattern similarity with word reading in native language. *Human Brain Mapping* 40: 98–109. https://doi.org/10.1002/hbm.24357.

Li, Le, Karen Emmorey, Xiaoxia Feng, Chunming Lu, and Guosheng Ding. 2016. Functional connectivity reveals which language the 'Control regions' control during bilingual production. *Frontiers in Human Neuroscience* 10: 616. https://doi.org/10.3389/fnhum.2016.00616.

Li, Jianhong, Wenjing Li, Junfang Xian, Yong Li, Zhaohui Liu, Sha Liu, Xiaocui Wang, Zhenchang Wang, and Huiguang He. 2012b. Cortical thickness analysis and optimized voxel-based morphometry in children and adolescents with prelingually profound sensorineural hearing loss. *Brain Research* 1430: 35–42. https://doi.org/10.1016/j.brainres.2011.09.057.

Li, Pei (李培), Junfei Liu (刘俊飞), and Yiming Yang (杨亦鸣). 2017. Longren hanyu cihui yuedu zhong de yuyin jiagong yanjiu 聋人汉语词汇阅读中的语音加工研究 [Phonological processing in Chinese word recognition by deaf readers]. *Yuyan Wenzi Yingyong* 语言文字应用 [*Applied Linguistics*] 2: 57–66.

Li, Weijun (李卫君), and Yufang Yang (杨玉芳). 2010a. Jueju yunlü bianjie de renzhi jiagong jiqi naodian xiaoying 绝句韵律边界的认知加工及其脑电效应 [The cognitive processing of prosodic boundary and its related brain effect in quatrain]. *Xinli Xuebao* 心理学报 [*Acta Psychologica Sinica*] 11: 1021–1032.

———. 2010b. Perception of Chinese poem and its electrophysiological effects. *Neuroscience* 168: 757–768. https://doi.org/10.1016/j.neuroscience.2010.03.069.

Li, Weijun (李卫君), Meng Liu (刘梦), Zhenghua Zhang (张政华), Nali Deng (邓娜丽), and Yushan Xing (邢钰珊). 2018. Kouchizhe jiagong hanyu qiyiduanyu de shenjing guocheng 口吃者加工汉语歧义短语的神经过程 [The neural process of processing Chinese ambiguous phrases in stutterers]. *Xinli Xuebao* 心理学报 [*Acta Psychologica Sinica*] 12: 1323–1335.

Li, Xiaoqing, Ximing Shao, Jinyan Xia, and Xu. Xiaoying. 2018. The cognitive and neural oscillatory mechanisms underlying the facilitating effect of rhythm regularity on speech comprehension. *Journal of Neurolinguistics* 49: 155–167. https://doi.org/10.1016/j.jneuroling.2018.05.004.

Li, Xiaoqing, Yuping Zhang, Jinyan Xia, and Tamara Y. Swaab. 2017. Internal mechanisms underlying anticipatory language processing: Evidence from event-related-potentials and neural oscillations. *Neuropsychologia* 102: 70–81.

Li, Yan (李艳), and Ming Li (黎明). 2019. Eryu shuiping dui zang-han shuangyuzhe xinlicihui yuyi biaozheng yingxiang de ERP yanjiu 二语水平对藏-汉双语者心理词汇语义表征影响的ERP研究 [An ERP study on the influence of second language level on the semantic representation of Tibetan – Chinese bilinguals]. *Waiyu Yanjiu* 外语研究 [*Foreign Languages Research*] 3: 42–48.

Li, Yanyan, Danling Peng, Li Liu, James R. Booth, and Guosheng Ding. 2014. Brain activation during phonological and semantic processing of Chinese characters in deaf signers. *Frontiers in Human Neuroscience* 8: 211. https://doi.org/10.3389/fnhum.2014.00211.

Li, Yanyan, Guosheng Ding, James R. Booth, Ruiwang Huang, Yating Lü, Yufeng Zang, Yong He, and Danling Peng. 2012c. Sensitive period for white-matter connectivity of superior temporal cortex in deaf people. *Human Brain Mapping* 33: 349–359. https://doi.org/10.1002/hbm.21215.

Li, Ying (李莹), Lei Mo (莫雷), Dapeng Shi (史大鹏), and John X. Zhang (张学新). 2016. Butong xingzhi hanyu yinyuju renzhi jiagong de fMRI yanjiu 不同性质汉语隐喻句认知加工的fMRI研究 [An fMRI Study on the cognitive processing of different type of Chinese metaphoric sentences]. *Zhejiang Daxue Xuebao (Renwen Shehui Kexueban)* 浙江大学学报 (人文社会科学版) [*Journal of Zhejiang University (Humanities and Social Sciences)*] 6: 33–45.

Li Yuming (李宇明) and Qianrui Chen (陈前瑞). 1997. Quanan errong de wenju lijie 群案儿童的问句理解 [Children's Understanding of Interrogative Sentences]. Huazhong shifan daxue xuebao (zhexue shehui kexueban) 华中师范大学 (哲学社会 科学版) [Journal of Central China Normal University (Philosophy and Social Sciences)]. 36(2): 77–84.

Liang, Baishen, and Yi Du. 2018. The functional neuroanatomy of lexical tone perception: An activation likelihood estimation meta-analysis. *Frontiers in Neuroscience* 12: 495. https://doi.org/10.3389/fnins.2018.00495.

Liang, Dandan (梁丹丹). 2004. Zhongguo shenjing yuyanxue de huigu yu qianzhan 中国神经语言学的回顾与前瞻 [Neurolinguistics in China: A retrospect and prospect]. *Dangdai Yuyanxue* 当代语言学 [*Contemporary Linguistics*] 2: 138–153.

Liang, Dandan (梁丹丹), Yiming Yang (杨亦鸣), Shiwen Feng (封世文), and Jiance Li (李建策). 2006. Hanyu ming, dong, xing chongdang mingci xiushiyu de fMRI yanjiu 汉语名、动、形充当名词修饰语的 fMRI 研究 [An fMRI study on modifiers of noun by nouns, verbs and adjectives in Chinese]. *Yuyan Wenzi Yingyong* 语言文字应用 [*Applied Linguistics*] 4: 90–97.

Liang, Jie (梁洁). 2002. Bingli yuyin shiyan fenxi jieguo: pingxuju de yudiao moshi 病理语音实验分析结果——平叙句的语调模式 [An intonation model of the statement in Putonghua: A phonetic analysis of an aphasian speech experiment]. *Dangdai Yuyanxue* 当代语言学 [*Contemporary Linguistics*] 2: 128–137.

Lin, Guhui (林谷辉), and Zhuoming Chen (陈卓铭). 1999. Hanyu chun shiduzheng 汉语纯失读症(附1例报告) [A Chinese pure alexia: A case study]. *Zhonguo Shenjing Jingshen Jibing Zazhi* 中国神经精神疾病杂志 Chinese Journal of Nervous and Mental Diseases [*Advances in Psychological Science*] 2: 18–22.

Lin, Liantong (林连通), and Shixi Gu (顾士熙). 2002. *Zhongguo Yuyanxue Nianjian (1995–1998) Shangjuan* 中国语言学年鉴 (1995–1998) 上卷 [*Yearbook of Chinese Linguistics (1995–1998) Volume 1*]. Beijing: Yuwen Chubanshe 北京: 语文出版社 [Beijing: Language Publishing House].

Liu, Junfei (刘俊飞). 2010. *Zhongguo Shouyu Yuyin Jiagong de Shenjing Jichu* 中国手语语音加工的神经基础 [*The neural basis of speech processing in Chinese sign language*]. Jiangsu Shifan Daxue Shuoshi Xuewei Lunwen 江苏师范大学硕士学位论文 [Master's thesis, Jiangsu Normal University].

Liu, Junfei (刘俊飞), Mengjie Xu (许梦杰), and Yiming Yang (杨亦鸣). 2018. Longtong zaoqi yuyan ganyu: Jiujing shi kouyu haishi shouyu? 聋童早期语言干预:究竟是口语还是手语? [Early language intervention for deaf children: Spoken language or signed language] *Nanjing Shifan Daxue Wenxueyuan Xuebao* 南京师范大学文学院学报 [*Journal of School of Chinese Language and Culture Nanjing Normal University*] 3: 10–16.

Liu, Meng, Yushan Xing, Liming Zhao, Nali Deng, and Weijun Li. 2018. Abnormal processing of prosodic boundary in adults who stutter: An ERP study. *Brain and Cognition* 128: 17–27. https://doi.org/10.1016/j.bandc.2018.10.009.

Liu, Tao (刘涛), and Huo Jiang (江火). 2016. Jufa yiwei de naoshenjing jiagong jizhi: Laizi hanyu beidongju de ERPs yanjiu 句法移位的脑神经加工机制——来自汉语被动句的 ERPs 研究 [Neural mechanisms of syntactic movement: An ERPs study of Chinese passive sentences]. *Yuyan Kexue* 语言科学 [*Linguistic Sciences*] 6: 612–624.

Liu, Tao (刘涛), Tongquan Zhou (周统权), and Yiming Yang (杨亦鸣). 2011. Zhuyu guanxi congju jiagong youshi de pubianxing-laizi hanyu guanxi congju ERP yanjiu de zhengju 主语关系从句加工优势的普遍性——来自汉语关系从句 ERP 研究的证据 [Universality of processing advantage for subject-relative clause: An ERP study of Chinese relative clauses]. *Yuyan Kexue* 语言科学 [*Linguistic Sciences*] 1: 1–20.

Liu, Tao (刘涛), Yiming Yang (杨亦鸣), Hui Zhang (张辉), Shanshan Zhang (张珊珊), Dandan Liang (梁丹丹), Jiexin Gu (顾介鑫), and Wei Hu (胡伟). 2008. Yufa yujing xia hanyu mingdong fenli de ERP yanjiu 语法语境下汉语名动分离的 ERP 研究 [Study on ERP of Chinese nomenclature separation in grammatical context]. *Xinli Xuebao* 心理学报 [*Acta Psychologica Sinica*] 6: 671–680.

Liu, Xiaojia (刘晓加), and Xiuling Liang (梁秀龄). 1996. Hanyu shixie jianchafa de zhiding he biaozhunhua 汉语失写检查法的制定和标准化 [Development and standardization of Chinese anaphorism test]. *Zhongguo Shenjing Jingshen Jibing Zazhi* 中国神经精神疾病杂志 [*Chinese Journal of Nervous and Mental Diseases*] 6: 331–333.

Lu, Chunming, Chuansheng Chen, Ning Ning, Guosheng Ding, Taomei Guo, Danling Peng, Yanhui Yang, Kuncheng Li, and Chunlan Lin. 2010a. The neural substrates for atypical planning and execution of word production in stuttering. *Experimental Neurology* 221: 146–156. https://doi.org/10.1016/j.expneurol.2009.10.016.

Lu, Chunming, Danling Peng, Chuansheng Chen, Ning Ning, Guosheng Ding, Kuncheng Li, Yanhui Yang, and Chunlan Lin. 2010b. Altered effective connectivity and anomalous anatomy in the basal ganglia-thalamocortical circuit of stuttering speakers. *Cortex* 46: 49–67. https://doi.org/10.1016/j.cortex.2009.02.017.

Luke, Kang-Kwong, Ho-Ling Liu, Yo-Yo Wai, Yung-Liang Wan, and Li Hai Tan. 2002. Functional anatomy of syntactic and semantic processing in language comprehension. *Human Brain Mapping* 16: 133–145. https://doi.org/10.1002/hbm.10029.

Luo, Hao, Jing-Tian Ni, Zhi-Hao Li, Xiao-Ou Li, Da-Ren Zhang, Fan-Gang Zeng, and Chen Lin. 2006. Opposite patterns of hemisphere dominance for early auditory processing of lexical tones and consonants. *Proceedings of the National Academy of Sciences of the United States of America* 103: 19558–19563. https://doi.org/10.1073/pnas.0607065104.

Luo, Yuejia (罗跃嘉), Jinghan Wei (魏景汉), Xuchu Weng (翁旭初), and Xing Wei (卫星). 2001. Hanzi shiting zairen de ERP xiaoying yu jiyi tiqu naojizhi 汉字视听再认的ERP效应与记忆提取脑机制 [ERP effects of recognition of Chinese spoken and written words and neural mechanism of Retrieval] *Xinli Xuebao* 心理学报 [*Acta Psychologica Sinica*] 6: 488–494.

Lyon, G. Reid, Sally E. Shaywitz, and Bennett A. Shaywitz. 2003. A definition of dyslexia. *Annals of Dyslexia* 53: 1–14.

Mei, Leilei, Gui Xue, Zhong-Lin Lu, Chuansheng Chen, Mingxia Zhang, Qinghua He, Miao Wei, and Qi Dong. 2014. Learning to read words in a new language shapes the neural organization of the prior languages. *Neuropsychologia* 65: 156–168.

Mei, Leilei, Gui Xue, Zhong-Lin Lu, Chuansheng Chen, Miao Wei, Qinghua He, and Qi Dong. 2015. Long-term experience with Chinese language shapes the fusiform asymmetry of English reading. *Neuroimag* 110: 3–10.

Mirman, Daniel, Qi Chen, Yongsheng Zhang, Ze Wang, Olufunsho K. Faseyitan, H. Branch Coslett, and Myrna F. Schwartz. 2015. Neural organization of spoken language revealed by lesion–symptom mapping. *Nature Communications* 6: 6762. https://doi.org/10.1038/ncomms7762.

Mo, Lei, Ho-Ling Liu, Hua Jin, and Ya-Ling Yang. 2005. Brain activation during semantic judgment of Chinese sentences: A functional MRI study. *Human Brain Mapping* 24: 305–312. https://doi.org/10.1002/hbm.20091.

Pan, Jinger, Hua Shu, Yuling Wang, and Ming Yan. 2015. Parafoveal activation of sign translation previews among deaf readers during the reading of Chinese sentences. *Memory and Cognition* 43: 964–972. https://doi.org/10.3758/s13421-015-0511-9.

Peng, Danling (彭聃龄), and Songlin Liu (刘松林). 1993. Hanyu juzi lijie zhong yuyi fenxi yu jufa fenxi de guanxi 汉语句子理解中语义分析与句法分析的关系 [The relationship between semantic analysis and syntactic analysis in Chinese sentence comprehension]. *Xinli Xuebao* 心理学报 [*Acta Psychologica Sinica*] 2: 22–29.

Peng, Danling (彭聃龄), Dejun Guo (郭德俊), and Sulan Zhang (张素兰). 1985. Zairenxing tongyi panduan zhong hanzi xinxi de tiqu 再认性同一判断中汉字信息的提取 [The extraction of Chinese character information in the recognition of identity]. *Xinli Xuebao* 心理学报 [*Acta Psychologica Sinica*] 3: 227–234.

Peng, Danling (彭聃龄), Yuan Deng (邓园), and Baoguo Chen (陈宝国). 2003. Hanyu duoyi danzici de shibie youshi xiaoying 汉语多义单字词的识别优势效应 [The dominant effect of polysemy word recognition in Chinese]. *Xinli Xuebao* 心理学报 [*Acta Psychologica Sinica*] 5: 568–575.

Perfetti, Charles A., and Li Hai Tan. 1998. The time course of graphic, phonological, and semantic activation in Chinese character identification. *Journal of Experimental Psychology-Learning Memory and Cognition* 24: 101–118. https://doi.org/10.1037/0278-7393.24.1.101.

Qiu, Weihong (丘卫红), Zulin Dou (窦祖林), and Guifang Wan (万桂芳). 2000. Yueyu shiyuzheng de pingjia yu yuyan zhiliao chutan 粤语失语症的评价与语言治疗初探 [Evaluation and treatment of cantonese aphasia]. *Zhongguo Kangfu Yixue Zazhi* 中国康复医学杂志 [*Chinese Journal of Rehabilitation Medicine*] 5: 278–281.

Schirmer, Annett, Siu-Lam Tang, Trevor B. Penney, Thomas C. Gunter, and Hsuan-Chih Chen. 2005. Brain responses to segmentally and tonally induced semantic violations in Cantonese. *Journal of Cognitive Neuroscience* 17: 1–12. https://doi.org/10.1162/0898929052880057.

Scovel, Thomas, Jiarong Li (李家荣), and Yunxing Li (李运兴). 1980. Shenjing yuyanxue yanjiu de zuijin qushi 神经语言学研究的最近趋势 [Trends in neurolinguistics] *Guowai Yuyanxue* 国外语言学 [*Contemporary Linguistics*] 4: 44–45.

Shen, Jiaxuan (沈家煊). 1989. Shenjing yuyanxue gaishuo 神经语言学概说 [Introduction to neurolinguistics]. *Waiyu Jiaoxue yu Yanjiu* 外语教学与研究 [*Foreign Language Teaching and Research*] 4: 23–28.

Shen, Jiaxuan (沈家煊), and Yao Le (乐耀). 2013. Cilei de shiyan yanjiu huhuan yufa lilun de gengxin 词类的实验研究呼唤语法理论的更新 [Emperical research on word category effect calls for new theories] *Dangdai Yuyanxue* 当代语言学 [*Contemporary linguistics*] 3: 253–267.

Sheng, Jingwei, Li Zheng, Bingjiang Lyu, Zhehang Cen, Lang Qin, Li Hai Tan, Ming-Xiong Huang, Nai Ding, and Jia-Hong Gao. 2019. The cortical maps of syntactic hierarchical linguistic structures during speech perception. *Cerebral Cortex* 29 (8): 3232–3240. https://doi.org/10.1093/cercor/bhy191.

Si, Xiaopeng, Wenjing Zhou, and Bo Hong. 2017. Cooperative cortical network for categorical processing of Chinese lexical tone. *Proceedings of the National Academy of Sciences of the United States of America* 114 (46): 12303–12308. https://doi.org/10.1073/pnas.1710752114.

Siok, Wai Ting, Charles A. Perfetti, Zhen Jin, and Li Hai Tan. 2004. Biological abnormality of impaired reading is constrained by culture. *Nature* 431: 71–76. https://doi.org/10.1038/nature02865.

Song, Yiqi, Zhongheng Jia, Shunhua Liu, and Dandan Liang. 2017. Discourse production of Mandarin-speaking children with high-functioning autism: The effect of mental and action verbs' semantic-pragmatic features. *Journal of Communicative Disorders* 70: 12–24. https://doi.org/10.1016/j.jcomdis.2017.10.002.

Su, Mengmeng, Jingjing Zhao, Michel Thiebaut de Schotten, Wei Zhou, Gaolang Gong, Franck Ramus, and Hua Shu. 2018. Alterations in white matter pathways underlying phonological and morphological processing in Chinese developmental dyslexia. *Developmental Cognitive Neuroscience* 31: 11–19. https://doi.org/10.1016/j.dcn.2018.04.002.

Sun, Lan (孙蓝). 2002. L2 cihui ciyi nengli fazhan chutan L2 词汇词义能力发展初探 [L2 vocabulary meaning ability development]. *Shandong Shida Waiguoyu Xueyuan Xuebao* 山东师大外国语学院学报 [*Journal of School of Foreign Languages, Shandong Normal University*] 1: 28–31.

———. 2004. Eryu shiyici chengxian fangshi ji gainian biaozheng de kaocah: Gean yanjiu fenxi 二语实义词呈现方式及概念表征的考察——个案研究分析 [A case study on the representation and conceptual representation of L2 content words]. *Yuyan Kexue* 语言科学 [*Linguistic Sciences*] 2: 34–41.

Sun, Youxia (孙友霞), Shanping Mao (毛善平), Meng Cai (蔡萌), Hua Zhao (赵华), Zhengfang Liu (刘正芳), and Li Xu (徐丽). 2007. Hnayu shiyu huanzhe fangwei jieci jiagong de chubu tancha 汉语失语患者方位介词加工的初步探察 [A study of processing characteristics of locative prepositions in patients with Chinese aphasia]. *Zhonghua Wuli Yixue yu Kangfu Zazhi* 中华物理医学与康复杂志 [*Chinese Journal Of Physical Medicine And Rehabilitation*] 12: 816–819.

Sun, Zhaocui (孙照翠). 2011. *Zhongguo Shouyu Yuyi he Jufa Jiagong de ERP Yanjiu* 中国手语语义和句法加工的ERP研究. [*ERP Study on Semantic and Syntactic Processing in Chinese Sign*

Language]. Jiangsu Shifan Daxue Shuoshi Xuewei Lunwen 江苏师范大学硕士学位论文 [Master's thesis, Jiangsu Normal University].

Tan, Li Hai, Chen Lin, Virginia Yip, Alice H.D. Chan, Jing Yang, Jia-Hong Gao, and Wai Ting Siok. 2011. Activity levels in the left hemisphere caudate-fusiform circuit predict how well a second language will be learned. *Proceedings of the National Academy of Sciences of the United States of America* 108: 2540–2544.

Tan, Li Hai, A. Spinks John, and Jia Hong Gao. 2000. Brain activation in the processing of Chinese characters and words: A functional MRI study. *Human Brain Mapping* 10: 16–27. https://doi.org/10.1002/(SICI)1097-0193.

Tang, Yi, Yi Xing, Zude Zhud, Yong He, Fang Li, Jianwei Yang, Qing Liu, Fangyu Li, Stefan J. Teipel, Guoguang Zhao, and Jianping Jia. 2019. The effects of seven-week cognitive training in patients with vascular cognitive impairment, no dementia (the Cog-VACCINE study): A double-blinded randomized controlled trial. *Alzheimer's & Dementia* 15: 605–614.

Tong, Xiuhong, Catherine McBride, Jason Chor Ming Lo, and Hua Shu. 2017. A three-year longitudinal study of reading and spelling difficulty in Chinese developmental dyslexia: The matter of morphological awareness. *Dyslexia* 23: 372–386. https://doi.org/10.1002/dys.1564.

Tzeng, Ovid J.-L., and William S.-Y. Wang. 1983. The first two R's. *American Scientist* 71: 238–243.

Tzeng, Ovid J.-L., Sylvia Chen, and Daisy L. Hung. 1991. The classifier problem in Chinese aphasia. *Brain and Language* 41: 184–202. https://doi.org/10.1016/0093-934X(91)90152-Q.

Wang, Dechun (王德春). 1997. *Shenjing Yuyanxue* 神经语言学 [*Neurolinguistics*]. Shanghai: Shanghai Waiyu Jiaoyu Chubanshe 上海：上海外语教育出版社 [Shanghai: Shanghai Foreign Language Education Press].

Wang, Lin, Marcel Bastiaansen, Yufang Yang, and Peter Hagoort. 2012a. Information structure influences depth of syntactic processing: Event-related potential evidence for the Chomsky illusion. *PLoS One* 7: e47917. https://doi.org/10.1371/journal.pone.0047917.

Wang, Lin, Peter Hagoort, and Yufang Yang. 2009. Semantic illusion depends on information structure: ERP evidence. *Brain Research* 1282: 50–56. https://doi.org/10.1016/j.brainres.2009.05.069.

Wang, Lin, Zude Zhu, and Marcel Bastiaansen. 2012b. Integration or predictability? A further specification of the functional role of gamma oscillations in language comprehension. *Frontiers in Psychology* 3: 187. https://doi.org/10.3389/fpsyg.2012.00187.

Wang, Peng (王彭), Shanshan Zhang (张珊珊), and Huo Jiang (江火). 2017. Hanyu houxuju zhong shenglue jiegou de shenjing jizhi yanjiu 汉语后续句中省略结构的神经机制研究 [A study on the neural mechanism of Chinese elliptical structure in subsequent sentence]. *Yuyan Wenzi Yingyong* 语言文字应用 [Applied Linguistics] 2: 46–56.

Wang, Suiping, Zude Zhu, John X. Zhang, Zhaoxin Wang, Zhuangwei Xiao, Huadong Xiang, and Hsuan-Chih Chen. 2008. Broca's area plays a role in syntactic processing during Chinese reading comprehension. *Neuropsychologia* 46: 1371–1378. https://doi.org/10.1016/j.neuropsychologia.2007.12.020.

Wang, Suiping (王穗苹), and Lei Mo (莫雷). 2001. Pianzhang yuedu lijie zhong Beijing xinxi de tongda 篇章阅读理解中背景信息的通达 [The understanding of background information in text reading comprehension]. *Xinli Xuebao* 心理学报 [*Acta Psychologica Sinica*] 4: 312–319.

Wang, Suiping (王穗苹), Lei Mo (莫雷), and Xin Xiao (肖信). 2001. Pianzhang yuedu zhong xianxing xinxi tongda de ruogan yingxiang yinsu 篇章阅读中先行信息通达的若干影响因素 [Some influencing factors of information access in text reading]. *Xinli Xuebao* 心理学报 [*Acta Psychologica Sinica*] 6: 508–517.

Wang, Tao (王涛), and Yiming Yang (杨亦鸣). 2017a. Wanqi gao shuliangdu Han Ying shuangyuzhe mingdong jiagong de fMRI yanjiu 晚期高熟练度汉英双语者名动加工的fMRI研究 [An fMRI study on neural substrates of nouns and verbs in late high-proficient Chinese-English bilinguals]. *Waiyu Xuekan* 外语学刊 [Foreign Language Research] 6: 60–66.

———. 2017b. Shuangyu kongzhi de shenjing jichu: Yili hanying shuangyu shikong de gean yanjiu 双语控制的神经基础———一例汉英双语失控的个案研究 [On the neural basis of bilingual language control: A case study of a Chinese-English bilingual losing control of languages]. *Yuyan Kexue* 语言科学 [*Linguistic Sciences*] 6: 561–572.

Wang, Xiaoyue, Suiping Wang, Yuebo Fan, Dan Huang, and Zhang Yang. 2017. Speech-specific categorical perception deficit in autism: An event-related potential study of lexical tone processing in Mandarin-speaking children. *Scientific Reports* 7: 43254. https://doi.org/10.1038/srep43254.

Wang, Xiaosha, Alfonso Caramazza, Marius V. Peelen, Zaizhu Han, and Yanchao Bi. 2015. Reading without speech sounds: VWFA and its connectivity in the congenitally deaf. *Cerebral Cortex* 25: 2416–2426. https://doi.org/10.1093/cercor/bhu044.

Wang, Xiaosha, Wei Wu, Zhenhua Ling, Yangwen Xu, Yuxing Fang, Xiaoying Wang, Jeffrey R. Binder, Weiwei Men, Jia-Hong Gao, and Yanchao Bi. 2018. Organizational principles of abstract words in the human brain. *Cerebral Cortex* 28: 4305–4318. https://doi.org/10.1093/cercor/bhx283.

Wang, Xiaolu (王小潞). 2009. *Hanyu Yinyu Renzhi yu ERP Shenjing Chengxiang* 汉语隐喻认知与 ERP 神经成像 [*Chinese metaphorical cognition and its ERP imaging*]. Beijing: Gaodengjiaoyu Chubanshe 北京: 高等教育出版社 [Beijing: Higher Education Press].

Wang, Xinde (王新德). 1994. Woguo hanyu shiyuzheng jiqi kangfu wenti 我国汉语失语症及其康复问题 [Chinese aphasia and its rehabilitation in China]. *Linchuang Shiyong Shenjing Jibing Zazhi* 临床实用神经疾病杂志 [*Journal of Clinical Practical Neurological Disorders*] 1: 13–15.

Wei, Jinghan (魏景汉), Peizi Kuang (匡培梓), Dongsong Zhang (张东松), and Zhuangtian Pan (潘垚天). 1995. Quanshiye hanzi ciyi lianxiang de ERP tezheng yu hanzi renshi de ERP zhenbie 全视野汉字词义联想的ERP特征与汉字认识的 ERP 甄别 [ERP characteristics of Chinese word meaning association and ERP discrimination of Chinese character recognition]. *Xinli Xuebao* 心理学报 [*Acta Psychologica Sinica*] 4: 413–421.

Wei, Zhiqiang (卫志强). 1994. Rennao yu renlei ziran yuyan: duofangwei yanjiu zhong de shenjing yuyanxue 人脑与人类自然语言——多方位研究中的神经语言学 [Human brain and human natural language: Neurolinguistics in multi-dimensional studies]. *Yuyan Wenzi Yingyong* 语言文字应用 [*Applied Linguistics*] 4: 43–49.

Wu, Sina (吴思娜), Hua Shu (舒华), and Yu Wang (王彧). 2004. 4–6 nianji xiaoxuesheng fazhanxing yuedu zhangai de yizhixing yanjiu 4~6 年级小学生发展性阅读障碍的异质性研究 [Heterogeneity of developmental dyslexia in primary school students in grades 4 to 6]. *Xinli Fazhan Yu Jiaoyu* 心理发展与教育 [*Psychological Development and Education*] 20(3): 46–50.

Wu, Yan Jing, and Guillaume Thierry. 2017. Brain potentials predict language selection before speech onset in bilinguals. *Brain and Language* 171: 23–30. https://doi.org/10.1016/j.bandl.2017.04.002.

Xia, Quansheng (夏全胜), Yong Lü (吕勇), and Feng Shi (石锋). 2012. 汉语名词和动词语义加工中具体性效应和词类效应的 ERP 研究 [An ERP study on the concreteness effect and grammatical class effect of Chinese nouns and verbs during semantic processing]. *Nankai Yuyan Xuekan* 南开语言学刊 [*Nankai Linguistics*] 1: 110–119.

Xia, Zhichao, Linjun Zhang, Fumiko Hoeft, Gu Bin, Gaolang Gong, and Hua Shu. 2017. Neural correlates of oral word reading, silent reading comprehension, and cognitive subcomponents. *International Journal of Behavioral Development* 42: 342–356. https://doi.org/10.1177/0165025417727872.

Xiao, Rong (肖蓉), Dandan Liang (梁丹丹), and Shanpeng Li (李善鹏). 2020. Hanyu Putonghua shengdiao ganzhi de laonianhua xiaoying 汉语普通话声调感知的老年化效应:来自 ERP 的证据 [*Effects of aging on the Mandarin lexical tone perception: Evidence from ERPs*] *Xinli Xuebao* 心理学报 [*Acta Psychologica Sinica*] 52(1): 1–11.

Xu, Lili (徐丽丽), Shanping Mao (毛善平), Meng Cai (蔡萌), and Youxia Sun (孙友霞). 2006. Shiyufa huanzhe dongci sunhai tedian ji jizhi yanjiu 失语法患者动词损害特点及机制研究 [The characteristics of verb impairment in Chinese agrammatism and its mechanism]. *ZhonghuaWuliyixue yu Kangfu Zazhi* 中华物理医学与康复杂志 [*Chinese Journal of Physical Medicine and Rehabilitation*] 9: 622–624.

Yan, Pei (晏培), and Surong Gao (晏培). 2000. Chuandaoxing shiyu huanzhe de fushu he langdu zhangai 传导性失语患者的复述和朗读障碍 [Repetition and reading aloud disturbance of the patients with conduction aphasia]. *Zhonghua Shenjingke Zazhi* 中华神经科杂志 [*Chinese Journal of Neurology*] 3: 141–143.

Yang, Xiaohong, Xiuping Zhang, Yufang Yang, and Nan Lin. 2018. How context features modulate the involvement of the working memory system during discourse. *Neuropsychologia* 111: 36–44. https://doi.org/10.1016/j.neuropsychologia.2018.01.010.

Yang, Yanhui, Bohan Dai, Peter Howell, Xianling Wang, Kuncheng Li, and Lu. Chunming. 2014. White and grey matter changes in the language network during healthy aging. *PLoS One* 9: e108077. https://doi.org/10.1371/journal.pone.0108077.

Yang, Yang, Fuyun Wu, and Xiaolin Zhou. 2015. Semantic processing persists despite anomalous syntactic category: ERP evidence from Chinese passive sentences. *PLoS One* 10: e0131936. https://doi.org/10.1371/journal.pone.0131936.

Yang, Yiming (杨亦鸣). 2002. *Yuyan de Shenjing Jizhi yu Yuyan Lilun Yanjiu* 语言的神经机制与语言理论研究 [*Research on the neural mechanism of language and language theory*]. Shanghai: Xuelin Chubanshe 上海: 学林出版社 [Shanghai: Xuelin Press].

———. 2007. Yuyan de lilun jiashe yu shenjing jichu—yi dangqian hanyu de ruogan shenjing yuyanxue yanjiu weili 语言的理论假设与神经基础——以当前汉语的若干神经语言学研究为例 [Theoretical hypotheses on language and their neurological basis]. *Yuyan Kexue* 语言科学 [*Linguistic Sciences*] 2007 (6): 60–83.

———. 2012. Shenjing yuyanxue yu dangdai yuyanxue de xueshu chuangxin 神经语言学与当代语言学的学术创新 [Neurolinguistics and academic innovation in modern linguistic studies]. *Zhongguo Yuwen* 中国语文 [*Studies of the Chinese Language*] 6: 548–560.

ang, Yiming (杨亦鸣), and Libo Geng (耿立波). 2008. Jiyu ERPs shiiyan de eryu cihui nengli xinli xianshixing yanjiu 基于 ERPs 实验的二语词汇能力心理现实性研究 [A study on the psychological reality of second language vocabulary competence based on ERPs experiment]. *Waiyu Jiaoxue yu Yanjiu* 外语教学与研究 [*Foreign Language Teaching and Research*] 3: 163–169.

Yang, Yiming (杨亦鸣), and Ming Cao (曹明). 1997. Hanyu pizhixia shiyu huanzhe zhudong jushi yu beidong jushi lijie, shengcheng de bijiao yanjiu 汉语皮质下失语患者主动句式与被动句式理解、生成的比较研究 [Comparative study on the comprehension and generation of active and passive sentence patterns in Chinese subcortical aphasia patients]. *Zhongguo Yuwen* 中国语文 [*Studies of the Chinese Language*] 4: 282–288.

———. 1998. Zhongwen danao ciku xing, yin, yima guanxi de shenjing yuyanxue fenxi 中文大脑词库形、音、义码关系的神经语言学分析 [Neurolinguistic analysis of the relationship among morphology, sound and semiotic code in Chinese brain lexicon]. *Zhongguo Yuwen* 中国语文 [*Studies of the Chinese Language*] 6: 417–424.

———. 2000. Jiyu shenjing yuyanxue de zhongwen danao ciku chutan 基于神经语言学的中文大脑词库初探 [A preliminary discussion of the Chinese mental lexicon based on neurolinguistics]. *Yuyan Wenzi Yingyong* 语言文字应用 [*Applied Linguistics*] 3: 92–99.

Yang, Yiming (杨亦鸣), and Tao Liu (刘涛). 2013. Hanyu huatiju zhong yuji de shenjing jizhi yanjiu 汉语话题句中语迹的神经机制研究 [Neural mechanisms of trace in Chinese topicalized constructions]. *Zhongguo Shehui Kexue* 中国社会科学 [*Social Sciences in China*] 6: 146–166.

Yang, Yiming (杨亦鸣), Dandan Liang (梁丹丹), Jiexin Gu (顾介鑫), Xuchu Weng (翁旭初), and Shiwen Feng (封世文). 2002. Ming dong fenlei: Yufa de haishi yuyi de — hanyu mingdong fenlei de shenjing yuyanxue yanjiu 名动分类:语法的还是语义的——汉语名动分类的神经语言学研究 [A neurolinguistic study on the classification of nouns and verbs in Chinese] *Yuyan Kexue* 语言学科 [*Linguistics Sciences*] 1: 31–46.

Yang, Yufang (杨玉芳). ed. 2015. *Xinli Yuyanxue* 心理语言学 [*Psycholinguistics*]. Beijing: Kexue Chubanshe 北京: 科学出版社 [Beijing: Science Press].

Ye, Caiyan (叶彩燕). 2004. Yueying shuangyu ertong zaoqi de yufa fazhan 粤英双语儿童早期的语法发展 [Early grammatical development in Cantonese and English bilingual children]. *Dangdai Yuyanxue* 当代语言学 [*Contemporary Linguistics*] 1: 1–18.

Ye, Zheng, Yue-jia Luo, Angela D. Friederici, and Xiaolin Zhou. 2006. Semantic and syntactic processing in Chinese sentence comprehension: Evidence from event-related potentials. *Brain Research* 1071: 186–196. https://doi.org/10.1016/j.brainres.2005.11.085.

Yin, Wengang (尹文刚). 1984. Naogongneng yicehua wenti de yanjiu 脑功能"一侧化"问题的研究 [Study on the 'lateralization' of brain function]. *Xinlixue Dongtai* 心理学动态 [*Journal of Developments in Psychology*] 4: 51–58.

———. 1990. Hanzi shidu de leixing yu yiyi 汉字失读的类型与意义 [Types and meanings of Chinese characters unreadable]. *Xinli Xuebao* 心理学报 [*Acta Psychologica Sinica*] 3: 297–305.

You, Zhijun (尤志珺), Shanping Mao (毛善平), Guojin Wang (王国瑾), and Xuefeng Feng (冯学峰). 2005. Zuobanqiu naocuzhong hou hanyu yufa queshi huanzhe gongnengci de lijie zhangai 左半球脑卒中后汉语语法缺失患者功能词的理解障碍 [Disturbance of understanding of functional morpheme in patients with Chinese agrammatism after left-hemisphere stroke]. *Zhongguo Linchuang Kangfu* 中国临床康复 [*Chinese Journal Of Clinical Rehabilitation*] 33: 4–5.

Yu, Mengxia, Ce Mo, You Li, and Lei Mo. 2015. Distinct representations of syllables and phonemes in Chinese production: Evidence from fMRI adaptation. *Neuropsychologia* 77: 253–259. https://doi.org/10.1016/j.neuropsychologia.2015.08.027.

Yu, Xi, Yanchao Bi, Zzaizhu Han, Caozhe Zhu, and Sam Po Law. 2012. Neural correlates of comprehension and production of nouns and verbs in Chinese. *Brain and Language* 122: 126–131. https://doi.org/10.1016/j.bandl.2012.05.002.

Zhang, Caicai, Kenneth R. Pugh, Einar W. Mencel, Peter J. Molfese, Stephen J. Forst, James S. Magnuson, Gang Peng, and William S.-Y. Wang. 2016. Functionally integrated neural processing of linguistic and talker information: An event-related fMRI and ERP study. *Neuroimage* 124: 536–549. https://doi.org/10.1016/j.neuroimage.2015.08.064.

Zhang, Hui, Long Jiang, Gu Jiexin, and Yiming Yang. 2013a. Electrophysiological insights into the processing of figurative two-part allegorical sayings. *Journal of Neurolinguistics* 26: 421–439. https://doi.org/10.1016/j.jneuroling.2013.01.004.

Zhang, Hui, Yiming Yang, Gu Jiexin, and Feng Ji. 2013b. ERP correlates of compositionality in Chinese idiom comprehension. *Journal of Neurolinguistics* 26: 89–112. https://doi.org/10.1016/j.jneuroling.2012.05.002.

Zhang, Hui (张辉), Hetao Sun (孙和涛), and Jiexin Gu (顾介鑫). 2012. Fei chengyu sizige cizu jiagong zhong yunlü yu jufa hudong de ERP yanjiu: shuyu biaozheng he lijie de renzhi yanjiu zhi liu 非成语四字格词组加工中韵律与句法互动的ERP研究——熟语表征和理解的认知研究之六 [ERP study on prosodic and syntactic interaction in processing of non-idiomatic four-character phrases: Cognitive study on the representation and comprehension of idioms]. *Waiyu yu Waiyu Jiaoxue* 外语与外语教学 [*Foreign Languages and Their Teaching*] 6: 6–11.

———. 2013. Chengyu jiagong zhong yunlü yu jufa hudong de shijian xiangguan dianwei yanjiu 成语加工中韵律与句法互动的事件相关电位研究 [Study on the event-related potential of prosodic and syntactic interaction in the processing of idioms]. *Waiguoyu* 外国语 [*Journal of Foreign Languages*] 36: 22–31.

Zhang, John X., Ching-Mei Feng, Peter T. Fox, Jia-Hong Gao, and Li Hai Tan. 2004. Is left inferior frontal gyrus a general mechanism for selection? *Neuroimage* 23: 596–603. https://doi.org/10.1016/j.neuroimage.2004.06.006.

Zhang, Mingxia, Chuansheng Chen, Gui Xue, Zhong-Lin Lu, Leilei Mei, Hongli Xue, Miao Wei, Qinghua He, Jin Li, and Qi Dong. 2014. Language-general and -specific white matter micro-structural bases for reading. *Neuroimage* 98: 435–441. https://doi.org/10.1016/j.neuroimage.2014.04.080.

Zhang, Ping (张萍). 2010. Zhongguo yingyu xuexizhe xinli ciku lianxiang moshi duibi yanjiu 中国英语学习者心理词库联想模式对比研究 [A comparative study of word association patterns in Chinese EFL learner's mental lexicon]. *Waiyu Jiaoxue yu Yanjiu* 外语教学与研究 [*Foreign Language Teaching and Research*] 1: 8–16.

Zhang, Qian, Hui Wang, Cimei Luo, Junjun Zhang, Zhenlan Jin, and Ling Li. 2019. The neural basis of semantic cognition in mandarin Chinese: A combined fMRI and TMS study. *Human Brain Mapping* 40: 5412–5423. https://doi.org/10.1002/hbm.24781.

Zhang, Qian (张强), and Yiming Yang (杨亦鸣). 2010. Hanyu binyu guanxicongju jiagong youshi:laizi shenjing dianshenglixue yanjiu de zhengju 汉语宾语关系从句加工优势：来自神经电生理学研究的证据 [Object preference in the processing of relative clause in Chinese: ERP evidence]. *Yuyan Kexue* 语言科学 [*Linguistic Sciences*] 4: 337–353.

Zhang, Qiang (张强), Xingan Shen (沈兴安), and Huo Jiang (江火). 2005. Woguo shenjing yuyanxue yanjiu de lilun he fangfa 我国神经语言学研究的理论和方法 [Theories and methods of neurolinguistics in China]. *Waiyu Yanjiu* 外语研究 [*Foreign Languages Research*] 6: 1–8.

Zhang, Shanshan (张珊珊), and Huo Jiang (江火). 2010. Liheci shi ci haishi duanyu?-Laizi shenjing dianshenglixue de zhengju 离合词是词还是短语？——来自神经电生理学的证据 [An ERP approach on whether split words are words or phrases]. *Yuyan Kexue* 语言科学 [*Linguistic Sciences*] 5: 486–498.

Zhang, Shanshan (张珊珊), Lun Zhao (赵仑), Tao Liu (刘涛), Jiexin Gu (顾介鑫), and Yiming Yang (杨亦鸣). 2006. Danao zhong de jiben yuyan danwei:laizi hanyu danyinjie yuyan danwei jiagong de ERPs zhengju 大脑中的基本语言单位:来自汉语单音节语言单位加工的ERPs证据 [Basic language unit in the brain: Evidence from ERPs study on monosyllabic units in Mandarin Chinese]. *Yuyan Kexue* 语言科学 [*Linguistic Sciences*] 3: 3–13.

Zhang, Wutian (张武田), and Ruixiang Peng (彭瑞祥). 1984. Danao jineng yicehua he biaoyi wenzi fuhao de rendu 大脑机能一侧化和表意文字符号的认读 [Brain function lateralization and ideographic sign recognition]. *Xinli Xuebao* 心理学报 [*Acta Psychologica Sinica*] 3: 275–281.

Zhang, Yang, Xiaoming Jiang, Henrik Saalbach, and Xiaolin Zhou. 2011. Multiple constraints on semantic integration in a hierarchical structure: ERP Evidence from German. *Brain Research* 1410: 89–100. https://doi.org/10.1016/j.brainres.2011.06.061.

Zhang, Yaxu, Jing Yu, and Julie E. Boland. 2010. Semantics does not need a processing license from syntax in reading Chinese. *Journal of Experimental Psychology: Learning, Memory, and Cognition* 36: 765–781. https://doi.org/10.1037/a0019254.

Zhao, Jing, Kerstin Kipp, Carl Gaspar, Urs Maurer, Xuchu Weng, Axel Mecklinger, and Su Li. 2014. Fine neural tuning for orthographic properties of words emerges early in children reading alphabetic script. *Journal of Cognitive Neuroscience* 26: 2431–2442.

Zhao, Jing, Urs Maurer, Sheng He, and Xuchu Weng. 2019. Development of neural specialization for print: Evidence for predictive coding in visual word recognition. *PLoS Biology* 17: e3000474.

Zhao, Jisheng (赵吉生), and Zhiqiang Wei (卫志强). 1987. (Trans.) *Shenjing Yuyanxue* 神经语言学 [*Basic Problems of Neurolinguistics*]. Beijing Daxue Chubanshe 北京大学出版社 [Peiking University Press]

Zhao, Lili (赵丽丽), Chengyan Li (李承晏), Zhijun You (尤志珺), and Xuefeng Feng (冯学峰). 2002. Hanyu yufa liangbiao de zhiding he biaozhunhua 汉语语法量表的制定和标准化 [Chinese agrammatism battery and its standardization]. *Cuzhong yu Shenjing Jibing* 卒中与神经疾病 [*Srtoke and Nervous Diseases*] 5: 296–298.

Zhao, Ming, Tao Liu, Gang Chen, and Feiyan Chen. 2015. Are scalar implicatures automatically processed and different for each individual? A mismatch negativity (MMN) study. *Brain Research* 1599: 137–149. https://doi.org/10.1016/j.brainres.2014.11.049.

Zhao, Ming (赵鸣), Zhiyuan Xu (徐知媛), Tao Liu (刘涛), Fenglei Du (杜锋磊), Yongxin Li (李永欣), and Feiyan Chen (陈飞燕). 2012. Yuyan leibi tuili de shenjing jizhi: laizi ERP yanjiu de zhengju 语言类比推理的神经机制:来自ERP研究的证据 [The neuromechanism underlying language analogical reasoning: Evidence from an ERP study]. *Xinli Xuebao* 心理学报 [*Acta Psychologica Sinica*] 44: 711–719.

Zhao, Rong, Rong Fan, Mengxing Liu, Xiaojuan Wang, and Jianfeng Yang. 2017. Rethinking the function of brain regions for reading Chinese characters in a meta-analysis of fMRI studies. *Journal of Neurolinguistics* 44: 120–133. https://doi.org/10.1016/j.jneuroling.2017.04.001.

Zheng, Qin, Zhongheng Jia, and Dandan Liang. 2015. Metaphor and metonymy comprehension in Chinese-speaking children with high-functioning autism. *Research in Autism Spectrum Disorders* 10: 51–58. https://doi.org/10.1016/j.rasd.2014.11.007.

Zhou, Guoguang (周国光). 2000. Ertong xide fuci de pianxiangxing tedian 儿童习得副词的偏向性特点 [Slant in the acquisition of adverbial word in modern Chinese]. *Hanyu Xuexi* 汉语学习 [*Chinese Language Learning*] 4: 22–27.

Zhou, Qiumin, Xiao Lu, Ying Zhang, Zhenghui Sun, Jianan Li, and Zude Zhu. 2018. Telerehabilitation combined speech-language and cognitive training effectively promoted recovery in aphasia patients. *Frontiers in Psychology* 9: 2312. https://doi.org/10.3389/fpsyg.2018.02312.

Zhou, Tongquan (周统权). 2007. Dongci peijia de liangxiaoying yu zhixiaoying: Laizi shiyuzheng yanjiu de zhengju 动词配价的量效应与质效应——来自失语症研究的证据 [The quantity effect and quality effect of verb valency: The evidence from the study of Chinese aphasia]. *Yuyan Wenzi Yingyong* 语言文字应用 [*Applied Linguistics*] 1: 102–110.

Zhou, Tongquan (周统权), and Siruo Zhou (周思若). 2018. Cong yuyan jiagong de yuqixing dao dingzhong jiegou jiagong de yuqixing 从语言加工的预期性到定中结构加工的预期性 [Prediction in language comprehension and in attributive-centred structure] *Dangdai Waiyu Yanjiu* 当代外语研究 [*Contemporary Foreign Language Studies*] 2: 18–24.

Zhou, Tongquan (周统权), Wei Zheng (郑伟), Hua Shu (舒华), and Yiming Yang (杨亦鸣). 2010. Hanyu binyu guanxicongju jiagong youshilun: laizi shiyuzheng yanjiu de zhengju 汉语宾语关系从句加工优势论:来自失语症研究的证据 [Preference of Chinese relative clause processing: Evidence from aphasia]. *Yuyan Kexue* 语言科学 [*Linguistic Sciences*] 3: 337–348.

Zhou Xiaolin (周晓林). 1997. Limit role of phonology in semantic access (语义激活中语音的有限作用). In Peng Danlin (彭聃龄, Hua Shu (舒 华) and Hsuan-Chih Chen (陈烜之), Cognitive Research of Chinese Language 汉语认知研究 (Hanyu Renzhi Yanjiu). Jinan:Shandong Jiaoyu Chubanshe 济南:山东教育出版社 [Jinan: Shandong Educational Press]. pp. 159–194.

Zhou, Wei, Zhichao Xia, Yanchao Bi, and Hua Shu. 2015. Altered connectivity of the dorsal and ventral visual regions in dyslexic children: A resting-state fMRI study. *Frontiers in Human Neuroscience* 9: 495. https://doi.org/10.3389/fnhum.2015.00495.

Zhu, Zude, Brian T. Gold, Chi-Fu Chang, Suiping Wang, and Chi-Hung Juan. 2015. Left middle temporal and inferior frontal regions contribute to speed of lexical decision: A TMS study. *Brain and Cognition* 93: 11–17. https://doi.org/10.1016/j.bandc.2014.11.002.

Zhu, Zude, Fengjun Yang, Dongning Li, Lianjun Zhou, Ying Liu, Ying Zhang, and Xuezhi Chen. 2017. Age-related reduction of adaptive brain response during semantic integration is associated with gray matter reduction. *PLoS One* 12: e0189462. https://doi.org/10.1371/journal.pone.0189462.

Zhu, Zude, John X. Zhang, Suiping Wang, Zhuangwei Xiao, Jian Huang, and Hsuan-Chih Chen. 2009. Involvement of left inferior frontal gyrus in sentence-level semantic integration. *Neuroimage* 47: 756–763. https://doi.org/10.1016/j.neuroimage.2009.04.086.

Zhu, Zude, Marcel Bastiaansen, Jonathan G. Hakun, Karl Magnus Petersson, Suiping Wang, and Peter Hagoort. 2019a. Semantic unification modulates N400 and BOLD signal change in the brain: A simultaneous EEG-fMRI study. *Journal of Neurolinguistics* 52: 100855. https://doi.org/10.1016/j.jneuroling.2019.100855.

Zhu, Zude, Peter Hagoort, John X. Zhang, Gangyi Feng, Hsuan-Chih Chen, Marcel Bastiaansen, and Suiping Wang. 2012. The anterior left inferior frontal gyrus contributes to semantic unification. *Neuroimage* 60: 2230–2237. https://doi.org/10.1016/j.neuroimage.2012.02.036.

Zhu, Zude, Wang Suiping, Xu Nannan, Li Mengya, and Yang Yiming. 2019b. Semantic integration declines independently of working memory in aging. *Applied Psycholinguistics* 40: 1481–1494. https://doi.org/10.1017/S0142716419000341.

Zhu, Zude, Xiaopu Hou, and Yiming Yang. 2018. Reduced syntactic processing efficiency in older adults during sentence comprehension. *Frontiers in Psychology* 9: 243. https://doi.org/10.3389/fpsyg.2018.00243.

Linguistic Typology in China

Zhengda Tang and Yue Wu

Contents

Introduction	116
The Introduction of Typology to China Through Translation	116
Reviews and Textbook Compilations	117
From Introductory Material to Typological Research	118
Typological Research into Sinitic Languages	120
Word Order Typology	121
Word Class Typology	124
Reference Grammar and Description of Languoids from Typological Perspectives	125
Theoretical Development and Linguistic Inventory Typology	127
Conclusion	129
References	129

Abstract

Since the seminal works of Greenberg and Comrie were translated into Chinese in the 1980s, linguistic typology has drawn increasing attention among Chinese linguists and has inspired researchers to investigate language "facts" through crosslinguistic comparisons, typological classifications, and generalizations of implicational universals. There has since been a steady stream of research into certain languages under the rubric of linguistic typology that are associated in one way or another with corresponding features in other languages that are genetically or geographically unrelated. In Chinese academic circles, however, most typological studies are still occupied by the description of Sinitic languages from typological perspectives. Crosslinguistically unbiased studies of categories or

---

Z. Tang (✉)
Institute of Linguistics, Chinese Academy of Social Sciences, Beijing, China
e-mail: tangzhd@cass.org.cn

Y. Wu
School of Humanities, Hangzhou Normal University, Hangzhou, China
e-mail: wuyuehznu@hznu.edu.cn

constructions based on data drawn from balanced samples remain underdeveloped; the similar is true for more finely grained studies of languoids other than the authors' native languages, especially those outside China. The recently emerged reference grammars, however, can present themselves to contribute to the theoretical and empirical aspects of linguistic typology. Additionally, Chinese typologists are also making efforts toward innovations in the development of typological theory and methodology.

Keywords

Linguistic typology · Crosslinguistic comparison · Reference grammar · Chinese dialects · Word order · Word class

Introduction

Since the 1980s, linguistic typology carried out by Chinese linguists has continued to develop at an accelerating rate, seeing a shift from introductory works to innovative ones, and an expansion in scope from research based on a limited sample of languages to ones based on large scale of corpora. This chapter seeks to outline this trajectory and provide a summary of the major features of the development of linguistic typology in China. Sections "The Introduction of Typology to China Through Translation" and "Reviews and Textbook Compilations" concern the introduction of typology to China by means of translations, book reviews, and textbook compilations. Section "From Introductory Material to Typological Research" outlines the transition between introductory and innovative works on typology. Section "Typological Research into Sinitic Languages," which constitutes the main body of the entry, discusses various studies carried out from the perspective of typology, including the typology of word order, parts of speech, and reference grammars within a typological framework. The next section provides a brief introduction of the latest theoretical developments in typology, followed by a conclusion.

The Introduction of Typology to China Through Translation

The emergence of linguistic typology in a modern sense is marked by the publication of Greenberg's (1963) seminal paper, "Some Universals of Grammar with Particular Reference to the Order of Meaningful Elements," and its translation into Chinese by Lu and Lu in 1984 marks the beginning of the studies and application of linguistic typology in China. Beyond increasing attention given to word order, several introductions to the general theoretical and methodological framework of linguistic typology followed. Among these is the translation of Bernard Comrie's *Language Universals and Linguistic Typology* (1981) by Shen, which continues to influence researchers and students of linguistics. The translation of monographs and papers on linguistic typology has continued since this period. Even with the increase of

Chinese scholars with a level of English language competency to read the original works in English, the Chinese translations of these important studies are still widely published and read. Examples of such include Croft (1991), translated by Gong Qunhu 龚群虎 et al (2010), and a collection of papers representative of recent developments in typology in the twenty-first century translated and compiled by Dai Qingxia 戴庆厦 and Wang Feng 汪峰 (2014) under the rubric of *The Primary Methodology and Theoretical Frameworks of Linguistic Typology*.

The translations of other works involving language comparisons and classifications – though not generally acknowledged as *linguistic typology proper* – have also been published and widely introduced to classrooms, such as those adopting genetic and geographic approaches. Mantaro Hashimoto's (1985) influential research in its Chinese translation by Yu Zhihong has fueled continuous interest and practices in the geologically based approach to Sinitic languages, including Sinitic dialects, and languages within and adjacent to China. Realizing that there could be possible confusion between contrastive linguistics and linguistic typology, Shen Jiaxuan 沈家煊 (1988) published his translation of Comrie's paper on the comparison between these two approaches. This and later translations highlight the necessity of cross-language comparison and the importance of well-balanced data building – thereby introducing new perspectives into Chinese academic circles. Chinese linguists have hitherto concentrated on explorations into individual languages or binary comparisons between Chinese and a certain foreign language. Considering the great number of dialect researchers in China, the introduction of linguistic typology in relation to dialectology is of special significance, as is demonstrated by the publication of Kortmann's (1999) *Typology and Dialectology* translated by Liu Haiyan 刘海燕 and Liu Danqing 刘丹青 (2004). The close link between typology and dialectology and the later concept of "languoid" (the language-like entity) has strengthened the involvement of dialect experts who have gone on to make noteworthy contributions to the theoretical development of linguistic typology as a whole.

Reviews and Textbook Compilations

Compared to translations of original monographs or papers, the continuous stream of reviews by generations of Chinese linguists has proven a more effective means to introduce typological perspectives and methodologies to Chinese academic circles. In this regard, Liao Qiuzhong's 廖秋忠 (1984) review of Comrie's (1981) monograph that appeared in the same year and same journal as the Chinese translation of Greenberg's paper noted above is worthy of note. Liao's review offers not only accurate chapter-by-chapter summaries of the findings detailed in the original work but also contains thought-provoking comments and criticisms, including methodological issues of data reliability and representability. As a heavily data-dependent discipline, these issues still loom large in linguistic typology. In 1991 William Croft published *Typology and Universals*. Shortly afterward, Shen Jiaxuan (1991) published his review of the work, in which his famous motto of "viewing Chinese, together with commonalities and exceptions, within but not beyond the range of language variations" has inspired and encouraged many later works in this area. Other important textbooks of general linguistic typology in twenty-first century have also been brought to the attention of Chinese linguists through book reviews, such as Guo Zhong 郭中 and Chu Zexiang's 储泽祥 (2012) review of the three-volume *Language Typology and Syntactic Description* edited by Timothy Shopen (2007), and Lin Zhong's 林忠 (2013) review of *The Oxford Handbook of Linguistic Typology* edited by Jae Jung Song in 2011.Through these reviews and other introductory works, the readers have glimpsed of not only the classic discipline, but also recent developments in linguistic typology. Outside of these, there are around ten journals with dedicated sections for reviews and introductions to more specific subjects associated with linguistic typology, such as categories, constructions, and other theoretical issues.

Textbooks on linguistic typology written in Chinese have also been published, such as *Linguistic Typology* by Liu Danqing 刘丹青 (2017), *What Is Linguistic Typology* by Jin Lixin 金立鑫 (2011), and *Course of Linguistic Typology* by Lu Bingfu 陆丙甫 and Jin Lixin 金立鑫 (2015). Jin's work (2011), in a question-and-answer format, consists of definitions, properties, and methodologies in linguistic typology, which is the so-called holistic typology, and specific typology concerning entities and categories. This book attaches considerable importance to formal features, logic procedures, and methodological rigidity. Written as a textbook designed for college students, Lu and Jin (2015) provides the basic features of linguistic typology and a large data sample. The textbook is also characterized by rich and specific references to Sinitic languages as both a canonical and peculiar language sample in typological theory. Liu's (2017) textbook is based on the classroom recordings of a course on linguistic typology and therefore facilitates students' understanding of otherwise more abstract concepts and operations in linguistic typology and serves to bridge the gap between principles and real-life language facts. Using witty generalizations and metaphors, Liu's textbook also presents his consideration of the role that linguistic typology has to play in relation to functionalist and formalist approaches, which can be characterised as "attesting, text, and testing." According to Liu, while the functionalist and formalist approaches focus on monolingual data, whether from existing texts or self-made sentences, typologists turn to more languages for the existence or absence of types or values. He also provides the basic concepts of "Linguistic Inventory Typology" which he has developed in the recent decade.

From Introductory Material to Typological Research

Neither linguists nor readers are content with translations and reviews of specific books or articles. Generalizations, reinterpretations, and other forms of extension are necessary for both research and learning. Jin Lixin 金立鑫 (2018) revisits the fundamentals of typology and classifies the evolution of linguistic typology into three stages: traditional, modern, and contemporary which can be specified by their own methodological tendencies, such as taxonomy, implicational correlations, and by multiple-field intervention of statistics, historical or geological information. He also emphasizes that linguistic typology is a set of methods or methodological systems rather than a set of principles or beliefs. This entry is representative of his insistence on having sufficient data and strict methodology. Jiang Yi 江轶 (2006) offers generalizations of linguistic typology through its 20 years of development, deeper interactions with other disciplines both within and beyond linguistics, extended fields and topics of research, and a large number of typological studies or studies that reference typology. The interaction of typology with other trends or disciplines is inevitable since the search for crosslinguistic universality is "innate" to various branches of linguistic pursuits. This is particularly true when formal and typological studies meet, as is found in the work of Wang Honglei 王洪磊. Wang (2019) points out that it is not sufficient to realize that formalists and typologists have more or less common goals for some sense of universals. Formalist typology, as it is termed by the author, is still controversial and under scrutiny in so far as the methodological and philosophical discrepancies between the formalist and typological approaches are concerned.

As linguistic typology has been regarded as a set of methodologies rather than a set of principles or theoretical prerequisites by many Chinese linguists, it is necessary to specify the "methods" which are specific and significant to typology. The "Western tradition," as is so termed, is believed to attach more importance to procedural or methodological matters. This can be shown by the considerable number of international publications "purely" on methodological issues, such as sampling and restricting universals. In contrast, Chinese typologists pay less attention to these "meta-knowledges" of typology itself but concern themselves with concrete case studies, such as of specific, discrete constructions and categories, in which typological methods are applied as operational devices. For example, few concepts can be more persuasively employed in typological researches than *markedness*, which was originally designed for phonology and later applied in almost all aspects of linguistic studies, such as lexicology, syntax, and semantics. Markedness soon proves to be a useful instrument especially in doing crosslinguistic comparison and finding correlations between language components, parameters, and values in typological research. Shen Jiaxuan 沈家煊 (1997) gives a panoramic view of the evolution and accommodation of the concept of markedness in relation to the late developed methods, such as implicational universals and multiparameter hierarchies. He also builds a connection between the well-known concept of cognitive prototype and markedness and predicts the application of markedness in the typological study of the Chinese language. Issues concerning data remain the core concern among linguists when considering methodological prerequisites in typological studies and the question of how to gloss raw data is as fundamental as the description and explanation of the facts. Widely acknowledged as the basic data glossing principles are the Leipzig Glossing Rules proposed by Comrie et al. (2008), which were introduced to Chinese academic circles in an article by Chen Yujie 陈玉洁 et al. (2014). This entry also discusses the application of these rules to data concerning Mandarin and dialects, some of which are quite uncertain against the background of the current glossing system, but nonetheless highly representative of meaningful elements in Chinese.

The introduction of the method and theory of semantic map to Chinese typologists around 2010 has had a great influence and led to a decade of research on the grammaticalization of multifunctional elements, crosslinguistic and cross-dialectal distribution of semantic primes, and methodological concerns. Zhang Min 张敏 (2010) offers a thorough introduction to the semantic map model and the related principles, operations, and the applications to the studies of multifunctional elements. Wu Fuxiang 吴福祥 (2012) emphasizes the relationship between semantic maps and grammaticalization. He elaborates three ways to "dynamicize" a conceptual space, that is, by crosslinguistic comparison based on functional implication relationship, by the use of grammaticalization paths, and through synchronic reconstruction by principles of grammaticalization. Researchers have also introduced the semantic map approach to both diachronically and synchronically varied languages. For example, Guo Rui 郭锐 of Peking University developed detailed illustrations and exemplifications of the primes and extensions of the semantic map model, as well as the application of this robust research model. See ▶ Chap. 6, "Development of the Semantic Map Model in China," by Zhang Ding 张定 on semantic map research in China in the present encyclopedia.

Typological research commits its findings, conclusions, and principles to the data– the very product of human linguistic activity. Even the pursuit of perfection concerning data-based methodology, such as the balanced representability among language families, as well as geological or demographic distributions, relies ultimately upon the reliability of data. In this sense, the investigation, presentation, gloss, and explanation of the data are of paramount importance. Therefore, fieldwork and language investigation constitute the very basis on which research into linguistic typology can be build. Comrie and Smith (1977) published a fine-grained questionnaire for descriptive language study, which has not drawn much attention among Chinese linguists until Liu Danqing's (2008) handbook was published. Liu (2008) provides a rich stock of interpretations, complementary remarks, and sufficient exemplifications based on the author's command of the 30 years of development of linguistic study, especially the study of linguistic typology. It also benefits from the formidable accumulations of knowledge concerning fieldwork on Sinitic dialects and other languages of China since the beginning of the twentieth century. This entry can be used for multiple purposes; it can be used as a detailed framework for both fieldworkers and writers of reference grammars, a guidebook on morphsyntactic features from the perspective of linguistic typology, or a dictionary of concepts, categories, and methodology closely related to linguistic typology. The influence of Liu's work is evident in the continuous emergence of hundreds of reference grammars on Sinitic and non-Sinitic languages.

Typological Research into Sinitic Languages

Following such introductory works on linguistic typology, innovative research has emerged that deals with linguistic issues from typological perspectives or with the typology itself, such as methodological issues, applicative issues, and principles and restrictions concerning typological variations and distributions. Among other approaches introduced from Western linguistics, linguistic typology has proved highly effective and has triggered profound thinking on the nature of Sinitic languages and their position in the typological framework, along with both descriptive and explanative issues concerning their seeming anomalies. Liu's (2003) paper entitled "Linguistic typology and Chinese research" offers a series of novel perspectives regarding both typology itself and typological approaches to Sinitic languages. Instead of identifying linguistic typology as one of the functionalist branches, Liu (2003) regards it as a full-fledged school of linguistics, as can be featured under the rubric of *attest*, to certify types/values by crosslinguistic classification and comparison, in contrast with *text* and *test*, to highlight what is essential of the methodology for functionalism and formalism, respectively. This chapter also critically comments on several well-acknowledged "characteristics of the Chinese language," such as mood particles, so as to explain why there is a lack of crosslinguistic perspectives. Liu (2003) also addresses general concerns about collecting large datasets and furnishes a finer-grained description of languages based on a relatively small corpus while still adopting crosslinguistic and typological perspectives. The acceptance of the multiplicity of typological approaches has fostered a great deal of work on exploring and revisiting Sinitic language including various Chinese dialects. Xu Jie's 徐杰 (2005) edited a volume on various constructional and categorical features of Chinese dialects and other languages in China contains studies based on crosslinguistic or cross-dialectal comparison and typological generalizations.

A summary of the characteristic properties of Sinitic languages is necessary here. Morphologically, Chinese has long been labeled as an inflectionally impoverished language in which lexicalization is mainly constituted by compounding, syntax by functional words and word order, and semantic interpretation is heavily dependent upon pragmatic and contextual factors. This basic "fact" has a series of profound effects on other features of Sinitic language, such as the robust use of prepositions, instead of case marking, for locating semantic roles, and of serial verb constructions and juxtaposition of "minor sentences," instead of subordinate measures for combining clauses, and sentence-final clitic-like particles, instead of various categories that are otherwise flagged by inflectional morphology such as tense, aspect, and modality, and even social deixis. From the perspective of word order, Sinitic languages can be characterized by the so-called "post verbal constraints," such as V-O order being prone to manifold constraints and even more conditions under which it is shifted to other alterative ordering such as topicalized OV or *ba* 把-OV. The other feature of word order worthy of note is within nominal phrases, for example, where all nominal modifiers (demonstratives, adjectives, and relative clauses) precede the nominal head. Such basic facts assist in shaping the typological features of Sinitic languages against the background of world languages.

Word Order Typology

Since the publication of Greenberg's seminal paper, the order of meaningful elements, in particular word order, has become one of the most prominent research topics among linguists. James H-Y Tai (1976) proposes that Chinese has experienced the word order change from SVO to SOV, with the latter correlated with the superset "modifier-modified" structure, emergence of postpositions, sentence-final question particles, and "verb-particle" construction. He additionally points out that this word order shift has been expanded from North to South due to the language contact of Chinese and the Northwestern Altaic languages (See in ▶ Chap. 14, "The Evolution of Chinese Grammar from the Perspective of Language Contact", Yonglong Yang and Jingting Zhang). Chauncy Chu (1984) addresses the basic word order in Chinese from both diachronic and synchronic perspectives. Chu draws three major conclusions about basic word order at clausal level and its evolution in Chinese. The first conclusion is that no transfer is observed and proved of basic word order from SVO to SOV even when there are more SOV in modern Chinese than in its old versions. The second is that the increased number of SOV clauses can be accounted for by three reasons: (i) increase of compound verbs and verb-complement combinations; (ii) the decrease of topic markers in historical Chinese; and (iii) the shift of basic word order may be correlated with some features, such as "Adj. – N" and sentence-final question markers, which by no means can be used reasonably as the criteria by which to judge the basic word order. Among many other efforts to further their research into word order issues, Jin Lixin 金立鑫 and Yu 于秀金 (2012) have different observations and revisit the word order type of Mandarin Chinese based on 15 pairs of grammatical elements correlated with VO-OV language types and seven other uncorrelated pairs. Jin and Yu's paper indicates that in ten correlated pairs, Mandarin Chinese has four pairs exhibiting both OV and VO word orders, three pairs tend to be OV order, and the other three VO order. Therefore, it concludes that Chinese is a language of mixed word order. In Sinitic languages or in Sino-Tibetan languages as a whole, the word order issue has attracted even more attention because of the apparent *sprachebund* features existing between Sinitic languages and other adjacent ones. Liu (2002) summarizes these features, such as the ordering concerning argument structure (the basic word order of SOV and SVO), preposition vs. postposition or circumposition, ordering concerning conjunctions, ordering of possessor relative to possessee, and ordering of modifiers (adjectives, relative clauses, classifiers, and demonstratives) relative to head nouns. He presents a series of features, especially those concerning the multiple modifiers preceding and following nominal heads, which cannot readily be generalized by certain Greenbergian universals, and calls for finer-grained parameter setting, evaluations, and explanations.

Since the multiple comparison of entities is a logic necessity for the realization of differences among them and therefore of the identities for each entity, typological perspectives and frameworks reasonably facilitate the studies of what is truly *characteristic* of Chinese, among which the most well-acknowledged are topic prominence, postverbal constraints, robust use of *ba*-construction with *disposal* meaning, complex system of classifiers, tonal and syllable prominent, verb-like adjectives or no independent adjectival category, widely used serial verb constructions, and prenominal relative clauses. Zhang Min 张敏 (2015, 2016) and several other conference papers endeavor to "crack" the puzzle of what seems to be typological "anomalies" of Chinese grammar, such as the division of labor of OV and VO orders, disposal constructions, and postverbal constraints (Chinese finds it difficult for adverbial elements to coexist with object when both occur in postverbal constructions). Zhang emphasizes the need to consider geological distribution and language contact in approaching these seeming anomalies. He assumes that some forms of contact may have occurred between Chinese and Mongolic, Tungus, Turkic, and Tibeto-Burman, and the contact is not of a once-and-for-all nature. This is evidenced by the scalar distribution pattern in which the tendential features toward VO or OV (less or more susceptible to postverbal constraints, respectively) are associated with, and gradually indicated by, where they are attested along the geological line from the North to the South, and from the West to the East. The more Northern, and within the vast Northern areas, the more Western where the language is spoken, the more OV features are attested. Zhang designs a multiple-parameter value system to evaluate these features to avoid the relative simplicity of interpretations in the previous studies, such as Hashimoto (1975, 1985) and Tai (1976). In his recent research, Zhang (2017) provides more language samples from South Asia to strengthen the cross-categorical correlation between classifiers and adjectives noticed by Rijkhoff (2000), according to whom *no* independent adjectives are found in languages that have classifiers. By evaluating several parameters Zhang argues that Chinese, from archaic period to the present, has never claimed an adjectival class independent of verbs.

Word order concerning relative clauses exemplifies the complexity and recursiveness of language, and is indicative of, and correlated with, the basic word order and orders of other elements. Based on the observation of sample data drawn from 189 languages on relative clauses, Tang (2006) tentatively proposes three principles concerning the word order issues in relativization, namely the Principle of Earliest Subject Head, Object Head Closest to Verb, and of Relative Construction Differing from Main Clause. The interaction of these three principles may account for some questions regarding the correlations between relativization-related word order and the basic word order, such as SVO and SOV. Specifically, these questions include those of why SOV languages fall into two subclasses with roughly balanced distribution between pre- and postnominal relative clauses, while the ratio of the two orders is almost all to none in SVO languages if it were not for Chinese in which the relative clauses precede the head nouns. The principles proposed in Tang (2006) also shed light on why a considerable number of genetically unrelated SOV languages produce their relative clauses without using any morpho-syntactically overt markers, whereas SVO languages predominantly find these markers quite indispensable. With these generalizations, an implication is drawn that typological distribution is in certain correlation with, and therefore at least partly accounts for, the factors facilitating or hindering processing of linguistic information. However, Tang also deems it unrealistic to suggest a once-and-for-all solution to issues concerning order of relativization, and he holds a stance that these principles work generally in statistical, rather than qualitative, manner. Tang (2006) is one of the few studies that distinguishes from other studies in that it bases its theoretical generalizations on data drawn from well-balanced sampling, instead of from randomly chosen languoids.

Besides a great number of studies concerning word order of specific categories or constructions, there are some studies that aim to develop more theoretical generalizations. Lu (2005) endeavors to further explore the Greenbergian concept of "dominance," which, as Lu proposes, is mainly motivated by an *identifiability* hierarchy. According to Lu, dominant values under certain parameters tend to be more "identifiable" in referentiality and thus have wider distribution among languages. He also offers a statement in the form of universal, that is, every other condition being equal, the more identifiable elements tend to precede those less identifiable. Lu applies the identifiability-precedence principle in many respects, including the word order concerning multiple modifiers within nominal phrases, possession constructions, and constructions expressing part-and-whole relationships. Useful as it may seem in explaining word order predominance and recession, the identifiability principle per se is still in want of a strict and operable definition, i.e., certain formal or morphosyntactic criteria are needed for how the principle works.

Word Class Typology

In Chinese, most verbs can be used as subjects or objects, and nouns can be used as predicates without any overt morphosyntactic alternations. This, among other various factors, may lead some to conclusions that Chinese has no categorical contrast or distinction of nouns and verbs. Wan Quan 完权 and Shen Jiaxuan 沈家煊 (2010) furnish a detailed introduction to the well-acknowledged crosslinguistic model of word class comparison proposed by Kees Hengeveld and his colleagues at the University of Amsterdam (Hengeveld et al. 2004). Based on this model, especially on its classification of languages into differentiated, flexible, and rigid types, the authors evaluate the impact of the model on the study of morphological and word order typology. After commenting on the typologization of the "Amsterdam Model," the article also identifies Chinese as one language with a flexible word class system due to what the authors regard as the asymmetry between nouns and verbs. They argue that Chinese possesses a word class of noun but not the case for a word class of verb, because, from their point of view, no lexical morphology or other devices in Chinese ensures a certain part of speech to serve only as the head of a predicate but not the head of a "reference," the prototypical function that a noun has to perform (cf. Hengeveld et al. 2004). This is one of the major reasons why Shen has proposed in many of his studies that Chinese has *no* such a verb-noun distinction, which has long been presupposed as basic to all languages. To be specific, Shen illustrates what he regards as one of most typologically characteristic paradigm in which verbs constitute the subcategory within the category of noun, as in Shen (2009). However, most of other research advocates or presupposes the distinction between nouns and verbs and contrasts the two independent word classes. Hu Jianhua 胡建华 (2013) believes that the noun-verb distinction is a basic lexical property for every natural language, Chinese being no exception, and that the only crosslinguistic difference lies in the degree of contrast between the two. He also argues that the uncommon neutralization of nouns and verbs by syntactic environments does not nullify the noun-verb contrast. On the contrary, Hu argues that such a distinction is derived from the "internalized noun-verb equilibrium" that holds at both the lexical and syntactic levels.

The introduction of the concept of scalarity or degree between the two basic categories leads to finer-grained discussions in this regard. Other studies obtain findings that have gone even further from the "verb-belonging-to-noun" paradigm. In his research based on a contrastive study between Chinese and English at various levels (speech act, sentences, subordinate clauses, phrases, word class shift, and of first language acquisition), Liu Danqing (2010) claims that typologically Chinese can be categorized roughly as a verb-oriented language, in comparison with English, a noun-oriented language. Respectively, the two languages serve to exemplify two linguistic types that are in sharp contrast in terms of grammatical prominence of word classes. Liu also provides explanations for a bunch of correlations between the values of being verb oriented and noun oriented and other features. Discussions of the noun-verb-associated controversies have led to more profound understandings concerning the scenario of Chinese parts of speech in the variations of world languages that has hitherto been neglected.

Reference Grammar and Description of Languoids from Typological Perspectives

Linguistic typology is built on the crosslinguistic comparison of formal features, which, in turn, is built on the basis of detailed description of languages. Reference grammar is one of the most frequently applied forms of language description. Chinese linguists have a long tradition and experience in the thorough description of languages including Chinese dialects and other languages in China. Ma Xueliang's 马学良 (1991) *An Introduction to Sino-Tibetan* provides elaborate yet concise descriptions and analyses of more than 40 languages belonging to several branches of the Sino-Tibetan language family mainly within the territory of China and a few in neighboring countries. These languages include Mandarin, Qiang, Jingpo and Lolo, Miao-Yao, and Zhuang-Dong (Kadai). This work offers many well-glossed examples and a relatively more descriptive, rather than prescriptive, framework for language research. The adequacy of categories and the prominence of features are combined. It also deals with cross-dialectal variations. Though introductory in nature, the work serves as one of the most cited reference grammars of Sino-Tibetan languages among Chinese linguists. Its publication can be regarded as one effort to bring about some changes to the well-established collegial curriculum system in which cross-languoid typological studies are lacking, as is pointed out in the preface of Ma's book by Zhu Dexi 朱德熙, an eminent linguist at Peking University.

We see reference grammars as essentially *typological* for two reasons. Firstly, a reference grammar must be structured so as to include categories, parameters, and values that lie within a more or less common and comparable framework to ensure further prototypical typological research. In this sense a reference grammar may resemble a precast, designed to be easy to handle and ready to use both in research and fieldwork. Secondly, a reference grammar differs diametrically from, say, traditional grammars as the framework of the latter does not usually consider typological comparison. Before the presence of reference grammars, most grammars of Sinitic languages and languages of Chinese ethnic minorities were written by observing Zhu Dexi's framework, which follows the structuralist and monolinguistic approach to single languages and often proves incongruous or even impossible for crosslinguistic comparison. For example, the well-established structural category of *pian-zheng* (modifier preceding modified) can mean AP-NP, VP-NP, Adv-VP, Adv-AP, or lexicalized AN, VN, AV, AdvV, AdvA, which are otherwise categorized into totally different classes within typological frameworks.

In recent years, more and more reference grammars have published, some of which are written in English. *A Grammar of Qiang: With Annotated Texts and Glossary*, by LaPolla and Huang in 2003, sets a high standard for a reference grammar from a typological perspective. It provides not only detailed descriptions within the category-based framework (that is, introduction, phonology, the noun phrase, the verbal complex, and the clause and complex structures), but also rich texts depicting various situations and genres of the native life and culture, with standardized morpheme-to-morpheme glosses. These texts are collected from the daily spontaneous talks which may reveal some properties that are difficult to be recalled from mere native intuition when speakers are asked to answer questions. Sheng Yimin 盛益民 (2014) is among the earliest efforts to offer detailed and systemic descriptions of a certain dialect from typological perspectives. Its framework is established with reference to Comrie and Smith (1977), Liu (2008), and other typological works, covering phonology, word class and derivation, nominal phrases, verbal phrases, simple clauses, complex sentences, and semantic categories. This reference grammar contains the specific features of the northern Wu dialect in the standardized, internationally well-acknowledged framework, providing more than 3000 examples at the clausal level, and over 50 tables and graphs, representing a valuable repository of every aspect of this particular Wu dialect. Matthews and Yip (1994/2011) provide a comprehensive grammar of Cantonese in their sizable volume that is structurally arranged so as to combine categorical comprehensiveness, systemic exhaustivity, and featural prominence, which can partially be seen from chapter or section titles, such as "Quantification and existential sentences," "Aspect and verbal particles," and "Sentence structure: word order and topicalization." What is special about Matthews and Yip's reference is that it discusses both the commonality and uniqueness of Cantonese. Besides the grammars of Sinitic languages written in English, there have been a great number of reference grammars written in Chinese of non-Sino-Tibetan languages. For example, most of reference grammars on Austronesian languages in Taiwan are written in Chinese, covering the Rukai, Tsou, Paiwan, Seediq, and Amis languages. This series of reference grammars are more concise than detailed and serve as suitable introductions for beginners. Among the monographs, dissertations, and language reports describing languages in China, the largest reference grammar ever published is *The Languages of China*, edited by Sun Hongkai 孙宏开 (2007). Spanning almost 3000 pages and covering 129 languages in China with contributions by over 90 authors, the work provides introductions and descriptions of languages belonging to Sino-Tibetan, Altaic, Austronesian, Austroasiatic, and Indo-European languages, along with mixed languages, such as Daohua, Tangwanghua, and Wutunhua. In terms of mixed languages, Atshogs (2004) has generated a great deal of discussion among Chinese linguists. The work contains detailed descriptions and reports of the languages that have a mixture of Chinese lexicon, Tibetan template for grammar, yet are agglutinative, ergative, and SOV languages. In addition, Atshogs (2004) also regards Daohua and several other mixed languages as independent languages, arguing that they all function as full-fledged languages by "branching" out from a proto-language. Atshogs also challenges some theories in historical comparative linguistics and genetic linguistics that have long underestimated or neglected the process of mixture as an important means to engender new languages.

Ironically, reference grammars under the typological framework for Chinese dialects are much less developed than those written for non-Chinese languages spoken in China. This is accounted for by the fact that hundreds of books on Chinese dialects have been written in a philological or structural manner, accumulated across generations. These grammars typically devote 40–50% of their contents to phonetics/phonology and lexicon, leaving only 10% of the contents to deal with grammar. And of the 10% of the content on grammar, it is typically of several discrete examples of what the author deems as to be peculiar and important, such as discussing about 50–100 pairs of sentences in Standard Mandarin and the dialect in question. On the other hand, most grammar works arrange their structure by loyally following Zhu Dexi's famous grammar system, which was written within the framework of structuralism. This situation led to a long period that lacked detailed grammars or accessible descriptions of facts compatible with the framework used for writing reference grammars, in order to facilitate crosslinguistic comparison. Compared with the stock of philological and structuralist descriptions, which are still produced today, typologically oriented and reference-grammar-framed descriptive publications are making steady progress in documenting the vast underdocumented languages in China.

Theoretical Development and Linguistic Inventory Typology

The largest proportion of all research under the typological rubric are fine-grained case studies of Sinitic languages including dialects from typological perspectives. These studies typically conclude or append a small-scaled crosslinguistic comparison under the parameters and values developed in the preceding body of detailed illustrations and analyses of a particular dialect. Some research offers a comparison of the target vernaculars with other geologically or genetically related languoids. Categorized as prototypical typological studies, these approaches are quite different from *typological* comparisons which take great effort in the pursuit of genetically and geographically well-balanced sampling and formal features. Systemic researches of the overall grammatical systems of natural languoids, discussion on principles and methodologies, and especially theoretical innovations are still far from proportionate, especially when more and more "anomalies" concerning Sinitic languages emerge through typological spectrums. Liu Danqing 刘丹青 (2011) and a series of other studies make efforts to develop a theoretical framework of typology known as linguistic inventory typology (LIT). Liu proposes and advocates LIT as an approach to, or a branch of, linguistic typology. It attempts to show how linguistic inventory, the total stock of encoding devices that a certain language has to offer, is employed to shape the typological scenario of the language. These devices may be used at morphological, lexical, or syntactic level. Liu proposes that some semantically or pragmatically prominent categories may override other categories, so that their corresponding formal devices may expand and "take over" the functions otherwise performed by other categories. For example, the topic prominence often manifests itself by "usurping" the formation of comparative structures, conditionals, most of transitive constructions, and even by exerting constraints over accessibility concerning relativization. In contrast to categorical prominence, lack of special devices may also have consequences; for instance, Chinese is believed to lack a tense category and therefore often applies aspectual, modal, or parasitic-in-lexicon devices plus contextual factors to encode temporal relations. These cross-category mismatches make the crosslinguistic form-meaning correspondence more complicated than it may appear. Besides reckoning differences and discrepancies in inventory, Liu also attaches varying degrees of significance to the categories or, in terms of semantic map approaches, semantic primes, so that the formal features and the overall typological scenario can be illustrated in a more dynamic and fine-grained manner. This dynamic and varying-weight approach adopted by LIT is different from the previous approaches, where values, parameters, and categories are assigned with default equal importance, as is symbolized by the evenly distributed semantic primes. Liu also points out that the LIT approach in typology mainly begins with semantic and pragmatic categories and focus on the examination of how these categories, prominent or parasitic, are denoted by formal devices. For example, syllable rather than phoneme prominence has had a profound influence on the shaping of Chinese phonological structures, such as the lack of cross-syllabic liaisons, almost one-to-one syllable-morpheme-word correspondence (except only for a tiny fraction of more-to-one cases), and even the necessity and emergence of tonal system. Other categories such as tense is no more than a potential value which can only be parasitically "fused" in many other categories or even lexicalized words. For example, the expressions *lianmang* and *ganjin* share the conceptional meaning of "immediately" but the former can only be used in modifying the past and therefore in "realis" situations and the latter solely in imperative or hortative expressions, thus with irealis and temporal denotation of future time. In the meantime, Liu also argues that more attention should be paid to universals behind the inventory diversity and the relationships between category expansion and the semantic map approach. The LIT approach to languages and to linguistic typology has drawn attentions in academic circles and inspired more and more studies in this regard, such as those on Chinese postverbal complements, serial verb constructions, and mixed word order of OV and VO, to name just a few (Wan 2014; Yu 2017; Wu 2018; Fang 2018; Guo 2018; Xu 2021).

Conclusion

As with other subdisciplines of linguistic research, typological studies in China have undergone a development from translations, summaries, and reviews of the state of research to consciously adopting typological frameworks, methodologies, and findings in individual case studies of certain languages. In terms of the target language, however, a large proportion of single-language descriptions or analyses with cross-linguistic comparison lack a comprehensive typological perspective. Feature-oriented or category-oriented studies based on well-balanced sampling are making steady progress but are still lacking in number. We call for more typologically oriented research, including reference grammars, which can provide an in-depth treatment of the differences, commonalities, and correlations among the features themselves– features that have long been regarded as the *core* of typology.

References

Atshogs, Yeshes Vodgsal (意西微萨·阿错). 2004. *Daohua Yanjiu* 倒话研究 *[Studies of Daohua]*. Beijing: Minzu Press.

Chen, Yujie (陈玉洁), de Sousa Hilario, Wang Jian (王健), Ngai Sing Sing (倪星星), Li Xuping (李旭平), Chen Weirong (陈伟蓉). 2014. Laibixi biaozhu xitong jiqi zai hanyu yufa yanjiuzhong de yingyong 莱比锡标注系统及其在汉语语法研究中的应用 [The Leipzig glossing rules and its application in the research of Chinese grammar]. *Fangyan* 方言 *[Dialect]* 1: 1–13.

Chu, Chauncy (屈承熹). 1984. Hanyu de cixu jiqi bianqian 汉语的词序及其变迁 [Chinese word order and word order change]. *Yuyan yanjiu* 语言研究 *[Studies in Language and Linguistics]* 1: 131–155.

Comrie, Bernard. 1981. *Language universals and linguistic typology*. Chicago: University of Chicago Press. Trans. Shen, Jiaxuan (沈家煊).*Yuyan Gongxing he Yuyan Leixing* 语言共性和语言类型, 1989. Beijing: Huaxia Publishing House.

———. 1986. Conditionals: A typology. In *On conditionals*, ed. Elizabeth Traugott et al. Cambridge: Cambridge University Press. Trans. Shen, Jiaxuan (沈家煊), 1988.

Comrie, Bernard, and Norval Smith. 1977. Lingua descriptive studies: questionnaire. *Lingua* 42 (1): 11–71.

Comrie, Bernard, Martin Haspelmath, and Balthasar Bickel. 2008. *The Leipzig glossing rules: Conventions for interlinear morpheme-by-morpheme glosses*. Department of Linguistics of the Max Planck Institute for Evolutionary Anthropology & the Department of Linguistics of the University of Leipzig. Retrieved January, 28, 2010. http://www.eva.mpg.de/lingua/resources/glossing-rules.php.

Croft, William. 1991. *Typology and universals*. Cambridge: Cambridge University Press. Trans. Gong, Qunhu et. al (龚群虎等). 2010. *Yuyan Leixingxue yu Yufa Gongxing* 语言类型学与语法共性, Shanghai: Fudan University Press.

Dai, Qingxia (戴庆厦), and Wang Feng (汪峰). 2014. *Yuyan leixingxue de jiben fangfa yu lilun kuangjia* 语言类型学的基本方法与理论框架 *[The primary methodology and theoretical frameworks of linguistic typology]*. Beijing: Commercial Press.

Fang, Di (方迪). 2018. Xiandai hanyu dongqushi-de xianhexing ji kuozhang xiaoying 现代汉语动趋式的显赫性及扩张效应 [Mightiness and expansion effect of verb-tendency construction in contemporary Chinese]. *Shijie Hanyu Jiaoxue* 世界汉语教学 *[Chinese Teaching in the World]* 32 (02): 229–240.

Greenberg, Joseph. 1963. Some universals of grammar with particular reference to the order of meaningful elements. In *Universals of language*, ed. Joseph Greenberg, 73–113. Cambridge, MA: MIT Press. Translated as "Moxie zhuyao gen yuxu youguan de yufa pubian xianxiang" 某

些主要跟语序有关的语法普遍现象, by Lu, Bingfu (陆丙甫) & Lu Zhiji (陆致极), *Guowai yuyanxue* 国外语言学 [*Overseas Linguistics*] 1984.2: 45–60.

Guo, Zhong (郭中). 2018. Lun Hanyu xiaocheng fanchou-de xianhexing jiqi leixingxue yiyi 论汉语小称范畴的显赫性及其类型学意义 [The mighty category of diminutive in Chinese from a typological perspective]. *Zhongguo Yuwen* 中国语文 [*Studies of the Chinese Language*] 2018 (02): 163–176.

Guo, Zhong (郭中), and Chu Zexiang (储泽祥). 2012. *Yuyan leixingxue he jufa miaoxie* (dier ban). 〈语言类型学和句法描写〉(第二版) [A review of *Language typology and syntactic description* (Second Edition)] *Waiyu jiaoxue yu yanjiu* 外语教学与研究 [*Foreign Language Teaching and Research*] 44(6): 954–958.

Hashimoto, Mantaro. 1975. Language diffusion on the Asia continent. Paper presented at the *8th international conference on Sino-Tibetan languages and linguistics*, Berkeley, California.

Hashimoto, Mantaro (桥本万太郎). 1985. *Yuyan dili leixingxue* 语言地理类型学 [A geographical typology of languages]. Trans. Yu, Zhihong (余志鸿), Beijing: Peking University Press.

Hengeveld, K., J. Rijkhoff, and A. Siewierska. 2004. Parts-of-speech systems and word order. *Journal of Linguistics*. 40 (3).

Hu, Jianhua (胡建华). 2013. Jufa duichen yu mingdong junheng – cong yuyi midu he chuanranxing kan shici 句法对称与名动均衡——从语义密度和传染性看实词 [Syntactic symmetry and noun-verb equilibrium: Lexical categories viewed from the latitudes of semantic density and virus infectivity]. *Dangdai Yuyanxue* 当代语言学 [*Contemporary Linguistics*] 1: 1–19.

Jiang, Yi (江轶). 2006. Guoji dangdai yuyan leixingxue fazhan dongtai 国际当代语言类型学发展动态 [Recent trends in the study of typological linguistics]. *Xiandai Waiyu* 现代外语 [*Modern Foreign Language (Quarterly)*] 29 (3): 302–308.

Jin, Lixin (金立鑫). 2011. *Shenme shi Yuyan Leixingxue* 什么是语言类型学 [*What is linguistic typology*]. Shanghai: Shanghai Foreign Language Education Press.

———. 2018. *Yuyan Leixingxue Tansuo* 语言类型学探索 [*Toward a typology of word order, tense, aspect and beyond*]. Beijing: Commercial Press.

Jin, Lixin (金立鑫), and Yu Xiujin (于秀金). 2012. Cong yu OV-VO xiangguan he buxiangguan canxiang kaocha Putonghua de yuxu leixing 从与 OV-VO 相关和不相关参项考察普通话的语序类型 [Word order type of Mandarin Chinese revisited on pairs of grammatical elements correlated and uncorrelated with OV-VO]. *Waiguoyu (Shanghai Waiguoyu Daxue Xuebao)* 外国语(上海外国语大学学报) [*Journal of Foreign Languages*] 35 (2): 22–29.

Kortmann, Bernd. 1999. Typology and dialectology. In *Proceeding of the 16th International Congress of Linguists, Paris*, ed. B. Caron. Amsterdam: Elsevier Science. Trans. Liu, Haiyan & Danqing Liu (刘海燕、刘丹青), Leixingxue yu Fangyanxue 类型学与方言学, *Fangyan* 方言 [*Dialect*], 2004.2: 148–157.

LaPolla, Randy & Chenglong Huang (黄成龙). 2013. *A grammar of Qiang: With annotated texts and glossary*. Berlin: Walter de Gruyter.

Liao, Qiuzhong (廖秋忠). 1984. *Yuyan de Gongxing yu Leixing* shuping 语言的共性与类型 述评 [A review of *language universals and linguistic typology: Syntax and morphology*]. *Guowai yuyanxue* 国外语言学 [*Overseas Linguistics*] 4: 1–15.

Lin, Zhong (林忠). 2013. *Niujin Yuyan Leixingxue Shouce* shuping 〈牛津语言类型学手册〉述评 [A review of *the Oxford handbook of linguistic typology*]. *Waiyu Jiaoxue yu Yanjiu* 外语教学与研究(外国语文双月刊) [*Foreign Language Teaching and Research (bimonthly)*] 5: 792–797.

Liu, Danqing (刘丹青). 2002. Hanzang yuyan de ruogan yuxu leixingxue keti 汉藏语言的若干语序类型学课题 [Aspects concerning word order of Sino-Tibetan languages]. *Minzu Yuwen* 民族语文 [*Minority Languages of China*] 2002: 1–11.

———. 2003. Yuyan leixingxue yu hanyu yanjiu 语言类型学与汉语研究 [Linguistic typology and Chinese research]. *Shijie Hanyu Jiaoxue* 世界汉语教学 [*Chinese Teaching in the World*] 4: 7–14.

———. 2008. *Yufa Diaocha Yanjiu Shouce* 语法调查研究手册 [*A handbook for grammatical investigation and research*]. Shanghai: Shanghai Educational Publishing House.

———. 2010. Hanyu shi yizhong dongcixing yuyan – shishuo dongcixing yuyan he mingcixing yuyan de leixing chayi 汉语是一种动词型语言——试说动词型语言和名词型语言的类型差异 [Chinese as a verby language: On typological differences between verby languages and nouny languages]. *Shijie Hanyu Jiaoxue* 世界汉语教学 *[Chinese Teaching in the World]* 1: 5–19.

———. 2011. Yuyan kucang leixingxue gouxiang 语言库藏类型学构想 [Linguistic inventory typology: A proposal for a new approach of linguistic typology]. *Dangdai Yuyanxue* 当代语言学 *[Comtemporary Linguistics]* 4: 289–303.

———. 2017. *Yuyan Leixingxue* 语言类型学 *[Linguistic typology]*. Shanghai: Zhongxi Book Company.

Lu, Bingfu (陆丙甫). 2005. Yuxu youshi de renzhi jieshi 语序优势的认知解释 [Word order dominance and cognitive explanations]. *Dangdai Yuyanxue* 当代语言学 *[Contemporary Linguistics]* 1: 1–15; 2005.2: 132–138.

Lu, Bingfu (陆丙甫), and Jin Lixin (金立鑫). 2015. *Yuyan Leixingxue Jiaocheng* 语言类型学教程 *[The course of linguistic typology]*. Beijing: Peking University Press.

Ma, Xueliang (马学良). 1991. *Hanzangyu Gailun* 汉藏语概论 *[An introduction to Sino-Tibetans]*. Beijing: Peking University Press.

Matthews, Stephen, and Yip, Virginia. 1994/2011. *Cantonese: A comprehensive grammar*. Abingdon: Routledge.

Pan, Chia-jung (潘家荣). 2012. *A grammar of Lha'alua: An Austronesian language of Taiwan*. Doctoral dissertation, James Cook University.

Rijkhoff, Jan. 2000. When can a language have adjectives? An implicational universal. In *Approaches to the typology of word classes*, ed. Vogel Petra Maria and Bernard Comrie, 217–249. Berlin: Mouton de Gruyter.

Shen, Jiaxuan (沈家煊). 1991. *Leixing he Gongxing* pingjie. 〈类型和共性〉评介 [A review of *Typology and universals*]. *Guowai yuyanxue* 国外语言学 *[Overseas Linguistics]* 3: 25–28.

———. 1997. Leixingxuezhong de biaoji moshi 类型学中的标记模式 [Patterns of marking in typology]. *Waiyu Jiaoxue yu Yanjiu* 外语教学与研究 *[Foreign Language Teaching and Research]* 1: 4–13.

———. 2009. Wokan hanyu de cilei 我看汉语的词类 [My view of word classes in Chinese]. *Yuyan Kexue* 语言科学 *[Linguistic Sciences]* 1: 5–16.

Sheng, Yimin (盛益民). 2014. *Wuyu Shaoxing Keqiaohua Cankao Yufa* 吴语绍兴柯桥话参考语法 *[A reference grammar of Shaoxing Keqiao Wu dialect]*. Doctoral Dissertation, Nankai University.

Shopen, Timothy. 2007a. *Language typology and syntactic description, Vol. 1: Clause structure*. 2nd ed. Cambridge: Cambridge University Press.

———. 2007b. *Language typology and syntactic description, Vol. 2: Complex constructions*. 2nd ed. Cambridge University Press.

———. 2007c. *Language typology and syntactic description, Vol. 3: Grammatical categories and the Lexicon*. 2nd ed. Cambridge University Press.

Song, Jae Jun (宋在晶), eds. 2011. *The Oxford handbook of linguistic typology*. Oxford: Oxford University Press.

Sun, Hong kai, Huang, Zengyi et al. (孙宏开、胡增益、黄行), eds. 2007. *Zhongguo de Yuyan* 中国的语言 *[Languages of China]*. Beijing: Commercial Press.

Tai, James H.-Y. (戴浩一). 1976. On the change from SVO to SOV in Chinese. In *Papers from the Parasessions on diachronic syntax, Chicago Linguistic Society*, ed. Sanford Steever, Carol Walker, and Salikoko Mufwene, 291–304.

Tang, Zhengda (唐正大). 2006. Yu guanxicongju youguande santiao yuxu leixing yuanze 与关系从句有关的三条语序类型原则 [Three word order principles regarding relativization: A typological perspective]. *Zhongguo Yuwen* 中国语文 *[Studies of the Chinese Language]* 5: 409–422.

Wan, Quan (完权). 2014. Fuci wenju-de yuyong gongneng 副词问句的语用功能 [Pragmatic functions of interrogative adverb]. *Hanyu Xuexi* 汉语学习 *[Chinese Language Learning]* 02: 11–19.

Wan, Quan (完权), Shen Jiaxuan (沈家煊). 2010. Kuayuyan bijiao de Amusitedan moxing 跨语言比较的"阿姆斯特丹模型" [A cross-linguistic comparison of parts of speech from Amsterdam model]. *Minzu Yuwen* 民族语文 *[Minority Languages of China]* 3: 4–17.

Wang, Honglei (王洪磊). 2019. Xingshi yuyanxue yu yuyan leixingxue de zhenghe: xingshi leixingxue shuping 形式语言学与语言类型学的整合：形式类型学述评 [Integration of formal linguistics and linguistic typology: Comments on formal typology]. *Jiefangjun Waiguoyu Xueyuan Xuebao* 解放军外国语学院学报 *[Journal of PLA University of Foreign Languages]*.

Wu, Fuxiang (吴福祥). 2012. Yuxu xuanze yu yuxu chuangxin – hanyu yuxu yanbiande guancha he duanxiang 语序选择与语序创新——汉语语序演变的观察和断想 [Choice and creation: Observations and thoughts of word order changes of Chinese]. *Zhongguo Yuwen* 中国语文 *[Studies of the Chinese Language]* 4: 61–69.

Wu, Jianming (吴建明). 2018. Yuyan leixingxue-de qianyan tansuo: Xunqiu 'kucang'-de yanguang 语言类型学的前沿探索——寻求"库藏"的眼光 [Frontiers in linguistic typology: Gaining insights from the 'inventory']. *Yuyan Jiaoxue yu Yanjiu* 语言教学与研究 *[Language Teaching and Research]* 02: 70–80.

Xu, Jie (徐杰). 2005. *Hanyu Yanjiu de Leixingxue Shijiao* 汉语研究的类型学视角 *[Studies in Chinese linguistics: A typological perspective]*. Beijing: Beijing Language and Culture University Press.

Xu, Zhongyun (徐中云). 2021. Kucang leixingxue shiye xia-de chaoxianyu youding fanchou biaoda yanjiu 库藏类型学视野下的朝鲜语有定范畴表达研究 [A study on definiteness expressions in Korean from the perspective of linguistic inventory typology]. *Minzu Yuwen* 民族语文 *[Minority Languages of China]* 2: 36–45.

Yu, Xiujin (于秀金). 2017. Kua-yuyan shi-ti-qingtai-de fanchouhua, xianhexing ji kuozhangxing: Kucang leixingxue shijiao 跨语言时-体-情态的范畴化、显赫性及扩张性——库藏类型学视角 [Categorization, mightiness and expansion of tense-aspect-modality across languages: A perspective from linguistic inventory typology]. *Zhongguo Yuwen* 中国语文 *[Studies of the Chinese Language]* 6: 670–692.

Zhang, Min (张敏). 2010. Yuyi ditu moxing: yuanli, caozuo ji zai hanyu duogongneng yufa xingshi yanjiu zhong de yunyong "语义地图模型":原理、操作及在汉语多功能语法形式研究中的运用 [The semantic map model: theory, methodology, and application in multifunctional forms in Chinese]. *Yuyanxue Luncong* 语言学论丛 *[Essays on Linguistics]* 42: 3–63. Beijing: The Commercial Press.

———. 2015. *The 23rd Annual Conference of the International Association of Chinese Linguistics (IACL), Hanyang University, Seoul, Korea, 26–28 August 2015*.

———. 2016. *The 4th Symposium of the Korean Association of Linguistic Typology, National Seoul University, Seoul, Korea, 11 January 2016*.

———. 2017. *Hanyu weishenme (haishi) meiyou dulide xingrongcilei – jianlun dongya dongnanya liangci yuyan junwu xingrongci* 汉语为什么(还是)没有独立的形容词类——兼论东亚东南亚量词语言均无形容词 *[Why does Chinese (still) have no independent adjectives: With peripheral views of southeast languages with classifiers and without adjectives]*. A speech at Institute of Linguistics, Chinese Academy of Social Sciences.

Development of the Semantic Map Model in China

Ding Zhang

Contents

Introduction	134
Research Domains	136
Research Methods	139
Diachronic Semantic Maps	143
"Second-Generation" Semantic Maps	147
Current Debates and Future Research Directions	147
Conclusion	149
References	150

Abstract

The chapter provides an overview of research employing the Semantic Map Model (SMM henceforth) in China, an emerging research area among Chinese scholars. It retraces the introduction of SMM into China, and demonstrates the various domains to which Chinese researchers have applied the semantic map. Moreover, it also analyzes the research methods employed by Chinese scholars doing SMM-based studies, with an account of how the Chinese academia responds to such state-of-art issues as diachronic semantic maps and "second-generation" semantic maps. This is followed by a discussion of current debates and future directions of SMM-based research in China.

Keywords

Semantic maps · Chinese languages · Semantic typology · Cross-linguistic comparison · Domains and methods

---

Translator: Wendi Xue The Australian National University

D. Zhang (✉)
Institute of Linguistics, Chinese Academy of Social Sciences, Beijing, China

Introduction

In natural languages, a linguistic form usually contains multiple meanings or functions, that is, polysemy or multifunctionality, which is an essential manifestation of linguistic economy. Remarkable achievements have been made in this field using traditional linguistic approaches. Links between the meanings within one linguistic form have been built by viewing them as results actuated by cognitive factors such as metaphor and metonymy, or discourse-pragmatics factors. However, traditional approaches fail to represent large-scale language data and therefore have limitations in revealing the universality and particularity of the multifunctional mode in human languages. It was the rise of cognitive linguistics and linguistic typology that motivated linguists to tentatively combine these two new approaches, which naturally led to the occurrence of the Semantic Map Model.

As a tool for analysis based on certain geometric patterns, the Semantic Map Model was initially utilized to represent the multifunctionality of grammatical forms in an attempt to reveal the variation and universality of the multifunctional nature in human languages via cross-linguistic comparison, in particular the cross-linguistic patterns underlying different multifunctional modes (Wu 2009). At present, this methodology has been most commonly recognized in linguistic circles as the "Semantic Map Model," while other terms like "Mental Map" (Anderson 1986) or "Cognitive Map" (Kormann 1997) have also been adopted by some scholars. Among others, Haspelmath (1997) termed it as the "Implicational Map," emphasizing the implicational universals encapsulated in these semantic graphs.

The origin of the Semantic Map Model can be traced back to research conducted by Lloyd B. Anderson in the 1980s. In late 1990s, studies using this model have ultimately come to fruition with substantial progress. The earliest classic achievements were concentrated in grammatical domains, including such categories as the aspect (Anderson 1982; Janda 2007), reflexives, and the middle voice (Kemmer 1993), intransitive predication (Stassen 1997), indefinite pronouns (Haspelmath 1997), modality (van der Auwera and Plungian 1998; van der Auwera et al. 2007), semantic roles (Haspelmath 2003; Rice and Kabata 2007), and adversative and contrast marking (Malchukov 2004). Furthermore, Croft (2001, 2003) has extensively applied this model to researches on part of speech, syntactic relations, and other aspects of language. After this, scholars such as Croft began to introduce the concept of Multidimensional Scaling (MDS henceforth), with representative works including Clancy (2006), Cysouw (2007), and Croft and Poole (2008).

Haspelmath (2003) proposed to apply this Semantic Map Model for cross-linguistic comparison in lexical semantics. In this chapter, he gives an example of conceptual space based on various lexemes for "tree/wood" in four European languages (Danish, German, French, and Spanish) based on Hjelmslev (1963). This involved constructing a one-dimensional semantic map representing five different functions ("tree," "firewood," "wood [stuff]," "small forest," and "large forest") in these four languages. François's (2008) study constructed a remarkable complete universal semantic map for "BREATHE" on the basis of 13 languages across eight language families. Providing a practical method for cross-linguistic comparison of semantic universals and typological tendencies concerning content words within the framework of semantic maps, this study by François has generally been accepted as a landmark in lexical and semantic typology.

Recently, scholars all over the world have extended the application of the Semantic Map Model to new fields of research, displaying promising prospects in regard to its general applicability.

Renzhi Li (2004) is the earliest Chinese scholar to pay attention to the Semantic Map Model. However, as his work was not published in China, few scholars were aware of it. In 2008, Min Zhang gave a detailed introduction to both the theory and application of the Semantic Map Model in several public lectures in Mainland China, which garnered the attention of Chinese academia. Fuxiang Wu (2009) marks the first publication of this model in Mainland China. His study examines the semantic changes of the polyfunctional morpheme "GET, ACQUIRE" in Southeast Asian languages, and discusses the grammaticalization path of the Chinese function morpheme "得 (de)." After this, research using the Semantic Map Model has gradually developed in China with the most noteworthy of these being Min Zhang (2008, 2010), Fuxiang Wu (2009, 2011, 2014), Rui Guo (2010, 2015), Ding Zhang (2010, 2016), Xiaolei Fan (2014, 2017), and Wenfang Sun (2018). In addition, the Singaporean scholar Chiew Pheng Phua has also published his research in China in such works as Phua (2011, 2015).

Meanwhile, special columns dealing with the Semantic Map Model emerged in several Chinese journals along with several collections of papers including research on the subject. For instance, *Essays on Linguistics* (2010, Vol. 42) includes four papers related to the Semantic Map Model. One of the papers found in this collection, *The Semantic Map Model and Its Application to the Study of Multifunctional Grams in Chinese* by Min Zhang, endeavors to provide a synthesis of previous research results and has had a wide-ranging and significant influence in Chinese academic circles. Another collection of papers *A Semantic Map approach to Chinese multifunctional forms* (Li et al. 2015) included 17 articles, covering various topics related to the Semantic Map Model from theoretical discussions to case studies. In 2017, the influential Chinese journal *Contemporary Linguistics* (4) initiated a special column for studies using semantic maps containing four articles.

This chapter seeks to provide a comprehensive introduction from various perspectives to research in China that employs the Semantic Map Model along with its prospects for future development. This chapter is organized into six sections. Section "Research Domains" focuses on the research domains involved using this model; Section "Research Methods" summarizes underlying methodologies; Section "Diachronic Semantic Maps" introduces diachronic semantic maps, a sub-model that can be promisingly combined with grammaticalization studies; Section "'Second-Generation' Semantic Maps" provides some brief information on "second generation" semantic maps that utilize new statistical scaling techniques; Section "Current Debates and Future Research Directions" discusses some current debates on the Semantic Map Model as well as future directions for the model, and section "Conclusion" provides a conclusion.

Research Domains

Similar to the development of semantic maps in international linguistic circles, after its introduction into Chinese linguistics, the Semantic Map Model was first applied to Chinese grammatical studies where it led to substantial achievements in regard to Chinese functional morphemes before its recent application to polysemy in Chinese content words.

The highlighted grammatical domain in this regard concerns such semantic roles as "pretransitive," "instrumental," "comitative," "passive," and their internal connections with each other. Haspelmath (2003) investigated the multiple functions of preposition and case markers in different languages and constructed a general conceptual space for instrumental and related functions (Fig. 1).

The preposition *with* in English expresses both the instrumental and comitative functions, while the German preposition *aus* represents both the source and the cause; French *par* can be used to denote both the passive agent and the cause, whereas German *aus* embraces such functions as the source and the cause. Thereupon, Haspelmath mapped the different boundaries of these grams in the conceptual space, resulting in a semantic map as per Fig. 2.

As shown in Fig.2, all the functions expressed by the specific languages noted above have occupied adjacent connected regions on the conceptual space, that is, not violating the contiguity/adjacency requirement of semantic maps. Therefore, the

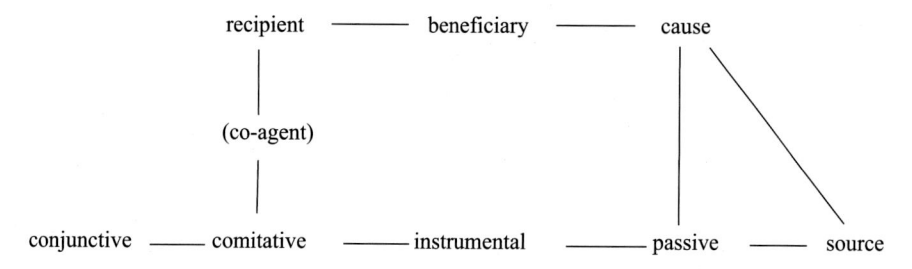

Fig. 1 Instrumental and related functions (Haspelmath 2003: 229)

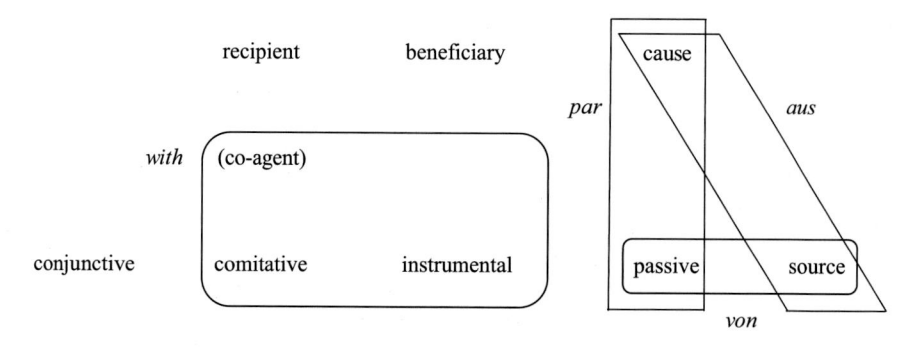

Fig. 2 The boundaries of some English, German, and French prepositions (Haspelmath 2003: 229)

assumed conceptual spaces have been verified in these languages. How then does this apply to the Chinese language?

By applying the instrumental conceptual space constructed in Haspelmath (2003) to modern Chinese dialectal data, Min Zhang (2008) has managed to verify, modify, and complement the Haspelmath's original semantic maps and presented his new version as shown in Fig. 3.

Zhang's revised map differs apparently from Haspelmath's. Firstly, the core of the conceptual space in Haspelmath (2003) is "instrumental," whereas Zhang (2008) places "pretransitive" at the center of the map, reflecting unique Chinese features. Moreover, Zhang has supplemented other new functions to the map, such as causative, manner, direction. The linking patterns between different functional nodes are also found to be different, such as the link added between "instrumental" and "source," which is not found in Haspelmath (2003).

Since Zhang (2008), semantic roles have remained the prevailing focus of Chinese research using the Semantic Map Model. Representative studies include Ding Zhang's (2010) conceptual space of "instrumental-comitative" markers, Wei Wang's (2015) semantic map of the spatial motion domain, Yawei Wang and Fuxiang Wu's (2017) map of concepts related to the combination domain, and Xiangnong Li and Yangyang Wei's (2019) study of the Chinese equative marker 和 (*he*).

At present, Chinese linguistic scholars have extended the Semantic Map Model to many other grammatical domains, including indefinite pronouns (Zhang 2010), modality (Zhang 2010; Fan 2014, 2017; Xia 2017), aspect (Chen 2010, 2016),

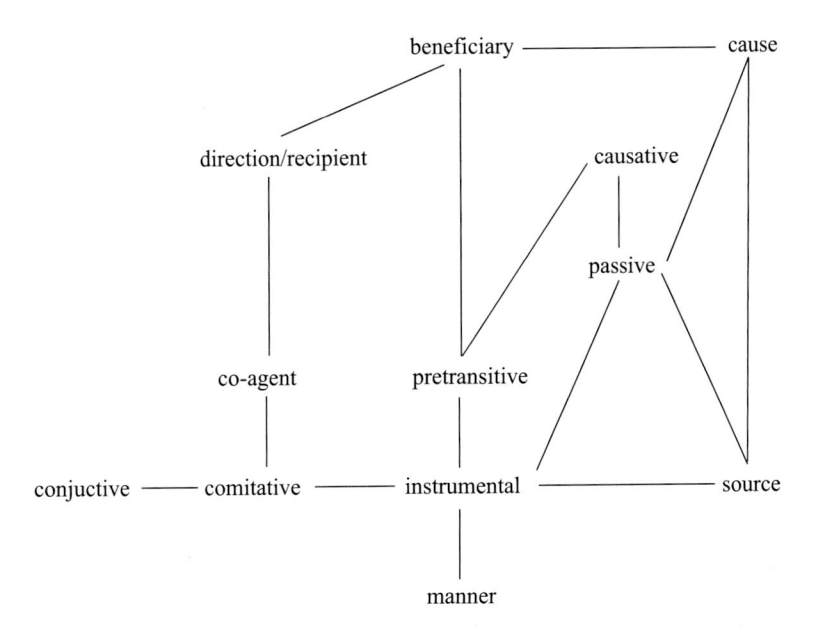

Fig. 3 The conceptual space for pretransitive and related functions (Zhang 2008)

numeral classifiers (Li 2011, 2016; Chen and Yue 2017; Zhang 2018), existential-possessive (Sun 2018), and reflexive possessive and intensifier (Ye 2019).

Inspired by Haspelmath (2003) and François (2008), some Chinese scholars have attempted to apply the Semantic Map Model to the study of lexical polysemy in content words. This has not only broadened the research scope of the model but has also advanced the development of lexical and semantic typology. Representative research in this regard includes Ding Zhang's (2016, 2017) semantic maps for "CHASE" (*zhuizhu*, 追逐) and "WEAR" (*chuandai*, 穿戴) respectively, Chiew-Pheng Phua and Kah Min Teoh's (2017) conceptual space for words related to the five flavors, and Chang Han and Jing Rong's (2019) semantic map for "SIT."

Ding Zhang (2016) constructs a 27-node verbal conceptual space for "CHASE" based on 50 verbal lexemes for CHASE from 30 languages/dialects. In this map, "RUN AFTER" (*zhuigan*, 追赶) and "DRIVE OFF" (*quzhu*, 驱逐) act as pivots from which most of the other 25 functions develop (except for "WHIP" [*bianda*, 鞭打] and "HUNT" [*dalie*, 打猎]). In line with the prominent element of each function as well as the semantic inheritance between the 25 functions and the 2 pivots, the 25 functions can be classified into four types, namely, the FORCE (*dongli*, 动力) type, the SEQUENCE (*shunxu*, 顺序) type, the AWAY (*likai*, 离开) type, and the GET (*qiuhuo*, 求获) type, as shown in Fig. 4.

Phua and Teoh (2017) build a semantic map for words related to the five flavors on the basis of related lexemes from six languages, as shown in Fig. 5.

It can be seen from these studies that the Semantic Map Model sheds new light on the polysemy of content words, though not as neat in regularity as the multi-functionality of grammatical forms. Many cross-linguistic lexical polysemy can be better described and explained using semantic maps.

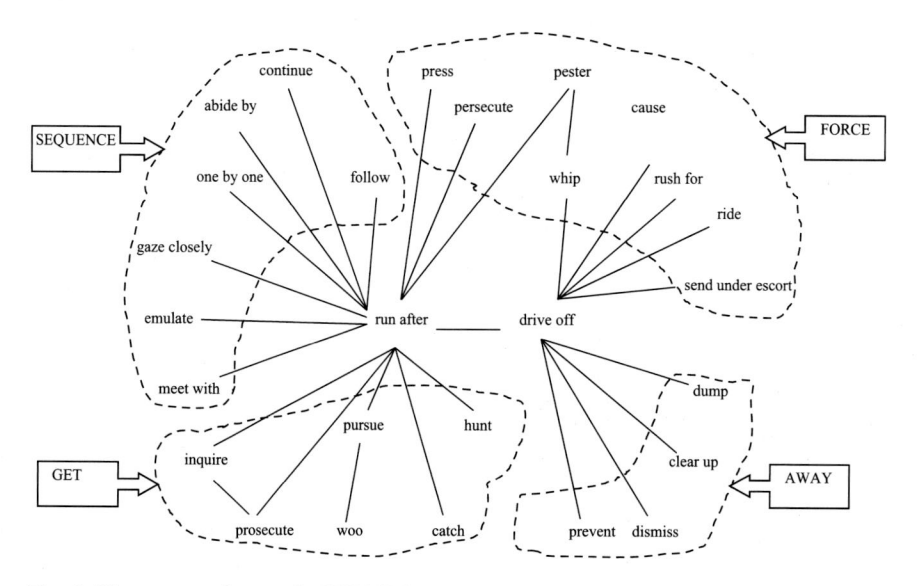

Fig. 4 The conceptual space for "CHASE"

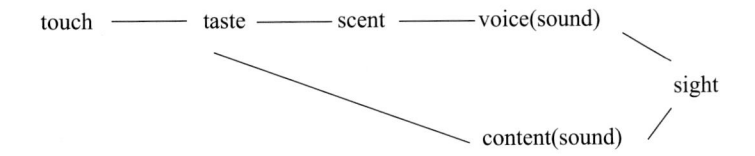

```
touch ——— taste ——— scent ——— voice(sound)
                                              \
                                               sight
                              content(sound)  /
```

Fig. 5 The conceptual space for words of the five flavors (Phua and Teoh 2017)

Research Methods

During the earlier stage of research using the model, Chinese linguists tended to verify, complement, and modify theoretical assumptions proposed by international general linguists who employ the Semantic Map Model in light of linguistic evidence from Mandarin or other Chinese dialects. For example and as noted previously, Min Zhang's (2008) research verifies and expands Haspelmath's (2003) conceptual space centered on the "instrumental." Subsequent to this, Ding Zhang (2010) and Xiaodan Hou (2017) studies revise Haspelmath's (1997) map of "Indefinite Pronouns" on the basis of Chinese evidence.

As the classic study in literature related to semantic maps, Haspelmath (1997) has attracted a great deal of attention to scholars working in typological studies (Croft 2007). The monograph has served as a role model for later studies along the same lines in terms of language sampling, selection of functions, construction of semantic maps, data analysis, and interpretation. Haspelmath examined 40 languages in depth along with 100 other languages in less depth before building a general conceptual space that includes nine functions and utilizes the semantic map to represent internal restrictions. His map for indefinite pronouns is shown in Fig. 6.

Indefinite pronouns in human languages tend to occur in a series. Haspelmath (1997: 307) summarizes the forms of indefinite pronouns in Modern Chinese. He suggests that the Modern Chinese bare interrogatives are used extensively as indefinite forms, and indefinite markers 都 (*dou*) and 也 (*ye*) have developed into series respectively; moreover, the determiner 任何 (*renhe*) constitutes a series by itself, and generic nouns such as 人 (*ren*, "people") are conventionally used as indefinite pronouns. Haspelmath (1997) managed to locate the functional distribution of different indefinite series in Chinese in light of relevant literature documentation and surveys of native Chinese speakers, and eventually produced the following semantic map (Fig. 7).

Haspelmath's research marked a breakthrough in Chinese linguistics, and yet while most of his findings conform to the linguistic facts of Modern Chinese, some conclusions are still open to discussion. Ding Zhang (2010) divides the indefinite morphemes in Mandarin Chinese into four major series, namely, unstressed interrogatives, stressed interrogatives, 任何 (*renhe*), and 某 (*mou*), the semantic map of which is shown in Fig. 8.

It is evident that Zhang (2010), though following the Haspelmathian framework and theoretical assumptions, has classified Mandarin indefinite pronouns into several

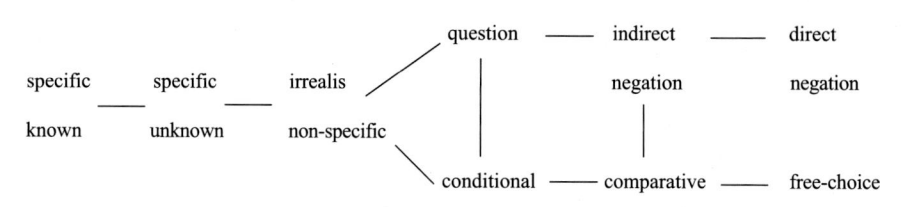

Fig. 6 The conceptual space for indefinite pronouns (Haspelmath 1997: 4, 64)

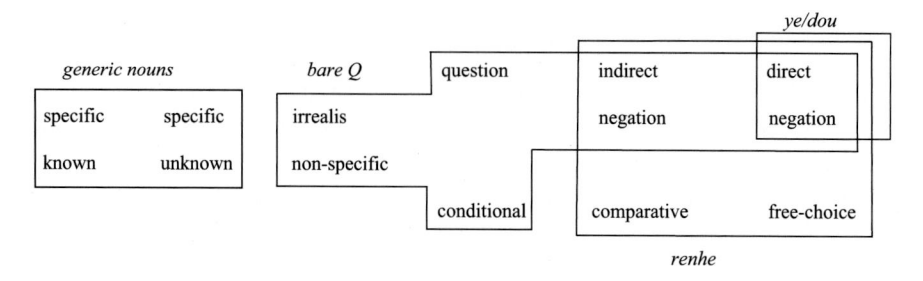

Fig. 7 A semantic map of Chinese indefinite pronouns (Haspelmath 1997: 307)

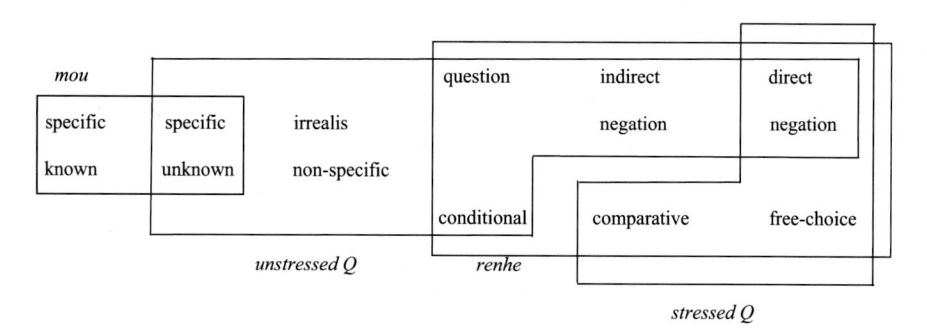

Fig. 8 A semantic map of the four indefinite series in Chinese

series quite distinct from Haspelmath's division and has modified the location of functional distribution.

Other noticeable studies featuring such verification and revision of previous semantic maps in light of Chinese data include Fangfang Tan (2016) and Xianqing Tang and Qiaoming Wang (2019). Tan (2016) has verified and supplemented some part of Malchukov's (2004) semantic map for adversative and contrastive markers on the basis of the typical examples of adversative and contrast relations found in both English and Chinese. Tang and Wang (2019) have examined the multiple functions of 是 (*shi*) in Chetian Renhua dialect of the Hmong (Miao) ethnic group in Guangxi Autonomous Region whereby the proposed semantic map in Sun (2018) was proved reasonable and supplemented as well.

On the other hand, with advances in research, Chinese linguists are no longer satisfied with the verification, supplementation, or modification of previously proposed semantic maps. Some researchers have attempted to construct new conceptual spaces and semantic maps based on linguistic evidence from Chinese dialects, minority languages in China, and other languages the world over.

Shanshan Weng and Xiaofan Li (2010) present an in-depth semantic analysis of the Mandarin complement morpheme 掉 (*diao*), in which 掉 (*diao*) is compared and contrasted with its semantic counterparts, that is, the *diao*-type lexemes, in other Chinese dialects. The study builds a conceptual space including five functions, which is later expanded to six functions with regard to the relevant senses found in the English counterpart *away*. Their semantic map is shown in Fig. 9.

To construct a general conceptual space or a semantic map, cross-linguistic comparison is an essential prerequisite and requires a large amount of language data. As a rule of thumb, a reasonable conceptual space can be built upon the basis of evidence from around 20 languages. With the increase of language samples, several new functions may be added to the map, or a few links between different nodes may be changed, but the overall layout of the map will remain roughly the same. However, with insufficient supporting language samples, the constructed conceptual space proves weak in its universality, and may not stand the verification based on other languages. In such cases, these maps often need to be altered drastically or even demolished. Among others, Fuxiang Wu (2009), Ding Zhang (2010), and Zhi-en Li (2011) stand out as representative of scholars that have created semantic maps using "big data" drawn from substantial language samples.

Wu (2009) constructed the conceptual space for "GET" based on the examination of "GET" morphemes in 31 minority languages in southern China and 17 Southeast Asian languages, as is shown in Fig. 10. This marks the first successful attempt in Chinese linguistic circles to directly build conceptual spaces using large amounts of language data.

Zhang (2010) scrutinized 35 instrumental-comitative prepositions from 30 Chinese dialects, and built a "instrumental-comitative"-pivoted semantic map for these prepositions in Chinese dialects, using a bottom-up approach. The map is shown in Fig. 11.

On the basis of Fig. 11, Zhang (2010) extended his survey scope to cover the other 100 multifunctional "instrumental-comitative" markers from 86 languages and eventually constructed, with reference to previous scholarship, a conceptual space for the instrumental-comitative markers in the world, as is shown by Fig. 12.

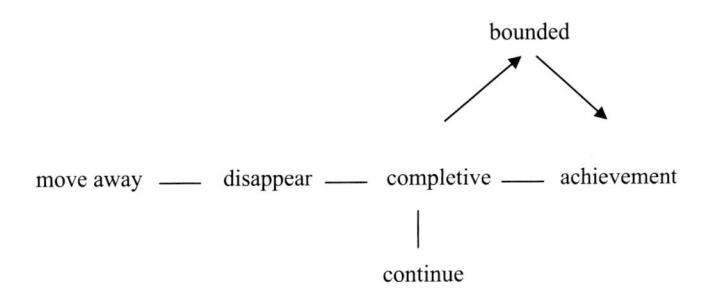

Fig. 9 The conceptual space for the *diao*-type words (Weng and Li 2010)

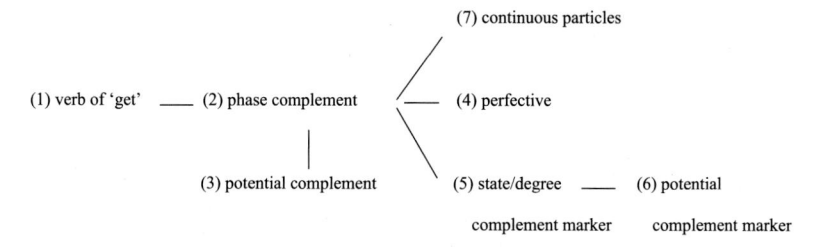

Fig. 10 The conceptual space for "GET" morpheme (Wu 2009)

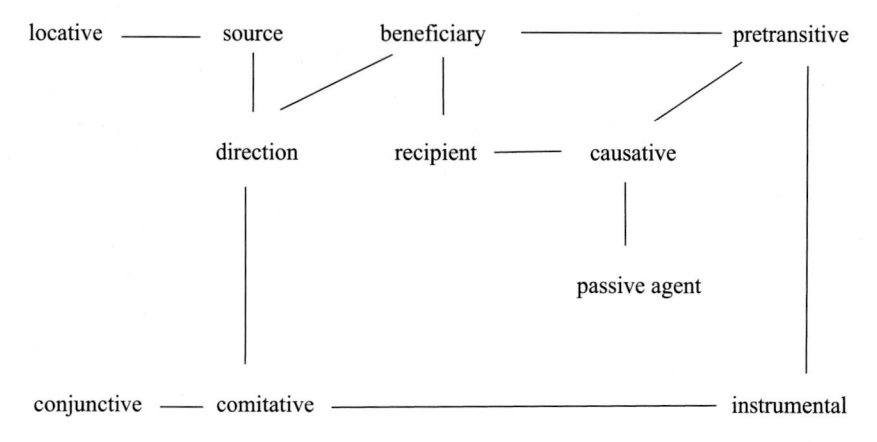

Fig. 11 The conceptual space for the instrumental-comitative prepositions in Chinese dialects

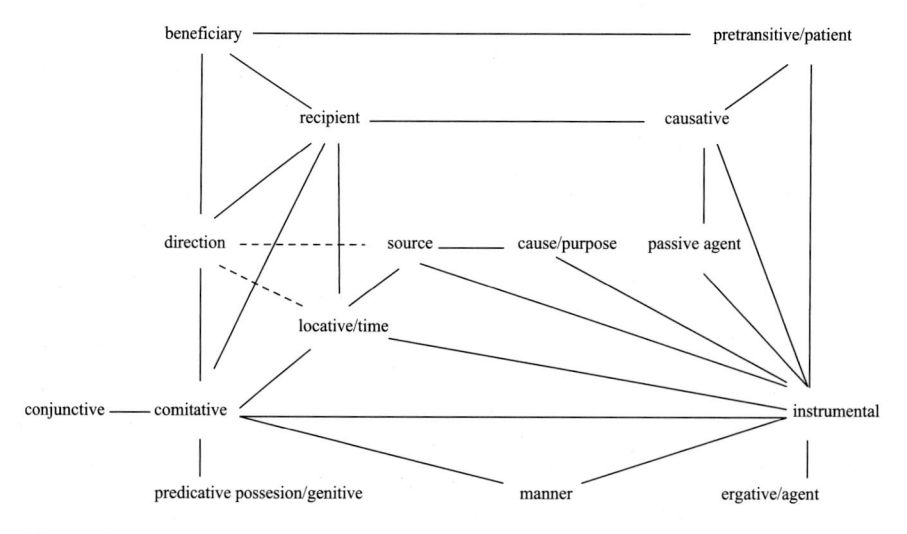

Fig. 12 The conceptual space for the instrumental-comitative markers in the world

Similar attempts include Zhi-en Li's (2011) semantic map of numeral classifiers, Wei Wang's (2015) map of the spatial motion domain, and Weifang Sun's (2018) map for "HAVE" (*you*, 有). Since language samples have become increasingly accessible, Chinese linguists now prefer constructing new semantic maps to verifying those proposed previously by other scholars. While it must be admitted that some defects such as sample selection, construction procedure, and explanatory depth still remain, new horizons have been opened up for semantic map-based research in Chinese linguistics.

Diachronic Semantic Maps

The Semantic Map Model can synchronically represent a series of implicational relations, and if we add the parameter of directionality to the map, changing the nondirectional interconnections between function nodes into directional arrows, it will become a dynamic conceptual space. This remains conventional practice in current research that employs diachronic semantic maps. For instance, Haspelmath (2003) transforms the stative, synchronic conceptual space of typical dative functions into a dynamic, directional one, based on the synchronic implicational relations and facts of related diachronic changes as is shown by Fig. 13.

According to Haspelmath (2003), some of the functions in the map are linked by nondirectional lines indicating correlation instead of arrows that imply evolutional directions for lack of related historical evidence. This indicates that the construction of semantic maps should be objectively based on the data.

There is also another approach to diachronic semantic maps that involves the direct construction of a dynamic conceptual space on the basis of existing diachronic linguistic findings instead of formulating the synchronic map before rendering it diachronic. A representative study of this approach can be found in the semantic map of modality proposed by van der Auwera and Plungian (1998). These researchers built a dynamic conceptual space for the modality category based on the series of grammaticalization paths proposed in Bybee, Perkins, and Pagliuca (1994) with the

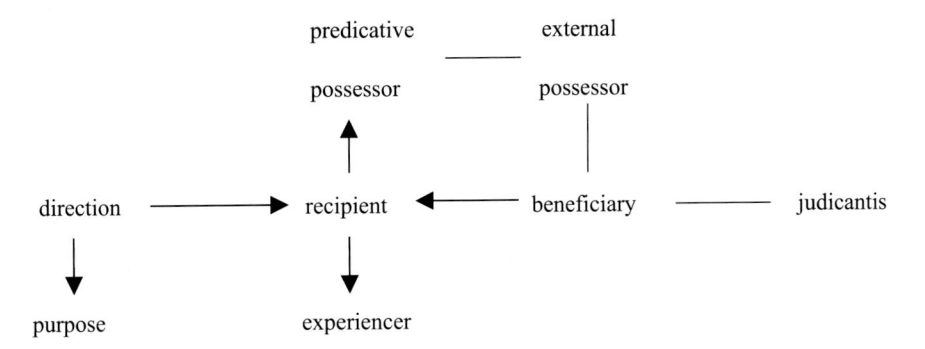

Fig. 13 A semantic map of typical dative functions, with directionality (Haspelmath 2003: 234)

integration of other relevant materials. The diachronic map (Fig. 14) they constructed can be used to predict the manner by which different languages encode the category of modality.

Led by scholars such as Fuxiang Wu and Bo Hong, the study of grammaticalization in Chinese has become a very significant field over the past two decades. Given the prominence of this field, the Semantic Map Model has been readily applied grammaticalization studies since its introduction into the Chinese academic sphere. At present, it is apparent that many findings that employ semantic maps in Chinese linguistic circles are relevant to grammaticalization. This means that many of these proposed maps can eventually be turned into dynamic diachronic ones. Most of these maps follow Haspelmath (2003) in constructing a synchronic semantic map before making it dynamic. However, there are several different methods of rendering synchronic maps into dynamic ones. Fuxiang Wu (2014) summarizes these methods and classifies them into three types: (a) cross-linguistic comparison based on functional implicational relations; (b) diachronic evidence of the path of grammaticalization and semantic change; (c) the degrees of grammaticalization and the principle of unidirectionality. Judging from existing studies, these three-way classifications remain effective as illustrated below:

(a) Cross-linguistic comparison based on functional-implicational relations: Wu (2009) provides such an example. As demonstrated by the semantic map for

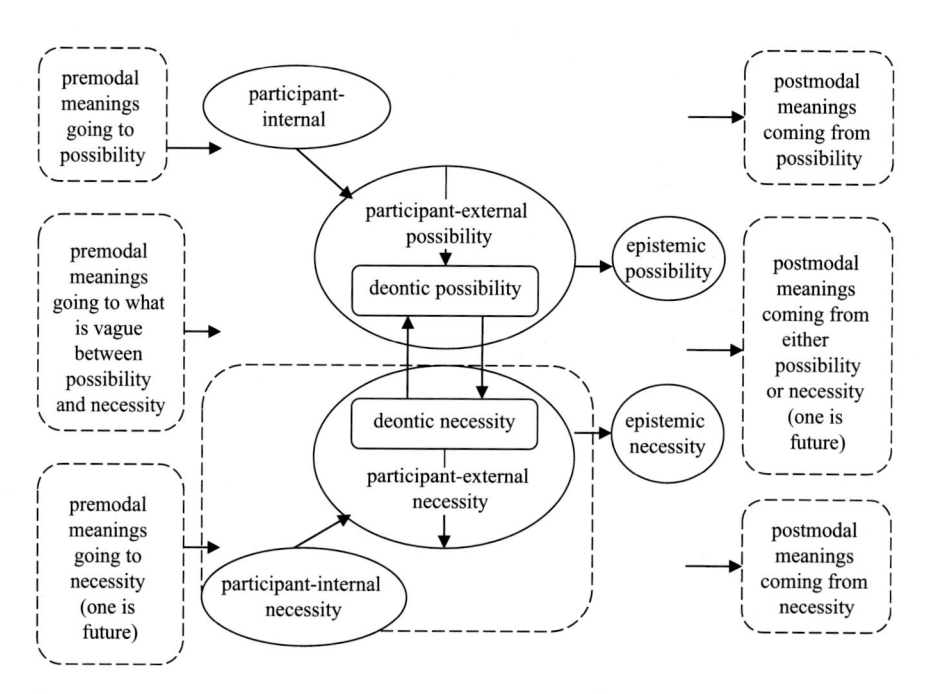

Fig. 14 A semantic map of modality (van der Auwera and Plungian 1998: 111; de Haan 2006: 47. With minor modifications)

"GET" (Fig. 10), it is obvious that there should be some sort of change between "GET" verbs, phase complement and potential complement. In light of cross-linguistic comparison of several Southeast Asian languages, Wu (2009) found that there are very clear implicational relations between the three aforementioned functions: (1) if a "GET" morpheme in a particular language can serve as a potential complement, it must at the same time function as a "GET" verb and a phase complement; (2) if a "GET" morpheme in a particular language can act as a phase complement, it must at the same time function as a "GET" verb, but it does not necessarily take on the function of potential complement. In other words, there exists an implicational relation "'GET' verb ⊃ phase complement ⊃ potential complement." From this implication, the direction of the change of the three functions can be inferred as "'GET' verb→phase complement→potential complement." The map in Fig. 10 can thereby be rendered dynamic as Fig. 15.

(b) Diachronic evidence of the path of grammaticalization and semantic change. Fuxiang Wu (2014) remarks that it is often difficult to judge the direction of change simply based on synchronic implicational relations. In such cases, diachronic evidence is usually needed to determine the direction, especially in regard to the paths of grammaticalization or the patterns of semantic change that repetitively occur cross-linguistically. For example, Zhang (2010) transforms his synchronic conceptual space for "instrumental-comitative" markers across the world into a diachronic one in light of previous historical studies, as is shown in Fig. 16.

(c) Degrees of grammaticalization and the principle of unidirectionality. According to Wu (2014), theoretical assumptions of grammaticalization can be employed to reconstruct the directions between different functions in the making of a dynamic conceptual space. The basic procedures are as follows: firstly, if two functions are directly connected, one can determine their degrees of grammaticalization using relevant parameters; secondly, one can reconstruct the evolution path in light of the principle of unidirectionality and the principle of gradualness. This can be illustrated by Fig. 3, the semantic map for pretransitive and related functions proposed in Zhang (2008). In this map, the path of change between "comitative" and "instrumental" should be "comitative" > "instrumental," because "comitative" is weaker than "instrumental" in terms of the degree of grammaticalization. Heine et al. (1991) argue that if the only difference between

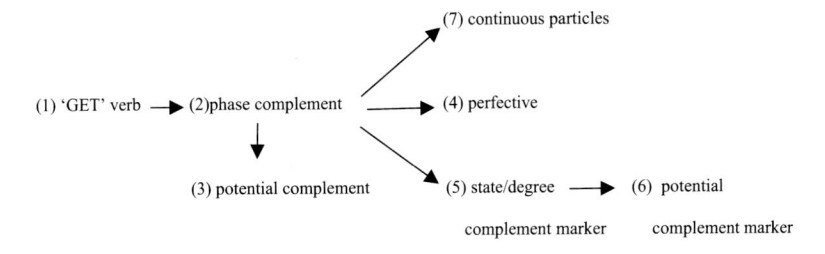

Fig. 15 The dynamic conceptual space for "GET" morpheme (Wu 2009)

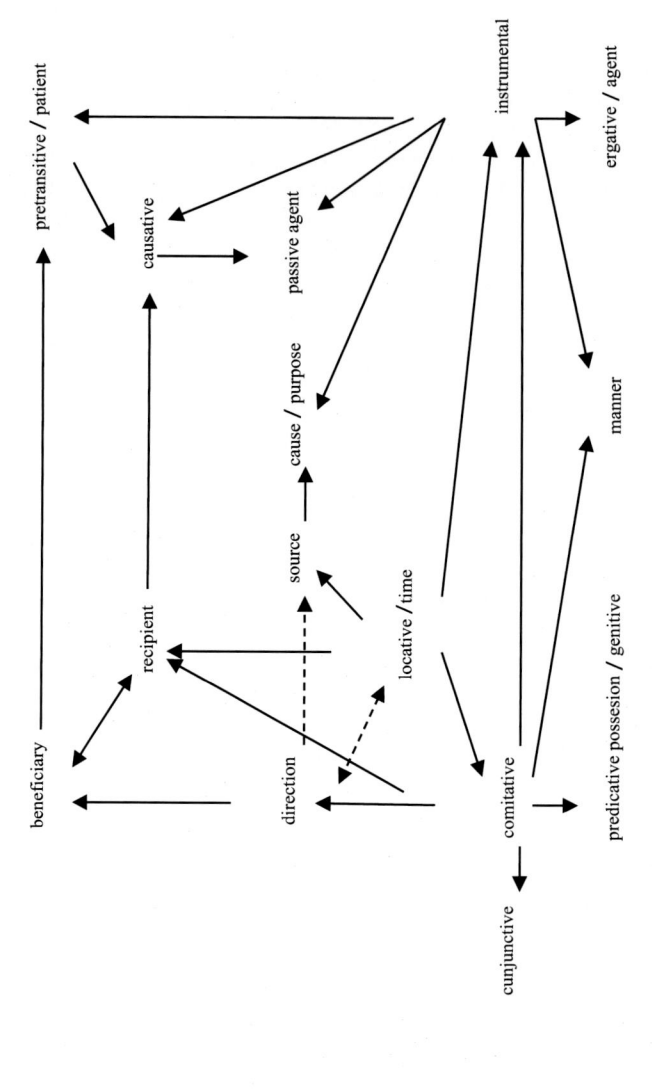

Fig. 16 The dynamic conceptual space for instrumental-comitative markers

two grammatical categories lies in the animacy of event argument, the one involving inanimate argument is always higher than the other with a human argument in terms of grammaticalization degrees. As is well known, "comitative" is a category that always involves human argument, while "instrumental" requires an inanimate argument. We can therefore propose and reconstruct relevant directions such as "instrumental>manner," "source>cause," "source>instrumental," "instrumental>pretransitive," "instrumental>passive," "passive>cause," "causative>passive," "instrumental>conjunction," etc.

The modes of turning a static conceptual space into a dynamic one involves only the grammatical domain, thus closely related to grammaticalization studies. Similarly, in studies of lexical polysemy for content words, we can also reconstruct the direction of semantic change in light of general rules governing semantic change, such as "concrete>abstract" and specialization, though such dynamic maps have little to do with grammaticalization.

"Second-Generation" Semantic Maps

In Chinese linguistic circles, classical or traditional semantic maps are often referred to as "first-generation" semantic maps. Despite their visual effectiveness, traditional semantic maps have their own limitations as noted by Clancy (2006), Wälchli (2007), Cysouw (2008), Croft and Poole (2008), and others. Such limitations include neglecting the difference in frequency, the lack of theoretical implications in the arrangement and distance between nodes, the limited number of functions presented on the conceptual space, and the difficulty in hand drawing maps. As a result, Clancy (2006), Cysouw (2007), and Croft and Poole (2008) introduced multidimensional scaling (MDS) to reinvigorate the Semantic Map Model, a transformation that was also readily accepted and followed by Chinese linguists, albeit with appropriate adjustments and improvements. These MDS-based semantic maps are referred to by some Chinese scholars as "second-generation" semantic maps.

In China at present, little research based solely on second-generation semantic maps has been carried out. Scholars tend to integrate the traditional semantic map with the second-generation model as is reflected in several significant studies. Inspired by Cysouw (2007), several studies have endeavored to complement MDS techniques with manual intervention, such as Rui Guo's (2010) conceptual space centered around the "resemble," "repeat," and "additive" functions, as well as Zhenyu Chen and Zhenning Chen's (2015) "weighted maps with least edges."

Current Debates and Future Research Directions

In the short period since its introduction, an increasing number of Chinese linguists have been attracted to the Semantic Map Model and several key controversies have been discussed at great length. These include such issues as how to identify connection patterns between different function nodes, whether or not these functions should be further divided, and how to ensure the reliability of data sources.

For instance, as for whether or not the represented functions in a conceptual space should be subdivided, de Haan (2004) proposed the term "primitive," arguing that any function on a semantic map should be primitive, that is, these functions must conform to the following "informal definition":

(a) A function X is "primitive" if it is not the case as shown below:

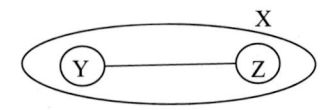

That means, if a function X can be subdivided into two or more functions expressed by two separate forms, it is not considered primitive. On the contrary, if the two postulated categories, Y and Z are never expressed by two different forms, they are not primitive and should be merged into a single category X.

For example, when constructing the conceptual space for the semantic roles of "instrumental-comitative," the maps cited in this chapter treat "direction" as only one single function/node. However, judging from the evidence of Chinese dialects, Chinese distinguishes the "animate direction" from the "inanimate direction." For instance, the preposition 跟 (*gen*) in the following Chinese sentence expresses the animate direction:

(b) 我   跟   他   汇报   了   一   件   事
    wo   gen   ta   huibao le   yi   jian   shi
    1SG PREP 3SG   report PART one CL   thing
    "I reported one thing to him."

In contrast, the preposition 向 (*xiang*) in the following example indicates the inanimate direction.

(c) 他   向   学校   走   去。
    ta   xiang xuexiao zou   qu.
    3SG   PREP   school   walk   go
    "He walked to the school."

Therefore, we can claim that the single node "direction" in the previously cited semantic maps is not a primitive function, since it can be subdivided into "animate direction" and "inanimate direction." A similar motivation saw Yimin Sheng (2015) subdivide "beneficiary" into "beneficiary" and "malefactive."

China has rich language resources: Sun et al. (2007) provide an overview of 129 languages in China, which fall under five language families. Such abundance provides a natural impetus for various linguistic discoveries. While significant achievements have been made in the field of Chinese grammatical studies that involve synchronic grammars for Standard Mandarin, other dialects, and historical grammatical research, the majority of these studies have been published in Chinese and therefore remain unknown to international typologists. This explains why a number of semantic maps proposed by linguists outside China have failed to reflect the linguistic facts of Chinese, and some researchers constructed semantic maps with the absence of related Chinese evidence. Many Chinese linguists who employ the Semantic Map Model have endeavored to incorporate the rich findings in Chinese grammatical studies when attempting to verify, supplement, and modify proposed semantic maps. The study of Chinese grammar has thereby been placed in a broader context of world languages, and the universals and particularities of Chinese have become more readily observed. In the meanwhile, some Chinese linguists have started to construct new semantic maps on the basis of linguistic data collected from Chinese, which has in turn enriched the studies that employ semantic maps and has expands research scope as a result.

Admittedly, in contrast to the profusion of studies dealing with Mandarin and other Chinese dialects, much remains to be done in regard to China's minority languages. Fortunately, shortly after the Semantic Map Model was introduced into China, scholars have attempted to include the minority languages of China in their studies. Examples of this include Jun Zhang's (2016) study of the polyfunctionality and grammaticalization path of mɑ33 in Lisu, a tonal Tibeto-Burman language spoken in Southwestern China, and Jing Wen's (2018) survey of the instrumental particle in Langsu (Maru or Lhao Vo), a Burmish language with only a few thousand speakers in China.

As we have previously noted, semantic maps can also be applied to the study of lexical semantic typology. Yet lexical typology has not yet been fully developed even in international linguistic circles, when compared with grammatical typology. Scholars have not yet reached a consensus over many related issues such as research methodology, rationale, and even a name for this new approach. Therefore, there are as yet very few studies that apply the Semantic Map Model to lexical and semantic typology.

Conclusion

In the decade or so since the introduction of the Semantic Map Model into Chinese linguistic circles, the general trend has gone from the verification and supplementation of semantic maps proposed by scholars outside China, to an attempt by Chinese scholars to construct their own original semantic maps. In other words, the past decade or so has witnessed a shift from imitation to innovation. Moreover, Chinese linguists have extended their research scope from grammatical categories to lexical semantics, and, insofar as second-generation semantic maps are concerned, scholars in China have caught up with their international counterparts. Indeed, the linguistic heritage of Chinese remains a rich source of materials for lexical typologists on account of its extensive and well-documented history. Moreover, traditional forms of exegesis applied to ancient texts and lexicological studies have left us with invaluable resources for contemporary researchers. These have contributed to significant findings of great relevance and the lexical and semantic complexities found in many Chinese dialects have started to garner the attention of many scholars. Given that the Semantic Map Model represents a new impetus for the study of lexical and semantic typology, it merits the increasing attention of linguists in China.

References

Anderson, Lloyd B. 1982. The 'perfect' as a universal and as a language-particular category. In *Tense-aspect: Between semantic and pragmatics*, ed. Paul J. Hopper, 227–264. Amsterdam: Benjamins.

———. 1986. Evidentials, paths of change, and mental maps: Typologically regular asymmetries. In *Evidentiality: The linguistic encoding of epistemology*, ed. Wallace Chafe and Johanna Nichols, 273–312. Norwood: Ablex.

Bybee, Joan, Revere Perkins, and William Pagliuca. 1994. *The evolution of grammar: Tense, aspect, and modality in the languages of the world*. Chicago: Chicago University Press.

Chen, Qianrui (陈前瑞). 2010. Nanfang fangyan "you" zi ju de duo gongneng xing fenxi南方方言"有"字句的多功能性分析 [On the Multifunctionality of "you(有)" in Southern Chinese Dialects]. *Yuyan Jiaoxue yu Yanjiu*语言教学与研究 [*Language Teaching and Language Studies*] 4: 47–55.

——— (陈前瑞). 2016. Wancheng ti yu jingli ti de leixingxue sikao完成体与经历体的类型学思考[Typological considerations on perfect and experiential aspects]. *Waiyu Jiaoxue yu Yanjiu (Waiguo Yuwen Shuangyuekan)*外语教学与研究(外国语文双月刊) [*Foreign Language Teaching and Research (bimonthly)*] 6: 803–814.

Chen, Zhuqin (陈祝琴), and Yue Xiuwen (岳秀文). 2017. Hanyu fangyan liangci "dou"delaiyuan yu xingcheng 汉语方言量词"兜"的来源与形成 [On the Origin and Formation of "兜(dou)" – A Quantifier in Chinese Dialects]. *Yuyan Yanjiu Jiukan (di shijiu ji)* 语言研究集刊(第十九辑) [*Bulletin of Linguistic Studies* (19)], 279–297.Shanghai: Shanghai Cishu Chubanshe 上海:上海辞书出版社 [Shanghai: Shanghai Lexicographic Publishing House].

Chen, Zhenyu (陈振宇), and Chen Zhenning (陈振宁). 2015. Tongguo ditu fenxi jieshi yufaxue zhong de yinxing guilǜ 通过地图分析揭示语法学中的隐性规律——"加权最少边地图" [Revealing covert laws in grammar with Semantic Map analysis: Weighted Maps with Least Edges]. *Zhongguo Yuwen*中国语文 [*Studies of the Chinese Language*] 5: 428–438.

Clancy, Steven J. 2006. The topology of Slavic case: Semantic maps and multidimentional scaling. *Glossos* 7: 1–28.

Croft, William. 2001. *Radical construction grammar*. Oxford: Oxford University Press.

———. 2003. *Typology and universals*. 2nd ed. Cambridge: Cambridge University Press.

———. 2007. Typology and linguistic theory in the past decade: A personal view. *Linguistic Typology* 11: 79–91.

Croft, William, and Keith T. Poole. 2008. Inferring universals from grammatical variation: Multidimentional scaling for typological analysis. *Theoretical Linguistics* 34 (1): 1–37.

Cysouw, Michael. 2007. Building semantic maps: The case of person marking. In *New challenges in typology: Broadening the horizons and redefining the foundations*, ed. Matti Miestamo and Bernhard Wälchli, 223–248. Berlin: Mouton.

———. 2008. Generalizing scales. In *Linguistische Arbeits Berichte*, ed. Marc Richards and Andrej L. Malchukov, vol. 86, 379–396. Leipzig: Universität Leipzig.

de Haan, Ferdinand. 2004. *On representing semantic maps*. Ms. Summer 2004 EMELD workshop, University of Arizona.

———. 2006. Typological approaches to modality. In *The expression of modality*, ed. William Frawley, 27–69. Berlin: Mouton.

Fan, Xiaolei (范晓蕾). 2014. Yi "xuke-renshi keneng" zhi queshi lun yuyi ditu de xingshi he gongneng zhi xifen——jian lun qingtai leixing xitong zhi xin jieding 以"许可—认识可能"之缺失论语义地图的形式和功能之细分——兼论情态类型系统之新界定 [Subdivision Principles of Forms and Function in Semantic Map Based on the Absence of Path 'Permission –Epistemic Possibility']. *Shijie Hanyu Jiaoxue* 世界汉语教学 [*Chinese Teaching in the World*] 1: 18–36.

———. (范晓蕾). 2017. Yuyi ditu de jiexidu ji biaozheng fangshi——yi "nengli yi wei hexin de yuyi ditu" wei li 语义地图的解析度及表征方式——以"能力义为核心的语义地图"为例 [Semantic Map: Resolution and Representation Mode]. *Shijie Hanyu Jiaoxue* 世界汉语教学 [*Chinese Teaching in the World*] 2: 194–214.

François, Alexandre. 2008. Semantic map and the typology of colexification: Intertwining polysemous networks across languages. In *From polysemy to semantic change: Towards a typology of lexical semantic associations*, ed. Martine Vanhole, 163–216. Amsterdam: John Benjamins.

Guo, Rui (郭锐). 2010. Fuci buchong yi yu xiangguan yixiang de yuyi ditu 副词补充义与相关义项的语义地图 [A Semantic map for adverbs with additive and other functions]. Zhongguo yuyan de bijiao yu leixingxue guoji yantaohui lunwen 中国语言的比较与类型学国际研讨会论文 [The International Symposium for Comparative and Typological Research on Languages of China], Xianggang Keji Daxue 香港科技大学 [The Hong Kong University of Science and Technology].

———. (郭锐). 2015. Yuyi ditu zhong gainian de zuixiao guanlian yuanze he guanliandu 语义地图中概念的最小关联原则和关联度 [The principle of minimum relatedness and the degree of association in semantic map]. *Hanyu duo gongneng yufa xingshi de yuyi ditu yanjiu* 汉语多功能语法形式的语义地图研究 [*A Semantic Map approach to Chinese multifunctional forms*], Li Xiaofan, Zhangmin, Guorui deng zhu 李小凡、张敏、郭锐等著 [Li Xiaofan, Zhang Min, Guo Rui, et al.], 152–172. Beijing: Shangwu Yinshuguan 商务印书馆 [Beijing: The Commercial Press].

Han, Chang (韩畅), and Rongjing (荣晶). 2019. Dongci "zuo" de cihui leixingxueyanjiu动词"坐"的词汇类型学研究[An analysis of Verb 'Sit' from the perspective of lexical semantic typology]. *Shijie Hanyu Jiaoxue* 世界汉语教学 [*Chinese Teaching in the World*] 4: 504–521.

Haspelmath, Martin. 1997. *Indefinite pronouns*. Oxford: Clarendon.

———. 2003. The geometry of grammatical meaning: Semantic maps and cross-linguistic comparison. In *The new psychology of language*, ed. M. Tomasello, vol. 2, 211–243. New York: Erlbaum.

Heine, Bernd, Ulrike Claudi, and Friederike Hünnemeyer. 1991. *Grammaticalization: A conceptual framwork*. Chicago: Chicago University Press.

Hou, Xiaodan (侯晓丹). 2017. Yuyitu moxing shijiao xia de Han Han yu buding biaodaduibi——yi "renhe" he "amu" wei li 语义图模型视角下的汉韩语不定表达对比——以"任何"和"amu"为例 [A comparison on Chinese and Korean indefinite expression under the semantic map model – A case study of "renhe" and "amu"]. *Hanyu Xuexi* 汉语学习 [*Chinese Language Learning*] 1: 56–69.

Janda, Laura A. 2007. Aspectual clusters of Russian verbs. *Studies in Language* 31 (3): 607–648.

Kemmer, Susan. 1993. *The middle voice*. Amsterdam: Benjamins.

Kortmann, Bernd. 1997. *Adverbial subordination: A typology and history of adverbial subordinators based on European languages*. Berlin: Mouton.

Li, Renzhi. 2004. *Modality in English and Chinese: A typological perspective*. Boca Raton: Dissertation.com.

Li, Zhi-en (李知恩). 2011. *Liangci de kua yuyan yanjiu* 量词的跨语言研究 [*Cross-linguistic study on numeral classifiers*]. Beijing Daxue Boshi Xuewei Lunwen 北京大学博士学位论文 [Ph.D. dissertation, Peking University].

Li, Xiaojun (李小军). 2016. Hanyu liangci "ge" de yuyi yanhua moshi 汉语量词"个"的语义演化模式 [The semantic evolution model of Quantifier "ge" in Chinese]. *Yuyan Kexue* 语言科学 [*Linguistic Sciences*] 2: 150–164.

Li, Xiangnong (李向农), and Wei Yangyang (魏阳阳). 2019. Hanyu "he" lei pingbibiaojide jianyong gongneng ji zai minzu yuyan de kuosan 汉语"和"类平比标记的兼用功能及在民族语言的扩散 [Poly-functionality of Chinese Equative Marker "*he*" and its Diffusion in Minority Languages]. *Hanyu Xuebao* 汉语学报 [*Chinese Linguistics*] 1: 47–56.

Li, Xiaofan (李小凡), Zhang Min (张敏), Guo Rui (郭锐), et al. 2015. *Hanyu duo gongneng yufa xingshi de yuyi ditu yanjiu* 汉语多功能语法形式的语义地图研究 [*A Semantic Map approach to Chinese multifunctional forms*]. Beijing: Shangwu Yinshuguan 商务印书馆[Beijing: The Commercial Press].

Malchukov, Andrej. 2004. Toward a semantic typology of adversative and contrast marking. *Journal of Semantics* 21 (2): 177–198.

Phua, Chiew-Pheng (潘秋平). 2011. Cong yuyi ditu kan Shanggu Hanyu de shuang jiwujiegou从语义地图看上古汉语的双及物结构 [Ditransitive constructions in Archaic Chinese: A semantic map methodology approach]. *Lishi Yuyanxue Yanjiu (di si ji)*历史语言学研究(第四辑) [*Research on Historical Linguistics* (4)], 129–159. Beijing: Shangwu Yinshuguan 商务印书馆[Beijing: The Commercial Press].

——— (潘秋平). 2015. Yuyi ditu he jushi duoyi 语义地图和句式多义[Semantic map and constructional polysemy]. *Hanyu duo gongneng yufa xingshi de yuyi ditu yanjiu* 汉语多功能语法形式的语义地图研究[*A Semantic Map approach to Chinese multifunctional forms*], Li Xiaofan, Zhangmin, Guorui deng zhu 李小凡、张敏、郭锐等著 [Li Xiaofan, Zhang Min, Guo Rui,*et al.*], 102–151. Beijing:Shangwu Yinshuguan 商务印书馆 [Beijing: The Commercial Press].

Phua, Chiew-Pheng (潘秋平), and TEOH Kah Min (张家敏). 2017. Cong yuyi ditu kanwuwei zhi ci de yuyi yanbian 从语义地图看五味之词的语义演变 [Semantic extension of words expressing the five cardinal tastes: A semantic map approach]. *Yuyanxue Luncong (di wushiwu ji)*语言学论丛(第五十五辑)[*Essays on Linguistics*(55)], 319–372. Beijing: Shangwu Yinshuguan 商务印书馆[Beijing: The Commercial Press].

Rice, Sally, and Kaori Kabata. 2007. Crosslinguistic grammaticalization patterns of the ALLATIVE. *Linguistic Typology* 11: 451–514.

Sheng, Yimin (盛益民). 2015. Yuyi ditu de bu lianxu he lishi yanbian 语义地图的不连续和历史演变[Unconnectivity in semantic map and historical change]. *Hanyu duo gongneng yufa xingshi de yuyi ditu yanjiu* 汉语多功能语法形式的语义地图研究 [*A Semantic Map approach to Chinese multifunctional forms*], Li Xiaofan, Zhangmin, Guorui deng zhu 李小凡、张敏、郭锐等著[Li Xiaofan, Zhang Min, Guo Rui, *et al.*], 333–349. Beijing: Shangwu Yinshuguan 商务印书馆 [Beijing: The Commercial Press].

Stassen, Leon. 1997. *Intransitive predication: An essay in linguistic typology*. Oxford: Oxford University Press.

Sun, Wenfang (孙文访). 2018. "You(have)"de gainian kongjian ji yuyitu "有(have)"的概念空间及语义图 [The conceptual space and semantic maps of "HAVE"]. *Zhongguo Yuwen* 中国语文 [*Studies of the Chinese Language*] 1: 15–36.

Sun, Hongkai (孙宏开), Hu Zengyi (胡增益), and Huang Xing黄行)(eds.). 2007. *Zhongguo de Yuyan* 中国的语言 [*Languages in China*]. Beijing: Shangwu Yinshuguan 商务印书馆 [Beijing: The Commercial Press].

Tan, Fangfang (谭方方). 2016. Ying Han "zhuanzhe" yu "duibi" de guanxi ji qi yuyi ditu jieshi 英汉"转折"与"对比"的关系及其语义地图解释 [The relation between English and Chinese adversativity and contrastiveness: Their explication in light of a semantic map]. *Waiyu yu Waiyu Jiaoxue* 外语与外语教学[*Foreign Languages and Their Teaching*] 3: 48–57.

Tang, Xianqing (唐贤清), and Wang Qiaoming (王巧明). 2019. Yuyitu shijiao xia Guangxi Chetian Renhua "shi" de duo gongneng xing yanjiu 语义图视角下广西车田苗族"人话""是"的多功能性研究[Research of the Multifunctional Word [ʐɿ33](BE) in the Renhua Dialect Used by Miao Nationality in Chetian, Guangxi from the Perspective of Semantic Map]. *Hunan Daxue Xuebao(shehui kexue ban)* 湖南大学学报(社会科学版) [*Journal of Hunan University (Social Sciences)*] 5: 81–86.

Van der Auwera, Johan, and Vladimir A. Plungian. 1998. Modality's semantic map. *Linguistic Typology* 2: 79–124.

van der Auwera, Johan, Peter Kehayov, and A. Vittrant. 2007. Acquistive modals. Ms. Cross-linguistic semantics of tense, aspect and modality. L Hogeweg, H. De Hoop and A. Malchukov (eds.), 271–302. Amsterdam: Benjamins.

Wälchli, Bernhard. 2007. Constructing semantic maps from parallel text data. Ms. https://www.eva.mpg.de/lingua/conference/07-SemanticMaps/pdf/waelchli.pdf, 29-Nov-2021.

Wang, Wei (王玮). 2015. Kongjian weiyi de yuyi ditu yanjiu 空间位移的语义地图研究[The semantic map of the spatial motion domain]. *Hanyu duo gongneng yufa xingshi de yuyi ditu yanjiu* 汉语多功能语法形式的语义地图研究[*A Semantic Map approach to Chinese multi-functional forms*], Li Xiaofan, Zhangmin, Guorui deng zhu 李小凡、张敏、郭锐等著 [Li Xiaofan, Zhang Min, Guo Rui, *et al.*], 302–332. Beijing: Shangwu Yinshuguan 商务印书馆 [Beijing:The Commercial Press].

Wang, Yawei (王娅玮), and Wu Fuxiang (吴福祥). 2017. Jiyu Hanyushi de yu lianjie fanchou xiangguan de gainian kongjian 基于汉语史的与连接范畴相关的概念空间 [The conceptual map for the Doamin of combination on the basis of the history of Chinese language]. *Dangdai Yuyanxue* 当代语言学[*Contemporary linguistics*] 4: 591–621.

Wen, Jing (闻静). 2018. Langsuyu gongju ge zhuci de duo gongneng xing ji yufahualujing浪速语工具格助词的多功能性及语法化路径[Polyfuntionality and grammaticalization path of the instrumental particle in Langsu]. *Minzu Yuwen* 民族语文 [*Minority Languages of Chinese*] 3: 59–69.

Weng, Shanshan (翁珊珊), and Li Xiaofan (李小凡). 2010. 从语义地图看现代汉语"掉"类词的语义关联和虚化轨迹The semantic connection and the grammaticalization process of *diao*(掉) words in the modern Chinese: From the semantic map. *Yuyanxue Luncong (di sishi'er ji)*语言学论丛(第四十二辑) [*Essays on Linguistics*(42)], 61–80. Beijing: Shangwu Yinshuguan 商务印书馆 [Beijing: The Commercial Press].

Wu, Fuxiang (吴福祥). 2009. Cong "de" yi dongci dao buyu biaoji 从"得"义动词到补语标记——东南亚语言的一种语法化区域[From the verb meaning 'GET,ACQUIRE' to the resultative construction marker: A kind of grammaticalization area in the Southeast Asian languages]. *Zhongguo Yuwen* 中国语文 [*Studies of the Chinese Language*] 3: 195–211.

——— (吴福祥). 2011. Duo gongneng yusu yu yuyitu moxing 多功能语素与语义图模型 [Multifuctional Morphemes and Semantic Map Model]. *Yuyan Yanjiu* 语言研究 [*Studies in Language and Linguistcis*] 1: 25–42.

——— (吴福祥). 2014. Yuyitu yu yufahua 语义图与语法化 [Semantic map and grammaticalization]. *Shijie Hanyu Jiaoxue* 世界汉语教学[*Chinese Teaching in the World*] 1: 3–17.

Xia, Liping (夏俐萍). 2017. Hunan Yiyang fangyan de"de" lei qingtai shi ——jian lun Xiangyu "de" lei qingtai shi yuyitu 湖南益阳方言的"得"类情态式——兼论湘语"得"类情态式语义图 [The *De*(得)Modality Construction in Yiyang County, Xaing Dialects: A Semantic Map of *De*(得)Modality in Xaing Dialects]. *Yuyanxue Luncong (di wushiliu ji)*语言学论丛(第五十六辑) [*Essays on Linguistics*(56)], 306–331. Beijing: Shangwu Yinshuguan 商务印书馆 [Beijing: The Commercial Press].

Ye, Jingting (叶婧婷). 2019. Fanshen lingshu yu qianghua de leixingxue kaocha 反身领属与强化的类型学考察Co-referential Processors and Intensifiers in World's Language: A Typological Perspective. *Waiguoyu* 外国语[*Journal of Foreign Languages*] 1: 25–38.

Zhang, Ding (张定). 2010. *Hanyu duo gongneng yufa xingshi de yuyitu shijiao* 汉语多功能语法形式的语义图视角[*Multifunctional grammatical forms in Chinese: A Semantic Map perspective*]. Zhongguo Shehui Kexueyuan Yanjiushengyuan 中国社会科学院研究生院博士学位论文 [Ph.D. dissertation, Graduate School of Chinese Academy of Social Sciences].

Zhang, Ding (张定). 2016. "Zhuizhu" dongci yuyitu "追逐"动词语义图[A Semantic Map for "CHASE"]. *Dangdai Yuyanxue* 当代语言学*Contemporary Linguistics*] 1: 51–71.

Zhang, Ding (张定). 2017. "Chuandai" dongci yuyitu "穿戴"动词语义图 [A Semantic Map for "WEAR"]. *Dangdai Yuyanxue* 当代语言学[*Contemporary Linguistics*] 4: 546–560.

Zhang, Jun (张军). 2016. Lisuyu ma³³ de duo gongneng xing yu yufahua 傈僳语ma³³的多功能性与语法化 [Polyfuntionality and grammaticalization of *ma³³* in Lisu]. *Minzu Yuwen* 民族语文 [*Minority Languages of Chinese*] 4: 26–37.

Zhang, Min (张敏). 2008. Kongjianditu he yuyi ditu shang de 'chang' yu 'bian':yi Hanyu beidong, shiyi, chuzhi, gongju, shouyizhe deng guanxi biaoji wei li 空间地图和语义地图上的'常'与'变':以汉语被动、使役、处置、工具、受益者等关系标记为例[Universals and particulars manifested on spatial maps and semantic maps: On passive, causative, pretransitive, instrumental, beneficiary markers across Chinese dialects]. 中国社会科学院语言研究所报告 [Handout of a talk, Institute of Linguistics, Chinese Academy of Social Sciences].

Zhang, Min (张敏). 2010. Yuyi ditu moxing: yuanli, caozuo ji zai Hanyu duo gongneng yufa xingshi yanjiu zhong de yunyong 语义地图模型:原理、操作及在汉语多功能语法形式研究中的运用[The Semantic Map Model and Its Application to the Study of Multifunctional Grams in Chinese]. *Yuyanxue Luncong (di sishi'er ji)*语言学论丛(第四十二辑)*Essays on Linguistics*(42)], 3–60. Beijing: Shangwu Yinshuguan 商务印书馆 [Beijing: The Commercial Press].

Zhang, Xu (张旭). 2018. Yingyu "lei liangci" yanjiu——jian yu Hanyu liangci zuo gongneng leixingxue duibi 英语"类量词"研究——兼与汉语量词作功能类型学对比[English quasi-numeral classifiers: A functional-typological perspective from Chinese]. *Waiyu Jiaoxue yu Yanjiu (Waiguo Yuwen Shuangyuekan)* 外语教学与研究(外国语文双月刊)[*Foreign Language Teaching and Research(bimonthly)*] 2: 173–185.

Advances in the Study of the Grammaticalization of Sinitic Dialects

7

Huayong Lin and Niting Yan

Contents

Introduction	156
The "Embryonic Period" (1994–2004)	158
The "Boom Period" (2005 to the Present)	160
Concurrent Phenomena and Grammaticalization	161
Research on Historical Documents and Grammaticalization	162
Research on Synchronic Materials and Grammaticalization	163
The Combination of Grammaticalization Theory and Other Research Methods	171
Grammaticalization Research on Pragmatics and Discourse	175
Theoretical Research on Grammaticalization Mechanism and Grammaticalization Patterns	176
The Current Situation and Prospects for Development in Grammaticalization Research	177
References	177

Abstract

After having been introduced into China almost 30 years ago, grammaticalization theory in the field of dialect research has undergone an "embryonic" and "boom" period and has already seen the development of a "mature" phase of its application. Based on the findings of research carried out by Chinese scholars, this entry reviews and examines the application of grammaticalization theory to Sinitic dialects from three research perspectives: materials, content, and methods, and offers suggestions for future research directions.

Keywords

Sinitic dialects · Grammaticalization · Chinese dialects

Translated by T. L. Edwards, The Australian National University.

H. Lin (✉) · N. Yan
Department of Chinese, Sun Yat-sen University, Guangzhou, China
e-mail: linhuay@mail.sysu.edu.cn

Abbreviations

1	first person
2	second person
3	third person
ATT	attributive marker
CLF	classifier
DEM	demonstrative
DIM	diminutive
DLM	dilimitative
DM	disposal marker
PERF	perfect
PFV	perfective
PROG	progressive
PRT	particle
REDUP	reduplicative
SCM	stative complement marker
SFP	sentence final particle
SG	singular
TOP	topic
TPF	temporal particle of first

Introduction

Traditional Chinese linguistics has a long history of research into the phenomenon of "turning 'content words' [*shici* 实词] into 'function words' [*xuci* 虚词]" (*shici xuhua* 实词虚化). As Yuan scholar Zhou Boqi 周伯琦 (1298–1369) points out in his *Liushu zheng'e* 六书正讹 [The Correct and the Spurious in the Six Character Types], "The 'function words' of today, are all 'content words' from antiquity 今之虚字, 皆古之实字." Outstanding historical studies on "bleaching" (*xuhua* 虚化) include Lu Yiwei's 卢以纬 *Yuzhu* 语助 [On Language Particles] from the late Yuan, and works by Qing authors such as Wang Yinzhi's 王引之 *Jingzhuan shici* 经传释词 [Explication of Terms in the Classics and Commentaries], Liu Qi's 刘淇 *Zhuzi bianlüe* 助字辨略 [An Outline of Particles], and Yuan Renlin's 袁仁林 *Xuzi shuo* 虚字说 [Explanation of Function Words]. Ma Jianzhong's 马建忠, *Mashi wentong* 马氏文通 [Master Ma's Rules of Grammar] of 1898 investigated the origin and evolution of several "function words" (*xuzi* 虚字) and while he did not employ modern technical terminology such as "syntactic context" (*jufa huanjing* 句法环境), Ma paid close attention to the relationship between "turning 'content words' into 'function words'" and its linguistic context. After this, linguists continued this research tradition by applying the theory to the study of particular syntactic structures such as Wang Li's *Hanyushi gao* 汉语史稿 [Script of History of Chinese Language] that looks at the mutual link between the evolution of the preposition *ba* 把 and *bei* 被 and the emergence and development of the disposal construction (*chuzhishi* 处置式) and passive construction (*beidongshi* 被动式).

Prior to the introduction of grammaticalization theory into China, Chinese research into "turning 'content words' into 'function words'" had already reached a stage of maturity. Li Yunhan (1981: 103) pays attention to the degree by which function words underwent bleaching and believes that "it also continues to evolve in pace with the developmental changes of grammatical structures." Xie Huiquan (1987: 213) points out that "the bleaching of content words should take their meanings as foundational and their syntactic status as the means by which this process occurs." These notions possess certain similarities with some views found in "grammaticalization" theory.

Chinese scholars have also applied the theory of grammaticalization to the study of the grammar of dialects. For example, Ben Liao (1983) utilizes data drawn from the Guiyang dialect 贵阳方言 to compare the use of the "*ba* 把" disposal construction in ancient Chinese and modern Chinese. While the entry's conclusion that "*ba* 把" is an "infix" remains open for debate, this research methodology of combining "ancient, dialectic, modern, and foreign" aspects of language and comparing the environment in which function words are distributed already exhibits the embryonic features of modern research on the grammaticalization of dialects. As such, it represents an attempt by Chinese scholars to apply the notion of "turning 'content words' into 'function words'" and grammaticalization theory in the study of dialectic grammar. It is evident, however, that the author is yet to distinguish between "bleaching" and "grammaticalization."

Shen Jiaxuan (1994) and Sun Chaofen (1994) introduced Western grammaticalization theory and its findings into China comparatively early. After this, Liu Jian, Cao Guangshun, and Wu Fuxiang (1995) carried out an examination of the mechanisms that induce grammaticalization of Chinese lexical items from a theoretical perspective. Chinese scholars have had a more comprehensive understanding of "grammaticalization" theory as applied to Sinitic language after Shen Jiaxuan's (1998a, 1998b) studies provided concrete examples from Chinese to illustrate a pragmatic approach to grammaticalization. Shen's works also provide an additional introduction to five types of "bleaching" mechanisms: "metaphor" (*yinyu* 隐喻), "inference" (*tuili* 推理), "harmony" (*hexie* 和谐), "generalization" (*fanhua* 泛化), and "absorption of context" (*xishou* 吸收). While the publication of the five articles noted above have laid the basic theoretical framework for grammaticalization research in China, this early stage of research does not strictly distinguish between "turning 'content words' into 'function words'" and "grammaticalization."

Later, as scholars' understanding of the theoretical concepts of "grammaticalization" deepened, they gradually learnt to distinguish the phrase "turning 'content words' into 'function words'" (*shici xuhua* 实词虚化), drawn from the ancient Chinese exegetical tradition, from the Western technical term "grammaticalization" (*yufahua* 语法化). The differences between these two notions as reflected in their main connotations can be summarized as follows: the notion of "turning 'content words' into 'function words'" is based predominately on semantics and focuses on the semantic changes of content words along with the generation of grammatical meaning; "grammaticalization," however, is not merely concerned with the evolution of content words to syntactic function words, but is also concerned with the evolution of structural patterns and pragmatics to syntactic forms. Although this notion of "turning 'content words' into 'function words'" only involves one aspect of grammaticalization theory, this local exegetic tradition has undoubtedly provided fertile ground for the rapid growth of grammaticalization theory in China and has had a profound impact on subsequent research.

With this background in place, research into the grammaticalization of Chinese dialects has developed rapidly. Since the introduction of grammaticalization theory into China, we can divide research on the grammaticalization of Chinese dialects into two periods: an "embryonic period" from 1994–2004 and a "boom period" from 2005 to the present. We can summarize the characteristics of these two periods as follows: the "embryonic period" focused on combining historical documents and synchronic dialect data along with a comparative approach to research into dialects, but the amount of research carried out during this stage was relatively limited. In the boom period," linguistic data from dialects served as the basis for research which greatly enhanced its scope and depth and placed emphasis on description, comparison, and reconstructing of the route of grammaticalization. In terms of methodology, grammaticalization theory became inseparably linked with a variety of methods including the semantic map model, language contact, pragmatics, cognition, and subjectivization. The following provides a review and outline of the results and advances during these two stages.

The "Embryonic Period" (1994–2004)

Grammaticalization research during this period can be further divided into two stages, 1994–1999 and 2000–2004. The first stage (1994–1999) saw the introduction of grammaticalization theory to China with scholars focusing on the connection between diachronic change and synchronic phenomena. Examples of this stage include Jiang Lansheng's (1995) use of historical documents such as the *Zhouli* 周礼 [The Rites of Zhou] along with examples drawn from modern dialects such as the Shandong and Gansu dialects to provide evidence for a hypothesis on the common origin of the character "么[mə^0]" in the interrogative pronoun "甚么[ʂən^{51} mə^0]," and the plural suffix "们[mən^0]" is the content word "物[mv-]," while also summarizing the characteristics of the grammaticalization of this word. Qian Nairong (1998) uses materials drawn from Ming and Qing literature and folk theater in the Wu dialect 吴语 to investigate the multiple usages of the measure word "个[kʌ]," and the demonstrative "介[kɑ]." Qian believes that semantic extension and conversion exist among these different usages and points out that these changes are common in the Wu dialect. Starting with the structure of "VP + (O) + 在里 [tsai^{51}li^{214}]/在[tsai^{51}]/哩[li^0]," Luo Ziqun (1999) combines materials from more than ten dialects including the Wu, Min, Gan, Xiang, and other dialects to discuss the varying states of the degree of bleaching in different dialects in terms of the preposition structure of sentence endings indicating continuity. Grounded in the inherent characteristics of Sinitic languages and the extensive use of dialect materials, the research methodologies this paper promotes serve as valuable reference materials.

There are two notable aspects that emerged during the first few years after the turn of the century. The first was that more attention was paid to accompanied language phenomena, such as the phonetic and semantic changes that are often displayed during the process of grammaticalization. Research conducted by scholars such as Jiang Lansheng, Shi Yuzhi, Liu Danqing, and Li Xiaojun are characteristic of this approach.

Rong Jing and Ding Chongming (2004) compare old and new streams of phonetic differences in the aspectual particle "着[tə^{44}/tʂɔ^{31}]" in the Kunming dialect 昆明话. Through the use of historical documents and other dialect materials, they observe that these phonetics differences are responsible for the emergence of different functions and look at the issue of the imbalance in the development of "着[tʂə^{0}]" in different dialects, and use this to explore the relationship between "diachronic specialization" (*lishi zeyi* 历时择一) and "synchronic balance" (*gongshi zhiheng* 共时制衡) as mechanisms of grammaticalization. Dai Zhaoming (2004) describes the process leading to the grammaticalization of notional words with nonchecked tones (*shusheng* 舒声) via semantic bleaching, phonetic reduction, and glottalization in Wu dialects. For example:

(1) Tiantai Wu Dialect 天台方言

a.	客		来了。
	k'a^{ʔ5}		lei^{224}-lau$_{21}$.
	guest		come-PFV
"Here comes the guest."			

b.	我	帮忙	来了。
	fiɔ^{214}	pɔ̃^{33}mɔ̃^{224} +	ləʔ-lau$_{21}$.
	1SG	help +	come-PFV
"I come here to help."			

In the Tiantai Wu dialect, "来 [lei^{224}]" in example 1a is an action verb. When "来 [lei^{224}]" is reduced to a directional marker, it becomes "来[ləʔ]" as per example 1b via phonetic lenition and glottalization. Their article considers phonetic lenition and glottalization to be the mechanisms through which the grammaticalization effect is solidified in phonetic form.

The second notable aspect during this period was a more mature approach to examining the path of grammaticalization from a diachronic perspective by using evidence drawn from dialect materials. By conductive a survey of "markers of stative complement" (*zhuangtai buyu biaoji* 状态补语标记) in a large number of southern dialects, Wu Fuxiang (2001, 2002, 2002) reaches a conclusion that markers of stative complement in southern dialects originate from the "perfect aspect marker" (*wanchengti biaoji* 完成体标记), i.e., "auxiliary words of perfective aspect" (完成体助词, or "phase complements" (*dongxiang buyu* 动相补语).

During this period, Chinese scholars have also paid attention to the grammaticalization of pragmatics. Through a synchronic description of the demonstratives *zhe* 这 ("this") and *na* 那 ("that") in the Beijing dialect, Fang Mei (2002) demonstrates that *zhe* 这 had already evolved into the definite article – a process whereby written usage gradually fixed it as a syntactic category. The entry also compares the different grammaticalization paths of demonstratives in southern dialects such as Wu and Yue 粤 and integrates grammaticalization theory and typological methods.

Similar uses of typological methods to carry out grammaticalization research include Wu Fuxiang's (2003) exploration of the evolutionary types of Chinese accompanying prepositions. By employing historical documents, synchronic dialects, minority languages, and crosslinguistic materials, this entry demonstrates that the two different evolutionary modes of comitative prepositions (*bansui jieci* 伴随介词) in SVO-type languages are caused by syntactic conditions and cognitive factors. Ding Jiayong (2003) also carried out an investigation into the sentence-final "着 [tʂə^0]" in Sinitic dialects from a typological perspective, and demonstrates that the aspectual particle "着[tʂə^0]" evolved from the modal "sentence-final particle" (SFP) that expresses the imperative mood.

The first decade since the introduction of grammaticalization theory to China can be grouped into three phases:

1. Assisted greatly by the long-standing exegetical tradition of "turning 'content words' into 'function words'," Chinese scholars were quick to absorb grammaticalization theory and apply it to local circumstances, seeing it gradually become a "hot research topic" in Chinese grammar. During this early phase, however, by focusing on traditional expressions such as "turning 'content words' into 'function words'," some scholars blurred the two terms with most still referring to the process as "bleaching."
2. While research into the grammar of Sinitic dialects has always been carried out with a focus on comparing historical documents, Standard Chinese, and other dialects, the introduction of grammaticalization theory during this second phase provided a more fruitful interface for integrating diachronic and synchronic aspects of grammaticalization.
3. After the turn of the century, based on the selection of more diverse materials and a greater richness of content, local research into the grammaticalization Sinitic dialects was characterized by a deepened exploration of its mechanisms. During this stage, the integration of pragmatics, typology, and grammaticalization theory could be seen as a period of "germination," one that provided a solid foundation for future scholars to carry out further research for a variety of perspectives.

The "Boom Period" (2005 to the Present)

When compared to the previous decade, the findings of the "Boom Period" of grammaticalization research made significant advances in terms of both the quantity and quality of research undertaken. In concrete terms, this can be seen in the exponential growth of "grammaticalization" as a keyword in research papers with the theory being used widely in multicategory and multiperspective research.

Wu Fuxiang (2005) integrates advances in foreign grammaticalization research with observations drawn from current research into Chinese to point out four fields that show great potential for Chinese grammaticalization research: the "grammaticalization of constructions" (*jiegoushi yufahua* 结构式语法化), "grammaticalization patterns" (*yufahua moshi* 语法化模式), "the grammaticalization of discourse markers" (*huayu biaoji yufahua* 话语标记语法化), and "semantic change with grammaticalization" (*yu yufahua xiangguan de yuyi yanbian* 与语法化相关的语义演变).

Using a large number of examples drawn from dialects, Liu Danqing (2009) provides of clear summary three areas that grammaticalization theory can reveal for research into the grammar of Sinitic dialects – grammatical units, grammatical categories, and grammatical means. Liu argues that dialect grammar is "dynamic and variable," which is mainly reflected in the "the coexistent phenomena of dynamics, variability, and heterogeneity in the characteristics of grammatical units," the "dynamicity and variability of grammatical category," and in the "dynamicity and variability of grammatical means." Zhang Yisheng (2016) provides a comprehensive summary of advances in Chinese function word research over the past 30 years from the perspective of "grammaticalization and subjectivisation." Zhang points out the inadequacies of former research and indicates directions for future research that involve adopting multiple theoretical interfaces, dialect contact, and the combination of multiple related research disciplines that reflect the thinking of Chinese scholars founded on the characteristics of Chinese along with the flexible application of Western theories and methods to Chinese research.

The articles noted above were a source of great inspiration for later studies of dialect grammar in the context of grammaticalization theory. Since 2005, domestic research on the grammaticalization of dialects has gradually deepened and we have seen it surge every five years or so. Research in these periods can be summarized under the following aspects.

Concurrent Phenomena and Grammaticalization

The process of grammaticalization often accompanies phonetic, semantic, and morphological change. When studying the grammaticalization of dialects, Chinese scholars focus on the interaction between concurrent phenomenon and grammaticalization and many studies have emerged with these phenomena as their starting points. In terms of research devoted to the linguistic features of Chinese itself, while there are numerous studies on phonetic change in Chinese, research into semantic change commenced relatively late and studies of morphological change remain quite rare.

Zhou Yimin (2005), Cao Fengfu (2006), Wu Zihui (2007), Zhang Zhanshan and Li Rulong (2007), and Zhao Rixin (2017) investigate phonetic change involving the relationship between grammaticalization and multitype phenomena such as the neutral tone, nasalization, glottalization, fusion, and vowel reduction. Investigations of diachronic and synchronic semantic change, multifunction words, and an account of their evolutionary mechanism and constraints are found in Jin Xiaodong and Wu Fuxiang (2016), Lei Rong (2017), and Liu Jinfeng (2017). By looking at phonetic materials drawn from dialects in Hebei, Henan, Shandong, and other regions, Chen Qianrui and Wu Jizhang (2019) analyze the tendency of semantic parallel weakening in the structure and meaning of the auxiliary word "*le* 了."

Research on Historical Documents and Grammaticalization

As Chinese historical documents are relatively abundant, research methods that utilize historical documents to investigate language evolution have also been extended to the study of Sinitic dialects. However, as the quantity of historical documents that have been preserved in various dialects differs, their use in the study of diachronic change has led to a situation where they are unevenly represented. Studies of grammaticalization that employ historical documents are more numerous in southern dialects such as Wu, Cantonese, and Southern Min, and several also touch on materials in dialects from the north and southwest. The majority of these studies are based on Ming and Qing documents. In addition to novels and dramatic works, several studies have also examined dictionaries, dialect textbooks, and translations of the Bible compiled by Christian missionaries.

Research in this area includes Lin Su'e's (2015) outline of the grammaticalization pathway of "等[təŋ^{35}]" that analyzes mid-nineteenth and early-twentieth-century missionary documents in the Ningbo dialect 宁波话 in comparison with documents in the Wu dialect. Yang Jingyu (2004) carries out an investigation of linguistic materials in the *Yueyu quanshu* 粤语全书 [Complete Writings in Cantonese], a late-Qing Cantonese textbook. Kwok Bit-Chee and Shin Kataoka (2006) combined 16 types of early Cantonese literature to explore the origin and evolution of the perfect aspect marker "*xiao* 哓."

Using materials drawn from Ming and Qing dramatical works, Zeng Nanyi and Li Xiaofan (2013) discuss the origin of the aspect marker "咧[lə?^{5}]/[lə^1]" (perfect, progressive, and continuous aspect) by investigating usage in the modern Quanzhou dialect 泉州方言, as per example (2).

(2) Quanzhou Min Dialect 泉州闽方言

a.

	我	咧写	论文。
	gua^{55}	lə?^{5}-sia$^{55\text{-}24}$	lun$^{31\text{-}11}$ bun^{24}.
	1SG	PROG-write	thesis

"I'm writing a thesis."

b.

	伊	在咧	煮食。
	i^{33}	tɯ$^{22\text{-}11}$lə?^{5}	-tsɯ$^{55\text{-}24}$·tsia?^{24}.
	3SG	PROG	-cook

"He's cooking."

In modern the Quanzhou dialect, the progressive aspect marker "咧[lə?^{5}]" can be used singularly as per Example (2a) and can also be reduced as per the structure "在 [tuɯ$^{22-11}$] +只/许处[tsit^5/hit^5 tə^{55}]" in Example (2b). Zeng's paper argues that the preposition structure in the Quanzhou dialect, after losing the demonstrative word "只/许[tsit^5/hit^5]," the locative word "处[tə^{55}]" gradually undergoes bleaching, which leads to phonetic change and finally changes into the aspect marker. Other scholar such as Chen Zeping, Li Wei, and Yiu Yuk-man also have integrated historical materials into dialect research.

Research on Synchronic Materials and Grammaticalization

After 2005, research on the grammaticalization of Sinitic dialects focused on the description and comparison of synchronic materials. No longer satisfied with a general description of the characteristics of grammar, Chinese scholars began to carry out more in-depth investigations to uncover more details on their development in the evolution of grammar. In terms of the objects of research during this period, there was an increasing trend for studies to focus on certain parts of speech or categories as they relate to form, construction, and pragmatics.

Correlations between Parts of Speech and Grammaticalization

Grammaticalization research into dialects touches upon "parts of speech" (*cilei* 词类) including verbs and verb phrases, nouns, pronouns, prepositions, particles, and modal particles.

Research on Verbs

Research on the grammaticalization of verbs and verb phrases in dialects started relatively early with the grammaticalization of speech verbs, giving verbs, directional verbs, and others attracting much attention.

Speech Verbs (*Yanshuo Dongci* 言说动词)

Shi Qisheng (1990) discusses the how the speech verb "咀 [tã^{213}]" is able to connect two predicate components to form a nonverbal usage in the predicate-object structure. Fang Mei (2006) analyzes the evolution of speech verbs in dialects and by the use of oral materials in the Beijing dialect demonstrates that *shuo* 说 ("to say") has been grammaticalized as complementizer and "irrealis clause marker" (*xuni qingtai congju biaoji* 虚拟情态从句标记).

Research on the grammaticalization of speech verbs in southern dialects reached a peak with studies produced by scholars such as Lin Huayong, Wang Jian, and Huang Yanxuan.

Han Weifeng and Shi Dingxu (2014) discuss the origin of the topic marker function of "伊讲 [ɦi^{23} kã^{34}]" in the Shanghai dialect 上海话. The article argues that it has successively undergone the following stages: "quotative clause (*zhiyin xiaoju* 直引小句) → reportative clause (*baogao xiaoju* 报告小句) → modal particle (*jumo zhuci* 句末助词) → topic-marking particle (*zhuti biaoji* 主题标记)" as per Example (3):

(3) Shanghai Wu Dialect 上海吴方言

a.	ɦi²³	kã³⁴	və⁷¹²hɔ³⁴.
	3SG	say	no

"He says NO!"

b.	və²¹hɔ³⁴,	ɦi²³	kã³⁴
	no	3SG	say

"He says NO!"

c.		və⁷¹²hɔ³⁴		ɦi²³ kã³⁴
		no		PRT

"Unexpectedly, there is a NO!"

d.	ɦi²³	ɦi²³kã³⁴,	kã³⁴	və⁷¹²hɔ³⁴
	3SG	TOP	say	no

"Unexpectedly, he says NO!"

Giving Verbs (*geiyu dongci* 给予动词)
The grammaticalization of giving verbs has also become a topic of interest for several scholars with representative studies carried out by Shen Ming, Cai Guomei, Huang Yanxuan, and Zhang Jingfen who have each discussed this phenomenon in the Jin, Xiang, Yue, and Min dialects.

Directional Verbs (*quxiang gondci* 趋向动词)
Directional verbs are also a hot topic in grammaticalization. By collating a large number of linguistic materials, Wu Fuxiang (2010a, b) summarizes the four types of grammaticalization patterns of directional verb:

1. "Directional verb → comparative marker (*bijiao biaoji* 比较标记)"
2. "Directional verb → dummy potential complement (*kuilei buyu* 傀儡补语) → functional particle (*nengxing zhuci* 能性助词)"
3. "Directional verb → complement marker (*buyu biaoji* 补语标记)"
4. "Directional verb → spatial/temporal proposition (*kongjian/shijian jieci* 空间/时间介词) → dative preposition (*yuge jieci* 与格介词)"

Lin Huayong and Kwok Bit-Chee (2010) construct the grammaticalization path of *lai* 来/ *qu* 去 "come/go" and observe the phenomenon whereby semantic bleaching brought about a convergence of the function of the verbs.

Other Verbs

Studies on the grammaticalization of other verbs in dialects include verbs of cognition (*renzhi dongci* 认知动词), verbs of arrival (*dazhiyi dongci* 达至义动词), verbs of holding (*chinayi dongci* 持拿义动词), and verbs of placing (*fangzhiyi dongci* 放置义动词). Based on an investigation of Southern Min operas from the Ming and Qing, Lien Chinfa (2007) believes that the experimental marker "八 [pat^4]" in Southern Min dialect has it origin in verbs of cognition. Yang Hsiu-fang (2014) discusses the syntactic context of "别 [pat^4]" in Taiwanese Southern Min and believes that the object followed by "别 [pat^4]" expanded from a noun to a verb. This structure caused the two verbs to juxtapose with the semantic focus shifting backward with the grammaticalization of "别 [pat^4]" to an adverb from a verb of cognition. For example, (4):

(4) Taiwan Min Dialect 台湾闽方言

a.	阿英	真	别	食。
	aliŋ1	tsin1	pat4	tsiaʔ8.[a]
	Aying	really	know	eat

"Aying really good at eating."

[a] In Yang Hsiu-feng (2014), examples 1, 2, 3, 4, 5, and 8 represent the "yinping tone" (*yinping* 阴平), "rising tone" (*shangsheng* 上声), "qu tone" (*qusheng* 去声), "yinru tone" (*yinru* 阴入), "yabgping tone" (*yangping* 阳平), and "yangru tone" (*yangru* 阳入).

b.	阿英	别	食。
	aliŋ1	pat4	tsiaʔ8.
	Aying	once	eat

"Aying has eaten before."

Some scholars look at semantic diversity and syntactic multifunctionality in materials from the Northern and Southern Wu dialects while others discuss the path of semantic change and grammaticalization for the verb of storage and placing.

Research on Prepositions

In research on the grammaticalization of prepositions in dialects, most attention has been paid to the usage of the preposition "disposal" and "passive" prepositions including disposal and passive markers. For example, Li Yuming and Chen Qianrui (2005) employ a quantitative method to investigate the passive sentence of *gei* 给 ("to give") in the Beijing dialect and argue that the passive use of the preposition *gei* in Beijing dialect was developed by "combining its own semantic basis and syntactic context" under the influence of the auxiliary word *gei* in a variety of sentence patterns. Other related studies on prepositions include Jiang Lansheng (2012), who investigates several prepositions used as conjunctions in Sinitic dialects. Jiang argues that "conjunction-prepositions" (*lian-jieci* 连-介词) have at least four sources including conjunctive verbs and causative verbs and illustrates three grammaticalization patterns.

Research on Particles and Markers

The origin of particles in dialects is a relatively complex phenomenon. Grammaticalization research touches on questions such as structure, dynamics, and stative particles which include the study of complement and aspect markers. Huang Xiaoxue (2006) points out that in the Susong dialect 宿松方言 in Anhui, the grammaticalization of "在那里" is derived from the sentence final stative particle "在." Kwok Bit-chee (2008) investigates the verbal particle "度[t^ho^{3B}]" in the Shaowu dialect 邵武话 and finds that it preserves a unique grammaticalization path from the "transference verb" (*chuandi dongci* 传递动词) → directional complement (*quxiang buyu* 趋向补语) → complement aspect marker (*buyuxing tizhuci zhuci* 补语性体助) → pure marker aspect marker (*chutti zhuci* 纯体助词) (The "3B" in "度[t^ho^{3B}]" refers to the "light departing tone" (*yangqudiao* 阳去调).).

The aspect particle or aspect marker is another hot topic in grammaticalization research, with many significant findings involving various aspect markers such as the perfect, progressive, and continuous aspect. Representative studies in this regard include Yang Yonglong (2005), who integrates diachronic literature with materials from Hakka, Wu, and Cantonese dialects and has unearthed a unique grammaticalization pathway seeing the evolution of adjectives meaning "stable" (*wenjin* 稳紧) in Chinese into continuous, progressive aspect particles.

Song Wenhui (2019) studies two variants of the suffix "了" in the Zhengding dialect 正定方言 in Hebei and demonstrates that there are differences in its expression of tense and form in regard to phonetic and syntactic structure and the context of its usage.

Research on Modal Particles

Modal particles have always been an issue that scholars have paid a great deal of attention to and applied grammaticalization theory relatively early to its study. Xing Xiangdong (2015) conducted a detailed analysis of modal particles in the Jin dialect that are also used as tense markers and separated the syntactic distribution between the pure indicative mood and the concurrent indicative mood in "来[$l_E^{44/0}$], 也[ia^0], 了[$lɛ^0$], 嘞[$lə?^0$]" of [$lɛ^0$] and 嘞[$lə?^0$]" and considers the indicative mood to be more fundamental with the function of the indicative mood derived from the grammaticalization of tense particles, which reveals the complex relationship between the tense component and tone component.

Li Xiaojun (2008) analyzes the distribution of modal particle "吵[sa~ṣa]" in the major southern dialects along with some northwestern dialects and argues that "吵[sa~ṣa]" is the phonetic change of "些[sia]" and that the modal particle is derived from the grammaticalization of quantifiers. Liu Cuixiang (2016) believes that in Shandong Qixia dialect 栖霞方言, the modal particles from the local structure have "唻儿[lɐr]" and "乜行儿[niə xɐr]/乜场儿[niə tɕiɐr]," but that their internal grammaticalization is different and the syntactic conditions of its appearance possess a differentiation between "causation" (*shiran* 使然) and "non-causation" (*feishiran* 非使然).

Research on Multifunction Words

Terms such as "multi-functional words" (*duogongneng xuci* 多功能虚词), "multi-functionality" (*duogongneng xing* 多功能性), and "multi-functional usage" (*duogongneng yongfa* 多功能用法) were used relatively late in dialect grammaticalization research but became more common after 2015. The characteristics of Chinese as an analytic language mean that morpheme often have multiple functions and as a result, research into multifunction words has developed rapidly. An early example of such research can be seen in Luo Xinru (2007), who on the basis of the description of the multifunctional usage of "滴[ti^{35}]" in Loudi dialect 娄底方言 of the Xiang dialect 湘语 points out that it can be used as a nonquantitative word, demonstrative, and structural auxiliary. Sheng Yimin (2010) argues that the multifunctional usage of the function word "作[tso^{ɔ}]/[tsoʔ$_{ɔ}$]" in the Shaoxing Keqiao dialect 柯桥话 was grammaticalized on the basis of the meaning of "to give" (*geiyu* 给与) and focuses semantic changes of the "marker of beneficiary (*shouyizhe biaoji* 受益者标记) → marker of comitive (*bansuizhe biaoji* 伴随者标记)" and "verbs of giving → marker of maleficiary (*shousunzhe biaoji* 受损者标记) → disposal marker (*chuzhi biaoji* 处置标记)/causative marker (*zhishi biaoji* 致使标记)."

Wang Fang and Liu Danqing (2011) employ a method of combining synchronic description and analysis of semantic changes to discuss the multifunctional word "的 [te]" in the Guangshan dialect 光山方言 of Henan and point out that it is a functional word with affixation, and multiple functions such as positional postposition, positional prepositions, attributive markers, and sentence markers are derived from "里 [lə]." Some scholars incorporated historical linguistic materials of early Southwestern Sichuan Mandarin to explore the connection between the multifunctional usage of "过[ko^{24}]" in the Zizhong dialect 资中方言 and reconstructed its grammaticalization process as follows: directional verb → preposition or prepositionalization; directional verbs → directional complement → Quasi-Iterative aspect → experiencial aspect.

Research on Substantive Words

The research on the grammaticalization of "substantive words" (*tici* 体词) in dialects mainly focuses on three aspects: pronouns, measure words, and "*zi* 子" nouns. Research into pronouns mainly touches on personal pronouns and demonstrative pronouns. Research into quantifiers includes general quantifiers and also "classifier-noun constructions" (*liangming jiegou* 量名结构). Research into nouns is mainly concentrated on diminutive *zi* nouns and there are also a small number of studies devoted to the grammaticalization of other nouns.

Pronouns

The singular and plural forms of personal pronouns and demonstrative pronouns are hot topics in grammaticalization research. Wang Huayun (2011) looks the forms of personal pronoun plural marker in Chinese dialects and argues that single syllable plural markers are the result of the omission of multiple syllables plural marks and that this represents the beginning of bleaching of content plural markers. Lin Huayong and Li Minying (2019) argue that the third person pronoun "佢" [k^{h}ei23] at the end of sentences in the Lianjiang dialect in Guangdong has been grammaticalized into the disposal marker. In Example (5), the function of "佢" is to express "subjective disposal in order to reach a certain result or state (non-current situation)."

(5) Lianjiang Cantonese Dialect廉江粤方言.

a.	佢			
	kʰei²³			
	3SG			
"he/she/it"				

b.	食	了	饭	佢。
	sek¹	liu²³	fan²¹	kʰei²³
	eat	finish	meal	3SG/DM
"Finish the meal!"				

c.	食	了	饭	嘚	佢。
	sek¹	liu²³	fan²¹	tɛ²¹	kʰei²³
	eat	finish	meal	–PFV	DM
"Finish the meal."					

In terms of demonstrative pronouns, Shi Xiuju and Shi Rong (2016) discuss the grammaticalization of the demonstrative pronouns "这[tʂ³¹]" and "兀[u³¹]/乃[nai³¹]" into structural auxiliaries in the Jiangxian dialect 绛县方言 in Shanxi and point out that as a result of the remains of their demonstrative function, they cannot completely replace true structural auxiliaries. Chen Weirong (2017) describes demonstrative words (the fusion form of referential structure) in Hui'an dialect 惠安方言 and believes that four sets of nominal demonstrative words in the Southern Min dialect can also serve as relativization clause markers (*guanxi congju biaoji* 关系从句标记).

Measure Words

Scholars have also paid attention quite early to grammaticalization as it relates to quantifiers. Peng Xiaochuan (2006) believes that the nonquantifier "改为:"啲[ti55] (些)"。55上标。" in Cantonese has been grammaticalized to become a structural auxiliary. Zhao Jianhua (2013) studies the structure and modal particle function of "个[ko³²⁴]" in the Changdu dialect 昌都话 of the Gan dialect. Other scholars such as Chen Zeping and Wang Huayun have also discussed this topic from the perspective of various dialects.

Nouns

There are comparatively fewer studies of the grammaticalization of nouns and they mainly focus on the diminutive combined with phonetic research. For example, Cao Fengfu (2006) observes the "grammaticalization cycle" (*yufahua lunhui* 语法化轮回) of Chinese nickname diminutives in data drawn from several dialects from the perspective of the relationship between semantics and phonetics, such as Example (6).

7 Advances in the Study of the Grammaticalization of Sinitic Dialects

(6) Mannan Dialect

a.		囝	
		kiã˦˥	
		child	
"child/children"			

b.		牛囝	
		gu˦˥ · a˦˥	
		cattle -DIM	
"calf/cattle"			

c.		牛 囝	囝
		gu˦˥ a˦˥	kiã˦˥
		cattle	-DIM
"calf"			

Lin Huayong and Ma Zhe (2008) investigate the grammaticalization process of the semantic morpheme "仔[tsɐi^{25}]" in Lianjiang Cantonese. Comparing it to surrounding dialects, they consider it to be a subordinate component that can be attached to nouns, temporary classifiers, predicate reduplication, adjective phrases, and other structures to indicate age or small size, capacity, low status, small quantity, and low degree. They suggest that the continuity of the nickname diminutive in the Lianjiang dialect reflects different degrees of grammaticalization, of which more special cases are predicate reduplications with reduplication and adjective phrases followed by the small predicate marker "仔[tsɐi^{25}]" as per Example (7).

(7) Lianjiang Yue Dialect 廉江粤方言

a.	佢	睇睇仔	电视	就	睡熟嘚。
	kʰei^{23}	tʰɐi^{25}tʰɐi^{25}-tsɐi^{25}	tin^{21}si^{21}	tsɐu^{21}	sui^{21}suk^{1} te^{21}.
	3SG	watch~REDUP-DLM	television	then	Sleep+well-PERF
"He fall asleep during watching TV."					

b.	你	个	面	红红仔。
	nei^{23}	kɔ^{33}	min^{21}	hoŋ^{21}hoŋ^{21}-tsɐi^{25}.
	2SG	ATT	face	red~REDUP-DIM
"Your face is getting a little red."				

c.	佢	生得	好高仔。
	kʰei^{23}	saŋ^{55}tɐk^{5}	hou^{25}+ kou^{55} -tsɐi^{25}.
	3SG	grow-SCM	quite+tall-DIM
"He is kind of tall."			

Related research results on the renewal, evolution, and exploration of the origins of the diminutive include Zhong Weiping and Kwok Bit-chee (2018) that investigate the diminutive suffix "儿[ni]" in Eastern Guangdong Min dialect 粤东闽语. There are also studies carried out from the perspective of semantic changes mechanisms and the motivation for grammaticalization such as Li Rulong (2005), Liu Xiuxue, and Shao Huijun.

The account above has provided an overview of research on lexical grammaticalization in Chinese dialects in recent years. In fact, in addition to the parts of speech noted above, the study of dialect grammaticalization also touches on parts of speech such as adjectives, adverbs, and auxiliary verbs, spanning many categories such as aspect, modality, and disposal. However, due to space constraints, these will not be discussed here.

Grammaticalization Research on Construction

As is widely recognized, grammaticalization in a broad sense includes not only the grammaticalization of a single word or morpheme, but also the grammaticalization of constructions composed of specific lexical items or morphemes. Wu Fuxiang (2005) points out that the "grammaticalization of construction" represents "a subject worthy of study." In terms of Sinitic dialects, research development on the "structural grammaticalization of construction" has been relatively slow to emerge but has become an increasing trend since 2010.

Related research in this area includes Wu Fuxiang (2005), who investigates the predicate-complement construction of Cantonese from the perspective of comparative dialectology, and finds that the predicate-complement construction of Cantonese negative words "Neg-V 得OC" and "Neg-V 得CO" structures represent a type of "analogy innovation" (*leitui chuangxin* 类推创新);

Qin Dongsheng (2011) discusses the predicate-complement construction in the Binyang dialect 宾阳话 in Guangxi and the distribution of aspect markers. Qin believes that "VOC" word order is preserved in modern Chinese and the "verb + object + aspect marker" word order is a result of its further development. Ding Chongming and Rong Jing (2013) describe eight special degree expressions in the Kunming dialect including the "V了[ləㄦ] V" and "V+了[ləㄦ]+不得得" and believe that they show differences in their degree of grammaticalization and that the reason for the rich syntactic form of these degree expressions in the Kunming dialect is the lack of lexical means.

The evolution of comparative sentences is also a topic of concern to scholars. Chang Song-hing and Kwok Bit-chee (2005) discuss the mechanism of the alternation of comparative sentence patterns ("过" and "比" phrases) in Hong Kong Cantonese and discuss their evolution process. Wu Fuxiang and Qin Fengyu (2010) describe the short disparity sentence in Nanning Cantonese 南宁粤语 and believe that it is due to the deletion of "Y" in the structural formula of "X + A + 过 + Y" which may have been affected by the Southern Zhuang dialect.

The Combination of Grammaticalization Theory and Other Research Methods

The sections above have described research into Sinitic dialect grammaticalization carried out in recent years in terms of its research stage and content. In the following section, we describe the use of grammaticalization in combination with other theoretical methodologies.

Grammaticalization and Linguistic Typology

The comparative method of typology has broadened the scope of the study of Sinitic dialects, examined the evolution of Sinitic language from the background of human languages, and observed its common and unique features. The main methods of typology include: crosslinguistic, comparative dialect research, regional typology research, and the use of semantic map models.

Scholars employing these methods to study Chinese dialects include Wu Fuxiang, Zhang Min, Chen Qianrui, and Fan Xiaolei. Wu Fuxiang (2010a, b) investigates the distribution of "X+A+过+Y" comparative construction in Cantonese from the perspective of comparative dialectology and regional typology and consideres it to be a significant feature that differentiates Cantonese other dialects. By comparing other southern dialects, southern subnational languages, and other Southeast Asian languages, Wu's paper demonstrates that the universality of "过" in comparative sentences in Southeast Asian languages is the result of regional diffusion caused by language contact, and that Cantonese is the model language and the source of diffusion.

The research method of employing the "semantic map model" (*yuyi ditu moxing* 语义地图模型) was mainly introduced into China by scholars such as Zhang Min and Wu Fuxiang with studies by Li Xiaofan, and Guo Zhang Min, Rui (2015) representing noteworthy achievements. Semantic maps provide the possibility for verifying the construction of the grammaticalization chain and have been widely used in the study of Sinitic dialects (see the entry on Semantic Map Model by Ding Zhang in the present encyclopedia).

Among the more influential studies, Zhang Min (2010) revises the semantic map of the dative functions of Haspelmath (2003) based on the investigation of language/dialect disposal and passive markers including Sinitic dialects. The introduction of this new method has triggered much fresh interest in the research of disposal and passive markers.

Chen Qianrui and Wang Jihong (2010) investigate "有" in front of verbs and adjectives in southern dialects and argue the usage of "asserting a state" (*queren zhuangtai shixian* 确认状态实现 in "有" phrases is a resultative aspect and that this usage in "asserting an event" (*queren Shijian shixing* 确认事件现实性) is a type of perfect aspect developed from the former and draws a semantic map of the "有" sentence. Based on the investigation of modal words in Chinese dialects, Fan Xiaolei (2015) classified the modal expression forms of more than 60 dialect spots from the perspective of form and meaning and drew up a semantic map of possibility modality in Sinitic dialects which caused great repercussions.

With the deepening of typological research, crosslingual and cross-dialect comparative research and grammaticalization have become more closely integrated with in-depth explorations of syntactic types, word order types, unidirectional implication, and common rules. Liu Danqing (2001) studies the materials from 12 dialect spots in the Wu dialect and compares their expression of different sentence patterns with the Beijing dialect concluding that "The Wu Dialect is a less typical SVO type and more typical topic priority type than the Beijing dialect." In contrast, postpositions in the Wu dialect are more developed with a lighter level of grammaticalization. Ding Jiayong (2003) investigates the sentence end modal particle *zhe* 着 in 22 Sinitic dialect spots. From the perspective of typology, Ding summarizes the semantics of *zhe* 着 at the end of sentences as modal and aspectual and demonstrates the evolution from modal particle toward aspectual particle. Fang Qingming and Sun Liping (2011) compare the predicate-complement construction with objectives in the northern dialect and point out that the northern dialect includes at least three word order types of predicate-complement constructions with objectives ("VC了/V不C," "V了(liao)了/V不了(liao)," and "V得了/V不得") and discusses their level of grammaticalization and their modes of evolution. Sheng Yimin (2017) uses samples drawn from 37 Sinitic dialects to distinguish between quasi-articles and quasi-demonstrative words that are used independently, modified, and restricted by different syntactic positions. After investigating other aspects, Sheng summarizes six commonalities in the structure of dialect definite referent quantity name. Other related studies include Wang Guosheng and Li Zhao on the question of "VP-Neg" in Sinitic dialects and Jiang Xiezhong (2018) on the phenomenon of overlap between dialect classifiers in various dialects.

In summary, the typological method has been relatively common in the study of dialect grammar – a fact that makes research into Sinitic dialect grammaticalization not only unique, but a field that shares many commonalities.

Grammaticalization and Language Contact

Dialects and minority languages in China have coexisted over a long period and there has been a long history of contact with non-Chinese languages such as Altaic and Southeast Asian languages. The influence of language contact on grammaticalization is another topic of note in the study of Sinitic dialects (see the entries on Contact-Induced Change in the Languages of Southern China by Fuxiang Wu and Yang Huang and the entry on the Evolution of Chinese Grammar from the Perspective of Language Contact by Yonglong Yang and Jingting Zhang in the present encyclopedia). Many scholars such as Wu Fuxiang, Kwok Bit-chee, Lin Huayong, Qin Fengyu, Qin Dongsheng, Yang Yonglong, and Zhang Ansheng have explored this aspect. Qin Fengyu and Wu Fuxiang (2009) discuss two special usages composed of "过[kɔ^{33}]" in Nanning Cantonese, and point out that the preposition "过 [kɔ^{33}]" derives from the lexicalized giving verb "给过[kɐi^{55}kɔ^{33}]." They argue that the evolution of the short-difference formula "X+A+过" is very likely to have been influenced by the Zhuang language. Other related research on contact with the

Zhuang-Dong language 壮侗语 includes Kwok Bit-chee's (2012) study of Nanning Cantonese adverbial words.

Qin Dongsheng and Qin Fengyu (2015) have studied "去" in Guangxi Cantonese 广西粤语. Lin Huayong and Li Yalun (2014) discuss the multifunctionality of "头先" and "正" in Lianjiang Cantonese. They compared early and modern Hakka and Cantonese materials respectively and argue that its multifunctionality comes from the "replication grammaticalization" of guest dialects and pay attention to the grammaticalization phenomenon caused by contact between dialects. Both "头先" and "正" are used a "temporary particle of first" as per Example (8).

(8) Lianjiang Yue Dialect 廉江粤方言

等	阿哥	回来	正!
tʰeŋ³³	a³³kɔ⁵⁵	hui²¹lɔi²¹	tseŋ³³
wait	elder brother	back.come	TPF
"wait your brother come back first."			

Other studies on grammatical evolution caused by interdialect contact also include Wang Chunling (2017). In addition, the contact between the northwest dialect and the Altai language has also been a prominent topic in recent years. Yang Yonglong (2014) discusses the origin of the multifunctional marker "哈" in Minhe Gangou dialect 民和干沟话 in Qinghai.

The question of language contact–induced grammaticalization is a complex one. Researchers often do not use a single method alone, but need to combine grammaticalization with semantic map model, crosslinguistic/dialect comparison, or other methodologies and use them flexibly and with great care. Kwok Bit-chee and Lin Huayong (2012) make a detailed description of the multifunction of the postposition component "倒[tou35]" in Lianjiang Cantonese from the perspective of language contact. On this basis, they combine historical documents, surrounding dialects including Cantonese and Hakka and consider that an aspect of its multifunctional reduplication derives from the neighboring Hakka dialects. This paper provides an example of how to differentiate "evolution caused by contact" and explains the mechanism of "replication."

This research method of combining methodologies is often used in areas with frequent language contact. The semantic map model can make the evolutionary relationships in different language/dialect systems more intuitive.

In addition, the reconstruction of the grammaticalization path cannot be inseparated from a full and accurate description. However, current discussion on how best to describe dialect grammar is insufficient. Lin Huayong and Li Minying (2019) take the Cantonese sentence end modal particle "佢" as an example and advocate using "minimal pairs" (*zuixiao chabi* 最小差比) to observe and test its syntactic meaning. By combining evidence from synchronic dialects and historical materials while reconstructing the grammaticalization path of "佢," this paper represents a helpful attempt on how best to test the description of dialect grammar.

Research on Grammaticalization, Subjectivity, and Subjectivization

The notion of subjectivity (*zhuguanxing* 主观性) and subjectivization (*zhuguanhua* 主观化) was first introduced to China by Shen Jiaxuan (2001) and is currently mainly applied to the study of modern Standard Mandarin with few examples of its application to Sinitic dialects. By combining subjectivization with grammaticalization research, Fang Mei (2007) points out that the evolution of the morphosyntactic function of Beijing dialect's erization includes the process of subjectivization. Biq Yung-O (2007) believes that the distant referents "那" and "在那边" in colloquial Taiwanese is due to the semantic extension of subjectivization with a "shift from referring to the essence of things to expressing the speaker's position on the things being discussed."

Shan Yunming (2009) studies the evolution of the verb "够" in the Guangzhou dialect and argues that its basic meaning is to express the satisfaction of objective quantity. When it develops into "达到" it exhibits subjectivity and becomes a verb to compare different items, such as Example (9). On the other hand, to a certain extent bleaching can also express similarity or a similar nature, which reduces the background information, such as Example (10).

(9) Guangzhou Yue Dialect 广州粤方言

a.	呢	条	绳	够	长	喇。
	nei^{55}	tʰiu^{22}	seŋ^{35}	kou^{33}	tʃʰɐŋ^{22}	la^{33}
	DEMprox	CLF	rope	enough	long	-SFP
"This rope is long enough."						

b.	你	唔 够	佢	(咁)	爱	我。
	nei^{24}	m^{21} kou^{33}	kʰɐi^{24}	kɐm^{24}	ɔi^{33}	ŋɔ^{24}
	2SG	Neg-like	3SG	such	love	me
"You don't love me like he does."						

(10) a.	咁	多	餸,	够	食	喇。
	kɐm^{24}	tɔ^{55}	soŋ^{33},	kou^{33}	sɪk^{1}	la^{33}
	such	mang	food	enough	eat	-SFP
"There is enough food (for us) to eat."						

b.	你	食	鲍鱼,	我	够	食	咯。
	nei^{24}	sɪk^{1}	pau^{55} y^{22},	ŋɔ^{24}	kou^{33}	sɪk^{1}	lɔk^{3}.
	2SG	eat	abalone	1SG	also	eat	-SFP
"Eating abalone is not a big deal, I can afford it too."							

Example (10b) means that "eating abalone" is not a big deal, and that "I" can afford it. Similar studies related to Cantonese include Leung Wai-mun's (2016) discussion on the Cantonese sentence end modal particle "啫." Liu Yongjing

(2016: 51) researches the phrase "又咱(又咋)" during the Ming and Qing in the Shandong dialect and argues that "the subjectivization of '又咋' is an extension of the semantic pragmatic environment" and also discusses the mechanism by which it emerged. Liu Danqing (2019) demonstrates the innovative grammaticalization path of sole complement function words (words which can only be complements) to create the scalar of objectives salient in the word "到[tɒ44/55]" in the Wujiang Tongli dialect 吴江同里话 from the perspective of tonal characteristics, semantic characteristics, and syntactic distribution.

Grammaticalization Research on Pragmatics and Discourse

Although Chinese linguists such as Liu Danqing and Tang Zhengda (2001) and Fang Mei (2002) took notice of the grammaticalization of Chinese pragmatics comparatively early, grammatical studies related to pragmatics and discourse have gradually become an area of attention since 2010. Based on a large number of Beijing spoken language materials, Li Xiaojun (2008) discusses discourse markers commonly used in Beijing spoken language from the perspective of grammaticalization and sociolinguistics. Tao Huan and Li Jialiang (2009) apply traditional rhetoric to the study of "伊讲" in the Shanghai dialect and believe that its grammaticalization process has experienced two stages: "afterthought forms" (*shihou zhuibuyu* 事后追补语) and "sentence final modal component" (*jumo yuqi chengfen* 句末语气成分), which are related to the "cooperative principle" (*hezuo yuanze* 合作原则) in the communication process.

Zhang Wei and Gao Hua (2012) also discuss the discourse marker "即係[zikhai]" in Cantonese natural conversation. Zhou Minli and Li Xiaojun (2018) explore the origin and function of the emphatic marker "别个" in Xinshao (Cunshi) dialect 新邵(寸石)话 in Henan.

Xing Xiangdong (2011) provides a detailed description and comparison of the phenomenon of "来" and "去" as topic markers in the Shenmu dialect 神木话 in Northern Shaanxi and argues that they derive from the grammaticalization of sentence final function words and also discussed its grammaticalization mechanism in combination with the surrounding words. Chen Shanqing (2015) describes the function of the "硬" topic marker in the Miluo dialect in Hunan province and outlined the origins of its grammaticalization. Guo Lixia (2017) believes that there are three sources of the topic markers in the Shanyin dialect 山阴方言 of the Jin dialect (modal particle, copula, and adverb). Based on this analysis of the syntactic distribution of these three topic markers, the paper also discusses the degree of their grammaticalization degree; Shi Xiuju and Hou Panjie (2018) also make a comparative analysis of topic markers in the Wenshui dialect 文水方言 in Shanxi. Wang Fang (2019) describes the semantic, grammatical, and pragmatic characteristics of the topic marker "不咋" in the Anyang dialect 安阳方言 in Henan.

From the perspective of pragmatics grammaticalization, another example is found in the work of Huang Lihe (2010) who compares subjunctive sentences in the Shanghai dialect with English and discusses the pragmatics motivation of the grammaticalization of subjunctive sentences.

Theoretical Research on Grammaticalization Mechanism and Grammaticalization Patterns

Grammaticalization mechanism and grammaticalization patterns are the most discussed aspects of grammaticalization theory in China. Current theoretical discussions, however, are based more on Standard Chinese and historical materials than dialects. The research on the grammaticalization theory of dialects is still in a stage of employing relevant theories to explain their phenomena and refine the details of their evolution. Zhang Ding (2009) investigates the syntactic expression and pragmatics limitations of the two echoic negative markers "不晓好" and "一大堆" in the Zongyang dialect 枞阳方言 and considered that the fundamental pragmatics motivation for their grammaticalization into echo negative markers is "verbal irony in echo utterances." Zhang's study also describes the "analogy" (*leitui* 类推) mechanism in their evolution step by step. Liu Chunhui (2009) and Wang Huayun (2017) also study this aspect. However, from a theoretical point of view, research on using dialect materials as supplementary evidence is still relatively limited.

On the other hand, from the perspective of dialect facts, scholars have put forward several evolutionary models in line with Chinese characteristics and advance several notions of grammaticalization theory. Wu Fuxiang (2005) points out that in Chinese, "a grammatical word or attached word does not further evolve into a inflectional affix," but usually integrates morphemes and adjacent components into a new lexical item. Zhao Rixin (2013) discusses the collocation function of "做" in both written and dialect materials and argues that its grammaticalization process is as follows: "verb → preposition → adverb → word forming morpheme." This syntactic operation is closely related to the semantic generalization of "做" and the focus shift in syntactic structure. Other studies also include Wang Huayun and Xiao Qingzhu's (2018) investigation on the origin and distribution of "叫莫" in the Huangxiao dialect 黄孝方言. In addition, the evolutionary mechanism with "Chinese characteristics" can also be seen in the evolutionary model of the diminutive. Cao Fengfu (2006) proposes the concept of "grammaticalization cycle" noted above in Example (6). Lin Huayong and Ma Zhe (2008) observe the evolution of the diminutive "仔" and believe that as its degree semantic bleaching deepens, "the affixing form develops from a small unit to a larger unit" which is not completely consistent with the "grammaticalization cline" model proposed by Givón.

The studies noted above are all attempts by Chinese scholars to further the field of grammaticalization theory as it relates to dialects. In general, however, theoretical research remains inadequate and it is one direction that deserves to be further developed in the future.

The Current Situation and Prospects for Development in Grammaticalization Research

From our current perspective, the study of dialect grammaticalization has gradually entered a "mature period." The features of this stage are as follows: First, the selection of dialects is richer and covers multiple dialect areas, dialect spots, and subdialect spots; second, the research objects are more diverse and, with the exception of morpheme, also pay attention to the grammaticalization of construction and pragmatics; third, a more detailed description of grammar, including multiple linked demonstrations of phonetics, semantics, and syntactic distribution; and fourth, there is a deeper level of theoretical discussion that involves notions such as the mechanism and motivation of grammaticalization.

There remain, however, several deficiencies in the study of the grammaticalization of dialects that deserve further attention in the future. First, there is a need for more discussion of single grammatical phenomenon and less on entire grammatical systems and categories; second, outside of the Beijing dialect, most findings have been in research into southern dialects dominated by Wu, Cantonese, and Min with relatively few studies on central and northern dialects; third, most studies deal with solitary examples and few are devoted to observing the details of evolution by making a comprehensive and systematic comparison between internal and adjacent dialects; and fourth, most studies deal with "hot issues" and few are devoted to systematic and more encompassing research.

Since being introduced into China almost 30 years ago, grammaticalization theory has become an important interface that links diachronic and synchronic aspects of dialect grammar. As China's language resources are particularly rich, synchronic evolution is bound to have richer details than diachronic evolution. The adept and rational use of these language resources will certainly benefit grammaticalization theory so that the study of the grammar of Sinitic dialects will not only reflect the commonalities of language evolution, but also enable us to clearly see their differences.

References

Ben, Liao (本了). 1983. Guiyang fangyan "ba" de shici xuhua 贵阳方言"把"的实词虚化 [The grammaticalization of "ba" in the Guiyang dialect]. *Guizhou minzu xueyuan xuebao* 贵州民族学院学报 *[Journal of Guizhou Minzu University]* 00: 97–99.

Biq, Yung-O (毕永峨). 2007. Yuanzhici "na" cichuan zai Taiwan kouyu zhong de cihuihua yu xiyuhua 远指词"那"词串在台湾口语中的词汇化与习语化 [Lexicalization and phrasalization of "na" collocates in spoken Taiwan Mandarin]. *Dangdai yuyanxu* 当代语言学 *[Contemporary Linguistics]* 2: 128–138.

Cao, Fengfu (曹逢甫). 2006. Yufahua lunhui de yanjiu 语法化轮回的研究——以汉语鼻音尾/鼻化小称词为例 [On the grammaticalization cycle]. *Hanyu xuebao* 汉语学报 *[Chinese Linguistics]* 2: 2–15.

Chang, Song-hing (张双庆), and Kwok, Bit-chee (郭必之). 2005. Xianggang yueyu Liangzhong chabiju de jiaoti 香港粤语两种差比句的交替 [Two types of comparative construction in Hong

Kong Cantonese: Competition and replacement]. *Zhongguo yuwen*中国语文 *[Studies of the Chinese Language]* 3: 232–238.

Chen, Qianrui (陈前瑞), and Wang, Jihong (王继红). 2010. Nanfang fangyan "you" ziju de duogongnengxing fenxi 南方方言"有"字句的多功能性分析 [On the multifunctionality of "you" in Southern Chinese dialects]. *Yuyan jiaoxue yu yanjiu* 语言教学与研究 *[Language Teaching and Linguistic Studies]* 4: 47–55.

Chen, Qianrui (陈前瑞), and Wu, Jizhang (吴继章). 2019. Cong fanyan yuyin kan "le" de gongneng yanhua 从方言语音看"了"的功能演化 [The functional evolution of particle "le"]. *Hanyu xuebao*汉语学报 *[Chinese Linguistics]* 2: 48–60.

Chen, Shanqing (陈山青). 2015. Hunan Miluo fangyan de huati biaoji zhuci "ying" jiqi yufahua laiyuan 湖南汨罗方言的话题标记助词"硬"及其语法化来源 [A study on the topic marker auxiliary "ying" in the Miluo dialect of Hunan Province of China and the origin of its grammaticalization]. *Zhongguo yuwen* 中国语文 *[Studies of the Chinese Language]* 1: 64–66.

Chen, Weirong (陈伟蓉). 2017. Fujian Hui'an fangyan de guanxi congju biaoji 福建惠安闽南方言的关系从句标记 [Relativization marker in Hui'an Southern-Min dialect, Fujian Province]. *Fangyan* 方言 *[Dialects]* 3: 352–359.

Dai, Zhaoming (戴昭铭). 2004. Ruohua, cuhua, xuhua he yufahua—Wufangyan zhong yizhong zhongyao de yanbian xianxiang 弱化、促化、虚化和语法化——吴方言中一种重要的演变现象 [On reduction, glottalization, semantic shift, and grammaticalization – An important evolution in the Wu dialect]. *Hanyu xuebao*汉语学报 *[Chinese Linguistics]* 2: 26–34.

Ding, Chongming (丁崇明), and Rong, Jing (荣晶). 2013. Kunming fangyan zhong de teshu Chengdu biaoda xingshi 昆明方言中的特殊程度表达形式 [On the special degree expressions in Kunming dialect in Yunnan Province]. *Zhongguo fangyan xuebao* 中国方言学报 *[Journal of Chinese Dialects]* 00: 124–130.

Ding, Jiayong (丁加勇). 2003. Hanyu fangyan jumo "zhe" de leixing kaocha 汉语方言句末"着"的类型学考察 [A typological study of the end of the sentence "zhe" in Chinese dialects]. *Changde shifan xueyuan xuebao* 常德师范学院学报*[Journal of Changde Teachers University]* 01: 100–103.

Fan, Xiaolei (范晓蕾). 2015. "Hanyu fangyan de nengxing qingtai yuyi ditu" zhi bulun "汉语方言的能性情态语义地图"之补论 [A supplement to "semantic maps of modality based on sinitic dialects"]. In *Hanyu duogongneng yufaxingshi de yuyi ditu yanjiu* 汉语多功能语法形式的语义地图研究 *[A study on semantic map of multifunctional grammatical forms in Chinese]*, ed. Li, Xiaofan 李小凡, Guo, Rui 郭锐, Zhang, Min 张敏, et al. Beijing: Shangwu chubanshe 北京: 商务出版社 [Beijing: The Commercial Press] 2015.

Fang, Mei (方梅). 2002. Zhishici "zhe" he "na" zai Beijinghua zhong de yufahua 指示词"这"和"那"在北京话中的语法化 [On the Grammaticalization of "zhe" in Beijing Mandarin: From the demonstrative to the definite article]. *Zhongguo yuwen* 中国语文 *[Studies of the Chinese Language]* 4: 343–356.

———. 2006. Beijinghua li "shuo" de yufahua——Cong yanshuo dongci dao congju biaoji 北京话里"说"的语法化——从言说动词到从句标记 [Grammaticalization of "shuo" in Beijing Mandarin: From lexical verb to subordinator]. *Zhongguo fangyan xuebao*中国方言学报 *[Journal of Chinese Dialects]* 1: 107–121.

———. 2007. Beijinghua erhua de xingtai jufa gongneng 北京话儿化的形态句法功能 [The Morpho-syntax of the non-syllabic "er" in the Beijing dialect]. *Shijie Hanyu jiaoxue* 世界汉语教学 *[Chinese Teaching in the World]* 2: 5–13.

Fang, Qingming (方清明), and Sun, Liping (孙利萍). 2011. Beifang fangyan nengxing shubu jiegou dai binyu de yuxu leixing kaocha—yibufen fangyandian wei li 北方方言能性述补结构带宾语的语序类型考察——以部分方言点为例 [An investigation into word order types with objects in the predicate-compliment Structre of Nnoerthern dialects—Using dialect spots as examples]. *Ningxia Daxue xuebao* 宁夏大学学报 *[Journal of Ningxia University]* 1: 45–48.

Guo, Lixia (郭利霞). 2017. Shanyin fangyan de youbiaoji huatiju 山阴方言的有标记话题句 [On the marked topics in the Shanyin dialect]. *Yuyan yanjiu jikan* 语言研究集刊 *[Bulletin of Linguistic Studies]* 1: 205–218.

Han, Weifeng, and Dingxu Shi. 2014. The evolution of *ɦi23 kã34* ('he says') in Shanghainese. *Language & Linguistics* 15 (4): 479–494.

Huang, Lihe (黄立鹤). 2010. Shanghai fangyan yu Yingyu xuniju zhi duibi kaocha 上海方言与英语虚拟句之对比考察——语用法的语法化视角 [A comparative study of the Shanghai dialect and English subjunctive sentences – Grammaticalization of language Useage]. *Huaxi yuwen xuekan* 华西语文学刊 *[Acta Linguistica et Litteraturaria Sinica Occidentalia]* 1: 50–56.

Huang, Xiaoxue (黄晓雪). 2006. Anhui Susong fangyan de shitai zhuci "zai" 安徽宿松方言的事态助词"在" [The stative particle "zai" in the Susong dialect of Anhui]. *Changjiang xueshu* 长江学术 *[Yangtze River Academic]* 3: 85–88.

Huang, Yanxuan (黄燕旋). 2016. Jieyang fangyan yanshuo dongci "da" de yufahua 揭阳方言言说动词"呾"的语法化 [Grammaticalization of the SAY verb "da" in the Jieyang dialect]. *Zhongguo yuwen* 中国语文 *[Studies of the Chinese Language]* 6: 686–694.

Jiang, Lansheng (江蓝生). 1995. Shuo "me" yu "men" de tongyuan 说"麼"与"们"的同源 [On the common origin of "me" and "men"]. *Zhongguo yuwen* 中国语文 *[Studies of the Chinese Language]* 3: 180–190.

———. 2012. Hanyu lian-jieci de laiyuan jiqi yufahua de lujing he leixing 汉语连-介词的来源及其语法化的路径和类型 [Origins, pathways, and types of grammaticalization of Chinese conjunction-prepositions]. *Zhongguo yuwen* 中国语文 *[Studies of the Chinese Language]* 2012 (4): 291–308.

Jiang, Xiezhong (蒋协众). 2018. Hanyu fangyan liangci chongdie de leixingxue kaocha 汉语方言量词重叠的类型学考察 [Typological study of classifier reduplication in sinitic dialects]. *Nankai yuyan xuekan* 南开语言学刊 *[Nankai Linguistics]* 2018 (1): 107–117.

Jin, Xiaodong (金小栋), and Wu, Fuxiang (吴福祥). 2016. Hanyu fangyan duogongneng xuxi "lian" de yuyi yanbian 汉语方言多功能虚词"连"的语义演变 [The semantic change of multi-functional word "lian" in sinitic dialects]. *Fangyan* 方言 *[Dialects]* 4: 385–400.

Kwok, Bit-Chee (郭必之). 2008. Shaowuhua dongtai zhuci "du" de laiyuan 邵武话动态助词"度"的来源——兼论邵武话和闽语的关系 [The origin of the verbal particle "du" in the Shaowu dialect]. *Zhongguo yuwen* 中国语文 *[Studies of the Chinese Language]* 2: 140–146.

———. 2012. Cong Nanning Yueyu de zhuangmaoci kan Hanyu fangyan yu minzu yuyan de jiechu 从南宁粤语的状貌词看汉语方言与民族语言的接触 [Language contact between Chinese dialects and minority languages: A case study of Ideophones in the Yue dialect spoken in Nanning]. *Minzu yuwen* 民族语文 *[Minority Languages of China]* 3: 16–24.

Kwok, Bit-Chee (郭必之), and Lin, Huayong (林華勇). 2012. Lianjiang yueyu dongci houzhi chengfen "dǎo" de laiyuan he fazhan——cong yuyan jiechu de jiaodu wei qierudian 廉江粵語動詞後置成分「倒」的來源和發展——從語言接觸的角度為切入點 [The origin and development of the post-verbal "dǎo" in Lianjiang Yue: A language contact perspective]. *Yuyan ji yuyanxue* 語言暨語言學 *[Language & linguistics]* 13 (2): 289–320.

Kwok, Bit-Chee (郭必之), and Shin Kataoka (片冈新). 2006. Zaoqi Guangzhouhua wanchengti biaoji "hiu" de laiyuan he yanbian 早期广州话完成体标记"晓"的来源和演变 [The origin and development of the perfective aspect marker "hiu" in early Cantonese]. *Zhongguo Wenhua Yangjiusuo xuebao* 中国文化研究所学报 *[Journal of Chinese Studies]* 2006 (46): 91–116.

Lei, Rong (雷容). 2017. Hanyu xiaocheng de yuyi yanbian jizhi 汉语小称的语义演变机制 [The mechanism of the semantic evolution of the Chinese diminutive]. *Hanyu xuebao* 汉语学报 *[Chinese Linguistics]* 2: 89–94.

Leung, Wai-mun (梁慧敏). 2016. Lun Yueyu jumo zhuci "tsɛ⁵⁵" de zhuguanxing 论粤语句末助词"啫"的主观性 [On the subjectivity of the sentence-final particle "$tsɛ^{55}$" in Cantonese]. *Zhongguo yuwen* 中国语文 *[Studies of the Chinese Language]* 3: 339–348.

Li, Rulong (李如龙). 2005. Minyu de "jian" jiqi yufahua 闽语的"囝"及其语法化 [On "jian" in the Min dialects and its grammaticalization]. *Nankai yuyan xuekan* 南开语言学刊 *[Nankai Linguistics]* 2: 1–8.

Li, Xiaofan (李小凡), Guo, Rui (郭锐), Zhang, Min (张敏), et al. 2015. *Hanyu duogongneng yufaxingshi de yuyi ditu yanjiu* 汉语多功能语法形式的语义地图研究 *[A study on semantic map of multifunctional grammatical forms in Chinese]*. Beijing: Shangwu chubanshe 商务出版社 [The Commercial Press].

Li, Xiaojun (李小军). 2008. Yuqici "sha" de laiyuan jiqi fangyan bianti 语气词"吵"的来源及其方言变体 [The generation of modal particles "sha" and its dialectal variants]. *Yuyan kexue* 语言科学 *[Linguistic Sciences]* 4: 398–405.

Li, Yuming (李宇明), and Chen, Qianrui (陈前瑞). 2005. Beijinghua "gei" zi beidongju de diwei jiqi lishi fazhan 北京话"给"字被动句的地位及其历史发展 [The status and development of passive "gei" construction in Beijing dialect]. *Fangyan*方言 *[Dialects]* 4: 289–297.

Li, Yunhan (黎运汉). 1981. Hanyu xuci yanbian de qushi chutan 汉语虚词演变的趋势初探 [A preliminary investigation into the evolutionary trends of Sinitic function words]. *Jinan Daxue xuebao* 暨南大学学报 *[Journal of Jinan University]* 4: 103–111.

Lien, Chinfa. 2007. Grammaticalization of Pat[4] in Southern Min: A cognitive perspective. *Language & Linguistics* 3: 723–742.

Lin, Huayong (林华勇), and Kwok, Bit-Chee (郭必之). 2010. Lianjiang Yueyu "lai/qu" de yufahua yu gongneng qujin xianxiang 廉江粤语"来/去"的语法化与功能趋近现象 [The grammaticalization of the directional verbs "qu" and "lai" and the convergence of their functions in the Lianjiang Yue dialect]. *Zhongguo yuwen* 中国语文 *[Studies of the Chinese Language]* 6: 516–525.

Lin Huayong(林华勇),and Li Minying(李敏盈). 2019. Cong Lianjiang fangyan kan Yueyu "qu"zi chuzhiju从廉江方言看粤语"佢"字处置句[The Cantonese disposal construction with "kheoi": Evident from the Lianjiang dialect].Zhongguo Yuwen中国语文[Studies of the Chinese Language]1:89–101.

Lin, Huayong (林华勇), and Li, Yalun (李雅伦). 2014. Lianjiang Yueyu "touxian" he "zheng" duogongnengxing de laiyuan 廉江粤语"头先"和"正"多功能性的来源 [The origins of multifunctional "touxian" and "zheng" in the Lianjiang Yue dialect]. *Zhongguo yuwen*中国语文 *[Studies of the Chinese Language]* 4: 317–325.

Lin, Huayong (林华勇), and Ma, Zhe (马喆). 2008. Guangdong Lianjiang fangyan "zi" yi yusu yu xiaocheng wenti 广东廉江方言的"子"义语素与小称问题 [The meaning morphemes "zi" and diminutive form in Lianjiang dialect of Yue Group]. *Yuyan kexue* 语言科学 *[Linguistic Sciences]* 6: 626–635.

Lin, Su'e (林素娥). 2015. Yibaiduo nianqian Ningbohua lian-jieci "deng" de yongfa jiqi laiyuan一百多年前宁波话连-介词"等"的用法及其来源 [A study of the function and origin of the conjunction and preposition "deng" in the Ningpo dialect more than 100 years ago]. *Yuyan kexue*语言科学 *[Linguistic Sciences]* 4: 417–428.

Liu, Chunhui (刘春卉). 2009. Henan Queshan fangyan zhong "gei" de yufahua jizhi kaocha 河南确山方言中"给"的语法化机制考察 [The grammaticalization of "gei" in the Henan Quanshan dialect]. *Yuyan kexue* 语言科学 *[Linguistic Sciences]* 1: 68–75.

Liu, Cuixiang (刘翠香). 2016. Shandong Qixia fangyan laiyuan yu chusuo jiegou de yuqici 山东栖霞方言来源于处所结构的语气词 [Modal particle originated from locative constructions in Qixia dialect in Shandong Province]. *Fangyan*方言 *[Dialects]* 3: 335–343.

Liu, Danqing (刘丹青). 2001. Wuyu de jufa leixing tedian 吴语的句法类型特点 [Two major typological features of Wu dialects]. *Fangyan*方言 *[Dialects]* 4: 332–343.

———. 2009. Yufahua lilun yu Hanyu fangyan yufa yanjiu 语法化理论与汉语方言语法研究 [Grammaticalization theory and grammatical research of Chinese dialect]. *Fangyan*方言 *[Dialects]* 2: 106–116.

———. 2019. "Dao" zi yufahua de xinquxiang: Wujiang Tonglihua de tiaojian biaoji ji zhuguan daliang biaoji "到"字语法化的新去向: 吴江同里话的条件标记及主观大量标记"到" [New directions in the Grammaticallization of "dao": The conditional marker and subjective quantity marker in the Tongli dialect of Wujiang]. *Yuwen yanjiu*语文研究 *[Linguistic Research]* 2: 1–7.

Liu, Danqing (刘丹青), and Tang, Zhengda (唐正大). 2001. Huati jiaodian mingan suanzi "ke" de yanjiu 话题焦点敏感算子"可"的研究 [A study on the adverb "kě" as a topical focus sensitive operator]. *Shijie Hanyu jiaoxue* 世界汉语教学 *[Chinese Teaching in the World]* 3: 25–33.

Liu, Jian (刘坚), Cao, Guangshun (曹广顺), and Wu, Fuxiang (吴福祥). 1995. Lun youfa Hanyu cihuihua de ruogan yinsu 论诱发汉语词汇语法化的若干因素 [On certain factors that trigger grammaticalization in sinitic lexical items], *Zhongguo yuwen* 中国语文 *[Studies of the Chinese Language]* 3:161–169.

Liu, Yongjing (刘永静). 2016. Zhuguanhua de kuozhan xiaoying——yi Ming-Qing Shandong fangyanci "youza" wei lie 主观化的扩展效应——以明清山东方言词"又咱(又咋)"为例 [The expansion effect of Subjectivisation: Using the subjective "youza" in the Shandong dialect during the Ming and Qing dynasty as an example]. *Guhanyu yanji* 古汉语研究 *[Research in Ancient Chinese Language]* 2: 51–57.

Lu, Xiaoyu (卢笑予). 2013. Linhai fangyan feiweiyu qianzhici de yufa duogongnengxing fenxi 临海方言非谓语前置词的语法多功能性分析 [An analysis of the multi-functionality of non-predicate prepositions in the Linhai dialect]. *Xiandai yuwen* 现代语文 *[Modern Chinese]* 5: 72–77.

Luo, Xinru (罗昕如). 2007. Xiangyu "[ti]" de duogongneng yongfa 湘语"滴"的多功能用法 [On the functions of "[ti]" in the Hunan dialect]. *Hanyu xuebao* 汉语学报 *[Chinese Linguistics]* 3: 9–15.

Luo, Ziqun (罗自群). 1999. Xiandai Hanyu fangyan "VP+(O)+zaili/zai/li" geshi de bijiao yanjiu 现代汉语方言"VP+(O)+在里/在/哩"格式的比较研究 [A Comparative Study of the "VP + (O) +zaili/zai/li" Structure in Modern Sinitic Dialects]. *Yuyan yanjiu* 语言研究 *[Studies in Language and Linguistics]* 2: 51–61.

Peng, Xiaochuan (彭小川). 2006. Guangzhouhua han fushuliang yiyi de jiegou zhuci "[ti]" 广州话含复数量意义的结构助词"啲" [Notes on the plural structural particle [ti] in Cantonese]. *Fangyan* 方言 *[Dialects]* 2: 112–118.

Qian, Nairong (钱乃荣). 1998. Wuyu zhong de "ge" he "jie" 吴语中的"个"和"介" [On "ge" and "jie" in the Wu dialect]. *Yuyan yanjiu* 语言研究 *[Studies in Language and Linguistics]* 2: 78–89.

Qin, Dongsheng (覃东生). 2011. Binyanghua de shubu jiegou he tibiaoji 宾阳话的述补结构和体标记 [The verb-compliment structure and aspect markers in the Binyang dialect]. *Baise Xueyuan xuebao* 百色学院学报 *[Journal of Baise University]* 2: 68–72.

Qin, Dongsheng (覃东生), and Qin, Fengyu (覃凤余). 2015. Guangxi Hanyu "qu" he Zhuangyu fangyan "pai¹" de liangzhong teshu yongfa——quyu yuyanxue shijiaoxia de kaocha 广西汉语"去"和壮语方言pai¹的两种特殊用法——区域语言学视角下的考察 [Two particular Useages of "qu" in Guangzi Mandarin and "pai¹" in the Zhuang lanuguage: A survey from the perspective of regional linguistics]. *Minzu yuwen* 民族语文 *[Minority Languages of China]* 2: 68–75.

Qin, Fengyu (覃凤余), and Wu, Fuxiang (吴福祥). 2009. Nanning baihua "guo" de liangzhong teshu yongfa 南宁白话"过"的两种特殊用法 [Two special usages of "kuo" in Nanning Cantonese]. *Minzu yuwen* 民族语文 *[Minority Languages of China]* 3: 16–29.

Rong, Jing (荣晶), and Ding, Chongming (丁崇明). 2004. Kunminghua de "zhe" zi jiqi yufahua guocheng zhong de lishi zeyi yu gongshi zhiheng wenti 昆明话的"着"字及其语法化过程中的历时择一与共时制衡问题 [The character "zhe" in the Kunming dialect and issues concerning diachronic specialization and synchronic balance in the grammaticalization process of Chinese]. *Zhongguo yuwen* 中国语文 *[Studies of the Chinese Language]* 3: 247–252.

Shan, Yunming (单韵鸣). 2009. Guangzhouhua dongci "gou" de yufahua he zhuguanhua 广州话动词"够"的语法化和主观化 [The grammaticalization and subjectification of the verb "gou" in Cantonese]. *Yuyan kexue* 语言科学 *[Linguistic Sciences]* 6: 633–640.

Shen, Jiaxuan (沈家煊). 1994. Yufahua yanjiu zongguan "语法化"研究综观 [A survey of studies on grammaticalization]. *Waiyu jiaoxue yu yanjiu* 外语教学与研究 *[Foreign Language Teaching and Research]* 4: 17–24.

———. 1998a. Shici xuhua de jizhi——"Yanhua erlai de yufa" pingjie 实词虚化的机制——《演化而来的语法》评介 [On the mechanism of the bleaching of content words: A review of "The evolution of grammar"], *Dangdai yuyanxu* 当代语言学 [*Contemporary Linguistics*] 3: 41–46.

———. 1998b. Yuyongfa de yufahua 语用法的语法化 [The grammaticalization of pragmatics]. *Fujian waiyu* 福建外语 [*Foreign Languages in Fujian*] 2: 1–8.

———. 2001. Yuyan de "zhuguanxing" he "zhuguanhua" 语言的"主观性"和"主观化" [A survey of studies on subjectivity and subjectivisation]. *Waiyu jiaoxue yu yanjiu* 外语教学与研究 [*Foreign Language Teaching and Research*] 4: 268–275.

Sheng, Yimin (盛益民). 2010. Shaoxing Keqiaohua duogongneng xuci "[tsoʔ₂]" de yuyi yanbian 绍兴柯桥话多功能虚词"作"的语义演变——兼论太湖片吴语受益者标记来源的三种类型 [The semantic change of multifunction function word "[tsoʔ₂]" in the Shaoxing Keqiao dialect]. *Yuyan kexue* 语言科学 [*Linguistic Sciences*] 2: 197–207.

———. 2017. Hanyu fangyan dingzhi "liangming" jiegou de leixing chayi yu gongxing biaoxian 汉语方言定指"量名"结构的类型差异与共性表现 [The definite classifier-noun constructions in Chinese dialects: Universality and diversity]. *Dangdai yuyanxu* 当代语言学 [*Contemporary Linguistics*] 2: 181–206.

Shi, Qisheng (施其生). 1990. Shantou fangyan de jiegou zhuci "dan" 汕头方言的结构助词"呾" [The structural particle "dan" in the Shantou dialect]. In *Yuyan wenzi lunji* 语言文字论集. Guangzhou: Guangdong renmin chubanshe 广州：广东人民出版社 [Guangzhou: Guangdong People's Publishing House], pp. 166–179.

Shi, Xiuju (史秀菊), and Hou, Panjie (侯盼洁). 2018. Shanxi Wenshui fangyan de huati biaoji 山西文水方言的话题标记 [Topic markers in the Wenshui dialect of Shanxi Province]. *Zhongbei Daxue xuebao* 中北大学学报 [*Journal of North China University*] 1: 8–13.

Shi, Xiuju (史秀菊), and Shi, Rong (史荣). 2016. Shanxi Jiangxian fangyan zhishi daici "zhai, wai, nai" yu jiegou zhuci "de" de yufa gongxing 山西绛县方言指示代词"这、乃、兀"与结构助词"的"的语法共性 [The similarities between demonstrative pronoun "zhai, wai, nai" structural particle "de" in the dialect of Jiangxian, Shanxi]. *Zhongbei Daxue xuebao* 中北大学学报 [*Journal of North University China*] 1: 60–64+70.

Song, Wenhui (宋文辉). 2019. Hebei Zhengding fangyan ciwei "le" liangge bianti de shiti yiyi 河北正定方言词尾"了"两个变体的时体意义 [The aspectual significance of two variants of the affix "le" in the Zhengding dialect of Hebei Province]. *Yuwen yanjiu* 语文研究 [*Linguistic Research*] 1: 57–65.

Sun, Chaofen (孙朝奋).1994. "Xuhualun" pingjie 《虚化论》评介, [A review of "on bleaching"]. *Guowai yuyanxue* 国外语言学 [*Contemporary Linguistics*] 4: 19–25.

Tao, Huan (陶寰), and Li, Jialiang (李佳樑). 2009. Fangyan yu xiuci de yanjiu jiemian——jianlun Shanghaihua "yijiang" de xiuci dongyin 方言与修辞的研究接面——兼论上海话"伊讲"的修辞动因 [Research on dialect and rhetoric – Also on the rhetorical motivation of "yijiang" in the Shanghai dialect]. *Xiuci xuexi* 修辞学习 [*Contemporary Rhetoric*] 3: 1–8.

Wang, Chunling (王春玲). 2017. Fangyan jiechu yinfa de yufa yanbian 方言接触引发的语法演变 [Grammatical change induced by dialect contact]. *Xinan Daxue xuebao* 西南大学学报 [*Journal of Southwest University*] 4: 147–158.

Wang, Fang (王芳). 2019. Anyang fangyan de huati biaoji "buza" 安阳方言的话题标记"不咋" [On the topic market "buza" in the Anyang dialect]. *Huazhong xueshu* 华中学术 [*Central China Humanities*] 2: 137–143.

Wang, Fang (王芳), and Liu, Danqing (刘丹青). 2011. Henan Guangshan fangyan laizi "li" de duogongneng xuci "de"——gongshi miaoxie yu yuyi yanbian fenxi 河南光山方言来自"里"的多功能虚词"的"——共时描写与语义演变分析 [The Multifunctional Word "de" originating from "li" in the Guangshan dialect: Synchronic description and semantic evolution]. *Yuyan yanjiu* 语言研究 [*Studies in Language and Linguistics*] 2: 8–17.

Wang, Huayun (汪化云). 2011. Shenglüe goucheng de rencheng daici fushu biaoji 省略构成的人称代词复数标记 [Plural markers of personal pronouns by ellipsis]. *Fangyan* 方言 [*Dialects*] 1: 20–25.

———. 2017. Huangxiao fangyan zhong "[ten]" de yufahua 黄孝方言中"等"的语法化 [Grammaticalization of "[ten]" in dialects of the Huangxiao cluster of Jianghuai Mandarin]. *Fangyan*方言 *[Dialects]* 2: 210–215.

Wang, Huayun (汪化云), and Xiao, Qingzhu (肖擎柱). 2018. Huangxiao fangyan zhong de "jiaomo" 黄孝方言中的"叫莫" ["Jiaomo" in the Huangxiao dialects]. *Yuyan kexue* 语言科学 *[Linguistic Sciences]* 3: 281–289.

Wang, Jian (王健). 2010. Suwan fangyan zhong "diao" leici de gongshi biaoxian yu yufahua dengji 苏皖方言中"掉"类词的共时表现与语法化等级 [The synchronic behaviors and the degrees of grammaticalization of "diao" type words in the dialects of Jiangsu and Anhui Provinces]. *Yuyan kexue*语言科学 *[Linguistic Sciences]* 2: 187–196.

Wu, Fuxiang (吴福祥). 2001. Nanfang fangyan jige zhuangtai buyu biaoji de laiyuan (yi) 南方方言几个状态补语标记的来源(一) [The origin of some markers of stative complements in southern Chinese dialects (one)]. *Fangyan*方言 *[Dialect]* 4: 344–354.

———. 2002. Nanfang fangyan jige zhuangtai buyu biaoji de laiyuan (yi) 南方方言几个状态补语标记的来源(二) [The origin of some markers of stative complements in southern Chinese dialects (two)]. *Fangyan*方言 *[Dialect]* 1: 24–34.

———. 2003. Hanyu bansui jieci yufahua de leixing yanjiu——jianlun SVO xing yuyan bansui jieci de liangzhong yanhua moshi 汉语伴随介词语法化的类型学研究——兼论SVO型语言中伴随介词的两种演化模式 [A typological study of grammaticalization of comitative preposition in Chinese language]. *Zhongguo yuwen*中国语文 *[Studies of the Chinese Language]* 1: 43–58.

———. 2005. Hanyu Yufahua de Dangqian Keti汉语语法化研究的当前课题 [Current issues in the study of grammaticalization of Chinese]. *Yuyan Kexue*语言科学 *[Linguistic Sciences]* 2: 20–32.

———. 2010a. Hanyu fangyan li yu quxiang dongci xiangguan de jizhong yuafahua moshi汉语方言里与趋向动词相关的几种语法化模式[Grammaticalization Patterns Related to the Directional Verbs in Chinese Dialects].Fangyan方言[Dialect]2:97–113.

———. 2010b. Yueyu chabishi "X+A+Guo+Y" de leixingxue diwei——bijiao fangyanxue he quyu leixingxue de shijiao 粤语差比式"X+A+过+Y"的类型学地位——比较方言学和区域类型学的视角 [The typological status of the comparative construction "X+A+guo(过)+Y" in Yue dialects: From the perspective of comparative dialectology and areal typology]. *Zhongguo yuwen*中国语文 *[Studies of the Chinese Language]* 3: 238–255.

Wu, Fuxiang (吴福祥), and Qin, Fengyu (覃凤余). 2010. Nanning Yueyu duanchabishi "X+A +Guo"de laiyuan 南宁粤语短差比式"X+A+过"的来源 [On origin of short disparity sentence "X+A+ Guo" in the Cantonese dialect in Nanning]. *Hefei Shifan Xueyuan xuebao*合肥师范学院学报 *[Journal of Hefei Normal University]* 2: 91–97.

Wu, Zihui (吴子慧). 2007. Shusheng cuhua zai Shaoxing fangyanzhong de yixie biaoxian 舒声促化在绍兴方言中的一些表现 [Glottalization of non-checked tones in the Shaoxing dialect]. *Zhejiang Jiaoyu Xueyuan xuebao* 浙江教育学院学报 *[Journal of Zhejiang Education Institute]* 5: 65–70.

Xie, Huiquan (解惠全). 1987. Tan shici de xuhua 谈实词的虚化 [On the bleaching of content words]. In *Yuyan yanjiu luncong* 语言研究论丛 第四辑 *[Linguistic studies: Volume 4]*, pp. 208–221. Tianjin: Nankai chubanshe 天津: 南开大学出版社 [Tianjin: Nankai University Press].

Xing, Xiangdong (邢向东). 2011. Shanbei Shenmuhua de huati biaoji "lai" he "qu" jiqi youlai 陕北神木话的话题标记"来"和"去"及其由来 [Topic markers "lai" and "qu" in the Shenmu dialect, Shaanxi Province]. *Zhongguo yuwen* 中国语文 *[Studies of the Chinese Language]* 6: 519–526.

———. 2015. Lun jinyu shizhi biaoji de yuqi gongneng——jinyu shizhi fanshou yanjiu zhiyi 论晋语时制标记的语气功能——晋语时制范畴研究之一 [A study on the modal function of tense markers in Jin dialects – The study of tense category in Jin dialects]. *Anhui Daxue xuebao* 安徽大学学报 *[Journal of Anhui University]* 4: 92–102.

Yang, Hsiu-fang (楊秀芳). 2014. Lun "bie" de xingtai bianhua ji yufahua 論「別」的形態變化及語法化 [On the morphological derivation and grammaticalization of the word "bie"]. *Qinghua Zhongwen xuebao*清華中文學報 *[Tsing Hua Journal of Chinese Literature]* 11: 5–55.

Yang, Jingyu (杨敬宇). 2004. *Qingmo Yuefangyan yufa jiqi fazhan yanjiu* 清末粤方言语法及其发展研究 *[Research on the grammar and development of the Cantonese dialect in the Late Qing]*. Guangzhou: Guangdong renmin chubanshe 广州：广东人民出版社 [Guangzhou:Guangdong People's Publishing House].

Yang, Yonglong (杨永龙). 2005. Cong wenjinyi xingrongci dao chixuti zhuci——Shishuo "ding", "wending", "lao", "shi", "wen", "jin" de yufahua从稳紧义形容词到持续体助词——试说"定"、"稳定"、"实"、"牢"、"稳"、"紧"的语法化 [From adjectives meaning "stable" to continuous aspect markers: The grammaticalization of "ding", "wending", "lao", "wen", "jin" in Chinese], *Zhongguo yuwen*中国语文 *[Studies of the Chinese Language]* 5: 408–417.

———. 2014. Qinghai Minhe Gangouhua de duogongneng gebiaoji "[xɑ]" 青海民和甘沟话的多功能格标记"哈" [The multi-functional case marker "[xɑ]" in the Gangou Chinese dialect in Minhe Hui and Monguor Autonomous County, Qinghai Province]. *Fangyan*方言 *[Dialects]* 3: 230–241.

Zeng, Nanyi (曾南逸), and Li, Xiaofan (李小凡). 2013. Cong Ming-Qing xiwen kan Quanzhou fangyan tibiaoji "ləʔ⁵/lə¹" de yufahua 从明清戏文看泉州方言体标记"咧"的语法化 [The gammaticalization of "ləʔ⁵/lə¹" in the Quanzhou dialect from the evidence of Quanqiang dramatic scripts from the Ming and Qing]. *Zhongguo yuwen* 中国语文 *[Studies of the Chinese Language]* 3: 205–214.

Zhang, Ding (张定). 2009. Zongyang fangyan liangge huisheng foudingci de yufahua枞阳方言两个回声否定词的语法化 [Grammaticalization of two echoic negative markers in the Zongyang dialect]. *Zhongguo yuwen*中国语文 *[Studies of the Chinese Language]* 4: 365–370.

Zhang, Min (张敏). 2010. "Yuyi ditu moxing": yuanli, caozuo jiqi zai Hanyu duogongneng yufa xingshi yanjiuzhong de yunyong "语义地图模型"：原理、操作及其在汉语多功能语法形式研究中的运用 [The semantic map model and its application to the study of multifunctional grams in Chinese]. In *Yuyanxue luncong*语言学论丛 *[Essays on Linguistics]*, vol. 42: 3–60. Beijing: Shangwu Yinshuguan北京：商务印书馆 [Beijing: The Commercial Press].

Zhang, Wei (张惟), and Gao, Hua (高华). 2012. Yueyu ziran huihuazhong "zikhai" de Huayu gongneng jiqi biaojihua粤语自然会话中"即係"的话语功能及其标记化 [The functions of "Zikhai" in Cantonese conversation and its grammaticalization]. *Yuyan kexue*语言科学 *[Linguistic Sciences]* 4: 367–395.

Zhang, Yisheng (张谊生). 2016. Sanshi nian lai Hanyu xuci yanjiu de fazhan qushi yu dangqian keti 30年来汉语虚词研究的发展趋势与当前课题 [The development trend of Chinese function word studies over the last 30 years and current issues of study]. *Yuyan jiaoxue yu yanjiu*语言教学与研究 *[Language Teaching and Linguistic Studies]* 3: 74–83.

Zhang, Zhanshan (张占山), and Li, Rulong (李如龙). 2007. Xuhua de zhongji: heyin——yi Yantai fangyan ruofan xuchengfen heyin wei li 虚化的终极:合音——以烟台方言若干虚成分合音为例 [The limits of bleaching: Using certain bleached elements from thethe Yantai dialect as examples]. *Ludong Daxue xuebao*鲁东大学学报 *[Ludong University Journal]* 2: 95–100.

Zhao, Jianhua (赵建华). 2013. Ganyu Duchanghua zhong "ge" zi de zhuci gongneng jiqi yufahua 赣语都昌话中"个"字的助词功能及其语法化 [The structure and modal particle function of "ge" in the Changdu dialect of the Gan dialect]. *Changzhou Xueyuan xuebao* 常州学院学报 *[Journal of Changzhou Institute of Technology]* 4: 73–77.

Zhao, Rixin (赵日新). 2013. "Zuo" de yufahua "做"的语法化 [Grammaticalization of "zuo"]. *Yuyan jiaoxue yu yanjiu* 语言教学与研究 *[Language Teaching and Linguistic Studies]* 6: 46–54.

Zhong Weiping(),and Kwok Bit-chee郭必之.Yuedong Minyu de xiaocheng houzhui "er":xingshi, gongneng,laiyuan ji yanbian粤东闽语的小称后"儿"：形式、功能、来源及演变[The Diminutive Suffix "er" in East Guangdong Min Dialects: Its Forms, Functions, Origin and Diachronic Development]. In *Yuyanxue luncong*语言学论丛 [Essays on Linguistics], vol. 57: 54–69. Beijing: Shangwu Yinshuguan北京：商务印书馆 [Beijing: The Commercial Press].

Zhou, Minli (周敏莉), and Li, Xiaojun (李小军). 2018. Hunan Xinshao (Cunshi) hua de qiangdiao biaoji "[piɛ˧˩ko˥]"——jianlun Hanyu fanyan tacheng daici dao qiangdiao biaoji de liangzhong leixing 湖南新邵(寸石)话的强调标记"别个"——兼论汉语方言他称代词到强调标记的两种类型 [The emphatic marker "[piɛ˦ko˥]" in the Xiang dialect in Cunshi town, Xinshao County, Hunan Province: Also on the two change types from the indefinite pronouns referring to others to emphatic markers]. *Fangyan* 方言 *[Dialects]* 4: 467–476.

Zhou, Yimin (周一民). 2005. Beijinghua de qingyin he yufahua 北京话的轻音和语法化 [Light syllable and Grammaticalization in Beijing dialect]. *Beijing shehui kexue* 北京社会科学 *[Social Science of Beijing]* 3: 148–151.

Recent Advances in the Study of Chinese Historical Syntax

Changcai Zhao

Contents

Grammaticalization in Chinese Historical Syntax	188
Theoretical Innovations Based on Case Studies of Grammaticalization in Chinese	189
Grammaticalization Research from a Typological Perspective	191
Important Monographs and Conferences on Grammaticalization	192
Construction Grammar in Chinese Historical Syntax	193
Recent Advances in Word Order Typology of the Chinese Language	194
The Basic Word Order of Chinese	194
On OV Word Order (Object-Fronting) in Archaic Chinese	195
Recent Advances in Verb Categorization in Chinese Historical Syntax	198
Verb Categorization Based on Transitivity	198
Unaccusative Verbs Versus Unergative Verbs	199
Verbs Categorization Based on Temporality	200
The Relations Between Nouns and Verbs	200
The Mechanisms for the Emergence of Chinese Aspect Markers	201
On Perfective Verbs, Perfective Markers, and Perfective Constructions	201
On Other Aspect Markers	202
Recent Advances in Important Syntactic Structures	202
On Verb-Complement Structure (Verb-Resultative Complement)	202
On Disposal Constructions	204
On Passive Constructions	205
Important Publications on Chinese Historical Syntax	206
The Future for Research on Chinese Historical Syntax	207
References	207

This chapter was originally written in Chinese by Changcai Zhao. The English version is a translation by Peng Xinjia 彭馨葭 and Zhao Lüyuan 赵绿原.

C. Zhao (✉)
Institute of Linguistics, Chinese Academy of Social Sciences, Beijing, China

> **Abstract**
>
> This chapter reviews research on Chinese historical syntax undertaken during the last 30 years. We will focus on the following topics: the application of grammaticalization theory and Construction Grammar, Chinese word order typology, verb categorization, the relations between nouns and verbs, the emergence of Chinese aspect markers and the mechanisms of their emergence, latest research development of a few important sentence structures, and finally a brief overview of the most important monographs on Chinese historical syntax. In conclusion, we aim at laying out potential prospects for future research directions and developments in Chinese historical syntax.

> **Keywords**
>
> Chinese historical syntax · Grammaticalization · Construction Grammar · Word order typology · Verb categorization · Aspect markers

Chinese historical syntax is probably the most vibrant research area in the field of Chinese linguistics. As scholars deepen their research, broaden their research focus and engage with a wider range of theoretical frameworks, we have seen tremendous breakthroughs in research on historical syntactic phenomena of the Chinese language. This chapter reviews recent progress in the following topic areas: grammaticalization theory, engagement with Construction Grammar, word order typology, verb categorization, the relation between nouns and verbs, aspect markers, and key sentence structures in the history of the Chinese language. We aim at providing a general overview of these areas by featuring the most influential studies carried out in the Chinese mainland over the last 30 years.

Following Jiang Shaoyu 蒋绍愚 (Jiang 2005, 2017), this chapter posits five major periods for the history of the Chinese language: *Yuangu Hanyu* 远古汉语 (Proto-Sinitic, the period prior to and including the Shang), *Shanggu Hanyu* 上古汉语 (Old Chinese, the period from the Western Zhou to the Western Han), *Zhonggu Hanyu* 中古汉语 (Middle Chinese, the period from the Eastern Han to the Sui), *Jindai Hanyu* 近代汉语 (Early Modern Chinese, the period from the Tang to the Mid Qing), and *Xiandai Hanyu* 现代汉语 (Modern Chinese, Late Qing up to the present). (Pre-Qin Chinese is another historical period that is traditionally considered linguistically significant).

Grammaticalization in Chinese Historical Syntax

Grammaticalization as a research paradigm emphasizes the integration of findings from diachronic data and synchronic phenomena with the goal of explaining linguistic variations on the synchronic plane with evidence from historical changes. Shen Jiaxuan 沈家煊 (1994, 1998a, 1998b), Chaofen Sun (1994), and Wu Fuxiang

吴福祥 (2004, 2005, 2013) provide a systematic introduction of grammaticalization theory to Chinese readers. Since its introduction, grammaticalization theory has contributed to our deepened understanding of historical change in the history of the Chinese language, as well as the syntactic system of the Chinese language in its current state.

Theoretical Innovations Based on Case Studies of Grammaticalization in Chinese

Starting in the late 1980s, many scholars in the Chinese mainland turned their attention to grammaticalization. The strong urge to apply grammaticalization theory to the study of Chinese historical syntax produced fruitful results. As research findings based on Chinese language data began to accumulate, these scholars also brought new insights to grammaticalization theory. Most noteworthy are studies by Xie Huiquan 解惠全 (Xie 1987), Liu Jian 刘坚, Cao Guangshun 曹广顺 and Wu Fuxiang (Liu et al. 1995), and Hong Bo 洪波 (Hong 1998), among others. Xie (1987, 213), for instance, points out "the grammaticalization of a content word takes its meaning as the basis of change, and the path of development is determined by its grammatical status. In most cases, the grammaticalization of a content word is only possible when it occurs in a slot that tends to mark certain grammatical function; such a syntactic distribution forms the basis for the bleaching of lexical meaning(s), which induces the stabilization of the slot as a grammatical one." Liu et al. (1995) study examines the emergence of grammatical words. Putting together case studies on 将 (*jiang*), 取 (*qu*), 得 (*de*), 个 (*ge*), 敢 (*gan*), 着 (*zhe*), 被 (*bei*), and 把 (*ba*), this chapter identifies several factors that induce the process of grammaticalization and how they affect the path of change (As the meanings of grammatical words are ambivalent and open to interpretation in the process of grammaticalization, the definition of grammatical words will not be given in this chapter). The factors examined here include syntactic distribution, semantic change, context, and the reanalysis.

Building upon these foundational works, scholars made great advancements in key issues in Chinese grammaticalization research. Based on empirical data from the Chinese language, new theoretical explanations were put forward to account for various phenomena in the process of grammaticalization, such as renewal, reinforcement, superposition, sound change, and antigrammaticalization. These theoretical advancements have in turn given new momentum to the study of grammaticalization in Chinese. The following highlights representative scholars and their research in this area.

Engaging with grammaticalization theory, Jiang Lansheng 江蓝生 conducted a series of empirical studies on Chinese historical syntax; based on her detailed description of language data, she made tremendous contributions to the theoretical discussions of grammaticalization. In one study (Jiang 1999a), the author discusses the relation between sound change and grammaticalization, taking the "V+$X_{prep.}$+$NP_{location}$" structure as an example. In vernacular literature during the Ming and Qing, the structure is usually written as "V+的+NP$_{location}$," and Jiang argues that 的 (*de*) in the X slot is in fact a weakened pronunciation of 着 (*zhe*). From this case study, Jiang discusses the relation between sound change and grammaticalization. She points out that sound change is a part of grammaticalization and is similarly a gradual process. While the grammaticalization of a content word usually follows the transition from content word > grammatical word > zero, sound change proceeds in a similar direction, toward a more and more reduced form until it becomes a zero form. Jiang's method of combining diachronic data (historical documents) and synchronic evidence (dialectal pronunciations) has proven successful in grammaticalization research. In another article, Jiang (1999b) shows a possible path of development from locative words to structural particles. Evidence comes from vernacular language data in Han, Wei, and Six Dynasties, where locative nouns 所 (*suo*) and 许 (*xu*) in the structure NP+NP$_{locative}$+NP takes the slot usually occupied by the structural particle 之 (*zhi*), and hence acquired the interpretation of a structural particle. When the locative nouns became fully grammaticalized, they began to be recognized as a structural particle through reanalysis. The proposed pathway provides an entirely new yet reasonable account for the origin of the structural particle 底 (*di*).

Cao Guangshun (1995) identifies a grammaticalization cline in the history of the Chinese language: V+V > V+C (verbal complement) > V+grammatical markers. This is a path of development that is testified on aspect particles such as 得 (*de*), 取 (*qu*), 将 (*jiang*), and 却 (*que*), and they have all gone through a stage where they would normally indicate the realization of an action, but given the right context, they could indicate the duration of an action.

Liu Danqing 刘丹青 (2001) proposes to consider renewal, reinforcement, and superposition as common processes in grammaticalization. Renewal refers to the replacement of a highly grammaticalized marker by a more independent linguistic unit to perform the same grammatical functions; an example would be the replacement of highly grammaticalized 于/於 (*yu*) in Archaic Chinese by content words such as 在 (*zai*), 对 (*dui*), 向 (*xiang*), 被 (*bei*), and 比 (*bi*). Reinforcement is a process whereby the meaning and syntactic role of an existing grammatical marker is reinforced by affixing a similar or related grammatical element. An example is the affixation of 物 (*wu*, "thing") to 何 (*he*, "what") in Middle Chinese, leading to the innovation of 何物 (*hewu*) as the new interrogative pronoun for things. These two processes are similar in that they produce polymorphic forms for a single grammatical function to counter the attritional forces of grammaticalization. On the contrary, superposition is a process whereby a single linguistic element becomes polysemic. It occurs with grammaticalization and represents a high level of grammaticalization.

Yang Yonglong 杨永龙 (2005) investigates the grammaticalization of durative and progressive markers in the history of Chinese language. Combining evidence from diachronic data in historical documents (the development of the durative marker 定 *ding*) and synchronic data from various Chinese dialects (稳定 *wending*, 牢 *lao*, 稳 *wen*, and 紧 *jin* are similar markers in Hakka, Wu, and Yue dialects), Yang proposes the grammaticalization of durative and progressive markers from a group of adjectives that mean "tight and stable" in Chinese and discusses the prerequisites, motivations, and mechanisms of change.

Engaging with discussions on the mechanisms of grammatical change and in particular Harris and Campbell (1995), Wu Fuxiang (2013) posits four major mechanisms for grammatical change: reanalysis and extension as internal mechanisms, grammatical borrowing and grammatical replication as external mechanisms. In discussing the internal mechanisms, Wu points out that the operation of reanalysis and extension is bound by language internal factors. The two mechanisms supplement each other and take turns to operate: Reanalysis operates on the underlying structure, which modifies the rules of a language, while extension works with the surface structure, which instantiates and extends rule change. Wu's argument is well supported by examples from grammatical changes in the Chinese language. In another article, Wu (2017) showcases four peculiar cases of antigrammaticalization (i.e., coordinating conjunction > comitative preposition, locative preposition > locative verb, dative preposition > benefactive verb, and comparative preposition > the LIKEN verb) and argues that such a path of development is largely due to the typological feature of Chinese. Cases of antigrammaticalization in Chinese challenge the unidirectionality hypothesis and provides a new insight into grammaticalization research in general, and grammaticalization in Chinese in particular.

The monograph by Ma Beijia 马贝加 (Ma 2014) is a systematic examination of grammaticalization from verbs into various grammatical markers (prepositions, adverbs, conjunctions, auxiliaries, and auxiliary verbs) in Chinese. Spanning across an extended period from pre-Qin to Modern Chinese, Ma's work is undoubtedly comprehensive yet still offers meticulous analysis for each case study, and it makes both empirical and theoretical contributions to research on the grammaticalization of verbs.

Peng Rui 彭睿 (2020) is another noteworthy monograph that deals with the theoretical aspect of grammaticalization. Peng takes grammatical changes in Chinese as data to review and evaluate existing theories, ultimately aiming at fine-tuning and completing theories on grammaticalization. The monograph provides in-depth discussion on context of change, cline of grammaticalization, and the relation between grammaticalization and the role of frequency, as well as the basis and methods for reconstructing the process of grammaticalization, while highlighting the validity of grammaticalization theory across languages and word categories.

Grammaticalization Research from a Typological Perspective

Imposing a typological assumption on Chinese grammaticalization research helps reveal the nature of grammatical change, whether the change is consistent with universal pattern, or is unique to the Chinese language. In this light, research on historical syntax could offer typological implications and contribute to an understanding of Chinese typology of grammatical change. In the last few decades, scholars have uncovered a large number of pathways of grammatical changes. Scholars make generalizations about the paths and patterns of Chinese grammaticalization from studies on grammatical changes in Chinese and attempt to explore more general typological features of grammaticalization in human languages.

Against the backdrop of such a new trend in grammatical research, research on Chinese prepositions and conjunctions are under careful scrutiny.

The majority of prepositions in Modern Chinese have a verb source, and in the process of grammaticalization, some of them have moved further down the cline and became conjunctions and clitics, while others stay somewhere in between. Approaching the source and path of grammaticalization of prepositions and conjunctions from a typological perspective thus helps reveal the mechanisms behind these grammatical changes.

Liu Jian (1989) and Yu Jiang 于江 (1996)'s respective studies examine the grammaticalization of grammatical words such as 与 (*yu*), 及 (*ji*), 共 (*gong*), 和 (*he*), 同 (*tong*), and 跟 (*gen*). These studies find that coordinating conjunctions in Chinese follow a common path of development as follows: "verb > preposition > conjunction."

Wu Fuxiang (2003a, b, c) also adopts a typological perspective to examine the grammaticalization of comitatives. Diachronic and synchronic data in Chinese reveal a path of "comitative verb > comitative preposition > coordinating conjunction." By comparing with data from Chinese dialects and other languages, Wu discusses possible typological implications of this path of development.

Jiang (2012) proposes the category "conjunction-prepositions." Jiang identifies four types of conjunction-prepositions by their verb source: (1) comitative verbs 和 (*he*), 跟 (*gen*), and 同 (*tong*); (2) verbs denoting calling and coercion such as 唤 (*huan*), 教 (*jiao*); (3) verbs of giving such as 与 (*yu*), 给 (*gei*); and (4) the dual-like compound 两个 (*liang-ge*) as in 我两个 (*wo liang-ge*). Jiang describes the motivation and path of development for these verb sources and distinguishes three types of grammaticalization.

Important Monographs and Conferences on Grammaticalization

"Grammaticalization, Lexicalization in Studies of the Chinese Language" is a book series showcasing the latest research findings on the topic of grammaticalization and lexicalization in Chinese. The series released by Xuelin Press in 2017 includes nine titles: *Motivations and Paths of Grammaticalization in the Chinese Language* (Jiang Lansheng), *Grammaticalization and Semantic Map* (Wu Fuxiang), *The Grammaticalization of Content Words and Structures* (Yang Yonglong), *Grammaticalization of Content Words in Chinese* (Li Zongjiang 李宗江), *A Study on Grammaticalization Theory and Grammatical Words in Chinese* (Zhang Yisheng 张谊生), *Lexicalization and Grammaticalization of High-frequency Disyllabic Words in Chinese* (Chen Changlai 陈昌来), *A Study on Grammaticalization and Aspect Markers in Chinese* (Chen Qianrui 陈前瑞), *Phenomena and Patterns of Lexicalization and Grammaticalization in Chinese* (Dong Xiufang 董秀芳), and *A Study on the Pragmatic Mechanisms of Grammaticalization and Grammatical Words in Chinese* (Shi Jinsheng 史金生).

The "Biennial International Conference on Grammaticalization in Chinese," organized by the Institute of Linguistics at Chinese Academy of Social Sciences starting from 2001, also led to the publication of another important research series on Chinese historical syntax. Ten proceedings of the conferences have been published by the Commercial Press (*Shangwu Yinshuguan*) under the title "Studies on Grammaticalization and Syntax."

Construction Grammar in Chinese Historical Syntax

Construction Grammar (CxG) as a research paradigm was first introduced to the linguistics community in China at the turn of the century. Zhang Bojiang 张伯江's (Zhang 1999) study on the ditransitive construction in Modern Chinese pioneers the application of CxG in the study of the Chinese language. Following Zhang, case studies applying CxG in the research of diachronic grammar of Chinese gradually increase. Informed by CxG, these studies offer better accounts of phenomena in Chinese syntax; these studies also serve as testing grounds for the application of CxG theory and methods in Chinese.

Jiang (2005) focuses on the "VP *de hao*" (VP 的好) construction and examines the motivations, processes, and features of its grammaticalization through changes in different domains: the demotion of grammatical functions, subjectification of semantics, and tightening of structure. She points out that the construction has two subtypes, each following a different path. One of the paths is morphologization, where patterns and lexical words develop into grammatical markers or morphological markers; the other path is syntacticization, where loose parataxis in a pragmatic mode of communication turns into a syntactic mode featuring tight syntax, or where an analytic language becomes agglutinative. During the grammaticalization of a construction, semantic, syntactic, prosodic, and pragmatic factors are all intertwined with each other and subject to the influence of other factors. Destabilizing any one factor would set off a chain reaction of changes, leading to the occurrence of morphologization and syntacticization.

In another article, Jiang (2007) discusses the principle of economy as a driving force of grammaticalization in Chinese. In this chapter, taking as examples (1) the reduplicative interrogative structure VP-NEG-VP in Chinese dialects and (2) several set phrases such as 爱怎怎 (*ai-zen-zen*, "whatever"), 爱谁谁 (*ai-shei-shei*, "whoever/whomever"), and 爱吃不吃 (*ai-chi-bu-chi*, "eat it or leave it") in Beijing dialect, Jiang shows the omission and compaction as both the forces and mechanisms of grammatical change. The explanation also applies to the innovation of "VP *bu* VP" (VP 不 VP) structure in the history of Chinese, as well as the emergence of interrogative "VP *bu*" (VP 不) construction in Archaic Chinese and "VP *ye wu*" (VP 也无) construction in Late Tang and Five Dynasties.

In Wu Fuxiang (2002), the author discusses the constructionalization of the Chinese potential verb-complement construction "V *de*/*bu* C" (e.g., 拿得动 /拿不动 *nadedong/nabudong* "able to lift/not able to lift") from the isomorphic construction that denotes the achievement of an action. The grammaticalization of "V *de* C / V *bu* C" and "V *de* OC / VO *bu* C" does not entail semantic, morphosyntactic, or phonological changes of any of the elements it contains; it was the constructional meaning that changed.

Yang Yonglong has two significant papers on CxG. In Yang (2011), He discusses the emergence of presupposed meaning and entailed meaning in the *"lian* X *dou* VP" (连 X 都 VP) construction. In Yang (2016), He reviews issues concerning the grammaticalization of constructions against traditional linguistic concepts such as grammatical words and grammatical structures. He focuses on the relationship between concept pairs, including: grammaticalization of content words/grammaticalization of structures, grammatical structures/constructions, constructional meaning/meaning of key grammatical words, and grammaticalization of structures/constructions.

In recent years, the linguistics community in Chinese mainland has also seen the publications of a handful of monographs on CxG, including *A Study on Construction Grammar in Chinese* (Zhu Jun 朱军 2010), *A Study on the Theory of Construction Grammar* (Niu Baoyi 牛保义 2011), *Theory of Construction Grammar* (Wang Yin 王寅 2011), *A Study on Construction Grammar* (edited by Liu Zhengguang 刘正光 2011), and *Grammaticalization, Diachronic Construction Grammar and Constructionalization* (Peng Rui 2016). The publications mentioned here have all provided systematic and in-depth introduction to the theoretical and applicational aspects of CxG while along their own critique. Some of these works also deal with case studies of constructional change in the history of the Chinese language. The Chinese translation of Adele Goldberg's *Construction at Work* by Wu Haibo 吴海波 has also been published early in 2013.

Recent Advances in Word Order Typology of the Chinese Language

The Basic Word Order of Chinese

Chinese is a morphologically impoverished language, and grammatical functions are marked by word order and grammatical words. Word order is thus a prominent linguistic feature in the discussion of Chinese syntax. It was generally believed that the basic word order of Chinese has been relatively stable throughout the history of the Chinese language, until the work of Greenberg and Lehmann on word order typology came out in the 1960s and 1970s. Scholars began to notice inconsistencies between the Chinese language and the proposed theory. A heated debate regarding the basic word order of Chinese and its diachronic change swept across the global community of Chinese linguistics. Chinese linguists focused their discussion in two areas: First, regarding the basic word order of contemporary Mandarin Chinese, is it SVO or SOV? Was there a shift from SVO in Classical Chinese to SOV in Modern Chinese? Second, was Archaic Chinese a purely VO language? If so, then how do we account for OV order (or object fronting) in Archaic Chinese? Was there a shift from SOV in Proto-Sinitic to SVO in Archaic Chinese?

Tai (1973, 1976) and Li and Thompson (1974) were among the earliest proposals that basic word order in Chinese is going through a transition from SVO to SOV, and Modern Chinese is still an incomplete SOV language. Tai (1976) argues that the OV word order in Chinese came from Altaic influence in the north, and as a result, OV features are more prominently seen in the northern dialects in China (see also Hashimoto 1985/2008). Contrary to Tai, Li and Thompson (1974) believe that because, culturally speaking, Chinese was a dominant language, it should not be susceptible to influences of neighboring languages; word order change can only be a result of internal factors.

Various scholars have since questioned and challenged the claim above (see discussions in Light 1979; Meng-chen Li 1980; Mei 1980; Sun and Givon 1985; Liu Danqing 2003 *inter alia*). Oppositions to the claim mainly focus on two aspects. First, proponents of Chinese as an OV language makes the assumption based on linguistic features such as "relative clause preceding the noun," "adverbial modifiers preceding the verb," and "preposition preceding the verb" following Greenberg's proposed word order universals, but they fail to recognize that the theory on word order universals is a description of tendency based on crosslinguistic evidence, and these features should not be taken as indicators of the basic word order. Second, those arguments have also failed to make a distinction between marked word order and unmarked ones. Structures that were presented as evidence of OV word order, such as the *ba* construction, the *bei* construction, and patient-topic sentences, are all marked constructions that serve specific pragmatic functions (such as contrast and emphasis) and should not be taken as part of the evidence.

Sun and Givon (1985) conducted a quantitative study on the word order in Chinese written and spoken texts and found that over 90% of the utterances exhibit VO word order, leaving the occurrence of OV word order to less than 10%, and when OV word order is observed, it is used to express pragmatic contrast or emphasis.

On OV Word Order (Object-Fronting) in Archaic Chinese

Word order in Archaic Chinese is one of the most debated issues in Chinese historical syntax. Scholars tend to approach this question from the perspective of regional shift of language typology and dialect geography and adopt a crosslinguistic typological methodology in their research, supported by detailed description of language data, many valuable conclusions have drawn from.

The Nature of OV Word Order (Object-Fronting)

It is widely conceived in the history of the Chinese language that the basic word order of Archaic Chinese is SVO. Some exceptions to the SVO word order, such as object fronting, have been pointed out. The *Mashi wentong* 马氏文通, an early work on Chinese grammar by Ma Jianzhong 马建忠 published in 1898, had located such a phenomenon and discussed the conditions of its occurrence. But the question arises: What is the nature of object fronting in Archaic Chinese?

As early as in 1958, Wang Li 王力 has put forward a hypothesis that pronominal object has a preverbal position in Proto-Sinitic. A handful of scholars took Wang's hypothesis one step forward and proposed that in Proto-Sinitic, all kinds of objects occur in a preverbal position, and the basic word order was SOV (Li and Thompson 1974; Yu Min 俞敏 1981; Zhang Qingchang 张清常 1989; Feng 1996) before transitioning into an SVO order in Archaic Chinese. Wei Peiquan 魏培泉 (1993) posits that Proto-Sinitic might contain both VO and OV word order as a result of language contact, but starting from Archaic Chinese (when earliest written records began to be available), VO became the dominant word order. All of the proposals above share the assumption that object fronting is a remnant of Proto-Sinitic. In other words, the existence of both the VO and OV order is in fact the heterogenous existence of two historic strata.

The hypothesis above is not without its oppositions (see Djamouri 1988; Shen Pei 沈培 1992; Peyraube 1994, 1996, 1997; Shi Yuzhi 石毓智 and Xu Jie 徐杰 2001; Aldridge 2012, 2015 and Yao Zhenwu 姚振武 2015, *inter alia*). On the one hand, in oracle bone inscription, the overall word order is a rather stable SVO, and the occurrence of OV is not more frequent than in Archaic Chinese. On the other hand, certain SOV patterns found in Archaic Chinese are completely absent in earlier texts. For instance, preverbal interrogative pronouns are not possible because there were no interrogative pronouns. Preverbal objects are also seen in sentences of negation; but in oracle bone inscriptions, preverbal objects in negation sentences are limited to person pronouns (mostly the first person pronoun 我 *wo*). The demonstrative pronoun 之 *zhi* does not occur preverbally either, unless led by the focus marker 唯 *wei*. Peyraube (1997) points out that object position as reflected by pronouns should not be taken as evidence of an earlier stratum, but it sets the standards for a new word order, which was only in its nascent form.

Currently, Chinese linguists are coming to agree that since there is no reliable evidence that puts OV as the basic word order of Proto-Sinitic, the occurrence of object fronting in Archaic Chinese should not be treated as remnant of an earlier OV word order, but rather, a peculiar phenomenon induced by language internal factors.

Explanations for Object Fronting in Archaic Chinese

The 1990s see increasing research interest in the occurrence of object fronting, and some attempts were made to account for its occurrence with modern linguistics theories. Researchers are concerned with the following questions: Is the occurrence of object fronting dictated by pragmatic rules or syntactic ones? What are the motivations and mechanisms for such a phenomenon? And some ensuing questions are: Why are pronouns allowed in a preverbal position, but not regular nouns? Why is preverbal interrogative pronouns compulsory, but not for pronouns in negation sentences? Why does object fronting cease to exist after Middle Chinese?

Among these inquiries, it is generally accepted that nominal objects take the preverbal position to be the focus of a sentence (He Leshi 何乐士 1989; see also Peyraube 1996; Ting 1997; Shi and Xu 2001). However, the author did not delve into the specific motivations and mechanisms for the respective cases of preverbal interrogative pronouns and pronouns in negation sentences; the author had only mentioned briefly that both cases are governed by syntactic rules.

Feng (1996) makes a distinction between the two cases. Taking evidence from synchronic data in Archaic Chinese, Feng points out that the two patterns have different underlying structures, and they follow separate paths in their emergence. In the case of negation, a pronominal object moves out of VP and attaches to the right of the negative marker as a clitic. An interrogative pronoun, on the other hand, undergoes focus-movement to the left edge of VP, which, if not blocked by a negative marker or an auxiliary, triggers cliticization to the verb. Djamouri (2000) even considers the preverbal pronominal object as a result of focus movement. He regards 不 (bu) as the negative copula at the clausal level, which triggers the movement of focus element – similar to the function of 唯 (wei) in oracle bone inscription. But there are two exceptions: If there is another linguistic focus (e.g., an interrogative), then the pronominal object cannot be moved to a preverbal position; the pronominal object in a subordinate clause does not go through movement either.

Shi and Xu (2001) argue that the occurrence of preverbal object is the result of focus movement and the behavior of object as clitics. The motivation of left-dislocation is to give focus to the object. Second, left-dislocation is restricted to pronouns; this is in accordance with Wackernagel's Law in diachronic linguistics, which states that enclitics tend to occur in second position. When objects occur in a postverbal position, it tends to be cliticized. They thus propose a rule for the word order shift in Archaic Chinese as follows: The focused enclitic (object pronoun) needs to be in second position of a sentence (following the subject and preceding the verb).

Edith Aldridge (2012) supposes that object fronting occurs as a result of syntactic movement. The left-dislocation of interrogative pronominal object and pronominal objects in negation sentences are different types of syntactic movement and should be analyzed separately.

In sum, proposals with regard to the fronting of object to preverbal position in Archaic Chinese consider both pragmatic and syntactic factors. The preverbal positions of interrogative pronouns and focus NP might be motivated by pragmatic factors, but the movement is subject to syntactic rules. The preverbal positions of pronominal object in negation sentences might be motivated entirely by syntactic factors. It is therefore valid to argue that preverbal positions in Archaic Chinese are a syntactic phenomenon. Liu Danqing (2004, 37–46) discusses the preverbal positions of various types of semantic patient as a result of "syntacticized word order." It occurs alongside the SVO order which is the dominant word order, and it is different from patient topicalization or the *lian* construction in Modern Chinese. Liu continues to argue that Archaic Chinese exhibits a mixed word order which is "dominantly VO and complemented by OV," versus a pure VO word order of Modern Chinese. Xu (2006) holds similar view but put the period when a pure VO order begins in Middle Chinese.

Other important works that discuss word order in Chinese include Yin Guoguang 殷国光 (Yin 1985), Zhang Cheng 张赬 (Zhang 2002, 2010), Matsue Takashi (Matsue 2010), and Wu Fuxiang (2012), *inter alia*. In these studies, Zhang (2002) focuses on the word order shift of prepositional phrase. Wu's (2012) study on the development of "V *de/bu* C" defines word order innovation in Chinese from the perspective of grammaticalization; he also discusses the nature of word order shift in the history of the Chinese language.

Recent Advances in Verb Categorization in Chinese Historical Syntax

The verb is the most basic word category in human languages. It is the pivot of a sentence. The syntax and semantics of verbs is highly complicated. Different approaches to analyzing verbs would result in completely different categorization of verbs. Early works on the history of the Chinese language tend to make such categorization based on the lexical semantics of verbs. Yang Bojun 杨伯峻 and He Leshi (1992), for instance, set up four categories of verbs: action verbs, mental activity verbs, verbs of existence, and copula verbs (such as 为 *wei* and 是 *shi*).

The lexical semantics of verbs is the basis of their grammatical functions and syntactic behavior. However, categorization of verbs basing entirely on lexical semantics, first of all, is not well grounded without the support of structural evidence; second, it does not account for the interaction between the semantic aspect and syntactic behavior of the verb. Therefore, such categorization is unable to reveal the systematicity of syntactic behavior of verbs. In the 1990s, scholars began to break away from categorization basing purely on lexical semantics, and they started to experiment with new approaches.

Verb Categorization Based on Transitivity

Transitivity has been the most basic and classic feature in verb categorization. Li Zuofeng 李佐丰 (Li 1994a, b) refers to the object of a transitive verb as "direct object," and the obliques of an intransitive verb "causative object" and "indirect object" to include certain NPs that denote locations, targets, recipients, and causes. Li further distinguishes transitive and intransitive verbs into four subtypes: (1) the real transitive, which is the equivalent of the typical transitive verbs; they usually take a direct object; some examples include 侵 (*qin*, "to invade"), 犯 (*fan*, "to offend"), 杀 (*sha*, "to kill"), and 袭 (*xi*, "to attack"); (2) the quasi-transitive, which includes transitive verbs that exhibit certain features of intransitivity, such as 赐 (*ci*, "to bestow"), 赏 (*shang*, "to bestow"), 夺 (*duo*, "to take away by force"), 问 (*wen*, "to question"), and 劝 (*quan*, "to persuade"); (3) the real intransitive, which covers the typical intransitive verbs that do not take an object, such as 卒 (*zu*, "to die"), 病 (*bing*, "to become sick"), 灾 (*zai*, "to spell disaster"), and 枯 (*ku*, "to wither"); and (4) the quasi-intransitive, which contains intransitive verbs that exhibit certain features of transitivity and usually do not take object or only allow causative object or indirect object, such as 怒 (*nu*, "to get angry"), 恐 (*kong*, "to fear" or "to threaten"), 来 (*lai*, "to come" or "to make someone come"), and 败 (*bai*, "to defeat" or "to be defeated").

Yin Guoguang's (1997) proposal is similar to Li's categorization, further dividing the two transitive-based categories into four subtypes. However, Yin's categorization of verbs is based on categorization of the secondary argument of the clause, which includes: (1) quasi-objects (relational objects and nonrelational objects such as causative objects, putative objects, and theme objects); (2) real objects (which are usually patient objects); and (3) other kinds of objects such as existential objects, equative objects, and semblative objects.

Guan Xiechu 管燮初 (1994, 152–153) was the first to propose a tripartite categorization for verbs in Archaic Chinese. Besides "external verbs" (外动词 *wai dongci*) and "internal verbs" (内动词 *nei dongci*), he proposed a third type of verbs which "sometimes takes an object but sometimes does not." Zhang Meng 张猛 (Zhang 1998) also proposes a similar tripartite categorization. Compared to verbs in Modern Chinese, verbs in Archaic Chinese alternate between different states of transitivity – this is another interesting research topic.

Unaccusative Verbs Versus Unergative Verbs

Cikoski (1978a, b) is among the earliest to propose an "ergative verbs vs. neutral verbs" dyad for verbs in Archaic Chinese: The semantic relation between the ergative verb and the subject changes as the verb takes or drops the object. Katsuya Ōnishi (2004) follows Cikoski's proposal and investigates the 40 ergative verbs found in *Shiji* 史记. Wei Peiquan (2001) also uses the term "ergative verbs" to refer to a type of intransitive verbs that have causative and putative usages – in this scenario, they are given the name "causative verbs"; Wei also analyzes more than a hundred ergative verbs that collocate with 弗 (*fu*) /勿 (*wu*).

Song Yayun 宋亚云 (Song 2014) also considered transitivity and ergativity in constructing a categorization system for verbs in Archaic Chinese. Song follows the tripartite categorization (transitive, intransitive, and ergative) and considers verbs that have both transitive and intransitive usages as transitive. He further divides the three categories into eight subtypes based on the strength of transitivity. He also points out that these categories are not marked by strict categorical boundaries, and these categories are mutually convertible given the right conditions. Song's innovative model has been applauded for how it reveals the different types of transitivity of verbs and the conditions for cross-categorical conversion.

Jiang Shaoyu (2017) discusses the identification of ergative verbs in Archaic Chinese and its path of further development. Jiang proposes only under the following conditions might a verb be considered ergative: The verb occurs simultaneously in two structures: "X+V," where V signifies the change of state of X, and "Y+V+X," where V is a causative. Jiang identifies three types of ergative verbs: action verbs, state verbs, and action-state verbs – with the last two categories, he proposes to consider only state verbs with causative usage (such as 亡 *wang* "to perish," 出 *chu* "to be out/to oust someone," and 劳 *lao* "to make someone to labor") and action-state verbs that highlight change-of-state (such as 灭 *mie* "to destroy/to be destroyed," 毁 *hui* "to destroy/to be destroyed," and 破 *po* "to brake/to be broken") as ergative. In terms of the path of development, Jiang discusses some significant factors, including the relative frequencies of the two predicative structures of V, their phonological realization, semantic change, etymology, and language data prior to the Spring and Autumn Period.

Verbs Categorization Based on Temporality

As the most typical linguistic element used in narration, temporality is the most outstanding feature of verbs. The categorization of verbs basing on temporality taps into the temporal structure (or the duration of the action and the type of the action) of an action. Three pairs of distinctive features – dynamicity, durativity, and telicity – are enlisted to capture the differences. What these features generate is also a tripartite categorization of verbs: state verbs, action verbs, and change-of-state verbs.

An early dichotomic categorization by Li Zuofeng (Li 1994b) shares some similarities with the trichotomy above. Li's proposal of action verbs (as an umbrella term for all "voluntary actions and activities") and state verbs (to refer to the rest of verbs, which are involuntary ones) is similar to action verbs and change-of-state verbs in the trichotomy. However, Li's categorization is based on actor intention rather than the temporal structure of verbs.

Ren He 任荷's (Ren 2020) monograph represents the latest progress in this area. Inspired by Van Valin (1997, 2005, 2006), Ren builds a categorization system of verbs in Archaic Chinese based on event structure. Four distinctive features (dynamicity, durativity, telicity, and causativity) are used to distinguish five types of verbs and event structure. Based on surveys conducted on a sample, the author identifies five types of verbs: state verbs (such as 有 *you* "to have," 爱 *ai* "to love"), activity verbs (such as 飞 *fei* "to fly," 食 *shi* "to eat"), change-of-state verbs (such as 死 *si* "to die," 入 *ru* "to enter"), accomplishment verbs (such as 戮 *lu* "to kill," 醢 *hai* "to smash someone to death"), and causative verbs (such as 毁 *hui* "to ruin," 纳 *na* "to accept").

By engaging with new theoretical approaches, scholars working on the history of the Chinese languages have opened up new possibilities in their research into verbs in Classical Chinese both on the semantic and syntactic plane. A typological perspective, characterized by the comparison of Chinese against other languages, also paves the path for the construction of a more scientific categorization system for verbs.

The Relations Between Nouns and Verbs

The relations between nouns and verbs, as well as their distinctions and convertibility, have garnered new research interest in recent years. Traditionally, nouns and verbs are considered to be categories exclusive to each other, but Shen Jiaxuan offers a new perspective which proposes nouns are inclusive of verbs.

Zhu Dexi 朱德熙's (1988) paper examines the relations between nouns and verbs in Archaic Chinese and directs particular attention to the "verbness" of nouns. Building upon Zhu's work, Shen Jiaxuan (Shen 2012) reexamines these two most important word classes in pre-Qin era and proposes the inclusive relation of nouns and verbs (for detailed introduction of Shen's proposal, see ▶ Chap. 3, "Cognitive Linguistics in China" by Quan Wan in the Handbook).

Other important works on the topic include Zhang Wenguo 张文国 (2005), Ren He (2014), and Yuan Jianhui 袁健惠 (2017). In particular, Wan Qun 万群 (2015) uses the *Guoyu* 国语 as the corpus for linguistic data and systematically compares the features of nouns and verbs and the convertibility of the two word classes in pre-Qin era.

The Mechanisms for the Emergence of Chinese Aspect Markers

On Perfective Verbs, Perfective Markers, and Perfective Constructions

In Modern Chinese, completion is marked by perfective markers and perfective constructions. Tsu-lin Mei (1981, 1999) traces the origin and development of this grammatical function back to its early occurrence in Archaic Chinese. Language data from texts such as the *Shiji* 史记, *Hanshu* 汉书, and *Zhanguo zonghengjia shu* 战国纵横家书 indicate that perfective structures such as "V (O) (Adv.) 已 + clause" were already in use in Archaic Chinese, and that this continues to develop through Middle Chinese into Late Tang. These data also show that 已 (*yi*)'s slot in the structure can also be taken by other verbs meaning completion (such as 讫 *qi*, 竟 *jing*, and 毕 *bi*). Mei's two studies outline the origin and the development for this important syntactic phenomenon.

Zhu Qingzhi 朱庆之 (Zhu 1993) and Seishi Karashima (2000) both find that the emergence of perfective marker 已 is under the influence of Buddhist sutra. Zhu argues that 已 is a Chinese translation of the past participle in Sanskrit, while Karashima proposes the source might be the absolute participle.

Jiang Shaoyu (2001) examines the usage of verbs of completion including 已 (*yi*), 竟 (*jing*), 讫 (*qi*), and 毕 (*bi*) and how they might contribute to the emergence of aspect particle *le* in original Chinese texts and translated Chinese texts. Jiang focuses on the structures V+O+已 and explicates its development in two stages. The first stage is prior to the translation of Buddhist sutra into Chinese; 已 was used only after durative verb (phrase) to denote the completion of a durative action. The second stage is marked by the translation of Buddhist sutra when translators used 已 as a counterpart for the absolute participle in Sanskrit, which then gave rise to a new usage of 已 denoting the accomplishment of an action. In other words, the new usage arises under the influence of Sanskrit. And because of its high-frequency occurrence, the new usage gradually becomes entrenched and takes hold in the colloquial language. But according to such an analysis, 已 is the only predecessor of 了 (le).

Other scholars have also explored the origin of the V+了+O structure and the perfective aspect marker *le* (see Wu 1998a; Cao 1999; Chen and Zhang 2007; Yang 2009).

Cao Guangshun's (1995) monograph provides a historical and systematic review of the perfective marker in Early Modern Chinese. Li and Shi (1997) approaches from a diachronic perspective and discusses the general mechanisms of emergence for Chinese aspect markers. Drawing upon theories on aspect, Yang (2001) conducts a comprehensive investigation on perfective markers in the *Zhuzi yulei* 朱子语类. Shuai Zhisong 帅志嵩 (Shuai 2014) takes a lexical-semantic approach to investigate the different means completion is marked in Middle Chinese while tracing their historical development.

On Other Aspect Markers

Other important works that deal with Chinese aspect markers include Liu Jian's "On aspect markers and VO-*guo*" (Liu et al. 1989), Peng Rui's "Mapping synchronic relation and diachronic development: The case of the historical evolvement of aspect particle *guo*" (2009), and Chen Qianrui and Zhang Man 张曼's (2015) paper on the emergence of experiential marker 过 (guo).

Studies that feature the origin and development of durative marker 着 (*zhe*) include Tsu-lin Mei (1989), Chaofen Sun (1997), Wu Fuxiang (2004), and Jiang Shaoyu (2006).

Recent Advances in Important Syntactic Structures

Verb-complement structure, disposal construction, and passive construction have always been the center issues in Chinese historical syntax. Some recent advances on these issues will be discussed in the following section.

On Verb-Complement Structure (Verb-Resultative Complement)

The emergence and development of verb-complement structure (verb-resultative construction, henceforth VR construction) is a topic attracting much research attention since the 1950s. In the past 30 years or so, a series of high-quality monographs have been published in this venue.

The most debated question on this topic concerns the earliest occurrence of verb-complement structure (VR construction) and the criteria for its identification. Dissatisfied with criteria basing entirely on semantics and intuition of Modern Chinese speakers, Wang Li, Tatsuo Ōta, Ryoji Shimura, Tsu-lin Mei, and Jiang Shaoyu have all offered their opinions on this issue.

Tatsuo Ōta (1958/2003) was the first scholar to propose using verb pairs that have been exclusively transitive or intransitive to determine the compounding relation within the verb-complement structure. For instance, the pair 杀 (*sha* "to kill") and 死 (*si* "to die") are a perfect candidate because they are related semantically, yet contrast with each other in terms of transitivity (number of event participants). Ōta's argument is as follows: Prior to Tang Dynasty, only usages following the V+杀+O pattern were in use, but beginning from Tang, we can see usages in V+死+O and V+死 in the passive voice. It is thus reasonable to deduce that verb-complement structure must have been in existence by Tang at the latest. Ryoji Shimura (1995), Tsu-lin Mei (1990), adopts similar method, but their study both put the time of emergence to as early as Six Dynasties.

Other scholars have questioned the validity of using the V杀-V死 pair as the formal marking of this structure. Wu Fuxiang (2000) points out that $S_{agent}+V_{transitive}+$杀$+O$ changes from a serial verb construction to a VR construction due to the intransitivization of the transitive verb 杀 (when 杀 is increasingly used as a complement denoting the results of "being dead"). Because of the assimilation of 杀 toward 死 in both its syntactic property (intransitive) and lexical meaning (animate object becoming lifeless), 死 began to take the slot of 杀 in the structure, leading to the structure $S_{agent}+V_{transitive}+$死$+O$. Similar discussions can be found in Song Shaonian 宋绍年 (1994), Yang Rongxiang 杨荣祥 (2002), and Liang Yinfeng 梁银峰 (2003), *inter alia*.

Jiang Shaoyu (1999a, b) provides an in-depth discussion on the period, path of emergence, and identification criteria of VR construction. Jiang points out that while semantic criteria is important, they should not be the only criteria. For instance, with the word 扑灭 (*pu-mie*, "to extinguish"), we should not come to conclusion based on the judgment that 扑 (*pu*, "to dash") is an action and 灭 (*mie*, "(for light and fire) to go out") is a result; it is equally insufficient to base our judgment of historical usage on modern-day interpretation. Jiang proposes that since the VR construction V1+V2 often evolves from the serial verb construction V1+V2, the relationship between the two verbs could be used as criteria of judgment. The first criterion taps into the semantic-syntactic behavior of V2 and the following object. When V2 is transitive or is a causative usage of an intransitive verb or adjective and V2 takes the following object as its argument, V1+V2 should be considered a serial verb construction, but when V2 becomes intransitivized or grammaticalized, or when the intransitive V2 does not take on causative usage and does not take the following object as its argument, then V1+V2 can be recognized as a VR construction. The second criterion looks at the semantic contribution of V1 and V2 in the structure. In a serial verb construction where V1 and V2 complement each other, the overall meaning is usually determined by V2. While in a VR construction, the overall meaning is usually determined by V1. In these two criteria, the first criterion has precedence over the second one.

Based on the principle of form-function mapping, Hu Chirui 胡敕瑞 (2005) deduces that VR constructions evolve from two structures in history: V+Complement+O (VCO) and V+O+Complement (VOC). Hu makes a distinction between three types of verbs in classical Chinese: (1) action verbs, which are typically transitive verbs that take an object (such as 杀 *sha* "to kill"); (2) state verbs, which are typically intransitive verbs that do not take an object (such as 熟 *shu* "to be fully cooked/mature"); and (3) action-state verbs, which have both transitive and intransitive usages (such as 破 *po* "to break"). Hu shows that action-state verbs are the main type that occupies the resultative complement slot; because it has the potential to be interpreted either way, the state interpretation slowly overtakes the action interpretation. The dominance of state interpretation of this type of verbs is thus thought to mark the emergence of VR constructions. And by this criterion, the earliest instances of the VR construction were found in VCO and VOC structures where the complement is filled by the third type of verbs, before other types of verbs were allowed in the slot.

Wei Peiquan (2000) discusses different types of causative constructions and how they might have an impact on the emergence of the verb-complement structure. Liu Chenghui 刘承慧 (2002) is a diachronic investigation of the development of verb-complement structures in Chinese. Liu sheds light on how causative construction might have contributed to the emergence of verb-complement structures. Departing from a synchronic perspective, Wu Fuxiang (1998b) examines syntactic, semantic features of verb-complement structures in Modern Chinese, while also tracing the origin, development, and factors of emergence of related structures. Other important works on the topic include Xu Dan 徐丹 (2001), Shengli Feng (2002), and Shi Yuzhi (2002).

There are in fact a vast array of subtypes for the verb-complement structure, each subtype following a distinct path and timeframe in its development. Tsu-lin Mei (1990), Jiang Shaoyu (1995, 2003), and Zhao Changcai 赵长才 (Zhao 2010) are several important papers that discuss their uneven development, focusing especially on the VOC type verb-complement structure.

All these subtypes of VR constructions reached full-blown development by Early Modern Chinese. Liang Yinfeng's (2006) monograph identifies five historical stages for the development of VR constructions as a whole while laying out the uneven development of each subtype in each period. Liu Ziyu 刘子瑜's (2008) monograph offers new insights into the categorization of subtypes of VR construction. Her conclusion was based on theoretical generalizations about the path of development, the driving forces, and mechanisms of change observed in *Zhuzi yulei*. Shi Huimin 石慧敏 (Shi 2011) introduces the theory of conceptual blending in her analysis. By combining diachronic and synchronic data, she analyzes the syntactic and semantic differences and prototypes of various VR constructions in terms of their scalarity and reviewed the driving forces and mechanisms of change including syntacto-semantic factors, disyllabification, and analogy, as well as the cognitive mechanisms of metaphor and metonomy.

On Disposal Constructions

The development of disposal constructions in Chinese has been a heated research topic since the 1990s. Scholars are interested in the categorization of disposal construction subtypes, the linguistic features of these subtypes, and how they evolved into the current state.

Disposal constructions only occur in one structure in Archaic Chinese: "以 *yi* +O_1+V+O_2," which is a general disposal construction. In Middle Chinese, other verbs meaning "to hold" (such as 取 *qui*, 持 *chi*, and 捉 *zhuo*) began to show up in this construction, and they possess usages that are both general and specific – these are also the predecessors of disposal constructions emerging in later periods led by 將 (*jiang*) and 把 (*ba*). Cao Guangshun and Yu Hsiao-jung 遇笑容 (Cao and Yu 2000) discuss disposal construction led by 取 (*qu*) in Middle Chinese in terms of its syntactic behavior and its origin. Zhu Guanming 朱冠明's (2002) analysis otherwise focuses on the construction led by 持 (*chi*).

Tsu-lin Mei (1990) argues that the disposal construction is a construction type that has diverse members. From a diachronic standpoint, the overall development of the disposal construction is the outcome of accumulative developments at different stages. In other words, the emergence of disposal construction does not necessarily follow a singular path or is motivated by a singular mechanism. It contains various historical strata, each having a different origin and going through different levels of grammaticalization. These seemingly similar constructions should not be united under a singular function/meaning but should each be attended to individually.

Wu Fuxiang (2003c) put forward a path of development for the Chinese disposal construction, which is as follows: serial verb construction > instrumental construction >general disposal construction > specific disposal construction > causative disposal construction. In the early stages of development (up until general disposal construction), the mechanism at work is reanalysis; at the later stages, it is extension at work.

Jiang Shaoyu (2005) made a distinction between disposal construction and instrumental construction. In the instrumental construction written as "$P_{instrument}$+O_1+V_2+O_2," the two objects are different concepts/entities; in the specific disposal construction "$P_{disposal}$+O_1+V_2+O_2," the two objects refer to the same entity/concept. In addition, in the instrumental construction "以+O_1+V," O_1 is not the semantic patient of the verb, but in the disposal construction "以+O_1+V," O_1 is. In a more recent article, Jiang (2008) deals with the general disposal construction, its origin, and its identification.

Guo Haoyu 郭浩瑜 (Guo 2010, 2012) distinguishes four types of disposal constructions: general, specific, causative, and suffering. She also offers discussions on their different processes of emergence.

Liu Ziyu (1995), Jiang Shaoyu (1997, 1999a, b), and Zhao Changcai (2010) all examine the various types of disposal constructions in Middle Chinese through Early Modern Chinese. A lot of discussion has also been devoted to whether Sanskrit influence contributed to the emergence of the disposal construction and how the disposal construction could be further categorized. These works, together with many others, have cleared the path for the research on Chinese disposal constructions.

On Passive Constructions

Starting in the 1950s and 1960s, scholars started to investigate the origin of the passive construction and passive markers. The research interest was rekindled in the 1980s as new theories were introduced and new language data from Archaic and

Middle Chinese were excavated. Important works on this topic include the following: Tang Yuming 唐钰明 and Zhou Xipo 周锡馥 (1985), Xie Huiquan and Hong Bo (1987), Tang Yuming (1987, 1988, 1991), Yao Zhenwu (1988, 1999), and Zhu Qingzhi (1995), among others.

He Hongfeng 何洪峰 (2004) discusses the emergence of the passive marker from the perspective of word order typology and topicalization. He hypothesizes that because Chinese has SVO word order and exhibits a topic-comment structure, when semantic patient becomes the topic of a sentence and is placed sentence-initial position, its semantic role is inconsistent with the one assigned by the structure and thus requires additional marking.

Shi Bing 时兵 (2008) focuses on the structure "$NP_{patient}+V+于+NP_{agent}$" in Archaic Chinese, which is a special construction expressing passivity. Shi posits that it originates from the structure $NP_{stimulus}+V+于+NP_{experiencer}$ found in oracle bone inscriptions and then became the $NP_{receiver}+V+ (NP_{object})+于+NP_{agent}$ structure in bronze inscriptions, which became a stable form during the Warring States period. Shi points out that both for the second stage of development ($NP_{stimulus}+V+于+NP_{experiencer}$) and choice of agent markers, there are typological universality and cognitive bases.

Song Yayun (2007) is a study that focuses on the patient-subject sentences. After examining the usage of 围 (*wei*, "to surround"), 诛 (*zhu*, "to kill"), 延 (*yan*, "to invite"), 就 (*jiu*, "to approach"), 剔 (*ti*, "to remove"), 剖 (*po*, "to cut open"), 烹 (*peng*, "to cook"), and 斩 (*zhan*, "to behead") (which are typical verbs in this sentence type) in Qin and Han texts, Song highlights the features and conditions of usage for patient-subject sentences and discusses the factors behind its rise and fall.

Another arena in the discussion of the passive construction is investigation on the nonstandard varieties of passive constructions led by 教 (*jiao*), 叫 (*jiao*), 让 (*rang*), 给 (*gei*), and 吃 (*chi*), which have implications for how passivity is related to causative and benefactive meanings. Some of the most outstanding papers include Jiang Lansheng (1989, 1999c), Jiang Shaoyu (2002, 2004), and Hong Bo and Zhao Ming 赵茗 (Hong and Zhao 2005).

Important Publications on Chinese Historical Syntax

Research on the History of Chinese Syntax in Early Modern Chinese, edited by Jiang Shaoyu and Cao Guangshun (published by the Commercial Press in 2005), is a comprehensive review of studies done on the historical syntax of Early Modern Chinese. The volume is organized into 14 chapters, each mapping out recent progresses in a certain topic in Chinese word classes and Chinese syntax.

An Outline for Early Modern Chinese by Jiang Shaoyu (revised version published by Peking University Press in 2017) is both a classic textbook and a research monograph. In the part on syntax, while reviewing some key issues in Early Modern Chinese research, the author offers insightful and original comments on these issues.

An Outline for Syntax in Archaic Chinese by Mei Kuang 梅广 (published by Taiwan Sanmin Book House in 2015, reprinted by Shanghai Education Press in

2018) is an in-depth investigation of syntax in Archaic Chinese in the generative framework. While the book is fundamentally theoretical, it also draws upon language data and provides a holistic description of the syntax of Archaic Chinese; it sheds light on the diachronic development of Archaic Chinese and has important implications on the typological changes of Chinese.

The Future for Research on Chinese Historical Syntax

In the future, more and more research will be conducted on the convergence of theoretical discussions and systematic description of linguistic facts. While digging into solid empirical data (historical written data, dialectal data, and language data from ethnic minority groups), we should also strengthen theoretical exploration and make summarizations, which might lead to the discovery of general linguistic patterns.

As regarding research methods, there should be a concerted effort to group together research in Archaic Chinese, Middle Chinese, Early Modern Chinese, and Modern Chinese, as well as to combine classic documents with newly excavated documents, written records with colloquial data from Chinese dialects. We should encourage typological studies that compare Chinese historical syntax with Chinese dialectal syntax, Chinese with neighboring languages of ethnic minorities, and Chinese with foreign languages.

References

Aldridge, Edith. 2012. Focus and Archaic Chinese word order. In *The proceedings of the 22nd North American Conference of Chinese Linguistics (NACCLS-22) and the 18th annual meeting of the International Association of Chinese Linguistics (IACL-18)*, ed. Lauren Eby Clemens and Chi-Ming Louis Liu, vol. 2, 84–101.

———. 2015. Pronominal object shift in Archaic Chinese. In *Syntax over time: Lexical, morphological and information-structural interactions*, ed. Theresa Biberauer and George Walkden, 350–370. Oxford: Oxford University Press.

Cao, Guangshun (曹广顺). 1995. Jindai Hanyu Zhuci 近代汉语助词 [*Auxiliary words in pre-modern Chinese*]. Beijing: Yuwen Chubanshe 北京:语文出版社 [Beijing: Language and Culture Press].

———. and Hsiao-jung Yu (遇笑容). 2000. Zhonggu yijingzhong de chuzhishi 中古译经中的处置式 [The disposal construction of translated Middle Chinese Buddhist scriptures]. *Zhongguo Yuwen* 中国语文 [*Studies of Chinese Language*] 2000.6: 556–563.

Cikoski, John S. 1978a. An outline sketch of sentence structures and word classes in classical Chinese. *Three essays on classical Chinese grammar*: I. Computational analyses of Asian & African Languages no.8.

———. 1978b. An analysis of some idioms commonly called "passive" in classical Chinese. *Three essays on classical Chinese grammar*: III. Computational Analyses of Asian & African Languages no.9.

Chou, Fagao (周法高). 1959. *Zhongguo Gudai Yufa: Chengdai Pian (Shang)* 中国古代语法·称代篇(上) [*Classical Chinese grammar: Substitution*]. Taibei: Zhongyanyuan Lishi Yuyan Yanjiusuo Zhuankan 台北:中研院历史语言研究所专刊 [Taipei: Monograph of IHP].

Djamouri, Redouane. 2000. Preverbal positions of the pronominal object in Archaic Chinese. Paper presented at *the 9**th* *International Conference on Chinese Linguistics*. The National University of Singapore.

Feng, Shengli. 1996. Prosodically constrained syntactic changes in early archaic Chinese. *Journal of East Asian Linguistics* 5: 323–371.

——— (冯胜利). 2002. Hanyu dongbu jiegou laiyuan de jufa fenxi 汉语动补结构来源的句法分析 [A formal analysis of the origin of VR-constructions in Chinese]. *Yuyanxue Luncong* 语言学论丛 [*Essays on Linguistics*] 26: 178–208. Beijing: Shangwu Yinshuguan 北京:商务印书馆 [Beijing: The Commercial Press].

Guo, Haoyu (郭浩瑜). and Rongxiang Yang (杨荣祥). 2012. Cong "kongzhidu" kan chuzhishi de butong yufa yiyi 从"控制度"看处置式的不同语法意义 [On the different grammatical meanings of the disposal constructions from the perspective of the control degree]. *Guhanyu Yanjiu* 古汉语研究 [*Research in Ancient Chinese Language*] 2012.4:50–55.

Hashimoto, Mantaro (桥本万太郎) 1985/2008. *Yuyan Dili Leixingxue* 语言地理类型学 [*A Geographical theory of linguistic typology*]. Beijing: Beijing Daxue Chubanshe 北京:北京大学出版社 [Beijing: Peking University Press].

He, Leshi (何乐士). 1989/2004. *Zuozhuan Xuci Yanjiu* 左传虚词研究 [*Study on function words in Zuozhuan*]. Beijing: Shangwu Yinshuguan 北京:商务印书馆 [Beijing: The Commercial Press].

Hu, Chirui (胡敕瑞) 2005. Dongjieshi de zaoqi xingshi jiqi panduan biaozhun 动结式的早期形式及其判断标准 [The early forms of resultative construction and the relevant criterion]. *Zhongguo Yuwen* 中国语文 [*Studies of Chinese Language*] 2005.3: 214–225.

Jiang, Lansheng (江蓝生). 1999a. Yufahua chengdu de yuyin biaoxian 语法化程度的语音表现 [Phonetic representation of degree in grammaticalization]. *Zhongguo Yuyanxue de Xin Tuozhan* 中国语言学的新拓展 [*Recent advances in Chinese linguistics*]. ed. by Feng shi (石锋) and Wuyun Pan (潘悟云). 195–204. Xianggang: Xianggang Chengshi Daxue Chubanshe 香港:香港城市大学出版社 [Hongkong: City of Hongkong University Press].

——— (江蓝生). 1999b. Chusuoci de lingge yongfa yu jiegou zhuci "di" de youlai 处所词的领格用法与结构助词"底"的由来 [The genitive use of the place words and the orgin of the structural auxiliary "di"] *Zhongguo Yuwen* 中国语文 [*Studies of Chinese Language*] 1999.2: 83–93.

——— (江蓝生). 2005. "VP de hao" jushi de liangge laiyuan: Jiantan jiegou de yufahua "VP 的好"句式的两个来源——兼谈结构的语法化 [Two sources of "VP de hao" combination: a case study on grammaticalization of a construction]. *Zhongguo Yuwen* 中国语文 [*Studies of Chinese Language*] 2005.5:388–398.

——— (江蓝生). 2012. Hanyu lian-jieci de laiyuan jiqi yufahua de lujing he leixing 汉语连—介词的来源及其语法化的路径和类型 [Sources, paths, and types of the grammaticalization of Chinese conjunction-prepositions]. *Zhongguo Yuwen* 中国语文 [*Studies of Chinese Language*] 2012.4: 291–308.

Jiang, Shaoyu (蒋绍愚). 2001. *Shishuoxinyu, Qiminyaoshu, Luoyangqielanji, Xianyujing, Baiyujing* zhong de "yi" "jing" "qi" "bi"《世说新语》,《齐民要术》,《洛阳伽蓝记》,《贤愚经》,《百喻经》中的"已","竟","讫","毕" ["yi" "jing" "qi" "bi" in *Shishuoxinyu, Qiminyaoshu, Luoyangqielanji, Xianyujing, Baiyujing*]. *Yuyan Yanjiu* 语言研究 [*Studies in Language and Linguistics*] 2001.1: 73–78.

——— (蒋绍愚). 2003. Weijin Nanbeichao de "shubinbu" shi shubu jiegou 魏晋南北朝的"述宾补"式述补结构 [The 'V-O-C' type verb-complement construction in Weijin and Nanbei dynasties]. *Guoxue Yanjiu* 国学研究 [*Sinological Studies*] 12: 293–322. Beijing: Beijing Daxue Chubanshe 北京:北京大学出版社 [Beijing: Peking University Press].

——— (蒋绍愚). 2017. Shanggu hanyu de zuoge dongci 上古汉语的作格动词 [Ergative verbs in old Chinese]. *Lishi Yuyanxue Yanjiu* 历史语言学研究 [*Studies on Historical Linguistics*] 11: 1–28. Beijing: Shangwu Yinshuguan 北京:商务印书馆 [Beijing: The Commercial Press].

Karashima, Seishi (辛嶋静志). 2000. Hanyi fodian de yuyan yanjiu 汉译佛典的语言研究 [Language study of Chinese Buddhist sutras]. *Wenhua de Kuizeng:Hanxue Yanjiu Guoji Huiyi Lunwenji* 文化的馈赠——汉学研究国际会议论文集 [*The gift of culture: Proceedings of*

Internetional Conference on Sinological Research]. 512–514. Beijing: Beijing Daxue Chubanshe 北京:北京大学出版社 [Beijing: Peking University Press].

Li, Charles N., and Sandra A. Thompson. 1974. An explanation of word order change: SVO-SOV. *Foundations of Language*. 12: 201–214.

Li, Charles N. (李讷). and Yuzhi Shi (石毓智). 1997. Lun hanyu tibiaoji dansheng de jizhi 论汉语体标记诞生的机制 [On the mechanism of the birth of Chinese aspect markers]. *Zhongguo Yuwen* 中国语文 [*Studies of Chinese Language*] 1997.2: 82–96.

Li, Zuofeng (李佐丰). 1994a. Xianqin de bujiwu dongci he jiwu dongci 先秦的不及物动词和及物动词 [Intransitive verbs and transitive verbs in pre-Qin Chinese]. *Zhongguo Yuwen* 中国语文 [*Studies of Chinese Language*] 1994.4:287–296.

——— (李佐丰). 1994b. *Xianqin Hanyu Shici* 先秦汉语实词 [*Notional words in pre-Qin Chinese*]. Beijing: Beijing Guangbo Xueyuan Chubanshe 北京:北京广播学院出版社 [Beijing: Beijing Broadcasting Institute Publishing House].

Liang, Yinfeng (梁银峰). 2006. *Hanyu Dongbu Jiegou de Chansheng yu Yanbian* 汉语动补结构的产生与演变 [*The origin and evolution of Chinese verb-complement constructions*]. Shanghai: Xuelin Chubanshe 上海:学林出版社 [Shanghai: Academia Press].

Liu, Chenghui (刘承慧). 1999. Shilun shichengshi de laiyuan jiqi chengyin 试论使成式的来源及其成因 [The historical development of the shicheng construction]. *Guoxue Yanjiu* 国学研究 [*Sinological Studies*] 6: 349–386. Beijing: Beijing Daxue Chubanshe 北京:北京大学出版社 [Beijing: Peking University Press].

——— (刘承慧). 2002. *Hanyu Dongbujiegou de Lishi Fazhan* 汉语动补结构的历史发展 [*Historical evolution of Chinese verb-complement construction*]. Taibei: Hanlu Tushu Chuban Youxian Gongsi 台北:瀚芦图书出版有限公司 [Taipei: Hanlu Book Publishing Co., Ltd].

Liu, Danqing (刘丹青). 2001. Yufahua zhong de gengxin, qianghua yu diejia 语法化中的更新、强化与叠加 [Renovation, reinforcement and superposisiton in grammaticalization]. *Yuyan Yanjiu* 语言研究 [*Studies in Language and Linguistics*] 2001.2: 71–81.

——— (刘丹青). 2004. Xianqin hanyu yuxu tedian de leixingxue guanzhao 先秦汉语语序特点的类型学观照 [Typological observations of word order characteristics in pre-Qin Chinese]. *Yuyan Yanjiu* 语言研究 [*Studies in Language and Linguistics*] 2004.1: 37–46.

Liu, Jian (刘坚). 1989. Shilun 'he'zi de fazhan, fulun 'gong'zi he 'lian'zi 试论"和"字的发展,附论"共"字和"连"字 [Discussion on the evolution of he, gong and lian]. *Zhongguo Yuwen* 中国语文 [*Studies of Chinese Language*] 1989.6: 447–453.

———., Guangshun Cao (曹广顺). and Fuxiang Wu (吴福祥). 1995. Lun youfa hanyu cihui yufahua de ruogan yinsu 论诱发汉语词汇语法化的若干因素 [Some causative factors in grammaticalization of Chinese lexicon]. *Zhongguo Yuwen* 中国语文 [*Studies of Chinese Language*] 1995.3: 161–169.

Liu, Zhengguang (刘正光). ed. 2011. *Ganshi Yufa Yanjiu* 构式语法研究 [*Studies on construciton grammar*]. Shanghai: Shanghai Jiaoyu Chubanshe 上海:上海教育出版社 [Shanghai Educational Publishing House].

Liu, Ziyu (刘子瑜). 2008. *Zhuziyulei Shubujiegou Yanjiu*《朱子语类》述补结构研究 [*Studies on predicate-complement construction in Zhuziyulei*]. Beijing: Shangwu Yinshuguan 北京:商务印书馆 [Beijing: The Commercial Press].

Ma, Beijia (马贝加). 2014. *Hanyu Dongci Yufahua* 汉语动词语法化 [*Grammaticalization of Chinese verbs*]. Beijing: Zhonghua Shuju 北京:中华书局 [Beijing: Zhonghua Book Company].

Matsue, Takashi (松江崇). 2010. *Guhanyu Yiwen Binyu Cixu Bianhua Jizhi Yanjiu* 古汉语疑问宾语词序变化机制研究 [*A study on word order change of interrogative pronouns as objects in classical Chinese*]. Dongjing: Haowen Chuban 东京:好文出版 [Tokyo: Kōben Publishing house].

Mei, Kuang. 1980. Is Modern Chinese really a SOV language? *Cahiers de linguistique-Asie Orientale*. 7: 23–45.

———(梅广). 2015. *Shanggu Hanyu Yufa Gangyao* 上古汉语语法纲要 [*An outline for syntax in Archaic Chinese*]. Taibei: Sanmin Shuju 台北:三民书局 [Taipei: Sanmin Book Co Ltd].

Mei, Tsu-lin (梅祖麟). 1981 Xiandai hanyu wanchengmao jushi he ciwei de laiyuan 现代汉语完成貌句式和词尾的来源 [The historical sources of perfective sentence pattern construction and perfective aspect marker]. *Yuyan Yanjiu* 语言研究 [*Studies in Language and Linguistics*]. 1981.1: 65–77.

——— (梅祖麟). 1990. Tang Song chuzhishi de laiyuan 唐宋处置式的来源 [The origin of the disposal construction in Tang and Song dynasties]. *Zhongguo Yuwen* 中国语文 [*Studies of Chinese Language*] 1990.3: 191–206.

——— (梅祖麟). 1991. Cong Handai de "dong, sha", "dong, si" lai kan dongbujiegou de fazhan 从汉代的"动、杀"、"动、死"来看动补结构的发展:兼论中古时期起词的施受关系的中立化 [The historical development of the 'verb-resultative complement' construction, with a note on the neutralization of the pre-verbal agent/patient distinction in Middle Chinese]. *Yuyanxue Luncong* 语言学论丛 [*Essays on Linguistics*].16: 112–136. Beijing: Shangwu Yinshuguan 北京:商务印书馆 [Beijing: The Commercial Press].

——— (梅祖麟). 1999. Xianqin Lianghan de yizhong wanchengmao jushi: Jianlun xiandai hanyu wanchengmao jushi de laiyuan 先秦两汉的一种完成貌句式——兼论现代汉语完成貌句式的来源 [A Chinese perfective sentence pattern in pre-Qin and Han dynasties, with a note on the orgin of perfective sentence pattern in Modern Chinese]. *Zhongguo Yuwen* 中国语文 [*Studies of Chinese Language*] 1999.4: 285–294.

Niu, Baoyi (牛保义). 2011. *Goushi Yufa Lilun Yanjiu* 构式语法理论研究 [*A study on theories of construction grammar*]. Shanghai: Shanghai Waiyu Jiaoyu Chubanshe 上海:上海外语教育出版社 [Shanghai: Shanghai Foreign Language Education Press].

Ōnishi, Katsuya (大西克也). 2004. Shishoutongci chuyi: Shiji de zhongxing dongci he zuoge dongci 施受同辞刍议——《史记》的"中性动词"和"作格动词" [On shī shòu tóng cí: neutral verbs and ergative verbs in *Shiji*]. ed. by Takashima K. (高岛谦一) and Shaoyu Jiang (蒋绍愚). *Yiyi yu Xingshi: Gudai Hanyu Yufa Lunwenji* 意义与形式——古代汉语语法论文集 [*Meaning and form: Essays in pre-modern Chinese grammar*]. 291–304. Muenchen: Lincom Europa.

Ōta Tatsuo (太田辰夫). 1958/1987. *Zhonguoyu Lishi Wenfa* 中国语历史文法 [*The historical grammar of Chinese*]. trans. by Shaoyu Jiang (蒋绍愚) and Changhua Xu (徐昌华). Beijing: Beijing Daxue Chubanshe 北京:北京大学出版社 [Beijing: Peking University Press].

Peng, Rui (彭睿). 2020. *Yufahua Lilun de Hanyu Shijiao* 语法化理论的汉语视角 [*Grammaticalization theory: a perspective from Chinese language*]. Beijing: Beijing Daxue Chubanshe 北京:北京大学出版社 [Beijing: Peking University Press].

Peyraube, Alain (贝罗贝). 2013. Shangu hanyu de yuxu 上古汉语的语序 [On word order in Archaic Chinese]. *Jingwai Hanyu Lishi Yufa Yanjiu Wenxuan* 境外汉语历时语法研究文选 [*Selected readings in Chinese historical syntax*]. ed. by Fuxiang Wu (吴福祥). 181–192. Shanghai: Shanghai Jiaoyu Chubanshe 上海:上海教育出版社 [Shanghai: Shanghai Educational Publishing house].

Ren, He (任荷). 2020. *"Mingci Dongyong" yu Shanggu Hanyu Mingci he Dongci de Yuyi Shuxing* "名词动用"与上古汉语名词和动词的语义属性 [*"Noun-verb" conversion in Archaic Chinese: from the perspective of lexical-semantic analysis*]. Beijing: Zhongguo Shehui Kexue Chubanshe 北京:中国社会科学出版社 [Beijing: China Social Sciences Press].

Shen, Jiaxuan (沈家煊). 1994. Yufahua yanjiu zongguan "语法化"研究综观 [A survey of studies on grammaticalization]. *Waiyu Jiaoxue yu Yanjiu* 外语教学与研究 [*Foreign Language Teaching and Research*] 1994.4: 17–24.

——— (沈家煊). 1998a. Yuyongfa de yufahua 语用法的语法化 [Grammaticalization of pragmatics]. *Fujian Waiyu* 福建外语 [*Foreign Language and Literature Studies*]1998.2: 1–8,14.

——— (沈家煊). 1998b. Shici xuhua de jizhi: Yanhua erlai de yufa pingjie 实词虚化的机制——《演化而来的语法》评介 [Mechanisms of grammaticalization: a review of *the evolution of grammar*]. *Dangdai Yuyanxue* 当代语言学 [*Contemporary Linguistics*] 1998.3: 41–46.

——— (沈家煊). 2012. Guanyu Xianqin hanyu mingci he dongci de qufen 关于先秦汉语名词和动词的区分 [On the distinction of nouns and verbs in pre-Qin Chinese]. *Zhongguo Yuyan Xuebao* 中国语言学报 [*Journal of Chinese Linguistics*] 15: 100–113. Beijing: Shangwu Yinshuguan 北京:商务印书馆 [Beijing: The Commercial Press].

Shen, Pei (沈培). 1992. *Yinxu Jiagu Buci Yuxu Yanjiu* 殷墟甲骨卜辞语序研究 [*A study on word order of oracle inscriptions in the Yin Ruins*]. Taibei: Wenjin Chubanshe 台北:文津出版社 [Taipei: Wen Chin Press].

Shi, Huimin (石慧敏). 2011. *Hanyu Dongjieshi de Zhenghe yu Lishi Yanbian* 汉语动结式的整合与历时演变 [*Integration and diachronic evolution of Chinese resultative construction*]. Shanghai: Fudan Daxue Chubanshe 上海:复旦大学出版社 [Shanghai: Fudan University Press].

Shi, Yuzhi (石毓智), and Jie Xu (徐杰). 2001. Hanyushi shang yiwen xingshi de leixingxue zhuanbian jiqi jizhi: Jiaodian biaoji "shi" de chansheng jiqi yingxiang 汉语史上疑问形式的类型学转变及其机制——焦点标记"是"的产生及其影响 [The process and mechanism of the typological change of the question form in the history of Chinese: The emergence of the focus marker shi and its effects]. *Zhongguo Yuwen* 中国语文 [*Studies of Chinese Language*] 2001.5: 454–465.

Shimura, Ryoji (志村良治). 1995. *Zhongguo Zhongshi Yufashi Yanjiu* 中国中世语法史研究 [*A study on historical grammar in middle Chinese*]. trans.by Lansheng Jiang (江蓝生) and Weiguo Bai (白维国). Beijing: Zhonghua Shuju 北京:中华书局 [Beijing: ZhongHua Book Company].

Song, Yayun (宋亚云). 2014. *Hanyu Zuoge Dongci de Lishi Yanbian Yanjiu* 汉语作格动词的历史演变研究 [*A study on the historical evolution of Chinese ergative verbs*]. Beijing: Beijing Daxue Chubanshe 北京:北京大学出版社 [Beijing: Peking University Press].

Sun, Chaofen, and Tamly Givon. 1985. On the so-called SOV word order in Mandarin Chinese: a quantified text study and its implications. *Language* 61 (2): 329–351.

Tai, James H.-Y. 1976. On the change from SVO to SOV in Chinese. In *Papers from the Parasession on diachronic syntax*, ed. S. Steever, C. Walker, and S. Mufwene, 291–304. Chicago: Chicago Linguistic Society.

Tang, Yuming (唐钰明), and Xipo Zhou (周锡馥). 1985. Lun Xianqin hanyu beidongshi de fazhan 论先秦汉语被动式的发展 [A discussion on the development of the passive construction in pre-Qin Chinese]. *Zhongguo Yuwen* 中国语文 [*Studies of Chinese Language*] 1985.4: 281–285.

——— (周锡馥). 1987. Han-wei Liuchao beidongshi luelun 汉魏六朝被动式略论 [A brief discussion of passive construction in Han-wei and six dynasties]. *Zhongguo Yuwen* 中国语文 [*Studies of Chinese Language*] 1987.3: 216–223.

Ting, Pang-hsin (丁邦新). 1997. Hanyu cixu wenti zhaji 汉语词序问题札记 [A note on word order of Chinese]. *Zhongguo Jingnei Yuyan ji Yuyanxue: Yuyan Leixing* 中国境内语言暨语言学:语言类型 [*Chinese Language and Linguistic IV: Language Typology*] 4: 155–162. Taipei: Zhongyanyuan Yuyan Yanjiusuo 台北:中研院语言研究所 [Taipei: Academia Sinica].

Wang, Li (王力). 1958. *Shangyushi Gao (Zhongce)* 汉语史稿(中册) [*Manuscript of Chinese language history (vol.2)*]. Beijing: Zhonghua Shuju 北京:中华书局 [Beijing: Zhonghua Book Company].

Wang, Yin (王寅). 2011. *Goushi Yufa Lilun* 构式语法理论 [*Researches on construction Grammar*]. Shanghai: Shanghai Waiyu Jiaoyu Chubanshe 上海:上海外语教育出版社 [Shanghai: Shanghai Foreign Language Education Press].

Wei, Peichuan (魏培泉). 2000. Shuo zhongu hanyu de shicheng jiegou 说中古汉语的使成结构 [The causative constructions in Middle Chinese]. *Zhongyanyuan Lishi Yuyan Yanjiusuo Jikan* 中研院历史语言研究所集刊 [*Bulletin of IHP*] 71–4: 807–856.

Wu, Fuxiang (吴福祥). 1998a. Chongtan "V+le+O" geshi de laiyuan he wanchengti zhuci 重谈"动+了+宾"格式的来源和完成体助词"了"的产生 [The further research of "V+le+O" structure's source and "le" as perfective particle's emergence]. *Zhongguo Yuwen* 中国语文 [*Studies of Chinese Language*] 1998.6: 452–462.

——— (吴福祥). 1998b. Shilun xiandai hanyu dongbu jiegou de laiyuan 试论现代汉语动补结构的来源 [The origin of Modern Chinese verb-complement construction]. *Hanyu Xianzhuang yu Lishi de Yanjiu* 汉语现状与历史的研究 [*Studies on the current situation and the history of Chinese*]. ed. by Lansheng Jiang (江蓝生) and Jingyi Hou (侯精一). 317–345. Beijing: Zhongguo Shehui Kexue Chubanshe 北京:中国社会科学出版社 [Beijing: China Social Sciences Press].

——— (吴福祥). 2000. Guanyu Dongbu jiegou "V si O" de laiyuan 关于动补结构"V死O"的来源 [On the formation of verb-complement "V si O"]. *Guhanyu Yanjiu* 古汉语研究 [*Research in Ancient Chinese Language*] 2000.3: 44–48.

——— (吴福祥). 2003a. Hanyu bansui jieci yufahua de leixingxue yanjiu 汉语伴随介词语法化的类型学研究 [A typological study of grammaticalization of the comitative preposition in Chinese Language]. *Zhongguo Yuwen* 中国语文 [*Studies of Chinese Language*] 2003.1: 43–58.

——— (吴福祥). 2003b. Nanfang fangyan nengxing shubu jiegou "V de/bu C" dai binyu de yuxu leixing 南方方言能性述补结构"V得/不C"带宾语的语序类型 [Types of word order of the potential verb-complement construction V+ de 得 +C/V+bu 不 +C with object in the southern Chinese dialect]. *Fangyan* 方言 [*Dialect*] 2003.3: 243–254.

——— (吴福祥). 2003c. Zailun chuzhishi de laiyuan 再论处置式的来源 [Further discussion on the origin of the disposal construction]. *Yuyan Yanjiu* 语言研究 [*Studies in Language and Linguistic*] 2003.3: 1–14.

——— (吴福祥). 2017. Hanyu fangyan zhong de ruogan niyufahua xianxiang 汉语方言中的若干逆语法化现象 [Antigrammaticalization in Chinese dialects]. *Zhongguo Yuwen* 中国语文 [*Studies of Chinese Language*] 2017.3: 259–279.

Xie, Huiquan (解惠全). 1987. Tan shici de xuhua 谈实词的虚化 [On the conversion of notional words into function words]. *Yuyan Yanjiu Luncong* 语言研究论丛 [*Linguistic Studies a Symposium*] 4: 208–227. Taijin: Nankai Daxue Chubanshe 天津:南开大学出版社 [Tianjin: Nankai University Press].

Xu, Dan. 2006. *Typological change in Chinese syntax*. Beijing: World Book Inc.

Yao, Zhenwu (姚振武). 1999. Xianqin hanyu shoushi zhuyuju xitong 先秦汉语受事主语句系统 [The system of accusative subject sentence in the pre-Qin Chinese]. *Zhongguo Yuwen* 中国语文 [*Studies of Chinese Language*] 1999.1: 43–53.

Yang, Bojun (杨伯峻), and Leshi He (何乐士). 1992. *Guhanyu Yufa jiqi Fazhan* 古汉语语法及其发展 [*Ancient Chinese grammar and its development*]. Beijing: Yuwen Chubanshe 北京:语文出版社 [Beijing: Language and Culture Press].

Yang, Yonglong (杨永龙). 2001. *Zhuziyulei Wanchengti Yanjiu*《朱子语类》完成体研究 [*Study on perfect aspect in Zhuziyulei*]. Kaifeng: Henan Daxue Chubanshe 开封:河南大学出版社 [Kaifeng: Henan University Press].

——— (杨永龙). 2012. Mudi goushi "VP qu" yu SOV yuxu de guanlian 目的构式"VP去"与SOV语序的关联 [The purpose construction "VP qu (去)" in Chinese and SOV order]. *Zhongguo Yuwen* 中国语文 [*Studies of Chinese Language*] 2012.6: 525–536.

Yin, Guoguang (殷国光). 1985. Xianqin hanyu dai yufa biaoji de binyu qianzhi jushi chutan 先秦汉语带语法标志的宾语前置句式初探 [A exploration of the Marked object-fronting lines in pre-Qin Chinese]. *Yuyan Yanjiu* 语言研究 [*Studies in Language and Linguistics*] 1985.2: 162–171.

——— (殷国光). 1997. *Lüshichunqiu Cilei Yanjiu*《吕氏春秋》词类研究 [*A study on word classes in Lüshichunqiu*]. Beijing: Huaxia Chubanshe 北京:华夏出版社 [Beijing: Huaxia Publishing House].

Yu, Jiang (于江). 1996. Jindai hanyu "he" lei xuci de lishi kaocha 近代汉语"和"类虚词的历史考察 [A survey of function word groups of 'he' in Modern Chinese]. *Zhonguo Yuwen* 中国语文 [*Studies of the Chinese Language*] 1996.6: 457–464.

Zhang, Cheng (张赪). 2010. *Hanyu Yuxu de Lishi Fazhan* 汉语语序的历史发展 [*The historical development of Chinese word order*]. Beijing: Beijing Yuyan Daxue Chubanshe 北京:北京语言大学出版社 [Bejing: Beijing Language and Culture University Press].

Zhang, Qingchang (张清常). 1989. Shanggu hanyu de SOV yuxu ji dingyu houzhi 上古汉语的SOV语序及定语后置 [SOV order in Archaic Chinese and the postpositive attributives]. *Yuyan Jiaoxue yu Yanjiu* 语言教学与研究 [*Language Teaching and Linguistic Studies*] 1989.1: 101–110.

Zhu, Dexi (朱德熙). 1988. Guanyu xianqin hanyu li mingci de dongcixing wenti 关于先秦汉语里名词的动词性问题 [Issues about the verbal characteristics of the nouns in pre-Qin Chinese]. *Zhongguo Yuwen* 中国语文 [*Studies of the Chinese Language*] 1988.1: 81–86.

Zhu, Jun (朱军). 2010. *Hanyu Goushi Yufa Yanjiu* 汉语构式语法研究 [*Construction grammar research of Chinese*]. Beijing: Zhongguo Shehui Kexue Chubanshe 北京:中国社会科学出版社 [Beijing: China Social Sciences Press].

The Sinicization of Buddhist Scriptural Language

Guanming Zhu

Contents

Introduction	214
The Eastern Spread of Buddhism, the Translation of Buddhist Classics, and Their Influence	214
The Initial Spread of Buddhism to China and Its Early Dissemination	214
An Overview of Buddhist Translation	216
The Flourishing of Buddhism and Its Significant Influence	217
Summary	219
The Characteristics of the Buddhist Chinese Language and the Reasons for Its Formation	219
The Characteristics of Buddhist Chinese	219
The Reasons Behind the Formation of the Characteristics of Buddhist Chinese	220
Summary	225
Research on "Sinicization" and Its Meaning Within Buddhist Studies	225
Research on Sinicization of Buddhist Language	225
On the Term "Sinicization" in Buddhist Studies	227
Summary	228
Modes of the Sinicization of Buddhist Language	228
The Origin of *Ziji* and Its Process of Sinicization	229
An Exegesis of *Shengse*, Its Origin, and Development	235
Historical Investigation of the Word *Xianzai*	239
Conclusion	246
References	247

Abstract

Soon after Buddhism was introduced to China during the Eastern and Western Han, a great number of Buddhist scriptures were translated into Chinese that possess a comparatively large number of unique characteristics when compared with locally produced literature. Just as Buddhism itself underwent a process of

---

Translated by T. L. Edwards, The Australian National University.

G. Zhu (✉)
Renmin University of China, Beijing, China
e-mail: zhuguanming@ruc.edu.cn

"Sinicization," many lexical and syntactic elements drawn from the language employed in the Buddhist classics were also gradually assimilated into the popular vernacular and brought about a "Sinicization" of Buddhist language. This chapter takes three terms, *ziji, shengse,* and *xianzai*, which exhibit differing degrees of lexical content and function to discuss concrete examples of how the spread of Buddhism affected the Chinese language. That is, the causes, processes, and channels by which Buddhist language was "Sinicized."

Keywords

Buddhism · Buddhist Chinese · Sinicization · *Ziji* · *Shengse* · *Xianzai*

Introduction

The language of the Buddhist scriptures refers to a type of language employed in Buddhist texts translated into Chinese that was produced when Buddhism was introduced into China following the large-scale translation of such texts. Over the passage of time along with the expansion of Buddhism's influence, a large number of lexical and syntactic elements drawn from this language were also gradually assimilated into Chinese and became part of the popular vernacular. This has not only enriched Chinese expressions but has also had a significant impact on the development and evolution of the Chinese language. This chapter discusses how the linguistic elements of this language were integrated into Chinese and the causes, processes, and channels by which Buddhist language was "Sinicized."

The Eastern Spread of Buddhism, the Translation of Buddhist Classics, and Their Influence

The Initial Spread of Buddhism to China and Its Early Dissemination

Buddhism is generally believed to have been introduced into China from India by way of Central Asia and Xinjiang during the period between the Eastern and Western Han in the first century CE. As far as we can see at present, the earliest unequivocal historical record of Buddhism's introduction into China can be found in Pei Songzhi's 裴松之 commentary on Yu Huan's 魚豢 *Xirong zhuan* 西戎傳 [*Account of the Xirong*] from the *Weilüe* 魏略 [*Brief History of Wei*] in the *Wuwan Xianbei Dongyi zhuan* 烏丸鮮卑東夷傳 [*Account of the Wuwan, Xianbei and Dongyi*] section of the *Sanguozhi* 三國志 [*Record of the Three Kingdoms*] which states:

> Previously, in the first year of the Yuanshou 元壽 era [2 BCE], during the reign of Emperor Ai of the Han 漢哀帝, Jing Lu 景盧, a student of the imperial academy received oral instructions from Yicun 伊存, an envoy of the king of the Great Yuezhi 大月氏 on the Buddhist *sūtra*s (*futu jing* 浮屠經) which say that this person is '*fuli*' 复立. 昔漢哀帝元壽

元年, 博士弟子景盧受大月氏王使伊存口授浮屠經, 曰復立者其人也. (*Sanguozhi*, Beijing: Zhonghua shuju, 1973; 30.854,)

According to Liu Xiaobiao's 劉孝標 commentary on the *Wen xue* 文學 [*Letters and Scholarship*] chapter in the *Shishuo xinyu* 世說新語 [*A New Account of the Tales of the World*], the term *fuli* 復立, as found in this text, should be written as *fudou* 復豆, which is a transliteration of "Buddha" (Gong Bin 2011: 412) (Kamata Shigeo (1980/1986: 17, 303) and Ren Jiyu (1981: 45, 90–91) both use material found in the *Weilüe* as evidence that Buddhism was introduced into China during the late Western Han and gradually became prevalent in society after the East Han. Zürcher (1990/1998), however, considers this material to be "of questionable historical value" and regards the edicts of Emperor Ming of the Han 漢明帝 (as discussed below) as being reliable evidence.). The influence of Buddhism at this time in China, however, was possibly not that great, and the *Shi-Lao zhi* 釋老志 [*Monograph on Buddhism and Taoism*] in the *Wei shu* 魏書 [*Book of Wei*] records the same account but adds the sentence, "the Central Lands [China] had heard of it [i.e. Buddhism], but did not yet believe in it 中土聞之, 未之信了也" (*Wei shu*, Beijing: Zhonghua shuju, 1977; 114.3025). To find figures in the historical record who believed in Buddhism, we need to look several decades after this. We find for example, that during the reign of Emperor Ming of the Eastern Han 漢明帝 (r. 57–75 CE), Liu Ying 劉英 the Prince of Chu 楚王 "fasted and performed sacrifices for the sake of the Buddha 為浮屠齋戒祭祀." In the eighth year of the Yongping 永平 era (65 CE) after Liu Ying offered fine silks to atone for a crime, Emperor Ming issued an edict declaring that the Prince of Chu, "recites the subtle words of the Yellow Emperor and Laozi and places great store in benevolent offerings to the Buddha. After purifying and fasting for three months, he has pledged an oath with the gods 誦黃老之微言, 尚浮屠之仁祠, 潔齋三月, 與神為誓" and that the silk offered to atone for his crime should be "used to aid the sumptuous entertainment of the *upāsaka* (*yipusai* 伊蒲塞) and *śramaṇa* (*sangmen* 桑門) 以助伊蒲塞、桑門之盛饌" ("Guangwu shi wang liezhuan" 光武十王列傳 ["Biographies of Ten Princes of Guangwu"] in *Hou Hanshu* 後漢書 [*Book of the Later Han*], Beijing: Zhonghua shuju, 1965; 42.1428). The *Hou Hanshu* 後漢書 [*Book of the Later Han*] also contains a record of Emperor Huan's 桓帝 faith in Buddhism by noting that he "erected a flowery canopy to offer sacrifice to the Buddha and Laozi 設華蓋以祠浮圖、老子" ("Xiao Huandi ji" 孝桓帝紀 ["Annals of Emperor Huan, the Filial"] in *Hou Hanshu*, 7.320). The above instances, however, are all examples of the Buddhist faith found in people from the court and upper levels of society. To find historical records of the Buddhist faith among the general populace, we possibly need to wait until the later part of the Eastern Han where the "Liu You zhuan" 劉繇傳 ["Biography of Liu You"] in the *Sanguozhi* contains the following account:

> Ze Rong 笮融, of Danyang 丹楊…then enlarged and erected a great Buddha shrine and used copper to construct a human [figure] with gold applied to the body and dressed in colored brocade. Nine layers of copper [candle] plates hung down and below there was a storied building with a covered walkway that could accommodate more than three thousand people who all received instruction and read the Buddhist sutras. So that Buddhist devotees

from within his realm and adjacent commanderies could hear the teachings, he exempted them from corvée labor to entice them. After this, over five-thousand households from near and afar arrived there. Each time the image of the Buddha was lustrated, wine and food were prepared, and mats were spread along the road for a distance of ten *li*. Around ten-thousand commoners arrived to observe and partake of the food offerings with the outlay amounting to hundreds of millions. 笮融者, 丹楊人, ... 乃大起浮圖祠, 以銅為人, 黃金涂身, 衣以錦採, 垂銅盤九重, 下為重樓閣道, 可容三千余人, 悉課讀佛經, 令界內及旁郡人有好佛者聽受道, 復其他役以招致之, 由此遠近前后至者五千余人戶 每浴佛, 多設酒飯, 布席於路, 經數十裡, 民人來觀及就食且萬人, 費以巨億計. (*Sanguozhi*, 49.1185).

During the late Eastern Han, Ze Rong 笮融 (?-195) conducted activities such as having people "receive instruction and read the Buddhist sutras 課讀佛經," and "lustrate the [image] of the Buddha 浴佛." From the fact that numerous people took part in Ze Rong's activities, we can see that Buddhism had already gained influence among the populace. The account given here of "using copper to construct a human [figure] 以銅為人" is the earliest reference to casting an image of the Buddha in the historical record (He Zhiguo (2005) provides a date for Ze Rong's construction of the Buddha image between the 4th year of the Chuping 初平 era (193 CE) and the 2nd year of the Zhixingping 至興平 era (195 AD) during the reign of Emperor Xian of the Han. For a detailed account of Ze Rong's Buddhist beliefs and activities including the construction of temples and statues and their role in the spread of Buddhism to the masses, see Lai Yonghai (2010a: 106–111).). In terms of images of the Buddha, the earliest excavated so far is a Buddha image on a bronze "Money Tree" unearthed in Fengdu 豐都 in Chongqing 重慶 and dates to the fourth year of the Yanguang 延光 era (125 CE) during the Eastern Han. The geographical extent of such images encompasses Shandong, Anhui, and Zhejiang in the east to Sichuan in the west. From the geographical distribution of such Buddha images by the mid- to late-Han, we can see that even though Buddhism was not simply confined to the court and had indeed spread among the populace, its influence was still relatively limited (Wang Suqi, 2007). For the influence of Buddhism to be properly felt in the central regions of China, we must wait until the period of Wei-Jin and Northern and Southern Dynasties.

An Overview of Buddhist Translation

Approximately fifty years after Buddhism was introduced into China, An Shigao 安世高 arrived in Luoyang 洛陽 from Parthia (Anxi guo 安息國) in the second year of the Jianhe 建和 era during the reign of Emperor Huan (148 CE). In Luoyang, he founded a translation bureau to translate Buddhist texts, thereby inaugurating a large-scale movement involving the translation of Buddhist texts from Indo-European languages (predominately Sanskrit) that lasted from the Eastern Han through to the Song (During the early period (Eastern Han, Wei, Jin, and Southern and Northern Dynasties), the source languages of Buddhist scriptures were mainly Middle Indic, the Northwest Indic prakrit (Gāndhārī), and Central Asian vernaculars such as Tocharian. Buddhist scriptures began to be "Sanskritized" during the first and second centuries CE which saw Sanskrit and Buddhist Hybrid Sanskrit serve as source languages. On the issue of Chinese Buddhist translations and their Indic source languages, see Karashima (2016: 13–17) and Spahr (1997).). According to Zhu Qingzhi's (1992a: 36) calculations, the Japanese edition of the Buddhist Canon, the *Taishō shinshū Daizōkyō* 大正新脩大藏經 [*The Newly Revised Tripiṭaka of the Taishō Era* (*Taishō Tripiṭaka*)] comprises over 2,206 sections (*bu* 部) and 8,899 traditional volumes (*juan* 卷) of Buddhist classics in classical Chinese amounting to approximately 70 million characters (including more than 50 sections of translation undertaken during several later dynasties including the Western Xia, Liao, Yuan, Ming, and Qing). Furthermore, according to Li Fenghua and He Mei's (2003: 37) introduction to the *Zhiyuan fabao kantong zonglu* 至元法寶勘同總錄 [*A Collated Catalogue of the Treasures of the Dharma of the Zhiyuan Era*], there were more than 190 primary translators working from the East Han to the late Song with thousands of other assistants involved in the work of translation.

The Flourishing of Buddhism and Its Significant Influence

The translation of Buddhist texts accelerated the propagation of Buddhism, quickly reaching its peak in the Northern and Southern dynasties and flourishing during the Tang. Outside of the sheer number of scriptures that were translated during this period, the degree of prosperity which Buddhism enjoyed from the Northern and Southern dynasties to the Tang can be seen in many features, including its level of official support, the construction of temples and grottoes, the number of monastic and lay followers, the establishment of various Buddhist lineages, and the economic development of temples and monasteries (For an account of the reasons and expressions of the flourishing of Buddhism during the Wei, Jin, Southern and Northern Dynasties, and Tang, please refer to the introduction and analysis of Zhang Wenbin and Li Shaolian (1987) and Guo Peng (1980: 381–402), respectively.). The following four accounts can give us an oversight of such developments:

1. **The renunciation of Emperor Wu of Liang** 梁武帝 **and his lectures on the Dharma**. The "Liang benji" 梁本紀 ["Basic Annals of the Liang"], in the *Nan shi* 南史 [*History of the Southern Dynasties*] states: "On a *guisi* day in the [ninth month in Autumn] of first year of the Zhongdatong 中大通 era (529 CE), [Emperor Wu] visited Tongtai temple 同泰寺 and held a great assembly that was open to all of the four groups of disciples [monks and nuns, laymen and laywomen]. [He] cast aside his imperial robes to don religious attire, carried out purification, and made the great renunciation. He used a side room for his chamber with a simple bed, earthenware vessels, rode in a small carriage, and had private persons as attendants and servants. [The next day], a *jiawu* day, he ascended to the Dharma seat in the hall and lectured on topics in the *Nirvāṇa sutra* 涅槃經 for the sake of the four groups. [Nine days later], on the *guimao* day, a large group of officials offered hundreds of millions in cash to redeem the

bodhisattva [i.e. Emperor Wu] from his great renunciation to which the monks tacitly acquiesced." 中大通元年…(秋九月)癸巳，幸同泰寺，設四部無遮大會。上釋御服，披法衣，行清淨大舍，以便省為房，素床瓦器，乘小車，私人執役。甲午，升講堂法坐，為四部大眾開涅槃經題。癸卯，群臣以錢一億萬奉贖菩薩大舍，僧眾默許 (*Nan shi*, Beijing: Zhonghua shuju, 1975; 7.206). The *Nan shi* records four instances when Emperor Wu renounced his role as sovereign. However, there are eight verifiable instances during this period where he personally lectured on the sutras and expounded Buddhist teachings with his largest audience numbering over 300,000 people (Lai Yonghai 2010b: 115).

2. **The establishment of mass popular Buddhist religious groups**. The Northern dynasties saw the foundation of a mass lay-focused Buddhist organization known as the Yiyi 義邑/邑義 which was founded with the main purpose of constructing Buddha images. While the majority of its members were nonreligious laypeople, there were also monks within the organization who carried out itinerant preaching. Han Lizhou's (2010) analysis of documents, mainly stone inscriptions, related to the construction of such images, shows that a large number of these images were the joint-production of hundreds of donors. The *Yiyi wubai ren zao xiang ji* 邑義五百人造像記 [*Record of Five Hundred People of the Yiyi Constructing Images*] dating to the first year of the Wuding 武定 era during the Eastern Wei 東魏 (541 CE) in Henan records the largest number of donors and lists each donor's name along with an account of the statue's production (Han Lizhou 2010: 593–597).

3. **Buddhist faith reflected in personal names**. During the Northern and Southern Dynasties, it was fashionable to name people after Buddhist personages or with terms that touched upon Buddhist themes. Lü Shuxiang (1988) has pointed out that "Buddhism has been propagated in China for close to two thousand years with the Northern and Southern dynasties representing the period in which it had already reached its greatest flourishing and had not yet lost its vitality. We can see just how pervasive and profound its influence was during this period by people's choice of personal names." We can count over forty personal names related to Buddhism found in the biographies of the official histories. For example, we find the name "Siddhārtha" (Xida 悉達) in people such as Liu Xida 劉悉達 (*Song shu* 宋書 [*Book of Song*], Beijing: Zhonghua shuju, 1974; 68.1807), Lu Xida 魯悉達 (*Chen shu* 陳書 [*Book of Chen*], Beijing: Zhonghua shuju, 1972; 13.198), and Wu Xida 吳悉達 (*Bei shi* 北史 [*History of the Northern Dynasties*], Beijing: Zhonghua shuju 1974; 84.2829), and Lu Xida 陸悉達 (*Bei Qi shu* 北齊書 [*Book of the Northern Qi*], Beijing: Zhonghua shuju, 1972; 39.518). Ji Xiuqin (1993: 92–93) lists almost fifty people from the Northern and Southern Dynasties with the character for "monk" (*seng* 僧) in their names and numerous others with the character for "Buddha" (Fo 佛) or the character "*dh*" (*tan* 曇) taken from the Sanskrit transliteration of the term for the Buddhist teaching ("*dharma*").

4. **Official prohibitions**. An edict from the seventh month of the second year of the Kaiyuan 開元 era (714 CE) during the reign of Emperor Xuanzong 唐玄宗 of the Tang states: "In light of having been informed that [there are people who] have opened shops in laneways and alleys to copy the sutras and brazenly cast [Buddha] images in public whose mouths consume liquor and flesh, and whose

hands are besmeared with the rank odor of mutton and raw meat, and that as a result, the way of veneration and respect has waned and feelings of contempt and insolence have arisen. The common people sometimes have cause to seek blessings but endure hunger and cold on account of this. When thinking of their foolish behaviour, I often feel mournful and sigh in sorrow. Little do they imagine that the Buddha is not something external, and the Dharma indeed lies in one's heart. If one knows to obtain it in one's own body or nearby, the way then is not far off. Because they are mired in these inveterate habits, it is obligatory to make a declaration. Henceforth from today, neither in village nor in town is it permitted to ever again to cast Buddha [images] or copy sutras as a profession." 如聞坊巷之內, 開鋪寫經, 公然鑄佛, 口食酒肉, 手漫羶腥, 尊敬之道既虧, 慢狎之心斯起。百姓等或緣求福, 因致飢寒, 言念愚蒙, 深用嗟悼。殊不知佛非在外, 法本居心, 近取諸身, 道則不遠。溺於積習, 實藉申明。自今以后, 村坊市等不得輒更鑄佛、寫經為業 (*Cefu yuangui* 冊府元龜 [*Prime Tortoise of the Record Bureau*], Nanjing: Fenghuang chubanshe, 2006; 159.1773). This prohibition reflects the fact that the demand for Buddhist scriptures and images among the populace was both great and pressing and that some people's profession involved casting images and copying scriptures. The fact that some people were willing to "endure hunger and cold" on account of their desire for Buddhist sutras and images serves to illustrate their devoutness.

Summary

The account above has briefly outlined the transmission of Buddhism to China, the translation of the Buddhist sutras, and its enormous influence in society from the imperial court to the marketplace. During the Tang, Buddhism had completed its process of "Sinicization" and had become an integral aspect of traditional Chinese thought and culture. It is precisely due to the large number of translated Buddhist scriptures and Buddhism's subsequent profound and long-lasting influence that this Buddhist language was used and understood by the Chinese populace and finally saw it become integrated into the language of the people as a whole.

The Characteristics of the Buddhist Chinese Language and the Reasons for Its Formation

The Characteristics of Buddhist Chinese

As the term implies, the "language of the Buddhist classics" (*Fodian yuyan* 佛典語言) refers to the specific type of language employed in Buddhist works and can be divided into broad and narrow categories. In a broad sense, the term refers to all texts closely related to Buddhism and includes the Buddhist sutras and canonical works translated into Chinese (*Hanyi Fodian* 漢譯佛典), domestic Buddhist works composed in China (*Zhongtu zhuanshu* 中土撰述, such as *Gaoseng zhuan* 高僧傳

[*Biographies of Eminent Monks*] and *Da Tang xiyu ji* 大唐西域記 [*The Great Tang Record of the Western Regions*]), Chan sayings (*Chanzong yulu* 禪宗語錄), and vernacular literature (*suwenxue* 俗文學, such as popular narrative literature *bianwen* 變文 and vernacular poetry *baihuashi* 白話詩). In a more narrow sense, the language of the Buddhist classics is limited to the Buddhist sutras and canonical works translated into Chinese whose language is characterized by explicit and obvious differences to the native Chinese language. When compared with the language the Buddhist sutras and canonical works translated into Chinese, other Buddhist works only exhibit a few of the specialized lexical and syntactic phenomena found in the translated works. When I discuss the "Sinicization" of the Buddhist language, it mainly refers to this narrow sense. (See Zhu Qingzhi (2001: 6) for a definition of the concept of "Buddhist Hybrid Chinese" (*Fojiao hunhe Hanyu* 佛教混合漢語); Zhu Qingzhi (2008) later changed the term "Buddhist Hybrid Chinese" to "Buddhist Chinese" (*Fojiao Hanyu* 佛教漢語).)

Many scholars have explored the characteristics of the language of Buddhist works translated into Chinese from a number of perspectives (Zhu Guanming 2013). Among these, Erik Zürcher's standpoint is comparatively early and representative of such scholarship. Zürcher argues that while the language employed in Buddhist translations during the Eastern Han represents "a somewhat formalized but nevertheless closer reflexion of the living language of second century Luoyang," these works also exhibit the following features (Zürcher 1977/1987):

1. It has been affected by the medium of literary Chinese.
2. The Indian original has exerted an influence.
3. A marked preference for four-syllable sentences.
4. The language of some texts is nonstandard, and some texts represent "a kind of Serindian Pidgin."

In fact, these four features outlined by Zürcher, along with the vernacular aspect of the "living language" previously discussed, are also applicable to Buddhist language after the Eastern Han period. Zhu Qianzhi (2008: 493) summed up the characteristics of Buddhist Chinese as "two processes of blending'" – one involves the mixture of a large number of linguistic elements from the original Indian language with Chinese, and the other involves the mixture of formal written Chinese, or classical Chinese (*wenyanwen* 文言文) with vernacular and nonstandard Chinese elements. I now will briefly discuss these two characteristics.

The Reasons Behind the Formation of the Characteristics of Buddhist Chinese

Based on previous studies, Zhu Qingzhi and Zhu Guanming (2006) analyzed the reasons behind the characteristics of Buddhist canonical language as having three main aspects: First, it is composed a mixture of written classical Chinese and vernacular dialects. Second, the emphatic use of four-character phrases. Third, it exhibits the infiltration of lexical and syntactic elements from the source language (Zhu Qingzhi and Zhu Guanming's (2006) original paper mainly discussed the causes behind the grammatical characteristics of Buddhist Chinese, but these are also applicable to vocabulary. The explanation here also adds the element of vocabulary.). In terms of the first of these, there are certainly some aspects of Buddhist canonical language that are due to the mixture of the two. For example, the inclusive adverb *dulu* 都盧 "all, the whole" was part of colloquial speech at the time, while a similar expression, the inclusive adverb *ci* 賜 was not, but was instead derived from the ancient meaning of "to use up completely" and was not in common use. These two were juxtaposed together to form the compound adverb *duluci* 都盧賜 "all, the whole, both": "Various humans and various non-humans, all (*duluci* 都盧賜) arrived here 諸人諸非人, 都盧賜來到是間" (*Daoxing bore jing* 道行般若經 [*The Sutra on Practicing the Way of* Prajñā; Skt. *Aṣṭasāhasrikā-prajñāpāramitā-sūtra*], T. vol. 8, no. 224, 2.435a2, Lokakṣema 支婁迦讖 trans., Eastern Han period). For further discussion of this compound adverb, see Karashima (2002). However, due to the intersection between the written and spoken language, along with the limitation of materials, it is very difficult to completely separate the written language of the period. Because of this, research in this field is as yet rather limited. Moreover, as both written classical Chinese and the vernacular spoken language are inherent aspects of Chinese itself, it is not possible, with this feature at least, to speak of "Sinicization." As such, I will discuss the "Sinicization" of Buddhist language from the following two aspects: the use of the "four-character phrase" form and the infiltration of elements of the source language. These two aspects brought linguistic elements along with their "Sinicization" that were different from domestic ones. In addition to this, Buddhist ideas and concepts also brought some particular elements to Buddhist language that are worthy of attention.

The Use of Four-Character Phrases

According to the research of Zürcher (1991/2001) and Yu Liming (1993: 29), the use of four-character phrases commenced in the Eastern Han and has played a dominant role ever since. In order to construct four-character phrases, translators would adopt a large variety of methods. These methods were even undertaken at the expense of splitting the text and its meaning and included "adding particles for euphony" (*tianjia chenyin ci* 添加襯音詞) like "in/at/on" (*yu* 於), "yet/and/but" (*er* 而), "it" (*zhi* 之), "doubling up synonyms" (*tongyici diejia* 同義詞疊加), "adding affixes" (*fujia cizhui* 附加詞綴), and other such strategies to increase the number of syllables, while "lexical economizing" (*cihui shengsuo* 詞匯省縮) and "syntactic elision" (*jufa shenglüe* 句法省略) were employed to decrease the number of syllables. These methods to increase or decrease the number of syllables in the Buddhist texts sometimes brought about lexical and syntactic forms that were entirely different to those found in local Chinese literature. For example, Wu Juan (2009) has proven that the interrogative phrase related to the length of time *jiuru* 久如 "how long" that is commonly seen in Buddhist texts, but not found in local Chinese ones, is an abbreviation of *jiujin ruhe* 久近如何 "approximately how long" and is used to construct four-character phrases. An example of this can be seen in the *Dafagu jing* 大法鼓經 [*The Great Dharma Drum Sutra:* Skt. *Mahābherīhāraka-parivarta sūtra*] translated by Guṇabhadra 求那跋陀羅 during the Liu Song: "At that time the holy king replied and asked the Buddha, "how long for me – should [it take] to attain Buddhahood? 爾時聖王-復問佛言:「我於久如-當得成佛?」" (*Dafalou jing, T,* vol. 9, no. 270, 9.294a12–13). An example of "syntactic elision" can be seen in the *Baiyu jing* 百喻經 [*Sutra of One Hundred Parables*] translated by Guṇavṛddhi 求那毘地譯 during the Xiao Qi 蕭齊 where the direct object pronoun *zhi* 之 "it" is omitted after the transitive verb "to kill" (*sha* 殺) in order to create a four-character phrase, "immediately, [the male dove] used [his] beak – to peck the female dove and kill [OMITTED 'it'] 即便以嘴-啄雌鴿殺" (*Baiyu jing, T,* vol. 4, no. 209, 4.557b1). Moreover, this placement of the transitive verb at the end of a sentence is not commonly found in traditional local literature (see Fang Yixin, 1994; Liang Yinfeng, 2003). This type of omission is the direct cause of the emergence of a "verb-result construction" that is still preserved in some dialects. The process of its emergence and its rise and fall and influence in later generation are precisely what forms the content of research into "Sinicization."

The Influence of Indic Source Languages

The infiltration of the source language into Buddhist language is related to their nature as works of translation. As such, these works cannot avoid influences from the source language, even though the form these influences take may be either implicit or explicit. With Sanskrit serving as the most representative of these source languages, there are enormous differences between the source language as the translator's native language and the Chinese language. Further to this, the translator's differing levels of proficiency along with their respect for the original texts invariably led to the blending of foreign and non-Chinese elements in terms of the lexical and syntactical aspects of Buddhist language. That is to say, those lexical and syntactical elements of the source language that differ from Chinese would sneak their way into Buddhist language under the guise of Chinese language.

Such infiltrations and borrowings are fairly obvious and mainly manifest as transliteration (*yinyi* 音譯) or free translation (*yiyi* 意譯) along with a large number of calques (fang*yici* 仿 譯詞), see Zhu Qingzhi (2002/2006). In addition to these, there is also a reasonably obscure mode of borrowing which has been termed "semantic/functional transfer" (*yizhi* 移植) by Zhu Guanming (2008). This refers to when a translator imposes a certain meaning or usage of a Sanskrit term onto a Chinese one on account of their psychological analogies. Various types of borrowings have caused many new words and expressions to appear in Chinese Buddhist language. How were these new words integrated into Chinese and what kind of changes occurred during this integration process in terms of morphology, semantics, and usage? An example of one such change can be seen in the Chinese transliteration of the Sanskrit word *bodhisattva* as *pusa* 菩薩, which is the commonly seen shortened version of *puti saduo* 菩提薩埵. Translator's note: The full transliteration of *puti saduo* 菩提薩埵 according to Edwin G. Pulleyblank's reconstruction of Early Middle Chinese (hereafter EMC) would have been pronounced "bɔ-dɛj sat-twa." The term is composed of two concepts, *bodhi* (*puti* 菩提; EMC bɔ-dɛj) or "enlightenment/awakening" and *sattva* (*saduo* 薩埵;) or "sentient being." The original meaning of the term was a "person destined for enlightenment, Buddha-to-be" (Edgerton 1953: 403) which in contemporary Chinese now refers to "a Buddha or certain divinities" or a "person with a charitable disposition" (*Xiandai Hanyu cidian* 現代漢語詞典 [*Dictionary of Modern Chinese*], 7th ed.) (Edgerton's (1953: 403) definition of *bodhisattva* is a "person destined for enlightenment, Buddha-to-be."). This word *bodhisattva* has also been transliterated into Chinese as *puti suoduo* 菩提索多 or as a free or liberal translation as *jueyouqing* 覺有情 "a sentient being [intent on] awakening," *kaishi* 開士 "a [commendable] person who opens [the way of enlightenment]," and *dashi* 大士 "a great [commendable] person." What facts led to the final choice of *pusa* 菩薩 to represent *bodhisattva* in Chinese and what were its developmental changes? These problems are also a focus of research into the "Sinicization" of Chinese Buddhist language.

In regard to the infiltration of syntactic elements from a modern perspective of linguistic typology, the differences between Sanskrit and Chinese are mainly displayed in the following aspects:

1. **Inflection**: Sanskrit is a typical inflected language with abundant morphological changes, while Chinese is an atypical, isolated language with no changes in morphology.
2. **Word order**: Sanskrit word order is flexible, but the object is generally placed before the verb (Object-Verb, or OV). In Chinese however, word order is Subject-Verb-Object (SVO).
3. **Function words**: The usage of some function words in Sanskrit is different to their counterparts in Chinese.

Therefore, theoretically speaking, these three aspects are most likely to have an impact on the syntax of Buddhist texts in Chinese translation. The following will provide an example for each of these three aspects:

1. **Inflection**: Sanskrit verbs have what is known as an "absolutive participle" or sometimes as an "independent" gerund. This "absolutive participle" is used in Sanskrit for the first of two successive actions by the same agent (or for several preceding actions in a multiple series of actions with the exception of the final one) and indicates "after ACTION was completed." Buddhist texts translated into Chinese add *yi* 已 "already" after verbs to indicate this in their translation. However, in domestic Chinese texts, *yi* could originally only follow nonpunctual verbs (*chixu dongci* 持續動詞), such as in the case of *shi yi* 食已 "after having eaten," and cannot follow punctual verbs (*shunjian dongci* 瞬間動詞). There are no such restrictions on the absolutive participle in Sanskrit, and we see new usages of *yi*. Examples of this include, "to have seen the Buddha already 見佛已" (Skt: *dṛṣṭvā*) from the *Lotus Sutra* (*Zhengfa huajing* 正法華經 [*The Lotus Sutra of the True Dharma;* Skt: *Saddharma-puṇḍarīka-sūtra*], trans., Dharmarakṣa 竺法護, *T*. vol. 9, no. 263, 4.90b16), with others such as *jueyi* 覺已 "awakened

already," *wen yi* 聞已 "heard already," and *si yi* 死已 "died already" in the *Xianyu jing* 賢愚經 [*Sutra on the Wise and Foolish*] (see *Xianyu jing*, trans., Huijue 慧覺, T. vol. 4, no. 202, 1.353a11; 1. 355a27; 4.379a18). This new usage further influenced the use of the particle *le* 了 and led to the emergence of *le* as an aspect particle (Karashima 2000; Jiang Shaoyu 2007).
2. **Word order**: Cao Guangshun and Yu Hsiaojung (2000) have discussed the origin of the disposal construction in Chinese. They have argued that the construction "take OV" (*qu OV* 取 OV), which omits the second object and originates from the consecutive verbal construction "take OV it" (*qu* OV *zhi* 取 OV之), is an influence of the OV word order in Sanskrit (We find examples of the "take OV" (*qu* OV) construction that omit the second subject in phrases such as "take the king and kill [him] 取父王殺" where second object "him" is elided (*Zengyi ahan jing*, 增壹阿含經 [*Ekōttarikāgama*], trans., Gautama Saṃghadeva 瞿曇僧伽提婆, T. vol. 2, no. 125, 2.762b28.). An example of the "take OV it" construction taken from the same source can be seen in the phrase "take the king and harm him 取父王害之" (*Zengyi ahan jing*, T. vol. 2, no. 125, 2.803b6).). They also argue that the "take OV" structure had a direct influence on the production of the *ba* 把 disposal construction (Zhu Guanming (2011) believes that the emergence of a new type of "Patient-Topic Clause" (*shoushi huati ju* 受事話題句) in Chinese after the Medieval Period (in the original article this is called "Patient-Subject Clause" *shoushi zhuyu ju* 受事主語句 according to Chinese convention) is also related to the influence of Sanskrit OV word order. The form and function of this new type of "Subject-Topic Clause" changed after entering Chinese and then also led to the emergence of "teach/give" (*jiao* 教/*gei* 給) causative and passive sentences in modern Chinese.).
3. **Function words**: This following text of this chapter will introduce the usage of the Chinese reflexive pronoun *zi* 自 "self, onself" which, due to the influence of the Sanskrit reflexive pronoun *sva*, had a new additional use as an attributive in the Buddhist classics. The emergence of the pronoun *ziji* 自己 "oneself" is also due to this new usage of *zi* (Compared with lexical borrowings, grammatical borrowings in Buddhist scriptures are much more ambiguous and require more in-depth observation and research in Sanskrit and Chinese uncover. At present, almost thirty items have been discovered that demonstrate the influence of Indic source texts on the language and grammar of Chinese Buddhist texts, see Zhu Qingzhi and Zhu Guanming (2006), Zhu Guanming (2013), and Yu Fangyuan and Zhu Guanming (2018).).

The Influence of Buddhist Concepts

In addition to the linguistic features of Buddhist language created by the internal factors of the language noted above, it cannot be ignored that there are some characteristic Buddhist doctrines and concept that are quite different from those native to China. Therefore, some attention must be paid to these new elements that were brought into the fold of Buddhist language. The emergence of these new elements is different to the borrowing of terms in the translation process. Here a new concept is borrowed from the source text where there are no words completely corresponding to Chinese. For example, Buddhist doctrine refers to vexations or afflictions as "fetters" or "knots" (*jie* 結, Skt. *bandhana*) and internal "mental afflictions" (*fannao* 煩惱) as "mental fetters" (*xinjie* 心結). This type of "fetter" or "knot" can be "untied" or "released" (*jeikai* 解開), and we see this usage in examples such as "Ānanda's mind was fettered (*xinjie* 心結) and the Buddha wished to release (*jie* 解) him 阿難心結, 佛欲解之" (*Zhong benqi jing* 中本起經 [*Middle-length Sutra on the Former Deeds {of the Buddha}*], trans., Tanguo 曇果 and Kang Mengxiang 康孟詳, *T*. vol. 4, no. 196, 2.163a19). When the "knot" is "untied," there are no mental afflictions, so there are many instances of the word *kaixin* 開心 ("happy," lit. "release/open/untie the heart") in Chinese Buddhist texts, and this is the origin of the word *kaixin* 開心 ("happy") in modern Chinese (see Zhu Guanming, 2007b/2009). Another example is the term *yingxiang* 影響 ("influence") in modern Chinese, which means "to act on the thoughts or actions of others." This word is also derived from the Buddhist notion that, "people's actions must be their retribution, like 'a shadow (*ying* 影) follows one's body, or an echo (*xiang* 響) responds to a sound'" (see Zhu Qingshi, 1992b).

Summary

The section above outlined the concept of "the language of the Buddhist scriptures" and discussed the characteristics and causes of this type of Buddhist Chinese. It should be noted that in comparison with the influence of the "four-character" literary style and the penetration of Buddhist concepts, there are evidently many more new structures found in the Chinese translation of Buddhist texts that are due to the infiltration of the Indic source language. Therefore, while this represents the starting point for us to investigate the Sinicization of the Buddhist language, the first two aspects also merit our full attention. The main focus of previous studies of Buddhist language was on the language itself, that is, its lexical and grammatical characteristics and their sources. However, how these characteristics became gradually integrated into Chinese after the Medieval Period, that is, the processes, reasons, and rules of the process of gradual "Sinicization," has as yet received scant attention.

Research on "Sinicization" and Its Meaning Within Buddhist Studies

Research on Sinicization of Buddhist Language

Wang Li (1958/1980: 519–522) was one of the first scholars to focus on the issue of Sinicization comparatively early and divides "Buddhist loanwords and translated terms" into three distinct types:

1. Specialized Buddhist terms that are generally used by a small number of people devoted to the study of Buddhist classics and are not part of the common

language of the whole population, such as *bore* 般若 ("wisdom"; Skt., *prajñā*), *youpose* 優婆塞 ("male lay practitioner"; Skt., *upāsaka*), or *youpoyi* 優婆夷 ("female lay practitioner"; Skt., *upāsikā*)
2. Words that have already become part of the common language but are recognizable as Buddhist terms, such as *heshang* 和尚 (Skt., *upādhyāya*; "preceptor," "master," and "teacher"), or *pusa* 菩薩 ("bodhisattva")
3. Words that have already penetrated into the lifeblood of the Chinese language to the extent that the general public are unaware of their origins and are no longer regarded as Buddhist terms, such as *shijie* 世界 ("the world"), *xianzai* 現在 ("now"), *jieguo* 結果 ("result"), or *yuanman* 圓滿 ("perfect")

As it is evident that the degree of "Sinicization" in these three types gradually deepens, we should focus on the last two of them when discussing the Sinicization of the language of the Buddhist classics – that is to explore how these loanwords became part of the language of the entire population, how they penetrated into the lifeblood of the Chinese language, and the regular patterns they exhibit in the process (The first of these three types however should not be completely excluded. While some of these terms did not enter into the common language, several did affect Chinese in some manner. For example, the word *youpoyi* 優婆夷 ("female lay practitioner"; Skt., *upāsikā*) was probably the origin of the modern word *ayi* 阿姨 ("auntie," a term of address for woman of the same generation as one's mother), see Zhu Qingzhi (2001). Zhu Guanming (2015) also notes that commonly used Buddhist words such as *jiemo* 羯磨 ("karma"), *bore* 般若, *jushou* 具壽 ("possessed of long-life," a term of respect), and *youqing* 有情 ("sentient being") are often seen in works that touch on Buddhist themes. While these are rarely used in the oral language, they form an integral aspect of formal written Chinese and despite only having undergone part way in the process of Sinicization are still deserving of our attention.).

Zürcher (1991/2001) points out that as early as the third century, the Chinese translation of Buddhist scriptures had developed a "scriptural style" that was different both from secular Chinese literature and the original Indic scriptures in terms of its terminology, style, and composition and that these represent a gradual process of "Sinicization." His main point, however, is that the style of the language used in translation itself became more Chinese and not that the language used in translation evolved to become a part of the language of the entire population. It is this last sense of the term "Sinicization" that the present chapter discusses.

As Zhu Qingzhi (2002/2006) expresses more clearly, "If the special components of Buddhist Chinese are only present in Buddhist scriptures, they cannot be said to have genuinely influenced the Chinese language – in other words, if these components are used by Chinese people in activities other than translating Buddhist scriptures, only then can they be regarded has having been genuinely assimilated into the Chinese language. Consequently, when discussing the influence of Buddhism on Chinese vocabulary, one must go a step further and discuss the spread of Buddhist terms drawn from translated scriptures to native Chinese literature and see what emerges from Buddhist or elements drawn from the translation of Buddhist scriptures found therein." Later, Zhu Qingzhi (2011) examines in detail the emergence of two terms, *diqing* 帝青 ("sapphire," or a "sky-blue precious stone") and *tianqing* 天青 ("reddish black") – both calques drawn from the Sanskrit word *indra-nīla* – in Buddhist Chinese and their later usage and development in modern Chinese. Such research objectives are identical with the "Sinicization" spoken of here.

Related research papers also include Chen Zhongyuan (2011) on the development of the word *ziji* 自己 after the Medieval Period, and Zhu Guanming (2015) on the spread of the word *weicengyou* 未曾有 ("marvelous, appearing rarely") after the Medieval Period and his work (2019) on the development and changes of the term *shengse* 生色 from the Tang to the Ming and Qing. Shi Guanghui (2011) devotes a chapter to the Sinicization of ten terms such as *pusa* 菩薩, *shanyou* 善友 ("virtuous friend"), *shijian* 是間 ("here"), and *zhongyang* 中央 ("middle"); Yang Tongjun (2011) also devotes a chapter to the evolutionary processes behind the "domestication" (*bentuhua* 本土化) of nine terms such as *jushi* 居士 ("householder"), *shangshou* 上首 ("most excellent," "chief"), *wuni* 五逆 ("five heinous crimes"), and *zuoye* 作業 ("effort," "work"); Zhang Yisan and Zhang Futong's (2013) volume is currently the only monograph on loanwords of Buddhist origin and provides a relatively rough outline of the Sinicization process of a great number of Buddhist terms.

On the Term "Sinicization" in Buddhist Studies

The outline above has provided an account of current research directly related to the Sinicization of the language of the Buddhist classics. These studies have predominately focused on lexical items and use content words as their basis. Research on the Sinicization process of function words and syntactic constructions is still rare. In regard to the actual term, as noted above, in addition to the term "Sinicization" (*Zhongguohua* 中國化, lit. "Chinization"), some have also called it *Hanhua* 漢化 (also "Sinicization," lit. "Han-ification") or "domestication" (*bentuhua* 本土化, lit. "native land-ification"). As we know, after being introduced to China, Buddhism underwent a process of adaptation to Chinese culture in terms of its doctrines, ceremonial regulations, and spiritual practices so as to become widely accepted in Chinese culture, found new schools of thought, and create a true Chinese Buddhism. In early studies, this process was known as "assimilation" (*tonghua* 同化) by Yao Baoxian (1936) and "domestication" by Hu Shih (1937) in his English language essay on the subject and subsequently termed *Huahua* 華化 (lit. "Hua-ification") by Liang Xiaohong (1996) or *Hanhua* 漢化 by Bai Huawen (2003). In recent years however, an increasing number of scholars employ the term "Chinization" (*Zhongguohua* 中國化) of Buddhism (Guo Peng 1980: 402; Fang Litian 1989; Li Shangquan 1989/1994; Xu Kangsheng 1990; Song Yubo 2012; Liu Huiqing 2019). While the connotation of these terms is basically identical, we follow the Buddhological convention and also use the phrase "Sinicization of the language of the Buddhist classics" (*Fodian yuyan de Zhongguohua* 佛典語言的中國化) to refer to the process whereby Buddhist Chinese became integrated into the language of the common people in China and define it as follows: Following Buddhism's increasing expansion and lasting influence in China, the lexical and grammatical components in Buddhist Chinese that differ from native literature underwent certain changes in form, meaning, and usage and were gradually integrated into the common language to become a part of the everyday speech of the Chinese people. This process is what we call the "Sinicization of the language of the Buddhist classics."

Summary

The section above has introduced previous research on the Sinicization of the language of the Buddhist classics along with the origin and definition of the term "Sinicization." Unlike previous research that attached importance to interpreting the progress of Buddhist Chinese and analyzing its origins and contributing factors, research into the "Sinicization" of the language of the Buddhist classics focuses on examining the development, evolution, and regular patterns of Buddhist Chinese since the Medieval Period. In short, earlier research focuses on the "causes" behind Buddhist Chinese language during the first half of its history while research into the "Sinicization" of such language deals with its "effect" in the common language during the latter half of its history. The following section will provide concrete examples of these "effects."

Modes of the Sinicization of Buddhist Language

The "Sinicization" of the language of the Buddhist classics is a process generally described as one that emerges from within the Buddhist classics and then enters the spoken language of the populace on account of the widespread influence of Buddhism within Chinese culture. The extent of its use then further expands bringing about changes in meanings and usage and then is finally assimilated into the "lifeblood" of the Chinese language. The primary task in researching this Sinicization is to describe this process as clearly as possible. Different lexical and syntactic forms exhibit distinct modes and channels of evolution. The following section employs the three terms *ziji* 自己 ("self," "oneself"), *shengse* 生色 ("outward expression," with an original meaning "gold" or "golden lustre"), and *xianzai* 現在 ("now") as examples, along with currently available relevant research (Zhu Guanming (2007a; 2019; 2021), Chen Zhongyuan (2011), to investigate the specific channels of their Sinicization. Of these, the reflexive pronoun *ziji* is a function word (*gongnengci* 功能詞), *shengse*, meaning of "gold" or "golden lustre," is a lexical word (*shici* 實詞), and the time word *xianzai* lies in between these two as a semifunctional word (*banshi banxu* 半實半虛), and it possesses the characteristics of both lexical and function words (Liu Yuehua et al. (2001: 50) clearly note that: "The grammatical functions of nouns that express position, location, and time are not identical to those of ordinary nouns and therefore must be introduced separately." Other works on grammar (such as Chao Yuanren 1979: 244–250; Zhu Dexi 1982: 43) also mainly handle such words in this manner.).

The Origin of *Ziji* and Its Process of Sinicization

The Origin of *Ziji*

Before discussing the origin of *ziji*, we must first speak of the transformation of the grammatical function of the word *zi* 自 ("oneself") in Buddhist classics during the Medieval Period. Prior to the Medieval Period, the word *zi* was a typical reflexive pronoun. It can only appear as an object before a verb or preposition (such as "unable to control himself" *bu neng zi ke* 不能自克, "Zhao gong shinian" 昭公十年, in *Zuo zhuan* 左傳 [*The Zuo Tradition {of the "Spring and Autumn Annals"}*]), or as an adverbial form to indicate emphasis ("the lord himself did not escort [her]," *Gong bu zi song* 公不自送, "Huan gong sannian" 桓公三年, in *Zuo zhuan*), and cannot be used as a subject or attributive (as a possessive term).

The word *ji* 己 ("oneself, 'one's own'"), however, is a common pronoun that can appear in all syntactic positions, such as the object after a verb together or a preposition ("Yuan Xuanzhong of Chen resented Shen Hou of Zheng for turning against him" *Chen Yuan Xuanzong yuan Zheng Shen Hou zhi fan ji* 陳轅宣仲怨鄭申侯之反己, "Xi gong wunian" 僖公五年, in *Zuo zhuan*), as a subject ("He himself was without faith" *ji ze wu xin* 己則無信, "Xiang gong wunian" 襄公五年, in *Zuo zhuan*), and as an attributive ("Zhuang Jiang took [him] as her own son" 莊姜以為己子, "Yin gong sannian" 隱公三年, in *Zuo zhuan*), yet cannot be used as an adverbial modifier. As a result, *zi* and *ji* are always found in different syntactic positions and cannot appear next to each other.

After the Medieval Period however, following the appearance of translations of Buddhist classics, a transformation of the usage of *zi* occurred in Buddhist works where the term began to appear in a large number of attributive positions such as the following examples:

Examples: 1
1.a

自	意	得	樂
Zi	Yi	De	Le
Own	Senses	Get	Joy

"...one's own senses gain joy."

Foshuo qichu sanguan jing 佛說七處三觀經 [*Sutra on the Seven Bases and Three Views as Spoken by the Buddha*], trans., An Shigao 安世高, T. vol. 2, no. 150, 1.878b14. Eastern Han 東漢 (25–220 CE)

1.b

或	見	自	字
Huo	Jian	Zi	Zi
One	See	Own	Name

"...one can see one's own name."

Daoxing bore jing 道行般若經 [*The Prajñā Sutra on the Practice of Enlightenment;* Skt. *Aṣṭasāhasrikā-prajñāpāramitā-sūtra*], trans., Lokakṣema 支婁迦讖, *T.* vol. 8, no. 224, 7.460b16. Eastern Han

1.c

鴿	踰	自	重
Ge	*Yu*	*Zi*	*Zhong*
Pigeon	Exceed	Own	Weight

"…the pigeon exceeded [the king's] its own weight."

Liudu ji jing 六度集經 [*The Sutra on Amassing the Six Perfections*], trans., Kang Senghui 康僧會, *T.* vol. 3, no. 152, 1.11c7. Wu 吳 (220–280 CE)

1.d

將	入	自	舍
Jjiang	*Ru*	*Zi*	*She*
Take	Into	Own	Room

"…[he] took [Ānanda] into [his] own chamber…"

Shisong lü 十誦律 [*Vinaya of Ten Recitations;* Skt. *Daśa-bhāṇavāra-vinaya*], trans., Puṇyatāra 弗若多羅 and Kumārajīva 鳩摩羅什, *T.* vol. 23, no. 1435, 3.26c7. Yao Qin 姚秦 (266–420 CE)

This type of use of *zi* as an attributive is first seen in Buddhist texts dating to the Eastern Han. While this usage is commonly seen in Buddhist works, it is extremely rare in domestic Chinese works from the same period. We believe that this new function of *zi* was produced during the translation of Buddhist texts. A comparative collation of Chinese and Sanskrit reveals that the use of *zi* in many of these attributive positions in translation corresponds to the Sanskrit reflexive pronoun *sva*. The following data is drawn from Unrai Ogiwara's, *Kan'yaku taisho Bonwa daijiten* 漢訳対照梵和大辞典 [*Sanskrit-Japanese Dictionary with Comparison to Chinese Translations*]:

Examples: 2

Sva-kāraṇa	*Zi yin* 自因 ("self-caused"), *zi xing* 自性 ("self-nature," "one's own nature")
Sva-lābha	*Zi caili* 自財利 ("one's own riches"), *zi suode* 自所得 ("self-gain")
Sva-kleśa	*Zi fannao* 自煩惱 ("one's own afflictions")
Sva-liṅga	*Zi xiang* 自相 ("one's own form")
Sva-guṇa	*Zi gongde* 自功德 ("one's own merit")
Sva-vāda	*Zi lun* 自論 ("one's own beliefs/theories"), *zi fa* 自法 ("one's own methods")
Sva-pakṣa	*Zi fen* 自分 ("one's own conditions"), *zi banlü* 自伴侶 ("one's own companions")
Sva-śakti	*Zi neng* 自能 ("one's own abilities"), *zi gongli* 自功力 ("one's own capabilities")

According to Monier-Williams (1899: 1275) and Speijer (1886: 198), the word *sva* is the Sanskrit reflexive pronoun, which when used as the object in a sentence is generally placed before the verb, but its main use is as an attributive. Since the Chinese *zi* has some of the same functions as the Sanskrit *sva*, the translators would impose the attributive function onto *zi* when translating the word *sva* (This is a common phenomenon in Buddhist translations that Zhu Guanming (2008) has termed "semantic/functional transfer" (*yizhi* 移植), or the means by which Buddhist translations influence the meanings and usage of Chinese vocabulary.).

As *zi* has an attributive function, it can be used together side by side with *ji* in the attributive position:

Examples: 3

3.a

內	省	自	己	軀
Nei	Xing	Zi	Ji	Qu
Inwardly	Examine	Own	Own	Body

"...inwardly observe one's own body."

Fo wubai dizi zishuo benqi jing 佛五百弟子自說本起經 [*Sutra on the Former Deeds of Five Hundred Disciples of the Buddha*], trans., Dharmarakṣa 竺法護, *T.* vol. 4, no. 199, 4.193b20. Western Jin 西晉 (266–420 CE).

3.b

到	自	己	境界
Dao	Zi	Ji	Jingjie
Arrive	Own	Own	Realm

"...[I] arrived at my own realm."

Zhong ahan jing 中阿含經 [*Madhyamāgama*], trans., Gautama Saṃghadeva 瞿曇僧伽提婆, *T.* vol. 1, no. 26, 1.625c29. Western Jin.

3.c

以	自	己	女	置	淫
Yi	Zi	Ji	Nü	Zhi	Yin
Took	Own	Own	Daughter	Install	Immoral

女	樓	上
Nü	Lou	Shang
Women	Building	On

"[he] took [his] own daughter and installed her in a house of ill repute"

Da zhuangyan lunjing 大莊嚴論經 [*The Sutra of Great Adornment Discourse*, Skt: *Sūtrālaṃkāra-śāstra*], trans., Kumārajīva 鳩摩羅什, *T.* vol. 4, no. 201, 4.272c22. Yao Qin

3.d

誤	偷	自	己	衣	缽
Wu	*Tou*	*Zi*	*Ji*	*Yi*	*Bo*
Mistake	Stole	Own	Own	Cassock	Alms bowl

"…[he] mistakenly stole his own cassock and alms bowl."

Mohe sengqi lü 摩訶僧祇律 [*Mahāsāṃghika-vinaya*], trans., Buddhabhadra 佛陀跋陀羅 and Faxian 法顯, *T*. vol. 22, no. 1425, 3.251b24. Western Jin.

3.e

長者	自	己	之	物
Zhangzhe	*Zi*	*Ji*	*Zhi*	*Wu*
Elder	Own	Own	GEN	Thing

"…the elder's own property."

Fo benxingji jing 佛本行集經 [*Sutra which Collects the Former Deeds of the Buddha*, Skt: *Buddha-carita-saṃgrāha*], trans., Jñānagupta 闍那崛多, *T*. vol. 3, no. 190, 3.861a10. Sui 隋 (581–618)

After having been frequently used together in the attributive position over a long period, *zi* and *ji* gradually solidified to become a compound word. According to Zhu Guanming (2007a), this solidification into the two-syllable word *ziji* 自己 ("oneself") came about during the Sui Dynasty (581–618) at the latest. At the beginning of its formation, the word *ziji* mainly appears in Buddhist texts where it is generally used as an attributive and rarely as the subject – a very different usage from the word *ziji* in modern Chinese. To understand how the function and usage of the word *ziji* gradually developed into its present state in modern Chinese is precisely the type of question that research into the "Sinicization" of such language seeks to answer.

The Development of the Function and Usage of *Ziji*

Chen Zhongyuan (2011) investigates the development of *ziji* since the Medieval Period in detail, and his views on this can be outlined as follows. In modern Chinese, the syntactic function of *ziji* is predominantly used as the subject, object, possessive phrases (i.e., attributive), adverb, and adjunct. In terms of whether or not these functions existed historically, *ziji* possesses all these functions by the late Tang and Five Dynasties at the latest; however, they all appear in translated Buddhist classics or works that touch on Buddhist themes. The following examples correspond to each of the respective five syntactic functions noted above:

Examples: 4

4.a (subject)

六	眾	苾芻	前	人	坐
Liu	*Zhong*	*Bichu*	*Qian*	*Ren*	*Zuo*
Six	Group	*Bhikṣu*	Before	Person	Sit

自己	立
Ziji	Li
Self	Stand

"…when others sit in front of them, a group of six *bhikṣu* (monks) themselves stand."

Genben shuoyiqie youbu pinaiye 根本說一切有部毘奈耶 [*Mūla-sarvāstivāda-vinaya-vibhaṅga*], trans., Yijing 義淨, *T*. vol. 23, no. 1442, 49.903c15. Tang (618–907)

4.b (object)

爭奈	自己	何?
Zhengnai	Ziji	He
What can be done	Self	What?

"What can be done about you yourself?"

Zutang ji 祖堂集 [*Anthology from the Hall of the Patriarchs*], eds., Sun Changwu 孫昌武, Kinugawa Kenji 衣川賢次, and Nishiguchi Yoshio 西口芳男, 2 vols, (Beijing: Zhonghua shuju, 2007), 381

4.c (possessive)

便是	自己	破弊	故	衣
Bianshi	Ziji	Pobi	Gu	Yi
Is just	Own	Shabby	Old	Robe

".. is just one's own shabby old robe.."

Genben shuoyiqie youbu pinaiye, 5.648b.29

4.d (adverb)

阿奴	來日	前朝	自幾(己)	宣	問
Anu	Lairi,	Qianchao	Ziji	Xuan	Wen
I	Tomorrow	Royal court	Myself	Declare	Ask

"I myself will ask and declare in the court tomorrow."

Han Qinhu huaben 韓擒虎話本 [*Vernacular Version of Han Qinhui*], in *Dunhuang bianwen jiaozhu* 敦煌變文校注, 1997, ed., Huang Zheng 黃征 and Zhang Yongquan 張涌泉校注, (Beijing: Zhonghua shuju, 1997), 299

4.e (adjunct)

如何	是	學人	自己	本分	事?
Ruhe	Shi	Xueren	Ziji	Benfen	Shi?
What	COP.	Learner	Own	Duty	Matter?

"What is a learner's own duty?"

Zutang ji, eds., Sun Changwu 孫昌武, Kinugawa Kenji 衣川賢次, and Nishiguchi Yoshio 西口芳男, 2 vols, (Beijing: Zhonghua shuju, 2007), 448

Although the use of *ziji* gradually increased from the Tang and Five Dynasties to the Ming, it is not until the Qing that there was an explosion in its frequency of use. From the Tang and Five Dynasties to the Song, *ziji* was used much more frequently in Buddhist works than domestic Chinese literature of the same period. These two trends can be seen in the following statistical data:

	Five dynasties	Song	Yuan	Ming	Qing
Instances per 10,000 characters	0.07	0.298	0.788	0.921	9.633
Domestic works/Buddhist works	0.01/0.45	0.15/1.52	–	–	–

The historical development of the five syntactic functions of *ziji* noted above can be summarized as follows:

1. *Ziji* as a possessive and subject were the functions that developed the earliest. These functions were in existence since the Tang and Five Dynasties and have continuously been its two most important functions.
2. The development of *ziji* as an object came about later than its function as a possessive or subject and is not frequently seen. This use as an object is earlier in Buddhist works than in domestic ones.
3. Its use as an adjunct (personal pronoun or personal name + *ziji*) is first seen in Buddhist literature, was rare in domestic Chinese works until the Song, and starts to be seen more during the Yuan.
4. Its adverbial use was the last to develop, and reliable examples of its use do not appear until the Ming. While this use developed slowly during the Ming and Qing, its frequency of use remained low.

Outside these syntactic functions, the use of *ziji* as a reflexive pronoun exhibits the strongest modern characteristics. While there is little difference in its use as an "anaphoric" or "generic reference" in Buddhist texts from the Tang to the Song, its emphatic usage (meaning its appositive or adverbial use) emerged earlier in Buddhist texts than domestic works. Its use as a "generic reference" has held dominant place in domestic Chinese literature, and it was not until the Yuan that this gave way to its anaphoric usage, which then declined sharply after the Ming. However, examples *ziji* allowing for "long-distance binding" (*chang juli yueshu* 長距離約束) do not appear until the Qing (such as "Xiren had known for some time that Grandmother Jia had already granted "she herself [i.e. Xiren]" (*ziji*) to Baoyu 襲人素知賈母已將自己與了寶玉的" from chapter six in *Hongloumeng* 紅樓夢 [*Dream of Red Mansions*]). The Yuan and Qing were important periods for the development of *ziji* with the Yuan seeing it undergo important changes in terms of function and usage and the Qing seeing an explosive increase in its frequency of use.

9 The Sinicization of Buddhist Scriptural Language

Summary

It is through Chen Zhongyuan's (2011) account of the development of *ziji* that we have a clearer understanding of its Sinicization. The word *ziji* was a product of the translation of Buddhist classics, is first seen in these texts, and by the Tang and Five Dynasties at the latest already possessed a relatively complete set of syntactic functions. These various functions and usages, however, did not occur at the same time within domestic Chinese literature and only began to be widely used during the Qing. Zhu Guanming (2007a) infers that *ziji* eventually became part of the ordinary vocabulary of the common language by the following means: that it emerged through the translation and dissemination of Buddhist texts, gradually gained currency among Buddhist clergy and lay followers, and through the influence of the propagation of Buddhism and Buddhist-inspired popular literature, finally became accepted as part of the language of the populace. Such an assertion is entirely consistent with the research noted above carried out by Chen Zhongyuan (2011) and also with Mair's (1994) argument along similar lines that Buddhism had a decisive influence on the emergence of modern Chinese. Dong Zhiqiao (2000) points out that the spread of Buddhist vernacular elements to domestic literature also underwent a similar process. As the following section hopes to show, the Sinicization process the word *ziji* underwent is certainly not the only means by which Buddhist terms entered into common Chinese parlance.

An Exegesis of *Shengse*, Its Origin, and Development

The Meaning of *Shengse* in the Poem "Qingong"

Opinions vary on how to interpret the term *shengse* in the following line from the Tang poet Li He's 李賀 (791–816 CE) poem, *Qingong* 秦宮:

> *In the inmost chamber lies a vast deep screen, a painting of 'golden hue' (shengse).*
> *Nei wu shen ping sheng se hua.*
> 內屋深屏生色畫。

Zhu Guanming (2019) investigates the origin of the term and its development in modern Chinese in detail. The word *shengse* is first seen in Chinese translations of Buddhist texts such as in the following examples:

Examples: 5

5.a

從	今日	不	聽	手	自
Cong	Jinri	Bu	Ting	Shou	Zi
PREP	Today	NEG	Let	Hand	Own

持	生色、	似色
Chi	Shengse	Sise
Hold	Gold	Silver

"From today, [you] will not allow [your] hand to hold gold or silver"

Mohe sengqi lü, *T.* vol. 22, no. 1425, 10.311a9

5.b

生色	者，	是	金	也
Shengse	Zhe	Shi	Jin	Ye
"Shengse"	PART	COP.	Gold	PART

"Shengse" means "gold"

Mohe sengqi lü, *T.* vol. 22, no. 1425, 10.311b

5.c

自	手	捉	生色、	似色
Zi	Shou	Zhuo	Shengse	Sise
Own	Hand	Grasp	Gold	Silver

"…one's own hand grasps gold or silver"

Mohe sengqi lü da biqiu jieben 摩訶僧祇律大比丘戒本 [*Greater Manual of Precepts for Bhikṣu from the Mahāsāṃghika-vinaya*], trans. Buddhabhadra 佛陀跋陀羅, *T.* vol. 22, no. 1426, X.551c10

Example 5.b from the *Mohe sengqi lü* 摩訶僧祇律 [*Mahāsāṃghika-vinaya*] clearly states that *shengse* means "gold" (*jin* 金). As the Sanskrit text for Example 5.c from the *Mohe sengqi lü da biqiu jieben* 摩訶僧祇律大比丘戒本 [*Greater Manual of Precepts for Bhikṣu from the Mahāsāṃghika-vinaya*] survives, we know that the Sanskrit original of *shengse* is *jāta-rūpa*, meaning "gold" (see Tatia (1976: 17). I would like to thank Professor Karashima for kindly supplying the materials employed here from his comparative collation of Chinese and Sanskrit texts.) The first part of the Sanskrit term, *jāta*, is the past participle of the verbal root √*jan* defined by Monier-Williams (1899: 410) as "to generate, beget, produce, create, cause, to cause to be born," which was translated into Chinese as *sheng* 生 ("to be born," "to beget"); the last half of the word, *rūpa*, is defined by Monier-Williams (1899: 885) as "any outward appearance or phenomenon or colour, form, shape, figure," which was translated into Chinese as *se* 色 ("color," "form"). From this, we can see that the word *shengse* is a loan word in its entirety and represents a foreign element in Chinese Buddhist language.

Li He was profoundly influenced by Buddhism and was very familiar with Buddhist texts. As a result, he quoted the Buddhist phrase *shengse*, signifying "gold" in his poem, meaning that the lofty screen in the inner chamber was a painting of gold to express its golden luster and splendor.

The Development of the Word *Shengse*

While the word *shengse* continues to appear in Buddhist literature after the Tang (including phonetic and semantic glosses on Buddhist scriptures and commentaries and explications of Buddhist monastic codes), there are few examples of its use and with Li He's poem perhaps the only instance of it in domestic Chinese literature. From the late Northern Song however, *shengse* begins to appear more frequently in domestic literature, and there is evidence to suggest that it had already become part of the common vernacular from the Song to the Ming.

During the late Northern Song, the word *shengse* begins to appear in the poetry and jottings of literati under three categories: (1) to represent painting, (2) to indicate clothing and forms of adornment, and (3) to describe picturesque scenery in poetry. The following three examples correspond to these three usages, respectively, with this general meaning also having changed from the metal "gold" itself to representing "golden" or "gold colored."

Examples: 6

6.a

During the Tang, General Li the Younger began creating landscape paintings in gold and green…if too much gold and green is employed like the situation nowadays with 'golden' (*shengse*) reticulated paintings that are entirely devoid of charm, what is there to be gained? "General Li the Younger" Xiao Li Jiangjun 小李將軍, refers to the Tang painter Li Zhaodao 李昭道 (*flor.* eighth century) who was renowned for his highly decorated landscape paintings in gold and green coloring that followed the style of his father, the painter Li Sixun 李思訓 (653–718), known as "General Li the Elder" Da Li Jiangjun 大李將軍.

唐小李將軍始作金碧山水...若多用金碧，如今生色罨畫之狀，而略無風韻，何取乎？

> "Jinbi shanshui" 金碧山水 ["Landscapes in Gold and Green"], in *Dongtian qinglu* 洞天清錄 [*A Pure Record of the Grotto Heavens*], by Zhao Xihu 趙希鵠, ed., Yi Yin (Hangzhou: Zhejiang renmin meishu chubanshe, 2016), 63. Northern Song

6.b

Emperor Zhenzong personally practiced thrift to transform all under Heaven, and so from His imperial quarters to the corners of the empire all clothing and ornaments that employed gold as embellishment were entirely forbidden…all households from his distaff relatives, down to civil and military officials, scholars, and commoners that had clothing or ornaments that were adorned with gold, including those that had 'golden color' (*shengse*) on the inside, had no fear of the law.

此蓋真宗皇帝躬行儉德，以化天下，故自中禁以及庶邦，凡衣服玩用以金為飾者，一切禁斷......其戚裡及臣僚士庶之家，衣服首飾並用銷金及生色內間金之類，並無避懼。

"Qing duan xiaojin deng shi" 請斷銷金等事 ["An Entreaty to Abandon Gold Ornamentation and Other Matters], in *Bao Zheng ji* 包拯集 [*The Collected Writings of Bao Zheng*], by Bao Zheng 包拯 (999–1062), (Beijing: Zhonghua shuju, 1963), 5.58. Song

6.c
The light of dawn declares the snow's retreat,
A 'golden hue' (shengse) floats about the wintery grass.
晨光報新霽，生色浮宿莽。

He Shanzhi you Fahua 和善之游法華, by Mou Yan 牟巘 (1227–1311), in *Quan Song shi* 全宋詩 [*Complete Poetry of the Song*], (Beijing: Beijing Daxue chubanshe, 1991–1998), 3511.41,927. Song

For paintings or adorned clothing to attain such shimmering golden luster, gold leaf or golden silk was certainly employed, so the term *shengse* is intimately related to its original meaning of "gold." In Example 6.c, this refers to the clear light of early morning when the sun shines on the perennial withered yellow winter grass in an entirely figurative sense that has nothing to do with "gold" itself.

I argue that the word *shengse* was a vernacular term from the Song to the Ming for the following reasons. First, there are many instances of its use, for example, it is found seven times in Meng Yuanlao's 孟元老 *Dongjing meng hua lu* 東京夢華錄 [*A Record of Dreaming of the Splendors of the past in the Eastern Capitol*] from the Song, and there are 83 instances of the word in the *Yuan shi* 元史 [*History of the Yuan*] compiled by Song Lian 宋濂 during the Ming. Second, since the Song, literati often use the word *shengse* to describe another uncommon expression *yanhua* 罨畫, meaning "variegated" or "brightly colored."

Examples: 7
7.a
I again asked him what other skills he had, he answered: 'I can also produce "yanhua".' Then I tested him with children's clothing, and I found it was just the 'shengse' we called today.
又問更有何藝, 曰:「亦能罨畫。」遂以小兒衣試之, 乃今之生色也。

Jiang Shaoyu 江少虞, "Fengsu zazhi" 風俗雜志 ["Miscellaneous Accounts of Customs"] in *Songchao shishi leiyuan* 宋朝事實類苑 [*A Categorized Garden of Historical Facts from the Song Dynasty*], (Shanghai: Shanghai guji chubanshe, 1981), 62.827. Song

7.b
The poem *Roaming Once More about the Temple on Mount Weiqu* has the following lines:
The mist over Magpie Crag dissolves, and the Jade Nest inclines,
The spring pool shines 'golden' (yanhua), and Venus sets.
The Yuan scholar Hao Tianting's commentary states: *"yanhua"* **means a painting of red-green** *"shengse."*
再游韋曲山寺:「鵲岩煙斷玉巢欹, 罨畫春塘太白低。」元人郝天挺注:「罨畫, 丹青生色圖畫也。」

Yuan Haowen 元好問, "Tan Yongzhi shi" 譚用之詩 ["On the Poetry of Tan Yongzhi"] in *Tangshi guchui pingzhu* 唐詩鼓吹評注 [*Fife and Drum Songs of Tang Poetry*], (Baoding: Hebei Daxue chubanshe, 2000), 9.484

Written records from the Ming period still use the word *shengse* to a large degree (such as the *Ming Taizu shilu* 明太祖實錄 [*Veritable Records of Emperor Taizu of the Ming*] compiled during the fourteenth and fifteenth centuries), but by the Qing period it disappears from the vernacular. One reason why it disappeared from the vernacular during the Qing is that it is only occasionally seen in literati writings from the period. Another perhaps more important reason is that Qing authors seemed to no longer understand the meaning of the word. For example, Wang Qi's 王琦 *Li He shige jizhu* 李賀詩歌集注 [*Collected Commentaries on the Poetry and Lyrics of Li He*] provides the following commentary on the term in Li He's poem; "a *shengse* painting" refers to a painting that is fresh and bright with lifelike colors 生色畫, 謂畫之鮮明, 色像如生者 (*Li He shige jizhu* 李賀詩歌集注, (Shanghai: Shanghai renmin chubanshe, 1977), 216.

Summary

Although also first seen in Buddhist texts from the Medieval Period as a foreign element, the word *shengse* took a different path in its Sinicization to the term *ziji*. While *ziji* infiltrated into domestic Chinese literature from Buddhist texts since the Tang, *shengse* is only found in Li He's poetry during the same period. While *ziji* developed steadily and saw its functions and usage gradually increase, *shengse* emerged suddenly during the late Northern Song and gradually entered the vernacular with many instances of its use during the Song, Yuan, and Ming. The Ming and Qing were key periods of development for *ziji* which was frequently used throughout the Qing and then became an integral part of modern Chinese. The Qing period, however, saw *shengse* begin to die out, and it completely fell out of vernacular use. Whether or not the word *shengse* existed in the vernacular during the Tang, why it erupted into use during the Song, and why it died out during the Qing are all questions to be solved by future research.

Historical Investigation of the Word *Xianzai*

The word *xianzai* is a commonly used noun of time in modern Chinese with two meanings: (1) the current period of time including a shorter or longer period of time around about the time of speaking; (2) the present moment, referring to the actual time of speaking (*Xiandai Hanyu cidian* 現代漢語詞典, 7th ed.). How did this word come about, and how did it become a commonly used word in modern Chinese? Zhu Guanming (2021) provides the following points on these questions.

The Origin of *Xianzai* and Its Use in Medieval Buddhist Texts

As previously noted, when discussing "Buddhist loanwords and translated terms," Wang Li (1958/1980: 521) specifically mentions that *xianzai* was translated from "a term in Sanskrit" and considers it to have "penetrated into the lifeblood of the Chinese language." Liang Xiaohong (1984) agrees with the view that *xianzai* has its origin in Buddhist texts, but she argues that it remained a technical Buddhist term representing one of the periods in the Buddhist concept of the "Three Preriods" (*sanshi* 三世), or "past" (*guoqu* 過去), "present" (*xianzai* 現在), and "future" (*weilai* 未來), and could not yet purely express time, i.e., the word *xianzai* was not yet a "time noun" in a grammatical sense. Zhu Qingzhi (2000) further points out that the three words "past," "present," and "future" are all translated from Sanskrit terms with *xianzai* being the translation of the Sanskrit word *pratyutpanna*. While these studies correctly point out the fact that *xianzai* is derived from its use in Buddhist texts as a calque, there are still several areas that require revision and supplementation.

The first thing to consider is the source term that the calque is modeled on. According to Seishi Karashima's (1998, 2001, 2010) three specialized Chinese-Sanskrit glossaries, there are numerous Sanskrit words that correspond to *xianzai* with some corresponding to individual Sanskrit words and others to combinations of multiple words. For example, there are three correspondences for *xianzai* in his glossary of the *Daoxing bore jing* 道行般若經: (1) *etarhi tiṣṭhanti dhriyante yāpayant*, "just now exist/at present reside"; (2) *pratyutpannā* "are present"; (3) *tiṣṭhanti dhriyante yāpayanti*, "reside"; and (4) *etarhi...tiṣṭhat~ dhriyamāna~ yāpayat~*, "at present reside" (Karashima, 2010: 534). In terms of its corresponding Sanskrit, *xianzai* is not the calque of a single Sanskrit word but uses two near-synonyms *xian* 現 "to appear" and *zai* 在 "to reside" in Chinese to translate a combination of multiple Sanskrit near-synonyms, such as in examples 1, 3, and 4. In these translation examples, after *xianzai* was created, it was then used to translate a single Sanskrit word as per example (2). In the other translation examples (1, 3, and 4), *xianzai* should be regarded as a parallel structured phrase, yet after it was used to translate a single Sanskrit word such as *pratyutpannā*, it can be regarded as a word.

The second thing to consider is whether *xianzai* can purely express time in Buddhist texts without being connected with the concept of the "Three Periods." The answer to this question is that it can as the Sanskrit words that correspond to the *xianzai* listed above all refer to ordinary time in Sanskrit and have no connection with the Buddhist notion of the "Three Periods." Moreover, we also investigated the use of *xianzai* in the *Chang a'han jing* 長阿含經 [*Dīrghāgama*]. In addition to the verbal meaning of "to presently exist" (*xianzai cunzai* 現在存在) and in juxtaposition with "past" (*guoqu* 過去) and "future" (*weilai* 未來) as one of the "Three Periods," there are several instances that represent unequivocal examples of its use as time nouns, such as in the following example:

Example: 8

8

At present, the *śramaṇa* ('monks') and elders [throughout] the Four Directions abound in wisdom and clearly explicate the *sutras* and *vinaya*.
現在四方沙門耆舊多智，明解經律。

Chang a'han jing 長阿含經 [*Dīrghāgama*], trans., Buddhayaśas 佛陀耶舍 and Zhu Fonian 竺佛念, *T.* vol. 1, no. 1, 4. 25c29. Yao Qin

The meaning of *xianzai* in this example conforms with definition (1) "the current period of time" noted earlier and definition, (2) "the present moment" is not yet seen in the translation of the *Chang a'han jing*.

The Usage of *Xianzai* in Medieval Domestic Literature

If we investigate the *Shishuo xinyu* 世說新語 [*A New Account of Tales of the World*], which was basically written at the same time as the *Chang a'han jing*, there are no instances of *xianzai* being used as a time noun, and words such as *jin* 今 ("now"), *fangjin* 方今 ("at present"), *jinri* 今日 ("today," "present day"), *dangnian* 當年 ("now"), and *dangjin* 當今 ("nowadays") are used to express the concept of present time instead.

Examples: 9

6.a

You always took delight in me doing the sound of a donkey, so 'now' (*jin* 今) I will do it for you.

卿常好我作驢鳴, 今我為卿作。

"Shangshi" 傷逝 ["Lamenting the Departed"], in *Shishuo xinyu* 世說新語 [*A New Account of Tales of the World*], by Liu Yiqing 劉義慶, SBCK ed., 17.10a. Liu Song 劉宋 (420–479 CE)

6.b

'At present' (*fangjin*), the officials who govern China...

方今宰牧華夏...

"Zhangshi" 政事 ["Political Affairs"], in *Shishuo xinyu*, 3.58b. Liu Song

In the examples above, Example (6.a) represents definition (2) "the present moment" and Example (6.b) represents definition (1) "the current period of time." The lack of examples of *xianzai* as a time word in the *Shishuo xinyu* indicates that the word was chiefly employed in Buddhist texts during the Medieval Period and had not yet spread to the domestic literature and the vernacular.

The Use of *Xianzai* Prior to the Mid-Ming Period

How then was *xianzai* as a time noun used after the Medieval Period? We have selected the Northern Song *Jingde chuandeng lu* 景德傳燈錄 [*The Jingde era Record of the Transmission of the Lamp*], the "Hundred Chapter" edition of the early Ming novel *Shuihu zhuan* 水滸傳 [*Heroes of the Water Margins*], and the mid-Ming Dynasty novel *Xiyouji* 西遊記 [*Journey to the West*] to carry out an investigation. The first of these are records of Chan Buddhist sayings, and the latter two are vernacular novels written by literati. As the vernacular elements in all three works are quite pronounced, they can be taken as representative examples to observe the spoken language of both Buddhists and the ordinary populace.

Similarly, the appearance of *xianzai* (also written as *xianzai* 見在) in these works is not as a time word. It is either a verb phrase meaning "to presently exist" (*xianzai cunzai* 現在存在) as per Example (10.a); or a combination of *xian* 現 "to appear" with the verb/preposition *zai* 在 "to reside" as in Example (10b.); or as per the notion of the "Three Periods" in Example (10.c):

Examples: 10

10.a

Bodhidharma nowadays 'presently exists' (*xianzai*)...
達磨如今現在。

Jingde chuandeng lu 景德傳燈錄 [*The Jingde era Record of the Transmission of the Lamp*], comp., Daoyuan 道原, *T.* vol. 51, no. 2076, 18.344c23. Song, 1004 CE

10.b

[The Iron Tablet] now resides in [my] home in Cangzhou...
見在滄州家裡...

Shuihu zhuan 水滸傳 [*Heroes of the Water Margins*], attr., Shi Nai'an 施耐庵, One Hundred Chapter ed., Ch. 52. 14th C

10.c

I pay homage (Skt., *namaḥ*) to the Buddhas of the Past, Future, and Present (*xianzai*).
南無過去未來現在佛。

Xiyouji 西遊記 [*Journey to the West*], attr., Wu Cheng'en 吳承恩, Ch. 100. Ming sixteenth C.

To express the time concept of "now," the *Jingde chuandeng lu* mainly uses such terms as *jin* 今 ("now"), *rujin* 如今 ("now"), *xian[you]* 見[有] ("appears [that there is]"), *shi[ren]* 時[人] ("contemporary person"), *jinshi* 今時, *jijin* 即今 ("right now"), and *xian* 現 ("appears"), and the *Shuihu zhuan* mainly uses terms such as *mujin* 目今 ("at present," "these days"), *xianjin* 見今 ("nowadays"), *jijin* 即今 ("right now"), *jimu* 即目 ("presently"), and *jinlai* 今來 ("present time") as per the following examples:

Examples: 11

11.a

I am now burdened by pain and wish to enter *nirvāṇa*.
吾今背痛, 欲入涅槃。

Jingde chuandeng lu, *T*, vol. 51, no. 2076, 1.205c7

11.b

At present [I have] the body of a nun, how could [you] not recognize [this]?
現是尼身, 何得不識。

Jingde chuandeng lu, *T*, vol. 51, no. 2076, 14. 313b29-313c1

11.c
These days the Son of Heaven's proclamations are to be read in public, how could we meet him [the Heavenly Master]?
目今天子宣詔, 如何得見。

Shuihu zhuan, Ch. 1

11.d
Yesterday it was early and cool when we left, why right now (*rujin*) we have to travel in the heat?
前日只是趁早涼走, 如今怎地正熱里要行?

Shuihu zhuan, Ch. 15

In the examples above, Examples (11.a) and (11.d) represent definition (2) "the present moment" and Examples (11.b) and (11.c) represent definition (1) "the current period of time." From this, we can see that from the Medieval Period to the mid-Ming, various words were used to express the concept of "now" in the vernacular, but instances of the typical use of the time word *xianzai* are not yet evident.

The Reappearance of *Xianzai* as a Time Word

We begin to see examples of the use of *xianzai* as a time word in literature from the late-Ming:
Examples: 12
12.a
Xu De is not always reunited with Mo Dajie. If she wants to seek another marriage partner now (*xianzai*), why not allow Yang Erlang to marry her and resolve the feud between the two families?
總是徐德不與莫大姐完聚了。現在尋人別嫁, 何不讓與楊二郎娶了, 消釋兩家冤仇?

Er ke Pai'an jingqi 二刻拍案驚奇 [*Slapping the Table in Amazement: Second Collection*], by Ling Mengchu 凌濛初, *j*. 38. Ming, 1632

12.b
I have no children now (*xianzai*) and I'll adopt and rear him [an abandoned baby] and will see what happens in eighteen years.
我今現在無子, 且收來養著, 到十八年後再看如何。

Er ke Pai'an jingqi, *j*. 30

In these two examples, *xianzai* is used adverbially before a verb phrase and cannot be understood as "now is doing 現在正在." They can only be time words as per the following: Example (12.a) according to the context of the story, Mo Dajie is not "presently" looking for someone to marry; Example (12.b) the verb "without" (*wu* 無) expresses a state, and it cannot be said to mean "without right now" [lit. "is withouting" 正在無].

These examples (12) represent some of the earliest instances of the use of the time word *xianzai* in vernacular language after the Medieval Period. This usage was rare during the late Ming but gradually increased in frequency during the Qing. We find sixteen instances of this in the first eighty chapters of *Hongloumeng,* such as the following examples:

Examples: 13

13.a

These days [we bring the pills] up north from the south, and [they] are now (*xianzai*) buried under the pear tree.

如今從南帶至北, 現在就埋在梨花樹底呢。

Hongloumeng, Ch. 7

13.b

His older brother Jin Wenxiang is 'now' (*xianzai*) the comprador at the venerable lady's establishment.

他哥哥金文翔, 現在是老太太那邊的買辦。

Hongloumeng, Ch. 46

In the examples above, Example (13.a) represents definition (2) "the present moment" and Example (13.b) represents definition (1) "the current period of time." When compared with the time word *xianzai* in Buddhist texts during the Medieval Period, definition (2) "the present moment" has been added in *Hongloumeng.*

In the Qing novel *Ernü yingxiong zhuan* 兒女英雄傳 [*Tales of the Heroism of Youths and Maidens*] we see instances of *xianzai zai* 現在在 ("presently located at"):

Example: 14

14.a

That foe, what type of fellow is he, and what location is [he] 'presently located' (*xianzai zai*)?

他這仇人又是何等樣人, 現在在甚麼地方?

Ernü yingxiong zhuan 兒女英雄傳 [*Tales of the Heroism of Youths and Maidens*], by Wen Kang 文康. Qing, pub., 1878

The appearance of *xianzai zai* 現在在 ("presently located at") indicates that the time word *xianzai* has become more closely integrated and that speakers have completely stopped using it as the detachable phrase *xian* + *zai* (as per Example 10b). We can see that during the mid-Qing, the status of *xianzai* as a "word" used in the spoken language was very stable. Therefore, during this period when the *Ernü yingxiong zhuan* was written, the usage of the word *xianzai* was entirely equivalent to that of contemporary Chinese (In contemporary Chinese, time words can also be used as adverbials of time after forming prepositional phrases with prepositions (Yu Dongtao, 2013). According to our research, this type of phrase *zai xianzai* 在現在 ("at present") can be found during the late-Qing and early Republic periods, such as: "The subscription for each car is one yuan, when "at present" (*zai xianzai*) it is provided, it truely feels painful." (Li Hanqiu 李涵秋, *Guangling chao* 廣陵潮 [Guangling Tide], Ch. 93, pub. 1936.).

Summary

From the reappearance of instances of *xianzai* in the *Er ke Pai'an jingqi,* published in 1632 during the late Ming Dynasty, to *Hongloumeng*, written around 1760 during the Qing, we see not only an increase in usage, but also additional meanings. By the time the *xianzai zai* appears in the *Ernü yingxiong zhuan,* written around 1850, it took more than 200 years for *xianzai* to complete its process of integration into Chinese where it has been retained as a commonly used word in contemporary Chinese.

While *xianzai* appeared as a translated term corresponding to a synonymous coordinate construction in the Indic Buddhist source texts and was used as a time word in the Medieval Buddhist texts, it remained silent for over a thousand years from the Medieval Period up to the end of the Ming, and only reappeared from the late-Ming onward in vernacular literature. This fact makes people wonder – is the term *xianzai* in modern Chinese actually a word borrowed from Buddhist texts? This is because during the Medieval Period, in Tang and Song dynasties where Buddhism influence was strongest, *xianzai* did not penetrate into the popular vernacular. Instead, during the Ming and Qing when Buddhist influence was weaker, it entered the spoken language and became a very common word.

It is unreasonable, however, to suggest that *xianzai* developed from the Chinese language itself, that is, it gradually underwent lexicalization from the phrase *xian/xian + zai* 現/見+在. This is because, as noted earlier, prior to reliable instances of the time word *xianzai* appearing during the late-Ming, there were few instances of the *xian/xian + zai* 現/見+在 type with other words such as *xian* 現 ("appears"), *rujin* 如今 ("now"), *and jijin* 即今 being used to express "now." In short, *xian/xian + zai* 現/見+在 lacks the necessary precondition for lexicalization – high-level frequency of usage. A more suitable explanation, perhaps, is that the time term *xianzai* was produced via the translation of Buddhist texts. Although it was at first only used in Buddhist texts, at any rate, it became a "reserve member" in the Chinese lexicon, and after a thousand years of silence, for reasons that remain unclear, it was reemerged during the late-Ming Dynasty and became integrated with several *xian/xian + zai* 現/見+在 type phrases to form the time word *xianzai* in modern Chinese (This situation, however, is not an isolated case. For example, the word *kaixin* 開心, which means "happy" in modern Chinese as mentioned above, follows a historical development that is also very similar to *xianzai*. Zhu Guanming (2007b/2009) points out that it comes from Buddhist scriptures and that there are many instances of its use in Buddhist texts from the Medieval Period referring to the untying of knots or fetters so that the mind is no longer troubled by such "vexations," miserliness, greed, jealousy, suspicion, and hesitation. However, according to Xu Shiyi (2007: 241–247) and Cui Xiliang (2009), *kaixin* 開心 is rarely seen in domestic Chinese literature after the Medieval Period and only begins to appear after the Qing after it had also "fallen asleep" for a thousand years.).

Conclusion

Since Buddhism was first introduced into China between the Western and Eastern Han, it has had a profound impact on all aspects of Chinese culture. The substantial number of Buddhist texts translated into Chinese brought fresh blood into the language, and the immense changes that occurred to Chinese were also largely related to Buddhism's long-lasting and deeply felt influence. As Mair (1994/2009) points out, Buddhism played a vital role in the emergence of national languages throughout East Asia including China, Japan, Korea, and Vietnam. Europe also underwent a similar process where religion influenced language. Fehling (1980) employed a substantial sample of linguistic data to prove that the spread of Christianity and Bible translation influenced the formation of modern European syntax. The distinctive syntactic features of Near Eastern languages entered European languages by means of biblical language, and the syntactic elements found in European languages that are not related to Latin or ancient Greek are without exemption drawn from those of Near Eastern languages (see Zhu Guanming 2011). What remains to be discussed is how Buddhist language entered step by step into vernacular and led finally to the current state of the Chinese language.

It is generally believed that the assimilation of the language of the Buddhist classics into Chinese underwent the following process: It emerged in the translation of Buddhist texts during the Medieval Period, was first used after this by clergy and lay followers in Buddhist-related literature, and then entered the common vernacular. However, if we consider the issues raised above by *ziji, shengse,* and *xianzai* and the differing degrees of content and function they exhibit, it is clear that this approach is only suitable for the term *ziji*. The path that *shengse* underwent can be summarized as follows: It emerged in the translation of Buddhist texts during the Medieval Period, was rarely seen during the Tang, was used through the Song, Yuan, and Ming, then underwent changes when it entered the vernacular, and finally died out during the Qing. The path that *xianzai* underwent was as follows: It emerged in the translation of Buddhist texts during the Medieval Period, was silent from the Tang through to the Yuan and Ming, started to be commonly used from the late-Ming to Qing, and gained additional meanings when it entered vernacular. In the Sinicization process of *shengse* and *xianzai*, we do not see monks and Buddhist laity playing a role in their dissemination and promotion. It is evident then that linguistic forms possess different modes of Sinicization.

This chapter began with an outline of Buddhist influence and the introduction of its scriptures into Chinese as the basic premise for the Sinicization of Buddhist language. This was followed by the routes this Sinicization took by using three terms to illustrate the complexity of these processes. However, many more questions requiring long-term research remain. For example, there are many unique forms found in Buddhist texts that differ from those found in domestic literature. Some underwent a journey of Sinicization and finally became integrated into the popular vernacular, some only went halfway, while others are still only found in Buddhist texts. What are the reasons and patterns behind these differences? Does the word *ziji* represent the Sinicization process of most linguistic forms found in Buddhist language as some have previously suggested? How many approaches to Sinicization are there, which can be used for each linguistic form, and what factors determine these various approaches? In reference to studies on the influence of biblical language on European syntax and the limited amount of research noted above, we believe that the unique syntactic forms found in Buddhist works have also had a substantial impact on Chinese syntax and have entered the spoken language. What are the particular circumstances behind this situation? These questions must be answered on the basis of further case studies.

References

Bai, Huawen (白化文). 2003. *Hanhua Fojiao yu fosi* 汉化佛教与佛寺 [*Sinicized Buddhism and Buddhist Temples*]. Beijing: Beijing chubanshe 北京:北京出版社 [Beijing: Beijing Press].

Cao, Guangshun (曹广顺) and Yu, Hsiaojung (遇笑容). 2000. Zhonggu yijing zhong de chuzhishi 中古译经中的处置式 [The disposal construction of translated middle Chinese Buddhist Sutras]. *Zhongguo yuwen* 中国语文 [*Studies of the Chinese Language*] 6: 555–563.

Chao, Yuenren (赵元任). 1979. *Hanyu kouyu yufa* (Lü Shuxiang yi) 汉语口语语法(吕叔湘译) [*A Grammar of Spoken Chinese (Translated by Lü Shuxiang)*]. Beijing: Shangwu yinshuguan 北京:商务印书馆 [Beijing: The Commercial Press].

Chen, Zhongyuan (陈中源). 2011. 'Ziji' zai Zhonggu yihou de fazhan "自己" 在中古以后的发展 [Development of 'ziji' in post-medieval Chinese]. *Hanyushi yanjiu jikan* 汉语史研究集刊 [*Collection of works on the history of the Chinese language*] 14: 54–66. Chengdu: Bashu Shushe 成都:巴蜀书社 [Chengdu: Bashu Press].

Cui, Xiliang (崔希亮). 2009. Shuo 'kaixin' yu 'guanxin' 说 "开心" 与 "关心" [On 'Kaixin' and 'Guanxin']. *Zhongguo yuwen* 中国语文 [*Studies of the Chinese Language*] Vol. 5: 410–418.

Dong, Zhiqiao (董志翘). 2000. *Gaosengzhuan* ciyu tongshi (1)—Jiantan Hanyifodian kouyuci xiang Zhongtu wenxian de kuosan 高僧传 词语通释(一)—兼谈汉译佛典口语词向中土文献的扩散 [Notes to the words of *biographies of eminent monks*, 1: On the spread of spoken words from Chinese Buddhist scriptures to non-Buddhist texts]. *Hanyushi yanjiu Jikan* 汉语史研究集刊 [*Collection of Works on the History of the Chinese Language*] Vol. 2: 251–266. Chengdu: Bashu shushe 成都:巴蜀书社 [Chengdu: Bashu Press].

Edgerton, Franklin. 1953. Buddhist hybrid Sanskrit grammar and dictionary. In *Dictionary*, vol. 2. New Haven: Yale University Press.

Fang, Litian (方立天). 1989. Fojiao Zhongguohua de lichen 佛教中国化的历程 [*Progress of the Chinization of Buddhism*]. *Shijie zongjiao yanjiu* 世界宗教研究 [*Studies in World Religions*] 3: 1–15.

Fang, Yixin (方一新). 1994. *Shishuo Xinyu* ciyu shigu 世说新语 词语拾诂 [Explaining some words in *a new account of Tales of the world*]. *Hangzhou Daxue xuebao* 杭州大学学报 [*Journal of Hangzhou University*] 1: 117–121.

Fehling, Detlev. 1980. The origins of European syntax. *Folia Linguistica Historica* 1 (2): 353–387.

Gong, Bin (龚斌). 2011. *Shishuo xinyu jiaoshi* 世说新语校释 [*Proofreading and Explanation of Shishuo xinyu*]. Shanghai: Shanghai guji chubanshe 上海:上海古籍出版社 [Shanghai: Shanghai Classics Publishing House].

Guo, Peng (郭朋). 1980. *Suitang fojiao* 隋唐佛教 [*Buddhism in Sui-Tang Dynasty*]. Jinan: Qilu shushe 济南:齐鲁书社 [Jinan: Qilu Press].

Han, Lizhou(韩理洲). 2010. *Quan Beiwei Dongwei Xiwei wen buyi* 全北魏东魏西魏文补遗 [*Addendum to the Whole Writing of the Northern, Eastern and Western Wei Dynasties*]. Xi'an: Sanqin chubanshe 西安:三秦出版社 [Xi'an: Sanqin Press].

He, Zhiguo (何志国). 2005. 'Xianfo moshi' he 'Xiwangmu + Fojiao tuxiang moshi' shuo shangque—zailun Fojiao chuchuan Zhongguo nanfang zhi lu "仙佛模式" 和 "西王母+佛教图像模式"说商榷–再论佛教初传中国南方之路 [Discussing the Views of 'Immortal & Buddha Model' and 'West Queen Mother + Buddhist Image Model'–Revisiting the Initial Propagative Road of Buddhism to Southern China]. *Minzu yishu* 民族艺术 [*National Arts*] 4: 96–105.

Hu, Shih (胡适). 1937. The Indianization of China: A case study in cultural borrowing. *Independence, convergence, and borrowing in institutions, thought, and art*. Cambridge: Harvard College.

Ji, Xiuqin (籍秀琴). 1993. *Zhongguo renming tanxi* 中国人名探析 [*Exploring Chinese Names*]. Beijing: Zhongguo guangbo dianshi chubanshe 北京:中国广播电视出版社 [Beijing: China Radio & TV Press].

Jiang, Shaoyu (蒋绍愚). 2007. Yuyan jiechu de yige anli – zaitan 'V(O)Yi' 语言接触的一个案例–再谈 "V(O)已" [A Case for Language Contact – Revisiting the Construction 'V(O)Yi']. *Yuyanxue luncong* 语言学论丛 [*Essays on Linguistics*] Vol. 36:268–285. Beijing: Shangwu yinshuguan 北京:商务印书馆 [Beijing: The Commercial Press].

Kamata, Shigeo (鎌田茂雄). 1980/1986. *Jianming Zhongguo Fojiaoshi* (Zheng Pengnian Yi) 简明中国佛教史(郑彭年译) [*Brief history of Chinese Buddhism (translated by Zheng Pengnian)*]. Shanghai: Shanghai yiwen chubanshe 上海:上海译文出版社 [Shanghai: Shanghai Translation Publishing House].

Karashima, Seishi (辛島靜志). 1998. A glossary of Dharmarakṣa's translation of the lotus sutra. Tokyo: The International Research Institute for Advanced Buddhology, Soka University.

——— (辛島靜志). 2000. Hanyi Fodian de yuyan yanjiu 汉译佛典的语言研究 [A Study of the Language of the Chinese Buddhist Translations]. Beijing Daxue Zhongguo chuantong wenhua yanjiu zhongxin, *Wenhua de kuizeng: Hanxue yanjiu guoji huiyi lunwenji* (yuyan wenxue juan) 北京大学中国传统文化研究中心编 文化的馈赠:汉学研究国际会议论文集 (语言文学卷) [Cultural Present: Proceedings of the International Symposium on Sinology Studies]. Volume of Linguistics and Literature: 512–514. Beijing: Beijing Daxue chubanshe 北京:北京大学出版社 [Beijing: Peking University Press].

——— (辛島靜志). 2001. A glossary of Kumārajīva's translation of the Lotus Sutra. Tokyo: The International Research Institute for Advanced Buddhology, Soka University.

——— (辛島靜志). 2002. *Daoxing bore jing* he 'yiyi' de duibi yanjiu – *Daoxingborejing* zhong de nanci 道行般若經 和「異譯」的對比研究—— 道行般若經 中的難詞 [Comparative Study of Lokakṣema's Translation of the *Aṣṭasāhasrikā Prajñāpāramitā* and Other Chinese Translations: Some Puzzling Words in Lokakṣema's Translation]. *Hanyushi Yanjiu Jikan 5* 漢語史研究集刊 5 [*Collection of Works on the History of the Chinese Language*] Vol. 5: 199–212. Chengdu: Bashu shushe 成都:巴蜀書社 [Chengdu: Bashu Press].

——— (辛島靜志). 2010. A glossary of Lokakṣema's translation of the Aṣṭasāhasrikā Prajñāpāramitā. Tokyo: The International Research Institute for Advanced Buddhology, Soka University.

——— (辛島靜志). 2016. *Fodian yuyan ji chuancheng* 佛典语言及传承 [*Languages and Transmission of Buddhist Texts*]. Shanghai: Zhongxi shuju 上海:中西书局 [Shanghai: Zhongxi Press].

Lai, Yonghai (赖永海). 2010a. *Zhongguo Fojiao tongshi* 中国佛教通史(第一卷) [*A General History of Chinese Buddhism (vol. 1)*]. Nanjing: Jiangsu renmin chubanshe 南京:江苏人民出版社 [Nanjing: Jiangsu People's Publishing House].

———(赖永海). 2010b. *Zhongguo Fojiao tongshi* 中国佛教通史(第二卷) [*A General History of Chinese Buddhism (vol. 2)*]. Nanjing: Jiangsu renmin chubanshe 南京:江苏人民出版社 [Nanjing: Jiangsu People's Publishing House].

Li, Shangquang (李尚全). 1989/1994. Jianlun Fojiao de Zhongguohua 简论佛教的中国化 [Brief Discussion on the Sinicization of Buddhism]. *Lanzhou xuekan* 兰州学刊 [*Lanzhou Academic Journal*] 4:72–76. / *Fojiao lun yi ji* 佛教论译集 [*Studies and Translated Works of Buddhism*]. 1994: 16–43. Lanzhou: Gansu renmin chubanshe 兰州:甘肃民族出版社 [Lanzhou: Gansu People's Publishing House].

Li, Fuhua (李富华) & He Mei(何梅). 2003. *Hanwen Fojiao Dazangjing yanjiu* 汉文佛教大藏经研究 [*Research on the Chinese Tripitaka*]. Beijing: Zongjiao wenhua chubanshe 北京:宗教文化出版社 [Beijing: China Religious Culture Publisher].

Liang, Xiaohong (梁晓虹). 1984. Fojing ciyu zhaji 佛经词语札记 [Notes on words in Chinese Buddhist scriptures]. *Nanjing Daxue xuebao* 南京大学学报 [*Journal Of Nanjing University*].2: 42–46.

———(梁晓虹). 1996. *Huahua Fojiao* 华化佛教 [*Sinicized Buddhism*]. Beijing: Beijing Yuyan Xueyuan chubanshe 北京:北京语言学院出版社 [Beijing: Beijing Language Institute Press].

Liang, Yinfeng (梁银峰). 2003. 'Zhuo cige sha' de 'sha' shi biao jieguo de bujiwu dongci ma? "啄雌鸽杀"的"杀"是表结果的不及物动词吗? [Is the 'sha' of 'Zhuo Cige Sha' an intransitive verb expressing the result?]. *Zhongguo yuwen* 中国语文 [*Studies of the Chinese Language*] 2: 181–182.

Liu, Huiqing (刘惠卿). 2019. *Fojiao Zhongguohua jincheng yu Jin-Tang wenyan xiaoshuo yanjin yanjiu* 佛教中国化进程与晋-唐文言小说演进研究 [*A Study of the Sinicization Process of Buddhism and the Evolution of Classical Chinese Novels from the Jin to Tang Dynasty*], Chengdu: Xinan Jiaotong Daxue chubanshe 成都:西南交通大学出版社 [Chengdu: Southwest Jiaotong University Press].

Liu, Yuehua (刘月华), Pan, Wenyu (潘文娱) and Gu, Wei(故韡). 2001. *Shiyong xiandai Hanyu yufa* 实用现代汉语语法 [*Practical Modern Chinese Grammar*]. Beijing: Shangwu yinshuguan 北京:商务印书馆 [Beijing: The Commercial Press].

Lü, Shuxiang (吕叔湘). 1988. Nanbeichao renming yu Fojiao 南北朝人名与佛教 [Personal names and Buddhism in the Southern and Northern dynasties]. *Zhongguo yuwen* 中国语文 [*Studies of the Chinese Language*] 4: 241–246.

Mair, Victor H. (梅维恒) 1994. Buddhism and the Rise of the Written Vernacular in East Asia: The Making of National Languages. *The Journal of Asian Studies*, Vol. 53, No. 3: 707–751. 汉译:佛教与东亚白话文的兴起:国语的产生 (王继红、顾满林译), 载 Zhu, Qingzhi (朱庆之) 编 *Fojiao Hanyu yanjiu* 佛教汉语研究 [*Studies of Buddhist Chinese*]. Beijing: Shangwu yinshuguan 北京:商务印书馆 [Beijing: The Commercial Press]. 2009: 358–409.

Monier-Williams, Monier. 1899. *A Sanskrit-English dictionary*. Oxford: Oxford University Press.

Ren, Jiyu (任继愈). 1981. *Zhongguo Fojiao shi (1)* 中国佛教史(第1卷) [*History of Buddhism in China*] vol.1. Beijing: Zhongguo shehui kexue chubanshe 北京:中国社会科学出版社 [Beijing: China Social Sciences Publishing House].

Shi, Guanghui (史光辉). 2011. *Donghan-Tang Hanwen chongyi Fojing cihui bijiao yanjiu* 东汉-唐汉文重译佛经词汇比较研究 [*Comparative Study of Lexicon in Various Translated Versions of the Buddhist Sutras from Eastern Han to Tang Dynasty*]. Beijing: Guojia shehui kexue jijin jiexian cailiao 北京:国家社会科学基金结项材料 [Beijing: Achievement Report of the Project Funded by National Social Science Fund of China].

Song, Yubo (宋玉波). 2012. *Fojiao Zhongguohua licheng yanjiu* 佛教中国化历程研究 [*Study of the Chinization of Buddhism Progress*]. Xi'an: Shanxi renmin chubanshe 西安:陕西人民出版社 [Xi'an: Shanxi People's Publishing House].

Spahr, Richard (徐真友). 1997. Guanyu Fodian yuyan de yixie yanjiu (Wan Jinchuan yi) 关于佛典语言的一些研究(万金川译) [Some Remarks on the Language of Buddhist Sutras (Traslated by Wan Jinchuan)]. *Zhengguan* 正观 [The Positive View] 1:68–83; also in Zhu, Qingzhi (朱庆之). 2009. Fojiao Hanyu yanjiu 佛教汉语研究 [Studies of Buddhist Chinese] 196–210. Beijing: Shangwu yinshuguan 北京:商务印书馆 [Beijing: The Commercial Press].

Speijer, J.S. 1886. *Sanskrit Syntax*. Leiden: Motilal Banarsidass.

Tatia, Nathmal. 1976. *Prātimokṣasūtram of the Lokottaravādimahāsāṅghika school*, Tibetan Sanskrit works series 16. Patna: Kashi Prasad Jayaswal.

Wang, Li (王力). 1958/1980. *Hanyu shigao* 汉语史稿 (修订本) [*A draft of the Chinese Language History*]. Beijing: Kexue chubanshe 北京:科学出版社 [Beijing:Science Press]/Beijing: Zhonghua shuju 北京:中华书局 [Beijing: Zhonghua Book Company].

Wang, Suqi (王苏琦). 2007. Handai zaoqi Fojiao tuxiang yu Xiwangmu tuxiang zhi bijiao 汉代早期佛教图像与西王母图像之比较 [Comparison between the Buddhist images in the early Han dynasty and those of the Western queen mother]. *Kaogu yu wenwu* 考古与文物 [*Archaeology and Cultural Relics*] 4:35–44.

Wu, Juan (吴娟). 2009. 'Jiuru' tanyuan "久如"探源 [Exploring the Origin of 'Jiuru']. *Hanyushi Xuebao 8* 汉语史学报 8 [*Journal of the Chinese Language History*] Vol. 8: 226–233, Shanghai: Shanghai jiaoyu chubanshe 上海:上海教育出版社 [Shanghai: Shanghai Education Publishing House].

Xu, Kangsheng (许抗生). 1990. Jianlun Fojiao de Zhongguohua 简论佛教的中国化 [Brief Remarks on Chinization of Buddhism]. *Zhongguo wenhua yu Zhongguo zhexue* 中国文化与中国哲学 [*Chinese Culture and Chinese Philosophy*] 1988: 361–375. Beijing: Shenghuo-Dushu-Xinzhi Sanlian shudian 北京:生活•读书•新知三联书店 [Beijing:SDX Joint Publishing Company].

Xu, Shiyi (徐时仪). 2007. *Hanyu baihua fazhan shi* 汉语白话发展史 [*Evolution of the Chinese Vernacular*]. Beijing: Beijing Daxue chubanshe 北京:北京大学出版社 [Beijing: Peking University Press].

Yang, Tongjun (杨同军). 2011. *Yuyan jiechu he wenhua hudong: Hanyi Fojing cihui de shengcheng yu yanbian yanjiu* 语言接触和文化互动:汉译佛经词汇的生成与演变研究 [*Language Contact and Cultural Interaction: A Study of the Formation and Evolution of the Words in the Chinese Translated Buddhist Scriptures*]. Beijing: Zhonghua shuju 北京:中华书局 [Beijing: Zhonghua Book Company].

Yao, Baoxian (姚宝贤). 1936. Fojiao ru Zhongguo hou zhi bianqian jiqi tezhi 佛教入中国后之变迁及其特质 [The Changing and Characteristics of Buddhism after Entering China], *Zhongshan wenhua jiaoyuguan jikan* 3.1 中山文化教育馆季刊 3.1 [*Zhongshan Cultural and Educational Museum Quarterly*] Vol 3.1: 267–272.

Yu, Liming (俞理明). 1993. *Fojing wenxian yuya n* 佛经文献语言 [*Buddhist Literature and Language*]. Chengdu: Bashu shushe 成都:巴蜀书社 [Chengdu: Bashu Press].

Yu, Dongtao (余东涛). 2013. Shuo xiandai Hanyu "jieci+shijianci" zuhe 说现代汉语"介词+时间词"组合 [On the 'preposition+temporal word' structure in mandarin Chinese], Hanyu xuebao 汉语学报 [Chinese Linguistics] 2: 72–77.

Yu, Fangyuan (于方圆) and Zhu, Guanming (朱冠明). 2018. Jin wunian Hanyi Fodian yufa yanjiu 近五年汉译佛典语法研究 [Review of the studies of the Buddhist Chinese grammar in recent five years], *Hanyushi Yanjiu Jikan* 25 汉语史研究集刊 25 [*Collection of works on the history of the Chinese language*] Vol. 25: 290–308. Chengdu: Bashu shushe 成都:巴蜀书社 [Chengdu: Bashu Press].

Zhang, Wenbin (张文彬) and Shaolian Li (李绍连). 1987. Shilun Wei Jin Nanbeichao Fojiao de xingsheng jiqi shehui yuanyin 试论魏晋南北朝佛教的兴盛及其社会原因 [A tentative remark on the prosperity of Buddhism in the Wei, Jin, and Southern and Northern dynasties and its social causes]. *Yindu xuekan* 殷都学刊 [*Journal of Yindu*] 3:43–47.

Zhang, Yisan (张诒三) and Futong Zhang (张福通). 2013. *Foyuan wailaici Hanhua yanjiu* 佛源外来词汉化研究 [*Study of the Sinicization of Buddhist Loanwords*]. Beijing: Zhongguo shuji chubanshe 北京:中国书籍出版社 [Beijing: China Book Publishing House].

Zhu, Dexi (朱德熙). 1982. *Yufa jiangyi* 语法讲义 [*Outline of Chinese Grammar*]. Beijing: Shangwu yinshuguan 北京:商务印书馆 [Beijing: The Commercial Press].

Zhu, Qingzhi (朱庆之). 1992a. *Fodian yu zhonggu Hanyu cihui yanjiu* 佛典與中古漢語詞彙研究 [*Chinese Buddhist Scriptures and the Study of Middle Chinese Vocabulary*]. Taibei: Wenjin Chubanshe 台北:文津出版社 [Taibei: Wenchin Press].

———(朱庆之). 1992b. 'Yingxiang' jinyi de laiyuan "影响"今义的来源 [The origin of the present meaning of the word 'Yingxiang']. *Wenshi zhishi* 文史知识 [*Chinese Literature and History*] 4: 99–100.

———(朱庆之). 2000. Fojing fanyi zhong de fangyi jiqi dui Hanyu cihui de yingxiang 佛经翻译中的仿译及其对汉语词汇的影响 [Calques in Buddhist Scripture Translation and Their Influence on Chinese Vocabulary]. *Zhonggu jindai Hanyu yanjiu* 中古近代汉语研究 [*Studies of the Middle and Pre-Modern Chinese*] Vol. 1: 247–262. Shanghai: Shanghai jiaoyu chubanshe 上海：上海教育出版社 [Shanghai: Shanghai Education Publishing House].

———(朱庆之). 2001. Fojiao Hunhe Hanyu chulun 佛教混合汉语初论 [A Preliminary Discussion on Buddhist Hybrid Chinese]. *Yuyanxue luncong* 24 语言学论丛 24 [*Essays on Linguistics*] Vol. 24: 1–33. Beijing: Shangwu yinshuguan 北京:商务印书馆 [Beijing: The Commecial Press].

———(朱庆之). 2002/2006. Lun Fojiao dui gudai Hanyu cihui fanzhan yanbian de yingxiang 论佛教对古代汉语词汇发展演变的影响 [On the Influence of Buddhism on the Evolution of Ancient Chinese Vocabulary]. (Hanguo) *Zhongguo yanjiu* (韩国)中国研究 [(South Korea) *Studies on China*] Vol. 30: 157–181. / *21 Shiji de Zhongguo yuyanxue* (2) 21世纪的中国语言学(二) [*Chinese Linguistics in the 21st Century*] Vol.2: 231–270. Beijing: Shangwu yinshuguan 北京:商务印书馆 [Beijing: The Commecial Press].

———(朱庆之). 2008. On some basic features of Buddhist Chinese. Journal of the International Association of Buddhist Studies, 31(1–2): 485–504.

———(朱庆之). 2011. Yige fanyuci zai gu Hanyu zhong de shiyong he fazhan 一个梵语词在古汉语中的使用和发展 [A case study of the use and development of a Sanskrit word in ancient Chinese]. *Zhongguo yuwen* 中国语文 [*Studies of the Chinese Language*] 4: 373–382.

———(朱庆之) and Guanming Zhu (朱冠明). 2006. Fodian yu Hanyu yufa yanjiu—20 Shiji guonei Fojiao Hanyu yanjiu huigu zhi er 佛典与汉语语法研究—20世纪国内佛教汉语研究回顾之二 [Buddhist scriptures and the study of Chinese grammar: Review of the study of Buddhist Chinese in China in the 20th century, part 2]. *Hanyushi Yanjiu Jikan* 9 汉语史研究集刊 9 [*Collection of works on the history of the Chinese language*] 9: 413–459. Chengdu: Bashu shushe 成都:巴蜀书社 [Chengdu: Bashu Press].

Zhu, Guanming (朱冠明). 2007a. Cong Zhonggu Fodian kan 'ziji' de Xingcheng 从中古佛典看"自己"的形成 [On the origin of the reflexive 'ziji' from the perspective of the medieval Chinese Buddhist scriptures]. *Zhongguo yuwen* 中国语文 [*Studies of the Chinese Language*] 5: 402–411.

———(朱冠明). 2007b/2009. Fodian ciyu kaoshi liu ze 佛典词语考释六则 [Textual Interpretation of Six Words in Buddhist Sutras], Disan jie Hanyushi xueshu yantaohui ji diliu jie zhonggu Hanyu guoji xueshu yantaohui 第三届汉语史学术研讨会暨第六届中古汉语国际学术研讨会 [The 3rd Symposium on the History of the Chinese Language & the 6th International Symposium on Middle Chinese]. Sichuan Daxue 四川大学 [Sichuan University]; Also in Zhongguo Renmin Daxue wenxueyuan bian *Yuyan Lunji* 6 中国人民大学文学院编 语言论集 第6辑 [Essays on Language and Linguistics, edited by the School of Liberal Arts, RUC] Vol. 6: 127–140. Beijing: Zhongguo shehui kexue chubanshe 北京:中国社会科学出版社 [Beijing: China Social Sciences Publishing House].

———(朱冠明). 2008. Yizhi: Fojing fanyi yingxiang Hanyu cihui de yizhong fangshi 移植:佛经翻译影响汉语词汇的一种方式 [Transfer: A Means of the Influence of the Translation of Buddhist Scriptures on Chinese Vocabulary]. *Yuyanxue luncong* 37 语言学论丛 37 [*Essays on Linguistics*] Vol. 37: 169–182. Beijing: Shangwu yinshuguan 北京:商务印书馆 [Beijing: The Commercial Press].

———(朱冠明). 2011. Zhonggu Fodian yu Hanyu shoushi zhuyuju de fazhan—Jiantan Fojing Fanyi Yingxiang Hanyu Yufa de Moshi 中古佛典与汉语受事主语句的发展—兼谈佛经翻译影响汉语语法的模式 [The medieval Chinese Buddhist scriptures and the development of the Chinese patient-subject clause: A case study of the pattern of the effect of the Buddhist sutra translation on Chinese grammar]. *Zhongguo yuwen* 中国语文 [*Studies of the Chinese Language*] 2: 169–178.

———(朱冠明). 2013. Hanyi Fodian yufa yanjiu shuyao 汉译佛典语法研究述要 [Concise Review of the Study of the Grammar of Chinese Buddhist Scriptures] in Jiang Shaoyu, Hu Chirui zhubian *Hanyi Fodian yufa yanjiu lunji* 载蒋绍愚、胡敕瑞主编 汉译佛典语法研究论

集 [*Collection of Works on Grammar of the Chinese Buddhist Scriptures*]. 1–45. Beijing: Shangwu yinshuguan 北京:商务印书馆 [Beijing: The Commercial Press].

———(朱冠明). 2015. Zhonggu yijing zhong de 'weicengyou' jiqi liuchuan 中古译经中的"未曾有"及其流传 [The word 'weicengyou' in medieval translated Buddhist scriptures and its evolution]. *Guhanyu yanjiu* 古汉语研究 [*Research in Ancient Chinese Language*] 2: 16–25.

———(朱冠明). 2019. Lihe Qingong shi 'shengse' kao 李贺 秦宫 诗"生色"考 [On Explaining the Word 'Shengse' in Lihe's Poem Qingong]. *Yuyanxue luncong* 59 语言学论丛 59 [*Essays on Linguistics*] 59: 140–165. Beijing: Shangwu yinshuguan 北京:商务印书馆 [Beijing: The Commercial Press].

———(朱冠明). 2021. Shijian mingci 'xianzai' de laiyuan ji Zhongguohua 时间名词"现在"的来源及中国化 [The origin and Sinicization of the temporal noun 'Xianzai']. *Hanyu xuebao* 汉语学报 [Chinese Linguistics] 1: 63–70.

Zürcher, Erik (许理和). 1977/1987. Late Han Vernacular Elements in the Earliest Buddhist Translations. *Journal of the Chinese Language Teachers Association*, Vol. 12, NO.3: 177–203. 汉译:最早的佛经译文中的东汉口语成分(蒋绍愚译), 语言学论丛 14 [Essays on Linguistics] Vol.14: 197–225. 北京:商务印书馆 [Beijing: The Commercial Press].

———.1990/1998. Handai Fojiao yu Xiyu (Wu Xuling yi) 汉代佛教与西域(吴虚领译) [Buddhism in the Han Dynasty and Western Regions (Translated by Wu Xuling)]. Guoji Hanxue 2 国际汉学 2 [International Sinology] Vol. 2: 291–310. Zhengzhou: Daxiang chubanshe 郑州:大象出版社 [Zhengzhou: Elephant Press].

——— 1991/2001 A New Look at the Earliest Chinese Buddhist Texts, in Koichi Shinohara and Gregory Schopen ed. *From Benares to Beijing: Essays on Buddhism and Chinese Religion*, Oakville-New York-London: Mosaic Press. 1991: 277–304. 汉译:关于初期汉译佛经的新思考(顾满林译), 汉语史研究集刊 4 [*Collection of Works on the History of the Chinese Language*] Vol. 4: 286–312. 成都:巴蜀书社 [Chengdu: Bashu Press].

Language Strategies in China

Xiaobing Fang and Yanhong Ge

Contents

Introduction	254
The Genesis of the Language Strategy Studies	256
Language Economy and Language Service	259
The Proposal on Developing Language Economy	261
The Notion of Language Service	263
Language Strategies as Chinese Innovations in LPP	265
Conclusion	269
References	269

Abstract

The studies of language strategy emerged as a new area of research in China in the twenty-first century. At the beginning, the studies were generally driven by the problem-solving needs of the country that was going through a fast process of modernization. Under such conditions, the application of the established theories in language policy and planning (LPP) was frequently implicit and unreflective while the pragmatic approach integrating knowledge from various sources seemed to be effective. The present chapter summarizes the studies of language strategies with their social effects in the theoretic perspective of LPP. The studies are seen to be outgrowth of the twentieth-century scholarship in LPP but seem to have elevated to a new level of multidisciplinary and cross-disciplinary development. The major issues the studies attempt to deal with are seen to be both theoretically important and broadly relevant beyond the Chinese situation. The economic approach by some of the language strategy studies is evaluated of their innovativeness in theoretic development and practical application.

X. Fang (✉) · Y. Ge
China Center for Linguistic and Strategic Studies, Nanjing University, Nanjing, China
e-mail: languefang@163.com; geyh@nju.edu.cn

Keywords

Language strategy · Language policy and planning (LPP) · Language resource · Language service · Language economy · Speech community planning

Introduction

There is a relatively rich literature on LPP in China, detailing its background, history of development, the standardization policy, policies on minority languages, foreign language education, and international education of the Chinese language and many other topics (Chen 陈平 1999; Spolsky 2014; Li 2015; Li and Li 李宇明和李嵬 2013, 2014, 2015; Zhou 周明朗 2004, 2019, Xu and Zhang 徐大明和张璟玮 2019, among others). However, to date, few dwell on the topic of "language strategy" although it is allegedly a Chinese innovation in LPP (Xu and Zhang 2019: 700–703). Possibly, it is just because of its innovative nature, the topic has been shied away to avoid controversy. The controversy, though being a minor one, seems to be inadvertently raised in the above assertion by Xu and Zhang (2019) in contrast to the omission or pass-over of the topic by others who summarize the LPP research in China in the twenty-first century (Spolsky 2014, 2019; Wang H. 王辉 2016, among others). Therefore, the present chapter furthers the introduction of language strategy study as a new area of study, while differing from Xu and Zhang (2019), less from the perspective of sociolinguistics as they do, more on its content and characteristics as research focus in LPP.

In a comprehensive introduction of Chinese language planning, Li 李宇明 (2015) has dealt with the topics of national language, language standardization, mother tongue, bilingualism, dialects, foreign languages, language corpus, lexicography, language in the information era, language protection, issues in language planning, theory of language life, report on language life, the international education of Chinese, the history of language planning practices in China, etc. The book is hailed that "[it] forms the basis for an understanding of language management in the PRC and a useful model for those undertaking a similar task elsewhere" (Spolsky 2019: 155).

In this 500-odd-page book by Li (2015), which is a collection of English translations of 30 papers written in Chinese by the author during the years 2001–2013 roughly, there are several sections about "language strategies": Sect. 2.1 of Chap. 4, "Formulate foreign language strategy as soon as possible" (Li 2015: 60–62); Sect. 2.2 of Chap. 4, "The scientific development of the international propagation strategy of the Chinese language" (Li 2015: 62–63); Sect. 2.1.1 of Chap. 6, "Field language policies and development strategies" (Li 2015: 93–93); Sect. 3 of Chap. 18, "China's language strategy in the information age" (Li 2015: 288–296) with Sect. 3.3.6 "Strategies for the international dissemination of the Chinese language" (Li 2015: 295–296); and Sect. 1.2 of Chap. 20, "National strategies of standardization" (Li 2015: 316–317).

To summarize the above, it can be seen that China's language strategies as recognized in Li (2015) are as follows: foreign language strategy, international education of Chinese strategy, field language development strategy, language in the information age strategy, and standardization strategy. Seen from the present perspective, all of the above are indeed important strategies for China's development, and they were timely proposed (Xu and Zhang 2019; Zhou 2019).

Although using the word "language strategy," Li (2015) does not seem to treat it as an important term in his work that "demonstrated dual emphasis on theory and practice" (Li 2015: v). One indication is that the term does not appear in the Index of the book (Li 2015: 489–490). There are 73 main items listed in the index and many of them also have subitems. For instance, the main item "language planning" comprises five subitems (the head term and four longer phrasal terms including the head term such as "field language planning," "macro-level language planning," etc.). The word "strategy" does not appear in any of the main items or subitems of the Index.

Unlike in Li 李宇明 (2015), in Zhou 周明朗 (2019), which is another up-to-date research on LPP in China but with a focus on "language ideology" and "language order," the word "strategy" does appear in the index of the book. However, the "strategy" there, as an entry subdivided into "global strategy," "grand strategy," and "publicity strategy," does not include "language strategy."

Li and Li 李宇明和李嵬 (2013, 2014, and 2015) is another important source on LPP in China. They are collections of English translations of selected chapters from a Chinese series called "The green cover book: Language situation in China" (中国语言生活报告 (China Language Life Report)) (SLA 2006–2011). Li and Li (2013) is a collection of 31 articles authored by over 40 scholars, mostly translations from the original texts in Chinese. It is regarded as "an in-depth account of the multifaceted and dynamic language situation in China" (Wang H. 王辉 2016: 99). The book comprises sections dealing with "government's efforts in the field of language management," "specific domains of language use," "hotly argued topics such as dialect craze, English craze, etc.," and "language situation and language policy of Hong Kong, Macau, and Taiwan," respectively (Wang H. 王辉 2016: 98). Li and Li (2014, 2015) are of the similar structure as Li and Li 李宇明和李嵬 (2013), the former with 33 chapters and the latter with 29 chapters, reporting on various aspects of "language life/situation" in China. These include the government's work and policies, their impacts on the masses in language issues, as well as some social debates on the issues.

Interestingly, the term "strategy" never appears in any of the titles of the chapters, forewords, or appendixes of the three volumes; neither does it in the books' indexes.

As will become clear in the following discussion, there are some important aspects of Chinese language planning that are covered in Li and Li (2013, 2014, and 2015), Li (2015), and Zhou (2019) but not described as language strategies. Moreover, in less lengthy works on the subject of Chinese LPP, language strategies are mentioned even less than Li (2015), sometimes not at all (Spolsky 2014, 2019;

Wang H. 2016; among others). Consequently, a discussion is prompted by this curious fact. Hopefully, the discussion will not only enlighten the question of why language strategy studies have been omitted in some works, but also enrich the theoretical framework of LPP.

According to Xu and Zhang (2019: 701), "sustainable multilingualism," "language economy," and "language service" were proposed as national language strategies in China. Notably, these are not among the strategies mentioned in Li (2015). Therefore, the present chapter will discuss these and other "strategies" that seem to miss in Li (2015) and others. As mentioned above, these language strategies will be examined for their innovativeness, apart from being evaluated for their importance.

The chapter is divided into five sections: Introduction, The genesis of language strategy, language service and language economy, language strategy studies as Chinese innovations in LPP, and conclusion.

The Genesis of the Language Strategy Studies

While it is relatively unknown to the English readership, there was a sudden surge of publications in Chinese with the term 语言战略 YUYAN ZHANLUE (language strategy) as a keyword in the last two decades. Research centers and institutes were also set up in China bearing names including the term. At the same time, new postgraduate programs and research projects were found in Chinese universities bearing similar names. Two journals were successively launched using the term in their titles, that is, 中国语言战略 China Language Strategies (Yiwen Press and Nanjing University Press 2012/2015) and 语言战略研究 (Chinese Journal of Language Policy and Planning 2016). As can be noted, the latter has a literally different title in English. The fact is worthy of some discussion because it bears on the question raised above.

However, in spite of the surge of studies in language strategy, only a few of the studies trace the line of the research in the perspective of theoretical development and many do not bother to define the newly introduced term or cite its source of introduction. The justification of the publications was apparently based on the significance of their social applications. However, there are a few studies published in Chinese that traced the origin and development of language strategy studies (Xu 2012; Wang X. 王晓梅 2016), which were briefly summarized in Xu and Zhang 徐大明和张璟玮 (2019). Accordingly, a more detailed account is in order.

The Chinese translations of the word "strategy" include 策略 CELUE and 战略 ZHANLUE, with the latter referring to "overall planning and directing in a war" or similar macro-level moves in business, politics, and other non-military operations, whereas the former has a meaning bordering on that of "tactics." For instance, for some time the English term "language strategy" was generally used in the meaning "strategies adopted by individuals in learning or using a (second) language" (see Meisel 1983, Oxford 1990, among others). These were generally translated as 策略 CELUE in Chinese. Similarly, the well-known book *Discourse Strategies* (Gumperz 1982) was translated as 会话策略 HUIHUA CELUE (约翰·甘柏兹 2001). In contrast, what the word refers to in the phrase "China language strategies" is clear in the context of LPP, especially in the meaning as "national language strategies." Therefore, 语言战略 YUYAN ZHANLUE is frequently translated as "national language strategies" in English, in order to avoid the possible interpretation of an individual's language strategy, but the translation unnecessarily narrows down the meaning in its exclusion of language strategies adopted by non-national level large organizations, such as a multinational company (see Bildfell 2012).

The phrase "grand strategy" in English represents perhaps the effort to stress the "overall" nature of certain strategies. When being translated literally into Chinese as "大战略 DA ZHANLUE," it sounds exaggerated or verbose at first blush. However, it is soon accepted and differentiated into a special meaning, something like "the grand ones among the 'big strategies'" (see Zhou 2019: 2, 50, 214, etc.).

At the turn of the century, China saw strategic planning in various fields and a proliferation of strategy studies (Zhou 2019: 2, 50). This prompted some Chinese language scholars to propose the term 语言战略 YUYAN ZHANLUE (language strategy) in Chinese to match up the strategy studies in other disciplines in the national wave of strategic planning at the time. According to Wang 王晓梅 (2014), when the term first appeared in Chinese publications, it referred to the National Security Language Initiative in the USA in 2006 (Wang 王建勤 2007). Wang (2007) calls it "a new language strategy of U.S. in the background of globalization" (Wang 2007: 11). At the same time, he points out, "The strategy certainly will threat and challenge China's national security of language and culture, but the research on China's national language strategy has lagged behind U.S. and European countries" (Wang 2007: 11).

Apart from the above paper, Wang (王建勤) subsequently published another paper on the same topic further clarifying his points (Wang 2010). In exemplifying the US language strategies, he identified two documents issued by the US government. The first is the National Security Language Initiative (NSLI) released on January 5, 2006, and the second is A Call to Action for National Foreign Language Capabilities (Call-to-Action) released on February 1, 2005. Both documents can be found on US government websites. The content of the two documents is quite similar, while the former may be seen as a response to the latter. As the titles of the documents show, they are the suggestion and action plan for strengthening foreign language education in the USA for the purpose of national security. Upon examination, in the 844-word text of NSLI, the word "strategy" is not used at all. Meanwhile, in the Call-to-Action, the word "strategy" occurs eight times in the 4577-word text of the document, practically all referring to "foreign language strategy." Specifically, the word is found in the phase "a/the national foreign language strategy" four times, in "that strategy," "a national strategy," "a national foreign language and cultural understanding strategy," and "a national strategy on foreign languages and cultural competency" each once. The phrases "a national foreign language and cultural understanding strategy," and "a national strategy on foreign languages and cultural competency" mean the same thing, pointing out that foreign language education should include cultural content as well. The phase "that strategy" or "a national strategy" refers anaphorically to "a/the national foreign language strategy" in the context.

As a response to calls such as Wang (王建勤 2007), the State Language Commission of the PRC decided to set up a research center for national language strategies. This is witnessed by a document released on the Internet in April 2007, The State Language Commission's (henceforth SLC) "Eleventh Five-Year Plan" for Research Work on Language Application. "National Language Strategy" is one of the key research directions listed in the "Eleventh Five-Year Plan" for SLC's Research Work on Language Application. It points out:

> Language strategy is an integral part of national development strategy. In recent years, some major countries in the world have been formulating their own language strategies, using languages to maintain the cultural security of the country, resolving social contradictions internally, uniting people's hearts, and spreading the ideas of their own countries and earn foreign exchange earnings. The current language life in our country is developing and changing rapidly. Various contradictions in language life are prominent. The types and methods of language services provided by the society are increasing day by day. The virtual space is expanding rapidly. The pace of Chinese going to the world has accelerated more than ever. Under such circumstances, China must study the macro language strategy in a timely manner, design an action plan to implement the language strategy, and propose scientific plans to deal with major language issues. (SLM 2007, translation by Baidu, with our revisions)

Consequently, the China Center for Linguistic and Strategic Studies (CCLASS) was set up at Nanjing University on November 9, 2007. The news released by the Ministry of Education on November 12, 2007, includes the following:

> The China Language Strategy Research Center is a substantive scientific research institution jointly established by the Ministry of Education's Department of Language and Information Management and Nanjing University. After the establishment of the center, it will carry out theoretical research on language policy and language planning, applied research on language national conditions, and China's international language strategy, and propose scientific plans to deal with important language issues at home and abroad, making them available for relevant government department for reference and basis for formulating national macro language strategies and language action plans.
>
> Language strategy is an organic part of national strategy. As China's urbanization and modernization process accelerate, language work faces many new problems and challenges. From the perspective of planning and constructing a linguistic life in a harmonious society, China's language policy and language are discussed. Planning is a topic that urgently needs to face. From a global perspective, some major countries are formulating their own language strategies to use languages to maintain national cultural security. The establishment of the Chinese Language Strategy Research Center has filled a gap in China's language strategy research institutions (Ministry of Education 2007, translation by Baidu, with our revisions).

The website of CCLASS was subsequently established and put out the following statement:

> China Language Strategy Research Center is engaged in the theoretical research of language policy and language planning and the applied research of national language strategy. The work objectives of the center are: first, to put forward new language planning theories that adapt to the current situation and China's national conditions; second, to disseminate new

achievements in language planning research and language strategy research and promote the transformation and application of relevant scientific research achievements; third, to carry out countermeasures research on major language problems at home and abroad, and to put forward policy suggestions and strategic plans to the State Language Commission, in providing strong support and guarantee for China's economic and social development (CCLASS 2007).

From the above, it can be seen that CCLASS recognizes the importance of both theoretical and applied research and links language strategy studies with LPP research explicitly. Both SLC and CCLASS define language strategy as part of the national development strategies. In doing so, the SLC was less concerned with theoretical issues in LPP but stressed the need for dealing with the domestic and international issues of the nation.

To summarize, the concept of "language strategy" seems to have undergone a change of meaning a few times. It may have started as an individual's language-learning strategy, transformed to include a state's initiative to develop its foreign language education, and further expanded to mean "any national strategies about language." It may have just been jargon for second language teaching and learning, a specific expression in one US document, and then become a special term in Chinese regularly used in government documents and academic research.

Language Economy and Language Service

The setting of "national language strategy" as a key research direction by the SLC, following up with the set-up of CCLASS and other research centers of SLA stimulated the research of LPP in China, and particularly the research on language strategies. The CNKI search shows that to date over 100 articles have been published on the topic.

As mentioned above, it is a Chinese invention to define language strategy as a national development strategy in any field and for any purpose, well beyond the field of foreign language education, and well beyond the purpose of national security. The US foreign language strategy was geared to national security, but the Chinese SLC's plan of research on national language strategies was aimed at a range of issues well beyond security and international conflicts. The security issue is certainly within the list of important issues Chinese language strategy studies have covered. The first language strategy mentioned in Li 李宇明 (2015: 60) is "foreign language strategy" for the country, which can be an answer to the question raised by Wang (Wang J. 王建勤 2007, 2010). This is also witnessed by the setting-up of the Research Center for Foreign Language Planning and Policy and the National Research Center for State Language Capacity by SLC in 2011 and 2014, respectively. However, as shown by the Eleventh-Five-Year Plan of the SLC on research (SLC 2007) cited above, the national language strategy studies are supposed to deal with many domestic and international issues, of which the majority seems to be domestic ones. This is shown by the following excerpts from the document:

> The current language life in our country is developing and changing rapidly. Various contradictions in language life are prominent. The types and methods of language services provided by the society are increasing day by day. The virtual space is expanding rapidly. (SLC 2007, translation by Baidu, with our revisions)

The later-released news report on the setting-up of CCLASS is more specific on the issues:

> As China's urbanization and modernization process accelerates, language work faces many new problems and challenges. From the perspective of planning and constructing a linguistic life in a harmonious society, China's language policy and language are discussed. (Ministry of Education 2007, translation by Baidu, with our revisions)

The following excerpts are from the official website of CCLASS:

> Since its establishment, China Language Strategy Research Center has made fruitful achievements. In terms of theoretical construction, the scholars of the center have initiated and developed language strategy theories such as "linguistic urbanization theory", "language exchange theory", "speech community planning theory", and applied these theories to solve social communication and social integration problems brought about by urbanization, to resolve for the minority people the apparent contradiction between the maintenance of ethnic language and culture and the economic development, and to provide for the social harmony in multilingual situations. In making policy suggestions, the center puts forward their suggestions to address national issues through reporting to superiors, editing and distributing internal communications, and publishing academic achievements and these include "sustainable multilingualism", "language trade as an international strategy" and "diversified second language education in the central and western regions" and other policy suggestions. In terms of discipline construction, the center has trained a large number of young and middle-aged scholars specializing in language survey, language planning and language strategy by means of degree courses, short-term training, cooperative research, etc., and they are currently active in the field of language planning and language strategy research at home and abroad. In addition, through academic publishing, academic exchange activities and social practice activities, the center has promoted the professional education, infrastructure construction and social application of language planning, explored and innovated the new research direction of "language strategic research", which has made international, interdisciplinary and social impacts (CCLASS 2019).

As shown by the above, the national language strategy studies carried out by CCLASS show the following characteristics: (a) They apply sociolinguistics and LPP theories to the resolution of current issues; and (b) they identify major issues of LPP in view of the country and the world. As reported by CCLASS in the above, the particular issues they have addressed include whether to support societal multilingualism, how to deal with other countries that have different languages, and what regional language education policies to adopt in the current situation. In deciding on the suggestions they put forward to the government and the society, they seemed to have some general strategies that they derive through the applications of the tested theories in sociolinguistics and in LPP research. As pointed out by Xu and Zhang 徐大明和张璟玮 (2019: 701), national language strategies proposed by CCLASS include "sustainable multilingualism," "language economy," and "language service." In the following, a more detailed account is given on each of the strategies.

The "sustainable multilingualism" strategy was first proposed in Chinese as DUOYU GONGCUN HE ER BUTONG 多语共存, 和而不同. Its literal translation in English would be "co-existence of many languages and harmony with differences," with the second part "harmony with differences" (other translations include "harmony without conformity," "harmony without uniformity," etc.) being a quote from Confucius (551–479 B.C.), educationist and philosopher in Ancient China. This is obviously the Chinese version of the internationally current point of view on the protection of linguistic diversity (see UNESCO 2001). With full awareness of different views on linguistic diversity and the associated controversy in LPP (see De Swaan 德斯万 2006, among others), the CCLASS researchers firmly recommended the policy of upholding linguistic diversity as a national language strategy (Xu 徐大明 2012; Wang X. 王晓梅 2016; Fang 方小兵 2013, 2014, 2015, among others). The effects of the acceptance of the recommendation are shown in the International Conference on Language held at Suzhou in June 2014 and its Suzhou Consensus, and the First World Language Resources Protection Conference held at Changsha in September 2018 that adopted the Yuelu Proclamation. The Chinese government has since become the most important partner of UNESCO in its drive for the protection of linguistic diversity in the world (Ministry of Education 2019).

The Proposal on Developing Language Economy

In the Chinese wave of language planning in the twenty-first century, the theoretical basis for the Chinese practices seems to be built on the notion of language as a resource (Li 李宇明 2018; Li 2015; Xu and Li 2002; among others). A particular interest seems to be that language is an economic resource (Li 2015: 9; Xu and Li 2002), and the strategic view that language resources should be developed to make economic gains (Xu 徐大明 2010; Li 李现乐 2010, 2014). Therefore, research on language economy was stimulated by such a point of view.

CCLASS played a key role in promoting the line of research on language economy. It was CCLASS that drew the attention of the Chinese scholars and SLC to the research of Grin (2009) on the contribution of multilingualism toward the GDP of Switzerland. The research report on the calculation of the multilingualism contribution on Switzerland's GDP in 2008 was released to the public in February 2009, and it immediately caught the attention of CCLASS. The researchers at CCLASS were actually looking for such research attentively with their belief that there should be an economic language strategy for China. This was prompted by two sources, one of the previous studies on language and economy (Marschak 1965, Coulmas 1992; Grin 1996, 2003; among others), the other of the understanding that economic development was a national priority for China and that economic construction was the central task of the Chinese government. With GDP goals being set for all levels of government, CCLASS found it intriguing that there were no concerns of some government agencies like SLC. CCLASS would like to draw the attention of SLC to the connection between language and economy. In doing so, it made use of the research report by Grin (2008). The research was first reported in Chinese by CCLASS in its internal communication newsletter SHIJIE YUWEN DONGDAI 世界语文动态 No. 1 Issue of 2009.

The advocating of developing language economy continued with the convening of the 2nd Summit Forum on National Language Strategies held at Nanjing in May 2009. The conference was organized by CCLASS and cosponsored by the Department of Language Information Administration of MOE, Language Commission of Jiangsu Province, and Nanjing University. Importantly, it was the first time that such conferences in the field of LPP fostered the theme Language Resources and Language Economy, connecting the theoretical studies of language resource with a national strategy of economic development, while being contextualized in the country's drive for GDP and SLC's demand for results of applied research. At the conference, a more specifically timed point of view was raised and subsequently widely reported was: "promoting language consumption and language employment can be measures for stimulating domestic demand and withholding the international impact of economic slowdown" (Nanjing University 2009).

A cross-disciplinary effect is mentionable also in respect of CCLASS's advocacy of language economy. The 1st Symposium of Chinese Language Economics was held at Ji'nan in October 2009, which was jointly organized by the Institute of Economics of Shandong University and CCLASS. The researchers in economics at Shandong University had long had an interest in language economics but were mainly confined to the theoretical reflections on the theme until it was prompted by CCLASS in the application-oriented research in language economy in China (Zhang and Liu 张卫国和刘国辉 2012). Since then, the Symposiums of Chinese Language Economics have been held regularly up to the present and an SLC-supported research center on language economics was established at Shandong University.

The research on language economy soon branched into "language industries" and "language services" while continuing in theoretical and applied research directions in macro-economics (Xu 徐大明 2010; Li 李现乐 2010, 2014; Zhang and Liu 张卫国和刘国辉 2012, among others). A CNKI search shows clearly that the rising trend of publications on language economy, language industry, and language service started in 2009. In promoting the understanding of language economy, CCLASS reports on language service industries, particularly in SHIJIE YUWEN DONGTAI. Meanwhile, the first PhD dissertation on language service was written by Li Xianle (Li 李现乐 2011), a researcher at CCLASS who since then became a prominent scholar in the field of language economy and language service in China.

In the last decade, "language industry" has become one of the most flourishing areas in language research in China, and there has been also economic gains correspondingly in the country (Li 李磊 2019). Reportedly, "in 2018, the number of enterprises engaged in language services or related services in China reached 72,500, and the industrial output value exceeded 280 billion Yuan." (China Language and Script Development Report, 2019). Practically, all reviews of the development of Chinese language industries refer to government support and the relevant research featuring Grin (2009) especially (Li 李磊 2019; Li 李现乐 2010, 2014; among others).

Associated with language industry is the concept of language technology, which is also classifiable into language economy in the strategic perspective of LPP. From the beginning, CCLASS has been alert to the impact on language life made by information and communication technologies (Zhang and Xu 张璟玮和徐大明 2019). The surveillance of international situation on such development reinforces the understanding. The first issue of SHIJIE YUWEN DONGTAI compiled by CCLASS has included a column on technology, including such reportages as "Google search engine using voice technology," "Customer-service talking machine recognizes speakers' gender and age and detects their mood," and "British scientists develop programs to translate ET languages." The technology column continued in the following issues of SHIJIE YUWEN DONGTAI. Meanwhile, CCLASS monitors the development closely. In 2016, the 13th International Conference on Urban Language Studies was held in Nanjing and the conference was co-organized by CCLASS. Following the decision of CCLASS, the conference adopted the theme of "language technology and language life." The decision had to be made in 2015 before sending out the call-for-papers and it was the time before the craze for language technology had infused the circles of applied linguistics and LPP in China.

To summarize, language economy, as inclusive of language services, language industry, and language technology, has become both a research area and an economic achievement in China ever since the introduction of the concept in LPP, especially with the promotion of CCLASS at its incipient period of dissemination in the circle of LPP as well as in the general public.

The Notion of Language Service

As mentioned above, the research on language services surged around 2009, apparently as a result of the promotion of language economy. However, it should also be noted that there was implicitly the notion of language service in LPP years ahead of those studies using the term as a key word. It is to be noted that the document "The 11th Five Year Plan of the State Language Commission on Research Work" has pointed out in the guiding ideology for language work was to "serve for promoting social harmony, economic development and the overall development of people" (SLC 2007). This being so, it follows that national language strategies are part of national development strategies (SLC 2007). The academic echoes to this national strategic planning can be found in such rationales that "language service is part of the service by the state to its people," and that "for the central government, language planning equals to the management of national language resources" (Xu 徐大明 2008: 14).

From the above, we can see that of the broad notion of "language service" that matches up practically to "language planning." However, the narrow notion of language service is as defined in the discussion on language economy. It refers to the commercialized language activities that realize their market values. An example is what is reported in the first issue of SHIJIE YUWEN DONGTAI "an average of 20% annual increase of U.S. language service companies in five consecutive years" (CCLASS 2008). However, what these companies do in translating, interpreting and language training can also be done by others without a commercial purpose. When the government or other organizations provide these as a public service, it makes the other half of language service not market-driven. Of course, government language service can be much more wide-ranged in including formulating and implementing language standards, guiding and providing language education, and other day-to-day management of language activities in social communication (Xu 徐大明 2008).

The advocacy of language economy and language service by Chinese LPP scholars fitted with the time when China was to enhance its service sectors of economy. The country has experienced a transition in the last decade from a small service-sectored economy into one that is mainly service-sectored economy. In the process, the language scholars who applied their expertise to LPP, apart from introducing the traditionally defined language services, had to adapt the concept of language service to the Chinese situation. Two major areas of work seem to have been done. One is to expand the range of "language industries," by adding "language exhibition," "language arts," etc., to the list (Li 李艳 2018). The other is to draw attention to the linguistic part of the service work that as a whole does not sell as a language product/service. It is pointed out that this part has a value-added function to the service in whole. An example of such language service includes a waitress at a restaurant striking up a pleasant conversation with a customer, or more essentially, her exchanges with the customer when taking the order of food. According to Li (李现乐 2010), there are differences and differentials in the awareness of the linguistic part of the services provided in the hospitality industry both on the part of the service providers and the part of the recipients. For instance, in the transitional process to a consumer economy, people's spending power correlates but varies to quite an extent as for the willingness to pay for the language part of restaurant service, as shown in a survey study by Li (Li 李现乐 2010). An analysis of chatting at work as a worker's welfare and its economic value illustrates the potential of language service and language economy (Xu 徐大明 2012).

The promotion of language service studies, now a common activity with various institutions nationwide, has also been closely connected with the work of CCLASS at the initial stage. In May 2011, CCLASS organized a meeting entitled "colloquium on language economy and language service," which was notably the first conference with the theme of language service. The conference took place at Beijing Language University, in conjunction with the dissertation defense meeting of two Nanjing University PhD candidates, with participants including such prominent figures as Prof Li Yuming, who now heads the Beijing Advanced Innovation Center for Language Resources, Prof. Qu Shaobing, who now heads the Research Center on Language Service at Guangzhou University, and Prof. 孙宏开 Sun Hongkai and Prof. 王远新 Wang Yuanxin, well-known Chinese linguists of anthropology, also with young participants including Dr. 李现乐 Li Xianle, the author of dissertation entitled "Language Service and Service Language" (Li 李现乐 2011). The Symposium on Language Service Studies was held in January 2012 at Guangzhou University, with CCLASS as one of the co-organizers. The Symposium has since then grown into a conference series and is sometimes conjointly organized with the Symposium of Chinese Language Industries or the Symposium of Chinese Language Economics.

Language economy and language services with minority languages have always been a concern for CCLASS, which has conducted the research project entitled "The Policy Option Study on the Salvage, Protection and Development of the Qiang Language in the Earthquake-Stricken Area of Wenchuan," notably the first international collaboration project on Chinese minority languages jointly sponsored by Chinese and Canadian governments (Xu 徐大明 2011; 李红杰 2011; Ge 葛燕红 2011). Later on, CCLASS organized The *Conference on Protection of Minority Languages and Economic and Social Development* in May 2012, with Chinese and Canadian participants including Prof. Li Hongjie, Deputy Director of Research Center on Ethnology and Anthropology, State Ethnic Affairs Commission of PRC, Prof. Ken Coates, Canada Research Chair in Regional Innovation, and many others (Ge 葛燕红 2011).

To summarize, the discussion on the realization of market value of language services raises the awareness of economic aspects of language life. The quality of language services has economic consequences and impacts service industry both with "language industries" as traditionally defined and other services that include language activity as a value-added part.

Language Strategies as Chinese Innovations in LPP

As mentioned above, language strategy studies are distinctively Chinese innovations. To recount what is already mentioned in Li Yuming 李宇明 (2015) and in Xu and Zhang 徐大明和张璟玮 (2019) in combination, we have the following strategies: "foreign language strategy, international education of Chinese strategy, field language development strategy, language in the information age strategy, and standardization strategy" (Li 李宇明 2015), "sustaining multilingualism, language economy, and language service as national language strategies" (Xu and Zhang 徐大明和张璟玮 2019). The above discussion has focused particularly on topics of language economy and language service. The reason for such choice is on the belief that the two language strategies are outstandingly innovative among all that has been proposed.

Language planning as a discipline has come along through a few different stages (Goundar 2017). The twentieth century saw the genesis of the discipline with successive milestones of research in building up the well-accepted theoretical framework now, including the foundational work by Haugen (1965), the distinction of status planning and corpus planning (Kloss 1969), the introduction of acquisition planning (Cooper 1989), and prestige planning (Haarmann 1990). The twenty-first century saw more of the reinterpretation, redefining and even challenges of the above framework, including also shifts toward language human rights issues and the notion of language as resources (Li 2015; Spolsky 2004, 2007, among others). Concepts have been introduced from other disciplines and adapted for new purposes, among which "language strategy" is one. The concept of strategic planning is appropriately traced to the national strategic planning field and it was then transferred to the field of LPP. However, disciplinary scholarships serve as a basis of innovations, whether or not with a cross-disciplinary approach. Ether the U.S. "foreign language strategy" or the Chinese "national language strategy" was defined firstly in the perspective of national needs and national development, in the "problem-solving" mode of early language planning practice (Goundar 2017). In this, we see a connection with the tradition of LPP. The theoretical reflections on many practical issues and their integration with the LPP framework were apparent a later development (Li 2015; Xu and Zhang 2019, among others), which perhaps started with the establishment of CCLASS when it claimed to be an LPP research institution.

As pointed out by Goundar (2017) and Veronica A. Razumovskaya and Yaroslav V. Sokolovsky (2012), the practices of language planning antecedents the establishment of the discipline of LPP, with the 1950s already saw vigorous activities of language standardization and policy-making, which became the conscious theme of linguistic nation-building of the discipline in the 1960s. Li (2015) actually traced the Chinese practices of language planning even further back to the beginning of the twentieth century when it was framed in the moves of modernization.

While it has been pointed out that the twentieth-century language planning tended to be limited to state and government acts (Goundar 2017, Spolsky 2004, among others), the language management theory introduced by Spolsky (2007) is a major attempt to explore the agentive dimension of LPP. Instead of a unidirectional top-down approach involving mainly national governments, language planning is redefined as a multi-level management practice in language choices motivated by different language ideologies and beliefs.

The language management theory has significantly expanded the theoretical basis of LPP, but the effort in replacing "language planning" with "language management" in terminology or conceptual framework has not always been successful (see Li (2015), Li and Li (2013) Wang H. (2016), Xu and Li 2001, among others). His article entitled "Language management in the People's Republic of China" (Spolsky 2014), his foreword to Li (2015) (Spolsky 2015), and his book review of Li (2015) (Spolsky 2019) show the continued interest in China by Bernard Spolsky and his consistent efforts to define what Li (2015) calls "language planning" as "language management." For instance, in evaluating Li (2015), Spolsky says,

> This collection of translated papers provides more than just data. By revealing the opinions of a scholar who was also long active in a management role, it forms the basis for an understanding of language management in the PRC and a useful model for those undertaking a similar task elsewhere. (Spolsky 2019: 155)

In mentioning of the author of the book as "a scholar who was also long active in a management role," Spolsky apparently refers to the fact that the author, Professor

Li Yuming of the Beijing Language University, had held the position of the Director of the Language Information Administration of the Ministry of Education, the PRC for over a decade starting in 2000. On the jacket of the book, Spolsky quotes the following sentence from the author of the book, "It is language life, instead of language per se, that a government should manage." Therefore, instead of the difference in calling the practices "language planning" or "language management," it is clear that Li (1015) does take the perspective of a government office in "language planning."

While not accepting explicitly the "language management" theoretical framework, Li is actually advocating a framework of "language life," which is seen as a Chinese contribution to LPP (Xu and Zhang 2019). The connection between the two is shown in the above quotation, as noted by Spolsky (2019).

It is already noted that some Chinese scholars are not comfortable with the English term "language strategy" while still entertaining the Chinese term YUYAN ZHANLUE, which is reflected in the absence of the reference to the term in the English rendition of the Chinese journal YUYAN ZHANLUE YANJIU (Xu and Zhang 2019). However, China Language Strategies (Nanjing University Press 2015), which does not shy away from the literal equivalence of the two languages in its title, defines clearly the relationship of language strategy studies with LPP as a new direction in the field. In contrast, there was some attempt in defining language strategy outside of the field of LPP (Cai 蔡永良 2016), which may have contributed to the qualms of those who were uncertain of its introduction to the international sphere of LPP.

Chinese language strategy studies can also be seen as works from a service perspective of language planning. With the understanding that language is not substance and that language service is usually embedded in other services, language management theories (Spolsky 2007, 2019) can be questioned. Li's 李宇明 (2013) position that the government should manage language life rather than language itself is seen by Spolsky (2019) as progress in the theoretical development of language management. However, whether the target of management is the symbolic system abstracted from human speaking activities or these speaking activities in action, the managers seem to be outside of them, distanced from the target. Moreover, they appear to be at an elevated and superior position. They have power and authority whereas the speakers under their management seem to be passive and powerless.

On October 13, 2020, the fourth national language conference since the founding of new China, also the first national language conference since the twenty-first century, was held in Beijing. The meeting summed up the achievements and experience of the language planning in the past, and the new situations and problems in the new era, and made clear the guidelines, objectives and tasks for the future, with language resource protection, national language capacity, and language AI as the core. It is a milestone in the development of China's language strategy.

With the understanding that language activities are just one aspect of the human activities that are more in focus and intentional, language as service and language planning as service seem to be a proper perspective to take. In contrast to the perspective of language management, this perspective has the advantage of being facilitative rather than imposing.

Language serves economy because economic activities are essential to human subsistence. Language planning becomes relevant and important if it facilitates economic activities. Traditional societies are made up of self-contained economic communities. The speech community is thus implicit in an economic community as its linguistic constituent (Hymes 1972).

The Chinese innovations in language strategy studies are twofold. The first is the integration of the concept of national development strategy from the field of strategic studies with the framework of LPP. The second, perhaps more importantly, in contextualizing the national language strategies of language economy and language service in China's economic development, Chinese scholars have developed the language-as-resource approach to LPP to a new level and changed LPP from a mainly politically motivated social engineering practice to a service of language science toward the general economic and social development of a nation.

The merits of the language strategy study are its integration with its sociopolitical context and its inclusion of the economic basis of human activities of which language plays a part. However, as it has achieved in the case of Chinese LPP, it has the limitations of taking mostly a national perspective and being biased toward the top-down approach, amiss of the insights from the language management theory in terms of its agency and domain dimensions. However, the non-discriminatingly instrumental role of the strategic type of LPP can be improved upon. The recent proposal of Speech Community Planning (Fang 方小兵 2018) can be a route toward this goal. The proposal applies the sociolinguistic theory of speech community. The theory defines languages as the facilities of speech communities and a speech community as a social organization of speaking individuals. More recent development of the theory includes the postulate that a speech community tends to dislocate from its original host community as a result of urbanization and immigration and that the speech community is a process of formation and transformation (Xu 徐大明 2015). In other words, individual speakers form a speech community by participating in situated speech interactions, and they are thus socially organized as a social dynamism.

In applying the sociolinguistic theory, the Chinese LPP researchers move away not only from the language-centered approach but also from just one angle of language resource management. In capturing the original insight of Haugen (1965) that the target and scope of language planning is a speech community, the new approach brings in the understanding that a speech community is no longer static or given. Therefore, the proposal for LPP is to have a vision of speech community construction. Consequently, the wisdom of strategic planning is also taken that particular policies are evaluated in the strategic perspective of community building.

Instead of serving a goal seemingly unrelated to language sciences, the approach of Speech Community Planning aligns any acceptable working objectives of LPP with the interest of a speech community. While it is unavoidable that any application work would fit with a socio-political and cultural context, the scientific stance of an LPP researcher can measure a policy option with a general goal of speech community construction. The sociolinguistic theory of speech community is built on the primary understanding that language is a social function and that language evolves in the direction of human civilization. Consequently, as an interventionist approach toward language, LPP may sacrifice non-human linguistic resources for human welfare but not vice versa.

Real-life situations are complex, and there are interwoven and intricate relationships between political, economic communities and speech communities. Therefore, following the line of speech community planning, new studies are analyzing the roles of existing speech communities in the process of visional speech community construction (Xu 2016; Fang 2020; Ge 2020; Zhang 2020).

Conclusion

The studies of language strategies are innovations of research in China in the twenty-first century. At first, the studies were driven generally by the problem-solving needs of the country which was involved in huge social transformations at an unprecedented scale and speed. Many of the studies under the name of "language strategies" may be appropriately defined as language policy studies in the traditional sphere, whereas others are drawn from other disciplines and non-academic sources. While the majority of the studies to date still concern them with immediate and practical goals in the fast-developing country that is China, several among them have transcended to the theoretical levels and become conscious of the theory-and-practice cycle.

In analyzing from the theoretical perspectives of LPP, those strategic studies on language planning in the Chinese situation have produced concrete results in language economy and language service, among other results, well rewarding their original motivation from the national development of the country. At the same time, the present analysis reveals the theoretical significance of the language strategy study in its expanding the horizon of LPP from basically a narrowly politically motivated to a both economically and socially driven approach.

References

Bildfell, Connor. 2012. Language strategies in China: An analysis and framework development for multinational companies. *Journal for Global Business and Community* 3 (2): 32–53.

CCLASS. 2016. Official website of the China Center for Linguistic and Strategic Studies. Available at www.chinalanguage.net. Accessed on 15 January 2018.

Chen, Ping. 1999. *Modern Chinese: History and sociolinguistics*. Cambridge: Cambridge University Press.

Chen, Zhangtai (陈章太). 2005. Yuyan guihua yanjiu 语言规划研究 [Studies on language planning]. Beijing: Shangwu yinshuguan 北京:商务印书馆 [Beijing: Commercial Press].

Connor, Bildfell. 2012. Language strategies in China: An analysis and framework development for multinational companies. *Journal for Global Business and Community* 2: 32–53.

Cooper, Robert. 1989. *Language planning and social change*. Cambridge: Cambridge University Press.

Coulmas, Florian. 1992. *Language and economy*. Oxford: Blackwell.

Van den Berg, Marinus, and Daming Xu. 2010. *Industrialization and the re-structuring of speech communities in China and Europe*. Newcastle: Cambridge Scholars.

Fang, Xiaobing (方小兵). 2013. Yuyan baohu de sanda zuoyandian: ziyuan,sentai yu quanli 语言保护的三大着眼点:资源、生态与权利 [Three focuses of language protection: Resources, ecology and rights]. *Minzu Fanyi* 民族翻译 [*Ethnic translation*] 4: 18–23.

——— (方小兵). 2014. Shuli yuyan yisi, youhua yuyan guih 梳理语言意识, 优化语言规划 [Combing language ideolody and optimizing language planning]. *Zhongguo shehui kexue bao* 中国社会科学报 [*Chinese Social Sciences Today*], 2014 /07/07.

——— (方小兵). 2015. Duoyu huanjin xia Muyu gainian de jieding: kunjin yuchulu 多语环境下"母语"概念的界定:困境与出路 [The definition of "mother tongue" in multilingual environment: Dilemma and solution]. 语言文字应用 [*Applied linguistics*] 24 (2): 77–86.

——— (方小兵). 2018. Cong jiating yuyan guihua dao shequ yuyan guihua 从家庭语言规划到社区语言规划 [From family language planning to community language planning]. Yunnan shifan daxue xuebao 云南师范大学学报 [*Journal of Yunnan Normal University* (Humanities and Social Sciences)] 50 (6):17–24.

——— (方小兵). 2020. Cong yuyan huoli dao yuyan renli: Yuyan sentai pinggu linian de youhua 从语言活力到语言韧力:语言生态评估理念的优化 [From language vitality to language resilience: The optimization of the concept of language ecological assessment]. Yunnan shifan daxue xuebao 云南师范大学学报 [*Journal of Yunnan Normal University* (Humanities and Social Sciences)] 52 (1):22–31.

Ge, Yanhong (葛燕红). 2005. Nanjingshi "xiaojie" chenhuyu de diaoca fenxi 南京市"小姐"称呼语的调查分析 [A survey of the appellation "Xiaojie" in Nanjing]. Zhongguo Shehui yuyanxue 中国社会语言学 [*Journal of Chinese Sociolinguistics*] (2): 196–206.

——— (葛燕红). 2011. Rescue, protection and development of the language of the Qiang ethnic minority after the Wenchuan Earthquake. Paper presented at the conference of Cultures of Humanitarianism: Perspectives from the Asia-Pacific held in the Australian National University (Canberra) on 2011-08-10.

——— (葛燕红). 2012. Nanda zhaokai shaoshu minzu yuyan baohu yantaohui 南大召开少数民族语言保护研讨会 [Seminar on minority language protection held at Nanjing University], Yangzi wanbao 扬子晚报 *Yangzi Evening News*, 2012-05-16(Column B15).

——— (葛燕红). 2020. Qiangyu huoli de shequ pinggu ji guihua celue 羌语活力的社区评估及规划策略 [Community-based assessment of the vitality of the Qiang language and its planning strategy]. Yunnan shifan daxue xuebao 云南师范大学学报 [*Journal of Yunnan Normal University* (Humanities and Social Sciences)] 52 (1):40–48.

Goundar, Prashneel Ravisan. 2017. *The characteristics of language policy and planning research: An overview*, in *Sociolinguistics: Interdisciplinary perspectives*, ed. Jiang Xiaoming (蒋晓鸣). Tech: Rijeka.

Grin, François. 1996. The economics of language: Survey, assessment and prospects. *International Journal of the Sociology of Language* 1: 17–44.

———. 2003. *Language policy evaluation and the European charter for regional or minority languages*. New York: Palgrave Macmillan.

———. 2008. Principles of policy evaluation and their application to multilingualism in the European Union. In *Respecting linguistic diversity in the European Union*, ed. Xabier Arzoz. Amsterdam: John Benjamins Publishing Company.

———. 2009. Promoting language through the economy: Competing paradigms. In *Language and economic development*, ed. J.M. Kirk and D.P.Ó. Baoill, 1–12. Belfast: Queen's University Press.

Haarmann, Harald. 1990. Language planning in the light of a general theory of language: A methodological framework. *International Journal of the Sociology of Language* 86: 103–126.

Haugen, Einar. 1965. *Language conflict and language planning: The case of modern Norwegian*. Cambridge, MA: Harvard University Press.

Heinrich, Patrick, and Christian Galan, eds. 2011. *Language life in Japan: Transformations and prospects*. New York: Routledge.

Hymes, Dell. 1972. Models of the interaction of language and social life. In *Directions in sociolinguistics*, ed. John J. Gumperz and Dell H. Hymes, 35–71. New York: Holt, Rinehart and Winston.

International Conference on Language. 2014. Conclusions of the International conference on language. Available at http://www.un.org/en/events/motherlanguageday/pdfs/suzhou_conclusions_en.pdf, Suzhou, China. Accessed 26 June 2016.

Meisel, Jurgen M. 1983. Transfer as a second-language strategy: An introduction. *Language and Communication* 1: 11–46l.

Kaplan, Robert B, and Richard B. Baldauf 1997. *Language planning: From practice to theory*. Clevedon: Multilingual Matters

Kloss, Heinz. 1969. *Research possibilities on group bilingualism: A report*. Quebec: International Center for Research on Bilingualism.

Li, Xianle (李现乐). 2010a. Yuyan ziyuan he yuyan wenti shijiaoxia d yuyan fuwu yanjiu 语言资源和语言问题视角下的语言服务研究 [A study of language services from the perspective of language resources and language problems]. Yunnan shifan daxue xuebao 云南师范大学学报 (社会科学版) [*Journal of Yunan Normal University* (Humanities and Social Sciences)] 5: 16–21.

Li, Yuming (李宇明). 2010b. Zhongguo yuyan guihua lun 中国语言规划论 [On language planning in China]. Beijing: Shangwu yinshuguan 北京:商务印书馆 [Beijing: Commercial Press].

Li, Yuming (李宇明). 2013. Understanding China's situation through its language life. In The language situation in China, 1, eds. Li Yuming and Li Wei, v–viii. Berlin/Beijing: De Gruyter Mouton, Commercial Press.

Li, Yuming, and Wei Li (李宇明, 李嵬). 2013. *The language situation in China* (Vol. 1). Berlin/Beijing: De Gruyter Mouton, Commercial Press.

——— (李宇明, 李嵬). 2014. *The language situation in China* (Vol. 2). Berlin/Beijing: De Gruyter Mouton, Commercial Press.

——— (李宇明, 李嵬). 2015. *The language situation in China* (Vol. 3). Berlin/Beijing: De Gruyter Mouton, Commercial Press.

Li, Xianle (李现乐). 2011. Yuyan fuwu yu fuwu yuyan 语言服务与服务语言 [Language service and serving language]. Nanjing daxue bosi lunwen 南京大学博士论文 [Nanjing University dissertation].

Li, Xianle (李现乐). 2014. Yuyan ziyuan yu yuyan jinji yanjiu 语言资源与语言经济研究 [Language resources and the studies of language economy]. Jinji wenti 经济问题 [*Economic Problems*] 9: 25–29.

Liang, Sihua (梁斯华). 2015. Language attitudes and identities in multilingual China: A linguistic ethnography. Heidelberg/New York: Springer.

Marschak, J. 1965. Economics of language. *Behavioral Science* 10 (2): 135–140.

Ministry of Education. 2016. Zhongguo yuyan wenziwang 中国语言文字网 [the website of Chinese languages]. Available at http://www.china-language.gov.cn/index.htm. Accessed 26 June 2016.

Oxford, Rebecca. 1990. *Language learning strategies: What every teacher should know*. New York: Newbury House Publishers.

Patrick, Peter L. 2002. The speech community. In *The handbook of language variation and change*, ed. Jack K. Chambers, Peter Trudgill, and Natalie Schilling-Estes, 573–589. Malden, Oxford: Wiley-Blackwell.

Razumovskaya, V.A., and Y.V. Sokolovsky. 2012. Modern tendencies of language policy and language planning in Russia and China: Comparative study. *Journal of Siberian Federal University* 5: 927–934.

Spolsky, Bernard. 2004. *Language policy*. Cambridge: Cambridge University Press.
Spolsky, B. 2007. Towards a theory of language policy. Working papers in educational linguistics. Retrieved from https://repository.upenn.edu/wpel/vol22/iss1/1
Spolsky, Bernard. 2014. Language management in the People's Republic of China. *Language Policy* 4: 165–175.
———. 2016. Language planning in China: Foreword. *Chinese Journal of Language Policy and Planning* 1: 89–90.
———. 2019. Book review: Li Yuming: Language planning in China. *Journal of language policy and Language planning* 1: 81–83.
State Language Commission. 2006–2011. The green cover book: Language situation in China. Beijing: Commercial Press.
Tseng, Chin-Chin, and Chen-Cheng Chun. 2019. Chinese language and new immigrants. In *The Routledge handbook of Chinese applied linguistics*, ed. Chu-Ren Huang, Zhuo Jing-Schmidt, and Barbara Meisterernst, 212–219. London: Routledge.
Van den Berg, Marinus. 2016. Restructuring Chinese speech communities: Urbanization, language contact and identity formation. *Special Issue of Journal of Asian-Pacific Communication* 26 (1): 1–173.
Wang, Xiaomei (王晓梅). 2014. The origin and development of linguistic strategic studies 语言战略研究的产生和发展. Zhongguo shehui yuyanxue 中国社会语言学 [*Journal of Chinese Sociolinguistics*] 1: 1–9.
Wang, Hui (王辉). 2016. Book review: Yuming Li and Wei Li (eds.): The language situation in China: Volume 1. Language Policy,15:97–99.
Xu, Daming (徐大明). 2004. Yanyu shequ lilun 言语社区理论 [Speech community theory]. Zhongguo Shehui yuyanxue 中国社会语言学 [*Journal of Chinese Sociolinguistics*], 1: 18–28.
——— (徐大明). 2006. Chengshi yuyan diaoca 城市语言调查 [Urban language survey]. Zhongguo Shehui yuyanxue 中国社会语言学 [*Journal of Chinese Sociolinguistics*], 2: 1–15.
——— (徐大明). 2010. The formation of a speech community: Mandarin nasal finals in Baotou. In Industrialization and the restructuring of speech communities in China and Europe, eds. Marinus Van den Berg and Daming Xu, 120–140. Cambridge: Cambridge Scholars Publishing.
——— (徐大明). 2012. The language strategies of China: Coexistence of multiple languages to achieve harmony in diversity "多语共存, 和而不同"的中国语言战略. In Aomen yuyan yanjiu sanshi nian 澳门语言研究三十年 [*Linguistic studies in Macau in the past thirty years*], eds. Xu Jie and Zhou Jian 徐杰, 周荐, 2–12. Macau: University of Macau.
———. 2015. Speech community and linguistic urbanization: Sociolinguistic theories developed in China. In *Globalizing sociolinguistics: Challenging and expanding theory*, ed. Dick Smakman and Patrick Heinrich, 95–106. New York: Routledge.
——— (徐大明). 2016. Speech community theory and the language/dialect debate. In *Restructuring Chinese speech communities: Urbanization, language contact and identity formation*, special issue of Journal of Asian Pacific communication, Vol. 26, ed. Marinus Van den Berg, 8–31. Amsterdam: +John Benjamins Publishing Company.
Xu, Daming and Jingwei Zhang (徐大明, 张璟玮). 2019. Chinese sociolinguistics. In *The Routledge handbook of Chinese applied linguistics*, eds. Chu-Ren Huang, Zhuo Jing-Schmidt, and Barbara Meisterernst, 691–708. London: Routledge.
Zhang, Jingwei (张璟玮). 2020. Yanyu shequ shijiaoxia de yuyan huoli: yi aomen tusheng puyu weili 言语社区视角下的语言活力:以澳门土生葡语为例 [Language vitality in the perspective of speech community: The case of Patua]. Yunnan shifan daxue xuebao 云南师范大学学报 [*Journal of Yunnan Normal University*] 52(1):32–39.
Zhang, Weiguo and Liu Guohui (张卫国,刘国辉). 2012. Zhongguo yuyan jinjixue yanjiu shulue 中国语言经济学研究述略 [A brief introduction to the linguistic economics in China]. Yuyan jiaoxue yu yanjiu 语言教学与研究 [*Language Teaching and Research*] 6:102–109.

Zhang, Jingwei, and Daming Xu (张璟玮,徐大明). 2008. Renkou liudong yu Putonghua puji 人口流动与普通话普及 [Population movement and the popularization of Putonghua]. Yuyan wenzi yinyong 语言文字应用 [*Applied Linguistics*] 18(3): 43–52.

——— (张璟玮, 徐大明). 2019. The impact of communication technology in the Chinese language life. In *The Routledge handbook of Chinese applied linguistics*, eds. Chu-Ren Huang, Zhuo Jing-Schmidt, and Barbara Meisterernst, 552–563. London: Routledge.

Zhou, Minglang. 2019. *Language ideology and order in rising China*. Singapore: Palgrave Macmillan.

Zhou, Minglang, and Hongkai Sun, eds. 2004. *Language policy in the People's Republic of China: Theory and practice since 1949*. Boston: Kluwer.

Internet Language Study in China

11

Minzhe Chen, Shuqing Liu, and Shuyu Zhang

Contents

Introduction	276
Studying Chinese Netspeak: Origins, Development, and Contemporary Research Trends	277
Pre-2000	279
2001–2010	280
2011–2019	284
Future Directions in Chinese Netspeak	289
Conclusion	292
References	292

Abstract

Research on Chinese netspeak took off just before the turn of the century. After almost thirty years, it is now a robust field of study that engages the next generation of Chinese linguistic researchers and draws attention from multiple disciplines. With an objective to identify research trends and relevant research interests for future Chinese netspeak studies, this chapter revisits previous research published in Chinese academic periodicals and sourced for the *Chinese Social Science Citation Index* (CSSCI) from 1994 – the first year China had access to the global Internet – to 2019.

An analysis of 504 studies published in major journals under the keyword "netspeak" (*wǎngluò yǔyán*, 网络语言) reveals a clear trajectory in research development. The earliest research, published before the year 2000, is

---

M. Chen (✉)
Hunan Normal University, Changsha, China

S. Liu
Guangdong Polytechnic, Foshan, China

S. Zhang
The Australian National University, Canberra, Australia

characterized by inquisitiveness about this new way of speaking and uncertainty over its implications on the Chinese language. In the first decade of the twenty-first century, the field shifted its focus to how netspeak as a social language interacts with netizens' psychology as well as Chinese society and culture. Between 2011 and 2019, the scope of research further expanded as netspeak became increasingly prominent in both official and public spheres. In the last decade, it has become an interdisciplinary subject that spans different fields of knowledge, including linguistics, education, law, communication, and cultural studies. This chapter concludes by calling for Chinese netspeak research to focus on strengthening its theoretical foundation, broadening its research scope, and enhancing its research focus to promote the future development of the field.

Keywords

Chinese Internet Language (CIL) · Netspeak · Chinese language · Meaning construction · Chinese linguistics

Introduction

The latest *47th Statistical Report on Internet Development in China* (China Internet Network Information Center 2021) recorded 989 million Internet users in China as of December 2020, an 43.7% increase in the past 5 years alone. Chinese Internet users account for about one-fifth of global Internet users, making them the world's largest digital society.

The dramatic increase of netizens in China has not only propelled the development of network technology and online media but also generated and popularized the Chinese Internet Language (henceforth "CIL"), which mainly refers to popular, conventionalized language expressions created by Internet users, such as the following:

(1)	A: 这是我收到的短信,说今晚要加班。
	zhèshì wǒ shōudào de duǎnxìn, shuō jīnwǎn yào jiābān
	"I got this text saying we are instructed to work overtime this evening."
	B: *How made winds.*

In (1), the underlined part is an echoism of the Chinese phrase *hǎoměi de wénzì* (好美的文字, "Such beautiful words"). It is used in mockery or as an act of venting in the face of a frustrating situation.

(2)	刘能换人演了,爷青结。
	Liú Néng huànrén yǎn le, yé qīng jiē

In (2), the underlined part *yé qīng jié* (爷青结) is a combination of: "grandpa" + "youth" + "has ended." The whole sentence means, "Liu Neng [a character in the Chinese television drama *Love Stories in the Countryside*] is being played by a different actor now—man, when did I get so old."

Owing to its rapid development and widespread circulation across the web, CIL has received significant attention from researchers. CIL is studied not only as a language variation in mundane communication but also as a research topic of sociolinguistic and cultural interest.

With an objective to identify research trends and relevant research interests for future CIL studies, this chapter revisits research on CIL published in Chinese academic periodicals and sourced for the *Chinese Social Science Citation Index* (CSSCI). It covers research published from 1994 to 2019, as it was in 1994 that China was officially and internationally recognized for its access to the global Internet. With no universally accepted label for this research topic, a variety of terms translated from equivalent notions in the English-speaking academic community are used in CIL-related studies. Common terms include *cyberspeak*, *Netspeak*, *Netlish*, *electronic discourse*, *Computer-Mediated Communication* (CMC), *Electronically-Mediated Communication* (EMC), and *Digitally-Mediated Communication* (DMC) (Crystal 2011: 92). While CIL may be called many different things, all these terms highlight language use characteristic of interaction on the Chinese Internet yet puzzling for laypeople or non-Internet users.

Researching CIL holds considerable significance for three key areas: the linguistic, cultural, and sociopolitical aspects of China and the Chinese language. Its linguistic significance resides in understanding the peculiarities of its lexicology, morphology, syntax, semantics, and so on. CIL also reflects rich cultural connotations that will serve as the heritage of future generations of Chinese language users. Gao Sai 高赛 (2010), for one, sees an equivalence between ancient Chinese poetry and contemporary Chinese pop songs, but to date, only the former has been studied extensively. Gao states that it is the social responsibility of scholars to be equally attentive to modern forms of language use, such as netspeak, for these may well "become the cultural treasures that our descendants endeavour to discover." Last, as discussed in section "Netspeak in the Media," CIL interacts with China's sociopolitical dynamics and is increasingly recognized and used in both public and authoritative discourse.

In short, a comprehensive and objective analysis of the origins, current status, and future development of Chinese netspeak research is very important. Section "Studying Chinese Netspeak: Origins, Development, and Contemporary Research Trends" of this chapter reviews existing CIL studies in chronological order, dividing the research into three distinct periods (pre-2000; 2001–2010; and 2011–2019) and identifying the major research trends of each one. This is followed by a discussion of the prospects for Chinese netspeak studies. In conclusion, the authors call for more work on the theoretical aspects of CIL and suggest the need for a broader research scope and sharper research focus on studies of CIL.

Studying Chinese Netspeak: Origins, Development, and Contemporary Research Trends

China did not have official access to the Internet until April 1994. At that time, the National Computing and Networking Facility of China (NCFC) put the 64K international special line into operation, allowing China to access the Internet through an

American company, Sprint (China Internet Network Information Center 2012). As a result, Chinese netspeak only emerged relatively recently. Compared to research on English netspeak (e.g., Crystal 2011), early CIL studies were several years behind and not as comprehensive or insightful in their analyses. However, as the number of Internet users in China has gradually increased, so has the volume and breadth of research on Chinese netspeak.

The data used in the current study was collected from the Chinese Academic Journals Full-text Database (CJFD) using the "advanced search" option. Administered by China National Knowledge Infrastructure (CNKI), CJFD compiles research published by Chinese scholars since 1996. In the current study, the publication time frame for the CJFD search was set from 1996 to 2019. There were no restrictions on the source of periodicals, but it was specified that either the study title or list of keywords includes the term *wǎngluò yǔyán* (网络语言, "netspeak"). The method of fuzzy matching was also adopted.

The search was conducted on June 30, 2019. A total of 3,314 data entries matching the search criteria were returned. Of these, 504 articles were published in key Chinese academic journals. The 504 articles were then categorized into three time periods: pre-2000, 2000–2010, and 2011–2019. The results show that the number of published research papers on Chinese netspeak has soared over the past two decades. Only 26 papers were published before the year 2000. By the end of the first decade of the twenty-first century, the number of published papers had jumped to 1,274, a 60-fold increase over a decade. The number of articles then doubled over the most recent decade to 2019, reaching a total of 2,014 articles. The handful of papers that were published prior to 2000 laid the foundation for Chinese netspeak studies. This foundation allowed the field to develop rapidly in the early twenty-first century. The decade since has been the boom era of Chinese netspeak research, a time in which the field has extensively drawn on the diverse disciplines that it encompasses, which includes linguistics, education, law, ethics, philosophy, communication, and cultural studies.

In what follows, section "Pre-2000" reviews the dawn of CIL research before the 2000s. The small number of studies on Chinese netspeak published during this time tended to be highly critical of the idiosyncrasy of this new language, contrasting it to a more "standardized" Chinese language. Section "2001–2010" then looks at changes researchers focus on and attitudes toward netspeak that occurred from 2000 to 2010, as the rise of the Internet in China helped to popularize netspeak. Increasingly, netspeak came to be studied as a social dialect in its own right, born and bred in its particular linguistic and cultural ecology within the online realm. The following decade from 2011 to 2019, in contrast, saw the blurring of the line between virtual reality and reality itself, a fact which came to be reflected in CIL research of this period. As reviewed in section "2011–2019," the widespread use of netspeak in media discourse, particularly by Chinese state-run media, has enriched netspeak and facilitated its reception among a much larger audience. This leads to the closing discussion of the interdisciplinary nature of CIL research.

Pre-2000

Before 2000, CIL received little scholarly attention. This is evidenced by the fact that only 26 papers on CIL were published in Chinese journals between 1994 and 2000. Of these, a mere six were published in key journals and three in journals listed in the Chinese Social Sciences Citation Index (CSSCI). Research of this period tended to focus on netspeak as a "problematic" development that challenges the standardized use of language. In this preliminary stage of research, Chinese scholars investigated and discussed the nature of netspeak, how netspeak words (henceforth, "net-words") are formulated, and its common figures of speech. Interestingly, although netspeak was predominately viewed in a negative light at that time, the field was nonetheless recognized for its research potential. The publication rate of CIL-related research in top-ranked key journals reflects this: the pre-2000 percentage of 23.08% surpasses the publication rate of later periods (15.62% for 2000–2010 and 14.85% for 2011–2019, respectively).

The first paper on Chinese netspeak was published by Li Xin 立鑫 (Li 1998). In his article, Li presented CIL as a threat to the standardized Chinese language and a source of language "pollution" and "violence," a common sentiment in CIL research of this period. Five papers published in 2000 in the journal *Language Construction* (《语言建设》) debated the definition, characteristics, and nature of Chinese netspeak. In one of the articles, Shan Xiong 闪雄 (2000) held that Chinese netspeak, full of "incorrect" Chinese characters, was destroying the purity of the Chinese language. In Shan's view, this could have a negative impact on future generations, as the majority of Internet users are young people. Shan further cautioned against the lack of legal regulations and moral restraints on the use of "incorrect" language in the online sphere, calling for authoritative regulations to be introduced to prevent such "pollutants" – products of the younger generation's desire to be unique and different – to cross over into people's everyday language.

Other scholars approached the novel language development of Chinese netspeak from a more descriptive point of view, deliberating on the nature of netspeak and its figures of speech. Jin Song 劲松 and Qi Ke 麒坷 (2000), for example, applied enumerative classifications to netspeak words and expressions. Jin and Qi then divided the words and expressions into seven types: graphic symbols; combinations of Chinese characters and graphic symbols; pidgin English; "Chinglish"; unrestricted abbreviation use; neologisms; and childish speech. Kuang Xia 邝霞 and Jin Zi 金子 (2000) then expanded the scholarly perspective on CIL, seeing Chinese netspeak as a "new social dialect used by a stable linguistic community" rather than a mere collection of buzzwords. In the main, however, CIL research before 2000 was largely preliminary and superficial in a theoretical sense. Of course, this is to be expected at a time when research findings were scarce and the Internet had yet to gain traction in China. The situation soon changed following the turn of the century. By the end of 2010, there were 457 million netizens in China and 60 times as many CIL-related research papers as had been published in the preceding decade.

2001–2010

Between 2001 and 2010, 1,274 papers on Chinese netspeak were published in Chinese journals. Of these, 199 papers were published in key journals and 134 papers in CSSCI journals. This decade-long research period can be further divided into two stages: 2001–2005 and 2006–2010. The first five years saw a subtle change in attitude toward Chinese netspeak and a shift in research focus from vocabulary to discourse as scholars continued to explore the nature of netspeak and how it is generated. It was also during this time that CIL came to be regarded as an integral part of modern culture. In the second half of the decade, the depth and breadth of research on Chinese netspeak continued to expand as scholars strove to understand CIL from a variety of sociocultural and psycholinguistic perspectives.

2001–2005

One of the most notable trends of CIL research published between 2001 and 2005 was a new recognition that netspeak is a social language and not a contravention of normal language use. During this period, Chinese scholars adopted a more open-minded perspective, and the focus of their research moved from netspeak vocabulary to discourse. Researchers started to theoretically analyze the nature, characteristics, and formation rules of netspeak and began to study interactions between netspeak and culture. This helped to engender a more systematic approach to the study of Chinese netspeak.

Much of the work in the early 2000s consisted of efforts to collate and classify Chinese netspeak words and expressions. The compilation of two dictionaries, the *Network Vogue Dictionary* (《网络时尚词典》) by Yi Wenan 易文安 (Yi 2000), and the *Chinese Netspeak Dictionary* (《中国网络语言词典》) by Yu Genyuan 于根元 (Yu 2001a), recorded the budding collection of new words and expressions. An overview of pre-2000 CIL research was also summarized in two books: *Introduction to Netspeak* (《网络语言概说》) by Yu Genyuan 于根元 (Yu 2001b), and *Netspeak* (《网络语言》) by Liu Haiyan 刘海燕 (Liu 2001).

However, the shift in scholars' perspectives of Chinese netspeak around the turn of the century came about moreover, the shift in scholars' perspectives of Chinese netspeak around the turn of the century was also influenced by David Crystal and his 2001 book, *Language and the Internet*. In the book, Crystal highlights the role of language on the Internet and the influence of the Internet on language (2001, viii). He recognizes that netspeak is born out of the diverse needs and communication settings of new and emerging social groups, with the Internet being just one example. In the book, Crystal also calls for greater appreciation of the innovation and adaptability that netspeak embodies, taking a critical view of attempts to intervene in the evolution of language simply to prevent any alteration of existing language "norms."

Crystal's initiative, together with the recognition that netspeak can in no way be eradicated from modern language, prompted Chinese scholars to adopt a more neutral stance toward CIL. Instead of denial and outright rejection of its legitimacy as a language form, researchers began to reflect on how Chinese netspeak could develop in the "right" direction without resorting to criticism of its characteristics. A growing number of scholars expressed the view that, instead of attempting to strictly uphold traditional written language norms in the hope of curbing netspeak altogether, it would be more realistic to accept the new status quo and adapt accordingly (Peng 2001; Zhong 2001; Qin 2003; Deng and Duan 2004). However, the ways in which netspeak affects Chinese culture and society remained a subject of debate. Some scholars studied netspeak through the prism of culture and literature. In one study (Xu 2004), Xu Zhongning 许钟宁 explored netspeak as a reflection of an emerging social culture, with Xu taking the view that the "cultural shock" that netspeak has on the wider community can result in different outcomes. Taking the opposing view, Chinese researcher Shen Fenfang 申芬芳 (Shen 2004) concentrated on the negative impacts that netspeak may have on the teaching of literature.

Between 2001 and 2010, there was considerably more research into the nature and word-formations of Chinese netspeak than had been conducted in the preceding decade. Notably, the scope of research began to extend from the lexical level to study netspeak in terms of discourse. Contextual specificity was afforded a greater role in explanations of the unique nature, formation, and characteristics of netspeak, in other words, in outlining how and why netspeak conforms to neither traditional written language nor everyday language.

For example, an article by Peng Yubo 彭育波 (Peng 2001) contrasted netspeak – which is considered a form of written language employed by Internet users – with traditional written language, that is, language used in print media such as newspapers, magazines, and books. Peng identified four distinctive features of netspeak, characterizing it as creative, visualizable, concise, and unstandardized. More specifically, the creativity of netspeak is its most prominent feature. Peng's characterization reinforces to some extent the findings of He Ziran 何自然 and He Xuelin 何雪林 (2003). Taking a memetic perspective, He and He found that netspeak is generated from memes or novel permutations of existing linguistic forms. The phrase *jūn nán* (菌男, "ugly men"), for instance, imitates the existing phrase *jùn nán* (俊男, "handsome men"). In his 2005 study on modes of CMC, Li Xuping 李旭平 (Li 2005) found that the characteristics of netspeak differ from everyday language. Meanwhile, in a departure from stylistic and systemic-functional linguistics approaches, respectively, Qin Xiubai 秦秀白 (2003) and Huang Guowen 黄国文 (2005) put forward the proposal that, as a new type of media discourse, netspeak is unique in its tone, domains, and modes of communication. Qiu and Huang additionally suggested that netspeak should be considered a hybrid of both spoken and written discourse.

In sum, CIL research carried out between 2001 and 2005 demonstrated a new approach that studies netspeak in terms of what it is – a social dialect characterized by the contextualized distinctiveness of the online community of Chinese speakers. Instead of seeing netspeak as a "deviation" from the traditional or standard language, researchers gradually came to recognize the social and cultural influences of netspeak. It was also during this period that netspeak came to be studied in more nuanced, in-depth, and systematic ways.

2006–2010

The second half of the decade to 2010 saw an increase in the number of CIL-related publications. Between 2006 and 2010, a total of 1,301 research articles were published. However, only 133 of these, approximately ten percent (10.22%) of the total, were published in key academic journals. This suggests that a greater number of research papers is not indicative of a greater understanding of Chinese netspeak. Indeed, the majority of research published during this period is similar to that conducted in the early 2000s. A number of studies looked to further investigate the nature (e.g., C. Shi 2010); stylistic features (e.g., He 2009); or lexical features (e.g., Y. Zhang 2007) of netspeak. Others continued to document figures of speech in netspeak (e.g., Zuo and Yao 2006), adopt a memetic perspective to study netspeak (e.g., Fu and Lu 2010), or describe the standard of netspeak (e.g., Zou 2007). Among this work, innovative findings and stronger theoretical frameworks were few and far between.

Nonetheless, CIL research took a major step forward in this period as the result of interdisciplinary collaboration. Efforts to understand netspeak from multiple perspectives, including sociopsychological and cognitive linguistics perspectives, concentrated scholars' attention on ways in which the formation of netspeak expressions relates to Internet users' psychology and cognition. During this time, several studies investigated the relationship between Chinese netspeak and its users, coming to the conclusion that Chinese netspeak reflects netizens' social and psychological needs in the Internet era (C. Liang 2006; H. Zhang 2006; Y. Liu 2009). The authors of these studies argued that the richness and diversity of linguistic forms manifested in netspeak – namely, the fact that Chinese netspeak not only makes use of Chinese characters but also creates novel combinations of English letters, numbers, graphic symbols, and pinyin – represents Internet users' response to the stress of their shifting social roles. Moreover, netspeak facilitates netizens' ability to craft a unique online persona for themselves, allowing them to fulfill a desire for recognition and acceptance by their peers.

The rise of cognitive linguistics perspectives on Chinese netspeak during this period also served to broaden the scope of research into the formation of Chinese netspeak, as researchers sought to unravel the cognitive mechanisms that informed the creation of new phrases and syntactic constructions. For example, Bai Jiehong 白解红 and Chen Minzhe 陈敏哲 (2010) examined how the "Agent-Action-Result" cognitive schema is activated during the online meaning construction of the Chinese netspeak compound X-kè (X客, "member of an online subgroup, e.g., blogger"). Similarly, Liu Dongqing 刘冬青 and Shi Jianping 施建平 (2010) looked to clarify the underlying cognitive basis that explains the revival of old and obscure Chinese characters in netspeak. As illustrated in Table 1, the glyph structures of several of these characters express emotion-related meanings, yet the characters simultaneously defamiliarize the reader on account of their unfamiliar visual forms. The humorous quality of these repurposed characters is also seen as important for an online audience. In another study, Wang Yanyan 王彦彦 (2010) analyzed the mechanisms underlying the creation of particular net-words from the perspective of cognitive metaphor. One of the net-words examined by Wang is *bēi jù* (杯具,

Table 1 Examples of repurposed glyphs in Chinese netspeak

Character	Pinyin (phonetic romanization)	Original meaning	Netspeak meaning and explanation
烎	yín	Brightness	To make one's blood boil (The character is made up of two components: 开 kāi "open" on top and 火 huǒ "fire" at the bottom)
囧	jiǒng	Brightness	Embarrassed, gloomy (The shape of the character resembles a downcast expression)
槑	méi	Plum blossom	Dull, stupid (The character consists of two 呆 dāi "dull")
氼	nì	To drown	Hello (The character has a similar pronunciation as 你 nǐ "you" in 你好 nǐhǎo "hello")
砳	lè	The sound of two stones clicking	Happy (The character has the same pronunciation as 乐 lè "joy, happiness")
叾	jiào	As long as (dialect)	As long as
嘂	jiào	Ancient instrument; To call out	To scream in excitement (The character is derived from 叫 jiào "to call out" by surrounding the original right-hand side component with four 口 kǒu "mouth")
玊	sú	Imperfect jade	Special individual

"cup"), which has come to mean "tragedy" in netspeak on account of it being a homophone of *bēi jù* (悲剧, "tragedy"). Wang also analyzed related derivative phrases, such as *xǐ jù* (洗具, "washware"), meaning "comedy" in netspeak as it is homophonous with *xǐ jù* (喜剧, "comedy"), and similarly, *cān jù* (餐具, "cutlery"), which is a homophone of *cān jù* ("惨剧", "disaster") and has acquired this meaning in netspeak as a result.

It was also during this period between 2006 and 2010 that CIL first appeared on the radar of China's language regulation and guidance authorities. In 2007, China's Ministry of Education (MOE) and the State Language Commission, a subordinate agency of MOE, jointly published the *Report on the Language Situation in China* (《中国语言生活状态报告》), a two-volume document that dedicated an entire chapter to popular netspeak expressions recorded in the previous year. This recognition of netspeak at a government level can be regarded as official acceptance of the fact that the novel words and expressions of Chinese netspeak are a new, continuously evolving, and integral part of the broader Chinese language. But in addition to this, publication of the report suggests that Chinese authorities were interested in incorporating netspeak into existing language planning and regulation frameworks, and factoring it into ideological and political education curricula. These developments will be discussed in more detail in the following section.

To summarize, despite the increase in the number of articles published between 2006 and 2010, the latter half of the 2000s was a relatively stagnant period for CIL-related research. And yet, although most publications continued to explore conventional research topics, this should not diminish the important contribution made by a handful of researchers during this time. The interdisciplinary efforts of these scholars significantly broadened the study of Chinese netspeak, revealing key psychological and cognitive factors that shape Chinese netizens' online language use.

2011–2019

A notable feature of CIL research published in the most recent decade to 2019 is its growing interest in the role of netspeak in both public and official domains. First of all, the acknowledgment of Chinese netspeak at an institutional level sanctioned the study of online phrases and expressions in a more positive and objective light. It also enabled netspeak to be embraced by mainstream (i.e., state-run) media in China, a domain in which it can continue to diversify, innovate, and proliferate among a much wider audience. These dynamic developments in Chinese society have subsequently informed recent research trends and points of scholarly debate related to Chinese netspeak. The field now boasts a solid multidisciplinary foundation, and there has been a tremendous improvement to the depth and breadth of research. Over this 10-year period, a total of 2,014 papers on netspeak were published in Chinese journals, among which 299 were published in key journals and 166 in CSSCI journals.

Rather than undertaking a chronological review, this section adopts a more holistic approach to summarize CIL-related research published between 2011 and 2019. Providing a holistic overview here accords with the fact that this period saw research findings from multiple disciplines integrated to yield new insights into Chinese netspeak. This interdisciplinary collaboration highlights the nature of CIL as a social dialect, one whose development has been increasingly propelled by social interactions in the age of the Internet and social media. Instead of netspeak constituting an insider language, Internet users now consciously assign new meanings to innovative netspeak coinages (i.e., netspeak neologisms). Hence, a purely ontological investigation will not be sufficient for understanding the many social, cultural, and occasionally political factors that have motivated the continual development and evolution of Chinese netspeak.

This section begins by discussing research into factors that have propelled recent innovations in CIL. In particular, it focuses on the important role played by the media (especially state-run media) in China. It then discusses two major implications for CIL research, ontological and social, which have come about as a result of the growing popularity and acceptance of netspeak in the broader community. Ontologically, the increased popularity and acceptance of netspeak has facilitated its conventionalization as a language phenomenon. In terms of its relationship to society, Chinese researchers have recognized that understanding both the positive and negative influences that netspeak has on the general public is becoming increasingly urgent, as the line between the virtual world and the real world begins to blur.

Netspeak in the Media

Chinese netspeak first entered the public consciousness through its promotion in the media, or, more precisely, state-run media. The extensive use of netspeak in the media has inspired several interdisciplinary research efforts in CIL research and media communication. While the disciplines involved may hold slightly different research objectives, their research interests are closely intertwined. For example, multiple disciplines are interested in studying the basis for netspeak's popularity in the media, how this maps onto the changing trends in public discourse, and the implications of netspeak on communication studies.

Netspeak words and constructions were introduced to a much larger community on November 10, 2010. An article entitled *Jiangsu "gěilì" to become a culturally rich province* (《江苏给力"文化强省"》) featured on the front page of the *People's Daily* (《人民日报》), an official newspaper of the Central Committee of the Chinese Communist Party (CCP) with a circulation of three million. The expression *gěilì* (给力, literally "give power/strength") can be roughly translated as "fantastic." Originating from a southern Chinese dialect (Hainan dialect), *gěilì* was later borrowed into Chinese netspeak as an expression of exclamation. By the end of 2012, a number of net-words had appeared in *People's Daily*, including *pīndiē* (拼爹, "to rely on one's family background to get ahead") and *diǎosī* (屌丝, "a person, usually male, who is deemed to have failed to meet society's expectations of them"). Both words speak to contentious social issues in contemporary China.

The introduction of netspeak neologisms, such as the examples above, has greatly enriched the linguistic ingenuity of Chinese state media. It also parallels the development of an increasingly complex and diversified media discourse in modern China. In a recent study, Shi Wei 施维 (Shi 2016) analyzed the language used in local newspapers in Chengdu since the beginning of the reform era in 1978. Shi observed two main changes in media discourse over the past forty years: (1) a shift from formulaic to complex expressions; and (2) a move away from uniform language toward diverse ways of speaking and writing. Shi proposed that widespread acceptance of innovative language among the younger generation has been a key factor in netspeak being adopted by the media. In this light, the conscious inclusion of net-words such as *gěilì* in official media is an astute move: The novelty and originality of net-words appeals to the communication needs of ordinary people (Zhou 2013). A similar case can be made for other Internet buzzwords that have been embraced by the general public, such as *diǎosī*, *yánzhí* (颜值, "attractiveness scale"), and *rènxìng* (任性, "capricious") (e.g., Y. Liu 2015; W. Shan 2015).

The field of communication studies has also been influenced by CIL research. This is because netspeak can be an effective tool for reflecting and changing public opinion. Researchers Zhang Ting 张挺 and Wei Hui 魏晖 have highlighted the need to study netspeak as an indicator of public opinion in the area of language policy, which could help inform decisions around language policy (Zhang and Wei 2011). A paper by Kang Qingye 康庆业 (Kang 2014) did well to encapsulate the ways in which studying Chinese netspeak benefits communication studies. As Kang noted, exploring netspeak not only opens up new avenues of research in network communication theory, it also deepens our understanding of people's online speech and behavior. In the longer term, CIL research also benefits the public. Its research findings help improve regulation of the Internet environment, something that is critical for today's information society.

In sum, the extensive use of netspeak in the Chinese media in recent years has seen it exposed to a wider audience than ever before. It has further demonstrated the importance of interdisciplinary collaboration in CIL research. The popularity of netspeak lies in its faithful reflection of the communication needs of ordinary people, who find its novel expressions both poignant and persuasive. Popular acceptance of netspeak has also made research in this area useful for understanding public opinion and guiding language policy.

The Grammar of Netspeak

As Chinese netspeak has continued to propagate through society with the help of the media, numerous language resources have emerged that have enabled CIL to be studied in terms of grammar. While the focus of CIL research remains studying neologisms, a growing number of studies have begun to explore the grammar embedded in these linguistic innovations, moving away from simple descriptions of word formation or meaning constructions. Recent ontological research is also notable for drawing more extensively on the latest theories and methodologies in linguistics, as can be seen in the increased adoption of corpus-based methodologies in CIL research.

With the theoretical foundation of netspeak largely formulated in the decade to 2010, the description of netspeak neologisms has gone one step further in the years since. Netspeak neologisms can now be described more comprehensively in terms of meaning, function, and evolution. More broadly, netspeak, no longer regarded as a "deviation" from conventional language, is at last discussed as an integral part of the Chinese language as a whole. For example, Jiang Haiyan 江海燕 and Zhai Huifeng 翟会锋 recently investigated the characteristics, functions, and changing usage of two Chinese modal particles *luo* (咯) and *da* (哒) (Jiang 2016) and two variants of *de* (的), *di* (滴), and *da* (哒) (Zhai 2017), respectively. Jiang and Zhai revealed the grammatical distinctiveness of these linguistic forms in netspeak by showing how netspeak usage differs from traditional usage as recorded in dictionaries and print media. Similarly, Ma Caixia 麻彩霞 (Ma 2018) pinpointed changes in the semantic and syntactic functions of the net-word *nèihán* (内涵, literally "connotation," and "to subtly badmouth" in netspeak) by comparing its netspeak usage to examples of conventional use given in the *Modern Chinese Dictionary* (《现代汉语词典》). In another study investigating the grammaticalization of the new net-word *zhēnxīn* (真心, literally "true heart" and "really, indeed" in netspeak), Liang Yonghong 梁永红 found that while *zhēnxīn* was mainly used as a noun or a descriptive adverb in the past, since 2010 it has also functioned as a comment adverb in netspeak (Y. Liang 2019).

CIL research carried out in the decade to 2019 is also notable for its more sophisticated application of linguistic theories. Many scholars have drawn on modern theories, such as memetics (e.g., Heylighen 1998; Blackmore 2000; Hyland and Tse 2004) and construction grammar (e.g., Ungerer and Schmid 2013), to analyze netspeak textually and syntactically. Fang Yan 方艳 (2005), for example, explicated the meaning and function of each component in the construction *Adj.+dào* (到, literally "to")+*VP* (e.g., *hǎochī dào kū* 好吃到哭, "so tasty that I'm gonna cry"). In his paper, Fang not only explored the meaning production process of this grammatical construction but also discussed the reasons for its prevalence. Similarly, Huang Ming 黄鸣 (2011) examined the *bèi*+XX construction (被 XX) from the perspective of memetics. Huang concluded that the mockery and pessimism entailed in this construction holds a pragmatic meaning particular to Chinese people and speaks to their rising awareness of social injustice.

In addition, recent CIL studies increasingly refer to the agency and creativity of Internet users as a major reason behind the dynamic interplay of netspeak and memes. One study by Lei Dongping 雷冬平 and Li Yaozhen 李要珍 (Lei and Li 2013) applied Hyland's (2005) canonical account of metadiscourse to an analysis of XX-*tǐ* (XX 体, "an Internet genre mocking the particular way of speaking that a TV character or person, usually of higher authority, is known for"). In their paper, Lei and Li discussed how these metadiscourses can continue to evolve through the circulation of memes. Wu Shuyan 吴术燕 (2014) also emphasize the importance of the receiver in determining the language style of XX-*tǐ*.

The task of compiling a grammar of netspeak has become far simpler in the past decade. First, the Internet itself is the single "field site" where new words and expressions can be generated. It is also a relatively free and open space for language contact. Second, the adoption of netspeak by the media has made it easier than ever before for novel expressions and constructions to reach a wide audience in a short period of time. As a result, there is now a huge amount of relevant information which can inform generalizations about the nature of Chinese netspeak. To take advantage of all the available data, corpus-based approaches to CIL research have become increasingly popular in the past few years. The data analyzed in such studies generally includes a set of net-words and expressions collected by the authors themselves as well as data taken from existing Chinese language corpora such as the CCL corpus managed by Peking University's Centre for Chinese Linguistics.

Chen Changlai 陈昌来 and Yong Qian 雍茜 (2011) created an online netspeak corpus using data collected from People's Daily Online, Shuimu BBS (a Tsinghua University-affiliated online discussion forum), and other search engines. Having access to this dataset allowed Chen and Yong to be among the first researchers to systematically analyze the semantic meaning, syntactic functions, and generative motivations of a net-word, in this case, the aforementioned net-word *gěilì*. Xing Nuhai 邢怒海 (2014) subsequently analyzed the net-word, *màosì* (貌似, "seemingly"), using the CCL corpus and Baidu search engine. In his paper, Xing discussed how the function and meaning of *màosì* have shifted as the word has moved from a literary context into colloquial and Internet language.

In terms of Chinese netspeak corpora more generally, significant progress was made on the construction of dedicated netspeak corpora between 2011 and 2019. A comprehensive review of the development of "instant dynamic corpora" was provided in a 2015 paper by Li Nan 李楠 (Li 2015). In the article, Li described the process by which an instant dynamic corpus was created using automatic recognition and collection of neologisms that appeared on five major Chinese news websites. Li then compared words featured in the instant dynamic corpus with those included in the CCL corpus. Shao Yanjun 邵燕君 (2018), on the other hand, took a more conventional approach to corpus compilation. Shao constructed a corpus of 245 net-words that she claimed were representative of contemporary Chinese culture and values. With a vision of communicating these net-words to posterity, each corpus entry is accompanied by a description of a related Internet cultural phenomenon, for example, ACG culture ("Anime, comics and games," sometimes called "two-dimensional space," 二次元宅文化), fan culture, female-oriented games, web novels (e.g., Wuxiaworld), computer games, and buzzwords.

In summary, ontological CIL research undertaken since 2010 has moved beyond basic description of net-words in an effort to reveal the underlying grammar embedded in Chinese netspeak. Greater sophistication in both theoretical and methodological approaches has contributed to a better understanding of netspeak grammar, and in particular, the social and cultural factors that motivate its linguistic forms.

Regulating Netspeak

There is no doubt that the Internet and netspeak have improved and enriched the lives of many people, especially university and college students. Nonetheless, many scholars have expressed concern that the transformative benefits of the Internet can only be fully realized once any potential risks that it poses to people's sense of morality and their regular language use in the real world have been understood and mitigated. Some also believe that stricter regulation and management of netspeak is necessary to counter any potential threats it may pose to online cultural security and netizens' well-being in cases of cyberbullying.

A number of Chinese researchers consider moral education among university and college students, particularly that of a political or ideological nature, to be critically important. They hold that universities, as institutions of higher education, should foster a "healthy and appropriate way of using language" among young people (Wang 2018: 43). Netspeak, some scholars have argued, can be used to enhance teaching and learning outcomes (e.g., Tian 2018). It is also seen to provide insight into students' beliefs and mindset (Xie & Xu 2018), at least when used appropriately. However, it is more common for scholars to express their reservations on the use of netspeak among young people. Some have contended that netspeak should be restricted on university campuses to some degree, if not banned outright. According to these scholars, the "lack of standardization" and "implicit negative connotations" of netspeak should be a cause for concern, especially when netspeak pertains to complex social issues (Wang 2018: 43). The net-word *diǎosī*, mentioned above, is one example. *Diǎosī* is a vulgar descriptive noun in and of itself (literally "dick hair"). But it also carries a connotation that society's measure of success centers on materialism and money worship, and not the "socialist core values" promoted in official discourse.

The academic journal *Language Construction* (《语言建设》) has published several papers on the relationship between netspeak and moral education that have stimulated a broader debate among Chinese researchers working on CIL-related topics (e.g., Tang 2013; Cheng and Jiang 2016). For example, in a study of university students' use of netspeak, Qi Meng 祁萌 and Li Zhengqi 李正琪 (2018) showed that netspeak can involve complex arrays of meaning. Qi and Li argued that the language must be strictly regulated as a result. On a similar note, Wang Zhijun 王志军 (2018) concluded that netspeak has both positive and negative influences on students, and called for measures to be implemented that would reinforce the positive impact of netspeak while minimizing deleterious effects on young people.

In addition, some scholars have expressed the view that use of netspeak should be controlled on account of its potential risks to online cultural security. Several Chinese researchers working in the field of Internet culture have advocated for greater monitoring of netspeak, which they see as not a language per se but a distinct culture (e.g., Yuan and Han 2018; Yu and Zhang 2019). In the name of safeguarding Chinese netizens' "virtual homes" and securing Chinese authorities' control over the Internet, these scholars maintain that close surveillance of Internet language is critical. Furthermore, they promote the need to vigilantly guard against the infiltration of netspeak into contemporary Chinese culture and society, not only in the education sphere but also in politics, the economy, and the media.

A more justifiable concern about the impact of netspeak pertains to its use as a vehicle for cyberbullying. Several researchers have called for more effective legal regulations to be introduced in order to prevent netspeak becoming a language of violence in cyberspace (e.g., Qiu and Ji 2013). Some go even further, arguing that cyberbullying through verbal abuse should be deemed a criminal offense and offenders should be liable for prosecution in accordance with the Criminal Law of the People's Republic of China (PRC) (e.g., Chen and Ma 2019). Other researchers have sought to examine the reasons why people engage in unprincipled use of language or even language violence online. Scholars including Xiao Feng 肖峰 and Dou Changyu 窦畅宇 (2016) and Zhou Bin 周彬 (Zhou 2018) have argued that this unethical behavior may be attributable to people's misunderstandings about the role of Internet language in contemporary society and their abuses of online "freedoms."

Overall, while there has been a growing acceptance that netspeak constitutes an integral part of the Chinese language, concerns persist that it may cause harm. This is particularly the case in contexts where netspeak is seen to be out of step with political and ideological education in China, or else is outright abusive and inappropriate.

Future Directions in Chinese Netspeak

Looking back over the past two decades, it is evident that CIL research has made considerable progress. Different research periods have focused on slightly different research topics, ensuring that knowledge of various aspects of Chinese netspeak has continued to accumulate. Early CIL research began by providing a pure description of what was then a newly emerging language phenomenon. These preliminary studies discussed the nature, characteristics, syntax, and figures of speech in netspeak, among other topics. This was followed by a period of deliberation during which researchers debated the "right" attitude toward netspeak, that is, whether Chinese netspeak should be viewed as a "deviant" or merely variant form of the Chinese language. In the late 2000s, netspeak became increasingly popular among the public, in large part due to its broader dissemination through state-run media. This prompted scholars to look more closely into the complex relationship between Chinese netspeak, netizens, and the society and culture in which they live. In recent years, CIL research has reached a mature stage of development, as it is now increasingly guided by theoretical considerations and advanced methodological approaches.

Chinese netspeak is a topic that continues to hold considerable interest for the next generation of Chinese researchers. To date, there have been eight doctoral dissertations and 405 master's theses on netspeak, which have already amassed a total of 368,004 citations on the CNKI database. CIL research also currently receives government support. For example, since 2006, the National Social Science Fund of China (NSSFC) has funded three research projects on netspeak: the construction and study of netspeak corpora (06BYY029), a study of factors contributing to the development of netspeak (07BYY021); and a multiperspective study of netspeak's influences on pragmatic language (08BYY022).

However, there is still much to be done in the study of Chinese netspeak. Three key research areas in particular warrant further exploration. These are the following:

(1) The collection of new expressions and constructions in Chinese netspeak together with a comprehensive explanation of how they emerge and evolve
(2) The establishment of Chinese netspeak linguistics as a distinct field
(3) A comparative analysis of Chinese netspeak and other major netspeak languages of the world

CIL research conducted over the past two decades has fully demonstrated the importance of the first topic listed above. The fleeting and rapidly evolving nature of netspeak makes it imperative that we continue to seek to add to and update our knowledge of this online language. This trend is likely to accelerate even further as new breakthroughs in computer science and network technology continue to emerge while, at the same time, social media integrates even further into people's lives. As shown in Table 2, the majority of Chinese netspeak neologisms coined in more recent years relate to contentious social issues. This marks a departure from the rich vein of satire and self-expression characteristic of early netspeak. Even existing net-words have continued to evolve. Take *gěilì* as an example. Its appearance in *People's Daily* has led to other derivative net-words such as *ungelivable, gelivable, geilivable, geliable,* and *geilivable.* As the Internet continues to permeate further into everyday life, it is simply a matter of time before such netspeak words and constructions find their way into popular discourse. Hence, an up-to-date understanding of netspeak helps maintain a harmonious coexistence between all the language varieties

Table 2 Examples of recent netspeak neologisms and their social contexts

Characters	Pinyin	Social context
躲猫猫死	*duǒ māomāo sǐ*	Originally referring to the children's game hide-and-seek, the expression *duǒ māomāo sǐ* has acquired a negative meaning in netspeak following widespread public skepticism over a police claim that the death of a 24-year-old man in custody occurred while the man was playing this game with his cellmates
被就业	*bèi jiùyè*	The phrase *bèi jiùyè* became popular after revelations that some college graduates were falsifying employment contracts in order to obtain their final graduation certificates, leading to artificially inflated employment rates for university graduates
蒜你狠	*suàn nǐhěn*	*Suàn nǐhěn* is homophonous with the conventional expression *suàn nǐhěn* (算你狠, "Fine, you win"). The netspeak version relates to a period of consumer panic in China after the price of garlic (*suàn* 蒜) soared

subsumed under the umbrella of modern Chinese; diversity makes a language more dynamic (Yao 2011).

On the second topic, netspeak already possesses a number of distinctive qualities that make it an attractive yet challenging new field for language researchers (see Crystal 2009). First of all, there is an abundance of netspeak text on the Internet. Information about netspeak is also more widely available as access to the Internet has become more common. Second, netspeak is a diverse language that spans different modes of communication, from emails to tweets. Appearing in various forms across online domains, there are multiple perspectives, qualities, strategies, and expectations of netspeak as a language. In addition, accurately tracing the origin and spread of netspeak words and expressions is relatively straightforward to do on the Internet. However, one challenge that the field continues to face is the fact that netspeak occurred in private communication (e.g., WeChat [the Chinese equivalent of WhatsApp and Messenger]), hence remains barely accessible to researchers. In short, netspeak is a new field with great potential and challenges to overcome, a field that can develop into an established discipline through the contributions of a new generation of CIL researchers.

The third key area which remains to be explored is the comparative study of world netspeaks. While Chinese netspeak has local characteristics and language tendencies that make it unique, it would also be insightful to study issues in CIL research from a broader, global perspective. Drawing on insights from netspeak studies conducted in other parts of the world is likely to help solve some of the problems and challenges that Chinese researchers face in their work on CIL. On the issue of netspeak regulation, for example, Chinese netspeak researchers can learn from the experiences of online communities who speak a language other than Chinese. For one, how netspeak can be used for the common good in an open yet regulated online environment has generated heated debate in English-language netspeak studies (e.g., Banks 2010). This debate has important implications which may help inform the research and development of netspeak regulations in Chinese cyberspace, as was discussed in section "Regulating Netspeak."

Conclusion

Chinese netspeak emerged in the wake of the Internet's establishment in China. It is a language that has continued to develop over almost three decades and continued to prosper thanks to the high Internet penetration rate and the prevalence of social media in contemporary China. As reviewed in this chapter, the study of Chinese netspeak has come along in leaps and bounds over the past few decades. It has inspired numerous interdisciplinary research efforts which pool the expertise from a range of disciplines including linguistics, education, law, ethics, philosophy, communication, and cultural studies. We should expect that our understanding of Chinese netspeak will deepen even further as these efforts continue. Moreover, the discipline will mature as it continues to strengthen its theoretical underpinnings, sharpen its focus, and broaden the scope of netspeak research. It is especially promising that researchers have begun to make progress on understanding the factors that inform the creation of Chinese netspeak words and constructions, and, furthermore, contextualizing these factors in light of netspeaks originating from other parts of the world. The next step is for Chinese language researchers to formally establish a dedicated field of Chinese netspeak linguistics and further our knowledge in this important area of study.

References

Bai, Jiehong (白解红) and Chen, Minzhe (陈敏哲). 2010. Hanyu wangluo ciyu de zaixianyiyi jian'gou yanjiu—yi "x ke" weili 汉语网络词语的在线意义建构研究——以"X客"为例 [A Study of Online Meaning Construction of Chinese Net-words——with "x客" as an Example]. *Waiyu Xuekan* 外语学刊 [*Foreign Language Research*] 2010(2): 25–30.

Banks, J. 2010. Regulating hate speech online. *International Review of Law, Computers & Technology 24* (3): 233–239.

Blackmore, S. 2000. *The meme machine*. Oxford Paperbacks.

Chen, Chunzhu (陈纯柱) and Ma, Shaoying (马少盈). 2019. Wangluo yuyan baoli de zhili kunjing ji lujing de xuanze 网络语言暴力的治理困境及路径选择 [On the Dilemma and Choices of Cyberbullying]. *Zhongguo Renmin Gong'an Daxue Xuebao (Shehui Kexue Ban)*《中国人民公安大学学报(社会科学版)》[*Journal of People's Public Security University of China (Social Sciences Edition)*] 2019(2): 138–147.

Chen, Changlai (陈昌来) and Yong, Qian (雍茜). 2011. Jiyu wangluo yuliaoku de "gěilì" yanjiu 基于网络语料库的"给力"研究 [A Study of gěilì based on Web Corpus]. *Dangdai Xiuci Xue* 当代修辞学 [*Contemporary Rhetoric*] 2011(03):82–88.

Cheng, Qian (成倩) and Jiang, Liping (蒋莉萍). 2016. Duoyuan huanjing xia wangluo yuyan dui daxuesheng sizheng jiaoyu de yingxiang 多元环境下网络语言对大学生思政教育的影响 [On the Impact of Net-language on Ideological and Political Education in a Multicultural Environment]. *Yuwen Jianshe*《语文建设》[*Language Construction*] 2016(9): 75–76.

China Internet Network Information Center. 2012. *The Internet Timeline of China 1986–2003*. Retrieved from https://www.cnnic.com.cn/IDR/hlwfzdsj/201306/t20130628_40563.htm on 22/08/2021.

———. 2021. *47th Statistical Report on Internet Development in China*. Retrieved from https://www.cnnic.com.cn/IDR/ReportDownloads/202104/P020210420557302172744.pdf on 21/08/2021.

Crystal, David. 2001. *Language and the Internet*. Cambridge: Cambridge University Press.

———. 2009. *The Future of Language*. Oxon: Routledge.
———. 2011. *Internet Linguistics: A Student Guide*. New York: Routledge.
Deng, Jun (邓军) and Duan, Huiru (段慧如). 2004. Lun wangluo yuyan yu guifan wenti 论网络语言与规范问题 [On Netspeak and Some Problems of Regular Regularization *suo* 求索 [*Seeker*] 2004(08): 119–120.
Fang, Yan (方艳). 2005. Xinxing wangluo yuyan "*A dào VP*" geshi tanxi 新兴网络语言"A到VP"格式探析 [An Analysis of the New Netspeak Construction "*A dào VP*"]. Suzhou Daxue Xuebao 苏州大学学报 [*Journal of Soochow University*] 2018(6): 175–183.
Fu, Fuying (傅福英) and Lu, Songlin (卢松琳). 2010. Lun wangluo yuyan de jinhua ji tese—yi moyinlun wei shijiao 论网络语言的进化及特色——以模因论为视角 [An Analysis of the Features of Computer-Mediated Language–from the Perspective of Memetics]. Nanchang Daxue Xuebao (Renwen Shehui Kexue Ban) 南昌大学学报(人文社会科学版) [*Journal of Nanchang University (Humanities and Social Sciences)*],41(04):158–161.
Gao, Sai (高赛). 2010. Wangluo yuyan nengfou zhuanzheng? 网络语言能否"转正" [Can Netspeak Become Formal Language?]. *Guangming Ribao* 光明日报 [*Guangming Daily*] 2010-12-26(5).
He, Wei (何巍). 2009. Qiantan wangluo yuti fengge tezheng 浅谈网络语体风格特征 [On the Stylistic Features of the Genre of Net-language]. *Shanxi Shifan Daxue Xuebao (Zhexue Shehui Kexue Ban)* 陕西师范大学学报(哲学社会科学版) [*Journal of Shanxi Normal University for Nationalities (Philosophy and Social Sciences)*] 2009(S1): 354–356.
He, Ziran (何自然) and He, Xuelin (何雪林). 2003. Moyinlun yu shehui yuyong 模因论与社会语用 [Memetics and Social Usage of Language]. *Xiandai Waiyu* 现代外语 [*Modern Foreign Languages*] 2003(02):200–209.
Heylighen, F. (1998). What makes a meme successful? Selection criteria for cultural evolution. *Proceedings of the 15th International Congress on Cybernetics (Association Internat. de Cybernétique, Namur)*, 418–423.
Huang, Guowen (黄国文). 2005. Dianzi yupian de tedian 电子语篇的特点 [Characteristics of Electronic Discourse]. *Waiyu yu Waiyu Jiaoxue* 外语与外语教学 [*Foreign Languages and Their Teaching*] 2005(12): 1–5.
Huang, Ming (黄鸣). 2011. Cong moyinlun shijiao tanxi qiangshi moyin *bèi+xx* 从模因论视角探析强势模因"被XX" [An Analysis of *bèi+xx* from the Perspective of Memetics]. *Waiguo Yuwen* 外国语文 [*Foreign language and Literature*] 2011(S1): 26–28.
Hyland, K., and P. Tse. 2004. Metadiscourse in academic writing: A reappraisal. *Applied Linguistics* 25 (2): 156–177.
Hyland, K. (2005). Stance and engagement: a model of interaction in academic discourse. Discourse Studies, 7(2), 173–192. https://doi.org/10.1177/1461445605050365
Jiang, Haiyan (江海燕). 2016. *Luo* and *Da* zheng liuxing "咯""哒"正流行 [On the Two Prevailing Net Words: *Luo* and *Da*]. *Yuwen Jianshe*《语文建设》 [*Language Construction*] 2016(1): 67–68.
Jing, Song (劲松) and Qi, Ke (麒坷). 2000. Wangluo yuyan shi shenme yuyan? 网络语言是什么语言 [What Kind of Language is Net-language?]. *Diannao Yuwen* 电脑语文 [*Computer Chinese*] 2000(11):13–14.
Kang, Qingye (康庆业). 2014. Chuanboxue shijiao xia de wangluo yuyan yanjiu 传播学视角下的网络语言研究 [A Study on Net-language from the Perspective of Communication]. *Yuwen Jianshe*《语文建设》 [*Language Construction*] 2014(24):39–40.
Kuang, Xia (邝霞) and Jin, Zi (金子). 2000. Wangluo yuyan—yizhong xin de shehui fangyan 网络语言———一种新的社会方言 [Netspeak—a New Social Dialect]. *Yuwen Jianshe*《语文建设》 [*Language Construction*] 2000(08):21.
Lei, Dongping (雷冬平) and Li, Yaozhen (李要珍). 2013. Yuanhuayu he wangluo yuyan zhong de *XX*体 de pianzhang goushi yanjiu 元话语和网络语言中"XX体"的篇章构式研究 [On the Textual Construction of Metadiscourse and the Construction *XX*体 on the Internet]. *Xiangtan Daxue Xuebao(Zhexue Shehui Kexue Ban)*《湘潭大学学报(哲学社会科学版)》 [*Journal of Xiangtan University (Philosophy and Social Sciences)*] 2013(3): 107–111.
Li, Xin (立鑫). 1998. Tantan wangluo yuyan de jiankang wenti 谈谈网络语言的健康问题 [On the Health Problems of Netspeak]. *Yuwen Jianshe*《语文建设》 [*Language Construction*] 1998(01): 3–5.

Li, Xuping (李旭平). 2005. Yuyu lilun moshi xiade wangluo jiaoji he wangluo yuyan 语域理论模式下的网络交际和网络语言 [The Application of Register Theory to the Analysis of CMC and Cyber-language]. *Waiyu Dianhua Jiaoxue* 外语电化教学 [*Technology Enhanced Foreign Languages Education*] 2005(5): 37–40.

Li, N. (李楠). (2015). Jiyu dongtai yuliaoku de xinciyu jiance 基于动态语料库的新词语监测 [Neologism Monitoring Based on Dynamic Corpus]. *Haiwai Yingyu* 海外英语 [*Overseas English*], 14, 200–201.

Liang, Caihua (梁彩花). 2006. Wangluo yuyan—wangluo shidai shehui xinli de zheshe 网络语言——网络时代社会心理的折射 [Netspeak—the Reflection of Social Psychology in the Cyber Age]. *Guangxi Yike Daxue Xuebao* 广西医科大学学报 [*Journal of Guangxi Medical University*] 2006(S1): 371–373.

Liang, Yonghong (梁永红). Wangluo xinci *zhenxin* de yufahua 网络新词"真心"的语法化 [The Grammaticalization of Internet Neologism "Zhenxin"]. *Hanyu Xuexi* 汉语学习 [*Chinese Language Learning*] 2019(1): 94–101.

Liu, Haiyan (刘海燕). 2001. Wangluo yuyan《网络语言》[Netspeak]. *Beijing: Zhongguo Guangbo Dianshi Chubanshe* 北京:中国广播电视出版社 [*Beijing: China Radio Film and TV Press*].

Liu, Yali (刘亚丽). 2009. Wangmin wangluo yuyan de xinli xinsu tanxi 网民网络语言的心理因素探析 [The Mental Implication of Network Language]. *Henan Daxue Xuebao(Shehui Kexue Ban)* 河南大学学报(社会科学版) [*Journal of Henan University (Social Science)*] 2009(5): 131–136.

Liu, Yan (刘妍). 2015. *Yanzhi* ye neng baobiao? "颜值"也能爆表? [Can *Yanzhi* over the limit as well?]. *Yuwen Jianshe*《语文建设》[*Language Construction*] 2015(07):72–73.

Liu, Dongqing (刘冬青) and Shi, Jianping (施建平). 2010. Shishuo wangluo shengpici—yi *yín* weili 试说网络生僻词——以"奀"为例 [A Study of Internet Uncommon Words——with "yín" as an Example]. *Yuwen Jianshe*《语文建设》[*Language Construction*] 2010(6): 34–35.

Ma, Caixia (麻彩霞). 2018. Neihan de yanhua "内涵"的演化 [The Evolution of *Neihan*]. *Yuwen Jianshe*《语文建设》[*Language Construction*] 2018(36): 65–67.

Ministry of Education of the People's Republic of China & State Language Commission. 2007. 2007 nian zhongguo yuyan shenghuo zhuangtai baogao 2007 年中国语言生活状况报告 [Report on Language Situation in China]. Retrieved from http://www.moe.gov.cn/s78/A19/A19_ztzl/baogao/201202/t20120210_130364.html on 23/08/2021.

Peng, Jiaqiang (彭嘉强). 2001. Zunzhong chuangxin jiangjiu guifan—tantan wangluo liuxingyu de guifan 尊重创新讲究规范——谈谈网络流行语的规范 [To respect Innovation and be Standard: a Study on Regularization of NRegularization *Jianshe*《语文建设》[*Language Construction*] 2001(08):16.

Peng, Yubo (彭育波). 2001. Lun wangluo yuyan de jige tedian 论网络语言的几个特点 [On the Characteristics of Netspeak]. *Xiuci Xuexi* 修辞学习 [*Rhetoric Learning*] 2001 (04): 12–13.

Qi, Meng (祁萌) and Li, Zhengqi (李正琪). 2018. Daxuesheng wangluo yuyan jidai jinghua 大学生网络语言亟待净化 [Netspeak used by College Students Need to be Cleansed]. *Renmin Luntan* 人民论坛 [*People's Tribune*] 2018(10): 68–69.

Qin, Xiubai (秦秀白). 2003. Wangyu he wanghua 网语和网话 [Cyberspeak and Cybertalk]. *Waiyu Dianhua Jiaoxue* 外语电化教学 [*Technology Enhanced Foreign Languages Education*] 2003(06):1–6.

Qiu, Yewei (邱业伟) and Ji, Lijuan (纪丽娟). 2013. Wangluo yuyan baoli gainian renzhi jiqi qinquan zeren goucheng yaojian 网络语言暴力概念认知及其侵权责任构成要件 [The Concept of Cyber Language Violence and the Elements of Its Tort Liability]. *Xinan Daxue Xuebao (Shehui Kexue Ban)*《西南大学学报(社会科学版)》[*Journal of Southwest University (Social Sciences Edition)*] 2013(1): 38–43.;173–174.

Shan, Xiong (闪雄). 2000. Wangluo yuyan pohuai hanyu de chunjie 网络语言破坏汉语的纯洁 [Netspeak is Damaging the Purity of Chinese Language]. *Yuwen Jianshe*《语文建设》[*Language Construction*]2000(10):15–16.

Shan, Wei (单威). 2015. *Renxingti* weihe liuxing "任性体"为何流行 [On the Prevalence of *rènxìng*-construction]. *Yuwen Jianshe*《语文建设》[*Language Construction*] 2015(07):70–72.

Shao, Yanjun (邵燕君). 2018. Pobi shu: wangluo wenhua guanjian ci 破壁书:网络文化关键词 [A Book to Break the Wall: Key Words about Network Culture]. *Shenghuo Shudian Chuban Youxian Gongsi* 生活书店出版有限公司 [*SDX Joint Publishing Company*] 2015(07):70–72.

Shen, Fenfang (申芬芳). 2004. Yuwen jiaoxue ruhe yingdui wangluo yuyan de fumian yingxiang 语文教学如何应对网络语言的负面影响 [What will We Do in Chinese Teaching in Response to the Negative Effects of Netspeak]. *Yuwen Jianshe*《语文建设》[*Language Construction*] 2004(05):30.

Shi, Chunhong (施春宏). 2010. Wangluo yuyan de yuyan jiazhi he yuyanxue jiazhi 网络语言的语言价值和语言学价值 [Web Language as a Language Variety and a Linguistic Issue]. *Yuyan Wenzi Yingyong* 语言文字应用 [*Applied Linguistics*] 2010(3): 70–80.

Shi, Wei (施维). 2016. Gaige kaifang zhijin chengdu bendi baozhi cihui bianyi yanjiu 改革开放至今成都本地报纸词汇变异研究 [On the Variations of the Vocabulary in the Local Newspapers of Chengdu since the Reform and Opening up]. *Zhonghua Wenhua Luntan* 中华文化论坛 [*Journal of Chinese Culture*] 2016(06):118–128.

Tang, Yizheng (唐益政). 2013. Dui daxuesheng wangluo yuyan he gaoxiao sixiang zhengzhi gongzuo guanxi yanjiu 对大学生网络语言和高校思想政治工作关系研究 [An Analysis of Relations between Netspeak used by College Students and the Ideological and Political Education in College]. *Yuwen Jianshe*《语文建设》[*Language Construction*] 2013(9): 53–54.

Tian, Yongfang (田永芳). 2018. Daxuesheng wangluo liuxing yu shiyong ji taidu xianzhuang de diaocha fenxi—yi shanxi daxue shangwu xueyuan wenke xuesheng wei li 大学生网络流行语使用及态度现状的调查分析——以山西大学商务学院文科学生为例 [On the Use of and Attitudes towards Netspeak—with Students of Arts at Business College of Shanxi University as an Example]. *Jishou Daxue Xuebao(Shehui Kexue Ban)*《吉首大学学报(社会科学版)》[*Journal of Jishou University (Social Sciences)*] 2018(12): 199–202.

Ungerer, F., and H.-J. Schmid. 2013. *An Introduction to Cognitive linguistics*. Routledge.

Wang, Yanyan (王彦彦). 2010. Wangluoyu *beiju* jiqi yansheng ciju de renzhi yanjiu 网络语"杯具"及其衍生词句的认知研究 [A Cognitive Study on the netspeak *bei ju* and its Derivatives]. *Dangdai Xiuci Xue* 当代修辞学 [*Contemporary Rhetoric*] 2010(1): 75–79.

Wang, Zhijun (王志军). 2018. Daxuesheng wangluo yuyan jinghua jizhi tanjiu 大学生网络语言净化机制探究 [Self-purifying Mechanism of Netspeak Use by College Students]. *Xuexiao Dangjian yu Sixiang Jiaoyu* 学校党建与思想教育 [*The Party Building and Ideological Education in Schools*] 2018(8): 42–44.

Wu, Shuyan (吴术燕). 2014. XXti yuyan fengge renzhi jiexi "XX体" 语言风格认知解析 [A Cognitive Analysis of the Language Style of *XXti*]. *Yuwen Jianshe* 语文建设 [*Language Construction*] 2014(30): 27–28.

Xiao, Feng (肖峰) and Dou, Changyu (窦畅宇). 2016. Wangluo shifan de zhexue fenxi 网络失范的哲学分析 [A Philosophical Analysis of Network Misconduct]. *Lilun Shiye*《理论视野》[*Theoretical Horizon*] 2016(1): 45–49.

Xie Qun (谢群) and Xu Jianjun (徐建军). 2018. Gaoxiao fudaoyuan wangluo sixiang zhengzhi jiaoyu huayuquan de jian'gou—jiyu wangluo yuyan shijiao 高校辅导员网络思想政治教育话语权的建构——基于网络语言视角 [The Construction of College Instructors' Discourse of Political and Ideological Education in the Network]. *Xiangtan Daxue Xuebao(Zhexue Shehui Kexue Ban)*《湘潭大学学报(哲学社会科学版)》[*Journal of Xiangtan University (Philosophy and Social Sciences)*] 2018(1): 148–153.

Xing, Nuhai (邢怒海). 2014. Qianlun wangluo ciyu "maosi" de tedian 浅论网络词语"貌似"的特点 [On the characteristics of netword màosì]. In *Jiaozuo shifan gaodeng zhuanke xuexiao xuebao* 焦作师范高等专科学校学报 [*Journal of Jiaozuo Teachers College*] 2014 (2):12–14.

Xu, Zhongning (许钟宁). 2004. Wangluo ciyu bianyi de yuyan wenhua jiexi 网络词语变异的语言文化解析 [A Study of Internet Language Variation on Chinese Language and Culture]. *Xiuci Xuexi* 修辞学习 [*Rhetoric Learning*] 2004(06): 58–60.

Yang, Jian (杨健). 2013. *Diaosi* dengtangshushi—qianxi yige wangluo reci ruhe jinru zhuliu meiti "屌丝"登堂入室——浅析一个网络热词如何进入主流媒体 [On the Reasons for the

Prevalence of Internet Buzzwords in Mainstream Media: A Case Study of the Chinese Internet Buzzword *Diaosi*]. *Bianji Xuekan* 编辑学刊 [*View on Publishing*] 2013(02):88–92.

Yao, Xiaodan (姚晓丹). 2011. Duoyanghua rang yuyan geng you huoli 多样化让语言更有活力 [Diversity: the Reason for a More Dynamic Language]. *Guangming Ribao* 光明日报 [*Guangming Daily*].

Yi, Wen'an (易文安). 2000. Wangluo shishang cidian 网络时尚词典 [Network Vogue Dictionary]. *Hainan: Hainan Chubanshe* 海南:海南出版社 [*Hainan: Hainan Publishing House*].

Yu, Genyuan (于根元). 2001a. Zhongguo wangluo yuyan cidian《中国网络语言词典》[Chinese Netspeak Dictionary]. *Beijing: Zhongguo Jingji Chubanshe* 北京:中国经济出版社 [*Beijing: Economic Press China*].

——— (于根元). 2001b. Wangluo yuyan gaishuo《网络语言概说》[Introduction to Netspeak]. *Beijing: Zhongguo Jingji Chubanshe* 北京:中国经济出版社 [*Beijing: Economic Press China*].

Yu, Ping (禹平) and Zhang, Xin (张鑫). Wenhuaxue shijiao xiade wangluo ciyu kaocha 文化学视角下的网络词语考察 [An Analysis of Netspeak from the Cultural Perspective]. *Xuexi yu Tansuo*《学习与探索》[*Study and Exploration*] 2019(6): 169–173.

Yuan, Zhoumin (袁周敏) and Han, Pugen (韩璞庚). 2018. Wangluo yuyan shiyu xiade wangluo wenhua anquan yanjiu 网络语言视域下的网络文化安全研究 [An Analysis of Cyber Cultural Security from the Perspective of Netspeak]. *Waiyu Jiaoxue*《外语教学》[*Foreign Language Education*] 2018(1): 39–43.

Zhai, Huifeng (翟会锋). 2017. *De* wangluo bianti jiqi renzhi dongyin kaocha "的"的网络变体及其认知动因考察 [An Exploration of Variants of *De* in Web Language and their Cognitive Motivations]. *Yuyan Jiaoxue yu Yanjiu* 语言教学与研究 [*Language Teaching and Linguistic Studies*] 2017(2): 104–112.

Zhang, Huijuan (张会娟). 2006. Wangluo yuyan de tedian jiqi dui wangmin xinli de tixian 网络语言的特点及其对网民心理的体现 [On the Characteristics of Netspeak and their Reflections in the psychology of netizens]. *Lilun Xuekan* 理论学刊 [*Theory Journal*] 2006(11): 125–126.

Zhang, Yunhui (张云辉). 2007. Wangluo yuyan de cihui yufa tezhen 网络语言的词汇语法特征 [On the Lexicogrammatical Characteristics]. *Zhongguo Yuwen* 中国语文 [*Studies of the Chinese Language*] 2007(6):531–535.

Zhang, Ting (张挺) and Wei, Hui (魏晖). 2011. Hulianwang huanjing xia yuyan wenzi yuqing jiance yu shizheng yanjiu 互联网环境下语言文字舆情监测与实证研究 [The Empirical Study of the Monitoring of the Public Sentiment about Language on the Internet]. *Yuyan Wenzi Yingyong*《语言文字应用》[*Applied Linguistics*] 2011(2): 6–12.

Zhong, Jiya (钟吉娅). 2001. Wangluo yuyan de tedian he guifan tanxi 网络语言的特点和规范探析 [An Analysis of the Characteristics and Regularization of Netspeak]. *HeRegularizationue Xuebao(Zhexue Shehui Kexue Ban)* 河南师范大学学报(哲学社会科学版) [*Journal of Henan Normal University (Philosophy and Social Sciences)*] 2001(06):126.

Zhou, Liying (周丽颖). 2013. Geili cheng meijie reci tanxi "给劲"成媒介热词探析 [An Analysis of the Medium Hot Word Geili]. *Zhongguo Chuban* 中国出版 [*China Publishing Journal*] 2013 (11):54-56.

Zhou, Bin (周彬). 2018. Wangluo changyu: wangluo yuyan、fuhao baoli yu huayu quan zhangkong 网络场域: 网络语言、符号暴力与话语权掌控 [Netdoms: Netspeak, Symbolic Violence, and controlling of discursive power]. *Dongyue Luncong*《东岳论丛》[*Dongyue Essays*] 2018(8): 48–54.

Zou, Lizhi (邹立志). 2007. Cong yuyan xitong benshen kan wangluo yuyan de guifan 从语言系统本身看网络语言的规范 [On the Regularization of Netspeak fromRegularizationem itself]. *Xiuci Xuexi* 修辞学习 [*Rhetoric Learning*] 2007(3): 61–64.

Zuo, Haixia (左海霞) and Yao, Ximing (姚喜明). 2006. Xiuci shijiao xiaode wangluo yuyan 修辞视角下的网络语言 [The Analysis of Net-language from Rhetorical Perspective]. *Waiyu Dianhua Jiaoxue* 外语电化教学 [*Technology Enhanced Foreign Language Education*]. 2006 (1): 27–31.

Part II
Interactions of Sinitic Languages

Interactions of Sinitic Languages: Introduction

12

Zhengdao Ye

Contents

Introduction	299
Chapter Summaries	300
References	302

Abstract

This chapter provides an overview of the selection of the chapters included in Section 3 of *The Palgrave Handbook of Chinese Language Studies*. The chapters present a broad picture of the dynamic interactions within Sinitic languages and of those between Sinitic languages and non-Sinitic languages, and explore how such interactions have influenced the formation of diverse Sinitic languages and of the languages with which they have been in contact.

Keywords

Sinitic languages · Language contact · Altaic languages · Mixed language · Chinese diaspora

Introduction

The selection of the five chapters in this section of the handbook aims to achieve two goals. First, it presents a broad picture of the dynamic interactions within Sinitic languages and of those between Sinitic languages and non-Sinitic languages. Sec-

---

Z. Ye (✉)
School of Literature, Languages and Linguistics, The Australian National University, Canberra, Australia
e-mail: Zhengdao.Ye@anu.edu.au

ond, it explores how such interactions have influenced the formation of diverse Sinitic languages and of the languages with which they have been in contact. In some cases, new languages were created, including those of a mixed form.

Chapter Summaries

The first two chapters provide overviews of language contact situations in China. While Fuxiang Wu and Yang Huang's chapter surveys contact-induced changes in the languages of Southern China, Yonglong Yang and Jingting Zhang's chapter focuses on how the grammar of Mandarin Chinese, especially its northern varieties, has evolved as a result of its contact with surrounding non-Sinitic languages. Wu and Huang's chapter begins with a sketch of the history of language contact research in China and a review of relevant studies in recent decades. The authors point out that, despite a large amount of literature on language contact situations in China and theoretical advancements in the study of contact-induced changes in the West, many aspects of the contact scenarios of Sinitic languages, particularly those concerning changes in grammatical structures and patterns, are still not fully understood or well accounted for. Wu and Huang then offer a detailed and comprehensive overview of different types of structural changes induced by language contact between Sinitic languages and a wide range of non-Sinitic languages (such as the Tai-Kadai, Miao-Yao, and Tibeto-Burman languages). In the light of the rich linguistic evidence presented, the authors evaluate the theory of contact-induced grammaticalization (e.g., Heine and Kuteva 2003 Matras 2009; Matthews and Yip 2009) as an alternative model for analyzing and accounting for changes in structural patterns triggered by language contact. The authors also consider the notion of "grammaticalization areas" in the broad context of language contact situations in the areas in focus, and discuss a number of important questions regarding future directions for studies in language contact in China.

Yang and Zhang's paper provides an overview of how the grammar of (Mandarin) Chinese, such as word order and postpositions, has evolved over time as a result of its direct or indirect contact with non-Sinitic languages. The authors stress the importance of a large-scale corpus of language materials to address two key research questions arising from language contact phenomena throughout Chinese history. One concerns the exact nature of contact-induced grammatical changes from ancient times to the present; the other is about understanding the processes behind this evolution. The authors identify four major periods as possessing rich linguistic materials, of which two periods have attracted much attention from Chinese scholars. The chapter provides detailed discussions of the evolution of a wide range of grammatical forms that took place during those two periods. One serves as an example of direct contact of Mandarin with northern Altaic languages, particularly during the Liao (916–1125 CE), Jin (1115–1234 CE), and Yuan (1271–1368 CE) Dynasties. During the Yuan Dynasty, a language with mixed Chinese-Mongolian elements known as "Han-er Yanyu" was also formed in Northern China. The other is an example of indirect contact with Sanskrit and Pali through

12 Interactions of Sinitic Languages: Introduction

the translations of Buddhist scriptures over a period of a thousand years from the late Eastern Han (202 BCE–220 CE) through to the Tang (618–907 CE) and Song (960–1279 CE) dynasties. The authors consider the role of language shift in the formation of Mandarin Chinese, and point to two periods of intense language contact – the Qing Dynasty (1644–1911 CE) and the late-Qing and early Republican period – that require further research.

The next chapter by Wilkinson Daniel Wong Gonzales is the first large-scale survey of the complex language contact situations in the Philippines, where many varieties of Sinitic languages co-exist with Philippine-based languages. It focuses on the dynamics between the former, the languages of (heritage and homeland) Chinese groups, and the latter, the languages of the historically indigenous population. Applying Haugen's (1971) framework on linguistic ecology and using oral and written data sampled from twelve linguistic varieties in three major cities across the archipelago, the chapter details the processes of Sinicization, Filipinization, and Sino-Philippine language creation, showing that languages co-existing in the same linguistic ecology can affect each other in different ways, depending on the socio-historical context in which such interactions take place. The languages that the chapter looks at closely include three varieties of Hokkien, two varieties of Cantonese, two varieties of Mandarin, Chinese Spanish Pidgin, and Chinese Tagalog Pidgin, as well as the mixed-language Philippine Hybrid Hokkien, known as *Lánnang-uè* (lit. "our people speech"). The four speaker groups that this chapter focuses on are Filipinos, Sangleys, Lannangs, and Mainland Chinese (new immigrants and sojourners). The chapter considers a number of factors that may account for the asymmetric ways the languages under discussion influence each other. They include differences in the dynamics between the groups in contact and social motivations.

The next two chapters look more specifically at languages belonging to the Yue and Min dialect groups. Cantonese is the representative variety of Yue Chinese and a major Sinitic language spoken among the Chinese diaspora. In his chapter on the expansion of Cantonese over the last two centuries, Hilário de Sousa looks into the formation of many what he calls "enclave" Cantonese varieties resulting from the migration of Cantonese speakers, in particular the varieties known as Nanning Cantonese and Hong Kong Cantonese. De Sousa examines a range of factors that caused the divergences amongst Cantonese varieties, including language policy, and points to the different language contact environment as the primary contributing factor. The author also discusses how written Cantonese varies from one jurisdiction to another and finds a continuum of registers.

Ruiqing Shen's chapter discusses interactions between Min and other Sinitic languages, with a particular focus on genetic inheritance and areal patterns. It first provides a general introduction to the Min group and its internal classification, before reviewing studies on interactions between Min and four other Sinitic groups, namely, Wu, Hakka, Waxiang, and Gan. The chapter shows that, while all four dialect groups share certain lexical and/or phonological features with Min, these commonalities are of a different nature. For example, the interactions between Min and Gan involve contact-induced areal patterns. However, features shared between Min and Waxiang are most likely due to shared retention, while those shared between Min and Southwestern Wu and between Min and Hakka include some innovations as well. The chapter also discusses implications of similarities and differences in patterns of interactions for the linguistic history of Min, and considers directions for future studies, particularly with respect to the interactions between Inland Min and Hakka and those between Coastal Min and Southwestern Wu.

All the chapters are rich in data, illustrating the dynamic and complex language contact situations and the diversity of Sinitic languages (see also Chappell 2001, 2015). They highlight the fact that the English term "Chinese" fails to capture the diversity of Sinitic languages.

References

Chappell, Hilary. 2001. Language contact and areal diffusion in Sinitic languages: Problems for typology and genetic affiliation. In *Areal diffusion and genetic inheritance: Problems in comparative linguistics*, ed. Alexandra Aikhenvald and R.M.W. Dixon, 328–357. Oxford: Oxford University Press.
———, ed. 2015. *Diversity in Sinitic languages*. Oxford: Oxford University Press.
Haugen, Einar. 1971. The ecology of language. *The Linguistic Reporter* 25 (25): 19–26.
Heine, Bernd, and Tania Kuteva. 2003. On contact-induced grammaticalizaion. *Studies in Language* 27 (3): 529–572.
Matras, Yaron. 2009. *Language contact*. Cambridge: Cambridge University Press.
Matthews, Stephen, and Virginia Yip. 2009. Contact-induced grammaticalization: Evidence from bilingual acquisition. *Studies in Language* 33 (2): 366–395.

Contact-Induced Change in the Languages of Southern China

13

Fuxiang Wu and Yang Huang

Contents

History of Research on Language Contact in China: Setting the Scene	304
The Changing Languages of Southern China	309
Contact-Induced Grammaticalization: An Alternative Model	310
Contact-Induced Changes from Chinese to Minority Languages of Southern China	311
Contact-Induced Changes from the Minority Languages of Southern China to Chinese	321
Areal Linguistics and Grammaticalization Area	322
Future Directions for Language Contact Study in China	324
References	327

Abstract

The aim of this chapter is to trace the development of contact-induced language change as a subfield of linguistic research in China. We first provide a brief introduction to the history of the field and present relevant research from recent decades. Drawing on data from existing language contact situations, we discuss contact-induced changes among languages of southern China. Findings and analytical approaches will contribute to future research on the contact-induced changes occurring between Chinese and the minority languages of southern/southwestern China.

Keywords

Language contact · Grammatical change · Languages of southern China

F. Wu (✉)
Beijing Language and Culture University, Beijing, China
e-mail: wufuxiang100@126.com

Y. Huang
Southwest Jiaotong University, Chengdu, China
e-mail: elvishuang@swjtu.edu.cn

© The Author(s), under exclusive licence to Springer Nature Singapore Pte Ltd. 2022
Z. Ye (ed.), *The Palgrave Handbook of Chinese Language Studies*,
https://doi.org/10.1007/978-981-16-0924-4_27

Abbreviations and Conventions

1sg	1st person single
1pl	1st person plural
2sg	2nd person single
2dl	2nd person dual
3sg	3rd person single
3pl	3rd person plural
ATTR	attributive
CL	classifier
COMP	complement
DIR	directional marker
DIST	distal
EXP	experiencer
EXH	exhaustion particle
GEN	genitive
GNO	gnomic evidential
IMPV	imperfective
INCL	inclusive
NEG	negative
NMLZ	nominalizer
NONPST	Non-past
PFV	perfective
PROX	proximal
PST	past
PTC	particle
Q	question marker
REL	relativizer
SEME	semelfactive
>	becomes/grammaticalizes to/develops into
A< > B	A corresponds to B

History of Research on Language Contact in China: Setting the Scene

Language change is simply a fact of life; it cannot be prevented or avoided (Campbell 2013: 3). When identifying the causes of language change, historical linguists generally focus on internal factors that highlight the relationships between very close languages and employ theories such as Schleicher's (1853) "tree model" (*Stammbaumtheorie*), or Schmidt's (1872: 27) "wave model" (*Wellentheorie*). Contact linguists, however, focus specifically on external factors such as borrowing, interference, metatypy, and other external causes (Weinreich 1953; Ross 1999; Thomason 2001). Since Weinreich's (1953) groundbreaking study, the field of language contact has witnessed profound theoretical and empirical developments (see Weinreich et al. 1968; Heath 1978; Thomason and Kaufman 1988; Aikhenvald 2003; Heine and Kuteva 2005; and Matras 2009). In the western world, the study of language contact draws on theoretical linguistics, second language acquisition, synchronic/typological description, contact-induced internal change, social motivation, and interdisciplinary methods that connect language evolution with ecology and biological evolution (Mufwene 2001).

Language contact is by no means an unfamiliar topic in the history of linguistic research in China. In the early 1920s, renowned sinologist Liang Qichao 梁启超 presented ten different uses of language apparent in Chinese Buddhist literature (Liang 1920). Following Liang's work, the famous Chinese scholars, Wang Jingru 王静如 and Lü Shuxiang 吕叔湘 (Yü et al. 2010) highlighted the effects of language contact on certain words in Middle Chinese and their work has a significant impact on subsequent studies of language contact in China. Zhu (1992), as well as Zhu and Zhu (2006), provide an extensive overview of the development of hybrid Buddhist Chinese. As a typical mixed language, this "hybrid" Buddhist Chinese is heavily influenced by Sanskrit, not only in its lexicon but also in its grammatical features. Studies of hybrid Buddhist Chinese have a great impact on the text-centric observation of language contact in ancient China.

It was not until recently that some western theories of language contact have attracted the attention of Chinese linguists. In Wang's (1969, 1979) papers, internal changes are understood as abrupt changes that require a long time span to diffuse; the gradualness of this diffusion process means that this process may be interrupted if one or more other changes compete against it and affect the same words during the course of diffusion. Despite several misunderstandings of some second-hand data, Wang's papers on lexical diffusion shed new light on the study of language contact in China.

Yue-Hashimoto (1976) observes an affinity between southern Chinese dialects and the Tai languages. She suggests that the present correlation between aspiration and tone in the Yue and dialects Min may be related to the phonetic situation in an old version of the Zhuang language, in which unaspirated stops and affricates occur in the so-called yang-tone categories (*yangsheng lei* 阳声类).

Yü (2004) discusses contact in medieval Chinese and introduces some interesting instances of contact during the Wei and Jin dynasty, while underlining the differences between genetic relations and language contact as they pertain to language change. The author examines a variety of contact mechanisms that gave rise to the hybridized grammatical structures of ancient Chinese. In a later collection (Yü et al. 2010), Yü brings together a range of articles by Chinese and international scholars devoted to the issue of language in medieval Chinese that look at a broad range of genres and texts. This collection includes a study by Ōta Tatsuo 太田辰夫 of the Han'er 汉儿 language (Yü et al. 2010: 1–23), a Chinese translation of Eric Zürcher's important early study of East Han vernacular elements in Chinese Buddhist texts (Yü et al. 2010: 24–48), Hashimoto Mantarō's 桥本万太郎 study of language contact in northern Chinese (Yü et al. 2010: 66–81), and a study by Seishi Karashima 辛岛静志 of Sanskrit-inspired Buddhist Chinese (Yü et al. 2010: 133–164). The volume also contains several focused case studies of particular issues related to language contact – such as a study by

Jiang Lansheng 江蓝生 of how contact shaped comparative constructions (Yü et al. 2010: 165–176), a study by Zu Shengli 祖生利 on multi-functional particles in Chinese inscriptions (Yü et al. 2010: 256–278), and research on the emergence of the "take" (*qu* 取) disposal form in ancient Chinese (Yü et al. 2010: 335–346). This collection is a pioneering work in the literature that has done much to illuminate how contact theory can be put into practice to analyze linguistic phenomena concerning ancient Chinese.

Ho (2004: 77–88) discusses the process of language contact whereby voiceless aspirated consonants become voiced aspirated consonants in modern Chinese dialects. Ho illustrates this process using data from the period of contact hundreds of years ago between the Yongxin dialect 永兴方言 and the Southwest Mandarin of Sichuan (SWM), yielding a process exemplified by the shift as pre-contact Yongxin *dz, zd* and the *tsh* > Yongxin *dzh*, and pre-contact Yongxin *b* and *ph* > Yongxin *bh*. A more general description of phonological contact among various Chinese dialects is provided by Wang (2005: 19–38). Wang distinguishes three types of contact scenarios based on data from earlier studies: contact between equally dominant dialects, the influence of dominant dialects on subordinate ones, and the influence of contact-induced varieties via bilingual speakers. The first type of contact scenario concerns the mutual influences of dialects on each other's isoglosses. For instance, the distinction between /n/~/l/ disappears in both the Jintan dialect 金坛方言 and the Liyang dialect 溧阳方言, as a result of Jianghuai Mandarin's 江淮官话 influence on the Wu dialect. The third scenario arises when bilingual speakers transfer aspects of their first language onto their imperfectly acquired second language. This is attested in the Linwu dialect 临武话, which has inherited the Mandarin phonemes /ŋ, ei, iau, yaŋ, yeŋ/ and the Tuhua 土话 phonemes /ã, ŋ, ĩ, iai, ũ, uã, ỹ/.

In addition to the changes in phonology that occur through language contact, a number of scholars have also paid great attention to the area of lexical borrowing. Norman and Mei (1976) use data from modern Austroasiatic languages to examine word correspondences between Proto-Austroasiatic and Old Chinese. They specifically focus on seven words from Old Chinese and Austroasiatic. They argue that these words in Old Chinese may bear some relationship to those in the Austroasiatic family (The arrow indicates the corresponding reconstruction available in Baxter and Sagart (2014): zhá札<>*tset "to die" → <>*s-qˤrət; sōu獀 <>*siô(g) "dog;" jiāng 江 <> *krong/kang/chiang "river" → *kˤroŋ; wèi蠚 <>*riwəi/jwi/wei "fly;" hǔ 虎 <>*kla(g)/χuo/hu "tiger"→ *qʰˤraʔ; yá牙 <> *ngra/nga/ya "tooth" → *m-ɢˤ<r>a; nǔ弩 <>*na/nuo/nu "crossbow").

In his classic paper, "Language Diffusion on the Asian Continent," Hashimoto (1976: 50) pioneers the study of language diffusion (areal diffusion) in China. He notes that: "[t]he structural diversity of Chinese languages can be best accounted for as the result of the diffusion of Altaic and Austroasiatic languages. Without consideration of these surrounding languages, these typological diversities of modern Chinese dialects must be hard to comprehend." Aside from tonal distinctions and syllable structure, Hashimoto also argues that the diffusion of dialects from north to south also shows a number of geographical characteristics, both morphological and syntactic.

Zhu (1990) examines the range of neutral interrogative constructions in Chinese dialects. He finds that Northern Chinese dialects use [VP-neg-VP] interrogative forms, in contrast to their Southern counterparts (e.g., the Yangzhou, Suzhou, and Shantou dialects), whose interrogative constructions are reported as either [ADV-VP] or [VP-neg-particle]. According to Zhu, the [VP-neg-VP] construction first emerged in the Tang dynasty. The [VO-neg-V] and [VO-neg-VO] structures are both used in today's Chinese dialects: [VO-neg-V] and [VO-neg-VO] are used in Beijing and Henan; the [V-neg-VO] and [VO-neg-VO] structures are used in Shandong and Fujian; and the [V-neg-VO] structure is used in Hubei. Zhu's paper was the first to investigate grammatical changes in negative constructions among various Chinese dialects from a contact/diffusion perspective.

T'sou (2001) concerns himself with the question of why word borrowing occurs. He begins by sketching a broad view of the lexical diffusion of Sino-Japanese loanwords (足 "foot" → *ashi*, 晚 "late" → *ban*, 袋 "bag" → *fukuro*, 东 "east" → *taba*, etc.) and Sino-Vietnamese morphemes. T'sou identifies two approaches to lexical importation: the "narrow approach" and the "broad approach." The narrow approach is supported by many scholars and primarily concerns phonetically adapted lexical items rather than semantically adapted items; the broad approach permits a much more broad-ranging analysis of cultural impact in language contact situations and suggests several useful indices with which to measure the extent and nature of such impact – an impact which is not easily quantified.

Inspired by new theories of contact-induced grammatical change, the twenty-first century has also witnessed a rapid expansion in the study of language contact in China. Some scholars have attempted to integrate these new models into their understanding of contact-induced grammatical changes between Chinese dialects and minority languages.

Chappell (2001) points out the need to take processes of diffusion and layering and other language-contact phenomena such as convergence, metatype, and hybridization into account, when reconstructing the history of the Chinese dialects. She proposes three main outcomes of language contact situations (stratification, hybridization, and convergence) for Chinese dialects and provides examples of particular contact scenarios that illustrate these three outcomes. Examples of morphological change in the Southern Min dialect in Taiwan (e.g., *lâng* 儂, *tai jîn* 大人) show stratificational distinctions in the lexicon between native colloquial morphemes and literary forms, with native and borrowed morphemes being hybridized into a single new form. Using examples drawn from the relative clause construction and the mixed comparative construction in Hong Kong Cantonese, the author demonstrates how these two syntactic structures can coexist in parallel use in different registers.

Matthews' (2006) paper focuses on the case of Cantonese and the minority linguistic stocks that have been in contact with it. He first lists a wide range on imported items in Cantonese that hint at a substrate influence from Tai languages on the Yue dialect (e.g., suffix -lou^2 "guy,", postverbal particle $saai^3$ "all," etc.). He then compares some specific features of Cantonese with the descriptions of Southeast Asian languages to illustrate that the inverted double object construction [V-DO-IO], the "surpass" comparative construction, the postverbal "acquire" modal, the productive "expressive" pattern, and the multifunctional classifiers in Cantonese may have much in common with the Tai language. These areal, typological characteristics mark Cantonese as being quite similar to the Tai-Kadai and Miao-Yao languages.

Language contact in China is prevalent, from the far north to the south and across the southwest (LaPolla 2001, 2010). Ordinary contact gives rise to convergent linguistic elements in different languages, while extensive contact accelerates the formation of hybrid phonological structures and mixed languages. Under Chinese language influences, the southern Qiangic variants (Tibeto-Burman) 南部羌语 have lost syllable stress and acquired four tones. Although this innovation is questioned in Sims (2020), tonogenesis in Southern Qiang is thought to have emerged through contact-induced reanalysis of older stress-accent patterns (Sun 1981: 7).

In another vein, some languages have lost their typological characteristics, shifting toward the structures of other languages with which they are in contact. The Huihui language 回辉话 has shifted from being an agglutinating language to an isolating type of language, and increased its number of monosyllables (Zheng 1997: 13). According to Chen (1982) and Li (1983), Wutun 五屯话 is significantly different from other members of the Chinese family and displays a fascinating mixture of Tibetan and Chinese features in every aspect of its structure. The language is strictly post-positional, lacks co-verbs, and its causative constructions are marked by a causative suffix -$k\partial$ instead of the resultative compound. Wutun phonology exhibits both Chinese and Bodic features, comprising a main vowel (V) surrounded by up to four optional, non-syllabic segments: an initial (C), a medial (M), a pre-initial (H), and a final (F); syllables take the form (CMVF) in Chinese lexical items, and (HCVF) in Tibetan lexical items (Janhunen et al. 2008: 25–26).

Waxiang 瓦乡话, an unclassified Chinese dialect found in Western Hunan, has attracted the attention of several linguists on account of some of its interesting features as a mixed language. On one hand, Waxiang has enjoyed a prolonged history of contact with Hmong-Mien 苗瑶语 and Tujia 土家语; while on the other hand, Southwest Mandarin has imposed a consistent influence on its language system, giving rise to a leveling of the diversity of linguistic features in these two dialects (Wu and Shen 2009).

Another mixed language, Daohua 倒话, found in Western Sichuan, is a hybrid of Southwest Mandarin and Tibetan. A-Tshogs (2003) reports that Daohua is spoken by the Han people living in the eastern part of the Qinghai-Tibetan plateau (in the western Sichuan Basin). Without any clear indication of its genesis, Daohua demonstrates a Chinese-style lexicon and a Tibetan-style grammatical structure. Almost all of the basic words in Daohua come from Chinese, while cultural words seem to be borrowed from Tibetan. Grammatically, Daohua resembles Tibetan both in word order and morphological coding strategy, and as a typical SOV language, it employs various markers to encode distinct grammatical categories. The linguistic profile of Daohua exemplifies the sort of mixed language that can result from profound contact between the Chinese and Tibetan strata.

As Huang (2005) states, despite the possibility of genetic relationships, the isomorphic features in most languages in China can be accounted for by areal influence and that numerous languages, both genetically affiliated and nonaffiliated, are prone to exhibiting linguistic convergence in a specific geographic area. Language contact is the mechanism that leads to this convergence.

The Changing Languages of Southern China

Apart from the general contact scenarios illustrated above, recent studies have provided insights into contact-induced changes by exploring new data from a wide range of languages spoken in southern China. Characterized by spectacular landscapes, charming natural scenery, and diverse ethnic customs, the region of southern China was originally inhabited by a mixture of tribal groups known to the Chinese as the Hundred Yue (Baiyue 百越), who were first absorbed into China during the Qin Dynasty (221–206 BCE). During the early Spring and Autumn period (770–476 BCE) and the Warring States 战国 (476–221 BCE), when the Chu state 楚国 was expanding its sovereignty to the border of the South China Sea, indigenous people in the south were in almost constant contact with Han migrants from the central plains.

While the linguistic profiles of some regions in southern China were described in a number of works before F. K. Li (1977), a comprehensive discussion of language contact did not come about until the 1970s. Xing (1979) compares the polyfunctional particles 了 -$l\varepsilon{:}u^4$ and 著 -ju^5 in the Tai group (especially Zhuang and Dong) with their counterparts in Chinese, concluding that these two particles reveal similar grammatical properties in Chinese and Zhuang-Tai Kadai, while noting that the cause of this isomorphic development is still unknown.

Ou'yang (1995) and Li (2000) provide examples of several similar phonological, lexical and grammatical traits in the Zhuang and Yue dialects in Guangxi, and posit that, with the spread of the Yue dialect from the Pearl River Delta to western Guangxi, these two languages have been in intensive contact ever since the Qing Dynasty.

Zeng (2003) presents a systematic survey of the Chinese loanwords in the Tai-Kadai languages of southwest China and offers new strategies for disentangling overlapping layers of various loanwords. Based on fieldwork data from Guizhou and Guangxi, Zeng identifies four historical layers of Chinese loanwords: (a) loans from Old Chinese, which are limited to a small number of basic words and morphemes; (b) loans from Middle Chinese, whose phonological properties in the Tai-Kadai languages correspond to the rhyme scheme as found in the *Qieyun* 切韵 dictionary; (c) contemporary loans in Sui水语, mostly borrowed from the Pinghua 平话 and Libo dialects 荔波话 (Southwest Mandarin); (d) modern loans, individually borrowed from the Libo and Guiyang dialects 贵阳话 or undergoing a transition (For example, 经济 "economy": Guiyang dialect $tein^1\ ci^4$ < > Libo dialect $tein^1 tci^4$ < > Shui $tin^3 ti^1$).

Based on historical records, demographic statistics, and traces of migration, Hong (2004: 104–120) proposes two contact categories – geographic adjacency and cultural interactions (education, matrimony, trade, etc.) – to explain the contact-induced change between Zhuang and Chinese. Periodic and frequent migrations from the central plains of China gave rise to complicated contact scenarios in Guangxi: contact among the Zhuang 壮语, Guangxi Yue 广西粤语, and Pinghua dialects falls under the former category, due to the geographic adjacency of these three speaker groups; contact between Zhuang and Southwest Mandarin belongs to the latter category, since most Zhuang speakers acquired Southwest Mandarin via schooling or skills training.

Lan (2005) compares Chinese loanwords and cognates in Zhuang. He first classifies the Zhuang-Chinese correspondences into two groups (old loans, which preserve overt Middle Chinese codas $-p$, $-t$, $-k$, $-m$; and new loans, which exhibit the phonological features of Southwest Mandarin). Second, he divides the old loans into two subcategories (Old Chinese layer vs. Middle Chinese layer). By dividing loans into distinct historical layers, Lan was able to distinguish between Chinese loans and Chinese cognates. Finally, he makes a statistical assessment of Chinese loanwords in more than twenty-three Zhuang dialects. Lan's work deepens our understanding of Chinese loanwords in Zhuang and shows conclusively that language contact between Chinese and Zhuang began in the ancient era of the Qin dynasty.

Qin (2007) compares the phonological forms and grammatical characteristics of the morphemes meaning *gei* 给 "give" in Pinghua, Yue, and Zhuang. He claims that the Pinghua $h\partial i^{55}$ 许, and the Nanning Yue 南宁粤语 pi^{35} 畀, are borrowed from the Zhuang hau^3, which in turn is derived from the Middle Chinese *jiǔ* 与 "to give." In the similar fashion, Qin and Wu (2009) suggest that the short comparative form [X-A-过:$_{CM}$] in Nanning Yue and Baise Yue 百色粤语 may have been copied from the southern Zhuang dialects in Guangxi.

In the past decades, a large amount of literature has sought to tackle the topic of language contact in China, but most studies to date have focused on the borrowing of phonological forms and loanwords rather than the replication of grammatical structures and patterns. Despite theoretical advances in the study of contact-induced changes in Western scholarship, the discussion of language change the Sinitic world has remained too general to account for a particular contact scenario. Previous studies on language contact in China may have failed to answer these questions:

(i) How can we tell whether language contact has occurred? What are the linguistic and socio-historical parameters?
(ii) Why and how does a particular innovation appear in some languages but not in others, even when the languages in question are geographically adjacent?
(iii) By what mechanisms have the languages in contact shared their syntactic, semantic, and pragmatic patterns?

Contact-Induced Grammaticalization: An Alternative Model

Soon after the contact-induced grammaticalization theory was initiated by Heine and Kuteva (2003: 533, 2005: 81–100) and further advocated in Matras (2009: 6) and Matthews and Yip (2009), this theory was first introduced into Chinese scholarship by Wu (2008), Phua (2009), Huang and Kwok (2013), and Huang (2014) and was employed to analyze grammatical patterns other than the transparent lexical borrowing triggered by language contact.

Inspired by this newly invented theory, many journal articles, dissertations, and conferences have provided detailed discussions of contact between languages in southern China. The majority of these can be classified under two aspects, such as "grammatical replication" and "constructional copying." We will now review the use of these methodologies in dealing with the contact phenomena in turn.

Contact-Induced Changes from Chinese to Minority Languages of Southern China

Wu (2007, 2009) provides a detailed investigation into contact-induced grammatical change among the languages in China. He first introduces the general theoretical frameworks espoused by Thomason (2001) and Heine and Kuteva (2005), analyzing the polyfunctionality of the "acquire" morpheme (Wu 2009) and the "dwell" morpheme (Wu 2010). Wu hypothesizes that the parallel developments of these morphemes in most Southeast Asian languages may be an outcome of language contact.

During the contact process, Chinese dialects, as the diffusional source, transferred their polyfunctional model and, in particular, the constructional model, to a wide range of minority languages. The author also discusses the word-order change/rearrangement of certain grammatical constructions in the Tai-Kadai, Hmong-Mein, Austronesian, and Austroasiatic languages in China, arguing that the [A-not-A] polar question form, fronting prepositional phrase, and [S-比:CM-St-A]/[S-A-比:CM-St] comparative configuration may have been replicated from parallel Chinese structures (Wu 2008, 2012).

Needless to say, the majority of Wu's work centers on coming to grips with a general picture of contact-induced grammatical change among the languages in China, rather than focusing on a specific manifestation of language contact occurring between two or three languages. Among these, certain contact exemplification found in China will extend the general issue of the contact theory proposed in Heine and Kuteva (2005). In terms of grammatical replication, the two authors argue that (The bold is highlighted by us.):

> Grammatical replication has the effect that the replica language (R) **acquires some new structure (Rx) on the model of another language (M)**...the new structure Rx is in most cases **not entirely new**; rather, it is built on some structure (Ry) that already existed in the replica language, and what replication then achieves is that it transforms Ry into Rx...What this suggests is that replication **does not start from nothing**; rather, it requires appropriate discourse patterns in the replica language to take place. (Heine and Kuteva 2005: 40–41)

This theory, based mainly on data found outside of China, casts a new perspective on understanding the process and outcome of grammatical replication. However, when it comes to explaining the language contact situation in southern China, this grammatical replication theory may be insufficient in part. Let us see certain instances of structural replication that have occurred in southern China.

Reordering

Heine and Kutava (2005: 112) state that "reordering (or restructuring) is a paradigm case of restructuring that concerns word order, where people rearrange the order of meaningful units in one language on the model of another language. For example, on the model of Indo-European Balkanic languages such as Macedonian and Albanian, speakers of these Turkish dialects have reversed the genitive and its head in possessive constructions."

"Reordering" is a common phenomenon found in the minority languages of southern China during the course of contact with Chinese. During language contact, a broad range of structures and orders in minority languages, such as genitive construction, relative clause order, locative phrase order, and order of verb complement, are rearranged according to the Chinese model, leading to a convergence of certain linguistic features from minority languages to Chinese. This replication *starts from something* that exists in the template of the modal language (Chinese), as shown below.

Genitive Construction

The genitive construction in Chinese usually precedes the head noun, giving rise to a [G-N] order. As a typical SVO typology, a number of minority languages in southern China demonstrate a [N-G] order of genitive construction. Yet, after contact with Chinese (1), these languages rearrange their model as a genitive-initial one. Examples (2)–(4) show the outcome of reordering in Dai, Biaohua, and Buyang (the original source of the data can be found in Wu 2014):

(1)	Chinese:	我　　的　　房子	[G-N]
		wǒ　　de　　fáng zǐ	
		1sg　GEN　house	
		"My house"	

(2)	Dai 傣语:	xɔ¹　　to¹xa³　/　məŋ²　tai²	[G-N]
		hoe　　1sg　　　place　Tai	
		"My hoe"　　　　"Tai's place"	

(3)	Biaohua 标话:	a¹tsɔ³　ken⁵　lək¹⁰　/　tsi³sy³　　　　tsu²　liak⁸	[G-N]
		uncle　CL　house　　branch secretary　CL　son	
		"Uncle's house"　　　"The branch secretary's son"	

(4)	Buyang 布央语:	mia¹¹tsɛ⁵⁴　ma⁰u:n⁵⁴　[N-G]　/　ma⁰u:n⁵⁴　ti³³　mia¹¹tsɛ⁵⁴ [G-N]
		aunty　　　child　　　　　　　child　　GEN　aunt	
		"The aunt's child"　　　　　"The child's aunt"	

Buyang has two orders – [N-G] and [G-N] – in its grammar, which is probably in the initial stage of language contact. It is evident that this group of minority languages shapes their new genitive construction on the Chinese model as indicated in Table 1

Table 1 Reordering of the genitive construction

	Initial stage	Outcome of contact
Chinese	**G-N**	G-N
Minority languages in southern China (excluding Tibeto-Burman languages)	N-G	**G-N**

Relative Clause

In a similar fashion, the relative clause in Chinese (5) tends to prepose the head noun in a [Rel-N] order. The adjacent minority languages (6–8) have copied this order and modified their default [N-Rel] template as the [Rel-N] one. To witness:

(5)	Chinese:	学生　　买　　过　　的　　书	**[Rel-N]**
		xué shēng mǎi guò de shū	
		student　bug　EXP　REL　book	
		"The book that the student bought"	

(6)	Dai 傣语:	*kun² au¹ pa¹*	[N-Rel]
		people catch fish	
		'People who catches fish'	
		kam² an² su¹tsau³ va⁶ nan⁴ di¹ tɛ⁴	[Rel-N]
		words REL 2sg say PTC good very	
		"The words that you said is very good"	

(7)	Biaohua 标话:	*jɔ² tsuŋ⁵ kɛ⁶ tau⁶*	[N-Rel]
		1pl.INCL raise REL bean	
		"The beans that we raise"	
		tsau² ma:i⁶ kɛ⁶ lan²	[Rel-N]
		catch fish REL people	
		"The people who catches fish"	

(8)	Maonan 毛南语:	*ai¹ zən¹ kwai¹ ʔja⁵ ka⁵*	[N-Rel]
		CL people cultivate land DIST	
		"That people who cultivates the land"	
		la:k⁸ce³ na⁴ ti⁰ ɦu⁴	[Rel-N]
		child eat REL rice	
		"The rice that the child ate"	

It is worth noting that all the minority languages from (6) to (8) negotiate with Chinese in terms of the word order of the relative clause, generating bipartite ambiguous structures that can be used in discourse together. Deep contact with Chinese will drive these languages to omit the [N-Rel] order at large. Table 2 is an illustration of such reordering:

Table 2 Reordering of the relative clause structure

	Initial stage	Outcome of contact
Chinese	Rel-N	Rel-N
Minority languages in southern China (excluding Tibeto-Burman languages)	N-Rel	Rel-N

Locative Phrase

Chinese features a preposition of the locative phrase, while the corresponding pattern in the minority languages in southern China is postposed. Rearrangement of the locative phrase order also occurs after a profound contact, as indicated from (9) to (12):

(9) Chinese:
他 [在屋里]$_{PP}$ 看书。 [PP-V]
tā　zài jiā lǐ　　kàn shū
3sg　in the house　read
"He is reading in the house."

(10) Dai 傣语:
kau⁶ het⁹ la³xɔŋ¹ [ti⁶　pə³tsin⁶]$_{PP}$. [V-PP]
1sg do work　be.at Beijing
"I am working in Beijing."

(11) Liyu 黎语:
kau⁶ het⁹ la³xɔŋ¹ [ti⁶　pə³tsin⁶]$_{PP}$. [V-PP]
1sg do work　be.at Beijing
"I am working in Beijing."
na¹ [du³ ŋa:i³ nom³]$_{PP}$ a:p⁷. [PP-V]
3sg be.at side river　shower
"He was taking a shower at the river bank."

(12) Biaohua 标话:
tsia¹ [ŋy⁴ mui¹ to³]$_{PP}$ phiam³ thu¹ [PP-V]
1sg　be.at DIST mountain cut　firewood
"I am cutting firewood on that mountain."

The Dai language preserves a relatively stable order in spite of the severe historical intrusion of Chinese. The scenario in other minority languages is in the opposite with Dai, where their relative phrase configuration has been reordered. Table 3 introduces this contact picture.

Verb-Complement Constructions

The Chinese language characterizes its verb-complement constructions. Typical cases of complements are free, predicative complements, as in 他走得很慢 tā zǒu de hěn màn "He walks is very slow," and bound verb-complement compounds, as in 我吃饱了 wǒ chī bǎo le "I have eaten full/ I am full of eating." As for the fact that most minority languages of southern China possess a corresponding construction as

13 Contact-Induced Change in the Languages of Southern China

Table 3 Reordering of the locative phrase structure

	Initial stage	Outcome of contact
Chinese	**PP-V**	PP-V
Minority languages in southern China (excluding Tibeto-Burman languages)	V-PP	**PP-V**

[VO-C], this construction is alternatively adjusted as [VC-O] after the contact with Chinese. This change can be demonstrated as follows:

(13)	Chinese:	他 射死 了 鸟。/ 他 关紧 了 门。	**[VC-O]**
		tā shè sǐ le niǎo tā guān jǐn le mén	
		3sg shoot.dead PFV bird 3sg close.tight PFV door	
		"He shot the bird to death." "He closed the door tightly."	

(14)	Dai 傣语:	xau^1 ju^2 sə^1 ta:i^1 to^1 nuŋ^6	[VO-C]
		3sg shoot tiger dead CL one	
		"He shot a tiger to death."	

(15)	Chadong 茶峒:	mən^2 wak^7 tai^1 tsi^4 hja^1 kok^8 tak^{8}nəm^4	[VC-O]
		3sg hit dead PFV two CL tiger	
		"He hit two tigers to death."	

(16)	Lajia 拉珈语:	lak^8 wak^8 na:ŋ^4 tiŋ^6 lie:u^3	[VO-C]
		3sg wash clothes clean PFV	
		"He cleaned the clothes to clean."	
		lak^8 wak^8 tiŋ^6 na:ŋ^4 lie:u^3	[VC-O]
		3sg wash clean clothes PFV	
		"He cleaned the clothes to clean."	

Note that the order of the complements in minority languages fits the Chinese [VC-O] model where the objects in general are included as a species of complements. Chinese, as a superstrate language, has diffused its particular construction to the substrate languages at large. Table 4 compares these two stages of contact separately.

Narrowing

Unlike the reordering process in which one template is cast and based on another model, narrowing concerns a selective procedure in which the replica language exhibits two (or more) structural options (say, A and B) to express one and the same grammatical function. When the model language has only one structure (A) for that function, restructuring may have the effect that speakers of the replica language narrow down these options to A, thereby establishing a one-to-one equivalence relation with the model language (Heine and Kuteva 2005: 114).

Table 4 Reordering of the verb-complement construction

	Initial stage	Outcome of contact
Chinese	VCO	VCO
Minority languages in southern China (excluding Tibeto-Burman languages)	VOC	VCO

Most students of language contact are aware of the narrowing process reported in Thomason (2001: 89) concerning the multiple word orders of the Kadiwéu language. Multiple word orders in Kadiwéu (i.e., OVS, VOS, SOV, OSV, VSO, and SVO) are narrowed down after contact with the single word-order language – Portuguese.

Narrowing is not a well-known process found in China, and we are yet to find any reports of narrowing triggered by contact among the languages spoken in southern China. However, it is not convincing to say that narrowing *never* happens. An endangered Tibeto-Burman language – Sabde Minyag 木雅语沙德话 – spoken in western Sichuan in southwestern China showcases a binary word-order of the question particles as [ʔæ=Clause] and [Clause=ʔæ/ʔa] (Minyag is an SOV language with an abundant vowel-harmony system. Vowels of the verbal affixes usually agree with those in the verb root in the neighboring environment. Vowels of verb roots attempt to keep a long-distance agreement with those in the sentence-initial subject, without a consideration of an insertion of a clause between the subject and verb.). Western Minyag is severely influenced by Khams Tibetan whose negative usually occurs in a [a=Clause?] configuration. When the Minyag language is in contact with Khams Tibetan, several Minyag-Tibetan speakers may figure out a fronting order of negatives as [ʔæ=Clause] in contrast to its original binary orders. This fronting order of question particles is seldom found in other Qiangic languages adjacent to the Minyag area. To see the examples:

i. The binary word orders of the question particle in Sabde Minyag (data are from our fieldwork)

(17)	ŋãtɕɤ khə-ɕɐ=ʔæ=pi ?
	lantern festival DIR-arrive=**Q**=IMPV$_3$
	"Is the Lantern Festival going to arrive?"

(18)	sɐde mtɕɔ́ yi-gɐ́=ʔæ=pi ?
	Sabde water DIR-rise=**Q**=IMPV$_3$
	"Will the river in Sabde flood (later)?"

(19)	vændæʁæ yi nɐ-rɐ́=pi ʔǽ=ndə ?
	old man sleep DIR-come=IMPV$_3$ **Q**=have
	"Has the old man come to sleep already?"

(20)	ta=ji ndzɔ́ yo-ndzi =si ʔǽ=ŋɐ=tí ?
	Tiger=ERG fox DIR-eat=PFV$_3$ **Q**=COP=GNO
	"The tiger has eaten the fox, right?"

ii. The word order of the question particle in (Dege) Khams Tibetan 德格藏语 (Geshang and Geshang 2002: 161)

(21)	tɕhø?	lɛ	ʑeʔpa	è	jĩ ?		
	2sg	DAT	cadre	**Q**	COP		
	"Are you (a) cadre?"						

(22)	khø	śama	lɛ	śi	è	ŋge ?	
	3sg	meal	DAT	PTC	**Q**	cook	
	"Is he cooking the meal?"						

iii. The word order of the question particle in Mebzang nDrapa 木绒扎坝语 (data are from our fieldwork)

(23)	ŋa	tɕhéto	jɪ =khə̆	a-zi-htṣə̂ =ra?	
	1pl	when	house=LOC	DIR-went-IMPV=**Q.PST**	
	"When were we going home?"				

(24)	pəhdzɔ́	tɕhéto	a-mnɛ́-htṣə̂ =a?	
	child	when	DIR-leave school-IMPV=**Q.NONPST**	
	"When does the child leave school?"			

While ʔæ and ʔa in (17) and (18) are suffixed to the predicates, they occur in front of the predicates in (19) and (20). The fronting pattern of question particles can also be found in the examples of the Dege Tibetan `e. In contrast, this fronting word order of question particles is not attested in other Qiangic languages like the nDrapa, the Qiang language in western Sichuan, or the Tangut language (see LaPolla with Huang 2003: 179 for Qiang; Gong 2011: 21 for Tangut). Although in Sabde Minyag, two possible word orders are available in daily conversation, the fronting ʔæ pattern takes up a majority of examples in our texts. It is reasonable then to hypothesize that Sabde Minyag may have narrowed down the word order options of question particles after contact with Khams Tibetan. A similar scenario can also be found in Guiqiong 贵琼语, another Qiangic language which has been in close contact with the Khams Tibetan language in western Sichuan. Guiqiong has a binary order of question particles (Jiang 2015: 304–306), one of which may have copied the Tibetan pattern. Geshiza rGyalrong 革什扎嘉戎语 also shows an enclitic tag question particle ʔæ= corresponding to the fronting Khams Tibetan pattern (Honkasalo 2019: 610–611). Note that both languages are grammatically similar to Tibetan in certain aspects.

Constructional Copying

Constructional copying differs from reordering in terms of the distinct patterns where grammatical replication takes place. Reordering is built on some existed models, while constructional copying is a holistic replication that requires no intimate or analogous template to exist in the model language. It is a *start-from-nothing* replication and is rare in world's languages.

Wu (2008) claims that minority languages in southern and southwestern China replicate some Chinese constructions that are not present in the grammars of those minority languages. These intruded constructions are essentially brand-new and have never been shared by minority languages before contact took place. The following discussion is built on this phenomenon.

"A/V-not-A/V" Polar Question

The "A-not-A" polar question is a Sinitic-specific constellation that has not been reported in the languages outside of China. The "A/V-not-A/V" structure displays areal distributions according to the geographic differences between south and north: [VP-neg] is distributed in the northwestern part of Jiangsu, and south of the Yangtze; [V-neg-VP] is distributed over a vast territory in the central southeast (Zhang 2000).

In addition to the Tai-Kadai and Miao-Yao languages, certain Tibeto-Burman languages (i.e., Lisu, Loloish) usually replicate the Chinese "A/V-not-A/V" model. To witness the examples from (25) to (29):

(25)	Chinese:	你 到底 走 不 走?	[V-not-V]
		nǐ dào dǐ zǒu bú zǒu	
		2sg at all go NEG go	
		"Will you go or not?"	

(26)	Zhuang 壮语:	*sou^1 yo^4 bou^3 yo^4 ne^6?*	[V-not-V]
		2dl know NEG know Q	
		"Do you know it or not?"	

(27)	Maonan 毛南语:	*nam^3 lɔ:t^7 kam^3 lɔ:t^7?*	[A-not-A]
		water hot NEG hot	
		"Is the water hot or not?"	

(28)	Younuo 优诺语:	*naŋ^{22}nɔ^{13} ne^{22}nɔ^{33} tɔ^{35} mɔ^{31} tɔ^{35} mai^{35}?*	[V-not-V]
		3pl today kill NEG kill pig	
		"Will they kill the pig or not today?"	

(29)	Lisu 傈僳语:	*e^{33}dʒɛ^{44} de^{42} ma^{31} de^{42}?*	[V-not-V]
		water drink NEG drink	
		"(Do you) drink the water or not?"	

"V-not-C" Negated Complement Structure

In most of the world's languages, modality can be expressed in terms of modal verbs, auxiliaries, particles, and affixes (Bybee et al. 1994). This notion can, in some cases, be encoded with a "V-not-C" construction where the negator is inserted between the main verb and its complement, yielding an overtone of situational possibility. This Chinese-specific construction has diffused to a wide range of minority languages as reported by Wu (2014):

(30)	Chinese:	小孩 拿 得 动, 大人 拿 不 动。	[V-not-C]
		xiǎohái ná dé dòng dà rén ná bú dòng	
		child take de COMP:move adult take NEG COMP:move	
		"This child can lift this (machine), but the adult cannot. (How strange!)"	

(31)	Bunu 布努语:	*ɲe³ tuŋ¹ hoŋ¹ nau³ tai⁵nau³ ɵu⁶ ma² caŋ⁴*	[V-not-C]
		one little work PROX today do NEG COMP:finish	
		"(We cannot even) finish the little work."	

(32)	She 畲语:	*ka³ khje⁴ kwei⁶ tsha⁴taŋ¹ kwa⁵ ha⁶ ŋu⁴*	[V-not-C]
		road very narrow car pass NEG COMP:go	
		"The road is so narrow that the car cannot pass through."	

(33)	Huihui 回辉:	*nau³³ sa³³ sien¹¹ lu³³ hu³³ zao²⁴ pu³³ phi⁵⁵*	[V-not-C]
		3sg GEN money much de count NEG COMP:finish	
		"He has too much money that it is countless."	

Reduplicated "V-V" Structure

Reduplication itself is not a rare phenomenon in the languages of the world, while the verbal reduplication is uncommon. The "VV" verbal reduplication in certain Asian languages implicates a change from lexical verbs to adverbial constructions, with a derivation of adverbial, auxiliary, irrealis, clause-linking, and particle functions (Hurch 2005). In Chinese, the "VV" reduplication denotes a semelfactive situation consisting of countable single-stage atelic events.

Likewise, this Chinese-specific constellation has also been copied by minority languages in southern China. Wu (2014) gives several interesting examples as follows:

(34)	Chinese:	这 本 书 我 看看 吧!	[V-V]
		zhè běn shū wǒ kànkàn ba	
		PROX CL book 1sg see.SEME PTC	
		"(Can) I try to read this book?"	

(35)	Zhuang 壮语:	*o:k⁷ yo:k⁸ pai¹ pja:i³ pja:i³*	[V-V]
		exit outside go walk walk	
		"(You) can try to walk outside."	

(36)	Jinghua 京话:	***tsuŋ⁵tai¹ di¹ja¹ tsai⁵ tsai⁵***	[V-V]
		1pl.INCL go out run run	
		"We can try to run outside."	

(37)	Huihui 回辉:	***ha³³ sioŋ²¹ sioŋ²¹ khiaŋ³²***	[V-V]
		2sg think think see	
		"You can try to think about it."	

In a nutshell, language contact in southern China resembles many other manifestations around the globe where reordering is a significant model in the course of the step-by-step replication. Moreover, a substantial number of minority languages of southern China replicate foreign constructions that are by no means pre-acquired in their grammar before contact. The Chinese language explicitly transfers its constructional model to these languages and abruptly marks new constructional settings in replica languages. In contrast to reordering, this replication process takes place suddenly. Copied constructions may be conventionalized so rapidly that most local residents are unaware of the intrusion of Chinese into their language profile.

On the basis of data drawn from the languages of southern China, the contact-induced transfer model proposed in Heine and Kuteva (2008: 37–59) can be extended (Wu 2012, 2014). Chart 1 describes this extended model:

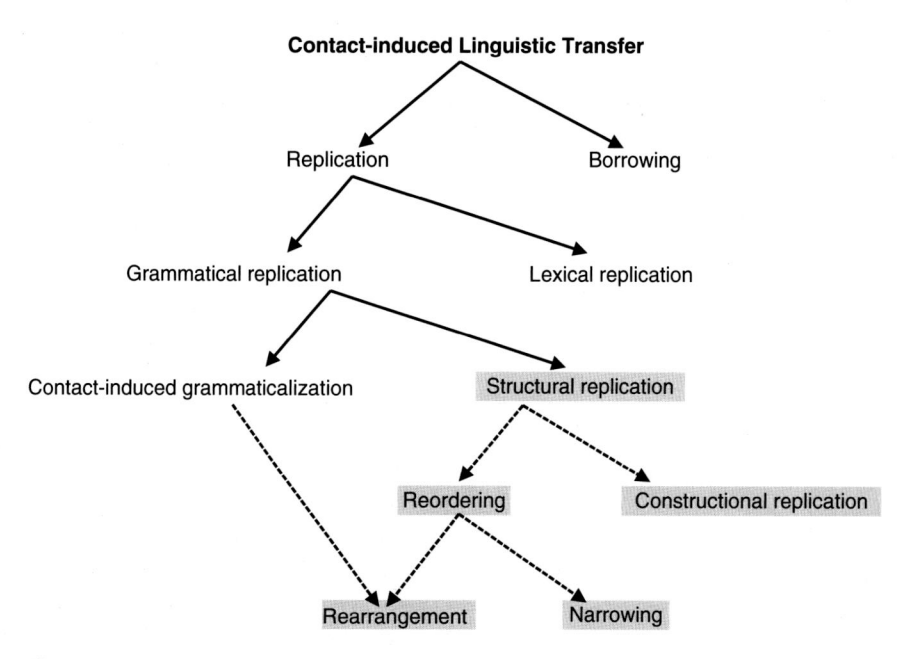

Chart 1 An extended model of contact-induced transfer

Contact-Induced Changes from the Minority Languages of Southern China to Chinese

In another vein, minority languages spoken in southern China usually diffuse their grammatical patterns to the abutting Chinese dialects that are deliberately learned by the minority people in order to facilitate communication with the dominant Han group.

Kwok's journal articles and monograph provide detailed discussions of contact in the Guangxi region of southern China. The author examines the functional change undergone related to three particular morphemes – the "acquire" verb (Kwok et al. 2011), the *na* 拿 "take" verb (Huang and Kwok 2013), and the *qu* 去 "go" verb (2014) – in the Chinese dialects and Tai-Kadai languages, and also investigates the contact-induced change in the [V-C-O] > [V-O-C] word order of the complement in Nanning Yue. Moreover, Kwok compares the expressives (ideophonic suffixes) in Nanning Yue and Cantonese, concluding that certain expressives in Nanning Yue may have developed through contact with Zhuang (Kwok 2012) when Zhuang speakers transferred the Zhuang exponents to their L2, Nanning Yue.

In one conference paper, Kwok and Lee (2013) offer an account of the grammaticalization of the postverbal Cantonese -*saai*^{3} based on a comparison with Nanning Yue. The two authors argue that Cantonese -*saai*^{3} differs from Nanning Yue -*saai*^{3} in its aspectual function. The functional extension in Nanning Yue is motivated by language contact between Zhuang and Nanning Yue.

Huang's (2014) dissertation is a comprehensive monograph that contains systematic analysis of the contact-induced change of languages in southern China. The author offers a detailed grammatical description of a target "finish" morpheme, depicting the polyfunctionality of the Nanning Yue -*łai*^{33} 晒, as well as its corresponding forms of $-le{:}u^4$, $-\theta o{:}t^7$, $-ju{:}n^2$, $-ja^5$, $-le{:}u^4$, $-thu{:}n^3$ in Zhuang and $-tɕ^hɐi^{21}$ 齐, $-liau^{55}$ 了 in the Pinghua dialect and Fusui Mandarin separately. After a functional comparison of these morphemes, Huang hypothesizes that the properties of Nanning Yue -*łai*^{33} 晒 are not found in other members of the Yue family. The Zhuang polyfunctional model has diffused to a wide range of languages in contact. Examples are shown from (38) to (42):

Polyfunctionality of the -łai^{33} 晒 "finish" in Nanning Yue

Universal quantifier:								
(38)	洞	里边	有	老虎，	村民	怕	晒	老虎。
	tu^{22}	*li*^{24}*pin*^{55}	*jɐu*^{24}	*lu*^{24}*fu*^{35}	*tɕʰyn*^{55}*mɐn*^{21}	*pʰa*^{33}	*łai*^{33}	*lu*^{24}*fu*^{35}
	cave	inside	have	tiger	villager	afraid	EXH:all	tiger
	"There is a tiger in the cave, thus all the villagers are afraid of it."							

Superlative:

(39)

一	笋	果	烂	晒。		
a^{55}	$lɔ^{21}$	$kɔ^{35}$	lan^{22}	$łai^{33}$		
one	CL	fruit	rotten	EXH:extremely		
i. "The fruits in the basket are extremely rotten."						
ii. "All the fruits in the basket are rotten."						

Completive:

(40)

大佬	啱啱	洗	晒	三	件	衫。
$tai^{22}lu^{35}$	$ŋam^{55}ŋam^{55}$	$łɐi^{35}$	$łai^{33}$	$łam^{55}$	kin^{22}	$ʃam^{55}$
brother	just	wash	EXH:completely	three	CL	clothes
i. "My brother washed all the three pieces of clothes just now."						
ii. "My brother finished washing the clothes just now."						

Postverbal aspect marker:

(41)

学校	修	晒	栋	楼。
$hɔk^{21}hau^{33}$	$łɐu^{55}$	$łai^{33}$	$tuŋ^{22}$	$łɐu^{21}$
school	build	FINISHED	CL	house
"The school has built a house."				

Conjunction:

(42)

佢	先	食	苹果，	晒，	再	食	沙梨。
$kʰy^{35}$	$łin^{55}$	$ʃek^{2}$	$pʰeŋ^{21}kɔ^{35}$	$łai^{33}$	$nɛ^{33}$	$tʃɔi^{33}$	$ʃek^{2}$ $la^{55}li^{21}$
3sg	first	eat	apple	THEN		will	eat pear
"He first ate the apple, then the pear."							

Polyfunctionality of the Diverse "Finish" Morphemes in Zhuang-Tai Kadai

Each language sample of the Zhuang group comprises two or three "finish" grams, which is a more complicated paradigm than that in Nanning Yue. Table 5 is a summary of the Zhuang model.

The evolution of -$łai^{33}$ 晒 involves an integrative contact procedure between Zhuang and Nanning Yue. Zhuang has diffused its linguistic exponents to a number of languages in the Central and Central Southern Guangxi Region. Not only does Nanning Yue, but also the other subset of the Chinese dialects in this region possesses the polyfunctional "finish" model of Zhuang-Tai Kadai (Huang 2014).

Areal Linguistics and Grammaticalization Area

The concept of "linguistic area" (*Sprachbund*) put forward by Boas (1920: 211), has been very effective at identifying instances of shared linguistic features that do not fit genetic classifications. This western theory was little understood by Chinese scholars

Table 5 The polyfunctional model of the "finish" grams in Zhuang variants

Function	Form	Universal quantifier	Superlative	Completive	Perfect marker	Linking verb	Connective 'so'	Connective 'but'
Liu jiang	$-\theta o:t^7$	−	−	+	−	−	−	−
	$-ju:n^2$	+	−	(+)	−	+	−	−
	$-le:u^4$	+	+	+	−	+	+	+
	$-le^{(6)}$	−	−	−	+	−	−	−
Ba ma	$-li:u^4$	+	+	+	−	+	+	+
	$-le$	−	−	−	+	−	−	−
Ma shan	$-\theta a:t^7$	−	−	(+)	−	−	−	−
	$-le:u^4$	+	+	+	−	−	−	−
	$-lo/\,le$	−	−	−	+	−	−	−
Jing xi	$-ja^5$	−	−	−	+	+	+	+
	$-le:u^4$	+	+	+	−	+	+	+
Long zhou	ja^5	−	−	−	+	+	+	−
	$thu:n^3$	+	+	+	−	−	−	−

until the early 1990s when Chen Baoya's 陈保亚 (1996) monograph *Lun yuyan jiechu yu yuyan lianmeng* 论语言接触与语言联盟 [*Language contact and language union*] was published. This monograph is concerned with the influx of loanwords borrowed from Chinese and incorporated into the Dai language in Yunnan. The author provides a tremendous amount of interesting examples of the Chinese influence on the minority languages in southern China (Chen 1996). Chen was not only trained as a linguist but also as an anthropologist. He spent several years living in Dai villages to observe the changes of loanwords both in the mixed Han-Dai and Dai-Han languages. According to his extrapolation, the complexity of language use results from a convergence of the languages spoken in this area. Chen's thought-provoking research has paved a new way to develop the study of language contact in China. However, Chen's research is not directed to any contact-induced change of grammatical patterns.

The notion of contact-induced sound change is a promising subject discussed by numerous scholars in China. A recent empirical- and theoretical-oriented study is presented in Shen (▶ Chap. 17, "Interactions Between Min and Other Sinitic Languages: Genetic Inheritance and Areal Patterns" this volume) to introduce the areal sound pattern in Min varieties. Shen claims that the development of three sounds features – implosives in Hainan Southern Min, voiced plosives in Cangnan Eastern Min, apical vowels in Far-western Min – correlates with the areal sound pattern hypothesis. Linguistic theories and the case study of Min can greatly benefit each other.

The tradition of language contact in China is on a par with phonological and lexical borrowing where some formal changes are easier to uncover. By contrast, the traits of some emergent linguistic areas must be explicitly understood via both formal and semantic perspectives. Huang and Wu's (2018) recent paper,

"Central-Southern Guangxi as a Grammaticalization Area," a step forward that helps establish a tentative language area after scrutinizing the diagnostics for a typical grammaticalization area. In this paper, the two authors deal with the ecological and linguistic context characterizing the Guangxi region. Numerous examples of area-specific linguistic phenomena (i.e., the polyfunctional grams *le: u¹* 晒 "finish," *pai¹* 去 "go," *hɔ:i³* 给 "give" and *ʔau¹* 攞 "take") from Sinitic (i.e., Nanning Yue, Pinghua dialect 平话, Fusui Mandarin 扶绥官话) and Tai-Kadai languages (i.e., southern and northern Zhuang 南壮和北壮, Dai) in this region are provided. Their concern is exclusively with four areal parallel grammaticalization chains that seem to be cross-linguistically rare, involving the verbs "finish," "go," "give," and "take". All of these replicated patterns are sufficient to constitute the Guangxi region of southern China as a part of the broad Mainland Southeast Asia linguistic area but different from other micro areas in exhibiting a set of rare areal features.

Future Directions for Language Contact Study in China

Rather than presenting a full-fledged introduction and analysis of all aspects of language contact, this chapter represents only the tip of the iceberg in understanding the full range of contact-induced grammatical changes that have occurred in the languages of China. Language contact is by no means an exotic topic compared with other fields that have been introduced into China only several decades ago. The unanswered questions raised in much of the relevant literature in China (or specifically written only in Chinese) provide a rich field for further inquiry. These include questions such as:

(i) *Is the theory of "language contact" a versatile model to account for the majority of changing phenomena in which two or more languages interact with each other and bear similar features?*

Our answer to this is negative. A number of linguists in China used to treat language contact as a universal and easily applied theory that could account for everything that historical and comparative linguistics is unable to provide. Concerning similarities shared by two or more languages, Aikhenvald and Dixon (2001: 1–3) propose a number of possible explanations: universal properties or tendencies, chance, borrowing or diffusion, genetic retention, and parallel development. Given the isomorphic properties of the Chinese language, there are Chinese linguists who are inclined to follow the methodology of language contact without an all-rounded consideration of other factors. This carelessness makes the theory of language contact a "universal key." There is a wide range of potential factors that may trigger cross-linguistic similarities. While language contact is a potential candidate that is involved in the expression of similarities, typological properties, chance, and genetic retention are also of equal importance.

(ii) *Does the contact-induced change that has taken place in the languages of southern China reflect a similar pattern in the languages of southwestern China with any precision?*

It does not have to be. In fact, as introduced in this chapter, contact repertoire in most languages of southern China reveals a pattern of contact-induced grammaticalization. In addition to the languages of southern China, a large percentage of languages spoken in southwestern China have a typology of SOV order opposite to their SVO counterparts. Grammatical borrowing is more widespread than contact-induced grammaticalization that is shared by most southern languages. Sun's (2019) conference paper outlines a prospect of language contact study of southwestern China. In contrast to the well-known contact situations from Chinese to minority languages, Sun focuses on an alternative contact situation where language contact takes place between minority languages (i.e., Dai influences Wa; Tibetan influences Horpa; rGyalrong influences Tibetan; and Qiang influences rGyalrong). Considering the emergence of sesquisyllabic structure, $[V_1-V_2]_{NMLZ}+V_{Head}$ verbal compound, and the verb root alternation, Sun argues that some unfamiliar tonal and morphological shifts are valid to take effect in language contact. Sun's paper sheds new light on further studies of contact in southwestern China in which borrowed phonology and morphology are cast in a large number of new linguistic components in most hybrid languages of this region. Besides, the process of grammatical borrowing is not a simple transfer of forms, but involves diverse negotiation and competition. Huang (forthcoming) reports that the nDrapa language 扎坝语 (Qiangic, Tibeto-Burman) spoken in the western Sichuan ethnic corridor used to borrow a copular -*ji* from its neighboring Khams Tibetan, and -*ji* loses its person-agreement and evidential character and occurs in the negative context where a compound copular configuration -*jirɛ* is used as in the [NEG-*ji-rɛ*]. This is not a simple occurrence of loanwords, it rather reflects a process of negotiation and competition when two particles compete with each other and eventually redistribute the functional paradigm. It is evident that from the nDrapa example that contact-induced grammaticalization has no say in capturing the mechanism of functional changes.

(iii) *Is contact-induced change a unilateral context or a bilateral context? What of the migration principle? Is it unreliable in most situations?*

The migration principle is important but is not applicable to all circumstances. On the basis of a vast survey of historical records from the Qin dynasty to the Qing dynasty, Hong (2004: 115–117) argues that wars during the Song dynasty (960–1279 CE) accelerated waves of migration to southern China. The indigenous Zhuang people were in frequent contact with urban residents as well as rural villagers, whose populations were made up of migrated Han soldiers and refugees, respectively. Contact between Pinghua 平话 and Zhuang-Tai Kadai was everywhere. Conversely, contact between Zhuang and Nanning Yue 南宁粤语 has only historically been limited to interactions during market trade (Ganxu 赶墟). Hong (2004) thus proposes a unilateral contact model where the Zhuang group had first came into contact with Pinghua speakers and sequentially the bilingual Zhuang-Pinghua speakers shifted to Nanning Yue over the course of two centuries.

However, a large number of cases reported by Kwok (2012, 2014) and Huang (2014) indicate that language contact in the Guangxi region of southern China is multidimensional. Different languages may have different contact processes, which may give rise to distinctive contact layers; in explaining the contact-induced functional extension, linguistic evidence carries more weight than migration evidence. In principle, however, it is not difficult to imagine rather straightforward situations in which linguistic migration theory would fail to produce reliable results (Campbell 2013: 430).

In other words, the migration principle is unreliable in most situations. Although it generally provides us with important information concerning the ins and outs of a putative historical contact picture, it is not applicable to all circumstances.

(iv) *How does the set of behavioral and conceptual habits in one language use influence another language in the course of contact?*

Behavioral habits and conceptual habits are equally significant. Historical linguists and sociolinguists are keen to describe the contact picture through familiar perspectives from their professions. This skewed viewpoint usually results in linguists ignoring some other parameters that are important incentives throughout the contact process. LaPolla (2009) classifies the effects of substratum, superstratum, and adstratum influence with reference to Tibeto-Burman languages, arguing that cognitive and behavioral causes for the manifested influence are the same as linguistic evidence in language contact. How a speaker represents a state of affairs reflects how that speaker conceives the state of affairs, and how one conceives a state of affairs depends on cultural norms.

With contact-induced changes, what happens is also related to cognition (e.g., conception of comparatives: stative locative in Rawang < > "surpass" type in Cantonese) and language habits (e.g., habit of thinking and behavior: sound segmentation: one phoneme [b] and [p] in Rawang < > two phonemes in Burmese). Language use is a set of behavioral and conceptual habits, and the habits acquired in speaking one language can influence the use of another language (LaPolla 2009).

Aside from linguistic evidence and social factors, the view of behavioral and conceptual habits has rarely been examined in previous studies of contact-induced changes in China. Therefore, a set of cognitive mechanism and behavioral principles should be considered to ensure a much more thorough investigation of language contact in China in future studies.

References

Aikhenvald, Alexandra Y. 2003. *Language contact in Amazonia*. New York: Oxford University Press.

Aikhenvald, Alexandra Y., and R.M.W. Dixon. 2001. Introduction. In *Areal diffusion and genetic inheritance: Problems in comparative linguistics*, ed. Alexandra Y. Aikhenvald and R.M.W. Dixon, 1–26. Oxford: Oxford University Press.

A-Tshogs, Yeshes vodgsal (阿错). 2003. *Zang hanyu zai daohua zhong de hunhe ji yuyan shendujiechu yanjiu* 藏汉语在倒话中的混合及语言深度接触研究 [Research on mixing of Tibetan and Chinese in Daohua and related languages: a study of deep language contact] (pre-published version). Tianjin: Nankai University.

Boas, Franz. 1920. The classification of American languages. *American Anthropologist* 22 (4): 367–376.

Bybee, Joan L., et al. 1994. *The evolution of grammar: Tense, aspect, and modality in the languages of the world*. Chicago: University of Chicago Press.

Campbell, Lyle. 2013. *Historical linguistics: An introduction*. 3rd ed. Edinburgh: Edinburgh University Press.

Chappell, Hilary. 2001. Language contact and areal diffusion in Sinitic languages: Problems for typology and genetic affiliation. In *Areal diffusion and genetic inheritance: Problems in comparative linguistics*, ed. Alexandra Aikhenvald and R.M.W. Dixon, 328–357. Oxford: Oxford University Press.

Chen, Naixiong 陈乃雄. 1982. Wutunhua chutan 五屯话初探 [A preliminary study of the Wutun dialect]. *Minzu yuwen* (民族语文) 1: 10–18.

Chen, Baoya (陈保亚). 1996. *Lun yuyan jiechu yu yuyan lianmeng* 论语言接触与语言联盟 [Language contact and language union]. Beijing: Yuwen Press.

Geshang, Jigmen (格桑居冕), and Yangchen Geshang (格桑央京). 2002. *Zangyu fangyan gailun* 藏语方言概论 [An introduction to the dialects of Tibetan]. Beijing: Minzu Press.

Gong, Hwang-cherng. 2011. *Tangut philology: Collection of papers by professor Hwang-cherng Gong*. Language and linguistics monograph series 46. Taipei: Academia Sinica.

Hashimoto, Mantarō J. 1976. Language diffusion on the Asian continent: Problems of typological diversity in Sino-Tibetan. *Computational Analyses of Asian and African Languages* 3: 49–66.

Heath, Jeffrey. 1978. *Linguistic diffusion in Arnhem Land*. Canberra: Australian Institute of Aboriginal Studies.

Heine, Bernd, and Tania Kuteva. 2003. On contact-induced grammaticalizaion. *Studies in Language* 27 (3): 529–572.

———. 2005. *Language contact and grammatical change*. Cambridge: Cambridge University Press.

———. 2008. Constrains on contact-induced linguistic change. *Journal of Language Contact (THEMA)* 2: 57–90.

Ho, Dah-an (何大安). 2004. *Guilue yu fangxiang bianqian zhong de yinyun jiegou* 规律与方向:变迁中的音韵结构 [Regulations and directions: The development of phonological structures]. Beijing: Beijing University Press.

Hong, Bo (洪波). 2004. Zhuangyu yu Hanyu de jiechushi ji jiechu leixing 壮语与汉语的接触史及接触类型 [The history and characteristic of language contact between Zhuang and Chinese]. In *Lezaiqizhong Wang Shiyuan jiaoshou qishi huadan qingzhu wenji* 乐在其中—王士元教授七十华诞庆祝文集 [The joy of research: A festschrift in honor of Prof. William S-Y. Wang's 70th birthday], ed. Feng Shi and Zhongwei Shen, 104–120. Tianjin: Nankai University Press.

Hongkasalo, Sami. 2019. *A grammar of eastern Geshiza: A culturally anchored description*. Doctoral dissertation submitted to the Faculty of Arts of the University of Helsinki. Helsinki: University of Helsinki.

Huang, Xing (黄行). 2005. Yuyanjiechu yu yuyan quyuxingtezheng 语言接触与语言区域性特征 [Language contact and areal features]. *Minzu yuwen* (民族语文) 3: 7–13.

Huang, Yang. 2014. *Synchronic variation, grammaticalization and language contact: The development of the FINISH morphemes in the Yue-Chinese and the Zhuang languages in the Guangxi region*. PhD thesis submitted to the Department of Chinese and History, City University of Hong Kong. Hong Kong: City University of Hong Kong.

———. 2017. *Revisiting the spatial orientation of the Muya language*. Paper presented at the 25th annual meeting of the International Association of Chinese Linguistics (IACL 25), Hungarian Academy of Sciences, Budapest, Hungary, June 25–27, 2017.

Huang, Yang (黄阳), and Bit-Chee Kwok (郭必之). 2013. Fangshizhuci zai Guangxi Hanyufangyan he Zhuangdongyu zhong de kuosan: yuantou guocheng ji qishi 方式助词在广西汉语方言和壮侗语中的扩散:源头、过程及启示 [Diffusions of the manner particles in Chinese dialects and Tai-Kadai languages of Guangxi: Their origins, processes and implications]. In *Eastward Flows the Great River: Festschrift in Honor of Professor William S-Y. WANG on this 80th Birthday* (大江东去:王士元教授八十岁贺寿论文集), ed. Gang Peng and Feng Shi, 521–540. Hong Kong: City University of HK Press.

Huang, Yang, and Wu. Fuxiang. 2018. Central Southern Guangxi as a grammaticalization area. In *New trends on Grammaticalization and language change*, ed. Sylvie Hancil, Tine Breban, and V.L. Jose, 105–134. Amsterdam/ Philadelphia: John Benjamins.

Hurch, Bernhard. 2005. *Studies on reduplication*. Berlin, New York: Mouton de Gruyter.

Janhunen, Juha, Marja Peltomaa, Erika Sandman, and Xiawu Dongzhou. 2008. *Wutun*. Muenchen: Lincom Europa.

Jiang, Li. 2015. *A grammar of Guiqiong*. Leiden, Boston: Brill.

Kwok, Bit-Chee (郭必之). 2012. Cong Nanning Yueyu de zhuangmaoci kan Hanyufangyan yu minzuyuyan de jiechu从南宁粤语的状貌词看汉语方言与民族语言的接触 [A case of language contact between the Chinese dialects and the ethnic minority languages: The ideophones in the Yue dialect spoken in Nanning]. *Minzu yuwen* (民族语文) 3:16–24.

——— (郭必之). 2014. Nanning diqu yuyan 'qu' yi yusu de yufahua yu jiechu yinfa de fuzhi 南宁地区语言"去"义语素的语法化与接触引发的"复制" [The morpheme GO in three languages of the Nanning region: Paths of grammaticalization and contact-induced 'replication']. *Language and Linguistics* 5: 663–697.

Kwok, Bit-Chee (郭必之), and Peppina Po-lun Lee (李宝伦). 2013. *Xianggang Yueyu dongci houzhi Chengfen saai de yufahua*香港粤语动词后置成分saai^{3}的语法化 [Grammaticalization of the Postverbal element saai^3 in Hong Kong Cantonese]. The 7th international symposium on the Grammaticalization in Chinese, Huazhong University of Science and Technology, Wuchang.

Kwok, Bit-Chee (郭必之), Andy Chin (钱志安), and Benjamin T'sou (邹嘉彦). 2011. Polyfunctionality of the preverbal 'acquire' in the Nanning Yue dialect of Chinese: An real perspective. Bulletin of the School of Oriental and African Studies (SOAS) 74.1: 119–137.

Lan, Qingyuan (蓝庆元). 2005. *Zhuang Han tongyuanci jieci yanjiu* 壮汉同源词借词研究 [A study of the cognates and loanwords between Zhuang and Han]. Beijing: The Minzu University Press of China.

LaPolla, Randy J. 2001. The role of migration and language contact in the development of the Sino-Tibetan language family. In *Areal diffusion and genetic inheritance: Problems in comparative linguistics*, ed. Alexandra Aikhenvald and R.M.W. Dixon, 225–254. Oxford: Oxford University Press.

———. 2009. Causes and effects of substratum, superstratum and adstratum influence, with reference to Tibeto-Burman languages. In *Issues in Tibeto-Burman historical linguistics*, Senro ethnological studies, ed. Nagano Yasuhiko, vol. 75, 227–237. Osaka: National Museum of Ethnology.

———. 2010. Language contact and language change in the history of the Sinitic languages. Selected papers of Beijing forum 2007. *Procedia – Social and Behavioral Sciences* 2: 6858–6868.

LaPolla, Randy, with Chenglong Huang. 2003. *A grammar of Qiang: With annotated texts and glossary*. Berlin, New York: Mouton de Gruyter.

Li, Fang-Kuei (李方桂). 1977. *A handbook of comparative Tai*. Hawaii: The University Press of Hawaii.

Li, Charles N. 1983. Languages in contact in Western China. *Papers in East Asian Languages* 1: 31–51.

Li, Jinfang (李锦芳). 2000. Yueyu xijian ji yu Zhuangdongyu jiechu de guocheng 粤语西渐及与壮侗语接触的过程 [The westward movement of the Yue dialect and its contact with Zhuang-Dong]. In *Proceeding of the 7th international symposium on Yue dialect*, ed. C. Y. Sin & Luke Kang Kwong, 62–75. Beijing: The Commercial Press.

Liang, Qichao (梁启超). 1920. Fanyiwenxue yu fodian 翻译文学与佛典 [Translated literatures and Buddhist texts]. In Qichao Liang (ed.), *Foxue yanjiu shibapian* 佛学研究十八篇 [Eighteen topics on the Buddhism]. Beijing: Zhonghua Book Company.

Matras, Yaron. 2009. *Language contact*. Cambridge: Cambridge University Press.

Matthews, Stephen. 2006. Cantonese grammar in areal perspective. In *Grammars in contact: A cross-linguistic typology*, ed. Alexandra Y. Aikhenvald and R.M.W. Dixon, 220–236. Oxford: Oxford University Press.

Matthews, Stephen, and Virginia Yip. 2009. Contact-induced grammaticalization: Evidence from bilingual acquisition. *Studies in Language* 33 (2): 366–395.

Mufwene, Salikoko S. 2001. *The ecology of language evolution*. Cambridge: Cambridge University Press.

Norman, Jerry, and Tsu-Lin Mei. 1976. The Austroasiatics in ancient South China: Some lexical evidence. *Monumenta Serica* 32: 274–301.

Ou'yang, Jüeya (欧阳觉亚). 1995. Liangguang yuefangyan yu zhuangyu de zhongzhong guanxi两广粤方言与壮语的种种关系 [The relation between Zhuang and the Yue dialects in Guangzhou and Guangxi]. *Minzu yuwen* (民族语文) 6: 49–52.

Phua, Chiew Pheng (潘秋平). 2009. *Cong fangyan jiechu he yufahua kan Xinjiapo huayu li de 'gen'* 从方言接触和语法化看新加坡汉语里的「跟」 [A study of (跟) in Singdarin: From a contact and grammaticalization perspective], In *Yufahua yu yufa yanjiu* 语法化与语法研究 [Grammaticalization Study], ed. Fuxiang Wu and Xiliang Cui 247–283. Beijing: The Commercial Press.

Qin, Yuanxiong (覃远雄). 2007. Pinghua Yueyu yu Zhuangyu 'gei' yi de ci 平话、粤语与壮语「给」义的词 [The 'GIVE' morphemes in Pinghua, Yue and Zhuang]. *Minzu yuwen* (民族语文) 5: 57–62.

Qin, Fengyu (覃凤余), and Fuxiang Wu (吴福祥). 2009. Nanning baihua 'guo' de liangzhong teshu yongfa 南宁粤语"过"的两种特殊用法 [two special usages of kuo (过) in Nanning Cantonese]. *Minzu yuwen* (民族语文) 3: 16–29.

Ross, Malcolm D. 1999. *Exploring metatypy: How does contact-induced typological change come about?* Conference talk notes, Australian Linguistic Society's Annual Meeting, Perth.

Schleicher, August. 1853. Die ersten Spaltungen des indogermanischen Urvolkes [The first splits of the indo-European people]. *Allgemeine Monatsschrift für Wissenschaft und Literatur* 3: 786–787.

Schmidt, Johannes. 1872. *Die Verwantschaftsverhältnisse der indogermanischen Sprachen* [On the genetic relations among the indo-European languages]. Hermann Böhlau.

Shen, Ruiqing. Forthcoming. Language contact in min: An areal perspective on sound patterns. In *The Palgrave handbook of Chinese language studies*, ed. Z. Ye. Berlin: Springer.

Sims, N.A. 2020. Reconsidering the Diachrony of tone in Rma. *Journal of the Southeast Asian Linguistics Society* 13 (1): 53–85.

Sun, Hongkai (孙宏开). 1981. *Qiangyu jianzhi* 羌语简志 [An introduction to the Qiang language]. Beijing: The Ethnic Publishing House.

Sun, Jackson T.-S (孙天心). 2019. *Yuyanjiechu dui yuyanjiegou de yingxiang yi Zhongguo xinandiqu weili* 语言接触对语言结构的影响:以中国西南地区为例 [Language contact and its influence on language structures: A case study of language contact in southwest China]. Unpublished paper presented in the 5th Workshop on Sino-Tibetan Languages of Southwest China, Nankai University, Tianjing, China, August 21–23, 2019.

T'sou, Benjamin K. T (邹嘉彦). 2001. Language contact and lexical innovation: Problems concerning the study of loan-words, lexical importation and cultural diffusion. In *New terms for new ideas. Western knowledge and lexical change in Late Imperial China*, ed. Michael Lackner et al., 35–56. Leiden: Brill.

Thomason, Sarah Grey. 2001. *Language contact*. Edinburgh: Edinburgh University Press.

Thomason, Sarah Grey, and Terrence Kaufman. 1988. *Language contact, creolization, and genetic linguistics*. Berkeley: University of California Press.

Wang, William S.-Y. 1969. Competing changes as a cause of residue. *Language* 45 (1): 9–25.

———. 1979. Language change: A lexical perspective. *Annual Review of Anthropology* 8: 353–371.

Wang, Futang (王福堂). 2005. Hanyu fangyan yuyin de yanbian he cengci 汉语方言语音的演变和层次 [Phonological developments and statifications of Chinese dialects]. Beijing: Language & Culture Press.

Weinreich, Uriel. 1953. *Language in contact: Findings and problems*. New York: Linguistic Circle of New York.

Weinreich, Uriel, William Labov, and Marvin I. Herzog. 1968. *Empirical foundations for a theory of language change*. Texas: University of Texas Press.

Wu, Fuxiang (吴福祥). 2007. Guanyu yuyan jiechu yinfa de yuyan yanbian 关于接触引发的语言演变 [On contact-induced language change]. *Minzu yuwen* (民族语文) 2: 3–23.

——— (吴福祥). 2008. Nanfang minzu yuyan chusuo jieciduanyu weizhi de yanbian he bianyi 南方民族语言处所介词短语位置的演变和变异 [The evolution and change of the position of preposition phrases in minority languages in Southern China]. *Minzu yuwen* (民族语文) 6: 3–18.

——— (吴福祥). 2009. Cong 'de' yi dongci dao buyu biaoji—Dongnanya yuyan de yizhong yufahua quyu 从"得"义动词到补语标记—东南亚语言的一种语法化区域 [From the verb meaning 'GET/ACQUIRE' to the resultative construction marker: A grammaticalization area in the Southeast Asian languages]. *Zhongguo yuwen* (中国语文) 3: 195–211.

——— (吴福祥). 2010. Dongnanya yuyan 'juzhu' yi yusu de duogongneng Moshi ji yufahua lujing 东南亚语言"居住"义语素的多功能模式及语法化路径 [On the polyfunctionality and grammaticalization of the 'DWELL/ LIVE' morphemes in the Southeast Asian languages]. *Minzu yuwen* (民族语文) 6: 3–18.

——— (吴福祥). 2012. Dongtaiyu chabishi de yuxu jiegou he lishi cengci 侗台语差比式的语序结构和历史层次 [The word orders and historical layers of the comparative constructions in the Tai-Kadai languages]. *Minzu yuwen* (民族语文) 1: 13–28.

——— (吴福祥). 2014. Yuxu chongzu yu goushi kaobei yufajiegou fuzhi de liangzhong jizhi 语序重组与构式拷贝:语法结构复制的两种机制 [Reordering and constructional copying: The two mechanisms in grammatical replication]. *Zhongguo yuwen* (中国语文) 2: 99–109.

Wu, Yunji (伍云姬), and Ruiqing Shen (沈瑞清). 2009. *Xiangxi Guzhang waxianghua diaocha baogao* 湘西古丈瓦乡话调查报告 [A fieldwork report of the Waxiang dialect in Guzhang, Xiangxi]. Shanghai: Shanghai Education Publishing House.

Xing, Gongwan (邢公畹). 1979. Xiandai Hanyu he Taiyu li de zhuci 'liao' he 'zhe' 现代汉语和台语里的助词"了"和"着" [The particle 'liao' and 'zhe' in modern Chinese and Tai-Kadai languages]. *Minzu yuwen* (民族语文) 2. 84–98.

Yü, Xiaorong (遇笑容). 2004. Hanyu yufashi zhong de yuyanjiechu yu yufa bianhua 汉语语法史中的语言接触与语法变化 [Language contact and grammatical change in the history of ancient Chinese studies]. *Hanyushi xuebao* (汉语史学报) 1: 27–34.

Yü, Xiaorong (遇笑容), Guangshun Cao (曹广顺), and Shengli Zu (祖生利). 2010. *Hanyushi zhong de yuyanjiechu wenti yanjiu* 汉语史中的语言接触问题研究 [A study of language contact in ancient Chinese]. Beijing: Yuwen Press.

Yue-Hashimoto, Anne (余霭芹). 1976. Southern Chinese dialects—The tai connection. Computational Analysis of Asian and African Languages 6: 1–9.

Zeng, Xiaoyu (曾晓渝). 2003. Lun Suiyu li de jin xiandai hanyu jieci 论水语里的近、现代汉语借词 [Chinese loanwords in the Sui language]. *Yuyan yanjiu* (语言研究) 23.2: 115–121.

Zhang, Min (张敏). 2000. Syntactic change in southeastern Mandarin: How does geographical distribution reveal a history of diffusion? In *Memory of Professor Li Fang-Kuei: Essays of linguistic change and the Chinese dialects*, ed. Pang-Hsin Ting and Yue-Hashimoto, 197–242, Language and Linguistic monograph series W-1. Taipei: Academia Sintica.

Zhu, Dexi (朱德熙). 1990. Dialectal distribution of V-neg-VO and VO-neg-V interrogative sentence patterns. *Journal of Chinese Linguistics (JCL)* 18.2:209–230.

Zhu, Qingzhi (朱庆之). 1992. *Fodian yu zhonggu Hanyu cihui yanjiu* 佛典与中古汉语词汇研究 [A study of Buddhist texts and vocabularies in medieval Chinese]. Taipei: Wenjin Press.

Zhu, Qingzhi (朱庆之), and Guanming Zhu (朱冠明). 2006. Fudian yu Hanyu yufa yanjiu— Ershi shiji guonei fojiao hanyu yanjiu huigu zhier 佛典与汉语语法研究—20世纪国内佛教汉语语法研究回顾之二 [Buddhist texts and the study of Chinese grammar: A review of Buddhist-Chinese studies in China (2)]. *Hanyushi yanjiu jikan* (汉语史研究集刊) 9. Chengdu: Bashu Book Company.

14

The Evolution of Chinese Grammar from the Perspective of Language Contact

Yonglong Yang and Jingting Zhang

Contents

Introduction	334
The Evolution of Old Chinese and Modern Chinese Word Order from the Perspective of Language Contact	336
Is Modern Chinese an SOV Language?	336
From SVO to SOV?	338
SOV Features and Language Contact	341
The Grammatical Evolution of the Translation of Buddhist Scriptures in Mediaeval Times from the Perspective of Language Contact	342
The Genesis of New Categories	343
The Emergence of New Functions	345
The Emergence of New Function Words	346
The Emergence of New Sentence Patterns	347
The Emergence of New Word Order	349
Increase in the Frequency of Usage	350
The Evolution of Chinese Grammar after the Liao and Jin Dynasties from the Perspective of Language Contact	351
Changes to Pronouns	353
Location Words Used as Case Markers or Postpositions	354
Auxiliaries Related to Verbs	356
Special Grammatical Phenomena in *Lao Qida*	359
Conclusion	360
References	361

Translators: Zhang Jingting 张竞婷 and Wan Quan 完权. Institute of Linguistics, Chinese Academy of Social Sciences, Beijing, China

Y. Yang (✉) · J. Zhang
Institute of Linguistics, Chinese Academy of Social Sciences, Beijing, China
e-mail: yangyl@cass.org.cn

© The Author(s), under exclusive licence to Springer Nature Singapore Pte Ltd. 2022
Z. Ye (ed.), *The Palgrave Handbook of Chinese Language Studies*,
https://doi.org/10.1007/978-981-16-0924-4_28

Abstract

Chinese language has been in contact with surrounding minority languages since ancient times. This chapter summarizes the evolution of Chinese grammar caused by contact based on previous studies. There are three main issues. First, the evolution of Chinese word order from the perspective of contact has been deeply discussed. Mandarin is not a typical SOV language, but it does have many features of SOV language, especially that the northwest dialects have almost all the word order features of SOV languages. From Old Chinese to Modern Chinese, the SOV features of northern Chinese gradually increase. The increase of these SOV features is related to the influence of the northern Altaic languages and other SOV languages. Second, the grammatical evolution of Buddhist scriptures translated into Chinese in the Six Dynasties has been studied from the perspective of contact. After the Eastern Han Dynasty, Buddhism was introduced into China, and the translation of Buddhist scriptures led to the indirect contact between Chinese and Sanskrit and Pali, which had a great impact on the Chinese language at that time. Third, the evolution of Chinese grammar after the Liao and Jin dynasties has also been developed from the study of the contact perspective. The Altai people ruled northern China in Liao, Jin, Yuan, and Qing Dynasties, which led to the direct contact between Chinese and varieties of the Altaic language. Finally, the Altai people who lived in the Central Plains switched to Chinese, and Chinese was also influenced by the Altaic language to a certain extent.

Keywords

Chinese grammar · Mandarin · Language contact · Buddhist scriptures · Altaic languages

Introduction

The evolution of language may be caused by either internal or external systemic factors. The most important of these external factors is language contact. The earliest written history of Chinese can be traced back to the oracle bone inscriptions of the Shang Dynasty (approx. 1600 BCE–1046 BCE). During the more than 3000 years since the Shang Dynasty, the development of Chinese has been accompanied by continuous contact with languages spoken by the ethnic groups that surround China. In addition to such contact, several of these predominately northern ethnic groups established dynasties that ruled over parts of China such as the Northern Wei (386–534 CE), Liao (916–1125 CE), and Jin (1115–1234 CE), or, in the case of the Yuan (1271–1368 CE) and Qing (1644–1911 CE), even came to rule over China in its entirety. As such, China's complex history of language contact raises many questions. What is the exact nature of the changes induced by such contact in Chinese grammar from the ancient period to the present? What were the processes behind this evolution? To answer these questions, a large-scale corpus of linguistic materials is required and there are four major periods of language contact throughout Chinese history that can be observed with abundant materials.

The first of such corpora can be found in linguistic materials related to indirect contact between Chinese and Indo-European languages such as Sanskrit that came into being through the translation of Buddhist scriptures. Over a period of a thousand years from the late Eastern Han (202 BCE–220 CE) to the Tang (618–907 CE) and Song (960–1279 CE) dynasties, groups of monks translated a large number of Buddhist scriptures from Sanskrit and other Indo-European languages into Chinese, which in turn had a substantial impact on Chinese culture and language.

The second is found in the corpus related to direct contact between Chinese and Mongolian peoples under Mongol rule during the Yuan when the Chinese language was in close contact with Mongolian. The degree of contact that occurred during this period was such that a language with mixed Chinese-Mongolian elements known as "Han-er Yanyu" 汉儿言语 formed in Northern China (Ota Tatsuo 1988/1991: 181–211). Materials from this period of Mongolian-Chinese contact are quite rich and include such sources as Mongolian-Chinese vernacular inscriptions and case records contained in the *Yuan dianzhang* 元典章 [*The Institutions of the Yuan Dynasty*].

The third group of materials relates to direct contact between Chinese and Manchu peoples during the Qing. Despite Manchu nobility ruling China for more than 200 years during the Qing, the Manchu language gradually disappeared in China and ethnic Manchu people turned to using the Chinese. At the same time, Chinese was also influenced by Manchu language to a certain extent. While we possess many materials that provide insights into the interaction between Manchu and Chinese such as the *Qingwen zhiyao* 清文指要 [*The Essentials of the Manchu Language*], research in this area remains relatively weak.

The fourth corpus involves the spread of western learning to China during the late-Qing and the early Republican periods, which witnessed the indirect contact between Chinese and European languages. During this period, a large number of western works were translated and published with some translated directly from European languages and other translated indirectly from Japanese. Mainly dealing with science, philosophy, politics, and history, these works appearing in translation profoundly influenced the thought and culture of the period and also influenced the movement toward the use of the written vernacular in modern Chinese.

At present, scholars have paid more attention to the first two of these four contact-period corpora, while research into the last two is still lacking, making it a subject worthy of further exploration. Current research on the influence of language contact on Chinese grammar is predominantly concerned with the following questions:

Firstly, in terms of word order type or overall structure, did Chinese undergo a process from SVO to SOV? Some scholars raised this issue as early as the middle of the last century and it is still being debated today.

Secondly, which grammatical categories and forms have been caused by language contact throughout the history of the Chinese language and what was the process behind this? There is a substantial amount of research in this field, especially on the grammar found in the translation of Buddhist scriptures during the Six Dynasties (222–589 CE) and on the Chinese grammar of the Yuan dynasty. As noted above, both periods are relatively rich in materials.

On the basis of previous research, this chapter features a general introduction to the evolution of Chinese grammar caused by language contact and is comprised of three sections: section "The Evolution of Old Chinese and Modern Chinese Word Order from the Perspective of Language Contact" focuses on the evolution of Old Chinese (*Shànggǔ Hànyǔ* 上古汉语, also known as "Archaic Chinese," approx. 1300–221 BCE) and Modern Chinese word order from the perspective of language contact; section "The Grammatical Evolution of the Translation of Buddhist Scriptures in Mediaeval Times from the Perspective of Language Contact" looks at the evolution in grammar that came about from the translation of Buddhist scriptures during the mediaeval period from the perspective of language contact; and section "The Evolution of Chinese Grammar after Liao and Jin Dynasties from the Perspective of Language Contact" summarizes the evolution of Chinese grammar after the Liao and Jin dynasties from the perspective of language contact.

The Evolution of Old Chinese and Modern Chinese Word Order from the Perspective of Language Contact

Has there been any evolution related to word order caused by language contact in the history of Chinese? Since the 1970s, many overseas scholars, such as Tai (1973, 1976), Li and Thompson (1974), and Hashimoto (1975), have paid attention to this question. Subsequently, it has also attracted the attention of scholars China, such as Jiang Lansheng 江蓝生 (1992) and Zhang Cheng 张赪 (2002). To answer this question, the following important questions need to be answered in advance. First, is Modern Chinese an SOV language? Second, is Old Chinese an SVO language, or does it have more SVO features than Modern Chinese? And third, is there any relationship between the evolution of word order and language contact?

Is Modern Chinese an SOV Language?

First of all, is the word order type of Modern Chinese SOV, or to what extent is it SOV? This question has been long-debated.

As Tai (1973 [1986: 341]) notes, "Chinese exhibits in surface structure both SVO and SOV word order" and that "Chinese has SOV as the underlying order," which is derived from SOV through a rule of NP-V inversion. Most linguists would agree that the basic word order of Modern Chinese is SVO, but according to Greenberg (1963) and other studies, languages with SOV as the dominant word order tend to have the following features of word order: (i) relative clause before noun, (ii) adjective before noun, (iii) genitive before the governing noun, (iv) adverbial before the main verb, (v) adverb before adjective, (vi) proper noun before common noun, (vii) final particle for yes-no question, (viii) postpositional, and (ix) standard before marker before adjective in comparative constructions. As Tai points out, "if we examine the word order in Chinese with these grammatical features," "we find that Chinese has all of the properties of a SOV language except that in most cases it has apparent prepositions rather than postposition." (Tai 1973 [1986: 347]).

There are also many scholars who do not agree with the premise that Modern Chinese is an SOV language. Light (1979) argues that Chinese is an SVO language, while Mei (1980) suggests that the *bǎ*-construction 把字结构 ("disposal construction") cannot be regarded as evidence that Chinese is a SOV language. Chu (1987 [1993: 142]) maintains that the view of Modern Chinese as an SOV language is not a tenable one, while according to Peyraube (1997: 17), Classical Chinese, Medieval, Modern, and Contemporary Chinese are all SVO languages.

As a result of investigating an increasing large sample of languages, linguistic typologists have come to several new understandings of word order correlations. Based on the materials drawn from 625 languages, Dryer (1992) sums up 16 word order correlation pairs. Following this, Jin Lixin 金立鑫 and Yu Xiujin 于秀金 (2012) point out that Dryer's research indicates that, in the correlated ten pairs, Mandarin Chinese has four pairs exhibiting both OV and VO word orders, three pairs tend to be OV order, and the last three have VO order and that as a result, Mandarin Chinese should probably be regarded as an OV-VO mixed word order type language. Furthermore, Dryer (2007) updated his previous list of correlation pairs. According to the revised correlation pairs, Dryer finds that although the basic word order of clauses in Mandarin Chinese is VO, it has more OV word order characteristics from the perspective of a larger range of word order harmonious matching (Yang Yonglong 杨永龙 2014).

In fact, Modern Chinese dialects are quite different in word order types. Hashimoto (1975) points out that the typological characteristics of modern Chinese dialects from north to south exhibit a continuum from Altaic to Thai. He illustrated that four of these syntactic features are related to word order.

1. a. Northern dialects: modifier-noun: *gōng-niú* 公牛 (male-cattle, "ox")
   b. Southern dialects: noun-modifier: *niú-gōng* 牛公 (cattle-male).
2. a. Northern dialects: adverb-verb: *nǐ xiān zǒu* 你先走 (you-first-go, "You go ahead.")
   b. Southern dialects (Cantonese): adverb-verb: *nǐ zǒu xiān* 你走先 (you -go-first).
3. a. Preferred in northern dialects: *tā dào Běijīng qù* 他到北京去 (he-to-Peking-go, "He is going to Peking.")
   b. Preferred in southern dialects: *tā qù Běijīng* 他去北京 (he-go-Peking).
4. a. Comparative object must precede the verb in northern dialects: *nǐ bǐ tā gāo* (你比他高, you-than-he-tall, "You are taller than him.")
   b. Comparative object can follow the verb in southern dialects: *li ke quan i* (你较悬伊, you-more-tall-he, "You are taller than him.")

Tai (1976 [1986: 431]) adds three subsequent observations. First, many *bǎ*-construction in Mandarin Chinese can be expressed only in SVO sentences in southern dialects. Second, while the aspect marker *le* 了 in Mandarin Chinese is a verb suffix, its equivalent in the Cantonese and Min dialects is an auxiliary verb *yǒu* 有 (have) preceding the main verb. Third, in cases where Mandarin Chinese allows either preverbal or postverbal positions for a locative phrase a without clear functional difference, southern dialects allow only postverbal positions.

Comrie (2008: 5) revisits this topic. Based on 20 structural features described by Haspelmath et al. (2005), including basic word order of clauses, prepositions/postpositions, and noun phrases, he conducted a comparative study of Chinese languages (Mandarin Chinese and Wu, Hakka, Cantonese, and Min dialects), Southeast Asia languages (Thai and Vietnamese), and North Asian languages (Yakut and Khalkha). In conclusion, Comrie notes that Chinese does indeed occupy an intermediate position typologically between North and Southeast Asia.

As discussed earlier, the conclusions we can draw from this are as follows:

First, Mandarin Chinese is not a typical SOV language, but it does have many features of an SOV language.
Second, different Chinese dialects have different SOV features. Generally speaking, southern dialects have more SVO features, while northern dialects, especially northwest dialects, have more SOV features.
Third, the dominant word order is SOV in the Chinese dialects of Hehuang 河湟 area, which are located in Northwest Gansu and Qinghai Provinces (Xu Dan 徐丹 2014; Yang Yonglong 2015, etc.). In some dialects in this area, SOV features predominate. For instance, Gangou 甘沟, a Chinese dialect in the Minhe 民和 County of Qinghai Province, has a series characteristic of SOV languages. Among the 16 word order correlation pairs that Dryer (2007) lists, 15 of them in Gangou Chinese demonstrate a strong correlation with verb-final languages, such as O-V, predicate-copula, NP-article, NP-Adposition, standard of comparison-adj., and final adverbial subordinator, etc. (Yang Yonglong 2015).

From SVO to SOV?

Secondly, has the Chinese language experienced an evolution from SVO to SOV? Or does Modern Chinese have more SOV features than Ancient Chinese?

Li and Thompson (1974) attempt to prove that Chinese has changed from the ancient SVO word order to the modern SOV word order concerning five aspects: (i) S+V+PP→S+PP+V, (ii) the emergence of the *bǎ*-construction, (iii) the emergence of *bèi*-construction 被字结构 ("passive construction"), (iv) the emergence of compounds, postpositions, and verbal suffixes, and (v) the general shift of Verb-Object constructions to Preposition-Object-Verb constructions. In addition, they note that Archaic Chinese possessed significant SOV characteristics even at the time when word order was strictly SVO, for prearchaic Chinese (before twelfth century BCE) was in SOV language long before archaic period (third to tenth century B.C.), and Chinese language has undergone the following cycle of changes during the past four or five millennia:

Pre-Archaic →Archaic →Modern
SOV →SVO →OV

Tai (1976 [1986: 429–432]) proposes a theory of word order changes in Chinese. "Archaic Chinese started as an SVO language like Thai. Through close and frequent contacts with Altaic languages in the north, it began to adopt the ordering principle of placing the specifier before the specified." The change started with "modifiers – head noun," and then "adverbials - main verb," and then most of the prepositional phrases shifted from the postverbal to the preverbal position. *Bèi*-construction (O *bei* S V) was produced in the second and first centuries BC, and *bǎ*-construction (S *ba* O V) was produced in the Tang Dynasty.

On the other hand, many linguists reject this view such as Light (1979), Chu (1987), and Sun (1996). Light (1979: 149) argues that there has been no change in the order of Verb and Object from Old Chinese to Modern Chinese but instead exhibits a change in the position of prepositional phrases and the relation of verbs to nouns and adverbs. Sun (1996: 81) notes that the *bǎ*-construction should properly be treated as a marker of high transitivity, and historically the emergence of the *bǎ*-construction in Middle Chinese does not symbolize the emergence of the SOV word order. While Chu (1987 [1993:169]) does not consider Modern Chinese to be a SOV language either, he also acknowledges that: (i) some sentences would be SVO in ancient Chinese, whereas they should be arranged according to SOV in modern Chinese; (ii) in terms of the number of SOV sentences, modern Chinese has many more than ancient Chinese.

The point that we are making here is as follows. Although modern Mandarin Chinese and most Chinese dialects have not developed into a typical SOV language, many features of SOV in Mandarin have gradually emerged in the history of its development, such as the position of prepositional phrases, the preference of postposition of localizers, the postposition of analogy *side* 似的 (–like), the hypothetical conjunctions *shí* 时 (when), *hòu* 后 (after), *hē* 呵 (a modal particle, "if"), *dehuà* 的话 (in the words of, 'if'), and *VP-qù* VP 去 (go to) expressing the purpose, etc.

Based on a large number of ancient linguistic materials, Zhang Cheng (2002) investigates the position of prepositional phrases in detail and analyzes the historical evolutionary process of locative prepositional phrases and instrumental prepositional phrases in Chinese from them being placed after verbs to being placed before verbs. Cheng finds that this evolution began in the Eastern Han Dynasty, became more noticeable during the Wei, Jin, and Northern and Southern Dynasties, and ended during the Tang and Five Dynasties, and by the Yuan and Ming dynasties, the position pattern of prepositions in Modern Chinese was fully formed. However, she does not agree the evolution of Chinese from SVO to SOV.

Modern Chinese monosyllabic localizers, such as *shang* 上 (above) and *li* 里 (in), are postpositions to a large extent as they exhibit the following properties: (i) presence after NPs; (ii) with a high degree of cliticization; (iii) cannot be used alone; (iv) cannot be modified by *de* 的 (genitive particle, nominalization) or adjectives; (v) bleaching of semantics; (vi) unstressed syllable pronounced without its original pitch; (vii) "NP + localizers" cannot act as an argument and has the characteristics of adverbial, and (vii) "NP + localizers" can express the location independently without another preposition. The postposition function of *shàng* and *lǐ* has gradually developed during the history of the Chinese language and has experienced a process of grammaticalization from nouns to localizers and from localizers to postpositions, which was complete by the Tang. Accordingly, the related structure has also undergone the evolution of "preposition + NP" > "NP + postposition," or "preposition + NP" > "preposition + NP + postposition" (Chen Changlai 陈昌来 2014). In the vernacular language of the Yuan Dynasty, which was deeply influenced by Altaic languages, postpositions derived from locative nouns, such as *shang*, *li*, *gēndi* 根底, and *gēnqian* 根前, have some case marking functions beyond location and are used to mark the ablative, instrumental, dative, and accusative case (see below). For example, in *mínguān héshàng-mei gēndi xiūjiào duàn zhě* 民官和尚每根底休教断者 extracted from *Yuán diǎnzhāng* 元典章, *gēndi* attached to the object *héshàng-mei* 和尚每 (monks) is postobjective case marker. The example sentence above means that administrative officials should not judge monks (Zu Shengli 祖生利 2003).

In Ancient Chinese, the predicate structure of the SVO type was mainly used to express analogy, such as the likening verbs *rú* 如 or *sì* 似 plus vehicle to form "*rú/sì* + N," for example, the phrase *yǒu nǚ rú yù* 有女如玉 ("There is a young lady like a gem") taken from the *Shijing* 诗经 [*The Book of Odes*]. In Middle Chinese, examples of SOV type "vehicle + postposition" as attributive or adverbial emerged in the translation of Buddhism sutras in the Six Dynasties. For example, *bǐchù fù yǒu huǒ xiàngsì shān* 彼处复有火相似山 ("There are also mountains similar to fire") was extracted from the *Zhèngfǎ niànchǔ jīng* 正法念处经 [*Saddharma-smrty-upasthana-sutra*]. After the Jin and Yuan dynasties, the postpositions *yěsì* 也似 (−like) and *sìde* 似的 (−like) to express analogy came into being as seen in such examples as *èrshísì suì huāzhī yěsì húnjiā* 二十四岁花枝也似浑家 ("a 24-year-old, flower-like wife") and *fēi yěsì pǎo dào jìnhún Zhāngyuánwài jiā* 飞也似跑到禁魂张员外家 ("flying to Mr. Zhang the penny pincher's house") extracted from the Vernacular Novel in Ming dynasty. As Jiang Lansheng (1992) points out, *yěsì* is composed of the verb *sì* 似 with the auxiliary word *yě* 也, which does not conform to the general rule of Chinese word formation. Therefore, its origin is very doubtful and was probably induced by the influence of Altaic grammar. Through further investigation, Yang Yonglong (2014) indicates that the emergence and development of "x + *xiāngsì* 相似 / *sì* 似 / *yěsì* 也似" and related postpositions are not only restricted by the structure of Chinese itself, but also influenced by language contact. Such linguistic influence is not limited to Mongolian but also includes the direct influence of different OV languages and the indirect influence in the translation of Buddhism sutras. After the Yuan Dynasty, the postposition *sì* / *yěsì* flourished, a feature closely related to the influence of Mongolian.

Chinese generally uses a prepositive conjunction to express a hypothetical condition. In pre-Qin Chinese, *ruò* 若 (if) and *rú* 如 (if) are generally used. During the Song and Yuan dynasties, *shí* 时 (when) and *hòu* 后 (after) used after the adverbial clauses of time were grammatically transformed into postpositive auxiliary particles of hypothesis. The postpositive auxiliary particle *he* 呵 ("if") appeared in the Yuan Dynasty (Ota Tatsuo 1958/2003). The prepositive conjunctional maker *dehuà* 的话 (in the words of, "if"), which is still very common in modern northern dialects, emerged in the Qing Dynasty (Jiang Lansheng 2004). Those postpositive auxiliary particles are all used at the end of the hypothetical clauses.

Hashimoto (1975) notices that the northern dialect preferred the form *Tā dào Běijīng qù* 他到北京去 (he-to-Peking-go, "He is going to Beijing"), while the southern dialect preferred *Tā qù Běijīng* 他去北京 (he-go-Peking). Lu Jianming 陆俭明 (1985) finds that in the syntactic structures of the purpose related to *qù* 去 (go), the northern dialect mainly uses "VP + *qù*," and the southern dialect mainly uses "*qù* + VP." Through the investigation of Chinese historical documents, Yang Yonglong (2012) finds that the earliest instance of "VP + *qù*" is found in the Chinese translations of Buddhist scriptures, after which they are commonly found in Buddhist classics or literature written under the influence of Altaic languages; in these materials, "VP + *qù*" coexists with "D(estination) *qù*," while "*qù* + VP" with "*qù* D"; the relationship between "VP + *qù*" and "D *qù*" is obvious and is actually a correlation to SOV languages.

To summarize the discussion thus far, the results of the research noted above show: (i) Ancient Chinese is not a typical SVO language, but it has more SVO word order features than modern Chinese; (ii) the SOV features of Modern Chinese have gradually increased in the history of Chinese; and (iii) it is true that Chinese has experienced a certain degree of evolution from SVO to SOV.

SOV Features and Language Contact

Does the evolution from SVO to SOV, or the increase in SOV features, have anything to do with language contact? This is a question that has puzzled researchers of the Chinese language for many years.

Many scholars maintain that the evolution of word order in the history of Chinese is motivated by language contact, such as, among others, Hashimoto (1975), Tai (1976), and Norman (1982), While other scholars hold different opinions, such as Li and Thompson (1974), Chu (1987/1993), etc.

Li and Thompson (1974: 206) clearly states that Chinese has changed from SVO to SOV without external influence and notes that, "any change observed in Chinese word order must be originated internally."

Hashimoto (1975), however, proposes that Chinese has been undergoing a consistent "Altaicization" since the very beginning of its history. Ancient Chinese is typologically much closer to the Thai language than modern Chinese is, and the typological characteristics of modern Chinese dialects from north to south exhibit a continuum from Altaic to Thai.

Tai (1976) answers the question why Chinese shows more SVO features in the south and more SOV features in the north and argues that this is because the change from SVO to SOV started in the north and extended to the south. Northern dialects started to change due to their contact with Altaic languages as northern China had been ruled by Altai ethnic groups several times throughout its history. Moreover, as early as the eleventh century BC, the Zhou Dynasty in Northwest China conquered the Shang Dynasty in the east and prior to their migration to the eastern regions, Zhou people had already undergone language contact with Altai people.

Both Jerry Norman (1982) and Hashimoto (1987) note the use of the same markers for both passive and causative constructions (i.e., *jiào* 叫, *ràng* 让) in northern Chinese. They suggest that *jiào* and *ràng* expressing both causative and passive are the result of the influence of the Manchu language, as Manchu also uses only one marker for both causative and passive constructions. Jerry Norman (1982) attempts to explain this process as follows: rather than suggesting this grammatical feature was directly borrowed from Manchu, it is more accurate to say that this started out as a speech habit of some Manchus who spoke both Manchu and Chinese. In the process of learning Chinese, they would naturally bring some features of their mother tongue into Chinese. Such a viewpoint, however, does not conform to historical facts. The Chinese word *jiào* 叫/教, which was used to express both causative and passive, appeared as early as the Tang Dynasty (Ota Tatsuo 1958/ 2003, Jiang Lansheng 1999a), has nothing to do with the influence of Manchu language. Norman (1982), however, distinguishes interference from borrowing and suggests that it was interference rather than borrowing that led to the evolution of word order. His view that the features of a speaker's mother tongue are brought into the target language in changes of word order is precisely what linguists latter called shift-induced interference.

The Grammatical Evolution of the Translation of Buddhist Scriptures in Mediaeval Times from the Perspective of Language Contact

From about the Eastern Han Dynasty to the Tang and Song dynasties, a large number of Indian Buddhist scriptures were translated into Chinese. Consequently, the translation of Buddhist scriptures and the spread of Buddhist culture impacted Chinese culture greatly and also led to indirect contact between Chinese and the original languages of Buddhist scriptures. The original languages of Buddhist scriptures during the initial stage of contact were Middle Indian, the Northwest Indian dialect (Gandhara), and Central Asian vernaculars such as Turura. In the first and second centuries, Buddhist scriptures began to be translated into Sanskrit, and the original language of Chinese Buddhist scriptures also included Sanskrit and Buddhist Hybrid Sanskrit (Karashima, 2016: 13–17). Most of the early translators were monks from Central Asia, India, and the Western Regions, such as Gobharaṇa 竺法兰 from Central India, Parthamasiris 安世高 from Parthia (now Iran), and Kumārajīva 鸠摩罗什 from Kucha (now Kuqa in Xinjiang, China). Most of these translators had no knowledge of Chinese at first and needed the assistance of Chinese scholars to translate Buddhist scriptures. Since the language of the original Buddhist scriptures and the mother tongue of the early translators were very different from Chinese, the Chinese translation of Buddhist scriptures was inevitably influenced by these foreign languages. Some of these influences disappeared in the development of Chinese language later, while others penetrated into Chinese language and become part of Chinese grammar.

Watters Thomas (1899) paid attention to the Chinese translations of Buddhist scriptures and analyzed the influence of Buddhism on Chinese, such as "*gù*" 故 expressing the reason and "*yǐ*" 已 expressing the past. The Chinese scholar Liang Qichao 梁启超 (1920) was the first to notice the linguistic features of these translated scriptures and their influence on the general literary language. He describes 10 grammatical and stylistic features found in translated scriptures, including features such as "the large amount of inverted syntax." Walter Liebenthal (1935a, b–1936a, b) introduced the method of comparative analysis of Sanskrit-Chinese versions and used it to study the grammatical aspects of translation such as "*suǒ*" 所 being used as perfect passive participant, and "*zài* 在/ *yú* 于......*zhōng* 中" being used to indicate the locative. Later, Wang Li 王力 (1937), Zhou Yiliang 周一良(1944), Ota Tatsuo (1958/2003), and Zürcher (1977/1987) were all interested in the study of the Chinese translation of Buddhist scriptures and revealed several important grammatical features (Zhu Qingzhi 朱庆之 and Zhu Guanming 朱冠明 2006). After the 1980s and especially since the turn of the century, more scholars have paid attention to the translation of scriptures during the Six Dynasties and the comparative analysis of Sanskrit-Chinese versions. A group of scholars, such as Wan Jinchuan 万金川, Zhu Qingzhi, Seishi Karashima, Yu Hsiao-jung 遇笑容, Cao Guangshun 曹广顺, Zhao Changcai 赵长才, Hu Chirui 胡敕瑞, and Zhu Guanming, have made great advances in analyzing and comparing some grammatical phenomena found in translations dating to the Six Dynasties period. In this regard, Zhu Qingzhi and Zhu Guanming (2006) provide summaries of their findings.

Based on previous research results, the grammatical evolution caused by language contact during the translation of Buddhist scriptures in the medieval period can be summarized under six aspects: the genesis of new categories, the emergence of new functions, the emergence of new function words, the emergence of new sentence patterns, the emergence of new word order, and the increase in frequency of use.

The Genesis of New Categories

1. The Genesis of Plural Markers

    It is generally believed that the singular and plural in pre-Qin Chinese were homomorphic, that there was no grammatical category of number, and that the plural marker "*men*" 们 appeared in Modern Chinese. However, prior to the appearance of "*men*," several plural markers had already emerged during the translation of Buddhist scriptures. In ancient Chinese, nouns indicating category such as "*chai*" 侪, "*cao*" 曹, "*bei*" 辈, and "*shu*" 属 were added to personal pronouns to form "*wuchai*" 吾侪 and "*ruoshu*" 若属, which mean "us" and "you." However, words such as ""*chai*" 侪 are not plural markers. Zhu Qingzhi (1993) points out that "*deng*" 等 and "*cao*" 曹 appeared frequently after pronouns or nouns referring to people and only indicated plural meanings in the translation of Buddhist scriptures in the period of Middle Chinese (*Zhōnggǔ Hànyǔ* 中古汉语, approx. 220-1279 CE). Besides, personal pronouns deliberately distinguished between singular and plural. Therefore, he believed that the emergence of plural markers in Chinese is related to the influence of Buddhist scriptures and that as such, it was the product of language contact, not the result of the development of Chinese itself.

2. The Genesis of Perfective Aspect

Mei Tsu-lin 梅祖麟 (1981,1999) suggests that the perfective "*le*"了 appeared after the Tang Dynasty and was originally used in "V + O + le, clause" and that the predecessor of this "*le*"了 were formerly the perfective verbs "*jing* 竟/*qi* 讫/*yi* 已/*bi* 毕" used in "V + O + perfective verb, clause" prior to the Six Dynasties. Jiang Shaoyu 蒋绍愚 (2001) finds that "*yi*" 已 is different from "*jing*竟/*qi*讫/*bi*毕" in the translation of Buddhist scriptures: "*yi*" 已 can only be used in a sentence or at the end of a clause, not at the end of a sentence, while "*jing*" 竟 can be used at the end of a sentence. Adverbs of time cannot be added before "*yi*" 已, while "*jing*" 竟 can be used. The verbs before "yi" 已 can be a continuous verb, instantaneous verb, or stative verb, while the verbs before "jing" 竟 and some others must be continuous. Jiang Shaoyu notes that while the use of "*yi*" 已 after continuous verbs has been an inherent feature in Chinese since the pre-Qin period, the use of "*yi*" 已 after instantaneous verbs first appears in Buddhist scriptures and represents a translation of the Sanskrit absolute participle. On this basis, he divides "*yi*" 已 into two classes: "*yi₁*" 已₁ and "*yi₂*" 已₂ with the second of these serving as the predecessor of the perfective marker "*le*"了. Jiang argues that this evolution from "*yi₁*" 已₁ to "*yi₂*" 已₂ was influenced by Buddhist Chinese owing to the fact the Sanskrit absolute participle can be placed after a continuous or noncontinuous verb and that given these features, when the Sanskrit absolute participle was translated as "*yi₁*" 已₁, the nature of "*yi₁*" 已₁ was transformed and this new function of "*yi₂*" 已₂ was added. Prior to this, Zhu Qingzhi (1993) and Karashima (1998) all suggest that the appearance of "*yi*" 已 in translated scriptures was influenced by Sanskrit. Zhu argues that "*yi*" 已 represents a translation of the Sanskrit past participle, while Karashima holds that it was a translation of the Sanskrit absolutive participle. By employing a comparative analysis of the Sanskrit and Chinese versions of texts, Wang Jihong 王继红 (2004) and Long Guofu (2007) find that "VP+*yi* 已" corresponds to both the Sanskrit absolutive participle and past passive participle. Regardless of whether it correspond to the absolutive participle or past passive participle, the extension of the usage of "*yi*" 已 from continuous to instantaneous verbs indicates the genesis of the Chinese perfective aspect and serves as the foundation for the emergence of the later "*le*" 了.

The Emergence of New Functions

The Interrogative Pronoun *"yún hé"* 云何 Does Not Express Interrogative Mood

As interrogative pronoun, *"yún hé"* 云何 is first found in the era of the *Shijing* 诗经 [*The Book of Odes*] and also appears in literature dating to the period of Middle Chinese and was used as a predicate or adverb in sentences to express interrogative or rhetorical mood. However, while there are few cases of its use in native Chinese literature, it is commonly found in the translation of Buddhist scriptures during the Middle Chinese period (Yu Hsiao-jung 2003). Yu Hsiao-jung (2003) notes that in Chinese Buddhist texts, *"yún hé"* 云何 can be used not only in wh-questions conveying interrogative mood, but also in yes-no questions, V-not-V questions, and alternative questions without the function of the interrogative mood and that removing *"yún hé"* 云何 does not affect the expression of the interrogative. Yu Hsiao-jung further notes that there are two usages of the interrogative pronoun *kim* in Sanskrit, with one bearing an interrogative function while the other does not and suggests that the noninterrogative mood of *"yún hé"* 云何 in the translation of Buddhist scriptures is the result of the translation of the Sanskrit *kim*.

The Case Marking Function of Location Words *"suǒ"* 所 and *"biān"* 边

The words *"suǒ"* 所 and *"biān"* 边 were originally nouns, mean place and vicinity, respectively, and later came the meaning of "for...(related objects)" (Ota Tatsuo 1988/1991: 12). On the basis of a large textual corpus, Zhao Changcai (2009) illustrates the usage of *"suǒ"* 所 and *"biān"* 边 in the translation of Buddhist scriptures and their relationship under the lens of language contact. He notes that *"suǒ"* 所 and *"biān"* 边 usually appeared in the structure of "Prep + N + *suǒ/ biān*" in Buddhist translations, which can be used to express location. For example, *"xíngdào fósuǒ"* 行到佛所 ("move towards the place where the Buddha lives") in the *Zhōng běnqǐ jīng* 中本起经 [*Madhyama-ityukta sutra*] and *"zài cǐchí biān"* 在此池边 ("beside this pond") in the *Bǎiyù jīng* 百喻经 [*Upamāśataka sūtra*]. It can also be used to express an object as the following two examples from the *Xiányú jīng* 贤愚经 [*Damamūka nidāna sūtra*]

"yú dàshīsuǒ búgǎn yǒu xī" 于大师所不敢有惜 ("won't be mean to the master.")
"bōpóqiélí suī hàiyú wǒ, wǒ yú qíbiān, yǒngwú chēnhèn" 波婆伽梨虽害于我，我于其边，永无嗔恨
("Although Pravaghāṭī hurt me, I would never have resentment to him")

There were even cases such as "N + *suǒ*" without a preposition, for example, *"shìgù bǐrénsuǒ, búyīng shēng bùwèi"* 是故彼人所，不应生怖畏 ("Therefore (we) should not be afraid of that person") in *Dàzhuāngyánlùn jīng* 大庄严论经 [*Kalpanāmaṇḍitikā*].

The development from locative to dative is a very common path of grammaticalization. However, the expansion of the function of location words in the translation of Buddhist scriptures may be related to the influence of translation from the original scriptures. Through investigating both Sanskrit and Chinese materials, Zhao Changcai finds that the Sanskrit locative marker can represent both location and object.

The Usage of the Coordinate Conjunction "*yì*" 亦

The word "*yì*" 亦 was used as a coordinate conjunction to connect nominal (sometimes predicate) components and is found in the translation of Buddhist scriptures from the Eastern Han Dynasty to the Six Dynasties. For example, "*jīmáo yì jīn*" 鸡毛亦筋 ("chicken feather and tendon") means "*jīmáo*" 鸡毛 and "*jīn*" 筋. Zürcher (1977/1987) notices this phenomenon and believes that it was related to the influence of the original text or the translator's mother tongue. Xu Zhaohong and Wu Fuxiang 吴福祥 (2015) point out that the "affinity adverb > coordinate conjunction" is a common evolution as a cross-language phenomenon, and the function of the coordinate conjunction of "*yì*" 亦 in the Chinese translation of Buddhist scriptures evolved from the function of an affinity adverb, but that this evolution was caused by language contact, instead of originating independently in Chinese. In the original Sanskrit scriptures, *ca* had the function of both affinity adverb and coordinate conjunction, which triggered the evolution of "*yì*" 亦 from an affinity adverb to a coordinate conjunction in the translation of Buddhist scriptures. However, the use of the coordinate conjunction of "*yì*" 亦 was only a temporary innovation. It did not fully expand and spread in Middle Chinese and died out later as a result.

The Emergence of New Function Words

The Generation of the Reflexive Pronoun "*zì jǐ*" 自己

Before the Buddhist scriptures of the Eastern Han, the distribution of reflexive pronouns "*zì*" 自 and "*jǐ*" 己 was different. "*zì*" 自 can only be used before a verb or a preposition and cannot be used as attributives, while "*jǐ*" 己 had no such restrictions. After the Han, the use of "*zì*" 自 gradually eliminated these restrictions and could later be used as an attributive. Lü Shuxiang 吕叔湘 (1984) was the first to notice that "*zì*" 自 could be used as a genitive pronoun in the *Sānguó zhì* 三国志 [*Annals of Three Kingdoms*]. Following up this clue, Mei Tsu-lin (1986) finds some use of "*zì*" 自 as a genitive pronoun in documents dating to the Six Dynasties, including in the translation of scriptures. Yu Liming 俞理明 (1989) and Dong Xiufang 董秀芳 (2002) hold that the genitive pronoun "*zì*" 自 was first found in the translated scriptures. Through a comparative analysis of Sanskrit and Chinese versions of texts, Zhu Guanming (2007) notes that "*zì*" 自 in translated scriptures is generally a translation of the Sanskrit *sva*. He also notes two further similarities between the use of *sva* and "*zì*" 自: first their similar use as a reflexive pronoun and second in that they are placed before the verb when it is an object. At the same time, *sva* can be used as possessor, such as *sva-kārana* (self-cause, self-nature), and *sva-lā hha* (self-profit, self-gain, self-possession). The usage of *"zì"* 自 as a possessor in the attributive position may have arisen under the influence of the translator's mother tongue, that is to say that usage was transplanted from Sanskrit *sva* and that this emergence of new functions was caused by the translation of Buddhist scriptures. On this basis, Zhu (2007) analyzes the evolution process of *"zì"* 自 and finds that owing to the new function of *"zì"* 自 under the influence of Buddhist scriptures, it became a synonym with *"jǐ"* 己 in the possessor position with the same meaning and function. Due to a proliferation in the trend of disyllabic words in Middle Chinese and the requirements of the four character style of Buddhist scriptures, *"zì"* 自 and *"jǐ"* 己 were often used in conjunction and as a result gradually merged to form a disyllabic word.

The Emergence of the Topic Shifting Adverb *"fùcì"* 复次

While the word *"fùcì"* 复次, which means repetition and is equivalent to *"yòu"* 又, can be found in the *Shǐjì* 史记 [*the Records of the Grand Historian*] dating to the Western Han, its usage is rare, yet is commonly found in the translation of Buddhist scriptures. It can be used before nouns and verbs to indicate repetition and at the beginning of a sentence to indicate a change of topic. Zhu Guanming (2005, 2008) finds that the Sanskrit counterpart to *"fùcì"* 复次 was *punar-apara*. *Punar* and *apara* were synonymous and both meant *"yòu"* 又 (again). *"fù"* 复 and *"yòu"* 又 were common adverbs that indicate repetition in Middle Chinese. The reason why translators chose *"cì"* 次 instead of *"yòu"* 又 was that both *apara* and *"cì"* 次 had the meanings of "later" and "secondary" and can also be used as conjunctions to connect words or sentences. This double correspondence made translators prefer *"cì"* 次 and *"fù"* 复 to form *"fùcì"* 复次 together to act as a topic shifting marker, thus producing the adverb *"fùcì"* 复次 with coordinate structure.

The Emergence of New Sentence Patterns

The Generation of the Passive Sentence *"suǒ* 所+V"

Zhu Qingzhi (1993, 1995) points out that passive sentences were widely used in the translation of scriptures and that such usage may be influenced by the original source texts. Among such structures, *"suǒ* 所+V" passive sentences exist in a large number of Chinese translated Buddhist scriptures, but were rarely seen in native Chinese literature. This may be the result of passive or passive participles in the translation of the original scriptures and as such is a product of the influence of the source language. Zhu Qingzhi (1995) proposes an explanation that non-Chinese native speakers often used "hard translation," that is a word-for-word or even a morpheme-to-morpheme approach in their translations of Buddhist sutras. Sanskrit has a specific grammatical marker attached to the predicate verb to express the passive, and most of these grammatical markers were translated into Chinese as *"suǒ"* 所, thereby forming a *"suǒ* 所+V" type passive sentence. The *"suǒ* 所+V" structure represents a full Chinese translation of the passive predicate verb in the original text, in which *"suǒ* 所" served as the Chinese translation of the passive marker.

Nevertheless, Long Guofu (2009) hold that structure of "*suǒ* 所+V" originated internally owing to the development of Chinese itself. This is a subject worthy of continued research.

The Generation of the Disposal Construction "*qǔ* 取+OV"

As the disposal construction functions by putting the object before the verb, many scholars believe that the "*bǎ*" 把 disposal construction is a kind of SOV sentence (Li & Thompson 1974, James Tai 1976). The "*bǎ*" 把 in the "*bǎ*" 把 disposal construction changed grammatically from the verb position V_1 of the serial verb construction "V_1+O+V_2" into a preposition during the Tang (Zhu Minche 祝敏彻 1957, Li & Thompson 1974). Prior to this, the "*chí*" 持, "*qǔ*" 取, and "*jiāng*" type disposal constructions had appeared in Chinese translations of Buddhist scriptures (Zurcher 1977, Ota Tatsuo 1988/1991). Cao Guangshun and Yu Hsiao-jung (2000) note that a large number of "*qǔ* 取+OV" sentence patterns in Buddhist scriptures were the result of the omission of the resumptive pronoun "*zhī*" 之 in the "*qǔ* 取 + OV + *zhī* 之" pattern. However, usually the ellipsis rule of Chinese serial verb constructions is to omit the previous object, that is, "*qǔ* 取 + OV + *zhī* 之" > "*qǔ* 取 + V + *zhī* 之." Why is the latter object omitted in the translation of scriptures, that is, "*qǔ* 取 + OV + *zhī* 之" > "*qǔ* 取 + OV "? Cao Guangshun and Yu Hsiao-jung (2000) propose that translators were influenced by their mother tongue. In Sanskrit and Pali, the native language of translators, the object was often placed in front of the verb. In the process of translation, the translator, driven by the mother tongue, may have chooses "*qǔ* 取+OV," a construction where the object is placed before the verb.

The Generation of the Determinative-Sentence "S, N *shì* 是"

A particular type of determinative sentence structure, "S, N *shì* 是," appeared in Buddhist translations in Middle Chinese and was used to express truth judgments. Jiang Lansheng (2003) points out that this type of sentence pattern differs from conventional word order and ellipsis rules in Chinese, but is consistent with judgment sentences in Sanskrit where the postposition verb "to be" expresses emphasis. Jiang therefore comes to the conclusion that this judgment sentence ending with "*shì* 是" found in Chinese translations of Buddhist scriptures may be a sentence pattern created under the influence of Sanskrit. Jiang Shaoyu (2009) disagrees with this opinion, argues against the influence of Sanskrit, and suggests that this structure was the inheritance of the sentence pattern of "NP_1, NP_2 (predicative) + *shì* 是 (copula) *yě* 也" that emerged after the Western Han. Zhu Guanming (2005, 2013) and Jiang Nan 姜南 (2008) support the view of Sanskrit influence and attempt to further demonstrate the relationship between this pattern and Sanskrit by means of the comparative analysis of Sanskrit and Chinese versions of texts. Jiang Nan (2008) notes that the "S *shì* N" sentence in Buddhist Chinese mainly corresponds to the simple judgment sentence in the original text, while "S, N *shì*" was mainly used in complicated sentence patterns in the original, which mainly appeared in such situations for emphasis of identity. Zhu Guanming (2013) holds that "NP_1, NP_2 + *shìyě* 是也" and "S, N *shì* 是" had different origins. The former, as Jiang Shaoyu

(2009) notes, was the direct inheritance of the Chinese determinative-sentences after the Western Han, while the latter, as Jiang Lansheng (2003) notes, came about under the influence of Sanskrit.

The Emergence of New Word Order

The Order of the Vocative

While there is no vocative marker in Chinese, vocative phrases have been placed at the beginning of a sentence since pre-Qin times. In addition to following such Chinese structures, some special uses of the vocative are found in the translations of Buddhist scriptures. Zhu Qingzhi (2001) was the first to note that the parenthesis was used to translate the vocative in Buddhist Chinese. Zhao Changcai (2015) made a thorough and detailed investigation of the order of the vocative in such translated scriptures and finds that the vocative can be inserted between the subject and predicate, the predicate and object, the object and object, continuous clauses, and between the relative conjunction of complex sentences and their guiding clauses and that these uses may sometimes lead to misunderstanding. For example, *"shèshǐ, xūpútí, bú biànjiàn fǎ, yǒusuǒshēngchù bú?"* 设使, 须菩提, 不遍见法, 有所生处不?("Subhūti, if we don't see the dharma completely, is there a dharma which is produced?") found in the *Dàoxíng bōrě jīng* 道行般若经 [*Aṣṭasāhasrikā Prajñāpāramitā*]. While "Subhūti" here is vocative, given the word order of Chinese it could also easily be considered to be the subject of the sentence. In fact, the subject is omitted, and it refers to you, me, or anyone. The reason why the vocative can appear in these special positions in translated scriptures is that in Sanskrit the vocative marker and the word order are relatively free, so the translator translates them directly.

Postpreposition Phrases

The combination of "*yǔ*" 与 and the companion object forms prepositional phrases which historically precedes the verb such as the sentence "*gōng yǔ sānzǐ rùyú jìshìzhīgōng*" 公与三子入于季氏之宫 ("The marquis and his three sons entered the palace of the Ji family") found in the *Zuozhuan* 左传 [*The Zuo Commentary (on the "Spring and Autumn Annals")*]. However, in the Buddhist translations, the phrase consisting of the companion led by "*yǔ*" 与 can be placed after the verb, as found in the following phrase from the *Shēng jīng* 生经 [*Jātaka Stories*]:

"*wén rúshì: yīshí fó yóu nánánguó bōhénàishù jiān, yǔ dàbǐqiūzhòngbǐqiū wǔbǎirén*" 闻如是：一时佛游那难国波和柰树间，与大比丘众比丘五百人 ("Thus I have heard: once Buddha traveled to Pravara of Nananda country, with about five hundred eminent monks").

According to the Chinese convention, "*yǔ* 与" here should be placed before the verb "*yóu*" 游. It is also possible to add "*jù*" 俱 after "*yǔ*" 与, thus forming the construction of "*yǔ......jù*" in the translation of Buddhist scriptures. Zhu Qingzhi

(2001) suggests that the sentence pattern "*yǔ......jù*" is a translation of the instrumental case in the original text. Zhu Guanming (2008) argues that the word "*jù*" 俱 in "*yǔ......jù*" can be used as a verb and that such usage is found in the *Shǐ jì* 史记 [*Records of the Grand Historian*] that predates the Buddhist translations. The construction "*yǔ......jù*" in the Chinese translations of Buddhist scriptures may indeed be the translation of the prepositional invariant word sā*rdham* with two nouns of the instrumental case both expressing accompany, or the translation of a single noun that expresses an instrumental case in Sanskrit. This translation, however, did not have a structural impact on Chinese syntax, nor was it "an unsuccessful Sinicization." However, "*yǔ* 与......" without the following verb "*jù*" 俱 can appear after the verb, which did not conform to the Chinese language habit, and was directly translated frow nouns of instrumental case expressing accompany in Sanskrit by the translator.

The Postpositional Conjunction "*gù*" 故

As we know, Chinese usually uses prepositional conjunctions. In Pre-Qin Chinese, the conjunction "*gù*" 故 expressing causality is generally placed at the head of the result clause on causational compound sentences, and the result clause is placed after the causal clause, forming the format of "......, *gù* 故......." In the translation of scriptures, the word "*gù*" 故 can be placed at the end of the causal clause to form the structure "......*gù* 故, ..." At the same time, the causal clause can also be postpositioned to form the structure of "......, ...*gù* 故." This usage of "*gù*" 故 is not an inherent component of Chinese, but is generated by the influence of the original Buddhist source texts in translation. As use of "*gù*" 故 as a postpositional conjunction is not found in native Chinese literature, Zürcher (1977) believes that this was brought about due to a direct translation of the causative ablative in Sanskrit. While Takasaki (1993) and Wang Jihong (2004) put forward a different view on the translation components of the original scriptures, they both agree that this usage of "*gù*" 故 was not inherent in Chinese, but resulted from the influence of the original scriptures through the process of translation.

Increase in the Frequency of Usage

By taking the patient-as-the-subject sentence as an example, an important feature of Chinese syntax, Sun Xixin 孙锡信 (1992) finds that even though the patient subjective sentence "P + A + V," such as in the example "*qiáncái tārén yòng*" 钱财他人用 ("other people use money") appeared during the Han Dynasty, it was still very rare. Yu Hsiao-jung (2008) notes that this sentence type appeared in large numbers in Buddhist translations and believes them to be related to the influence of Sanskrit. Zhu Guanming (2005, 2011) demonstrates this in detail. With abundant examples taken from his comparative analysis of Sanskrit and Chinese versions of texts, Zhu illustrates how such sentences were more common in translated scriptures but rare in native Chinese literature composed during the same period. On one hand, the reason for their high frequency in translated works was a result of the development of Chinese itself. On the other hand, owing to case markers in Sanskrit, the position of the patient subject was relatively free and was often placed before the verb. During the translation process, the translator, influenced by Sanskrit, copied the Sanskrit word order and placed the patient subject first, thereby forming the patient-as-the-subject sentences of Chinese.

Of the grammatical variations noted previously, some innovations were not preserved in later periods such as the noninterrogative usage of "*yún hé*" 云何, "*yì*" 亦 for coordinate relations and "*fùcì*" 复次 for topic shifting, postposition of prepositional phrases. Some innovations, such as the plural category and the perfect category, however, were retained and developed further. Although the specific expression forms have changed, these two categories have taken root in Chinese and continue to this day. Another example is the postpositional conjunction "*gù*" 故. Although it was no longer used after the Tang Dynasty, the lexical category of postpositional conjunction remained, and the direct successors later is "...... *de yuángù*"......的缘故. Outside of this, forms such as, "...... *shí* 时," "...... *de shíhòu* 的时候," "...... *dehuà* 的话" are also in the same vein. Some have been retained and are still used even now, such as the reflexive pronoun "*zìjǐ*" 自己 and the patient-as-the-subject sentence.

Yu Hsiao-jung (2008) summarizes the theoretical issues related to grammatical evolution caused by language contact via translation. Many points of view in this paper are noteworthy. First, that scriptural translation is the result of second language acquisition with Chinese as the target language operating under contact between Chinese and the original source languages and that the grammatical phenomena found in these translations are variations due to translator's incomplete acquisition influence by their mother language. Second, that interference and borrowing are the modes of grammatical change caused by language contact. Such forms of borrowing are not easily retained, and interference serves as the basic cause of linguistic change. Third, interference can change the appearance and developmental direction of Chinese. For example, the emergence of the disposal construction in Middle Chinese and V in "VO+ *yǐ* 已" can be an instant verb. In these two cases, one represents a structural development interfered by language contact, and the other is where contact leads to the disappearance of semantic constraints. Interference can also increase the frequency of the use of phenomena inherent in Chinese such as patient-as-the-subject sentences. From the perspective of historical processes, interference due to contact usually exerts an influence on the development process and direction of Chinese within a short period of time, which makes the motivation of Chinese development diversified and the development path complicated.

The Evolution of Chinese Grammar after the Liao and Jin Dynasties from the Perspective of Language Contact

Prior to Mongol rule over the entirety of China during the Yuan Dynasty, the Liao, and Jin dynasties established by the northern Altai people ruled parts of northern China for over 300 years. During this period, the language of northern China underwent profound contact with the Khitan and Jurchen languages. However, due to the limited materials that have come down to use, we know very little about the changes in Chinese grammar caused by language contact during this period. During the Yuan when Mongolian peoples established a dynasty that ruled the entire country, the Mongolian language came into even deeper contact with northern Chinese. This contact eventually saw the formation of a variant of northern Chinese with significant Mongolian characteristics known as Han-er Yanyu 汉儿言语. This variant prevailed in northern China where it had already undergone influence from the Altaic language contact from the period of Middle Chinese onward (Ota Tatsuo 1953). As opposed to the lack of sources from the Liao and Jin periods, an abundance of linguistic sources and documents have come down to use from the Yuan that reflect the features of Chinese language of the time provide the necessary conditions for us to observe the history of language contact between Chinese and Mongolian and evaluate the evolution of Chinese grammar engendered by such contact.

The linguistic corpora reflecting such contact during the Yuan can be broadly divided into two types. The first type consists of Mongolian imperial edicts, government decrees, and documents translated into Chinese. The vernacular inscriptions of the Yuan Dynasty and metaphrastic documents such as *Yuán diǎnzhāng* 元典章 [*The Institutes of the Yuan Dynasty*] and *Tōngzhì tiáogé* 通制条格 [*A Newly Revised Compilation of Yuan Dynasty Legal Documents*] are the representative of this type. The second type consists of materials similar to the everyday spoken vernacular such as Yuan drama, conversation books for learning Chinese, and confessions of prisoners during interrogation. Works such as the conversation book known as the *Lao qida* 老乞大 which was designed for Korean people to learn Chinese can be regarded as representative documents of this type.

After the Yuan and Ming dynasties, the Qing Dynasty saw the Manchu people rule the entirety of China, and as a result the Chinese language once more underwent a period of profound contact with the Manchu language of the Altaic family. Zu Shengli (2013) uses the Chinese-Manchu conversational textbooks *Qīngwén qǐméng* 清文启蒙 [*Rudiments of the Manchu Language*], *Qīngwén zhǐyào* 清文指要 [*Essentials of the Manchu Language*], and *Xùbiān jiān Hàn Qīngwén zhǐyào* 续编兼汉清文指要 [*A Continuation of "Rudiments of the Manchu Language" with concurrent Chinese text*] compiled during the middle of the Qing as his main linguistic corpus, and systematically investigates the characteristics of the influences of Manchu among Chinese bannerman, which he further compares with the Mongolian Chinese of the Yuan. Zu Shengli (2011: 221) argues that while most of the special grammatical phenomena found in Mongolian Chinese under the influence of Mongolian grammar also occur in the language used by Chinese bannerman during the Qing, their forms of expression and syntactic components related to their corresponding source languages were different. Despite having had contact with Chinese during different historical periods, as both Mongolian and Manchu are Altaic languages, the results of this contact have several common characteristics.

Based on currently available research, the following section provides a brief summary of the evolution of Chinese grammar brought about by language contact after the Liao and Jin dynasties (mainly during the Yuan), which we suggest falls into four aspects.

14 The Evolution of Chinese Grammar from the Perspective of Language Contact

Changes to Pronouns

The Emergence of Inclusive and Exclusive First Personal Pronouns in the Northern Dialect

Prior to the Song, there was no distinction between the inclusive and exclusive plural form of the first person pronouns. However, from the twelfth century onwards, there was a contradiction between inclusive and exclusive pronouns in northern Chinese and this feature has continued to the present day in northern dialects. Liu Yizhi 刘一之 (1988) finds that in *Liú Zhīyuǎn zhūgōngdiào* 刘知远诸宫调 [*Liu Zhiyuan's Tune*], a text dating to the Jin, the exclusive form is clearly distinguished from the inclusive form with "*zán*" 咱 used as an inclusive form appearing 11 times, and "*ǎn*" 俺 used as exclusive form appearing 18 times. Furthermore, in this text, these two pronouns "*zán*" 咱 and "*ǎn*" 俺 are used solely for their respective conditions and are never conflated. Zhang Qingchang 张清常 (1982) maintains that the opposition between inclusive and exclusive forms in northern Chinese was due to the influence of Altaic. Mei Tsu-lin (1988) further points out that this opposition between inclusive and exclusive forms exists in Mongolian, Manchu-Tungusic, and some Turkic languages of the Altaic Family. The Khitan language is a Mongolian language and Jurchen is a Manchu-Tungusic language. Therefore, Northern Chinese introduced the distinction between inclusive and exclusive forms under the influence of Jurchen language or Khitan language. In addition, the continuation of this opposition in northern Mandarin was partly due to the later influence of Mongolian and Manchu.

The Special Usage of Plural Suffix "*měi*" 每

Zu Shengli (2002a, 2003) finds several usages of the plural suffix "*měi*" 每 in metaphrastic documents that are distinct to those found in the Chinese of the Yuan period. The main usage of "*měi*" 每 includes the following: (i) "N *měi* +N *měi* +N *měi*......," that is, "*měi*" 每 is added after the juxtaposition plural nouns one by one; (ii) "N *měi*," which can be modified with a quantifier before it; (iii) "*zhòng* 众 (*duō* 多) N *měi* 每," which is a combination of both "*zhòng* 众 (*zhū* 诸) N" and "N *měi* 每"; (iv) nouns not referring to a person can be followed by "*měi*" 每, that is to say, "*měi*" 每 can be used after the noun of inanimate things; (v) "V/VP *de měi* 的每" representing the identity of the character, which is the result of translating the plural form of adjective verbs referring to human in the Mongolian language; (vi) "*měi*" 每 used after the demonstrative personal pronouns "*zhè de* 这的(底)" and "*nà de* 那的(底)," with the meaning of "they."

In Middle Mongolian (*Zhōnggǔ Měnggǔ* 中古蒙古), nominal words such as nouns and pronouns have the grammatical feature of number. The singular is expressed in the stem form and the plural is expressed in the stem followed by the additional elements, or the inflection of the stem. The special usage of the plural suffix "*měi*" 每 (i.e., "*men*" 们) in the metaphrastic documents is closely related to the additional elements that indicate the plural in Mongolian.

Other Special Pronoun Usage

Zu Shengli (2001b, 2003) also discusses other special uses of pronouns in the vernacular inscriptions of the Yuan Dynasty and the "Xíngbù" 刑部 ["Compilation of Legal Instruments"] chapter found in the *Yuán diǎnzhāng* 元典章 [*The Institutes of the Yuan Dynasty*]. First, Zu Shengli notes that the genitive form of the personal pronoun being used as an attributive and being placed after the head of the noun is the result of the literal translation of Mongolian. Second, that demonstrative pronouns can also be used as third person pronouns. This is because there are no independent and specialized third person pronouns in Middle Mongolian, which are often expressed by demonstrative pronouns. Third, that interrogative pronouns are often used in arbitrary reference and can be used with unconditional conjunctions such as "*bùjiǎn*" 不拣 and "*bùyǐ*" 不以 to indicate arbitrary reference. The expression of "unconditional conjunctions + interrogative pronouns" translates the arbitrary reference usage of interrogative pronouns in Mongolian. And fourth, special rhetorical question "*bù......nàshénme*" 不......那甚麼 is the translation of a rhetorical question sentence pattern in the Middle Mongolian.

Location Words Used as Case Markers or Postpositions

The Localizer "*háng*" 行 Used as the Case Marker

Yu Zhihong 余志鸿 (1983, 1987) looks at the syntactic function of the postposition "*háng*" 行 in Yuan drama and the *Ménggǔ mìshǐ* 蒙古秘史 [*Secret History of the Mongols*] and believes that this postposition was the translation of an auxiliary word in the accusative, dative, locative, ablative, and genitive case in Mongolian. In this regard, Jiang Lansheng (1998) puts forward a new view. She distinguished sentences with the postposition word "*háng*" 行 in literature dating to the Song, Yuan, and Ming into two categories, namely, type-A; "V/Prep +N *háng* (+VP)" (such as "*xiàng zánháng* 向咱行 'to us'" and "*xiàng shuíháng sù* 向谁行宿 to + whom + stay 'where to stay'") and type-B; "N *háng* +VP" (such as "*jūnwángháng zòu* 君王行奏 to the throne + memorialize 'memorialize to the throne'"). Jiang Lansheng points out that only type-A existed during the Song and suggests that this type originated from within Chinese itself. She further suggests that type-B was a new sentence pattern in Chinese during the Yuan Dynasty, that this pattern was influenced by the word order of Mongolian, and that it gradually disappeared after the end of the Yuan Dynasty. In terms of its origin, she argues that "*háng*" 行 was not a simple translation of the Mongolian case auxiliary word, and its pronunciation had nothing to do with the case auxiliary word. Moreover, the original word of "*háng*" 行 was "*shàng*" 上, which was originally a Chinese word. It is an issue open to further investigation.

The Localizers "*shàng /shàngtou*" 上/上头 Used as the Causal Postposition

In the metaphrastic documents, the localizers "*shàng/shàngtou*" 上/上头 were often used after causal adverbials, which was equivalent to "......*deyuángù*" 的缘故

[reason] or "*yīnwéi*" 因为 [because]. "*Shàng/shàngtou*" 上/上头 was often used with the causal conjunctions "*wèi*" 为 and "*yīn*" 因 to form the structure of "*wèi/yīn......shàng/(de)shàngtou.*" After investigating the translations of the *Ménggŭ mìshĭ* and *Huáyí yìyŭ* 华夷译语 [*Translation of Chinese and Foreign Languages*] from the early Ming, Zu Shengli (2003, 2004) finds that the usage of "*shàng/shàngtou*" 上/上头 as the causal postposition was mainly the result of the translation of *tula* and *ar/bar* from Middle Mongolian with *tula* as the postposition expressing cause and *ar/bar* as an additional element of the instrumental case of adjectival verbs. As they had the same grammatical position and similar meaning and function, they could be used in translation for each other. Under the influence of the Mongolian causal postposition, many cases of the causal postposition "*shàng/shàngtou*" 上/上头 can be found in the nonmetaphrastic documents of the Yuan period. During the Ming Dynasty, with the decline of the influence of language contact with Mongolian, traditional prepositions in Chinese regained their dominant position causing the causal postposition "*shàng/shàngtou*" 上/上头 to gradually disappear. However, during the Qing, due to the influence of the word order of Manchu, "*shàng/shàngtou*" 上/上头 was often used as a causal postposition once more.

Other Location Words Used as Postpositions

Yu Zhihong (1992) points out that Chinese had a well-organized postposition system during the Yuan. In addition to the multifunctional "*háng*" 行, "*shàng/shàngtou*" 上/上头 indicating the cause, and "*hē/shí*" 呵/时 indicating the hypothesis, there were also the postposition words "*gēnqián/gēndĭ*" 根前/根底 used as case markers for the locative and dative and postposition words "*lĭ/chù*" 里/处 indicating place and origin. These postpositions were often used in parallel with the inherent prepositions in Chinese, thus forming the PRE-POS structure. Yu Zhihong (1992) believes that this was the result of contact between two languages with different structures.

Based on specific analysis of the case marking function of location words in the vernacular inscriptions of the Yuan Dynasty and the "Xíngbù" chapter of the *Yuán diănzhāng*, Zu Shengli (2001a, 2003) points out that the reason for the translation of these words was due their same grammatical position and similar function. Mongolian is an agglutinative language, and its nominal words, such as nouns, pronouns, numerals, and adjectives (including verbs), have the grammatical category of case, which indicates the various grammatical relations between nominal words and verbs. Mongolian nominatives bear zero case marking and the declensions of other cases are marked by various additional elements after the root or stem, while Chinese, as an analytical language, mainly expresses grammatical meaning of case by using function words and changes to word order. In the Mongolian-Chinese translation materials from the Yuan, many declensions of Mongolian nominal words were omitted, combined, and simplified, and the Chinese location words "*lĭ*" 里 [inner], "*gēndĭ*" 根底 [front], "*háng*" 行 [place], and "*chù*" 处 [place] were used to translate these case markers, thereby giving these location words new grammatical functions. These usages tended to disappear gradually during the Ming. However, due to the influence of the Manchu language during the Qing Dynasty, some location words regained their function as case markers (Zu 2013).

Postposition Hypothetical Conjunction "*he*" 呵

In the metaphrastic documents of Yuan Dynasty, the modal particle "*he*" 呵 was frequently used in a variety of ways, the most common of which was used as a hypothetical marker at the end of the preceding clause in a hypothetical sentence with "*he*" 呵 as a translation of the additional elements of the hypothetical adverbial verbs in Mongolian. "*He*" 呵 was often used with hypothetical conjunctions such as "*ruò*" 若 and "*rú*" 如 to form "*ruò/ rú......he.*" These phenomenon have been discussed in many works including Li Taizhu 李泰洙 (2000a) and Zu Shengli (2002b, 2003, 2013).

By means of comparison, "*shí*" 时 and "*hòu*" 后, which were derived from the grammaticalization of time nouns during the Tang, were the postpositional conjunctions (generally called auxiliaries) that appeared earlier in Modern Chinese to express hypotheses. "*He*" 呵 emerged during the Song and was used in or at the end of a sentence to express various moods, such as pause, hypothesis, interrogative, rhetorical, imperative, or exclamation.

Influenced by Mongolian, the usage of "*he*" 呵 at the end of a hypothetical clause to indicate hypothesis was widely used during the Yuan. After the middle of the Ming, the hypothetical auxiliary "*he*" 呵 gradually disappeared and was replaced by "*shí*" 时 as the main posthypothesis marker. The reason why the metaphrastic documents chose "*he*" 呵 instead of "*shí*" 时 or "*hòu*" 后 as the main translation form of *asu/basu* is probably that "*shí*" 时 or "*hòu*" 后 as a hypothetical auxiliary word still had lexical meaning to some extent, and at the same time they were two common nouns, which were also used to translate corresponding nouns in Mongolian. In contrast, "*he*" 呵 was a full-time modal particle, which originally indicated the interval of mood, so it can be seen in a variety of tenses and sentence patterns, producing a variety of mood effects, and it was also close to *asu/esu* phonetically.

Auxiliaries Related to Verbs

While Chinese lacks morphological changes after the pre-Qin period, the surrounding Altai languages were rich in morphological changes, including grammatical categories, such as states, form, aspect, tense, and person, as well as morphemes of adverbial verbs and adjective verbs, auxiliary verbs, and quotation verbs, which are quite different from Chinese. Influenced by Mongolian, some auxiliary words related to the form of verbs appeared in the metaphrastic documents of the Yuan. Their grammatical functions were obviously different from the previous usage, for example, "*zhe*" 着 [a particle for continuous] was used as the conjunctive component of verbs, "*yǒu*" 有 [to have] was provided with multiple functions at the end of a sentence, and "*zhě*" 者 [nominalization marker] expressed the imperative mood at the end of a sentence.

"*Zhe*" 着 as the Conjunctive Component of Verbs

The word "*zhe*" 着 has been used as a continuous particle since the late Tang. Zu Shengli (2002b), however, finds that in the metaphrastic documents of the Yuan

Dynasty, the dynamic auxiliary "*zhe*" 着 was often used between two verbs to construct serial verbs "V$_1$ + *zhe* + V$_2$," or constructed "V$_1$ *zhe* (O$_1$) + V$_2$ *zhe* (O$_2$) … V$_N$ (O$_N$))" structure to connect coordinated/serial verbs, or used to translate the marker of successive converbs or preparative converbs in Mongolian, so that "*zhe*" 着 can not only express continuity, but also used as the linker of verbs. Li Chongxing 李崇兴 (2005) further points out the influence of the translation of "*zhe*" 着 on Chinese grammar. One was the emergence of the manner-expressing usage of "*zhe*" 着. The addition of "*zhe*" 着 after the verbs in the Yuan was necessary for translating Mongolian adverbs. The translation expanded the application of "*zhe*" 着 and changed its grammatical function accordingly. Therefore, the way-expressing usage of "*zhe*" 着 appeared, and "*zhe*" 着 even became a grammatical marker of way-expressing. The other is preposition with "*zhe*" 着, such as "*yīnzhe*" 因着 [because], "*yīzhe*" 依着 [in accordance with]. In the literal translation, "*zhe*" 着 was often used to translate the marker of successive converbs or preparative converbs in Mongolian. In the structure of "*yīn/yī* + O + V," "*yīn*" 因 and "*yī*" 依 were very similar to preparative converbs in Mongolian. Therefore, it was possible to make "*yīn*" 因 and "*yī*" 依 carry the word "*zhe*" 着 by following the convention of translating preparative converbs.

In the Chinese of the bannermen of the Qing Dynasty, the structure of "V + *zhe* + V" was also a reflection of the form of successive converbs in Manchu, which was consistent with the correspondence of "*zhe*" in the metaphrastic documents (Zu 2013). The rationale for such translation is that the "*zhe*" in "V$_1$*zhe* (O$_1$) V$_2$(O$_2$)" in Song and Yuan Chinese was similar to the marker of Mongolian successive converbs and preparative converbs in syntactic position and function, as well as in pronunciation.

Postpositioned Saying Verbs and their Usage as Quotation Markers
In Yuan period documents, "*dào*" 道 (speak) and "*medào*" 么道 (speak) can be used as postpositioned quotation verbs, such as "'*bìng*' *medào tuīcízhe*" "'病' 么道推辞着 (refuse because of illness) in the "*Xíngbù*" chapter of the *Yuán diǎnzhāng*. It can be further used before another speaking verb, which was equivalent to a quotation marker as per the following example from the same work, "*cóng chéngjísī huángdì dàojīn, 'jiāo sēngrén zhùshòu zhě' medào dàolái.*" 从成吉思皇帝到今,'教僧人祝寿者'么道道来 (From the Emperor Genghis Khan to the present, "let the monks celebrate birthday" said). Zu Shengli (2003) analyzes this problem and considers it to be the result of translating the Mongolian quotation verb *ke.e-*. Zu Shengli (2004) further believes that "*medào*" 么道, as a translation form of *ke.e-*, was not pure spoken Chinese vocabulary of the Yuan Dynasty but was likely to be used in the Mongolian-Chinese spoken by the Mongolians, rather than simply as a written form found in translation. On the other hand, due to the influence of language contact, the quotation verbs which belong to Mongolian may have also infiltrated into the Chinese of the Yuan period. In Chinese-Manchu conversational materials of the Qing Dynasty, we can also see the usage of "*shuō*" 说 (speak) and "*dehuà*" 的话 (saying) corresponding to the Manchu quotation verb *se-*.

The Sentence-Final *"yǒu"* 有 with Multiple Functions

In Chinese, the verb *"yǒu"* 有 [to have] is a notional verb indicating possession or existence. In the metaphrastic documents from the Yuan Dynasty, *"yǒu"* 有 had a series of special usages at the final in a sentence besides the conventional usages. Here *"yǒu"* 有 can be used as a verb, but its meaning is equivalent to *"zài"* 在 which means existence, or *"shì"* 是 which means judgment. When indicating existence, *"yǒu"* 有 can be used independently, for example, *"Tiēmùzhēn jiālǐ yǒuyěwú?"* 帖木真家里有也无 ("Is Temujin at home") in the translation of *Yuáncháo mìshǐ* 元朝秘史 [*The Secret History of Yuan Dynasty*]. The word *"yǒu"* 有 can also be collocated with *"zài"* 在 as per the following from the same work, *"Tiēmùzhēn āndá de bǎixìng zài wǒzhèlǐ yǒu."* 贴木真安答的百姓在我这里有 ("The people of the Temujin brother are here with me"). It can also be used independently or be collocated with *"shì"* 是 for expressing judgment, for example, *"'nà dálǔhuāchì shì shènmerén yǒu?' medào shèngzhǐ wènhe, huízòu: 'xìngcuī de hànrrén yǒu'."* "那达鲁花赤是甚么人有?" 么道圣旨问呵,回奏:"姓崔的汉儿人有。" ("Who is that Darughachi?" the Emperor asked, replied: "It's a Chinese surnamed Cui.") found in the "Xíngbù" chapter of the *Yuán diǎnzhāng*. As the final of a sentence, *"yǒu"* 有 can also be used as an auxiliary to indicate grammatical meanings such as tense and aspect, for example, *"zǎizhe yángmáo yǒu."* 载着羊毛有 ("Carrying the wool") found in the *Yuáncháo mìshǐ*.

Zu Shengli (2003) makes a detailed analysis of this phenomenon and argues that the usage of *"yǒu"* 有 in the metaphrastic translation documents of the Yuan are mainly the result of copying and transplanting the special verbs *a-* and *bü-* in Middle Mongolian. Since *a-* and *bü-* had substantial meaning and can be used as auxiliary verbs, and there was no equivalent grammatical component in Chinese, therefore they are literally translated as the common verb *"yǒu,"* resulting in some special uses of *"yǒu."* These special uses of *"yǒu"* existed both in the ancient text *Lao Qida* 老乞大 and in the Chinese of Han bannermen during the Qing Dynasty. Lee Tae Soo (2000c) classifies and describes the auxiliary *"yǒu"* at the final in a sentence in *Lao Qida*. In the structure *"yǒu/méi......yǒu"* 有/没......有, former *"yǒu"* is a verb, and the latter *"yǒu"* is an auxiliary. The *"yǒu"* at the final in a sentence can also be used in determinative, existential, declarative, and interrogative sentences. In modern Qinghai Chinese, there is a similar such a sentence with *"yǒu"* at the end of sentence and Jia Xiru 贾晞儒 (2015) suggests that its grammatical function here is basically the same as that of *"yǒu"* found at the end of vernacular sentences during the Yuan and that this represents the legacy of Mongolian influence on Yuan Chinese in Qinghai Chinese.

The Sentence-Final Particle *"zhě"* 者 as an Imperative Marker

In the literal translations from the Yuan period, the modal particle *"zhě"* 者 was widely used as the final in a sentence to express the imperative mood, for example, *"shālerénde, qiāolezhě."* 杀了人的,敲了者 (those who kill others shall knock them down) in the "Xíngbù" chapter of the *Yuán diǎnzhāng*. Prior to this, the sentence-final particle *"zhě"* 者 appeared very early in Chinese, and *"zhě"* 者 was widely used in Song and Yuan Chinese and can express many types of imperative moods, such as will, command, hope, request, and warning. From the functional point of view, it was quite similar to the imperative verb in Mongolian. Meanwhile, from the perspective of syntactic position, because of the SOV word order, the marker of the imperative verb in Mongolian was always at the end of the sentence, which was completely consistent with the position of the modal auxiliary particle *"zhě"* 者 in Chinese sentences. Therefore, Zu Shengli (2003) concludes that the expression of imperative by *"zhě"* 者 in Yuan Dynasty was the direct translation of the imperative marker of the Mongolian verbs, and the chosen of *"zhě"* 者 in the translation had both functional and syntactic position motivation. However, the nature of *"zhě"* 者 changed in metaphrastic documents in Chinese. The Chinese *"zhě"* 者 and Mongolian imperative marker were essentially different grammatical components. The former was a kind of postpositive function word, which can be omitted, and the corresponding imperative meaning can also be expressed by means of vocabulary and intonation, while the latter was a suffix, which must be attached to the stem of the verb or other additional elements, and the imperative meaning of the sentence should be reflected by the imperative form of the verb. In the Manchu-Chinese comparative materials of the early to middle Qing period, the modal particle *"shìne"* 是呢 expressing imperative meaning at the final in a sentence corresponded to the imperative suffix of Manchu verbs, which was the interference feature of Manchu. *"Shìne"* 是呢 and *"zhě"* 者 in the Yuan Dynasty had a type relationship of contact influence (Zu Shengli 2013).

Special Grammatical Phenomena in *Lao Qida*

Many of the important variations of Chinese grammar after the Liao and Jin (mainly during the Yuan) have been summarized above according to the subcategories of syntax and semantics. The following section looks at this from another perspective and focuses on grammatical phenomenon caused by contact during the Yuan and Ming as reflected in the *Lao Qida* 老乞大. The *Lao Qida* is a conversational textbook for learning Chinese written during the Goryeo (918–1392 CE) and Joseon (1392–1897 CE) periods in Korea. Ota Tatsuo (1953) notes that it faithfully reflects the spoken language of that time. There are four versions of *Lao Qida* in existence: (i) The original *Lao Qida*, a newly discovered version found in 1998 that may reflect northern Chinese during the Yuan; (ii) The *Lǎo Qǐdà yànjiě* 老乞大谚解 annotated by Cui Shizhen during the reign of Joseon Jungjong in Korea (1506–1515) and roughly reflects the northern Chinese of the early Ming; (iii) The *Lǎo Qǐdà xīnshì* 老乞大新释 published in 1761 and was heavily revised on the basis of previous editions and reflects the situation of northern Chinese during the Qing; and (iv) The *Chóngkān Lǎo Qǐdà* 重刊老乞大 published in 1795 also mainly reflects the appearance of northern Chinese during the Qing. In the past, more attention was paid to the *Lǎo Qǐdà yànjiě* but since the beginning of this century, scholars have generally concentrated on the ancient edition the *Lao Qida* as the basis for their research. Lee Tae Soo (2000a, 2000b, 2000c) and Zu Shengli (2011) mainly discuss some special grammatical phenomena arising from language contact in the *Lao Qida* which mainly correspond to the contents discussed in the previous three sections above:

   (i) The special use of the plural suffix "*měi*" 每
   (ii) The plural first person pronouns strictly distinguishing between inclusive and exclusive
  (iii) The use of location words as case markers
  (iv) "*shàng/shàngtou*" 上/上头 used as the causal postposition
   (v) The use of the sentence-final particle "*he*" 呵 extensively as a hypothetical clause marker
  (vi) The substantial use of the answer phrase "*nàbānzhě(hē)*" 那般者(呵) [that's all right]
 (vii) The "V$_1$+*zhe*着+V$_2$" structure
(viii) The abnormal word order of some adverbs
  (ix) A large degree of SOV word order
   (x) The extensive use of the sentence-final auxiliary word "*yǒu*" 有
  (xi) The widely used imperative sentence-final particle "*zhě*" 者
 (xii) The use of the determined modal particle "*yězhě*" 也者 at the end of sentences
(xiii) The rhetorical question of "*bú*+VP+*nàshènme*" 不 VP 那甚么 [not-VP SFP do-what]

Conclusion

China's vast territory, long history, and precious language resources represent a cultural heritage and diverse linguistic environment that provide us with unique conditions for the study of language contact. Many scholars have been keenly aware of the enormity of this field and have made great achievements in its advancement. A thorough study of the phenomena of language contact in Chinese, its history, conditions, contact types, and its mechanism of evolution can not only increase our knowledge of the Chinese language and its history, but also has great value for enriching linguistic theories in general.

Over the several thousand years of contact between the Chinese and northern Altaic languages, the results have shown that on the one hand, the ethnic minorities who entered China eventually abandoned their original languages and shifted to using Chinese, yet on the other hand, the Chinese language has also undergone a series of changes under the influence of language contact.

While the evolution of specific grammatical forms and meanings in the history of the development of Chinese may be very complex and the specific grammatical phenomena, evolution, duration, and final outcome of such evolution may also be very varied, there remain some common features – that the evolutionary process of Chinese grammar is not only forged by internal factors and inherent structural constraints, but that this evolution also came about through contact-induced changes such as the evolution of word order and postposition system described above. In brief, the influence of language contact, specifically language shift, plays an important role in the evolution of grammar and cannot be ignored. The influence of language contact on Chinese grammar occurred mainly through the transference of other languages in the process of language shift. That is to say, the original ethnic minorities in northern China gradually gave up their mothers tongue and shifted to Chinese after they settled in China. In the process of language shift, some features of the mother tongues of the original ethnic minorities were brought into the Chinese language, which led to the evolution of the northern Chinese language. The morphological and syntactic evolution of the transferring features that were absorbed into Chinese slowly shaped and changed the linguistic type of the Chinese language. This evolution took place under the synergistic effect of internal development patterns and external contact influences.

A certain grammatical component or structure in Chinese language may be the result of a comprehensive effect under the influence of multiple foreign language contacts in different historical periods. Some types of evolution caused by contact may have occurred repeatedly and intensified throughout history. For example, the usage of case markers of localizers, the postposition "(yě) sì" (也)似 expressing analogy, and "VP+qù 去" expressing purpose, all appeared successively in the translation of scriptures during the Six Dynasties and were further intensified during the Jin and Yuan.

Some features of evolution caused by contact have survived and entered the Chinese grammatical system, while some have only lasted for a short period of time and were not finally retained in the Chinese language. Whether or not such features were retained in Chinese is not only related to the depth and duration of language contact, but also to the structural constraints of the Chinese language itself.

References

Cao, Guangshun (曹广顺) and Yu Hsiao-jung (遇笑容). 2000. Zhonggu yijing zhong de chuzhishi 中古译经中的处置式 [The disposal construction of translated Middle Chinese Buddhist Sutras]. *Zhongguo Yuwen* 中国语文 [*Studies of the Chinese Language*] 6: 555–563,576.

Chen, Changlai (陈昌来). 2014. *Hanyu "jieci kuangjia" yanjiu* 汉语"介词框架"研究 [*Study on Chinese "Preposition Frame"*]. Beijing: Shangwu Yinshuguan 北京: 商务印书馆 [Beijing: Commercial Press].

Chu, Chauncey Cheng-His. 1987. *Historical syntax: Theory and Application to Chinese*. Taipei: Crane Publishing Company//1993. *Lishi yufaxue lilun yu hanyu lishi yufa* 历史语法学理论与汉语历史语法 [*The theory of historical grammar and Chinese historical grammar*]. Beijing: Beijing Yuyuan Xueyuan Chubanshe 北京:北京语言学院出版社 [Beijing: Beijing Language Institute Press].

Comrie, Bernard. 2008. The areal typology of Chinese: Between North and Southeast Asia. In Redouane Djamouri, Barbara Meisterernst & Rint Sybesma (eds.) *Chinese linguistics in Leipzig*. Collection des Cahiers de Linguistique Asie Orientale 12. Paris: Ecole des Hautes Etudes en Sciences Sociales, Centre Recherches Linguistiques sue L'Asie Orientale. 1–21.

Dong, Xiufang (董秀芳). 2002. Guhanyu zhong de "zi" he "ji"—xiandai hanyu "ziji" de teshuxing de laiyuan 古汉语中的"自"和"己"——现代汉语"自己"的特殊性的来源 ["Zi"(自) and "Ji"

(己) in classical Chinese—The origin of the particularity of "ziji" in Modern Chinese]. *Guhanyu Yanjiu* 古汉语研究 [*Research in Ancient Chinese Language*] 1: 69–75.

Dryer, Matthew S. 1992. The Greenbergian word order correlations. *Language* 68: 81–138.

———. 2007. Word order. In *Language typology and syntactic description*, ed. Timothy Shopen, vol. 1. Cambridge, UK: Cambridge University Press.

Greenberg, Joseph H. 1963. Some universals of grammar with particular reference to the order of meaningful elements. In *Universals of language*, ed. Joseph H. Greenberg, 73–113. Cambridge, MA: MIT Press.

Hashimoto, J.M. 1975. Languge diffusion on the Asia continent: Problems of typological diversity in Sino-Tibetan. *Computational Analyses of Asian and African Language.* 3: 49–65.

Haspelmath, Martin, Matthew S. Dryer, David Gi, and Bernard Comrie, eds. 2005. *The world atlas of language structures.* Oxford: Oxford University Press.

Hsiao-jung Yu (遇笑容). 2003. Shuo "yunhe" 说"云何" [A discussion on the interrogative word "yunhe"]. In *Kaipian* 开篇(第22期)[*The opening,* vol. 22], 48–50. Dongjing: Haowen Chuban 东京:好文出版 [Tokyo: The Kohbun Press].

——— (遇笑容). 2008. Lilun yu shishi: yuyan jiechu shijiao xia de zhonggu yijing yufa yanjiu 理论与事实:语言接触视角下的中古译经语法研究 [Theory and Fact – A Study of the translated Buddhist Sutras of the Mediaeval period from the perspectives of language contact]. In *Hanyushi Xuebao* 汉语史学报(第七辑) [*Journal of Chinese language history*, vol. 7], 121–127. Shanghai: Shanghai Jiaoyu Chubanshe 上海:上海教育出版社 [Shanghai: Shanghai Educational Publishing House].

Jia, Xiru (贾晞儒). 2015. Mengguyu yu hanyu jiechu guocheng zhong de "you" de xiaozhang—yuandai baihua jumo "you"zi tanqi 蒙古语与汉语接触过程中的"有"的消长——元代白话句末"有"字谈起 [A study of Chinese "you" through inspecting of the contact between Mongolian and Chinese languages – Talk about "you" at the end of the sentence in vernacular Chinese in Yuan Dynasty]. *Xibu Menggu Luntan* 西部蒙古论坛 [*Journal of the Western Mongolian Studies*] 1: 61–69,127.

Jiang, Lansheng (江蓝生). 1992. Zhuci "shide" de yufa yiyi jiqi laiyuan 助词"似的"的语法意义及其来源 [The grammatical meaning and origin of the auxiliary word "shide"]. *Zhongguo Yuwen* 中国语文 [*Studies of the Chinese Language*] 6: 445–452.

——— (江蓝生). 1998. Houzhici "hang" kaobian 后置词"行"考辨 [A Study of the postposition word "Hang"]. *Yuwen Yanjiu* 语文研究 [*Linguistic Research*] 1:1–10,15.

——— (江蓝生). 1999a. Hanyu shiyi yu beidong jianyong tanyuan 汉语使役与被动兼用探源 [On the origin of the dual use of causation and passivity in Chinese]. In Honor of Mei Tsu-Lin: Studies on Chinese historical syntax and morphology (A Peyraube and Sun Chaofen, eds.). Paris: Ecole des Hautes Etudes en Sciences Sociales.

——— (江蓝生). 1999b. Cong yuyan shentou kan hanyu binishi de fazhan 从语言渗透看汉语比拟式的发展 [The development of analogy in the Chinese language as seen from language infiltration]. *Zhongguo Shehui Kexue* 中国社会科学 [*Chinese Social Sciences*] 4:169–179,208.

Jiang, Shaoyu (蒋绍愚). 2001.《Shishuoxinyu》、《Qiminyaoshu》、《Luoyangqielanji》、《Xianyujing》、《Baiyujing》zhong de "yi"、"jing"、"qi"、"bi"《世说新语》、《齐民要术》、《洛阳伽蓝记》、《贤愚经》、《百喻经》中的"已"、"竟"、"讫"、"毕" [The "yi", "jing", "qi" and "bi" in Shishuoxinyu(A new account of the tales of the world), Qiminyaoshu (Essential techniques for the peasantry), Luoyangqielanji(Record of Buddhist Temples in Luoyang), Damamūka nidāna sutra and Upamāśataka sutra]. *Yuyan Yanjiu* 语言研究 [*Studies in Language and Linguistics*] 1: 73–78.

Jiang, Lansheng (江蓝生). 2003. Yuyan jiechu yu yuanming shiqi de teshu panduanju 语言接触与元明时期的特殊判断句 [Language contact and special determinative-sentences in the Yuan and Ming Dynasties]. In *Yuyanxue Luncong* 语言学论丛(第二十八辑) [*Essays on linguistics*, vol. 28], 43–60. Beijing: Shangwu Yinshuguan 北京:商务印书馆 [Beijing: The Commercial Press].

——— (江蓝生). 2004. Kuaceng feiduanyu jiegou "dehua" de cihuihua 跨层非短语结构"的话"的词汇化 [About the lexicalization of the cross-structural combination "de hua (的话)"]. *Zhongguo Yuwen* 中国语文 [*Studies of the Chinese Language*] 5: 387–400,479.

Jiang, Shaoyu (蒋绍愚). 2009. Yetan hanyi fodian zhong de "NP1, NP2+shiye/shi" 也谈汉译佛典中的 "NP1, NP2+是也/是" [Talking about "NP1, NP2+shiye/shi" in the Chinese translation of Buddhist scriptures]. In *Zhongguo Yuyanxue Jikan* 中国语言学集刊(第三卷第二期) [*Collection of Chinese Linguistics*, vol. 3(2)], 29–44. Beijing: Zhonghua Shuju 北京: 中华书局 [Beijing: Zhonghua Book Company].

Jiang, Nan (姜南). 2010. Hanyi fojing "S, N shi" ju feixici panduanju 汉译佛经"S, N是"句非系词判断句 [The determinative construction "S,N shi(是)" in Chinese Buddhist scriptures]. *Zhongguo Yuwen* 中国语文 [*Studies of the Chinese Language*] 1: 59–66, 96.

Jin, Lixin (金立鑫) and Yu, Xiujin (于秀金). 2012. Cong yu OV-VO xiangguan he bu xiangguan canxiang kaocha putonghua de yuxu leixing 从与OV-VO相关和不相关参项考察普通话的语序类型 [Word order of Mandarin Chinese revisited in pairs of grammatical elements correlated and uncorrelated with OV-VO]. *Waiguoyu* 外国语 [*Journal of Foreign Languages*] 2: 22–29.

Karashima Seishi. 1998. Hanyi fodian de yuyan yanjiu(er) 汉译佛典的语言研究(二) [Linguistic research on the Chinese translation of Buddhist scriptures (2)]. *Suyuyan Yanjiu* 俗语言研究 [*Study of folk language*] 5:47–57.

———. 2016. *Fodian yuyan ji chuancheng* 佛典语言及传承 [*Buddhist Scripture language and inheritance*]. Shanghai: Zhongxi Shuju 上海:中西书局 [Shanghai: Publishing house of Chinese and Western].

Lee, Tae Soo (李泰洙). 2000a.《Laoqida》sizhong banben congju juwei zhuci yanjiu《老乞大》四种版本从句句尾助词研究 [A comparative study of post subordinate-clause particles in the four different editions of Lao Ch'i-ta]. *Zhongguo Yuwen* 中国语文 [*Studies of the Chinese Language*] 1: 47–56, 94.

——— (李泰洙). 2000b. Guben yanjieben《Laoqida》li fangweici de teshu gongneng 古本、谚解本《老乞大》里方位词的特殊功能 [The special functions of localizers in the ancient version and the translation of Lao Ch'i-ta]. *Yuwen Yanjiu* 语文研究 [*Linguistic Research*] 2: 30–38.

——— (李泰洙). 2000c. Guben《Laoqida》de yuzhuci "you" 古本《老乞大》的语助词"有" [The auxiliary word "you" in the ancient version of Lao Ch'i-ta]. *Yuyan Jiaoxue yu Yanjiu* 语言教学与研究 [*Language Teaching and Linguistic Studies*] 3:77–80.

Li, Chongxing (李崇兴). 2005. Lun yuandai mengguyu dui hanyu yufa de yingxiang 论元代蒙古语对汉语语法的影响 [The influence on the Chinese grammar caused by the Mongolian in the Yuan dynasty]. *Yuyan Yanjiu* 语言研究 [*Studies in Language and Linguistics*] 3: 77–81.

Li, Charles N., and Sandra A. Thompson. 1974. An explanation of word order change SVO→SOV. *Foundations of Language* 12: 201–214.

Liang, Qichao (梁启超). 1920/2001. Fanyi wenxue yu fodian 翻译文学与佛典 [Translation Literature and Buddhist scriptures]. In *Foxue yanjiu shiba pian* 佛学研究十八篇 [*Eighteen Buddhist Studies*], 165–201. Shanghai: Shanghai Guji Chubanshe 上海:上海古籍出版社 [Shanghai: Shanghai Ancient Books Press].

Liebenthal, Walter.1935a–1936a. On Chinese –Sanskrit Comparative Indexing. *Monumenta Serica* (华裔学志) 1:173–185.

———.1935b–1936b. The Problem of a Chinese-Sanskrit Dictionary. *Monumenta Serica* (华裔学志) 1:168–172.

Light, Timothy. 1979. Word order and word order change in Mandarin Chinese. *Journal of Chinese Linguistics*. 7: 149–180.

Liu, Yizhi (刘一之). 1988. Guanyu beifang fangyan zhong diyirencheng daici fushu baokuoshi he paichushi duili de chansheng niandai 关于北方方言中第一人称代词复数包括式和排除式对立的产生年代 [On the generation time of the contrast between the inclusive and exclusive plural forms of the first person pronoun in the Northern dialects]. In *Yuyanxue Luncong* 语言学论丛(第十五辑) [*Essays on linguistics*, vol. 15], 92–140. Beijing: Shangwu Yinshuguan 北京: 商务印书馆 [Beijing: The Commercial Press].

Long, Guofu (龙国富). 2007. Hanyu wanchengmao jushi he fojing fanyi 汉语完成貌句式和佛经翻译 [The Chinese perfective construction and Buddhist translations].*Minzu Yuwen* 民族语文 [*Minority Languages of China*] 1:35–44.

——— (龙国富). 2009. Zhonggu hanyi fojing beidongshi yu fojing fanyi 中古汉译佛经被动式与佛经翻译 [Research on the passive in Chinese Buddhist Scriptures and their translations]. In

Lishiyuyanxue Yanjiu 历史语言学研究(第二辑) [*Study of historical linguistics*, vol. 2], 147–157. Beijing: Shangwu Yinshuguan 北京: 商务印书馆 [Beijing: The Commercial Press].

Lü, Shuxiang (吕叔湘). 1984. Du《sanguozhi》读《三国志》[Reading the Three Kingdoms Chronicles]. In *Yuwen Zaji* 语文杂记 [*Chinese Miscellaneous*], 23–25. Shanghai: Shanghai Jiaoyu Chubanshe 上海: 上海教育出版社 [Shanghai: Shanghai Educational Press].

Lu, Jianming (陆俭明). 1985. Guanyu "qu+VP" he "VP+qu" jushi 关于"去+VP"和"VP+去"句式 [On sentence pattern of "qu +VP" and "VP +qu"]. *Yuyan Jiaoxue yu Yanjiu* 语言教学与研究 [*Language Teaching and Linguistic Studies*] 4:18–33.

Hashimoto Mantaro. 1987. Hanyu beidongshi de lishi·quyu fazhan 汉语被动式的历史·区域发展 [History of Chinese passive·regional development]. *Zhongguo Yuwen* 中国语文 [*Studies of the Chinese Language*] 1: 36–49.

Mei, Kuang. 1980. Is modern Chinese really a SOV language? *Cahiers de linguistique-Asie Orientale* 7: 23–45.

Mei Tsu-lin (梅祖麟). 1981. Xiandai hanyu wanchengmao jushi he ciwei de laiyuan 现代汉语完成貌句式和词尾的来源 [The origin of the perfect form sentence pattern and suffix in modern Chinese]. *Yuyan Yanjiu* 语言研究 [*Studies in Language and Linguistics*] 00: 65–77.

——— (梅祖麟). 1986. Guanyu jindai hanyu zhidaici———du lüzhu《jindai hanyu zhidaici》关于近代汉语指代词———读吕著《近代汉语指代词》[About modern Chinese demonstrative pronouns – Reading Lü's "modern Chinese demonstrative pronouns"]. *Zhongguo Yuwen* 中国语文 [*Studies of the Chinese Language*] 6: 401–412.

——— (梅祖麟). 1988. Beifang fangyan zhong diyirencheng daici fushu baokuoshi he paichushi duili de laiyuan 北方方言中第一人称代词复数包括式和排除式对立的来源 [On the origin of the contrast between the inclusive and exclusive plural form of the first person pronoun in the Northern dialects]. In *Yuyanxue Luncong* 语言学论丛(第十五辑) [*Essays on linguistics*, vol. 15], 141–145. Beijing: Shangwu Yinshuguan 北京: 商务印书馆 [Beijing: The Commercial Press].

——— (梅祖麟). 1999. Xianqin lianghan de yizhong wanchengmao jushi———jianlun xiandai hanyu wanchengmao jushi de laiyuan 先秦两汉的一种完成貌句式———兼论现代汉语完成貌句式的来源 [A perfect form sentence pattern in the Pre-Qin and Han dynasties – And a discussion of the origin of the perfect form sentence pattern in modern Chinese]. *Zhongguo Yuwen* 中国语文 [*Studies of the Chinese Language*] 4: 285–294.

Norman, Jerry. 1982. Four notes on Chinese-Altaic linguistic contacts. *Tsing Hua Journal of Chinese Studies* 1–2: 243–247.

Peyraube, Alain. 1997. On word order in archaic Chinese. *Cahiers de linguistique-Asie orientaie*. 26: 3–20.

Sun, Xixin (孙锡信). 1992. *Hanyu lishi yufa yaolue* 汉语历时语法要略 [*Outline of Chinese diachronic grammar*]. Shanghai: Fudan daxue Chubanshe 上海:复旦大学出版社 [Shanghai: Fudan University Press].

Sun, Chaofen. 1996. *Word order change and grammaticalization in the history of Chinese*. Stanford: Stanford University Press.

Tai, James H-Y. 1973. Chinese as an SOV language. In *The Ninth regional meeting of Chicago linguistic society*, 659–671. Reprinted in *Readings in Chinese transformational syntax*, Shou-hsin Teng, (ed.), 1986. Taiwan: Crane Publishing Company, 340–354.

———. 1976. On the change from SVO to SOV in Chinese. *Parasessions on Diachronic Syntax, Chicago Linguistic Society*, 291–304. Reprinted in *Readings in Chinese Transformational Syntax*, Shou-hsin Teng, (ed.), 1986. Taiwan: Crane Publishing Company, 428–444.

Tatsuo Ota. 1958/2003. *Zhongguoyu lishi wenfa* 中国语历史文法 [*A historical grammar of modern Chinese*]. Jiang, Shaoyu 蒋绍愚 and Xu, Changhua 徐昌华. (trans.) Beijing: Beijing Daxue Chubanshe 北京:北京大学出版社 [Beijing: Peking University Press].

———. 1988/1991. *Hanyushi tongkao* 汉语史通考 [*A general study of Chinese language history*]. Jiang, Lansheng 江蓝生 and Bai, Weiguo 白维国. (trans.) Chongqing: Chongqing Chubanshe 重庆:重庆出版社 [Chongqing: The Chongqing Press].

Thomas, Watters. 1899. *Essays on the Chinese language*. Shanghai: American Presbyterian Mission Press.

Wang, Li (王力). 1937. Zhongguo wenfa zhong de xici 中国文法中的系词 [Copula in Chinese grammar]. *Qinghua Daxue Xuebao(Ziran Kexue Ban)* 清华大学学报(自然科学版) [*Journal of Tsinghua University (Natural Science Edition)*] 1: 1–67.

Wang, Jihong (王继红). 2004. *Jiyu fanhanduikan de fojiao hanyu yufa yanjiu—Yi《apidamojushelun·fenbiejiepin》weili* 基于梵汉对勘的佛教汉语语法研究——以《阿毗达摩俱舍论·分别界品》为例 [*A study of Buddhist Chinese grammar based on the comparative analysis of Sanskrit-Chinese versions – Take "Abhidharmakosa-sastra·Distinguishment" as an example*]. Beijing Daxue Boshi Xuewei Lunwen 北京大学博士学位论文 [Ph.D. dissertation, Peking University].

Xu, Dan (徐丹). 2014. *Tangwanghua yanjiu* 唐汪话研究 [*Study on the Tang Wang Dialect*]. Beijing: Minzu Chubanshe 北京:民族出版社 [Beijing: The Ethnic Publishing House].

Xu, Zhaohong (徐朝红) and Wu, Fuxiang (吴福祥). 2015. Cong leitong fuci dao binglie lianci: zhonggu yijing zhong xuci "yi" de yuyi yanbian 从类同副词到并列连词:中古译经中虚词"亦"的语义演变 [From additive adverbs to coordinative connectives: Semantic change of the function word yi (亦) in the medieval Chinese Buddhist scriptures].*Zhongguo Yuwen* 中国语文 [*Studies of the Chinese Language*] 1: 38–49, 96.

Yang, Yonglong (杨永龙). 2012. Mudi goushi "VP qu" yu SOV yuxu de guanlian目的构式"VP去"与SOV语序的关联 [The purpose construction "VP qu (去)" in Chinese and the SOV order]. *Zhongguo Yuwen* 中国语文 [*Studies of the Chinese Language*] 6: 525–536, 576.

——— (杨永龙). 2014. Cong yuxu leixing de jiaodu chongxin shenshi "X+xiangsi/si/yesi" de laiyuan 从语序类型的角度重新审视"X+相似/似/也似"的来源 [Revisiting the origin of simile-making postpositions si/xiangsi/yesi (似/相似/也似) from the perspective of word order typology]. *Zhongguo Yuwen* 中国语文 [*Studies of the Chinese Language*] 4: 291–303,383.

——— (杨永龙). 2015. Qinghai minhe gangouhua de yuxu leixing 青海民和甘沟话的语序类型 [The word order of the Gangou Chinese dialect in Minhe Hui and Monguor county of Qinghai province]. *Minzu Yuwen* 民族语文 [*Minority Languages of China*] 6:15–30.

Yu, Zhihong (余志鸿). 1983. Yuandai hanyu zhong de houzhici "Hang" 元代汉语中的后置词"行" [The postpositive word "Hang" in Chinese of the Yuan Dynasty]. *Yuwen Yanjiu* 语文研究 [*Linguistic Research*] 3:48–50,63.

——— (余志鸿). 1987. Yuandai hanyu "-hang" de yufa yiyi 元代汉语"-行"的语法意义 [The grammatical meaning of Chinese "-Hang" in the Yuan Dynasty]. *Yuwen Yanjiu* 语文研究 [*Linguistic Research*] 2:16–20.

Yu, Liming (俞理明). 1989. Cong fojing cailiao kan zhonggu hanyu renji daici de fazhan 从佛经材料看中古汉语人己代词的发展 [Viewing the development of reflexive pronouns in middle ancient Chinese from Buddhist scriptures]. *Sichuan Daxue Xuebao(Zhexue Shehui Kexue Ban)* 四川大学学报(哲学社会科学版) [*Journal of Sichuan University (Philosophy & Social Sciences Edition)*] 4: 61–67.

Yu, Zhihong (余志鸿). 1992. Yuandai hanyu de houzhici xitong 元代汉语的后置词系统 [The postpositional system of Chinese in the Yuan Dynasty]. *Minzu Yuwen* 民族语文 [*Minority Languages of China*] 3:1–10.

Zhang, Qingchang (张清常). 1982. Hanyu "zamen" de qiyuan 汉语"咱们"的起源 [The origin of Chinese "zamen"]. In *Yuyanxue Yanjiu Luncong* 语言研究论丛(第二辑) [*Collection of Linguistics Research*, vol. 2], 91–95. Tianjin: Tianjin Renmin Chubanshe 天津:天津人民出版社 [Tianjin: The Tianjin People's Press].

Zhang, Cheng (张赪). 2002. *Hanyu jieci cizu yuxu de lishi yanbian* 汉语介词词组语序的历史演变 [*The historical evolution of the word order of Chinese prepositional phrases*]. Beijing: Beijing Yuyan Wenhua Daxue Chubanshe 北京:北京语言文化大学出版社 [Beijing: Beijing Language and Culture University Press].

Zhao, Changcai (赵长才). 2009. Zhonggu hanyi fojing zhong de houzhici "suo" he "bian" 中古汉译佛经中的后置词"所"和"边" [The postpositions suo (所) and bian (边) in medieval translations of Buddhist scriptures]. *Zhongguo Yuwen* 中国语文 [*Studies of the Chinese Language*] 5: 438–447, 480.

——— (赵长才). 2015. Zhonggu yijing you yuandian huge de duiyi suodailai de jufa yingxiang 中古译经由原典呼格的对译所带来的句法影响 [The study of the Buddhist influence on Chinese syntactic structure: the vocative case]. In *Lishiyuyanxue Yanjiu* 历史语言学研究(第九辑)

[*Studies in historical linguistics*, vol. 9], 11–24. Beijing: Shangwu Yinshuguan 北京: 商务印书馆 [Beijing: The Commercial Press].

Zhou, Yiliang (周一良). 1944. Zhongguo de fanwen yanjiu 中国的梵文研究 [Sanskrit studies in China]. *Sixiang yu shidai* 思想与时代 [*Thoughts and Times*] 35: 10–13.// 1963. In *Weijinnanbeichao shi lunji* 魏晋南北朝史论集 [*A collection of essays on the history of Wei, Jin, Southern and Northern Dynasties*], 323–338. Beijing: Zhonghua Shuju 北京: 中华书局 [Beijing: Zhonghua Book Company].

Zhu, Minche (祝敏彻). 1957. Lun chuqi chuzhishi 论初期处置式 [On the initial disposal construction]. In *Yuyanxue Luncong* 语言学论丛(第一辑) [*Essays on linguistics*, vol. 1], 17–33. Beijing: Shangwu Yinshuguan 北京: 商务印书馆 [Beijing: The Commercial Press].

Zhu, Qingzhi (朱庆之). 1993. Hanyi fodian yuwen zhong de yuandian yingxiang chutan 汉译佛典语文中的原典影响初探 [A preliminary study of the influence of original texts in the Chinese translation of Buddhist scriptures]. *Zhongguo Yuwen* 中国语文 [*Studies of the Chinese Language*] 5: 379–385.

——— (朱庆之). 1995. Hanyi fidian zhong de "suo V" shi beidongju jiqi laiyuan 汉译佛典中的"所V"式被动句及其来源 [The "suo V" type passive sentence and its source in the Chinese translation of Buddhist scriptures]. *Guhanyu Yanjiu* 古汉语研究 [*Research in Ancient Chinese Language*] 1: 29–31,45.

——— (朱庆之). 2001. Fojiao hunhe hanyu chulun 佛教混合汉语初论 [A preliminary study on Buddhism mixed Chinese]. In *Yuyanxue Luncong* 语言学论丛(第二十四辑) [*Essays on linguistics*, vol. 24], 1–33. Beijing: Shangwu Yinshuguan 北京: 商务印书馆 [Beijing: The Commercial Press].

Zhu, Guanming (朱冠明). 2005. *Zhonggu hanyi fodian yufa zhuanti yanjiu* 中古汉译佛典语法专题研究 [*A study of the grammar of medieval Chinese Buddhist scriptures*]. Beijing Daxue Boshihou Gongzuo Yanjiu Baogao 北京大学博士后工作研究报告 [Postdoctoral research report, Peking University].

——— (朱冠明). 2007. Cong zhonggu fodian kan "ziji" de xingcheng 从中古佛典看"自己"的形成 [On the origin of the Chinese reflexive ziji(自己) from the perspective of the medieval Chinese Buddhist scriptures]. *Zhongguo Yuwen* 中国语文 [*Studies of the Chinese Language*] 5: 402–411,479.

——— (朱冠明). 2008. Fanhanben《amituojing》yufa zhaji 梵汉本《阿弥陀经》语法札记 [Notes on some grammatical elements in two Chinese translations of the Land of Bliss and their Sanskrit original: A comparative study]. In *Lishiyuyanxue Yanjiu* 历史语言学研究(第一辑) [*Studies in historical linguistics*, vol. 1], 108–119. Beijing: Shangwu Yinshuguan 北京: 商务印书馆 [Beijing: The Commercial Press].

——— (朱冠明). 2011. Zhonggu fodian yu hanyu shoushi zhuyuju de fazhan—Jiantan fojing fanyi yingxiang hanyu yufa de moshi 中古佛典与汉语受事主语句的发展——兼谈佛经翻译影响汉语语法的模式 [The medieval Chinese Buddhist scriptures and the development of the Chinese patient subject clause: A case study on the pattern of the effect of the Buddhist scripture translation on Chinese grammar]. *Zhongguo Yuwen* 中国语文 [*Studies of the Chinese Language*] 2: 169–178,192.

——— (朱冠明). 2013. Hanyi fodian yufa yanjiu shuyao 汉译佛典语法研究述要 [A summary of studies of the grammar of Chinese Buddhist scriptures]. In *Hanyi fodian yufa yanjiu lunji* 汉译佛典语法研究论集 [*On the grammar of Chinese Buddhist scriptures*], ed. Shaoyu Jiang (蒋绍愚) and Chirui Hu (胡敕瑞), 1–45. Beijing: Shangwu Yinshuguan 北京: 商务印书馆 [Beijing: The Commercial Press].

Zhu, Qingzhi, (朱庆之) and Zhu, Guanming (朱冠明). 2006. Fodian yu hanyu yufa yanjiu—20 shiji guonei fojiao hanyu yanjiu huigu zhier 佛典与汉语语法研究——20世纪国内佛教汉语研究回顾之二 [Buddhist scriptures and Chinese grammar research – Review 2 of domestic Buddhist Chinese studies in the 20th century]. In *Hanyushi Yanjiu Jikan* 汉语史研究集刊(第九辑) [*Studies on the History of Chinese Language*, vol. 9], 413–459. Chengdu: Bashu Shushe 成都: 巴蜀书社 [Chengdu: The Bashu Press].

Zu, Shengli (祖生利). 2001a. Yuandai baihua beiwen zhong fangweici de gebiaoji zuoyong 元代白话碑文中方位词的格标记作用 [The use of marking Mongolian nominal cases of localizers in the vernacular inscriptions of the Yuan dynasty]. *Yuyan Yanjiu* 语言研究 [*Studies in Language and Linguistics*] 4: 62–75.

——— (祖生利). 2001b. Yuandai baihua beiwen zhong daici de teshu yongfa 元代白话碑文中代词的特殊用法 [The special usage of pronouns in the vernacular inscriptions of the Yuan Dynasty]. *Minzu Yuwen* 民族语文 [*Minority Languages of China*] 5:48–62.

——— (祖生利). 2002a. Yuandai baihua beiwen zhong ciwei "mei" de teshu yongfa 元代白话碑文中词尾"每"的特殊用法 [The special usage of suffix "mei" in the vernacular inscriptions of the Yuan Yanjiu]. *Yuyan Yanjiu* 语言研究 [*Studies in Language and Linguistics*] 4: 72–80.

——— (祖生利). 2002b. Yuandai baihua beiwen zhong zhuci de teshu yongfa 元代白话碑文中助词的特殊用法 [The special usages of particles in the vernacular inscriptions of the Yuan dynasty]. *Zhongguo Yuwen* 中国语文 [*Studies of the Chinese Language*] 5: 459–472,480.

——— (祖生利). 2003. 《Yuandianzhang·Xingbu》zhiyiti wenzi zhong de teshu yufa xianxiang 《元典章·刑部》直译体文字中的特殊语法现象 [The special grammatical phenomena in the metaphrastic translation documents of "Compilation of legal instruments, The institutes of the Yuan Dynasty"]. In *Menggushi Yanjiu* 蒙古史研究(第七辑) [*Mongolian history studies*, vol.7],138–190. Huhehaote: Neimenggu Daxue Chubanshe 呼和浩特: 内蒙古大学出版社 [Huhehaote: The Inner Mongolia university Press].

——— (祖生利). 2004. Yuandai zhiyiti wenxian zhong de yuanyin houzhici "shang/shangtou" 元代直译体文献中的原因后置词"上/上头" [The special usage of causal postposition of Shang&Shangtou in the translated documents of the Yuan Dynasty]. *Yuyan Yanjiu* 语言研究 [*Studies in Language and Linguistics*] 1: 47–52.

——— (祖生利). 2011. Guben《Lao Qida》de yuyan xingzhi 古本《老乞大》的语言性质 [Linguistic features of the oldest edition of "Lao Ch'i-ta"]. In *Lishiyuyanxue Yanjiu* 历史语言学研究(第四辑) [*Study of historical linguistics*, vol. 4], 17–53. Beijing: Shangwu Yinshuguan 北京: 商务印书馆 [Beijing: The Commercial Press].

——— (祖生利). 2013. Qingdai qiren hanyu de manyu ganrao tezheng chutan—yi《Qingwen qimeng》deng sanzhong jianhan manyu huihua jiaocai wei yanjiu de zhongxin 清代旗人汉语的满语干扰特征初探—以《清文启蒙》等三种兼汉满语会话教材为研究的中心 [Manchu interference features in the Manchurian Mandarin during the Qing Dynasty: Based on three Manchu conversation textbooks with Mandarin paraphrases]. In *Lishiyuyanxue Yanjiu* 历史语言学研究(第六辑) [*Study of historical linguistics*, vol.6],187–227. Beijing: Shangwu Yinshuguan 北京: 商务印书馆 [Beijing: The Commercial Press].

Zürcher, E. 1977. Late Han vernacular elements in the earliest Buddhist translations. *Journal of the Chinese Language Teacher's Association* 12: 177-203.//Jiang, Shaoyu (蒋绍愚). (trans.) 1987. Zuizao de fojiao yiwen zhong de donghan kouyu chengfen 最早的佛教译文中的东汉口语成分. In *Yuyanxue Luncong* 语言学论丛(第十四辑) [*Essays on linguistics*, vol. 14],197-225. Beijing: Shangwu Yinshuguan 北京: 商务印书馆 [Beijing: The Commercial Press].

Interactions of Sinitic Languages in the Philippines: Sinicization, Filipinization, and Sino-Philippine Language Creation

15

Wilkinson Daniel Wong Gonzales

Contents

Introduction	371
Terminology	372
The Philippines in a Nutshell	373
Chinese Heritage Groups and Their Languages	374
Colonial Era	376
Contemporary	376
Sinitic Languages and Their Interactions	379
Data	379
Orthography	379
Annotations	379
Overview	379
Hokkien	380
Early Manila Hokkien	381
Philippine Hokkien	382
Mainland Hokkien Variety in Manila	388
Cantonese	389
Influences on Local Languages	390
Influences from Local Languages	390
Mandarin	391
Influences on Local Languages	392
Influences from Local Languages	393
Sino-Philippine Contact Languages	394
Pidgins	394
Mixed Languages	396

W. D. W. Gonzales (✉)
University of Michigan, Ann Arbor, MI, USA
e-mail: wdwg@umich.edu

Inter-clausal Code-Switching 399
Discussion 400
  Summary 400
  Patterns of Asymmetric Contributions 402
Summary 404
Conclusion 404
References 406

Abstract

This chapter explores the interactions between Sinitic and Philippine-based languages in the Philippine context. It focuses on the complex dynamics between the languages of the historically indigenous population and those of the (heritage and homeland) Chinese groups. Using oral and written data sampled from 12 linguistic varieties in three major Philippine cities across the archipelago, the chapter features the processes of Sinicization, Filipinization, and language creation. It shows that languages that co-exist in the same linguistic ecology can actually affect each other differently, depending on the sociohistorical context in which such interactions take place. Overall, this descriptive overview chapter hopes to highlight the intricacies of the relationship between the Sinitic and Philippine-based languages and attempts to provide a holistic characterization of the Sino-Philippine and, consequently, broader Philippine linguistic landscape.

Keywords

Sino-Philippine linguistic varieties · Language ecology · Language contact · Linguistic interactions · Sociolinguistics

List of Abbreviations

1	First person
2	Second person
3	Third person
ADV	Adverb
CLF	Classifier
CLI	Clitic
COM	Comparative
COND	Conditional
CONJ	Conjunction
COP	Copula
DEM	Demonstrative
DET	Determiner
GEN	Genitive
INC	Inclusive
INF	Infinitive
INT	Interjection

LNK	Linker
LOC	Locative
NEG	Negative marker
NOM	Nominative case
PFV	Perfective
PL	Plural
POL	Politeness marker
PRT	Particle
Q	Question marker
REL	Relativizer
SG	Singular
UV	Undergoer voice

Introduction

This chapter is mainly concerned with the dynamics of Sinitic languages (e.g., Hokkien) in relation to non-Sinitic languages used in the Philippines (e.g., Tagalog). Up to now, most linguistic studies that focus on multilingual interactions in the Philippines have stressed the dynamics between historically indigenous languages and Western languages such as English and Spanish (Schuchardt 1883; Toribia 1963; Llamzon 1969; Sobolewski 1980; Bautista 2004). Some of these studies include work on the local English (e.g., Bautista 2000 for English in Manila), creoles (e.g., Lesho 2013 for Cavite Chabacano), and bilingual code-switching between English and a regional language (e.g., Sobolewski 1980; Bautista 2004 for Tagalog-English; Abastillas 2015 for Cebuano-English). Although these works were successful in examining aspects of dynamic linguistic contact phenomena in the Philippines, they become inadequate if one seeks to gain a fuller understanding of the linguistic landscape of the Philippines and of its complexities. This is because earlier research tends to downplay, if not ignore, the role other non-Western languages (apart from historically indigenous ones) play in the Philippine "language ecology" – a term that refers to the set of interactions between languages and their environment (Haugen 1971).

There are several non-Western ethnic groups that have historically resided in the Philippines and have interacted with the Filipinos in varying degrees (e.g., Koreans, Imperial 2016; Chinese, Wickberg 1965; Doeppers 1986; See and Teresita 1990; Gonzales 2017a). This suggests that "Philippine-based" languages – linguistic varieties that are historically indigenous to the Philippines (e.g., Tagalog) and nativized varieties of non-indigenous languages (e.g., Philippine Englishes, Gonzales 2017b) – have also come into contact with the languages of these ethnic groups, and not just the Western languages. Yet, very little scholarly attention has been given to these ethnic groups and to the codes they use.

As a minority group in the Philippines, the Chinese were engaged in trade long before the arrival of the Spanish (Wickberg 1965), have historically been considered economic pillars of Philippine society, and dominate several major industries (Tan 1993: 77). But despite their influence and contributions to early and contemporary Philippine society, studies focusing on Sinitic language(s) and the dynamics of these languages in relation to other Philippine-based languages continue to be eclipsed by research and reference works that focus solely on Western-local interactions.

Despite the fact that overseas Chinese trade and migration to the Philippines and greater Southeast Asia have always been a crucial part of Chinese history (Wickberg 1965), there has been very little work on how the Sinitic languages used by the local Chinese communities interact with other languages spoken in the Philippines. This is in contrast with the many studies that have focused on descriptions of the Sinitic languages without accounting for the interactive dynamics (of language contact) among them (see Li and Thompson 1989 for Mandarin; MacGowan 1883, Bodman 1987 for Amoy Hokkien; Chappell 2019 for Southern Min, etc.).

To encourage the development of dynamic-based linguistic research in both Philippine and Sinitic scholarly circles, this chapter introduces and explores the nexus of Sinitic and Philippine linguistics. It aims to bridge the gaps between fields by highlighting the complex, dynamic nature of the interactions between the Sinitic languages and Philippine-based languages. By doing so, it provides a more holistic and, consequently, more faithful characterization of the (Sino-)Philippine linguistic landscape that could potentially serve as a springboard for further, much-needed research in the field.

Specifically, this chapter will survey several linguistic codes that are related to Sinitic languages. It will illustrate and discuss the varied consequences of contact between Sinitic and Philippine-based languages as reflected in the general processes of Filipinization, Sinicization, and Sino-Philippine language creation. The chapter is intended to be a *descriptive* overview of Sino-Philippine interactions, focusing on three major Sinitic languages spoken in the Philippines – Southern Min (Hokkien), Cantonese, and Mandarin. It also aims to showcase new languages resulting from contact between Sinitic and Philippine-based languages. The four speaker groups that this chapter will focus on are the Filipinos, the Lannangs, Mainland Chinese (new immigrants and sojourners), and the Sangleys.

The rest of the chapter is organized as follows: the next section will cover the operational definitions of select terms in language contact studies. The two sections that follow this will set the stage by providing a general overview of the Philippines, the Chinese heritage and homeland groups in the Philippines, and their languages. This is succeeded by explorations of Sino-Philippine linguistic interactions. A general discussion that accounts for these interactions and concluding notes follow these sections.

Terminology

In this chapter, "code-switching" will be used to refer to situations when a speaker completely shifts from one language to another – whether it is at the word, phrase, clause, or sentence level. A speaker code-switches when they switch from a monolingual (unilingual) to a bilingual language mode; this is typically socially motivated. For example, a speaker can code-switch as a mark of inclusion of another bilingual speaker, or to exclude an outsider conversation, or to assert their identity.

Code-switching is different from "borrowing," which is operationally defined here as the integration of the word from one language into another (Grosjean 2010). In contrast to code-switching, borrowing involves phonological adaptation. In code-switching, a given word, for instance, is clearly flagged as originating from another language based on its phonology. In contrast, in borrowing, the loaned word is considered as fully integrated to the recipient language.

While some scholars may view borrowing as being either lexical, functional, or structural (see Thomason and Kaufman's 1988 borrowing scale) in this chapter, I limit myself to using "borrowing" to designate only lexical borrowing or loan words. Functional and structural borrowing will be referred to as "language transfer" or simply "transfer" here. For example, it is uncommon to say that a language "borrows" the pronominal system from another language. For the present purposes, we will say instead that the pronominal system is "transferred" from the source language to the recipient language (Odlin 1989; Pavlenko and Jarvis 2008).

The Philippines in a Nutshell

The Philippines is an archipelagic nation in Southeast Asia that consists of around 7,641 islands (Fig. 1).

The population is roughly 106 million (The World Bank 2018), and most people belong to one of the seven major indigenous ethnic groups – the Tagalogs, Cebuanos, Ilocanos, Visayans, Ilonggos, Bikolanos, and Warays. Table 1 shows the breakdown of the population by ethnicity.

As mentioned in the introduction, the Philippines has had a long history of trade with other non-indigenous groups, such as the Chinese (Patanne 1996), and had been occupied by several foreign powers, such as the Spaniards in the 1500s, the Americans around the 1900s, and, briefly, the British (1762–1764, Manila) and the Japanese (1942–1945) (Agoncillo and Guerrero 1970).

The interaction between the indigenous and non-indigenous populations throughout history is one reason why the Philippines is generally thought of as a hotbed of languages. According to Simons and Fennig (2020), it has 184 living languages, comprising 175 indigenous languages and 9 non-indigenous ones. Of these codes, Filipino – operationally defined in this chapter as a standardized Tagalog variety with overt non-Tagalog lexical mixing – functions as the national language. Alongside English, Filipino is also an official language, that is, both English and Filipino are used in official domains, such as in government documents. Both English and Filipino are also used as the primary media of instruction in school, whereas the other languages are mainly used for wider communication within their respective ethno-geographic groups.

Figure 2 shows the predominant indigenous languages in the Philippines by language family.

Chinese Heritage Groups and Their Languages

Historically, those of Chinese descent in the Philippines comprise three major groups – the Sangleys, the Lannangs, and the Mainland Chinese.

Fig. 1 Map of the Philippines with Manila, Iloilo, and Cebu highlighted with red stars

Table 1 Population by ethnicity according to the 2010 Census of Population and Housing conducted by the Philippine Statistics Authority (2010) (latest data not available)

Ethnicity	Population	%
Tagalog	22,512,089	24.44
Cebuano	9,125,637	9.91
Ilocano	8,074,536	8.77
Bisaya/Binisaya	10,539,816	11.44
Hiligaynon/Ilonggo	7,773,655	8.44
Bikol	6,299,283	6.84
Waray	3,660,645	3.97
Chinese (Lannang?)[a]	1,410,000	1.53
Other foreign ethnicity	63,017	0.07
Others	22,632,850	24.57
Not reported	6,450	0.01
Total	92,097,978	100

[a]The population estimates are based on Uytanlet (2014: 3)

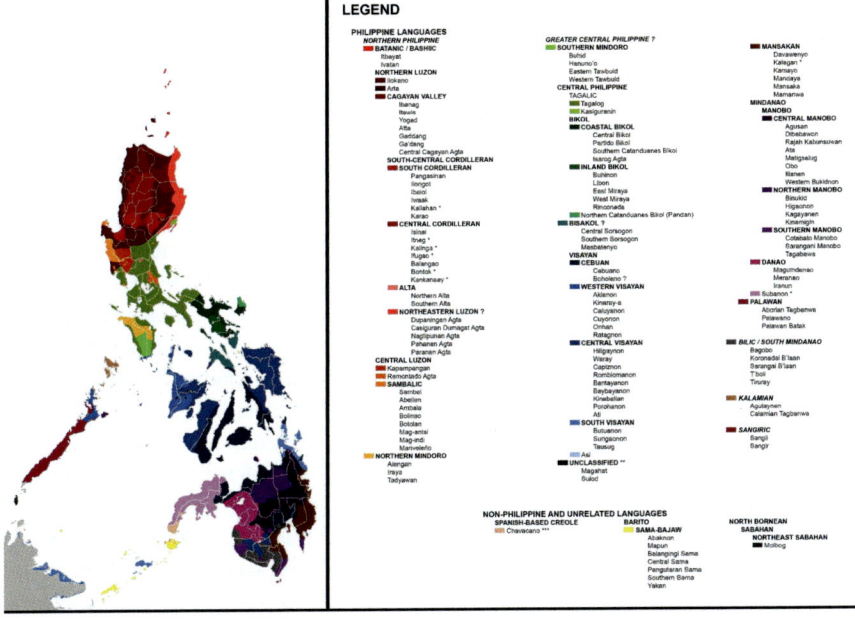

Fig. 2 Linguistic map of the Philippines based on indigenous and/or predominant language groups. (Map by Reddit user u/*pansitkanton*, based on data from Ethnologue in 2016, Simons and Fennig 2020)

Colonial Era

Sangleys

The Sangleys were Southern Chinese merchants and traders during the Spanish colonial era of the Philippines. They functioned as the "middlemen" between the indigenous population and the Spanish colonizers (Uytanlet 2014: 48). The Sangleys were engaged in trade among other professions (e.g., bakers, cooks, barbers, vendors, blacksmiths) that sought to fill in the occupational vacuum that the natives and the colonizer left (Uytanlet 2014): the natives, at that time, generally only practiced subsistence agriculture, while the Spaniards were not too keen on engaging with hard labor (Uytanlet 2014).

The Sangleys, who had Hokkien (Klöter 2011) and Chinese Spanish Pidgin (Fernández and Sippola 2017) in their linguistic repertoire, interacted frequently with the local population. Some, if not most, Sangleys intermarried the natives, having offspring who eventually assimilated to mainstream Philippine society at the turn of the nineteenth century (Wickberg 1965: 237). Although most Sangleys were concentrated in Manila, they had also established trade in other Philippine ports such as those in Iloilo, Zamboanga, and Jolo (Wickberg 1965: 5).

Contemporary

Lannangs

The Lannangs (derived from Hokkien *lân láng* "our people") generally consist of Southern Chinese who emigrated from China around the late nineteenth to early twentieth century and their descendants. They are individuals with Chinese ancestry (but not necessarily "pure" Chinese) who spent the majority of, if not their entire, life in the Philippines. Many Lannangs have Filipino citizenship, but there are some who were unable to acquire it despite the mass naturalization decree issued in the 1970s (Tan 1993). The Lannangs also go by many names, depending on their political alignment, citizenship, and personal preference. Some of these include Filipino-Chinese, *Huīdīpīn Huakiaú* "Philippine Overseas Chinese," Tsinoy/Chinoy, or Chinese Filipino (see Gonzales 2021 for a comparison). Regardless of race, citizenship, or designation, the Lannangs all share the experience of a mixed Chinese and Filipino culture. They are generally more oriented toward the Philippines.

Exposed to a multicultural environment, most Lannangs are multilingual in both Sinitic and Philippine-based languages. They are knowledgeable in Philippine Hokkien, Mandarin, and Cantonese (if they are of Cantonese heritage) and can also communicate in the dominant regional language (e.g., Hiligaynon, if in Iloilo City). Like the Filipinos without Lannang heritage (henceforth, Filipinos), the Lannangs also have the dominant languages Filipino and English in their linguistic repertoire, as they have also been schooled using a national English-Filipino curriculum.

In terms of location, the Lannangs are dispersed all across the Philippine islands (Doeppers 1986). In the National Capital Region of Metropolitan Manila (see

Fig. 3), the Lannangs tend to reside in either the historic Chinese enclave areas of Binondo and Santa Cruz (see Fig. 4) or in the Banawe area in Quezon City, although a large number of Lannangs do not live in any of these areas and reside in areas such as Tondo as well as San Juan City and Makati City.

Fig. 3 A map of Metropolitan Manila

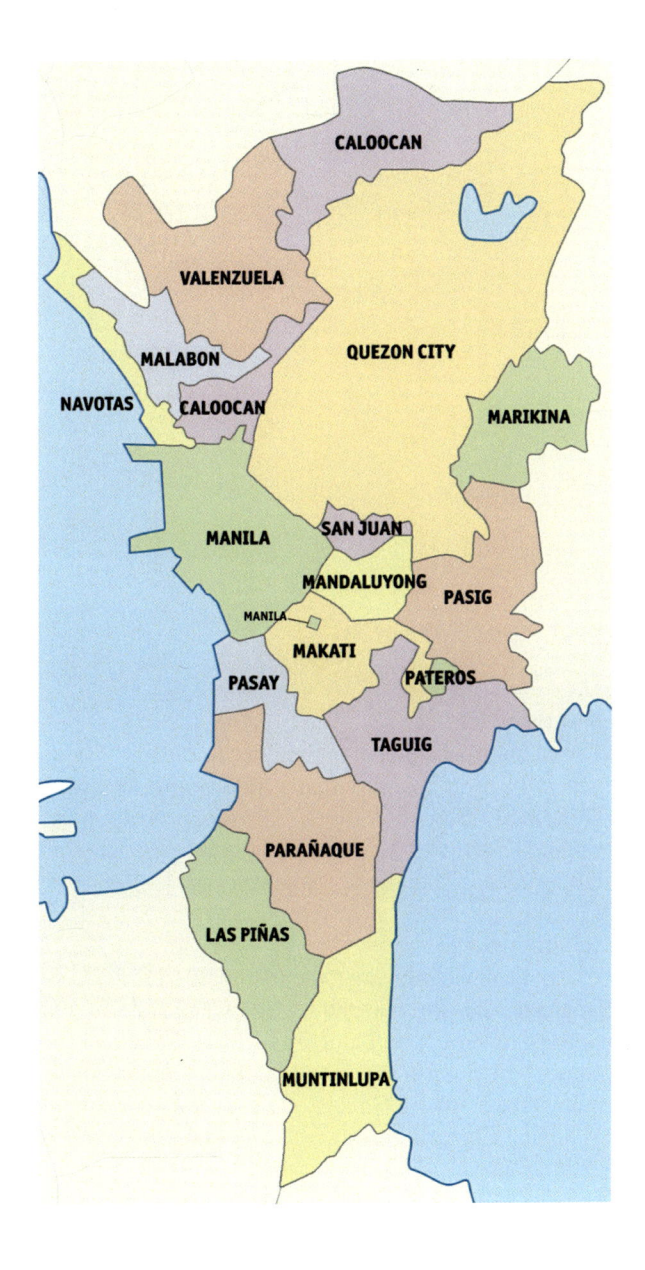

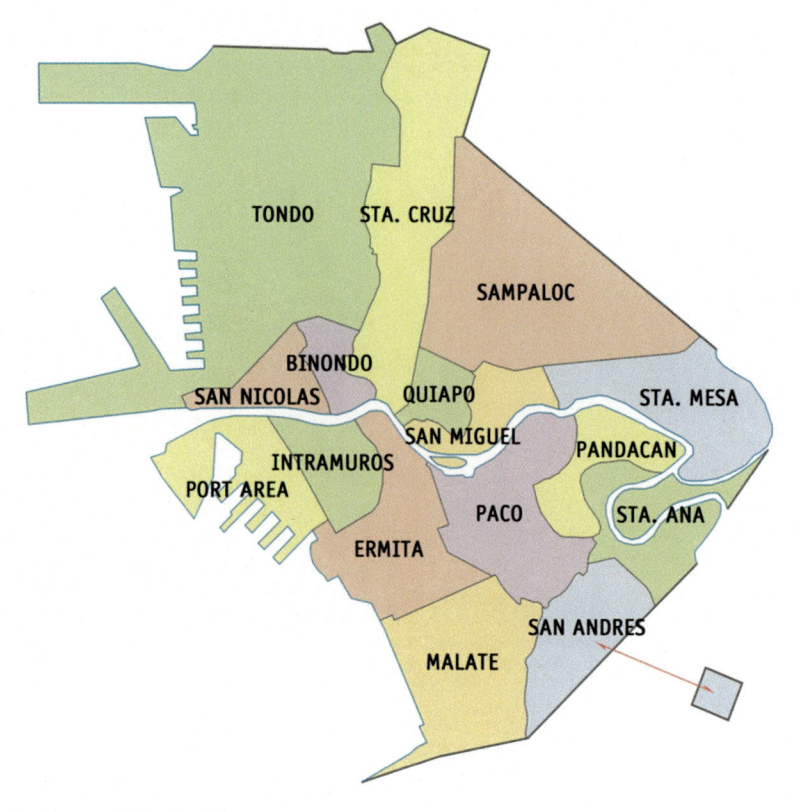

Fig. 4 A map of Manila City

Mainland Chinese

The Mainland Chinese – called the *Taīdiōkâ*s "Mainlander" by many Lannangs – generally refer to individuals of Chinese ancestry who began arriving in the Philippines from the 1990s. They consist of new immigrants from Southern China as well as other Chinese groups that have no intention of staying in the Philippines (henceforth, sojourners). Unlike the Lannangs, the Mainland Chinese are more oriented toward their homeland in China compared to the Philippines.

Unlike most Lannangs, the Mainland Chinese have Mandarin as their dominant language and are also not as knowledgeable of Philippine-based languages, in contrast to those with Lannang heritage. Mainland Chinese with Southern Chinese heritage also have Hokkien as another dominant language. Most of these new immigrant Southern Chinese are concentrated in the Binondo area, while the other Mainland Chinese groups reside in less predictable areas compared to the immigrant group.

Sinitic Languages and Their Interactions

This section explores how the Sinitic languages of the heritage and homeland groups – particularly Southern Min (henceforth, Hokkien), Cantonese, and Mandarin – interact with major Philippine-based languages. The section also features new languages created from these interactions.

Data

The observations and analysis of these interactions draw from two main sources: (a) examples from published sources and (b) a data bank consisting of transcribed fieldwork recordings, elicitations, narratives, and interviews collected in Metropolitan Manila between 2017 and 2019. A subset of the data bank, particularly that collected from Lannangs (approximately 155,000 words), will hereafter be referred to in this chapter as the Lannang Corpus (LC).

Orthography

With the exception of Pinyin-using Mandarin, the latter data source had been transcribed (and will be presented) in Lannang Orthography or the orthographic conventions of The Lannang Archives (The Lannang Archives 2020). Linguistic data from the literature review will be presented as is.

Annotations

Under each linguistic example, I use the following labelling convention: (Reference, if applicable; date of utterance/writing, location, type of data, linguistic variety, participant ID/information)

Overview

Overall, the succeeding sections will feature the following languages and varieties:

- Hokkien
  - Early Manila Hokkien
  - Mainland Hokkien: Manila
  - Philippine Hokkien (Lannang Hokkien): Manila
- Cantonese
  - Lannang Cantonese: Manila
  - Lannang Taishanese: Manila

- Mandarin
  - Mainland Mandarin: Manila
  - Lannang Mandarin: Manila
- Chinese Spanish Pidgin
  - Manila variety
- Chinese Tagalog Pidgin
  - Manila variety
- Lánnang-uè (Philippine Hybrid Hokkien)
  - Manila variety
  - Cebu variety
  - Iloilo variety

Using these codes as case studies, the following sections feature three broad linguistic processes – Filipinization, Sinicization, and Sino-Philippine language creation. Filipinization is where Philippine languages influence Sinitic ones; Sinicization is where the reverse occurs; Sino-Philippine language creation is where new languages not genetically traceable to the source languages are created.

Each of the sections on linguistic interactions will begin with an overview of the highlighted language. Seminal research on the language along with relevant sociohistorical information will then be outlined. This will then be followed by descriptions of individual varieties within that language and discussions of how these varieties contribute to the processes of Sinicization and Filipinization. A section featuring new languages created out of Sino-Philippine language contact will follow this.

Hokkien

No survey of Sinitic languages in the Philippines is complete without a discussion of Hokkien, arguably the oldest and most widely used Sinitic language in the Philippines. The earliest systematic study of this language in the Philippine context, to my knowledge, is a Spanish missionary grammar entitled *Arte de la lengua Chio Chiu* (Klöter 2011). One of the earliest (if not the earliest) concrete linguistic evidence of Hokkien use, on the other hand, can be found in translations of Christian catechisms in the local Hokkien vernacular – the *Doctrina Christiana en letra y lengua china*, published in 1604. It dates Hokkien use in the Philippines to at least the seventeenth century, although historical references suggest that Hokkien use predates Spanish colonization (Van der Loon 1966, 1967; Chirino 1604). In the absence of records of other Sinitic languages, such as Mandarin or their speakers in the Philippines prior to the 1600s, one could reasonably assume that Hokkien is the oldest Sinitic language in the Philippines. It is also claimed to be widely used among the contemporary Philippine Chinese communities as a heritage language, even if some of them have non-Hokkien heritage (e.g., Cantonese). This is because the Hokkien-speaking population has traditionally formed the bulk of the Philippine Chinese population

(90%, by Ang See's 1990 estimation), a situation that continues up to the present. Its established and dominant use in the community has most likely caused the minority Chinese population – the Standard Cantonese and Taishanese-speaking Cantonese – to assimilate with the majority Hokkien-speaking Chinese by acquiring Hokkien as one of their dominant languages.

This section does not attempt to survey all historical varieties of Hokkien in the Philippines and will focus only on three salient Hokkien varieties during two different eras: Early Manila Hokkien is discussed for the earlier era (see Klöter 2011); and Philippine Hokkien and Mainland Hokkien in Manila are discussed for the contemporary era.

Early Manila Hokkien

The earliest attested variety of Hokkien in the Philippines is Early Manila Hokkien, used by the Sangleys. The literature on the variety is scarce: the most extensive study of it, to my knowledge, is Van der Loon's (Van der Loon 1967) analysis and grammar reconstruction, which focused on the pronunciation, morphology, and syntax of this variety. Following this is Klöter's (2011) investigation of the variety in the Spanish-era *Arte*, where he argues that Early Manila Hokkien is distinct from the other Hokkien varieties spoken in mainland China (i.e., Amoy, *Zhāngzhōu, Quánzhōu*, and *Cháoshàn*) based on lexical and phonological evidence (see also ▶ Chap. 17, "Interactions Between Min and Other Sinitic Languages: Genetic Inheritance and Areal Patterns," by Ruiqing Shen's in this volume). The new variety, he claims, had emerged as a result of dialect levelling – a standardization process in which dialectal differences are "reduced" (Klöter 2011: 162). He uses the innovative second-person plural *lun* as an example and claims that it does not appear in any of the other Hokkien dialects (Klöter 2011). But while Klöter posits dialect emergence as a result of dialect contact here, he did not find interactions with Tagalog (Klöter 2011). Klöter's omission of Tagalog influence on Early Manila Hokkien is somewhat surprising, as its speakers, the Sangleys, were known to intermarry local women and interact with locals (Wickberg 1965).

Influences from Local Languages

Van der Loon (1967) did not highlight linguistic interactions in his study, but the data he described offers some insights into the situation of language contact involving Early Manila Hokkien. While he did not pinpoint explicit Tagalog influence, he documented loanwords from Spanish in Early Manila Hokkien. For example, in (1), the Spanish word for God, *Dios*, was borrowed into Hokkien as *Liosi/Diosi*. Religious terms, such as the word for Jesus Christ (i.e., *Sesu*) or that for the Holy Spirit (i.e., *Si Piritu Santo*), were also borrowed from Spanish into Hokkien (Van der Loon 1967). Despite the lack of oral data in Early Manila Hokkien, Van der Loon's study shows that borrowings from Spanish were commonplace. An initial look at the same data indicates no evidence of Tagalog borrowings, structural or functional transfer, confirming Klöter's initial observations (Klöter 2011: 157).

(1) ***Diosi*** *u   chap   si   kia   su-sit.*
    God   have   four ten   CLF   reality
    "There are fourteen facts about God."
    (Van der Loon 1967: 148; 1604, *Doctrina Christiana en letra y lengua china*, translation, Early Manila Hokkien, unknown)

Influences on Local Languages

Despite there currently being no evidence of Tagalog borrowings into Early Manila Hokkien, there is some evidence of the influence of Early Manila Hokkien on Tagalog. Residuals of Early Manila Hokkien on Tagalog exist in Filipino (Manila), traditionally characterized as a variety of standardized Tagalog with Spanish, Hokkien, and English influence, although historically, Filipino is a newly developed language with Tagalog as its base. It was formed as a response to Philippine nationalism in the 1930s (Thompson 2003: 28).

A large number of words in Filipino for cookery, cutlery, and trade-related expressions are considered to be Tagalog origin by locals, despite their origins in Hokkien (Chan-Yap 1980). In (2), for example, the word for a particular rice porridge dish cooked with ox tripe, referred to as *goto* "ox tripe," originates from Hokkien *gu to*. Other words include *taho* "bean curd" from Hokkien *taūhù* and *bilao* "device for winnowing rice" from *bí laū* (Chan-Yap 1980).

(2) *Ano   bang   pinagkaiba   ng   lugaw,   **goto**   at   arroz-caldo?*
    what PRT   difference   LNK   porridge tripe   and   arroz caldo
    "What is the difference between rice porridge, ox tripe, and arroz-caldo?
    (Magbanua 2018; 2018, tabloid, text, Mainstream Filipino, unknown)

Some posit that these loanwords entered Tagalog in the seventh century (Chan-Yap 1980: 2). Others like Manuel (1949: 94) propose a timeline slightly before the period of Spanish colonization in the 1500s. Regardless of the exact time of introduction, what is clear is that there was already interaction between Hokkien and the local languages, Spanish, and Tagalog (at least in Manila), even in the early period – Hokkien was influenced by Spanish through the borrowing of Spanish religious lexicon; on the other hand, Tagalog was influenced by Hokkien through lexical borrowing of cookery and food-related terminology. Despite there being bi-directional contact between early Sinitic Hokkien and Philippine-based languages, there seems to be no evidence of structural or functional transfer of Spanish or Tagalog onto Early (Manila) Hokkien or vice versa.

Philippine Hokkien

The second variety of Hokkien to be discussed in this chapter is Philippine Hokkien or *Huīdīpīn Hōkkiên-uè*. It is a dialect of Hokkien used by the Lannangs which has drawn influence from other Southern Chinese Hokkien dialects, such as *Ēmúng*/Amoy/Xiamen and *Tsînkāng*/Jinjiang Hokkien. It is sometimes referred to as

Lánnang-uè "Our People speech," but I will reserve the use of this term to refer to a different language in this chapter.

Philippine Hokkien is typically used by the Lannangs in religious and cultural domains (e.g., traditional opera, funeral rituals, temple rituals, sermons, lineage, or clan association meetings) (Nicolas 2016; Uytanlet 2014). In terms of spheres of usage, the situation is complex: a small percentage of the Lannangs, particularly those in their 90s and above, use it across domains (e.g., home, peers, community, religion). Younger Lannangs are known to use Philippine Hokkien in restricted domains (e.g., religion). There are, however, some Lannangs who use this variety as their dominant and native language.

Generally, Philippine Hokkien is viewed by the community as an ancestral language and is held in high regard. Many Lannangs pride themselves in being able to speak the language fluently, even if some of them (particularly younger ones) cannot actually speak it, instead using a mixed language with Hokkien characteristics, a point I will return when I discuss Lánnang-uè later.

Research done on Philippine Hokkien is largely inadequate. One of them is Dy's (1972) description of Philippine Hokkien syntactic structures, perhaps the earliest work on the variety. The other is that of Tsai (2017), whose description of Philippine Hokkien is perhaps the most comprehensive. In her dissertation, she explored the phonology of this dialect and conducted surveys focusing on the language situation of the Lannang community in Manila. Her study, for example, found that the phonological system (e.g., tone) of Philippine Hokkien is distinct from other Hokkien varieties.

In the following sections, I describe two general processes of interaction – one where Hokkien influences other Philippine-based languages and the other where it is influenced by other local languages.

Influences on Local Languages

Mainstream Tagalog and Filipino (henceforth, Tagalog for simplification purposes) in Manila have borrowed words for relatively modern Chinese paraphernalia from Philippine Hokkien. The word *āngpau* "red packets with money," for example, has been borrowed into Tagalog as *ampao*, with the velar nasal sound [ŋ] phonologically adapted as a bilabial nasal sound [m] through the process of assimilation, which is a sound process where two neighboring sounds become more similar to each other, and the loss of contrastive tone, as shown in (3).

(3) *Binigyan ako ni kuya ng* **ampao**
    give   1.SG   NOM   brother   LNK   red-packet
    "My brother gave me a red packet."
    (December 2019, unknown, elicited speech, Tagalog, Filipino)

The claim of borrowing from Philippine Hokkien is based on the fact that *ampao* did not appear in Chan-Yap's (1980) comprehensive list of early Hokkien borrowings into Tagalog. This section, thus, assumes that the borrowing is from Philippine Hokkien, and not from earlier Hokkien varieties.

As shown in this subsection, borrowing from Philippine Hokkien (henceforth, Hokkien) is present in Tagalog. There has so far been no evidence of linguistic transfer from Hokkien to it, though. There is, however, transference into the Tagalog used by the Lannangs within the community (henceforth, Lannang Tagalog).

Not only did the Lannangs borrow vocabulary from Hokkien and English into their Tagalog; their Tagalog was also influenced by the structural features of Hokkien. For example, apart from the borrowing of "basic" Hokkien words like *takpai* ("always") and "technical" English words such as *protective*, there is the tendency for speakers not to use special clitic rules or morphological contrasts.

In Tagalog, the special second-person pronominal *-ka* is a clitic that should always succeed the first element of the sentence, such as *lagi* ("always") in (4).

(4) *Lagi   -ka   nalang   protective.*
    always   2.SG   PRT   protective
    "You're always protective"
    (2019, home, translation by native speaker, Tagalog, Filipino)

However, in Lannang Tagalog, *ka* can be located at the end of the sentence (5), potentially due to structural transfer or interference from Hokkien, a variety that does not allow the insertion of *ka* between adverb-adjective modifiers.

(5) *Takpai   protective   nalang   **ka.***
    always   protective   PRT   2.SG
    "You're always protective."
    (2017, Lannang home recording # 001- LC Manila, spontaneous speech, Lannang Tagalog, Lannang)

Similarly, in mainstream Tagalog, the special politeness clitic *po*, if used, should typically be the second element of the sentence (Anderson 2008). In (6), *po* succeeds the first constituent *mahal na mahal* ("really loves").

(6) *Mahal   na   mahal -po   niya   tayo.*
    love   LNK   love POL   3.SG   1.PL.INC
    "He really loves us."
    (2020, home, translation by native speaker, Tagalog, Filipino)

However, in Lannang Tagalog, there is a tendency to place this politeness clitic at the end of the clause, perhaps due to a Hokkien-triggered reanalysis of the clitic as a particle (7).

(7) *Mahal   na   mahal niya   tayo   **po.***
    love   LNK   love 3.SG   1.PL.INC POL
    "He really loves us."
    (2019, LC Manila, spontaneous speech, Lannang Tagalog, 55-year-old female Lannang)

It is also common for speakers of the variety to ignore the subtle distinctions of Tagalog morphology marking. For example, mainstream Tagalog makes a distinction between nonfinite and finite verbs by using different morphological markers. However, there is a tendency for Lannang Tagalog speakers to drop the distinction, since Hokkien does not have this morphological feature but uses sentence-final particles to relay information of verb finiteness. In (8), the perfective *nag-*, rather than the *mag-* prefix in Tagalog, is used as a morpheme marking a nonfinite state.

(8) *Pag   nag-   start   na   ako   nag - thaktsheh this year...*
    COND   PERF   start   PRT   1.SG   INF   study   this year
    "When I start to study this year..."
    (2017, LC Manila, spontaneous speech, Lannang Tagalog, PC0152)

Apart from interacting with Tagalog, the Hokkien spoken in Manila also influenced (Philippine) English. I particularly focus on the variety used by some Lannangs – a Lannang variety of Philippine English used in Manila (hereafter, Manila Lannang English, MLE) (Gonzales and Hiramoto 2020).

Borrowings from Hokkien are present in this variety of English. For instance, in (9), the Hokkien word *chiong guan*, instead of *jackpot*, was used.

(9) *I like Dr. [omitted]'s suggestion of a laptop or Terabyte as the ultimate chiong guan **than** a TV.*
    (Gonzales and Hiramoto 2020; 2017, LC Manila, E-mail text, Manila Lannang English, PC0002)

There are also structural influences from Hokkien on this variety of English. Two of these are the lack of auxiliary inversion in *wh*-questions, as shown in (10), and the use of plain comparative *than*, as is the case in (9).

(10) *Why you did not answer?*
     (Gonzales and Hiramoto 2020; 2017, LC Manila, text message, Manila Lannang English, PC0002)

Both features have been attributed to Hokkien influence (Gonzales and Hiramoto 2020): the absence of an auxiliary inversion system for asking questions and optional use of a bipartite comparative marker in Hokkien (i.e., *pí + khâ*) in Hokkien seem to have influenced the English of the Lannangs, making both features innovations in the variety.

So far the discussion has focused solely on interactions between Hokkien and regional languages within the Manila context. However, there are also Hokkien-related interactions in other Philippine cities and provinces where Lannang and Filipino interactions are present, such as the central Philippine city of Iloilo. In addition to Tagalog, the Lannangs in Iloilo also use Hiligaynon, the regional language. Hokkien has also influenced the Hiligaynon that they speak, referred to here as Lannang Hiligaynon.

Lexical borrowings are observed in this variety of Hiligaynon. Example (11) shows the borrowing of Hokkien *boksu* ("pastor").

(11) *Daw   may   komplikasyon   bala   **boksu,***
    PRT   have   complication   PRT   pastor

   teh   bata   niya   nag-   takeover   sa   [omitted].
   so   child   3.SG.GEN   PERF   takeover   LOC   [omitted]

"I heard that they had a complication, pastor, so their child took over there."
(2018, LC Iloilo, spontaneous speech, Lannang Hiligaynon, PC0143)

Just as Lannang Tagalog has structural influences from Hokkien, Lannang Hiligaynon is also influenced by the structure of Hokkien. Whereas mainstream Hiligaynon typically follows the VSO constituent order (Wolfenden 1975: 69), Lannang Hiligaynon can follow the canonical SVO constituent order exhibited in Hokkien and English, as in (11) and (12). The phrase *în* "3.PL" in (12) also seems to have been transferred from Hokkien to Hiligaynon. Along with other features that will not be discussed here, the innovative word order and pronominal system show that the Lannangs in Iloilo both borrow and transfer structure from Hokkien onto Hiligaynon.

(12) *În   disagree   gid   ya   sa   ubra   nila.*
    3.PL disagree   PRT   PRT   LOC   work   3.PL
    "They disagree with what they did."
    (2018, LC Iloilo, spontaneous speech, Lannang Hiligaynon, PC0143)

There is also an influence of Hokkien on the English variety the Iloilo Lannangs use (henceforth, Iloilo Lannang English), just as there is a Hokkien influence in Manila Lannang English. This is unsurprising, given that the Lannangs in Iloilo are knowledgeable in English, as well as in Hokkien and Hiligaynon. Structural transfer from Hokkien to English, for instance, is observed in the cases of Hokkien-sourced polysemy and in pro-dropping – or the dropping of the subject – in English below.

For example, the phrase *saw good* in (13) can be interpreted as "see well" in English, whereas in Hokkien, this phrase is ambiguous. This is because "to see" or *khuà* in Hokkien can mean "to read" or "to comprehend." In (13), it appears that the speaker intended to use *read well* but instead used the word *see*, as how they would in Hokkien, making the extended use of *see* – to read – an innovation in Iloilo Lannang English.

(13) *My face is Chinese but I cannot saw good in Chinese. I cannot speak in Chinese.*
    "My face is Chinese but I cannot read Chinese well. I cannot speak Chinese."
    (2018, LC Iloilo, spontaneous speech, Lannang English, PC0143)

Furthermore in (14), the subject can be dropped, like in Hokkien, whereas this is not considered grammatical in "standard" English.

(14) *Always change mind.*
    "He keeps changing his mind."
    (2018, LC Iloilo, spontaneous speech, Lannang English, PC0143)

Influences from Local Languages

The local languages have also influenced Hokkien. In (15), for example, a Lannang pastor facilitating a committal funeral service, a domain where the "unmixed" Hokkien is used, is observed to "code-switch" to English, as indicated by the lack of phonological adaptation (Grosjean 2010). He could have used Hokkien *āntsòng* [an^{33} tsoŋ^{52}] ("committal"); instead, he used English *committal* [koˈmital]. When asked why he switched to English, he explained that he did so to accommodate the audience who happened to be more dominant in Tagalog and English. He switched for the sake of those who could not understand the Hokkien word.

(15) *Diēn-aū   dân   tsiū   ū   committal.*
     after   1.PL   ADV   have   committal
     "After this, we will have a committal."
     (2019, LC Manila, Cosmopolitan Memorial Chapels and Crematory, spontaneous speech/ funeral speech, Philippine Hokkien in Manila, PC0002)

Apart from code-switching, local influence on Hokkien is also reflected in loan words. Place names in the Philippines that cannot be originally found in Hokkien, for example, are borrowed from the regional language(s). For instance, in the speech of the same pastor, the term used to refer to the city of Manila, Manila [maˈ.ni.la], is borrowed into Hokkien with suprasegmental and segmental modifications – [ma^{33}.ni^{33}.laʔ^{35}] (16). Older Lannangs loan the term differently, as [bien^{33}.ni^{33}.laʔ^{35}].

(16) *Mānīlá   kaûhuè*
     Manila   church
     "Church in Manila"
     (2020, home, spontaneous speech, Philippine Hokkien in Manila, PC0002)

There is also evidence of English borrowing into Hokkien (e.g., Pōló "Paul" and "Mádī-ā "Mary") in the domain of religion (e.g., sermons).

English is the source of a number of words here because there is evidence that American and British scholars were deeply involved in the publication of Hokkien dictionaries and ultimately in the Amoy Hokkien Bible in the 1800s that was used by Christian immigrants to the Philippines (Uayan 2014) in the 1900s. Some of these immigrants had the mission of establishing churches in the country.

However, unlike the borrowing of place names from regional languages, English borrowing in Hokkien does not seem to be a consequence of interactions within the Sino-Philippine language ecology. Instead, the aforementioned historical evidence suggests that religious terms in English were borrowed into (Amoy) Hokkien before its entry into the Philippines – before it competed with other Hokkien dialects during the formation of Philippine Hokkien.

Overall, only code-switching and regional language borrowing – not English borrowing – are innovations resulting from Sino-Philippine interactions. No instances of structural transfer from local languages to Hokkien were observed.

Mainland Hokkien Variety in Manila

This section focuses on the other co-existing variety of Hokkien of the contemporary Philippine Chinese population – the Mainland Hokkien variety in Manila, primarily used by the Mainland Chinese (the homeland Chinese).

Mainland Hokkien has linguistic features that distinguish it from Philippine Hokkien. Comparing the phonological systems, for instance, there is socially conditioned variation in the use of voiced alveolar onsets in some words. In Mainland Hokkien, the voiced alveolar lateral approximant [l] is generally used. In Philippine Hokkien, the voiced alveolar stop [d] is also used, particularly for words that have a high vowel succeeding the alveolar onset such as *di/li* "2.SG" and *duwe/luwe* "woman." Despite the variation, speakers of Philippine Hokkien can understand Mainland Hokkien speakers, and vice versa.

Unlike Philippine Hokkien that is used restrictively, Mainland Hokkien is used by its speakers in many other domains (e.g., home, with peers) and is used by both young and old generations ubiquitously, who regard it as their native tongue. Some Lannangs explicitly characterize the Mainland variety as "too Chinese" or *Intsik* "chink/Chinese."

To my knowledge, Mainland Hokkien in the Philippines has not been investigated by scholars, perhaps because of an assumption that this variety has not been influenced by the local languages. The assumption seems reasonable, at first glance, particularly for those who came from Southern China in the 2010s, as these Mainland Chinese may not have been extensively exposed to the local language; nor have they been educated in Tagalog or English, unlike the Lannangs. This does not necessarily mean, however, that they were not in contact with the Filipinos – the Mainland Chinese actually frequently interact with locals (e.g., with caretakers, with market vendors, etc.). With evidence of interactions such as this, it is problematic to discount the possibility of language contact and the role Mainland Hokkien plays in the Philippine language ecology.

Influences on Local Languages
So far, this Manila Mainland variety of Hokkien has not been observed to influence any of the local languages. No borrowing or instances of structural or functional transfer from Manila Mainland Hokkien to local languages can be observed.

Influences from Local Languages
There is evidence showing that Mainland Hokkien, as used by the Southern Chinese immigrants in Manila (henceforth, Manila Mainland Hokkien), has been influenced by local languages.

For instance, in (17), a 26-year-old male immigrant, who came to the Philippines when he was 21, borrowed the word *kābātū-ān* to refer to the town Cabatuan with phonological adaptations similar to Philippine Hokkien.

(17) *Guâ   tsētsùn tī   **kābātū-ān.***
     1.SG   now   LOC Cabatuan.

"I am now in Cabatuan."
(December 2019, WeChat, spontaneous speech, Manila Mainland Hokkien, 35-year old male)

This is slightly different from the strategy used by some immigrant children. In (18), for example, a sentence spoken by the 7-year-old daughter of a Mainland Chinese immigrant in Binondo located in Manila shows the girl borrowed the English word *Shoppers* "Shoppers Mart" into her Hokkien, with some tonal modifications. The usage of *Shoppers* is similar to the use of *Shoppers* in Lánnang-uè used by the Lannangs.

(18) **Shoppêrs** lî –ē thau à!
    Shoppers 2.SG GEN head PRT
    "I don't agree that we should go to Shopper's Mart."
    (December 2019, Ongpin Street, spontaneous speech, Manila Mainland Hokkien, 7-year-old girl)

The use of *lî* ("2.SG") in (18) indicates that the borrowing is into Manila Mainland Hokkien and not into Philippine Hokkien, because of the lack of the *lî* [li^{55}] feature in the latter, which uses *dî* [di^{55}] for the second-person singular instead (Dy 1972:75).

Apart from English words, commonly used native Tagalog words, such as *yaya* ("female domestic helper"), are also borrowed into Manila Mainland Hokkien. Similar to the case of Early Manila Hokkien and Philippine Hokkien, cases of transfer from the local languages are, likewise, not observed in this variety of Hokkien.

Having looked at the interactions of Hokkien varieties with Philippine-based languages, I now turn to the dynamics of Cantonese varieties with the local languages.

Cantonese

Cantonese has been in the Philippines since at least the1850s (Wickberg 1965: 38). It is one of the dominant languages of the Cantonese heritage groups in the Philippines, whose ancestors have historically engaged in hard labor (e.g., making roads) but eventually shifted to commercial and retail business (i.e., bakeries, restaurants, tailor shops) (Wickberg 1965; oral tradition, third-generation 57-year-old Taishanese descendant, 2019). It is not, however, a prominent language of the greater Philippine Chinese community because of the tendency of the Cantonese heritage groups to immigrate to Western countries, leaving only a relatively small, stable population of heritage speakers behind.

In contemporary Philippine society, Cantonese – particularly the Standard Cantonese and Taishanese dialects – is used in domains such as the home, in restaurant associations (*kong^{55} -ong^{44} than^{11} kon^{22} woi^{11} kon^{44}*), the Cantonese Association (i.e., *kong^{55}-ong^{44} woi^{11} kon^{44}*), ancestral lineage and locality associations, as well as work and profession associations; it is regarded as identity markers of their Cantonese heritage. In the Lannang communities of Cantonese heritage, the Cantonese varieties are used alongside Philippine Hokkien and Lánnang-uè.

Although Cantonese presence in the Philippines is well-known in Philippine scholarly circles (Wickberg 1965; See and Teresita 1997), Cantonese heritage languages do not enjoy the same attention. The fact that Cantonese speakers have historically remained a minority population of Philippine Chinese society might have further discouraged work on them (See and Teresita 1997), even if they possess innovative features akin to other dominant Philippine languages.

Influences on Local Languages

Anecdotal observations show that Cantonese influence on contemporary English, Tagalog, and Hokkien used in the Philippines is limited to food item and phrase borrowings. For instance, Chinese New Year greetings in the Filipino and Lannang spheres typically are in Cantonese-origin *kung hei fat choi* ("Congratulations and wishing [you] prosperity"), rather than Hokkien *kiōng hî huât tsaí* or Mandarin *gōng xǐ fā cai*. Filipinos and Lannangs have, for example, also loaned Cantonese *hakaw* ("shrimp dumplings") into Tagalog and Hokkien, respectively.

Influences from Local Languages

Borrowing is common in the Manila variety of Standard Cantonese (henceforth, Lannang Cantonese), as well as in the Manila variety of Taishanese (henceforth, Lannang Taishanese). For instance, in (19), *TV* [ti^{55}vi^{55}] is borrowed from English into Lannang Cantonese, similar to how English *sorry* [so^{55}ɻi^{21}] is borrowed in Lannang Taishanese (20).

(19) Ngó m thaí **TV** lò.
   1.SG NEG see television PRT
   "I don't want to watch television anymore."
   (2019, LC Manila, elicited speech, Lannang Cantonese, PC0002)

(20) *Ay,* khui kong, "**Sorry**, ngoi hai m o lo."
   INT 3.SG say   sorry   1.SG see NEG CLI PRT
   "Oh, it said, 'Sorry, I can't see it anymore.'"
   (2019, frog story – LC Manila, narrative, Lannang Taishanese, PC0002)

Functional transfer of Tagalog particles, such as *nga*, into Lannang Taishanese is also present, as seen in (21).

(21) *Ni* hiak fan lo **nga.**
   2.SG eat   rice PRT PRT

"Just eat rice."
(2019, LC Manila, elicited speech, Lannang Taishanese, PC0002)

Given that the speakers are also proficient in Lánnang-uè, a language with the clause-final *nga* feature, it is plausible that the feature has been borrowed from Lánnang-uè rather than from Tagalog or Filipino. But given the lack of data, I assume that the *nga* particle has been transferred from Tagalog. Likewise, it is also likely that the code described as Lannang Taishanese here is a Cantonese counterpart of the mixed language Philippine Hybrid Hokkien (Lánnang-uè). However, in this chapter, I refrain from using the term "Philippine Hybrid Cantonese" due to lack of evidence.

Mandarin

Of the three languages, Mandarin is arguably the latest addition to the Sino-Philippine linguistic ecology. Although the time of its exact introduction to Philippine society is unknown, due to the lack of linguistic documentation, socio-historical records show that it was most likely introduced after Hokkien and Cantonese (Wickberg 1965). Before the early 1900s, there were no existing records of Mandarin users – there seem to be only those of the Hokkien and Cantonese-speaking Chinese (Wickberg 1965). Compare this to accounts of modern Philippine Chinese society (1900s onward), where Mandarin, Hokkien, and Cantonese users are mentioned (Poa 2004). Both past and present accounts collectively provide evidence for the late entry of Mandarin in the Philippines.

While Mandarin is used in the Philippines, its use in the Philippine Chinese population is not homogeneous; that is, not everyone uses Mandarin frequently and proficiently. This is in part because three different populations use it – the Mainland Chinese immigrants, the Mainland Chinese sojourners, and the Lannangs. Each of these groups does not share the same relationship with Mandarin.

For most of the Lannangs, Mandarin is not their native language. In contrast to Hokkien and Cantonese, Mandarin is widely regarded as a school language only and is rarely used beyond the academic domain. It was said to be first introduced to the mainstream Lannang school curriculum around the 1950s (Poa 2004), making it one of the four languages used in school (i.e., Mandarin, Hokkien, English, and Tagalog). Despite Mandarin's inclusion in the curriculum and its status as a global language in modern Lannang society, the Lannangs, in general, did not adopt Mandarin as the community lingua franca (Poa 2004). And this continues to be the case even up to the present, although the domain of Mandarin use has slowly been expanding due to the increasing influence of China in recent years.

Proof of its increasing dominance can be found in evangelical church services held by a subgroup of Lannangs, which cater to elderly and middle-aged members. For example, Chinese hymns are now sung in either Hokkien or Mandarin, whereas they were previously only sung in Hokkien. There are also some reports of Catholic churches using Mandarin. These mostly cater to the new immigrants and sojourners.

Broadcasting companies (e.g., ChinoyTV, Chinatown TV) targeting the Philippine Chinese use Mandarin instead of Hokkien. Regardless of this, the use of Mandarin is still relatively restricted in the contemporary Lannang community.

However, this is not the case for the new immigrant and sojourner groups. Since they are from post-Cultural Revolution China – a society with Mandarin as its common language – Mandarin is one of their native languages. Unlike the Lannangs, the immigrant group can use Mandarin proficiently and use it ubiquitously among their peers and family members. However, most of the time, when conducting business or with peers, they choose to use Hokkien or a Tagalog Pidgin, which will be discussed later.

The Mainland Chinese sojourners, like the immigrant group, also have Mandarin in their linguistic repertoire. Mandarin is also their native language, as they have been exposed to Mandarin in formal schooling in China. However, in contrast to the Hokkien-using immigrants, the sojourners use Mandarin in most, if not all, domains of language use.

It is clear that the use of Mandarin is dominant among the Mainland Chinese, but not among the Lannangs. This is not to say, however, that no Lannang households would use Mandarin. Some families that have at least one Chinese parent from other Mandarin-speaking regions (e.g., Taiwan) do use Mandarin as their native tongue and home language. They do, of course, represent the minority in Lannang communities.

To my knowledge, published works that focus on the linguistic structure of Mandarin varieties in the Philippines do not exist, despite the increasing presence of Mandarin in contemporary Philippine society.

Influences on Local Languages

Mandarin in the Philippines has been observed to interact with Hokkien and English by functioning as a contributor of select (not necessarily basic) lexicon. In Lannang Mandarin, for example, Mandarin *pàng-pàng* ("fat") occurred in the Hokkien utterance of a young female Lannang speaker that is more proficient in Mandarin than Hokkien (22).

(22) *ū   tsīgē   pàng pàng*
 have one   fat   fat
 "There is a fat kid."
 (2017, LC Manila, interview, Philippine Hokkien, PC0155)

The word for *frog* in Mandarin, *qīngwā*, is also seen in the Lannang English of an old Lannang speaker in Manila (23).

(23) *The   boy   is   looking   at   qīngwā.*
 The   boy   is   looking   at   frog
 "The boy is looking at the frog."
 (2019, frog story-LC Manila, narrative, Manila Lannang English, PC0125)

In both examples, the words are code-switched (Grosjean 2010) into the local languages because Hokkien-like tone was not applied to *pàng-pàng* in (22), and Mandarin tone was not modified in *qīngwā* in (23). Specifically, in the former, Hokkien-sourced phonological dissimilatory processes (e.g., tone sandhi) should have been observed in the word (e.g., *phângphàng* or *phāngphàng*) just as such a process can be observed in Lánnang-uè and Philippine Hokkien. For the latter, *qīngwā* should have acquired English-like stress.

Overall, the influence of Mandarin on the Philippine-based languages is observed to be limited to code-switching only. Borrowing and transfer from Mandarin to the local languages have not been noticed in my fieldwork.

Influences from Local Languages

Apart from affecting the local languages, Mandarin in the Philippines is observed to be influenced by these languages as well. The Mandarin used by the Mainland Chinese in Metropolitan Manila (henceforth, Manila Mainland Mandarin), for example, borrowed the technical or culture-specific Tagalog word *Pasay* [pa^{55}saj^{11}] ("Pasay City") (24).

(24) *Zài* **Pasay**...
     at   Pasay
    "At Pasay."
    (2019, house, spontaneous speech, Mainland Chinese Mandarin, 40-year-old male)

The local languages' influence on Mandarin can also be observed in the non-technical vocabulary present in the Mandarin variety used by the Metropolitan Manila Lannangs (henceforth, Lannang Mandarin). For example, in (25), the word *beībǐ* "baby" [peɪ^{55} pi^{21}] was used in place of *yīnger* ("infant").

(25) *Kěnéng   jiùshì   tāmen de   xiǎo   **beībǐ**.*
     possible   that   3.PL GEN small baby
    "It is possible that that is their small baby."
    (2019, frog story- LC Manila, spontaneous speech, Lannang Mandarin, PC0096)

In (26), the word *ōwl* [aʊl^{55}] ("owl") was used instead of *māotóuyīng* with English stress removed and Mandarin stress imposed, as in borrowing.

(26) *Nà   gě   jiào   **ōwl**   ma?*
     DEM CLF call   owl   Q
    "Is that called an owl?"
    (2019, frog story- LC Manila, spontaneous speech, Lannang Mandarin, PC0125)

There is also code-switching between Mandarin and English, such as in (27), where the word *technology* [tekˈnoloˌdʒi] not only has English-like prosodic features (unlike borrowing) but is also used intentionally with English features to fulfill a pragmatic function – to mark the word for translation.

(27) ***Technology*** shì   shénme   yìsi?
     technology COP what   meaning.
     "What does 'technology' mean?"
     (2019, LC Manila, elicited speech, Lannang Mandarin, PC0068)

In summary, the influence of the local languages on Mandarin in the Philippine context is not only limited to intra-clausal code-switching. Technical and non-technical lexical borrowing from the local languages exists. But while there is code-switching and borrowing in Mandarin, neither functional nor structural transfer from the local languages to Mandarin was observed.

Sino-Philippine Contact Languages

What has been presented so far has showed how the Chinese heritage and homeland speakers' languages were influenced by Philippine-based languages. Furthermore, it showed how the local languages also influenced their Sinitic languages. However, there are also cases where speakers create new (contact) languages, using the linguistic resources they have in their repertoire. This section focuses on two types of such languages – pidgins and mixed languages.

Pidgins

Chinese Spanish Pidgin

"Chinese Pidgin Spanish" (henceforth, Chinese Spanish Pidgin) is a language that has a Spanish lexifier and a Hokkien substrate (Fernández 2018: 137). Spanish provides most if not all of the vocabulary of the language, whereas Hokkien contributes to other non-lexical aspects of the language, such as the syntax.

Chinese Spanish Pidgin, first attested in 1718, was primarily used by the Sangleys (Chinese merchants) in business transactions with their customers (Fernández and Sippola 2017; Fernández 2018). Two distinctive features of this language include the subject pronouns *mia* ("1.SG"), *suya* ("2.SG") (28), and lack of preverbal marking, which are not present in Spanish, Hokkien, or Chabacano, a Spanish creole (Fernández and Sippola 2017; Fernández 2018). The use of the voiced alveolar approximant [l], rather than the Spanish alveolar tap [ɾ] in intervocalic condition, is another feature of this language, as seen in the use of *pala* [pala] in (28) (cf. *para* [paɾa]).

(28) *Suya tiene ba   comision   pala pidi pasapote con   mia*
     2.SG have Q   commission for   ask   passport with 1.SG

     y   pala   jase   buluca   aqui   na   calle?
     CONJ   for   make   problem   here   REL   street

"Do you have a commission to ask for my passport and to make problems here on the street?"

(Tombo 1860: 284, in Fernández and Sippola 2017; 1860, article, text, Chinese Spanish Pidgin, unknown)

Chinese Tagalog Pidgin

Chinese Tagalog Pidgin is a pidgin with a Tagalog lexifier and Sinitic substrates (most likely Hokkien and Mandarin), primarily used by the contemporary Mainland Chinese immigrants. It emerged out of a need of a language to communicate with locals, particularly with customers and domestic helpers who speak Tagalog. To the best of my knowledge, it has not yet been documented by any scholar.

Chinese Tagalog Pidgin seems to have several features that are distinct from Tagalog. These include the absence of inflectional affixes, the verbal negation marker *wala*, and the generalized third-person pronoun *siya*. For example, in (29), Tagalog Pidgin speakers do not use the "undergoer voice" suffix *-in* (Latrouite 2011: 148), instead opting to use the bare verb *pindot* ("press"). Contrast this with (30) where Tagalog speakers used the suffix *-in*.

(29) *Ako   din   kala   ko   kasi   siya   wala   pindot   ako.*
    1.SG ADV thought 1.SG CONJ 3.SG   NEG   press   1.SG
    "Me too, that's what I thought because they didn't help me press the button."
    (December 2019, Ongpin Street in Manila, spontaneous conversation, Chinese Tagalog Pidgin, 7-year-old boy)

(30) *Ako   rin, kala   -ko   kasi   hindi -niya   ako   t<in>ulung-an*
    1.SG ADV though   1.SG CONJ   not   3.SG   1.SG   <PFV>help-UV

   pindut-in   yung   button
   press   DEM   button
   "Me too, that's what I thought because they didn't help me press the button."
   (July 2020, Manila, translation, Tagalog, native speaker of Tagalog)

Mainstream Tagalog speakers would also not use *wala* (29), a determiner typically used to modify a head noun; they would instead use the verbal negation marker *hindi* (30). Another difference between Mainstream Tagalog and Chinese Tagalog Pidgin is the use of pronouns. In Mainstream Tagalog, the pronominal and verbal inflection systems are tightly intertwined: the *-niya* "3.SG" in (30) should be used instead of *siya* because the verb has an undergoer voice affix. In Tagalog Pidgin, the pronominal system appears to be simpler and does not seem to interact with the verb phrase due to the Pidgin's lack of verbal inflectional affixes. In (29) and in other Tagalog Pidgin speech samples, *siya* is used in contexts where *niya* is supposed to be used (in Tagalog).

With regard to its stability in the new immigrant community, my preliminary observations from my fieldwork in Binondo in 2019 reveal that the features described earlier are used not just by one speaker but by other new immigrants as well. This suggests some degree of conventionalization for the pidgin.

Mixed Languages

So far, there is only one documented Sino-Philippine mixed language in the Philippines – Lánnang-uè "Our People Speech" or Philippine Hybrid Hokkien. Lánnang-uè is a language that has features consistent with what Matras and Bakker (2003:1) refer to as "mixed languages," which are languages that systematically combine elements or (sub)systems from the source languages (Gonzales and Starr 2020). Lánnang-uè sources its vocabulary and grammar from the regional languages, English, and Hokkien in a systematic manner, as will be illustrated in this section.

Lánnang-uè does not seem to be mutually intelligible with Hokkien. Speakers of Lánnang-uè without proficiency in Philippine Hokkien report not being able to understand Mainland Hokkien speakers. Likewise, Mainland Hokkien users with no proficiency in Lánnang-uè cannot understand or produce Lánnang-uè utterances even with knowledge of the source languages. In other words, the grammar of Lánnang-uè (or the "proper" way of mixing) must be acquired.

Although Lánnang-uè behaves like a language, many of its native speakers – the Lannangs – perceive it as an "adulterated" variety of Hokkien (See and Teresita 1990: 14) or a failed attempt to acquire Hokkien (Gonzales 2021). Those who claim that it is an independent language from Hokkien are in the minority. Because some perceive it as Hokkien, it shares the name "Lánnang-uè" with Philippine Hokkien (Tsai 2017). In this chapter, the term "Lánnang-uè" refers exclusively to the mixed language.

Based on a preliminary analysis of interview data with old and young Lannangs, Lánnang-uè most likely emerged from code-switching between Hokkien, English, and the regional languages (e.g., Tagalog). During the interviews, the older, Hokkien-dominant Lannangs (those in their late 80s and early 90s) code-switched from Hokkien to English and with the regional languages. When asked why they switched from one language to another, they said that it was to accommodate to the younger Lannangs who were more dominant in English and the regional languages. Contrast this with the responses of the younger Lannangs who generally did not have any specific motivation to switch to English or the regional language(s). Among some of the words they used to characterize the switches or, rather, mixing were "regular," "normal," and, for some, even "conventionalized." This suggests that it is possible that Lánnang-uè originated from code-switching between the languages mentioned above.

Lánnang-uè is used in at least three cities – Iloilo, Cebu, and Manila – where Lannangs are known to be dominant (Doeppers 1986). The regional Lánnang-uè varieties all have linguistic features from Philippine English, Philippine Hokkien, and at least one regional language.

Lánnang-uè: Manila (Luzon Island, Luzon Region)

Manila Lánnang-uè has a composite lexicon that originates systematically from Hokkien, Tagalog, English, and Mandarin. 46.1% of its "basic" vocabulary exclusively comes from Hokkien, 3.7% from Tagalog, and 17.8% from English (Swadesh 1972; Gonzales and Starr 2020). "Technical" words can be sourced from these languages as well. Examples include *distinguîsh* [dis^{33}tiŋ^{33}guıʃ^{55}] and *bâse* [beıs^{55}] in (31).

(31) *Dî   tsiûwâ distinguîsh   na   î   sī   Taīdióklâng   a*
    2.SG how   distinguish REL   3.SG COP   Chinese   PRT

    from   bâse   on   î   –e   láng?
    from   base   on   3.SG   GEN   person

"How do you distinguish that he is Chinese based on his personhood?"
(December 2019, LC, interview, Manila Lánnang-uè, PC0068)

Words of English and Tagalog origin undergo phonological modifications in Lánnang-uè. They have acquired tone-like properties, as exemplified above.

Manila Lánnang-uè also has a composite structure or grammar, based on initial observations of the Lannang Corpus. Like its lexicon, many of its grammatical properties can be systematically traced back to the four languages. The aspect, negation, copula, and pronominal system, for example, are generally of Hokkien origin. The language's plurality marking, complementizer, and derivational affixation (e.g., like *pāng-khùn* "for sleeping" or *tagā-taīdiók* "person from Mainland China") systems are generally from Tagalog (Gonzales 2018). Phrasal conjunctions (e.g., *ānd sō* "and so," *āt leâst* "at least") in Manila Lánnang-uè originate from English, while the *wh*-question system of the language seems to be derived from Mandarin (Table 2).

Lánnang-uè: Cebu (Cebu Island, Visayas Region)

Outside the context of Metropolitan Manila, explorations on Lánnang-uè have historically not been given emphasis in linguistic research. This certainly does not mean however that Lánnang-uè is not used by the Lannangs outside Manila. Like the Manila Lannangs, the Lannangs of Cebu also use Lánnang-uè (henceforth, Cebu Lánnang-uè).

Similar to the Manila variety, Cebu Lánnang-uè has a composite vocabulary and grammar. Cebu Lánnang-uè words like *cûte* ("cute") in (32) and *squirrèl* ("squirrel") in (33), for instance, are sourced from English, whereas words like *kaû* ("dog") in (33) are from Hokkien. The conjunction *nya* in (33) is sourced from Cebuano, whereas the preposition *tī* ("at") is sourced from Hokkien.

(32) *Yá   cûte   mân.*
    very   cute   PRT
    "It's cute."
    (2019, frog story - LC Cebu, narrative, Cebu Lánnang-uè, PC0019)

Table 2 Distribution of selected Lánnang-uè elements by source language

Linguistic components		Hokkien	Tagalog	English	Mandarin
Lexicon (basic)	Origin	70.8%	21.0%	42.5%	0%
	Exclusive	46.1%	3.7%	17.8%	0%
Grammar		• Aspect system • Negation • Copula • Pronominal system	• Plurality marker • Complementizers • Most derivational affixes • Approximators • Conjunctions • Adverbials • Interjections • Yes-no question system	• Phrasal conjunctions • Most prepositions	• *Wh-*question system (?)

(33) *Nya   tsî   tsiâh   kaû   guāntsaī   tī   tsiá   tsâ   squirrèl   tī-tsiá.*
 then DEM CLF   dog   still   PREP here then   squirrel   here.
"Then, the dog continued to be here and the squirrel is here."
(2019, frog story - LC Cebu, narrative, Cebu Lánnang-uè, PC0019)

Lánnang-uè: Iloilo (Panay Island, Visayas Region)

The Lánnang-uè spoken in Iloilo City (henceforth, Iloilo Lánnang-uè) has a lexicon sourced from multiple languages. But instead of having Hokkien, English, and Cebuano as primary source languages, Iloilo Lánnang-uè has the regional language Hiligaynon, English, and Hokkien. (34) shows an example where English-sourced *equipmênt* ("equipment") and Hokkien-sourced *huāngtshiā* ("car") are used in Iloilo Lánnang-uè.

(34) *Taīdiók   -e   nā   sī   ū   huāngtshiā o,   khâ   bo   siâ*
 China   GEN   if   COP have   car   PRT COM NEG   as

 kwan bala   ho   pero   taīdiók –e   mga   heavy equipmênt   o,
 PRT   PRT   PRT   but   China –GEN   PL   heavy equipment   PRT

 tsîtsùn   malakàs.
 now   strong
"If we are talking about China's car industry, they are not as good; however, their heavy equipment business is strong."
(2018, car - LC Iloilo, spontaneous speech, Iloilo Lánnang-uè, PC0144)

The grammar of Iloilo Lánnang-uè also seems to be systematically sourced from its source languages. For example, the pluralizer *mgā* ("PL") in (35) is acquired from Hiligaynon, whereas the pronominal system originates from Hokkien (e.g., *î* "3.SG"). Iloilo Lánnang-uè, like the variety of Cebu, also has pragmatic particles sourced from the regional language Hiligaynon, as seen in particles like *bala* in (34).

(35) Pero   î   –e   mga   dust   ūm   tsaī   tolóh   a
     but   3.SG   GEN   PL   dust   NEG   know   where   PRT
     "But [I] don't know where his/her remains are."
     (2018, car – LC Iloilo, spontaneous speech, Iloilo Lánnang-uè, PC0143)

Inter-clausal Code-Switching

The discussion so far seems to characterize the interactions of Sinitic languages and Philippine-based languages (including contact languages) as happening only within the clause, but interactions of this nature also happen at levels higher than the clause, such as via inter-clausal code-switching (Gonzales 2016).

Several scholars have highlighted the existence of such macro-level interactions in the Lannang community (Chuaunsu 1989; Zulueta 2007; Gonzales 2016). In these types of interactions, a Lannang speaker may opt to switch languages within a single utterance, usually when a new phrase or clause is introduced. For instance, in a conversation in Manila, a speaker switches among four languages – Hokkien, Tagalog, English, and Lánnang-uè (36).

(36) hindi   niya   b<in>ayad,   (Tagalog)
     NEG   3SG   <PFV>pay

     î   tsiāgēh   diāpdì   ū   siá   cheke   hó   (Lánnang-uè)
     3.SG   January   twenty two   have   write   cheque   PRT

     î   lāktsāp   kuí   tshiēng   huán   i   kô   ó   (Hokkien)
     3.SG   sixty   around thousand   return   3.SG   PRT   PRT

     It's   because   (English)
     It's   because

     nag-   bigay   siya   ng   clearance.   (Tagalog)
     PERF   give   3.SG   PRT   clearance

     "She/he didn't pay. he wrote a check on January returning him/her 60,000.
     It is because she/he gave him/her clearance."
     (2017, house - LC Manila, spontaneous speech, code-switching, PC0153)

In Iloilo, the switching among English, Hiligaynon, and Lánnang-uè is observable (37).

(37) At   least   you   celebrate   Father's Day with [omitted] (English)
     At   least   you   celebrate   Father's Day with [omitted]

     hò   khâ   siammíh   khâ   hose   na   nah.   (Lánnang-uè)
     yes   COM   what   COM   good   PRT   PRT

     Oo   gane.   (Hiligaynon)

yes PRT

Bīnná	*nalang*	*boksu*	*e*		(Lánnang-uè)
Tomorrow	PRT	pastor	PRT		

Boksu,	*maniláh*	*tsādít*	*lohhō*	*bá?*	(Lánnang-uè)
pastor	Manila	yesterday	rain	Q	

"At least you celebrated Father's Day with [omitted]. It is better that way. Yeah. Let's do tomorrow. Pastor, did it rain in Manila yesterday?"
(2018, car – LC Iloilo, spontaneous speech, code-switching, PC0143)

The exact nature of the inter-clausal mixture depends on the situation and the region where the code-switched utterances are used.

Discussion

Summary

The previous sections have explored Hokkien, Cantonese, and Mandarin and highlighted the interactions among these three languages and other neighboring languages in the Philippine context. Using Haugen's (1971) ecolinguistic framework, these sections bring forth the interrelationship among these languages (organisms) in the Philippine environment. In summary, three broad processes were explored – Filipinization, where the local languages influenced the Sinitic ones instead (e.g.. English code-switching in Philippine Hokkien); Sinicization, where Sinitic languages influenced the local languages (e.g., food-related borrowings in Filipino from Early Manila Hokkien); and language creation, where independent new codes were formed (e.g., Lánnang-uè).

From the languages and varieties that emerged as a result of these three main processes, three general contact mechanisms were identified and discussed: code-switching (inter-clausal vs. intra-clausal) (Grosjean 2010), borrowing (Thomason 2001; Grosjean 2010), and transfer (Siegel 1999, 2003).

A summary of most of the languages discussed in the previous section in relation to the said contact mechanisms is provided below. To facilitate comprehension, this is split into two tables – the first highlights the Sinitic languages and the influence from other languages (Table 3).

The second puts historically indigenous languages at the center and focuses on how they are influenced by other Philippine-based languages, including Sinitic ones (Table 4). Tables 3 and 4 do not summarize the contact languages that emerged as a result of Sino-Philippine interactions.

The presence of these mechanisms and interactions – as manifested in the cornucopia of varieties and languages in Tables 3 and 4 – indicates that the role that Sinitic languages play is not a static one, that is, Sinitic languages can take the active role of a feature or lexical contributor, but can also be receivers of linguistic elements. The present survey also shows that the Filipinization, Sinicization, and

Table 3 Chapter summary of Sinitic languages/varieties and interactions with other languages

Language	Variety	Code-switching (lexical)	Borrowing (technical)	Borrowing (basic)	Transfer
Mandarin	Mainland (Manila)		T		
	Lannang (Manila)	E		E	
Cantonese	Lannang (Manila)		E		
	Lannang Taishanese (Manila)		E		T/L
Hokkien	Mainland (Manila)	M	ET	T	
	Early Manila Hokkien		S		
	Lannang (Manila)	E	T		

(H, Hokkien; C, Cantonese; E, English; T, Tagalog; G, Hiligaynon; Ce, Cebuano; L, Lánnang-uè; M, Mandarin; S, Spanish; H*, Early Manila Hokkien)

Table 4 Chapter summary of indigenous languages/varieties and interactions with other languages

Language	Variety	Code-switching (lexical)	Borrowing (technical)	Borrowing (basic)	Transfer
English	Lannang (Manila)	M	H		H
	Lannang (Iloilo)				H
Tagalog/Filipino	Mainstream (Manila)		H*H		
	Lannang (Manila)		E	H	H
Hiligaynon	Lannang (Iloilo)		H		H

H, Hokkien; C, Cantonese; E, English; T, Tagalog; G, Hiligaynon; Ce, Cebuano; L, Lánnang-uè; M, Mandarin; S, Spanish; H*, Early Manila Hokkien

language creation processes co-exist in the Sino-Philippine linguistic ecology and demonstrates the complexity and dynamism of the Sino-Philippine linguistic landscape.

A natural question arises: what caused these interactions? Based on the descriptions of the Chinese heritage and homeland groups earlier, it is clear that the processes of Filipinization, Sinicization, and language creation are likely to have been triggered by contact between the Sinophone and Philippine language-speaking groups. However, as hinted earlier, the consequences of contact are asymmetric, that is, language varieties in the Sino-Philippine ecology do not necessarily need to "impose" and "receive" the same degree of Sinitic or Philippine influence from each other. Speaker groups also do not necessarily need to create new (contact) languages.

Patterns of Asymmetric Contributions

Three general patterns of asymmetric contributions pertaining to language interactions in the Sino-Philippine sphere can be identified, each of which can be accounted for by the dynamics between the speaker groups involved, as well as specific social motivations.

Sinitic Over Philippine

The first pattern of asymmetry relates to cases where Sinitic influence is greater than local Philippine influence. Influence here is measured by how much a language contributes to another language.

One such case involves Early Manila Hokkien used by the Sangleys. The current data indicate that Early Manila Hokkien provided technical lexicon for Tagalog (and modern Filipino) but not the other way around. Sinitic influence was greater perhaps because of the Sangleys' dominant role as economic brokers in early Philippine society. Considered the middlemen between the *Indios* (locals) and the Spaniards, the Sangleys dominated the economy. The economic power they wielded is reflected in the Filipinos' tendency to borrow lexical items that relate to their business or trade activities (e.g., food-related terms). The Hokkien used by the Sangleys might have been influenced by the *Indios*' Tagalog, but currently there is no evidence to support this. Regardless, the lack of foreign influence on the Sangleys' Early Manila Hokkien might be due to this power dynamic: in other words, there is no need or pressure for them to "compromise" their language, as they are in a position of power.

Philippine Over Sinitic

A reversed pattern is also observed, one in which Philippine influence is more salient than Sinitic influence. Influence here is measured by how much a language is "resistant" to foreign influences.

The pattern of asymmetric contribution, for one, is evident in the Lannang and Filipino community mixed codes – Lánnang-uè and Filipino, respectively. While Lánnang-uè has vocabulary and features almost equally sourced from both Sinitic languages and regional languages like Tagalog and English, Filipino generally does not source linguistic elements from Sinitic languages, apart from select lexical items. This is not surprising given the Lannangs' historical attempt to assimilate to larger Philippine society – an action that requires one to be "Filipinized" by learning the languages and associated norms of the community they are trying to assimilate to, not the reverse. As the majority of the Philippine population, the Filipinos have no motivation to assimilate to the Lannangs.

Note that, unlike the Sangleys, the Lannangs are not economic brokers. They are dominant in retail and other businesses, but do not monopolize them and, as such, do not hold the kind of power that would pressure locals to assimilate or adjust to them. Another crucial point is the fact that Lannangs identify more as Filipino rather than Chinese, in contrast to the Sangleys. So even if they did monopolize businesses, they used other languages apart from Hokkien; they employed Tagalog and English (in Manila). In both situations, there was simply no incentive for locals to assimilate to the Lannangs, as seen in the lack of Hokkien features in Filipino. For the Lannangs, however, there is a need to retain aspects of their heritage but also identify with Philippine society. A result of this negotiating act, I argue, is reflected in their mixed language Lánnang-uè with features sourcing from both Philippine-based and Sinitic languages.

Yet another case where Philippine influence is noticeably greater than Sinitic influence is in the genesis of pidgins. Here, a language is more influential if it gets chosen as a lexifier. As indicated in the chapter, the Mainland Chinese immigrants (in Manila, at least) have created Chinese Tagalog Pidgin, which has a Tagalog lexifier. However, no such counterpart of a pidgin, a Hokkien-lexifier pidgin (e.g., "Philippine Hokkien Pidgin"), can be observed among the locals. The reason is probably due to an economic need on the part of the Mainland Chinese immigrants. In order to establish a successful business in the Philippines, the Mainland Chinese immigrants have to cater to, and in some cases rely on, Tagalog language-speaking customers. This meant having to accommodate them by communicating in a created Tagalog-based code that does not necessarily have to be identical to mainstream Tagalog. Contrast this with the Filipinos who do not rely on the Hokkien-speaking Mainland Chinese immigrants for their needs and, consequently, have less of a need to communicate in Hokkien (or a Hokkien Pidgin). This one-sided need explains the asymmetry found in pidgin emergence. In both cases of asymmetry, pidgin genesis and feature selection in mixed languages, the pattern is clear – Philippine influence is greater than Sinitic influence.

Southern Over Non-southern

Within the Sinitic languages, a different pattern of asymmetric contributions has emerged. The grammar of Southern Sinitic languages like Hokkien tends to either influence or be influenced by Philippine-based languages. This contrasts with non-southern Sinitic Mandarin, whose grammar appears to neither influence nor be influenced by them. While the Mandarin used by Lannangs is only limited to lexical borrowing and code-switching ("lower-order" mechanisms), there is evidence of southern Taishanese, for example, also being involved in other "higher-order" contact mechanisms such as transfer (e.g., transfer of the discourse particle system). Non-southern Mandarin also does not seem to influence the regional languages' grammar as much as southern Hokkien.

Perhaps the chronological entry of Sinitic languages into the Philippine linguistic ecology can account for this imbalance in contributions or influence. As pointed out earlier, Mandarin is the latest addition among the Sinitic languages discussed in the chapter. This is in contrast to the southern Sinitic languages that have been in the Philippines during the Spanish colonial time (or even before that period in the case of Hokkien). Since Mandarin was introduced later than Hokkien and Cantonese (as a school language), it is not surprising that Mandarin is not as explicitly involved in the observable interactions compared to Hokkien or Cantonese.

In the Lannang context, another possible reason why Mandarin does not interact as much with the regional languages compared to Southern Sinitic languages could have something to do with heritage. Mandarin was never a heritage language in the Philippines, whereas Hokkien and Cantonese are. In other words, the languages that mark the Philippine Chinese identity are Hokkien and Cantonese (Chuaunsu 1989), the latter particularly for the Cantonese heritage Lannang community.

Mandarin's arguably late entrance into the linguistic ecology and its lack of an index for group identity are two possible reasons for the asymmetric influences. The small number of Mandarin speakers (e.g., Mainlanders with non-Southern Chinese heritage) in relation to Hokkien speakers, as well as relative social distance between Mandarin speakers and other groups that speak Philippine-based languages, could also account for this other pattern of asymmetric contributions.

Summary

Overall, what is emerging from this discussion is that dynamic interactions between speakers of Sino-Philippine languages display asymmetric contributions. There are cases where Sinitic influence is greater than local influence, some where the reverse is true, and other cases where Southern Chinese influence prevails over non-Southern influence. And we see that these asymmetries can be partially accounted for if we direct our attention to the dynamics between different social groups (Table 5).

It goes without saying that one should not simply rely on a single sociohistorical account of language contact phenomena (like the ones offered earlier) and use it to predict the outcomes of similar contact situations. It is naturally impossible to account for all social factors and conditions affecting contact-induced change (Thomason 2000). That is, other unreported or non-observable factors can also affect the result of such interactions. For instance, innovative features may be deliberately inhibited or introduced in a linguistic variety due to speaker attitudes (Thomason 2007). Given this, the accounts and social factors used to explain the patterns of asymmetric contributions in this chapter are not predictive at all. This discussion is only meant to help us understand observable linguistic interactions within the Sino-Philippine language ecology.

Conclusion

With the goal of highlighting the complex and dynamic nature of Sinitic languages in the Philippine context, this chapter has primarily explored three Sinitic languages. It has demonstrated how their speakers interact with speakers of Philippine-based languages through the processes of Filipinization, Sinicization, and language creation. This study has not only shown the complex dynamics between Sinitic languages. It has also illustrated that the ways in which these languages influence one another are not always symmetric, partly due to differences in the dynamics between the groups in contact. It could also be partly due to other social motivations. Beyond that, this chapter has shown the relevance of these linguistic varieties to the broader Philippine society. It does so by identifying (new) Philippine languages that have

Table 5 Summary of patterns of asymmetric contributions Asterisks (*) here mean 'not'

Pattern	Asymmetry	Social group(s) involved	Motivation
Sinitic > Philippine	*Borrowing* Early Manila Hokkien → Tagalog *Tagalog → Early Manila Hokkien	Sangleys versus *Indios* (locals)	Economic power
Philippine > Sinitic	*Feature selection* Philippine + Sinitic → community mixed code (Lánnang-uè) Philippine + *Sinitic → community mixed code (Filipino)	Lannangs versus Filipinos	Minority assimilation
	Pidgin genesis Chinese Tagalog Pidgin *Philippine Hokkien Pidgin	New immigrants versus Filipinos	Economic need
Southern Sinitic > non-southern Sinitic	*Higher-level interactions* Philippine Hokkien, Cantonese *Mandarin	Lannangs versus mainlanders with non-Southern Chinese heritage	Chronological order, heritage, identity

emerged out of Sino-Philippine contact. The chapter also highlights the ways in which Philippine features have enriched Sinitic languages.

It should be clear that the present discussion is obviously an oversimplification of the complex linguistic ecology in which these languages evolve and continue to develop. Throughout this chapter, I have had to simplify some categories present in the contact varieties under study, although we know all too well that language categories are not always clear-cut (Thomason 2001; Matras and Bakker 2003). For example, Lánnang-uè was characterized in this chapter as a mixed language, but it also shares some characteristics with "indigenized varieties" (Winford 2020: 4). This could potentially challenge the notion of contact languages fitting into specific "types" and support instead the theory of a continuum of contact varieties (Thomason 2001; Baptista 2015; Winford 2020). Future research may consider investigating more linguistic features for each variety with the aim of testing whether these languages can be viewed as "types" or rather exemplify a continuum of contact varieties.

Outside of these issues, I have also not discussed interactions beyond the second order (defined here as interactions involving the contact varieties themselves) in detail. There are some cases of transfer in Manila Lannang English, for example, that were attributed to Tagalog and Hokkien influence in the discussion but can theoretically be attributed to Filipino and Lánnang-uè as well (Gonzales and Hiramoto 2020). Such cases of second-order transfer are, indeed, possible, but fall beyond the scope of this chapter, which aims to provide a *descriptive* overview of the dynamics between Sinitic and Philippine-based languages in the Philippines. Other potential interactions that involve the contact varieties described in this chapter could be further explored in future work.

The overall objective of this chapter is modest: it is meant to provide scholars and interested individuals a snapshot of the Sino-Philippine language ecology. Being the first large-scale survey and description of its kind in the field of Sino-Philippine linguistics, spanning varieties past and present, this chapter takes a much-needed step in providing readers with a more holistic introduction to Sino-Philippine and, consequently, to the broader Philippine linguistic landscape.

References

Abastillas, Glenn. 2015. *Divergence in Cebuano and English code-switching practices in Cebuano speech communities in the Central Philippines*. Unpublished M.A. thesis, Georgetown University.
Agoncillo, Teodoro A., and Milagros C. Guerrero. 1970. *History of the Filipino people*. Quezon City: Malaya Books.
Anderson, Stephen R. 2008. Second-position clitics in Tagalog. In *The nature of the word: Studies in honor of Paul Kiparsky*, ed. K. Hanson and S. Inkelas. Cambridge: MIT Press.
Baptista, Marlyse. 2015. Continuum and variation in Creoles: Out of many voices, one language. *Journal of Pidgin and Creole Languages* 30 (2): 225–264.
Bautista, Maria Lourdes S. 2000. *Defining standard Philippine English: Its status and grammatical features*. Manila: De La Salle University Press.
———. 2004. Researching English in the Philippines: Bibliographical resources. *World Englishes* 23 (1): 199–210.
Bodman, Nicholas Cleaveland. 1987. *Spoken Amoy Hokkien*. New York: Language Services.
Chan-Yap, Gloria. 1980. *Hokkien Chinese borrowings in Tagalog*. Canberra: Pacific Linguistics.
Chappell, Hilary. 2019. Southern Min. In *The Mainland Southeast Asia linguistic area*, ed. A. Vittrant and J. Watkins. Berlin: De Gruyter Mouton.
Chirino, P. Petrus [Pedro]. 1604. Dictionarium Sino Hispanicum. Ms. 60 [= RMS], 83 ff. Biblioteca Angelica, Rome. /Ms. Chinois 9276 [= PMS], Bibliothèque nationale de France (BnF), Paris.
Chuaunsu, Rebecca Shangkuan. 1989. *A speech communication profile of three generations of Filipino-Chinese in Metro Manila: Their use of English, Pilipino, and Chinese languages in different domains, role-relationships, speech situations and functions*. Unpublished M.A. thesis, University of the Philippines.
Doeppers, Daniel F. 1986. Destination, selection and turnover among Chinese migrants to Philippine cities in the nineteenth century. *Journal of Historical Geography* 12 (4): 381–401.
Dy, Carmen J. 1972. The syntactic structures of Amoy as used in the Philippines. *Philippine Journal of Linguistics* 3 (2): 75–94.
Fernández, Mauro. 2018. El pidgin chino-español de Manila a principios del siglo XVIII. *Zeitschrift für Romanische Philologie* 134 (1): 137–170.
Fernández, Mauro, and Eeva Sippola. 2017. A new window into the history of Chabacano: Two unknown mid-19th century texts. *Journal of Pidgin and Creole Languages* 32 (2): 304–338.
Gonzales, Wilkinson Daniel Wong. 2016. Trilingual code-switching using quantitative lenses: An exploratory study on Hokaglish. *Philippine Journal of Linguistics* 47: 109–131.
———. 2017a. Language contact in the Philippines: The history and ecology from a Chinese Filipino perspective. *Language Ecology* 1 (2): 185–212.
———. 2017b. Philippine Englishes. *Asian Englishes* 19 (1): 79–95.
———. 2018. *Philippine Hybrid Hokkien as a postcolonial mixed language: Evidence from nominal derivational affixation mixing*. Unpublished Master's thesis, National University of Singapore.

Gonzales, Wilkinson Daniel Wong. 2021. Filipino, Chinese, neither, or both? The Lannang identity and its relationship with language. Language & Communication 76, Elsevier.
Gonzales, Wilkinson Daniel Wong, and Mie Hiramoto. 2020. Two Englishes diverged in the Philippines? A substratist account of Manila Chinese English. *Journal of Pidgin and Creole Languages* 35 (1): 125–159.
Gonzales, Wilkinson Daniel Wong, and Rebecca Starr. 2020. Vowel system or vowel systems? Variation in the monophthongs of Philippine Hybrid Hokkien in Manila. *Journal of Pidgin and Creole Languages* 35 (2): 253–292.
Grosjean, François. 2010. *Bilingual*. Cambridge: Harvard University Press.
Haugen, Einar. 1971. The ecology of language. *The Linguistic Reporter* 25 (25): 19–26.
Imperial, Rowland Anthony. 2016. *Speech production and sociolinguistic perception in a 'non-native' second language context: A sociophonetic study of Korean learners of English in the Philippines*. Unpublished M.A. thesis, National University of Singapore.
Klöter, Henning. 2011. *The Language of the Sangleys: A Chinese vernacular in missionary sources of the seventeenth century*. Leiden/Boston: Brill.
Latrouite, Anja. 2011. *Voice and case in Tagalog: The coding of prominence and orientation*. Unpublished Ph.D. dissertation, Heinrich Heine University Düsseldorf.
Lesho, Marivic. 2013. *The sociophonetics and phonology of the Cavite Chabacano vowel system*. Unpublished Ph.D. dissertation, The Ohio State University.
Li, Charles N., and Sandra A. Thompson. 1989. *Mandarin Chinese: A functional reference grammar*. Berkeley: University of California Press.
Llamzon, Teodoro A. 1969. *Standard Filipino English*. Manila: Ateneo University Press.
MacGowan, John. 1883. *English and Chinese dictionary of the Amoy dialect*. London: Trubner & Co. 57 Ludgate Hill.
Magbanua, Djan. 2018. 7 spots for your goto/lugaw cravings. Retrieved from https://libre.inquirer.net/5487/7-spots-for-your-goto-lugaw-cravings
Manuel, Arsenio. 1949. The origin of the Tagalog language and the Chinese contributions to its growth. Fookien Times Yearly.
Matras, Yaron, and Peter Bakker. 2003. The study of mixed languages. In *The mixed language debate: Theoretical and empirical advances*, ed. Y. Matras and P. Bakker, 1–22. Mouton de Gruyter: Berlin/New York.
Nicolas, Arsenio. 2016. Chinese music in the Philippines: History and contemporary practices. In *The folk performing arts in ASEAN*, ed. Narupon Duangwises and Lowell D. Skar. Princess Maha Chakri Sirindhorn Anthropology Centre: Bangkok.
Odlin, Terrence. 1989. *Language transfer: Cross-linguistic influence in language learning*. Cambridge: Cambridge University Press.
Patanne, Eufemio P. 1996. *The Philippines in the 6th to 16th centuries*. San Juan: LSA Press.
Pavlenko, Aneta, and Scott Jarvis. 2008. *Crosslinguistic influence in language and cognition*. New York/London: Routledge.
Philippine Statistics Authority. 2010. The 2010 census of population and housing reveals the Philippine population at 92.34 Million. Retrieved from https://psa.gov.ph/content/2010-census-population-and-housing-reveals-philippine-population-9234-million
Poa, Dory. 2004. From quadrilingual to bilingual: On the multilingual teaching in the Chinese schools in the Philippines. In *Bilingual studies*, ed. Dai Qingxia, vol. 2. Beijing: Minzu Chubanshe.
Schuchardt, Hugo. 1883. *Kreolische Studien IV: über das Malaiospanische der Philippinen*. The Rosetta Project: A Long Now Foundation Library of Human Language.
See, Ang, and Teresita. 1990. *The Chinese in the Philippines: Problems and perspectives*. Vol. 1. Manila: Kaisa Para sa Kaunlaran.
———. 1997. *The Chinese in the Philippines: Problems and perspectives*. Vol. 2. Manila: Kaisa Para sa Kaunlaran.
Siegel, Jeff. 1999. Transfer constraints and substrate influence in Melanesian Pidgin. *Journal of Pidgin and Creole Languages* 14 (1): 1–44.
———. 2003. Substrate influence in creoles and the role of transfer in second language acquisition. *Studies in Second Language Acquisition* 25 (2): 185–209.

Simons, Gary F. and Charles D. Fennig (eds.). 2020. Ethnologue: Languages of the World, 23rd edition. Dallas, Texas: SIL International. Online version http://www.ethnologue.com. (consulted 1 May 2020).

Sobolewski, Frank Andrew. 1980. *Some syntactic constraints in Tagalog-English language mixing.* Unpublished Master's thesis, University of Hawaii.

Swadesh, Morris. 1972. What is glottochronology? In *The origin and diversification of language*, ed. Joel F. Sherzer, 271–284. London: Routledge.

Tan, Susan Villanueva. 1993. *The education of Chinese in the Philippines and Koreans in Japan.* Unpublished thesis, University of Hong Kong, SAR.

The Lannang Archives. 2020. Lannang orthography. Retrieved July 16, 2020 from https://www.lannangarchives.org

The World Bank. 2018. Population, total – Philippines. Retrieved May 1, 2020 from https://data.worldbank.org/indicator/SP.POP.TOTL?locations=PH

Thomason, Sarah Grey. 2000. On the unpredictability of contact effects. *Estudios de Sociolingüística* 1 (1): 173–182.

———. 2001. *Language contact: An introduction.* Washington, DC: Georgetown University Press.

———. 2007. Language contact and deliberate change. *Journal of Language Contact* 1 (1): 41–62.

Thomason, Sarah Grey, and T. Kaufman. 1988. *Language contact, creolization, and genetic linguistics.* Berkeley: University of California Press.

Thompson, Roger M. 2003. *Filipino English and Taglish: Language switching from multiple perspectives.* Amsterdam/Philadelphia: Benjamins.

Tombo, Juan Manuel. 1860. Juancho. *Ilustración Filipina II* 23&24: 283–285.

Toribia, Mano. 1963. *The Zamboangueno Chabacano grammar.* The Rosetta Project: A Long Now Foundation Library of Human Language.

Tsai, Huiming. 2017. A study of Philippine Hokkien language. Unpublished Ph.D. dissertation, National Taiwan Normal University.

Uayan, Jean. 2014. The "Amoy Mission": Lessons and reflections. Retrieved from https://bsop.edu.ph/the-amoy-mission-lessons-and-reflections/

Uytanlet, Juliet Lee. 2014. The hybrid tsinoys: Challenges of hybridity and homogeneity as sociocultural constructs among the Chinese in the Philippines. Ph.D. dissertation, Asbury Theological Seminary.

Van der Loon, Piet. 1966. The Manila incunabula and early Hokkien studies (part 1). *Asia Major* 12: 1–43.

———. 1967. The Manila incunabula and early Hokkien studies (part 2). *Asia Major* 13: 95–186.

Wickberg, Edgar. 1965. *The Chinese in Philippine life, 1850–1898.* New Haven: Yale University Press.

Winford, Donald. 2020. Theories of language contact. In *The Oxford handbook of language contact*, ed. Anthony Grant. Oxford: Oxford University Press.

Wolfenden, Elmer. 1975. *A description of Hiligaynon syntax.* Norman: Summer Institute of Linguistics.

Zulueta, Johanna. 2007. I speak Chinese but. . . : Code-switching and identity construction among Chinese-Filipino youth. *Caligrama* 3 (2).

The Expansion of Cantonese Over the Last Two Centuries

16

Hilário de Sousa

Contents

Introduction	410
Earlier History of Yue Chinese	411
Cantonese Since the First Opium War, and the Notion of "Cantonese"	412
Nanning Cantonese	416
Cantonese Under Different Jurisdictions	426
Written Cantonese	430
Conclusion	434
References	435

Abstract

Cantonese is the representative variety of Yue Chinese. Since the end of the First Opium War (1839–1842), a large number of Cantonese people have emigrated from the heart of the Pearl River Delta, thereby creating many "enclave" varieties of Cantonese elsewhere in far southern China and overseas. This chapter, descriptive in nature, looks into the formation of these enclave Cantonese varieties, concentrating on Nanning Cantonese and Hong Kong Cantonese. The primary factor that caused the variation among the Cantonese varieties is the difference in their language contact environments. Being spoken in so many different countries and territories has also increased the variation among the Cantonese varieties, with the difference in language policy being one of the factors. Also discussed in this chapter is Written Cantonese; in the Cantonese world, one finds a continuum of written registers from Standard Written Chinese to Written Cantonese. Being used in different jurisdictions also means that Written Cantonese has evolved slightly differently in the different jurisdictions.

H. de Sousa (✉)
EHESS – CRLAO, Paris, France
e-mail: hilario@bambooradical.com

Keywords

Cantonese · Yue Chinese · Guangzhou · Hong Kong · Macau · Nanning · Overseas Chinese · Language contact

Introduction

Cantonese, the representative variety of Yue Chinese, is one of the better-known Sinitic languages. In this chapter, some aspects of the development of Cantonese will be discussed. The chapter is not so much about the linguistic changes in Standard Cantonese; it is mainly about the development of the various Cantonese varieties away from the heart of the Pearl River Delta, i.e., the Guangzhou area, where Cantonese originated. This chapter is primarily descriptive in nature.

Despite not being particularly widely spoken within China, outside China Cantonese is one of the best-known Chinese varieties besides Mandarin. What has contributed to the prominence of Cantonese? Part of it is due to its diversity: massive emigration by Cantonese people from the heart of the Pearl River Delta since the 1840s created many enclaves of Cantonese speakers elsewhere, both within far southern China and overseas, causing Cantonese to be spoken in many different countries and territories. Cantonese has received favorable treatment with the language policies in some of them. These varieties of Cantonese spoken outside the heart of the Pearl River Delta are referred as "enclave Cantonese" varieties in this chapter.

One important theme, from as early as the formation of Yue Chinese to the emergence of the modern Cantonese varieties, is language contact. While Standard Cantonese in Guangzhou and the enclave Cantonese varieties elsewhere have remained highly mutually intelligible, there are some variations. The variations among the Cantonese varieties are often the result of the differences in their language contact environments. Some of the enclave Cantonese varieties are still very close to the Cantonese of Guangzhou, e.g., Hong Kong Cantonese (e.g., Zhān et al. 2002: 213–218; Cheung 2007; Cheng 1999). Others have become more divergent from Guangzhou Cantonese. In this chapter, as an illustration of a more divergent enclave Cantonese variety, the case of Nanning Cantonese will be discussed.

Data on Standard Cantonese are drawn from both literature and the present author's first-language knowledge. Data on Nanning Cantonese are primarily drawn from Lín and Qín (2008), and also from knowledge acquired by the present author based on his fieldwork on Nanning Pinghua (e.g., de Sousa 2013, 2017, forthcoming-a; Lǐ 2000; Qín 2000, 2007), another Sinitic language spoken in Nanning.

This chapter follows the English linguistic convention of treating speech varieties that are not mutually intelligible as separate languages. (See Mair (1991) on the Western linguistic concept of *language* versus *dialect*, and the Chinese concept of *yǔyán* 语言 versus *fāngyán* 方言, which are not identical. The Western and Chinese approaches are simply two different ways of classifying speech varieties; both have their merits and limitations. See also Cheng and Tang (2014) on the issue of languagehood from the perspective of Hong Kong Cantonese.) Based on this English convention, Cantonese, Hakka (Kejia 客家), Teochew (Chaozhou 潮州), Mandarin, etc. are separate languages, and the family of languages that descend from Old Chinese is called the Sinitic language family (e.g., Mair 2013; Chappell 2015a; Handel 2015).

Earlier History of Yue Chinese

Cantonese is the representative variety of the Yue dialect group. Having an understanding of what "Cantonese" is and of what "Yue dialect group" is are each essential in understanding what the other is. When speakers of Sinitic languages talk about *Yuèyǔ* (粵語 $jyt^2\ jy^{13}$), "the Yue language," they are most usually referring to Standard Cantonese. However, the notion of the "Yue dialect group" is much wider than the notion of "Cantonese." There are many Yue dialects which are of very low intelligibility to speakers of Standard Cantonese without pre-exposure. (However, due to exposure to Cantonese media, many speakers of other Yue dialects understand Standard Cantonese.) Here the earlier history of the Yue dialect group will be briefly outlined, before the notion of "Cantonese" is discussed in the next section.

The Yue dialects are primarily spoken in the Pearl River basin, plus the many small river basins in Guangdong and Guangxi, south of the Pearl River basin between Macau and the border with Vietnam. The Pearl River basin is situated to the south of the Yangtze River basin. The Yangtze and Pearl River basins are separated by the Nanling 南岭 mountains. The following is a summary of Lǐ (2002: 121–134) on the migration history of Yue speakers, and the interaction that Yue speakers had with indigenous people in the Pearl River basin (see also de Sousa (forthcoming-b)). During the Qin Dynasty (221–206 BCE), the Lingqu 灵渠 Canal was built (in modern day Xing'an 兴安 County in Guangxi near Hunan), linking the Yangtze River system and the Pearl River system. Before then, Chinese political structures existed only in the Yangtze, Huai, and Yellow River regions to the north. With the opening of the Lingqu Canal, for the first time Chinese political structures were set up in the Pearl River region. For the next millennium or so, the number of Han people in the Pearl River region was small in relation to the indigenous population. In the eighth century CE, during the Tang Dynasty (618–690, 705–907 CE), the Plum Pass Road (*Méiguāndào* 梅关道) was built (in modern day Nanxiong 南雄 City in Guangdong near Jiangxi), greatly improving the accessibility of the Pearl River delta from the Yangtze region to the north. That sped up the migration of Han people from the north into Guangdong. Within decades of the opening of the Plum Pass Road, the number of Han people (preexisting population and new migrants) in the Pearl River Delta was so great that reports of indigenous people in the Pearl River Delta had become infrequent. Within Guangdong, Han people gradually spread from the Pearl River Delta, primarily in a westward direction (as the west was relatively lightly populated), forming the Yue dialect group.

(The areas north and east of the Pearl River Delta were already relatively heavily populated by Han settlers; later on, these areas became primarily Hakka speaking.) On the way, they encountered indigenous people, and also pockets of other Han Chinese people who had settled in the region earlier. During the Northern Song Dynasty (960–1127 CE), there were still many reports of indigenous people in western Guangdong. However, by the Yuan Dynasty (1271–1368 CE), there were already very few reports of indigenous people in western Guangdong; most indigenous people had assimilated into the Yue-speaking Han communities. Yue language continued to spread westward from western Guangdong to eastern Guangxi. By the eighteenth century, in the middle of the Qing dynasty (1636–1912 CE), there were already few reports of indigenous people in the southeastern third of Guangxi. The history of Cantonese since the middle of the eighteenth century will be discussed in the next section. See also Yóu (2000: 106) on Northern Chinese migration to Guangdong around the Northern Song Dynasty, and Wáng (2009) for a model of the formation of the various Sinitic dialects groups from a historical phonological perspective.

Linguistically, the most significant influence on Yue was the Middle Chinese introduced by Northern Chinese migrants in about tenth-century CE (during later parts of the Tang Dynasty (618–907) and the Five Dynasties period (907–979); Wáng 2009) and Early Mandarin in the thirteenth-century CE (towards the end of the Song Dynasty (960–1279); Lau 2001). Other influences include the earlier Sinitic varieties in the Pearl River region (e.g., Kwok 2004), and also the indigenous languages in the region. The indigenous languages that Yue was in contact with were primarily Kra-Dai languages (also known as Tai-Kadai, or *Dòng-Tái* 侗台 or *Zhuàng-Dòng* 壯侗 in Chinese). Yue has a strong Kra-Dai substratum; some Kra-Dai influences are present throughout the Yue area, while others are more restricted towards the west, where contact between Yue and Kra-Dai languages lasted till more recently, or is still ongoing. There have been many studies on the Kra-Dai influence on Yue; some examples are Bái (2009), Bauer (1996), Chappell (2017), Huang (1997), Huang and Wu (2018), Lǐ (2002), Liú (2006), Matthews (2006), Peyraube (1996), de Sousa (2015, forthcoming-b), and Yue-Hashimoto (1991). See also discussions below on section "Nanning Cantonese" and Chap. 13, ▶ "Contact-Induced Change in the Languages of Southern China."

Cantonese Since the First Opium War, and the Notion of "Cantonese"

Looking at the distribution of the subtypes of Yue dialects, it is clear that their distribution is not entirely caused by a gradual spread of population from the Pearl River Delta towards the west (as described in the preceding section); the Yue dialects do not simply form an east-west dialect chain along the Pearl River. There are many enclaves of Cantonese in far southern China and overseas that have remained linguistically quite close to Standard Cantonese. In the context of the Yue area in western Guangdong and eastern Guangxi, these Cantonese dialects are noticeably different from the pre-established Yue dialects that surround them. What has caused this pattern?

A major starting point for this new pattern is the cessation of the centuries-long maritime prohibitions (*hǎijìn* 海禁), after which a large number of Cantonese speakers started migrating by boats directly from the heart of the Pearl River to further away places along the waterways and coast of far southern China, and also overseas. Between the fourteenth and the first half of the nineteenth century (spanning the Yuan, Ming, and the first half of Qing Dynasty), most of the time there were restrictions on civilian maritime traffic; civilian seafaring was prohibited, and navigation on domestic rivers was not totally free. (There were already some Yue migrants overseas, legally or illegally, before the First Opium War; many were in foreign lands, for instance dealing with financial transactions between China and foreign countries.) After the First Opium War (1839–1842), however, China was forced to end its centuries-long maritime prohibitions; there were no longer restrictions on civilian watercraft ownership and maritime movements. From the heart of the Pearl River Delta, a large number of Cantonese people, especially merchants, migrated by boats in all sorts of directions, bringing with them the Cantonese language to new places. (There were also speakers of other Yue dialects, and other southern Sinitic languages, that migrated at around the same time, but they are outside of the scope of this chapter.) Some Cantonese speakers went up the Pearl River system to localities across Guangdong and Guangxi. Others went along the coast to Hong Kong, Macau, west along the Guangdong and Guangxi coasts, and then to Vietnam and further. Many went across the ocean to the other continents. The emigration has not stopped since, with the number of emigrants spiking during turbulent times. Often, through the commercial prowess of the Cantonese people, Cantonese became the dominant Sinitic variety in many cities and towns. There are many of these "enclave" varieties of Cantonese scattered across far southern China and overseas. These Cantonese varieties are inevitably in contact with the languages that surround them. The level of influence that these enclave Cantonese varieties receive from their surrounding languages varies. Some factors involved are the number of Cantonese migrants versus the other linguistic groups, the socioeconomic power that each language group has, the level of multilingualism, and language shift. Another factor which influences the linguistic features that an enclave Cantonese variety has is the type of Cantonese spoken by the initial settlers: Which part of the Pearl River Delta did they come from? or whether they spoke yet another enclave Cantonese variety to start off with. (For example, the Cantonese of Hekou 河口 in Yunnan was mostly formed by speakers from Cantonese enclaves in Guangxi like Baise 百色 and Nanning 南宁, plus some later Cantonese migrants who moved up the Red River from Northern Vietnam (Lǐ 2002: 132–133).) Some enclave Cantonese varieties remain highly similar to the Cantonese spoken in Guangzhou; Hong Kong Cantonese is an example. Others have been more strongly influenced by the surrounding languages. In the next section, the case of Nanning Cantonese will be discussed.

Before continuing, the notion of "Cantonese" has to be defined first. There is Standard Cantonese, and the other Cantonese varieties. Standard Cantonese in this chapter refers to the language of Canton, a Western name for Guangzhou (广州 $kʷɔŋ^{35}$ $tsɐu^{53}$), the capital of Guangdong province. (A competing standard is Hong Kong Cantonese, but Hong Kong Cantonese is minimally different from Guangzhou Cantonese.) Beyond the speech of Guangzhou, the speech varieties that descended from the Cantonese spoken by people who emigrated from the Guangzhou area since the First Opium War (1839–1842) are also considered Cantonese in this chapter. "Guangzhou area" is the area traditionally referred to as Sanyi (三邑 sam^{55} $jɐp^{5}$) "three counties": the historical counties of Nanhai 南海, Panyu 番禺, and Shunde 顺德. (The historical Panyu County included the central districts of Guangzhou.) Away from the Guangzhou area, some examples of enclave Cantonese varieties within China are Hong Kong 香港, Macau 澳门, Shaoguan 韶关, Wuzhou 梧州, Beihai 北海, and Nanning 南宁. As for the overseas distribution of Yue dialects, map B-16 in the first edition of the Language Atlas of China (Wurm et al. 1987/1989) classifies overseas Yue dialects into three groups: Sanyi 三邑 "three counties," Siyi 四邑 "four counties" (the historical counties of Taishan 台山, Kaiping 开平, Enping 恩平, and Xinhui 新会), and Zhongshan 中山. "Cantonese" here corresponds with Sanyi Yue. Some examples of Chinatowns overseas that are traditionally Cantonese dominant are Hanoi, Kuala Lumpur, Sydney, Vancouver, and London. There are some vocabulary differences among these various Cantonese varieties, but their phonologies are minimally different from that of Guangzhou Cantonese. A phonologically oriented definition of this notion of Cantonese is presented below.

The definition of Cantonese outlined above is perhaps a somewhat narrow definition of Cantonese. Ideas vary about the range of Yue dialects that is encompassed by the label of Cantonese. In the widest sense, the term Cantonese is sometimes applied to the entire Yue dialect group. However, this wide approach is not recommended: there are many Yue dialects that are of rather low intelligibility to speakers of Standard Cantonese without prior exposure. This is often the case with the Yue dialects that are not spoken in the Pearl River basin, for instance the Yue dialects of Taishan 台山 and Yangjiang 阳江 in Guangdong, and Bobai 博白 (Dilao dialect 地佬话) and Hepu 合浦 (Lianzhou dialect 廉州话) in Guangxi. "Cantonese" is literally the language of Canton/Guangzhou city: applying the term Cantonese to the entire Yue dialect group is akin to applying the term "Shanghainese" or "Suzhounese" to the entire Wu dialect group, including highly divergent Wu varieties like Wenzhou 温州. Just as it would be misleading to call Wenzhou Wu "Shanghainese," it would conjure up the wrong impression if the term "Cantonese" were applied to Yue dialects as divergent as, e.g., the Lianzhou dialect of Hepu.

(Also notice that the definition of "Cantonese" cannot simply be the language spoken by descendents of people from the Guangzhou area, because they have not necessarily maintained the Cantonese language. For instance, there was a significant Cantonese community in Liuzhou 柳州. While Cantonese had a strong influence on the Southwestern Mandarin of Liuzhou (e.g., Liú 1995; Táng 2012), most Cantonese speakers have shifted to Liuzhou Mandarin.)

Table 1 Tones in Guangzhou Cantonese (Zhān et al. 2002: 292)

	*A	*B	*C	*DL	*DS
*voiceless	55 ~ 53	35	33		55
*voiced	21	13	22		

Table 2 Tones in Hong Kong Cantonese (Matthews and Yip 1994: 22)

	*A	*B	*C	*DL	*DS
*voiceless	55	35 ~ 25	33		55
*voiced	21 ~ 11	13 ~ 23	22		

Table 3 Tones in Beihai Cantonese (Chén and Chén 2005: 7)

	*A	*B	*C	*DL	*DS
*voiceless	55	35	33	3	5
*voiced	21	13	22	2	

It is beyond the scope of this chapter to present a detailed study of the features of the various Cantonese varieties, or the internal classification of the Yue dialects more generally. Given the brief history of migration out of the heart of the Pearl River Delta, the phonologies of the various Cantonese varieties (as per the definition of Cantonese adopted in this chapter) have remained very similar to each other. There are some slight segmental differences (i.e., differences in the consonants and vowels), but the tones have remained remarkably similar. As a quick demonstration of the uniformity of the phonological systems, only the tonal systems of some Cantonese varieties are shown here. Tables 1, 2, 3, and 4 show the tonal systems in four Cantonese varieties: Guangzhou, Hong Kong, Beihai, and Nanning. In the tables: *A/B/C/D are the tonal categories of *píng* 平/*shǎng* 上/*qù* 去/*rù* 入 in Middle Chinese, L/S refer to the "long" and "short" vowels in the modern Yue dialects, and *voiceless/*voiced refer to the voicing of the initial consonant of a syllable in Middle Chinese (i.e., "Yin"/"Yang" tones, respectively, in Chinese historical linguistics). The cells in the middle show the pitch values of the tones: 5 is highest pitch

Table 4 Tones in Nanning Cantonese (Lín and Qín 2008: 14)

	*A	*B	*C	*DL	*DS
*voiceless	55	35	33	3	5
*voiced	21	24	22	2	

and 1 is lowest pitch. (The tilde "~" indicates free variations.) As can be seen in the tables, the tones are basically the same across these Cantonese varieties; the variation shown here can be viewed as mere notational differences.

Macau Cantonese also has the same tones as Guangzhou and Hong Kong, except that some people do not, or cannot easily, distinguish the two rising tones (the two tone B's) (Bauer and Benedict 1997), probably a trait related to the Zhongshan-type of Yue that used to be spoken in Macau. See also, e.g., Bauer, Cheung, and Cheung (2003), and Zhang (2019), on recent tone mergers in Hong Kong and Macau. In Beihai, Xiǎn (2018 ms), and Chén and Lín (2009: 3) report that among younger speakers, *voiced A and *voiced C have merged to become [21], and the two tones *B's have merged to become [13].

Based on the definition of Cantonese adopted in this chapter, some Yue varieties geographically close to Guangzhou are not considered Cantonese here. Examples are Zhongshan 中山 Yue and Dongguan 东莞 Yue. (Before the arrival of Cantonese, Macau Yue was similar to Zhongshan Yue (Zhān et al. 2002: 196–202), while the majority of indigenous Yue varieties in Hong Kong are similar to Dongguan Yue (Zhān et al. 2002: 188–195; Chang et al. 1999).) As can be seen in Tables 5 and 6, the tones in Zhongshan and Dongguan are noticeably different from those in Cantonese.

Nanning Cantonese

To give an example of a more divergent Cantonese variety, some features of Nanning Cantonese will be discussed below. Most of these are the results of the language contact environment in the Nanning area.

Nanning is the capital of Guangxi Zhuang Autonomous Region. The city is divided into a northern and southern half by the Yong River 邕江, a tributary of the West branch of the Pearl River (i.e., upriver from and west of Guangzhou). The city center lies on the northern bank, and is dominated by Nanning Cantonese. In the surrounding suburbs, Nanning Pinghua is spoken. In the surrounding rural areas, the indigenous Zhuang languages are spoken: roughly speaking, Northern Zhuang north of the river and Southern Zhuang south of the river. There are also two types of Mandarin in Nanning: Old Nanning Mandarin and New Nanning Mandarin. Old Nanning Mandarin (*Yōngzhōu Guānhuà* 邕州官话) is a type of Southwestern Mandarin that used be spoken across a few blocks in the city center. Old Nanning Mandarin is now moribund in the city center, but it is still spoken in several villages

Table 5 Tones in Zhongshan Yue (Zhān et al. 2002: 294)

	*A	*B	*C	*DL	*DS
*voiceless	55	213	33		55
*voiced	51				

Table 6 Tones in Dongguan Yue (Zhān et al. 2002: 295)

	*A	*B	*C	*DL	*DS
*voiceless	213	35	32	22 / 224	44
*voiced	21	13		22	

south of the river. New Nanning Mandarin (*Nánníng Pǔtōnghuà* 南宁普通话, or *NánPǔ* 南普) is Nanning's version of modern Standard Mandarin, strongly influenced by the local languages. The indigenous Zhuang languages are Tai languages (the branch of the Kra-Dai language family that also includes major languages like Thai and Lao), while Pinghua, Cantonese, and Mandarin are Sinitic. (Today, few young people under 20 speak anything other than New Nanning Mandarin.)

Cantonese first arrived in Nanning in the middle of the nineteenth century. During the early days of the Republic of China (the 1910s), Old Nanning Mandarin was still spoken by half of the population in Nanning's city center (Zhōu et al. 2006). However, as more Cantonese people arrived, Cantonese gradually replaced Old Nanning Mandarin as the dominant language in the city center. Nanning Cantonese has been heavily influenced by the local languages, especially from the indigenous Zhuang languages. So much so that Nanning Cantonese, which has been spoken in Nanning for less than 200 years, is at times even more Zhuang influenced than Nanning Pinghua is, despite Pinghua having been spoken in the area for more than one millennium (see de Sousa 2013, 2015).

The phonology of Nanning Cantonese is recognizably Cantonese. The tones are the same as Standard Cantonese. Table 7 below lists the tones in Nanning Cantonese (repeated from Table 4 above); compare this with the inventory of tones in the other Cantonese varieties shown above (Tables 1, 2 and 3).

Nanning Cantonese is noticeably different from the surrounding languages. Tables 8 and 9 illustrate the tonal systems of the surrounding Sinitic languages.

(There are different accents of Pinghua in the various suburbs of Nanning; their tones, and their phonologies in general, differ slightly. Southern Pinghua, of which

Table 7 Tones in Nanning Cantonese (Lín and Qín 2008: 14; repeated from Table 4)

	*A	*B	*C	*DL	*DS
*voiceless	55	35	33	3	5
*voiced	21	24	22	2	

Table 8 Tones in Nanning Weizilu Pinghua (de Sousa forthcoming-a)

	*A	*B	*C	*D
*voiceless aspirated	53	33	35	3
*voiceless unaspirated			55	
*voiced sonorant	21	13	22	23
*voiced obstruent				2

Table 9 Tones in Old Nanning Mandarin (Zhōu et al. 2006)

	*A	*B	*C	*D
*voiceless	35	54	13	31
*voiced sonorant	31			
*voiced obstruent				

Nanning Pinghua is a dialect, is on a dialect continuum with the non-Cantonese Yue dialects in Guangxi. The migration of Cantonese speakers from the Guangzhou area to the Nanning area was, roughly speaking, the migration of people directly from the eastern end of the dialect continuum to the western end of the dialect continuum along the Pearl River. See de Sousa (2015, forthcoming-a, forthcoming-b).)

The segments of Nanning Cantonese are largely the same as Standard Cantonese. Some features of earlier Cantonese (as seen in the eighteenth and nineteenth century Cantonese sources, e.g., the rime book *Fēnyùn Cuōyào* (分韵撮要 $fɐn^{55}$ $wɐn^{13}$ $tsʰyt^{3}$ jiu^{33}) and Western documentations of Cantonese) can still be seen in Nanning Cantonese. For instance, the diphthongization of high vowels is absent in Nanning Cantonese, e.g., 机 *jī* "machine," Nanning Cantonese ki^{55}, Standard Cantonese kei^{55}. In the earlier Western documentations of Cantonese, for the affricates and fricatives, there were two coronal places of articulation: alveolar and post-alveolar, rendered <ts/tsʻ/s> vs. <ch/chʻ/sh> or the like (a distinction that is preserved to a degree by the British spelling of Hong Kong place names; see, e.g., Bauer 2005; Kataoka and Lee 2008). Standard Cantonese merged these two series within the last century. In Nanning Cantonese, while the two sets of affricates have merged more recently (vestiges are still maintained by some of its oldest speakers (Lín and Qín 2008: 11)), the two fricatives are still largely distinct. The articulation of the two fricatives is notable: *ɬ* and *f*. The lateral fricative *ɬ* (or dental fricative *θ* in some areas) is an areal feature in Guangxi and parts of Guangdong. One possibility is that Cantonese speakers acquired the lateral fricative after arriving in Guangxi; looking at other Sinitic languages in the region, most Guangxi Hakka and Southern Min varieties also arrived in Guangxi within the last two hundred years or so, and many have also acquired the lateral fricative within this short period of time. Another possibility is that the ancestors of Nanning Cantonese started off having the lateral fricative in the Pearl River Delta. Although there is no evidence that the lateral fricative existed in Guangzhou, at present the lateral fricative is still found in some Yue dialects not too far away from Guangzhou: in the Siyi 四邑 region to the southwest of Guangzhou, e.g., Taishan 台山, and also in Fogang 佛冈 to the north of Guangzhou (Mài 2010).

Zhuang, Nanning Pinghua, and Cantonese have the same consonantal codas of *-m -n -ŋ -p -t -k*; these languages are usually rather conservative with them. Old Nanning Mandarin has fewer codas: *-n -ŋ*, plus a few cases of *-m -p -t -k* in loanwords. Interestingly, there are some cases in Nanning Cantonese, and sometimes also in Nanning Pinghua, where certain syllables ended up having the "wrong" coda. This is probably caused by speakers of Old Nanning Mandarin hypercorrecting when they speak Cantonese, and then shifting en masse to Cantonese, causing these "errors" to become mainstream. Nanning Cantonese has subsequently influenced Nanning Pinghua. In particular, cases of $*-n > -m$ are extraordinarily rare in Chinese historical phonology (the overwhelmingly dominant direction of change is $*-m > -n$). Some examples are:

- 典 "scripture," Middle Chinese ten^B:
  - Old Nanning Mandarin $tien^{54}$, Standard Mandarin *diǎn*, Standard Cantonese tin^{35}; but
  - Nanning Cantonese tim^{35}, Nanning Pinghua tim^{33};
- 演 "act," Middle Chinese jen^B:
  - Old Nanning Mandarin ien^{54}, Standard Mandarin *yǎn*, Standard Cantonese jin^{35}; but
  - Nanning Cantonese jim^{35}, Nanning Pinghua im^{33};
- 建 "build," Middle Chinese $kjon^C$:
  - Old Nanning Mandarin $kien^{13}$, Standard Mandarin *jiàn*, Standard Cantonese kin^{33}; but
  - Nanning Cantonese kim^{33}, Nanning Pinghua kim^{55}.

One example of the more common sound change of $*-m > -n$ is:

- 鐮 "sickle," Middle Chinese *ljem⁴*:
  - Standard Cantonese *lim²¹*, Nanning Pinghua *lim²¹*; but
  - Nanning Cantonese *lin²¹* (cf. the regular reflexes in Old Nanning Mandarin *lien³¹*, and Standard Mandarin *lián*).

There are also some synchronic phonetic loans from Mandarin, i.e., direct loaning through contemporary phonetics and not through historical sound correspondences. For instance, for the verb "give," in Nanning Cantonese there is the native Cantonese verb 畀 *pi³⁵* (Standard Cantonese *pei³⁵*), and also the Mandarin phonetic loan 给 *kei⁵⁵* (< Liuzhou Mandarin 给 *kei⁵⁴*, Old Nanning Mandarin *kei⁵⁴*; the regular pronunciation of 给 in Cantonese is *kʰɐp⁵*, from Middle Chinese *kipᴰ*). The traditional term for "corn" 包粟 *pɛu⁵⁵ ɬuk⁵ ~ pau⁵⁵ ɬuk⁵* in Nanning Cantonese (Standard Cantonese 粟米 *sʊk⁵ mɐi¹³*) has been replaced by the term 玉米 *jy²² mɐi²⁴*, which is a partial loan from Mandarin: the whole word is in Mandarin (cf. Old Nanning Mandarin 玉米 *iu³¹ mi⁵⁴*, Liuzhou Mandarin *y²⁴ mi⁵⁴*, Standard Mandarin *yùmǐ*), the segments of the first syllable are Mandarin like, while the tone is in Cantonese (cf. Cantonese 玉 *juk²* "jade"); the second syllable 米 *mɐi²⁴* is in Cantonese. Nanning Cantonese *jy²² mɐi²⁴* has in turn been loaned via normal sound correspondences into Nanning Pinghua as *ɲəi²² mɐi¹³*.

Standard Cantonese already has a noticeable number of lexical items from Kra-Dai language, and Nanning Cantonese has even more Zhuang loanwords. Examples of Zhuang words that are found in Nanning Cantonese but not in Standard Cantonese include *mɐp²* "hit with thing," *nɐm⁵⁵* "unsophisticated," *kʰɐm²¹* "concave," cf. Northern Zhuang *moeb [mop³]* "hit," *numq [num³⁵]* "slow," and *gumz [kum³¹]* "concave."

One also sees transfer of Zhuang grammatical patterns into Nanning Cantonese. When comparing the grammars of Nanning Cantonese and Nanning Pinghua, sometimes there is a curious case of Nanning Cantonese, or sometimes even Standard Cantonese, resembling the indigenous Zhuang more than Nanning Pinghua does. This is despite Pinghua having been spoken in Nanning for at least one millennium, whereas Cantonese has only been in Nanning for less than 200 years. This can perhaps be explained by a general lack of social inhibitions among Cantonese people when it comes to intermarrying and interacting with Zhuang people (as well as the socioeconomic power of Cantonese speakers), causing a huge number of Zhuang people to shift to Cantonese, to the extent that many second-language Cantonese features among Zhuang speakers have become mainstream in Nanning Cantonese (see, e.g., Kwok 2019, on Zhuang-like grammatical patterns in Nanning Cantonese). Pinghua has also been strongly influenced by Zhuang. However, until recently, there was some social distance between Pinghua and Zhuang speakers, leading to fewer opportunities for mainstream Pinghua to be influenced by the variety of Pinghua spoken by Zhuang people. Possibly yet another factor is how Nanning Pinghua people, who strongly identify with their Northern Chinese origin, might have been more receptive to the linguistic influences of Old Nanning Mandarin or Guangxi Mandarin in general, or to any influence from Hunan or further north. These points are discussed in more detail in de Sousa (2015, forthcoming-b). Here, the following grammatical features of Nanning Cantonese, Nanning Pinghua, and Northern Zhuang are briefly discussed: negation, the degree modifier "too," attributive possession, [ADJ + CLF + N] phrases, lone classifiers, the position of resultative complements, and the grammaticalization of "go" as an imperative marker.

Sinitic languages differ in the way that they express negation. Mandarin has two commonly used negators: 不 *bù* and 没 *méi* ~ 没有 *méiyǒu*. The differences between these two are complex (M. Li 1999; Hsieh 2001; Lin 2003; Xiao and McEnery 2008, among others); here, in an oversimplified manner, 不 *bù* is called a nonperfective negator, and 没 *méi* ~ 没有 *méiyǒu* is called a perfective negator. An example of the nonperfective 不 *bù* is 明天我不去 *míngtiān wǒ bú qù* (tomorrow I NEG.NPFV go) "tomorrow I will not go," and an example of the perfective 没 *méi* ~ 没有 *méiyǒu* is 昨天我没去 *zuótiān wǒ méi qù* (yesterday I NEG.PFV go) "yesterday I did not go." The verb of existence 有 *yǒu* (e.g. "there is X"), which also indicates predicative possession (e.g. "I have X"), calls for special attention: in Mandarin, the negative form of 有 *yǒu* is always the perfective 没 *méi* ~ 没有 *méiyǒu*, e.g., 我没钱 *wǒ méi qiǎn* (I NEG.have money) "I have no money."

Standard Cantonese functions similarly; there are the nonperfective negator 唔 m^{21}, and the perfective negator 冇 mou^{13}: e.g., 听日我唔去 $t^h\text{ɪŋ}^{55} jɐt^2 ŋɔ^{13} \underline{m^{21}} hɵy^{33}$ (tomorrow I NEG.NPFV go) "tomorrow I will not go," versus 琴日我冇去 $k^hɐm^{21} jɐt^2 ŋɔ^{13} \underline{mou^{13}} hɵy^{33}$ (yesterday I NEG.PFV go) "yesterday I did not go." Existence and possession are similarly negated by the perfective 冇 mou^{13}, e.g., 我冇钱 $ŋɔ^{13} \underline{mou^{13}} ts^h in^{35}$ (I NEG.have money) "I have no money." (See also Law 2014 on Guangdong Yue negation.)

On the other hand, the Sinitic languages in Nanning follow a pattern that is used in most modern Tai languages: not distinguishing nonperfective and perfective negation, and using an analytic expression for "not exist/have." For instance, in contrast to the 唔 m^{21}/冇 mou^{13} distinction in Standard Cantonese, Nanning Cantonese uses 冇 mu^{24} for both: 听日我冇去 $t^h eŋ^{55} jɐt^2 ŋɔ^{24} \underline{mu^{24}} hy^{33}$ (tomorrow I NEG go) "tomorrow I will not go," and 琴日我冇去 $k^hɐm^{21} mɐt^2 ŋɔ^{24} \underline{mu^{24}} hy^{33}$ (yesterday I NEG go) "yesterday I did not go." In contrast to Standard Cantonese where "not exist/have" is simply 冇 mou^{13}, the same meaning in Nanning Cantonese has to be formed analytically by a negator 冇 mu^{24} followed by the verb 有 $jɐu^{24}$ "exist/have": 我冇有钱 $ŋɔ^{24} \underline{mu^{24} jɐu^{24}} tʃ^h in^{21}$ (I NEG have money) "I have no money." (As the analytic construction "negator + have" is also found in Standard Cantonese as late as the nineteenth century (Law 2014), it is most likely a retention in Nanning Cantonese. The retention of this construction in in the Zhuang Language may have also influenced its retention in Nanning Cantonese.) Nanning Pinghua similarly uses 冇 mi^{13} (NEG) and 冇有 $mi^{13} jɔu^{13}$ (NEG have) in the same manner. This is the pattern that most modern Tai languages have; for instance Northern Zhuang uses *mboux* (NEG) and *mboux miz* (NEG have) (e.g., Wéi and Qín 2006), and Thai has ไม่ *mâi* and ไม่มี *mâi mīi* (e.g., Smyth 2002: 138–152). (Pittayaporn, Iamdanush, and Jampathip (2014) reconstruct a Mandarin-type 不 *bù* versus 没 *méi* negator distinction for Proto-Tai, but the distinction is kept in only one Tai variety in Vietnam among the 64 modern Tai varieties in their survey, about two-thirds of which are Zhuang varieties in China. Attestation of this distinction is also found in Thai documentations from the fifteenth and sixteenth centuries.)

In Standard Cantonese, the degree modifier "too" is expressed by a normal Sinitic pre-adjectival 太 t^hai^{33} (cf. Mandarin 太 *tài*), e.g., 太热 $t^hai^{33}\,jit^2$ (too hot) "too hot," 太冻 $t^hai^{33}\,toŋ^{33}$ (too cold) "too cold" (ambient temperature). On the other hand, Northern Zhuang has a post-adjectival *lai* "many/much" for this function, e.g., *hwngq lai* (hot much) "too hot," *nit lai* (cold much) "too cold." Nanning Pinghua has calqued this post-adjectival "much." In Nanning Weizilu 位子渌 Pinghua (data collected by the present author), either the Sinitic pre-verbal "too" or the Tai post-verbal "much," or both, can be used, e.g. 太热/○多 $t^hai^{25}\,ɲit^{23}$ / $jən^{53}$ (too hot/cold), 热/○多 $ɲit^{23}$ / $jən^{53}\,tɔ^{53}$ (hot/cold much), 太热/○多 $t^hai^{25}\,ɲit^{23}$ / $jən^{53}\,tɔ^{53}$ (too hot/cold much) "too hot/cold." On the other hand, Nanning Cantonese and some Nanning Pinghua varieties like Tingzi 亭子 (Qín et al. 1997: 71) only have the post-adjectival "much" construction from Zhuang, e.g., Nanning Cantonese 热多 $jit^2\,tɔ^{55}$ (hot much) "too hot," 冻多 $tuŋ^{33}\,tɔ^{55}$ (cold much) "too cold." They do not use the Sinitic pre-adjectival degree marker.

Attributive possession is usually conveyed in Mandarin and Nanning Pinghua through a modifier marker (MOD; i.e., 的 *de* in Mandarin). A modifier marker marks the preceding constituent as a noun modifier. For example, in Mandarin "my pig" and "my book" are expressed as 我的猪 *wǒ de zhū* (1SG MOD pig) "my pig," 我的书 *wǒ de shū* (1SG MOD book) "my book." In Nanning Pinghua, the two expressions are 我个猪 $ŋa^{13}\,kə^{55}\,tʃəi^{53}$ (1SG MOD pig) "my pig," 我个书 $ŋa^{13}\,kə^{55}\,ɬai^{53}$ (1SG MOD book) "my book." Cantonese (both Standard and Nanning Cantonese) also has a modifier marker 嘅 $kɛ^{33}$ that can be used in this environment. However, a more common strategy (for nonabstract possessums) is to use the classifier of the possessum instead, e.g., Nanning Cantonese 我只猪 $ŋɔ^{24}\,tʃɛk^3\,tʃy^{55}$ (1SG CLF pig) "my pig," 我本书 $ŋɔ^{24}\,pun^{35}\,fy^{55}$ (1SG CLF book) "my book." Northern Zhuang is similar: it also has a possessive marker *duh* (Wéi and Qín 2006: 203–204; functionally narrower than the Sinitic modifier marker), but the more common strategy is to use the classifier of the possessum. However, unlike the possessor–possessum word order in Sinitic languages, most Zhuang varieties have the possessum–possessor word order: *duz mou gou* (CLF pig 1SG) "my pig," *bonj saw gou* (CLF book 1SG) "my book."

Continuing on the syntax of classifiers, there are some classifier constructions in Nanning Cantonese that are reminiscent of Zhuang, but are not found in either Standard Cantonese or Nanning Pinghua. One such construction is the adjective + classifier + noun [ADJ + CLF + N] construction. In Standard Cantonese and Nanning Pinghua, the only adjectives that can immediately precede a classifier are the size adjectives, e.g., Standard Cantonese 大间屋 $tai^{22}\,kan^{55}\,ok^5$ (big CLF house) "big house," Nanning Pinghua 大间屋 $tai^{22}\,kan^{53}\,ok^3$ (big CLF house) "big house." It is ungrammatical with other types of adjective (e.g., Standard Cantonese *空间屋 *$hoŋ^{55}\,kan^{55}\,ok^5$ (empty CLF house) and Nanning Pinghua *空间屋 *$hoŋ^{53}\,kan^{53}\,ok^3$ (empty CLF house) are ungrammatical). One can instead have the adjective between the classifier and the noun [CLF + ADJ + N], e.g., Standard Cantonese 间空屋 $kan^{55}\,hoŋ^{55}\,ok^5$ (CLF empty house) "the empty house," Nanning Pinghua 个间空

屋 $kə^{55}$ kan^{53} $hʊŋ^{53}$ $ʊk^3$ (this CLF empty house) "this empty house." (Cantonese allows classifier-initial noun phrases; Nanning Pinghua does not allow classifier-initial noun phrases except when the noun phrase is after a verb, similar to Mandarin.) Alternatively, one can put the adjective into a relative clause, e.g., Standard Cantonese 空嗰間屋 $hʊŋ^{55}$ $kɔ^{35}$ kan^{55} $ʊk^5$ (empty that CLF house) "the house that is empty," Nanning Pinghua 空个間屋 $hʊŋ^{53}$ $kə^{55}$ kan^{53} $ʊk^3$ (empty this CLF house) "the house that is empty."

On the other hand, in Nanning Cantonese, [ADJ + CLF + N] noun phrases are very common, and any adjective can go into the ADJ slot, e.g., 空間屋 $hʊŋ^{55}$ kan^{55} uk^5 (empty CLF house) "the empty house." The following are some other examples. (In Chinese linguistics, a distinction is often made between *xíngróngcí* 形容词, for the "verby" type of adjectives, as in 高 ku^{55} "tall" in 2 below, and *fēnbiécí* 分别词, for the "nouny" type of adjectives, as in 黃色 $wɔŋ^{21}$ $ʃek^5$ "yellow" in 1 below. This distinction is ignored here.)

Nanning Cantonese

1. 黃色支笔冇写得哂, 黑色支重得。
   <u>$wɔŋ^{21}$</u> <u>$ʃek^5$</u> $tʃi^{55}$ $pɐt^5$ mu^{24} $lɛ^{35}$ $tɐk^5$ lai^{33}, <u>$hɐk^5$</u> <u>$ʃek^5$</u> $tʃi^{55}$ $tʃuŋ^{22}$ $tɐk^5$.
   yellow color CLF pen NEG write can PRF black color CLF still can
   "The yellow pen is unusable, the black one can still be used." (Lín and Qín 2008: 278)
2. 妈糊高只男崽好呖嘅。
   <u>$ma^{55}wu^{21}$</u> ku^{55} $tʃɐk^3$ nam^{21} $tʃɐi^{35}$ hu^{35} $lɛk^5$ $kɛ^{33}$.
   quite tall CLF male child very capable MOD
   "The rather tall boy is very capable." (Lín and Qín 2008: 277)

The [ADJ + CLF + N] construction in Nanning Cantonese is analogous to the [CLF + N + ADJ] construction in Zhuang, with [ADJ] and [CLF + N] reordered following the head-initial noun phrase order in Zhuang.

Northern Zhuang

3. <u>Diuz[-]buh moq gou</u> deng nou   haeb baenz   congh.
   CLF-clothes new 1SG PASS mouse bite complete hole
   "My new shirt was ruined by a mouse." (Wéi and Qín 2006: 242)

In Nanning Cantonese, a lone classifier can be used as an anaphor. This usage is not found in Nanning Pinghua or Standard Cantonese (or Standard Mandarin). In example 4 below from Nanning Cantonese, the classifier 只 $tʃɐk^3$ (the general classifier for animals) on its own functions as an anaphor. In each instance the classifier refers to one dog, and the referent is determined by the context (in this case probably by pointing). Example 5 below in Standard Cantonese is a translation of example 4; Standard Cantonese requires at least a demonstrative in front of the classifier in this case.

Nanning Cantonese

4. 啲狗我中意只, 冇中意只, 只难睇多。
   ti^{55} $kɐu^{35}$ $ŋɔ^{24}$ $tʃuŋ^{55}ji^{33}$ $tʃɛk^3$, mu^{24} $tʃuŋ^{55}ji^{33}$ $tʃɛk^3$, $tʃɛk^3$ $nan^{21}tʰei^{35}$ $tɔ^{55}$.
   CLF.mass dog 1SG like   CLF   NEG like   CLF   CLF ugly   too
   "The dogs, I like (this) one, I do not like (that) one, (that) one is too ugly." (Lín and Qín 2008: 277)

Standard Cantonese

5. 啲狗我中意呢只, 唔中意嗰只, 嗰只太难睇。
   ti^{55} $kɐu^{35}$ $ŋɔ^{13}$ $tsʊŋ^{55}ji^{33}$ ni^{55} $tsɛk^3$, m^{21} $tsʊŋ^{55}ji^{33}$ $kɔ^{35}$ $tsɛk^3$,
   CLF.mass dog 1SG like   this CLF   NEG like   that   CLF
   $kɔ^{35}$ $tsɛk^3$ $tʰai^{33}$ $nan^{21}tʰei^{35}$.
   that CLF   too   ugly
   "The dogs, I like this one, I do not like that one, that one is too ugly."

The lone classifier construction is also found in Zhuang, also functioning as an anaphor.

Northern Zhuang

6. mwngz dawz duz ma de   daeuj hawj gou, gou cawz duz.
   2SG   take CLF dog that come give 1SG 1SG buy CLF
   "[Y]ou bring that dog to me, I'll buy it[.]" (Sio and Sybesma 2008: 191; Qín 1995: 83)
7. mwngz bi   bi   ndaem faex, go baenzlawz ha?
   2SG   year year plant tree CLF how   Q
   "[Y]ou plant trees every year, how are they doing?" (Sio and Sybesma 2008: 191; Qín 1995: 83)

With verb phrase syntax, the Sinitic languages in the Nanning area have also calqued many patterns from the Zhuang languages. For instance, Nanning Cantonese has the word order [verb + object + resultative complement], e.g., 食饭饱 $ʃek^2$ fan^{22} $pɐu^{35}$ (eat rice be.full) "having eaten and being full." This [verb + object + resultative complement] order is more common than the normal Sinitic word order of [verb + resultative complement + object], e.g., Standard Cantonese 食飽飯 sɪk² pau³⁵ fan²² (eat be.full rice), Mandarin 吃饱饭 chī bǎo fàn (eat be. full rice) "having eaten and being full" (Kwok 2010). The Nanning Cantonese pattern is a Tai pattern, cf. Northern Zhuang *gwn haeux imq* (eat rice be.full) (Wéi and Qín 2006: 203), Lao *khòòj5 kin3 makø-muang1 qiim1 lèèw4* (I eat CLF-mango be.full PRF) "I've eaten my fill of mangoes" (Enfield 2008: 412). (Nonetheless, see also Qín, Qín, and Tián 2016: 385–391 on their critique of Kwok 2010; they argue that the pattern in Nanning Cantonese is from Pinghua and not Zhuang.)

Another example is the grammaticalization of the verb "go" to an imperative marker. ("Go" also has a range of other grammaticalized meanings in this region.) This is a development led by Zhuang, and subsequently calqued into the Sinitic languages (see, e.g., Kwok 2014, 2019; Huang and Wu 2018: 115–118; see also ▶ Chap. 13, "Contact-Induced Change in the Languages of Southern China," by Wu and Huang's in this volume or whatever is appropriate).

Nanning Cantonese

8. 拧铰剪剪断苊绳去。
   nɛŋ⁵⁵ kɛu³³ tʃin³⁵ tʃin³⁵ tʰyn²⁴   tɐu⁵⁵ ʃeŋ²¹ hy³³.
   take scissors cut   be.severed CLF string IMP (<go)
   "Take scissors and cut the string!" (Lín and Qín 2008: 340)

Northern Zhuang

9. *Rumz baek rem   lai,   gven   aen[-]cueng   bae.*
   wind north strong much close CLF-window   IMP (<go)
   "The north wind is too strong, close the window!" (Wéi and Qín 2006: 208)
10. *Gwn vanj haeux liux   bae,*
    eat   bowl rice   finish IMP (<go)
    "Eat up the bowl of rice!" (Wéi and Qín 2006: 208)

It is beyond the scope of this chapter to present a detailed account on the language contact situation in Nanning. In this section we have seen some examples of an enclave Cantonese variety: Nanning Cantonese. All varieties of Cantonese, including Standard Cantonese in Guangzhou, are affected by their local language contact environments to some degree. Nanning Cantonese is a Cantonese variety that has diverged relatively strongly from Standard Cantonese. Its lexicon and grammar have been strongly influenced by the other languages in the Nanning area. Nonetheless, its phonology is still recognizably Cantonese, and Nanning Cantonese is still quite highly intelligible to speakers of Standard Cantonese.

Cantonese Under Different Jurisdictions

When we look at the variation among the Cantonese varieties, there is one sociopolitical aspect of Cantonese that makes it stand out among the Sinitic languages: Cantonese is one of the few Sinitic languages that are spoken in large numbers across many different jurisdictions. What has caused Cantonese to be spoken in so many different jurisdictions? Another question is that, with Cantonese easily being one of the best-known Sinitic languages in the West, what caused its prominence, especially when we consider that it is – relatively speaking – not widely spoken in China?

Both of these questions can be answered through a number of interrelated factors: the prosperity of the Port of Guangzhou, the dominance of the Hong Kong entertainment industry, Cantonese being used in an official capacity in Hong Kong and Macau, and the dominance of Cantonese in many Chinatowns overseas. In what follows, each of these factors will be briefly discussed.

The prominence of Cantonese began with the prosperity of the Port of Guangzhou. During the time of the maritime prohibitions, Guangzhou and Macau were some of the very few ports in China where foreign traders were allowed to conduct business. Between 1757 and the end of the First Opium War in 1842, Guangzhou was the only port in China where international trading was allowed. The intermediaries were mostly Cantonese speakers. Macanese Creole developed in Macau (e.g., Batalha 1985; de Senna Fernandes and Baxter 2004; Wong 2007), and Chinese Pidgin English developed around the Guangzhou area (e.g., Baker and Mülhäusler 1990; Ansaldo et al. 2010). Both Macanese Creole and Chinese Pidgin English contain many Cantonese/Yue elements, and both are products of the language contact that occurred in the Pearl River Delta between Cantonese and European languages.

The commercial importance of Guangzhou attracted European colonization on the coast of Guangdong. The Portuguese arrived in Macau in 1557; Britain-annexed Hong Kong in 1842; and France-annexed Kouang-Tchéou-Wan 广州湾 (i.e., Zhanjiang 湛江/Fort-Bayard) in 1898. Hong Kong, in particular, and also Macau, to a smaller degree, formed a link between Mainland China and the foreign world. The intermediaries were mostly Cantonese speakers. After European colonization, many people from the Guangzhou area migrated to Hong Kong, Macau, and Zhanjiang. Cantonese became the dominant language in those places.

The second factor has to do with the dominance of the Hong Kong entertainment industry. In the earlier decades of the twentieth century, the Chinese entertainment industry was centered in Shanghai. During the wars of the 1940s, many people who were involved in the entertainment industry fled from Shanghai to Hong Kong, which significantly enriched the Hong Kong entertainment industry. In the 1950s, 1960s, and 1970s, the Hong Kong entertainment industry was cut off from the Mainland Chinese market, as Mainland China closed itself off from the rest of the world. The Hong Kong entertainment industry remolded itself to suit Chinese audiences overseas, thereby pushing Hong Kong Cantonese popular culture to the world, with the largest market being the Chinese diaspora in Southeast Asia. Hong Kong's popular culture had influence in general in many parts of Southeast Asia (e.g., Thomas 2002; Heryanto 2013). The Hong Kong entertainment industry continued to flourish. Today, in consideration of the dominance of the Cantonese television media from Hong Kong, Mainland China has one Cantonese satellite television channel, TVS2 (of Guangdong Radio and Television), one of the very few satellite television channels in Mainland China that broadcast exclusively in a language other than Mandarin.

The third factor is the use of Cantonese in an official capacity in Hong Kong and Macau. The Hong Kong and Macau SAR governments primarily function in Cantonese, making Cantonese one of the very few Sinitic languages with official status. Officials speak Cantonese at all sorts of occasions, including the most formal. This has given Cantonese exposure to the world unmatched by other Sinitic languages except Mandarin.

The fourth factor is the spread of Cantonese speakers around the world. Since the end of the First Opium War (1839–1842), a large number of Cantonese (and other Yue) people migrated overseas. Wǔ (2007) estimates that there are more than 8.5 million Yue speakers outside China. Yue is not as prominent as Min and Hakka in many parts of Southeast Asia. However, Yue dominates many Chinatowns in Europe, Africa, the Americas, and Oceania (see, e.g., T'sou and Yóu 2003). Hence, traditionally, the Chinese culture that people in the West are familiar with is often Yue culture, and the Chinese language that they hear is often Cantonese. This is another factor which has contributed to the prominence of Cantonese outside China.

How has being spoken in many different jurisdictions affected the development of the various Cantonese varieties? Some issues related to the development of Cantonese across different jurisdictions will be discussed below. Two common themes are the difference in the language contact environments, and the difference in the language policies of the various countries and territories.

Hong Kong Cantonese is the best-known enclave Cantonese variety. Cantonese is not indigenous to Hong Kong: before the arrival of Cantonese, indigenous Hong Kongers spoke a number of different Yue, Hakka, and Southern Min varieties. The majority spoke a Yue variety that was similar to the indigenous Yue varieties in nearby Shenzhen and Dongguan. (Many of these varieties are now moribund; see Chang, Wán, and Zhuāng (1999) for a survey of the indigenous speech varieties in the New Territories of Hong Kong.) In the 1950s, the Cantonese-speaking population had not yet surpassed 50% of the population in Hong Kong, and Cantonese speakers were concentrated in the urban areas in Hong Kong Island and Kowloon Peninsula. However, with the socioeconomic dominance of Cantonese, there was a massive shift towards Cantonese by indigenous and nonindigenous Hong Kongers who spoke other speech varieties. Prominent groups of non-Cantonese-speaking migrants to Hong Kong include Hoishanese (Taishan and other Siyi Yue varieties), Hakka (Kejia), Teochew (Chaozhou), Hokkien (Southern Min), Shanghainese, and various South Asian groups. Apart from other groups shifting to Cantonese, there was also a large number of newer Cantonese-speaking migrants from the Guangzhou area. Since the 1970s, the percentage of Cantonese speakers in Hong Kong has risen to about 90%, while the percentage of other Sinitic varieties has continuously dropped, except for Mandarin. (See, T'sou and Yóu 2003; Lau 2004a, b on the formation of Hong Kong Cantonese and the changes in the linguistic demographics in Hong Kong. See Ding 2010 on the influences that the other Sinitic varieties and English have on the phonology of Hong Kong Cantonese.) The situation in Macau was similar; Macau also had Yue, Hakka, and Southern Min speakers; the majority spoke a Yue variety that was similar to that of nearby Zhongshan (Zhān et al. 2002: 196). However, the old Yue of Macau was supplanted by Cantonese, with only some traces of the former variety of Yue left. (See, e.g., Wong 2007, on the linguistic situation in Macau.)

Hong Kong Cantonese is known to have many English loanwords, and Hong Kongers often code mix or code switch between Cantonese and English (e.g., D. Li 1999; Wong et al. 2009; Chan 2019). With English being an official language of Hong Kong, and with the history of colonization by Britain, English is well established in Hong Kong society. As an illustration of how unaware Hong Kongers can be of their use of English loanwords, there is a memeified phrase in Hong Kong: $t^h\upsilon\eta^{21}$ $ts\textit{e}m^{22}$ $t\!f^h\varepsilon k^5$-ha^{13}, uttered in a (serious) television period drama by the role of the last Ming Emperor Chongzheng 崇禎 (seventeenth century). English loanwords sound so natural to Hong Kongers that no one noticed the anachronism during the entire production process of the drama: the "Ming Emperor" said 同朕再 check 吓 $t^h\upsilon\eta^{21}$ $ts\textit{e}m^{22}$ $ts\upsilon i^{33}$ $t\!f^h\varepsilon k^5$-ha^{13} (for 1SG.emperor again check-DELIMITATIVE) "check for me again," with an English loanword included.

Macau, heavily influenced by Hong Kong, follows Hong Kong in most respects, including having basically the same set of English loanwords. Portuguese remains one of the official languages of Macau SAR. However, Portuguese has never had the same level of penetration among the general public in Macau as English has in Hong Kong. Some Portuguese loanwords are still used in Macau Cantonese, but many such loanwords and words in Macau Cantonese, in general, are being replaced by words from Hong Kong Cantonese. For example, in Macau, "tuna" is traditionally 亞東 a^{33} $t\upsilon\eta^{55}$ (< Portuguese *atum*), but this has largely been replaced by 吞拿 $t^h\textit{e}n^{55}$ na^{21} (< English *tuna*). Similarly, 阿刁 a^{33} tiu^{55} "uncle" (< Portuguese *tio* "uncle") and 阿窩 a^{33} $w\upsilon^{55}$ "grandmother/old woman" (< Portuguese *avó* "grandmother") are no longer commonly used; these days people usually say 阿叔 a^{33} $s\upsilon k^5$ "uncle" in Cantonese, or even *uncle* in English, and 阿婆 a^{33} $p^h\upsilon^{21}$ "grandmother/old woman" in Cantonese.

In contrast to the prevalence of English loanwords in Hong Kong and Macau, many expressions in Guangzhou Cantonese are cognates of those found in Mandarin, the official language. For instance, the verb for sending things electronically is often $s\varepsilon n^{55}$ in spoken Hong Kong Cantonese (< English *send*), whereas it is 发 fat^3 in Guangzhou (< Mandarin 发 $f\bar{a}$ "distribute") (an alternative for both is 寄 kei^{33} "send (letter)"). Lexical semantics in Guangzhou Cantonese is also more observably affected by Mandarin. For instance, the verbs 闩 san^{55} "close (door/window)" and 熄 $s\iota k^5$ "switch off (lights/electrical appliances)" are both often replaced by 关 k^wan^{55} in Guangzhou (< Mandarin 关 $gu\bar{a}n$). Similarly, 截的士 $tsit^2$ $t\iota k^5si^{35}$ "hail taxi" and 搭的士 tap^3 $t\iota k^5si^{35}$ "ride taxi" are both commonly replaced by 打的 ta^{35} $t\iota k^5$ in Guangzhou (< Mandarin 打的 $d\check{a}$ $d\bar{\imath}$. While the noun 的士 $t\iota k^5si^{35}$ was loaned from Cantonese to Mandarin as *dìshì*, the phrase 打的 *dǎ dī* was loaned back from Mandarin to Cantonese as *ta³⁵ tık⁵*; traditionally Cantonese did not use the verb 打 *ta³⁵* "hit" for means of transportation). (Not all European loanwords are gone in Guangzhou; Guangzhou has kept many of the older European loanwords, e.g., 波 *pɔ⁵⁵* "ball" (< English *ball*), 麻甩 *ma²¹lɐt⁵* "pervert" (< French *malade* "sick").)

In Southeast Asia, Cantonese is on the whole less prominent than other Sinitic languages such as Hokkien, Teochew, and Hakka. Nonetheless, a few larger Chinatowns are Cantonese dominant, e.g., Hanoi, Ho Chi Minh City, and Kuala Lumpur. (Kuala Lumpur in particular has an active Cantonese television industry.) Even in non-Cantonese dominant areas, Chinese people often have some familiarity with Cantonese from Hong Kong popular culture, and/or having lived in the big cities with Cantonese-dominant Chinatowns. Naturally, these overseas Cantonese varieties are also influenced by their local linguistic environments. For instance, Malayan Cantonese (e.g., Chén 2003; Sin 2009) has many linguistic elements from English and Malay, e.g., in Kuala Lumpur Cantonese *sık³nə⁵⁵* "signal" (< English *signal*), *kɐm³³pɔŋ⁵⁵* "village" (< Malay *kampung*), 食风 *sık² fʊŋ⁵⁵* (eat wind) "travel" (< Malay *makan angin* (eat wind) "travel"). There are also loans from other Sinitic languages that are commonly encountered in Malaysia, e.g., Malayan Cantonese *tsʰin⁵⁵tsʰai⁵⁵* "any/whatever," from Hokkien 清彩 *tɕʰin⁵³tsʰai⁵³* (the equivalent in Standard Cantonese is 求其 *kʰɐu²¹kʰei²¹* or 是但 *si²²tan²²*). See Chén (2013) on borrowings among the various Sinitic languages in Southeast Asia, and Tan (this volume) on the contact among the Sinitic languages and English in Malaysia.

Cantonese varieties in different Anglophone countries have many English loan words. Nonetheless, their forms are not necessarily the same in different countries. For instance, "apartment" is 雅柏文 *ŋa¹³pʰak³mɐn²¹* in Australia and New Zealand (see Chén 2012 on Sydney Cantonese). On the other hand, this term has evolved to just 柏文 *pʰak³mɐn²¹* is US and Canada.

Not all differences are due to language contact; for instance, many words are simply coined differently in different countries and territories. For instance, "social housing" is 组屋 *tsou³⁵ ʊk⁵* (combination house) in Singapore, 人民组屋 *jɐn²¹mɐn²¹tsou³⁵ ʊk⁵* (people combination house) in Malaysia, 社屋 *sɛ¹³ ʊk⁵* (social house) in Macau, 公屋 *kʊŋ⁵⁵ ʊk⁵* (public house) in Hong Kong, and 经适房 *kıŋ⁵⁵sık⁵fɔŋ²¹* (economy suitable house) in Mainland China. (Also notice the use of 屋 *ʊk⁵* for "house," which is more common in Cantonese and Hakka, versus 房 *fáng* for "house," which is more common in Mandarin.) In another example, a power bank (USB external battery) is 充电宝 *tsʰʊŋ⁵⁵ tin²² pou³⁵* (charge electricity treasure) in Guangzhou (< Mainland Mandarin 充电宝 *chōng diàn bǎo*), but commonly 尿袋 *niu²² tɔi³⁵* (urine bag, i.e., urostomy bag) in Hong Kong (a metaphor of people walking around with cables/tubes leading out of their bodies). Macau often sides with Hong Kong when it comes to lexical choices, but in some cases it sides with Guangzhou. For instance, an eraser is 胶擦 *kau⁵⁵ tsʰat³* (rubber scrub) in Guangzhou and Macau, but 擦胶 *tsʰat³ kau⁵⁵* (scrub rubber) in Hong Kong.

There are huge differences among the legal systems of Hong Kong, Macau, and Mainland China. Legal practitioners in Hong Kong and Macau often coin legal terms in Chinese that bear a stronger resemblance to Classical Chinese than the ones in Mainland China. For instance, "property tax" is 差餉 $tsʰai^{55}$ $hœŋ^{35}$ (police wage) in Hong Kong Cantonese, and 業鈔 jip^2 $tsʰau^{55}$ (property banknote) in Macau Cantonese, both more classical sounding and less semantically transparent than the term 房产税 $fɔŋ^{21}$ $tsʰan^{35}$ $søy^{33}$/fáng chǎn shuì (house estate tax) used in Mainland China. Another example is the Classical Chinese-sounding term 入禀 $jɐp^2$ $pɐn^{35}$ (rù bǐng, enter report): in Hong Kong 入禀 $jɐp^2$ $pɐn^{35}$ is to file a lawsuit; in Macau 入禀 $jɐp^2$ $pɐn^{35}$ is to file a lawsuit, or to submit an application for driving test. Ho (2012) discusses some differences in the Chinese legalese in Macau, Hong Kong, Mainland China, and Taiwan.

Written Cantonese

The culture of writing also varies in different parts of the Cantonese world. One obvious difference is the use of Simplified versus Traditional Chinese characters. In Mainland China, Simplified Chinese is near-universal. Simplified Chinese is also more common in Malaysia and Singapore. In Hong Kong, Macau, and most other Cantonese communities overseas, Traditional Chinese is dominant. (Although currently, with the increased mobility of people from Mainland China, Simplified Chinese has become more commonly seen in Hong Kong, Macau, and overseas).

In addition, there are the different registers of writing, on a continuum from Modern Standard Written Chinese to Written Cantonese. Formal written communications are mostly conducted in Standard Written Chinese, which is based on Standard Mandarin. However, what people consider "Standard Written Chinese" differs slightly in different parts of the Sinitic world (similar to how Standard Written English differs slightly in different parts of the Anglophone world). In Cantonese societies, there can be conscious or subconscious admixtures of Cantonese linguistic features in people's Standard Written Chinese. For instance, instead of using the "compare" comparative construction (e.g., 甲比乙好 kap^3 pei^{35} jyt^2 hou^{35} (A compare B good) "A is better than B"), which is the construction used in Standard Written Chinese, Cantonese-influenced Standard Written Chinese might use the "surpass" comparative construction (e.g., 甲好过乙 kap^3 hou^{35} $kʷɔ^{33}$ jyt^2 (A good surpass B) "A is better than B"), which is the dominant pattern in Cantonese (see Chappell 2015b on comparative constructions among Sinitic languages). Scholarly discussions on written *Gǎngshì Zhōngwén* 港式中文 "Hong Kong-style Chinese," or the broader *Yuèshì Zhōngwén* 粤式中文 "Yue-style Chinese," include Shí (2006), Shí, Shào, and Chu (2014), and Tin (2008).

On the other side of the spectrum is Written Cantonese. The distinguishing feature of Written Cantonese is the use of Cantonese grammatical words like 係 $hɐi^{22}$ "be," 佢哋 $kʰøy^{13}tei^{22}$ "they," the negators 唔 m^{21} and 冇 mou^{13} (see section on "Nanning Cantonese" above), instead of Written Chinese equivalents like 是 si^{22} "be," 他們 $tʰa^{55}mun^{21}$ "they," the negators 不 $pɐt^5$ and 沒 mut^2 (< Mandarin 是 *shì*, 他們 *tāmén*, 不 *bù*, 沒 *méi*). Within Written Cantonese, there is a continuum between what can be called "high" Cantonese and "low" Cantonese. While Cantonese grammatical words are used in both types, "high" Cantonese utilizes more words that are reminiscent of Literary Chinese, and Mandarin-like grammatical constructions. The formal Spoken Cantonese used in high school Cantonese oral exams in Hong Kong (e.g., Lee and Leung 2012), and in news broadcasts, can be considered the spoken equivalent of "high" Written Cantonese. In Cantonese oral exams, pupils would be instructed to use literary-sounding lexical items like 認為 $jɪŋ^{22}wɐi^{21}$ "consider" (Mandarin 认为 rènwéi) instead of colloquial Cantonese equivalents like 諗 $nɐm^{35}$ "think." Traditionally, newscasters receive their texts in Written Chinese, and they translate them orally into "high" Spoken Cantonese. (Translational errors are sometimes heard during news broadcasts. For instance, the modifier marker 的 $tɪk^5$ in Written Chinese (Mandarin 的 de) has to be translated into Colloquial Cantonese 嘅 $kɛ^{33}$. However, there were unfortunate instances where newscasters misapplied this rule to cases where 的 $tɪk^5$ was *not* a modifier marker, and ended up saying, e.g., 波羅嘅海 $pɔ^{55}lɔ^{21}$ $kɛ^{33}$ $hɔi^{35}$ (pineapple MOD sea) "Sea of Pineapple" when they saw the text 波羅的海 $pɔ^{55}lɔ^{21}tɪk^5$ $hɔi^{35}$ "Baltic Sea.")

Written Cantonese is stigmatized to a degree. For instance, Written Cantonese is heavily suppressed by the education systems in all jurisdictions. Chinese written works are expected to be in Standard Written Chinese, and Cantonese influences in students' Chinese writings are considered inappropriate in an education setting. Nonetheless, Written Cantonese can be easily found in Hong Kong and Macau, for instance in advertisements, and in the "gossipy" sections of mainstream newspapers and magazines. In their "serious" sections, some newspapers leave the direct quotes in Written Cantonese instead of translating them into Modern Written Chinese. Online discussions by younger people are primarily in Written Cantonese. Headlines of (less formal) government public announcements are sometimes in Written Cantonese. There is also the interesting case of news.gov.hk, the Hong Kong SAR government's news outlet: the Chinese press releases on their website are in Standard Written Chinese, but the posts on their social media accounts are entirely in Written Cantonese. Recently, there has been a slight decrease in the stigma towards Written Cantonese in Hong Kong and Macau.

Despite the stigma, the tradition of vernacular Cantonese literature has never been broken since the first written representation of colloquial Cantonese in the seventeenth century (towards the end of the Ming Dynasty). Written Cantonese has never been standardized; people sometimes find ad hoc ways to represent Cantonese-specific words, including using Roman characters. For instance, the mass classifier ti^{55} is written 啲, or sometimes with the Roman letter *D*. (The mass classifier denotes a mass, whereas a normal classifier denotes an individual; for instance, compare 啲狗 ti^{55} $kɐu^{35}$ (CLF.mass dog) "the dogs," 啲沙 ti^{55} sa^{55} (CLF.mass sand) "the sand," versus 只狗 $tsɛk^3$ $kɐu^{35}$ (CLF dog) "the dog," 粒沙 $nɐp^5$ sa^{55} (CLF sand) "the grain of sand.") In another illustration of this ad-hoc-ness, in days before Unicode, Hong Kong and Macau computer users used the Big-5 Chinese character encoding standard developed in Taiwan (instead of the GB standard of Mainland China). However, the Chinese character sets developed in Taiwan did not have most of the

Cantonese-specific characters in them. (The Hong Kong and Macau governments did publish extended character sets for the Cantonese characters, but not all users bothered installing them. In addition, the Chinese input methods from Taiwan could not necessarily handle these extended character sets.) The informal solution in Hong Kong and Macau for rendering the mass classifier 啲 ti^{55} was "o的," with the Roman letter o substituting the mouth radical □, followed by the normal Chinese character 的. Similarly, other Cantonese characters with the mouth radical like 唔 m^{21} (NEG) and 嘅 $kε^{33}$ (MOD) were rendered "o吾" and "o既" respectively.

In addition to the aforementioned written registers, since the early nineteenth century, there has been a register called Saam Kap Dai 三及第 sam^{55} $k^hɐp^2$ $tɐi^{35}$, which is a mixture of Classical Chinese, Modern Standard Chinese, and Written Cantonese (Wong 2002; Snow 2004: 127). In the middle of the twentieth century this register was popular in the newspapers in Hong Kong and Macau. Earlier it was also popular in Guangzhou. While it is still possible to find younger people who can write reasonable Classical Chinese, the art of mixing Classical Chinese, Standard Written Chinese, and Written Cantonese is now moribund.

In addition to the issues outlined above, there is yet another issue that caused a difference in how Cantonese is written in different places: the differences in the development of computing culture. In the early days of computing, Hong Kong and Macau looked towards Taiwan for Chinese language computing. The education systems in Hong Kong and Macau are relatively poor when it comes to teaching the phonological principles of Mandarin and English, and even poorer for Cantonese. Hence, instead of the pronunciation-based input methods that are popular in Taiwan and Mainland China, Hong Kong and Macau have mostly gravitated towards the shape-based input methods of Chinese Characters, e.g., Cangjie 倉頡, "Stroke" method 笔划. Each key on a keyboard corresponds to a shape component of a Chinese character. With these shape-based input methods, (for competent typists) there is – especially now with Unicode – no problem in rendering the traditional Cantonese characters that were used across the Cantonese world.

Computing culture evolved separately in Mainland China. While there are also shape-based input methods in Mainland China (e.g., Wangma Wubi 王码五笔), the vast majority of people in Mainland China uses Mandarin Pinyin-based input methods. When people in Mainland China type in Cantonese, most of them use Mandarin Pinyin-based input methods to come up with Cantonese-specific characters. Cantonese characters are often replaced by characters that are quicker to type with Mandarin-based input methods. Sometimes these Cantonese-specific characters do not appear in Mandarin-based input methods. At other times, they do appear, but appear at the "bottom of the list" when a particular Mandarin syllable is typed in, and these lists of characters can be very long, as there are many homophones in Mandarin. People are thus more inclined to use a character that appears earlier in the list as a substitute, instead of scrolling to the bottom of the list for the "correct" Cantonese character. Also, sometimes people do not know the pronunciation of the Mandarin cognate of these Cantonese characters, if a cognate exists at all. An

16 The Expansion of Cantonese Over the Last Two Centuries   433

example is the rendition of the modifier marker $kɛ^{33}$ in Cantonese (functionally similar to 的 *de* in Mandarin; it marks the preceding constituent as a nominal modifier). Traditionally, the most common way of rendering $kɛ^{33}$ is 嘅; the character 嘅 $kɛ^{33}$ has a "mouth" radical 口 indicating that it is "colloquial," and 既 kei^{33} as the phonological component. The character 嘅 is still commonly used in Hong Kong and Macau. The following is an example of 嘅 from Macao Daily News, the best-selling newspaper in Macau. (News articles there are mostly written in Standard Written Chinese; this sentence in Written Cantonese is a direct quote from a member of the Legislative Assembly of Macau.)

Macau Written Cantonese

11. 睇唔到有人講嘅解決唔到嘅嘢。
    $t^hei^{35} m^{21} tou^{35}$ $jɐu^{13}$ $jɐn^{21}$ $kɔŋ^{35}$ $\underline{kɛ^{33}}$ $kai^{35}k^hyt^{3}$ m^{21} tou^{35} $\underline{kɛ^{33}}$ $jɛ^{13}$.
    cannot:see   exist   people   say   MOD   solve   NEG can   MOD   thing
    "(I) cannot see the things that some people say that cannot be solved."
    (www.macaodaily.com/html/2018-08/05/content_1285391.htm; accessed 11 Feb 2020)

In Mainland China, on the other hand, $kɛ^{33}$ is nowadays often rendered 噶: the traditional 嘅 is not, or not easily, typable with Mandarin-based input methods, whereas 噶 is easily typable using Mandarin-based input methods with its Mandarin pronunciation *gé*, which sound somewhat like Cantonese $kɛ^{33}$. (This usage of 噶 is not formed through Cantonese phonology: the phonological component 葛 is $kɔt^{3}$ in Cantonese, rather divergent from $kɛ^{33}$.) The following is an example of 噶 from the official Xinhua news website, in an article about learning Cantonese.

Guangzhou Written Cantonese

12. 你咁论尽嘎, 咁重要噶嘢都可以整唔见。
    nei^{13} $kɐm^{33}$ $lɵn^{22}tsɵn^{22}$ ka^{33}, $kɐm^{33}$ $tsoŋ^{22}jiu^{33}$ $\underline{kɛ^{33}}$ $jɛ^{13}$ tou^{55} $hɔ^{35}ji^{13}$ $tsɪŋ^{35}$
    2SG   so   clumsy   SFP such important MOD   thing even can   make
    $m^{21}kin^{33}$
    be.lost
    "You are so careless, you even manage to lose such an important thing."
    (www.gd.xinhuanet.com/newscenter/2018-02/02/c_1122346832.htm; accessed 6 Aug 2018)
    [SFP: sentence final particle]
    (嘎 ka^{33} is another character that Hong Kong and Macau readers might be less familiar with; ka^{33} is usually rendered 㗎 in Hong Kong and Macau.)

Conclusion

Cantonese is the representative variety of Yue Chinese. People's definitions of "Cantonese" vary; in this chapter, Cantonese is the language of Canton/Guangzhou, and also the Yue varieties that descended from the ones spoken by migrants from the Guangzhou area since the end of the First Opium War (1839–1842). Throughout its history, the development of Yue Chinese has been intimately tied to language contact, from the interactions with the indigenous languages in the Pearl River basin (which is still ongoing on the western edge of the Yue-speaking area), to the interactions with the European merchants, missionaries, and colonizers (Portuguese, British, and French) that arrived in Guangdong in the last few centuries, as well as the myriad of languages that Cantonese migrants encounter in the many Chinatowns overseas that they find themselves in.

This chapter is primarily descriptive in nature; some aspects of the development of selected Cantonese varieties were discussed in this chapter. Guangzhou has been a prosperous city for more than one millennium; before the maritime restrictions ended at the end of the First Opium War, Guangzhou was one of the very few ports, or at times the only port, where foreign traders could conduct business in China. Since the lifting of the maritime restrictions, millions of Cantonese people emigrated from the heart of the Pearl River Delta. Some went up the Pearl River to places like Wuzhou and Nanning, while others went out towards the sea to places like Hong Kong, Macau, Fort Bayard (Zhanjiang), and further to many foreign countries. Cantonese dominates many Chinatowns overseas. Cantonese enclaves can be found in many parts of Far Southern China and around the world.

One enclave Cantonese variety discussed in this chapter is Nanning Cantonese. Nanning is the capital of Guangxi Zhuang Autonomous Region. Nanning Cantonese started taking shape less than 200 years ago. Within these 200 years, Nanning Cantonese has acquired a great deal of linguistic influence from the other languages in Nanning: Old Nanning Mandarin, Nanning Pinghua, Northern Zhuang, and Southern Zhuang. While Nanning Cantonese is still largely intelligible to speakers of Standard Cantonese, many second-language features from speakers of the other Nanning languages have become mainstream in Nanning Cantonese. With Zhuang being the most divergent from Cantonese, features from Zhuang are especially observable in all areas of Nanning Cantonese, from phonetics and morphosyntax to discourse practice, lexical forms, and semantics. The greater social engagement between Cantonese and Zhuang speakers (in contrast to the slight distance that Pinghua and Zhuang speakers kept with each other in the past) means that occasionally Nanning Cantonese resembles Zhuang more than Nanning Pinghua does, despite Pinghua having been spoken in the Nanning area for about one millennium, whereas Cantonese has been spoken in the area for less than two centuries.

Another enclave Cantonese variety discussed in this chapter is Hong Kong Cantonese. Cantonese is not indigenous to Hong Kong SAR; indigenous Hong Kongers spoke a range of other Yue dialects, and also some Hakka and Southern Min varieties. The special status of Hong Kong and Macau, and the commercial success of the Cantonese migrants, resulted in Cantonese being favored in the language policies there, and Cantonese being used at an official capacity in the two SARs. The dominance of Hong Kong media and popular culture helped spread Cantonese worldwide. All Cantonese varieties are influenced by their local linguistic environments. For instance, the Cantonese varieties spoken in Hong Kong and in many Anglophone countries contain many English loanwords. The Cantonese of Kuala Lumpur and other places in Malaya has calqued many Malay expressions (in addition to English expressions), and also loanwords from other Sinitic languages commonly encountered in Malaya (▶ Chap. 36, "Chinese Languages and Malaysian English: Contact and Competition," by Siew Imm Tan's in this Handbook).

In the Cantonese world, there is a continuum of written registers from Standard Written Chinese to Written Cantonese. While stigmatized, the tradition of Written Cantonese has never been broken, and its stigma has slightly decreased recently. The separate evolution of computing culture in Hong Kong/Macau and Mainland China has created differences in the choice of characters used in rendering Cantonese words, beyond the distinction of Traditional versus Simplified Chinese characters.

All Sinitic languages are important components of the Chinese heritage. Research on Cantonese not only enhances people's understanding of Cantonese and the wider Yue dialect group, it also enriches studies of the other Sinitic languages. Research on Cantonese provides a similar, yet nonidentical, perspective with which one could compare and contrast research on the other languages in China and Southeast Asia.

References

Ansaldo, Umberto, Stephen Matthews, and Geoff Smith. 2010. China coast pidgin: Texts and contexts. *Journal of Pidgin and Creole Languages* 25 (1): 63–94.
Bái, Yào Tiān 白耀天. 2009. Yuèyǔ yǔ Zhuàngyǔ 粤语与壮语 [Yue and Zhuang]. *Guǎngxī Mínzú Yánjiū* 广西民族研究 2009 (4): 120–126.
Baker, Philip, and Peter Mülhäusler. 1990. From business to pidgin. *Journal of Asian Pacific Communication* 1 (1): 87–115.
Batalha, Graciete Nogueira. 1985. Situação e perspectivas do Português e dos crioulos de origem Portuguesa na Ásia Oriental (Macau, Hong Kong, Malaca, Singapura, Indonésia). *Congresso sobre a Situação Actual da Lingua Portuguesa no Mundo: Actas* 1: 287–304.
Bauer, Robert S. 1996. Identifying the Tai substratum in Cantonese. In *Pan-Asiatic linguistics – Proceedings of the fourth international symposium on languages and linguistics, V*, 1806–1844. Bangkok: Institute of Language and Culture for Rural Development, Mahidol University.
———. 2005. Two 19th century missionaries' contributions to historical Cantonese phonology. *Hong Kong Journal of Applied Linguistics* 10 (1): 21–46.
Bauer, Robert S., and Paul Benedict. 1997. *Modern Cantonese phonology*. Berlin: Mouton de Gruyter.
Bauer, Robert S., Kwan-hin Cheung, and Pak-man Cheung. 2003. Variation and merger of the rising tones in Hong Kong Cantonese. *Language Variation and Change* 15: 211–215.
Chan, Ka Long Roy. 2019. Trilingual code-switching in Hong Kong. *Applied Linguistics Research Journal* 3 (4): 1–14. https://doi.org/10.14744/alrj.2019.22932.
Chang, Song Hing 張雙慶, Bō Wàn 萬波, and Chū Shēng Zhuāng 莊初昇. 1999. Xiānggǎng Xīnjiè fāngyán diàochá bàogào 香港新界方言調查報告 A study of the geographic distribution of dialects in the New Territories before urbanization. *Journal of Chinese Studies* 中國文化研究所學報 39: 361–396.

Chappell, Hilary M. 2015a. Introduction: Ways of tackling diversity in Sinitic languages. In *Diversity in Sinitic languages*, ed. Hilary M. Chappell, 3–12. Oxford: Oxford University Press.

———. 2015b. Linguistic areas in China for differential object marking, passive, and comparative constructions. In *Diversity in Sinitic languages*, ed. Hilary M. Chappell, 13–52. Oxford: Oxford University Press.

———. 2017. Languages of China in their East and South-East Asian Context. In *The Cambridge handbook of areal linguistics*, ed. Raymond Hickey, 196–214. Cambridge: Cambridge University Press.

Chén, Hǎi Lún 陈海伦, and Yì Lín 林亦 (eds.). 2009. *Yuèyǔ Pínghuà Tǔhuà fāngyīn zìhuì Dì yī biān Guǎngxī Yuèyǔ, Guìnán Pínghuà bùfèn* 粤语平话土话方音字汇第一遍广西粤语、桂南平话部分 *[Dialectal syllabary of Yue, Pinghua, and Tuhua Volume 1 Part on Guangxi Yue, Southern Guangxi Pinghua]*. Guǎngxī Dàxué Yǔyánxué Cóngshū 广西大学语言学丛书 [Guangxi University Linguistics Series]. Shanghai: Shanghai Educational Publishing House 上海教育出版社.

Chén, Xiǎo Jǐn 陈晓锦. 2003. *Mǎláixīyà de sāngè Hànyǔ fāngyán* 马来西亚的三个汉语方言 *[Three Chinese dialects of Malaysia]*. Beijing: China Social Science Press 中国社会科学出版社.

——— 陳曉錦. 2012. Xīní Yuèfāngyán Guǎngfǔhuà 悉尼粤方言廣府話 [Sydney Yue Cantonese]. *Estudos de Cantonense* 粵語研究 12: 20–26.

——— 陳曉錦. 2013. Dōngnányà Huárén shèqū xiōngdì Hànyǔ fāngyán de hùjiècí 東南亞華人社區兄弟漢語方言的互借詞 Loan words among Chinese dialects of Chinese Societies at Southeast Asia. *Estudos de Cantonense* 粵語研究 13: 76–82.

Chén, Xiǎo Jǐn 陈晓锦, and Tāo Chén 陈滔. 2005. *Guǎngxī Běihǎishì Yuèfāngyán Diàochá Yánjiū* 广西北海市粤方言调查研究 *[Investigative studies of the Yue dialects in Beihai City, Guangxi]*. Beijing: China Social Science Press 中国社会科学出版社.

Cheng, Ting Au 郑定欧. 1999. Xiānggǎng Yuèyǔ yǔ Guǎngzhōu Yuèyǔ zhī bǐjiào 香港粤语与广州粤语之比较 [A comparison of Hong Kong Yue and Guangzhou Yue]. In *Shuāngyǔ shuāngfāngyán yǔ xiàndài Zhōngguó* 双语双方言与现代中国 *[Bilingualism, bidialectalism and modern China]*, ed. Ēn Quán Chén 陈恩泉, 405–415. Beijing: Beijing Language and Culture University Press 北京语言大学出版社.

Cheng, Siu-Pong, and Sze-Wing Tang. 2014. Languagehood of Cantonese: A renewed front in an old debate. *Open Journal of Modern Linguistics* 4 (3): 389–398. https://doi.org/10.4236/ojml.2014.43032.

Cheung, Samuel Hung-nin 張洪年. 2007. *Xiānggǎng Yuèyǔ yǔfǎ de yánjiū (zēngdìngbǎn)* 香港粤語語法的研究(增訂版) *A grammar of Cantonese as spoken in Hong Kong*, revised ed. Hong Kong: The Chinese University Press 中文大學出版社.

de Senna Fernandes, Miguel, and Alan Baxter. 2004. *Maquista Chapado: Vocabulary and expressions in Macao's Portuguese creole*. Macau: Instituto Internacional de Macau.

de Sousa, Hilário 蘇沙. 2013. Nánníng Shàngyáo Pínghuà de yīxiē míngcí duǎnyǔ xiànxiàng duìbǐ yánjiū 南宁上尧平话的一些名词短语现象对比研究 [Comparative studies of some noun phrase phenomena in Nanning Shangyao Pinghua]. In *Hànyǔ fāngyán yǔfǎ yánjiū de xīnshìjiǎo – Dì wǔ jiè Hànyǔ fāngyán yǔfǎ guójì xuéshù yántǎohuì lùnwénjí* 汉语方言语法研究的新视角 – 第五届汉语方言语法国际学术研讨会论文集 *[New viewpoints in the studies of the grammar of the Chinese Dialects – Proceedings of the fifth international conference on the syntax of Chinese dialects]*, eds. Dān Qīng Liú 刘丹青, Léi Zhōu 周磊, and Cái Dé Xuē 薛才德, 141–160. Shanghai: Shanghai Educational Publishing House 上海教育出版社.

de Sousa, Hilário. 2015. Language contact in Nanning: Nanning Pinghua and Nanning Cantonese. In *Diversity in Sinitic languages*, ed. Hilary M. Chappell, 157–189. Oxford: Oxford University Press.

———. 2017. Pínghuà 平話 dialects. In *Encyclopedia of Chinese language and linguistics*, ed. Rint Sybesma, Wolfgang Behr, Yueguo Gu, Zev Handel, C.-T. James Huang, and James Myers, vol. 3, 425–431. Leiden: Brill.

de Sousa, Hilário. forthcoming-a. *A grammar of Nanning Pinghua*. Sinitic Languages of China Series. Berlin/Boston: De Gruyter Mouton.

———. forthcoming-b. On Pinghua, and Yue: Some historical and linguistic perspectives. Submitted to *Crossroads: An Interdisciplinary Journal of Asian Interactions*.

Ding, Picus Sizhi. 2010. Phonological change in Hong Kong Cantonese through language contact with Chinese topolects and English over the past century. In *Marginal dialects: Scotland, Ireland and beyond*, ed. Robert McColl Millar, 198–218. Aberdeen: Forum for Research on the Languages of Scotland and Ireland.

Enfield, N.J. 2008. *A grammar of Lao*. De Gruyter Mouton: Berlin/Boston.

Handel, Zev. 2015. The classification of Chinese: Sinitic (The Chinese language family). In *The Oxford handbook of Chinese linguistics*, ed. William S.-Y. Wang and Chaofen Sun, 34–44. Oxford: Oxford University Press. https://doi.org/10.1093/oxfordhb/9780199856336.001.0001.

Heryanto, Ariel. 2013. Popular culture for a new Southeast Asian studies? In *The historical construction of Southeast Asian studies: Korea and beyond*, ed. Victor T. King and Seung Woo Park, 226–262. Singapore: Institute of Southeast Asian Studies.

Ho, Pan 何斌. 2012. Liǎng'àn sìdì fǎlǜ zhōng de tèyǒu yòngyǔ jǔyú 兩岸四地法律中的特有用語舉隅 [Examples of terms specific to the legal systems in the four jurisdictions across the strait]. In *Àomén yǔyán yánjiū sānshínián: Yǔyán yánjiū huígù jì qìngzhù Chéng Xiánghuī Jiàoshòu Àomén cóngyán cóngjiào sānshínián wénjí* 澳門語言研究三十年:語言研究回顧暨慶祝程祥徽教授澳門從研從教三十周年文集 *Macau language research review*, eds. Jié Xú 徐傑 and Jiàn Zhōu 周荐, 110–130. Macau: Universidade de Macau 澳門大學.

Hsieh, Miao-Ling. 2001. *Form and meaning: Negation and question in Chinese*. PhD dissertation, Los Angeles: University of Southern California.

Huang, Yuan Wei. 1997. The interaction between Zhuang and the Yue (Cantonese) dialects. In *Comparative Kadai: The Tai branch*, ed. Jerold A. Edmondson and David B. Solnit, 57–76. Dallas: Summer Institute of Linguistics & the University of Texas at Arlington.

Huang, Yang, and Fuxiang Wu. 2018. Central southern Guangxi as a grammaticalization area. In *New trends in grammaticalization and language change*, ed. Sylvie Hancil, Tine Breban, and José Vicente Lozano, 105–134. Amsterdam: John Benjamins Publishing Company.

Kataoka, Shin, and Cream Yin-Ping Lee. 2008. A system without a system: Cantonese Romanization used in Hong Kong place and personal names. *Hong Kong Journal of Applied Linguistics* 11 (1): 79–98.

Kwok, Bit-Chee 郭必之. 2004. Cóng Yú Zhī liǎngyùn "tèzì" kàn Yuè fāngyán gēn Gǔ-Jiāngdōng fāngyán de liánxì 從虞支兩韻「特字」看粵方言跟古江東方言的聯繫 Evidence for an Ancient Jiangdong layer in the Yue dialects as revealed in "special words" of the Yu and Zhi rhymes. *Language and Linguistics* 語言暨語言學 5 (3): 583–614.

——— 郭必之. 2010. Yǔyán jiēchù zhōng de yǔfǎ biànhuà: Nánníng Yuèyǔ "shúyǔ + bīnyǔ + búyǔ" jiégòu de láiyuán 語言接觸中的語法變化:南寧粵語「述語 + 賓語 + 補語」結構的來源 [Grammatical change in language contact: On the origin of the "verb + object + complement" structure in Nanning Yue]. In *Lìshǐ yǎnbiàn yǔ yǔyán jiēchù: Zhōngguó Dōngnán fāngyán* 歷時演變與語言接觸—中國東南方言 *Diachronic change and language contact: Dialects in South East China*, eds. Hung-nin Samuel Cheung 張洪年 and Song Hing Chang 張雙慶, 201–216. Journal of Chinese Linguistics Monograph Series Number 24. Hong Kong: The Chinese University of Hong Kong 中文大學出版社.

——— 郭必之. 2014. Nánníng dìqū yǔyán 'qù'-yì yǔsù de yǔfǎhuà yǔ jiēchù yǐnfā de "fùzhì" 南寧地區語言「去」義語素的語法化與接觸引發的「複製」 The Morpheme GO in three languages of the Nanning Region: Paths of grammaticalization and contact-induced "replication". *Language and Linguistics* 語言暨語言學 15 (5): 663–697. https://doi.org/10.1177/1606822X14528640.

——— 郭必之. 2019. *Yǔyán jiēchū shìjiǎo xià de Nánníng Yuèyǔ yǔfǎ* 語言接觸視角下的南寧粵語語法 *[Grammar of Nanning Cantonese under the viewpoint of language contact]*. Beijing: Zhonghua Book Company 中華書局.

Lau, Chun-Fat 刘镇发. 2001. Xiàndài Yuèyǔ yuányú Sòngmò yímín shuō 现代粤语源於宋末移民说 [Theory on the end-of-Song origin of modern Yue]. In *Dìqījiè Guójì Yuè Fāngyán Yántǎohuì lùnwénjí* 第七届国际粤方言研讨会论文集 [*Proceedings of the 7th International Conference on Yue Dialects*], eds. Chow Yiu Sin 单周尧 and K.K. Luke 陆镜光, 76–83. Beijing: The Commercial Press 商务印书馆.

——— 劉鎮發. 2004a. Xiānggǎng liǎngbǎiniánlái de yǔyán shēnghuó yǎnbiàn 香港兩百年來的語言生活演變 [Changes in the linguistic life in Hong Kong in the last two hundred years]. In *Táiwān yǔ Dōngnányà Huárén dìqū yǔwén shēnghuó yántǎohuì lùnwénjí* 台灣與東南亞華人地區語文生活研討會論文集 [*Proceedings of the conference on the linguistic life in Taiwan and Chinese areas in Southeast Asia*], 128–143. Hong Kong: Oi Ming Publishers 靄明出版社.

——— 劉鎮發. 2004b. Cóng Guǎngzhōuhuà dào Xiānggǎng Yuèyǔ – Xiānggǎng Yuèyǔ de xíngchéng 從廣州話到香港粵語—香港粵語的形成 [From Guangzhou dialect to Hong Kong Yue – The formation of Hong Kong Yue]. In *Táiwān yǔ Dōngnányà Huárén dìqū yǔwén shēnghuó yántǎohuì lùnwénjí* 台灣與東南亞華人地區語文生活研討會論文集 [*Proceedings of the conference on the linguistic life in Taiwan and Chinese areas in Southeast Asia*], 144–165. Hong Kong: Oi Ming Publishers 靄明出版社.

Law, Paul. 2014. The negation mou5 in Guangdong Yue. *Journal of East Asian Linguistics* 23: 267–305. https://doi.org/10.1007/s10831-013-9116-0.

Lee, Kwai Sang, and Wai Mun Leung. 2012. The status of Cantonese in the education policy of Hong Kong. *Multilingual Education* 2: 2. https://doi.org/10.1186/10.1186/2191-5059-2-2.

Li, D. C-S. 1999. Linguistic convergence: Impact of English on Hong Kong Cantonese. *Asian Englishes* 2 (1): 5–36. https://doi.org/10.1080/13488678.1999.10801017.

Li, M 1999. *Negation in Chinese*. PhD thesis, Manchester: University of Manchester.

Lǐ, Lián Jìn 李連進. 2000. *Pínghuà yīnyùn yánjiū* 平話音韻研究 [*Studies in Pinghua phonology*]. Nanning: Guangxi People's Publishing House 廣西人民出版社.

Lǐ, Jǐn Fāng 李锦芳. 2002. *Dòng Tái yǔyán yǔ wénhuà* 侗台语言与文化 [*Tai-Kadai language and culture*]. Beijing: The Ethnic Publishing House 民族出版社.

Lin, Jo-Wang. 2003. Aspectual selection and negation in Chinese. *Linguistics* 41 (3): 425–459. https://doi.org/10.1515/ling.2003.015.

Lín, Yì 林亦, and Fèng Yú Qín 覃风余. 2008. *Guǎngxī Nánníng Báihuà yánjiū* 广西南宁白话研究 [*Studies on Nanning Cantonese of Guangxi*]. Guilin: Guangxi Normal University Press 广西师范大学出版社.

Liú, Cūn Hàn 劉村漢. 1995. *Liǔzhōu fāngyán cídiǎn* 柳州方言詞典 [*Liuzhou dialect dictionary*]. Nanjing: Jiangsu Educational Press 江蘇教育出版社.

Liú, Shū Xīn 刘叔新. 2006. *Yuèyǔ Zhuàngdǎiyǔ wèntí* 粤语壮傣语问题 [*The Yue Zhuangdai question*]. Beijing: The Commercial Press 商务印书馆.

Mài, Yún 麦耘. 2010. Yuèyǔ de xíngchéng, fāzhǎn yǔ Yuèyǔ hé Pínghuà de guānxì 粤语的形成、发展与粤语和平话的关系 [The formation and development of Yue, and the relationship between Yue and Pinghua]. In *Yánjiū zhī lè – Qìngzhú Wáng Shìyuán Xiānshēng qīshíwǔ shòuchén xuéshù lùnwénjí* 研究之乐 – 庆祝王士元先生七十五寿辰学术论文集 *The joy of research II – A festschrift in honor of Professor William S-Y. Wang on his seventy-fifth birthday*, eds. Wù Yún Pān 潘悟云 and Zhōng Wěi Shěn 沈钟伟, 227–243. Shanghai: Shanghai Educational Publishing House 上海教育出版社.

Mair, Victor H. 1991. What is a Chinese "dialect/topolect"? Reflections on some key Sino-English linguistic terms. *Sino-Platonic Papers* 29: 1–31.

———. 2013. The classification of Sinitic languages: What is "Chinese"? In *Breaking down the barriers: Interdisciplinary studies in Chinese linguistics and beyond*, ed. Guangshun Cao, Hilary M. Chappell, Redouane Djamouri, and Thekla Wiebusch, 735–754. Taipei: Academia Sinica.

Matthews, Stephen. 2006. Cantonese grammar in areal perspective. In *Grammars in contact: A cross-linguistic typology*, eds. A.Y. Aikhenvald and R.M.W. Dixon, 220–236. Oxford: Oxford University Press.

Matthews, Stephen, and Virginia Yip. 1994. *Cantonese: A comprehensive grammar*. London/New York: Routledge.

Peyraube, Alain. 1996. Le Cantonais est-il du Chinois ? *Perspectives Chinoises* 34: 26–29.

Pittayaporn, Pittayawat, Jakrabhop Iamdanush, and Nida Jampathip. 2014. Reconstruction of Proto-Tai negators. *Linguistics of the Tibeto-Burman Area* 37 (2): 151–180.

Qín, Fèng Yú 覃凤余, Dōng Shēng Qín 覃东生, and Chūn Lái Tián 田春来. 2016. Xiàpiān: Pínghuà de yǔfǎ yánjiū 下篇平话的语法研究 [Lower section: Grammatical studies of Pinghua]. In *Guǎngxī Pínghuà yánjiū* 广西平话研究 [*Studies on Guangxi Pinghua*], ed. Jǐn Yú 余瑾, 281–426. Beijing: China Social Science Press 中国社会科学出版社.

Qín, Xiǎo Háng 覃晓航. 1995. *Zhuàngyǔ tèshū yǔfǎ xiànxiàng yánjiū* 壮语特殊语法现象研究 [*Studies of special syntactic phenomena in Zhuang*]. Beijing: The Ethnic Publishing House 民族出版社.

Qín, Yuǎn Xióng 覃远雄. 2000. *Guìnán Pínghuà yánjiū* 桂南平话研究 [*Studies on Southern Guangxi Pinghua*]. PhD thesis, Guangzhou: Jinan University 暨南大学.

——— 覃远雄. 2007. Pínghuà he Tǔhuà 平话和土话 [Pinghua and Patois]. *Fāngyán* 方言 2007 (2): 177–189.

Qín, Yuǎn Xióng 覃遠雄, Shù Guān Wéi 韋樹關, and Chéng Lín Biàn 卞成林. eds. 1997. *Nánníng Pínghuà cídiǎn* 南寧平話詞典 [*Nanning Pinghua dictionary*]. Nanjing: Jiangsu Educational Press 江蘇教育出版社.

Shí, Dìng Xǔ 石定栩. 2006. *Gǎngshì Zhōngwén liǎngmiàn dì* 港式中文兩面睇 [*Looking at Hong Kong style Chinese bothways*]. Hong Kong: Sing Tao Publishing Ltd 星島出版.

Shí, Dìng Xǔ 石定栩, Jìng Mǐn Shào 邵敬敏, and Chi Yu Chu 朱志瑜. 2014. *Gǎngshì Zhōngwén yǔ Biāozhǔn Zhōngwén de bǐjiào* 港式中文與標準中文的比較 [*Comparison of Hong Kong Chinese and Standard Chinese*], 2nd ed. Hong Kong: Hong Kong Educational Publishing Company 香港教育圖書有限公司.

Sin, Ka Lin 冼偉國. 2009. "Mǎláixīyà de sāngè Hànyǔ fāngyán" zhōng zhī Jílóngpō Guǎngdōnghuà yuètán《馬來西亞的三個漢語方言》中之吉隆坡廣東話閱譚 A review on Kuala Lumpur's Cantonese in part of the three Chinese dialects in Malaysia. *New Era College Academic Journal* 新纪元学院学报 6: 83–131.

Sio, Joanna Ut-Seong 蕭月嬌, and Rint Sybesma 司馬翎. 2008. The nominal phrase in Northern Zhuang – A descriptive study. *Bulletin of Chinese Linguistics* 中國語言學集刊 3 (1): 175–225.

Smyth, David. 2002. *Thai: An essential grammar*. London/New York: Routledge.

Snow, Don. 2004. *Cantonese as written language – The growth of a written Chinese vernacular*. Hong Kong: Hong Kong University Press.

T'sou, Benjamin K. Y. 鄒嘉彥, and Rǔ Jié Yóu 游汝杰. 2003. *Hànyǔ yǔ Huárén shèhuì* 漢語與華人社會 *Chinese language and society*. Hong Kong: City University of Hong Kong Press 香港城市大學出版社.

Táng, Qī Yuán 唐七元. 2012. Cóng cíhuì jiǎodù kàn Yuè fāngyán duì Liǔzhōu fāngyán de yǐngxiǎng 从词汇角度看粤方言对柳州方言的影响 [Looking at the influence of Yue dialect on Liuzhou dialect from a lexical viewpoint]. *Journal of Changchun University* 长春大学学报 2012 (7): 815–818.

Thomas, Mandy. 2002. Re-orientations: East Asian popular cultures in contemporary Vietnam. *Asian Studies Review* 26 (2): 189–204. https://doi.org/10.1111/1467-8403.00022.

Tin, Siu-lam 田小琳. 2008. Gǎngshì Zhōngwén jí qí tèdiǎn 港式中文及其特点 Hong Kong Chinese and its characteristics. *Journal of College of Chinese Language and Culture of Jinan University* 暨南大学华文学院学报 2008 (3): 68–79.

Wáng, Hóng Jūn 王洪君. 2009. Jiān'gù yǎnbiàn, tuīpíng hé céngcì de Hànyǔ fāngyán lìshǐ guānxì móxíng 兼顾演变、推平和层次的汉语方言历史关系模型 A historial relation model of Chinese dialects with multiple perspectives of evolution, level and stratum. *Fāngyán* 方言 2009 (3): 204–218.

Wéi, Jǐng Yún 韦景云, and Xiǎo Háng Qín 覃晓航. 2006. *Zhuàngyǔ tōnglùn* 壮语通论 [*General studies of Zhuang*]. Beijing: Central University for Nationalities Press 中央民族大学出版社.

Wong, Chung Ming 黃仲鳴. 2002. *Xiānggǎng Sānjídì wéntǐ liúbiànshǐ* 香港三及第文體流變史 [*Evolutionary history of the Saam Kap Dai literary style in Hong Kong*]. Hong Kong: Hong Kong Writers Association 香港作家協會.

Wong, Yee 黃翊. 2007. *Àomén yǔyán yánjiū* 澳门语言研究 [*Studies on Macau languages*]. Beijing: The Commercial Press 商务印书馆.

Wong, Cathy Sin Ping, Robert S. Bauer, and Zoe Wai Man Lam. 2009. The integration of English loanwords in Hong Kong Cantonese. *Journal of the Southeast Asian Linguistics Society* 1: 251–266.

Wǔ, Wèi 伍巍. 2007. Yuèyǔ 粤语 On Yue group. *Fāngyán* 方言 2007 (2): 167–176.

Wurm, S.A., Rong Li, et al., eds. 1987/1989. *Language atlas of China*. 1st ed. Hong Kong: Longman Group (Far East) Ltd.

Xiǎn, Yáng 冼洋. 2018ms. Běihǎi Báihuà cídiǎn 北海白话词典 [Dictionary of Beihai Cantonese]. Beihai.

Xiao, Richard 肖忠华, and Tony McEnery. 2008. Negation in Chinese: A corpus-based study 汉语中的否定:基于语料库的研究. *Journal of Chinese Linguistics* 中国语言学报 36 (2): 274–330.

Yóu, Rǔ Jié 游汝杰. 2000. *Hànyǔ fāngyánxué dǎolùn* 汉语方言学导论 [*Introduction to Chinese dialectology*], Revised ed. 修订本. Shanghai: Shanghai Educational Publishing House 上海教育出版社.

Yue-Hashimoto, Anne. 1991. The Yue dialect. In *Languages and dialects of China*, ed. William S.Y. Wang, 292–322. Journal of Chinese Linguistics Monograph Series Number 3.

Zhān, Bó Huì 詹伯慧, Xiǎo Yàn Fāng 方小燕, Yū Ēn Gān 甘于恩, Xué Qiáng Qiū 丘学强, Choi Lan Tong 汤翠兰, Jiàn Shè Wáng 王建设, and Qí Zhōng 钟奇, eds. 2002. *Guǎngdōng Yuè fāngyán gàiyào* 广东粤方言概要 *An outline of Yue dialects in Guangdong*. Guangzhou: Jinan University Press 暨南大学出版社.

Zhang, Jingwei. 2019. Tone mergers in Cantonese: Evidence from Hong Kong, Macao, and Zhuhai. *Regional Chinese in Contact* 5 (1): 28–49.

Zhōu, Běn Liáng 周本良, Xiáng Hé Shěn 沈祥和, Píng Lí 黎平, and Yù Juān Wéi 韦玉娟. 2006. Nánníng Xiàguójiē Guānhuà tóngyīn zìhuì 南宁下郭街官话同音字汇 A list of homonyms of Xiaguo street dialect in Nanning. *Journal of Guilin Normal College* 桂林师范高等专科学校学报 20 (2): 1–8.

Interactions Between Min and Other Sinitic Languages: Genetic Inheritance and Areal Patterns

17

Ruiqing Shen

Contents

Introduction to Min	442
Classification of Min	443
Interaction Between Min and Other Sinitic Languages	444
Wu and Min	445
Hakka and Min	446
Xianghua and Min	448
Gan and Min	448
Discussion	451
Appendix: Data and Sources	452
References	452

Abstract

This chapter discusses interaction between Min and other Sinitic languages, with a focus on genetic inheritance and areal patterns. The first section is a general introduction to the Min group and its internal classification. The second section reviews studies on interactions between Min and four other Sinitic groupings, namely, Wu, Hakka, Waxiang, and Gan. It is demonstrated that while all four share certain lexical and/or phonological features with Min, these commonalities are of different kinds. The interaction between Min and Gan involves contact-induced areal patterns. In the other cases, while genetic inheritance is a factor, features shared between Min and Waxiang are most likely due to shared retention, while those shared between Min and Southwestern Wu and between Min and Hakka include some shared innovations as well. The last section discusses implications for the linguistic history of Min and directions for future studies.

Keywords

Min · Genetic · Language contact · Areal patterns

R. Shen (✉)
Department of Chinese Studies, National University of Singapore, Singapore, Singapore

© The Author(s), under exclusive licence to Springer Nature Singapore Pte Ltd. 2022
Z. Ye (ed.), *The Palgrave Handbook of Chinese Language Studies*,
https://doi.org/10.1007/978-981-16-0924-4_33

Introduction to Min

The Min varieties, part of the larger Sinitic language family, are mainly spoken in the southeastern coastal regions of Greater China, predominantly in Fujian, Guangdong, Hainan, and Taiwan. Min varieties are also spoken by large communities of overseas Chinese in Southeast Asia and other countries including Malaysia and Singapore (see Gonzales' chapter and Tan's chapter in this Handbook) (for linguistic introductions to Min, see Branner 2001; Lien 2015).

Historically, Fujian province is the homebase of the Min group. Physiographically, Fujian is often described as "eight-tenths mountain, one-tenth water, and one-tenth field" (八山一水一分田). The reality is if anything more extreme: composing some 87% of total land area, mountains and highlands dominate the province's topography. The Wuyi mountains 武夷山 in the north, rising to elevations above 2000 m, serve as an imposing barrier between Fujian and the interior (Yeung and Chu 2000). The Min 閩 river and other waterways, key means of transport and travel down to modern times, are largely confined within provincial borders (Norman 1991: 325). These geographical characteristics have made Fujian what Johanna Nichols calls a linguistic "residual zone," relatively unaffected by the periodic waves of immigration from the North (Nichols 1992: 13–24).

Since Karlgren (1954), Min has been regarded as the most archaic branch of the Sinitic family, differing significantly from other members. Many scholars argue that Min split from mainstream Chinese in the Qin-Han period (221 B.C.–220 A.D.; see Norman 1979; Ting 1983; Baxter 1995, among others). Some recent studies (Handel 2010; Nohara and Akitani 2014; Akitani and Nohara 2019) contend that Min may preserve some archaic features belonging to pre-Qin-Han Old Chinese, indicating an even earlier division. However, migration history indicates a much later period for the formation of Min. Fujian had been inhabited by non-Chinese people, known as Bai Yue 百越, before Chinese migrants began to enter Fujian from the first century onward (Bielenstein 1959; Lapolla 2001: Table 1).

According to Norman (1991), the Min family is recognizable as a highly distinctive subgroup of Sinitic based on both phonological and lexical evidence. Norman (1991: 349) provides a list of lexemes uniquely shared by different varieties of Min, a few of which are presented in Table 1.

Four varieties have been selected as representative of the various branches of Min (see Table 2 for details). All of these words, while rarely found in other Sinitic

Table 1 A lexical list for identifying Min. (Based on Norman 1991: 349; see appendix for data and source)

	Glossary	HP	JO	FZ	CZ
骹	"foot(+leg)"	k^hau^1	k^hau^1	k^ha^1	k^ha^1
戍	"house"	$tɕ^hio^5$	$ts^hiɔ^5$	$ts^huɔ^5$	ts^hu^5
喙	"mouth"	(ts^hui^5)	ts^hy^5	ts^hui^5	ts^hui^5
渐食	"insipid"	$t^hæm^4$	$tsian^3$	$tsian^3$	$tsiã^3$
囝	"son"	kin^7	$kyen^3$	$kian^3$	$kiã^3$
–	"to wear"	$ɕion^6$	$tsœyn^6$	$søyn^6$	ts^hen^6
–	"firewood"	t^hau^7	ts^hau^5	ts^ha^2	ts^ha^2

languages, show regular sound correspondences among Min varieties. Therefore, these items are regarded as Min diagnostic words.

Note also that Table 1 includes several basic words, such as "foot(+leg)" and "mouth." Not all of these words have corresponding Chinese characters. Even when Chinese characters do exist, they are not found as common words in Chinese texts. Rather, many of them first appear in the *Ji Yun* 集韻 (*Collected Rhymes*), a rhyme book published in A.D. 1037 which collected quite a few words of dialectal origins.

Classification of Min

Within the Min group, at least six branches have been identified so far (Norman 1991). The hierarchical relationships between these six branches are shown in Table 2.

As shown, the primary division in Min is between Inland and Coastal branches. This determination is based on lexical differences (Norman 1991: 350, Chen and Li 1991: 89–93), as well as comparative phonology (Coblin 2018). In the second divisions, Inland Min splits into Northwestern and Central branches, while Coastal Min splits into Southeastern and Eastern branches. In the final divisions, both Northwestern Min and Southeastern Min undergo a further split, yielding six branches in total. Each branch may involve several mutually unintelligible varieties (Inoue 2018). Some of these varieties are shown in the last column of Table 2.

Today, five out of the six Min branches are still mainly spoken in Fujian Province, China, while Southern Min is now widely spoken in other areas of China, as well as in Southeast Asia and other countries by a large number of overseas Chinese (Ding 2015). The geographic distribution of Min branches and their linguistic neighbors is presented in Fig. 1.

Table 2 Classification of Min (based on Norman 1991; Branner 2000: 43; Kwok 2018: 16. See also Zheng 2018: Fig. 6. All the varieties listed in the last column are mutually unintelligible)

1st division	2nd division	3rd division	MU varieties
A. Inland Min	1. Northwestern Min	1.1 Northern Min (N.Min)	Jianou (建甌) Jianyang (建陽) Huangkeng (黃坑)
		1.2 Far Western Min (FW.Min)	Heping (和平) Gaotang (高唐)
	2. Central Min	Central Min (C.Min)	Yong'an (永安)
B. Coastal Min	1. Southeastern Min	1.1 Southern Min (S.Min)	Xiamen (廈門) Chaozhou (潮州) Haikou (海口)
		1.2 Puxian Min (PX.Min)	Putian (莆田)
	2. Eastern Min	Eastern Min (E.Min)	Fuzhou (福州) Ningde (寧德) Cangnan (蒼南)

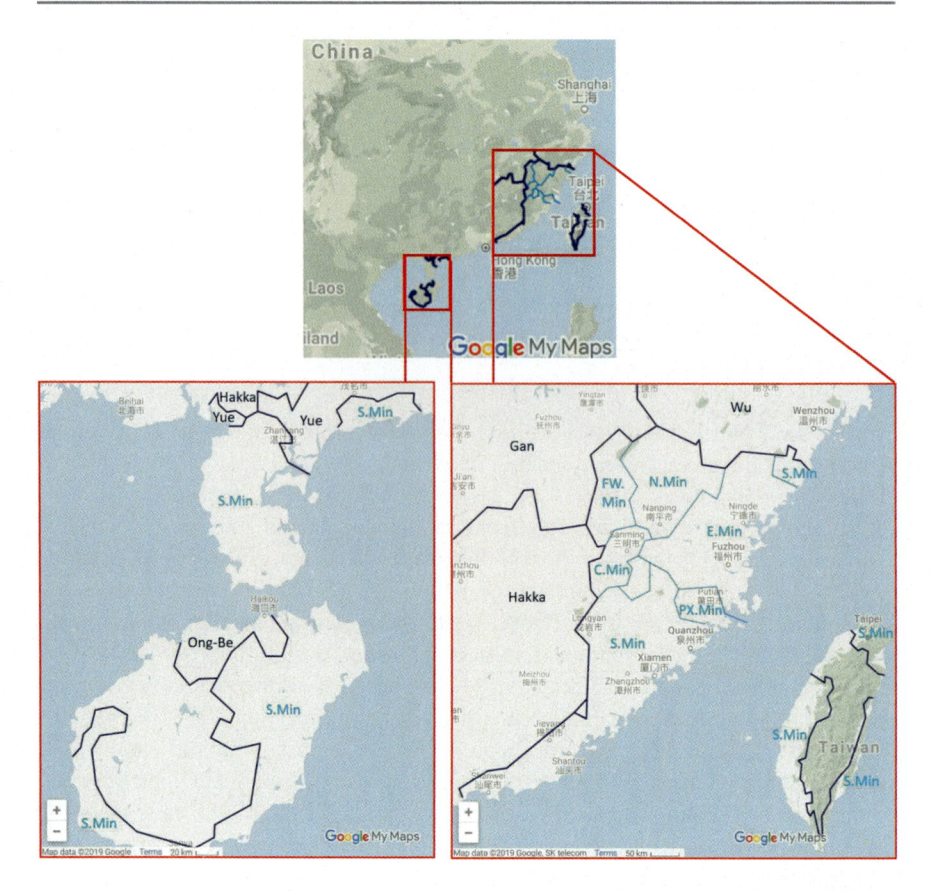

Fig. 1 Geographic distribution of Min branches and their linguistic neighbors in Mainland China and Taiwan. (Boundaries are reproduced from Wurm et al. (1987) with curve details omitted and a few modifications based on my knowledge)

Among the six branches of Min listed in Table 2, the affiliation of Far Western Min, spoken in the areas of Shaowu 邵武 and Jiangle 将乐, is still controversial. Since Norman (1982) proposed the "Shaowu hypothesis," arguing that Shaowu and nearby varieties belong to the Min family, there has been a long debate between scholars who support this hypothesis (see, e.g., Long 2010; Akitani 2013; Coblin 2018; Shen 2018) and those who instead argue for a Gan affiliation for Shaowu (see, e.g., Lei 1984; Chen 1993; Chang and Wan 1996). The nature of interaction between Far Western Min and neighboring Gan will be discussed below.

Interaction Between Min and Other Sinitic Languages

No language is totally isolated. Min is not an exception. Besides interactions with other Sinitic languages, Min varieties also feature substrata of non-Sinitic origin. However, it is beyond the scope of this chapter to evaluate different proposals regarding these substrata. Past proposals have referenced Hmong-Mien (Ballard 1981), Kra-Dai (Yue-Hashimoto 1976), and Austroasiatic (Norman and Mei 1976).

This section, instead, discusses interactions between Min and four other Sinitic groupings, namely, Wu, Hakka, Waxiang, and Gan. We will focus our discussion mainly on lexical and phonological evidence since very few grammatical descriptions have been published beyond those concerning the most widely known Southern Min varieties (e.g., Xu 2007).

Wu and Min

The lexical similarities between Southern Wu and Min have long been noticed (see, e.g., Ting 1988; Ballard 1992; Pan 1995; Zhengzhang 2002). If we focus on Southwestern Wu (i.e., Chuqu Wu), even more shared lexicon with Min can be found (Norman 1990; Akitani 2000). Below are some of the cognates between Min and Southwestern Wu which are rarely found in other Sinitic languages (Table 3).

Note that the first three items are also listed in Table 1 as Min diagnostic words. Interestingly, the last three items were recorded as Lower Yangtze region dialect words (*Jiangdong fangyan* 江东方言) by Guo Pu 郭璞 (A.D. 276–324), a scholar of the Jin dynasty (more such items can be found in Norman 1983; Li 2002). This has led some scholars to argue that both Min and Southern Wu are the descendants of an early Lower Yangtze dialect (see Ting 1988, 2006; Mei 2001, 2015 for further discussion).

Besides shared lexicon, Min and Southwestern Wu also share more than a dozen phonological features. Some of these features are listed in Table 4.

Most of the Table 4 features (A1–A6) are retentions, meaning they were extant in Old Chinese and are preserved in Min, while some (A7–A9) are innovations, meaning they were not present in Old Chinese. Some, such as (A1)(A4)(A5)(A8), are also shared by Southeastern Wu (Oujiang Wu; see Ting 1988; Pan 1995; Zhengzhang 2002; Mei 2001, 2015). Furthermore, some scholars (Zheng 2015; Tao 2018) argue that Northern Wu also shares a few lexical and phonological features with Min, indicating genetic closeness between Min and Wu in general. They argue that the current marked differences between Northern Wu and Min were caused by migration in later periods, as a result of which most Wu dialects acquired

Table 3 Lexical items shared by Min and Southwestern Wu (based on Norman 1990; Akitani 2000; see appendix for data and sources. X indicates an additional morpheme of which phonetic information is omitted; forms in brackets indicate the form is extracted from a word with different semantic meaning)

	Glossary	Min		Southwestern Wu	
		HP	FZ	SC2	QY
骹	"foot(+leg)"	k^hau^1	k^ha^1	$k^hɐɯ^1$	$k^hɒ^1$
厝	"house"	$tɕ^hio^5$	$ts^huɔ^5$	$tɕ^hyɤ^5$	$tɕ^hye^5$
喙	"mouth"	(ts^hui^5)	ts^hui^5	$tɕ^hyeʔ.X$	–
䘼	"sleeve"	–	$uoŋ^3$	$X.əŋ^3$	$X.iəŋ^3$
薸	"duckweed"	p^hieu^7	p^hiu^2	$biɐɯ^2$	$piɒ^2$
夥	"many"	uai^4	(uai^6)	–	$uɑ^4$

Table 4 Phonological features shared by Min and Southwestern Wu. (Based on Norman 1990; Akitani 1999, Old Chinese reconstruction from Baxter and Sagart 2014)

Phonological features	Example
A1) Retention of Old Chinese *-ai and *-oi (Rhyme Ge 歌部)	破 "broken," 蟻 "ant"
A2) Retention of Old Chinese *-ak (Rhyme Duo 铎部)	石 "stone," 惜 "love; comfort (v.)"
A3) Retention of Old Chinese *-u (Rhyme You 幽部)	扫 "sweep," 草 "grass"
A4) Retention of Old Chinese *g- as velar plosives	厚 "thick," 汗 "sweat"
A5) Old Chinese *l- (Initial Yi 以母) realized as affricative/fricative	痒 "itch," 蝇 "fly(n.)"
A6) Old Chinese *s-t- (Initial Shu 书母) realized as affricative	水 "water," 舂 "pestle(v.)"
A7) Merger between the rhyme of 放 "put" and 篷 "sail(n.)"	
A8) Merger between the rhyme of 病 "ill" and 彭 "surname Peng"	
A9) Merger between the rhyme of 船 "boat" and 春 "spring"	

Table 5 Lexical items shared by Min and Hakka. (Based on Norman 1986: 336–337; see appendix for data and sources)

		Min		Hakka
	Glossary	JY	FZ	MX
瀾	"saliva"	luen^5	lan^3	lan^1
脝	"heel, elbow"	tian^1	tan^1	tsan^1
–	"thin, watery"	loin^9	tsin^1	tsin^1
沕	"to dive"	me^6	mei^6	mi^5
蛤	"chicken louse"	loi^9	tai^2	ts^{h}i^2
–	"to wipe away"	tsui^8	souk^8	ts^{h}ut^8

later layers of Northern Chinese. Southwestern Wu, on the other hand, has largely retained its earlier forms due to the region's isolated, mountainous geography.

Hakka and Min

Norman (1986) listed 20 lexical items shared by Min and Hakka including some basic words, some of which are reproduced in Table 5.

Note that most of these items are colloquial words lacking Common Sinitic origins, even though they may have corresponding Chinese characters recorded in rhyme books or dictionaries.

In terms of phonological features shared by Min and Hakka, Norman (1986) finds that the two contrasting series of sonorants, absent from Middle Chinese, must be recognized for both Min and Hakka (O'Connor 1976). Furthermore, the two series show striking similarities in terms of the lexical items to which they pertain (Table 6).

In the Table 6, words in the left-hand column are reconstructed with voiceless sonorants in Proto-Min and upper register tones (*Yin diao* 阴调, indicated by odd numbers) in Proto-Hakka, while words in the right-hand column are reconstructed with voiced sonorants in Proto-Min and lower register tones (*Yang diao* 阳调, indicated by even numbers) in Proto-Hakka. In addition, Akitani (1993, 1995) found two features shared by Min and Hakka: (1) different rhymes for 飞 "to fly" and 肥 "fat(adj.)" and (2) an unusual rhyme for 泉 "spring."

Recent studies on the varieties of Sinitic spoken in Liancheng 连城, a county located at the border between the Min and Hakka regions, seem to be of crucial importance in understanding the nature of the relationship between these two groups. Akitani (1996) discusses the genetic position of the Wenheng 文亨 variety, finding that while it shares many phonological features with Hakka, it does not show the distinctive Hakka features outlined in Norman (1986). Meanwhile, Wenheng also shares many lexical and phonological features with Min. Therefore, he concludes that Wenheng should be regarded as a sister language of Proto-Hakka and a "missing link" between Hakka and Min. This relationship is represented in Fig. 2:

Besides Wenheng, there are many mutually unintelligible varieties spoken in the Liancheng area (Branner 1995, 1996), as well as in nearby areas such as Wan'an 万安 (Branner 2000). Branner (1995, 1996, 2000) argues that these varieties clearly belong to Min, although Hakka influences can be found. Through an investigation of Min features in 25 Liancheng varieties, Yan (2002) argues that all of these Liancheng varieties were originally Min varieties. While most of them have been significantly influenced by Hakka to the west, the Min features are relatively well preserved in the northeastern part of the region. It seems that a better understanding of the relationship between Min and Hakka awaits further exploration of the Liancheng varieties, which is just at its beginning.

Table 6 A comparison between Norman (1973) and O'Connor (1976), with some forms added from Norman (1986)

	Glossary	Proto-Min	Proto-Hakka		Glossary	Proto-Min	Proto-Hakka
毛	"hair"	*mh	*m-1	嫲	"female suffix"	*m	*m-2
面	"face"	*mh	*m-5	卖	"sell"	*m	*m-6
目	"eye"	*mh	*m-7	麦	"wheat"	*m	*m-8
聋	"deaf"	*lh	*l-1	来	"come"	*l	*l-2
六	"six"	*lh	*l-7	落	"descend"	*l	*l-8

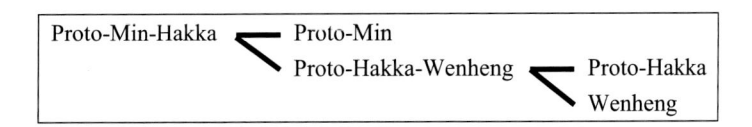

```
Proto-Min-Hakka  ←  Proto-Min
                 ↘  Proto-Hakka-Wenheng  ←  Proto-Hakka
                                         ↘  Wenheng
```

Fig. 2 Proposed relationship between Proto-Min, Proto-Hakka, and Wenheng (Akitani 1996)

Xianghua and Min

Xianghua, also known as the Waxiang dialects, is an unclassified Sinitic language spoken in western Hunan, on the boundaries of counties and in mountainous areas where the Miao (Hmong) and Tujia ethnic minorities live (Wu 2006). They seem to share lexical items with Min which are rarely found in other Sinitic languages, some of which are shown below in Table 7.

All of these items can be traced back to Old Chinese (some with semantic shift, such as "year old"). Besides lexical similarities, Min and Xianghua also share certain phonological features, as shown below.

Note that most of the features in Table 8 are retentions rather than innovations. The only innovative feature is the phonological feature B1. It could be a parallel development, as it is also found in Yue (Kwok 2004). Therefore, shared lexical and phonological features indicate that both Min and Xianghua are conservative and have preserved many archaic features inherited from Old Chinese.

Gan and Min

As mentioned earlier, the affiliation of Far Western Min, spoken in the Shaowu and Jiangle areas, is still controversial. Since Norman's (1982) "Shaowu hypothesis," in which he argues that the Shaowu dialect belongs to the Min family, there has been a long debate among scholars.

Table 7 Lexical items shared by Min and Xianghua (see Wu and Shen 2010: 30–39; see appendix for data and source)

	Glossary	Min HP	Min FZ	Xianghua GZ2
啼	"cry"	hi^7	t^{h}ie^2	lie^{13}
犬	"dog"	–	k^{h}ein^3	k^{h}uai^{25}
晬	"year old"	tsui^5	–	tsua^{33}
炙	"warm by fire"	tɕio^7	–	tso^{41}

Table 8 Phonological features shared by Min and Xianghua (see Wu and Shen 2010: 25–29, Old Chinese reconstruction and Middle Chinese notation in italic forms from Baxter and Sagart 2014)

Phonological features	Example
B1) Merger between the rhyme of 柱 "pillar" and 抽 "to draw"	
B2) Retention of Old Chinese *-ak (Rhyme Duo 铎部)	石 "stone," 尺 "a unit of length"
B3) Separate rhyme for Middle Chinese *i* (Rhyme Zhi 之韵)	箕 "dustpan," 起 "get up"
B4) Middle Chinese *tr-* series (Initial Zhi group 知组) realized as plosives (e.g. /t/)	锤 "hammer," 重 "heavy"
B5) Middle Chinese *hj-* (Initial Yun 云母) realized as fricative /h/	雨 "rain," 远 "far"
B6) Old Chinese *s-t- (Initial Shu 书母) realized as unaspirated affricative (e.g., /ts/)	水 "water," 舂 "pestle (v.)"

As it is very hard to find a list of lexical items exclusively shared by Far Western Min and Gan, the argument for the Gan affiliation of Far Western Min has largely been based on shared phonological features. One often-mentioned feature is the devoicing process in which voiced stops become voiceless aspirated (hereforth "aspirated devoicing"). It has been argued that aspirated devoicing is a shared innovation by Gan and Far Western Min. However, aspirated devoicing is a natural sound change that could occur independently at different times and places. Alternatively, this common mode of devoicing could be due to language contact. For example, Ho Ne (She), a Hmongic language (Ratliff 1998), may have undergone aspirated devoicing due to close contact with Hakka (Sagart 2002; Nakanishi and Kwok 2009), and yet no one argues for a genetic relationship between Ho Ne and Hakka. The aspirated devoicing affecting Gan, Hakka, and Far Western Min could be regarded either as parallel development or as an areal sound change that spread across Jiangxi Province and adjoining regions. In neither case does it indicate any particularly close genetic relationship between Gan, Hakka, and Far Western Min.

Another feature shared by Gan and Far Western Min is the chain shift *t^h>h followed by *ts^h>t^h, discussed by many scholars (Sagart 1993: 241–260; Sagart 2000; Zheng 2002: Chapter 3; Jiang 2003: 85–98; Shen 2019). This change occurred in a geographically continuous area, involving some Gan and Far Western Min varieties as well as some Northern Min varieties, as shown in Table 9.

In Table 9, two varieties from Gan, Far Western Min, and Northern Min, respectively, are represented. For each group, one variety has undergone the change, while the other hasn't. Note that the change affects different words in each group. For example, the word "bean" underwent *t^h>h in Gan and Far Western Min, but not in Northern Min. While scholars are still debating where this change initiated (see Sagart 2000; Shen 2019), there is a general consensus that this change occurred in recent times and does not indicate any especially close genetic relationship between Gan, Far Western Min, and Northern Min.

The Gan varieties geographically close to Far Western Min, such as the Lichuan 黎川 variety, possess a very productive phonological process which utilizes tonal alternation to express diminutive meaning (Yan 1993). Some scholars (e.g., Chen 1993; Chang and Wan 1996) argue that a "diminutive tone" like that of Gan is the underlying cause of unusual tonal splits in Far Western Min. However, this argument

Table 9 Examples of *t^h>h and *ts^h>t^h in Gan, FW.Min, and N.Min (adapted from Shen 2019: Table 1)

Glossary	Gan		FW.Min		N.Min	
	NC	LC	SC	GZ	JO	JY
桃 'peach'	t^h	h	t^h	h	t^h	h
豆 'bean'	t^h	h	t^h	h	t	t
茶 'tea'	ts^h	t^h	ts^h	t^h	t	t
菜 'vegetable'	ts^h	t^h	ts^h	t^h	ts^h	t^h

is problematic upon further investigation. Although some diminutive forms are indeed found in Far Western Min, the quantity of such forms varies substantially across different varieties, as shown in Table 10.

As can be seen in Table 10, Guangze (GZ), the geographically closest variety to the Gan-speaking area, possesses a large amount of forms featuring diminutive tones (highlighted by grey), while Shunchang (SC), the furthest variety from Gan, possesses almost no such forms. Therefore, the amount of diminutive forms is strongly correlated with geographical location: the closer to the Gan-speaking area, the more forms are found. Furthermore, these forms do not form regular correspondences across different Far Western Min varieties. This is very different from the pattern of tonal splits found consistently in all Far Western Min varieties, as shown below.

All the words in Table 11 belong to the Middle Chinese Level Tone (*Ping sheng* 平声) and have voiced initials. Therefore, they belong to the same tonal category in most modern Sinitic varieties. However, in all Far Western Min varieties, such as

Table 10 Diminutive forms in Far Western Min varieties. (Adapted from Shen 2018: Table 8; The diminutive forms are highlighted by grey)

Glossary		GZ	HP	GT	SC
蚁	'ant'	nie.7	nie.7	ŋie.7	ŋe.5
虾	'shrimp'	ha.7	ha.7	ha.7	ha.5(?)
饼	'cake'	piaŋ.7	piaŋ.7	piaŋ.7	piaŋ.3
龟	'turtle'	kuei.7	-	ky.1	ky.1
鸟	'bird'	tiau.7	tieu.3	tiu.3	-
衣	'raincoat'	i.7	i.1	i.1	i.1

Table 11 Tonal splits in FW.Min and the corresponding forms in Northern Min. (Based on Akitani 2013; see appendix for data and source)

	Glossary	FW.Min		N.Min		
		HP	GT	LJ	SB	JY
A	瓶 'vase'	pʰen²	pʰãi²	bĩ²	baiŋ²	vaiŋ⁹
	长 'long'	hoŋ²	tʰoŋ²	daõ²	dɔŋ²	loŋ⁹
	球 'ball'	kʰiu²	kʰiu²	giu²	giu²	kiu⁹
B	爬 'to climb'	pʰa²	pʰa²	ba²	pa⁵	pa²
	藤 'vine'	hen²	tʰãi²	dẽ²	taiŋ⁵	taiŋ²
	晴 'clear (weather)'	tʰaŋ²	tsʰiaŋ²	dzã²	tsaŋ⁵	tsaŋ²
C	藻 'duckweed'	pʰieu⁷	pʰiau⁷	pʰiao⁵	pʰiau⁵	pʰio²
	虫 'insect'	hŋ⁷	tʰuŋ⁷	tʰəŋ⁵	tʰəŋ⁵	huŋ²
	- 'field'	tʰen⁷	tsʰãi⁷	tsʰẽ⁵	tsʰaiŋ⁵	tʰaiŋ²

Heping (HP) and Gaotang (GT), a tonal split AB vs. C is found. Note that this tonal split is clearly different from the sporadic forms with diminutive tone shown in Table 10. Also note that the AB/C tonal split is different from the A/BC split found in the majority of Northern Min varieties, as represented by Shibei (SB) and Jianyang (JY). Importantly, the Linjiang (LJ) Northern Min variety newly reported by Akitani (2013) points to an important intermediate stage that links the two tonal split patterns found in Far Western and Northern Min, respectively: Akitani argues that the A/BC pattern in Linjiang (LJ) may be seen as a development from an earlier Shibei-type stage, one in which unaspirated voiceless stops have merged into voiced stops in terms of both VOT (voicing onset time) and tonal category (*p>b, *Tx>T2). Subsequently, this Linjiang-type A/BC pattern in Linjiang (LJ) further developed into the parallel A/BC pattern of Far Western Min via aspirated devoicing (*b>p^h). Therefore, the tonal split of Far Western Min can be explained in terms of a two-stage sound change from the pattern seen in Northern Min, but not by Gan-style diminutive tones.

Discussion

In the previous sections, we have examined four possible candidates which could in theory bear a close genetic relationship to Min: Wu (particularly Southwestern Wu), Hakka, Xianghua, and Gan. It has been demonstrated that while all of them share certain lexical and/or phonological features with Min, they are of different natures.

The shared features between Gan and Min have been found to involve contact-induced influence within a geographically continuous area, where changes implicate Gan, Far Western Min, and sometimes Northern Min varieties as well. Therefore, these features should be regarded as constituting an "areal sound pattern" (Blevins 2017). The lexical features shared by Min and Xianghua are all of Old Chinese origin, and phonological features are most likely due to retention rather than innovation. Such commonalities only indicate that both groups are conservative and have preserved many archaic features inherited from Old Chinese. On the other hand, the lexical and phonological features shared by Min and Southwestern Wu, and also by Min and Hakka, cannot be attributed solely to inheritance from Old Chinese. At least part of these features are shared innovations.

However, the picture becomes much more complex if we consider that most of the lexical and phonological features shared by Min and Southwestern Wu are different from those shared by Min and Hakka. This makes a representation of the interrelationship between the three groups very difficult, if not entirely impossible. One of the possible solutions is related to the "convergence hypothesis" of Min proposed by Norman and Coblin (Coblin 2018), which argues that Inland Min and Coastal Min are originally two separate genetic groupings which became similar due to convergence at a later stage. This hypothesis is proposed based on evidence that Inland Min and Coastal Min show considerable differences in terms of lexicon and comparative phonology (see the section "Classification of Min" and references therein).

The "convergence hypothesis" is supported by the demographic historical study of Bielenstein (1959), which shows that the early Chinese-speaking population colonized Fujian by various routes. The first wave was from Southwestern Zhejiang into Northern Fujian (191–205 A.D.); the second wave was from Eastern Jiangxi into Far Western Fujian (ca. 260 A.D.); and the third wave was from the Wenzhou area to Eastern Fujian (post-280 A.D.). A map illustrating these migration routes can be found in Coblin (2018: 111). Ge (1995) provides additional evidence in support of Bielenstein's (1959) migration routes and argues that most of these immigrants were from the nearby Zhejiang and Jiangxi regions rather than remote areas of Northern China. According to these migration routes, the isolation of the coast from the inland was not broken until the seventh century (Bielenstein 1959: 109), indicating separate linguistic histories for Inland Min and Coastal Min before that time.

If Inland Min and Coastal Min indeed have gone through different linguistic histories, we should expect that Inland Min bears a close relationship with Hakka, now spoken in Southern Jiangxi and nearby areas, while Coastal Min bears a close relationship with Southwestern Wu, now spoken in Southern Zhejiang. Whether this is the case needs to be explored in future studies.

Appendix: Data and Sources

Abbreviation	Location		Source
CZ	Chaozhou	潮州	Beijingdaxue (2005)
FZ	Fuzhou	福州	Beijingdaxue (2005)
GT	Gaotang	高唐	Norman (p.c.)
GZ2	Guzhang	古丈	Wu and Shen (2010)
HP	Heping	和平	Norman (1992)
JO	Jianou	建瓯	Beijingdaxue (2005)
JY	Jianyang	建阳	Norman (1969, 1986)
LJ	Linjiang	临江	Akitani (2013, p.c.)
MX	Meixian	梅县	Beijingdaxue (2005) & Norman (1986)
QY	Qingyuan	庆元	Cao et al. (2000)
SB	Shibei	石陂	Akitani (2008)
SC2	Suichang	遂昌	Cao et al. (2000)

References

Akitani, H. (秋谷裕幸). 1993. 客家語における微韻唇音字 [Rhyme Wei with bilabial initials in Hakka]. *Chuugoku Gogaku* 中国語学 [Bulletin of the Chinese Language Society of Japan] 240: 11–20.

———. (秋谷裕幸). 1995. 閩·客家語の「泉」音について [On the pronunciation of 'Spring' in Min and Hakka]. *Chuugoku Bungaku Kenkyuu* 中國文學研究 [Bulletin of the Chinese Literature Studies] 21: 68–83.

———. (秋谷裕幸). 1996. 福建省連城縣文亨鄉崗尾村方言の系統論上の位置 [The genetic position of Wenheng Gangwei dialect in Liancheng, Fujian]. *Chuugoku Gogaku* 中国語学 [Bulletin of the Chinese Language Society of Japan] 243: 153–162.

———. (秋谷裕幸). 1999. Ye tan Wuyu Chuqufangyan zhong de Minyu chengfen 也谈吴语处衢方言中的闽语成分 [Another look at Min features in Chuqu Wu dialects]. *Yuyan Yanjiu* 语言研究 [Studies in Language and Linguistics] (1): 114–120.

———. (秋谷裕幸). 2000. Wuyu Chuqufangyan zhong de Minyu ci 吴语处衢方言中的闽语词——兼论处衢方言在闽语词汇史研究中的作用 [Min words in Chuqu Wu: On the use of Chuqu dialects in the study of Min lexical history]. *Yuyan Yanjiu* 语言研究 [Studies in Language and Linguistics] (3): 99–106.

———. (秋谷裕幸). 2008. *Minbeiqu sanxianshi fangyan yanjiu* 閩北區三縣市方言研究 [Studies on Northern Min dialects: Shipi, Zhenqian, and Dikou]. Language and linguistics monograph series A12–2. Taipei: Institute of Linguistics, Academia Sinica.

———. (秋谷裕幸). 2013. Minbeiqu Pucheng Linjiang fangyan he Shaojiangqu Guangzezhaili fangyan de guzhuopingsheng fenhua 闽北区浦城临江方言和邵将区光泽寨里方言的古浊平声分化 [Tonal split of Middle Chinese Ping Sheng in Pucheng Linjiang Dialect of Northern Min and Guangzezhaili Dialect of Shaojiang]. 太田斎・古屋昭弘两教授还历纪念中国语学论集 ed. 中国語学論集刊行会 Chūgoku Gogaku Ronshū Kankōkai, 310–319. Publisher: 好文出版, Tōkyō : Kobun Shuppan.

Akitani, H. (秋谷裕幸) and M. Nohara (野原将挥). 2019. Shanggu chunhuayuanyin jiashuo yu minyu 上古唇化元音假说与闽语 [The rounded-vowel hypothesis in Old Chinese and Min dialects]. *Zhongguo Yuwen* 中国语文 [Studies of the Chinese Language] 1: 15–26.

Ballard, W.L. 1981. Aspects of the linguistic history of South China. *Asian Perspectives* 24 (2): 163–185.

———. 1992. Lexical variation in Southern wu dialects and the Princeton hypothesis revisited twenty years later. *Acta Linguistica Hafniensia* 24 (1): 5–32.

Baxter, W.H. 1995. Pre-Qieyun distinctions in the Min dialects. In *Papers from the first international symposium on languages in Taiwan*, ed. International Symposium on Languages in Taiwan, 393–406. Taipei: Crane Publishing.

Baxter, W.H., and L. Sagart. 2014. *Old Chinese: A new reconstruction*. New York: Oxford University Press.

Beijingdaxue. (北京大学中文系语言学教研室). 2005. Hanyu fangyan cihui 汉语方言词汇 [The lexicon of Chinese dialects]. Beijing: Yuwen Chubanshe.

Bielenstein, H. 1959. The Chinese colonization of Fukien until the end of the T'ang. In *Studia Serica Bernhard Karlgren Dedicata*: Sinological studies dedicated to Bernhard Karlgren on his seventieth birthday, ed. Søren Egerod and Else Glahn, 98–122. Copenhagen: Ejnar Munksgaard. Bielenstein

Blevins, J. 2017. Areal sound patterns: From perceptual magnets to stone soup. In *The Cambridge handbook of areal linguistics*, ed. A. Grant, 88–121. Cambridge: Cambridge University Press.

Branner, D.P. 1995. A Gutyan Jongbao dialect notebook. In *The Yuen Ren society treasury of Chinese dialect data I*, 243–338. Seattle: Yuen Ren Society.

———. 1996. A Gerchuan Juyu dialect notebook. In *The Yuen Ren society treasury of Chinese dialect data II*, 289–349. Seattle: Yuen Ren Society.

———. 2000. *Problems in comparative Chinese dialectology: The classification of Miin and Hakka*. Berlin/New York: Mouton de Gruyter.

———. 2001. Min. In *Facts about the world's major languages*, ed. Jane Garry and Carl Rubino, 151–157. New York/Dublin: H. H. Wilson Company.

Cao, Z., Akitani, H., Ôta, I., and R. Zhao. 2000. *Wuyu Chuqu fangyan yanjiu* 吳語處衢方言研究 [Studies on the Chuqu subgroup of Wu dialect]. Tokyo: Kohbun.

Chang, S.H. (张双庆), and B. Wan (万波). 1996. Cong Shaowu fangyan jige yuyan tedian de xingzhi kan qi guishu 从邵武方言几个语言特点的性质看其归属 [The affiliation of Shaowu dialect: Evidences from some linguistic features]. 语言研究 Studies in language and linguistics (1): 3–17.

Chen, Z.M. (陈忠敏). 1993. Shaowu fangyan rushenghuazi de shizhi 邵武方言入声化字的实质 [The nature of Rushenghua in Shaowu dialect]. 中央研究院历史语言研究所集刊 [Bulletin of the Institute of History and Philology] (63.4): 815–830.

Chen, Z.T. (陈章太), and R.L. Li (李如龙). 1991. *Minyu yanjiu* 闽语研究 [A study of the Min dialects]. Beijing: Yuwen Chubanshe.

Coblin, W.S. 2018. Convergence as a factor in the formation of a controversial common min phonological configuration. *Yuyan yanjiu jikan* 语言研究集刊 [Bulletin of Linguistic Studies] 21: 79–122.

Ding, P.S. 2015. *Southern Min (Hokkien) as a migrating language: A comparative study of language shift and maintenance across national borders*. Singapore: Springer.

Ge, J.X. (葛剑雄). 1995. Fujian zaoqi yiminshi bianzheng 福建早期移民史实辨正 [A reexamination of early migration history of Fujian]. 复旦学报(社会科学版) [Fudan Journal (Social Science Edition)] (3): 165–171.

Handel, Z. 2010. Old Chinese and Min. *Chuugoku Gogaku* 中国語学 [Bulletin of the Chinese Language Society of Japan] 257: 34–68.

Inoue, F. 2018. Continuum of Fujian language boundary perception: Dialect division and dialect image. *Dialectologia: revista electrònica* 20: 147–180.

Jiang, M.H. (江敏華) 2003. *Kegan fangyan guanxi yanjiu* 客贛方言關係研究 [A study of the relationship between Hakka and Gan]. 國立臺灣大學中國文學研究所博士論文 [Ph.D. Dissertation, Department of Chinese Literature, National Taiwan University].

Karlgren, B. 1954. *Compendium of phonetics in ancient and archaic Chinese*. Göteborg: Elanders Boktryckeri Aktiebolog.

Kwok, B.C. (郭必之). 2004. Cong Yu Zhi liangyun tezi kan Yuefangyan gen Gujiangdongfangyan de lianxi 從虞支兩韻"特字"看粵方言跟古江東方言的聯繫 [Evidence for an ancient Jiangdong layer in the Yue dialects as revealed in "special words" of the Yu and Zhi rhyme]. *Language and Linguistics* 5: 583–614.

———. 2018. *Southern min: Comparative phonology and subgrouping*. Abingdon: Routledge.

LaPolla, R.J. 2001. The role of migration and language contact in the development of the Sino-Tibetan language family. In *Areal diffusion and genetic inheritance: Case studies in language change*, 225–254. Oxford: Oxford University Press.

Lei, B.C. (雷伯长). 1984. Shuo Shaowu fangyan 说邵武方言 [On Shaowu dialect]. 语言研究 [Studies in Language and Linguistics] 2: 144–147.

Li, R.L. (李如龙). 2002. Minfangyan zhong de Guchuyu he Guwuyu 闽方言中的古楚语和古吴语 [Old Chu and Old Wu preserved in Min dialects]. 福州: 闽语国际学术讨论会论文.

Lien, C. (连金发). 2015. Min languages. In *The Oxford handbook of Chinese linguistics*, 160–172. Oxford: Oxford University Press.

Long, A.L. (龙安隆). 2010. Fujian Shaowu fangyan Zhuopingruhua de xingzhi 福建邵武方言浊平入化的性质 [The nature of Zhuopingruhua in Fujian Shaowu dialect]. 方言 [Dialect] 4.

Mei, T.L. (梅祖麟). 2001. Xiandai Wuyu he 'Zhizhi Yuyu, gongweibuyun' 现代吴语和 "支脂鱼虞, 共为不韵" [The survival of two pairs of Qieyun distinctions in Southern Wu dialects]. *Zhongguo Yuwen* 中国语文 [Studies of the Chinese Language] 1: 3–15.

———. (梅祖麟). 2015. Shishi "Yanshijiaxun" li de 'Nanran Wu Yue, bei za Yi Lu' 試釋《顏氏家訓》裡的『南染吳越, 北雜夷虜』——兼論現代閩語的來源 [the "Wu dialect" of southern dynasties and the origin of modern Min; Plus an exegesis of Yan Zhitui's dictum, "the south is tainted by Wu and Yue features, and the North is intermixed with barbaric tongues of Yi and Lu"]. *Language and Linguistics* 16(2): 119–138.

Nakanishi, H., and B.C. Kwok. 2009. Evolution of the initial consonants in the She language induced by contact with Hakka. *Journal of Chinese Linguistics* 37 (2): 207–226.

Nichols, J. 1992. *Linguistic diversity in space and time*. Chicago: University of Chicago Press.

Nohara, M. (野原将挥), and H. Akitani (秋谷裕幸). 2014. Yetan laizi shanggu *ST- de shumuzi 也谈来自上古*ST- 的书母字 [A discussion on the shu initial derived from Old Chinese *ST-]. *Zhongguo Yuwen* 中国语文 [*Studies of the Chinese Language*] 4: 340–350.

Norman, J.L. 1969. *The Kienyang dialect of Fukien*. Ph.D. dissertation, University of California, Berkeley.

———. 1973. Tonal development in Min. *Journal of Chinese Linguistics* 1: 222–238.

———. 1979. Chronological strata in the Min dialects. *Fang yan* 方言 [Dialect] 4: 268–274.

———. 1982. The classification of the Shaowu dialect. 中央研究院歷史語言研究所集刊 [Bulletin of the Institute of History and Philology] (53): 543–583.

———. 1983. Some ancient Chinese dialect words in the Min dialects. *Fangyan* 方言 [Dialect] (3): 202–211.

———. 1986. *What is a Kejia dialect? Paper presented at the second international conference on sinology*. Taipei: Academia Sinica.

———. (罗杰瑞). 1990. Jiangshan fangyanzhong leisi Minyu de chengfen 江山方言中类似闽语的成分 [The Min-like features in Jiangshan Dialect]. *Fangyan* 方言 [Dialect] (4): 7–10.

———. 1991. The Mǐn dialects in historical perspective. *Journal of Chinese Linguistics Monograph Series* 3: 323–358.

———. 1992. A glossary of the Herpyng dialect. In *The Yuen Ren society treasury of Chinese dialects*, 1–107. Seattle, WA: Yuen Ren Society for the Promotion of Chinese Dialect Fieldwork.

Norman, J.L., and T.L. Mei. 1976. The Austroasiatics in ancient South China: Some lexical evidence. *Monumenta Serica* 32 (1): 274–301.

O'Connor, K.A. 1976. Proto-Hakka. アジア・アフリカ言語文化研究 [Journal of Asian and African Studies] 11: 1–64.

Pan, W.Y. (潘悟云). 1995. Wen Chu fangyan he Minyu 温、处方言和闽语 [Wenzhou Wu, Chuqu Wu and Min]. In *Wuyu he Minyu de bijiao yanjiu* 吴语和闽语的比较研究 [A comparative study of Wu and Min], 100–121. Shanghai: Shanghai Educational Press.

Ratliff, M. 1998. Ho Ne (She) is Hmongic: One final argument. *Linguistics of the Tibeto-Burman Area* 21 (2): 97–109.

Sagart, L. 1993. Les dialectes gan. *Revue bibliographique de sinologie* 11: 261–262.

———. 2000. Two Jiangxi sound changes in spatial perspective: The case of th- > h- and tsh- > th-. *In Memory of Professor Li Fang-Kuei: Essays of Linguistic Change and the Chinese Dialects*. Language and linguistics monograph series W1. Taipei: Academia Sinica.

———. 2002. Gan, Hakka and the formation of Chinese dialects. In *Dialect variations in Chinese*, pp.129–154. Taipei: Academia Sinica.

Shen, R.Q. (沈瑞清). 2018. Zaoqi Shaojiang fangyan xiangyin shengmu de shengdiao yanbian: Chongfang Luojierui xiansheng de Shaowu jiashuo 早期邵将方言响音声母的声调演变——重读罗杰瑞先生的"邵武假说" [The tonal change of voiceless sonorants in Early Shaojiang]: Jerry Norman's 'Shaowu hypothesis' revisited]. 语言研究集刊(第二十一辑) [Bulletin of Linguistic Studies (Volume 21)], 140–174.

———. 2019. Spatial diffusion and local motivation of sound change, paper presented at the 1st conference of Geolinguistic Society of Japan, Aoyama Gakuin University, Tokyo, Japan, 5–6 Oct 2019.

Tao, H. (陶寰). 2018. Wu Min Yu Yun, Xia Mu de duyin he Minyu quanzhuoshengmu de qinghua 吴闽语云、匣母 的读音和闽语全浊声母的清化 [The pronunciation of the Yun (云) and Xia (匣) initials in Wu and Min dialects and the devoicing of Middle Chinese voiced initials in Min dialects]. *Zhongguo Yuwen* 中国语文 3: 335–350.

Ting, P.H. 1983. Derivation time of colloquial Min from Archaic Chinese. *Bulletin of the Institute of History and Philology* 54 (4): 1–14.

———. (丁邦新). 1988. Wuyu zhong de Minyu chengfen 吳語中的閩語成分 [A Min substratum in the Wu dialects]. *Bulletin of the Institute of History and Philology Academia Sinica* 59(1): 13–22.

———. (丁邦新). 2006. Cong lishi cengci lun Wu Min guanxi 从历史层次论吴闽关系 [Historical Strata in the Wu and Min dialects]. *Fang yan* 方言 [Dialect] (1): 1–5.

Wu, Y. (伍云姬). 2006. The evolution of negative forms in the Hunan Waxiang dialects. Evolution, 52, 000117-1.

Wu, Y. (伍云姬), and R.Q. Shen (沈瑞清). 2010. *Xiangxi Guzhang Waxianghua diaocha baogao* 湘西古丈瓦乡话调查报告 [A linguistic survey of Guzhang Waxianghua in Western Hunan]. 上海教育出版社.

Wurm, S.A., R. Li, T. Baumann, and M.W. Lee, eds. 1987. *Language atlas of China*. Hong Kong: Longman.

Xu, H.L. 2007. *Aspect of Chaozhou grammar: A synchronic description of the Jieyang variety*. Hong Kong: The Chinese University of Hong Kong Press.

Yan, S. (颜森). 1993. *Lichuan fangyan yanjiu* 黎川方言研究 [A Study of Lichuan dialect]. 社会科学文献出版社.

Yan, X.H. (严修鸿). 2002. Liancheng fangyan yunmu yu Minyu xiangtong de cengci tezheng 连城方言韵母与闽语相同的层次特征 [The shared rhyme features between Liancheng dialects and Min]. In *Minyu yanjiu jiqi yu zhoubianfangyan de guanxi* 闽语研究及其与周边方言的关系 [The study of Min dialects and its relationship with other peripheral dialects]. Hong Kong: The Chinese University of Hong Kong.

Yeung, Y.M., and D.K. Chu. 2000. *Fujian: A coastal province in transition and transformation*. Hong Kong: Chinese University Press.

Yue-Hashimoto, A. 1976. Southern Chinese dialects: The Tai connection. *Computational Analysis of Asian and African Languages* 6: 1–9.

Zheng, X.F. (鄭曉峰). 2002. *Fujian Kuangze fangyan* 福建光澤方言 [The Guangze dialect of Fujian]. 新竹: 清華大學語言學研究所博士論文.

Zheng, W. (郑伟). 2015. The historical relationship between Northern Wu and Min dialects from the perspective of comparative phonology. *Journal of Chinese Linguistics*, 43(1): 119–149.

Zheng, Z. 2018. A new proposal for Min subgrouping based on a maximum-parsimony algorithm for generating phylogenetic trees. *Lingua* 206: 67–84.

Zhengzhang S.F. (郑张尚芳). 2002. Minyu yu Zhenanwuyu de shenceng lianxi 闽语与浙南吴语的深层联系 [A deep connection between Min and Southern Wu]. In *Minyu yanjiu jiqi yu zhoubianfangyan de guanxi* 闽语研究及其与周边方言的关系 [The study of Min dialects and its relationship with other peripheral dialects]. Hong Kong: The Chinese University of Hong Kong.

Part III

Meaning, Culture, and Translation

Meaning, Culture, and Translation: Introduction

18

Zhengdao Ye

Contents

Introduction 460
Chapter Summaries 460
Conclusion 462
References 463

Abstract

This chapter provides an overview of Section 3 of *The Palgrave Handbook of Chinese Language Studies*. The selection of seven studies investigates how the Chinese meaning universe intersects and interacts with other linguistic and cultural traditions through a range of media, such as subtitling, and across various domains, such as political, legal, and literary. Placing Chinese languages in a global context and examining how their meaning and acts of meaning-making interact with other linguacultural spheres through the lens of translation, the studies in this section broaden our understanding of the possibilities and challenges in the global circulation of texts, ideas, and signs.

Keywords

Meaning and culture · Cultural keywords · Kinship terms and kinship relations · Translation of children's literature · Legal translation · Subtitling and multilingual cinema

Z. Ye (✉)
School of Literature, Languages and Linguistics, The Australian National University, Canberra, Australia
e-mail: Zhengdao.Ye@anu.edu.au

© The Author(s), under exclusive licence to Springer Nature Singapore Pte Ltd. 2022
Z. Ye (ed.), *The Palgrave Handbook of Chinese Language Studies*,
https://doi.org/10.1007/978-981-16-0924-4_36

Introduction

While Part I and Part II deal with the structural and sociolinguistic aspects of Sinitic languages, the seven chapters in this section investigate ways in which the Chinese meaning universe intersects and interacts with other linguistic and cultural traditions. It does so through a range of media, such as subtitling, and across various domains, such as political, legal, and literary.

Chapter Summaries

It is widely held among students of language and culture that one of the most fruitful ways of understanding a people's culture is through systematic analysis of the meaning of the keywords of that culture (e.g., Williams 1983; Wierzbicka 1997, 2006; Cassin 2014; Levisen and Waters 2017). The first two chapters represent two different approaches to analyzing cultural keywords. The chapter by James Underhill and Mariarosaria Gianninoto is firmly grounded in the ethnolinguistic tradition associated with Wilhelm von Humboldt (1767–1835), while the chapter by Paweł Kornacki adopts the Natural Semantic Metalanguage (NSM) approach and shows how it can tackle the issue of untranslatability.

Underhill and Gianninoto show how words are part of worldviews (*Weltansicht*) and how worldviews tend to change when new words are adopted into a linguistic community. What they call "migrating meanings" introduces a new and dynamic way of looking at words and meanings when they enter into another culture. More specifically, this chapter examines the ways in which the concept of "the citizen" develops within the Chinese linguacultural sphere over time. Based on a careful examination of collocational patterns, phraseological units, and idiomatic forms across a large collection of examples drawn from varied texts, the authors follow the new meanings that emerge, and the associations that *gōngmín* 公民 takes on. They argue that, in many respects, the Chinese words *gōngmín* 公民 and *guómín* 国民 are colored by foreign meanings. This study challenges the idea that words and ideas are simply translated from one language to another or simply transplanted into a new culture; cultures constantly evolve and reinvent themselves through their contact with foreign ideas. However, this study also shows that scrutinizing migrating meaning across cultures can potentially help pinpoint what concepts are shared by all people.

By definition, cultural keywords, being situated at the heart of a culture, are resistant to translation. Kornacki's chapter represents another method that deals with the issue of untranslatability, the Natural Semantic Metalanguage (NSM) approach. It first undertakes a review of the works on Chinese cultural keywords in sinology, cultural anthropology, and cultural and cross-cultural psychology. It then demonstrates how the method can be employed to render and represent culture-specific meanings in a neutral, culture-independent way with two illustrative case studies: *xīngfú* 幸福 and *rènao* 热闹, the former relating to the domain of emotion and the latter to sociality. The author shows that once the meaning is spelled out, the underlying cultural values and beliefs also become apparent. The chapter also considers directions for future studies in relation to a series of questions raised by Arthur Kleinman about the "quest for meaning" (Kleinman 2011).

Wendi Xue and Zhengdao Ye's chapter turns to one of the most important domains in Chinese culture – kinship relations. It provides the first major review of literature on Chinese kinship studies in linguistics and anthropology and discusses the commonality and differences in approaches to kinship semantics between Chinese and Anglophone scholarship. While giving primary focus to distinctive approaches to the meaning of kinship terms in Sinitic languages over the span of two and half-thousand years of Chinese history, the authors also situate their discussions within the broad context of cross-linguistic, typological research on kinship terms, with two methodological questions in mind. One concerns building semantic databases of kinship terms in Sinitic languages. The other relates to the issue of translatability, that is, how Chinese kinship terms can be precisely and authentically represented in comparative studies of world kinship systems. The authors point out that these two questions remain a priority in future research of Chinese kinship semantics, and that they require a combination of methods and tools.

The next two chapters concern two areas of translation, the highly specialized area of legal translation and the translation of children's literature. It is commonly acknowledged that legal translation, especially between Chinese and English, is difficult; it is a complex and special type of linguistic activity involving mediation and crossover between languages and cultures, and above all between different legal systems. In her chapter on legal translation between Chinese and English, Deborah Cao explains what legal language entails and what makes Chinese legal language distinct. It explores and highlights some of the sources of difficulty and peculiar challenges in this type of translation activity, such as systemic differences in law and linguistic and cultural differences, and discusses their implications for practice with illustrative examples. The author maintains that, complex though it is, legal translation is a specific skill that can be acquired.

The chapter by Minjie Chen and Helen Wong is the first comprehensive study of the history of Chinese children's literature translated into English from the late Qing dynasty to the present. In the first part of the chapter, the authors examine the types of texts selected for translation, analyze the fluid relationship between the source and target text, and discuss ideological incongruity as a major barrier to importing children's literature from China to the Anglo world. It then reviews the breakthrough of Chinese children's literature in English translation during the first two decades of the twenty-first century, highlighting major authors, illustrators, titles, and international recognition. In the second part, the authors offer a survey of commercial and non-commercial agents that have facilitated an international network of authors, illustrators, translators, and publishers. They highlight international children's literature organizations, libraries, festivals, book fairs, academic institutions, translators' professional communities and initiatives, and the most active figures who have played important roles in raising the visibility of Chinese-language children's literature, promoting high-quality translations, and professionalizing the field of translation. The chapter offers insights into what it takes for Chinese literature to cross linguistic and cultural boundaries in the global circulation of ideas and texts.

The final two chapters deal with the fast-growing areas of audio-visual translation (AVT), a multimodal form of translation from spoken to written language. In her chapter on translating films about ethnic minorities in China into English, Haina Jin provides a rare insight into how the translation of multilingual cinema is dealt with when it involves Mandarin-Chinese and non-Sinitic languages spoken by China's ethnic minorities. The author provides an overview of the films produced in China about its ethnic minorities. As she notes, many of these films have been screened in international festivals or distributed outside China. However, as the author also points out, before the 1980s, films about Chinese ethnic minorities were made in Mandarin, which masked the multilingual reality. With increasing awareness of multilingualism and openness in filmmaking policies, filmmakers began to experiment in making films with or in minority languages. Multilingual cinema naturally poses challenges for translation. The author illustrates these with a Mongolian and Mandarin film with Mandarin and English subtitles. She investigates how bilingualism is reflected in the English translation, discusses strategies used by the translator, and analyzes the effects of such strategies in translating bilingual scenes. Jin argues that English translations of films about China's ethnic minorities afford an opportunity to broaden the current understanding of film translation, which calls for translators to be more aware of the languages and cultures of ethnic minorities, in order to develop effective solutions to preserve the multilingual features of those films.

Dingkun Wang's chapter surveys research and practices regarding the subtitling of television programs and films, primarily in the context of Mainland China. It defines the meaning and function of subtitling and provides a brief overview of the development of subtitling in the history of audiovisual translation (AVT). It also considers how technological progress constantly poses new challenges for those who perform subtitling work – for instance, they must comply with the spatial and temporal constraints of the audiovisual medium, while striving to render the information emanating from the multimodal convergence of signs in the source text in an effective and efficient way. The author examines how subtitlers recontextualize source materials for cross-cultural consumption by walking the line between covert translation (for comprehensibility) and overt translation (for authenticity). Finally, the chapter discusses directions for future research, including the potential for studies of the translation of music and *danmaku*-driven participatory subtitling work to explore the commonalities among the modes, types, and strategies of translation of audiovisual programming.

Conclusion

Two decades ago, the literary critic Lydia Liu wrote in the introduction to her edited book *Tokens of Exchange: The Problem of Translation in Global Circulations* that there was a new expectation growing out of the then-theoretical work on translation. She formulated the expectation as a series of questions as follows:

How do signs and meanings travel from place to place in global circulations? Can a theory of translation illuminate our understanding of how meaning-value is made or unmade between languages? Is translatability a value in itself or a product of repeated exchange and negotiation in the translation process? What do we stand to gain or lose when we take up a position for or against the commensurability of verbal or nonverbal signs in multilinguistic or multisemiotic situations? (Liu 1999, p. 2)

Placing Chinese languages in a global context and examining how their meaning and acts of meaning-making interact with other linguacultural spheres through the lens of translation, the chapters in this section, in one way or another, address these questions and broaden our understanding of the possibilities and challenges in the global circulation of texts, ideas, and signs.

References

Cassin, Barbara (ed.) 2014. *Dictionary of untranslatables: A philosophical lexicon*. Trans. and Ed. Emily Apter, Jacques Lezra & Michael Wood. Princeton: Princeton University Press.
Kleinman, Arthur. 2011. Quest for meaning. In *Deep China: The moral life of the person: What anthropology and psychiatry tell us about China today*, ed. Arthur Kleinman, Yunxiang Yan, Jing Jun, Sing Lee, Everett Zhang, Tianshu Pan, Wu Fei, and Jinhua Guo. Berkeley: University of California Press.
Levisen, Carsten, and Sophia Waters, eds. 2017. *Cultural keywords in discourse*. Amsterdam: John Benjamins.
Liu, Lydia H. 1999. Introduction. In *Tokens of exchange: The problem of translation in global circulations*, ed. L.H. Liu, 1–12. Durham/London: Duke University Press.
Wierzbicka, Anna. 1997. *Understanding cultures through their key words*. Oxford: Oxford University Press.
———. 2006. *English: Meaning and culture*. Oxford: Oxford University Press.
Williams, Raymond. 1983. *Keywords: A vocabulary of culture and society*. Oxford: Oxford University Press.

19

Migrating Concepts in Chinese

James Underhill and Mariarosaria Gianninoto

Contents

Migrating Meanings	466
Keywords and Keyword Approach	467
Ethnolinguistics and Worldviews	469
Methodology	471
A Case Study of Migrating Concepts	471
The Chinese Citizen: Trajectories of *Gōngmín* and *Guómín*	475
The Origins of the Chinese Words	475
Rival Meanings	478
The Return of the Citizen?	480
A Comparative Perspective	485
References	488

Abstract

This chapter introduces an ethnolinguistic account of keywords and how they transform the languages into which they are introduced. Relying on recent work in philology and philosophy, it shows how words are part of worldviews and how worldviews tend to change, when new keywords are adopted. However, the authors stress that keywords are interesting for this very reason; they help us unlock cultures. Migrating meanings introduces a new way of understanding borrowed words. The cultures that welcome new keywords redefine them, just as they are transformed by them. More specifically, this chapter shows the ways in which the concept of the citizen develops within the Chinese language. In many respects, the forms *gōngmín* 公民 and *guómín* 国民 are colored by foreign

---

J. Underhill (✉)
Rouen University, Rouen, France
e-mail: james-william.underhill@univ-rouen.fr

M. Gianninoto
Université Paul Valéry, Montpellier, France
e-mail: mariarosaria.gianninoto@univ-montp3.fr

© The Author(s), under exclusive licence to Springer Nature Singapore Pte Ltd. 2022
Z. Ye (ed.), *The Palgrave Handbook of Chinese Language Studies*,
https://doi.org/10.1007/978-981-16-0924-4_37

meanings. The authors adopt a Keyword Approach and investigate electronic corpora to analyze how words related to citizenship are assimilated into Chinese and reinvented in the Chinese worldview. Based on more than a thousand examples taken from academic articles, corpora, databases, and dictionaries, the words related to citizenship are considered from a historical perspective. This study challenges the idea that words and ideas are simply translated from one language to another or simply transplanted into a new culture. Citizen, in English, becomes a migrating meaning, and the authors strive to follow the new meanings that emerge and the associations that *gōngmín* 公民 takes on in Chinese.

Keywords

Migrating meanings · Keyword approach · Ethnolinguistics · Worldviews · Cross-linguistic research · Cross-cultural studies · Citizen · *gōngmín* 公民, *guómín* 国民

Migrating Meanings

It is common knowledge, of course, that words often come from other languages. Whole dictionaries are devoted to words of foreign origin. And philosophers, philologists, and linguists often find that the history of ideas is difficult to imagine without many keywords that are borrowed from abroad. In this sense, understanding who we are, where we are going, and what we are doing in society together is often tied up with an ongoing dialogue with other cultures and languages. For this reason, keywords that come from elsewhere – migrating meanings – are important keys to understanding culture.

The fate of migrating meanings is always fascinating for two reasons: the curious ways concepts must be adapted to make them understood and the way keywords take on new usages which partly redefine their original meanings. In one sense, migrating meanings are "borrowed." However, the very metaphor of "borrowing" begins to break down as soon as this process gets underway, because these migrating concepts set sail for a journey with no return. They are "taken on," but they can't be "given back."

This takes us to the core of comparative semantics and cross-lingual research. When we study other cultures, we ask what the world means for them. But when we study keywords that are borrowed, trying to understand these migrating meanings involves trying to grasp what they come to mean for us. Because in the course of migrating, and after they become established, they often take on unique new meanings in our language. In this sense, words build bridges: it's important to look across the bridge to the other side, to understand the word's origins. But then, it's important to look back at our own side and see how our own perspectives have been altered. In studying languages and cultures, it's crucial to understand people with their perspectives, their culture, their history, and their worldview. This is an encounter with otherness. But once keywords migrate, we begin to understand that our own perspectives, culture, history, and worldview are somewhat different from the lives of generations that lived without those keywords. Migrating meanings can entail semantic and cultural mutations.

When words are adopted, they take root in a very different worldview, and they take on very different associations. Lydia Liu sees this more as a process of reinvention than appropriation: "when knowledge passes from the guest language to the host language, it inevitably takes on new meanings in its new historico-linguistic environment; the translation remains connected with the original idea as no more and, perhaps no less, than a *trope of equivalence*. Everything else must be determined by the users of the host language" (Liu 1995: 60).

Often borrowed words, or words invented to translate foreign keywords, displace other rival words or force speakers to redefine or reorganize the definitions they have of existing concepts. As it will become clear in this case study, this was the case of the various terms adopted to define citizen and citizenship in Chinese.

Citizen may be a migrating meaning in Chinese: the people, on the other hand, is not. The people (*mín* 民 and later *rénmín* 人民) was central in Chinese political tradition, but Marxist-Leninist discourse introduced a German and a Russian version of "the people" to the Chinese worldview, adding new connotations. So these new meanings contributed to redefining how existing keywords were used; In what contexts? By what groups? For what reasons? Who was for the people? And who set themselves against the people? Who was part of the people? Who became an enemy of the people? It is clear that these new meanings are not being "borrowed"; they are being invented, created by a creative people rethinking how they should understand the relations between people and the relations between the people and the state and between the people of their nation and other peoples. Meanings migrate, but what those new definitions come to mean remains inevitably something of a mystery to the speakers of those cultures from which they are borrowed.

Investigating "migrating meanings" entails comparing and contrasting broadly similar concepts in several languages and then asking what happens to words that move from one linguistic tradition to another. As the keyword migrates, it takes on a new role in a new environment, and by competing with other words in that new context, the keyword is transformed and reconfigured. Once they leave home, keywords reinvent themselves.

Keywords and Keyword Approach

When we try to understand our own culture, when we try to explain it to other peoples, and when we try to translate what we do, and who we are, we need words, of course. And when we speak about what we do together in society, we use words. When we speak of ourselves as a people, as individuals, and as citizens of a given nation, we use cultural keywords. But as translators know, words are rarely innocent. They have their traditions, their origins, and their associations. And these often prove very difficult to translate.

This highlights the importance of keywords. We need concepts to understand society, but as we move from society to society, from culture to culture, and from language to language, we find that the definitions given to keywords vary. Keywords are therefore to be understood in two very different lights. Keywords can be studied as cultural, political, or sociological terms; or they can be considered as key concepts that enable us to gain insight into a given culture. The first understanding of keywords has a long tradition. For over a century, lexicographers and specialists of various fields have been proposing specialist dictionaries and glossaries. And this tradition arguably culminates in Raymond Williams' seminal text, *Keywords: A Vocabulary of Culture and Society* (Williams 1983).

Raymond Williams' *Keywords* inspired a generation of research in English-speaking cultures by demonstrating that keywords such as "nature," "private," and "democracy" are highly political (Williams 1983). These are the concepts that powerful players in society argue over and seek to appropriate for their own interests. Williams was a critical thinker and a Marxist. But his influence went far beyond political writing and transformed the way we think about the words we think with and the way those words change over time. Around the world, linguists have renewed their efforts to understand semantics and word use within politics and within communities working with the paradigm of keywords, Anna Wierzbicka, among many others.

Anna Wierzbicka is engaged in a critique of what she calls "the Anglo Worldview" (Wierzbicka 2014). Many dominant cultures have come to feel that their words and their concepts, their ideas, and their ideals are universal. This was the case of the French during the eighteenth century, and it is, in many respects, true of the North Americans today. But by examining a wide range of languages around the world, Anna Wierzbicka and her colleague Cliff Goddard have provided considerable evidence to support their conclusion that the number of conceptual universals does not exceed around 70 key terms (including pronouns, such as "I" and "you," and locations, such as "there" and "far," and quantifiers, such as "big" and "small"; for Mandarin Chinese, see Chappell 1994 and 2002). This is an important point for translation studies. This forces us to radically revise the idea we have of what cultures share and what we can exchange. But translators, interpreters, linguists, and anthropologists have long known that words do not coincide when we compare and contrast how they work in the lexicons of any two languages.

Another approach takes keywords to be fragments of the worldview that makes up a language and a culture. Studying how a people uses these keywords can help us to understand how a people understands itself, its role in the world, and the organization of its society, its members, and their relations. One of the highpoints in this tradition was Barbara Cassin's *Dictionary of Untranslatables: A Philosophical Lexicon*, published in the United States in 2014 and adapted from the original French edition published in 2004. Cassin's authors are concerned with understanding how the Greeks, and the Romans, and the French and the Germans understand the mind, the memory, madness, the state, the nation, the people, the individual conscience, and the collective will, among other concepts. Cassin's encyclopedia opens up a whole range of traditions and helps us understand how concepts are translated from one context to another as the various versions of the Western culture emerge and evolve.

Cassin's work combines a philological and philosophical approach and represents the culmination of a tradition in the history of ideas in France. This is why her encyclopedia has been translated into many languages and is making a considerable impact in the French and English editions. Lexical research has been transformed by the IT revolution, however, and the new technology that has been made available for studying language. And this means we have to rethink the resources we use for establishing what words mean to incorporate electronic linguistic corpora and other electronic texts and tools.

Originally, Williams and the scholars of his time during the second half of the twentieth century relied on dictionaries, encyclopedias, and glossaries of specialized terms, in addition to their own vast experience of the history of ideas and their erudition in discourse and social theory. Williams considered concepts first and words second. Words for the sociologist and the politician are the "means" of expression. But what if we inverse this and admit that without words, we cannot conceive of ideas? This is the challenge that the German philosopher Wilhelm von Humboldt (1767–1835) and ethnolinguistics face us with (see, e.g., Underhill 2009, 2011, 2012). Conceptualization depends upon formulating, defining, and negotiating the shared meanings we think with. Discourse analysis means studying the words we adopt in our discourse strategies and debates. This forms part of the semantic approach to the history of ideas. Rather than seeing words as terms for preexisting concepts, semantic studies are careful to place words within contexts. And ethnolinguistics places keywords within worldviews.

Ethnolinguistics and Worldviews

Ethnolinguistics is a field of study that investigates words and worldviews. In one sense, all words are keys to understanding a culture, not only the key terms Williams had in mind. But ethnolinguists and cultural thinkers like Williams tend to limit their studies of keywords to fundamental concepts that change over time. This enables them to use keywords as ways of unlocking what is different in other cultures and what changes over time. There is a long tradition in ethnolinguistics and linguistic anthropology in America inspired by W. Humboldt, including Franz Boas, Edward Sapir, Benjamin Whorf, and Dell Hymes. And Humboldt's ethnolinguistic tradition has stimulated a great deal of scholarship in Germany and France and around the world (see Trabant 2012; Underhill 2009, 2011, 2012, and the Rouen Ethnolinguistics Project, especially Trabant 2015).

Wilhelm von Humboldt (1767–1835), the great German philosopher of language, saw words as concepts that formed part of the dynamic fabric of the worldview of a people. The worldview (*Weltansicht*) evolves over time, as individuals work together to develop their minds and their shared consciousness as a people, a linguistic community (what Humboldt called *ein Nation*). Humboldt was equally interested in the destiny of nations, the shared consciousness of the linguistic community, and the expressive capacities of individuals. He was both a linguist and a politician. But he did not believe that the individual imposes his will on politics and politics does not impose itself on the people. Humboldt believed that language was at the center of all cultural enterprises and that we think and feel together in language or thanks to it. We develop together as a people and as individuals, and for this to happen, Humboldt maintained that intellectual freedom was crucial. For this reason, after the liberation from Napoleon's invading army in 1815, Humboldt accepted to set up the Prussian university to defend culture and free thinking. And above all, he was concerned with defending the mother tongue of the nation, German. Humboldt was one of the first ethnolinguists (Underhill 2009). And because he invented a philosophy of language, he asked questions that contemporary linguists are often unwilling to engage in: How does the mind develop in language? How does a linguistic community develop an identity and defend it in its language? How does translating works and borrowing words transform our ways of seeing things? These are the questions of migrating meanings that are fundamental to the approach developed in the present chapter (see also Underhill 2013; Underhill and Gianninoto 2019).

One of the most exciting and tenacious schools in contemporary ethnolinguistics today can be found in Poland, in Lublin, where the keywords of various European and Slavic languages are studied and compared. Jerzy Bartmiński's and Maciej Abramowicz's study of the people, for example, verified and established that French and Polish provide a largely analogous concept. As these two scholars point out (1998: 16–17), the words *lud* and *peuple* cover three main meanings: (a) a great group of humans that can be gathered together because they inhabit the same area or share the same laws, customs, or religion; (b) a mass of people, a crowd, or multitude; and (c) the lower social orders or classes of people without rights or privileges. Both French and Polish provide a concept that can be characterized as sharing four fundamental characteristics in the imagination: *lud-peuple* is (a). impersonal, (b). united, (c) passive, and (d). laborious.

However, the Lublin ethnolinguists were perplexed to come across a large number of contemptful references to the "peuple" in French that contrasted with a certain romantic idealism associated with *lud*, in the Polish worldview. In both languages, "the people" were often represented as an "object." "The people" was not made up of "individuals" or "citizens" in many French sources. The people formed an indivisible mass; it was a whole deprived of individual characteristics. This coincides with certain of our own findings concerning "the people" in English and "peuple" in French (Underhill and Gianninoto 2019). "The people" is often perceived from without. It is observed or looked down upon. Very often, the people are feared. The Lublin ethnolinguists were equally perplexed by a paradox in their findings: "the people" is often considered to have no will proper to itself. And yet, in both French and Polish, the people can be considered the source of great works. What this underlines is that keywords are cultural values, always complex, and ever-changing.

Like J. Bartmiński and M. Abramowicz, R. Williams, B. Cassin, and A. Wierzbicka are all aware of the degree to which keywords evolve over time. They understand that the people in the streets, in cafes, and at work will use keywords in ways very different from the way poets and politicians use them. This is part of the dynamic interactive process of discussing and debating meanings that forms part of social life and which makes each keyword specific to a given culture. There is little disagreement on this point in translation studies, linguistic anthropology, and ethnolinguistics. Meanings are shared and social; ultimately no one, or no group, owns them.

Methodology

In our own research (Underhill and Gianninoto 2019), it proved useful to combine the ethnolinguistic perspective with the keyword approach. And to enrich their marriage, findings from electronic corpus-based research were brought to bear. Furthermore, it appeared necessary to combine a synchronic and a diachronic perspective in order to explain what keywords mean at any one time, by following the trajectories of our migrating meanings and the way their usages change over time.

The migrating meanings study relied on six sources of information to enable us to form a reasonable idea of the way keywords were being used and the way their usage was being transformed.

- Dictionaries and encyclopedias
- Research articles
- Electronic corpora and databases
- Newspaper articles in paper and online form
- Dialogue and discussion with linguistic specialists
- Dialogue with native speakers of the languages we studied

These sources provided us with up to 1000 usages for each keyword in each of the languages studied (English, French, German, Czech, and Chinese), and this greatly nourished our reflection and inspired our research in comparing how our migrating meanings form keywords in different linguistic and cultural traditions focusing primarily on the socio-political lexicon.

According to Kipnis (2006), three dimensions are essential in "a keyword approach": the "linguistic history," the "genealogy of related discourses," and the "analysis of the contemporary socio-political context." Similarly, in our study of migrating meanings, we combined diachronic and synchronic perspectives, in order to retrace the trajectories of words and concepts and to compare contemporary usages of specific words in various discursive contexts.

A Case Study of Migrating Concepts

In what follows, we focus on the Chinese terms used for what is usually translated into English as the "citizen" within the socio-political lexicon. Unlike "the people," "citizen" is a representative example of a "migrating concept." As we shall see, many dimensions of the contemporary usage of keywords related to citizenship in China have been colored by European influence. Citizenship emerges as one mode of perceiving persons, individuals, and members of society. The contemporary words for citizen can – it is true – be found early in Chinese. Nonetheless, the meanings that migrate from Europe began to take on their present meaning in the late Qing period, under Japanese and Western influences (Guo 2011, 2013; Zarrow 1997: 3; Liu 1995: 25). In this case study, we will be analyzing the contemporary word for "citizen," starting with the early meanings attributed to this word, and we shall trace the way this word and its meanings develop. This will enable us to analyze contemporary usages of the word *gōngmín* 公民 ("citizen").

What Chinese sources were consulted? In order to gain a comprehensive overview of word trajectories and usages, we consulted various corpora and databases, for classical and modern Chinese. To gain an idea of the general usage of our keywords, we studied how our keywords were used in:

- The Chinese Text Project database (*Zhōngguó zhéxué shū diànzǐhuà jìhuà* 中国哲学书电子化计划, http://ctext.org/zh), developed by Donald Sturgeon. This is "the largest database of pre-modern Chinese texts in existence," bringing together "over thirty thousand titles and more than five billion characters." The sources included in this database cover many centuries and are extremely useful for identifying the first usages of the Chinese word forms.
- The 100 million-character corpus of Classical Chinese (*Gǔdài hànyǔ yǔliàokù* 古代汉语语料库 by the State Language Commission Corpora (国家语委语料库, http://www.cncorpus.org/ACindex.aspx)).
- The Modern Chinese Scientific Terminology database (hereafter abbreviated as MCST), collecting words coined in the nineteenth and early twentieth century "as equivalents of European and/or Japanese terms used in many different branches of knowledge" (http://mcst.uni-hd.de/helpMCST/wscdb.lasso).

For more recent usages, we relied upon the:

- Beijing Language and Culture University Corpus Center (BCC) corpora of Modern Chinese (*Běi yǔ hànyǔ yǔliàokù* 北语汉语语料库, http://bcc.blcu.edu.cn/hc). In particular, we consulted the 2 billion character newspaper corpus (报刊), the 3 billion character literary corpus (文学), the 3 billion character microblogging corpus (微博), and the 1 billion character general (综合) corpus.
- Peking University Center for Chinese Linguistics (CCL) corpus of Modern Chinese (http://ccl.pku.edu.cn:8080/ccl_corpus/). The Peking Centre Corpus with 581 million characters was compiled by the Institute of Linguistics of Peking University (*Běijīng dàxué hànyǔ yǔyán xué yánjiū zhōngxīn* 北京大学汉语语言学研究中心).

Corpus research has obvious advantages, enabling us today to gather together a vast amount of usages and to compile and study them in a systematic manner. This facilitates generating commonly used phrases and collocations and enables us to perceive more clearly the contexts in which keywords are used and the way they are harnessed in dialogue and debate. We can get a firm grasp on the words and ideas, impressions, and feelings associated with keywords in this manner.

Nonetheless, corpora have their shortcomings. They are assembled by groups with specific aims and interests: they are not neutral. The Corpus of Contemporary American (COCA), for example, provides a fantastic user-friendly corpus of up-to-date written and spoken American sources. But it will often provide examples from radical neoliberal sources or sources closely linked to the Republican Party, such as the Christian Monitor or Fox News. This can give us a skewered perspective and prevent us from gaining an accurate impression of the way many Americans express themselves and the way they think. If we want to understand words and worldviews, it therefore seems a good idea not to give up the more traditional means of analyzing discourse and political rhetoric. For this reason, we also set up our own corpora of texts and read widely to gain a more accurate all-round view of how our keywords work (Underhill and Gianninoto 2019: 16–17).

We wanted to ensure we had a large spectrum of Chinese sources in order to do justice to the diversity of ways in which keywords were used. For this reason, we made great use of the website of an influential Chinese newspaper, the *People's Daily* (*Rénmín wǎng* 人民网, http://www.people.com.cn), and a database for human and social sciences academic writings (http://www.nssd.org/). These resources have the advantage of enabling researchers to observe how words work in different contexts and the way words are used within ongoing debates.

Research on written texts is, of course, vital, but it does not replace discussion. Our initial comparative study of the keyword "people" was influenced by the Lublin School of Ethnolinguistics. The Lublin School distinguishes itself from other schools in cultural linguistics by working in a three-tier methodology. Its researchers work in a wide variety of Slavic and other European languages, with dictionaries and encyclopedias, texts, interviews, and questionnaires.

For our study, we had recourse to native informants and to specialists of the Chinese language and culture in both Asia and Europe, to verify our findings. Besides discussions with various native speakers, exchanges with specialists as Huáng Yáng 黄阳 from the School of Humanities of Southwest Jiaotong University in Chengdu (China), Giorgio Casacchia from Naples Oriental University (Italy), Yuán Zhōngjùn 袁中隽 (PhD student, Institute of Oriental Languages and Civilizations, Paris, France), Zhū Bīng 朱冰 (PhD student) from the Department of Applied Linguistics of Nagoya University (Japan), and Hilary Chappell from the School of Advanced Studies in the Social Sciences (Paris, France) all proved especially enlightening for our research on keywords.

The various angles of investigation in our methodological approach should not be seen as a series of steps: our research involved constantly alternating between various sources. We generated working hypotheses, and tried them out, and revised them as our various sources enabled us to refine our emerging impressions. We would present our findings to specialists, and in turn, we would return to corpora, texts, and encyclopedias to verify the affirmations of our native speakers. So, despite taking on new forms of interactive and electronic research, the traditional means of verifying meanings were systematically exploited. Dictionary entries were also taken into account, to form a general idea of common definitions and shared perceptions of words as they emerge over time in specific periods. We tried to retrace the trajectories of words and meanings by following dictionary definitions, consulting the reference etymological dictionary *Cíyuán* 辭源, and comparing the different editions of two reference dictionaries for Modern Chinese, the *Xiàndài hànyǔ cídiǎn* 现代汉语词典 and the *Xīnhuá cídiǎn* 新华词典. In this respect, we adopted the methodology of the Lublin school of "cognitive ethnolinguistics" (e.g., Bartmiński 2009:25). Dictionaries were studied in order to establish the key importance of:

- Collocations
- Phraseological units
- Idiomatic forms

Our research was also inspired by the previous research into Chinese keywords, undertaken from various perspectives. Many scholars have drawn inspiration from the kind of approach that was founded upon Williams' *Keywords* and applied it to Chinese. The detailed analysis of the word *sùzhì* 素质 ("quality") investigated by Andrew Kipnis (2006) explicitly adopts a keyword approach inspired by Williams. The volume *Words and Their Stories: Essays on the Language of the Chinese Revolution*, edited by Wang (2011), analyzes the keywords and key expressions of the Chinese revolution (such as "revolution," "rectification," and "socialist realism"), making reference to Williams as a source of inspiration (Wang 2011a: 1–2). Working within the framework of Anna Wierzbicka's and Cliff Goddard's Natural Semantic Metalanguage, Zhengdao Ye proposes an innovative "corpus-based lexical-conceptual analysis" to elucidate how meanings work in Chinese (Ye 2018). Other similar keyword approaches can be found in Ye (2013, 2014), Goddard and Ye (2016), and Kornacki (2017) among others, while words and expressions of Sinitic languages other than Mandarin Chinese have also been investigated, for instance, in Shanghainese (Ye 2012) and in Cantonese (Leung 2017).

There are also noteworthy publications that deal with specific words that are crucial in this specific historical context, although they do not adopt a keyword approach. One highly significant contribution to Chinese studies was made by the project *Language and Politics in Modern China* launched by Jeffrey Wasserstrom in the 1990s. Among the "working papers" published between 1993 and 1996 within the framework of this project are found essays on the specific concepts and significative words or expressions: on *guójiā* 國家 "nation," *yúlùn* 輿論 "public opinion" in late Qing China (Judge 1994), and the changing terminology used to designate state administrators in Republican China, from "lettered officials" (*wénguān* 文官) to "cadres" (*gànbù* 干部) (Strauss 1995). Similarly, in the two volumes edited by Lackner, Amelung, and Kurtz (2001) and Lackner and Vittingoff (2004) devoted to the lexical change in late Qing China, we can find essays on the trajectories of key socio-political concepts: "liberty," "president," "right," and "power." For our own study of migrating meanings, Guo's study on the notion of "citizen" (Guo 2013), and Huang's study on the translation of "individualism" (Huang 2004), provided remarkable insights into those keywords (see Underhill and Gianninoto 2019).

The Chinese Citizen: Trajectories of *Gōngmín* and *Guómín*

The Origins of the Chinese Words

At the turn of the century in 2000, a series first appeared on Chinese television, *Gōngmín liángxīn* 公民良心, that might be translated as either "citizen conscience" or "civil conscience." It was directed by Wú Zǐniú 吴子牛 and Xīn Jiārùn 辛嘉润. And it told the story about a common citizen and his courageous struggle to uncover the scandalous corruption related to a flood control embankment. That citizen considered it his vocation to protect the lives of his fellow citizens. The title of this successful TV series contains the contemporary word *gōngmín* 公民 "citizen" and demonstrates clearly the positive connotations associated with this word. Citizenship is, therefore, not only a political concept, a question of state governance, and a crucial question for individuals in society, but it forms part of the imagination and the entertainment of contemporary people in China, it would seem.

How is the keyword formed and what are its roots? This form, *gōngmín* 公民 (formed with *gōng*, meaning "public," and *mín* "the people" or "one of the people"), has a long history and can be found in early sources. It appears in the chapter "The Five Vermins" (*Wǔ dù* 五蠹) of the *Hán Fēizi* 韩非子 (III BCE), in which we can read:

> 私安则利之所在, 安得勿就?是以公民少而私人眾矣。[1]
> If they can obtain anything so profitable as private security, how can you expect them not to resort to such measures? Hence men who are concerned with public welfare grow fewer, and those who think only of private interests increase in number. (Watson transl. 2003: 116)

In this passage, *gōngmín* 公民 is translated as "men who are concerned with public welfare," as opposed to *sīrén* 私人 "those who think only of private interests." For Hán Fēizi, citizenship is a question of caring for the public interest rather than focusing on self-interest. Judge (1994: 3) claims that the word *gōng* 公 is "rich in classical resonances and multiple historical meanings" and that it is "conceived in a dichotomous relationship with *si* (meaning privateness and selfishness, wickedness or injustness [*buzheng*])." The word *gōng* 公 evokes notions of "openness and publicness." Selfishness and self-interest are doubtless universal. But *gōngmín* 公民 is a Chinese concept with Chinese roots. Goldman and Perry (2002: 4) add that "the origins of the term *gongmin* lie in the Confucian celebration of the public

---

[1] Chinese text retrieved from the Chinese Text Project database. http://ctext.org/hanfeizi/wu-du/zh

service." These semantic connotations still resonate within the contemporary word *gōngmín* 公民 "citizen."

However, the disyllabic word *gōngmín* 公民 was used to designate the citizen only in the late Qing period. Here we are dealing with a concept that becomes harnessed by new forces and remodeled when external influences begin reinventing an indigenous keyword. Guo Zhonghua (2013) states that "modern citizenship is to be considered the product of the European specific political culture" (现代公民身份被看作欧洲特殊政治文化的产物). Zarrow (1997:3) concurs with this view, claiming that: "thinking precisely and explicitly in terms of citizenship did not emerge until the late nineteenth century under Western influence." For this reason, the word *gōngmín* 公民 has been listed among the *wàiláicí* 外来词 "words of foreign origin" by Gao and Liu (1984: 121). A Chinese word comes to seem external when meanings migrate and the word resonating within the worldview comes to appear foreign to those who use it or encounter it.

How are we to understand the paradoxical status of migrating meanings? It would surely be legitimate to define *gōngmín* 公民 as a "return graphic loan" (Masini 1993), that is to say a form found in Classical Chinese, which was adopted in Japanese to designate a new concept and reintroduced with new connotations into Chinese. We are dealing with a complex process of migration that filters one culture through another before the keyword returns to take on new roots in the soil it has taken flight from. We find numerous such terms and concepts that can be ascribed to Western thought migrating between the end of the nineteenth and the beginning of the twentieth century. In fact, the Japanese language played a crucial role in East-West semantic exchanges and became a medium for introducing Western knowledge into China (Masini 1993: 104).

This is the first lesson we drew from considering how meanings migrate between Europe and China. Binary oppositions do little to help explain the subtlety of exchanges. When meanings migrate, we are invariably dealing with complex origins, a complicated passage, and complex repercussions. Keywords migrate, but worldviews must reintegrate keywords into existing structures and patterns of thinking and feeling. And often in passing from one worldview to another, the voyage transforms the voyager.

The transformation of *gōngmín* 公民 and the introduction of new facets of meaning were part of the massive process of the introduction of Western knowledge that took place in the late Qing period, a period in which China experienced a deep political, economic, and social crisis. Certain periods favor the migration of meanings, and the direction of migration depends on the powerplay of geopolitical and historical forces. During the Qing period – a period of transition – new disciplines and theories were introduced as tools to be used in an attempt to rescue the country (Hsü 1970). Conflicts created the need to find new means of defense, and semantics and keywords, like knowledge and ideas, are part of this process. As Lackner, Amelung, and Kurtz (2001: 2) underscore, within only a few decades, "the Chinese language absorbed, or indeed 'devoured' the nomenclatures of the most diverse branches of Western knowledge, the formation of which had taken millennia – including several periods of cross-cultural translation – in the occident."

As much as technology, words and works proved to be the crucial resources that were being absorbed. In particular, a massive process of translating Western works related to the social sciences (Tsien 1954: 327) took place at the turn of the century. It was in this context that several key terms from Western political thought were introduced to China.

Hence, the disyllabic form *gōngmín* 公民 acquired the new meaning of "citizen" as a graphic loan from the Japanese word, *kōmin* 公民. This word was used by the scholar Kang Youwei 康有为 (1858–1927), one of the leaders of the 1898 Hundred Day Reform Movement. Kang composed an essay entitled *Gōngmín zìzhì piān* 公民自治篇 "On Self-government by the Citizenry" (published in 1902). In that essay, he wrote: "人人有议政之权, 人人有忧国之责, 故命之曰公民" (see Fang and Wang 2006: 141) ("Since all the people have the right to participate in assemblies and they all have the responsibility to be concerned about their country, they are called citizens," transl. by Goldman 2005: 11).

The second lesson to be learned on contemplating migrating meanings is that when words are imported or transformed, they remain in competition with other keywords. They cover similar meanings, and for this reason, we consider these similar terms to be "rival synonyms." This proved to be the case for "citizen": the current word for citizen, *gōngmín* 公民, was not the only term in use, nor did it appear to be the most important word used to define citizenship in late Qing China. Both of these points have to be borne in mind when we compare the Chinese concepts with the French, German, or English counterparts that these Chinese keywords were used to translate (*citoyen*, *Bürger*, and citizen, respectively).

What were these rival synonyms? The word *guómín* 国民 (non-simplified form 國民) was widely used between the end of the nineteenth century and the beginning of the twentieth century, according to Guo (see Guo 2011: 145, 148). This word was formed with *guó* meaning "country" and *mín* "the people" or "one of the people," and it was strongly associated with a concern for Chinese sovereignty and national rescue, as Goldman and Perry underscore:

> in adopting the term *guomin*, late Qing and Republican-era elites revealed their preoccupation with asserting China's position vis-à-vis the foreign imperialist powers, rather than with ensuring the rights of individuals vis-à-vis the state. (Goldman and Perry 2002: 6)

There was no immediate consensus about which term would dominate. It was the word *guómín* that "entered state-sponsored textbooks [...] around the time of the 1911 revolution" (see Goldman and Perry 2002: 4). The form *guómín* is still used today to mean "citizen" or "national": besides, it already existed in ancient Chinese sources. This form was found in the *Chūnqiū Zuǒzhuàn* 春秋左传 "Zuǒ Qiūmíng's 左丘明 Commentaries [to the Spring and Autumns Annals]" (in the section dedicated to the thirteenth year of the duke Zhao; see Liu 1995: 308). In this passage, the form *guómín* 国民 was used to designate "the people of the state," according to J. Legge's translation (1872: 650):

私欲不違, 民無怨心, 先神命之, 國民信之

> He has not, to gratify himself, gone against the people. They have no feeling of animosity against him, and the Spirits formerly gave the appointment to him. The people of the State believe in him.

How was *guómín* 國民 colored by the forces that were acting upon *gōngmín* 公民? In some respects, the two keywords followed similar courses. As Liu highlights (1995: 308), *guómín* 國民 can also be considered a return graphic loan. Like *gōngmín* 公民, it was a Chinese form which acquired a new meaning in Japanese (*kokumin* 國民), before being reintroduced with this new meaning into Chinese. However, both Japanese and Chinese were influenced by this process of migrating meanings. The form *guómín* 國民, like the Japanese *kokumin* 國民, assumed various connotations during the nineteenth century under Western influence. The Japanese word *kokumin* 國民 was used in Japanese to designate the people, the nation, and the citizen, during the Meiji era (Burtscher 2012: 65).

In much the same way, *guómín* 國民 was progressively used to designate "citizen," even though the meanings of "the people," "people of a nation," and "nation" were also current throughout the nineteenth and early twentieth centuries, as sources clearly demonstrate according to our findings based upon the MCST database. Scholars confirm the importance of these new facets of citizenship and the importance of translations. Bastid-Bruguière (1997: 230–231) underlines that the form *guómín* 國民 was used by Liang Qichao to express J. K. Bluntschli's (1808–1881) German notion of "*Volk*" (people) in the essay "The doctrine of the political scientist Bluntschli" (*Zhèngzhìxué dà jiā Bólúnzhīlǐ zhī xuéshuō* 政治学大家伯伦知理之学说), published in 1903. Furthermore, Liu (1995: 47) points out that this word was used to translate the notion of "national character." Liu quotes two expressions for "national character" formed with *guómín*: *guómínxìng* 国民性 (*xìng* meaning "nature" or "quality") and *guómín de pǐngé* 国民的品格 (*pǐngé* meaning "moral qualities").

At this stage, it becomes clear that lexicon and word formation are crucial dimensions of cultural exchanges. Words are not uprooted and transplanted. Meanings do not grow like that. When meanings migrate, they must build new nests in their adopted homes, and that means making use of the resources at hand. The central term of "citizen" is caught up in attempts to counter foreign influence and consolidate the Chinese society and culture as a people and a nation. For this reason, in her work, Liu underlines that Liang Qichao reflected on the Chinese national character to identify "the cause of the evils responsible for the deplorable state of the Chinese *guomin* (citizen)" in the late Qing period (Liu 1995: 48).

Rival Meanings

The rival synonyms for citizenship are created within the context of these debates and negotiations. Some neologisms and keywords assert themselves and gain wide usage. Others disappear without a trace. Migrating meanings and the offspring they bear play their role in cultural exchanges and struggles, and their fates are unpredictable. It is worth pointing out, for example, that various other forms were used in the nineteenth and early twentieth century as equivalents of citizen: two forms used to designate the people, *mín* and *rénmín*, were also used to designate the citizen, but in the course of time, this meaning was lost, and the contemporary word *rénmín* now designates the people but not the citizen.

The form *mín* (the ancient word for "the people," meaning "one of the people") was used as an equivalent of "citizen" by W. A. P. Martin (1827–1916), in the *Wànguó gōngfǎ* 万国公法 (1864), his translation of the *Elements of International Law* (1936) by Henry Wheaton (see Wheaton, Martin trans. 2002: 25).

Rénmín 人民, the current word for "the people," and *píngmín* 平民 ("common people") were also considered equivalents of citizen in Médard's *Vocabulaire francais-chinois des sciences morales et politiques* (1927), and the word *chénmín* 臣民, litt. "subject," figures among the equivalents of "citizen" in the MCST database.

Creativity and overlapping meanings are part of language use and innovative attempts to refine and redefine terms and keywords. Polysemy, semantic instability, and overlexicalization – the coexistence of several rival synonyms – are not surprising consequences in this context of cultural transformation and reform. This is a context in which we can observe the evolution of knowledge and the adoption of radically new ideas and expressions. Keywords and terminology are part of this process of transformation. But there is something else at stake here.

These improvised new keywords reflect the complexity of the process of adopting and introducing new words. Above all, this highlights the difficulties involved in "negotiating equivalence" (Amelung, Lackner, and Kurtz 2001). Liu speaks of this in as "negotiating commensurability" (see Liu 2004: 131). These authors are concerned with negotiating commensurability between the Western-inspired notions and Chinese conceptualizations in the formative period of modern Chinese terminologies. So rather than a simplistic model of borrowing something, an exchange between two clearly defined partners, we are dealing with the intellectual, cultural, and communicative strategies that must create a third space, a new emerging space in which migrating meanings can mean something and make sense. When those new keywords begin to make sense, we have clearly moved into a new period in which we think and feel about our society and our culture and our language slightly differently.

Migrating meanings is a dynamic process. It involves individuals, authors, peoples, nations, political actors, economic actors, and the media. As such, it would be simplistic to think in terms of simple transformations: befores and afters. Western influence did arguably "transform" the concept of citizenship for the Chinese between the end of the nineteenth and the beginning of the twentieth centuries. But that transformation was in no way permanent or static. A whole new set of forces was to be set in play, and no one in the West or in the East, in Europe, or in China, could have foreseen the dramatic ways those new forces would affect society and culture and the ideas and cultural keywords that came to play a major role in how cultures see themselves.

The words for "citizen," which played an important role in reformist circles and in the intellectual debates between the end of the nineteenth century and first decades of the twentieth century, were destined to be overshadowed in the first years of the People's Republic of China, when the emphasis shifted from the citizen to the people and the masses (Keane 2001: 1). Here we see an entirely different form of competition in language and discourse. In this context, the rival concepts risked ousting the concept of citizenship as it had only recently come to be understood. The new roots of citizenship were weak, and it would prove relatively easy to uproot those newly coined meanings. It came to be considered that the concepts of "citizen" and "citizenship" evoked negative connotations in the new society. Because of their Western and bourgeois legacies, these concepts were to be downgraded and were confined for the most part to legal contexts (Tang 1986: 276). Consequently, the words for citizen became marginal in official and political discourse.

This represents a radical shift in meaning, and it goes hand in hand with the transformation of the Chinese society. Nonetheless, this new way of understanding peoples and individuals was not permanent. And it would be more accurate to consider citizen to be a dormant than a dead concept in this period. It had been marginalized, but it could make a comeback. Between the end of 1970s and the 1980s, a renewed interest in the concept of the citizen emerged, reflecting China's political and social changes. In the post-Mao era, the notions of citizen and citizenship are stressed more and more, and in particular it is the word *gōngmín* that is increasingly used.

The Return of the Citizen?

In the reference dictionary for contemporary Mandarin Chinese, the *Xiàndài hànyǔ cídiǎn* 现代汉语词典 "Dictionary of Modern Chinese" (2015, sixth edition), the word *gōngmín* 公民 "citizen" is defined as: "a person having or obtaining the nationality of a given country, enjoying rights and assuming obligations according to the Constitution and the laws of that country" (具有或取得某国国籍, 并根据该国宪法和法律规定享有权利和承担义务的人). In this lemma, we also find a reference to the other word for "citizen," *guómín* 国民, but there appears to be little difference, and the dictionary simply affirms that "in our country, [the words] *gōngmín* 公民 and *guómín* 国民 are similar in meaning" (在我国, 公民与国民含义相同). In the same dictionary (*Xiàndài hànyǔ cídiǎn*, 2015), the definition of *guómín* is, however, shorter than *gōngmín*. The definition refers only to nationality: "a person having the nationality of a country is a citizen [or a 'national'] of this country" (具有某国国籍的人是这个国家的国民).

It is worth stressing that in the 1997 edition of the *Xīnhuá cídiǎn* 新华词典 (hereafter abbreviated as XHCD) "Dictionary of new China," the entry *gōngmín* 公民 contained a reference to the different rights and obligations of the citizens of different countries: "According to the various countries' social systems, the determination of the rights and obligations of the citizen show substantial differences" (不同社会制度的国家,对公民权利和义务的规定,有本质的不同 (XHCD

1997: 297). However, this reference was removed from the 2001 edition of the same dictionary (XHCD 2001: 326). In this edition of the *Xīnhuá cídiǎn* 新华词典, the entry *guómín* 国民 is contrasted with the word *gōngmín* 公民, the prevalent term used to designate the citizen, and it is compared with this form's use in Japanese, a language that uses the same characters in its writing system. Thus, the first meaning given for *guómín* 国民 is: "Citizen (*gōngmín* 公民). As a legal term, most countries use *gōngmín* 公民, only a few countries use *guómín* 国民, like Japan" (公民。作为法律名词，多数国家用公民,仅少数国家用国民，如日本。See XHCD 2001: 361). Which keyword won out in the rivalry between competing synonyms? These quotations clearly show that *gōngmín* 公民 has come to be commonly accepted as the main term used to designate the citizen in contemporary Chinese. Despite the predominance of *guómín* at the turn of the twentieth century, this keyword has been relegated a secondary term in Chinese.

This shift becomes evident if we compare the frequency diagrams (频次图) of the BCC corpus for the two words (BCC现代汉语语料库, http://bcc.blcu.edu.cn/, consulted January 2019). The diagram of *gōngmín* shows an increase in the usage of this word since the end of the 1970s. Why is this period significant? Because it coincides with the launching of the "reform and opening up" policy. After a period of consolidation in which the people was considered to be of the utmost importance for the unity of China, we move into a period in which individual citizenship begins to grow as a social concern.

Study of corpora and frequency certainly appears to confirm this impression. There are some peaks (for instance, in 1989 and 2002–2003), and then frequency becomes stable at a relatively high level as of 2011. Otherwise, the frequency diagram (频次图) for *guómín* shows a strong decrease in the early 1950s (after the foundation of the People's Republic) and frequency gradually falls off as of the 1980s. This corresponds to the increasing use of *gōngmín*.

This will appear difficult for Europeans to understand perhaps: since English does not distinguish between two keywords in conceptualizing what a "citizen" is and since French, with "citoyen," and German, with "Bürger," hold together various meanings within a single keyword. But where two rival synonyms are in play, as in Chinese, changing economic, social, political, and cultural forces will inevitably have an impact upon which one gains ground and which one is marginalized. It is *gōngmín* that has become the legal term used to define citizen. It was adopted in the Constitution of the People's Republic of China, for example, in "Chapter II," which is entitled "The Fundamental Rights and Duties of Citizens" (第二章 公民的基本权利和义务). This appears to reflect a more general trend in the way this keyword has spread in usage, and we find *gōngmín* widely used in different discursive contexts.

Who uses *gōngmín*? Where? And why? Is it – like "citizen" in English and "citoyen" in French – restricted to certain spheres of language use? Or is it used within a wide variety of spheres of life? The word *gōngmín* 公民 is frequently used in the influential Chinese newspaper, *People's Daily* (online version *Rénmín wǎng* 人民网, http://www.people.com.cn/, consulted January 2019). For our purposes, it proved interesting to analyze the frequency with which the word is used in specific fields. Of a total of 228,794 occurrences, the highest percentage of examples are to be found in the sections "International" (*guójì* 国际, 16,642 occurrences) and "Opinions" (*guāndiǎn* 观点, 11,168 occurrences), followed by 7229 occurrences related to contemporary political affairs (*shízhèng* 时政), 6964 in the section "society" (社会), and 5453 related to the "rule of law" (*fǎzhì* 法治).

What are we to conclude from these statistics? It would appear that the frequency with which the word is used in the section related to contemporary political affairs demonstrates that *gōngmín* is largely represented in official political discourse. On the other hand, current debates on the topic of citizenship could explain the high frequency of the occurrences in the section "Opinions," "society," and "rule by law."

It is, however, interesting to note that in the "[President] Xi Jinping's important speech Database" (习近平系列重要讲话数据库, consulted January 2019) on *People's Daily* website (http://jhsjk.people.cn/), we obtained 195 results for this word. Hence, the word *gōngmín* 公民 does indeed form part of political discourse today in China. Nonetheless, although *gōngmín* is gaining ground, it is certainly not displacing other more established keywords, like "the people": with less than 200 references above, and 6608 results for the word *rénmín* 人民, it is clear that *gōngmín* is not competing with "the people." It would be fairer to consider that *gōngmín* as a keyword is being invited back into the arena of Chinese political discourse.

Elsewhere, *gōngmín* is gaining ground. It is interesting to compare the above statistics with the frequency of this word in the BCC corpora (BCC 现代汉语语料库, http://bcc.blcu.edu.cn/, consulted January 2019). The word *gōngmín* is much more frequent in the newspaper corpus with 57,315 occurrences. The general corpus provided 51,041 results. The keyword would appear to be taking root in these spheres much more than in others, judging from both the microblogging and the literary corpus, which provided, respectively, 9120 and 2128 occurrences.

What phrases does *gōngmín* appear in? Which are its most frequent collocations? The following themes appear to be among the main ones associated with this keyword:

- Virtue and morality of citizens
- Citizens' voting and referendums
- Citizens' quality (*sùzhì* 素质)
- Socialist and good citizens
- Civic or citizens' awareness
- Citizens' or civil society

The following expressions have been selected among the most frequent collocations for the word *gōngmín* in the BCC general corpus and will be discussed in turn: *gōngmín dàodé* 公民道德 (citizens' or civic virtue), *gōngmín sùzhì* 公民素质 (citizen's "quality"), *hǎo gōngmín* 好公民 (good citizen), *gōngmín yìshí* 公民意识 (citizens' or civic awareness), and *gōngmín shèhuì* 公民社会 (civil society).

Dàodé 道德 – "virtue" or "morality" – appears as the most frequent collocation for *gōngmín* in the general corpus and in the microblogging corpus: 3249 and 381 occurrences were found for *gōngmín dàodé* 公民道德 "citizens' or civic virtue."

Dàodé "virtue" or "morality" is also the second most frequent collocation in the newspaper corpus. In this newspaper corpus, the most frequent collocation is *tóupiào* 投票 "vote" in the expression *gōngmín tóupiào* 公民投票 "referendum" (lit. "citizens' vote," 3359 occurrences).

Among the collocations for *gōngmín*, *sùzhì* 素质 also figures highly. This word has more than 30 different translations, but it is usually rendered by "quality" (Kipnis 2006): *sùzhì* 素质 can be found among the frequent collocations throughout the general corpus, but it is also found both in the newspaper corpus and in the microblogging corpus. Is this a sign of the times? This concern for "quality" – *sùzhì* 素质 – appears to be part of a general trend, and the word seems to be asserting itself as a new keyword, with the emergence of what has come to be called a "*suzhi* discourse" (see Kipnis 2006). Kipnis (2006: 295) claims, for example, that: "Reference to *suzhi* justifies social and political hierarchies of all sorts, with those of 'high' *suzhi* being seen as deserving more income, power and status than those of 'low' *suzhi*."

Among the examples of the expression *gōngmín sùzhì* 公民素质, which can be rendered as "citizen's quality," in the *People's Daily Online* (*Rénmínwǎng* 人民网), we can find "improving citizens' quality" (*tíshēng gōngmín sùzhì* 提升公民素质) and "the overall improvement of citizens' quality and the degree of social civilization" (*gōngmín sùzhì hé shèhuì wénmíng chéngdù quánmiàn tíshēng* 公民素质和社会文明程度全面提升).

The concern for civic responsibility and the moral obligations of the citizen are certainly not a thing of the past in Chinese society, judging from the findings in our study of *gōngmín*. The word *gōngmín* is widely used in expressions related to "socialist citizens" and "good citizens." "Good citizens" has become a "topic of public discourse," as Hooper points out (2005: 3). This expression was found throughout the corpus we consulted, but it must be admitted that it is not particularly frequent. We found only 302 occurrences of the expression *hǎo gōngmín* 好公民 ("good citizen") in the BCC general corpus and 215 in the newspaper corpus. However, 1324 results were found in the *People's Daily Online*. The following expressions were representative of the examples we found in the CCL corpus for the expression "good citizen" and "socialistic citizen":

- "Be a good party member and a good citizen" (好党员好公民).
- "The education of the 'three goods': Be a good child at home, be a good student in school, be a good citizen in society" (在家做个好孩子、在校做个好学生、在社会上做一位"好公民"的三好教育).
- "Cultivate socialist citizens with ideals, moral qualities, culture and discipline" (培养有理想、有道德、有文化、有纪律的社会主义公民).

Among the frequent collocations for *gōngmín*, we found *gōngmín yìshí* 公民意识 "citizens' or civic awareness." This proved to be the fifth most frequent collocation in the newspaper corpus and the sixth most frequent collocation throughout the general corpus. The expression *gōngmín shèhuì* 公民社会, one of the forms used to designate "civil society," is the second most frequent collocation in the microblogging corpus and the sixth most frequent collocation in the BCC corpus. Moreover, 11,684 occurrences of the expression *gōngmín shèhuì* 公民社会 "civil society" can be found in the *People's Daily Online*. It would appear that *gōngmín yìshí* 公民意识 "citizens' or civic awareness" is central to a series of questions in debates related to the dynamics and tensions involving citizens living in today's Chinese society. Among the 164 occurrences found in the CCL corpus of Modern Chinese (consulted in May 2016), we found examples revealing the complexity of developing civil society. Such examples included:

- 在十多年前的中国,"公民社会"还是一个十分敏感的话题
  "In the China over the last ten years, 'civil society' has still been a very sensitive topic."
- 改革开放后,一个相对独立的公民社会已经在中国迅速崛起
  "After the 1978 reforms, a relatively independent civil society has been rapidly rising in China."
- 中国公民社会的发展也面临着许多困境
  "The development of civil society in China is also facing many difficulties."
- "公民社会"是改革开放后引入的对civil society的新译名
  "Civil society [*gōngmín shèhuì*] is a neologism, a translation of 'civil society' [*in English in the text*] introduced in China after the 'Reform and Opening up' policy."

These examples clearly indicate a complex multi-faceted concept of *gōngmín* 公民, one that is becoming increasingly central as civil society emerges and transforms itself. While some dimensions are highlighted, others fade into the background, but without actually being abandoned. In recent decades, a new dimension regarding the rights, standards, and aspirations of citizens has come to the fore in debates.

Academics and specialists appear to concur on the importance of redefining the roles of citizens and their significance for society as a whole. The "civil society" (*gōngmín shèhuì*) can be considered one of the main topics of academic publications related to "citizens" (*gōngmín*), according to the results of the National Social Sciences Database (http://www.nssd.org/, accessed January 2019). Currently, the word *gōngmín* 公民 is mainly used in writings related to "civil society" (*gōngmín shèhuì* 公民社会, 1790 papers), "citizens' rights" (*gōngmín quánlì* 公民权利, 1648 papers), "citizens' education" or "civic education" (*gōngmín jiàoyù* 公民教育, 1460 papers), and "citizens' participation" or "civic participation" (*gōngmín cānyù* 公民参与, 1436 papers). Broadly speaking, the word *gōngmín* is to be found primarily in academic publications in the fields of politics and law (政治法律); 13,460 such articles were found.

Taken together, this broad overview of Chinese language and discourse reveals that the keyword *gōngmín* 公民 is widely used throughout various discursive contexts. Well-represented in newspapers and academic literature (in particular in the fields of politics and law), the idea and ideal of citizenship have clearly become an integral part of official discourse. But *gōngmín* 公民 is not limited to these written or political spheres of life; it has also forged a place for itself in the minds of individuals, as we can see in its usage in less formal discursive forms like microblogging. What does this tell us? It suggests that the citizen is a cultural keyword that has taken root throughout language and life in China. From the point of view of cultural semantics and ethnolinguistics, we can consider *gōngmín* 公民 part of the contemporary Chinese worldview. And if we wish to understand that worldview from the perspective of other cultures, other languages, and other worldviews, we first need to understand that *gōngmín* 公民 is not the exact equivalent of "citizen," "citoyen," or "Bürger." Just as individuals from other cultures struggle to understand what we mean by our concepts for citizen, and the values and associations we attach to the citizen, so Europeans will have to grasp the way the form *gōngmín* 公民 is used in Chinese.

A Comparative Perspective

If citizens are to understand one another and work together for the greater good, they must agree on what it means to be a citizen, who forms part of the citizens of a country, and what those citizens are doing to develop and promote the greater good of civic society. This would appear to be a fundamental question for any society. However, our study of migrating meanings (Underhill and Gianninoto 2019) leads us to conclude that when we debate ideas, we cannot escape keywords and keywords are defined within language. The definitions of concepts change over time. As history and politics change over time, presenting new circumstances, people react differently to ideas and ideals.

During periods of political and national crisis, often it is the people and the nation that come to the forefront. In France during the revolution, this was the case. And once more, in the post-war period following the liberation of France in 1945, the people, the nation, and their destiny became key concerns, for the French.

In China at the turn of the twentieth century, it was the word *guómín*, with its explicit reference to the country and its nationalistic appeal, that became the prevalent term for "citizen." Similarly during the first decades of the twentieth century, the words for individual and individualism, central in intellectual debates, were associated with reformist, anti-traditionalist movements and campaigns, calling for national salvation and China's modernization (Huang 2004: 46; Liu 1995: 10; Zarrow 2012: 87).

In the first decades of communist China, after years of the Sino-Japanese war and civil war, the political discourse celebrated the people, and the concepts of citizen and individual were overshadowed. Only later, in times of peace and prosperity, when social harmony had been assured, the words for individual and citizen came to the foreground. Nevertheless, it is the word for "the people" which is still pivotal in Chinese political discourse today.

Cultures adopt different attitudes to individuals, citizens, and the people as a whole. In countries such as England, the people has invariably been considered as "the common people." The leaders of England have rarely presented themselves as being men or women of the people. Their leaders do at times speak the language of the people and try to demonstrate they are close to the people. The walkabouts of the Royals and their visits abroad form part of this practice or policy. But although these leaders may shake hands with the people, they do not work hand and hand with the people. They do not tend to consider the people as the force that will create the destiny of the nation. The leaders lead the people, but do not belong to them.

What can be said about citizens in this respect? Citizens can be considered to be individuals. But it is essential to understand that this is not necessarily the case. At various stages in history in various languages, individualism has been considered an evil or an excess, something that does not coincide with the will of the people, its aims, or its well-being. Individual and individualism were criticized in the 1950s, in the Czechoslovak Communist Republic, just as it was mistrusted in the first decades of Mao's presidency in China (Underhill and Gianninoto 2019). But neither was individualism particularly welcome in many periods of English history. In Victorian England, individualism was often denounced just as much. Individualism was an ethical question in such periods. For religious and political thinkers, individualism was considered tantamount to selfishness and self-centeredness; it goes against a sense of civic duty and solidarity.

China and Western countries have influenced each other. The relations fluctuate as geopolitical and cultural powerplays evolve. However, one thing is constant; influence is always complex. When cultures meet and mingle and when keywords take flight and enter another worldview, there are three spheres of complexity that come into play: the origin, the destination, and the transfer. We must remember that in their place of origin, the meaning and status of Western keywords are hotly debated. Keywords are translated into a similarly complex context. And finally, keywords must make their own place in language by displacing other keywords.

When new keywords are introduced, they often force speakers to reorganize the way key concepts work together. This accounts for the complexity of the contexts of origins and destinations. The third sphere is the sphere of transition: often a new complex context will be set up in which a series of keywords begin to transform key questions of social importance. Do meanings migrate? Do we manage to translate words and worldviews? Certainly, but meanings are translated out of worldviews and they are transformed in the process of transfer. The worldviews that they are translated into are dynamically changing over time. And often the transfer of meanings between worldviews creates a new situation.

The example of the Japanese prism through which Western concepts enter into the modern Chinese worldview is a case in point. Meanings do migrate, then, but the meanings that settle often prove to be very different from the meanings that take flight. Transfer transforms them. So *gōngmín* 公民 clearly takes on a role and associations unlike the keywords for "citizen" in play in European languages. As the host language's meanings begin to dynamically reorganize the way keywords are structured, often their meaning become unrecognizable, incomprehensible even, to the speakers from whose language they have been translated. If our study has demonstrated one thing, it is that it would be a mistake to think that English speakers have understood what *gōngmín* 公民 means for the Chinese, once they have translated it back into English as "citizen." If meanings are to migrate, they must take root in language. And if researchers are to recognize what these meanings now mean, we have to understand the soil they grow in and the climate that enables them to grow. This is certainly the case for keywords such as "individual," "citizen," and "people."

Gōngmín "citizen" and *guómín* "national, citizen" are both forms found in ancient Chinese sources: meanings that have been adopted to designate foreign-inspired notions. In this regard, they can be considered emblematic examples of migrating concepts and meanings. However, as Liu (1995: 26) underlines, "meanings (...) are not so much 'transformed' when concepts pass from the guest language to the host language as invented within the local environment of the latter." This is undoubtedly the case of the words designating citizen and citizenship whose trajectories are strictly tied to Chinese political and social history.

Guómín, with its explicit reference to the country and its nationalist appeal, was central between the end of the nineteenth and the beginning of twentieth century in China. The Communist Revolution in China shifted the emphasis from citizens to the people, from groups of individuals to the collective masses, marginalizing the words *gōngmín* for some decades. In recent decades, we have seen a partial comeback of the words for citizen, with a particular emphasis on the word *gōngmín*, a "return graphic loan" whose components have deep roots in the Chinese cultural and linguistic tradition. In corpora and databases, texts, and online discussions, *gōngmín* 公民 is far less frequent than *rénmín* 人民 "the people." Nevertheless, various trends over the last hundred years have not erased *gōngmín* 公民 from the Chinese language, and this keyword can be found throughout a large spectrum of discursive contexts, from academic literature to newspapers and from blogs to TV series. In this sense, *gōngmín* 公民 is very much part of the cultural mindset of the Chinese people.

Meanings that migrate are transformed: the emerging meanings that begin to pattern the contemporary Chinese worldviews are not exotic alien concepts imposed upon Chinese society. When meanings migrate, they follow trajectories and paths that pattern the Chinese cultural traditions and that are specific to the history China and the Chinese people. For this reason, our work on migrating meanings takes us beyond the West, beyond the East, and beyond the East-West opposition. What do we do in comparing keywords? We enter into new worlds and seek to understand what we share. We ask what we owe to other cultures, and we try to see how our cultures adopt and recreate new meanings to help us grasp what we share together as individuals, as citizens of the world, and as speakers of different languages. Whatever language we speak, the words we use often come to us from other languages and other cultures. Understanding that can help us understand who we are and how we have become what we are. This does not tie us to the past but shows us how cultures are constantly reinventing themselves using existing resources. Meanings migrate, but they make sense to us, even when they take on very new meanings. What words will migrate in the coming decades? How will their meanings change, and how will we change with them? These are the questions we focus on in studying cultural keywords and migrating meanings.

References

Abramowicz, Maciej, and Jerzy Bartmiński. 1998. *Langues et peoples d'Europe Centrale et Orientale dans la culture française*. Paris: Institut d'Études Slaves.

Bartmiński, Jerzy, 2009. *Aspects of cognitive ethnolinguistics*, London: Equinox.

———. 2012. *Aspects of cognitive ethnolinguistics*, trans. Adam Głaz, ed. Jörg Zinken, London: Equinox.

Bastid-Bruguière, Marianne (Bāsīdì 巴斯蒂). 1997. 'Zhōngguó jìndài guójiā guānniàn sùyuán——guānyú Bólúnzhīlǐ « guójiā lùn » de fānyì' 中国近代国家观念溯源——关于伯伦知理〈国家论〉的翻译 [The origins of the concept of the state in modern China: On the translation of Bluntschli's *The theory of the state*]. *Jìndàishǐ yánjiū* 近代史研究 [Modern Chinese History Studies], (4): 221–232.

Burtscher, Michael. 2012. A nation and a people? Toward a conceptual history of the terms Minzoku 民族 and Kokumin 國民 in early Meiji Japan. *Journal of Political Science and Sociology*, (16): 47–106. http://koara.lib.keio.ac.jp/xoonips/modules/xoonips/download.php?file_id=64793(consulted March 2015).

Casacchia, Giorgio, and Mariarosaria Gianinnoto. 2012. *Storia della linguistica cinese*. Venise: Libreria Editrice Cafoscarina.

Cassin, Barbara, ed. 2004. *Vocabulaire européen des philosophes, Dictionnaires des intraduisibles*. Paris: Robert.

——— (ed.) 2014. *Dictionary of untranslatables: A philosophical Lexicon*. Trans. and Ed. Emily Apter, Jacques Lezra & Michael Wood. Princeton: Princeton University Press.

Chappell, Hilary. 1994. Mandarin semantic primitives. In *Semantic and lexical universals: Theory and empirical findings*, ed. Cliff Goddard and Anna Wierzbicka, 109–148. Amsterdam/Philadelphia: John Benjamins.

———. 2002. The universal syntax of semantic primes in Mandarin Chinese. In *Meaning and universal grammar – Theory and empirical findings*, ed. Cliff Goddard and Anna Wierzbicka, vol. 1, 243–322. Amsterdam/Philadelphia: John Benjamins.

Fang, Zhiqin 方志钦, and Jie Wang 王杰. 2006. *Kāng Yǒuwéi yǔ jìndài wénhuà* 康有为与近代文化 [Kang Youwei and modern culture]. Kāifēng: Hénán dàxué chūbǎnshè.

Gao, Mingkai 高名凯, and Zhengtan Liu 刘正埮. 1984. *Hànyǔ wàiláicí cídiǎn* 汉语外来词词典 [Dictionary of Chinese loanwords]. Shànghǎi: Shànghǎi císhū chūbǎnshè.

Gianninoto, Mariarosaria. 2016. Da *xiǎojǐ* 小己 a *gèrén* 个人: Alcune riflessioni sui termini per individuo e individualismo in cinese' [From *xiǎojǐ* 小己 to *gèrén* 个人: Some thoughts on the terms for individual and individualism in Chinese]. In *Linguistica cinese: Tendenze e prospettive [Chinese linguistics: trends and perspectives]*, ed. Clara Bulfoni, 89–101. Milan: Unicopli.

Goatly, Andrew. 2007. *Washing the brain: Metaphor and hidden ideology*. Amsterdam/Philadelphia: John Benjamins.

Goddard, Cliff, and Zhengdao Ye. 2016. Exploring "happiness" and "pain" across languages and cultures. In *"Happiness" and "pain" across languages and cultures*, ed. Cliff Goddard and Zhengdao Ye, 1–18. Amsterdam/Philadelphia: John Benjamins.

Goldman, Merle. 2005. *From comrade to citizen: The struggle for political rights in China*. Cambridge, MA: Harvard University Press.

Goldman, Merle, and Elizabeth J. Perry. 2002. Introduction: Political citizenship in modern China. In *Changing meaning of citizenship in modern China*, ed. Merle Goldman and Elizabeth J. Perry, 1–22. Cambridge, MA: Harvard University Press.

Guo, Taihui 郭台辉. 2011. Zhōng-rì de "guómín" yǔyì yǔ guójiā gòujiàn ———Cóng míngzhì wéixīn dào xīnhài gémìng 中日的"国民"语义与国家构建——从明治维新到辛亥革命 [Meanings of "guomin" and state-building in early modern China and Japan: From the Meiji Japan to China's revolution]. *Shèhuìxué yánjiū* 社会学研究 [Sociological Studies] (4): 137–245.

Guo, Zhonghua 郭忠华. 2013. Jìndài gōngmín gàiniàn yǔ fānyì de xiàndàixìng 近代公民概念与翻译的现代性 [The modern concept of citizenship and the modernity of translation]. *Zhōngguó shèhuì kēxué bào* 中国社会科学报 [Chinese Social Sciences Today].

Hooper, Beverly. 2005. The consumer citizen in contemporary China. Working papers in contemporary Asian Studies, Centre for East and South-East Asian Studies, Lund University. http://www.ace.lu.se/images/Syd_och_sydostasienstudier/working_papers/Hooper.pdf

Hsü, Immanuel C.Y. 1970. *The rise of modern China*. New York/Oxford: Oxford University Press.

Huang, Kewu 黄克武. 2004. Gèrénzhǔyì de fānyì wèntí: cóng Yán Fù tánqǐ,「個人主義」的翻譯問題:從嚴復談起 [The translation of 'Individualism' in Chinese: About Yan Fu], *Èrshíyī shìjì* 二十一世纪 [21st century]. (84): 40–51.

Humboldt, Wilhelm von. 1995. *Schriften zur Sprache*. Stuttgart: Reklam, Universal-Bibliothek.

———. 1999. *On language: On the diversity of human language construction and its influence on the mental development of the human species* (1836). Trans. Peter Health, ed. Michael Losonsky. Cambridge: Cambridge University Press.

———. 2000. *Sur le caractère national des langues et autres écrits sur le langage*. Trans. Denis Thouard. Paris: Seuil.

———. 2003. *Über die Verschiedenheit des menschlichen Sprachbaues/Über die Sprache*. Berlin: Fourier verlag.

Judge, Joan. 1994. Key words in the late Qing reform discourse: Classical and contemporary sources of authority. In *Indiana East Asian working paper series on language and politics in modern China, paper 5*, 1–33. Bloomington: Indiana University East Asian Studies Center.

Keane, Michael. 2001. Redefining Chinese citizenship. *Economy and Society* 30 (1): 1–17.

Kipnis, Andrew. 2006. Suzhi: A keyword approach. *The China Quarterly* 186: 295–313.

Kornacki, Paweł. 2017. Rènao: What does it mean to have a good time the Chinese way? An ethnopragmatic exploration of a Chinese cultural keyword. In *East-Asian and Central-European encounters in discourse analysis and translation*, ed. Anna Duszak, Arkadiusz Jabłoński, and Agnieszka Leńko-Szymańska, 57–82. Warsaw: Institute of Applied Linguistics.

Lackner, Michael, and Natascha Vittinghoff. 2004. *Mapping meanings: The field of new learning in late Qing China*. Leiden: Brill.

Lackner, Michael, Amelung Iwo, and Joachim Kurtz, eds. 2001. *New terms for new ideas. Western knowledge and lexical change in late imperial China*. Leiden: Brill.

Legge, James. 1872. The Ch'un ts'ew, with the Tso chuen. In The Chinese classics, with a translation, critical and exegetical notes and copious indexes by J. Legge, vol V, part II, containing Dukes Seang, Ch'au, Ting, and Gae, with Tso's appendix; and the indexes. Hong Kong/London: Lane, Crawford/Trübner. https://babel.hathitrust.org/cgi/pt?id=uc1.b000970399;view=1up;seq=254. Accessed July 2016.

Leung, Helen Hue Lam. 2017. Cantonese "mong4". In *Cultural Keywords in Discourse*, ed. C. Levisen and S. Waters, 183–210. Amsterdam/Philadelphia: John Benjamins.

Liu, Lydia H. 1995. *Translingual practice. Literature, national culture, and translated modernity–China, 1900–1937*. Stanford: Stanford University Press.

Masini, Federico. 1993. *The formation of modern Chinese lexicon and its evolution toward a national language: The period from 1840 to 1898*. Journal of Chinese linguistics: Monograph VI. Berkeley: University of California, Project on Linguistic Analysis.

Rouen Ethnolinguistics Project (forum for international online conferences founded and directed by James Underhill), http://rep.univ-rouen.fr/

Sapir, Edward. 1949. *Language: An introduction to the study of speech (1921)*. New York: Harcourt, Brace& World.

———. 1985. In *Selected writings in language, culture, and personality (1949)*, ed. David G. Mandelbaum. Berkeley: University of California Press.

Schoenhals, Michael. 1994. 'Non-people' in the People's Republic of China: A chronicle of terminological ambiguity. In *Indiana East Asian working paper series on language and politics in modern China, Paper 4*, 1–48. Bloomington: East Asia Studies Center, Indiana University.

Tang, Tsou. 1986. *The cultural revolution and post-Mao reforms: A historical perspective*. Chicago: University of Chicago Press.

Trabant, Jürgen. 2012. *Weltansichten*. München: C. H. Beck.

———. 2015. *The Jürgen Trabant Wilhelm von Humboldt Lectures*, eight hours of on-line conference lectures, Rouen. http://rep.univ-rouen.fr/content/films-trabant
Tsien, Tsuen-hsuin. 1954. Western impact in China through translation. *Far Eastern Quarterly* 13 (3): 305–327.
Underhill, James W. 2009. *Humboldt, worldview, and language*. Edinburgh: Edinburgh University Press.
———. 2011 *Creating worldviews: metaphor, ideology and language*, Edinuburgh: Edinburgh University Press.
———. 2012. *Creating worldviews: Metaphor, ideology, & language*. Edinburgh: Edinburgh University Press.
———. 2013. *Ethnolinguistics and cultural concepts: Truth, love, hate, & war*. Cambridge: Cambridge University Press.
Underhill, James, W. 2016. *Individuals or peoples for Europe?* On-line conference paper, Prague. https://slideslive.com/38897474/jednotlivci-nebo-narody-pro-evropu
Underhill, James, and Mariarosaria Gianninoto. 2019. *Migrating meanings: Sharing keywords in a global world*. Edinburgh: Edinburgh University Press.
Wang, Ban. 2011a. Understanding the Chinese revolution through words: An introduction. In *Words and their stories: Essays on the language of the Chinese revolution*, ed. Ban Wang, 1–14. Leiden: Brill.
———, ed. 2011b. *Words and their stories: Essays on the Language of the Chinese Revolution*. Leiden: Brill.
Watson, Burton. (trans.) 2003. *Han Feizi: Basic writings*. New York: Columbia University Press.
Wheaton, Henry. 1866. *Elements of international law*. London: Sampson Low, Son.
Wheaton, Henry (Huì Dùn 惠顿), Martin, William A. P. (trans.) 2002. (Dīng Wěiliáng 丁韪良), *Wànguó Gōngfǎ* 万国公法 [*Elements of International Law*]. Shànghǎi: Shànghǎi shūdiàn chūbǎnshè.
Whorf, Benjamin Lee. 1984. In *Language, thought and reality: Selected writings (1956)*, ed. John B. Caroll. Cambridge, MA: M.I.T. Press.
Wierzbicka, Anna. 1997. *Understanding cultures through their key words*. Oxford: Oxford University Press.
———. 2014. *Imprisoned in English: The hazards of English as a default language*. Oxford: Oxford University Press.
Williams, Raymond, 1986. Keywords: A Vocabulary of Culture and Societyn (1976), Oxford: Oxford University Press.
Ye, Zhengdao. 2012. Eating and drinking in mandarin and Shanghainese: A lexical-conceptual analysis. In *Quantitative approaches to problems in linguistics: Studies in honour of Phil Rose*, ed. Cathryn Donohue, Shunichi Ishihara, and William Steed, 265–280. Munich: Lincom Europa.
———. 2013. Understanding the conceptual basis of the "old friend" formula in Chinese social interaction and foreign diplomacy: A cultural script approach. *Australian Journal of Linguistics* 33 (3): 365–385.
———. 2014. The meaning of "happiness" (xingfu) and "emotional pain" (tongku) in Chinese. *International Journal of Language and Culture* 1 (2): 194–215.
———, ed. 2017. *The semantics of nouns*. Oxford: Oxford University Press.
———. 2018. On Chinese "Happiness": A semantic perspective. Paper given at a conference held at the École des hautes études en sciences sociales (Paris). 16 Jan 2018.
Zarrow, Peter G. 1997. Introduction: Citizenship in China and in the West. In *Imagining the people, Chinese intellectuals and the concept of citizenship 1890–1920*, ed. Joshua A. Fogel and Peter G. Zarrow, 3–38. New York/Londres: M. E. Sharpe.
———. 2012. *After empire: The conceptual transformation of the Chinese state, 1885–1924*. Stanford: Stanford University Press.

Chinese Cultural Keywords

20

Paweł Kornacki

Contents

Introduction: Cultural Keywords from Williams to Wierzbicka	492
Approaching the Studies of Chinese Cultural Keywords	495
Selected Western Sinological Studies	495
Cultural Anthropology of Chinese Society	496
Cultural and Cross-Cultural Psychological Studies of Chinese People	496
The Natural Semantic Metalanguage Approach	497
Two Case Studies of Cultural Keywords in Chinese Language and Culture	498
Introduction	498
Case Study 1: *xìngfú* 幸福 ("Happiness") – Chinese Notion of "Happiness"	499
Case Study 2: *rènao* 热闹 ("Hot and Noisy") – Having a Good Time with Other People	503
Food and Chinese Cultural Value of *rènao* 热闹	505
Conclusions and Prospects for Future Studies	507
References	509

Abstract

This chapter examines the idea of cultural keywords conceived as symbolic focal points which lexicalize salient values, attitudes, and concerns in Chinese language and culture. After a brief review of the origins of the idea in British cultural studies and its manifestations in philological, anthropological, and psychological research, it introduces a linguistic approach to the study of cultural keywords, known as the Natural Semantic Metalanguage (NSM) approach to meaning, offering arguments in favor of employing it to render and represent culturally specific meanings in a non-ethnocentric fashion, with two case studies of Chinese cultural keywords. They pertain to the domains of Chinese emotions (the word *xīngfú* 幸福 "happiness") and sociality (the word *rènao* 热闹 "lively; hot and

---

P. Kornacki (✉)
Institute of English Studies, University of Warsaw, Warsaw, Poland
e-mail: p.kornacki@uw.edu.pl

© The Author(s), under exclusive licence to Springer Nature Singapore Pte Ltd. 2022
Z. Ye (ed.), *The Palgrave Handbook of Chinese Language Studies*,
https://doi.org/10.1007/978-981-16-0924-4_43

noisy"). While highlighting various aspects of Chinese culture, the presented analyses of Chinese cultural keywords are couched within the Natural Semantic Metalanguage framework, situating them within a developing field of cultural keyword studies.

Keywords

Chinese language · Chinese culture · Chinese social interaction · Chinese cultural keywords · Cultural keywords · Natural Semantic Metalanguage

Introduction: Cultural Keywords from Williams to Wierzbicka

What are cultural keywords? Informally put, they are words that offer a particularly revealing insight into how a culture works. According to one recent scholar of the subject (see Goddard 2005, p. 281), a cultural keyword (or key word) can be defined as "a highly salient and deeply culture-laden word which acts as a focal point around which a whole cultural domain is organized, a word which designates a culturally important concept." As such, cultural keywords are of interest not only to academics the in humanities and social sciences, such as anthropologists, linguists, psychologists, and scholars in cultural studies, but also anyone who has a professional interest in Chinese society and Chinese culture and who can benefit from a deepened understanding of Chinese words and their meanings.

The English expression "cultural keyword" used in the aforementioned sense is usually traced to a notable book *Keywords: A Vocabulary of Culture and Society* authored by British (Welsh) Marxist social historian Raymond Williams (1921–1988). In the book, Williams (1983) defines a significant field of study in Western humanities by demonstrating that not all words are equal, as far as their cultural and historical significance is concerned. The "key and lock" metaphor contained in the title of his book conveys an intuitively plausible idea that some words in English might open up a particularly engrossing view of its speakers' lifeworld, their institutions, and social practices, while other words might not fit this task as well.

Originally, Williams' book contained over a hundred short essays, each one of them devoted to one English word referring to some essential aspect of modern-era English society, e.g., *art, communication, country, democracy, elite, man, science, work*, etc. While it seems that Williams did not originally envisage his work to have any cross-cultural and comparative dimensions, his writings drew considerable interest from scholars researching non-Western cultural traditions as well (see, e.g., Barlow 1991 on Māori language and culture). However, an explicitly comparative perspective extending Williams' critical keyword studies approach to Chinese culture is contained in Gentz (2009), Gentz's work purports "to critically reflect upon Western theoretical approaches towards culture through an examination of the application of Western theories to the non-Western culture of China" (p. 5), by focusing on broad interpretive categories of cultural studies and their varied significances in the multiple contexts of Chinese traditions and Chinese-Western cultural translation.

The book's expressed aim is "to stimulate sustained reflection on and systematic exploration of terminology employed in the study of culture, especially Chinese culture" (Gentz 2009:13). Its chapters are devoted to such broad subjects as "Culture/Wenhua, Orientalism/Occidentalism, Postmodernism, Historiography and their significance in Western Anglophone countries as well as China." Significantly, Gentz points to similarities and differences between traditional Chinese collections of keywords such as "The Literary Mind and the Carving of Dragons" *Wénxīn diāolóng* 文心雕龙 by Liú Xié 刘勰 (465–522 CE), which was an early work organized according to a set of 49 keywords of literary criticism, and "Master North Stream's Meanings of Words" *Běixī zìyì* 北溪字义 written by Chén Chún 陈淳 (1159–1223 CE), which has served a similar purpose in the unfolding of Neo-Confucianism, consisting of a selection of 26 central analytical terms used in the then newly emerging Neo-Confucian doctrine (Gentz 2009).

Broad issues of cross-cultural, philological, and philosophical interpretation of Chinese cultural keywords are of paramount importance in another major recent work in the domain of Chinese studies and cultural translation, which is the collection of studies by Li and Pines (2019) *Keywords in Chinese Culture: Thought and Literature*. According to the blurb of the book: "The authors of the volume explore different keywords and focus on different periods and genres, ranging from philosophical and historical texts of the Warring States period (453–221 BCE) to late imperial (ca. 16th–18th centuries CE) literature and philosophy." The book showcases studies dealing with such key Chinese cultural words as *hé* 和 "harmony, harmonize," *míng* 名 "name," and *xiào* 孝 "filial piety." It also includes a highly relevant methodological chapter by Christoph Harbsmeier, which highlights interpretive dangers of "subsumption of the Chinese evidence under our conceptual scheme, or tediously repetitive diagnosis of conceptual and cognitive deficiencies in the Chinese conceptual system" (see Harbsmeier 2019, p. 397).

The historical-comparative analysis of key Chinese categories of textual practice, such as that carried out by Gentz (2009), and the comparative philosophical perspectives adopted in Li and Pines (2019), built upon and extended Williams' original domain of keywords, which was conceived by the latter pair of scholars as "pivotal terms of political, ethical, literary and philosophical discourse" (Li 2019, pp. ix–x). Another conceptual dimension of research which fed into the modern studies of keywords has been a helpful cultural anthropological distinction between "experience-near" and "experience-distant" concepts. This is clearly articulated in Geertz's (1973) influential essay:

> An experience-near concept is, roughly, one that someone – a patient, a subject, in our case an informant – might himself naturally or effortlessly use to define what he or his fellows see, feel, think, imagine, and so on, and which he would readily understand when applied by others. An experience-distant concept is one that specialists of one sort or another – an analyst, an experimenter, an ethnographer, even a priest or ideologist – employ to forward their scientific, philosophical, or practical aims. (Geertz 1973, p. 57)

Arguably, the distinction between "experience-near" and "experience-distant" concepts seems to be particularly relevant to any approach to the subject of cultural keywords which accrues due significance to usage-based linguistic evidence examined in its proper cultural environment. The challenge of describing and clarifying the actual meanings of "experience-near" concepts in different cultures and languages has become one of the major research tasks of the Natural Semantic Metalanguage (NSM; see below) approach to keywords studies. In *Understanding Cultures Through Their Key Words*, Anna Wierzbicka characterizes cultural keywords as words which are "particularly important and revealing in a given culture... and salient in the collective psyche of a society, and their meanings resonate with meanings of other linguistic units and cultural practices" (Wierzbicka 1997, pp. 15–16). She further adds that cultural keywords "commonly denote values, attitudes, speech acts, social categories, among others" (Wierzbicka 1997, p. 16).

Of central importance to the NSM approach to keyword studies is the issue of their translatability. The fact that cultural keywords manifestly defy conventional translation has been acknowledged by many scholars. For example, Geertz (2000, p. 209), who cites Wierzbicka's work, calls for "extended glosses, sample uses, contextual discriminations, behavioral implications, alternate terms" in order to overcome the translatability problem. In his view, failing to heed the problems of cross-cultural translation in a semantically adequate fashion will result in "words ethnocentrically, tendentiously, or merely lazily translated from one language to another into English, as those affective clichés 'guilt' and 'shame'". To address the translatability issue in a semantically illuminating way which moves beyond the mere enumeration of single-word glosses, NSM researchers aim to delve in what Geertz (1973, p. 7) called "thick description" of culturally significant local vocabularies which yield insights into the "stratified hierarchy of meaningful structures" characteristic of cultural categories in general. In order to address such semantic requirements, NSM methodology offers descriptions of the meaning of a word in the form of paraphrases composed of simple and universal words which make up the metalanguage (for a detailed discussion of the NSM program, see Goddard and Wierzbicka 1994, 2002, 2014).

As a brief illustration of NSM explicative methodology, consider the following paraphrases (referred to in NSM terminology as explications) which aim at capturing the essential meanings of two culturally significant words: English *mind* and Chinese *xīn* 心 ("heart/mind").

mind
 one of two parts of a person
 one cannot see it
 because of this part, a person can think and know (based on Wierzbicka 2016, p. 458)

xīn 心 ("heart/mind")
 one of two parts of a person
 because of this part, this person can think and know
 because of this part, this person can feel good things
 because of this part, this person can feel bad things
 because of this part, this person can be a good person
 because of this part, this person can be a bad person (based on Kornacki 1995, Chapter 2)

Such paraphrases, composed of simple and cross-translatable words, allow one to render examined meanings of culture-specific concepts in sufficient detail. They spell out and highlight the relevant cultural features of key verbal items observable in linguistic and cultural data. For example, one conspicuous feature of the Chinese concept of *xīn* 心 "heart/mind," rendered semantically transparent in the above explication, is to reflect what many scholars (see, e.g., Munro 1985, Ed.) used to term as "holism" – inseparability of cognition, affect, as well as moral knowledge of what is good and what is bad – which is perceived as a salient mode of thinking in Chinese culture. This particular feature of the concept of *xīn* 心 makes it meaningfully different from the modern English (Anglo) cultural keyword *mind*, which is predominantly focused on cognitive activities (for more on the subject of heart and soul-like linguistic constructs from the NSM perspective, see Peeters 2019, Ed.).

Generally speaking, linguistic evidence examined in order to arrive at a semantic explication subsumes collocational data which contains the studied lexical item (or items). The explication considers standard linguistic tests of collocation, which indicate changes in meaning (e.g., *clear mind, inquisitive mind* vs. older English *happy mind, fiery mind*; see Wierzbicka 1992, p. 44), substitutability, and contrast in context (e.g., consider different implications of syntactically parallel English sentences *X lost his* **mind** vs. *X lost his* **soul**). Initial semantic hypotheses (e.g., such that modern English noun *mind* is currently predominantly linked with thinking and knowing, whereas in earlier English it also had affective and conative dimensions) need to be further examined against more linguistic data mined from language corpora and/or relevant cultural texts which contain and depend on the examined word. Both of the NSM-based Chinese keyword studies introduced in this chapter make use of such varieties of linguistic and cultural evidence to arrive at their proposed semantic explications, formulated in terms of universal semantic primes.

Approaching the Studies of Chinese Cultural Keywords

How can one study Chinese cultural keywords? How have key Chinese cultural meanings been studied? Which Chinese words qualify as suitable objects of such research? Given that Chinese words and the realities that they represent are of clear interest to several distinct academic disciplines, each with its own research priorities and methodologies, the issue of deciding on one or the other approach in the study of Chinese cultural keywords is of major consequence.

Tentatively, one can identify four broad research approaches which accommodate a close attention to Chinese words as a part of their general methodology, although they do so in distinct and sometimes complementary ways.

Selected Western Sinological Studies

Traditionally, Western sinological studies have been largely devoted to a philological exegesis of textual sources of Chinese cultural traditions. While they did *not* make verbatim use of the expression "cultural key words," they traced and examined the history of some of the most notable Chinese lexical items in painstaking detail. The detailed philological studies of Michael Carr 1992, 1993; Mark Elvin 1989; or Halvor Eifring (1999), to mention but a few, offer rich insights into the historical and cultural significance of meanings of key cultural terms and expressions appearing in the textual traditions of China. A large number of studies exist that select and describe the essential *zì* 字 (words/characters) of Chinese culture. For the sake of illustration, two of them might still be mentioned as directly relevant to the "keywords" approach – Ivor Armstrong Richards' (1932) theoretically significant study *Mencius on the Mind*, which is a meaning-based exploration of the keyword *xīn* 心 ("heart/mind") in the classical philosophical text *Mengzi*, and Peter Boodberg's (1953) paper entitled *The Semasiology of Some Primary Confucian Concepts*.

Cultural Anthropology of Chinese Society

The second group of studies originates from within the American cultural anthropology tradition. While focusing on the characteristic elements and qualities of social interaction in Chinese societies, these studies pay close attention to the language of participants and their communicative symbolic exchanges taking place in this context. One particularly insightful study in this tradition is Mayfair Mei-hui Yang's (1994) monograph *Gifts, Favors and Banquets: The Art of Social Relationships in China*, which is a detailed ethnographic and meaning-oriented exploration of the cultural practice of *guānxì* 关系 (roughly, "human relations") in modern China. The book addresses some of the key related cultural concepts, such as *rénqíng* 人情 ("favor"), *gǎnqíng* 感情 ("attachment, feelings"), and *miànzi* 面子 ("face"), in numerous examples of their contextual usage among Chinese people of twentieth-century Mainland China. As these culturally distinct meanings of Chinese keywords are central to the Chinese social intercourse studied by Yang, they are also among the staple objects of research of a variety of psychological approaches to the study of Chinese society, addressed in the following section.

Cultural and Cross-Cultural Psychological Studies of Chinese People

The third set of research methods and procedures, which has some affinity with the cultural anthropological approach, could be said to trace its scholarly lineage to studies in cultural and cross-cultural psychology. This third set focuses on the research questions of communication and cultural change in China (see Chu and Ju 1993), and the suitability of Western analytical concepts in making sense of Chinese social and individual behavior (see, e.g., Yang 2006). Since studies of indigenous psychologies, by definition need to rely on indigenous (i.e., local language and culture-specific) concepts, they furnish students of keyword studies with a valuable pool of "experience-near" verbal and conceptual items. See e.g., Bond and

Hwang (1986) or Bond (1991), Ho (1976), and Yang and Ho (1988) for an introduction to and exemplification of this psychological approach. The Chinese words as addressed in their research included such culturally significant notions as that of *xiào* 孝 "filial piety," *miànzi* 面子 "face," and *rénqíng* 人情 "favor."

The Natural Semantic Metalanguage Approach

The fourth approach to the studies of Chinese cultural keywords focuses on the methodological issues identified by Anna Wierzbicka and Cliff Goddard, the two linguists most closely associated with the NSM research approach. According to Wierzbicka, a word might be considered a keyword if it meets the following criteria: (1) it has a relatively high frequency in the semantic domain it belongs to, (2) it is at the center of a phraseological cluster, and (3) it might be part of common sayings or expressions (Wierzbicka 1997, p. 16). Goddard (2003a, pp. 15–16) further clarifies that "Terms for values, social categories, ethnopsychological constructs, and ethnophilosophical terms have proved particularly fruitful sources, but cultural key words can also turn up in unexpected places, as with the Australian English swearword and discourse marker *bloody* (Wierzbicka 2002b)" (Goddard, ibid.).

Wierzbicka (1997) effectively introduces a culture-independent, universal metalanguage for cross-cultural comparison of meanings. The metalanguage is known as Natural Semantic Metalanguage (NSM). One of the characteristic features of the NSM methodology lies in its principled way of combining (one might say, harnessing) basic meaning units that are common among languages (i.e., universal semantic primes) to describe (or explicate, in NSM terminology) the meanings of words and expressions that make each human language unique (or at least very special to its speakers).

As Goddard (2002) argues, NSM allows for developing a maximally culture-free method of representing linguistic meaning. Semantic explications of linguistic and cultural symbols in the form of semantic scripts written in cross-translatable words offer yet another avenue to better understand native words and ideas, and reflect an experience-near perspective. The advantages of a descriptive linguistic methodology based on NSM include its attention to lexical detail and focus on actual usage-based examples, semantic analysis techniques, and cross-linguistic translatability. Thus, for students of cultural keywords who wish to ground their research in the actual linguistic usage of the examined culture and avoid the pitfalls of "conceptual ethnocentrism," NSM-based studies of cultural vocabularies may constitute an appealing alternative (see, e.g., Goddard 2002, 2006a).

Yet, given that cultural keywords are, by their nature, expressive of the culturally important aspects of a given society, acknowledging a verbal item as a "cultural keyword" seems to be an inescapably social act; it defies any attempts to derive a list of such words by a quantitative procedure alone. The non-reducible quality of their meaning has been accentuated by Wierzbicka and other scholars engaging with the NSM research framework. For example, in their recent review of Wierzbicka's approach to meaning across languages and cultures, Gladkova and Larina (2018, pp. 718–720) highlight the relevant points:

> Wierzbicka (1997a, p. 16) argues that 'there is no finite set of such words in a language, and there is no "objective discovery procedure" for identifying them. To show that a particular word is of special importance in a given culture, one has to make a case for it.' According to Wierzbicka, a word might be claimed to be a keyword if it meets the following criteria: (1) it has a relatively high frequency in the semantic domain it belongs to, (2) it is at the center of a phraseological cluster and (3) it might be part of common sayings or expressions.

Why do linguists want to study cultural keywords? For Goddard (2003a, pp. 15–16), "cultural keywords" are the first item on his list of the chief types of linguistic evidence which shed light on what anchors communicative interaction in a specific language and culture. What kinds of words does Goddard have in mind?

> Terms for values, social categories, ethnopsychological constructs, and ethnophilosophical terms have proved particularly fruitful sources, but cultural key words can also turn up in unexpected places, as with the Australian English swearword and discourse marker *bloody* (Wierzbicka 2002b). (Goddard, ibid.)

What may strike one about Goddard's description is its certain affinity with Geertz's notion of *experience-near concepts*, so that in the former scholar's usage the expression "cultural key word" effectively comes to signify something close to a linguist's take on the classic psychological idea of "experience-near" concept in the anthropological sense.

While each one of the briefly described approaches to the study of Chinese cultural keywords tends to focus on somewhat different groups of culturally revealing expressions, all of them underscore a need to render and communicate their meanings.

The next section will showcase two examples of studies which apply the linguistic semantic techniques of the NSM approach to two prominent Chinese cultural keywords.

Two Case Studies of Cultural Keywords in Chinese Language and Culture

Introduction

To date, the research framework of Natural Semantic Metalanguage has been employed by some scholars to probe into the meanings and broader significance of several Chinese cultural keywords. For example, Goddard (2005, pp. 84–89) proposed semantic explications of the key Chinese cultural value terms *xiào* 孝 "filial piety" and *rěn* 忍 "perseverance/forbearance," exemplifying their connection with prevailing cultural concerns of Chinese societies. Li, Ericsson, and Quennerstedt (2013) examined the significance of the cultural keyword *xīn* 心 "heart/mind" in the Chinese health maintenance system of *Qìgōng* 气功. The aim of this section is to present two Chinese cultural keywords and their more recent studies, in order to illustrate different uses of linguistic evidence compatible with the overarching methodology of NSM. The two keywords under discussion are:

- The prominent Chinese emotion word *xìngfú* 幸福 ("happiness") relevant in the field of cross-cultural emotion studies
- The attitudinal adjective/verb *rènao* 热闹 ("hot and noisy," "lively") which sheds light on some characteristically Chinese aspects of human sociality

The above lexical items draw from studies undertaken by Ye (2016) and Kornacki (2017). While each of these studies deals with a different aspect of Chinese culture, the authors of these papers rely on the common semantic methodology of the Natural Semantic Metalanguage approach in order to reflect a culture-internal perspective.

These studies of keywords have been selected to reflect two distinct domains of research and some particular aspects of Chinese culture. Case Study 1 addresses a persistent translation problem plaguing cross-cultural psychological studies of emotions. Case Study 2 takes a meaning-driven NSM-based approach to Chinese sociable interaction, making it relevant to the studies of cultural anthropology and sociality. At the same time, these case studies also illustrate some of the prominent strands of current research within the NSM framework such as the examination of emotion vocabularies, study of social concepts, and investigations into culture-internal accounts of communicative practices. Both case studies make use of somewhat different (yet ultimately complementary) varieties of linguistic and cultural evidence. The first study relies on usage and frequency data derived mostly from Chinese language corpora coupled with native speaker's insights. The second study focuses on a close reading of some everyday Chinese cultural texts.

Case Study 1: *xìngfú* 幸福 ("Happiness") – Chinese Notion of "Happiness"

Zhengdao Ye (2016) conducted a detailed study of the Chinese emotion word *xìngfú* 幸福 (roughly, "happiness"), arguing that this prominent emotion word encapsulates a notable amount of cultural knowledge conceived as normative expectations and culturally desirable affects.

Placing her study in the broad context of ethnopsychology, Ye (2016) argues that the newly emerging "Science of Happiness" seems to miss (if not to neglect) an important linguistic and cultural issue: "What are the indigenous 'happiness' concepts in Chinese? And what do they mean?" As noted by Ye (p. 68), the linguistic optimism of many "happiness" researchers quoted in the first part of her study is hardly warranted. Having scrutinized Chinese psychological terminology employed to translate the conceptual framework of international (i.e., English-based) "happiness research," she casts doubts as to the conceptual adequacy of the employed social science research method, which consists of collecting and comparing questionnaire data based on self-report of subjective happiness and subjective well-being

(SWB). Indeed, as one of the scholars quoted by Ye rather frankly admits, the scientific value of such questionnaire and self-report-based psychological studies hinges on the supposed equivalence of meanings of the key words compared across languages in such studies:

> Of course one could question whether the word 'happy' means the same thing in different languages. If it does not, we can learn nothing by comparing different countries. (Layard 2003, p. 16, quoted in Ye 2016, p. 66)

As of now, the central word in the contemporary Chinese "happiness discourse" appears to be *xìngfú* (幸福), which is used to translate such terms as Gross National Happiness (GNH) (*guómín xìngfú zǒngzhí* 国民幸福总值) and the National Happiness Index (NHI) (*guómín xìngfú zhǐshù* 国民幸福指数). This particular Chinese word has been chosen to gloss the English language expression "well-being" as well and to render the specialist psychological term "subjective well-being" (SWB) and its derivatives, as employed in Chinese language research on what is called in Chinese *zhǔguān xìngfúguān* (主观幸福观), literally "subjective sense of *xìngfú*" (Ye 2016, p. 67, n.d. 2).

However, what appears to be "conspicuously missing" from the studies of comparative sociology and psychology of happiness and subjective well-being (SWB) in the Chinese context is precisely what anthropologists and linguists, such as Clifford Geertz and Anna Wierzbicka, have argued for, namely, "knowledge and understanding of the native concepts with which local people talk and think about 'happiness'." To satisfy such a demand requires both the appropriate linguistic data and a proper semantic methodology which is able to account for and explain the meanings (and meaning differences) of the relevant linguistic expressions in a non-arbitrary and cross-culturally illuminating way.

According to Ye, Chinese people's discourse on "happiness," both at the official level and in people's daily experience, is centered around the concept of *xìngfú* 幸福 (provisionally glossed by her as "happiness"). The word appears noticeably often in TV series and popular songs (e.g., as a part of their titles). Its salience and frequency are evidenced in Chinese language corpora, four of which are mined in her study for the frequency data on *xìngfú*, its synonym *kuàilè*, its opposite *tòngkǔ*, and a borrowed English word *happy*, which is strikingly popular among young Chinese Internet users (cf. Ye 2016, pp. 70–71).

While the status of *xìngfú* as "a key and salient concept in contemporary Chinese society" is quite clear, the issue of its actual semantic meaning requires elucidation. Given that most Chinese-English bilingual dictionaries simply pair it with English words "happy" (adj.) and "happiness" (n.), it is essential to select linguistic examples which would enable one to formulate and test initial hypotheses about the meaning of the word in question.

Ye relies on three kinds of sources to collect the relevant usage-based data that would enable her to arrive at an NSM semantic explication of *xìngfú*. They are:

- Excerpts from Chinese autobiographical writing

- Posts from Sina Weibo (the Chinese equivalent of Twitter)
- Frequency-of-usage information obtained from Chinese language corpora

Several types of frequency data demonstrate that *xìngfú* is undeniably common in the large samples of linguistic data contained in four major corpora: Leiden Weibo Corpus (LWC), Beijing University's Modern Chinese Corpus (CCL PKU), Lancaster Corpus of Written Chinese (LCMC2), and the UCLA Written Chinese Corpus (UCLAv2). Yet, while the quantifiable statements of high frequencies suggest that the concept is highly salient among native users of Chinese, frequencies simply do not deliver (and they are not meant to deliver) any answer as to the exact meaning of *xìngfú*. Such an answer can only be arrived at through a careful and rigorous examination of the range of contexts in which the word is used. Sources like autobiographical writings which contain affective scenarios and descriptive passages are helpful in formulating a hypothesis about the meaning of *xìngfú*. Initially, Ye looks at nine short fragments (each about 20–100 characters in length) which highlight some salient meaning components. Two examples are presented here:

(1) 此时你是多么幸福, 你同你所爱的人在一起, 在蓝天阔野中跑, 在碧波白 浪中游, 你会是怎样的幸福! (Shi 2000, p. 48)
At this moment, how happy (*xìngfú*) you are. You are with the person you love, running in the open field under the blue sky and frolicking in the green sea and white waves. What happiness (*xìngfú*) you have!
(2) 听说, 婚后很不幸福, 两个人常常打架。 (UCLA2)
Apparently, she was not happy after they got married. They were often in fights.

According to Ye, such examples point to "a deep sense of connection with others." The others are "typically family members, or those one is in love with." It thus makes sense that dysfunctional interpersonal relations are tellingly described as *bú xìngfú* (which can be literally translated as "not happy/no happiness").

These examples also show that the meaning of *xìngfú* is "markedly different" from English words *happy* and *happiness*, because the notion of *xìngfú* is "earthly happiness that is anchored in an interpersonal relationship." Described in an informal way, Ye characterizes it as "an expansive and gratifying feeling that stems from the belief that one is cared for and loved."

Apart from observing patterns of usage of the word in question, collocational tests are also conducted to determine the boundary of usage. For example, while a "happy monk" sounds natural in English, it is unacceptable if the term *xìngfú* is used to describe monks in Chinese, precisely because *xìngfú* places emphasis on filial bond.

On the basis of this and similar informative linguistic examples, a distinctive semantic theme is proposed which clarifies the meaning differences between the Chinese *xìngfú* and English *happiness/happy*, as reflected in the following semantic explication formulated in NSM:

[A] Semantic explication of *xìngfú* ("the belief that one is cared for and loved")
(a) it can be like this:
    someone feels something very good for some time

because this someone thinks like this at this time:
(b) "I know that I can be with someone at many times
I feel something very good when I am with this someone
I feel something very good when I think about this someone
(c) at the same time, I know that it is like this:
"this someone feels something very good towards me
this someone often thinks about me
this someone wants to do good things for me
(d) I want it to be like this'"
(e) when this someone thinks like this, this someone feels something
very good for some time, like people feel at many times when they think like this
(f) it is very good for this someone if it is like this

In this explication, *xìngfú* is formulated in the form of a prototypical cognitive scenario which describes a sequence of thoughts that go with the specified feeling and which distinguishes it from other emotions. It attempts to capture a state of inner security accompanied by the experiencer's thought that they are cared for and loved by those they are attached to. Structurally speaking, Ye says that the hypothesized "thoughts are presented in three bundles" with the initial two introduced with the phrase "I know." Thus, the first bundle, represented in (a), specifies "feelings towards the other party, trust and deep attachment," from which one's subjective belief in and expectations of the other party originate. The second bundle, which Ye calls "my belief and expectations," is expressed by the component (c), in order to capture the "relational and mutual" dimensions of meaning encoded in *xìngfú*, as well as its "other-dependent" nature, which "relies not on one's own volition, but on what the other party does."

Yet, what may be most interesting for the student of cultural key words is the way the concept of *xìngfú* throws light on other areas of cultural psychology embedded in Chinese vocabulary. There are two aspects of human interaction where the idea of *xìngfú* is of paramount importance:

- The first is the caring element of *xìngfú* and the characteristically Chinese idea of love, which emphasizes "action and deeds over words" (Ye 2016, p. 75). Characteristically, this is the case for *both* familial and romantic love.
- The second is that as for *xìngfú* and the Chinese idea of love, it is the caring element that is crucial. This situation is related to the enduring Chinese value of *xiào* 孝 "filial piety" (see also, e.g., Chu and Ju 1993, pp. 199–203; Goddard 2005, pp. 84–87) and the significant cultural value of caring deeds performed by children towards their parents, "consistent with the general Chinese cultural ethos" (Ye 2016, p. 76).

Ye characterizes *xìngfú* as "relational and deeply earthly," yet "passive and fatalistic," owing to its "other-dependent nature." Moreover, because of this cultural perception, it seems to "lie beyond one's individual volition and is subject to external forces," with such other key Chinese cultural notions as, for example, *mìng* 命 "life, fate, destiny" (ibid.). This is evidenced by a number of linguistic expressions which conceptualize *xìngfú* in non-volitional terms. This points to a radically different conceptualization behind common Chinese expressions such as 幸福来敲门 *xìngfú lái qiāomén* "when *xìngfú* knocks on the door," which actually has a fatalistic ring to it, when contrasted with the culturally significant English phrase "pursuit of happiness."

Methodologically speaking, NSM-based keyword studies enable researchers to connect to other related disciplines. For example, Ye locates her own work within the tradition of ethnopsychological studies of emotions and value concepts, finding affinity with work of such cultural anthropologists as Kirkpatrick and White (1985) or Lutz (1995). The latter scholars highlighted the hard-to-eliminate dangers of cultural and cognitive biases inherent in the researcher's own language, arguing for an appropriate interpretive framework that is both culture-sensitive and have the descriptive flexibility to make sense of local cultural vocabularies. However, these ethnopsychological studies face translatability issues. With its repertoire of cross-linguistically identified and tested conceptual primes (see Goddard and Wierzbicka 1994, 2002, 2014), NSM offers a solution. As Ye points out, "the exponents of semantic primes in any language can be regarded as constituting the most basic 'experience-near' concepts within a language and culture" while "at the same time making that perspective accessible to culture-outsiders" (Ye 2016, p. 69).

The explication of *xìngfú* shows that the culturally important nuance of the meaning of the Chinese keyword is lost in its typical English glosses. An explication formulated in NSM can make such nuanced meanings more accessible to cultural outsiders.

Case Study 2: *rènao* 热闹 ("Hot and Noisy") – Having a Good Time with Other People

The purpose of the second case study is to probe into the chief uses and symbolic significance of the Chinese cultural keyword *rènao* 热闹. Routinely rendered in everyday English with its literal morpheme-by-morpheme gloss of "hot and noisy," *rènao* has been seen as crucial for making sense of Chinese social behavior across a variety of contexts by both Chinese and Western scholars. Like the previous study, this section shows how the NSM approach can be used to explore the broader meaning of rènao in culture.

Kornacki's (2017) study shows a different approach to Chinese language data. It purports to offer a semantic insight into the meanings which routinely show up during Chinese sociable (rather than simply social) interaction. In order to access a local Chinese language perspective on such social behavior, the study examines two cultural texts – a Chinese Internet media report from a local temple festival and a fairly recent controversy over two different styles of feasting, both of which frequently appeal to this salient cultural notion. The formula *crowds*, *events*, *noise* that is sometimes offered in the psychological literature dealing with this Chinese social value (see, e.g., Warden and Chen 2009) is essentially confirmed by Kornacki's linguistic and cultural data. Kornacki argues that paying close attention to the meaning and form of descriptive language used by Chinese speakers yields valuable insights into their cultural viewpoints. Significantly, the notion of *rènao* 热闹 turns out to be closely intertwined with some other prominent (and well-acknowledged in scholarly research) Chinese cultural concerns, including the idea of *rénqíngwèi* 人情味 (lit. "flavor of human feelings"; for an in-depth discussion of its sociocultural grounding, see Yang 1994, pp. 119–123), Chinese cultural identity, Chinese language, and a particularly complex culinary culture described in the anthropological literature (see, e.g., Simoons 1990).

A few quotations from sociological and anthropological research attests to the culturally prominent status of this Chinese concept:

- The literal translation is hot and noisy which also is often literally true and leads Western visitors (...) to mistake *renao* locations as chaotic and out-of-control – problems to be solved. For Chinese, however, *renao* is so ordinary it is often cognitively taken for granted. (Warden and Chen 2009, p. 217)
- To the Chinese people, *renao* is a key concept and an important feature of any successful celebration. It is considered embarrassing or foreboding to hold a wedding, banquet, or in some cases, a funeral that lacks *renao* – the emotion that transforms formal occasions into warm and interactive events. (Yu 2004, p. 138)

The social and psychological relevance of *rènao* 热闹 ("hot and noisy") in Chinese social interaction is captured particularly aptly in Yu's (2004) description of the not-so-official aspects of Chinese social behavior:

- being in a *renao* place (like a night market) creates a subjective feeling of safety. (...) A physical and emotional sense of relaxation leads to what many informants believe is undisciplined behaviour. (...) Dressing down is perhaps more frequent than any other aspect of relaxed behaviour – wearing slippers instead of shoes and shorts instead of pants is part of what night market strolling is about (Yu 2004, p. 140).

One might note that Yu draws attention to the likely psychological dynamics underlying such observed ways of behaving – he points out that other frequently noted manifestations of such relaxed behavior involve "eating while walking along a street, eating food in an inappropriate manner, talking and laughing freely in public place, bargaining without considering one's own status, rummaging through merchandise, and littering carelessly." As the Chinese scholar perceptively concludes, "All of these are considered to be actions of people from lower social and economic classes – except when they are done in night markets" (Yu 2004, p. 140).

Thus, as argued by the researchers quoted in the first part of this paper, the word *rènao* tends to evoke a number of popular, socially salient, and "close-to-the-experience" cultural practices, objects, activities, and values. Such values and practices can be also accessed by studying vocabulary expressing them, as is done in Kornacki's (2017) paper.

Chinese linguistic expressions routinely used to describe various aspects of social behavior perceived as *rènao* are valuable cues which allow cultural outsiders to gain insight into everyday conceptualizations of Chinese social interaction. In order to examine them in their natural context, Kornacki's paper focuses on meaningful descriptive features of a journalistic report of a *rènao* event (temple festival, *miàohuì* 庙会). The title of the report makes use of this culturally significant Chinese lexical item twice, and reads:

(3) 烟台玉皇顶庙会热闹开场, 好看好玩好热闹。
Yāntái Yùhuángdǐng miàohuì rènao kāichǎng, hǎokàn hǎowán hǎo rènao
"Temple festival at the Yúhuángdǐng temple in Yāntái begins with *rènao*, it's good to watch, good fun, and good *rènao*."

As the paper shows (for further discussion, see pp. 65–67), salient features of the event are identified with Chinese descriptive phrases which focus on the

Performative (active) nature of the event [which] is highlighted with phrase such as *mínsú biǎoyǎn* 民俗表演 ("folk performance"), *gāoqiāo biǎoyǎn* 高跷表演 ("walking on stilts performance"), *yāogǔ biǎoyǎn* 腰鼓表演 ("waist-drum performance"), and *huájījù biǎoyǎn* 滑稽剧表演 (comic talk performance).

Large numbers of participating people are important, and this is explicitly conveyed with nominal expressions referring to very large groups of people, rather than individuals: *shàngwàn* 上万 (lit. "over ten thousand" – i.e., very numerous), *rénshānrénhǎi* 人山人海 (lit. "people-mountains-people-seas" – huge crowds), etc.

Tasting and sampling of local delicacies *měishí* 美食 and "snacks" *xiǎochī* 小吃 is typical of social events rich in *rènao* 热闹 atmosphere (see also Yu 2004).

Engaged spectatorship of the attending public who not only *guānkàn* 观看 ("watch, view") and *cānyù* 参与 ("participate") but also share their appreciation and interest are implied by symbolic expressions such as *bùduàn jiàohǎo* 不断叫好 ("unceasing applause") or taking pictures, described as *liúxià měihǎo shùnjiān* 留下美好瞬间 ("preserving beautiful moments").

Food and Chinese Cultural Value of rènao 热闹

Much has been written about the uniquely rich symbolic significance of food and food consumption in Chinese culture (see Simoons 1990 for a standard reference on this subject). While the cultural anthropology of food is an established subject of research (see Pilcher 2012), it can be argued that the Chinese cultural concept of *rènao* 热闹 provides a way to explore social aspects of food in its communicative context. In order to highlight some relevant linguistic evidence indicative of Chinese speakers' attitudes towards feasting, the second part of Kornacki's paper (pp. 70–77) looks at a minor controversy surfacing in Chinese Internet media which ostensibly consisted of a debate over the advantages and disadvantages of two styles of feasting. These two styles of feasting are referred to in Chinese as:

- *fēncānzhì* 分餐制 "separate/individual meal system"
- *jùcānzhì* 聚餐制 "together/group meal system"

Due to the risk of the SARS epidemic affecting parts of China at the time of reporting (roughly, 2004), the authorities decided to implement the former system more broadly, so that individual patrons would consume their dishes strictly from their own plates. The essence of the ensuing problem is pinpointed in the title of one of the posts, which read:

(4) 春节分餐制叫好不叫座降低热闹气氛人情味不足
Chūnjié fēncānzhì jiàohǎo bù jiàozuò jiàngdī rènao qìfēn rénqíngwèi bùzú
"Separate meal system at the Spring Festival is praised but unappealing: the atmosphere of *rènao* is lowered and *rénqíngwèi* is insufficient."

While many Chinese patrons complained about the "individual meal system" arrangements, they articulated their dissatisfaction by referring to two prominent Chinese cultural keywords – *rènao* 热闹 and *rénqíngwèi* 人情味 (roughly, "the flavor of human feelings"). For example, one of the Chinese customers phrased his opinion in the following terms:

(5) 今天来的客人全是家宴，几乎每一桌都有老人和小孩。按照咱们中国人的传统家宴讲的是丰盛气氛和热闹, 所以没有客人要求分餐。
Jīntiān lái de kèrén quán shì jiāyán, jīhū měiyī zhuō dōu yǒu lǎorén hé xiǎohái. Ànzhào zánmen Zhōngguórén de chuántǒng, jiāyàn jiǎngde shi fēngshèng, qìfēn he rènao, suǒyǐ méiyǒu kèrén yāoqiú fēncān.
"The guests who have come today all take family feasts, there are old people and children at almost every table. According to our Chinese tradition, family feasts should be sumptuous, (rich in) atmosphere and *rènao*, therefore there are no guests who demand 'separate style' meals."

Another illuminating comment is provided by a Chinese customer who noted that:

(6) 春节之前机关团体连环餐，朋友聚会餐 时曾有客人要求分餐之, 前几天有三个老外就是一人点一个菜，个人吃自己盘子里的菜，但在春节合家团圆之际，分餐制自然会使餐桌的热闹气氛降低人情味不足
Chūnjié zhīqián jīguān tuántǐ liánhuāncān, péngyou jùhuìcān shí céng yǒu kèrén yāoqiú fēncānzhì, qián jǐ tiān, yǒu sānge lǎowài jiùshì yī rén diǎn yīgè cài, gèrén chī zìjǐ pánzi lǐ de cài, dàn zài Chūnjié héjiā tuányuán zhījì, fēncānzhì zìrán huì shǐ cānzhuō de rènao qìfēn jiàngdī, rénqíngwèi bùzú.
"In the time before the Spring Festival, among the institutional group-meals and groups of friends getting together, there had been guests requesting a separate-style meal, but in the last couple of days there were just three foreigners requesting one meal per one person, with each person eating one's dish from one's own plate. However at the Spring Festival when the families reunite and get together, the separate eating style can only diminish the *rènao* atmosphere of the festive table, and create insufficient *rénqíngwèi*." (See p. 71, example 3)

The key point of such interactions at a festive time is the very special sociable atmosphere which gives a unique flavor to the event itself. According to the words of the Chinese reporter:

(7) 记者注意到聚餐的客人大多是一家人或亲朋好友围坐一桌, 亲密无间, 热热闹闹的聚餐, 几乎没有实行分餐制的

Jìzhě zhùyìdào, jiùcān de kèrén dàduō shi yījiārén huò qīnpéng-hǎoyǒu wéizuò yī zhuō, qīnmì-wújiān, rèrènaonao de jùcān, jīhū méiyǒu shíxìng fēncānzhìde.

"The journalist has noticed that the majority of the consumers are people from one family or close friends, sitting around the table, dining together (*jùcān* 聚餐) in a very *renao* manner. They haven't practiced the 'separate system' (分餐制 *fēncānzhì*) at all." (See p. 73, example 6)

Crucially, as further examples in the original paper illustrate (see, e.g., pp. 73–75), it is the "*rènao* atmosphere" (热闹气氛, *rènao qìfēn*), arising from and cementing positive feelings among Chinese people who are in relationships of *shúrén* 熟人 ("acquaintances"), *péngyǒu* 朋友 ("friends"), and *qīnrén* 亲人 ("one's family members"), that sustains and nourishes these significant interpersonal relations in Chinese society.

In order to account for such dynamic interpersonal imagery, Kornacki proposes the following semantic explication for "*rènao* atmosphere" (热闹气氛, *rènao qìfēn*) in Chinese social interaction.

[B] Semantic explication for *rènao qìfēn* ("*rènao* atmosphere")
(a) it is like this:
(b) many people are in this place now
(c) because many things are happening in this place now, they want to see these things
(d) when they are in this place, they can see many things, they can hear many things, they are with many other people at the same time in one place
(e) because of this, they feel something very good
(f) at the same time, they think like this:
"we can do some things at this time not like we can do things at other times
we can do some things in this place not like we can do things in other places
this is good"
(g) people think like this: "it is good if it is like this"

With components (a-b) clarifying a typical background of a bustling scene where many things happen at a given moment, lines (c-d) signal intensified sensory stimulation (forms, sounds, color, taste), while the elements (e) and (f) contain a reference to the characteristic social mood generated by the "*rènao* atmosphere," namely, the feeling of "togetherness" and a certain loosening of social norms, so aptly characterized in Yu (2004). On the cultural semantic level, one might emphasize here how the study of cultural keywords (such as *rènao* 热闹) feeds into the study of cultural scripts. It yields culturally revealing interactive and procedural dimensions of their social significance, which tend to be reflected in Chinese descriptive vocabulary depicting the sensory nature of such social events (see Kornacki 2017, pp. 65–67 for more examples).

Conclusions and Prospects for Future Studies

In her book *The Structure of Chinese Values: Indigenous and Cross-Cultural Perspectives*, Chinese cross-cultural psychologist Li Jiajun (2016: p.1) describes the motivation behind her study as follows: "For a changing China to clarify what

Chinese values are and how Chinese values change is an important step towards improving inter-/cross-cultural communication with the culturally different other." Increasingly, clarifying the meanings of key cultural words and expressions has been identified as an essential task in the area of cross-cultural contact. This seems particularly desirable for the long term, as comprehensive understanding of the interacting people's communicative intentions is of prime importance and useful to all involved. A further statement arguing for the necessity of acknowledging *meaning* as an indispensable research area in studies of Chinese culture comes from the eminent American medical anthropologist and renowned scholar of Chinese cultural psychology Arthur Kleinman. He concludes his book *Deep China* with a chapter titled, significantly, "Quests for Meaning." Among other issues, he formulates two related questions in this chapter which lend cross-cultural significance to the project of explicating Chinese cultural keywords (Kleinman 2011, p. 213):

- How do ordinary Chinese make sense of their experiences?
- What kind of quests for meaning do they embark on?

As demonstrated in the case studies presented in this chapter, the NSM approach has much to offer in elucidating how Chinese words and their culture-specific meanings can be approached so as to render their significance in a semantically illuminating way. The key advantage of employing NSM in keyword studies can be seen in its distinct focus on meaning and its attention to usage-based linguistic evidence. NSM techniques of semantic explications reflect a culture-internal perspective on local meanings by rendering their content with simple and cross-linguistically tested words. While originally conceived as a program of research in linguistic semantics, NSM offers a unique blend of distinct advantages to keyword studies: a theory-driven method of dealing with terminological ethnocentrism; a considerable body of existing research in cross-cultural semantics; and a distinctly articulated research focus which balances a linguistic search for semantic primes and universals with detailed semantic studies of culture-specific domains and language-specific vocabularies of social experience. For more examples of recent NSM research on cultural keywords in a number of languages and cultures, see Goddard and Ye 2014; Levisen and Waters 2017; Peeters 2019.

The Chinese cultural keywords presented in this chapter belong to a group of words which are maximally language- and culture-specific (see Wierzbicka 2016), even though in some psychological research they have been used to gloss assumed universals of human experience. The case study of the key Chinese emotion word *xìngfú* 幸福 ("emotional happiness") exposes methodological shortcomings of such semantically myopic universalism where the English word *happiness* is taken as a universal yardstick of human experience. The second case study of *rènao* 热闹 offers a focused look at the salient linguistic exponents and descriptors of Chinese sociable (as distinct from plainly social) interaction. Approaching the subject of Chinese cultural concepts such as *xìngfú* 幸福 (roughly, "happiness") and *rènao* 热闹 (roughly, "hot and noisy") from a cross-linguistic point of view, one cannot help but notice that they resist standard translation, while their makeshift glosses usually hide more than they reveal (for more on the persistence of semantically inadequate translations of cultural vocabularies, see Geertz 2000, pp. 203–217).

What is particularly encouraging in this regard is that a major advantage of applying NSM in the context of Chinese language and culture lies in its "consistent effort to model a local meaning system so that a local perspective can be framed in culture-neutral terms, an attempt made possible by the use of these universal primes. It is clear that if we want to have a deep understanding of China and its people, knowing the 'what' and 'how' of Chinese social behaviour is not sufficient. It is crucial that we fully understand the question of **why**" (Ye 2013, p. 369) (emphasis added).

Future research and study within NSM will assist students of Chinese language and move towards clarifying the objectives stated by Kleinman, in that it will focus students' attention on specific lexical expressions and their salient cultural meanings. In particular, three current avenues of research using the NSM framework – namely the investigation of emotion words, personhood constructs (words like *xīn* 心 *heart* or *mind*), and social relation words (see Ye (2017) for more examples) – could enrich future studies of Chinese language in ways which allow for better cross-linguistic comparability and accessibility in both formal and informal cross-cultural communicative encounters.

References

Barlow, Cleve. 1991. *Tikanga Whakaaro. Key concepts in Māori culture*. Auckland: Oxford University Press.
Bond, Michael Harris. 1991. *Beyond the Chinese face. Insights from psychology*. Hong Kong: Oxford University Press.
Bond, M.H., and K.-k. Hwang. 1986. The social psychology of Chinese people. In *The psychology of the Chinese people*, ed. M.H. Bond, 213–266. New York: Oxford University Press.
Boodberg, Peter A. 1953. The semasiology of some primary Confucian concepts. *Philosophy East and West* 2 (4): 317–332. https://doi.org/10.2307/1397493.
Carr, Michael. 1992. Chinese "face" in Japanese and English (Part I) Otaru University of Commerce 人文研究 (1992), 84: 39–77. http://hdl.handle.net/10252/1737. Accessed June 2001.
———. 1993. Chinese "face" in Japanese and English (Part II) Otaru University of Commerce 人文研究 (1993), 85: 69–101. http://hdl.handle.net/10252/1585. Accessed June 2001.
Chu, Godwin, and Yanan Ju. 1993. *Great Wall in Ruins. Communication and cultural change in China*. Albany: State University of New York Press.
Eifring, Halvor, ed. 1999. *Minds and mentalities in traditional Chinese literature*. Beijing: Culture and Art Publishing House.
Elvin, Mark. 1989. The tales of xin and shen. Body-person and heart-mind in China during the last 150 years. In Feher et al., eds., Fragments for a history of the human body. Part 2. 266–350. New York: Zone.
Feher, M., R. Naddaff, and N. Tazi, eds. 1989. *Fragments for a history of the human body. Part 2*. New York: Zone.
Geertz, Clifford. 1973. *The interpretation of cultures*. London: Fontana Press.
———. 2000. *Available light. Anthropological reflections on philosophical topics*. Princeton: Princeton University Press.
Gentz, Joachim. 2009. *Keywords re-oriented*. interKULTUR, vol. IV, Universitätsverlag Göttingen, Göttingen.

Gladkova A.N., Larina T.V. 2018. *Anna Wierzbicka, words and the world*. In Studies in semantics: For Anna Wierzbicka's anniversary, Vol 22, No 3, 499–520. http://journals.rudn.ru/linguistics/article/view/19344. https://doi.org/10.22363/2312-9182-2018-22-3-499-520

Goddard, Cliff. 2002. Overcoming terminological ethnocentrism, 27–28. IIAS Newsletter, Leiden: International Institute for Asian Studies.

———. 2005. *The languages of East and Southeast Asia: An introduction*. Oxford: Oxford University Press.

———. 2006a. Ethnopragmatics: a new paradigm. In *Ethnopragmatics. understanding discourse in cultural context*, ed. Cliff Goddard, 1–30. Boston/Berlin: De Gruyter Mouton.

———, ed. 2006b. *Ethnopragmatics. Understanding discourse in cultural context*. Boston/Berlin: De Gruyter Mouton.

Goddard, Cliff, and Wierzbicka, Anna (Eds.) 1994. Semantic and lexical universals: Theory and empirical findings. Amsterdam: John Benjamins. https://doi.org/10.1075/slcs.25

Goddard, Cliff, and Wierzbicka, Anna (Eds.) 2002. *Meaning and universal grammar: Theory andd empirical findings*. 2 volumes. Amsterdam: John Benjamins. https://doi.org/10.1075/slcs.60 (vol. 1), https://doi.org/10.1075/slcs.61 (vol. 2).

Goddard, Cliff, and Anna Wierzbicka. 2014. *Words and meanings: Lexical semantics across domains, languages & cultures*. Oxford: Oxford University Press.

Goddard, Cliff, and Zhengdao Ye. 2014. Exploring "happiness" and "pain" across languages and cultures. *International Journal of Language and Culture* 1: 131–148. https://doi.org/10.1075/ijolc.1.2.01god.

Harbsmeier, Christoph. 2019. Philological reflections on Chinese conceptual history: Introducing Thesaurus Linguae Sericae. In Li, W., & Pines, Y. (2019). Keywords in Chinese culture: Thought and literature. Hong Kong: The Chinese University of Hong Kong Press, pp.381–404.

Ho, David Y.F. 1976. On the concept of 'face'. *American Journal of Sociology* 81: 867–884.

Kirkpatrick, J., and G.M. White. 1985. Exploring ethnopsychologies. In *Person, self and experience: Exploring pacific ethnopsychologies*, ed. G.M. White and J. Kirkpatrick, 3–34. Berkeley: University of California Press.

Kleinman, Arthur. 2011. Quests for meaning. In *Deep China: The moral life of a Person*, ed. Kleiman et al., 263–290. Berkeley, Los Angeles, London: University of California Press.

Kleinman, Arthur, Yunxiang Yan, Jing Jun, Sing Lee, Everett Zhang, Tianshu Pan, Fei Wu, and Jinhua Guo. 2011. *Deep China: The moral life of the person*. Berkeley/Los Angeles/London: University of California Press. https://doi.org/10.1525/j.ctt1pnb7k.

Kornacki, Paweł. 1995. *Heart & face: Semantics of Chinese emotion concepts*. PhD thesis, Australian National University. Open access

———. 2017. *Rènao*: What does it mean to have a good time the Chinese way? An ethnopragmatic exploration of a Chinese cultural keyword. In *East-Asian and Central-European encounters in discourse analysis and translation*, ed. Anna Duszak, Arkadiusz Jabłoński, and Agnieszka Leńko-Szymańska, 57–82. Warsaw: Institute of Applied Linguistics. PDF (open access).

Layard, Richard. 2003. Happiness: has social science a clue. Lecture. 1.

Levisen, Carsten, and Sophia Waters, eds. 2017. *Cultural keywords in discourse*. Amsterdam: John Benjamins.

Li, Jiajun. 2016. *The structure of Chinese values: Indigenous and cross-cultural perspectives*. Reading: Paths International Ltd., UK.

Li, W., and Y. Pines. 2019. *Keywords in Chinese culture: Thought and literature*. Hong Kong: The Chinese University of Hong Kong Press. muse.jhu.edu/book/73729.

Li, Jing, Christer Ericsson, and Mikael Quennerstedt. 2013. The meaning of the Chinese cultural keyword xin. *Journal of Languages and Culture* 4 (5): 75–89. https://doi.org/10.5897/JLC12.054. / Open access.

Lutz, Catherine. 1995. Need, nurturance, and the emotions on a Pacific atoll. In *Emotions in Asian thought: A dialogue in comparative philosophy*, ed. J. Marks and R.T. Ames, 235–252. Albany: State university of New York Press.

Munro, Donald J., ed. 1985. *Individualism and holism: Studies in Confucian and Taoist values*. Ann Arbor: Center for Chinese Studies, University of Michigan.

Peeters, Bert, ed. 2019. *Heart- and soul-like constructs across languages, cultures, and epochs*. New York: Routledge.

Pilcher, Jeffrey M., ed. 2012. *The Oxford handbook of food history*. Oxford. ISBN: 9780199729937 Published online: Nov 2012: Oxford University Press. https://doi.org/10.1093/oxfordhb/9780199729937.001.0001.

Richards, Ivor A. 1932. *Mencius on the mind: Experiments in multiple definition*. London: Routledge.

Simoons, Frederick J. 1990. *Food in China: A cultural and historical inquiry*. London: CRC Press.

Warden, Clyde A., and Judy F. Chen. 2009. When hot and noisy is good: Chinese values of renao and consumption metaphors. *Asia Pacific Journal of Marketing and Logistics* 21 (2): 216–231.

Wierzbicka, Anna. 1992. *Semantics, culture, and cognition. Universal human concepts in culture-specific configurations*. New York, Oxford University Press.

———. 1997. *Understanding cultures through their key words: English, Russian, Polish, German, and Japanese*. Oxford: Oxford University Press.

———. 2016. Two levels of verbal communication, universal and culture-specific. In *Verbal communication*, ed. Andrea Rocci and Louis de Saussure, 447–481. Berlin: De Gruyter Mouton.

Williams, Raymond. 1983. *Keywords: A vocabulary of culture and society*. London: Fontana.

Yang, Mayfair Mei-hui. 1994. *Gifts, favors & banquets. The art of social relationships in China*. Ithaca/London: Cornell University Press.

Yang, K.S., and David Y.F. Ho. 1988. The role of *Yuan* in Chinese social life: A conceptual and Empirical analysis. In *Asian contributions to psychology*, ed. A.C. Paranjpe, David Y.F. Ho, and R.W. Rieber, 263–281. New York: Praeger Publishers.

Ye Zhengdao. 2013. Understanding the Conceptual Basis of the 'Old Friend' Formula in Chinese Social Interaction and Foreign Diplomacy: A Cultural Script Approach. *Australian Journal of Linguistics* 33 (3): 365–385. https://doi.org/10.1080/07268602.2013.846459

Ye, Zhengdao. 2016. The meaning of "happiness" (xìngfú) and "emotional pain" (tòngkǔ) in Chinese. *International Journal of Language and Culture* 1 (2): 194–215. https://doi.org/10.1075/ijolc.1.2.04ye.

———. 2017. The semantics of social relation nouns in Chinese. In *The semantics of nouns*, ed. Zhengdao Ye, 63–88. Oxford: Oxford University Press. https://doi.org/10.1093/oso/9780198736721.003.0003.

Yu, Shuenn Der. 2004. Hot and noisy – Taiwan's night market culture. In *The minor arts of daily life: Popular culture in Taiwan*, ed. David K. Jordan, Andrew D. Morris, and Marc L. Moskowitz, 129–149. Manoa: The University of Hawai'i Press.

The Semantics of Kinship in Sinitic Languages

21

Wendi Xue and Zhengdao Ye

Contents

Introduction	514
The Broad Context of Kinship Studies	515
Early Period	516
Contemporary Period	519
Survey of Chinese Kinship Studies in English Scholarship	520
A Lexicographic Tradition in Chinese Scholarship	524
Recent Development	530
Documenting Kinship Terms of Chinese Dialects	531
Research on Specific Kinship Terms	532
Methodological Issues in Typological Research	533
Future Directions	535
References	536

Abstract

This chapter provides an overview of scholarly efforts to collect, codify, and explain kinship terms in Chinese history and research on kinship terms in Sinitic languages (Chinese dialects) in general and focuses primarily on distinctive Chinese approaches to the semantics of kinship terms and recent developments in Chinese kinship research. It places discussions of Chinese kinship semantics within the broad context of crosslinguistic, typological research of kinship semantics in contemporary anthropological and linguistic research. The chapter starts with a description of some influential approaches to kinship terms within these two disciplines. Some of these

---

W. Xue (✉)
Australian National University, Canberra, Australia
e-mail: Wendi.Xue@anu.edu.au

Z. Ye
School of Literature, Languages and Linguistics, The Australian National University, Canberra, Australia
e-mail: Zhengdao.Ye@anu.edu.au

approaches have influenced recent studies of Chinese kinship semantics. It then discusses general approaches to kinship semantics in both Chinese and Anglophone scholarship, highlighting both their commonality and the distinctive aspects of Chinese approaches. This is followed by an account of recent developments in studies of Chinese kinship semantics, and a discussion of future directions.

Keywords

Meaning and culture · Chinese kinship terms · Kinship Semantics · Kinship lexicography · Translatability of kinship vocabulary · Kinship typology · Kinship system · Kinship database · Linguistic normativity · Lexical typology · Psychological reality

Introduction

Kinship plays a vital role in Chinese culture. It underlies Chinese social structure and guides social interaction. As early as the Western Zhou dynasty (founded in the eleventh century BCE), several kinship terms were employed as titles of nobility known as *wǔděng jué* 五等爵 ("five aristocratic peerage ranks"). The highest rank, equivalent to a duke, was entitled *gōng* 公 (originally "father," present-day "grandfather"); a count, *bó* 伯 (originally "elder brother," present-day "father's elder brother"); and a viscount, *zǐ* 子 (originally "child," present-day "son"). The normative role of kinship relations in Chinese society was reinforced in Confucian canons. For example, the *wǔ cháng* 五常 ("Five Constant Virtues"), as posited in the Confucian classic *Shàng Shū* 尚书 ("Canon of History"), suggests that a father should possess the virtue of *yì* 义 ("justice"); a mother, of *cí* 慈 ("affection"); an elder brother, of *yǒu* 友 ("friendliness"); a younger brother, of *gōng* 恭 ("respect"); and a son, of *xiào* 孝 ("filial piety"). These virtues were formulated to regulate the normativity of kinship relations inside a traditional Chinese family, and the ideal structure of all forms of social organization is based analogously on this normativity. That is why in the *Analects*, Confucius claims that the essence of governance is to "let the ruler be a ruler, the subject a subject, the father a father, and the son a son" (*jūn jūn chén chén, fù fù zǐ zǐ* 君君臣臣, 父父子子). The relation between a ruler and subjects is clearly modelled on the relationship between the father and the son within a family. From a Confucian standpoint, social harmony stems in part from every member knowing their own position and enacting their respective role in the hierarchical social organization. Kinship was chosen as the bedrock of such a social order. Therefore, greater precision in the system of descriptive terminology used to distinguish different kinship relations became essential in Chinese culture.

Socially regimented kinship relations define interactional behavior between members of a society and enact the core Confucian value of *lǐ* 礼 ("rites; norms of appropriateness"). The fundamental role that kinship plays in Chinese society has given rise to an elaborated and extensive vocabulary for referring to and talking about kinship relationships. It is not surprising then that there is a long tradition of the study of kinship terms in Chinese scholarship dating back to the *Erya* (尔雅), the first Chinese dictionary and encyclopedia written approximately during the fourth to second century BCE. Alongside this lexicographic tradition, the Chinese scholarly tradition has engaged in systematic and meticulous efforts to collect, codify, and explain kinship terms throughout its history.

The objective of this chapter is to provide an overview of these efforts to record Chinese kinship terms and research kinship terms in Sinitic languages (Chinese dialects) in general. It will focus primarily on distinctive Chinese approaches to the semantics of kinship terms and outline recent developments in Chinese kinship research. In doing so, we place our discussion within the broad context of crosslinguistic, comparative studies of the meaning and codification of kinship terms carried out in anthropology and linguistics.

Two methodological issues are of particular interest to our discussion. One concerns **building semantic databases of kinship terms** in Sinitic languages. This includes collecting Chinese kinship terms in as many varieties as possible and provides systematic definitions of these terms. The other relates to **the issue of translatability**, i.e., how can Chinese kinship terms be precisely and authentically represented in comparative studies of world kinship systems? Both issues relate to meaning, and the second is concerned with meaning in a crosslinguistic context. Having these two issues as our central concerns for this overview also necessitates that the literature covered in the chapter is selective.

It should be stressed that kinship terms are used both referentially and vocatively in the Chinese context. A kinterm (a shorthand for kinship terms) is used vocatively when the speaker addresses the hearer directly; but is used referentially when the term is employed to refer to a relative other than the addressee. In this chapter, we cover literature dealing with both usages. As we shall see, the two types of usage for kinship terms are often dealt with differently by Chinese scholars on account of their different linguistic and social functions. These two types – referring to one's own or to others' relatives versus addressing interlocutors, are both fundamental to defining proper social relationships and regulating appropriate behaviours in social interaction, and therefore occupy important places in the history of Chinese scholarship on kinship research.

Given that this review places the discussion of Chinese kinship semantics in the broad context of crosslinguistic, typological research of kinship semantics across anthropology and linguistics in recent times, the chapter will start with a description of some influential approaches to kinship terms in these two disciplines. Some of these approaches have also influenced recent studies of Chinese kinship semantics. It will then discuss general approaches to kinship semantics in both Chinese and Anglophone scholarship, highlighting both their commonality and the distinctive features of Chinese approaches. This will be followed by an account of recent developments in studies of Chinese kinship semantics. The final section discusses future directions.

The Broad Context of Kinship Studies

The celebrated British anthropologist Robin Fox famously said, "Kinship is to anthropology what logic is to philosophy or the nude is to art" (Fox 1967:10). According to J.A. Barnes (1980) and others, the study of kinship was an unquestionably core focus for social and cultural anthropologists in the early half of the twentieth century. In light of this we will commence our review of seminal anthropological works from this period first, and then turn to more recent works and discuss trends in kinship research in linguistics over the last two decades, a period often referred to by kinship researchers as a renaissance of kinship studies.

Early Period

L. H. Morgan in the late nineteenth century, who is often regarded as the initiator of the anthropological study of kinship, proposed two macro-categories of kinterms found in different societies: the "classificatory" and the "descriptive" systems (Morgan 1871, 1877; Peletz 1995:344; Dziebel 2007:3). In a classificatory system, relatives of different kintypes are often grouped together under a general kinship term, while a descriptive system tends to describe each and every individual relative with distinctive specific terms. Morgan proposed three "systems of consanguinity": the Malayan system, the Turanian or Ganowanian system, and the Aryan system, with the first two primarily organized on a classificatory basis while the last on a descriptive one (cf. Shen, 1989).

Morgan's (1871) work touches upon Chinese kinterms. This is rare among works from this early period that aim to classify world kinship systems. He classifies the Chinese kinship system as a descriptive system for different individual relatives in Chinese tend to be described with specific terms. By extension, Chinese kinship terms are of an Aryan system. We will discuss Morgan's work on Chinese kinship terms in the next section.

Morgan's classification of kintypes with reference to specific cultures and societies has had an influence on subsequent anthropological kinship research. This is exemplified in George Murdock's (1949) proposal of six kinship systems: (a) Hawaiian, (b) Eskimo, (c) Iroquois, (d) Omaha, (e) Crow, and (f) Sudanese. This taxonomy, or typology, is by far the most comprehensive and is thus adopted by many scholars in their analysis of kinship systems as well as kinterms. In the same vein of Murdock's work, Schusky's (1965) manual for kinship analysis is a comprehensive and user-friendly guide to this traditional paradigm, with the major typological framework based on Murdock's six kinship systems. For a recent critique of the model, see Read (2013).

In today's terms, these early anthropological efforts, by Morgan and Murdock, could be described as building kinterm databases while attempting macro-taxonomy. It seems, however, that, despite the fact that typology or taxonomy can only be based on some sort of kinship databases, scholars of that era did not really pay much attention to building large-scale kinship databases or compiling kinship dictionaries. This is markedly different from the Chinese tradition, which will be discussed in the next section.

Nor were scholars of the early period particularly concerned with the translatability issue. They use English for describing kinship relations and assume interpretation based on the English language is applicable to all languages. For example, Murdock (1949) proposes a convenient notation for kinship semantic analysis, using the first two letters from eight English-based primitive kinterms as the notation symbols: **Fa** for father, **Mo** for mother, **Br** for brother, **Si** for sister, **So** for son, **Da** for daughter, **Hu** for husband, and **Wi** for wife. All the other terms are supposed to be composed of these primitives, e.g., a paternal uncle is **FaBr**. As can be seen, translation was certainly not their focus. But they did recognize the importance of kinterm meaning representation, and gave more attention to detailed analysis of individual kinship terms.

Early anthropologists, despite their strong interests in macro-taxonomy of kinship systems and indifference to crosslinguistic translatability between varied kinship lexicons, did regard the description of different kinterms as one of their core tasks, for only by that can a specific kinship system be placed in the right position of the universal kinship classification. In return, the taxonomic findings also provide tools of description and analysis for the study of kinship terminology in different ethnic groups, known as the "traditional kin-type designation" method, as summarized and detailed in Wallace and Atkins's (1960) paper "The Meaning of Kinship Terms," a landmark in early kinship semantics. In that vein, every kinterm in a non-English language can be semantically analyzed via a "primitive English term" or the assembly of several such primitive terms (ibid.:58). For instance, Chinese *mā* 妈 can be represented by "mother," but *jiù* 舅 should be analyzed by two such terms, as "mother's brother"; however, English *aunt* requires juxtaposition of such multiple-term compounds, including "father's sister" and "mother's sister." This method, clearly inspired by Murdock (1949)'s ad hoc notation scheme, is especially important regarding the translatability issue, in that it has intuitively proposed some reductionist framework and attempts to analyze and describe each kinterm by a limited number of "primitive" elements. This tradition not only paved way for the advent of the structuralist method of componential analysis but also, to some extent, inspired the later Natural Semantic Metalanguage school.

It was not until 1950 when structuralism in linguistics started to influence the anthropological circle that scholars began to embark on large-scale analysis of individual kinship terms. "Componential analysis" (CA henceforth) as a tool of semantic analysis of individual kinship terms was then introduced chiefly by Goodenough (1956) and Lounsbury (1956). Although CA was apparently built upon the previous kin-type notation tradition, it marks a great stride forward toward the internal organization logic of the target kinship system since it endeavors to find out the fundamental dimensions shared by a kinterm system that constitute a logical semantic network with all terms occupying their respective slots. Therefore, each kinterm in that system can be defined by these parameters indicating minimal dimensions (termed as "components"), such as sex of relative, generation, and lineality. For example, English "grandfather" and "grandmother," in the vein of CA, are described as "$a_1b_1c_1$" and "$a_2b_1c_1$," respectively, in which "**a**" stands for the sex distinction (a_1 for male and a_2, female), "**b**" for generation (b_1 indicates "two generations above ego"), and "**c**" for lineality (c_1 for "lineal"). Therefore, if the two components a_1 and c_1 in "grandfather" are kept while b_1 is replaced by b_2 ("one generation above ego"), this is the semantic structure of the English "father"; if a_1 and b_2 of "father" are kept but "c_1" is replaced by "c_2" ("co-lineal"), the componential combination generates an English "uncle" (cf. Wallace and Atkins 1960). Obviously the "components" in CA are more logically organized and justified, and more abstract, than the former primitive "kin-type" symbols intuitively borrowed from English. Nevertheless, these "components" are, at a deeper level, still rooted in the English (or rather, European) cognitive mindset of scientific analysis.

Although CA had, since its proposal, become a dominant descriptive tool in the analysis of kinship terms, it also faced criticism. For example, Wallace and Atkins (1960), as mentioned previously, examined the newly-emergent CA and discussed its assorted "assumptions and procedures." On that basis, the authors identified five problems of CA, including (a) the problem of homonyms and metaphors; (b) the problem of definition versus connotation; (c) the problem of complementarity, paradigms, and spaces; (d) the problem of noncommutative relational concepts; and (e) the problem of psychological reality, social-structural reality, and the indeterminacy of semantic analysis.

From a semantic point of view and for the purpose of the discussion of kinship semantics in Chinese (see below), the problems of "psychological reality" and ethnocentrism are perhaps the most important, the latter Wallace and Atkins (1960: 64) identify as involving "an apparent indeterminacy in analysis which in turn raises a number of basic theoretical problems concerning the purposes of componential analysis, the nature of model-building in anthropology, and the constraints on purpose imposed by the technical demands of a convenient model." Goodenough (1956), as well as other anthropologists that follow, all strive to achieve an insider-based description of kinterms which can reflect the individual insider's cognitive reality, but the issue of ethnocentrism, among others, remains. This is because culture-specific terms are, to a large extent, explained via a defining vocabulary forged in the vein of specific ethnic languages, especially English. Inevitably, such a methodology, which is based on an outsider's cognitive structure, cannot authentically represent the meanings of the target terms or represent what is psychologically real for the insiders.

David Schneider initiated the shift of American anthropologists' focus from describing the unfamiliar kinship systems in remote indigenous communities to re-analysing the English kinship terminology (Schneider 1968). Among others, Robbins Burling noticed that analysis of kinterms at the time had relied heavily on various abstract symbols and mathematical formalism (i.e., CA-style), neglecting native speakers' own interpretation and the diversity in daily usage of the terms, which, combined together, did not result in the claimed objective of representing the psychological structure of the insiders (Burling 1970). To counter this, Burling resorted to some very plain, or "banal" as he stated it, but reasonable ways, such as basing his analysis on the native speakers' verbal definition of the term, as well as on the natural sequence of the acquisition of usage of the kinterm from children to adults, and tried to work out a more cognitively real representation of the American kinship terms (ibid.). For instance, an "uncle" in American English is thus explained as "**parent's brother**" (for adults and children who have learned the cover term "parent") or "**mother's or father's brother**" (for children who have not yet learned "parent"), instead of the CA-based definition as "male (a_1), co-lineal (c_2), of the first ascending generation (b₂)." In other words, Burling prefers an approach of kinterm analysis that is understandable and learnable to native children (cf. Wallace and Atkins 1960, Burling 1970). However, Burling merely relies on the speakers' native natural languages for such psychologically real analysis instead of proposing some universal and intertranslatable metalanguage.

Contemporary Period

Although research on kinship has declined since the 1970s (e.g., Keen 2014), there has been a resurgence of interest in kinship from both anthropologists and linguists, as well as from scholars in other disciplines (e.g., Peletz 1995; Moravcsik 2013) due to the irreplaceable role of kinship as the foundation of all human societies (Dash 2004:88–99; Good 2010:396). In the words of Patrick McConvell and others, researchers of different fields are now "turning once more toward kinship" (McConvell et al. 2013:2).

In this renewed interest in kinship terms, anthropological linguists have played an important role. They aim to document kinship terms in as many languages as possible, aided by the greater knowledge researchers now have on many underdocumented languages. At the same time, they aim to work out a typology of kinship systems encoded in diverse languages.

Several trends are particularly noticeable in these endeavors. The first trend is the greater importance attached to the grammar and pragmatics in linguistic studies of kinship terms, such as Zeitlyn (1993), Dziebel (2007), McConvell et al. (2013), and Moravcsik (2013). The second one is the building of kinship databases. A great deal of work along this line of research has been undertaken by scholars working on Australian Aboriginal languages, such as the Australian Research Council-funded AustKin project (its online database was launched on March 3rd 2016 by P. McConvell, H. Koch, L. Dousset, J. Simpson, R. Hendery, J. Bell, P. Kelly, and other scholars). Building empirical databases of kinship terms is also intrinsically tied to their typological studies and to a deeper understanding of the foundations of human society and the diversity of human cognition (e.g., Dousset et al. 2010). Researchers are keenly aware that documenting kinship terms is not just about database building; it also relies on well thought-out semantic tools that can make crosslinguistic comparison possible. Thus, a third trend can be found in greater attention being given to the coding language. There have been continuous and lively debates among scholars about the most suitable metalanguage for defining kinship terms in crosslinguistic research (cf. Evans 2010; Wierzbicka 1992, 2013, 2016, 2017). In regard to this trend, the study of meaning and the notation system formulated by Anna Wierzbicka aims to provide a neutral metalanguage which can simultaneously enable fine-grained meaning analysis from an "emic," that is, an insider's, perspective. In her view, neither the traditional kin-type taxonomy nor the classical CA methods can guarantee a psychologically real "emic" exploration of the native speakers' cognitive world. In arguing against Burling's (1969) pessimistic viewpoint that the semantic analysis of anthropologists can only work out an artificial rule system to explain what he has observed, Wierzbicka (1992:329–370) claims that researchers are able to restore the "psychological reality" of the target speakers by adopting Natural Semantic Metalanguage (NSM) (Wierzbicka 1992: 329–370). We can take the English term "father" as an example. "Father" is defined in the classical CA tradition as being composed of the following components:

1. "male (a_1)
2. of the first ascending generation (b_2)
3. lineal (c_1),"

Such semantic components were criticized by Burling (1970: 23) as a "fancy," as being "esoteric," and unnatural in the native culture. In contrast, the NSM-based explication of English "father" is represented as follows:

Someone's father
a man
some time before this someone was born
   this man did something with a part of the body to a woman's body
something happened inside this woman's body because of this
some time after this this someone was born because of this

All of these trends discussed above are highly relevant to the present review of Chinese kinship semantics, especially with respect to recent developments in the study of Chinese kinship semantics, a topic we will address later in the chapter.

It should be pointed out that the development of kinship studies in anthropology and linguistics has, in large, stemmed from researchers' work in pre-industrial societies that do not have extensive written records. This context is very different to the context where research into the Chinese kinship system has been carried out under, either within China or outside China, or in Chinese or English. In the next two sections, the reader will come to appreciate how a long and unbroken written history has exerted its influence on approaches to the study of Chinese kinship semantics.

Survey of Chinese Kinship Studies in English Scholarship

In this section, we provide a survey of those studies in the English literature, mostly conducted by scholars based outside China, which specifically look at Chinese kinship terms and their meanings. We first delineate those studies in a chronological order and then point out some distinctive features of this line of study against the background of kinship studies outlined in the previous section. It should be noted that a combination of various Chinese Romanisation systems will be used hereafter: if a scholar uses a particular system, we follow their convention; otherwise, the contemporary *pinyin* system for Standard Mandarin is used as per the previous sections.

In *Chinese Kinship*, Stuart Thompson remarks that kinship analysis has become the "most developed domain in the anthropology of Chinese society" (Thompson 1985: 344). To many scholars, western-based anthropological studies of Chinese society are much less fruitful, or even appear to be "caught in parochial eddies and theoretical backwaters," perhaps due particularly to the seemingly insurmountable philological, or Sinological obstacles, as pointed out, for example, by P. Steven Sangren (1982:628). However, anthropologists and linguists have been attracted by the distinctive Chinese system of kinship terminology since very early times. For example, Kroeber, one of the most influential American anthropologists, sees the Chinese kinship system as "the most perfect system conceivable to mankind" for its "richness and exactitude" (cited in Lévi-Strauss 1969:328).

As noted earlier, one of the earliest works on Chinese kinship terms in the English literature is Morgan's (1871) *Systems of Consanguinity and Affinity of the Human Family* and includes a pioneering survey of the Chinese kinship system in Anglophone academia. However, the survey served only as supporting evidence of his macro-kinship taxonomy, and his material was based on one single Englishman's experience in China, of which the reliability may be questioned.

Ching-chao Wu's (1927) "The Chinese Family: Organization, Names, and Kinship Terms" presented the first panoramic sketch of the Chinese kinship terms. However, Wu's data were a mixture of colloquial and formal forms in Mandarin, with obvious influence from the author's own local accent. For instance, the nasal initial "n-" in *nǚ* 女 "daughter" and *nèi* 內 "inner" were misspelled as *Lü* and *Lei*, and formal terms like *Shu-Mu* 叔母 (colloquial *shěnshen* 婶婶) "wife of father's younger brother" co-occur with colloquial terms like *Ko* 哥 (formal *xiōng* 兄) "elder brother."

T. S. Chen and J. K. Shryock's (1932) "Chinese Relationship Terms" marks the beginning of monographic research on Modern Chinese kinterms. They described kinship terminology in Classical, as well as in Modern Chinese, based on two Chinese dictionaries, the *Erh-ya* (i.e., *Erya* 尔雅, the earliest attested Chinese lexicon) and the *Tsu-yuan* (辞源, a 1915 sourcebook of Chinese vocabulary). They also explained the many and various additional morphemes, such as *t'ang* (堂 pinyin: *táng*, "paternal cousin with the same surname") and *piao* (表 pinyin: *biǎo*, "maternal cousin or paternal cousin with a different surname") for different cousin relations, which could combine with the major kinship terminological roots such as *hsiung* (兄 pinyin: *xiōng*, "elder brother") and *ti* (弟 pinyin: *dì*, "younger brother") to achieve clearer designation (e.g., *t'ang-ti* "father's brother's son younger than oneself" and *piao-ti* "father's sister's son or mother's sibling's son younger than oneself"). This evolved later into a tradition of exploring the inner "grammar" of the Chinese kinterms, especially in regard to their word-formation mechanism. We will come to this point in the next section.

Kroeber (1933) proceeds to analyze the "grammar" or "morphology" underlying Modern Chinese kin terms, demonstrating that the present-day "rich" and "descriptive" Chinese system is actually an independent development from an originally primitive "classificatory" or "non-descriptive" system by gradual development and application of various added descriptive elements through the ages. In Kroeber's view, Chinese kinship terms start as what Lowie (1928, cited in Kroeber 1933) terms as a "bifurcate-collateral" system found in many indigenous peoples, particularly in western America, e.g., the Cocopa's kinship system. In such a non-descriptive system, only two sets of distinctions were crucial, viz. male-line versus female-line kin, and older versus younger siblings. Later on, Chinese kinship terms developed into a rich and precise descriptive system on the original "classificatory" base. Kroeber (ibid.: 156) suggests that this might have resulted from the fact that Chinese people "obviously remain interested in kinship, whereas we [English people] want to refer to it as sketchily as possible." H. Y. Fêng's PhD dissertation is the first English monograph entirely devoted to the systematic study of the Chinese kinship (Fêng 1937). Drawing upon social anthropological research in kinship studies such as Morgan (1871) and later theories, Fêng examined the diachronic development of Chinese kinship terms in the past 2500 years, on the basis of evidence culled mainly from historical documents. In the monograph, in addition to the principles of terminological composition such as the nuclear terms (e.g., *hsiung* 兄 "elder brother" and *ti* 弟 "younger brother") and basic modifiers (e.g., *t'ang* 堂 "paternal cousin") which had been touched on to some extent by Chen and Shryock (1932), he also expounds on the structural principles, i.e., the principles that Chinese kinship structure is based upon, including the lineal vs. collateral differentiation, and generation stratification (obviously CA-inspired). The former is a vertical segmentation, such as the distinction between one's own brothers (*hsiung ti* 兄弟, lineal line) and one's father's brother's sons (*t'ang hsiung ti* 堂兄弟, collateral line), while the generation stratification is a horizontal segmentation, distinguishing kin of one generation from another by various modifiers such as *tsu* 祖 for the generation immediately above one's father's lineal and collateral relatives, and *sun* 孙 for the generation immediately below one's son's lineal and collateral kin.

Fêng (ibid.) also discusses the factors affecting the system, where he attributes the "slow but persistent" tendency found in Chinese kinship terms toward a more systematic and descriptive system to such "systematizing forces" as the *tsung-fa* 宗法 "law of kindred" and the *sang-fu* 丧服 "mourning system." Fêng (ibid.:172) defines *tsung-fa* as "sib organization," where sib, or *tsung-tsu* 宗族, refers to "a group of people possessing a common sibname (patronym), descended from a common male ancestor, no matter how remote, and characterized by a feeling of relationship" (ibid.:142). Two *tsung-fa* factors are considered by Fêng (ibid.:174) as having "direct influence" on the kinship alignment, i.e., patrilineal descent and exogamy, which led to the distinction of *tsung ch'in* 宗亲 "sib relatives" from non-sib relatives. As for the *tsung-fa*-based mourning system, it is well-known for the *wu-fu* 五服 "five grades of mourning," assigned according to different degrees of relationships. Since *sang-fu* application stops at the fourth collateral line, the Chinese kinship system accordingly highlights the sharp distinction between the first four collateral lines and all other remote collaterals outside *wu-fu*. Fêng (ibid.:183) opines that it is the influence of the sophisticated mourning system that consequently drove and transformed the original classificatory Chinese kinship system into a descriptive type. Perceptively, Fêng (ibid.:159) also notices that in Chinese kinship exists a third category of kinship terms, the "supernumerary terms," in addition to the universally acknowledged "referential *versus* vocative" dichotomy. These supernumerary terms are for sacrificial or literary use, and not to be found in vocative addressing or referential designation of natural conversation. To exemplify that, *k'ao* 考 for "father" in epitaphs belongs to this category.

Ruey (1949) represents the first monographic study focusing on a specific group of Chinese kinterms. He demonstrated that three "uncle"-type kinship words, i.e., elder paternal uncle (*po* 伯), younger paternal uncle (*shu* 叔), and maternal uncle (*chiu* 舅), and two "aunt"-type terms, namely paternal aunt (*ku* 姑) and maternal aunt (*i* 姨), had all undergone drastic semantic shift in history, and a tentative account for the underlying causes was also given.

Yuen Ren Chao (1956) is the first study to examine the Chinese address system (including kinterms) from a linguistic perspective, and the first to have paid due attention to the kinship terminological system based on spoken Modern Chinese, generally known as Mandarin. Admitting that "excellent descriptive and historical studies of Chinese kinship terms have been made" previously, Chao demonstrates his insight into the complicated variable forms of the Chinese kinterms and the actual interpersonal condition under which a particular term is decided for use. Evidently, Chao's division of kinterms into vocative, designative and learned forms follows Fêng's (1937) threefold categorization, but his observation on the conditions of kinterm use is unprecedented. For instance, he describes in detail what distinctive forms a speaker should be using when referring to a relative of a higher-than-ego generation (despite age seniority), or to an elder kinsman of the same generation, or to a younger one of the same generation, etc. Interestingly, he argues that in reference to a younger kinship member of the same or lower-than-ego generation, the name of the designated relative would conventionally replace the specific kinterm in conversation.

Chao's work is thought provoking and can be regarded as a pioneering study of the pragmatics of kinship terms, since the role of contexts in the study of kinship terms has not yet been heeded at the time. This tradition, however, does not attract many followers. Wu (1990) and Sandel (2002) are among the few. In such studies, the contextual usage of kinterms is highlighted, but it seems that little effort has been made to incorporate such usages into defining the semantic structure underlying the Chinese kinterms, a point we will return later when discussing key methodological issues in recent studies of Chinese kinship semantics.

While most of the studies available in the English literature focuses on Mandarin, Hugh D. R. Baker's (1968) study of a Chinese lineage village in Hong Kong touches on a slight smattering of the Cantonese kinterms, such as *a-kung* 阿公 (lit. "grandfather") for "ancestor," and *mui-tsai* 妹仔 (lit. "little younger sister") for "female servant," a non-kin use of the original kinterm. However, the work does not provide any description or analysis on the kinship terminology, except for an observation that "the use of an extensive kinship terminology is dying" especially among the young locals (Baker 1968:209).

A striking feature which sets these early studies of Chinese kinship terms apart from those studies outlined in the previous section, is that generally speaking, the former attach great importance to written documents while the latter attaches importance to native speakers' oral discourse. This difference might be attributed to the inaccessibility of fieldwork in China. It is possible that Yuen Ren Chao was able to give attention to kinterms used in spoken Mandarin because of the fact that he was as a native speaker and a descriptive linguist at the same time (Chao 1956). But it also reflects the enduring influence of Classical Chinese orientation in the study of kinship terms, that is, the literature-based lexicographic tradition (the topic of next section).

Apart from Morgan (1871), most early works on Chinese kinship terms in the English scholarship treat them as a self-contained domain. The terms were studied not as part of a typological panorama. That is, they were not taken to compare with kinship terms in other languages. However, recent years have witnessed more endeavours in the English academia to include Chinese kinship terms as a significant part of the crosslinguistic contrastive studies, the representative of which is Nikolayeva's (2014) *Typology of Kinship Terms*. Unfortunately, the Chinese kinship data used in Nikolayeva's research are far from the kinterms actually used by modern Chinese speakers. For instance, the term for son's wife is *zifu* 子妇 in her work, which, though literally understandable, is never heard in a real-life Chinese conversation. Therefore, the typological findings deduced from such flawed Chinese kinship data remain dubiously problematic. That also explains why building reliable kinship databases is of vital importance before we embark on any typological analysis.

We can also see from these early studies that most efforts went to Mandarin and that Chinese kinship terms are usually treated as a whole system in discussion, and only some of them, like Ruey (1949), dwell on some specific kintype groups (e.g., the "uncle"-type group). This may not be surprising, given the written sources with which the scholars worked. In some aspects, these early approaches to Chinese kinship terms, championed in Anglophone literature, are congenial to the traditional, lexicographically-based approach to kinship terms in Chinese scholarship. This point will become clearer when we discuss how kinship terms have been studied in Chinese scholarship.

A Lexicographic Tradition in Chinese Scholarship

In this section, we sketch studies of Chinese kinship terms available in the Chinese language. We begin with the longest and most enduring lexicographic tradition of kinship terms in Classical Chinese, and then survey kinship-related specialized dictionaries emerged in the late Qing Dynasty (1840–1912), as well as in the period between 1988 and 2007, which represents a fertile period for modern Chinese kinterm dictionaries.

The systematic documentation of Chinese kinship terms can be traced back to the *Erya* 尔雅, the earliest classificatory dictionary in China, compiled in the third or second century BCE (Fêng 1937:144; Yan 2012:26; Karlgren 1931 argues that the major part of *Erya*'s glosses probably dates from the third century BCE). The author of the *Erya* is still unknown, though ancient Chinese scholars often attributed this dictionary to Duke Zhou (c. eleventh century BCE), or to Confucius' disciples in the early Warring States period (fifth to fourth century BCE), while modern researchers tend to opine that the compilation was between the late Warring States period (third century BCE) and the early Western Han (second century BCE) by anonymous authors (Hu & Fang 2004). The title, consisting of *er* 尔 (a phonetic loan character of the homophone *er* 迩 "to approach") and *ya* 雅 (lit. "refined, elegant," hence "the Standard language"), is generally interpreted as "approaching what is correct, proper, refined" (Coblin 1993), implying that the major aim of this general dictionary is to promote the correct and proper usage of assorted Chinese vocabulary for educated Chinese.

The *Erya* contains 19 sections in total, and the fourth section, *Shiqin* 释亲 (literally "explaining relatives") deals exclusively with kinship terms and aims at promoting the correct and proper usage of such terms for an educated Chinese of the period. The terms contained in the section are by far the earliest description of the Classical Chinese kinship system, in which 96 kinterms were explained, involving 61 relations, e.g. *kao* 考 ("father"), *bi* 妣 ("mother") and *shufu* 叔父 ("younger paternal uncle") under the *zongzu* 宗族 ("paternal clan") category, *jiu* 舅 ("maternal uncle") and *waiwangfu* 外王父 ("mother's father") under the *mudang* 母党 ("maternal relatives") category, *waijiu* 外舅 ("wife's father") and *yi* 姨 ("wife's sister") under the *qidang* 妻党 ("wife's relatives") category, and *xionggong* 兄公 ("husband's elder brother") and *nügong* 女公 ("hubsand's elder sister") under the *hunyin* 婚姻 ("marriage") category.

The 96 kinterms, which are still in use to this day, constitute the core of the Chinese kinship vocabulary. This lexicographic effort to collect and explain Chinese kinship terms not only reflects the importance of kinship terms in the Chinese lexicon, but also marks the start of China's unique lexicographic tradition of documenting kinship terms. The *Erya* not only established the important status of kinship terms in Chinese lexicon, but also the canonical approach to kinship studies in Chinese scholarship, in which terms are recorded and explained in terms of a prescriptive manner (though in the guise of a descriptive style) and terms of converse relations are also named. Below we give a few examples of the *Erya* entries (Hu & Fang 2004: 194, 203, translated by the present authors).

> Of Father's brothers, those born before Father are addressed or referred to by the term *shifu* 世父, and those born after Father, by the term *shufu* 叔父. [父之晜弟，先生为世父，后生为叔父。]
> A male born before another male is addressed or referred to by the latter as *xiong* 兄, and a male born after another male is addressed or referred to by the latter as *di* 弟. [男子先生为兄，后生为弟。]
> A female born before a male is addressed or referred to by the male as *zi* 姊, and a female born after a male is addressed or referred to by the male as *mei* 妹. [男子谓女子，先生为姊，后生为妹。]
> If a male relative address me by the term *jiu* 舅, I will address him by the term *sheng* 甥. [谓我舅者，吾谓之甥也。]

We see this explanatory mode as "prescriptive in the guise of a descriptive style" because the purpose of the *Shiqin* section in the *Erya* is not to describe what terms existed in the timespan when the lexicon was written, but to prescribe, as is implied by the dictionary title, the "correct and proper" way of addressing different kintypes, despite the somewhat apparently descriptive explanation style with no "should" or "should not" expressions.

The *Erya* belongs to the category of "general dictionaries" which are aimed at "covering the whole vocabulary for the 'general' user" (Hartmann & James 2002:129) and is not a specialized dictionary that concentrates on a restricted domain of the lexicon (ibid.). The most important legacies of the *Erya* are twofold. Firstly, it established a lexicographic norm throughout Chinese history for general dictionaries to include kinship terms as an essential component section; secondly, it spurred the compilation of specialized dictionaries covering kinship terms in later days.

There are two types of specialized dictionaries related to kinterm documentation in China. The first type is the specialized dictionary dedicated exclusively to kinship terms, while the other is the dictionary of address terms including kin terminologies, because addressing people with kinship terms has been an essentially significant social practice throughout Chinese history. Thus, kinship terms are included and treated in three types of Chinese dictionaries: general reference dictionaries, specialized kinterm dictionaries, and specialized dictionaries of address terms. The two types of specialized dictionaries differ from each other in their purposes for compilation, a point that will become clearer when specific examples are given below. We will focus on two periods, the late Qing Dynasty (1840–1912) when the two earliest attested specialized dictionaries of such kinds were published, and the period between 1988 and 2007 that witnessed the flowering of kinterm-related specialized dictionaries.

During the late Qing period, there was renewed interest among scholars, particularly some Confucian philologists who systematically collected extant kinship terms, compiled specialized dictionaries with large inventory of entries, and developed a unique system for categorizing kinship terms. The first attested specialized kinship dictionary, Zheng Zhen's 郑珍 *Compendium of Kinship Terms* (*Qinshuji* 亲属记, lit. "Notes on Relations") was finalized in 1860 and first printed in 1886. Zheng passed the Provincial Examination of Imperial China and obtained the *juren* degree in 1837, but did not take any official positions in his life (Schmidt 2013). He was regarded by his contemporaries as an accomplished universal Confucian scholar, and by researchers as one of the greatest poets in China's history (ibid.).

The dictionary, small in size, is prescriptive in nature in that the entries were selected in line with Confucian rites and ethics. For example, the popular kinterms, such as *gumu* 姑母 (one's father's sister) and *jiufu* 舅父 (one's mother's brother), were rejected by Zheng, because he believed that these terms blurred the clear-cut Confucian borderline between parents and their siblings by "misapplying" the morphemes *fu* 父 "father" and *mu* 母 "mother" to the collaterals (Zheng 1886). For that reason, *Qinshuji* includes a very limited range of kinship relations, but often with dense commentaries. The lexicon is divided into two sections, one for the various terms in reference to one's parents, and the other for the assorted words denoting grandparents, marriage-related kintypes, and descendants. As can be seen, only terms for kintypes of lineal consanguinity are admitted by Zheng in his lexicon, which differs greatly from the *Erya* quadruple categorization. However, for each relation, Zheng endeavors to display the myriad terms used in ancient classics and histories. For instance, the book includes 22 terms for "father," of which 9 can only be used for one's deceased father, and 13 for one's living father (with a few obviously from Chinese dialects or even non-Sinitic languages, e.g., *mohe* 莫贺 "father" from Serbi <*Xianbei*> and *langba* 郎罢 "father" from Min Chinese). Zheng's lexicon also features many "supernumerary terms" (Fêng 1937) as introduced in the previous section, which is very typical in Chinese dictionaries of kinship terms, since these terms bear more ritual functions in Chinese society than mere conversational reference or addressing.

The prescriptive nature of *Qinshuji* is clearly indicated in its preface, in which Zheng explains three reasons for compiling the dictionary. Firstly, the work is to collect and present all correct kinship terms from ancient records and present-day use. Secondly, it is to provide the readers with the semantic evolution process of Chinese kinship terms. Thirdly, it is to demonstrate that certain terms are used to colloquially conform in principle to Confucian teachings on rites, while other terms that would be deemed inappropriate from a Confucian perspective are excluded. The prescriptive approach of *Qinshuji* obviously differs from the descriptive approach of *Chengweilu*, to be discussed next.

As the first attested specialized dictionaries of address terms, Liang Zhanglu's 梁章钜 *Collection of Terms of Address* (*Chengweilu* 称谓录) was finalized in 1848 and printed in 1884. Liang (1775–1849 CE) was a famous Confucian scholar-official, holding such offices as the Surveillance Commissioner of Shandong Province and the Governor of Guangxi and Jiangsu Provinces in the late Qing dynasty.

The dictionary contains over 5000 terms. The first 11 chapters (approximately one third of the total 32 chapters) are chiefly about kinship terms. The kintypes included in these 11 chapters are listed as follows:

1. Terms for one's father, grandparents and other remote ancestors
2. Terms for one's father's various spouses in the polygamous context
3. Uncle- and aunt-related terms, including granduncles and grandaunts
4. Terms for brothers, their spouses, their children and other related kintypes
5. Terms for husbands and wives
6. Terms for sons, daughters and grandchildren
7. Terms for marriage-related kintypes, or in-laws
8. Terms for sisters, their spouses, their children, and other related kintypes, as well as terms for children's spouses and their children/children-in-law
9. Terms for parents and grandparents and spouses of the Emperor
10. Terms for the Emperor's sons, grandsons and their spouses
11. Terms for other kintypes in the royal family

In the preface, Liang explains two reasons for compiling the dictionary. One is to facilitate the reading of Classical Chinese texts. In his view, the Chinese system of address terms is inherited from Western Zhou Dynasty (founded around 1045 BCE), and has grown increasingly complex with passage of time, but no dictionaries have been compiled to represent this large reservoir of terms after the *Erya*. The other reason is to expand the vocabulary of writers, since written Chinese at that time was composed in literary Chinese and an educated person had to command a large vocabulary to write well. It follows that the entries in the dictionary are terms used in existing Classical Chinese texts, to the exclusion of colloquial or dialectal terms that merely occur in oral communication, with no literary attestation. However, if a colloquial or dialectal kinterm is found in Classical Chinese texts, it will still be culled into the *Chengweilu* corpus. For instance, we can find the colloquial term *diedie* 爹爹 ("dad") in *Chengweilu*, since it is recorded in the works of literati in the Song dynasty (960–1279 CE). Liang's explanation style is also descriptive in nature, simply indicating a certain term is used by people of some period/region to refer to some kintype, as is recorded in some classics. The significance of Liang's dictionary lies precisely in this respect, that is, it started a descriptive tradition in the lexicographical treatment of Chinese kinship terms.

The lexicographical tradition has persisted to contemporary period, as reflected in the plethora of specialized kinship dictionaries and dictionaries of address terms compiled and published in a period spanning from the late 1980s through early 2000s. During that time, two large kinship dictionaries were published. They are Bao Haitao 鲍海涛 and Wang Anjie's 王安节 1988 *Dictionary of Kinship Address Terms* (*Qinshu Chenghu Cidian* 亲属称呼词典) and Liu Chaoban's 刘超班 1991 *A Dictionary of Kinship Terms in China* (*Zhonghua Qinshu Cidian* 中华亲属辞典). The former includes 3500 kinship entries, while the latter collects approximately 4000 kinterms.

Although Bao and Wang (1988) is exclusively devoted to Chinese kinship terms and include a large number of entries (Yan 2012), it is marred by its oversimplified definitions, scarce illustrative examples, and insufficient treatment of terms in modern Sinitic languages. In this regard, the dictionary far from represents the rich panorama of Chinese kinterms. However, it employs a subcategorization framework of Chinese kinship distinct from the previous traditions such as the *Erya* or *Chengweilu* tradition. In Bao and Wang (1988), ten categories are found:

1. Father-side terms, incl. Father and other paternal progenitors/ancestors
2. Mother-side terms, incl. Mother and other maternal progenitors/ancestors
3. Husband-side terms, incl. Husband and other kintypes related to the husband
4. Wife-side terms, incl. Wife, concubines and other kintypes related to them
5. Sibling-side terms, incl. Brothers, sisters and their spouses
6. Descendant-side terms, incl. Descendants (e.g. children, grandchildren) and their spouses
7. *Tang*-collateral-side terms, incl. Uncles, aunts, cousins and related kintypes sharing the same family name with the ego
8. *Biao*-collateral-side terms, incl. Uncles, aunts, cousins and other related kintypes not sharing the same family name with the ego
9. Collective terms in reference to two or more kintypes, e.g. *erfu* 二父 "father and his younger brother"
10. General terms in reference to the kinship relations, e.g. *zongzu* 宗族 "patrilineal clan"

This subcategorization is more detailed and user-friendly than the previous divisions.

In contrast, Liu (1991) adopts an alphabetical order in arranging the kinship terms instead of the semantic subcategorization. In his dictionary, there are slightly more colloquial or dialectal terms of modern Chinese, most of which are paired with textual attestations from contemporary fictions or dialect dictionaries, but kinship terms from non-Han ethnic groups are also included, such as *ata* 阿塔 the "term for father in Chinese Tajik" (Authors' note: *ata* is a term for father used in many Turkic-speaking ethnic groups, and the Chinese Tajik counterpart *ato* is a term borrowed from neighboring Turkic people). Despite the text-based descriptive approach inherited from Liang (1884), the kinship data of modern languages collected in these two dictionaries are not very reliable as no fieldwork investigation or verification was conducted.

During the same period, at least eight dictionaries of address terms were compiled which contain kinship terms, such as Wang Huo 王火 and Wang Xueyuan's 王学元 (1988) *Dictionary of Chinese Terms of Address* (*Hanyu Chengwei Cidian* 汉语称谓词典), Lu Ying's 陆瑛 (1989) *Concise Dictionary of Address Terms* (*Jianming Chengwei Cidian* 简明称谓词典) and Han Xingzhi's 韩省之 (1991) *Unabridged Dictionary of Address Terms* (*Chengwei Da Cidian* 称谓大辞典). An examination of the prefaces to these dictionaries indicates two motives for compiling such dictionaries. Firstly, they were compiled in order to promote the correct use of Chinese kinship terminology in writing, as well as in everyday communication. It appears that in the 1980s, the misapplication of kinship terms was very common. The surge of specialized dictionaries of address terms in the late 1980s and early 1990s was probably not surprising after the disruptive and destructive Cultural Revolution when social order was in disarray. The rejuvenation of China's education system after the Cultural Revolution called for promotion of "correct and refined national language." Furthermore, traditional cultural values started to be restored and once again valued by Chinese people. This concentrated publication of dictionaries of address forms during the period further attests to the normative role of kinship terms in Chinese social interaction, and in society in general.

Secondly, they were meant to help ethnic Chinese, both in and outside China, particularly the young, to regain familiarity with the multilayered kinship system and to present a panorama of Chinese kinterms, including a thesaurus of various register synonyms for each kintype. For example, Bao and Wang (1988) lists 113 terms for "father" (incl. stepfathers and sworn fathers, and many "supernumerous terms" used only in sacrificial or literary contexts), and a large portion of them are register synonyms, such as *fu* 父 "father (used oft. in letters)," *laozi* 老子 "father (colloquial, only as a term of reference, never as an address term; can be used by a father to refer to himself in the face of the son," *jiafu* 家父 "my father (lit. "house + father," a respectful term referring to the speaker's own father)," *jiayan* 家严 "my father (lit. "house + strict," a respectful term in reference to one's own father in written Classical Chinese)," *lingchun* 令椿 "your father (lit. "your honorable + long-living tree," a respectful term referring to the hearer's father in ancient China)," etc.

This effort could be also be seen as being in tandem with the promotion of standardized *Putonghua* ("the common language"; generally known in English as Mandarin), since Mandarin kinship terms account for the majority of the corpus, and dialectal terms are only accepted, for the large part, when they are present in one form or another in the written literary texts, a principle inherited from *Chengweilu*. (For monographs and dictionaries on Chinese dialects see the next section.)

Of the eight dictionaries, what is noteworthy is Zhu Jiancheng's 朱建成 (2007) bilingual *Chinese-English Dictionary of Terms* (*Han-Ying chenwei cidian* 汉英称谓词典), since it is perhaps the first in China to pair Chinese kinterms and their possible English counterparts in a specialized dictionary of terms, touching upon the translatability issue in the Chinese lexicographic tradition of kinterm documentation. The kinship terms are found in the 28th section of the lexicon, titled "Terms for addressing or referring to common interpersonal relations." However, it is apparent that many of "translated" English terms are problematic. For example, *tangdi* 堂弟 "a younger male cousin on father's side" is paired in Zhu's dictionary with "younger paternal brother," while "paternal brother" is often used to denote a half-brother sharing only the male parent with the ego, which is quite different.

Regarding authors and compilers of these dictionaries published in the late twentieth to early twenty-first century period, not much is known in terms of their biographic details, and we can only surmise from the prefaces or postscripts that the majority are modern Chinese language teachers in universities (e.g., Wang Huo and Wang Xueyuan), or researchers of modern Chinese literature (e.g., Liu Chaoban). That explains why written Chinese kinship terms were preferred and colloquial dialectal terms largely escaped their lens. Additionally, the lack of theoretical linguistic participation also undermined the reliability of the database.

The specialized Chinese dictionaries of kin terminologies surveyed in this section reflect the persistently important role of kinship relations in Chinese society across time. They also constitute valuable records of the type of kinship relations named and codified in the Chinese language. Through these dictionaries, scholars have been able to trace the continuation and changes in kinship relations in China.

The development of the unique lexicographic tradition of documenting kinship is by no means accidental. There are deep cultural reasons behind such a development, such as the essential role of kinship in Confucianism. Over time, the diachronic accumulation and juxtaposition of kinship terms, resulting from an unbroken written record, also adds to the complexity of the multilayered nature of Chinese kinship terms. Without the continuous lexicographic pursuit in Chinese kinship semantics, it might be difficult for modern researchers to straighten out the entangled enigma underlying the sophisticated Chinese kinship systems. In that sense, this lexicographic tradition can be regarded as an important contribution that Chinese scholars have made to the contemporary study of kinship terms across different languages and cultures.

Recent Development

While the lexicographical tradition in China continues to this day, in recent times, many scholars in China have also started to incorporate methods for kinships studies developed in anthropology and give more attention to dialectal terms and usages, resulting in a plethora of kinship lexicons (as part of general dictionaries), as well as many studies of dialectal kinship terms (independently or as part of larger descriptive Sinitic language studies). In this section, we provide an account of these recent developments, focusing on efforts to document kinship terms in Chinese dialects and account of the methodological issues involved.

Documenting Kinship Terms of Chinese Dialects

Although the majority of literature on Chinese kinterms is based on Standard Chinese, or the Mandarin variety, there is in fact one tradition of documenting/studying the kinship system within a specific dialect region of China (Yan, 2012), such as Liu's (2001, 2010) articles on kinterms in Hainan Chinese dialects. This tradition stems from modern Chinese linguistics dating back to the Republic of China era (1912–1949) when renowned linguists like Yuen Ren Chao advocated for fieldwork investigation and documentation of the many and various Chinese dialects. This then is another tradition that differs from the time-honored lexicographic one, and the academic achievements of this linguistic tradition are not specialized dictionaries of kinship or address terms for promoting the standard language, but often take the form of general dictionaries that document a particular Chinese dialect, with a specific portion dedicated to the local kinship terms; additionally, monographs to describe and analyze the kinterm system of a specific Chinese dialect are not a rarity.

For the first type, i.e., the general dialectal dictionaries with a specific kinship database within, the representative is the famous series *Great Dictionary of Modern Chinese Dialects* (*Xiandai Hanyu Fangyan Da Cidian* 现代汉语方言大词典), with Li Rong 李荣 as the chief editor, published from 1992 to 2002. The individual dictionaries of this series cover 42 Chinese dialects, including 16 Mandarin varieties (Harbin, Jinan, Muping, Luoyang, Wanrong, Xi'an, Xining, Xuzhou, Urumqi, Yinchuan, Chengdu, Guiyang, Liuzhou, Wuhan, Nanjing, and Yangzhou), two Jin dialects (Taiyuan, Xinzhou), eight Wu dialects (Chongming, Danyang, Hangzhou, Jinhua, Ningbo, Shanghai, Suzhou, and Wenzhou), one Hui dialect (Jixi), three Gan dialects (Lichuan, Nanchang, Pingxiang), two Xiang dialects (Changsha, Loudi), five Min dialects (Fuzhou, Haikou, Jian'ou, Leizhou, and Xiamen), two Yue varieties (Dongguan, Guangzhou), two Hakka varieties (Meixian, Yudu), and one Pinghua dialect (Nanning). These dictionaries were compiled by linguists specializing in the target Chinese dialects, and the coverage of words for each individual dictionaries varies from 7000 to 10,000 entries, with a uniform "semantically-based index" that sorts the whole lexicon into 30 semantic categories, of which "kinship" is the 10th category.

As for the second type, the literature is vast, and, to illustrate that, we herein chiefly focus on the monographs of some southern dialects, such as Cantonese and Teochew, and discuss their strengths and weaknesses.

Among monographs and articles on Cantonese kinterms, Li Qiwen's (1989) "Analyzing the System of Kinship Terms in Guangzhou Dialect" (in Chinese) is the first systematic description of modern Cantonese kinship terminology, based on the author's investigation of the Cantonese variety spoken in contemporary

Guangzhou. Li made a comparison between Mandarin and Guangzhou Cantonese, and tried to explain the differences between them, but the methodology adopted is still the traditional lexicographic one.

On the contrary, Hung-nin Samuel Cheung's (1990) "Terms of Address in Cantonese" uses the anthropological framework similar to Wallace and Atkins's (1960) denomination using certain semantic primitives. For example, *baak* 伯 is analyzed as "ego's father's elder brother," and *syūn* 孙 as "ego's son's son."

Cheung's (1990) is also the first research devoted to the use of relationship terms (including kinship words) in the Cantonese variety spoken in Hong Kong. Cheung attempted to generalize the phonological and semantic features of the Hong Kong Cantonese kin terms which had never been covered before.

The previous research on Teochew kinterms is even less extensive than that on Cantonese. Wu (2007) is the first and only monograph on the system of Teochew kinship terminology based on the Teochew varieties spoken in the City of Chaozhou. The thesis covers such issues as the synchronic description of the terminological system, the word-formation patterns, the comparison and contrast with the Amoy Hokkien, and so on. It is by far the most comprehensive in this field, but the author relied on the anthropological approach to semantic analysis, and the presupposition that kinship terms within the whole Sinitic group are of no significant distinction in between is also found therein, similar to Li (1989).

From our point of view, methodologically, this presupposition that Sinitic kinship terms are homogeneous rather than diverse has been a persistent problem among many dictionaries of kinship terms in Chinese dialects. Take Hu Shiyun's 胡士云 (2007) *Study of Chinese Kinship Terms* (*Hanyu Qinshu Chengwei Yanjiu* 汉语亲属称谓研究) as another example. This monograph is by far the most comprehensive piece of research on Chinese kinship terminology, covering 23 Sinitic dialects in scope and discussing the evolution of kin terms across 2500 years. Hu's framework of semantic analysis is something similar to the Chinese lexicographic tradition, and his cross-dialectal comparison and contrast seem to treat the non-Mandarin kinterms as absolute synonyms of their Mandarin counterparts, with mere difference in form. For instance, when he discusses the various terms for "father" in different Chinese dialects, Hu holds that the forms can be divided into six subgroups: *fuqin* 父亲 group, *ba* 爸 group, *die* 爹 group, *ye* 爷 group, *laozi* 老子 group, and the miscellaneous group. In Hu's opinion, these terms share the same semantic content in kintype designation, despite the formal disparity. This indicates that the author presupposed absolute homogeneity among all Sinitic varieties, a clearly false but prevalent impression. While his system is compatible with the anthropological tradition of kinship documentation, how to reflect an insider's psychological reality remains a sore methodological issue, an issue that will be discussed later in this section.

Research on Specific Kinship Terms

While dictionaries and typological studies deal with kinship terms as a whole domain, there are many studies devoting to subordinate kinship categories with specific analysis. This is because by elaborating on specific kinterms or kinterm groups, it is more likely that the minute subtle shades underlying the apparently similar terms across languages are discovered when examined in a more macro perspective.

Among such studies, Huang (1994) and Gong (2010) both addressed the semantic evolution of *bó* (伯, father's older brother) and *shū* (叔, father's younger brother) in Chinese, but their conclusions are based on speculation without any reliable methodology.

Shen (2014) is the only paper to discuss the two Mandarin terms for a father's sister's husband: *gūfu* (姑夫) and *gūfù* (姑父) from a pan-chronic perspective. The author argued that *gūfu* appeared later than *gūfù* in history and suggested some possible reasons for the gradual replacement of *gūfù* with *gūfu* in contemporary Mandarin.

The only scholar who has probed specifically into the "uncle"-type terminologies in non-Mandarin Sinitic languages is Liu (2001, 2010). His study focused on several Sinitic languages of Hainan Island, particularly the features in usage of the "uncle"-type term *jiù* (舅, mother's brother), as well as the aunt-type terms *gū* (姑, father's sister) and *yí* (姨, mother's sister). According to Liu, the varieties in Hainan all distinguish a mother's younger brother from her older brother, and a father's or mother's younger sister from his or her older sister, which is rarely found in most Sinitic languages. The author attributed this phenomenon to long-term language contact between the Hainan Chinese communities and speakers of Kam-Kadai languages on the same island.

Methodological Issues in Typological Research

Many scholars see the methodological issue of translatability as being central to the study of Chinese kinship terms in the broad context of crosslinguistic, typological research. In comparative studies on kinship terms within Sinitic languages, Cao"s (2008) *Linguistic Atlas of Chinese Dialects: Lexicon* constitutes a remarkable effort to provide a cross-dialectal, typological overview of 15 kinship terms, including paternal and maternal grandparents, parents, maternal aunt, husband, wife, son, daughter, son-in-law, daughter-in-law, daughter's son, and sister's son. By examining the semantic intersection and differences between the terms for a given kintype found in different Sinitic dialects, it is possible to create a lexical semantic map for a certain kintype or for a specific kinterm, though Cao (2008) has not advanced in this pursuit.

When Sinitic languages are compared with non-Sinitic languages, or when Mandarin kinterms are contrasted with non-Mandarin Chinese terms, the translatability issue is brought to the fore. CA or other anthropological approaches can be useful in initial documentation of kinship terms, but they do not reflect the psychological reality of the native speakers. In fact, even the current prevailing mainstream linguistic typological approach to kinship semantics, which features quantitative analysis of "big data" related to kinship terms on a crosslinguistic basis, can be categorized as largely "etic" in nature. In striving for a culturally neutral panorama of kinship terminology, this current approach pays scant attention to the native speaker's inherent culture-based semantic interpretation of the kinterms in question. In this sense, the Natural Semantic Metalanguage approach, which is sometimes also regarded as belonging to the category of lexical and semantic typology, has forged a quite different "emic" approach in complementing the mainstream "etic" pathway of typology, since it attaches significant importance to the issue of psychological reality in crosslinguistic semantic analysis and endeavors to represent a balance between cognitive universalism and the cultural diversit that underlies linguistic typology.

Xue's (2016) thesis "The Semantics of 'Uncle'-type Kinship Terms in Cantonese (Guangzhou) and Teochew (Jieyang)" marks an important step in attempting to explore, using the Natural Semantic Metalanguage approach, the psychological reality of a Chinese local insider when he or she is to interpret a kinship term against the local linguistic and cultural context. In her work, she proposes four culture-specific semantic molecules in explicating the uncle-type kinship terms in the Cantonese and Teochew dialects (for discussion on semantic molecules, see Goddard 2016; conventionally, semantic molecules will be represented by [m] in NSM explications, as can be seen in the following parts). The four proposed molecules are *ago* [阿哥, Cantonese]/*a-hia^n* [阿兄, Teochew] for older brothers, *sailou* [细佬, Cantonese]/*a-ti* [阿弟, Teochew] for younger brothers, *jehje* [姐姐, Cantonese]/*a-che* [阿姐, Teochew] for older sisters, and *muihmui* [妹妹, Cantonese]/*a-mue* [阿妹, Teochew] for younger sisters.

The use of these four molecules secures not only a simple and succinct explication but also a logical and psychologically real one, consistent in principle with the Chinese tradition of kinterm explanation starting from *Erya*. Take the explication of *tua-peh* 大伯 ("the oldest paternal uncle") in Teochew as an example. Without the proposed culture-specific molecules, *tua-peh* is explicated the way as:

someone's *tua-peh* [大伯]

(a) A man[m], this someone can say about this man[m]: "this is my tuā-peh"
(b) Someone can say this about a man[m] if it is like this:
(c) "This man[m] is a brother[m] of this someone's father[m]
(d) This man[m] was born[m] before this someone's father[m]
(e) If there are other sons[m] of this man's father[m]
(f) They were all born[m] after this man[m]"

But we can apply to this explication the proposed molecule of *a-hia^n*, whose explication looks like this:

someone's *a-hia^n* [阿兄]

(a) Someone can say about a man[m]: "this is my *a-hia^n*"
(b) If this someone can think about him like this:
(c) "His mother[m] is my mother[m], his father[m] is my father[m]
(d) He was born[m] before I was born[m]"
(e) A child[m] can say the same about another child[m]

(f) If after sometime this other child[m] can be a man[m]

After the application of this molecule, the explication for *tua-peh* becomes the following structure:

Someone's *tua-peh* [大伯]

(a) A man[m], someone can say about this man[m]: "this is my *tuā-peh*"
(b) Someone can say this about a man[m] if it is like this:
(c) "This man[m] is the *a-hia^n*[m] of this someone's father[m]
(d) This man[m] doesn't have an *a-hia^n*[m]"

Apart from the culture-specific molecules proposed to approach the native speaker's psychological reality, Xue's thesis highlights another group of kinship terms that merit academic attention: the uncle-type kinterms. The "uncle"-type kin terms constitute a distinctive group of the Chinese kinship terminology (cf. Ruey, 1949), which differs starkly from the more classificatory pattern in European and American speech communities. An emic probe into the cognitive world underlying this phenomenon can help culture-outsiders understand some of the psychological foundations of traditional, as well as modern Chinese society, though previous research in this regard is very limited, and even Cao's (2008) atlas lacked investigation data for any forms of "uncle." Thus, the introduction of NSM approach to the study of "uncle"-type kinship terms in Chinese dialects is both significant and necessary in that it cannot only fill the lacunae of previous researches such as Cao (2008), but also contribute to the typological exploration of Chinese kinship semantics.

Future Directions

More than three decades ago, Thompson (1985) made a remark on Chinese kinship analysis that, in spite of progress in this area, "much remains to be done." This statement is applicable to the present situation. Although previous studies of Chinese kinship terms appear to be plentiful, most of them follow a lexicographic tradition. This kind of linguistic documentation has been carried out largely for the purposes of preserving cultural heritage and for linguistic codification in line with Confucian ethics. Furthermore, such documentation has not been oriented primarily toward a non-Chinese audience. In the resurgence of kinship studies in anthropology and linguistics, when Chinese and its various dialects have much to contribute in both empirical data and ways of organizing and codifying kinship semantic typology, the challenges of bringing a Chinese perspective to an international audience remains an obstacle and requires solutions that can speak to the general data pool while at the same time can reflect the "psychological reality" of Chinese native speakers. In achieving this, a combination of methods and tools seems unprecedentedly pressing for the whole academic field.

References

Baker, H. 1968. *A Chinese lineage village: Sheung Shui*. London: Frank Cass. & Co., Ltd.
Bao, H., and A. Wang. 1988. *Qinshu chenghu cidian [Dictionary of kinship address terms]*. Changchun: Jilin Education Press.
Barnes, J. 1980. Kinship studies: Some impressions of the current state of play. *Man* 15 (2): 293–303.
Burling, R. 1969. Cognition and componential analysis: God's truth or hocus-pocus? *Cognitive anthropology*, 419–428. New York: Holt, Rinehart and Winston.
———. 1970. American kinship terms once more. *Southwestern Journal of Anthropology* 26 (1): 15–24.
Cao, Z. 2008. *Linguistic atlas of Chinese dialects: Lexicon*. Beijing: Commercial Press.
Chao, Y.R. 1956. Chinese terms of address. *Language* 32 (1): 217–241.
Chen, T., and J. Shryock. 1932. Chinese relationship terms. *American Anthropologist* 34 (4): 623–664.
Cheung, H.S. 1990. Terms of address in Cantonese. *Journal of Chinese Linguistics* 18 (1): 1–43.
Coblin, W.S. 1993. Erh Ya. In *Early Chinese texts: A bibliographical guide*, 94–99. Berkeley: Society for the Study of Early China; Institute of East Asian Studies, University of California Berkeley.
Dash, K. 2004. *Invitation to social and cultural anthropology*. New Delhi: Atlantic.
Dousset, L., R. Hendery, C. Bowern, H. Koch, and P. McConvell. 2010. Developing a database for Australian indigenous kinship terminology: The AustKin project. *Australian Aboriginal Studies* 1: 42.
Dziebel, G. 2007. *The genius of kinship*. Youngstown, N.Y: Cambria Press.
Evans, N. 2010. Semantic typology. In Song, J. J. (ed.), *The Oxford handbook of linguistic typology*, 505–534. Oxford: OUP.
Fêng, H. Y. 1937. *The Chinese kinship system*. PhD University of Pennsylvania.
Fox, R. 1967. *Kinship and marriage*. Harmondsworth: Penguin Books.
Goddard, C. 2016. Semantic molecules and their role in NSM lexical definitions. *Cahiers de Lexicologie. Special Issue on 'Lexical Definition'* 2016 (109): 13–34.
Gong, M. 2010. Discussion on the change of 'Bo (伯)' and 'Shu (叔)' from order words to kinship terms. *Journal of Kaili University* 28 (5): 96–99.
Good, A. 2010. Kinship. In Barnard, A. & Spencer, J. (ed.), *The Routledge encyclopedia of social and cultural anthropology*, 2nd ed., 396–404. London: Routledge.
Goodenough, W. 1956. Componential analysis and the study of meaning. *Language* 32 (1): 195–216.
Han, X. 1991. *Chengwei da cidian [Unabridged dictionary of address terms]*. Beijing: New World Press.
Hartmann, R.R.K., and G. James. 2002. *Dictionary of lexicography*. London: Routledge.
Hu, S. 2007. *Hanyu qinshu chengwei yanjiu [A study of Chinese kinship terms]*. Beijing: The Commercial Press.
Hu, Q. & Fang, H. 2004. Erya yizhu [Erya: A modern Chinese translation with commentary]. Shanghai: Shanghai: Shanghai Guji Chubanshe.
Huang, R. 1994. Shi 'Shu' 'Bo' [explanations on 'Shu' and 'Bo']. *Chinese Literature and History* 02: 88–90.
Karlgren, B. 1931. The early history of the *Chou Li* and *Tso Chuan* texts. *Bulletin of the Museum of Far Eastern Antiquities* 3: 1–59.
Keen, I. 2014. Language in the constitution of kinship. *Anthropological Linguistics* 56 (1): 1–53.
Kroeber, A. 1933. Process in the Chinese kinship system. *American Anthropologist* 35 (1): 151–157.
Lévi-Strauss, C. 1969. *The elementary structures of kinship*. Trans. J. von Sturmer, J. Bell & R. Needham. Boston: Beacon Press.
Li, Q. 1989. Guangzhou Fangyan Qinshu Chengweici Xitong Fenxi [an analysis of the system of kinship terminology in Guangzhou dialect]. *Social Sciences in Guangdong* 1: 165–173.
Li, R. 2002. *Great dictionary of modern Chinese dialects*. Nanjing: Jiangsu Educational Press.
Liang, Z. 1884. *Chengwei lu [Collection of terms of address]*. Hangzhou: Jiajingwenzhai.

Liu, C. 1991. *Zhonghua qinshu cidian [A dictionary of kinship terms in China]*. Wuhan: Wuhan Publishing House.
Liu, J. 2001. On the appellation of relatives in the Chinese dialects of Hainan. *Humanities & Social Sciences Journal of Hainan University* 19 (2): 25–29.
———. 2010. Cong 'Gu' 'Yi' 'Jiu' lei Chengwei Kan Hainan Yuyan Jian de Xianghu Yingxiang [language contact and the peculiarity of kinship terms 'Gu' 'Yi' and 'Jiu' as used in Hainan Chinese communities]. *Minority Languages of China* 2: 50–56.
Lounsbury, F.G. 1956. A semantic analysis of the Pawnee kinship usage. *Language* 32 (1): 158–194.
Lu, Y. 1989. *Jianming chengwei cidian [concise dictionary of address terms]*. Nanning: Guangxi Nationality Publishing House.
McConvell, P., I. Keen, and R. Hendery. 2013. *Kinship systems: change and reconstruction*. Salt Lake City: The University of Utah Press.
McConvell, P., Koch, H., Dousset, L., Simpson, J., Bell, J., Kelly, P., McGrath, M., & Hendery, R. The AustKin Project [Internet]. ACT (Australia): Australian Research Council (with contributions from EHESS). 2016- [cited 2020 May 1]. Available from: http://www.austkin.net/.
Moravcsik, E.A. 2013. *Introducing language typology*. New York: Cambridge University Press.
Morgan, L. 1871. *Systems of consanguinity and affinity of the human family*. Washington: Smithsonian Institution.
———. 1877. *Ancient society*. Tucson, Ariz: University of Arizona Press.
Murdock, G. 1949. *Social structure*. New York: Macmillan.
Nikolayeva, L. 2014. *Typology of kinship terms*. Frankfurt am Main: Peter Lang.
Peletz, M. 1995. Kinship studies in late twentieth-century anthropology. *Annual Review of Anthropology* 24 (1): 343–372.
Read, D. 2013. A new approach to forming a typology of kinship terminology systems: From Morgan and Murdock to the present. *Structure and Dynamics* 6 (1).
Ruey, Y. 1949. On the Chinese kinship terms *Po, Shu, Ku, Chiu* and *I*: A study of Chinese kinship system and its development. *Bulletin of the Institute of History and Philology Chinese Academy of Sciences* 14: 151–211.
Sandel, T. 2002. Kinship address: Socializing young children in Taiwan. *Western Journal of Communication* 66 (3): 257–280.
Sangren, P. 1982. Recent studies in Chinese kinship: *Chinese family and kinship*, by Hugh Baker; *The politics of marriage in contemporary China*, by Elisabeth Croll; *Marriage and adoption in China, 1845–1945*, by Arthur P. Wolf and Chieh-shan Huang. Reviewed in: *American Anthropologist* 84 (3): 628–632.
Schmidt, J.D. 2013. *The poet Zheng Zhen (1806–1864) and the rise of Chinese modernity*. Leiden: Brill.
Schneider, D. 1968. *American kinship: A cultural account*. Chicago: Chicago University Press.
Schusky, E. 1965. *Manual for kinship analysis*. New York: Holt, Rinehart and Winston.
Shen, X. 1989. Lun Jiegou Renleixue de Yuyan Zhexue Ji qi Qinshu Chengwei Yanjiu [language philosophy of structural anthropology and its study of kinship terminology]. *Journal of Xinjiang Normal University (Social Science)* 03: 75–82.
Shen, S. 2014. 'Gūfū' he 'Gùfù' – Jianlun Xiandai Hanyu Qinshu Chengweici de Yimingtongzhi Xianxiang [a preliminary study of the absolute synonymy in kin terms of modern Chinese]. *Journal of Guilin University of Aerospace Technology* 03: 279–283.
Thompson, S. 1985. *Chinese kinship*, by Paul Chao. Reviewed in: *The China Quarterly* 102: 344–345.
Wallace, A., and J. Atkins. 1960. The meaning of kinship terms. *American Anthropologist* 62 (1): 50–58.
Wang, H., and X. Wang. 1988. *Hanyu chengwei cidian [dictionary of Chinese terms of address]*. Shenyang: Liaoning University Press.
Wierzbicka, A. 1992. *Semantics, culture, and cognition*. New York: Oxford University Press.
———. 2013. Kinship and social cognition in Australian languages: Kayardild and Pitjantjatjara. *Australian Journal of Linguistics* 33 (3): 302–321.

———. 2016. Back to 'Mother' and 'Father': Overcoming the Eurocentrism of kinship studies through eight lexical universals. *Current Anthropology* 57 (4): 408–429.

———. 2017. The meaning of kinship terms: A developmental and cross-linguistic perspective. In *The semantics of nouns*, ed. Zhengdao Ye, 19–62. London: OUP.

Wu, C. 1927. The Chinese Family: Organization, Names, and Kinship Terms. *American Anthropologist* 29 (3): 316–325. new series.

Wu, Y. 1990. The usages of kinship address forms amongst non-kin in mandarin Chinese: The extension of family solidarity. *Australian Journal of Linguistics* 10 (1): 61–88.

Wu, J. 2007. *Chaozhou Fangyan Qinshu Chengwei Yanjiu [the research on the kinship terms of Chaozhou (Teochew) dialect]*. Master: Jinan University.

Xue, W. 2016. *The semantics of 'uncle'-type kinship terms in Cantonese (Guangzhou) and Teochew (Jieyang)*. ANU: Unpublished Master Thesis.

Yan, Y. 2012. *Hongloumeng qinshu chengweiyu de Ying yi yanjiu [a study of the English translation of kinship terms in Hong Lou Meng]*. Shanghai: Shanghai Foreign Language Education Press.

Zeitlyn, D. 1993. Reconstructing kinship, or, the pragmatics of kin talk. *Man* 28 (2): 199.

Zheng, Z. 1886. *Qinshu ji [compendium of kinship terms]*. Guiyang: Wulan Yinguan.

Zhu, J. 2007. *Han-Ying chenwei cidian [Chinese–English dictionary of terms]*. Guangzhou: Guangdong World Publishing Corporation.

Translating Legal Language Between Chinese and English

22

Deborah Cao

Contents

Introduction	540
Sources of Difficulty in Translating the Legal Language Between Chinese and English	541
Translating Legal Terms Between Chinese and English	545
Conclusion	549
References	550

Abstract

It is commonly acknowledged that legal translation, especially legal translation between Chinese and English, is difficult. It is a complex and special type of linguistic activity involving mediation and crossover between different languages and cultures, and above all different legal systems. Special skills, knowledge, and experience on the part of the translator are required to produce such translation. This chapter focuses on legal translation between Chinese and English. It explores and highlights some of the sources of difficulty and peculiar challenges in this type of translation activities.

Keywords

Chinese legal language · Chinese law · Legal translation · Legal terminology

---

D. Cao (✉)
Griffith University, Brisbane, Australia
e-mail: D.Cao@griffith.edu.au

© The Author(s), under exclusive licence to Springer Nature Singapore Pte Ltd. 2022
Z. Ye (ed.), *The Palgrave Handbook of Chinese Language Studies*,
https://doi.org/10.1007/978-981-16-0924-4_41

Introduction

In June 2017, the Chinese Foreign Ministry spokesman had this to say during a routine press briefing:

> 我想强调,香港是中国的特别行政区,香港事务属于中国内政。1984年的《中英联合声明》就中方恢复对香港行使主权和过渡期有关安排作了清晰划分。现在香港已经回归祖国怀抱20年,《中英联合声明》作为一个历史文件,不具有任何现实意义,对中国中央政府对香港特区的管理也不具备任何约束力。英方对回归后的香港没有主权,没有治权,也没有监督权。希望上述人士认清现实。(https://www.fmprc.gov.cn/web/wjdt_674879/fyrbt_674889/t1474476.shtml)
> I wish to stress that Hong Kong is China's Special Administrative Region (SAR). Hong Kong affairs are China's domestic affairs. The 1984 Sino-British Joint Declaration made clear delineation regarding China resuming the exercise of sovereignty over Hong Kong and the relevant transitional arrangements. It has been 20 years now since Hong Kong has returned to the motherland. The Sino-British Joint Declaration, as a historical document, does not have any practical significance, and it is no longer binding for the Chinese central government's management over Hong Kong. The UK has no sovereignty, no power to rule and no power to supervise regarding Hong Kong after the handover. It is hoped that those people alluded to earlier will see and recognise the reality. (This translation is my own)

Understandably, reactions from the UK were immediate as the Chinese foreign affairs spokesman apparently said that the Sino-British Joint Declaration is no longer binding. "The Sino-British Joint Declaration remains as valid today as it did when it was signed over 30 years ago," a British Foreign Office spokeswoman reportedly said, and "It is a legally binding treaty, registered with the UN and continues to be in force. As a co-signatory, the UK government is committed to monitoring its implementation closely." (https://uk.reuters.com/article/uk-hongkong-anniversary-china/china-says-sino-british-joint-declaration-on-hong-kong-no-longer-has-meaning-idUKKBN19L1IR) It is generally accepted (including China) that the Joint Declaration is a legally binding document, recognized under international law, deposited with the UN, and it cannot be revoked unilaterally by China or the UK.

The Chinese official may have intended to be provocative, or may have misspoken, or simply used language too loosely; it is difficult to ascertain. However, what is also interesting and relevant for the purpose of this chapter is the Chinese Foreign Ministry's translation of the statement into English which leads to a different interpretation. The translation from the Chinese Foreign Ministry reads:

> I want to stress that Hong Kong is China's SAR, and Hong Kong affairs belong to China's domestic affairs. The Sino-British Joint Declaration (1984) clearly marks the transitional period off from China resuming the exercise of sovereignty over Hong Kong. It's been 20 years now since Hong Kong's return to the motherland, and **the arrangements** during the transitional period prescribed in the Sino-British Joint Declaration **are now history and of no practical significance, nor are they binding on the Chinese central government's administration of the Hong Kong SAR**. The British side has no sovereignty, no power to rule and supervise Hong Kong after the handover. It is hoped that relevant people will come around to this (bold added). (http://www.china-embassy.org/eng/fyrth/t1474637.htm)

If the Chinese Foreign Ministry spokesman had spoken too loosely, the Foreign Ministry's official translation was done even more loosely, and perhaps it was done after the strong international reactions. In the Chinese Foreign Ministry official translation, the words 《中英联合声明》作为一个历史文件,不具有任何现实意义 (The Sino-British Joint Declaration, as a historical document, does not have any practical significance) were translated as "the arrangements during the transitional period prescribed in the Sino-British Joint Declaration." This translation largely changed the original words and meaning that had caused so much angst in the first place. The original Chinese is unambiguously clear and does not allow the kind of alterations as found in the official translation of the Chinese Foreign Ministry.

Although this is not a case of translating legal texts, it can nevertheless illustrate some of the translation problems. In real life situations, it is not uncommon that a translator is asked to omit words or change the meanings of words to suit particular purposes and agenda, especially for political and commercial reasons. (for a brief discussion of ethics and legal translators, see Cao (2014)). As the above example shows, translation, especially in diplomatic and legal contexts, can carry significant consequences. After all, one may never know whether the translation of the Chinese word *yi* 夷 or *manyi* 蛮夷 as "barbarians" or "foreign barbarians," as opposed to "foreigners," contributed to the start of the Opium Wars (1839–1860) between China and Western countries (Liu 1999). This chapter discusses translating legal language between Chinese and English, with specific reference to translating legal terminology, and some of the practical issues and challenges.

Sources of Difficulty in Translating the Legal Language Between Chinese and English

To begin with, what it is meant by the Chinese legal language? In this chapter, legal language refers to the language of and related to law and legal process. This includes language of the law, language about law, and language used in other legal communicative situations (cf. Kurzon 1998, who distinguishes language of the law and legal language which is language about law). Legal language is a type of register, that is, a variety of language appropriate to specific occasions and situations of use, and in this case, a variety of language appropriate to legal situations of use, language for legal purposes. Thus, the Chinese legal language refers to the language of and related to laws and legal processes of a Chinese jurisdiction such as China.

Furthermore, four major variants or subvarieties of the legal language in the written form can be distinguished: (1) legislative language, e.g., domestic statutes and subordinate laws, international treaties and multilingual laws, and other laws produced by lawmaking authorities; (2) judicial language produced in the judicial process by judicial officers and other legal authorities; (3) legal scholarly texts produced by academic lawyers or legal scholars in scholarly works and commentaries; and (4) legal language used in private legal documents, for example, contracts, leases, wills, and litigation documents (see Cao 2007). (Cf. Jean-Claude Gémar (1995 cited in Sarcevic 1997: 11) identifies six subdivisions of legal language: they are the language of the legislator, judges, the administration, commerce, private law, and scholarly writings.) For the Chinese legal language, these subtext types have their own peculiarities as well as commonalities. As noted, legal language is not homogeneous, not just one legal discourse, but "a set of related legal discourses" (Maley 1994: 13). Legal language does not just cover the language of law alone, but all communications in legal settings.

A related linguistic feature of the Chinese legal language is that the modern Chinese legal language has evolved into three variations used in three legal speech communities, that is, mainland China, Hong Kong, and Taiwan, and each has its own linguistic and legal characteristics (for the evolution of modern Chinese legal language and translation, see Tian and Li (2000), Lackner et al. (2001), Wang (2001), Yu (2001), Qu (2013), and Cao (2018)). Hong Kong now is a bilingual jurisdiction with English and Hong Kong Chinese as its official legal languages in a Common Law jurisdiction, and its legal Chinese is mainly a translated language from legal English with influence from its indigenous dialect or language of Cantonese, very different from mainland China or Taiwan in terms of the legal system and language use. The written language used in mainland China and Taiwan is largely the same writing system, but there are many variations due to historical, political, and other reasons. The Chinese characters used in Taiwan are in the traditional form, and its laws are traced to the beginning of the twentieth century. The mainland Chinese legal language began in the 1950s when the People's Republic of China (PRC) started to enact law and build legal institutions and uses the simplified Chinese characters. There are also substantive differences in the actual laws of two jurisdictions. Linguistically, the legal language in mainland China is less formal and more modern, using more plain language, while Taiwan's statutes are much more formal and have retained the classical style, usage, and terminology from the previous age. Consequently, one must not assume that meanings of identical words from the laws in mainland China, Taiwan, and Hong Kong are the same. In fact, very often, the meanings of legal terms in the three jurisdictions differ considerably deriving their meanings from their respective laws. This sociolinguistic phenomenon also contributes to some of the difficulties in translating between Chinese and English. This chapter focuses on the legal language used in mainland China.

Next, one may ask: What is special about the Chinese legal language and Chinese legal translation, and why is it difficult to translate legal texts between Chinese and English or other Western languages?

There are a number of sources of difficulty. First, with regard to legal translation in general, the nature of law and legal language contributes to the complexity and difficulty in legal translation. This is compounded by complications arising from crossing two languages and legal systems in translation. Major sources of difficulty in legal translation for most languages include the systemic differences in law, and cultural and linguistic differences, to be briefly explained next.

Law as an abstract concept is universal as it is reflected in written laws and customary norms of conduct in different countries. However, legal systems are peculiar to the societies in which they have been formulated. Each society has different cultural, social, and linguistic structures developed separately according to its own conditioning. Legal concepts, legal norms, and application of laws differ in each individual society reflecting the differences in that society. Legal translation involves translation from one legal system into another. Unlike pure science, law remains a national phenomenon. Each national law constitutes an independent legal system with its own terminological apparatus, underlying conceptual structure, rules of classification, sources of law, methodological approaches, and socio-economic principles (Sarcevic 1997: 13). Due to the differences in historical and cultural development, the elements of the source legal system cannot be simply transposed into the target legal system (Sarcevic 1997: 13).

Thus, one major challenge to the legal translator is the incongruency of legal systems in the source language and the target language. Furthermore, when one translates legal texts between different legal systems and different languages, the degrees of difficulty may vary. There are the following scenarios according to the affinity of the legal systems and languages according to de Groot (1988: 409–410): (1) When the two legal systems and the languages concerned are closely related, for instance, between Spain and France, the task of translation is relatively easy; (2) when the legal systems are closely related, but the languages are not, this will not raise extreme difficulties, for instance, translating between Dutch laws in the Netherlands and French laws; (3) when the legal systems are different but the languages are related, the difficulty is still considerable, and the main difficulty lies in *faux amis*, for instance, translating German legal texts into Dutch, and vice versa; and (4) when the two legal systems and languages are unrelated, the difficulty increases considerably, for instance, translating the common law in English into Chinese, and vice versa. In short, the degree of difficulty in legal translation is related to the degree of affinity between the legal systems and languages in question (de Groot 1988: 410). Such systematic gaps are manifested in both the law and language, that is, in both Chinese and Western laws and Chinese and Western languages, making translation between them particularly arduous.

Another source of difficulty in legal translation is cultural differences. Language and culture or social contexts are closely integrated and interdependent. In this connection, a legal culture is meant those "historically conditioned attitudes about the nature of law and about the proper structure and operation of a legal system that are at large in the society" (Merryman et al. 1994: 51). Law is an expression of the culture, and it is expressed through legal language. As pointed out, "[e]ach country has its own legal language representing the social reality of its specific legal order" (Sarcevic 1985: 127). Legal translator must overcome cultural barriers between the source language and target language societies when reproducing a target language version of a law originally written for the source language reader. Relevantly, Weston (1983: 207) writes that the most important general characteristic of any legal translation is that an unusually large proportion of the text is culture-specific.

The existence of different legal cultures and traditions is a major reason why legal languages are different from one another, and will remain so. It is also a reason why legal language within each national legal order is not and will not be the same as ordinary language.

Legal language is a technical language, but legal language is not a universal technical language but one that is tied to a national legal system (Weisflog 1987: 203), very different from the language used in pure science, say mathematics or physics. Law and legal language are system-bound, that is, they reflect the history, evolution and culture, and above all, the law of a specific legal system. Examples of such linguistic difficulties are found at the lexical, grammatical, syntactical, and textual levels. For instance, legal documents, in both English and Chinese, are often long and complex. Complicated syntactical structures can cause comprehension difficulties for the reader, including the translator. This also makes the rendering into the target language difficult. Another instance is lexical difficulty, in terms of legal concepts, jargon, and other terminology associated with law, the focus for the rest of the chapter (for discussions of the other areas, see Cao 2007).

Applicable to the translation of most legal languages, three major terminological challenges can be identified. They are: (1) legal conceptual issues and the question of equivalence and nonequivalence of legal concepts in translation; (2) legal jargon and usage that are bound to the legal institutions, personnel, and areas of law; and (3) legal language as a technical language in terms of ordinary vs legal meanings, and legal synonyms (see Cao 2007). Due to the fact that a basic linguistic difficulty in legal translation is the absence of equivalent terminology across different languages, this requires constant comparison between the legal systems of the source language and target language. As appropriately stated:

> The absence of an exact correspondence between legal concepts and categories in different legal systems is one of the greatest difficulties encountered in comparative legal analysis. It is of course to be expected that one will meet rules with different content; but it may be disconcerting to discover that in some foreign law there is not even that system for classifying the rules with which we are familiar. But the reality must be faced that legal science has developed independently within each legal family, and that those categories and concepts which appear so elementary, so much a part of the natural order of things, to a jurist of one family may be wholly strange to another. (David and Brierley 1985: 16)

For Chinese/English legal translation, Chinese legal terminological challenges and peculiarities include (1) nonequivalence of legal concepts and terms; (2) legal and institutional differences resulting in nonequivalence of terms, especially in light of and against the backdrop of the inherent nature and characteristics of the Chinese legal language and how it has evolved in modern times. Even for seemingly basic words such as "law" (see Cao 2004, 2018), "rights" (see Cao 2004, 2017), "justice" (see Cao 2018), and "court" (see Cao 2018), there are considerable differences when these words and concepts are translated or understood in Chinese and English respectively.

Translating Legal Terms Between Chinese and English

Yan Fu (1854–1921), one of the most influential Chinese modern thinkers who was a leading translator, in his Chinese translation of Montesquieu's *De l'esprit des lois* published in 1913, warned his readers about the difference between the Chinese *fa* (law) and Western "law" this way:

> In the Chinese language, objects exist or do not exist, and this is called *li* [order in nature, things as they are, or the law of nature]. The prohibitions and decrees that a country has are called *fa* [human-made laws]. However, Western people call both of these 'law'. Westerners accordingly see order in nature and human made laws as if they were the same. But, by definition, human affairs are not a matter of natural order in terms of existence or non-existence, so the use of the word 'law' for what is permitted and what is prohibited as a matter of law of nature is a case in which several ideas are conveyed by one word. The Chinese language has the most instances in which several ideas are expressed by one word, but in this particular case the Chinese language has an advantage over Western languages. The word 'law' in Western languages has four different interpretations in Chinese as in *li* [order], *li* [rites, rules of propriety], *fa* [human-made laws] and *zhi* [control]. Scholars should take careful note. (cited in Cao 2007: 1)

These words were written about a 100 years ago, but they are still true today despite the much closer encounter and frequent interactions between China and the West in the intervening years. Although today "law" and *fa* are used as linguistic equivalents, how these words are understood, construed, and, above all, practiced in the respective societies is not the same.

Words are the building blocks of language. As mentioned earlier, one distinctive feature of legal language is the complex and unique legal vocabulary. Legal terminology is the most visible and striking linguistic feature of legal language as a technical language, and it is also one of the major sources of difficulty in translating legal documents between English and Chinese. The legal vocabulary in a language, including legal concepts, jargon, and other legal usage, is extensive. It is resulted from and reflects the law of the particular legal system that utilizes that language. Words matter. In law, words often become points of legal contention. In translation, due to the systemic differences in law, many legal words in one language do not find ready equivalents in another, causing both linguistic and legal complications. For instance, the English legal concept and law of "equity" in common law does not have an equivalence in Chinese language or Chinese law. A new word had to be created in Chinese, *hengpingfa* 衡平法 (equity or law of equity). Many Chinese neologisms were created in the late nineteenth and early twentieth century associated with Western laws, including *zhuquan* (主权 sovereignty), *minquan* (民权 civil right), *fayuan* (法院 court), *zeren* (责任 responsibilities, liabilities, and duties), *liyi* (利益 interest), *renmin* (人民 people), *peichang* (赔偿 compensation), *zizhi* (自治 autonomy), *xuanju* (选举 election), *sifa* (司法 judiciary), *zhengduan* (争端 dispute), *xianzhi* (宪制 constitutional government), *lingshi* (领事 consul), *guanxia* (管辖 jurisdiction), *guohui* (国会 parliament), *zongtong* (总统 president) *budongchan* (不动产 real estate), *dongchan* (动产 movable property or chattel), *jicheng* (继承 succession or inheritance), *qinquan*

(侵权 tort), and *buzuowei* (不作为 omission), among many others (for further discussions of new legal terms in Chinese and Chinese legal translation around the turn of the nineteenth century, see Tian and Li (2000), Lackner et al. (2001), Wang (2001), Yu (2001), Svarverud (2001), Qu (2013), and Cao (2004)). Today, creating neologism is not common in legal translation, although it sometimes is still necessary.

Directly relevant to this is the question of how to translate. Translation methods have always been a topic of interest and contention since ancient times, in both the West and China. One of the perennial themes in the practice and study of translation is the discussions about literal and free translation and how much constraint or freedom the translator has in the translating process. For translating legal texts, the approaches were sometimes stipulated as written rules. For instance, the first codified rule on the translation of legislative texts is said to be that from Emperor Justinian in the *Corpus juris civilis*. Justinian issued a directive explicitly permitting only translations into Greek that reproduced the Latin texts word for word to preserve the letter of the law and to prevent distortion of his monumental codifications (Sarcevic 1997: 24). In other words, it meant to stipulate that the words of the source text were to be translated literally into the target language, a strict literal translation approach in legal translation. In the more recent past, starting from the 1970s, in order to implement the principle of equality, translators in Canada, among other bilingual or multilingual jurisdictions, were given freedom to produce idiomatic translation of legislation instead of literal translation (Sarcevic 1997: 46). For the Chinese, although there has never been any legal stipulation as to how to translate, the most popular and widely cited translation criterion in modern China is embodied in the dictum of *xin* (faithfulness), *da* (accuracy), and *ya* (elegance) from Yan Fu for translation in general. As the target-reader-oriented and communicative approach gains more currency, the literal and free translation debates have not been as prominent in the recent decades in the study of translation. Nevertheless, for scholars of translation, translation practitioners, and end users of translation, the question of how close or accurate a particular translation presents or represents in the final product of translation, or to put it in another way in more colloquial terms, how much is lost in translation, never really goes away.

In contemporary discussions of translation, of particular interest to the discussion here, two general translation strategies have been identified for dealing with foreign words, concepts, and texts in terms of the degree to which the translator makes a text conform to the target culture. One is foreignization – that is, the form of translation deliberately foregrounds the cultural other so that the translated text can never be presumed to have originated in the target language, and the final product may seem strange and unfamiliar (Bielsa and Bassnett 2009: 9). In contrast, domestication or acculturation is the strategy of making the translated text closely conform to the culture of the target language, which may involve the loss of information from the source text (Venuti 1995). In domestication, the translated text is adapted to suit the norms of the target culture as the signs of its original foreignness are erased (Bielsa and Bassnett 2009: 9). The domesticating strategy is said to "violently" erase the cultural values, and thus create a text that appears as if it had been written in the target language and that follows the cultural norms of the target reader (Venuti 1995). Furthermore, Venuti (1992, 1995) is of the view that the dichotomy between domestication and foreignization is an ideological one. According to Venuti (1995), every translator should look at the translation process through the prism of culture, which refracts the source language cultural norms. It is the translator's task to convey them, preserving their meaning and their foreignness, to the target language audience. There is a tendency in English translation of foreign texts of overdomestication, the tendency to minimize the foreignness of the target text and reduce the foreign cultural norms to English language and cultural values.

For legal translation, there is no rule as for what translating approach is preferable, although it is generally accepted that the translation of legal concepts should try to preserve as much as possible the original tenor and style. Legal translation, generally speaking, should be source language and culture oriented, especially in terms of translating legal concepts and institutional terms, that is, to preserve and convey the foreignness of the legal meanings in the source language. For translating legal terms from Chinese into English, including institutional terms, particularly from classical Chinese into English, there is a tendency to domesticate – that is, target language and culture-oriented, making the Chinese words, the Chinese legal processes, and practices sound more like English legal words and laws, often erasing the Chineseness. Sometimes, this is also true in translation from English into Chinese, making Western legal terms sound like traditional Chinese. A story was told from the 1940s in China. An American official delegation was visiting China. A Chinese host in conversation asked about the American "Emperor" to the amazement of the visitor. It turned out that the Chinese host was under the impression that the USA was an empire and it had an emperor as the head of state because at the time the word "President" in English had been mistakenly translated into Chinese as "Emperor." It is a case of domestication, that is, making the translated foreign words sound more familiar for the Chinese audience.

An example of domestication in English translation from Chinese is the indigenous legal and institutional term *xing bu* 刑部 used in imperial China, no longer used today. Xing bu was an imperial Chinese government branch, literally "Punishments Department" or "Board of Punishments," but it is mostly translated as 'Ministry of Justice." Such a translation was oped, presumably because the "Board of Punishments" would sound too foreign for the English readers, and if without added explanation, the meaning may be lost. However, not to translate *xing bu* as Board of Punishments because of its foreignness and for the lack of such an English term and to opt for a more familiar term in English would also give one the wrong impression that the Chinese *xing bu* was the same or similar to the English ministry of justice. It would fail to convey the meanings of the Chinese words and the role of law and what law and the legal process were like in imperial China. If we take the source language and culture-oriented approach, *xing bu* would be translated as the "Board of Punishments," retaining the original meaning and connotation, despite its foreignness.

A contemporary example is how *xuanchuan bu* 宣传部 should be translated into English. Is it "Ministry of Propaganda" or "Ministry of Information"? In modern

Chinese, *xuanchuan* means propaganda, information, communication, publicity, and promotion and can be used as a noun or a verb, and it does not have a negative connotation by itself, or one can say that its meanings are often and largely neutral. The meanings of *xuanchuan* vary depending on the context and actual use. *Xuanchuan bu* is an internal division of the Chinese Communist Party and the Chinese government in charge of ideological work, the dissemination of information, and control over the media. Before the 1990s, *xuanchuan bu* was often translated as Department of Propaganda by the Chinese government in its English materials. However, as China opened up and Chinese translators became more proficient and culturally aware in relation to the English language, they started to question whether the translation should be changed to Department of Information or Communication instead of the word "propaganda." They started to realize that "propaganda" in English is often used as a derogatory term with negative connotations, for instance, the propaganda department of the Nazi regime. Then the Chinese started to use neutral words in English for *xuanchuan*, such as publicity, information, and communication. The Chinese Communist Party Central Propaganda Department officially changed its English name to the Central Publicity Department in 1998, while its Chinese name was unchanged, retaining the word *xuanchuan*. Originally, *xuanchuan* means "announce" or "publicize." In modern Chinese, apart from the original neutral meanings of "publicize" and "disseminate," *xuanchuan* is also often associated with ideological or political communication, that is, propaganda, for instance, *xuanchuanyuan* (political propaganda worker), *xuanchuan huodong* (publicity or promotion activities), which can refer to political or ideological promotion and also ordinary promotion activities, and *fazhi xuanchuan* (promoting law or publicizing or popularizing information about the law or legal knowledge). Even in political context, the word does not normally carry derogatory connotation to the Chinese language users in China, but the word "propaganda" in English is negative. If the original usage in Chinese does not carry negative connotation, then is it appropriate for the translator to add the derogatory meanings, changing a neutral word into negative one? There may not be one single correct answer to these situations, but for the translator, these are legitimate factors for consideration.

A recent contemporary example is the legal term *xunxin zishi* 寻衅滋事, literally "picking quarrels and provoking troubles." Under Article 293 of China's Criminal Law (1997 amended), the offense of *xunxin zishi* can attract 5 years up to 10 years' imprisonment depending on the seriousness of the acts, and it may be applied to a broad and unspecified range of offenses. *Xunxin zishi* charge, along with several similar ones (Articles 290–293 of the Criminal Law (1997), is sometimes referred to more generally as "disturbing social order." Previously, *xunxin zishi* was applied narrowly to actions disrupting public order, involving actual assault or property damage. In recent years, the application of this charge has been extended. It has been used to detain, charge, and convict people with a wide range of acts from ordinary assaults and public brawls, to public drunken and disorderly conduct, petition, assaulting medical staff in hospitals, destroying a roadside clothes donation bin, and spreading rumors or false information on the Internet, to journalists and rights lawyers (see Cao 2018). Given the uncertain and broad meanings of *xunxin zishi*, how to translate the Chinese term into English cannot be easily answered. A question arises as to whether it should be translated into English using familiar English legal terms such as "public nuisance," "disorderly conduct," or "disturb the peace." In this case, it seems a source language-oriented translation method would be preferable, retaining the original Chinese meanings even though it may sound foreign and strange in English. *Xunxin zishi* is indeed a very Chinese legal term and offense with its Chinese uniqueness, in both the word and legal substance. It is not at all the same as "public nuisance" or "disorderly conduct" in English laws.

In 2019, heated debates and considerable controversy erupted in Chinese legal circles, especially among Chinese students studying law in English-speaking countries. It involves the translation of the law degree, J.D., in the USA, Australia, Canada, and other countries (J.D. – Juris Doctor). Until recently, the term had been translated into Chinese as 法律博士学位 literally "law doctorate." However, public notices from the Chinese Ministry of Education were issued in mid-2019 stating that after investigation and consideration, in order to avoid confusion to potential employers and in the society caused by the "doctorate" title in the name of the J.D. degree certificate, the Ministry of Education's Study Abroad Service Centre has decided to adjust the Chinese translated name of the J.D. certificate from the USA, Canada, and Australia to "Professional Law Diploma." It was also noted in the official notices that the J.D. from such programs in the USA will be classified as equivalent to a professional doctorate degree, J.D. from Australia as equivalent to a professional diploma degree, and J.D from Canada as equivalent to a bachelor's degree. These official notifications of changes in the Chinese translation of the degree title caused a great deal of uproar and consternation among Chinese law students outside China. Such a change may very well change the lives and careers of these future students without the word "doctorate" in the naming of the degrees in Chinese. After much protests and discussions, it was reported that the Ministry of Education of late withdrew its original notices and reverted to the original Chinese translation. Relevantly, it is apparent whether the term is translated literally as it has been or translated as an explanatory term entails serious consequences in real life.

Conclusion

It is important to bear in mind that legal translators are not lawyers. Likewise, bilingual lawyers are not translators. The legal translator's job is not to provide legal advice and solve legal problems, but to translate and facilitate communication across linguistic, cultural, and legal barriers through the medium of language. The legal translator's skills and tasks are very different from the lawyer's. The legal translator does not read and interpret the law the way a lawyer does. The legal translator does not write the law either. However, the legal translator needs to know the language used in law, how lawyers, including judges and lawmakers, think and write and why they write the way they do, and at the same time, to be sensitive to the intricacy, diversity, and creativity of language, as well as its limits and power.

Translating legal texts between Chinese and English presents unique challenges as well as sharing similarities with other types of translation. Legal translation, as other types of translational activities, is a norm-governed human and social behavior, a text-producing act of legal communication. Legal translation is complex, but it is a skill that can be acquired. Learning to translate law and legal language means the acquisition of the relevant competence, in languages, legal and general knowledge, and translation skills, among others, required to select and apply those norms and rules that will produce legitimate translations (Cao 2007).

References

Bielsa, Esperanca, and Susan Bassnett. 2009. *Translation in global news*. New York: Routledge.
Cao, Deborah. 2004. *Chinese law: A language perspective*. Aldershot: Ashgate.
———. 2007. *Translating law*. Clevedon: Multilingual Matters.
———. 2014. Afterword: Trials and tribulations of legal translation. In *Ashgate handbook on legal translation*, ed. A. Wagner et al., 313–315. Aldershot: Ashgate.
———. 2017. On the universality of rights: Absence and presence of rights in the Chinese language. *Intercultural Pragmatics* 14 (2): 277–292.
———. 2018. *Chinese language in Law: code red*. Lanham: Lexington Books.
David, René, and John Brierley. 1985. *Major legal systems in the world today*. London: Stevens.
De Groot, G.R. 1988. Problems of legal translation from the point of view of a comparative lawyer. In *XIth world congress of FIT proceedings: Translation – Our future*, 407–421. Maastricht: Euroterm.
Kurzon, Dennis. 1998. Language of the law and legal language. In *Special language: From humans thinking to thinking machines*, ed. L. Christer and M. Nordman, 283–290. Clevedon: Multilingual Matters.
Lackner, Michael, Iwo Amelung, and Joachim Kurtz, eds. 2001. *New terms for new ideas: Western knowledge and lexical change in late Imperial China*. Leiden: Brill.
Liu, Lydia H., ed. 1999. *Tokens of exchange: The problem of translation in global circulations*. Durham/London: Duke University Press.
Maley, Yon. 1994. The language of the law. In *Language and the Law*, ed. J. Gibbons, 11–50. New York: Longman.
Merryman, John Henry, Clark, Avid S. and Haley, John O. 1994. *The civil law tradition: Europe, Latin America, and East Asia*. Virginia: The Michie Company.
Qu, Wensheng. 2013. *Cong cidian chufa (Starting from Dictionaries)*. Shanghai: Shanghai renmin chubanshe.
Sarcevic, Susan. 1985. Translation of culture-bound terms in laws. *Multilingua* 4 (3): 127–133.
———. 1997. *New approach to legal translation*. The Hague: Kluwer Law International.
Svarverud, Rune. 2001. The nations of 'power' and 'rights' in Chinese political discourse. In *New terms for new ideas: Western knowledge and lexical change in late Imperial China*, ed. M. Lackner, I. Amelung, and J. Kurtz, 125–146. Leiden: Brill.
Tian, Tao, and Zhuhuan Li. 2000. Qingmo fanyi waiguo faxue shuji pingshu (Commentary on the Chinese translations of Foreign legal works in the late Qing dynasty). Zhongwai faxue (Peking University Law). *Journal* 12 (3): 355–371.
Venuti, L., ed. 1992. *Rethinking translation: Discourse, subjectivity, ideology*. London: Routledge.
———. 1995. *The translator's invisibility: A history of translation*. London: Routledge.
Wang, Jian, ed. 2001. *Xifa dongjian: Waiguoren yu zhongguo fa de jindai bianqe (Western law going East: Foreigners and the reform of modern Chinese Law)*. Beijing: Zhongguo zhengfa daxue chubanshe.
Weisflog, W.E. 1987. Problems of legal translation. Swiss reports presented at the XIIth international congress of comparative law, pp. 179–218. Zürich: Schulthess.
Weston, M. 1983. Problems and principles in legal translation. *The Incorporate Linguist* 22 (4): 207–211.
Yu, Jiang. 2001. Jindai Zhongguo faxue yuci de xingcheng yu fazhan' (Formation and development of modern Chinese legal language and terms). In *Zhongxi falü chuantong (Chinese and Western legal tradition)*, vol. 1, 24–66. Beijing: Zhongguo zhengfa daxue chubanshe.

Chinese Children's Literature in English Translation

23

Minjie Chen and Helen Wang

Contents

Introduction	552
Part I – A Historical Overview	553
Pioneer Work by Missionaries During the Late Nineteenth Century	554
Importing Chinese Folk and Supernatural Tales During the Twentieth Century	558
English Translations Made in China	565
A Bumpy Journey to the West	567
Crocoducks and Republican Bunnies Leaping to the West: Progress in the New Century	570
Part II – International Interest and Translation Initiatives in Chinese Children's Books Since the 1980s	574
Commercial and Noncommercial Interest in Chinese Children's Books	574
(A) Noncommercial Interest	575
(B) Commercial Interest	577
Cultural Shifts Conducive to the Translated Book Industry	579
Translator-Led Initiatives	580
Conclusion	589
Appendix 1: A List of Chinese Folk Literature Adapted or Illustrated by Ed Young	590
Appendix 2: Notable Children's and Young Adult Books by Contemporary Chinese Authors in English Translation (1995–2020)	592
Appendix 3: Chinese Children's Books Selected as White Ravens, 2014–2019	594
References	597

Abstract

This chapter outlines a history of English translations Chinese children's literature from the late Qing dynasty to the present. Part I examines the types of text selected for translation, analyzes the fluid relationship between the source and

M. Chen (✉)
Princeton University Library, Princeton, NJ, USA

H. Wang
The British Museum, London, UK
e-mail: hwang@britishmuseum.org

© The Author(s), under exclusive licence to Springer Nature Singapore Pte Ltd. 2022
Z. Ye (ed.), *The Palgrave Handbook of Chinese Language Studies*,
https://doi.org/10.1007/978-981-16-0924-4_45

target text, and reveals how the text served shifting religious, political, educational, cultural, and commercial interest. It also discusses ideological incongruity as a major barrier for importing children's literature from China to the West. It then reviews the breakthroughs of Chinese children's literature in English translation during the first two decades of the twenty-first century, highlighting major authors, illustrators, titles, and international recognition. Part II offers a survey of commercial and noncommercial agents that have facilitated an international network of authors, illustrators, translators, and publishers. It highlights international children's literature organizations, libraries, festivals, book fairs, academic institutions, translators' professional communities and initiatives, and the most active figures that have played important roles in raising the visibility of Chinese-language children's literature, promoting high-quality translated works, and professionalizing the field of translation.

Keywords

Chinese children's literature · Translations into English · Chinese folk tales · Chinese picture books · Chinese-to-English translators · Translation initiatives · Missionaries in China

Introduction

If we apply the conventional definition of "children's literature," that is, reading materials tailored for the entertainment of youth, to Chinese books, then this type of publication barely existed in premodern China, where children's traditional texts were limited to primers, the Confucian canon, and educational contents that youth were supposed to learn, read aloud, and memorize. *The Twenty-Four Paragons of Filial Piety* 二十四孝, first compiled during the Yuan dynasty (1271–1368), was one of the few texts with redeeming entertainment value for a young listener, viewer, or reader. Its apparent appeal to youth lies in the agency exhibited by child paragons of filial piety for problem solving, the fantastical elements in the miracles seen as rewards for their extraordinary virtue and sacrifice, and the illustrations that commonly accompany the text. Its engagement is relative and contextual, though. Lu Xun 鲁迅, writing in the 1920s, famously decried the alienating effect of some of the 24 paragon stories, finding "Kuo Chu Buries His Son," which thoroughly disempowers the child in the story, alarming and demoralizing (Lu 1976, 34).

Chinese children's reading materials began to transform during the last quarter of the nineteenth century. Pursuing agendas ranging from evangelical to nationalistic, Western missionaries and Chinese intellectuals imported both the practice and content of children's literature from abroad, and expanded publications for Chinese youth to text and illustrations that were more attuned to the interests and cognitive capacities of young minds. Translations have had an indelible impact upon Chinese children's materials and, consequently, translated works are an integral part of the scholarship of Chinese children's literature (e.g., Li 2010, 2017; Song 2015; Chen 2019). In contrast, the history of Chinese children's literature that has been exported and translated into other languages is little studied. This chapter contains two parts. Part I delineates the history of Chinese children's reading materials available in the English language from the eighteenth century to the present. Part II provides an overview of translation initiatives and activities of contemporary Chinese children's literature in the English-speaking world.

Part I – A Historical Overview

As we shall see, the early practice of translating Chinese texts into English-language children's books was not always aligned with modern process. The relationship between the target text and source text ranged from a "free rendering" to the more faithful. The source text might be haphazardly documented and hard to trace; or it was directly taken from an oral narrative and did not exist in a written Chinese version; or the English text was translated from an intermediary language; or the source text was not necessarily intended as "children's literature" in China. Regardless, a "prehistory" of Chinese children's literature in English translation constitutes a substantial part of how English children's literature absorbed Chinese culture and literature even before Chinese children's literature came into being by the late nineteenth century and began to thrive during the Republic period.

Chinese topics made only rare appearances in English-language juvenile books between the eighteenth century and the First Opium War (1840–1842), which ushered in modern China. These earliest texts project an admiring view of Chinese, reflecting the legacy of a historically favorable perception of Chinese civilization held in Western society, a perception which would soon turn critical and contemptuous after the late eighteenth century (Chen 2004b). Solomon Sobersides' holiday gift books – titled *Christmas Tales* for the 1780 London edition and *A Pretty New-Year's Gift* for the 1786 American edition – include several stories "from oriental and Chinese sources." The American edition, with the subtitle *Entertaining Histories, for the Amusement and Instruction of Young Ladies and Gentlemen, in Winter Evenings*, contains an anecdotal story about a Chinese emperor called Hamti – the description of whom roughly matches that of Emperor Wu of the Han dynasty, considered one of the greatest emperors in Chinese history. Hamti is portrayed as a wise king who cares about the disabled and the most vulnerable of his subjects (Fig. 1).

An anonymous work of fiction, with the self-explanatory title *Chun and Si-ling: An Historical Romance: In Which Is Introduced Some Account of the Customs, Manners, and Moral Conduct of the Chinese*, was published in London in 1811. An admirer of Chinese culture, the author wished to teach young readers "how to acquire and maintain equanimity of soul" (*Chun* 1811, "Introduction") as modeled by the fictional Chinese protagonists.

Fig. 1 A pageant led by the Chinese King Hamti, "followed by the blind, the maimed, and the strangers of the city," as imagined in an uncredited engraving in Solomon Sobersides's *A Pretty New-Year's Gift*. Worcester, Massachusetts: Printed by Isaiah Thomas, 1786 (Sobersides 1786). (page 113)

Pioneer Work by Missionaries During the Late Nineteenth Century

As a result of the two Opium Wars, ending in 1842 and 1860, respectively, the disgraced Qing government was forced to open up the entire country to missionary activity, attracting Christian missionaries of diverse denominations from Britain, America, and Europe to the Middle Kingdom to preach the Gospel. Missionaries produced some of the earliest English translations of Chinese texts for children. In order to better understand the minds of Chinese people and thus to communicate more effectively, missionaries learned Chinese language(s) and culture(s). The same skills and knowledge prepared the Christian proselytizers for the role of cultural ambassadors. By exporting Chinese children's culture and introducing it to young people in the English-speaking world, these bicultural writers hoped to cultivate interest and empathy in Chinese people, and to bolster support for the mission.

The Twenty-Four Paragons of Filial Piety was among the earliest texts that found their way into English-language children's reading. Claudia Nelson (2014) analyzed examples of the 24 paragons of filial piety that appeared in American children's literature between 1845 and the 1920s, and pointed out that, to the American readership, self-sacrificing filial piety was taken as a hallmark of Chinese culture, and filial devotion symbolized both Chinese "virtue and stagnation, a regard for the past that put a stop to progress" (3). Of the 24 examples, Nelson found "He Fed Mosquitoes with His Blood" to be the most-often retold and most familiar to non-Chinese readers (1).

The persistent prominence of filial themes is demonstrated in the 1867 edition of *Peter Parley's Annual*, which was marketed as a holiday gift book for children. It contains a piece titled "Boat Life in China (1867)." Seemingly a description of life and culture in Canton (Guangzhou), much of it, upon close reading, is old wine in a new bottle. Part of the piece refashions the mosquito story – originally published as "Filial Affection of the Chinese" in the June 1845 issue of *Robert Merry's Museum* – into a narrative about a boy in Canton; another part recycles an even older text, a historical story of filial piety, fraternal loyalty, and family duties about Lu Nanjin 陆南金, who lived in the eighth century. The story was recorded in the *Old Book of Tang* 旧唐书 (945). The earliest English version, translated from French Jesuit historian Jean-Baptiste Du Halde's *Description de la Chine* (1736), appeared as "Example of Filial Piety in China" in the December 1736 issue of the *Gentleman's Magazine*. The version in *Peter Parley's Annual* is similar but for a slightly different spelling for Lu's name (Loo-nan-kin versus Lou Nan-kin).

Adele M. Fielde (1839–1916), a Baptist missionary who did extraordinary Bible work with local women during her tenure at the port city of Swatow (Shantou, Guangdong Province) from 1873 to 1889, provided a free English rendering of *The Twenty-Four Examples of Filial Piety* and published what appears to be the earliest Chinese-English bilingual illustrated version of the stories at the Swatow Printing Office Company in 1879.

Chinese Stories for Boys and Girls and Chinese Wisdom for Old and Young, first published in 1880, includes 31 stories widely used to acculturate Chinese children to the moral principles of Confucianism, filial piety, and fraternal duties. Arthur E. Moule (1836–1918), an English missionary to Zhejiang province, selected and translated them from a popular book he acquired in Hangzhou and had ten of the wood-block printed illustrations reproduced in the English edition. The purpose of Moule's (1881) book was threefold: First, to amuse young readers in England (10); second, to increase their interest in Chinese people (34); thirdly and most importantly, to sustain support for the Chinese mission (Fig. 2). Moule wrote, with an eloquent quote from Confucius,

> [T]here is much, as I said before, to admire in the Chinese; and I hope that these stories will interest my readers as showing how the Chinese *think*, and how *human* they are; and therefore how well worth the while it is for English Christians to spend their money and their lives in doing good to this great people. "Within the four seas" – that is, everywhere, said Confucius, "all men are brethren." (18)

It is to be determined who published the first Chinese folktale to English-speaking children and when. Andrew Lang (1892) compiled *The Green Fairy Book*, an illustrated collection of fairy tales, and referred to one of the entries, "The Story of Hok Lee and the Dwarfs," as Chinese (ix, 229). The plot of the story closely resembles "The Lump-Removing Old Man" こぶ取り爺, a folktale widely known in Japan but which also circulated in Korea and evidently in China. "Demons remove and return lumps," which is close to the plot of "The Story of Hok Lee and the Dwarfs," has been indexed as a Chinese folk tale type (Jin 2002, 713). It is unclear if Lang used a Chinese source or if this was a case of imprecise attribution of cultural origins. Based on linguistic clues, Mikyung Bak (2014) speculated that Lang adapted a Korean version of the tale. Confusing East Asian cultures would be found to be a reoccurring error in English-language children's books – Arlene Mosel's picture book *Tikki Tikki Tembo* (1968), allegedly a Chinese tale but traced to Japanese folk culture, being the best-known example.

Fig. 2 Illustration of "He Cried and the Bamboo Sprouted," one of the stories from *The Twenty-Four Paragons of Filial Piety*, in *Chinese Stories for Boys and Girls and Chinese Wisdom for Old and Young*. London: Seeley, Jackson, & Halliday, 1881. (Unpaginated plate between page 42 and 43)

Bibliographical records suggest that the earliest Chinese folktale collection in English translation, with a degree of authenticity professed by the translator, was by Adele M. Fielde. Having returned to the United States from Shantou, she published *Chinese Nights' Entertainment: Forty Stories Told by Almond-Eyed Folk Actors in the Romance of the Strayed Arrow* in 1893, and prefaced the book with valuable contextual information of the collection:

> These tales have been heard or overheard by the writer, as they were told in the Swatow vernacular, by persons who could not read. They and their kind have furnished mental entertainment for her during many nights when travelling in a slow native boat, or sitting in a dim native hut, with almond-eyed women and children...She is not aware that any of these stories have before been rendered into a European tongue. (Fielde 1893, v)

Fielde's revelation allows us to appreciate that the stories were told by illiterate Chinese from the oral lore of Shantou, and, true to the tradition of folktales before they were appropriated by children's literature, were meant to entertain "folks" regardless of age. However, her English-language publication was oriented for a youth readership. The 40 entries are framed within a story, in which a teacher and his pupils share the stories for entertainment – the teacher emphasizes that he is going to tell a story that has "no moral" (5).

Fielde in effect played the double-role of folklorist/translator in Shantou. *Chinese Nights' Entertainment* captured Chinese voice, especially the rare voice of illiterate women, at a time when folklore was not yet established as a discipline in China – the systematic collection and documentation of Chinese oral tales did not begin until the early twentieth century (Li 2008). In the second edition of the book, Fielde (1912) referred to herself as the one who recorded and translated the tales, and rightly speculated that they were not to be found in Chinese books. For example, variant versions of the first story, "The Five Queer Brothers," did not appear in Chinese publications until the early 1930s, according to the folklorist Wolfram Eberhard ("Variations" 1977, 5). Interestingly, Fielde's book was translated back into Chinese and published as *Chao Shan ye hua* 潮汕夜话: 潮汕老古一八七三 in Hong Kong in 2016. The book also treated readers with authentic Chinese art: the 25 illustrations that accompany the stories were done, under the direction of Fielde, by local Chinese artists (Fig. 3).

Fielde's intention is made clear in the second edition, which had a new title, *Chinese Fairy Tales*. For Fielde (1912), these traditional tales revealed "the Chinese minds" (viii), presented a favorable image of Chinese behavioral traits and values, and encouraged an optimistic outlook for their future.

Chinese Mother Goose Rhymes, published in 1900, is a compilation of about 130 nursery rhymes presented in Chinese-English bilingual form. Isaac Taylor Headland (1859–1942), the translator, also prepared all the illustrations, mostly of photographs of Chinese children by themselves or with their caregivers. He arrived in China as a missionary under the auspices of the American Methodist Episcopal Mission in 1890 ("Isaac" 2016) and was teaching at Peking University when the book was published. According to the preface, he and his colleagues collected more than six hundred nursery rhymes mainly from two provinces. Headland expressed two motives for compiling the book, which resemble those of Moule's *Chinese Stories*. First, the rhymes were intended to entertain English-speaking children. Partly for this reason, as the translator disclaimed in the preface, he made no attempt at regularity in the meter of the rhymes. Below is an example of the extent of the liberty that Headland took with Chinese rhymes in his English rendition, which prioritizes appeal to English readers over fidelity to the original text (Table 1).

Second, by presenting "a new phase of Chinese home life" (5), Headland hoped to "lead the children of the West to have some measure of sympathy and affection for the children of the East" (6). Dated October 1900, shortly after the siege of Beijing in the midst of the anti-Christian Boxer Uprising, Headland's message was all the more significant (Fig. 4).

Fig. 3 Captioned "The teacher expounded a classic at eventide under a tree," the frontispiece of *Chinese Nights' Entertainment* illustrates a scene from the frame story of the book. New York: G.P. Putnam's Sons, 1893

Importing Chinese Folk and Supernatural Tales During the Twentieth Century

In the wake of the First World War, there was a keenly felt need in the United States to bring up a new generation of citizens as respectful and friendly towards people of other nations and races (Chen 2016, 186). Folktales translated from Chinese as well as fantasy stories that feature Chinese characters were published in the 1920s, a welcome addition to counter previous evil images of Chinese found in the likes of *The Insidious Dr. Fu Manchu* (1913). Several decades earlier, at the height of American hostility against Chinese people – the Chinese Exclusion Act was passed

Table 1 An excerpt from *Chinese Mother Goose Rhymes* translated by Isaac Taylor Headland

Source text	Gloss	Translated text
老頭子出來拄拐棍兒	The old man comes out leaning on a stick	The old man comes out With a great deal of trouble
老婆子出來就地兒擦	The old lady comes out bent to the ground	His wife hobbles after, Her body bent double
看家的狗兒三條腿	The guard dog has three legs	Their three-legged dog Is as thin as a rail
避鼠的狸貓短个尾巴	The rat-avoiding civet cat is short of a tail	And their rat-fearing cat Is minus a tail

(Headland 1900, 150–151)

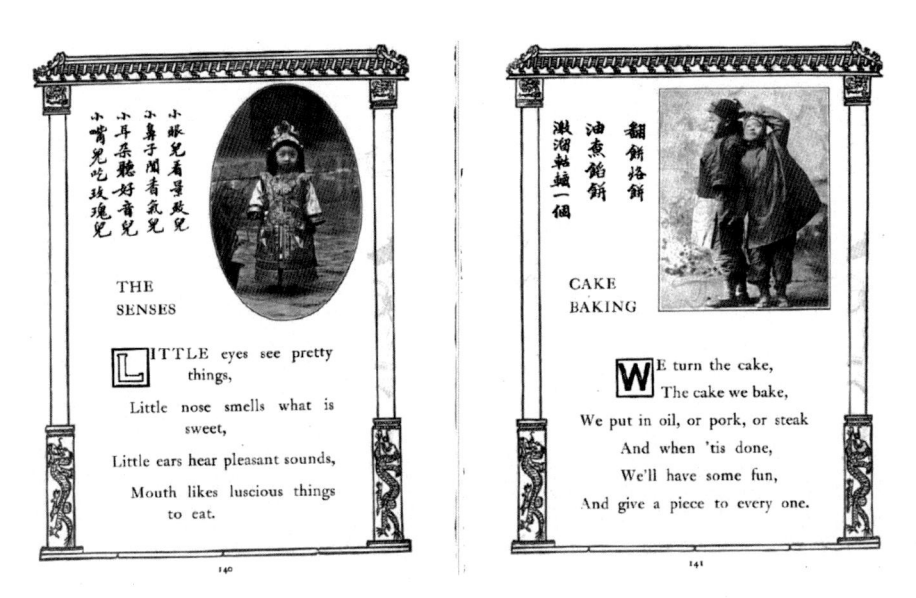

Fig. 4 *Chinese Mother Goose Rhymes*, translated and illustrated by Isaac Taylor Headland. New York: Fleming H. Revell Co., 1900

in 1882 to bar the immigration of Chinese laborers to the United States – Chinese deities and supernatural beings would have been condemned as part of Chinese heathen beliefs and superstitions. Now they were embraced as fantastic tales suitable for entertaining young readers in English-speaking countries. The trajectory of Chinese folktales to the English-language world, through translations and adaptations, blurs the customary demarcation between "international" and "multicultural" children's literature, revealing cases where one morphs into the other.

The Chinese Fairy Book, first published in New York in 1921 and then in London in 1922, is a collection of Chinese folktales and legends. Its title page credits Dr. R. Wilhelm as editor and Frederick H. Martens as translator. Wilhelm was sent as a pastor by the General Evangelical Protestant Mission, a German missionary

Fig. 5 Cover and the frontispiece of *The Chinese Fairy Book*, which includes six illustrations in color by George W. Hood (1869–1949). New York: Frederick A. Stokes, 1921

society, to Qingdao in 1899, when the city was a German colony ("Wilhelm, Richard" 2011). He lived in China for 25 years and became a sinologist and an influential translator of ancient Chinese philosophy.

Martens (1921) expressed confidence about the appeal of these Chinese stories. He compared them to *One Thousand and One Nights* and predicted that "[t]here is no child who will not enjoy their novel colour, their fantastic beauty, their infinite variety of subject" (v). "They have been retold simply, with no changes in style or expression beyond such details of presentation which differences between oriental and occidental viewpoints at times compel" (vi). The translator's statement raises no suspicion that the text might not have been directly translated from Chinese. The English edition was actually selected and translated from German from Richard Wilhelm's *Chinesische Volksmärchen*, first published in Jena in 1914 (Fig. 5).

The Chinese Fairy Book covers a remarkable extent of significant names from Chinese culture and history. Among the 73 stories in the collection are familiar tales about animals, deities, ghosts, and other supernatural beings, including Chang'e 嫦娥 ("The Lady of the Moon"), the Cowherd and the Weaver Girl 牛郎织女 (with an illustration in the frontispiece), the Eight Immortals 八仙, Nezha 哪吒, and the Monkey King 孙悟空 ("The Ape Sun Wu Kung"). They also include legends about historical figures such as Confucius, Laozi, Guan Yu 关羽 ("The God of War"), Yang Guifei 杨贵妃 (an imperial consort, she is considered one of the "Four Beauties of China"), and Ximen Bao 西门豹 ("How the River God's Wedding Was Broken Off").

The book was "translated after original sources" according to its title page, although they are either rarely cited or attributed to oral tradition. One of the unattributed entries, "The Frog Princess," appears to have been adapted from "Qing wa shen" 青蛙神 in Pu Songling's (蒲松龄, 1640–1715) *Strange Stories from a Chinese Studio* 聊斋志异. "Rose of Evening" is a sanitized version of "Wan xia" (晚霞, Cloud of Sunset), also from Pu's book. An exception is *Sin Tsi Hia* 新齐谐, a collection of supernatural stories by Yuan Mei (袁枚, 1716–1797), which is cited as the source of three tales in *The Chinese Fairy Book*.

Another anthology of Chinese folktales translated from the intermediary language of German is *Chinese Fairy Tales and Folk Tales* (1937) by Wolfram Eberhard (1909–1989), who began collecting folklore from China in 1934 and published the original German edition, *Chinesische Volksmärchen* (with an identical title to Wilhelm's work), in Leipzig in 1936. Neither the German nor English edition was oriented toward young readers; the work was of a scholarly nature, deemed a landmark in the history of Chinese folktale studies and probably the most frequently reprinted collection of Chinese folktales in America (Dorson 1965; "Variations" 1977).

The Five Chinese Brothers (1938), written by Claire Huchet Bishop, is one of the earliest titles that retell a Chinese folktale in the format of a stand-alone picture book. It is remarkable how filial and fraternal themes from previous Chinese stories coalesce in this new work. In the aforementioned story about Lu Nanjin, after he was imprisoned for committing a crime for which he might face a death penalty, Lu's younger brother turned himself in and made a false confession. He hoped to rescue the more capable elder brother, who could bury their deceased mother and arrange their sister's marriage. When the judge found out that the latter was innocent, and learned that the younger brother was willing to die to ensure filial and family duty could be fulfilled, the case was reported to the emperor, who pardoned the criminal, suggesting the virtue of filial piety was so esteemed as to potentially redeem other wrongdoings. In *The Five Chinese Brothers*, when the First Brother faces execution, the Second Brother literally sticks his neck out to save the elder brother's life – born with an iron neck, the Second Brother cannot be killed by decapitation. As in the case of Lu Nanjin, filial piety is highly respected by the authorities, and each time a brother entreats to go home and pay one last visit to his aged mother on the eve of his execution, the wish is granted by the judge.

It is intriguing how Bishop, who was raised in France, learned about the Chinese tale. Except for a slight difference in ending, *The Five Chinese Brothers* is largely the same as "The Five Queer Brothers," recorded by Adele M. Fielde from the oral culture of Shantou during the late nineteenth century. Bishop knew the oral French version of the story and cited her father, a China buff, as the possible source of her knowledge ("Variations" 1977). The elder Bishop could possibly have learned the story directly or indirectly from Fielde's book – both the first and second edition of her folktale collection are registered in the catalogues of libraries in European countries, including France, although that is not definitive evidence of how widely the copies had circulated among the public. Bishop recalled in an interview that she initially told the story to French children in a library, and later prepared the English translation in writing in order to entertain American children at the New York Public Library (Hile 2007). At any rate, brothers looking alike and each bestowed with a

Fig. 6 *Five Brothers* in the Chinese comic book format (Shanghai People's Fine Arts Publishing House, 1955 (Fu et al. 1955))

Fig. 7 *Ten Brothers* in the Chinese comic book format (Shanghai People's Fine Arts Publishing House, 1958 (Hong and Zheng 1958))

unique magic power is a popular tale type circulating in the oral culture of many regions of China (Liu 2002, 463); adaptations into Chinese children's materials have been made since the 1950s, including the paper-cut animation TV series *The Calabash Brothers* 葫芦兄弟 (1986–1987) (Figs. 6, 7, and 8).

Despite its controversial visual representation of Chinese, *The Five Chinese Brothers* has remained a children's classic and is still in print. It was even translated into Chinese and published by Zhejiang People's Fine Arts Publishing House in 2018 (Bishop and Wiese 2018). The illustrations, done by Kurt Wiese, have drawn vehement criticism in the United States for what is perceived as racist depiction of Chinese skin color in bright yellow pigment, among other offensive Chinese stereotypes. For complex reasons, such visual features, if admittedly not realistic, are apparently not a source of insult for Chinese in China. For one thing, in the Chinese language "yellow" is employed as the formal term for describing the skin tone of the

Fig. 8 *Calabash Brothers*, a comic book spin-off of the 2008 movie remake of the original animation TV show (Shanghai Bookstore Publishing House, 2008 (Hu lu xiong di 2008))

Han Chinese – true insult would have come from rejecting the label – making it an unusual case of a welcome "loss in translation."

The narrative space for Chinese stories expanded and contracted in American society during the twentieth century, contingent on political climate, social movements, race relations, the international relationship between China and the United States, and so on. Stories set in premodern China and content with a focus on traditional Chinese culture, however, have endured the vicissitudes of political and social tensions and maintained their presence in English-language children's literature. During the Cold War, in particular, tales and legends of ancient China increasingly dominated American children's books about Chinese. Folktales and stories about traditional Chinese social life and customs offered double advantages. First, these topics were safely removed from the conflicts and chaos under McCarthyism and Maoism; second, by choosing to celebrate Chinese traditions, American authors, especially ethnic Chinese writers, disassociated themselves from Communist China, which was known to be waging a crusade against heritage and tradition, particularly during the Cultural Revolution (1966–1976) (Chen 2016, 199–200).

Writers and illustrators of Chinese or non-Chinese heritage have both contributed to the introduction of Chinese folktales to the English-speaking world. Notable names include Frances Carpenter (1890–1972), Laurence Yep, Demi, married couple Raymond Chang and Margaret Scrogin Chang, Zhang Song Nan 张颂南, among others. The Tianjin-born artist Ed Young 杨志成 is arguably the most prominent author and illustrator of picture books that expose American children to Chinese folktales and culture. His long career of making children's books spans more than half a century, from 1962 to the present. Chinese folktales and legends make up a substantial part of his illustrious list of works. The first translated Chinese folk literature he worked on was *Chinese Mother Goose Rhymes* (1968), which revives 41 rhymes from Isaac Taylor Headland's (1900) compilation by the same title and another source, Guido Vitale's *Pekinese Rhymes* (1896), with new translations and color illustrations. Young's paintings have preserved the clothing style and children's games of the late Qing dynasty as captured by Headland's black-and-white photographs. Young's *Lon Po Po: A Red-Riding Hood Story from China* (Philomel, 1989) is based on "Grandmother Wolf," a household story in China. *Yeh-Shen: A Cinderella Story from China* (Putnam, 1982), retold by Ai-Ling Louie and illustrated by Young, reproduces the Chinese source text from Duan Chengshi's *Miscellaneous Morsels from Youyang* 酉阳杂俎, dated around 850 of the Tang dynasty.

A century of importing traditional Chinese tales into the reservoir of English-language children's literature culminated in two remarkable books: Gene Luen Yang's graphic novel *American Born Chinese* (2006), winner of a Will Eisner Comic Industry Award, and Grace Lin's *Where the Mountain Meets the Moon* (2009), recipient of a Newbery Honour. In *American Born Chinese*, Yang masterfully reshaped *Journey to the West* into a sophisticated metaphor for Asian Americans' painful struggle with racial stereotypes, social isolation, and internalized sense of inferiority. In Yang's retelling, the Monkey King is freed from five hundred years' imprisonment beneath a mountain of rock only when he embraces himself and stops wishing to be one of the deities, paralleling the Chinese American boy protagonist's trajectory from self-rejection to self-acceptance of his ethnic identity.

A fantasy-adventure story, *Where the Mountain Meets the Moon* retells numerous Chinese folktales, employs an extensive cast of figures from folktales and legends, introduces surprising twists to age-old plots, and fuses what once were unrelated tales into one coherent grand narrative. In the storyland constructed by Lin, the Weaver Girl spins the red threads that the Old Man of the Moon uses to tie people's destinies together; a peach grove, which the girl protagonist must pass, is guarded by raucous monkeys, recalling the Garden of Immortal Peaches under the mischievous Monkey King's charge in *The Journey to the West*; an entire house full of residents disappears into thin air overnight, reminiscent of the eerie way handsome mansions of ghosts and fox spirits evaporate at dawn in *Strange Stories from a Chinese Studio*; examples of morphing elements from Chinese tales are too many to enumerate. Lin followed the book with two award-winning installments, *Starry River of the Sky* (2012) and *When the Sea Turned to Silver* (2016) (a National Book Award Finalist), stringing yet more retold Chinese folktales like pearls in a necklace.

Another example of the influence of Chinese literature upon English children's literature was *The Gruffalo* (1999) by Julia Donaldson and illustrated by Axel Scheffler.

The author was inspired by a fable about a fox tricking a tiger (狐假虎威), which was found in the *Strategies of the Warring States* 战国策 compiled two millennia ago. *The Gruffalo* has since been translated into more than 50 languages, including a Chinese version titled *Gu lu niu* 咕噜牛 (2005), translated by Ren Rongrong 任溶溶.

English Translations Made in China

Educators and librarians in the United Kingdom and the United States have repeatedly pointed out the dearth of translated children's literature, especially those from countries outside Europe and Japan. Quantitatively speaking, however, there is more Chinese children's literature in English translation than the public is aware of. Driven by motives that range from disseminating Communist ideology, Chinese culture, and Chinese language to diffusing political soft power, Chinese publishers have been the major force in producing translations or bilingual versions of Chinese children's literature since the 1950s.

The publishing houses of the China Foreign Languages Publishing Administration have been responsible for the bulk of Chinese children's literature translated into English. The Foreign Languages Press, the main and oldest publisher of the Administration, was established in 1952 as a state organ to specialize in foreign language editions of Chinese works. Bibliographical research suggests that FLP, along with Zhaohua (朝华) Publishing House, another affiliate to the Administration, published no fewer than 282 English translations of Chinese children's literature or folktale collections between 1954 and 1987, and probably many more. Dolphin Books, previously the children's literature editorial department of FLP, split off as a separate house in 1986 (Hua and Shi 2020; "Ji gou jian jie" n.d.). Overseen and controlled by the Communist Party of China, Dolphin Books is an official organ charged with the international promotion of Chinese children's literature, although other Chinese presses have diversified since the 1990s and have also participated in the production of English and bilingual children's books (Table 2).

1. The incomplete tally of 282 titles was based on bibliographical information in WorldCat as of September 2020 and its accuracy is contingent on the granularity, consistency, and comprehensiveness of the metadata shared by member libraries

Table 2 Chinese children's literature translated into English by the China Foreign Languages Publishing Administration between 1954 and 1987

Years	Foreign Languages Press (1952–)	Zhaohua (1982–)
1954–1959	45	n.a.
1960–1969	42	n.a.
1970–1979	52	n.a.
1980–1987	109	34
Total	248	34

globally. The title list can be found at https://www.worldcat.org/profiles/minjiec_oclc/lists/3969535 curated by Minjie Chen.
2. The reduced output of the 1960s reflects the disruption of the Cultural Revolution, which nearly paralyzed China's book publishing until the early 1970s.

Spanning three decades, the FLP list encompasses poetry, folktales, legends, fables, supernatural tales adapted from Chinese classical literature, and contemporary Chinese children's literature. The years between 1966, when the Cultural Revolution broke out, and 1978 witnessed a substantial increase in stories about Communist and military heroes, the Sino-Japanese War (1937–1945), and conflicts between the (virtuous) poor and the (evil) rich. Folktales and adaptations from traditional literature were banned as part of "the Four Olds," or Old Customs, Old Culture, Old Habits, and Old Ideas. Jiang Qing (1967), Chairman Mao Zedong's wife, delivered an influential speech in 1964, condemning popular opera stories for portraying "emperors, princes, generals, ministers, scholars, beauties, ghosts and monsters" and failing to represent workers, peasants, and soldiers (1). Folk tales and supernatural stories made a comeback to children's literature in the 1980s. In fact, adaptations of *Journey to the West*, the mesmerizing saga of the Monkey King's adventures and redemption, and eerie accounts of fox fairies and monsters from *Strange Stories from a Chinese Studio* dominate FLP's output of children's titles between 1984 and 1987. Overall, nearly one out of every five or six titles in FLP's list can be attributed to either classic.

The period from 1949 through the 1980s, except for the decade-long, detrimental chaos caused by the Cultural Revolution, is widely considered a thriving time for Chinese children's literature. FLP translated some of the best contemporary Chinese literary works for youth into English and other foreign languages. Examples of major titles are Xu Guangyao's 徐光耀 *Little Soldier Chang Ka-tse*, Ye Shengtao's 叶圣陶 *The Scarecrow*, Hong Xuntao's 洪汛涛 *Ma Liang and His Magic Brush*, Li Xintian's 李心田 *Bright Red Star*, Sun Youjun's 孙幼军 *The Adventures of a Little Rag Doll*, as well as works by Ge Cuilin 葛翠琳, He Yi 贺宜, Jin Jin 金近, Ke Yan 柯岩, Lin Songying 林颂英, brothers Ren Daxing 任大星 and Ren Dalin 任大霖, and Yan Wenjing 严文井. FLP's list also includes works by the finest children's illustrators, cartoonists, and artists in China, for example, Cheng Shifa 程十发, Feng Zikai 丰子恺, He Yanrong 何艳荣, He Youzhi 贺友直, Hua Junwu 华君武, Miao Yintang 缪印堂, Yan Dingxian 严定宪, Yang Yongqing 杨永青, and Zhan Tongxuan 詹同渲. One misleading difference between the translations published by FLP and what was read by average Chinese children is that the former was mostly illustrated in color. Due to cost, Chinese children's access to books with full-color illustrations was limited before 1980. Comic books with black-and-white illustrations were the most affordable and commonly consumed format until the mid-1980s (Chen 2016, 24, 33).

Among FLP's in-house translators were Yang Hsien-Yi 杨宪益 and Gladys Yang 戴乃迭, two formidable figures and a married couple. Gladys Yang (neé Tayler) was born in Beijing in 1919 to a British missionary family and educated in Oxford. The interracial couple was best known for collaboratively translating masterpieces such as *A Dream of Red Mansions* 红楼梦 and Lu Xun's works into English. Their collaborations also include *Ancient Chinese Fables* (1957) and *Stories About Not Being Afraid of Ghosts* (1961), potentially accessible to young readers. Gladys Yang translated Zhang Tianyi's 张天翼 *Big Lin and Little Lin* (1958) and *The Magic Gourd* (1959), the latter being the basis of *The Secret of the Magic Gourd*, a Disney movie adaptation released in 2007. In Zhang's fantasy novel, the titular Magic Gourd represents the shamelessly exploitative class, which lives off other people's labor like thieves, as well as the bureaucratic ruling class, which steals other people's credit for contributions. The Gourd is manipulative and selfish, a seductive and toxic force against altruism and socialism. Aside from the socialist theme, which would make *The Magic Gourd* an uncomfortable story in capitalist countries, its imaginative plot and narrative art are among the best of Chinese children's literature. The boy protagonist Wang Bao is a flawed, and ultimately likeable character, with a weakness that children can easily relate to. His first-person narrative is at times defiant, unreliable (but young readers know better), and self-reflective. His moral battle against the Gourd's evil influence is tense and engaging. Gladys Yang was no fan of footnotes as far as children's books were concerned. In her fluent translation of *The Magic Gourd*, street food *tanghulu* 糖葫芦 (candied Chinese hawberry) becomes "toffee-apples" (Chang 1959, 22); *budaoweng* 不倒翁 (tilting doll) becomes "jack-in-the-box" (19); and *bawangbian* 霸王鞭 (sullu spurge) (22), used in a simile, is decidedly dropped from the English text. Her translation is faithful to the narrative tone and the vivid characterization in Zhang's writing, but takes liberty with specificities inconsequential to the universal course of a child's moral development.

A Bumpy Journey to the West

The great majority of the translations from Chinese publishers remain unknown to young readers in the English-speaking world. At least two barriers were found between the Chinese books on offer and the intended audience – the distribution mechanism and antagonistic ideologies. As Thomas A. Zaniello (1974) pointed out, there was a problem in accessing these English translations made in China. Unlike works whose translation rights are sold to foreign publishers, they were not distributed through conventional channels in target countries and were thus largely absent in libraries. During the height of the liberation movements of the 1970s, researchers reported getting hold of copies via a bookstore in Hong Kong or the mailing service of China Books and Periodicals (e.g., Krasilovsky 1973; Zaniello 1974). The latter was established in the midst of the Cold War, in 1960, by Henry Noyes, who was born to a Presbyterian missionary family in Guangzhou in 1910. His bookshop was the sole distributor of books, magazines, and newspapers from mainland China in North America, serving a niche market of China scholars, Marxist and socialist political groups, and overseas Chinese in North America (Üngör 2009).

On the rare occasion where a book did enter an American or British school during the Mao era, it could provoke controversy and furor among adult gatekeepers. For instance, a British school protested about the "politically subversive literature" when it received a copy of *Little Sisters of the Grassland* among a pack of "Third World children's books for 'multiracial' schools" (Stones 1977, 8). The book was published by Foreign Languages Press in 1973 and relates the story of Longmei and Yurong, Inner Mongolian sisters who nearly lost their lives tending the commune sheep in a blizzard. An entire generation of Chinese children was instructed to look up to the self-sacrificing girls as role models. Their heroic deeds were told in news reports, picture books, comic books, and an animated film, and commemorated in songs and *pipa* concertos. Of all juvenile books by FLP, *Little Sisters of the Grassland* was probably the title that circulated the widest. Anecdotal evidence suggests that it even left an impression on the minds of some American children, whose leftist or Maoist parents purchased the English edition for them. The majority of Western society, however, would perceive the story to be Communist propaganda.

The relative success of Chinese folktales adapted and published in the United States and Britain, in contrast to the obscurity of children's titles translated by FLP, offers insights into the challenge of translated children's books in English-speaking countries. To borrow ideas from scientific experiments, the parallel bodies of literature – translated in and outside of China respectively – function almost as a control and an experimental group, teasing out the weight of "independent variables" that influence receptions. Common factors that are cited to explain the scarcity of translated books in the United States include the appeal of style and content (Lystad 1974; Hearne 1991), expense, language, the "national character" (Garrett 2006, 11) of Americans to reject foreign books, and, more specifically, Americans' intolerance of foreign cultures and values that deviate from the American normalcy. The nearly imperceptible influence of FLP books upon English-speaking children, however, illuminates the foremost significance of political and cultural values in the reception of translated literature. After all, as a state organ FLP was under no economic pressure to turn a profit and was able to employ expert translators. Yet even Gladys Yang's first-rate translation could not help Zhang Tianyi's *The Magic Gourd*, which condemns economic exploitation and promotes socialism, to circumvent gatekeepers to reach child readers in the West during the Cold War.

A rare controversy surrounding a Chinese folktale published in the United States threw into relief the key factor to the success of this body of texts. *The Voice of the Great Bell* (1989), illustrated by Ed Young, was retold by Margaret Hodges from Lafcadio Hearn's "The Soul of the Great Bell." Hearn did not know Chinese but relied on the works of European Sinologists as his source of information ("The Soul" 2016). The legend of the Great Bell, one of the tales that appear in his *Some Chinese Ghosts* first published in 1887, is remarkably similar to versions that circulate in China, echoing the motif of a maiden's sacrifice for the perfection of craftsmanship in Chinese folklore. In the legend, a man is ordered by the emperor to make a giant bell. The emperor threatens to kill the man should he fail to make the bell. The man's daughter learns from an astrologer that the flesh and blood of a virgin must be mixed in for the metals to blend. To save her father's life, she throws herself into the foundry and ensures the casting of the bell. Ed Young's impressionistic illustrations soften the harrowing tale and won the book a starred review in the influential *School Library Journal*. The picture book subsequently invited criticism from a librarian for the portrayal of female martyrdom for the benefit of a tyranny (Denham and Gregory

1990, 65). The reviewer, Helen Gregory defended her rating, arguing that the evaluation was based on the quality of the writing and the art, not on the disagreeable message of the story. She went further to champion children's exposure to "varied value systems" (68). That controversies like this rarely happen to folktales adapted by authors in America, however, indicates that generally only those tales *perceived* to be congruent with mainstream values are selected for publication – a filtering process that FLP books did not undergo.

As a telling contrast, Ed Young's other picture book from the same year, *Lon Po Po: A Red-Riding Hood Story from China* (1989), won a Caldecott Medal. The tale type of "Grandmother Wolf" circulates widely in Chinese oral culture with varying details, but the main plot typically includes a youngest child being eaten by the wolf before older children finding a way to defeat the fake grandmother (Liu 2002, 105). Young's adaptation (though he is listed as translator on the title page) spares the death of the child and portrays three sisters who do not wait to be rescued and, with wit and courage, take fate into their own hands. Mulan is another successful case of folktale adaptation. In Chinese tradition it is a story about a filial daughter who feels obligated to relieve her aging father from military service, whereas in Western adaptations it has been shaped into and received as a tale about a woman warrior shattering traditional gender roles.

Technically outside the scope of translated literature, another category of Chinese stories having been heartily embraced in the West is works written directly in the languages of their adopted countries by Chinese diaspora writers. Prominent among their works is a body of memoirs and novels based on family and personal history in the tumultuous twentieth-century China, particularly ordeals and trauma they coped with during the Cultural Revolution before moving to democratic countries and thriving as immigrants. Notable titles for young readers include Jiang Ji-li's *Red Scarf Girl* (1997) and Li Cunxin's *Mao's Last Dancer* (2003). Their stories were rarely published in mainland China – suppressed on grounds of ideological incongruity – but found welcoming narrative space in English owing to geopolitical circumstances of the West. As Lingchei Letty Chen (2008) pointed out in her study, aptly titled "Translating Memory, Transforming Identity," the extensive popularity of memoirs about the Cultural Revolution serve to comfort the moral and emotional vulnerability felt in the post-Cold War and post-9/11 West, reminding the society of the value of freedom and democracy it has advocated and fought for (25).

The limitation of the FLP model – "English translations made in China" – is why the Chinese government now exhorts, with the catchphrase "Go out" (*zou chuqu*), an alternative approach to the state project of exporting Chinese culture and soft power: selling translation rights (Hua and Shi 2020). The China Writers Association began in 2013 to administer funds that support the translation of Chinese contemporary literature; qualified recipients must have secured publishing contracts with foreign publishers (Abrahamsen 2014; "2019 nian du" 2019). Chinese publishers have also registered companies overseas and published translations of Chinese children's books in those countries. Their publications are seldom available for purchase, and the very small print runs suggest that they serve rather as samples to help sell the rights or for publicity purpose.

Crocoducks and Republican Bunnies Leaping to the West: Progress in the New Century

Mainland China initially experienced a slump in the output of children's literature in the 1990s, one of the reasons why no publisher was able to replicate FLP's scale of achievement in translating contemporary Chinese literary works for youth. The symptoms and causes of the publishing ennui have been the subject of scholarly discussions, educators' laments, and hand-wringing journalist reports. Often-cited factors include an education system driven by high-stakes testing that discouraged time spent on leisure reading, publishers' trying to transition from a planned economy to a market economy (Tang 2006), and competition from television. (It should be pointed out that Chinese children did not so much stop reading for fun as developed variant tastes. Romance novels from Taiwan, kung fu novels from Hong Kong, and manga from Japan – reportedly in rampant pirated editions (Yue 2003; Liu 2009) – enjoyed a huge readership among late teens in the 1990s, but were not acknowledged as "proper reading" by adults.) Critics observed a renaissance of children's literature in mainland China since the 2000s, citing the "Harry Potter effect" (Tang 2006, 22), a boost to the fantasy genre, and a re-kindled interest in the format of picture books.

Picture books, though not new to Chinese children's literature, were prohibitively expensive for most Chinese families and little appreciated until the second half of the 2000s. Thanks to the efforts of publishers, parent advocates, and literacy scholars, the significance of parent-child shared reading, the practice of literacy enrichment programs (library story time), and classical titles (picture books like *Where the Wild Things Are* and *The Runaway Bunny* had been practically unknown to Chinese toddlers) were introduced to a new generation of middle-class Chinese families blessed with growing disposable income.

Japan and Taiwan, with a much longer history with picture books, provided culturally affinitive models for publishing and reading promotion in the early days. Long segregated by political regimes since the end of the First Sino-Japanese War, Taiwan, and mainland China developed distinct bodies of children's literature (see Lin Wenbao and Qiu Gerong (2018) and You Peiyun (2007) on the history of Taiwanese children's literature). The Taiwan-based Hsin-Yi Foundation, an early-childhood education organization established in 1977, opened an offshoot in Nanjing in 2004. The Nanjing Hsin-yi Children Cultural Development Company adapted Taiwan's practice with creating picture books and promoting reading, and has been a major player in the field. A Taiwanese investor launched the "House of Aesop's Fables" in Shanghai in 2006 and offered story time for families, heralding numerous privately-run "picture book houses" that popped up in Chinese cities (Chen and Hearne 2012). Japanese publisher Matsui Tadashi 松居直, whose theory of picture books was introduced to China by children's author and scholar Peng Yi 彭懿, is a major influence on Chinese appreciation of the format.

Dozens of bestselling and award-winning works created by children's authors from both Taiwan and mainland China were purchased by American, British, Canadian, Australian, and New Zealand publishers between 2000 and 2020, earning critical acclaim and accolades abroad. To begin with, Taiwanese author Chen Chih-Yuan's 陳致元 works received unprecedented popular attention thanks to a title directly translated from Chinese into English. His picture books *On My Way to Buy Eggs* (2001) and *Guji Guji* (2004a) were published by Kane/Miller in the United States. The latter, an ALA Notable Children's Book, was a phenomenal success with both children and critics. The endearing and triumphant story of a crocodile raised by a mother duck celebrates a child's agency and provokes thinking on the questions of identity, race, and family. Another Taiwanese picture book maker who broke through to the English world is Jimmy Liao 幾米, whose multiple works were published with Little, Brown and Company between 2005 and 2009.

The influx of translated picture books not only entertained Chinese children, but also had a profound impact on mainland Chinese writers and artists, who began to make picture books that incorporate visual narrative techniques observed in masterpieces into their own stories. Some of the best Chinese titles have since been brought to English-speaking countries. Brothers Xiong Liang (熊亮, aka Kim Xiong) and Xiong Lei 熊磊 are collaborators on children's books and considered the first authors to pioneer and experiment with the fresh format of picture books in mainland China since 2000. Kim Xiong's picture book *The Little Stone Lion* was published in separate Chinese and English editions in 2005 by Heryin Books – a Taiwan-based organization which also published Chen Chih-Yuan's works in English. This first exposure of contemporary picture books from mainland China was followed by several remarkable titles.

No! That's Wrong! (Kane/Miller, 2008) by Ji Zhaohua 姬炤华 and Xu Cui 徐萃, a married couple, won a glowing review from Betsy Bird in her influential blog hosted by *School Library Journal*. Dramatically juxtaposing the elegant visual style of Chinese ink-wash painting, giggle-inducing underpants comics for toddlers, and the bold application of "breaking down the fourth wall" (Bird 2008), the hilarious book encloses a sophisticated message about defiance of authority and independent thinking. *No! That's Wrong!* technically may not be considered a title translated from China, but part of a new phenomenon of international publishing that elides traditional categorization. *After* the success of its English version overseas, the title, by the Beijing-based authors and illustrators, was published in Chinese first in Taiwan (2009) and later in mainland China (2011).

The decade between 2011 and 2020 witnessed the biggest international recognition of Chinese children's literature. *A New Year's Reunion* (Candlewick, 2011) written by Yu Liqiong 余丽琼 and illustrated by Zhu Chengliang 朱成梁 was named one of the ten *New York Times* Best Illustrated Children's Books of 2011, a singular success for a children's book translated from China. The book is both familiar and refreshing to an American audience. It fits into the tradition of multicultural holiday books and reiterates the universal theme of family and love in a new setting – early twenty-first-century China. The book presents a bittersweet story of how, after an entire year's separation, a girl named Maomao reconnects with her father during the New Year's celebration before his all-too-soon departure. Maomao is your average stubborn pre-schooler, and her changing feeling toward her father is depicted in subtlety that invites close reading. She is another memorable girl character in a picture book about the Chinese New Year since the titular figure in Thomas Handforth's Caldecott-winning *Mei Li* (1938). Another author worthy of a mention is Liu Cixin 刘慈欣, some of whose science fiction enjoys a readership across youth and adults. His *The Three-Body Problem* (Tor Books, 2014), translated by Ken Liu, won the prestigious Hugo Award for Best Novel in 2015, although none of his works suitable for young readers is available in English yet (Wu and Mallan 2006, 42; Shi 2019).

The year 2016 was a historic moment for the international visibility of Chinese children's literature. First, Taiwanese writer Fang Suzhen's 方素珍 *Grandma Lives in a Perfume Village* (NorthSouth Books, 2015) was named an honor title by the Mildred L. Batchelder Award, a first for a Chinese work since the establishment of the award for translated children's literature half a century before. The success of the book exemplified the dividend of collaboration that crosses political differences and cultural barriers. The title was originally published in Taiwan in 2007. A new edition, published in Beijing seven years later, hired Sonja Danowski, a German artist, as its illustrator. Danowski's exquisite artwork drastically transformed Fang's book and must take due credit for the charm of the otherwise melancholy story with the difficult topic of death (Fig. 9).

Second, also in 2016 the Hans Christian Andersen Award for children's writing was won by a Chinese author, Cao Wenxuan 曹文轩, for the first time since it began 60 years before. Cao's middle-grade novel, *Bronze and Sunflower*, translated by Helen Wang, was cited in many best books lists, including *The New York Times* Notable Children's Books of 2017. More of Cao's works have been purchased by foreign houses since his award, including *Feather* (Elsewhere Editions, 2017), *Summer* (Imprint, 2019), and *Dragonfly Eyes* (Walker Books, 2021).

Fig. 9 Cover images of *Grandma Lives in a Perfume Village*. Left: illustrated by Jiang Binru 江彬如 (Taipei, Taiwan: Children's Publications Co., Ltd., 2007 (Fang and Jiang 2007)). Right: illustrated by Sonja Danowski (New York: NorthSouth Books, 2015; translated from the 2014 edition by the China Juvenile and Children Publishing House)

Helen Wang's translation of *Bronze and Sunflower* suggests that a translator sometimes has to do a little subtle editing. In an interview she discussed the difference in the style of storytelling in Chinese and English literature, pointing out that Chinese stories often provide more information and more repetition than the English reader expects or can tolerate (Toft 2015; Wang 2015a, b). Her rendition of the original text smoothed the journey of a Chinese-language work to the English-speaking world.

The number of Chinese children's books in English translation was greatly boosted by the Candied Plums Publishing House. Since its establishment in 2016, that publisher has released 16 picture books in English or bilingual version, winning warm reviews and awards. Candied Plums was a project of the China Educational Publications Import & Export Corporation ("Guo ji" 2017), a state-owned book trade corporation. By embracing a diverse team of bilingual and bicultural editors, translators, and children's literature experts based in China, the United States, and Britain, Candied Plums avoids the pitfall of FLP's operation in isolation. It also stands out from the American and UK publishing industry, which overall lacks diversity and foreign-language expertise (Lee & Low 2020; Roxburgh 2006, 7; CILIP 2019). Candied Plums is not without its challenges though. Being subject to fickle administrative support from its parental state organ, it lacks sustained resource for marketing and continual content development (Li Xiaocui, text message to author, July 9, 2020). Its sales record is disproportionate to the quality of its works, and the publisher has not announced any new titles since 2017.

The titles selected by Candied Plums for translation are not only among the best of contemporary Chinese picture books, but also gauged to be interesting and suitable for an overseas market. In an interview, Roxane Feldman, a New York-based middle school librarian who was engaged as the publication consultant of Candied Plums, revealed how she helped the publisher navigate around cultural differences and select titles that would be a good fit. "A lot of Chinese books don't shy away from the harsh truths of life," she observed, and cited as an example Cao Wenxuan's environmentally themed picture book *The Last Leopard* 最后一只豹子 (Tomorrow Publishing House, 2010) (Kantor 2017). It tells a tale of extinction, with what she perceived to be a disturbing ending – one that does not sit well with American conventional understanding of what constitutes appropriate reading for children. In American juvenile fiction about a topic as deadly as the Holocaust, for example, writers are most often compelled to let child protagonists survive. Even dystopian novels for young readers cannot resist a hopeful ending. Feldman's success with cultural intermediary drives home the metaphor that translated literature opens a *window* – in lieu of an unblocked view – into another culture.

Among the offerings of Candied Plums, Gan Dayong's *Little Rabbit's Questions* (2016) was cited in USBBY's 2018 Outstanding International Books List; Wang Zaozao's *An's Seed* (2017) won an Honourable Mention of the 2017 Freeman Book Awards for children's literature, for "contribut[ing] meaningfully to an understanding of East and Southeast Asia" ("Freeman" n.d.). Both works were translated by Helen Wang.

Little Rabbit's Questions is a series of dialogue between Mama Rabbit and Little Rabbit, reminiscent of the loving verbal exchange between Little Nutbrown Hare and its parent in *Guess How Much I Love You*, but appropriating the sinister Q and A between Little Red Riding Hood and the Wolf with a radical twist. Once again like *A New Year's Reunion*, *Little Rabbit's Questions* celebrates the familiar theme of family bonds in a new cultural setting with a refreshing aesthetic style. Gan's ink painting recalls the brush-pen cartoon art of Feng Zikai (1898–1975), after whom the Chinese Children's Picture Book Award is named. Feng, who was most active during Republican China, applied Chinese brush painting to cartoon work, breathing liveliness into the tradition of high art and injecting a distinct Chinese flavor into a format that was introduced from the West. Gan paid homage to Feng Zikai by setting the bunnies' story in Republican China, portraying a *qipao*-donning Mama Rabbit and furnishing the Little Rabbit's bedroom with a plain pendant lamp from that bygone era.

Part II – International Interest and Translation Initiatives in Chinese Children's Books Since the 1980s

Although there has long been an interest in Chinese children's books, until recently it has been difficult to satisfy that interest. Children's book publishing in English-speaking countries is usually commercially driven, and publishers are looking for appropriate content that will be commercially viable. For reasons already discussed, for much of the twentieth century Chinese children's books rarely met Western publishers' criteria. Adaptations, or recreations, of traditional Chinese stories continued to be more successful.

Commercial and Noncommercial Interest in Chinese Children's Books

The situation began to change in the late 1980s and 1990s, as the Chinese children's book world started to become more commercial. However, it was the noncommercial children's literature world that first followed new developments in Chinese children's books and this deserves attention before we move on to the commercial world. Two institutions in particular, the International Youth Library (IYL), in Munich, and the International Board for Books for Young People (IBBY) were instrumental in introducing Chinese children's books, writers, and illustrators to an international audience. The Asian Festival of Children's Content (AFCC), an annual event, also merits attention.

(A) Noncommercial Interest

The International Youth Library (IYL), Munich
The International Youth Library was founded in 1949 by journalist and author Jella Lepman (1891–1970). At the end of World War II, Lepman was a consultant to the US army, and in 1946 she organized the first international exhibition in post-war Germany, the *Internationale Jugendbuchaustellung* (International Youth Book Exhibition), which displayed 2000 books from fourteen countries. The exhibition was shown in several cities in Germany, and the books became the founding collection of the International Youth Library. Lepman was convinced that books could offer children hope for the future, and it was thanks to Lepman that a conference about international understanding through children's books was held in 1952, which led to the founding of the nonprofit organization, the International Board for Books for Young People (IBBY).

Since 1993, the International Youth Library has selected outstanding books from countries around the world for recommendation in its annual White Ravens catalogue. Between 1993 and 2019, a total of 127 books published in the Chinese language (the selection is by language, rather than by country) have been selected as White Ravens, with bibliographic details and a synopsis given in English for each book. The first list (1993) included books published in the late 1980s, and thereafter the White Ravens catalogue uniquely offers an almost annual survey of the best Chinese children's books as selected and recommended by a European library. The Chinese books selected as White Ravens in the last five years are listed in the Appendix below; details of the previous years can be found on the IYL website (White Ravens).

The International Board for Books for Young People (IBBY)
The International Board for Books for Young People was founded in Zurich, Switzerland, in 1953. IBBY is "a non-profit organization which represents an international network of people from all over the world who are committed to bringing books and children together" (International Board). It is a nongovernmental organization with an official status in UNESCO and UNICEF, and has a policy-making role as an advocate of children's books. It is committed to the principles of the International Convention on the Rights of the Child, ratified by the United Nations in 1990 [and by China in 1992]. One of its main proclamations is the right of the child to a general education and to direct access to information, and it is thanks to IBBY's insistence that the resolution includes an appeal to all nations to promote the production and distribution of children's books. IBBY cooperates with many international organizations and children's book institutions around the world and exhibits at the International Children's Book Fair in Bologna, and other international book fairs.

While IBBY and the IYL provide a framework through which to bring children's books from around the world to international attention, they do not normally

Table 3 IBBY China's nominees for the Hans Christian Andersen Awards

Year	Author nominee	Illustrator nominee
1988	–	Ke Ming 柯明
1990	Sun Youjun 孙幼军	Qiu Zhaoming 裘兆明
1992	Jin Bo 金波	Yang Yongqing 杨永青
1994–2000	–	–
2002	Qin Wenjun 秦文君	Wu Daisheng 吴带生
2004	Cao Wenxuan 曹文轩	Wang Xiaoming 王晓明
2006	Zhang Zhilu 张之路	Tao Wenjie 陶文杰
2008	Qin Wenjun	–
2010	Liu Xianping 刘先平	–
2012	–	–
2014	Yang Hongying 杨红樱	Xiong Liang 熊亮
2016	**Cao Wenxuan**	Zhu Chengliang 朱成梁
2018	Qin Wenjun	Xiong Liang
2020	Huang Beijia 黄蓓佳	Zhu Chengliang

commission or publish translations of children's books, nor do they market or distribute them. However, they provide essential noncommercial opportunities for people in the children's book world, including cultural organizations, libraries, charities, academics, writers, illustrators, and translators to interact and to share information about highly regarded authors, illustrators, and books. A good example of this is the Hans Christian Andersen Awards organized by IBBY.

The Hans Christian Andersen Awards are made every other year to a living author (since 1956) and a living illustrator (since 1966) for their "lasting contribution to children's literature." The author's prize is nicknamed "the little Nobel Prize in literature" on account of its international prestige. Nominations are made by the IBBY organization in each country, which prepares a full dossier about each nominated author and illustrator in English that is then shared internationally in IBBY publications and online, and at the bi-annual IBBY conferences. Several Chinese authors and illustrators have been nominated for the Hans Christian Andersen Awards since 1988 (Glistrup 2002; IBBY). Cao Wenxuan was the first Chinese nominee to win the award (Table 3).

The Asian Festival of Children's Content (AFCC)

Special mention should also be made of the Asian Festival of Children's Content (AFCC), established in 2000 (AFCC 2020). The AFCC is an annual literary festival with an Asian focus rather than a commercial bookfair. It is organized by the National Library Board of Singapore and the Book Development Council of Singapore "to promote the habit of lifelong learning and to create and provide opportunities for the creative and joyous learning of Asian content and culture among children." The focus is on Asian content, rather than the perpetuation of Western content. China was the guest of honor at the 2015 AFCC.

Academic Conferences

There are, of course, academic conferences on children's literature, such as the International Symposium for Children's Literature & the Fourth US-China Symposium for Children's Literature, held at the Cotsen Children's Library, Princeton, in 2018 (Chen and Wang 2018), and the Biennial Congress of the International Research Society for Children's Literature (IRSCL), last held in Stockholm in 2019 (Wang 2019). However, academic conferences are more concerned with original content and theory, rather than translated books. As mentioned earlier in this chapter, the field of children's literature is growing in China, but knowledge of Chinese children's literature outside of China remains weak. On a practical level, attending conferences is expensive, and while there is much to learn at academic conferences, literary translators are more likely to go to conference-type events at bookfairs and festivals where there is more interaction with authors, readers, editors, publishers, and agents.

(B) Commercial Interest

The business of buying and selling rights takes place in the commercial world: for example, at international book fairs, most notably at the annual specialist children's book fairs, such as Bologna and Shanghai. These large international book fairs also schedule events featuring authors, illustrators, editors, and translators. Teri Tan has been reporting on the Chinese children's publishing industry for *Publisher's Weekly* for over a decade, and produces an annual summary volume (Tan 2009–2020).

The Bologna Children's Book Fair

The Bologna Children's Book Fair (*La fiera del libro per ragazzi*) was established in 1963. It takes place annually over four days in April, and is the leading professional fair for children's books in the world. Most participating publishers are commercial, and go to Bologna to promote their books and products and to buy and sell international rights. Authors, illustrators, translators, critics, academics, librarians, media representatives, and translators also attend, and there is a lively program of events, including a Translators' Centre. International organizations such as the International Book on Board for Young People (IBBY) and the International Youth Library (IYL) also attend and display their respective selections of the best children's literature from around the world. IBBY displays books produced by the Hans Christian Andersen Award nominees along with their nomination dossiers, the IBBY Honour Books, and books for children with disabilities. The International Youth Library displays its latest selection of outstanding books, known as White Ravens. China was Guest of Honour at the Bologna Children's Book Fair in 2018 ("Bologna").

The Bologna Children's Book Fair Translators' Centre: The 2021 Bologna Children's Book Fair website introduces its Translators' Centre as catering "for professional translators and those wishing to specialise in literature for children . . . an information, debate and international exchange hub for sector professionals meeting each year in Bologna." It highlights professional skills, specialization, and networking; the Translators' Café (a seated area for a program of talks, workshops, and seminars), the World Directory of Children's Book Translators (an online database, linked to the UNESCO Index Translationum website, where listed translators must have translated at least two children's books), Update for Professionals (a special seminar on translation-geared technologies); special focus on translator training (in collaboration with the School of Languages, Literature, Translation and Interpreting, and the University of Bologna); "In Altre Parole" (a translation competition for translators working in Italian); networking (with partners such as the UK children's reading charity Book Trust); and the Centre for Books and Readership (a databank for translators working in Italian). The BCBF also welcomes translators by offering reduced price entry tickets.

In 2018, the year in which China was Guest of Honour, the Translators' Centre program included for the first time, thanks to children's literature scholar Junko Yokota, events that specifically addressed Chinese themes. The first, "Translation to/from Asian languages: issues, ideas and inspiration" was a panel with Junko Yokota, Center for Teaching through Children's Books, National-Louis University; Helen Wang; Lucia Obi, International Youth Library, Munich; and Frances Weightman, the Leeds Centre for New Chinese Writing. All talked about their work, Yokota on the need for more discussion about Asian books at international events; Wang about translating and increasing visibility of translated books; Obi about the IYL and the White Ravens; and Weightman about the children's literature component of the Leeds Centre for New Chinese Writing. The second, titled "Translating Cao Wenxuan," was a discussion about the translation of Cao's novel *Bronze and Sunflower*, with the Italian translator Paolo Magagnin, Ca' Foscari University, Venice; his editor Alessandra Gnecchi Ruscone, at Giunti, and the English translator Helen Wang. Frances Weightman also spoke about the reception of Cao Wenxuan via the Leeds Centre's school bookclubs. Questions were raised about the titles – the English title *Bronze and Sunflower* followed the Chinese original, while the Italian was shortened to *Girasole* (Sunflower) – and who has the final decision over a translation – the translator or the publisher? (Magagnin forthcoming). The third event was "In Altre Parole," a Chinese-to-Italian translation workshop with Paolo Magagnin, and Silvia Pozzi, translator, University of Milano-Bicocca.

Twenty-six Chinese writers, scholars, and reading promoters attended the fair, along with 24 Chinese illustrators. The authors included Cao Wenxuan, Han Yuhai 韩毓海, Xue Tao 薛涛, Gerelchimeg Blackcrane 格日勒其木格·黑鹤, Tang Sulan 汤素兰, and Qin Wenjun. Illustrators included Yu Hongcheng 于虹呈, Xiong Liang, Hei Mi 黑眯, Yu Rong 郁蓉, and Zhu Chengliang. Scholars included Fang Weiping 方卫平 (Fang 2018) and Zhao Xia 赵霞, who spoke with Lucia Obi at the event "Chinese Children's and Youth Literature: Recommendations and Trends," organized by the International Youth Library. Also at the exhibition was a display of 150 works of illustration and books by 30 Chinese illustrators ("Dream" 2018).

The China Shanghai International Children's Book Fair (CCBF)

The China Shanghai International Children's Book Fair was established in 2013. Initially organized by Shanghai Xinhua Distribution Group Co., Ltd., China Education Publishing & Media Group Ltd., and China Universal Press & Publication Co., Ltd., it has, since 2018, been co-organized with Ronbo BolognaFiere Shanghai Ltd. since 2018 (China Shanghai). The CCBF takes place annually over three days in November. The annual Chen Bochui International Children's Literature Awards are announced at the Fair.

The Chen Bochui International Children's Literature Awards 陈伯吹国际儿童文学奖 are the longest continuously running literary awards in China. They are given for promoting excellence in children's publishing and cultural diversity, and are named after Chen Bochui (1906–1997), who translated Pushkin's *Children's Tales*, *The Wizard of Oz*, and *Don Quixote* into Chinese in the 1940s, and donated his entire life savings to establish the award. Originally called the Children's Literary Garden Prize 儿童文学园丁奖, the first awards were made in 1981 and thereafter biennially. In 1988 the awards were renamed as the Chen Bochui Children's Literature Awards, and since 2014 as the Chen Bochui International Children's Literature Awards, with awards made annually.

The CCBF arranges a wide range of events bringing together international authors, illustrators, publishers, children's book specialists, and reviewers. Key to their success is Carolina Ballester, the CCBF's International Programme Manager, who also leads the CCBF Publishers Fellowship Programme and organizes exhibitions.

Cultural Shifts Conducive to the Translated Book Industry

The surveys by CILIP (2019) in the UK and Lee & Low Books (2020) in the United States highlighted the need for diversity in the children's book world in two major English-speaking countries. Fortunately, general knowledge of, and familiarity with, China and Chinese language(s) and culture(s) is improving. The past 20 years have seen a growing number of native English speakers learning Chinese at all levels. Significant numbers of Chinese nationals have studied overseas, and some have settled and raised families overseas, many of whom are bilingual. Today, Mandarin is more widely taught to children in mainstream schools in English-speaking countries than it was in the 1980s and 1990s when the majority of learners of Mandarin were university students and adults. There has also been a growth in translation training courses in universities internationally, and thus, there is a growing pool of translators of different backgrounds and experiences. In previous decades, translations of Chinese literature in English-speaking countries were mostly done by academics and published by university presses, often with a high retail price for a specialist audience and university libraries. Today, there is a much greater diversity of translators and publishers, and a wider range of genres of Chinese books are being published in translation.

The impact of all these developments is that many more people in English-speaking countries now have at least some familiarization with Chinese people and China generally. This is significant because children's books are usually purchased by adult family members and friends, teachers and librarians, rather than by children themselves. A growing awareness, knowledge, and understanding among adults contributes to a positive interest in the reading materials they choose to give to children.

Even when that familiarization is superficial or nascent, there is at least some understanding and empathy that comes from going to school together, or living and working together. After centuries in which China was regarded as exotic in both positive and negative senses, as well as decades of mutually antagonistic political propaganda, the greater familiarization and increased understanding and empathy, are very significant indeed.

Also contributing to a receptive culture to Chinese literature is an expanding corpus of works by diaspora Chinese in English-speaking countries. As Nicky Harman commented when she was translator-in-residence at the Free Word Centre, in London (UK), Chinese diaspora writers straddle two or more worlds and English readers (and publishers) often find their writing more accessible (and marketable). Indeed, their works, frequently about China, their Chinese heritage, and immigrant experience, have won critical acclaim and prestigious prizes, and made best sellers lists. A prime example of this is Jung Chang's *Wild Swans* (Harper Collins, 1991), which was "the biggest grossing non-fiction paperback in publishing history," selling more than 10 million copies worldwide and translated into 30 languages (Allardice 2005). Writing directly in the English language, writers like Jung Chang and Li Yiyun, as well as a burgeoning group of successful children's and young adult authors such as Gene Luen Yang, Grace Lin, Jean Kwok, Kelly Yang, Emily X.R. Pan, Sue Cheung, Teresa Robeson, and Justine Laismith, "translate" Chinese stories, concepts, and vocabulary into idiomatic English, further acquainting the English readership with Chinese history and culture.

Translator-Led Initiatives

The early twenty-first century has seen a growth in literary translation in English-speaking countries, largely due to increased activity by literary translators and small independent presses campaigning for more foreign language literature to be published in English. Certain factors have facilitated this: increased international travel; technological changes, particularly digital communication and publishing, and training and mentoring. In particular, developments in the digital world have enabled translators to communicate with other translators with unprecedented ease and at low-cost. The professionalization of the field, and the generosity of established literary translators toward newcomers, sometimes through formal teaching or mentoring, but often through informal advice and sharing of experiences has helped the next generation of translators to understand better the process and business of translation from author to reader. The chain can include author

\> publication in the home country > the sale of rights, sometimes through an agent > translator + editor(s) > production > publication > marketing > reviewing > distribution > bookseller + library > purchaser > reader.

The professionalization of literary translation and the publishing market has been discussed more fully elsewhere (e.g., Constantine 2018). The following is a snapshot of what has been happening in the UK, with particular reference to Chinese-to-English translation. There are, of course, similar initiatives, short courses, residencies, and events in the United States, Canada, Australia, and other English-speaking countries.

The translator-led initiatives of the early twenty-first century have been remarkable. By training, supporting, and mentoring newcomers, established translators have, in effect, helped to nurture a community of literary translators who are better connected and informed about the industry. They not only translate for publication but also join forces and participate in literary translation events. For example, Nicky Harman is not only an award-winning literary translator, she has also been extremely dynamic in promoting Chinese literature in translation in the UK. Harman studied Chinese at Leeds University in the 1970s (China was closed at the time and Harman did not visit China until the 1980s). While teaching Chinese translation at Imperial College London, she started translating Chinese fiction in her free time, and became a full-time literary translator in the 2010s. Harman joined the Translators Association (a members' group of the Society of Authors), and was for several years on the committee and, for three years, a co-chair. While on the TA committee she helped to devise and organize events for the Literary Translation Centre at the London Book Fair (annually, in March/April), and International Translation Day events (annually, in September/October). She also tutors at translation summer schools, for example, at Norwich (British Centre for Literary Translation), London (Translate in the City) (see below), and Warwick (Warwick Translates).

In 2011 Harman was one of the first translators in residence, sponsored by the Gulbenkian Foundation, at the Free Word Centre, London. At the same time as she was forging links with the wider literary translation community, Harman was also an early member of the Chinese-to-English translator collective Paper Republic. She attended an important literary translation event at Moganshan, China, in 2008, and received an Arts Council award to develop the Paper Republic database with Eric Abrahamsen. Paper Republic translators also contributed translations to the Chinese arts and literature journal *Tiannan/Chutzpah*, which was the brainchild of editor, artist, and film-maker Ou Ning 欧宁. Thus, Harman was part of both the UK literary translation community and the international Chinese-English literary translation community. In 2010 she and Helen Wang founded the China Fiction Book Club in London, which welcomed translators and readers to read and discuss Chinese fiction; this was initially a real life book club with events, but transformed to a Twitter information feed (@cfbcuk) in 2012.

Harman has written numerous articles, reviews, and given interviews to journals (printed and digital), and in 2012 persuaded the books editor at *The Guardian* newspaper to publish a Chinese short story each day in the week leading up to the London Book Fair at which China would be guest of honor. In 2016, Harman proposed a more ambitious project, Read Paper Republic, to publish a Chinese short story, poem, or essay each week for a year on the Paper Republic website, drawing on authors and translators around the world to contribute (see below, "Paper Republic"). In 2014, Harman was invited to join the University of Leeds "Writing Chinese" project as a consultant. This project subsequently developed into the Leeds Centre for New Chinese Writing. In 2019, Paper Republic transformed and registered as a UK charity, with Harman on the board of trustees. At the same time, Harman continues to translate, to write blogs and articles, to judge translation competitions, and, crucially, to support other Chinese to English translators. Thus, in the first two decades of the twenty-first century, largely thanks to Harman's enthusiasm and energy, a strong Chinese-English literary translation network has been established in the UK.

Harman's enthusiasm rubbed off on others, notably Helen Wang. Twenty years after her first translations of short stories and essays were published in the 1990s, Wang started translating short stories again. When Harman told members of the China Fiction Book Club that Egmont UK was looking to translate two Chinese children's books, Wang decided to try her hand. Egmont invited short, unpaid, samples of the beginnings of the two books, and from these commissioned about six translators to submit longer, paid, samples. Wang won the contract for Shen Shixi's *Jackal and Wolf*, and Petula Parris Huang won the contract for Wu Meizhen's *An Unusual Princess* (Egmont UK, 2012). Egmont's plan was for relay translation: the two middle-grade novels would be translated from Chinese to English, and then from English into seven other languages, enabling Egmont to launch eight foreign language editions simultaneously (eight being a lucky number) at the London Bookfair in 2012, when China would be guest of honor.

In 2013, Walker Books commissioned Wang to translate Cao Wenxuan's *Bronze and Sunflower*. This was Walker Books' first translation of a Chinese children's book, there was no one in the company who could read Chinese or knew much about China, and they had not worked with Wang before. They did, however, have staff who had read the French translation of the novel, *Bronze et Tournesol* (Editions Philippe Picquier, 2010). Wang suggested that she translate the first chapter keeping very close to the original Chinese, so that she could see how they would edit it before translating the second chapter. That way, she could adjust, and they would know what to anticipate in terms of editing. They also agreed that Wang should flag up any concerns for discussion during the editing stage (Sullivan 2018). *Bronze and Sunflower* was published in the UK in April 2015 and, after some editing for the American edition, in the United States in 2017. Cao was awarded the Hans Christian Andersen Award in April 2016 (the first Chinese author to win this award). Wang's translation won the Marsh Christian Award for Children's Literature in Translation in January 2017 (the first Chinese book to win this award). In November 2017, Wang was also awarded the Chen Bochui International Children's Literature Award, at the China Shanghai Children's Bookfair. Since then, Wang has translated an additional two middle-grade novels (*The Ventriloquist's Daughter* by Lin Man-Chiu; and *Dragonfly Eyes* by Cao Wenxuan, along with several picture books).

Like Harman, Wang has also been active in the broader translation community. She joined the Translators Association, served on the committee, started

@TranslatedWorld on Twitter, and initiated the hashtag #namethetranslator. Keen that the Paper Republic blog should be buzzing in the run-up to the London Bookfair in 2012, she offered to write a blog post every day for a month. Her posts were simple but the activity drew attention to the blog and website.

The developments mentioned at the beginning of this section created possibilities for new ways of trying, doing, improving, and collaborating. A great deal has been achieved by dynamic individuals working collaboratively, with a committed but flexible approach, albeit with very small funds. It is perhaps worth pointing out here that few people can earn a living from literary translation alone. However, the diversity of experience of today's literary translators is encouraging, as is the range of genres that are being translated. It is very clear that translations of novels, excerpts, short stories, and poetry paved the way for translations of children's books, and that experienced literary translators have been commissioned for recent translations of Chinese picture books and middle-grade novels (e.g., Nicky Harman [2020b], Jennifer Feeley [*White Fox*, 2019], Chloe Garcia Roberts [*Feather*, 2017]). A brief description of key translator-led initiatives and events is given below.

Paper Republic (2007–Present)

Paper Republic has been the most influential Chinese-English translation initiative. It was founded in 2007 by Eric Abrahamsen, then in Beijing, as an online forum for translators of Chinese literature to share information about Chinese books and authors, and discuss how to get them translated and published abroad (Paper Republic). Run as a collective of volunteers, it soon expanded to include translators around the world, some of whom attended the Moganshan Literary Translation Residency in Zhejiang, China, in 2008 (see below). Since its founding, Paper Republic has served as an information site, with a database of authors, translators, publications, and publishers, and has worked collaboratively or consultatively with other organizations. Activities include devising the program for the Beijing International Book Fair; working on *Peregrine*, the English-language edition of *Tiannan/Chutzpah!* between 2011 and 2013 (Tiannan/Chutzpah n.d.), and working with People's Literature Publishing House (Renmin wenxue chubanshe) to produce *Pathlight*, a journal of Chinese Literature in Translation (Laughlin 2013).

In December 2012, Paper Republic began publishing an annual list of Chinese-English translations (books only), and has done so ever since. In 2015–2016, a Paper Republic team of four editors (Nicky Harman, Helen Wang, David Haysom, and Eric Abrahamsen) undertook to publish a short story, essay, or poem in translation, one a week for a year – the Read Paper Republic series was free-to-view, and was designed as a way for readers to "dip their toe" into contemporary Chinese writing. Some of the stories were later selected for use in speed-book-club events organized by Paper Republic members at the Free Word Centre, in London (Mezzanotte 2017). A total of 53 pieces were published on the Paper Republic website, with the kind permission of the authors and translators, and edited by the four mentioned above. The impact of Read Paper Republic was felt across the translation community, and in 2016, Paper Republic was shortlisted for the Publisher's Weekly Literary Translation Initiative Award, at the London Book Fair (PW Awards 2016). The Read Paper Republic translation activity continued, with two mini-series entitled "Afterlives" and "Bare Branches," and led, in 2018, to a collaboration with *OWMagazine* (*Dandu*) and the *LA Review of Books*' China Channel to translate literary nonfiction (LARB China Channel), and in 2020 to the mini-series "Epidemic," including a collaborative translation by 124 translators from 20 countries on five continents (Harman 2020a).

In 2017 Paper Republic collaborated with the newly formed organization Global Literature in Libraries (see below), and produced 28 blog-posts, one a day for the month of February, again inviting translators and specialists to introduce specific areas of interest, including children's books, to non-China specialists ("The GLLI-PR" 2017). During the activities of 2015–2018 the four editors of Read Paper Republic were located across three continents, their collaboration owing much to digital communication, in particular, email and Slack. In February 2019, Paper Republic was relaunched as a UK-registered charity. The Paper Republic website remains a rich and active resource for translators, publishers, librarians, and readers. A 2020 initiative, from translator Jack Hargreaves, is #SundaySentence: every Sunday, a sentence or two from a novel or short story is posted, together with some background to the author and work. Readers are invited to "have a go" and submit their translation as a comment within the week. Initiatives, competitions, and events such as translation slams (Marven 2016) are designed to be inclusive and welcoming, and to enable everyone to enjoy both the processes and products of translation.

The Moganshan Literary Translation Residency, 2008

The first Sino-British Literary Translation Course took place in Moganshan, Zhejiang province, in March 2008. Organized by Jo Lusby, of Penguin China, and Kate Griffin, of Arts Council England, in collaboration with China's General Administration of Press and Publications, it brought together four professional translators (Howard Goldblatt, Bonnie S. McDougall, Zhang Chong and Shi Zhikang), and over 40 translation students, and four authors (Li Er 李洱, Tie Ning 铁凝, Hari Kunzru, and Bernardine Evaristo) whose work the students were to translate. The aim was to forge links between Chinese and British writers, translators, and publishers (Swift 2008). One of the many outcomes of the Moganshan Residency was a successful application by Eric Abrahamsen and Nicky Harman to the UK's Arts Council for a grant to develop the Paper Republic website (see above) and its database of authors, translators, publishers, and publications. The Moganshan Literary Translation Residency was a formative or confirmatory experience for many translators on the residency.

In the UK, there are two key annual events for literary translators of all languages: the London Bookfair (March) and International Translation Day (30 September). The programs for these events are devised by a group of organizations which are concerned with writers and literature from across the world, including the Translators' Association, and English PEN. These are ticketed public events and are open to all.

The London Book Fair – Literary Translation Centre (2010–Present)
The Literary Translation Centre (LTC) was a new initiative in 2010. In its first year, the Book Fair was badly affected by the volcanic ash from Iceland, which caused many flights to be cancelled, and resulted in an unusually low attendance. Fortunately, the organizers of the LTC – the British Council, English PEN, the Translators Association and others – were able to keep the center lively and interesting. It has since become an established part of the London Bookfair – indeed, it is described on the LBF website as a "nerve center" of the bookfair – with a wide range of panels and talks, and a natural meeting place. It is usually packed, with standing room only. Many of the sessions are recorded, and can be viewed on YouTube (search for London Bookfair Literary Translation Centre).

China was the country of honor at the London Book Fair in 2012. A large number of authors came to London for the fair and to tour the UK afterwards. Among them were the well-known children's authors Shen Shixi 沈石溪 ("China's King of Animal Stories") and Wu Meizhen 伍美珍.

International Translation Day (2011–Present)
In 2017 the UN designated 30 September, the feast day of St Jerome, patron saint of translators, as International Translation Day ("International" n.d.). Like the Literary Translation Centre, representatives of several organizations jointly produce a program that brings together speakers from the translation world and the publishing world. The aim is to understand the business of literary translation better. It also provides a professional development and networking opportunity for translators, both new and experienced.

China Fiction Book Club (2010–Present)
The China Fiction Book Club was founded in 2010 by Nicky Harman and Helen Wang as an informal book club that met in London roughly every 6–8 weeks for about two years to share interesting books, and news about Chinese literature. In 2012, they decided to end the book club meetings, but to open a twitter account (@cfbcuk) to continue to share news. The twitter account was launched on the day that Mo Yan won the Nobel Prize for Literature.

Literary Translation Summer Schools
Since the late twentieth century there has been an expansion of postgraduate training in translation, and translation to and from Chinese (usually Mandarin) is now offered on many MA courses in the UK. Special mention should be made of the British Centre for Literary Translation (BCLT) founded in 1989 by W.G. Sebald (1944–2001) at the University of East Anglia (UEA) to promote and encourage the translation of literary works, to support practicing translators and to foster the study of translated literature (REF 2014). Many of the current leading literary translators in the UK have some association, direct or indirect, knowingly or unknowingly, with the BCLT, having taught or studied at the BCLT, or simply by being part of the translation community in

the UK and benefiting from the knowledge and experience of others. The impact has been a growth in literary translation courses, opportunities for continued professional development, mentoring, summer schools, international training days, public events, translation events at book festivals, and the Emerging Translators Network (ETN). These activities have a cumulative and positive impact, promoting understanding and improving best practice not only in process of translating texts but also in the working relationships with authors, editors, publishers, events programmers, and others. Many translators work extremely hard to promote the books and authors they translate. There are also other summer schools that offer Chinese in the UK, for instance, City University's Translate in the City (now defunct) and Warwick University's Warwick Translates.

China in Britain: Myths and Realities Project, 2012–2014

China in Britain: Myths and Realities was an AHRC [UK Arts and Humanities Research Council]-funded research project led by Anne Witchard, a specialist in Victorian literature at the University of Westminster, to investigate changing conceptions of China and Chineseness in Britain (China in Britain). It was an important and timely project with its ultimate aim "to contribute towards the ongoing reformulation of both British and Chinese cultural understandings in the context of a multicultural Britain still structured by racialised inequalities and Orientalist stereotypes." The project acknowledged the UK government's interest in commercial engagement with China: "In the light of China's emerging global profile as a country of major economic and political impact, Chinese visibility is undergoing significant transformations. Interest in Chineseness has seen an upsurge concomitant with that of our interest in fostering economic relationships with China." And it sought to understand the British-Chinese community better: "Although the British-Chinese voice has been marginal in mainstream cultural and political life it *is* beginning to make itself heard." *Translating China* (as the project was also known) sought to "explore the current and potential future course of Chinese-British relations and the place of history within it" (Translating China).

The project came under the AHRC's "Translating Cultures" theme. It is interesting that "translate" (like "curate") was a fashionable term at precisely the same time that literary translation was on the rise, and was used to convey a sense of exploring and re-framing. "In a world seen to be increasingly characterized by transnational and globalized connections, the need for understanding and communication within, between and across diverse cultures is stronger than ever. The Translating Cultures theme addresses this need by studying the role of translation, understood in its broadest sense, in the transmission, interpretation, transformation and sharing of languages, values, beliefs, histories and narratives." Witchard's team organized six study days: (1) Film (10 May 2012), (2) Film (31 May 2012), (3) Performance/Theatre and Music (18 July 2012), (4) Aesthetics: Visual and Literary Cultures (8 December 2012), (5) Archiving (27 April 2013), and (6) Diasporic Translations (24 May 2013). Although this project did not focus on literary translation specifically, it was important in highlighting new developments, and provided opportunities for interdisciplinary networking, both academic and nonacademic.

The Leeds Centre for New Chinese Writing (2014–Present)

The Leeds Centre for New Chinese Writing developed out of the 2014 AHRC-funded project "Writing Chinese." ("Leeds" n.d.) The Centre works in collaboration with a number of partners, including Paper Republic, the Mandarin Teachers Network (based at the Institute of Education's Confucius Institute, London), along with an increasing number of publishers, notably Balestier Press, Penguin China and Alain Charles Asia. The aim is to make new Chinese writing accessible to everyone, and to that end they run a series of author talks, readings, workshops, symposia, and other activities in Leeds, which they make as inclusive and open as possible. They have a monthly featured author in their bookclub, run a book review network dedicated to contemporary Chinese literature in translation, and hold an annual Chinese-English literary translation competition, as well as masterclasses and residential weekends for the book-review network. The Centre also provides a hub for research into all aspects of Chinese literature and Chinese-English literary translation. To date, they have featured over 50 Chinese writers as "author of the month," including four children's authors: Cao Wenxuan, Meng Yanan 孟亚楠, Lin Man-Chiu 林滿秋, and Huang Beijia.

The annual translation competition run by the Leeds Centre has been very successful not only in encouraging people to "have a go" but also in boosting the translation careers of the winners. The winning translations are published, and the first few winners were fortunate enough to be offered a free place at the Translate in the City summer school, tutored by Nicky Harman and leading translators working in other languages. Some have been invited back to judge subsequent competitions and have won further acclaim, commissions and awards, notably Natascha Bruce, a joint winner of the first competition. Two of the translation competitions have been for school students (under 18/19 years). The first was Meng Yanan's picture book *Happy Mid-Autumn Festival*, selected partly because the content was about a festival, and festivals feature in the primary school curriculum. The winning translation by Jasmine Alexander was published by Balestier Press in 2018. Alexander was mentored by Helen Wang between winning the competition and publication, and was later invited to Leeds to talk to younger students about her experience of learning Chinese and translating. The winner of the second competition for school students, Izzy Hasson, will see her translation of the picture book *Sleepy Sleepy New Year*, also by Meng Yanan, published by Balestier Press in 2020.

The Leeds Centre advises and develops the teaching of Chinese in UK schools – in particular, the government-funded Mandarin Excellence Programme (MEP), which provides intensive tuition in Mandarin for secondary school pupils in selected schools throughout England from the age of 11. Under the UK government's new criteria for language teaching, schools that offer Mandarin classes are now required to include Chinese literature in their coursework. The Centre works regularly with Katherine Carruthers (a key figure in Chinese teaching in schools, and in teacher-training), her team at the Institute of Education and the Confucius Institute to develop materials that would be appropriate for school-age students. It runs series of school bookclubs for pupils to discuss and reviews translated works of Chinese fiction, and hosts resources for teachers including a range of short stories in translation (selected from the Read Paper Republic database) suitable for class discussions, along with simple Chinese language literary texts.

In summer 2016, the Centre organized a symposium for teachers of Chinese to learn about children's books, and invited Minjie Chen, a specialist of Chinese children's literature, Anna Gustafsson Chen, a prolific Chinese-Swedish translator, and Helen Wang as speakers. Wang had worked with Carruthers and her team before, having given presentations at the annual Chinese Teachers Day, initially in her capacity as a curator at the British Museum. She supported events at the Writing Chinese project at Leeds, has been a judge of their translation competition three times, and contributed suggestions on ways to meet the government requirement on pupils' exposure to literature in the language they were learning. It was at this event in 2016 that Wang invited Chen and Chen to join her and co-write the blog *Chinese Books for Young Readers*.

Chinese Books for Young Readers

Chinese Books for Young Readers is an independent blog-website to share books and information about Chinese books. While contributing to the Paper Republic website, Helen Wang had included some information about Chinese children's books, and thought this subject merited a site of its own. Minjie Chen writes on the curatorial blog of the Cotsen Children's Library, Princeton University. Anna Gustafsson has her own blog (*Bokberget*, in Swedish) on Chinese books, including children's books. The new site was to be a dedicated space to post in English about Chinese books for young readers, offering a variety of reviews, interviews, industry knowledge, and insight into the world of Chinese children's books ("Chinese Books" n.d.). The site also has a twitter feed (@cb4yr) for news, and a Pinterest page (https://www.pinterest.co.uk/cb4yr/boards/).

Global Literature in Libraries Initiative (GLLI)

Global Literature in Libraries is another translator-led initiative, seeking to increase collaboration between librarians and translators. Librarians wishing to source more diverse books do not always have time to research in-depth. Meanwhile, literary translators often have this knowledge readily available. The GLLI website was created by German-English translator Rachel Hildebrandt. The Global Literature in Libraries Initiative "strives to raise the visibility of world literature for adults and children at the local, national and international levels" (Global). Helen Wang was a founding member of the board, and for the month February 2017 edited the GLLI site, focusing on Chinese literature, and thanks to support from colleagues at Paper Republic, posted every day throughout that month ("The GLLI-PR" 2017).

When Wang left the board, she introduced David Jacobson, a Japanese-English translator, who remains on the board. Jacobson conducted a survey of translations from Chinese, Japanese, and Korean and presented his findings on a panel titled "Asian American Experiences in Children's Books" at the 12th IBBY Regional Conference at the University of Washington, Seattle, in October 2017. His work highlighted the difficulty in collecting data about translated books, and concluded that "there is considerably more children's literature available in translation from East Asia than the CCBC (Cooperative Children's Book Center, at the University of Wisconsin) and USBBY (US Board on Books for Young People) surveys suggest" (Jacobson 2017). He counted 123 translations from Japanese since 2001, 64 from Chinese since 2005, and 38 from Korean since 2002. The contribution of a handful of publishers who have emphasized literature in translation is striking – nearly 60% of the Chinese titles came from just three small publishers: Balestier (UK), Starfish Bay (Australia), and Candied Plums (US). Jacobson updated and expanded his survey in 2019 (Jacobson 2019).

Jacobson was the recipient of a grant from the International Youth Library in 2017 to examine primary materials relating to Jella Lepman. Two years later, Jacobson, Jongsun Wee, who was a fellow researcher he met at the IYL, English-Japanese translator Rieko Nakaigawa Lee, and Minjie Chen co-authored an article titled "The Japan-China-Korea Peace Picture Book Project," which they published in two parts on Jacobson's guest series on *A Fuse #8 Production*, Betsy Bird's children's book review blog, and *Chinese Books for Young Readers*. Their article brings the Japan-China-Korea Peace Picture Book Project to the attention of English-language readers. Responding to rising animosity between Japan and her East Asian neighbors over the legacy of the Pacific War, the book project produced a series of 11 picture books on the themes of war and peace from the three countries and made most of them available in all three languages (Jacobson et al. 2020; Chen et al. 2020).

Conclusion

Chinese children's literature received unprecedented international attention during the first two decades of the twenty-first century. However, the breakthroughs, significant and welcome, did not appear overnight. Chinese texts have taken a winding path entering English-language children's literature, ushered by missionaries, sinologists, folklorists, translators, and Chinese diaspora writers since the late nineteenth century. Children's books portraying non-white experiences are often examined under the binary criteria of cultural authenticity/inauthenticity. A historical overview corroborates Nelson's (2014) insight that Chinese texts translated for English readers are a reflection as much of Chinese culture as of Western perception of that culture. The Chinese children's texts that have been published or well received in the West confirm how English-speaking children are exposed to a selective array of Chinese contents that, for their perceived representations of China and Chinese, serve changing utilities. The earliest text during the late eighteenth century about a wise Chinese ruler reflected the lingering favorable perception of China as an admirable model. Texts from Chinese classics and folklore, selected by missionaries for child readers back in their home countries, are perceived to reveal the humanity, decent values, and positive characteristics of Chinese, a people worthy of empathy, affection, the Christian mission, and a hopeful political future. Folktales and fantastical stories translated and retold by sinologists, folklorists, and, later, children's authors fill the need for international and multicultural literature as progressive educators, influenced by world wars and the Civil Rights Movement, put out urgent calls for cultural diversity in children's books to combat bias and bigotry, and to raise the visibility and esteem of minority groups.

Multiple factors combined to win reception and recognition for selected Chinese children's literature in English translation during the twenty-first century. First, foreign works notwithstanding, these books are perceived to reinforce the mainstream values of the target countries, partly explaining why family bonds and children's agency are prevalent themes. Chinese children's literature has come a long way from treating children as a means to another group's agenda to treating them as an end in itself. Its concern shifted from shaping the child audience into a filial son or daughter for the welfare of parents, to a patriotic citizen for the interest of the nation, to a selfless commune member for the advancement of Communism, to an obedient, well-behaving child approved by parents and teachers, finally to a human being entitled to safety, emotional health, and freedom of thought. As Chinese and Western societies find more common grounds in values concerning children, childhood, and child-rearing, they dislodge the main roadblock between Chinese children's literature and the Western market.

Second, contemporary Chinese children's literature, having benefited from the cross-fertilization of the tradition of Chinese literature and art and Western bookmaking for young readers, brings distinct aesthetic experience that is both familiar and fresh to the English-language world.

Thirdly, commercial and noncommercial interest in Chinese literature, a shifting culture that embraces Chinese topics and the Chinese language, the talent and skills of a growing pool of Chinese-English literary translators from a diverse background, and the myriad activities of their initiatives are all indispensable to the production and consumption of high-quality translated works. When contemporary stories became more appealing to English-language publishers – enough to risk investing in the cost and effort of translating and marketing otherwise unknown authors and their books (even when they are favorites and bestsellers in China), it set in motion a benign cycle of a growing demand and supply of more diverse and translated children's books.

Appendix 1: A List of Chinese Folk Literature Adapted or Illustrated by Ed Young

Title (and Parallel Title)	Common Chinese title	Text	Publisher	Year
The Emperor and the Kite	(An original tale inspired by a Chinese legend)	Jane Yolen	World Publishing	1967
			World Publishing	1968

(continued)

Title (and Parallel Title)	Common Chinese title	Text	Publisher	Year
Chinese Mother Goose Rhymes = 孺子歌圖		Selected and edited by Robert Wyndam		
Eight Thousand Stones: A Chinese Folktale	曹冲称象	Diane Wolkstein	Doubleday	1972
Cricket Boy: A Chinese Tale	促织	Feenie Ziner	Doubleday	1977
The Terrible Nung Gwama: A Chinese Folktale		Ed Young	Collins	1978
White Wave: A Chinese Tale	田螺姑娘	Diane Wolkstein	Crowell	1979
High on a Hill: A Book of Chinese Riddles		Ed Young	Collins	1980
Yeh-Shen: A Cinderella Story from China = 葉限		Al-Ling Louie	Putnam	1982
Eyes of the Dragon	画龙点睛	Margaret Leaf	Lothrop	1987
Lon Po Po: A Red-Riding Hood Story from China	狼外婆	Ed Young	Philomel	1989
The Voice of the Great Bell	铸钟娘娘	Lafcadio Hearn; retold by Margaret Hodges	Little Brown	1989
Seven Blind Mice	盲人摸象	Ed Young	Philomel	1992
Red Thread	月下老人	Ed Young	Philomel	1993
Cat and Rat: The Legend of the Chinese Zodiac	猫鼠结仇	Ed Young	Holt	1995
Donkey Trouble	抬驴	Ed Young	Atheneum Books for Young Readers	1995
Night Visitors	黄粱一梦	Ed Young	Philomel	1995
Mouse Match: A Chinese Folktale	老鼠嫁女	Ed Young	Silver Whistle	1997
The Lost Horse: A Chinese Folktale	塞翁失马	Ed Young	Silver Whistle	1998
The Hunter: A Chinese Folktale	猎人海力布	Mary Casanova	Simon & Schuster	2000
Monkey King	美猴王	Ed Young	HarperCollins	2001
The Sons of the Dragon King: A Chinese Legend	龙生九子	Ed Young	Atheneum	2004

Appendix 2: Notable Children's and Young Adult Books by Contemporary Chinese Authors in English Translation (1995–2020)

Author	Title	Title in Chinese	Illustrator	Translator	Publisher	Year	Format/Genre
Jicai Feng 冯骥才	Let One Hundred Flowers Bloom			Christopher Smith	Viking	1995	Juvenile fiction
Chen Chih-Yuan (Taiwanese) 陳致元	Guji Guji	Same as English	Same as author		Kane/Miller	2004	Picture book
Chunshu 春树	Beijing Doll	北京娃娃		Howard Goldblatt	Riverhead Books	2004	Young adult fiction
Kim Xiong 熊亮	The Little Stone Lion	小石狮	Same as author		Heryin Books	2005	Picture book
Ma Yan 马燕	Diary of Ma Yan: The Struggles and Hopes of a Chinese Schoolgirl	马燕日记		Lisa Appignanesi	HarperCollins	2005	Diary
Jimmy Liao (Taiwanese) 幾米	The Sound of Colors: A Journey of the Imagination	地下鐵	Same as author	Sarah L. Thomson	Little, Brown	2006	Picture book
Ji Zhaohua & Xu Cui 姬炤华, 徐萃	No! That's Wrong!	天啊! 错啦!	Same as author		Kane/Miller	2008	Picture book
Yang Hongying 杨红樱	Mo's Mischief series	淘气包马小跳			HarperCollins	2008	Juvenile fiction
Yu Liqiong 余丽琼	A New Year's Reunion	团圆	Zhu Chengliang 朱成梁		Candlewick	2011	Picture book
Shen Shixi 沈石溪	Jackal and Wolf	红豺		Helen Wang	Egmont	2012	Juvenile fiction
Gerelchimeg Blackcrane 黑鹤	Black Flame	黑焰		Anna Holmwood	Groundwood Books	2013	Juvenile fiction
Cao Wenxuan 曹文轩	Bronze and Sunflower	青铜葵花	Meilo So	Helen Wang	Walker Books Candlewick	2015 2017	Juvenile fiction
Fang Suzhen (Taiwanese) 方素珍	Grandma Lives in a Perfume Village	外婆住在香水村	Sonja Danowski	Huang Xiumin 黄秀敏	NorthSouth Books	2015	Picture book

(continued)

Author	Title	Title in Chinese	Illustrator	Translator	Publisher	Year	Format/Genre
Guojing 郭婧	The Only Child	独生小孩	Same as author	N.A.	Schwartz & Wade	2015	Wordless picture book
Bao Dongni 保冬妮	Who Wants Candied Hawberries?	冰糖葫芦,谁买?	Wu Di 吴翟	Adam Lanphier	Candied Plums	2016	Picture book
Chen Xiaoting 陈晓婷	Father and Son Go Fishing	和爸爸一起去海边	Huang Ying 黄缨	Duncan Poupard	Candied Plums	2016	Picture book
Dong Yanan 董亚楠	Express Delivery from Dinosaur World	恐龙快递	Same as author	Helen Wang	Candied Plums	2016	Picture book
Fu Wenzheng 符文征	Buddy Is So Annoying	我讨厌宝弟	Same as author	Adam Lanphier	Candied Plums	2016	Picture book
Gan Dayong 甘大勇	Little Rabbit's Questions	小兔的问题	Same as author	Helen Wang	Candied Plums	2016	Picture book
Lin Songying 林颂英	Borrowing a Tail	借尾巴	Zhang Le 张乐	Duncan Poupard	Candied Plums	2016	Picture book
Wang Xiaoming 王晓明	The Peanut Fart	花生米样的屁	Same as author	Adam Lanphier	Candied Plums	2016	Picture book
Wang Yimei 王一梅	Rory the Rabbit	兔子萝里	李春苗, 何萱 Li Chunmiao & He Xuan	Adam Lanphier	Candied Plums	2016	Picture book
Wang Zaozao 王早早	An's Seed	安的种子	Huang Li 黄丽	Helen Wang	Candied Plums	2016	Picture book
Xia Lei 夏蕾	Who Ate My Chestnut?	谁吃了我的毛栗子	Wang Chao 王超	Duncan Poupard	Candied Plums	2016	Picture book
Xiao Mao 萧袤	CeeCee	西西	李春苗, 张彦红 Li Chunmiao & Zhang Yanhong	Helen Wang	Candied Plums	2016	Picture book
Xiao Mao 萧袤	The Frog and the Boy	青蛙与男孩	陈伟, 黄小敏 Chen Wei & Huang Xiaomin	Helen Wang	Candied Plums	2016	Picture book
Zhou Xu 周旭	Picking Turnips	拔萝卜	Same as author	Adam Lanphier	Candied Plums	2016	Picture book

(continued)

Author	Title	Title in Chinese	Illustrator	Translator	Publisher	Year	Format/Genre
Zhu Chengliang 朱成梁	Flame	火焰	Same as author	Helen Wang	Candied Plums	2016	Picture book
Cao Wenxuan	Feather	羽毛	Roger Mello	Chloe Garcia Roberts	Elsewhere Editions	2017	Picture book
Lin Man-Chiu (Taiwanese) 林滿秋	The Ventriloquist's Daughter	腹语师的女儿		Helen Wang	Balestier Press	2017	Fantasy fiction
Meng Yanan 孟亚楠	Happy Mid-Autumn Festival	中秋节快乐	Same as author	Jasmine Alexander	Balestier Press	2018	Picture book
Nie Jun 聂峻	My Beijing: Four Stories of Everyday Wonder	老街的童话	Same as author	Edward Gauvin	Graphic Universe	2018	Comics (graphic works)
Chen Jiatong 陈佳同	White Fox	白狐迪拉	Viola Wang	Jennifer Feeley	Chicken House Books	2019	Fantasy fiction
Gerelchimeg Blackcrane 黑鹤	The Moose of Ewenki	鄂温克的驼鹿	Jiu Er	Helen Mixter	Greystone Books	2019	Picture book
Yan Ge 颜歌	White Horse	白马	James Nunn	Nicky Harman	Hoperoad	2019	Young adult fiction

(A searchable database of translated children's literature is maintained by the Global Literature in Libraries Initiative at https://www.librarycat.org/lib/GLLI)

Appendix 3: Chinese Children's Books Selected as White Ravens, 2014–2019

Year	Title (cited with synopses in White Ravens catalogue, not translations)
2014	Sister Rabbit's Gigantic Pumpkin Jiu, Er 九儿. 2012. *Meimei de nangua* 妹妹的南瓜. Beijing: Lianhuanhua chubanshe.
2014	A Boy's Honour Li, Donghua 李东华. 2014. *Shaonian de rongyao* 少年的荣耀. Taiyuan: Xiwang chubanshe.
2014	I Saw a Bird Liu, Bole 刘伯乐, 2014. *Wo kanjian yi zhi niao* 我看见一只鸟. Jinan: Mingtian chubanshe.
2014	Gezi's Book of Time Lu, Mei 陆梅. 2013. *Gezi de shiguang shu* 格子的时光书. Nanning: Jieli chubanshe.
2014	The Most Terrible Day Tang, Muniu 汤姆牛. 2013. *Zui kepa de yi tian* 最可怕的一天. Beijing: Beijing lianhe chuban gongsi.

(continued)

Year	Title (cited with synopses in White Ravens catalogue, not translations)
2014	Tour of Beijing Sun, Xinyu (Joyce Sun) 孙心瑜. *Beijing you* 北京游. Taibei: Xiaolu wenhua (Tien-Wei Publishing Company).
2015	The Sun Was Once a Little Boy Li, Shanshan 李姗姗. 2015. *Taiyang xiao shihou shi ge nanhai* 太阳小时候是个男孩. Chengdu: Sichuan haonian ertong chubanshe.
2015	Pop-up Taiwan Lau, Kit (Sze Kit Lau) 劉斯傑. 2014. *Taiwan danqi* 台灣彈起. Hong Kong: Zhuonian chuangzuo youxian gongsi (SK Ronex Ltd). Parallel edition: Taipei: Shibao wenhua (China Times Publ.)
2015	The Boys and the Sea Zhang, Wei 张炜. 2014. *Shaonian yu hai* 少年与海. Hefei: Anhui shaonian ertong chubanshe.
2015	Granny Mian Couldn't Sleep Liao, Xiaoqin 廖小琴 (text) and Zhu, Chengliang 朱成梁 (illus.). 2014. *Mian popo shuibuzhao* 棉婆婆睡不着. Jinan: Mingtian chubanshe.
2015	The Brand Cao, Wenxuan 曹文轩. 2015. *Huoyin* 火印. Beijing: Tiantian chubanshe.
2015	The Excuse for Being Late Yao, Jia 姚佳. 2014. *Chidao de liyou* 迟到的理由. Jinan: Mingtian chubanshe.
2016	Saving "The Fisher Boy" Zhao, Lihong 赵丽宏 (text) and Feng, Xueqing (illus.). 2015. *Yu tong* 渔童. Fuzhou: Fujian shaonian ertong chubanshe.
2016	Summer Cao, Wenxuan (text) and Yu Rong 郁蓉 (illus.). 2015. *Xiatian* 夏天. Nanchang: Ershiyi shiji chubanshe.
2016	The Generals' Hutong Shi, Lei 史雷 (text), Lin, Xian and Wang, Yue (illus.). 2015. *Jiangjun hutong* 将军胡同. Beijing: Tiantian chubanshe.
2016	Children Blowing the Whale Whistle Zhou, Rui 周锐. 2016. *Chui jing shao de haizi* 吹鲸哨的孩子. Dalian: Dalian chubanshe.
2016	Children of the Bajau Peng, Yi 彭懿. 2016. *Baoyao ren de haizi* 巴夭人的孩子. Jinan: Mingtian chubanshe, 2016.
2017	Through the Eyes of a Child Huang, Beijia 黄蓓佳. 2017. *Tong mou* 童眸. Nanjing: Jiangsu fenghuang shaonian ertong chubanshe.
2017	The Fantastic Book Yang, Sifan 杨思帆. 2016. *Qimiao de shu* 奇妙的书. Guilin: Guangxi shifan daxue chubanshe.
2017	Each Grain in Your Dish Yu, Hongcheng 于虹呈. 2016. *Pan zhong can* 盘中餐. Beijing: Zhongguo shaonian ertong chubanshe.
2017	The Propitious Time Zhang, Zhilu 张之路 (text) and Du, Lingyun (illus.). 2016. *Jixiang shiguang* 吉祥时光. Beijing: Zuojia chubanshe.

(continued)

Year	Title (cited with synopses in White Ravens catalogue, not translations)
2017	Above Ground, Under Ground. An Illustrated Account of the Insect Microcosm Through the Four Seasons Qiu, Chengzong 邱承宗. 2016. *Di mian di xia* 地面地下. Taipei: Xiaolu wenhua shiye gufen youxian gongsi.
2017	The Postman from Buluo Town Guo, Jiangyan 郭姜燕 (text) and Chen, Shu (illus.). 2016. *Buluo zhen de youdiyuan* 布罗镇的邮递员. Shanghai: Shaonian ertong chubanshe.
2018	*Welcome to the H(e)aven Town* Ander (Ye, Ande 葉安德) (text/illus.). 2017. *Huanying guanglin tiantang xiaozhen* 欢迎光临天堂小镇. Xinzhu: Heying wenhua (Hsinchu: Heryin Books).
2018	Playing with Lanterns Wang, Yage 王亚鸽 (text) and Zhu, Chengliang 朱成梁 (illus.). 2017. *Da denglong* 打灯笼. Beijing: Lianhuanhua chubanshe, 2017. 32 pp.
2018	A Lian's Story Tang, Sulan 汤素兰 (text) and Zhang, Wenqi (illus.). 2017. *A Lian* 阿莲. Changsha: Hunan shaonian ertong chubanshe.
2018	The Pagoda Qin, Wenjun 秦文君 (text), various artists (illus.). 2017. *Baota* 宝塔. Jinan: Mingtian chubanshe, 2017.
2018	Nic is Me Chang, Xingang 常新港 (text), WANG Jue (illus.). 2018. *Nike daibiao wo* 尼克代表我. Beijing: Tiantian chubanshe.
2018	The Flight of a Bullet Bai, Bing 白冰 (text) and Liu, Zhenjun 刘振军 (illus.). 2018. *Yi ke zidan de feixing* 一颗子弹的飞行. Beijing: Zhongguo shaonian ertong xinwen chubanshe.
2019	The Summer of Pigeons Liu, Haiqi 刘海栖 (text), Wang, Zumin (illus.), and Wang, Ying (illus.). 2019. *You ge gezi de xiatian* 有鸽子的夏天. Jinan: Shandong jiaoyu chubanshe.
2019	The Wild Child Dong, Hongyou 董宏猷 (text), and Cao, Qing (illus.). 2018. *Ye wazi* 野娃子. Nanchang: Ershiyi shiji chubanshe.
2019	The Toil of the Red Headscarves. The story of the female construction workers in Singapore Lin, Denan 林得楠 (text) and Yu, Guangda 余广达 (Patrick YEE) (illus.). 2018. *Xinku le hong toujin* 辛苦了红头巾. *Lao shi cheng jianzhu nügong de gushi*. Singapore: Ling zi chuan mei si ren youxian gongsi (Lingzi Media).
2019	The Horses in Grandma's House Xie, Hua 谢华 (text), Huang, Li 黄丽 (illus.). 2019. *Waipo jia de ma* 外婆家的马. Zhengzhou: Haiyan chubanshe.
2019	Plastic, Plastic, Everywhere Fung, Michelle Kuen Suet (Feng, Juanxue 馮捲雪). 2017. *Tin Hong Gaau Jyu (Tiankong jiaoyu* 天空膠雨*)*. Hong Kong: Sanlian shudian youxian gongsi.
2019	Sand and Stardust Xue, Tao 薛涛 (text), Hui shi guang wenhua and Liu Huang (illus.). 2019. *Shali yu xingchen* 沙粒于星尘. Hefei: Anhui shaonian ertong chubanshe.
2019	Mat Jenin. A Malaysian Folminjiecktale Maniniwei (Lin Wanwen) (text & illlus.); Lim Wan Kee (Lin Wanqi) (trans.). 2018. *Mareni* 马若尼. *Malai minjian gushi / Mat Jenin. Cerita rakyat Melayu*. Xinbeishi: Bubu chubanshe. Chinese and Malaysian.

(continued)

Year	Title (cited with synopses in White Ravens catalogue, not translations)
2019	Don't let the sun fall! Guo, Zhenyuan 郭振媛 (text) and Zhu, Chengliang (illus.) 朱成梁. 2018. *Bie rang taiyang diaoxialai* 别让太阳掉下来. Beijing: Zhongguo heping chubanshe.

For fuller details, see "White Ravens: Chinese titles, 1984–2020," Chinese Books for Young Readers (no.111), 8 October 2020. https://chinesebooksforyoungreaders.wordpress.com/2020/10/08/111-white-ravens-chinese-titles-1984-2020/

References

Literature

Bishop, Claire Huchet, and Kurt Wiese (illus.). 1938. *The five Chinese brothers*. New York: Coward-McCann.
Bishop, Claire Hucher 毕肖普, and Kurt Wiese 维泽 (illus.). 2018. 中国五兄弟 *(The five Chinese brothers)*. Trans. Fangli Fei. 杭州: 浙江人民美术出版社.
Boat Life in China. 1867. In *Peter Parley's annual: A Christmas and New Year's present for young people*, ed. William Martin, 254–259. London: Darton.
Chang, Tien-yi. 1959. *The magic gourd*. Trans. Gladys Yang. Peking: Foreign Languages Press.
Chen, Chih-Yuan. 2004a. *Guji Guji*. La Jolla: Kane/Miller.
Chun and Si-ling: An historical Romance, in which is introduced some account of the customs, manners, and moral conduct of the Chinese. 1811. London: Printed for the author.
Eberhard, Wolfram. 1937. *Chinese fairy tales and folk tales*. Trans. Desmond Parsons. London: Kegan Paul & Co.
Example of Filial Piety in China. 1736. *Gentleman's Magazine* 6(December 1): 738.
Fang, Suzhen, and Sonja Danowski (illus.). 2015. *Grandma lives in a perfume village*. Trans. Xiumin Huang. New York: NorthSouth Books.
Fang, Suzhen 方素珍, and Binru Jiang 江彬如 (illus.). 2007. 外婆住在香水村 *(Grandma lives in a perfume village)*. 台北市: 青林國際出版股份有限公司.
Fielde, Adele M., trans. 1879. *Twenty four examples of filial piety*. Swatow: Printed by the Swatow Printing Office Company.
———. 1893. *Chinese nights' entertainment, forty stories told by almond-eyed folk actors in the romance of the strayed arrow*. New York: G.P. Putnam's Sons.
———. 1912. *Chinese fairy tales: Forty stories told by almond-eyed folk*. 2nd ed. New York/London: G.P. Putnam's Sons.
Fu, Runhua 傅润华, Zhihao 之浩 (illus.), and Lushan 鹿山 (illus.). 1955. 五兄弟 *(Five brothers)*. 上海: 上海人民美术出版社.
Gan, Dayong. 2016. *Little rabbit's questions*. Trans. Helen Wang. Seattle/Washington: Candied Plums.
Headland, Isaac Taylor, trans. and illus. 1900. *Chinese Mother Goose rhymes*. New York/Chicago: Fleming H. Revell Co.
Hodges, Margaret, Ed Young (illus.), and Lafcadio Hearn. 1989. *The voice of the great bell*. Boston: Little, Brown and Company.
Hong, Xuntao 洪汛涛, and Jiasheng Zheng 郑家声 (illus.). 1958. 十兄弟 *(Ten brothers)*. 上海: 上海人民美术出版社.
Hu lu xiong di (The Calabash Brothers). 2008. 上海: 上海书店出版社.
Ji, Zhaohua, and Xu. Cui. 2008. *No! That's wrong!* La Jolla: Kane/Miller.
Lang, Andrew. 1892. The story of Hok Lee and the dwarfs. In *The green fairy book*, ed. Andrew Lang, 229–233. New York: A.L. Burt Company.

Lin, Grace. 2009. *Where the mountain meets the moon*. New York: Little, Brown and Co.
Louie, Ai-Ling, and Ed Young (illus.). 1982. *Yeh-Shen: A Cinderella story from China*. New York: Philomel Books.
Moule, Arthur E. 1881. *Chinese stories for boys and girls: And Chinese wisdom for old and young*. London: Seeley, Jackson, & Halliday.
Shanghai dian ying zhi pian chang Little Sisters of the Grassland Compiling and Drawing Group. 1973. *Little sisters of the grassland*. Peking: Foreign Languages Press.
Sobersides, Solomon. 1786. History of Hamti, emperour of China. In *New-Year's gift; or, entertaining histories, for the amusement and instruction of young ladies and gentlemen, in winter evenings*, ed. A. Pretty, 111–113. Worcester: Printed by Isaiah Thomas.
Wilhelm, Richard. 1921. *The Chinese fairy book*. Trans. Frederick Herman Martens. New York: Frederick A. Stokes.
Yang, Gene Luen. 2006. *American born Chinese*. New York: First Second.
Young, Ed, trans. and illus. 1989. *Lon Po Po*. New York: Philomel Books.
Yu, Li-Qiong, and Cheng-Liang Zhu (illus.). 2011. *A new year's reunion*. Somerville: Candlewick Press.

Secondary Sources

2019 nian du Zhongguo dang dai zuo pin fan yi gong cheng zheng ji gong gao (Announcing the year 2019 contemporary Chinese literary works translation program). 2019. 中国作家网. http://www.Chinawriter.com.cn/n1/2019/0912/c403937-31350961.html
Abrahamsen, Eric. 2014. China writers association translation fund – Some news. *Paper Republic*. https://paper-republic.org/pers/eric-abrahamsen/China-writers-association-translation-fund-some-news/
Allardice, Lisa. 2005. Inside story: Jung Chang's novel wild swans smashed best-selling records worldwide. *The Guardian*, May 26, 2005. https://www.theguardian.com/books/2005/may/26/biography.china
Asian Festival of Children's Content (AFCC). 2020. https://afcc.com.sg/. Accessed 19 July 2020.
Bak, Mikyung. 2014. The folktale 'Hokpuri Yongkam' and the visual representation of the Korean Dokkaebi. In *Proceedings of the 9th conference of the international committee for design history and design studies*, 231–236. Aveiro: Editora Edgard Blücher. https://doi.org/10.5151/despro-icdhs2014-0028.
Bird, Elizabeth. 2008. Review of the day: No! That's wrong! *A Fuse #8 Production* (blog). March 18, 2008. http://blogs.slj.com/afuse8production/2008/03/18/review-of-the-day-no-thats-wrong/
Bokberget (blog). http://baodaobooks.blogspot.com/
Bologna Children's Book Fair. http://www.bookfair.bolognafiere.it/en/home/878.html. Accessed 19 July 2020.
Chen, Jeng-Guo S. 2004b. The British view of Chinese civilization and the emergence of class consciousness. *The Eighteenth Century* 45 (2): 193–205.
Chen, Lingchei Letty. 2008. Translating memory, transforming identity: Chinese expatriates and memoirs of the cultural revolution. *Tamkang Review: A Quarterly of Literary and Cultural Studies* 38 (2): 25–40.
Chen, Minjie, and Betsy Hearne. 2012. New house on the block: Private children's libraries in China. *The Horn Book Magazine* 88 (5): 52–57.
Chen, Minjie. 2016. *The Sino-Japanese war and youth literature: Friends and foes on the battlefield*. London/New York: Routledge.
Chen, Minjie, and Lydia Qiuying Wang. 2018. Cotsen conference report: The international symposium for children's literature & the fourth US-China symposium for children's literature. *Cotsen Children's Library* (blog). June 25. https://blogs.princeton.edu/cotsen/2018/06/symposium-2/

Chen, Minjie, Jongsun Wee, David Jacobson, and Reiko Nakaigawa Lee. 2020. War and peace in China-Japan-Korea picture books. *Chinese Books for Young Readers* (blog). March 12. https://chinesebooksforyoungreaders.wordpress.com/2020/03/12/cjk-peace/

Chen, Shih-Wen. 2019. *Children's literature and transnational knowledge in modern China: Education, religion, and childhood*. Singapore: Palgrave Macmillan.

China in Britain. http://translatingcultures.org.uk/awards/research-networking-awards/china-in-britain-myths-and-realities/

China Shanghai International Children's Book Fair. http://www.ccbookfair.com/en. Accessed 19 July 2020.

Chinese Books for Young Readers. n.d. https://chinesebooksforyoungreaders.wordpress.com/

CILIP. 2019. CILIP Carnegie and Kate Greenaway Children's Book Awards Independent Diversity Review. https://www.cilip.org.uk/page/CKGDiversityReviewFinal#downloads

Constantine, Peter. 2018. Professionalisation of literary translation and the publishing market. In *The Routledge handbook of literary translation*, ed. Kelly Washbourne and Ben Van Wyke. London: Routledge. (Chapter 6).

Denham, Isabel Y., and Helen Gregory. 1990. The voice of the great bell. *School Library Journal* 36 (2): 65, 68.

Dorson, Richard M. 1965. Foreword to *Folktales of the world*, ed. Wolfram Eberhard. Rev. ed., v–xxxi. Chicago: University of Chicago Press.

Dream. 2018. *Dream. Illustrations from China*. (Catalogue of the Bologna Book Fair Guest of Honor Illustration Exhibit – hosted by State Administration of Press, Publication, Radio, Film and Television of The People's Republic of China. Organised by China Children's Press & Publication Group [CCPPG], Beijing Fanglue Bohua Culture Media Co Ltd. Executive Curator: Guo Yawen 郭亚文).

ETN – Emerging Translators Network. https://emergingtranslatorsnetwork.wordpress.com/

Fang, Weiping 方卫平. 2018. *Four decades of Chinese children's literature: A historical overview*. Trans. Yuehong Huo 霍跃红. Ed. Alida Allison. Beijing: Zhongguo shao nian er tong chu ban she.

Freeman Book Awards. n.d. *The National Consortium for Teaching about Asia* (blog). http://nctasia.org/freeman-book-awards/. Accessed 23 June 2020.

Garrett, Jeffrey. 2006. Of translations and tarantulas: What's at stake when American children read books from other countries. In *Crossing boundaries with children's books*, ed. Doris Gebel, 10–14. Lanham: Scarecrow.

Glistrup, Eva. 2002. *The Hans Christian Andersen awards 1956–2002*. Trans. Patricia Compton. Copenhagen: IBBY.

Global Literature in Libraries Initiative. https://glli-us.org/. Accessed 19 July 2020

Guo ji chu ban (International publishing). 2017. 中国教育图书进出口有限公司 (China Educational Publications Import & Export Corporation Ltd.). http://www.cepiec.com.cn//index.php?m=content&c=index&a=lists&catid=81

Harman, Nicky. 2020a. How Paper Republic ended up leading what is possibly the world's biggest collaborative translation. *Asian Books Blog*. May 27. http://www.asianbooksblog.com/2020/05/how-paper-republic-ended-up-leading.html#more

———. 2020b. I want to be good. Nicky Harman tells us about Huang Beijia's novel. *Chinese Books for Young Readers* (blog). June 18. https://chinesebooksforyoungreaders.wordpress.com/2020/06/18/97-i-want-to-be-good-nicky-harman-tells-us-about-huang-beijias-novel/

Hearne, Betsy. 1991. Coming to the states: Reviewing books from abroad. *Horn Book Magazine* 67 (5): 562–568.

Hile, Kevin S. 2007. Claire Huchet Bishop. In *Gale literature: Contemporary authors*. Farmington Hills: Gale.

https://hub.londonbookfair.co.uk/lbf-international-excellence-awards-2016-shortlist-revealed-us-china-and-australia-lead-the-way/

Hua, Meng 花萌, and Qiong Shi 石琼. 2020. 从对外传播到版权输出: 中国儿童文学的译介之路: 海豚出版社策划总监梅杰访谈录 (From global propaganda to selling rights). *昆明学院学报* 42 (2): 1–5.

International Board on Books for Young Readers. https://www.ibby.org/. Accessed 19 July 2020.

International Translation Day. n.d. 30 September. United Nations. https://www.un.org/en/events/translationday/. Accessed 19 July 2020.

Isaac Taylor Headland. 2016. The Freer Gallery of Art and Arthur M. Sackler Gallery, Smithsonian Institution. https://asia.si.edu/wp-content/uploads/2017/09//Headland-Isaac-Taylor.pdf

Jacobson, David. 2017. Survey of translations of children's and YA literature translated from Chinese, Japanese and Korean. Chinese Books for Young Readers (no. 51), 16 October 2017. https://chinesebooksforyoungreaders.wordpress.com/2017/10/16/david-jacobsons-survey/

Jacobson, David. 2019. Beyond 3 percent: Translated children's literature in the U.S. *A Fuse #8 Production* (blog). November 12. http://blogs.slj.com/afuse8production/2019/11/12/guest-post-beyond-3-percent-translated-childrens-literature-in-the-u-s/

Jacobson, David, Minjie Chen, Reiko Nakaigawa Lee, and Jongsun Wee. 2020. The Japan-China-Korea peace picture book project. *A Fuse #8 Production* (blog). March 12. http://blogs.slj.com/afuse8production/2020/03/12/guest-post-what-were-missing-gems-of-world-kid-lit-2/

Ji gou jian jie (A brief introduction to the institution). n.d. 中国外文出版发行事业局 (The China Foreign Languages Publishing Administration). http://www.cipg.org.cn/node_1006430.htm. Accessed 22 June 2020.

Jiang, Qing 江青. 1967. 谈京剧革命：一九六四年七月在京剧现代戏观摩演出人员的座谈会上的讲话 (On the Revolution in Peking Opera). 人民日报, May 10, 1.

Jin, Ronghua 金荣华. 2002. 中国民间故事集成类型索引(一)简目 (An index to Chinese folk tale types, pt. 1, abridged). In 中国民间故事类型研究, ed. Shouhua Liu, 709–24. 武汉: 华中师范大学出版社.

Kantor, Emma. 2017. Candied plums launches with bilingual picture book list. *Publishers Weekly*. https://www.publishersweekly.com/pw/by-topic/childrens/childrens-industry-news/article/73195-candied-plums-launches-with-bilingual-picture-book-list.html

Krasilovsky, Phyllis. 1973. What Chinese children read: A morality tale. *Publishers Weekly* 203 (9): 100–101.

LARB China Channel. *Los Angeles review of books China Channel*. https://chinachannel.org/

Laughlin, Charles. 2013. The new translators and contemporary Chinese literature in English: A review of the journals 'Chinese Literature Today, Pathlight', and 'Chutzpah!/Peregrine'. *Chinese Literature: Essays, Articles, Reviews (CLEAR)* 35: 209–214. www.jstor.org/stable/43490168

Lee & Low Books. 2020. Where is the diversity in publishing? The 2019 diversity baseline survey results. *The Open Book Blog*. Jan 28. https://blog.leeandlow.com/2020/01/28/2019diversitybaselinesurvey/

Leeds Centre for New Chinese Writing. n.d. https://writingchinese.leeds.ac.uk/

Li, Jing. 2008. Chinese Tales: History of Chinese folktales and storytelling. In *The Greenwood encyclopedia of folktales and fairy tales*, ed. Donald Haase, vol. 1, 194–200. Westport: Greenwood Press.

Li, Li 李丽. 2010. 生成与接受 (Production and reception: a study of translated children's literature in China, 1898–1949). Wuhan Shi: Hubei ren min chu ban she.

Li, Wenjie. 2017. *A history of translation and interpretation: The Chinese versions of Hans Christian Andersen's Tales*. Odense: University Press of Southern Denmark.

Lin, Wenbao 林文寶, and Gerong Qiu 邱各容. 2018. 臺灣兒童文學史 (A history of children's literature in Taiwan). 臺北市: 萬卷樓圖書股份有限公司.

Liu, Shouhua 刘守华, ed. 2002. 中国民间故事类型研究 (A study of Chinese folktale types). 武汉: 华中师范大学出版社.

Liu, Ting 刘婷. 2009. 中少社引进日本漫画遇到难题 (The China Children's Press encountered challenges in importing Japanese comics). 北京晨报, March 26, 2009, A13-A13.

Lu, Xun. 1976. The picture-book of twenty-four acts of filial piety. In *Dawn blossoms plucked at dusk*. Trans. Xianyi Yang and Gladys Yang, 26–35. Beijing: Foreign Languages Press.

Lystad, Mary H. 1974. *A child's world: As seen in his stories and drawings*. National Institute of Mental Health. Rockville. Appendix B: A Comparative Note on Early American, Modern Russian and Chinese books for children, 119–124.

Magagnin, Paolo. (forthcoming). Chinese children's literature and the challenges of internationalization. Cao Wenxuan's *Qingtong kuihua* in Italian. In *Humanities, different traditions and methodologies. Multicultural perspectives in Chinese language and literature*, ed. Xiaoming Chen and Tiziana Lippiello. Beijing: Peking University. https://core.ac.uk/download/pdf/223179717.pdf

Martens, Fredrick H. 1921. Preface to *The Chinese fairy book*, ed. Richard Wilhelm, v–vi. New York: Frederick A. Stokes.

Marven, Lynn. 2016. University hosts translation slam. University of Liverpool's *Modern Languages and Cultures blog 2011–2017*. https://soclas.wordpress.com/2016/11/11/university-hosts-chinese-translation-slam/

Mezzanotte, Marinella. 2017. On a cold December evening I headed to the Free Word Centre in London to do something I had never tried before. *Global Literature in Libraries Initiative* (blog). Feb 28. https://glli-us.org/2017/02/28/on-a-cold-december-evening-i-headed-to-the-free-word-centre-in-london-to-do-something-i-had-never-tried-before-by-marinella-mezzanotte/

Nelson, Claudia. 2014. Adapting Chinese children's texts for U.S. readers: Three case studies. In *Second US-China children's literature symposium, Columbia, South Carolina, June 2014*. https://www.academia.edu/16703447/

Paper Republic – Chinese literature in translation. https://paper-republic.org/. Accessed 19 July 2020.

PW Awards 2016 – LBF International Excellence Awards 2016: Shortlist revealed US, China and Australia lead the way. *London Book Fair* (blog). March 7.

REF 2014 – Impact case study (REF3b): University of East Anglia (29) English language and literature – Literary translation: Theory into practice, research excellence framework, 2014. https://ref2014impact.azurewebsites.net/casestudies2/refservice.svc/GetCaseStudyPDF/1444

Roxburgh, Stephen. 2006. Sie Müssen Den Amerikanischen Sektor Verlassen: Crossing boundaries. In *Crossing boundaries with children's books*, ed. Doris Gebel, 5–10. Lanham: Scarecrow.

Shi, Jingnan 史竟男. 2019. 刘慈欣科幻作品推出少儿版 (Liu Cixin's science fiction released in juvenile edition). Xinhua Net. http://www.xinhuanet.com/politics/2019-09/02/c_1124951701.htm. Accessed 27 Sept 2020.

Song, Lihua 宋莉华. 2015. 近代来华传教士与儿童文学的译介 (Translation and introduction of children's literature by missionaries in modern China). Shanghai: Shang hai gu ji chu ban she.

Stones, Rosemary. 1977. An introduction to children's books in the People's Republic of China. *Bookbird* 15 (1): 8–18.

Sullivan, Julie. 2018. Interview with Helen Wang. *Words and Pictures, the SCWBI British Isles online magazine* [SCWBI = Society for Children's Book Writers and Illustrators]. https://www.wordsandpics.org/2018/10/translation-helen-wang.html

Swift, Rebecca. 2008. Crossing literary continents. *The Guardian* (Books Blog), 25 March 2008. https://www.theguardian.com/books/booksblog/2008/mar/25/crossingliterarycontinents

Tan, Teri. 2009–2020. List of her articles on Chinese children's book publishing in Publisher's Weekly. https://www.publishersweekly.com/pw/article_authors/74.html

Tang, Rui. 2006. Chinese children's literature in the 21st century. *Bookbird* 44 (3): 21–29.

The GLLI-PR Collaboration on Chinese Literature, Feb 2017 (List of All 28 Posts). 2017. *Global Literature in Libraries Initiative* (blog), Feb 28. https://glli-us.org/2017/02/28/the-glli-pr-collaboration-on-chinese-literature-feb-2017-list-of-all-28-posts/

The Soul of the Great Bell. 2016. *Story of the Week* (blog). http://storyoftheweek.loa.org/2016/06/the-soul-of-great-bell.html

Tiannan/Chutzpah 2011–2013 (16 issues). n.d. https://paper-republic.org/publishers/chutzpah/

Toft, Zoe. 2015. An interview with the translator of Bronze and Sunflower by Cao Wenxuan. *Playing by the Book* (blog). http://www.playingbythebook.net/2015/04/27/an-interview-with-the-translator-of-bronze-and-sunflower-by-cao-wenxuan/

Translating China. http://translatingchina.info/

Üngör, Çağdaş. 2009. Reaching the distant comrade: Chinese communist propaganda abroad (1949–1976). Ph.D., State University of New York at Binghamton.

Variations of a Folk Tale. 1977. *Interracial Books for Children Bulletin* 8(3): 5.

Wang, Helen. 2015a. Sherlock Holmes in Chinese. *Paper Republic*. Feb 6. https://paper-republic.org/pers/helen-wang/sherlock-holmes-in-chinese/

———. 2015b. Translating children's books. *Books from Taiwan*. March 26. https://booksfromtaiwan.tw/latest_info.php?id=24

———. 2019. International Research on Chinese Children's Literature (IRSCL 2019). *Chinese books for young readers*. Sept 12. https://chinesebooksforyoungreaders.wordpress.com/2019/09/12/86-international-research-on-chinese-childrens-literature-irscl-2019/

White Ravens. Munich: International Youth Library.

1993–2007. http://www.childrenslibrary.org/servlet/WhiteRavens

1996–2013. http://www.iylmuc.org/files/HM_4/whiteravens.htm

2012–2019. http://whiteravens.ijb.de/list

Wilhelm, Richard. 2011. In *Deutsche Biographische Enzyklopädie*. Berlin: K. G. Saur. https://db.degruyter.com/view/DBE/_10-1083

Wu, Yan, and Kerry Mallan. 2006. Children's science fiction in China. *Bookbird* 44 (3): 40–47.

You, Peiyun 游珮芸. 2007. 日治時期台灣的兒童文化 (Children's literature in Taiwan under Japanese rule). 臺北市: 玉山社.

Yue, Hua 粤华. 2003. 台湾言情小说出版商进入大陆 (Taiwanese publishers of romance novels enter mainland China). 出版参考 (31): 23.

Zaniello, Thomas A. 1974. Heroic quintuplets: A look at some Chinese children's literature. *Children's Literature* 3 (1): 36–42. https://doi.org/10.1353/chl.0.0441.

Translating Films About Ethnic Minorities in China into English

24

Haina Jin

Contents

Introduction	604
Multilingual China and the Implications for Films About Ethnic Minorities in China	606
Translating Multilingual Films into English	609
Jail	611
Rehearsal	612
On Stage	613
Interview	614
Conclusion	614
References	615

Abstract

China has produced a significant number of films on its ethnic minorities, which provide great opportunities for international audiences to get to know the multilingual and multicultural China. Before the 1980s, films on Chinese ethnic minorities were made in Mandarin, which masked the multilingual reality. With increasing awareness of multilingualism and openness in filmmaking policies, filmmakers began to experiment in making films with or in ethnic minority languages. Multilingualism poses great challenges for English translation. This chapter will use *Hajab's Gift* (Tianzhi enci, dir. Chen Liming, 2012), a Mongolian and Mandarin film with Mandarin and English subtitles to investigate how bilingualism is reflected in the English translation. It will discuss strategies used by the translator and analyze the effects of such strategies in translating bilingualism scenes.

H. Jin (✉)
School of Foreign Languages and Cultures, Communication University of China, Beijing, China
e-mail: jinhaina@cuc.edu.cn

Keywords

Chinese films · Subtitling · Audio-visual translation · Multilingualism · Bilingualism · Ethnic minorities

Introduction

China has produced a significant number of films on its ethnic minorities, and many of them have been screened in international festivals or have been distributed outside China. Films about ethnic minorities in China provide great opportunities for international communities to get a glimpse of China as a multilingual and multicultural country. It is worth noting that, over the years, scholars have used different terms to refer to films about ethnic minorities in China, such as "non-Han films" (Frangville 2012, 2016) and "minority nationality film" (Berry 2016; Lo 2015). Some propose a narrow definition of it. Wang (1997) proposes that ideally films about ethnic minorities are directed by ethnic minority directors on ethnic minority life to reflect their distinctive ethnic culture. Rao (2011) advocates a broader definition, which includes films that encourage interactions between Han and ethnic minorities in a transcultural approach. In this chapter, the sense in which the term "films about China's ethnic minorities" is used is in line with Rao's, which is used to emphasize that the topic of the films concerned is ethnic minorities in China.

This chapter will discuss the translation of feature films made in the PRC about ethnic minorities. These include both domestic productions and transnational co-productions. The directors of these films can be Han Chinese, non-Han Chinese, or foreign nationals. In this sense, the films falling into the definition range from *The Romantic History of Yao Mountain* (Yaoshan yanshi, dir. Yang Xiaozhong, 1933), a Chinese silent film, to *Wolf Totem* (Lang tuteng, dir. Jean-Jacques Annaud, 2015), a Sino-French co-production, and *Jinpa* (Zhuangsile yizhi yang, dir. Pema Tseden, 2018), a Tibetan-language film.

Since the 1990s, films about ethnic minorities in China have begun to draw more and more academic attention. Berry (1992) and Gladney (1995) discuss the differences between Tian Zhuangzhuang and earlier Chinese filmmakers and point out that one of the differences lies in Tian's use of ethnic minority languages in his films. Zhang (1997) addresses questions of nationhood and ethnicity in Chinese cinema and examines changes in their ideological construction, culture, and politics from the 1950s to the 1980s. Lo (2015) analyzes how Hollywood style was reflected in minority nationality films made in the PRC in the 1950s and 1960s. In 2016, a special issue of *Journal of Chinese Cinemas* was devoted to the Tibetan cinema of Pema Tseden, following the event of "Transgressing Tibet: International Symposium on Pema Tseden's Films, Fictions and Translations" held in Hong Kong in October 2014. Although the special issue covers interesting topics, including road movies, minor cinema, Buddhist film, palimpsests, and contested landscapes in analyzing Pema Tseden's Films (Lo and Yeung 2016), the translation of his films is unfortunately not discussed. Frangville (2017) contends that non-Han films have become a new focus of China's strategies for soft power. Films about ethnic minorities in China show their languages and cultures and pose great challenges for translators, especially when they are multilingual or bilingual. It is of great academic and practical importance to address this topic.

There are different modes of film translation: subtitling, dubbing, voiceover, and interpretation. Subtitling and dubbing are the most widely employed in China. Díaz Cintas and Remael (2007) define subtitling as a translation practice that consists of presenting a written text, generally on the lower part of the screen, that aims to translate the original dialogue, as well as any other information that might be conveyed in the source language on the soundtrack, and any discursive elements that appear on the screen. By way of contrast, dubbing, following Chaume's definition (2012: 1), consists of replacing the original audio track which contains a film's (or any audiovisual text) source language dialogue with another track in which translated dialogue is recorded in the target language. Both subtitling and dubbing have their advantages and disadvantages. Dubbing does not depend on literacy and requires less focus, so it can reach a larger audience, but it costs more, takes more time to produce, and is technically difficult, due to the challenge of maintaining lip-synchronization. Because it completely replaces the original audio track, dubbing also necessarily sacrifices a degree of authenticity. Subtitling is relatively inexpensive, can be done far more rapidly, and allows the original voices to be retained, which, among other advantages, can facilitate language learning. However, subtitling cannot reach an illiterate audience and may obscure the image on the screen. As for the translation of Chinese films into English in China, subtitling is the most used method, preferred by commercial film companies for two reasons. First, it is much more economical than dubbing. Second, it preserves the original dialogue (Jin and Gambier 2018). Dubbing is used in the state-sponsored audiovisual translation projects. These are well funded which can afford the more expensive method, and since they aim to cover as many target audiences as possible, dubbing is preferred. Such projects are initiated by the State Administration of Radio, Film and Television to promote Chinese films and television abroad. For example, the Sino-African Film and Television Cooperation Project was initiated in 2012. Under this project, 10 Chinese television dramas and 52 Chinese films are translated into seven different African languages each year. The aim is to give people in Africa a deeper understanding of Chinese culture, project a modern image of China, and develop an African market for Chinese film and television (Jin 2020). A number of films about ethnic minorities in China were selected for this project. *A Postman in Shangri-La* (Xiangbala xinshi, dir. Yuzhong, 2007), *A Tibetan Love Song* (Kangding qingge, dir. Jiang Ping, 2010), *Whiter than Snow* (Meili xueshan, dir. Xu Hong 2007), *My Shangri-La* (Wode xianggelila, dir. Zhang Liming, 2007), *My Mongolian Mother* (Eji, dir. Ning Cai, 2010), and *The Silent Mani Stones* (Jijingde Manishi, dir. Pema Tseden, 2006) were dubbed into English and French. *Looking for Nadam* (Xunzhao Nadamu 2009) was subtitled in English and French.

This chapter aims to fill a gap by examining the translation into English of films about ethnic minorities in China. It will first discuss the linguistic situation in China and its implications for films about ethnic minorities. It will trace the development of such films and illustrate how minority languages are represented in them. It will use *Hajab's Gift* as a case study to explore the translation into English of a bilingual film. It will identify the strategies adopted by the translator and discuss the effects of such strategies.

Multilingual China and the Implications for Films About Ethnic Minorities in China

The People's Republic of China (PRC) is a multiethnic and multilingual country. There are 56 officially recognized ethnic groups. The vast majority of the population belongs to the Han. According to the Seventh Population Census of the PRC, carried out in 2010, the 1.286 billion Han constitute 91.11% of China's total population. The remaining 8.89% of the population of mainland China, amounting to 125.47 million people, belongs to one of the 55 ethnic minority groups (National Bureau of Statistics of China [NBS] 2021). According to researchers from the Chinese Academy of Social Sciences, 129 languages are spoken in total in the PRC (Sun et al. 2007). Mandarin is the official language of the PRC, but certain ethnic minority languages also have official status within certain autonomous regions, such as Tibetan in Tibet and Mongolian in Inner Mongolia. According to the Ministry of Education of the PRC (2021), 80.72% of the Chinese population can speak Mandarin, which means that 19.28% of the population, around 271 million people, cannot. A great number of people from ethnic minorities fall into the latter category, especially those who live in rural areas. Some of the ethnic minorities speak both Mandarin and their native language, while others speak only their native language. This diverse and complex language reality in China was first masked and then represented in films about ethnic minorities. From the 1930s to 1980s, only Mandarin was used in dialogue for public screening, and all the characters spoke fluent Mandarin, regardless of their ethnicity. It was not until the 1980s that the native languages of ethnic minorities were used in film dialogue.

Chinese filmmakers began to make films about ethnic minorities in the silent film era. In 1933, *The Romantic History of Yao Mountain* (Yaoshan yanshi, dir. Yang Xiaozhong) was made by the Yilian Film Company. It depicts Huang Yuhuan, a Han teacher who goes to Guanxi and falls in love with the daughter of the Yao leader. Huang helps the leader to effect social and economic development in the Yao region. In 1940, *Storm on the Border* (Saishang fengyun, dir. Ying Yunwei) showed how the Han and Inner Mongolians fought together against the Japanese invasion. *The Romantic History of Yao Mountain* is believed to have been lost, while *Storm on the Border* can still be watched today. The director, the screenwriter, and the cast of the two films were all Han Chinese. In *Storm on the Border*, both Mongolian and Han actors speak Mandarin and even Mongolian songs are sung in Mandarin.

After the establishment of the PRC in 1949, films about ethnic minorities became an important genre largely due to their political importance. In the first few years of the PRC, the Ministry of Culture was given responsibility for governing and promoting China's film industry. The newly established government considered film a powerful and effective medium for mass communication. Zhou Yang (1951), then Deputy Minister of Culture, argued that "film is the most powerful art form, and should therefore be a priority for our efforts with regard to culture and the arts." Clark (1987) observes that, after the PRC was established, film was a vital tool in the remolding of new socialist citizens. Domestically, film was used as a tool for publicity and to educate the people about the revolution, the newly founded nation, its policies, and guidelines and to encourage them to participate in the construction of socialism. Internationally, film was used as a tool to project the image of the newly established China. Lo (2015) points out that films about ethnic minorities became an important genre in China after the establishment of the PRC and particularly served as "a means of territorial inclusion," mapping new boundaries, new landscapes, and showing a newly imagined Chinese people under the socialist regime. Therefore, a certain percentage of film production was devoted to this genre under the planned economy.

The first film about ethnic minorities made in the PRC was *The Victory of the People in Inner Mongolia* (Neimengu renmin de shengli, dir. Yu Xuewei, 1950). It shows how the Han Chinese and the Mongolians united to defeat the common enemy under the PRC's national policy. The director and the screenwriter were Han Chinese, while the actors were Han and Mongolian Chinese. The film was made in Mandarin for nationwide screening. Later, it was dubbed into Mongolian by the Inner Mongolian Film Studio and Changchun Film Studio for audiences in Inner Mongolia (Jin 2020). In June 1951, the film was exhibited at the sixth Karlovy Vary International Film Festival. In October 1951, a Chinese film week was held in the 16 capitals and other major cities of the Soviet Union to celebrate the second anniversary of the establishment of the PRC. A dozen Chinese films were shown, including *The Victory of the People in Inner Mongolia*.

From the establishment of the PRC in 1949 to the beginning of the Cultural Revolution in 1966, there was a surge of films about ethnic minorities. In total, 45 such films were made, showing 18 ethnic minority groups, including Mongolian, Tibetan, Uyghur, and Miao. Apart from those mentioned above, notable ones included musicals such as *Five Golden Flowers* (Wuduo jinhua, dir. Wang Jiayi, 1959), *Third Sister Liu* (Liu Sanjie, dir. Su li, 1960), and *Ashima* (dir. Liu Qiong, 1964), as well as thrillers such as *Visitors on the Icy Mountain* (Bingshan shang di laike, dir. Zhao Xinshui 1963). *Five Golden Flowers* was probably the most shown overseas. Xia Yan, a renowned writer and the then Deputy Minister of Culture, directly participated in the screenwriting. He asked the production team to make a film that could be distributed abroad to present a new image of China and increase the influence of Chinese socialism. Released in 1959, the film was shown in 46 countries and regions (Rao 2011). Films were used to instill revolutionary ardor for the building of a new nation and educate and mobilize the masses for the successful completion of the political task of national unification under socialist auspices. The directors in this period were all Han Chinese, though the screenwriters and actors sometimes included ethnic minorities. During shooting, actors from ethnic minorities spoke their native languages in some films. However, the films were universally dubbed into Mandarin for initial release. In a small number of films, songs sung in the native languages of the minorities were used, such as *Victory of the People in Inner Mongolia* and *Dawn of Menghe* (Menghe de liming, dir. Lu Ren, Zhou Danxi, 1955). There are two possible reasons for the practice of dubbing in Mandarin. First, in the early years, most of the population were illiterate. Second, the use of Mandarin by different ethnic groups was a symbol of national unity and full integration.

In the 1980s, filmmakers were more conscious of the languages they used in films. *On the Hunting Ground* (Liechang zhasan, dir. Tian Zhuangzhuang, 1984) witnessed a different approach in dealing with the question of language. In this film, Tian Zhuangzhuang did not replace Mongolian dialogue with Mandarin. Instead, he kept the Mongolian dialogue while adding a monotone male voiceover in Mandarin, with a time-lag of several seconds. His method revealed a long-masked reality, the multilingual situation in China. Tian Zhaungzuang insisted on using Tibetan actors and the Tibetan language when shooting his next film, *The Horse Thief*. On completion, he added Chinese subtitles produced by a Tibetan group. However, his method did not pass its review by the China Film Bureau, and the film was sent to the Shanghai Film Translation Studio to be dubbed into Mandarin. Both Mandarin and Tibetan copies were kept by the China Film Company. Only the Mandarin version was screened in Mainland China. When a French distributor wanted to purchase the film, and learned that there was a Tibetan version, he decided to purchase that version. Thus, for French release, the film was in Tibetan, while in Mandarin for release in other markets. In 1988, the film was broadcast on Channel Four in the UK and introduced by Tony Raynes as the film for the next generation. It was widely released in the USA in the early 1990s. The film director Martin Scorsese listed the film as one of his favorites from the 1990s (Ebert 2000). In 2018, a Blue-ray restoration was released with both Tibetan and Mandarin dialogue and two sets of English subtitles, one translated from the Tibetan subtitles and the other translated from the Mandarin. Similarly, In *Sacrifice of Youth* (Qingchun ji, dir. Zhang Nuanxing, 1985), when the Dai people speak to each other, Dai language is used. When the Dai people speak to the Han people, the language is Mandarin. The experiments language-use carried out by Tian Zhuangzhuang and Zhang Nuanxing inspired more directors to make films with or in the languages of ethnic minorities.

In 2003 and 2004, the State Administration of Radio, Film Television issued new policies of filmmaking. Under these policies, private companies are allowed to make films, which resulted in an increase in the production of films about ethnic minorities. Rao (2011) estimated that there were over 350 films about ethnic minorities produced from 1949 to 2010, and nearly 100 were produced after 2000. Under the new policies, dialogue in minority languages is no longer required to be dubbed into Mandarin, though Chinese subtitles are still required for public release.

A number of Han and non-Han directors still use Mandarin as the main language in their films about ethnic minorities. For example, *Mai Mai Ti's 2008* (Maimaiti de liangqianlingba, dir. Xierzati Yahrfu, 2008) tells the story of a group of Uyghur boys in a village in Hotan, Xinjiang, who play football and dream of attending the Olympic opening ceremony in Beijing. The director and the most of cast are Uyghur, as are most of the characters. However, the dialogue is in

Mandarin for public screening. There are practical reasons for using Mandarin for the dialogue. Since most of the Chinese audience speak Mandarin, a film using Mandarin is more accessible for most of the audience and may potentially bring more box office revenue.

Meanwhile, an increasing number of filmmakers from ethnic minorities began to make films in their native languages. Among them, Pema Tseden's achievements were the most remarkable. Pema Tseden is a Tibetan director and screenwriter. His films, including *The Silent Mani Stones*, *Soul Searching* (Xunzhao Zhimei Gengdeng 2008), *Old Dog* (Lao gou 2011), *Tharlo* (Taluo 2015), *Jinpa* (Zhuansile yizhi yang 2018), and *Balloon* (Qiqiu 2019), were about the Tibetan people's experience in modern society. The main language in his films is his mother tongue: Tibetan. His films are often screened with dialogue mostly in Tibetan accompanied by both Mandarin and English subtitles. However, even in his films, Mandarin is not absent in the dialogue. The occasional use of Mandarin shows the encounter of two languages and two cultures and the life of the Tibetan people in modern society.

Apart from directors from ethnic minorities, Han Chinese directors also made films using minority languages, such as the Miao Language in *La La's Gun* (Gun Lala de qiang, dir. Ning Jingwu, 2008) and the Lisu language in *Deep in the Clouds* (Biluo xushan, dir. Liu Jie, 2010). For most of the Chinese audience, watching a film whose dialogue is in a minority language rather than Mandarin would be like watching a foreign film. Often, they must rely on Mandarin subtitles to understand the film. At the same time, they are more exposed to the language and culture of the ethnic minority and become more aware of the cultural and linguistic diversity. Directors of Chinese films now have a high awareness of international audiences and international markets. Most Chinese films are produced with English and Chinese bilingual subtitles and screened that way, even in China (Jin and Gambier 2018).

With the increasing use of native languages in the films about ethnic minorities in China, it is worth noting that the English subtitles are often translated from the Chinese subtitles rather than the native languages due to the lack of translators who are able to work directly from those languages. According to Wang Lei, a producer of Pema Tseden's films, the English subtitles of Pema Tseden's films are translated from the Chinese subtitles rather than from the Tibetan dialogue (Personal Interview, 2020, March 13).

The following section uses *Hajab's Gift* to illustrate the complex issues involved in translating a film about an ethnic minority. The film contains mixed Mongolian-Mandarin dialogue. It is directed by Chen Liming, a Han-educated female Mongolian director, and the English subtitles are translated by Linda Jaivin from the Chinese subtitles.

Translating Multilingual Films into English

Hajab's Gift is a biographical film about Hajab, a renowned singer, known as the King of Urtiin duu in Inner Mongolia. He was born into a poor nomadic family. Due to his singing talents, he was adopted by a Mongolian baron. In order to learn Urtiin duu, a traditional Mongolian lyrical chant, he left the aristocratic family for the steppes. By the time of the establishment of the PRC, he was well known as a singer in Inner Mongolia. He performed for Mao Zedong four times and was honored with the title of People's Vocalist. In the Cultural Revolution, he was prosecuted and sent to jail for ten years. After the Cultural Revolution, he regained his freedom and continued to sing. The Chinese name of the film is Tianzhi enci, meaning "gift from Heaven." Before Hajab was born, the local Lama foretold that he would be an extraordinary person and gave him the name Hajab. The Chinese title of the film plays on the meaning of his name, "Tianzhi enchi," while the English title, *Hajab's Gift*, refers to his talent.

Chen Liming, the director, a Chinese-educated Mongolian, wrote the script in Mandarin. Her partner, the editor and cinematograph director Bai Long, translated it into Mongolian. Chen Liming's first language was Mandarin, and Mongolian her second. Bai Long spoke Mongolian as his first language. The cast was a mixture of monolingual nomads and bilingual professionals from the ethnic minority class of the performance major at the Shanghai Theater Academy. The actor who played Hajab in old age was Dashdondong, a renowned singer from Mongolia. The English subtitles of the film are by Linda Jaivin, an Australian writer, translator, and sinologist, who has been translating Chinese films since the 1990s. She has translated many award-winning Chinese films including *Farewell, My Concubine*, *Hero*, and *The Grandmaster*. She does not speak Mongolian and had to work from the Chinese subtitles. She often checked with the director as well as the Mongolian translator, to make sure which line of dialogue was being spoken. When she felt that a Chinese line did not really reflect what was being said in Mongolian, she would discuss it with them, seeking a more suitable Chinese version. The filmmakers did not speak English, so the three must had to rely on Mandarin to communicate (Personal interview, Jaivin, September 27, 2013). Though not speaking Mongolian, the translator shows a high awareness of the Mongolian language and culture. The names of the characters are translated into English according to the English spelling of Mongolian names, instead of using pinyin. For example, the leading character's name is translated as Hajab, instead of Hazhabu, the pinyin form of his Chinese name.

The bilingual soundtrack was rendered into English and Chinese subtitles. The color and the font of the English subtitles remain the same throughout the film, and no indication is given of language-switching in the original soundtrack. While the English subtitles convey the meaning of both Mongolian and Mandarin, they sometimes obscure the code-switching, which is a fundamental part of the Han-Mongolian interaction. Code-switching is an essential multilingual procedure that illustrates "how linguistic choices are socioculturally symbolic" (Baldo 2009). Li (2018) thinks that as the characters move between languages, they call upon different languages to signal their cultural and in-group identities, to express their emotions and cultural attitudes, or to exert authority over one another. Such usage underlines the complex interplay between language, culture, and power in code-switching. Although film is an audiovisual medium, international audiences who have little knowledge of Mandarin or Mongolian may not notice the code-switching.

Just as Tortoriello (2012) has observed, an audience who is not fluent in the languages spoken in the film may have difficulty in detecting any linguistic variations from the soundtrack. Mongolian is the dominant language in the film, and Mandarin is used when a Mongolian is communicating with a Han Chinese who cannot speak Mongolian. There are several bilingual characters in the film. Hajab himself speaks Mongolian and Mandarin. When communicating in Mandarin, Hajab speaks with a Mongolian accent, which indicates that Mandarin is not his native language. A troupe leader also speaks both languages. However, when he speaks Mongolian, he shows unfamiliarity with Mongolian culture, which indicates that his Mongolian is an acquired language.

As Meylaerts and Serban (2014) point out, multilingualism makes translation and communication issues more visible, and it can highlight internal tension within cultures. In this film, when Mongolian replaces Mandarin, tension often arises. However, in the English translation, this tension is not always maintained, as may be seen in the following four scenes.

Jail

The first scene where Mandarin is used occurs in a jail in Hohhot, the capital of Inner Mongolia, in 1968. When two guards mock and torture Hajab, they speak to him in Mandarin.

Character	Chinese subtitle	English subtitle
Guard 1 (Mandarin in Beijing dialect)	来 给学段驴叫呗	Teach us to bray like a donkey!
Guard 2 (Mandarin with Mongolian accent)	咳 听见了没	Hey, you listening?
Guard 2 (Mandarin with Mongolian accent)	你聋啦还是哑啦	Your deaf? Or just dumb?
Hajab (Mandarin with Mongolian accent)	我是人民的歌唱家，你不知道吧	Don't you know I am a People's Vocalist?
Hajab (Mandarin with Mongolian accent)	我唱的是长调	I sing "long song".
Hajab (Mandarin with Mongolian accent)	不是驴叫	It is not braying.
Guard 2 (Mandarin with Mongolian accent)	吆 长调	Huh? Long song?
Guard 2 (Mandarin with Mongolian accent)	长调是啥 我看跟驴叫一样	What's that? Sounds like donkey braying to me.
Guard 2 (Mandarin with Mongolian accent)	来 给好好学一个	C'mon, teach us one.

In this scene, all the characters speak in what can be loosely described as Mandarin, though their forms of Mandarin are different. Guard 1 speaks a Beijing dialect, which indicates that he might be a Red Guard from Beijing, given the time, 1968. Guard 2 speaks with a Mongolian accent, suggesting that he is from Inner

Mongolia and has become a Red Guard. The translator uses a marker to indicate that they are speaking in Mandarin. The marker is the translation of changdiao. Changdiao (长调, Mongolian: Urtiin duu, English: long song) is one of the central elements of the traditional music of Mongolia. When the Mongolians speak to each other about changdiao, the translator uses Urtiin duu to translate it. When the Mongolian and the Han Chinese talk about it, it is rendered as "long song" to show that the Mongolian wishes to explain it to the Han Chinese, who is not familiar with it. In this first scene, Hajab uses Long Song to explain to the Red Guards and defend the music he loves.

Rehearsal

At rehearsal of the singing troupe, while Hajab is walking in, a group of five male singers in different clothes are singing a song in Mandarin to celebrate the revolution. Their dress represents different ethnic groups, including Han and Mongolian, as well as different trades, including soldiers, farmers, herdsmen, and intellectuals. The fact that they all sing in Mandarin indicates that people of different ethnic groups and walks of life all embrace the revolution and are fully integrated into the PRC. However, this is not shown in the English subtitles. Hajab then has a conversation with the new troupe leader in Mongolian.

Character	Chinese subtitle	English subtitle
Troupe leader (Mongolian)	什么节目?你是干什么的?	Performing? Who are you?
Hajab (Mongolian)	我是唱长调的	I sing Urtiin duu.
Troupe leader (Mongolian)	长调?	Urtiin duu?
Troupe leader (Mongolian)	什么长调?	What is that?
Hajab (Mongolian)	我是哈札布, 是唱长调的演员。	I am Hajab, a singer of Urtiin duu, long song.

When the new troupe leader, who might be a Han Chinese who has acquired Mongolian, asks Hajab what he does, Hajab replies that he sings Urtiin duu. The new leader apparently does not understand and asks what Urtiin duu is. Hajab answers, "I am Hajab, a singer of Urtiin duu, long song." In the Chinese subtitle, there is only one word: changdiao. In the English subtitle, the translator uses both "Urtiin duu" and "long song" to show that Hajab is trying to help the new troupe leader understand, which indicates that the new leader is probably not a native Mongolian. He cannot find Hajab's name on the list of performers and suggests that he go to ask the leader of the Workers' Propaganda Team:

Character	Chinese subtitle	English subtitle
Troupe leader (Mongolian)	你去工宣队长办公室问一下吧	Go ask at the office.

The literal translation of the Chinese subtitle of the troupe leader is "Go ask at the office of the leader of the Workers' Propaganda Team." The Workers' Propaganda

Team is a shorthand term for the Workers' Mao Zedong Thought Propaganda Team, which was introduced in 1968 to implement the leadership of the working class in educational and cultural institutions. The Chinese subtitle has a strong political coloration, which a Chinese audience will associate with the Cultural Revolution. The English "Go ask at the office" lacks any political suggestion. This translation may be due to constraints of time and space in subtitling. As Díaz Cintas and Remael (2021) point out, for TV, cinema, and DVD, standard practice allows a maximum of 37–39 characters per line. The literal translation of the Chinese line is 70 characters, which is too long. A suggested translation would be "Ask the leader of the Propaganda Team." This conforms to accepted subtitling practice and retains the political coloration.

A subsequent conversation between Hajab and the leader of the Propaganda Team is in Mandarin. The leader speaks fluent Mandarin, with much political jargon of the Cultural Revolution, while Hajab speaks in accented Mandarin and sometimes struggles to understand. The English translation reduces this dramatic effect.

Character	Chinese subtitle	English subtitle
The Workers' Propaganda Team Leader	无产阶级专政是不会放过你们这些牛鬼蛇神的	The People's Dictatorship won't let you off.

The phrase "*niugui sheshen* (牛鬼蛇神)," literally "cow-ghosts and snake-gods," is not translated in the English subtitle. In Chinese mythology, *niugui sheshen* are evil spirits who take human form to perform evil tricks, but when unmasked revert to their ghostly forms. The term was popularized by Mao Zedong in 1957 and widely used during the Anti-Rightist campaign to describe intellectuals who pretended to support the Party, only to be unmasked when they launched a barrage of criticism when allowed to speak out during the Hundred Flowers campaign (Ji 2004: 195). In the Chinese subtitle, the worker's team leader refers to Hajab as one of the "cow-ghosts and snake-gods," a serious political accusation, which could lead to severe punishment. The English sentence has only "you," which greatly reduces the severity of the accusation. Again, this reduction might o be due to the constraints of subtitling.

On Stage

The third scene is in a gala performance. The Mongolian announcer first speaks in Mongolian and then in Mandarin with a Mongolian accent to introduce Hajab, who will perform a song.

Character	Chinese subtitle	English subtitle
Announcer (in Mongolian)	现在有请人民歌唱家哈札布上场	We now invite the People's vocalist Hajab
Announcer (in Mongolian)	为大家演唱蒙古长调走马	To perform the Urtiin duu "Mongolian Pony"

(continued)

Character	Chinese subtitle	English subtitle
Announcer (in Mandarin)	现在有请人民歌唱家哈札布上场	We now invite the People's vocalist Hajab
Announcer (in Mandarin)	为大家演唱蒙古长调走马	To perform the Urtiin duu "Mongolian Pony"

Although the Chinese and English subtitles appear to be the same and convey the same meaning, the effect is different. The Chinese audience can distinguish the Mandarin soundtrack from the Mongolian. However, the English-speaking audience might not notice the change of language. When they see the English subtitle repeat, they may wonder about the reason. An alternative translation could be: "We now invite the People's vocalist Hajab to perform the long song 'Mongolian Pony,'" instead of a mere repetition. The suggested translation would better signal the code-switching and indicate that the intended audience is Han Chinese.

Interview

The fourth scene occurs when a female journalist from Beijing visits Hajab in the Xinlin Gol steppe in Inner Mongolia in 2000 and conducts an in-depth interview with him. She speaks in Mandarin, and Hajab answers in Mandarin and Mongolian alternately. When he recalls Beijing, he sings a song in Mandarin. The seemingly smooth conversation hides the fact that this kind of conversation will probably not happen in real life, because the journalist would need a translator to understand Hajab's Mongolian utterance. The Chinese audience might be aware of this illusion. However, the journalist's Mandarin speech and Hajab's Mongolian and Mandarin are translated into English with no marker, which obscures the bilingual situation and facilitates the smoothness of conversation.

To summarize, the English translator of *Hajab's Gift* shows a high awareness of Mongolian language and culture. However, the English subtitles do not indicate any linguistic features, and the color and the font in the subtitles are the same throughout. There is no indication of the language used in the original. The translator does manage to indicate the changes of languages such as by using different words for changdiao. On some occasions, the English subtitles fail to show the bilingual nature of the film and do not reveal the tension behind the code-switching.

Conclusion

Translating films dealing with China's ethnic minorities into English has offered English-speaking viewers a glimpse of China's diversity of cultures. With the increasing use of native languages, multilingual or bilingual films pose great challenges to the translators. Heiss (2004) and Zabalbeascoa and Voellmer (2014) emphasize the importance of not depriving the target audience of the linguistic diversity of the source text. Bartoll (2006) suggests the use of some features taken from subtitling for the hard of hearing, such as different colors or paratextual information in brackets. These, however, are seldom used, because they are highly distracting and may confuse the audience. The analysis of the translation of *Hajab's Gift* shows that code-switching in the source film may be maintained by strategies such as using different terms in the target text to translate one term in the source text. Meylaerts (2006:5) points out that multilingualism should be regarded more as an opportunity than a problem for translation studies, because it stretches the limits of translation by showing that it cannot be simply the "full transposition of one (monolingual) source code into another (monolingual) target code for the benefit of a monolingual target public". The English translation of films about China's ethnic minorities also stands as an opportunity to broaden the current understanding of film translation, which calls for translators to be more aware of the languages and cultures of ethnic minorities and develop effective solutions to preserve the multilingual features of those films.

References

Baldo, Michela. 2009. Subtitling multilingual films. The case of lives of the saints, an Italian Canadian TV screenplay. In *Translating regionalised voices in audiovisuals*, ed. Federico Federici, 117–135. Rome: Aracne.
Bartoll, Eduard. 2006. Subtitling multilingual films. In *Proceedings of the Marie Curie Euroconferences MuTra: Audiovisual Translation Scenarios*, ed. Mary Carroll, Heidrun Gerzymisch-Arbogast and Sandra Nauert. Copenhagen: MuTra. https://www.euroconferences.info/proceedings/2006_Proceedings/2006_Bartoll_Eduard.pdf
Berry, Chris. 1992. "Race (minzu)": Chinese film and the politics of nationalism. *Cinema Journal* 2: 45–58.
Berry, Chris. 2016. Pema Tseden and the Tibetan Road Movie: Space and Identity beyond the "Minority Nationality Film". *Journal of Chinese Cinemas* 10(2): 89–105.
Chaume, Frederic. 2012. *Audiovisual translation: Dubbing*. London, New York: Routledge.
Clark, Paul. 1987. *Chinese cinema: Culture and politics since 1949*. Cambridge: Cambridge University Press.
Díaz Cintas, Jorge, and Aline Remael. 2007. *Audiovisual translation: Subtitling*. Manchester: Routledge.
———. 2021. *Subtitling: Concepts and practices*. London, New York: Routledge.
Ebert, Roger. 2000. Ebert & Scorsese: Best films of the 1990s. https://www.rogerebert.com/rogers-journal/ebert-and-scorsese-best-films-of-the-1990s
Frangville, Vanessa. 2012. Translated by Elizabeth Guil. The Non-Han in socialist cinema and contemporary films in the PRC. *China Perspectives*. https://doi.org/10.4000/chinaperspectives.5877.
Frangville, V. 2016. Pema Tseden's The Search: The Making of a Minor Cinema. Journal of Chinese Cinema 10(2): 106–119.
Frangville, Vanessa. 2017. Going to Hollywood with Non-Han Films: A Potential Soft Power Synergy. In Screening China's soft power. eds. Paola Voci and Luo Hui. 73–90. London and New York: Routledge.
Gladney, Dru C. 1995. Tian Zhuangzhuang, the fifth generation, and minorities film in China. *Public Culture* 1: 161–175.
Heiss, Christine. 2004. Dubbing multilingual films: A new challenge? *Meta* 49 (1): 208–220.
Ji, Fengyuan. 2003. *Linguistic engineering: Language and politics in Mao's China*. Honolulu: University of Hawaii Press.

Ji, Fengyuan. 2004. Language and Violence During the Chinese Cultural Revolution. *American Journal of Chinese Studies* 11(2): 93–117.

Jin, Haina. 2020. Film translation into ethnic minority languages in China: A historical perspective. *Perspectives* 4: 575–587.

Jin, Haina, and Yves Gambier. 2018. Audiovisual translation in China: A dialogue between Yves Gambier and Jin Haina. *Journal of Audiovisual Translation* 1: 26–39.

Li, Dang. 2018. Chinese-dialect film and its translation: A case study of The World (2004). *Journal of Chinese Cinemas* 12(3): 267–284

Lo, Kwai-Cheung. 2015. (Un)Folding Hollywood and new Chinese subjectivity through PRC's minority nationality films in the 1950s and 1960s. In. *American and Chinese-Language Cinemas: Examining cultural flows*. Eds. Lisa Funnell and Man- Fung Yip. 71–85. London, New York: Routledge.

Lo, Kwai-Cheung and Jessica Wai Yee Yeung. Eds. 2016. Special issue on the Tibetan cinema of Pema Tseden. *Journal of Chinese Cinemas* 10(2).

Meylaerts, Reine. 2006. Heterolingualism in/and translation: How legitimate are the other and his/her language? An introduction. *Targets* 1: 1–15.

Meylaerts, Reine, and Adriana Şerban. 2014. Introduction: Multilingualism at the cinema and on stage: A translation perspective. *Linguistica Antverpiensia* 13: 1–13.

———. 2021. The Seventh Communiqué of the National Bureau of Statistics of PRC on major figures of the 2020 population census. Retrieved from http://www.stats.gov.cn/tjsj/tjgb/rkpcgb/qgrkpcgb/202106/t20210628_1818821.html

Rao, Shuguang. 2011. Zhanguo Shaoshu minzu dianying shi, [A history of films about ethnic minorities in China], zhongguo dianying chubanshe.

Sun, Hongkai, Zengyi Hu, and Xing, Huang. 2007. *Zhongguo de yuyan* [Languages of China]. Beijing: The Commercial Press.

The Ministry of Education, PRC. 2021. 80.72% of the Chinese population can speak Mandarin. Retrieved from http://www.news.cn/2021-08/23/c_1127787835.htm

Tortoriello, Adriana. 2012. Lost in subtitling? The case of geographically connotated language. In *Audiovisual translation across Europe: An ever-changing landscape*. eds. Silvia Bruti and Elena Di Giovanni. 97–112. Oxford: Peter Lang.

Wang, Zhimin. 1997. Shaoshu minzu dianying de ganian jieding. (On the definition of the concept of ethnic minority films). In *Luo Zhongguo Shaoshuminzu dianying*, edited by China writers' association. Beijing: China Film Press.

Zabalbeascoa, Patrick, and Elena Voellmer. 2014. Accounting for multilingual films in translation studies: Intratextual translation in dubbing. In *Media and translation: An interdisciplinary approach*, ed. Dror Abend-David, 25–52. New York: Bloomsbury.

Zhang, Yingjin. 1997. From "minority film" to "minority discourse": Questions of nationhood and ethnicity in Chinese cinema. *Cinema Journal* 3: 73–90.

25

Subtitling in Chinese: Reflections on Emergent Subtitling Cultures in Contemporary China

Dingkun Wang

Contents

Introduction	618
A Very Short Historical Overview	619
Subtitling in Practice	621
Multimodality	622
Constraints	624
The Impact of Technology	625
Subtitling in Theory: Recontextualization and Intercultural Mediation	626
Future Research	630
Concluding Remarks	631
References	632

Abstract

This chapter surveys research and practices regarding the subtitling of television programs and films, primarily in the context of Mainland China. It defines the meaning and function of subtitling and provides a brief overview of the development of subtitling in the history of audiovisual translation (AVT). It also reflects on how technological progress constantly poses new challenges for those who perform subtitling work – for instance, they must comply with the spatial and temporal constraints of the audiovisual medium while striving to render the information that emanates from the multimodal convergence of signs in the source text in an effective and efficient way. It examines how subtitlers recontextualize source materials for cross-cultural consumption by walking the line between covert translation (for comprehensibility) and overt translation (for authenticity). Finally, this chapter discusses directions for future research, includ-

D. Wang (✉)
University of Hong Kong, HKSAR, Hong Kong, China
e-mail: wdingkun@hku.hk

ing the potential for studies on the translation of music and *danmaku*-driven participatory subtitling work to explore the commonalities among the modes, types, and strategies of translation of audiovisual programming.

Keywords

Chinese subtitling · Constraints · Multimodality · Recontextualization · Research

Introduction

Audiovisual translation (AVT) refers to any form of intercultural and/or interlingual mediation that involves the convergence and interaction between language and audiovisual footage (e.g., film, television programs, and video games). To date, a plethora of translational modes have been developed to both cope with the accelerating global dissemination of audiovisual products and cater to ever more diverse audiences in and across different linguacultural contexts. Captioned audiovisual text, often known as subtitles, involves "the addition of text onto or next to the screen, in both interlingual and intralingual cases of transposition" (Chaume 2018a, 84). In subtitling, "snippets of written texts [are] superimposed on visual footage [to] convey a target language version of the source speech" (Pérez-González 2014, 15–16). Subtitling is less complicated than dubbing or voicing over, which is performed with great sophistication atop the piece's original soundtrack and involves more complex procedures for reproducing dialogue in the same or different languages. Although subtitling is "quicker and a lot cheaper than dubbing" (Díaz-Cintas and Anderman 2009, 4), it is no less subtle or evocative. Subtitles also translate and convey visual and vocal discursive elements – such as letters, graffiti, inscription, songs, and voices off camera – thus revealing the complex relationship between the verbal and the visual.

AVT studies have a relatively long history. According to Díaz-Cintas (2009), the earliest AVT research was conducted in the 1930s. This research was extended by various scholars in the late 1950s and early 1960s (Gottlieb 1994). In this early period, AVT research was not merely academic – it was also published in lay writing in newspapers, magazines, and tabloids. Scholarship was developed and shared privately through the social connections of professionals and academics. Diaz-Cintas (2009, 1–2) points out that this informal circulation makes it difficult to determine the authorship of some of these earlier materials, leaving the historiography of AVT research incomplete.

AVT-related research began to flourish in the 1990s, when the gap between the academic and social impacts became obvious. For instance, some established theories of translation did not apply well to AVT (Díaz-Cintas 2004). In addition, debates raged as to whether AVT was within the scope of translation or even worthy of academic attention. Scholars such as Whitman-Linsen responded to the latter by stressing the need to "dispel the disdain of literary intelligentsia, who seem to dismiss film translating and the degree of difficulty involved in it as not worthy of their attention" (Whitman-Linsen 1992, 17). While studies completed in the late 1990s and the early 2000s compared the advantages and disadvantages of different

AVT modes for linguistic and cultural mediation, later research tends to explore these modes' distinctive capacities to cater to different audiences and viewing contexts (see Pérez-González 2019). For instance, Chinese subtitling is among the most popular subjects of PhD dissertations in translation studies. Such dissertations have dealt with a wide range of topics related to the linguistic, cultural, and technological aspects of AVT-enabled media exchanges between China and the rest of the world, especially in the context of globalization (Li 2015; Liang 2017; Wang 2014; Yuan 2016; Zhang 2015a, etc.).

However, AVT-related research and pedagogy are still in their early stages in China. Although scholars research interesting and meaningful applications and reflections on AVT – for example, what is at stake in AVT-enabled interlingual mediation, how AVT-derived cultural and linguistic representations are unique, how audiences understand and respond to AVT products, and the extent to which AVT can promote intercultural literacy – today's scholars tend to insist on maintaining rigorous distinctions between existing modes of AVT:

> The challenges raised by the shift from speech to written text are typically encountered in subtitling but not in dubbing; the transfer of discourse markers, exclamations and interjections plays a crucial role in the perceived naturalness of the dubbed exchanges but not so much in subtitling; the nature and recurrence of the translation strategies activated in dubbing and subtitling vary greatly and whilst condensation and deletion can be considered pivotal to subtitling, they are not so pervasive in dubbing; the ineffability of linguistic variation in written subtitles can be easily overcome by dubbing actors' voice inflection; and the cohabitation of source and target languages in the subtitled version straightjackets potential translation solutions in a way that does not happen in dubbing. (Díaz-Cintas 2019, 183)

This chapter will focus on how subtitling in contemporary China acts as a form of intercultural mediation. Because Chinese subtitling has yet to be recognized as a proper intellectual pursuit in Chinese translation and language studies, this chapter will also survey the historical, practical, and theoretical aspects of subtitling in the Mainland Chinese context. It will look at how subtitlers render the multimodality and constraints of audiovisual texts while adapting to new challenges brought by technological progress. It discusses the translation of discourse markers, nonstandard usage and linguistic varieties, and sexual references within the theoretical framework advanced by Juliane House (2001, 2006), who makes distinctions between overt and covert translation in the overarching process of recontextualization. Finally, this chapter will detail why future researchers should focus on the translation of music, digital game localization, and *danmaku*-driven participatory subtitling to renegotiate differences and commonalities between various modes of AVT.

A Very Short Historical Overview

According to Markus Nornes (2007, 149), the practice of subtitling began with Herman Weinberg's translation of the German operetta *Zwei Herzen im Dreiviertel-Takt* (*Two Hearts in Waltz Time*) in 1930. However, Ivarsson (1995) noted that subtitles had already been applied to film translation in 1909 when M.N. Topp registered his patent on a "device for the rapid showing of titles for moving pictures other than those on the film strip" (Ivarsson 1995, 295). The first attested exhibition of a subtitled sound film was in Paris, where the film *The Jazz Singer* was shown with French subtitles in 1929 (Ivarsson 1995). The earliest recorded use of film translation in China was published in an 1897 article in *The Game Newspaper*, which mentioned the use of live interpretation during the screening of American films (Jin and Gambier 2018, 27). According to Jin (2018), Chinese subtitling practices began in 1923 when Peacock Film Company, the first Chinese film studio, hired Shuren Cheng to produce Chinese subtitles for American films. Ten years later, the Republican government made translation compulsory for all foreign films screened in China as a demonstration of respect for China and its people. Major international studios, such as Paramount, began to add Chinese subtitles to films for screenings in major Chinese cities such as Shanghai and Peking (Beijing) around 1936. These practices continued until 1949 when the People's Republic of China was founded. Dubbing prevailed as the dominant form of AVT for the next four decades (Du 2015).

Dubbing maintained its popularity among Chinese audiences through the mid-1990s but gradually lost its appeal after the turn of the century. Subtitling began to replace dubbing in the twenty-first century, due to the demand for rapid and global distribution of audiovisual products, the widescale retirement of senior dubbing artists, increased workloads, insufficient funding support, and a lack of talent and enthusiasm among new dubbing artists (Du et al. 2013). The proliferation of audiovisual content online has also made it easier for international audiences to view popular foreign telefiction (Gao 2012, 65).

At present, more than 50 foreign films are imported into Mainland China every year for cinematic exhibition, generating billions in box office revenue. The movie channel of the national television network China Central Television 6 has broadcast 400 foreign films annually every year since it was launched in 1996. In addition, more than 1000 films have been streamed online by Internet corporations such as Tencent Video (腾讯视频) and iQiyi (爱奇艺). The translation of films for cinematic exhibitions is carried out by the four state-owned film production and translation studios – Shanghai Film Translation and Production Studio (上海电影制片厂), August First Film Studio (八一电影制片厂), Changchun Translation and Production Studio (长春电影制片厂), and China Film Group Corporation (中国电影集团) – in collaboration with private translation agencies such as Chong Er Chuan Qi Creative Culture LLC (虫二创意文化有限公司).

Not all AVT work in China is done by professionals. Chinese amateurs and hobbyists come together in online translation communities and work together to share and translate their favorite audiovisual products. In doing so, they strive to secure their "increased control over the production and distribution of media content" (Turner 2010). Indeed, fansubbing – subtitling done by fans for fans (Díaz-Cintas and Muñoz Sánchez 2006) – is something of a popular-culture phenomenon. Dedicated consumers who have little or no previous training or experience in translation devote themselves to subtitling and circulating audiovisual content online. Fansubbing has, therefore, become a focus in some of the recent and relevant literature (see Luis Pérez-González 2014, 2017 – see also Kung 2016; Li 2015; and Wang 2017 for works that focus on the Chinese context in particular).

Fansubbing communities appeared in China as soon as construction began on the national broadband network in the late 1990s. Active Internet users and enthusiastic audiences met at the intersection between shadow economies, global media exchange, and voluntary online social activity. Nationwide Internet access enabled these individuals to expand those affinity networks centered on foreign pop culture and to nurture and cultivate such networks among the diverse digital audiences of the Chinese-speaking world (Hu 2018). Fansubbers "effectively [act] as self-appointed translation commissioners" (Pérez-González 2007, 71). They translate a great variety of genres, languages, and media. Some fansubbers provide very high quality fansubs and even become professionals. For instance, the fansubbing community *Ren Ren Ying Shi* (人人影视) created a successful business venture – the online subtitling platform *Yi Shi Jie* (译视界; http://www.1sj.tv/), which is devoted to automated AVT work. It has since become a commissioned translation agency of the Shanghai International Film Festival, and one of its senior members was invited to be a keynote speaker at an AVT symposium at Shanghai Jiao Tong University organized by the China Foreign Languages Publishing Administration, a state-owned publishing agency that translates and disseminates up-to-date information about the Chinese government. Indeed, the fansubbing community's expansion into the media and language industries has seen it develop good relations with the Chinese party-state (Chu 2012; Wang and Zhang 2017; Wu 2017).

Suffice it to say that subtitling is a vast frontier of translation practice in China. A great number of people are making significant contributions to transnational media exchange in a wide variety of languages. Many of them are volunteers and amateurs who work outside the officially sanctioned distribution channels, with or without official translation qualifications. Thus, the boundary between amateur and professional AVT practices is blurred, if not totally absent, in the Chinese context. Given this context, this chapter's reflections on AVT for Chinese audiences will draw on examples from both official and fansubbed translations.

Subtitling in Practice

Subtitling challenges the fundamental concepts of meaning and equivalence in translation because these concepts have largely been theorized with reference to printed text. Subtitling places audiovisual narratives within a restricted time and space by maneuvering multi-semiotic and meta-linguistic clues. Subtitling is not only about giving a literal translation of audiovisual content – it is also about "a moment in the growth of the original" (Derrida 1985, 188). That is, subtitling actualizes the source information and allows the original to survive in a target context; subtitles are legible only when they synthesize the spoken (and sometimes) written information text from the audiovisual original. In that sense, effective translation is woven into the multimodal web of visual and acoustic signs that make up any audiovisual product.

Multimodality

Kress (2010) advanced the multimodal approach as a means to analyze communication processes and performances. In his view, multimodality stands for the nature and outcome of the assemblage of distinct modes of communication (e.g., speech, writing, gestures, images, and sounds) to represent and communicate meaning. Under this framework, AVT can be considered a culturally sensitive, socially motivated, and medially conscious practice that integrates different verbal and nonverbal modes for the realization of specific meanings in transnational media exchanges.

Dirk Delabastita (1989) observed that meaning is generated in audiovisual narratives via the collaboration of verbal acoustic signs (e.g., dialogue), nonverbal acoustic signs (e.g., music and sound effects), verbal visual signs (written information on screen, including subtitles), and nonverbal visual signs (other visual elements). This means not only that verbal and nonverbal information combine in various ways but that the source information emerges from combinations of various signifying channels, each of which exerts a different degree of influence on the realization of meaning in the target language.

As Remael (2001, 17) observes, some parts of the original message are transmitted more efficiently and effectively by the audiovisual content itself, and other parts are transmitted better through text, but there is no clear division regarding what parts of a message are best suited to which mode. Some forms of meaning-making rely more on one specific channel of communication than on others. TV programs, for example, rely more upon verbal signs than on nonverbal signs. In the German comedy *Knaller Frauen* (2011–2015), Frauen asks her father if he has had any difficulty using the iPad she bought for him. Her father demonstrates his confidence in handling the new technology by using the iPad as a chopping board.

A more recent example is found in *Crazy Rich Asians* (2018). In one scene, the film's protagonist, Rachel Chu, joins the Goh family for lunch in their extravagantly decorated mansion. Mrs. Goh says to Rachel, "This is simple food lah" (https://www.youtube.com/watch?v=syg3WKOnuGY). Here, a cliché of hospitality is coated in Singlish (as signified by *lah*) (See ▶ Chap. 35, "The "Chineseness" of Singlish" of the Handbook) to differentiate the speaker and her family from their Chinese-American guest. The subtitle in this scene (see Table 1) bridges the language, prosody, and image on screen. This enhances the meaning of Mrs. Goh's remark, which was intended to flaunt her family's wealthy lifestyle in front of Rachel – a young academic who grew up in a middle-class immigrant household in New York – by referring to the veritable feast on the table as "点粗茶淡饭" (roughly, "a little unrefined tea and food") with a

Table 1 Crazy Rich Asians (2018)

Source information	Cinematic subtitle	Back translation into English
This is simple food lah.	就是一点粗茶淡饭啦	This is simply a little harsh tea and tasteless food!

condescending gesture and gaze. Thus, the combination of speech, subtitles, and visuals can not only assist the target audience's comprehension of the situation depicted on the screen but also aid their comprehension of subtle meanings that might be lost without the combination of all of these features.

Today, transnational filmmaking tends to purposefully select and organize visual and vocal resources and leave audiences to interpret the largely unmediated – and, in many cases, highly complex – ensemble of images, gestures, and sounds. Pérez-González (2014, 188–190) describes how a scene from the 2007 Dutch-Spanish film *Alle is Liefe* (*Love is All*) demonstrates a less conventional way of representing the meaning within the original audiovisual text. Early in the film, the film's protagonist, Jan, is hitchhiking in a truck delivering oranges from Valencia to Amsterdam. The truck encounters an accident with a minivan carrying a group of actors who are travelling to the same destination to perform in a St. Nicolas Day celebration. The actor who was supposed to play Sinterklaas (St. Nicolas, or Santa Claus) died of a heart attack in the accident, so the actors enlist Jan to play the role. Although these bits of spoken information are expressed and encoded in the audiovisual original, the culturally specific connections between them might escape the attention of Chinese viewers who are not acquainted with:

- The truck's colors (which are the colors of the Spanish national flag)
- The painted sign on the truck, which reads *Naranjas Valencianas* ("Oranges from Valencia")
- The reference to St. Nicolas, who travels from Spain every year to deliver presents to children
- Oranges, which are mentioned in Dutch stories about St. Nicolas
- Jan's role as a modern-day St. Nicolas

Ideally, AVT combines subtitled verbal exchanges and explanatory textual inserts which help viewers make sense of cultural and/or contextual information which might escape some target viewers. However, the fast pace of cinema makes this a difficult task: subtitles often appear the moment a character begins to speak and disappear when the character stops speaking. Consequently, AVT audiences must attend to both the subtitles and audiovisual material simultaneously (Nornes 2007, 162) without the opportunity to go at their own pace or re-read a passage, like they would with written text. Subtitlers are not obliged to consider these factors because their task is to adapt the subtitles to the audiovisual medium rather than manipulate the pace of the original to facilitate translation. They must also format their subtitles carefully, given the limited space on the screen. To date, subtitling work still conforms to the space-time constraints that have been prescribed and adopted by global audiovisual producers since the 1930s (O'Sullivan and Cornu 2019). Having looked at the multimodal nature of AVT, we will now turn to some of the space-time constraints that are specific to subtitling.

Constraints

Chinese subtitles are traditionally displayed in white text and centered at the bottom of the screen, in a "safe zone" (Karamitroglou 2000) at least 1/12th of the total screen height from the bottom and 1/12th of the total screen width on both sides to stimulate eye movement and make for easier reading. Subtitling software (e.g., WinCAPS, SWIFT, and EZTitles) enable subtitlers to customize the placement of subtitles and perform highly precise and complex operations. For example, such software displays the length of the source video in hours, minutes, seconds, and microseconds, enabling subtitlers to mark the entry and exit points of source-language speech, and therefore provide subtitled text, with great precision (Schwarz 2002). The program then automatically attaches the translation after the input is completed. This allows subtitlers to judge the suitability of the translation and troubleshoot when or if the translation breaches the original's own spatial constraints.

According to Kuo (2017), subtitles should last a minimum of 20 frames and should ideally last at least two seconds per full line of text, to a maximum of six or eight seconds. Most Chinese audiences can read four Chinese characters per second, so by Kuo's reckoning a two-second subtitle should include eight Chinese characters at the most. Multiple lines are used when a sentence that exceeds the maximum number of characters in one line cannot be further condensed or segmented, when two bits of dialogue are closely related, or when an on-screen conversation moves quickly. It is generally recommended that two-line subtitles consist of a shorter first line and then a full second line. Two-line subtitles should not be presented in quick succession, so that the audience can follow the action on screen without reading too much text. Furthermore, two-line subtitles are not often used when a speaker speaks slowly or when important information is revealed gradually through dialogue – in these cases, it is better to use single lines of subtitles to not disrupt the flow of the narrative. Chinese subtitles display incomplete sentences only when necessary – for example, when mirroring a speaker's pace and cadence. Subtitlers always seek the most comprehensive and simple expressions. While Chinese uses punctuation sparingly, no consensus on the use of punctuation marks has been reached within the present proto-industrial network of subtitlers.

Subtitles rarely, if ever, perform perfect, literal, and/or exact linguistic and cultural mediation due to the complex and conflicting parameters of space and time in the audiovisual original. According to Adami and Pinto (2019), these parameters include but are not limited to:

- The genre and the rhythm of the original
- The complexity of language

Table 2 The Big Bang Theory: Season 5, Episode 18 (2011; https://subhd.tv/do0/25845393)

Explanation by *Ren Ren Ying Shi*	Back translation into English
托马斯.斯特恩特.艾略特 著名美国诗人 诺贝尔文学奖得主	Thomas Stearns Eliot, a renowned American poet and Nobel Laureate for Literature
这句话改自他的作品《空心人》原文为: '世界就是这样告终的 不是砰的一声而是一声抽泣'	These sentences are adapted from his [Eliot's] *The Hollow Men*. The original is "This is the way the world ends – Not with a bang, but with a whimper."

- The density of information presented on screen
- The audience's familiarity with the subject matter
- The audience's level of education and language proficiency

Despite these limitations, fansubbers creatively and effectively mediate culturally loaded source materials. As mentioned above, fansubs often provide audiences with more assistance than conventional subtitles by adding extra cultural and contextual explanations. For example, in *The Big Bang Theory* (Episode 18, Season 5, 2011), the fansubs provided by *Ren Ren Ying Shi* displayed two explanatory notes on the top of the screen when Sheldon performed a verbal parody of T.S. Eliot's *The Hollow Men*. The first note appeared when T.S. Eliot was first mentioned, and the second appeared with the last line of the parody (displayed in Table 2).

Despite this good and creative work on the part of fansubbers, recent studies have expressed doubt about the extent to which these contextual and cultural explanations can be trusted, due to their subjectivity (e.g., Díaz-Cintas 2018; Dwyer 2017, 2018).

The Impact of Technology

While digital media technologies continue to enable grassroots participation in the translation and spread of audiovisual productions, the deepening engagement of amateur AVT workers (such as fansubbers) has fundamentally transformed the professional AVT work (Díaz-Cintas 2020). To meet the challenges posed by the increasingly amateur and technological nature of their work, AVT professionals must be flexible and adaptable and the industry at large should nurture amateur AVT workers' investment in their work rather than aim to automate their operations (Díaz-Cintas and Massidda 2019, 267–268).

Despite technological advances, human labor remains crucial to every stage of subtitling work (Georgakopoulou 2019), not least because new technologies often pose new challenges that cannot be met by artificial intelligence or automation. For instance, AVT work in 3D films requires the adaptation of new workflows, skills, and production practices. Subtitlers working on 3D films must insert subtitles in places that do not jeopardize or confuse the audience's comprehension of 3D images. Furthermore, they must keep in mind that viewing a 3D film with subtitles "requires a very fast and constant swapping of targets (subtitle/scene) while our eyes converge always to the same distance" (González-Zúñiga et al. 2013, 18). Improperly placed subtitles, then, can give audiences a headache, make them dizzy and fatigued, and cause nausea (Hoffman et al. 2008).

3D subtitlers' creative input, therefore, goes beyond bridging linguacultural barriers – they create a 3D audiovisual environment in which images and subtitles consolidate the original's meaning without ruining its immersive and spectacular visuals (Wan et al. 2013). However, subtitling practices continue to be criticized for altering or spoiling original works, even though such practices open original works up to wider viewership (O'Sullivan 2019). In this respect, future research should investigate subtitling practices for 3D films in more detail in order to examine how subtitles supplement original works and how dynamic technological developments affect subtitling practices and the accompanying work of cultural mediation.

Subtitling in Theory: Recontextualization and Intercultural Mediation

Bakhtin (1986) states that film dialogue is a secondary speech genre that maintains some of the primary functions of daily conversation based on different fictional contexts. Whereas everyday conversation can be formulaic and banal, film dialogue must be pertinent, dramatic, or intriguing to keep an audience's attention (Pavesi 2018). Film dialogue is "written to be spoken as if not written" – that is, it aims to maintain a degree of spontaneity (Gregory and Carroll 1978, 42). The audience should feel like they are eavesdropping on on-screen characters rather than feel that they are being addressed by these characters directly (Hatim and Mason 1997, 82). In this sense, subtitling work is an intervention – subtitlers address the target audience directly by mediating between the audience and the original. To do this effectively, subtitlers must analyze the multimodality of meaning conveyed through film and preserve the pragmatic style of original dialogue as much as possible (Pettit 2005). Although translation should honor a target language's literary styles and traditions (Creagh 1995; in Venuti 2000, 471–472), subtitles are concerned not with the original text (in this case dialogue) but with the original audiovisual narrative.

Subtitling is therefore "a form of cultural ventriloquism, and the focus must remain on the puppet, not the puppeteer" (Egoyan and Balfour 2004, 85). This view echoes Juliane House's notion of covert translation, which considers translation as being concerned with contextual and pragmatic equivalence rather than "formal, syntactic and lexical similarity" (House 2001, 247–248). Covert translation is a pragmatic application of recontextualization, by which the ideational and interpersonal function of an original can be realized in a particular "context of situation" of using the target language (House 2001, 248). However, in a subtitled version, if subtitlers incline to recreate an equivalent (or nonequivalent) speech event by manipulating the source text, their preference for target-oriented rhetorical styles, linguistic patterns, or cultural references will be in contrast to the overt presence of the source information.

Despite the need to assimilate the audience into the original narrative, it is crucial to maintain a distance between the world in the original and the world presented by the subtitles so that the difference between the two can be preserved. In the example presented in Table 3, the subtitler completely alters the original narrative through their translation. The subtitle refers to a popular Chinese Internet celebrity called Phoenix Sister who is popular for her controversial comments on social affairs, undesirable appearance, and exaggerated manners. The reference to Phoenix Sister in the subtitle corresponds to the speaker's mention of ugliness, and provides the Chinese audience with a small joke, but it changes and contrasts with the actual reference made by the speaker.

Diaz-Cintas and Remael (2007) emphasize that good subtitles should leave target audiences with the impression that the characters on screen are speaking the target language. However, AVT cannot completely erase the original contents of a film; the mere presence of subtitles brings the audience's attention to the various differences between the original and the mediated, subtitled work. This means that target-language audiences may perceive subtitled audiovisual media with a different focus than source-language audiences.

Subtitles deliver both familiar and unfamiliar information to audiences. Familiar information might include universal meanings and concepts that are inaccessible for the target audience due to language differences. Unfamiliar information includes meanings and concepts that are alien to the target-language audience because of irreconcilable cultural differences or gaps between the source and target languages. Hence, subtitling work entails recontextualizing original work so that it can be comprehended fully and easily by audiences across different cultural contexts. In short, although subtitles are designed to translate dialogue from "one sub-code (the seemingly unruly spoken language) to another (the more rigid written language)" (Gottlieb 1994, 106), subtitlers can give better translations when they thoroughly understand the context and cultural meanings behind the source information. For example, the target culture's norms determine the quantity and quality of discursive markers translated by subtitlers (Mattson 2006, 7). Chinese subtitlers often strive to convey cognitive and emotive values in the source-language dialogue in order to fill the functional gap between Chinese and English (Fong 2009). For example, in the example shown in Table 4, "Oh, yeah" was translated as "对啊" to reflect the function of discourse markers in information management.

AVT also affects audience comprehension through omission. Audiences strive to understand performances by linking ideas and units of speech (Hatim and Mason 1990, 194); therefore, subtitlers need to be cautious when omitting parts of the source meaning because their decision to omit may (a) change the meaning of the original and/or (b) not correspond to the audience's expectations of the audience.

Table 3 Madagascar 3 (2013)

Source information	Cinematic subtitles	Back translation into English
That is one ugly, mug-ugly lady!	这不是麻辣凤姐吗?	Isn't this the hot and spicy Phoenix Sister?

Table 4 Two and a Half Men: Season 5, Episode 1 (2007)

Source information	DVD subtitles	Back translation into English
Alan: Uh, Jake? Jake, where are you?	杰克?你在哪里?	Jake? Where are you?
You're still on the bus.	你还在校车上啊	You're still on the school bus.
Funny, I don't hear the other kids.	真奇怪, 我没听见其他小朋友的声音	Really strange, I can't hear the voices of the other kids.
Jake: Oh, yeah, it was a tough day. Everybody's tired.	对啊, 今天很辛苦大家都累了	That's right, today is a tough day, everybody is tired.

Table 5 Two and a Half Men: Season 5, Episode 3 (2007)

Source information	DVD subtitles	Back translation into English
Berta: Wow, I never thought I'd see Charlie Harper on his knees.	我没想过我会看见查理哈波卑躬屈膝	I didn't think I could see Charlie Harper bow and kneel.
You know, figuratively.	那只是一种比喻	That's merely a metaphor.

This is a large consideration for AVT workers, because "[i]t is difficult for the TL [target-language] audience to retrieve interpersonal meaning in its entirety. In some cases, they may even derive misleading impressions of characters' directness or indirectness" (Hatim and Mason 1997, 96). Subtitlers choose to either translate discourse markers literally or omit them – rather than, say, adopt a range of equivalents – and must therefore be aware of the risks of omission. In Table 5, the omission of "you know" disconnected the two parts of the source-language speech, while the speaker on screen was referring to what she had just said as shared knowledge between herself and her addressee.

By omitting sociolinguistic elements of dialogue, subtitlers both translate the essential semantic meaning from the source text and convert the source information from one register (dialect, informal) to another (standard, formal). Sarah Kozloff (2000) maintains that nonstandard linguistic varieties in works of fiction are often stereotypical mimeses rather than genuine or accurate portrayals of linguistic phenomena in real-life contexts. Here, mimeses refer to the recreation of what has been used in previous film and literary texts in accordance with the aesthetic, diegetic, stylistic, and functional objectives of the author. The degree of (in)tolerance toward creative mimeses of nonstandard discourse varies among different reception contexts. As a result, source and target context audiences may respond differently to the representation of sociolinguistic elements in the original (if preserved in the translated version), the means of mediation, and their associated sociocultural implications (Lane-Mercier 1997). For example, the film *Lust, Caution* (2007) begins with a scene depicting four women playing Mah-jong and chatting in Shanghainese. The dialogue is difficult to understand for the majority of Chinese audience members, who rely on subtitles in Simplified Written Chinese (SWC) or official "common speech," known as *putonghua*. In such situations, it is difficult for subtitlers to use SWC to represent pre-scripted, dialect-driven speech without creating a counterforce that may challenge the status of the national lingua franca. Although the use of *fangyan Yan* demonstrates an effort to protect language and cultural diversity, this example shows the difficulty of using existing subtitling strategies to represent Han ethnic multilingualism (Zhang 2015a; see also ▶ Chap. 24, "Translating Films About Ethnic Minorities in China into English," by H. Jin's in this handbook).

Subtitlers may also utilize target-language expressions and colloquialisms when the corresponding expressions in the original are linguistically and culturally universal. This can be difficult when translating taboo or coarse language, because translators are pressured to eliminate or tone down the offensiveness of the source information in the target context (Ávila-Cabrera 2015; Trupej 2019). For example, in *Guardians of the Galaxy Vol. 2* (2017) Rocket embarrasses a galactic pirate with the alias "Taser Face" by comparing his face to a "scrotum hat." The subtitler Bowen Fu (2017) disguised the explicit reference by the double entendre 裹蛋皮 (*guo dan pi*; "thin flatbreads for wrapping eggs"). This translation creates a pun because it is phonetically identical to 果单皮 (*guo dan pi*), a sweet and sour snack made of thornapples and beetroots. In another example from *Sex and the City* (Episode 3, Season 1, 1998), Lisa uses the word "pussy" to criticize her husband's lack of masculinity. This word is translated as 多愁善感 ("sentimental"), which totally neutralizes Lisa's aggression. Some researchers have suggested that this translation reflects the Chinese social order by "preventing Chinese viewers from accessing an ideology in which men are perceived to be soft and somewhat feminine, weaker than women" (Yuan 2016, 162).

In contrast, fansubs are more liberal in their use of bad language. In the example shown in Table 6, the fansub tones down the swearing from the original by replacing the taboo Chinese character 屄 ("cunt") with a neutral one that has the same pronunciation (逼, "force"). Euphemisms and homonyms are often used this way in Chinese fansubbing to avoid subbing taboo or crass Chinese characters. Nevertheless, some target audiences are against the replacement of source-language swearing with corresponding target-language elements because they feel that this might overshadow important aspects of the original and distort the accuracy of the translation (Greenall 2015, 51).

This kind of euphemization poses two challenges. First, it can be difficult for the subtitler to find suitable substitutes or homonyms for vulgar Chinese characters. Second, the target audience may not fully understand the process and may be confused when applied. Despite rising English literacy in China (Fong 2013), swear words and taboo references in films may still escape the awareness of target audiences who are unfamiliar with colloquial English (Hu 2005; Xu and Michael Connelly 2009). Thus, subtitlers must have a proper understanding of the offensiveness of the source-language swearing in order to perform an effective and appropriate translation.

In summary, subtitled films demand that viewers read and take in translations alongside the on-screen action that demands their primary attention (Hillman 2011,

Table 6 Southpaw (2015; http://www.rrys2019.com/subtitle/45456)

Source information	Fansubs by YYeTs	Back translation into English
Bitch-ass motherfucker!	他妈的蠢逼玩意儿!	His-mother's dumb cunt-like thing.

380). This necessitates that AVT workers adjust their practices so that their linguistic and cultural mediations fit the expectations, abilities, and needs of their audiences, rather than aiming at direct or literal translations. Hence, subtitling both enhances a target audience's comprehension of source information and reminds them that they are viewing a mediated work. Given the above, this paper will now reflect on directions for future AVT research.

Future Research

Although Chinese subtitling is part of many Western universities' AVT courses and programs, it is only offered at one university in China – the Communication University of China. However, the landscape of AVT research in China is beginning to change. For example, the Baker Centre for Translation and Intercultural Studies at the Shanghai International Studies University is beginning to conduct interdisciplinary AVT research. Given subtitling's role in the international media landscape, this section recommends two avenues for future AVT research: the translation of song lyrics and examinations of participatory subtitling on online streaming platforms.

Instead of maintaining the division between individual translation strategies, types, and processes, Marais and Meylaerts (2019, 1–3) stressed the importance of studying how these elements can be effectively synergized within a single audiovisual product to address the complex and often paradoxical processes of mediation. To this end, future researchers should explore the cross-cultural mobility of song lyrics. Potential research questions in this vein might include the following:

- How does subtitling culturally and cross-culturally render the meaning of song lyrics in ways that assist target audiences' appreciation of the meaning, rhythm, and rhyme of (for example) music videos? (Bosseaux 2015; Low 2005, 2017; Susam-Sarajeva 2008; West 2019)
- How can subtitles visualize the vocal aspects of lyrics and be integrated with translated music videos? (Di Giovanni 2008; Johnson 2019)
- What might we expect the cumulative societal and cultural impacts of subtitled music videos to be in target cultures? (Susam-Saraeva 2019)

Such research could apply data-driven, algorithmic processing to explore the user-driven extension of a transmedia IP-engine – that is, "a single source that is transformed across multiple formats, platforms, and products" (Yecies et al. 2019) – of digitally distributed musical texts. Potential findings might illuminate how users accommodate new music production and reception technologies and explore the affective elements of music transposed across languages, cultures, and digital platforms (see also Desblanche 2019; Hamilton 2019; Nowak 2016).

Another emerging area of inquiry is *ad hoc* participatory translation on streaming platforms, where the *danmaku* function allows users to post comments on videos streaming in real time. Named after the military term for a barrage, *danmaku*-based participatory media were initially launched in Japan. They have since spread to

China and other Asian countries and engendered a new form of grassroots translation and remediation of audiovisual content. *Danmaku* comments scroll across the screen from right to left in two or three seconds and can obscure almost the entire streamed image. In this way, *danmaku* synthesizes private viewership and collective anonymous online texting (Bachmann 2008a, b). Rather than reducing the original video to a multimodal palimpsest, *danmaku*-mediated communication opens "an interstitial space of captioning that does not necessarily engage narration-like content" (Johnson 2013, 302). Instead, the chunks of textual exchange are blended with counter-transparent typographic plays featuring typos, symbols, and other forms of image-oriented writing in the virtual process of meta-interactive sociality between viewers.

In the Chinese context, *danmaku* enables viewers to access foreign media through user-derived translations. When translation is initiated, the bottom of the screen is reserved for *danmaku*-derived subtitles rather than comments. Volunteer translators within the community are rewarded with appreciation from other viewers and are encouraged to be prudent and responsible with their translations (Yang 2021). *Danmaku*-derived translations accommodate the ongoing co-occurrence of multiple renditions which proceed nonlinearly through stages of negotiation, knowledge-sharing, and correction (Yang 2020). Some scholars are already beginning to focus on how to theorize and integrate the study of *danmaku* into the literature (Zhang and Cassany 2019a, b). Future AVT research might draw on the French literary theoretical framework of genetic criticism – which examines how a particular text comes into being by uncovering "the many divergences operating inside its transformation" (De Biasi 2004, 38) – to study the diachronic, polyphonic, (re)constructive process of translation represented by *damanku*. Such research might help researchers think through how *danmaku* reconfigures the traditional spatial and temporal constraints of subtitling work, *danmaku*'s potentially democratizing effects, and what new uses of and meanings within Chinese are invented and negotiated in the process of *danmaku*-mediated communication? By asking and answering these questions, scholars can invite a new and productive dialogue between Chinese language studies and translation and media studies.

Concluding Remarks

This chapter has described some of the linguistic, cultural, and technological aspects of contemporary subtitling practices in China and suggests that Chinese language studies might make a healthy contribution to the subtitling literature in the future, especially as subtitling practices evolve online. It has pointed out how subtitlers mediate between verbal and visual modes of communication within the constraints of the audiovisual medium, and how these constraints and the semiotic complexity of audiovisual media demand that AVT is not limited to language but instead encompasses the entirety of an audiovisual production. However, meaning writ large can be "happily lost in translation" (Desilla 2019) when there is more source information than can be conveyed in neat sentences, and subtitlers' compliance with the usual norms of text translation varies across target linguistic and cultural norms.

Subtitlers are always expected to strike a balance between the source information and the target culture and to swing the pendulum of audience experience between full and no access to the source information (besides paralinguistic features, such as intonations). Bearing this in mind, future research should explore the widespread, emergent forms of participatory subtitling to identify various elements of cross-cultural interaction and distinguish between those features of such communication that appear only in these specific contexts.

References

Adami, Elisabetta, and Sara Ramos Pinto. 2019. Meaning-(re)making in a world of untranslated signs: Towards a research agenda on multimodality, culture, and translation. In *Translation and multimodality: Beyond words*, ed. Monica Boria, Ángeles Carreres, María Noriega-Sánchez, and Marcus Tomalin, 71–93. Abingdon: Routledge.

Ávila-Cabrera, Javier. 2015. Subtitling Tarantino's offensive and taboo dialogue exchanges into European Spanish: The case of pulp fiction. *Revista de Lingüística y Lenguas Aplicadas* 10: 1–11.

Bachmann, Götz. 2008a. Wunderbar! Nico Nico Douga Goes German – and Some Hesitant Reflections on Japaneseness (pt 1). *Metagold: A Research Blog about Nico Nico Douga*. Retrieved from: http://d.hatena.ne.jp/metagold/20080722/1216744317. Last accessed on 16 Oct 2019.

———. 2008b. 'Part 2 about Japaneseness – and Some Scenarios for Nico Nico Douga in the West' *Metagold: A Research Blog about Nico Nico Douga*. Retrieved from: http://d.hatena.ne.jp/metagold/20080727/1217109922. Last accessed on 16 Oct 2019.

Bakhtin, M. Mikhail. 1986. *Speech genres and other late essays*. Trans. Vern W. McGee, ed. Caryl Emerson and Michael Holquist. Austin: University of Texas Press.

Bosseaux, Charlotte. 2015. *Dubbing, film and performance: Uncanny encounters*. Bern: Peter Lang.

Chan, Sin-wai, James Minett, and Florence Li Wing Yee, eds. 2016. *The Routledge encyclopaedia of the Chinese language*. Abingdon: Routledge.

Chaume, Frederic. 2018a. Is audiovisual translation putting the concept of translation up against the ropes? *The Journal of Specialised Translation* 30: 84–104.

———. 2018b. An overview of audiovisual translation: Four methodological turns in a matural discipline. *Journal of Audiovisual Translation: Inaugural Issue*, 40–63. Retrieved from: http://jatjournal.org/index.php/jat/article/view/43

Chu, Donna S.C. 2012. Fanatical labor and serious leisure: A case of fansubbing in China. In *Frontiers in new media research*, ed. Francis L.F. Lee, Louis Leung, Jack Lunchuan Qiu, and Donna S.C. Chu, 259–277. New York: Routledge.

Coblin, W. South. 2000. A brief history of mandarin. *Journal of the American Oriental Society* 120 (4): 537–552.

Creagh, Patrick. (trans.) 1995. *Declares pereira: A true account*. London: Harvill.

De Biasi, Pierre-Marc. 2004. Towards a science of literature: Manuscript analysis and the genesis of the work. In *Genetic criticism. Texts and avant-textes*, ed. Jed Deppman, Daniel Ferrer, and Michael Groden, 36–68. Philadelphia: University of Pennsylvania Press.

DeFrancis, John. 1984. *The Chinese language: Fact and fantasy*. Honolulu: University of Hawaii Press.

Delabastita, Dirk. 1989. Translation and mass-communication: Film and TV translation as evidence of cultural dynamics. *Babel* 35 (4): 193–218.

Derrida, Jacques. 1985. From Des Tours de Babel. In *Difference in translation*, ed. J.F. Graham, 165–207. New York: Cornell University Press.

Desblanche, Lucile. 2019. *Music and translation: New mediations in the digital age*. Basingstoke: Palgrave Macmillan.

Desilla, Louisa. 2019. Happily lost in translation: Misunderstandings in film dialogue. *Multilingua*, 38 (5): 601–618.
Di Giovanni, Elena. 2008. Translation, cultures and the media. *European Journal of English Studies* 12 (2): 123–131.
Díaz-Cintas, Jorge. 2004. Subtitling: The long journey to academic acknowledgement. *The Journal of Specialised Translation* 1: 50–70.
———. 2009. *New trends in audiovisual translation*. Bristol: Multilingual Matters.
———. 2012. Clearing the smoke to see the screen: Ideological manipulation in audiovisual translation. *Meta: Translator's Journal* 57 (2): 279–293.
———. 2018. 'Subtitling's a carnival': New practices in cyberspace. *Journal of Specialised Translation* (30): 127–149.
———. 2019. Audiovisual translation in mercurial mediascapes. In *Advances in empirical translation studies: Developing translation resources and technologies*, ed. Meng Ji and Michael Oakes, 177–197. Cambridge: Cambridge University Press.
———. 2020. Audiovisual translation. In *The Bloomsbury companion to language industry studies*, ed. Erik Angelone, Maureen Ehrensberger-Dow, and Gary Massey, 209–230. London: Bloombury Academic.
Díaz-Cintas, Jorge, and Gunilla Anderman. 2009. *Audiovisual translation: Language transfer on screen*. Basingstoke: Palgrave Macmillan.
Díaz-Cintas, Jorge, and Serenella Massidda. 2019. Technological advances in audiovisual translation. In *The Routledge handbook of translation technology*, ed. Minako O'Hagan, 255–270. Abingdon: Routledge.
Diaz-Cintas, Jorge, and Pablo Muñoz Sánchez. 2006. Fansubs: Audiovisual translation in an amateur environment. *The Journal of Specialised Translation* (6): 37–52.
Diaz-Cintas, Jorge, and Aline Remael. 2007. *Audiovisual translation: Subtitling*. Manchester: St. Jerome.
Du, Weijia. 2015. Beyond the ideology principle: The two faces of dubbed foreign films in PRC, 1949–1966. *Journal of Chinese Cinemas* 9 (2): 141–158.
Du, Zhifeng, Yao Li, and Gang Cheng. 2013. *Basic literacy in AV translation & research*. Hangzhou: Zhejiang University Press.
Dywer, Tessa. 2017. *Speaking in subtitles*. Edinburgh: Edinburgh University Press.
Egoyan, Atom, and Ian Balfour. 2004. *Subtitles: On the foreignness of film*. Cambridge, Mass., London: MIT.
Fong, Gilbert. 2009. The two worlds of subtitling: The case of vulgarisms and sexually-oriented language. In *Dubbing and subtitling in a world context*, ed. Gilbert C.F. Fong and Kenneth K.L. Au, 39–61. Hong Kong: Chinese University Press.
Fong, Tsz Yan Emily. 2013. *English in China: Language, identity and culture*. Ph.D. thesis, The Australian National University.
Fu, Bowen. 2017. *Man Wei Xi Lie Dian Ying Fan Yi An Li Jie Xi* (Translation the Marvel Cinematic Universe: With Reference to Mainland Chinese Versions). Keynote speech during The Advanced Forum on Screen Translation, 16–18 June 2017. Shanghai Jiao Tong University, China.
Gao, Yang. 2012. *TV Talk: American television, Chinese audiences, and the pursuit of an authentic self*. Ph.D. thesis, Vanderbilt University, Tennessee
Georgakopoulou, Panayota (Yota). 2019. Technologisation of audiovisual translation. In *The Routledge handbook of audiovisual translation*, ed. Luis Pérez-González, 516–539. London: Routledge.
González-Zúñiga, Diego, Jordi Carrabina, and Pilar Orero. 2013. Evaluation of depth cues in 3D subtitling. *Online Journal of Art and Design* 3 (1): 16–29.
Gottlieb, Henrik. 1994. Subtitling: People translating people. In *Teaching translation and interpreting 2: insight, aims, visions*, ed. Cay Dollerup and Annette Lindegaard, 261–274. Amsterdam: John Benjamins.
Greenall, Annjo K. 2015. Translators' voices in Norwegian retranslations of Bob Dylan's songs. *Targets* 27 (1): 40–57.

Gregory, Michael, and Susanne Carroll. 1978. *Language and situation: Language varieties and their social context*. London: Routledge.

Hamilton, Craig. 2019. Popular music, digital technologies and data analysis: New methods and questions. *Convergence: The International Journal of Research into New Media Technologies* 25 (2): 225–240.

Hatim, Basil, and Ian Mason. 1990. *Discourse and the translator*. London/New York: Longman.

———. 1997. *Translator as communicator*. London: Routledge.

Hoffman, David M., Girshick, Ahna R., and Akeley, Kurt. 2008. Vergence-accommodation conflicts hinder visual performance and cause visual fatigue. *Journal of Vision* 8 (3). www.ncbi.nlm.nih.gov/pmc/articles/PMC2879326/pdf/nihms202260.pdf.

House, Juliane. 2001. Translation quality assessment: Linguistic description versus social evaluation. *Meta: Translators' Journal* 46 (2): 243–257.

Hillman, Roger. 2011. Spoken word to written text: Subtitling. In *The oxford handbook of translation studies*, ed. Kristijan Malmkjear, and Kevin Windle, 379–393. New York: Oxford University Press.

———. 2006. Text and context in translation. *Journal of Pragmatics* 38 (3): 338–358.

Hu, Guangwei. 2005. English language education in China: Policies, progress, and problems. *Language Policy* 4 (1): 5–24.

Hu, Kelly. 2018. Between informal and formal cultural economy: Chinese subtitle groups and flexible accumulation in the age of online viewing. In *The Routledge handbook of East Asian popular culture*, ed. Koichi Iwabuchi, Eva Tsai, and Chris Berry, 45–54. Abingdon: Routledge.

Ivarsson, Jan. 1995. The history of subtitles. *FIT Newsletter-Nouvelles de la FIT* 74 (3–4): 294–302.

Jin, Haina. 2018. Introduction: The translation and dissemination of Chinese cinemas. *Journal of Chinese Cinemas* 12 (3): 197–202.

Jin, Haina, and Yves Gambier. 2018. Audiovisual translation in China: A dialogue between Yves Gambier and Haina Jin. *Journal of Audiovisual Translation* 1 (1): 26–39.

Johnson, Daniel. 2013. Polyphonic/pseudo-synchronic: Animated writing in the comment feed of Nicovideo. *Japanese Studies* 33 (3): 297–313.

Johnson, Rebecca. 2019. Audiovisual translation and popular music. In *The Routledge handbook of audiovisual translation*, ed. Luis Pérez-González, 418–435. Abingdon: Routledge.

Karamitroglou, Fotios. 2000. *Towards a methodology for the investigation of norms in audiovisual translation*. Amsterdam/Atlanta: Rodopi.

Kozloff, Sarah. 2000. *Overhearing film dialogue*. Berkeley: University of California Press.

Kress, Gunther. 2010. *Multimodality: A social semiotic approach to contemporary communication*. London: Routledge.

Kung, Szu-Wen. 2016. Audienceship and community of practice: An exploratory study of Chinese fansubbing communities. *Asia Pacific Translation and Intercultural Studies* 3 (3): 252–266.

Kuo, Arista Szu-Yu. 2017. Subtitling quality beyond the linguistic dimension. In *The Routledge handbook of Chinese translation*, ed. Chris Shei and Zhao-Ming Gao, 415–431. Abingdon: Routledge.

Lane-Mercier, Gillian. 1997. Translating the untranslatable: The translator's aesthetic, ideological and political responsibility. *Targets* 9 (1): 43–68.

Li, Dang. 2015. *Amateur translation and the development of a participatory culture in China: A Netnorgraphic Study of the Last Fantasy Fansubbing Group*. Ph.D. thesis, University of Manchester.

Liang, Lisi. 2017. *Subtitling English-language films for a Chinese audience: Cross-linguistic and cross-cultural transfer*. Ph.D. thesis, Cardiff University.

Low, Peter. 2005. The pentathlon approach to translation. In *Song and significance. Virtues and vices of vocal translation*, ed. Dinda L. Gorlée, 185–212. Amsterdam/New York: Rodopi.

———. 2017. *Translating song: Lyrics and texts*. Abingdon: Routledge.

Marais, Kobus, and Reine Meylaerts. 2019. Introduction. In *Complexity thinking in translation studies: Methodological considerations*, ed. Kobus Marais and Reine Meylaerts, 1–18. Abingdon: Routledge.

Mattson, Jenny. 2006. *Linguistic variation in subtitling – the subtitling of swearwords and discourse markers on public television, commercial television and DVD*. Paper presented at the EU High Level Scientific Conference Series, Copenhagen.

Nornes, Markus. 2007. *Cinema babel: Translating global cinema*. Minneapolis: University of Minnesota Press.

Nowak, Raphaël. 2016. When is a discovery? The affective dimensions of discovery in music consumption. *Popular Communication* 14 (3): 137–145.

O'Sullivan, Carol. 2019. "A splendid innovation, these English titles!": Source of evidence for early subtitling practice.

O'Sullivan, Carol, and Jean-François Cornu. 2019. History of audiovisual translation. In *Routledge handbook of audiovisual translation*, ed. Luis Pérez González, 15–30. Abingdon: Routledge.

O'Sullivan, Carol. 2019. "A splendid innovation, these English titles!": Sources of evidence for early subtitling practice. In *The translation of films, 1900–1950*, ed. Carol O'Sullivan and Jean-François Cornu. Oxford: Oxford University Press for the British Academy.

Pavesi, Maria. 2018. Reappraising verbal language in audiovisual translation: From description to application. *Inaugural issue of Journal of Audiovisual Translation* 1: 101–121.

Pérez-González, Luis. 2007. Fansubbing anime: Insights into the "Butterfly Effect" of globalization on audiovisual translation. *Perspectives: Studies on Translation Theory and Practice* 14 (4): 260–277.

———. 2014. *Audiovisual translation: Theories, methods and issues*. Abingdon: Routledge.

———. 2017. Investigating digitally born amateur subtitling agencies in the context of popular culture. In *Non-professional subtitling*, ed. David Orrego-Carmona and Yvonne Lee, 15–36. New Castle upon Tyne: Cambridge Scholar Publishing.

———. 2019. Rewiring the circuitry of audiovisual translation: Introduction. In *The Routledge handbook of audiovisual translation*, ed. Luis Pérez-González, 1–12. Abingdon: Routledge.

Pettit, Zoë. 2005. Translating register, style and tone in dubbing and subtitling. *The Journal of Specialised Translation* 4 (4): 49–65.

Remael, Aline. 2001. Some thoughts on the study of multimodal and multimedia translation. In *(Multi) Media translation*, ed. Yves Gambier and Henrik Gottlieb, 13–22. Amsterdam: John Benjamins.

Schwarz, Barbra. 2002. Translation in a confined space – film subtitling with special reference to Dennis Potter's "Lipstick on Your Collar". *Translation Journal*. https://translationjournal.net/journal/22subtitles.htm

Susam-Saraeva, Şebnem. 2019. Interlingual cover versions: How popular songs travel round the world. *The Translator* 25 (1): 42–59.

Susam-Sarajeva, Şebnem. 2008. Translation and music. *The Translator* 14 (2): 187–200.

Trupej, Janko. 2019. Avoiding offensive language in audiovisual translation: A case of study of subtitling from English to Slovenian. *Across Languages and Cultures* 20 (1): 57–77.

Turner, Graeme. 2010. *Ordinary people and the media: The demotic turn*. New York: SAGE.

Venuti, Lawrence. 2000. Translation, community, Utopia. In *Translation studies reader*, ed. Lawrence Venuti, 468–488. New York: Routledge.

Wan, Shuai, Bo Chang, and Fuzheng Yang. 2013. Viewing experience of 3D movie with subtitles: Where to put subtitles in a 3D movie?. 2013 Fifth international workshop on Quality of Multimedia Experience (QoMEX), 3–5 July 2013, Klagenfurt am Wörthersee.

Wang, Dingkun. 2014. *In search of principles for Chinese subtitling: The application of Lu Xun's 'Hard Translation' in Modern Media*. Ph.D. thesis, The Australian National University.

———. 2017. Fansubbing in China – With reference to the Fansubbing Group YYeTs. *The Journal of Specialised Translation* 28 (2017): 165–188.

Wang, Dingkun, and Xiaochun Zhang. 2017. Fansubbing in China: Technology-facilitated activism in translation. *Targets* 29 (2): 301–318.

West, David. 2019. Introduction: The challenges of the song lyric. *Language and Literature* 28 (1): 3–6.

Whitman-Linsen, Candace. 1992. *Through the dubbing glass. The synchronization of American motion pictures into German, French and Spanish.* Frankfurt & Main: Peter Lang.

Wu, Zhiwei. 2017. The making and unmaking of non-professional subtitling communities in China: A mixed-method study. In *Non-professional subtitling*, ed. Davide Orrego-Carmona and Yvonne Lee, 115–144. Newcastle upon Tyne: Cambridge Scholars Publishing.

Xu, Shijing, and F. Michael Connelly. 2009. Narrative inquiry for teacher education and development: Focus on English as a foreign language in China. *Teaching and Teacher Education* 25 (2): 219–227.

Yang, Yuhang. 2020. The danmaku interface on Bilibili and the recontextualised translation practice: A semiotic technology perspective. Social Semiotics 30 (2): 254–273.

Yang, Yuhong. 2021. Danmaku subtitling: An exploratory study of a new grassroots translation practice on Chinese video-sharing websites. *Translation Studies* 14 (1): 1–17.

Yecies, Brian, Jack (Jie) Yang Aegyung Shim, and Peter Yong Zhong. 2019. Global transcreators and the extension of the Korean Webtoon IP-engine. *Media, Culture & Society*. Published on line on 30 Sept 2019: https://doi.org/10.1177/0163443719867277.

Yuan, Long. 2016. *The subtitling of sexual Taboo from English to Chinese.* Ph.D. thesis, The Imperial College London.

Zhang, Xiaochun. 2015a. Cinematic multilingualism in China and its subtitling. *Quaderns: Revista De Traducció* 22: 385–398.

Zhang, Leticia, and Daniel Cassany. 2019a. 'Is it always so fast?': Chinese perception of Spanish through Danmaku video comments. *Spanish in Context* 16 (2): 217–242.

Zhang, Leticia Tian, and Daniel Cassany. 2019b. The 'danmu' Phenomenon and Media Participation: Intercultural Understanding and Language Learning Through 'The Ministry of Time'. Comunicar 58. Retrieved from https://eric.ed.gov/?id=EJ1201217. Last accessed on 16 Oct 2019.

Part IV

New Trends in Teaching Chinese as a Foreign Language

New Trends in Teaching Chinese as a Foreign Language: Introduction

26

Li Yu

Contents

Introduction 640
Chapter Summaries 641
Conclusion 644
References 645

Abstract

This section introduction describes the common themes of the seven chapters included in the volume "New Trends in Teaching Chinese as a Foreign Language." Teaching Chinese as a Foreign Language (TCFL) emerged as an interdisciplinary field of study in the 1980s and early 1990s in both the Chinese speaking world and the United States. In the past three decades, the field has undergone exponential growth which is manifested in a drastic increase in the number of Chinese language learners worldwide as well as rapid progress made in the research area of Chinese language pedagogy. This chapter identifies three new trends of learner needs in the TCFL field, namely, "starting young" 低龄化, "reaching high" 高级化, and "more-diverse and individualized needs" 多样化个别化. To address these emerging learning needs, the TCFL field must innovate its teaching practices to apply the most recent findings in related research disciplines and design new curricula that would reflect learners' interests. The seven chapters included in this volume report the most up-to-date findings in four research areas – teacher training, instructional technology, assessment, and curriculum and instructional design. The collective wisdom shared by all of these articles is that TCFL is a field in which pedagogical theories must be translated into actual,

---

L. Yu (✉)
Department of Asian Studies, Williams College, Williamstown, MA, USA
e-mail: lyu@williams.edu

© The Author(s), under exclusive licence to Springer Nature Singapore Pte Ltd. 2022
Z. Ye (ed.), *The Palgrave Handbook of Chinese Language Studies*,
https://doi.org/10.1007/978-981-16-0924-4_1

effective, and efficient classroom practice. The more prepared our teachers and the better designed and streamlined our curricula, the shorter the time required for students to start from zero background in Chinese and proceed to reaching the truly advanced level.

Keywords

Teaching Chinese as a Foreign Language (TCFL) · New trends of learner needs · Pedagogical approach

Introduction

Although the teaching of Chinese to speakers of other languages has a relatively long history (Yao and Yao 2010; Ling 2018; Chao 2020), the field of Teaching Chinese as a Foreign Language (TCFL) is relatively young. The 1980s and early 1990s saw the emergence of TCFL as a unique academic discipline or, more precisely, an interdisciplinary field, in both the Chinese-speaking world and the United States. In Mainland China, a bachelor's and then graduate degrees in *duiwai hanyu* 对外汉语 (Chinese as a Foreign Language), later officially renamed in 2012 as *hanyu guoji jiaoyu* 汉语国际教育 or Teaching Chinese to Speakers of Other Languages (TCSOL), were established during this period initially at a few designated universities in Beijing, Shanghai, and Chongqing and then more widely at a large number of universities across the country (Pan 2004; Shei 2020). In Taiwan, the first program to grant a master's degree in *huayuwen jiaoxue* 华语文教学 (Teaching Chinese as a Foreign Language) was founded in 1995 at the National Taiwan Normal University. A doctoral degree and a bachelor's degree followed in 2003 and 2006, respectively, and similar degree programs sprang up across the region. In the United States, the 1996 publication of the volume *Chinese Pedagogy: An Emerging Field* marked the early formation of this interdisciplinary field in the English-speaking world (McGinnis 1996), followed by the awarding of the first doctoral degree in Chinese language pedagogy in 2003 at The Ohio State University. Coincidentally, 2004 became a defining year for the Chinese field when more than 2000 US schools expressed interest in offering the Chinese Advanced Placement (AP) Course and Examination. Since then, both the US and PRC governments have invested heavily in the field through a variety of initiatives, stimulating an exponential growth of the field in the ensuing years (Wang 2012).

More than three decades have passed since TCFL grew from an unrecognized and underappreciated profession to an active and fertile field that crosses the disciplines of linguistics, Second Language Acquisition (SLA), psychology, cognitive science, and education. In the meantime, Chinese learning has enjoyed increasing popularity worldwide, thanks to China's rise as a world power and fast-growing economy. In US higher education institutions, enrollment in Chinese language courses peaked in 2013 at 61,084 (Looney and Lusin 2019, 32). In the past 5 years, the total number of college students studying Chinese has continued to hover between 50,000 and

60,000 nationally. More students are now starting their Chinese learning journey earlier, beginning in their high school, middle school, or sometimes even elementary school years. According to a 2010 survey conducted by the Center for Applied Linguistics, the number of Chinese programs grew dramatically during the decade between 1997 and 2008 at the K-12 levels (Rhodes and Pufahl 2010). Together with this first trend of "starting young," more students are reaching the advanced level and need to take more advanced-level courses in colleges and universities. This second trend of "reaching high" was captured in the ratio of introductory to advanced undergraduate enrollments data reported by the Modern Language Association (Looney and Lusin 2019, 10). In 2016, this ratio was 3:1 for Chinese, meaning 25 percent of undergraduate enrollments were in advanced-level courses. This represents a small step forward compared to a ratio of 4:1 (20 percent of total enrollments in advanced-level courses) in 2013 and 2009 (Looney and Lusin 2019, 50). For previous generations, the main purpose of learning Chinese was for missionary, academic, and military reasons. This has been replaced by a more-diverse range of personal and professional interests and motives (Ling 2018; Li et al. 2014). The needs of Chinese learners have thus changed and diversified dramatically over the past three decades, constituting the third trend of "more-diverse and individualized needs."

To meet these new learning needs, the field of TCFL must keep reinventing itself, innovating its teaching practices to reflect the most recent findings in many related disciplines and research areas, as well as designing new curricula that would reflect learners' interests. This section on the new trends in teaching Chinese as a foreign language was envisioned to achieve this goal. The seven articles included in this section represent the most recent advances made in the areas of teacher training, instructional technology, assessment, and curriculum and instructional design for Chinese language programs located in a non-target cultural environment and address various issues that are central to the further development of the TCFL field in the coming decades. Collectively, they provide an insight into the future directions of the field of TCFL.

Chapter Summaries

Jianhua Bai's chapter addresses a pressing issue in TCFL – teacher training. Given the increasing number of students enrolled in the K-16 system in the United States, there are still not enough qualified Chinese language teachers to teach in these programs. All too often, many students who go through years of Chinese learning from kindergarten to high school find they cannot function properly in the language when entering college. Novice teachers who have undergone training in Mainland China or Taiwan usually have no firsthand teaching experiences or training in actual classroom teaching techniques. They do not fully understand the institutional culture of the Western educational system and thus find it extremely challenging to build a Chinese program from scratch, let alone maintain it. The various teacher training programs, especially the one spearheaded by the Middlebury Chinese School described in Bai's chapter, provide some good models for future TCFL teacher preparation and education.

An equally important issue is assessment. Two national surveys have found that Chinese language teachers tend to not prioritize the use of standardized testing to assess the learning outcomes of their students, whereas program administrators tend to think otherwise (Walker and Li 2003; Li et al. 2014). Experts in the field have advocated establishing programmatic goals and using systematic assessment tools to measure learning outcomes and adjust program goals accordingly (Li et al. 2014). Teacher trainers have also emphasized the importance of standards-based assessment in teacher preparation and development (Everson 2012). Dan Nie and Qifeng Zhao's chapter provides a historical overview of the development of assessment tools for Chinese as a Second Language and considers the future of Chinese proficiency testing in both China and elsewhere. They point out that proficiency tests that cater to learners at a more-advanced level and better customized and computerized tests powered by AI technologies are the new trends in the realm of TCFL assessment. Their overview and discussions of the future of Chinese proficiency testing will be useful to teachers who want to take a more-systematic approach to designing a holistic language program with smooth articulation between the different levels.

While Chinese language teachers may disagree on what assessment tools to use, one thing they would all agree on is that technology plays an important role in language teaching and learning, and if well integrated, technology can make teaching and learning more effective, efficient, and enjoyable (Zhang 1998; Yao 2009). The role of instructional technology became even more prominent when the 2019 pandemic hit the globe and jump-started remote learning in 2020 worldwide. Shijuan Liu and Jun Da's chapter provides a comprehensive overview of the technologies used in Chinese language teaching and learning from 1900 to 2020 and offers suggested categorization, frameworks, and guidelines for understanding instructional technologies and their application to teaching Chinese as a foreign language in the past and future. Although technology has evolved over the years, it has always worked as a supporting tool for language learning and teaching and cannot replace the teacher. However, the teacher's role has been redefined and made even more challenging and "high-end" by the advancements made in instructional technology. It is more urgent than ever for teachers to skillfully select and use the various tools of instructional technology.

As technology helped shoulder some of the responsibilities of a traditional teacher, course design and curricular innovation play an even-larger part in redefining the teacher's role. Innovative curricular design is the key to success for any course or language program. The next three chapters present three sets of pedagogical and curricular design ideas. Junqing Jia and Zhini Zeng's chapter proposes a new curricular design for beginner-level reading and writing classes. One of the most active yet futile fields of research within TCFL in the past three decades is character instruction. Many Chinese language teachers have asked how to help students learn to read and write Chinese characters in an efficient manner. Jia and Zeng's answer to this problem is that we may have been asking the wrong question all along. Reading Chinese is not as difficult as assumed if teachers recognize the primacy of speech and the secondary role of writing systems to speech. Jia and Zeng's innovative design of a beginner-level literacy curriculum is based on the pedagogical concept that reading is equated not with the concept of character recognition, decoding, or character recall, but with the idea of doing things with texts in the target sociocultural environment.

Yang Wang's chapter deals with yet another pressing issue in TCFL, but from the other end of the spectrum: how to help advanced-level learners improve their academic writing skills in Chinese. For years, the development of advanced writing skills has been given the lowest priority due to both time constraints of most language programs and limited student need. In the past, even if a student began studying Chinese in their freshman year, they would still only reach the level of intermediate-low or at best intermediate-high in terms of writing by the time they graduated. Writing skills consistently lagged behind students' listening, speaking, and reading skills. Many government-run programs did not emphasize the development of writing skills because of the lack of real-world needs for diplomats and military personnel to acquire such skills. However, thanks to the "starting young" and "reaching high" trends mentioned earlier as well as improvements made in language pedagogy in the past three decades, more students are able to reach a truly advanced level when they graduate from college. This level is not complete without more advanced-level writing skills. Wang's chapter makes a convincing argument for including academic writing skill training in advanced-level curricula and offers a detailed description of the instructional design of such a curriculum.

As students' personal and intellectual interests broaden, they are also increasingly attracted by courses that go beyond the domains of Chinese literature, a mainstay discipline well established in many university and college Chinese programs in the United States. Yu Li's chapter on the inclusion of calligraphy education in a Chinese curriculum demonstrates the power of thinking out of the box to introduce elements of Chinese achievement culture other than literature to satisfy the intellectual curiosity of a general student body who may not be as literature bound as previous generations. In addition, in this digitized world, calligraphy is perhaps one of the last few culturally authentic tasks that would require handwriting of Chinese characters. Li's curricular design shows that a more liberal way of designing new courses that pique students' interests and cater to their learning and intellectual needs would be a new direction for language professionals to consider when building and maintaining a successful program.

Making language learning more meaningful, rewarding, and efficient for our students would require rethinking our pedagogical approach. Since the 1970s, the Communicative Approach has dominated the field of foreign language education, yet it reached a dead end in recent years (Klapper 2003; VanPatten 2017). Presentation, Practice, and Production (PPP) and Task-Based Instruction (TBI) have failed to highlight the central role of culture in language learning. The recent grassroots movement of TPRS (Teaching Proficiency through Reading and Storytelling) does not discuss the role of culture in language learning either. In the old paradigm, culture is deemed to be something outside of language or only as a small component of language. However, more and more language professionals have realized the importance of bringing intercultural communicative competence to language learning in our ever-more diversified and globalized world (Byram 1997; Spring 2020). They have also realized the importance of centralizing the humanistic side of language learning rather than dwelling on its instrumental value (MLA 2007). Yet how do we do that? Li Yu's chapter on the Performed Culture Approach describes a revolutionary, paradigm-shifting pedagogical approach that would help us achieve the goal of seamlessly integrating culture (especially behavioral culture) instruction with language instruction. Developed in the field of East Asian language pedagogy and combining the wisdom of Chinese educational philosophy with Western sciences, the Performed Culture Approach presents a promising pedagogical framework that would effectively bring different cultures and worldviews together.

Interdisciplinarity has been the hallmark feature of the field of TCFL since its inception. The seven chapters in this section are authored by ten frontline Chinese language teachers and researchers from various disciplinary backgrounds. The collective wisdom shared by all of these articles is that TCFL is a field in which pedagogical theories must connect with actual, effective, and efficient classroom practice. These authors share a vision that the more prepared our teachers are and the better designed and streamlined our curricula are, the shorter the time required for students to start from zero background in Chinese and proceed to reaching the truly advanced level. Since a minor in Chinese, which usually requires only 2 or 3 years of language study, still appears to be the predominant form of academic credentials in the United States (Li et al. 2014, 11), there is still a long way to go for Chinese language professionals to make Chinese language learning a successful and rewarding experience for the majority of learners who embark on this journey.

Conclusion

Although Chinese has grown from a Less Commonly Taught Language (LCTL) to be one of the top ten most studied languages in the United States, it remains as one of the less successfully taught languages. This is due to a lack of common understanding of principles among the ranks and files of frontline teachers as to the most effective and efficient methods of translating pedagogical and SLA theories and research findings into effective classroom practices. In addition, the majority of Chinese language courses in the United States are staffed by nontenure-stream faculty (full-time or part-time lecturers and adjuncts) or graduate teaching assistants (Li et al. 2014, 16). Although many of them are qualified and experienced teachers, most have a labor-intensive workload and lower status in the administrative structure of language programs and departments, a condition that limits their access to the time, resources, and power required to conduct action research, receive continued professional training and development, and make key program-wide decisions. A 2003 report on a survey of East Asian language programs finds that both frontline teachers and language program supervisors place teacher training, instructional technology, and improved materials as the top priorities for improving the study of East Asian languages (Walker and Li 2003). In the larger field of foreign language education, language professionals have pushed for more standards-based assessment and drafting of learning objectives based on can-do statements (Li et al. 2014, 49). This section has responded to these visionary calls and made a small, yet significant, step forward in the field. To prepare us for the upcoming decades for TCFL, Chinese language teachers need to take a liberal perspective in examining Chinese learning and teaching in a global context (Zhang 2018; Chen 2020). Critical modifications need to take place not only in the area of teacher education and training for pre-service instructors (Kubler 2021) but also in the way of thinking for many veteran teachers in the field. Only through some transformative changes in our pedagogical thinking and practices are we able to make fundamental, positive changes to the field of TCFL and help every learner who aims to reach the advanced level in Chinese to achieve this goal within a reasonable time frame.

References

Byram, Michael. 1997. *Teaching and assessing intercultural communicative competence*. Philadelphia: Multilingual Matters.
Chao, Der-lin. 2020. The beginning of Chinese professorship and Chinese language instruction in the United States: History and implications. In *The Routledge handbook of Chinese language teaching*, ed. Chris Shei, Monica E. McLellan Zikpi, and Der-Lin Chao, 47–63. London: Routledge.
Chen, Guangyan. 2020. *Performed culture in Chinese language education: A culture-based approach for U.S. collegiate contexts*. Lanham: Lexington Books.
Everson, Michael. 2012. The preparation and development of Chinese language teachers: The era of standards. *Journal of the Chinese Language Teachers Association* 47 (3): 7–18.
Klapper, John. 2003. Taking communication to task: A critical review of recent trends in language teaching. *Language Learning Journal* 27 (1): 33–42.
Kubler, Cornelius. Forthcoming, 2021. Transformative aspects of teacher education and training for preservice instructors of Chinese and Japanese. In *Transformative language learning and teaching*, ed. Betty Lou Leaver, Dan E. Davidson, and Christine Campbell. Oxford: Cambridge University Press.
Li, Yu, Xiaohong Wen, and Tianwei Xie. 2014. CLTA 2012 survey of college-level Chinese language programs in North America. *Journal of the Chinese Language Teachers Association* 49 (1): 1–49.
Ling, Vivian. 2018. *The field of Chinese language education in the U.S.: A retrospective of the 20th century*. London: Routledge.
Looney, Dennis, and Natalia Lusin. 2019. Enrollments in languages other than English in United States institutions of higher education, summer 2016 and fall 2016: Final report. *The Modern Language Association of American Web Publications*. https://www.mla.org/content/download/110154/2406932/2016-Enrollments-Final-Report.pdf. Accessed 12 Aug 2020.
McGinnis, Scott. 1996. *Chinese pedagogy: An emerging field*. Columbus: Foreign Language Publications, The Ohio State University.
MLA Ad Hoc Committee on Foreign Languages. 2007. Foreign languages and higher education: New structures for a changed world. *Profession* 2007: 234–245.
Pan, Wenguo 潘文国. 2004. Lun duiwai hanyu de xuekexing 论对外汉语的学科性 [On the academic discipline of Chinese as a foreign language]. Shijie hanyu jiaoxue 世界汉语教学 [Chinese Teaching in the World] 1:11–19.

Rhodes, Nancy C., and Ingrid Pufahl. 2010. Foreign language teaching in U.S. schools: Results of a national survey. Center for Applied Linguistics, National K-12 Foreign Language Survey. Web publication. https://www.cal.org/what-we-do/projects/national-k-12-foreign-language-survey. Accessed 12 Aug 2020.

Shei, Chris. 2020. From "Chinese to foreigners" to "Chinese international education": China's efforts in promoting its language worldwide. In *The Routledge handbook of Chinese language teaching*, ed. Chris Shei, Monica E. McLellan Zikpi, and Der-Lin Chao, 32–46. London: Routledge.

Spring, Madeline. 2020. Intercultural communication competence in CFL language curricular. In *The Routledge handbook of Chinese language teaching*, ed. Chris Shei, Monica E. McLellan Zikpi, and Der-Lin Chao, 336–357. London: Routledge.

VanPatten, Bill. 2017. *While we're on the topic: BVP on language, acquisition, and classroom practice*. Alexandria: The American Council on the Teaching of Foreign Languages.

Walker, Galal, and Minru Li. 2003. Report on the survey of East Asian language programs. National East Asian Languages Resource Center at The Ohio State University. Web publications. https://nealrc.osu.edu/assessments/survey. Accessed 12 Aug 2020.

Wang, Shuhan C. 2012. Sustaining the rapidly expanding Chinese language field. *Journal of the Chinese Language Teachers Association* 47 (3): 19–41.

Yao, Tao-chung. 2009. The current status of Chinese CALL in the United State. *Journal of the Chinese Language Teachers Association* 44 (1): 1–23.

Yao, Tao-chung, and Kuang-tien Yao. (2010). Meiguo hanyu jiaoxue lishi huigu yu xianzhuang美国汉语教学历史回顾与现状 [Chinese language instruction in the United States: A look at its history and current status]. In *Beimei Zhongguoxue: Yanjiu Gaishu yu Wenxian Ziyuan*北美中国学:研究概述与文献资源 [Chinese studies in North America: Research and resources], ed. Haihui Zhang, Zhaohui Xue and Shuyong Jiang, 773–784. Beijing: Zhonghua shuju.

Zhang, Zhengsheng. 1998. CALL for Chinese: Issues and practices. *Journal of the Chinese Language Teachers Association* 33 (1): 51–82.

———. 2018. My never-ending education 活到老学到老. In *The field of Chinese language education in the U.S.: A retrospective of the 20th century*, ed. Vivian Ling, 281–289. London: Routledge.

Teacher Training in the Field of Teaching Chinese as a Foreign Language in the United States

27

Jianhua Bai

Contents

Introduction	648
Teacher Training Programs in the United States	649
In-service Teacher Training: The Case of Middlebury Chinese School	649
Summer Professional Development Programs: The cases of STARTALK and CARLA	657
On-Demand Training: The Case of AP Workshops	660
Essential Competencies for CFL Teaching Excellence	662
A Call for a Systematic and Integrated Approach to Teacher Training	664
Conclusion	672
References	673

Abstract

This chapter deals with teacher training in the field of teaching Chinese as a foreign language in the United States. It starts with a summary and critical analysis of some models of teacher training and professional development, and then continues to discuss what constitutes CFL (Chinese as a Foreign Language) teaching excellence. The chapter states that great progress has been made in Chinese language pedagogy research, but notes the need for more effective and systematic models of teacher training in the CFL field. A systematic and integrated approach to CFL teacher training is proposed and discussed through the description of a teacher training framework and the important elements of the framework that are designed to develop the essential competencies of CFL teaching excellence. It addresses not only what successful CFL teachers should know, but also what they should be able to do, applying a strong knowledge base into successful and effective teaching practice.

J. Bai (✉)
Kenyon College, Gambier, OH, USA
e-mail: bai@kenyon.edu

Keywords

Teacher training · Teacher professional development

Introduction

The field of teaching and learning Chinese as a foreign language (CFL) has come a long way. From the seventeenth to the early twentieth centuries the learning of Chinese was mostly the interest of missionaries in the west. According to Walton (1997) "the first wave of intense national interest dates back to World War II, as our rather modest and haphazard dealings with China moved toward more pressing pragmatic concerns. The second wave arose during the 1980s after the normalization of relations between the People's Republic of China and the United States." Great changes have occurred in the past three decades. First, we witnessed fast growing interest in learning Chinese during the first 10 years of the twenty-first century. The MLA 2002 survey showed that 34,153 students were enrolled in Chinese courses nation-wide in US institutions of higher education, a 20% increase since the last MLA survey in 1998 (Welles 2004). The instruction of Chinese at secondary schools also developed rapidly since the Geraldine Dodge Foundation generously supported 60 high schools to teach Chinese in the 1980s (CLASS website). In 2003 the College Board decided to add the course of Chinese Language and Culture to their Advanced Placement (AP) program. The Chinese AP program has had a very positive impact on the learning of CFL and definitely improved the K-16 articulation. It has helped enhance the national visibility and provided more support for professional development, curriculum development, and instructional resources. Most importantly, it has encouraged CFL learners to start their learning of Chinese in the early grades, which is essential for producing the urgently needed professionals who can function professionally in CFL. A recent survey (Davidson 2017) found the number of young Americans studying Mandarin (K-12) had doubled from 2009 to 2015. The increasing number of students learning Chinese at all levels reflects the general recognition of the growing importance of acquiring the Chinese language and culture competence for cross-cultural understanding and effective communication in world events.

As the importance of learning Chinese language and culture is widely recognized, we notice that the field of teaching CFL has become more and more professional. It has developed from a marginal field with less developed pedagogy to a strong profession. We witness great progress in Chinese language pedagogy research, curriculum design, and material development. Despite the fact that the interest in learning Chinese is steadily growing and the field is becoming more professional, US schools and higher education institutions still face the need for effective models of teacher training in the CFL field. Asia Society convened a meeting in April 2005 with the leaders in the field and concluded that, in order to build the infrastructure to support a K-16 pipeline of Chinese language learners to meet national needs, three critical issues must be addressed: "creating a supply of qualified Chinese language teachers; increasing the number and quality of school programs; and developing appropriate curriculum, materials, and assessments, including technology-based delivery system (Asia Society and College Board 2008)." In order to understand the specific needs and provide a more systematic model in the area of CFL teacher training, the next section will first provide a summary and critical analysis of some models of teacher training and professional development.

Teacher Training Programs in the United States

As the importance of learning Chinese language and culture is widely recognized we have seen the increase of number of students at all levels. Teacher training has also received more attention in the field of teaching Chinese as a foreign language. In addition to the graduate programs for Chinese language pedagogy, we have also seen the growth of different teacher training and professional development programs in the field of teaching Chinese as a foreign language such as the in-service training of various summer programs, short-term workshops such as STARTALK, SPEAC, ALLEX, and the AP summer institutes. These programs provide opportunities for teachers to apply what they learn from the books into actual teaching. They are valuable additions to what the conventional graduate programs of universities offer for CFL teachers because most of the traditional university graduate programs do not provide sufficient guided teaching practicum to help develop CFL teachers' teaching excellence. This section is a summary and critical analysis of the teacher training and professional development programs that the author is familiar with.

In-service Teacher Training: The Case of Middlebury Chinese School

One form of teacher training is the in-service training common among some of the well-established intensive immersion programs of Chinese. According to my observation and/or active participation in the teacher training at the Chinese School of Middlebury College, Associated Colleges in China (ACC), the CET China programs and the overseas US Flagship programs, there are some common elements shared by these programs: learning to assess students' learning needs and determining the learning objectives, improving the base knowledge about the Chinese language and the relevant disciplines such as second language acquisition and instructional technology, learning about the recent theoretical framework of foreign language pedagogy such as the standards-based instruction and learning, and developing effective teaching techniques for students of different proficiency levels. Most of the examples below are from the Chinese School of Middlebury College where I played an important role in teacher training and supervision.

One important component of the in-service training is to learn to understand learner needs, based on which we identify the learning objectives. This part of the teacher training is often done collaboratively among the director and the lead teachers of the programs. One of the major tasks for the weekly meetings of the director and lead teachers at the Chinese School of Middlebury College is to discuss and examine the appropriateness and difficulty level of the learning material and instruction. Based on our observation, reflection, and examination of the results of regular teaching evaluations we try to make constant adjustments to meet the needs of students of various levels. The theoretical framework that guides our efforts in meeting the needs of students is the Backward Design approach (Wiggins and McTighe 1998): identify the learner needs and learning objectives, determine acceptable evidence, plan teaching and learning activities, and consider how formative and summative testing can be integrated. The following procedure is what we use to guide the training on curriculum design for different levels.

- Identify the specific needs of the group of learners that we teach at a particular level, e.g., advanced Chinese for college level students of various majors whose proficiency level ranges across the intermediate range on the Oral Proficiency Interview (OPI) scale.
- Identify instructional objectives by asking what students should know and be able to do communicatively, considering linguistic, pragmatic, and strategic competencies when determining learning objectives.
- Determine acceptable/expected learning outcome by asking what our students can do at the end of each learning unit and at the end of the course.
- Select instructional material, both the textbook and supplemental instructional material and plan for instructional and learning activities.
- Make testing/assessment an integral part of instruction and learning: pre-assessment, formative assessment, and summative assessment, all aligned with the learning objectives identified at the beginning of planning.

The Backward Design approach to curriculum planning for various levels of CFL courses has worked well. According to Buehl (2000), advantages of this approach include:

- Students are less likely to become so immersed in the factual detail of a unit that they miss the whole point for studying the topic.
- Instruction focuses on global understanding and not on daily activities; daily lessons are constructed with a clear vision of what the overall "gain" from the unit is to be.
- Assessment is designed before lesson planning, so that instruction drives students toward the essence of what they need to know.

Once the lead teachers and the director reach an agreement on the curriculum design, they work collaboratively to prepare for the teacher training activities. In order to help teachers become well informed about the curriculum of their level and the content of the pre-service teacher training, we create a dynamic web site to facilitate the teacher training process. In addition to the course description and syllabi for each level, the website also contains recommended readings that help teachers develop the essential competencies of CFL teaching excellence. In order to help the teachers prepare for the level they are assigned to teach, syllabi for all levels are posted on the website to help faculty members learn about the textbooks and the instructional objectives of different levels of the curriculum.

The initial training workshop at the start of the program last 2 or 3 days that address five major components: the philosophy and pedagogical principles of the program, theme-based approach to language teaching and learning, the teaching of grammar and other linguistic elements such as Chinese characters in communicative contexts, effective lesson planning that leads to engaging, challenging, and intellectually stimulating learning activities, and familiarize faculty with proficiency-based assessment and testing procedures and how to make testing an integral part of the curriculum. The initial training workshop is to help teachers achieve a better understanding of the current relevant theories on language teaching and learning with a focus on the development of Chinese language proficiency. Second language acquisition theories and effective Chinese language pedagogy are always discussed in the context of actual teaching of various levels of classes of the program. Training sessions are connected with specific lesson planning of the actual classes that they teach. Specific examples are used to guide the design of CFL curriculum, learning materials, and specific learning activities. The following chart is what I made to guide teachers for their weekly team lesson planning:

备课会的一些注意事项 (a guide for effective team lesson planning)
- 熟读课文, 找出重要的语言点及功能项并想出讲解方法及练习活动

  Read the text to be taught and identify the important points, linguistic and functional, to be covered in teaching, and consider ways to teach and practice the identified points.
- 记下重点(包括重点词汇、句型、句段、语用功能等等)

  Write down the identified points (vocabulary, grammar patterns, paragraph structures, and pragmatic elements) to bring to team lesson planning.
- 设计课堂活动, 例如:情景、角色、讨论题、讲和练的方法

  Design classroom teaching and learning tasks such as contextualized practice, role-play, and group discussion tasks.
- 对所教的学生要有充分的了解, 做到因人而异的个性化教学

  Understand the different strengths and needs of the students and prepare to deliver individualized and differentiated instruction.
- 了解学生的学习情况及出现的问题, 并找出解决问题的措施

  Understand the specific difficulties of the students and prepare methods and techniques to help students overcome their difficulties.

教学目标 Learning goals (功能 functional)	例如: 描述能力、论述能力 e.g., the ability to describe or explain
教学重点 Elements to teach (语言点、语法、语段结构、重点词汇、修辞等等, important vocabulary and grammar points, text structures, and rhetorical devices etc.)	
如何热身? how to warm-up 如何讲解? how to explain	

(continued)

如何以练代讲? how to practice 使用什么例句? modeling to use	
练习方法: 用什么情境出句型? design context 替换练习? substitution practice 转换练习? transformation practice 完成句子? sentence completion 设计交际情景和活动? designing communicative tasks 语段练习? sentence to paragraph level practice 角色扮演? role-play	
布置作业 homework design	
考什么? 怎么考? 考试的结果怎么用来完善教学? formative testing: what to test, how to test, and how test results are used to improve teaching and learning	

The chart above is designed to help teachers learn to develop effective lesson plans. Teachers of each level meet twice weekly under the leadership of the lead teacher of that level. The chart follows the Backward Design approach and serves as a guide for the team lesson planning meeting. Before the meeting, teachers are required to read the text of the lesson carefully, take notes on important areas of the lesson to cover and prepare a draft of the lesson plan. During the meeting, they will exchange ideas about the teaching of the lesson and collaboratively determine the following items: (1) learning objectives of the lesson, linguistic and functional, aligned with the overall learning objectives of the course for the students they teach, (2) highlighted linguistic forms and their functions in lesson planning, (3) methods to explain/present the new content (warm-up and pre-assessment activities, how to provide structured input and modeling etc.), (4) techniques to use for students to practice the new content (step-by-step learning activities for both controlled and automatic processing), (5) assignments and formative assessment to use to reinforce teaching and learning, and (6) summative assessment (weekly tests) to be used to assess students learning outcome. Part of the team lesson planning is devoted to a discussion on common errors and strategies for error correction at both group and individual levels.

Hands-on training experience is the major component of the workshop. Research has shown that the provision of the necessary knowledge about language can be successful in changing the conception of language teaching, but the acquisition of relevant knowledge and the changing of conceptions of teaching alone does not appear to allow full and consistent transfer of knowledge to successful and effective teaching of the language (Bartels 2005). In other words, teachers have to develop effective teaching competence through sufficient hands-on training dealing with the particular group of students they teach. Bartels (2005) argues that the factors that help teachers develop teaching excellence are concrete information, a focus on using the relevant knowledge on specific teaching activities, and time spent on such practice tasks. Each part of the training sessions held at the Chinese School contains a hands-on component either in the form of demo teaching or actual lesson planning.

Following the demo teaching there is always a reflection and discussion session, which consists of teacher's self-assessment and peer critiques. Micro-teaching sessions often last 15 min for each new teacher. For the micro-teaching, the teachers use the lesson plans prepared by themselves with help from their lead teachers. The follow-up critique section includes reflection by the teachers themselves and other teachers' comments as well.

Teachers are informed what textbooks to use in advance and, in some cases, teachers are required to select one lesson and bring one lesson plan they had written and then guided to address the following questions: are the lesson and teaching techniques aligned with the goals and objectives of the curriculum written for their students? What can be done to enhance them? How do we use the materials most effectively in different types of class sessions? How do we make the material accessible and understandable for students? How do we use them as models of effective communicative competency for the students to follow? What specific questions and learning activities should we develop in order to help students to understand authentic materials at the appropriate level and how can we help students develop productive skills/competencies? After addressing these questions, teachers leave the workshop with better understanding of the process of selecting material that meets the need of their own students; they also leave with a completed and enhanced lesson plan that they can use in their own classes.

The training component on proficiency-based testing and assessment is designed to align testing with the proficiency learning goals, making sure that testing is an integral part of the curriculum. We often invite one of the teachers who is a certified OPI tester to conduct a session on OPI testing in order to familiarize the teachers with the criteria and procedures of OPI testing so that they can apply the principles of proficiency-based language assessment in their own teaching and testing. Understanding of proficiency levels is also valuable for setting the learning goal adequately according to learners' proficiency levels, using the test results formatively for enhanced teaching, and documenting test results to show program effectiveness.

We strongly believe that effective teacher training needs to be ongoing. Like learning a foreign language, learning to become a successful CFL teacher takes time and repeated practice. As a follow-up to the initial training workshop, we have ongoing training activities. First, we often put in place a professional teacher trainer who spends time to observe and videotape new teachers in action during the first week of instruction, and then meets individually with the teachers and provides specific and constructive suggestions to help teachers improve their teaching. Then, an additional in-service teacher training workshop is conducted by the teacher trainer and the program director, incorporating examples from the class observations. Rigorous teacher training and close supervision at the beginning of the program is essential to ensure excellent teaching performance throughout the program. Results of the regular teaching evaluations also lead to another round of class observation and training to address the areas that need improvement. The director holds meetings with each lead teacher to discuss the teaching evaluation results and seek ways to improve and enhance teaching and learning. When necessary, new teachers may be assigned mentors who help the new teachers in need through class observation and/or preparing lesson plans together. Teacher training continues to take place throughout the program. There are several ongoing activities such as mini workshops that address specific issues such as innovative approaches to the teaching of grammar, material development, action research by teachers, developing advanced level Chinese competency, and integrating computer technology into the CFL language curriculum. The presenters are selected by the director based on proposals submitted by the lead teachers.

Another important ongoing teacher training activity is the team lesson planning. Lead teachers of each level meet twice a week not only to prepare lessons, but also to assess their students' performance and seek ways to improve instruction and learning. Each lead teacher often supervises 3–8 teachers. All teachers are experienced CFL teachers, but some of them need more on-the-job training. Lead teachers observe the less experienced instructors' classes and give feedback and suggestions for improvement. Their hard work and dedication to help the new teachers grow is very impressive and noted repeatedly by students on evaluations. Successful teacher training needs a team of extremely conscientious and experienced "master" teachers to help new "apprentice" teachers develop teaching excellence.

Another effective approach to in-service teacher training is through our ongoing collaborative action research projects. We have worked collaboratively on theme-based approach to material development, grammar pedagogy, and other projects. Action research is a systematic process of inquiry that helps teachers understand more about the teaching and learning process, reflect on the effectiveness of their work, explore the underlying principles of practice, and then make informed and enhanced instructional decisions. For instance, we spent three summers working collectively on the grammar pedagogy project. The project explored effective approaches and techniques to develop students' grammatical awareness in task-based communicative contexts. We started the project by identifying the important and difficult grammar points that we need to teach and then distribute the grammar points among the teachers. Before starting the development of teaching procedures of the grammar points we conducted workshops that deal with how second language acquisition and foreign language pedagogy research findings inform effective teaching of grammar and then reach an agreement on the framework. After the teaching procedures of the grammar points are completed by individual teachers, the director and the lead teachers meet to discuss the completed work and offer suggestions for revising by the teacher who originally completed the writing of the grammar teaching procedures. After 3 years of hard work, our book on Chinese grammar pedagogy was published by Yale University Press in 2009 (Bai 2009). The book has been well received by the field of teaching Chinese as a foreign language. Most importantly, the process of completing the project was a valuable element for effective teacher training, helping teachers understand the process of learning and the procedures of effective pedagogical scaffolding to help students learn Chinese grammar in meaningful and engaging contexts. For each of the grammar points, teachers first need to consult useful references of Chinese pedagogical grammar to summarize the syntactic, semantic, and pragmatic characteristics of the grammar point. For example, the following is a paragraph on what teachers need to know (教师须知) about the grammatical structure of 无论……都…… and the common errors made by learners are described.

> "无论"是连词，与副词"都"构成"无论……都……"句型。"无论"又作"不论、不管"，"不管"常用于口语。本语言点表达的是不管出现什么情况，结果都是一样的，或前提条件不同，结果不变，强调没有例外。学习者在该语言点上容易出现的问题:a) 主语用在"都"后，如:无论怎么样都他不该打人。B)在"选择疑问短语"中用"或者"取代"还是"，例如，无论中国人或者美国人都喜欢打篮球。(Bai 2009, p. 227)
> Translation: "无论 (no matter) is a Chinese conjunction. With the adverb "都", it forms the "无论 (or 不论、the more colloquial term 不管) ……都……structure. It is used to mean "no matter what happens, the results will remain the same, no exceptions." Students often make two types of errors: a) that the adverb "都" is misplaced before the subject of a sentence, e.g., 无论怎么样都他不该打人; b) that "或者" is misused instead of the correct word "还是" in the alternative question phrase, e.g., 无论中国人或者美国人都喜欢打篮球。(Bai 2009, p. 227)

The main part for the teaching of each grammar point is the instructional procedures (课堂操作程序) that include the interactive method to present the grammar point, and learning activities – from teacher-controlled to open-ended – to help students learn the grammar point. The following is an example to illustrate the innovative approach to present and practice the new grammar point 无论……都……:

> 课堂操作程序 (Teaching Procedures)
> <u>语言点导入举例 Example for Introducing Grammar Items</u>
> 问答导入式(课前板书语言点"无论……都……")
> Introducing and explaining the structure interactively
> 师:学生能不能在教室里吸烟?
> Teacher: Can students smoke in the classroom?
> 生:不能。．
> Student:No.
> 师:老师能不能在教室里吸烟?
> Teacher: Can teachers smoke in the classroom?
> 生:……
> 师:校长呢?
> Teacher: How about the principle/director of the school?
> 生:
> 师:老师指着板书示范:无论谁都不能在教室里吸烟。老师再领说。
> Teacher:Pointing at the structure written on the board: No matter who you are, you cannot smoke in the classroom.
> <u>操练: 句子转换(主题＝校园生活)</u>
> Practice: sentence transformation
> 在明德暑校学生上午、下午、晚上都得上课。
> Students go to classes in the morning, in the afternoon and in the evening as well.
> 在学校学生上午、下午、晚上都可以上网。
> Students can get online all the time, morning, afternoon, and evenings.
> 张老师、刘老师、李老师说话，我们都听得很清楚。
> We understand all the teachers clearly, Professors Zhang, Liu, and Li.
> 这里的课, 语言课、文学课……我都喜欢。
> I like all the classes here: language, literature …

操练 (open-ended):看图说话(写故事):认真的李英
Practice: Story-telling based on the pictures given, the diligent Li Ying

小英是个用功的好学生 (Xiao Ying is a highly diligent student),

(去上学, go to school)。

放学之后 (after school), (复习, study), ……

The purpose of this open-ended learning activity is for students to practice the grammar point repeatedly in a meaningful theme and produce a paragraph such as the following:

> 小英是个用功的好学生,无论晴天或雨天,她都会准时去学校上课。放学之后,不论是哪一个科目,她都会认真复习,所以每一次考试,不管是数学还是国语,小英都能考一百分。小英在家里也是个好孩子,无论是爸爸还是妈妈说的话,她都会仔细地听。吃晚饭的时候,妈妈无论做什么菜,小英都会把它吃完。小英的朋友无论有什么困难,小英都会帮助他。所以不论是谁,都称赞小英是好孩子。(Bai 2009, p. 228)
> Translation: Xiao Ying is a diligent student. She goes to school on time no matter the weather is good or bad. After class, she studies and reviews all subjects of the school, so she gets perfect scores on all her tests, math, Chinese, etc. Xiao Ying is a good child at home. She listens to her parents all the time. It does not matter what her mother cooks, she is always happy to finish her meals. Xiao Ying is always ready to help her friends, no matter what difficulties they encounter. All people around her praise her and say that she is a good child.

Although most of the examples cited are from the Chinese School of Middlebury College, the components of teacher training discussed in this section are also found in other intensive CFL programs mentioned earlier. These intensive programs have made great contributions to CFL teacher training and professional development. CFL teachers from these programs have become some of the best teachers in teaching CFL in the US colleges and universities. They report positively about their experience in participating in the teacher training sessions discussed in this section. However, we notice some challenges and limitations in these programs such as the lack of time we can spend on teacher training and the diverse professional backgrounds of the teachers. Most programs require heavy teaching load in order to ensure sufficient level of practice for their students both in and out of classrooms. Therefore, it is hard for teachers to have enough time and energy to participate fully and actively in all the teacher training activities. The diverse professional backgrounds of the teachers also make it difficult for the programs to offer training workshops that appeal to all teachers. There needs to be a more systematic structure to offer differentiated approach to individual needs and a way to ensure sustained learning, practice, and reflection on best practices.

Summer Professional Development Programs: The cases of STARTALK and CARLA

One of the well-received summer teacher training programs is STARTALK that was created in 2006 and funded by the National Security Agency and administered by the National Foreign Language Center at the University of Maryland (https://startalk.umd.edu/. It supports teacher training across the United States. It also supports summer camps for students to learn Chinese at different levels. Its website contains information about the number of programs and the strength and focus of each program. It has funded teacher training workshops to provide professional development for world language teachers including CFL teachers. The teacher training workshops are often 2 or 3 weeks long and support world language programs and teachers to improve their program development and teaching excellence. The workshop participants are mostly current teachers of Chinese from pre-college levels, whose academic backgrounds are various such as Chinese language pedagogy, Chinese literature, East Asian Studies, or other content areas of humanities and social sciences.

The Chinese School of Middlebury College received several grants from STARTALK to establish training programs for CFL teachers. Workshop participants investigate how current relevant theories relate to local decisions on organizing and planning for instruction in a CFL program. Through lectures, discussion, and hands-on experiences, participants learn to understand the process of learning Chinese as a foreign language, to identify critical issues in CFL pedagogy, and to enhance their ability to develop appropriate teaching materials and to plan and implement effective classroom activities that help students develop proficiency in the target language across the three communicative modes (interpersonal, interpretive, and presentational), in the five goal areas (communication, cultures, connections, comparisons, and communities) as outlined in the Standards for Foreign Language Learning in the twenty-first Century (ACTFL 2012).

The Middlebury STARTALK is an intensive training program. Workshop participants meet for five hours daily (9–3:00 with a one-hour lunch break) and are expected to speak only the target language during the period of training. They are required to eat and socialize "in the language" and participate in the cocurricular activities and cultural presentations that are offered in the Chinese School. Unlike the conventional university Chinese graduate courses which are mostly lectures or discussions on theoretical and practical issues, the workshops create a learning-by-doing environment. Participants receive training in developing, organizing, and delivering effective curriculum. They are required to observe and critique classroom procedures, strategies and teaching techniques and evaluated on special projects, teacher portfolios, reflection papers on their observation of classroom instruction, and micro-teaching demonstrations.

At the conclusion of the workshop participants were able to do the following:

- Describe the national standards and explain how they can incorporate the standards in their instruction
- Plan activities that align with student performance outcomes

- Align assessment pieces, both formative and summative, with the activities they design
- Apply their knowledge while developing thematically based instructional activities that they can use in their classrooms, and
- Design activities that would accommodate the diverse needs of different student populations

We had participants who teach CFL to either college-level students or K-12 students. The exposure to the national standards is certainly not as prominent for teachers from post-secondary institutions. However, our participants can incorporate some of the standards-based principles into their own teaching and testing, making them more effective. College teachers also have a better understanding of what is going on at the K-12 level and from what learning environments their students are coming. The training program enabled better K-16 articulation.

At the end of the workshop, participants presented, in front of a panel of four professors of Chinese pedagogy and their classmates, standard based thematic units and sample activities and materials they created during the workshop. Suggestions were made during this stage of the workshop to facilitate final revisions. Participants submitted final copies of their materials by the end of the workshop. These materials incorporated activities and tasks that reflected components of the national standards. They are also required to do micro-teaching followed by critiques and suggestion from their professors and other workshop participants. Our workshop participants also developed a resource network among themselves with online resources and websites that can be useful to their students and colleagues at their home institutions. Workshop participants agreed to make their materials available online (https://startalk.umd.edu) to the community of CFL teachers who would like to use them. Participants reported that they used the thematic units they developed during the training in their actual teaching when they returned to their schools.

Since its start in 2006, the program has played an important role in professionalizing the CFL field, especially in the K-12 teaching and learning of Chinese as a foreign language. It has enabled the creation of a systematic curriculum framework aligned with standards-based world language teaching. CFL program developers and teachers can get online and access rich resources that deal with the following important areas of the CFL curriculum (https://startalk.umd.edu/public/principles/additional-resources):

- Implementing a Standards-Based and Thematically Organized Curriculum
- Conducting Performance-Based Assessment
- Integrating Culture, Content, and Language
- Using the Target Language and Providing Comprehensible Input
- Facilitating a Learner-Centered Classroom
- Adapting and Using Age-Appropriate Authentic Materials

The program has also enabled the creation and sharing of a variety of instructional activities designed to keep learners engaged as they process and produce language.

All of the activities include a short overview description, step-by-step directions, as well as helpful tips and suggestions for differentiation, which are arranged by the three communicative modes: interpretive, interpersonal, and presentational (http://teach.nflc.umd.edu/startalk/classroom-activities).

The workshops are short, intensive training programs. Most of the participants had no knowledge of standards-based material design. This meant they had to learn, process, and apply a good amount of new information in a very short period of time. Workshop organizers had to tighten the schedule and requirements to accommodate this constraint. We need to allow more time for discussion and experimenting with the standards-based instruction as well as for collaboration and design. The workshops would be more valuable if they were longer than 2 or 3 weeks. A system can be created so that teachers can participate in a number of well-articulated and sequential programs that help teachers reinforce their learning and develop the required competencies systematically in a series of well-sequenced workshops. Another drawback of the workshops is that most of them are designed to develop pedagogical competencies; few of them are designed to develop linguistic and/or cultural competencies. Many CFL teachers at the K-12 levels need metalinguistic training such as Chinese pedagogical grammar or other aspects of Chinese linguistics.

In addition to STARTALK, there are other summer workshops such as CARLA's summer institutes (https://carla.umn.edu). The institutes are all graduate-level courses and participants can get graduate credits. However, like other short-term teacher training programs, they are very intense and short. According to my conversations with past participants, they do learn a lot in a week and, if they are taking the institute courses for credits, they will have a final project and will normally be given more time to do after the institute ends. These summer institutes are for different language teachers, and participants of different languages may be placed together. Therefore, the training is conducted in English and there is no learning opportunity to improve Chinese language skills. But, on the other hand, since these institutes are for all kinds of language teachers, participants have the opportunity to learn from colleagues of other languages about best practices in teaching world languages. The following are some very thoughtful comments from a CARLA participant (Wei 2019):

- It really brings in the latest practices in language teaching, so you feel like you are well-informed.
- I also met some Chinese teachers teaching kindergarten from immersion schools in San Francisco as well as College ESL teachers from Chinese university in Beijing.
- "Creativity in the Language Classroom" was extremely helpful and just what I needed because my class tends to be a little "serious" and my activities are not very varied and fun. This institute was very well organized and very informative.
- Some drawbacks are that some institutes may indicate it is for K-12, but may not have too much for kindergarten teachers. I also think the pace was very fast and you really hope that the institute could be longer so the instructors can go more in

details and you have more time to explore the topic. My brain was not functioning at the end, so taking three institutes at a time is a lot and you do need time to let what you learned sink after each institute.
- You may feel overwhelmed or even confused during the institute, but after it ends, you may feel clearer. If you actually practice putting what you learned into practice after the institute, you may find yourself learning even more and are clearer about what they talked about during the institute.

In this section, we described two examples of summer teacher training programs that have contributed greatly to the professional growth of many CFL teachers. These summer programs are well received by CFL teachers. However, the programs are short and do not provide sufficient amount of time for participants to practice implementing what they learn from the courses they take. In the next section, we will discuss another type of teacher training workshops that are available throughout the year.

On-Demand Training: The Case of AP Workshops

In addition to the in-service training model and the summer programs described above there are other teacher training workshops throughout the year. One example is the AP workshops conducted by College Board Consultants in different parts of the USA and other countries. The information about the times and places of these workshops are published on the College Board AP central website.

The AP Chinese Language and Culture Course was first offered in the fall semester of 2006 and the first exam was administered in 2007. It provides students with ongoing and varied opportunities to develop their communicative competence and their understanding of the Chinese culture. The AP Chinese course parallels the level of a fourth semester college level course. The exam is organized according to the three communicative modes articulated in the National Standards and includes listening, speaking, reading, and writing. The College Board offers a broad range of professional development opportunities for teachers and administrators such as one-day workshops, information sessions with the AP Chinese Development Committee at the AP National Conference, weeklong summer institutes, Pre-AP world languages strategies, and Vertical Teams workshops. These workshops are offered throughout the year and during the summer as well.

The AP Chinese workshop (half day or 1 day long), and the summer institute (a week-long event, 30 contact hours) are designed to provide professional development opportunities for AP Chinese teachers and others who are interested in teaching other levels of Chinese as well. During the workshop College Board endorsed trainers who offer interactive lectures and hands-on experience for workshop participants. The lectures and hands-on learning activities are to help participants understand the standards-based course framework, the proficiency-based exam aligned with the standards-based AP course, best pedagogical practices for teaching Chinese as a foreign language, and how the new AP resources provided by College

Board can be utilized to enhance the teaching and learning of Chinese as a foreign language. The new resources include the theme-based course description that are based on the six universal themes and eight skill sets developed by a group of world language education professionals under the leadership of the College Board, instructional guides for the thematic units, personal progress checks, and the AP question bank. The most well-received component of the workshop is the discussion and application of effective instructional strategies to develop student' communicative competence in all three modes of communications (interpersonal, interpretive, and presentational). Participants also like the hands-on experiences in analyzing the sample student responses to the AP test items by using the rubrics/scoring guidelines designed for assessing writing and speaking competencies.

In addition to the workshops, many CFL teachers also find the College Board webinars and the on-line material of AP Central useful for their professional development. Due to the impact of Covid-19, the College Board has offered many online professional development opportunities. The following are some of the recent webinars:

- Guiding Principles and High-Leverage Teaching Practices for Designing Oral Activities (speaker: Cecilia Chang)
- Steps to Building a Successful School to School Partnership (speakers: Bob Davis, Cleopatra Wise, and Tom Martineau)
- Focus on Proficiency: Coach for Performance (speaker: Laura Terrill)
- Supporting Activities Beyond the Classroom to Enhance Language and Culture Learning (speaker: Marybeth Fuller)
- Proactive Management through Classroom Engagement (speaker: Lise Olsen-Dufour)
- The Communicative Approach to Teach Culture and Structures (speaker: Maggie Chen)

In addition to the aforementioned professional development opportunities, AP teachers can access regularly developed training material such as the interactive online modules, e.g., Building an Effective Program, Diagnosing Student Progress, Interpersonal Communication: Developing Speaking abilities and many more at the resources page on their website. The newly developed training material such as the AP Classroom includes new instructional resources that provide daily support for the teaching and learning of AP Chinese. The teachers of AP Chinese can access Unit Guides that include the outline of the content and skills covered on the exam and suggestions to help teachers integrate material into their courses. They can also use Personal Progress Checks, learning how to use formative tests to measure student progress through each unit and throughout the year. College Board has conducted both in-person workshops and webinars to train teachers making use of these resources. The instructional materials and training workshops provided by the College Board are significant for CFL teachers' professional development. For example, the universal themes and skill sets were identified as essential after careful study by the College Board. However, the limitations of the training workshops designed for AP Chinese language and culture are similar to the other short-term training programs, i.e., workshop participants do not have sufficient time to practice applying the principles into actual teaching through hands-on experience. For instance, the universal themes are important content areas for the AP Chinese language and culture course to cover, but it requires much more work and consideration on how to adapt these themes appropriately to different levels of AP and pre-AP classes.

In this section, we have examined some examples of teacher training and professional development. It is obvious that much progress has been made in teacher training in the area of teaching Chinese as a foreign language, but, we notice a lack of systematic planning for teacher training in many Chinese programs that we have observed. The training workshops offered are discrete in nature, each workshop focusing one area such as standards-based assessment or literacy development. They are randomly done without a structure to systematically address the essential elements of effective teacher development. In the next section we discuss the essential competencies of CFL teaching excellence and then, in the last section of this chapter, we provide a framework and some examples for a systematic and integrated approach to teacher training.

Essential Competencies for CFL Teaching Excellence

There has been continuous research and discussion on what constitutes teaching excellence in the field of teaching Chinese as a foreign language (Lü 1989, Kubler 1997, 2006, Christensen and Noda 2002, Chang 2006, Sung 2006, Zhang 2006a, b, Yu 2007, Lu and Ma 2016, Lu 2017). The conclusions and recommendations about essential elements of teaching excellence can be summarized into three categories: knowledge about Chinese language and language teaching, knowledge about the relevant subject areas such as Chinese literature and psycholinguistics, a set of skills of Chinese language pedagogy, and psychological readiness. On the knowledge base about Chinese language for CFL teachers, Lu (2017) argues for the inclusion of Chinese phonetics and phonology, morphology and character formation, lexicon and syntax. According to Lu (2017) CFL teachers should have a general understanding of (1) Chinese phonetics such as how sounds and tones are made through different places and manners of sound articulation and how sounds and tones are produced differently according to the phonological context, (2) Chinese character formation (六书), the historical development of Chinese characters and frequent errors made by CFL learners, (3) Chinese lexicon such as word frequency, the denotative and connotative meanings of words and the implications on teaching and learning of Chinese as a foreign language, and (4) Chinese syntax such as parts of speech of words, the characteristics of commonly used functional words, and various grammatical structures and their uses in communication. On the skill sets for CFL teachers, Lu (2017) considers the following skills essential for CFL teachers to develop: strong Chinese and foreign language skills, classroom management and teaching skills to keep students highly motivated and actively engaged in their learning of CFL, and research skills and strong skills to interact with students and their parents successfully and effectively. In addition to the knowledge base and skill sets necessary for CFL teachers, Lu (2017) also values an important psychological factor, the sense of responsibility and high level of dedication to CFL learners. Competent CFL teachers should genuinely care for their students and are willing to devote time and energy to learning about the learning styles, strengths, and needs of their students and adjust their teaching accordingly.

The NFLC Guide for Basic Chinese Language Programs (Kubler 1997) provides a general and practical framework for teacher training. The recommendation is similar to what has been discussed above in that it calls for teachers to develop understanding of the nature and defining characteristics of the Chinese language such as the four tones, the Chinese characters, the new grammatical concepts such as topic/comment, verb aspect, and classifiers, the transcription systems such as Pinyin, and the difference between spoken and written Chinese that is heavily influenced by classical Chinese, both lexically and syntactically. Part II of the NFLC Guide for Basic Chinese Language Programs offers the following recommendations: adopting learner-centered approach to teaching, teaching learners how to learn in order to become life-long learners, integrating culture in language learning, paying special attention to listening development, both interactive and interpretive listening, distinguishing speech and written representation, regarding fluent communicative competence as the goal of developing speaking competence, teaching vocabulary in context, developing reading competence after speaking and listening, developing reading skill in meaningful context and integrate authentic materials, learning characters in contexts instead of individually, developing writing competence after the development of other three skills, distinguishing productive and receptive characters, developing tests that reflect learning goals, and providing feedback after testing. In addition to the importance of knowing the Chinese language and teaching skills, the NFLC Guide for Basic Chinese Language Programs also emphasizes the importance of the following skills: becoming familiar with and taking advantage of information and available support systems, participating in ongoing professional development activities, recruiting students and sustaining students' motivation in learning Chinese, increasing the amount of language practice beyond the classrooms, and integrating technology and distance learning resources into teaching and learning.

In summary, a successful and effective CFL teacher training program should deal with the elements of teaching excellence discussed above: a solid knowledge base relevant to the successful teaching and learning of CFL such as knowledge about the Chinese language, Chinese meta-linguistic competencies, and about theories on Chinese language acquisition and pedagogy, a set of strong pedagogical skills and the psychological readiness to be a successful CFL teacher. Sufficient ongoing hands-on training under close supervision by expert teacher trainers are essential to help teachers transfer knowledge to effective teaching practice in actual classrooms. Tsui (2003) argues that expert teachers should:

- Exercise more autonomy, not dictated by models
- Be more aware of and more responsive to the contextual factors
- Have in store more well-established routines (sets of classroom procedures)
- Be more able to integrate their knowledge of the subject, the learner, the classroom management etc.

A Call for a Systematic and Integrated Approach to Teacher Training

It is essential for the CFL training program to provide learning opportunities for teachers to learn to integrate their knowledge about the Chinese language and effective language teaching, and develop and implement effective teaching methods that lead to successful learning outcome. The teacher trainees should acquire adequate knowledge and teaching skills, and be able to make informed pedagogical decisions that work well for their own students in their local contexts, knowing what works and what does not in their own classes. In the remaining part of this chapter we attempt to propose a framework for a systematic and integrated approach to teacher training and professional development. We will use the framework and the main integral elements of the graduate program that we developed at the Chinese School of Middlebury College in 2007. Since its start in 2007, the program enrolls 25–30 students each summer and has achieved excellent results, i.e., students becoming employed or returning to their teaching positions as high quality CFL teachers. One major difference between this program and other university graduate programs is that it is well integrated into a residential immersion program that offers all levels of Chinese as a foreign language. The graduate students observe CFL students of all levels learning Chinese on a daily basis.

The graduate program at the Chinese School is designed to produce effective CFL teachers for both pre-college and college level Chinese programs. Most participants are current CFL teachers from the pre-college levels. They come for four summers to earn an MA degree in Chinese Language Pedagogy. There are also CFL teachers who come to the program for further professional development instead of earning an MA degree. In order to obtain the MA degree, students are required to take 12 courses and a language teaching practicum during the final summer session. The 12 courses consist of three important areas: Chinese linguistics, second language acquisition and pedagogy, and literature and culture courses; they are designed to help students/teacher trainees to build the knowledge base for effective teaching of Chinese as a foreign language. Students/teacher trainees benefit greatly from the unique Chinese-only learning environment where all courses are taught in Chinese and there is also a strictly enforced Chinese language pledge. This is another major difference from other US university graduate programs where for most courses of Chinese language pedagogy, Chinese linguistics or other related subjects are taught in English. The first type of required courses at the graduate program of the Chinese School of Middlebury College is linguistics such as Introduction to Chinese Linguistics, Chinese Phonetics, Chinese Semantics, Sociolinguistics, Discourse Analysis, and Pedagogical Grammar. The following are three examples of courses in this area. These courses are designed to develop the teachers' meta-linguistic competencies. Many CFL teacher certification programs require a Chinese proficiency level such as Intermediate High, but it is not sufficient just to maintain a high level of Chinese language proficiency. Being a native speaker of Chinese does not mean that he or she has the necessary meta-linguistic competencies to be an effective CFL teacher.

Introduction to Linguistics: This is a graduate course on the basic foundation and principles of linguistics. It combines the teaching of language with research application, explanation, and discussion on how to understand linguistic phenomenon, and how to understand linguistic rules. Central content includes: the definition of language, linguistic concepts, sub-division of linguistics such as phonetics, morphology, semantics, systemized syntax, and other related issues, the communicative function of language, the relationship between language and thought; regional and societal variations in language, the relationship between language and speech, the relationship between language and writing, and the nature of language and language acquisition.

Phonetics: This course uses the actual phonetic situation of Mandarin Chinese to explore basic concepts and research methods in phonetics. The specific content in this course includes the biology of phonetics, the physics of phonetics, psychological foundations, and an introduction to the real manifestations of Chinese phonetics' syllabic structure, tones, vowels, consonants, rhythm, and intonation. The instructor will detail the procedures for using phonetic experiments to analyze tones, vowels, consonants, and intonation. In addition, the course will also cover the process of acquisition of Chinese phonetics for both first- and second-language learners, contact and development between languages and dialects, and other topics, building extensions upon the foundation of basic research.

Sociolinguistics: The object of sociolinguistic research is language in society. Language's history, language's current situation, and language's development are all inseparable from groups of people who use language, and all of these have an intimate relationship with humanity and society. Language is the tool of communication for people. Language exists only within society; once one leaves society, there is no language. In both studying and researching language, one cannot leave societal life aside. The content of sociolinguistics includes hierarchies in language and society, ethnic communities and languages, gender differences in language, linguistic and speech situations, communication and interaction between language and society, language's ethnic variants, and geographic conditions and language. Sociolinguistics is a branch of linguistics, yet it is also its origin. This course gives an account of sociolinguistic theory and research methods, and through case study analysis, allows students to design their own research proposals, engage in real linguistic studies, and connect theory with reality.

The second area of the graduate curriculum relates to second language acquisition and foreign language pedagogy such as Principles and Practices of Teaching Chinese as a Foreign Language, Second Language Acquisition, Methods & Materials

Development, Action Research for CFL Teachers, Teaching Vocabulary in Chinese as a Foreign Language, Reading Development, and Language Testing and Assessment. These courses are designed to help students understand the process of learning a foreign language and the pedagogical scaffolding necessary for effective learning of a Chinese as a foreign language. The following are three examples of courses in this area.

> *Theory and Practice in Teaching Chinese as a Second Language*: This course introduces students to the principles of second language acquisition (SLA), a field of study that investigates how people learn a second language and provides a basis for understanding the SLA research related to learning and teaching Chinese as a second language. Theoretical issues to be covered include what it means to know a language and how one becomes proficient in a foreign language, factors that affect the learning process, and the role of one's native language in the process of second language acquisition. We will also examine what SLA research has discovered about teaching grammar, pronunciation, vocabulary, and writing. The goal of this course is to explore ways in which SLA theories can be applied to facilitate acquisition of Chinese as a second language both in terms of classroom teaching and syllabus design.
> *Language and the Brain*: Language is an important feature separating humans from the animal kingdom. Starting with a basic physiological and psychological understanding of neuroscience from the perspective of language, the course will cover the cognitive aspects in the relationship between language and the brain's structure and function. Using the most current scientific research and drawing from various disciplines, students in this course will explore the mysteries of language and the brain.
> *Methods in Teaching Chinese as a Foreign Language*: This course instructs second language teaching methodology and classroom teaching skills, including the theoretical foundations in linguistics, pedagogy, psychology, and culture study; main schools and the background of second language teaching methodology; the teaching methodology of different language elements (pronunciation, characters, words, and grammar); the teaching methodology of different language skills (listening, speaking, reading, and writing), the teaching consciousness and behaviors of second language instructors, etc. The purpose of this course is to give students more opportunities to know details of second language teaching theories and methodology, furthermore, master the process, methods, and skills of Chinese classroom teaching. At the end of this course, students will be able to use appropriate methodology in different circumstances.

There are also courses that combine the content of the first and second areas of courses such as Pedagogical Grammar or the Teaching of Chinese Vocabulary. In addition to the conventional lecture and discussion formats there are also assigned class observations and micro-teaching. These courses are valuable and well received because they help teacher trainees apply what they have learned from theoretical linguistics courses to the teaching of various aspects of Chinese language and culture. The following are three examples of the courses of this type.

Teaching Chinese Vocabulary: This course combines teaching practice and acquisition theory, and introduces to students vocabulary teaching and acquisition in CFL. Students are expected to understand the basic contents and main features of modern Chinese vocabulary system research findings of Chinese semantics and lexicon as well as some achievements and problems of vocabulary acquisition in Second Language Acquisition. Emphasis will be on the key points and difficulties in Chinese vocabulary teaching. Through lectures, discussions, and micro-teaching activities, students will enhance their abilities to select appropriate strategies to teach vocabulary, predict and avoid common vocabulary errors, and develop learning modules and activities for effective teaching of Chinese vocabulary.

Pedagogical Grammar: This course stems from practical applications, leading students to a full understanding of the fundamentals of Chinese grammar as well as the principles and methods of teaching Chinese grammar. Through lectures, discussions, and micro-teaching activities, students will take the first steps toward using theory in the practical application of teaching Chinese grammar. Equipped with a higher level of active knowledge of grammar, students will enhance their abilities to predict and avoid grammatical errors, resolve grammatical problems, and put together grammar-teaching exercises. The main content of this course is separated into three units. The first unit will introduce the basic principles of teaching grammar. The second unit will cover the basic content of Chinese grammar to be taught. The third unit will focus on honing students' skills and methods for teaching Chinese grammar. The second unit is the centerpiece of the course; it is not a comprehensive overview of modern Chinese grammar, but rather, was developed based upon the actual needs and requirements of teaching. For this unit, the instructor has chosen difficult grammar points, and will specifically discuss and analyze Chinese content words, words with grammatical meaning only, complements, sentence patterns, word order, compound phrases, and other such important, difficult, or easily mistaken points and educational tactics for overcoming them. Through micro-teaching activities, students will connect theoretical knowledge with the actual practice of teaching. Instructors will analyze, explain, and comment upon students' recording teaching practice sessions to guide them toward improvement in their grammar-teaching practice.

Sociolinguistics and Teaching CFL: This course is designed to introduce the subject of sociolinguistics and deal with its implications for teaching CFL. To build sociality, human beings inevitably use languages as a social artifact to conduct their transactions in the world they live. Foreign language learners, in order to be proficient in a target language, are reasonably required to develop their awareness of socio-cultural aspects of language structures they employ to communicate.

Their development of the socio-cultural awareness is contingent upon their interactions with their language instructors, for whom this course is designed. The goal of the course is to provide a systematic introduction to the nature and dynamics of the intricate relationship between language, culture, and society. Special emphases will be put on basic theories, research methods, and research outputs in the topics of interactional sociolinguistics, ethnography of communication, language socialization and language acquisition, linguistic variation analysis, and sociology of language. Upon completion of the course, students are expected to be able to (1) define different research methods in sociolinguistics studies, (2) to summarize key findings in sociolinguistics outputs and to provide to critiques against and offer suggestions for the outputs, (3) integrate observable sociolinguistic phenomena in Chinese speech communities in their designs of curriculums for Chinese teaching.

In addition to the foundation courses in Chinese linguistics, language acquisition and pedagogy, and research methodology, teacher trainees are also required to take the third area of courses on literature, film, fiction, contemporary issues of China and so on. Course offerings in this area include Chinese literature and cultural studies such as Masterpieces of Chinese Literature, Chinese Cinema, Cultural Topics in Contemporary China, Women in Chinese Society, Modern Chinese Novel and Culture, Social Changes Reflected in Contemporary Chinese Literature, Chinese Thought & Culture: Traditional Patterns, Modern Manifestations. These courses are designed to help CFL teachers to gain adequate understanding of various aspects of the Chinese culture and society and the fundamental knowledge about Chinese literature and films. These courses are also valuable to sharpen the students' analytical and critical thinking skills and their creative writing ability as well. What they learned from these courses can help them be more effective in integrating culture into their Chinese language classes and also enable them to teach content area courses of Chinese studies for advanced-level CFL students. The following are four examples of the courses in this area.

Chinese Cinema: This course examines the cinemas of Mainland China, Taiwan, and Hong Kong, with a special focus on a number of topics such as the birth of the nation-state, history and memory, gender and sexuality, social migration, and urban culture. We will consider how Chinese films between and beyond the two fin-de-siécles have created the spectacles of "China," narrated its oftentimes traumatic history, and reflected its ongoing social and cultural transformation. Students will learn to appreciate Chinese films as artifacts of specific historical contexts and develop critical perspectives in understanding modern China through films. Class meetings are conducted in Chinese and combine lectures, discussions, and students' presentations.

Classical Chinese Poetry: Classical genres, themes, and poets are introduced and examined through poetry recitation and construction. Students will understand and appreciate the linguistic and aesthetic features of Chinese language in poetic form. The course integrates linguistic mastery with poetry study, introduces the formal structure of Chinese classical poetry, and examines its stylistic variations within historical contexts.

The Quest for Spontaneity in Ancient Chinese Thought: This course examines a fascinating issue that attracts the attention of all the major Chinese thinkers in ancient times, that is, the freedom of acting without calculation or conscious effort – a state of being that can be best summarized as ziran (self-so) in Chinese, or "spontaneity" in English. Through close readings of selected passages from the original texts by such big names as Confucius, Mencius, Mozi, Xunzi, Liezi, and especially Laozi and Zhuangzi, students will learn to detect and analyze the differences – and similarities, if any – between the varied understandings and interpretations of this "spontaneity" from the perspectives of different schools of thought.

Dream as a Literary Mode: Through a survey of the "dream" stories in various genres, this course examines how a recurrent literary theme evolves into a powerful mode of expression and narration, a convenient and effective tool for authors from different ages to represent the ethos of their times. This "case study" of a literary mode provides a vivid example of how a literary tradition reinvents and revitalizes itself though its development. Students can also expect to become more sensitive to the distinctive generic features as well as the conventions of the major genres in Chinese literature.

The course descriptions are constructed by faculty and the director of the Chinese School of Middlebury College collaboratively. The three areas of courses discussed above are required for completing the graduate program; they are designed to help teacher trainees to build the necessary knowledge base for successful CFL teachers. New courses are continuously added as we learn more about what the teachers should know to be successful CFL teachers. For instance, some students suggested that a course be created to help them understand the US pre-collegiate school system and its culture. This is particularly true for the graduate students who are new to the United States. Issues such as K-12 classroom management and advocacy can become great obstacles to Chinese program development and effective language instruction. In response to the newly identified need, a one-day workshop was added for all graduating students in response to this suggestion. For a complete list of the courses and their descriptions one can find them on the website of the Chinese School of Middlebury College (https://www.middlebury.edu/language-schools/languages/chinese/graduate/courses).

Throughout the program, students not only improve their Chinese language proficiency and acquire the necessary knowledge for teaching Chinese as a foreign language, but also received instruction and class assignments that guide students to transfer the knowledge they acquire into successful teaching competencies. Most of the courses contain elements that help students learn how to develop and refine teaching techniques and strategies, to learn to use tools specifically targeted to the challenging areas of Chinese language instruction, and address national trends in standards-based world language teaching and assessment. In order to enable the CFL teacher trainees to develop the ability to transfer knowledge to successful teaching skills, we make the hands-on training an integral element of the curriculum through class observation, material development, collaborative team lesson planning, and micro-teaching. Students are encouraged to take advantage of the available learning opportunities of the Chinese School of Middlebury College. For instance, they can observe some of the best Chinese language teachers in action; they can observe the CFL learners both in the classroom and in informal situations; we also arrange for them to attend the team lesson planning of different levels of the language program of the Chinese School. At the end of the program, each teacher trainee is required to take the course of Language Teaching Practicum, which provides the ideal opportunity to synthesize the Middlebury learning experiences and apply them meaningfully into their own teaching and research. From observation to instructional design to actual teaching, students apply what they have learned in an authentic experience.

Unlike other courses where trainees focus on the learning of one element of the required areas that build a strong base knowledge for CFL teachers, the course of Language Teaching Practicum requires trainees to synthesize and integrate what they have learned to develop teaching excellence. In this course, students have the opportunity to reflect upon the content from the three areas of courses that they have already taken and to improve their ability to design and effectively implement classroom teaching practices. For the first part of the course students work collaboratively to lead each of the sessions on the following topics in a seminar style: Standards-based Language Teaching and Learning, Developing Teaching Excellence through Effective Class Observation and Critique, Backward Design and Standard-base Curriculum Design and Implementation, Building Thematic Units and Effective Lesson Planning, Standards-based/Performance-oriented Testing and Assessment in CFL, and Building and Sustaining a Successful CFL Program. Students' assignments include: readings on TCFL, reflections on the questions and problems brought up in readings, classroom observation, directed course design, class observation and reflection papers, micro-teaching activities, and a comprehensive portfolio, which helps them synthesize and integrate what they have learned to their developing of teaching excellence through reflection and application. The portfolio should include their term papers from other courses, their class observation notes and reflection papers, their micro-teaching lesson plans and reflection papers on their teaching, the thematic learning units they developed and plan to take home to use in their own teaching. This course is an integral component of a rigorous and intensive graduate curriculum in which teacher trainees build a solid bridge from theory to application and practice.

One of the hands-on learning tasks that help trainees to transfer their knowledge base to effective teaching is the class observation and critique papers on what they observe. Students are arranged to observe classes of their choice with the approval and assistance of the lead teacher of the Chinese School. The following are the major questions that guide the students' class observation and critique papers: What did the instructor try to accomplish? What did the students learn today? What communicative competencies did they develop? What classroom tasks and learning activities contributed to their learning? What topics, activities would you have done differently if you had been the instructor? What went well? What went as planned? What went not as planned and how did the instructor handle the unplanned happenings? In order to answer these questions, students are required to record the specific examples and exact descriptions of what happened in the classes. Rubrics for assessing the quality of their class observation and critique papers are distributed to the students in advance as a learning guide for this hands-on learning task.

The major hands-on learning task that helps teacher trainees to transfer their knowledge base to effective teaching is the micro-teaching assignment. For this task they work collaboratively in pairs to co-plan and co-instruct a lesson in four sessions, each teaching two classes and acting as a peer-coach and co-teacher for the partner (s). Following the teaching, they are required to write a formal critique of the process and the experience, using specific examples from the taped micro-teaching classes. The goal of this task is to provide an opportunity to link theoretical knowledge with standards-based, pedagogically sound lesson design and teaching practice, to provide an opportunity to design and teach meaningful lessons in an authentic environment to an authentic audience (student volunteers from the language classes of the Chinese School), to provide an opportunity to observe and evaluate their own teaching as well as that of a partner, and to learn how to share observations constructively with others, to provide participants with specific suggestions to improve the effectiveness of lesson design and implementation, and to provide an opportunity to adapt lessons in order to make them developmentally appropriate and applicable in their own teaching situations.

The topic(s) of the lesson for their micro-teaching can be chosen from a list generated by the lead teachers of the level that the teacher trainees have decided to teach. The teacher trainees choose the focus of the lesson (e.g., teaching contextualized grammar points 语言点 such as the Chinese ba-sentence or a functional item 功能项 like how to apologize in Chinese), plan it, and put the elements of effective language teaching into practice. Teacher trainees are reminded to pay particular attention to active student participation by considering ways that they can actively involve students in their teaching of the lesson. They are also supposed to include appropriate and ongoing formative and summative assessment techniques to identify what their students have learned and whether they meet the learning objectives. In addition to classroom learning, the teacher trainees should also design and give independent practice tasks as homework assignment. The teacher trainees should also learn to identify what materials will be needed for the micro-teaching such as equipment, materials, visuals, handouts, etc. Prior to their teaching their professors are available to go over their plans and answer questions that the teacher trainees may have. Their professors provide them with suggestions and feedback, and the teacher trainees have the opportunity to revise the lesson plans, if necessary, before they teach the lesson.

For each lesson, one of the paired trainees will be the "lead teacher," and the other partner will provide support. While the lead teacher presents most of the "input," the support partner(s) may actively participate as well. She or he may help with modeling, questioning, assisting with guided practice, while also observing specific student and teacher behaviors, taking notes, helping look for evidence of learning, or focusing on a specific aspect of the lesson requested by the lead teacher. To prepare the reflection papers, the trainees view the taped classes, take notes, discuss, and critique their own teaching using the principles discussed in the seminar sessions. The self-reflection paper should contain specific references to the lesson planning, the teaching, and the initial impressions. It should not be a summary of what is taught, but a meta-cognitive reflection of the process. The summary and the reflection should help improve future teaching practices.

As mentioned above, teacher trainees are also required to complete two other hands-on learning tasks that help them learn to transfer their knowledge base into teaching excellence: the composing of a portfolio and development of a thematic learning unit that they are to take home to use for their own students. The portfolio includes their term papers from other courses, their class observation notes and reflection papers, their micro-teaching lesson plans and reflection papers on their teaching, and the thematic learning units they developed.

The thematic unit should contain clearly stated instructional objectives that are connected to a particular theme, appealing to their own students. It should include a complete and well-sequenced unit with learning objectives for each lesson, aligned with the unit learning goals. Teacher trainees should demonstrate their understanding and command of the standards-based world language instruction and learning as well as the techniques of the Backward Design approach.

To summarize, this section describes a unique graduate program of Chinese language pedagogy to illustrate a systematic and integrated approach to teacher training. It provides learning opportunities for teachers not only to develop the knowledge base required of CFL teachers, but also to learn to apply the knowledge base effectively in curriculum design, material development, and classroom teaching and assessment through sufficient amount of guided learning-by-doing activities. The program is well integrated into a residential immersion program that offers all levels of Chinese as a foreign language. The learning environment enables the graduate students to observe CFL students of all levels learning Chinese on a daily basis and makes it easy and convenient to arrange for teaching practicums. Other teacher training program may not have this unique learning environment, but the systematic and integrated approach to teacher training and its integral elements proposed in this chapter may apply in other programs.

Conclusion

As the importance of learning Chinese language and culture is widely recognized, the field of teaching Chinese as a foreign language has developed from a marginal field with less developed pedagogy to a stronger profession. We witness great progress in Chinese language pedagogy research, curriculum design and material development, and teacher training and professional development. However, a review of the current available programs and practices of teacher training reveals the need for more effective and systematic models of teacher training in the field of teaching Chinese as a foreign language. In this chapter, we proposed a systematic and integrated approach to teacher training and professional development that is viable for effective teacher training.

The graduate program at the Chinese School of Middlebury College has worked well in terms of developing competent CFL teachers that are well recognized in the field. We have found that it is essential for an effective teacher development program to offer balanced courses in three areas of Chinese linguistics, foreign language acquisition, and Chinese language pedagogy as well as Chinese literature and cultural studies. The program needs to ensure that the teacher trainee/graduate students have a solid knowledge base relevant to the successful teaching and learning of CFL. The program needs to provide the teacher trainees with sufficient on-going hands-on training through guided class observation and reflection and micro-teaching sessions under close supervision by expert teacher trainers to help teachers transfer their knowledge about teaching CFL to effective teaching practice in the actual classroom. There may be limitations and challenges to apply this model in other contexts that are different from the intensive summer immersion program where the graduate program is situated. But the framework and proposed training content and activities can be adapted to fit different situations. For instance, instead of a formal degree program, a CFL program can create a series of systematic and integrated workshops that enable CFL teachers to develop the required essential elements of teaching excellence and offer these workshops in multiple years. The CFL teachers can create a plan in the form of a portfolio that document the required training sessions that address what teachers should know and be able to do in order to become successful and effective teachers of Chinese as a foreign language.

References

ACTFL (American Council on Teaching of Foreign Languages). 2012. *Proficiency Guidelines*. https://www.actfl.org/publications/guidelines-and-manuals/actfl-proficiency-guidelines-2012. Accessed 28 June 2020.

Asia Society and College Board. 2008. Chinese in 2008: An expanding field. https://asiasociety.org/sites/default/files/C/Chinesein2008.pdf. Accessed 28 June 2020

Bai, Jianhua, ed. 2009. *Chinese grammar made easy: A practical and effective guide for teachers*. New Haven: Yale University Press.

Bartels, Nat. 2005. *Applied linguistics and language teacher education*. New York: Springer.

Buehl, Doug. 2000. Backward design; Forward thinking. *Education News*. https://web.archive.org/web/20130327141424/http://www.weac.org/news_and_publications/education_news/2000-2001/read_backwards.aspx. Accessed 28 June 2020.

CARLA Summer Institutes. https://carla.umn.edu. Accessed 28 June 2020.

Chang, C. (张曼荪). 2006. Mingde zhongwen shuxiao 2006 nian shizi peixun jishi yu pingshu 明德中文暑校2006年师资培训纪实与评述 [The Teacher-Training Workshop of the 2006 Middlebury Summer Chinese School: Innovations and Reflections]. Guoji Hanyu Jiaoxue 国际汉语教学 [International Chinese Language Teaching and Learning] 2006. No. 4.

Christensen, Matt, and Mari Noda. 2002. *A performance-based pedagogy for communicating in cultures: Training teachers for East Asian languages*. Columbus: National East Asian Languages Resource Center, The Ohio State University.

CLASS (Chinese Language Association of Secondary-Elementary Schools). https://www.classk12.org. Accessed 29 June 2020.

Davidson, Dan. 2017. The National K-12 Foreign Language Enrollment Survey Report. https://www.americancouncils.org/sites/default/files/FLE-report-June17.pdf. Accessed 28 June 2020.

Kubler, Cornelius. 1997. *NFLC guide for basic Chinese language programs*. Washington, DC: National Foreign Language Center.

———. 2006. *NFLC guide for basic Chinese language programs*. 2nd ed. The Ohio State University Foreign Language Publications.

Lü, Bisong (吕必松). 1989. Guanyu duiwai hanyu jiaoshi suzhi de jI ge wenti 关于对外汉语教师素质的几个问题 [On the fundamental requirements of CFL teachers]. Shijie hanyu jiaoxue 世界汉语教学 [Chinese Teaching in the World] 1:1–17.

Lu, Jianming (陆剑明). 2017. Hanyu jiaoshi peiyang zhi wo jian 汉语教师培养之我见 [On Chinese Teacher Development]. Guoji Hanyu Jiaoyu 国际汉语教育 [International Chinese Language Education] 2: 3 (第二卷第三期).

Lu, Jianming, and Zhen Ma (陆剑明、马真). 2016. *Hanyu jiaoshi yingyoude suzhi he jibengong* 汉语教师应有的素质和基本功 [Fundamental qualities of Chinese teachers]. Beijing: Waiyu Jiaoxue yu Yanjiu Chubanshe 外语教学与研究出版社 [Foreign Language Teaching and Research Press].

STARTALK. https://startalk.umd.edu/public. Accessed 28 June 2020.

Sung, Juyu. 2006. *Teaching Chinese as a second language: A practice-oriented approach*. Taipei: Xiuwei Publishing.

Tsui, Amy. 2003. *Understanding expertise in teaching*. New York: Cambridge University Pres.

Walton, R. 1997. Preface. In: *NFLC guide for basic Chinese Language Programs*. Washington, DC: National Foreign Language Center.

Wei, Binbin. 2019. Personal email communication.

Welles, Elizabeth. 2004. Foreign language Enrollments in United States Institutions of Higher Education, Fall 2002. *ADFL Bulletin* 35: 7–26.

Wiggins, Grant, and Jay McTighe. 1998. *Understanding by design*. Alexandria: Association for Supervision and Curriculum Development.

Yu, Li (虞莉). 2007. Meiguo daxue zhongwen jiaoshi shizi peiyang moshi fenxi 美国大学中文教师师资培养模式分析 [An analytical study on the models of Chinese teacher development in US universities]. *Chinese Teaching in the World* (世界汉语教学) 1: 114–121.

Zhang, Hesheng (张和生). 2006a. Duiwai hanyu jiaoshi suzhi yu peixun yanjiu 对外汉语教师素质与教师培训研究 [Research on CFL teacher quality and teacher training]. Beijing: Shangwu Yishuguan 商务印书馆 [The Commercial Press].

——— (张和生). 2006b. Duiwai hanyu jiaoshi suzhi yu peixun yanjiu de huigu yu zhanwang 对外汉语教师素质与培训研究的回顾与展望 [Research on CFL teacher quality and teacher training: past, present and prospect. *Journal of Beijing Normal University* [北京师范大学学报] 3: 108–113.

The History and Development of Assessment of Chinese as a Second Language

28

Dan Nie 聂丹 and Qifeng Zhao 赵琪凤

Contents

Introduction	676
Historical Development of CSL Assessment	676
Foundational Phase (1980–1990): Starting from Scratch	676
Expansion Phase (1990–2000): Constructing a Comprehensive Examination System	677
Improvement Phase (2000–2010): Enhancing the HSK's Student Support Services	678
Diversification phase (2010–): Meeting the Needs of the Times	679
An Overview of Contemporary CSL Assessments	682
Contemporary CSL Assessments in China	682
International Chinese Language Proficiency Tests	689
Summary of CSL Assessments	691
Issues and Future Directions in CSL Assessment	691
Candidate Differentiation in Current CSL Assessment	691
Examination Categorization in Current CSL Assessment	692
Technology for Assessment Implementation	693
Future Directions for CSL Assessment	693
Conclusion	698
References	698

Abstract

This chapter discusses assessments of Chinese as a second language (CSL), tracing the development of a comprehensive system to assess non-native speakers' proficiency in spoken and written Chinese. The chapter begins with a historical overview of CSL testing and then reviews contemporary trends. In conclusion, it considers the future of Chinese proficiency testing and offers suggestions for improvement.

Translated by Janet Davey, The Australian National University

D. Nie 聂丹 (✉) · Q. Zhao 赵琪凤
Beijing Language and Culture University, Beijing, China
e-mail: donna0923@163.com; qifengspring@hotmail.com

Keywords

Chinese proficiency test · Historical development · Future directions · Chinese as a second language

Introduction

Studying Chinese as a second language (CSL) proficiency tests has been a major focus of the language assessment field. But while the research topic has valuable insights to offer, to date there has been no comprehensive overview of the history and development of CSL testing, leaving a lacuna to be filled by Chinese language assessment researchers. This chapter seeks to address the sizeable gap in language assessment literature by tracing the development of CSL assessment from its origins to the present day, exploring the directions the field has taken. More broadly, the overview presented here acts as a reference point for further developments in Chinese language testing and promotion of the Chinese language around the globe.

Historical Development of CSL Assessment

Chinese language proficiency testing is an important part of international Chinese language education. Proficiency tests are used for several purposes, including the assessment, selection, and evaluation of students. ("Chinese language" or "Chinese" refers to Mandarin Chinese unless specified otherwise.) Research on this topic is crucial to the development of international Chinese language education (Zhao 2020). In the 1980s, the formal designation of *duiwai Hanyu* ("Chinese as a foreign language") as a distinct field, and subsequent emergence of proficiency tests for non-native learners, had a formative impact on the teaching of Chinese as a foreign language (TCFL). Tests of Chinese as a second language (CSL) have continued to be devised, developed, and improved ever since. Drawing on Sun's (2009) definitive history of the *Hanyu Shuiping Kaoshi* (HSK), China's main Chinese language proficiency test, this section presents a timeline of CSL assessment development in China.

Foundational Phase (1980–1990): Starting from Scratch

In the early 1980s, TCFL – a form of foreign (or second) language teaching – had yet to develop into a mature academic discipline. At that time, there was little in the way of basic research on teaching methods and a dearth of language proficiency tests. The global introduction of the "Test of English as a Foreign Language" (TOEFL), an English language proficiency test, had a significant influence on China's foreign language teaching community. It inspired a vision for some pioneering Chinese academics to develop an equivalent CSL standardized proficiency test. In December

1984, an HSK design group was established at Beijing Language Institute (now Beijing Language and Culture University or BLCU). The group, comprised of academic Liu Xun and five others, released the first set of HSK test papers in 1985. Moving from test design to development, the team then launched the HSK as a distinct Chinese "brand"; determined where it would fit within a broader language assessment system; established a systematic theoretical foundation underpinning the test; and conducted a large number of trials to verify its reliability and validity.

In terms of the HSK's place in the language assessment system, Liu, Guo, and Wang (1988, 110) described it as "not only a multipurpose proficiency test, but also a standardized examination with fixed and accurate assessment criteria." The HSK's multipurpose nature is reflected in several aspects. First, HSK is the main assessment scale to measure the true Chinese proficiency of examination candidates. Second, HSK data is an important tool for examining and evaluating both teaching outcomes and teaching quality. Third, the macro-level guidance that it provides to TCFL also promotes improvements in instruction methods and revision of teaching materials. In addition, the HSK is used to specify the level of Chinese proficiency required for international students to (1) attend a university in China; (2) satisfactorily complete a Chinese preparatory education course; (3) be assigned to the appropriate level of a Chinese language program; and (4) obtain a certificate of general Chinese language proficiency (Liu et al. 1988).

The existence of a standardized language proficiency test is an important indicator of a country's language teaching capability. The successful development of the HSK in the late 1980s filled a gap in China's system of language assessment. It was also a sign that TCFL had taken a significant step in its development as a discipline.

Expansion Phase (1990–2000): Constructing a Comprehensive Examination System

In its second decade of development, the HSK expanded to become a comprehensive examination system with a full set of assessment criteria and grading guidelines. During this phase, the HSK development group continued to study and analyze the HSK at a theoretical level.

As the HSK became more influential, the number of candidates sitting the test gradually increased. This prompted the development group to begin contemplating and mapping out the next steps for the examination system. When the test was officially launched in 1989, it had been limited to candidates with an elementary or intermediate level of Chinese proficiency and had no means of assessing those at either an advanced or beginner level. Following several years of research and development, the HSK Testing Center (formerly of BLCU) introduced "HSK (Advanced)" in 1993 and renamed the original exam "HSK (Elementary-Intermediate)." This was supplemented by "HSK (Basic)" in 1997. Today, the HSK is a comprehensive examination system made up of the HSK (Basic), HSK (Elementary-

Intermediate), and HSK (Advanced) formats. With this set of exams, the HSK is able to cater to learners of Chinese regardless of their proficiency level, providing suitable proficiency tests for beginners, intermediate learners, and advanced students alike.

Examination procedures and grading guidelines were then formulated to support the HSK examination framework. While work to devise the different exams was still in progress, the HSK development group also began to draft a series of guidelines for TCFL assessment. In 1988, the group published a set of "Chinese Proficiency Assessment Criteria and Grading Guidelines." This was followed by the "HSK (Advanced) Guidelines for China" in 1994 and the "HSK (Basic) Guidelines for China" in 1998.

During the HSK's decade-long expansion phase, examination researchers began to study the HSK at greater depth and with a broader scope, especially at a theoretical level. By proceeding from broader issues – such as proposed theoretical frameworks of second language proficiency, studies of Chinese language abilities, and concept validity research – researchers were able to explore some of the acknowledged hard problems of linguistics and psychology while contributing valuable findings of their investigations and analysis to what was then only an emerging field in China (Chen 1996, 1997). Investigation into these fundamental theories then became the prerequisite for and foundation of subsequent work to improve the design of the HSK and would ensure that it offered a meaningful measurement of language proficiency. At the same time, exploring fundamental theories also led scholars and research students involved in language assessment to consider major issues in the field and promote practical, applied research.

Improvement Phase (2000–2010): Enhancing the HSK's Student Support Services

By the turn of the millennium, the HSK had become a well-known Chinese language proficiency test with almost 20 years of development behind it. Nearly one million candidates had completed the test, including members of China's ethnic minorities (Sun 2009). But if the HSK was to reap further benefits for society, it needed to be able to meet the needs of as many Chinese learners as possible. During this time, the HSK development group combined in-depth studies of language cognition with an evolving concept of language assessment in order to improve the HSK and to develop well-targeted Chinese language tests.

This informed the "Scheme for a Revised Version of the Chinese Proficiency Test (HSK)," published by the BLCU HSK Testing Center in 2007. According to the scheme, the improved HSK would be characterized by the following: (1) the nature and applications of its assessment would remain unchanged; (2) equal consideration would be given to examining communication skills and knowledge of the Chinese language; (3) its marking system, interpretation of scores, and grade divisions would be fair and reasonable and guided by scientific evidence; (4) each of the Basic, Elementary/Intermediate, and Advanced level formats would provide a comprehensive assessment of learners' listening, speaking, reading, and writing skills and include both written and oral test components;

and (5) the test paper would have a logical structure, and question design would reflect principles of language acquisition and instruction (HSK Testing Center 2007).

During this period the BLCU HSK Testing Center also drew on survey responses and user feedback from overseas markets to develop two new exam formats. In 2006, it introduced "HSK (Entry Level)" for beginners with little prior knowledge of Chinese. It also developed a second set of examinations, collectively known as the "Test of Practical Chinese" (C.TEST). Intended for people working in business or trade who require Chinese language communication skills, the C.TEST assesses Chinese proficiency in everyday contexts by means of a spoken interview and a written paper.

It is evident that ongoing development has allowed the HSK to advance with the times. This has ensured that the reliability and scientific rigor of the test continue to improve and that the HSK retains the ability to adapt to changing market demands and candidates' needs, draw on diverse perspectives to develop its examinations, and enhance student support services.

Diversification phase (2010–): Meeting the Needs of the Times

In 2012, China's Ministry of Education (MOE) published a "List of Undergraduate Majors at Regular Higher Education Institutions." In the list, what had previously been known as the TCFL Program was renamed "International Chinese Language Education" and became a second-tier discipline within the more general "Chinese Language and Literature" program (An and Han 2018). The reclassification highlights the fact that contemporary international Chinese language education involves teaching Chinese not only to non-native learners in their home countries and regions but also in China itself (Li 2014). Accordingly, the development and implementation of a Chinese language assessment system have become increasingly multifarious since 2010.

The trend toward diversity is reflected in the range of candidates and test purposes catered for in the contemporary language assessment system. There are now examinations designed for particular groups of Chinese learners. The "New HSK," for example, assesses the ability of non-native speakers to communicate in Chinese in their daily, academic, and professional lives. It is run by the Office of Chinese Language Council International 国家汉语国际推广领导小组办公室 (also known as Confucius Institute Headquarters), an organization affiliated with the Ministry of Education. The Office of Chinese Language Council International is more commonly known as *Hanban* 汉办. The C.TEST and "Business Chinese Test" (BCT) assess the Chinese proficiency of non-native speakers working in sectors that rely on international communication, such as business, trade, education, and the arts. The "Chinese Language University Entrance Examination for International Students" (CLUEE-IS) – which is targeted at specific academic disciplines, international student markets, and tertiary institutions in China – identifies talented students for admission to undergraduate programs. The "Youth Chinese Test" (YCT) assesses primary- and middle-school students whose first language is not Mandarin on their ability to use Chinese in everyday and school settings. It aims to enhance the students' self-confidence and sense of achievement when communicating in Chinese. The "Chinese Language Proficiency Test for China's Ethnic Minorities" (MHK) is a national standardized examination designed to assess the proficiency of Chinese people from an ethnic minority background whose first language is not Mandarin. Finally, the "Overseas Chinese Proficiency Test" (HSC) is targeted toward Chinese people living abroad and assesses their ability to use their heritage language.

The diverse nature of China's contemporary language assessment system also extends to the range of assessment methods and applications of different examinations. For example, while the "C.TEST Spoken Interview," the "BCT Speaking Test" or "BCT (Oral iBT)," and the "Spoken Chinese Test" (SCT) are all designed to assess oral proficiency, each exam has a different format. As the name implies, the C.TEST Spoken Interview is an interview. The BCT (Oral iBT) is conducted online using a computerized adaptive testing (CAT) methodology, while the SCT employs automated speech processing technology. Among the numerous Chinese language tests, the main body of examinations is still closely associated with China's national priorities and the needs of society. The following paragraphs will discuss two of these exams in detail, the New HSK and CLUEE-IS.

The New HSK: Integrating Testing and Teaching

The New HSK was officially launched in November 2009 and fully introduced in 2010. Following a principle of "integrating testing and teaching," the New HSK was designed to align with associated teaching materials and reflect current trends in international Chinese language teaching. The test aims to "enhance teaching through testing" and "enhance learning through testing" and is used to help stimulate greater interest in learning Chinese.

The New HSK consists of a written test and an oral test, which are completed separately. The written test can be taken at one of six levels: HSK 1, HSK 2, HSK 3, HSK 4, HSK 5, or HSK 6. The oral test, known as the HSKK (an acronym of *Hanyu Shuiping Kouyu Kaoshi*), has three levels: HSKK Beginner, HSKK Intermediate, and HSKK Advanced. The HSKK is a recorded exam. The number of New HSK candidates for the years 2012–2016 is given in Table 1.

As the figures above illustrate, the number of people sitting the New HSK – the current general-purpose Chinese language proficiency test – nearly doubled in 5 years, increasing from 250,000 in 2012 to 470,000 examinees in 2016. Annual

Table 1 New HSK candidates (2012–2016)

Year	Candidates
2012	250,000
2013	300,000
2014	350,000
2015	400,000
2016	470,000

growth in the number of candidates remained above 15% across this period, reaching approximately 21% at its peak. This shows that the New HSK remains the leading examination among current Chinese language proficiency tests.

Because the New HSK is a new type of international standardized test of Chinese language proficiency, associated research is still in its infancy. Studies have been limited; the few that have been conducted have tended to be exploratory in nature. In a review paper, Tang (2016) identified a total of 144 research papers on the New HSK published between 2010 and 2015, which were retrieved from the China Knowledge Resource Integrated Database (CNKI). The research papers were roughly divided by topic into four major areas: the development, implementation, and global promotion of the New HSK; assessment tools used in the New HSK; the relationship between teaching, exam preparation, and the New HSK; and comparisons of different language tests. Results from their statistical analysis showed that from 2010 to 2015, research on the New HSK concentrated mainly on teaching and exam preparation; such studies accounted for 58.33% of the research articles retrieved from the CNKI database. The next most common topic was New HSK assessment tools – the subject of 24.31% of the 144 articles examined. Articles on the New HSK's development, implementation, and global promotion, along with comparative studies of different language assessments, were relatively small in number, accounting for 6.94% and 10.42%, respectively.

Chinese Language University Entrance Examination for International Students (CLUEE-IS): Selection of Students

The number of international students applying for formal education programs in China has maintained a high growth rate in recent years. International students come to China not merely in order to learn Chinese. Many wish to complete a full tertiary education in China, at an undergraduate or postgraduate level, and graduate with a diploma or degree in a chosen discipline.

The "Chinese Government Scholarship Undergraduate Preparatory Education Examination for International Students" (CGS UPEE-IS), currently conducted at the 17 tertiary institutions that offer preparatory education courses for international students, was developed and implemented by the Institute of International Student Education Policy and Evaluation Research at BLCU in 2013. Three years later, the examination became the official standard used to assess completion outcomes of Chinese Government Scholarship (CGS) preparatory education courses for incoming international undergraduates. It also became a "threshold" examination required for CGS-funded international students to commence undergraduate studies in China.

The university admission system for international students includes a comprehensive language examination – assessing conversational and academic Chinese proficiency – as well as a test of students' knowledge of mathematics, physics, and chemistry. The language examination, CLUEE-IS, differs from other language examinations for international students in two main ways. First, general language examinations for international students have a wide range of applications. They are not designed for specific tertiary institutions, academic disciplines, or student markets, but rather used to broadly assess whether the candidate's language proficiency meets the minimum requirement for them to undertake an academic program at any university in the host country. Conversely, CLUEE-IS is targeted toward particular higher education institutions, academic disciplines, and student markets so that it reflects the educational philosophy and assessment standards of these institutions. Second, general language examinations for international students typically only examine language skills, whereas CLUEE-IS offers a comprehensive assessment of a range of abilities, encompassing not only language proficiency but also the knowledge and basic skills the student would require to study their chosen discipline (Wang et al. 2016).

An Overview of Contemporary CSL Assessments

While remaining cognizant of the history of international Chinese language education, universities and examination authorities in China and overseas have recently begun to focus on new frontiers in Chinese language assessment. Chinese language testing within China is currently undergoing a process of diversification and refinement. A range of Chinese language tests have been developed to suit different groups of candidates. For example, there are specific tests for non-native Chinese speakers (such as international students), overseas Chinese, and members of ethnic minority groups. Development of specialized oral tests, which come in a variety of assessment formats, has also facilitated a more targeted assessment of language capabilities than was possible through general tests. The next section provides an overview of the most representative CSL assessments currently conducted in China or overseas, focusing on the group of candidates catered for by each test.

Contemporary CSL Assessments in China

This section introduces the most representative Chinese language tests conducted in China and aimed at one of the three main examinee groups: non-native Chinese speakers (mainly international students), overseas Chinese, and members of China's ethnic minorities. It also discusses the focus of each test in terms of the specific language skills it assesses (i.e., whether it is broad or specialized).

Chinese Language Tests for Non-Native Speakers

New HSK: A Comprehensive Examination
The New HSK is an international standardized test of Chinese proficiency, which focuses on assessing CSL learners' ability to communicate in Chinese in their daily, academic, and professional lives. It was formally announced in November 2009 and fully introduced in 2010.

The New HSK consists of separate written and oral tests. The written test can be taken at one of six levels: HSK 1, HSK 2, HSK 3, HSK 4, HSK 5, or HSK 6. The oral test, the HSKK, has three levels: HSKK Beginner, HSKK Intermediate, and HSKK Advanced. The HSKK is a recorded exam.

The New HSK follows a principle of "integrating testing and teaching." Uniting language teaching and language learning, the exam was devised with reference to relevant teaching materials and current trends in international Chinese language teaching. This design reflects its aims to "enhance teaching through testing" and "enhance learning through testing."

Examination results for the New HSK are recognized worldwide for an extended period of time and can be used for the following purposes: (1) university admission, class allocation, course exemption, and awarding of academic credits; (2) to hire, train, and promote staff through employment agencies; (3) to assist Chinese learners in understanding and improving their language proficiency; and (4) to evaluate the quality of Chinese language teaching or training at teaching institutions or training organizations. However, for international student applications to Chinese tertiary institutions, a New HSK certificate can only be used as proof of Chinese proficiency for a period of 2 years after completing the exam.

Chinese Language University Entrance Examination for International Students (CLUEE-IS)

Preparatory education courses are the foundation of a university's teaching curriculum. Preparatory education for international students in China began in the 1950s. Since the start of the twenty-first century, the number of students coming to China has increased every year, as have the range of disciplines studied. Now, preparatory education for international students is moving into a new phase of development.

At present, CGS UPEE-IS is conducted at the 17 tertiary institutions that offer preparatory education courses for international students. This examination is the sole official standard used to assess completion outcomes of CGS preparatory education courses for international undergraduate students.

The international student admission system includes a comprehensive language examination (CLUEE-IS), which assesses students' proficiency in conversational and academic Chinese, and also a test of their knowledge of mathematics, physics, and chemistry. The value of the language examination is illustrated by the fact that CLUEE-IS assesses proficiency in both everyday and academic Chinese. It also differs from other language examinations for international students in two major ways. First, general language examinations for international students are used for various purposes; they are not specially designed for particular tertiary institutions, fields of study, or student markets. Instead, they are mainly used to conduct a broad assessment as to whether the candidate's language proficiency meets the minimum admission requirement for an academic program at any given university in the host country. In contrast, CLUEE-IS is targeted toward specific higher education institutions, disciplines, and student markets to ensure that it reflects the educational philosophy and assessment standards of those institutions. Second, general language examinations only assess language proficiency, whereas CLUEE-IS provides a comprehensive assessment of different abilities, including international students' academic knowledge and overall learning capabilities in addition to their Chinese proficiency.

Test of Practical Chinese (C.TEST): A Comprehensive Examination

The C.TEST examination series was expressly designed by BLCU and tailored to meet the needs of potential candidates and other stakeholders of this new Chinese language proficiency test. The C.TEST has been in use for more than a decade, having been officially launched in 2006. It is used to test non-native speakers' ability to communicate in Chinese in their everyday professional and social lives. The main purpose of the C.TEST is to assess the Chinese proficiency of people who work in areas that involve international communication – such as the business, trade, cultural, or education sector – and to provide an authoritative certification of their practical language abilities. The C.TEST was developed with the aim of producing a comprehensive assessment of a candidate's listening, speaking, reading, and writing proficiency in Chinese.

The C.TEST examination series includes both a written test and a spoken interview. The written paper has seven test levels, which are grouped into three categories: C.TEST (Level G) is for beginners, C.TEST (Levels E–F) is for elementary learners, and C.TEST (Levels A–D) assesses intermediate to advanced learners. The question design on the written paper is noteworthy in a number of ways. First of all, the listening comprehension component of the intermediate-advanced C.TEST (Levels A–D) includes a new picture-based question task. This requires examinees to look at a picture while listening to recorded questions and then choose the correct answer by relating the question to the picture. Thus, only by comprehending both the picture and recorded question will the candidate be able to answer correctly; visual comprehension without listening comprehension or vice versa would not be enough. In addition, because the C.TEST (Levels A–D) is an intermediate to advanced level exam, it also features a more complex listening-based summarizing task, which requires candidates to take brief notes on the main points of a class lecture they have just heard. As it comprehensively assesses students' listening, reading, and note-taking abilities, this type of question is highly effective for making finer distinctions in students' level of proficiency.

More broadly, the C.TEST emphasizes the importance of helping and supporting students on their journey to language proficiency. This distinctive feature is reflected in the following aspects. First, the examination materials are disclosed in full. Test materials (including the recordings for the listening component and the written paper) are released after each C.TEST, and candidates are free to take the exam paper away with them for future reference. Second, C.TEST grade reports provide examinees with detailed, useful information. The report notes the correct answer and correct response rate for every question together with the examinee's chosen responses. This allows the student to analyze their answers and compare them with the correct answers in order to identify their strengths and weaknesses. It therefore facilitates improvement in their level of proficiency. Third, the C.TEST examination certificate not only includes a test score and level of certification but also provides a diagnostic evaluation of the student's actual Chinese proficiency.

Finally, the C.TEST applies modern measurement theories and technologies to standardize examination scores. This ensures that results are consistent and comparable for candidates sitting the C.TEST at different times.

Test of Chinese as a Foreign Language (TOCFL): A Comprehensive Examination

In August 2001, Taiwan began planning for the introduction of its own Chinese language proficiency test. Its "Test of Chinese as a Foreign Language" (TOCFL) – initially called "Test of Proficiency–Huayu" (TOP) – was developed in a joint project led by three research teams from the Mandarin Training Center, the Graduate Institute of Teaching Chinese as a Second Language, and the Psychological Testing Center at National Taiwan Normal University. Since the first examination was held in December 2003, TOCFL has assessed candidates from over 60 countries. TOCFL has developed in two stages. The original exam – TOP – ran from 2003 until 2008. It was then superseded by TOCFL, which was developed according to theories of communicative language testing. Whereas TOP was a single comprehensive assessment of listening, speaking, reading, and writing abilities, TOCFL is divided into separate listening, speaking, reading, and writing tests.

The TOCFL grading system follows the international standard of language ability outlined in the Common European Framework of Reference for Languages (CEFR). There are six levels grouped into three proficiency bands, with two levels in each band. (TOCFL now also has a fourth "Novice" band with another two levels at the lower end of the grading scale, giving eight proficiency levels in total.) From lowest to highest proficiency, Band A covers beginner and basic level Chinese, Band B spans intermediate to high-intermediate proficiency, and Band C corresponds to fluency and then superior proficiency in Chinese.

Business Chinese Test (BCT): A Comprehensive Examination

The BCT is another international standardized test of Chinese language proficiency. Officially introduced in 2013, the BCT assesses the ability of examinees whose first language is not Mandarin to use Chinese to communicate in a real-world business or workplace setting. The test features separate written and oral components. The written exam has two levels – BCT (A) and BCT (B) – while BCT (Oral iBT) is a computer-based speaking test run automatically with CAT.

The BCT was introduced with the following objectives in mind: First is to create a broad-based exam that meets the practical needs of foreigners studying business Chinese or working in a Chinese language environment. As part of this, the BCT aims to increase foreigners' interest in Chinese and their confidence in communication. Second is to provide Chinese language teachers and self-taught students with teaching resources and suggested assessment tasks in order to facilitate targeted, tangible improvements in non-native speakers' communication skills in real-life business and workplace situations. And third is to evaluate candidates' ability to complete language-based tasks and offer constructive feedback for further improvement through assessment tasks modeled on typical business scenarios.

The two levels of the BCT written test, BCT (A) and BCT (B), are separate examinations that differ in test structure. Both BCT (A) and BCT (B) have listening and reading sections, but while BCT (A) assesses writing with simple tasks, BCT (B) asks for a longer, structured written composition.

Youth Chinese Test (YCT): A Comprehensive Examination

The YCT is an international standardized test of Chinese proficiency that assesses primary- and middle-school students whose first language is not Mandarin on their ability to use the Chinese language in everyday and school settings. The YCT was formally introduced in 2010 with the aim of enhancing students' self-confidence and sense of achievement when communicating in Chinese. The YCT includes a written test and an oral test, which are scored separately. The test paper and assessed content of the written exam differ for each of its four levels.

Like the New HSK, the New YCT advocates "integrating testing and teaching." The examination was designed to align with current international primary- and middle-school teaching materials and education levels. It promotes effective Chinese language teaching and learning as it "enhances teaching through testing" and "enhances learning through testing."

The New YCT also advocates that learning be "stress-free, lively and interesting." It strives to make learning Chinese practical and enjoyable so that students develop an affinity for the language and continue on the pathway to proficiency. This principle is threaded through the New YCT, from its development to its question design. Development and promotion of the YCT have stimulated interest in learning Chinese among primary- and middle-school students. It has also fostered young learners' Chinese language thinking skills, enhanced their "language sense," and strengthened their self-confidence and pride in learning Chinese. Together, these benefits lay a strong foundation for them to further their Chinese language studies while also encouraging rich language use. The test reflects a belief that students' mastery of Chinese will be largely manifested through communication skills, through their ability to use a finite vocabulary to express an infinite number of concepts and big ideas, and will not occur simply as a result of mastering mechanical grammatical structures and fixed linguistic knowledge. Examination scores from the YCT can be used as a reference point to help students assess and improve their Chinese proficiency and also to evaluate teaching outcomes in schools and other educational institutions.

The C.TEST Spoken Interview: A Specialized Examination

The C.TEST Spoken Interview was the first interview-style oral examination in China expressly designed to assess the proficiency of non-native Chinese speakers. The live interview is completed either in person or online using a standardized examination procedure and scoring process. The interview runs for 10–15 min in total, progressing through three stages: "Warm Up"; "Level Checks" and "Probes"; and "Wind Down." The combined Level Checks and Probes stage is itself subdivided into four phases: "estimation," "floor," "positioning," and "ceiling." (The terms used to describe the interview stages mimic those of the ACTFL Oral

Proficiency Interview (OPI). In the Level Checks and Probes stage, the examiner starts with an estimate of the candidate's proficiency based on the Warm Up (*estimation*) and proceeds to assess the lower and upper limits of that proficiency in order to pinpoint their true speaking ability (*positioning*). Questions that enable the examiner to determine the highest level of performance the speaker can sustain (the *floor*) are known as Level Checks, while questions that establish the performance level the speaker is unable to consistently sustain (the *ceiling*) are termed Probes.) In addition to its standardized procedure, the C.TEST Spoken Interview is also notable for being realistic (featuring everyday scenarios and conversation topics), focused on communication (through face-to-face dialogue and interaction), and evidence-based (with a verified test structure; rigorous and effective interviewer training; objective assessment process; and impartial review of all interviews and results).

Spoken Chinese Test (SCT): A Specialized Examination

The SCT was developed by Peking University in partnership with the US education company Pearson. It aims to assess non-native speakers' Chinese proficiency through automated speech processing technology.

The SCT is most noteworthy for its convenience, adaptability, and computerized scoring system. Because it assesses speech automatically, candidates can take the test at any place or time, either over the phone or on the computer, and receive their results immediately.

The duration of the exam is 25 min, there are eight types of questions, and the candidates are asked 80 questions in total. Once the test concludes, the system calculates an overall proficiency score, along with five diagnostic scores for each of grammar, vocabulary, fluency, pronunciation, and tone accuracy.

Research on the SCT shows that it can successfully determine an examinee's speaking proficiency and it has both validity and reliability as an oral examination (Li and Li 2014).

Chinese Oral Proficiency Test (HKC): A Specialized Examination

The Chinese Oral Proficiency Test (HKC) is a standardized test of Chinese proficiency. The HKC was designed by the MOE and the National Spoken and Written Chinese Language Working Committee with three main aims: first, to become part of its comprehensive system of language assessment in China; second, to respond to international demand for Chinese language learning and communication; and finally, to assess the oral proficiency of non-native Chinese speakers and overseas Chinese. From its format to its content, the HKC is a test that fully reflects everyday life in China and real-life scenarios. An HKC certificate provides official documentation of an examinee's spoken Chinese proficiency (Chai 2018b). The HKC is principally focused on pronunciation, comprehension, and expression in spoken communication. It assesses the accuracy of pronunciation and the use of vocabulary and grammatical structures. The HKC grading system has three proficiency bands with three levels in each band, making a total of nine assessment levels. Levels 1–3 form the Elementary Band, Levels 4–6 the Intermediate Band, and Levels 7–9 the Advanced Band.

Chinese Proficiency Tests for Ethnic Minority Groups: MHK

Since early 2001, the MOE has led the development of the MHK, a national standardized examination specifically devised to assess the proficiency of Chinese people who are members of an ethnic minority group and whose first language is not Mandarin. The MHK has four interconnected levels, which cater to candidates with ethnic minority backgrounds at primary school, middle school, high school, and university graduate level, respectively. Students who pass the examination at a particular grade level are awarded the corresponding certificate. As students progress from Level 1 to Level 4, improvements in their Chinese proficiency are reflected not only in the accumulation and expansion of their linguistic knowledge but also in their communication skills. The four components of the exam – listening comprehension, reading comprehension, written expression, and oral expression – test different language skills and so together ensure that the MHK provides a thorough assessment of examinees' ability to communicate in Chinese.

The design of the MHK test paper not only has made Chinese language examinations more informative evaluations of language proficiency but has also actively guided Chinese language teaching. Each level of the examination includes a speaking test. The format of the speaking test is intended to reflect examinees' actual ability to express themselves in Chinese and is adapted to accord with the language skills that could reasonably be expected for speakers from an ethnic minority background to participate in Chinese society. Furthermore, the MHK provides certification of the candidate's Chinese proficiency, which can be drawn upon by relevant bodies to guide decisions about student admission, job recruitment, and staff appointment. Schools and other institutions teaching Chinese also use the test results when evaluating students' and teachers' Chinese proficiency (Peng 2005).

Chinese Proficiency Tests for Overseas Chinese: HSC

The HSC is a system of standardized language proficiency examinations. It was created to assess the language proficiency of Chinese people overseas, to cater for the diverse backgrounds and experiences of overseas Chinese and CSL speakers more generally. Introduction of the HSC was a flagship project of the Overseas Chinese Affairs Office of the State Council and part of a major research program at Jinan University's College of Chinese Language and Culture, which also oversaw the development and implementation of the exam through its Overseas Chinese Testing Center (Zheng 2019).

In its current form, the HSC acts as a proficiency test for young people; it aims to assess overseas Chinese adolescents whose knowledge of and experience with the Chinese language lie between that of a child and an adult. Accordingly, its grading scale takes into account different stages of young people's cognitive development and uses five proficiency levels that essentially correspond to year levels in the basic education system (Wang 2018). The grading system is outlined in Table 2.

Table 2 HSC grades with corresponding age groups)

HSC level	Age of candidate[a]	Test structure
Level 1	7–8	Written exam: assessing listening, reading, and writing
Level 2	9–10	Oral exam (optional)
Level 3	11–12	
Level 4	13–15	
Level 5	16–18	

[a]From a Chinese-speaking home environment

International Chinese Language Proficiency Tests

This section briefly introduces Chinese language proficiency tests currently in use outside China. In particular, it discusses two tests conducted in the USA, the "Advanced Placement (AP) Chinese Language and Culture Exam" and the "SAT Chinese with Listening Subject Test."

AP Chinese Language and Culture Exam

The AP Chinese Language and Culture (CLC) course outline stipulates that the exam includes a broad range of questions designed to assess students on their interpersonal, interpretative, and presentational communication skills in Chinese. Whether open-ended or close-ended, questions on the AP CLC Exam focus on students' ability to use the Chinese language in the real world. The question scenarios recognize the importance of real-life language practice and so tend to be closely related to common aspects of students' social lives, such as extending an invitation, providing an introduction, making an apology, or saying thank you. Written assessment tasks often take the form of e-mails, letters, posters, advertisements, and news articles. Dialogues, broadcasts, announcements, phone messages, etc. may be used for oral assessment.

The writing section of the AP CLC Exam focuses on students' ability to produce written compositions in Chinese based on their understanding, analysis, and interpretation of materials provided. It includes two writing tasks, "Story Narration" and "Reply to a Letter." The Story Narration task requires examinees to write a detailed description of a set of interrelated pictures based on their understanding of the significant elements and overarching story expressed in the pictures. The Reply to a Letter task requires students to read and fully comprehend a letter provided before writing a reply that gives suitable responses to the letter's enquiries. The reply letter must also demonstrate a writing style and logical structure appropriate to Chinese.

The speaking section of the AP CLC Exam has two main tasks, "Conversation" and "Cultural Presentation." The computerized Conversation task assesses the examinee's ability to converse with a virtual dialogue partner. When it is their turn to speak, the candidate is required to give a timely, appropriate, and complete response within 20 s. In the Cultural Presentation task, the student is asked to speak in Chinese about a specified topic within a time limit. The presentation topic may relate to a certain aspect of Chinese culture, such as Chinese festivals, cuisine, art, or a famous or historical figure. The candidate is given 4 min to prepare and 2 min to deliver their presentation.

The AP CLC course uses rubric-based scoring to assess the above tasks. Examiners assess the degree to which the student has satisfactorily demonstrated competency in each of "task completion," "delivery," and "language use" and provide a mark for these three aspects of student performance accordingly. The final grade is a score from 0 to 6, with 6 corresponding to the highest grade. An example rubric for the highest grade is given below (Zhou and Feng 2010) (Table 3).

SAT Chinese with Listening Subject Test

US examinations similar to "China's National College Entrance Examination" (more commonly known as the "*Gaokao*") include the ACT and SAT (originally abbreviations of "American College Testing" and "Scholastic Assessment Test," respectively). The ACT and SAT are representative of US college entrance examinations more generally, and student results are used as key selection criteria for college admissions and scholarships. The US College Board introduced SAT Chinese – later renamed "SAT Chinese with Listening" (SAT CL) – as one of its Language Subject Tests in 1994, while the AP CLC Exam has been offered to students since 2007.

Table 3 AP CLC Exam writing section scoring guidelines for grade of "6: Excellent" (College Board 2020)

Criterion Task	Task completion	Delivery	Language use
Story narration	Narration includes a thorough and detailed beginning, middle, and end that tell a logical and complete story consistent with the stimulus Well organized and coherent, with a clear progression of ideas; use of appropriate transitional elements and cohesive devices; well-connected discourse of paragraph length	Consistent use of register appropriate to situation	Rich and appropriate vocabulary and idioms, with minimal errors Wide range of grammatical structures, with minimal errors
Cultural presentation	Presentation addresses all aspects of prompt with thoroughness and detail Well organized and coherent, with a clear progression of ideas; use of appropriate transitional elements and cohesive devices; well-connected discourse of paragraph length Cultural information is ample, accurate, and detailed	Natural pace and intonation, with minimal hesitation or repetition Accurate pronunciation (including tones), with minimal errors Consistent use of register appropriate to situation	Rich and appropriate vocabulary and idioms, with minimal errors Wide range of grammatical structures, with minimal errors

The language knowledge and skills assessed by the SAT CL test are not directly related to any particular textbook. All of the exam questions are given in both simplified characters with *Hanyu Pinyin* and traditional characters with phonetic symbols (*Zhuyin*). The questions are designed to focus on student's actual language abilities and make use of advertisements, letters, diaries, notes, notices, etc. The total duration of the test is 1 hour, divided into three 20-min periods for the listening, grammar, and reading sections. There are 85 questions in total.

Summary of CSL Assessments

In keeping with recent trends in the development of CSL assessments in China and overseas, the New HSK, CLUEE-IS, C.TEST, HSC, BCT, YCT, AP CLC, and SAT CL are all designed to assess the four language skills of listening, speaking, reading, and writing. This is reflected in the way that the test paper is structured and the exam administered in each case. Regarding the main objective and configuration of questions, all of the tests are oriented toward providing a comprehensive assessment and thorough evaluation of a candidate's language abilities. In this sense, they differ from specialized tests of speaking skills, such as the C.TEST Spoken Interview and SCT. The C.TEST Spoken Interview and SCT can also be distinguished from one another by their different formats and purposes. The interview-based assessment of the C.TEST Spoken Interview has the distinct advantage of being realistic and interactive, while the SCT's computer-assisted examination and grading method have improved the efficiency of oral exams.

Issues and Future Directions in CSL Assessment

Candidate Differentiation in Current CSL Assessment

The New HSK is currently the main examination of Chinese language proficiency. The initial development of its grading system was guided by a recognition of the importance of promoting the Chinese language around the world and an appraisal of the market for Chinese language testing. This resulted in an exam that aims to encourage students to learn Chinese; map out a pathway to Chinese proficiency; refute the outdated notion held by many beginners that Chinese is difficult to learn; and create a global evaluation system for Chinese language teaching – one that is centered on a trifecta of teaching, learning, and testing. These goals have driven the development of HSK examinations for low levels of Chinese proficiency, namely, the Level 1, Level 2, and Level 3 exams. At the same time, the level of difficulty of the intermediate and advanced examinations has been reduced. Thus, the overall difficulty of the New HSK has decreased over time.

As the world has changed and the Chinese language has become more influential in recent years, so has it become easier for international students to come to China to study. But according to feedback from stakeholders, the HSK (Level 6) exam no longer meets the needs of current candidates, educational institutions, and employers in China as an assessment of high-level language proficiency. It is a particular concern for students about to begin higher degree research programs and for people commencing employment in a teaching, research, or translation-related area in China (Chai 2015). After consulting with experts and conducting market surveys, the New HSK Research Unit recently concluded that a new exam above HSK (Level 6) should be introduced. This is an indication that the efficacy of higher-level exams has become a major issue for the HSK.

Examination Categorization in Current CSL Assessment

Customized Examinations

The "New HSK Examination Series Development Plan" published in 2006 includes several types of HSK exams designed to cater to different target users (e.g., children) and different language needs (e.g., business, tourism, administrative work). Surveys conducted over the past 2 years have indicated that there is also considerable demand for – and a pressing need to develop – a Chinese proficiency test for the medical profession. In the future, Chinese proficiency tests could become even more highly targeted, through the introduction of a Chinese language and culture exam, for example. Therefore, niche targeting of Chinese proficiency tests will become a key focus of research in the next phase of assessment development.

Discipline-Specific Academic Aptitude and Language Testing

The number and demographics of international students in China have undergone numerous changes over the past decade. This has generated urgent demand from China's educational institutions, which require a way to assess international undergraduates' academic aptitude as part of the student admission process. Apart from the various academic Chinese proficiency examinations independently developed by individual institutions, most tertiary institutions in China do not yet have dedicated entrance examinations for different academic disciplines. In other words, among the suite of international student entrance examinations currently used in China, there are no universal or standardized examinations that assess students' academic aptitude to study a chosen discipline in the Chinese language.

The criteria for international student admission as outlined in "Document No. 668," issued by China's State Education Commission (now the MOE) on December 26, 1995, are largely based on a system of measurement through Chinese proficiency tests. Such a system, however, is not entirely suited to the task of assessing whether today's international students have the requisite Chinese proficiency to commence studies in a particular discipline. An examination that assesses international students' academic aptitude – and hence, their suitability for admission to undergraduate studies – is sorely needed in China. As Chai (2015, 44) notes:

This gap in the coverage of China's undergraduate admission examinations, firstly, leads to confusion among international students applying to study at a university in China. Secondly, university admission offices lack the means to accurately gauge the character and learning abilities of potential students. This may in turn create challenges for maintaining education quality and graduation standards.

In sum, there are currently a number of issues in the design and development of CSL assessments. These are problems that educators and examination developers alike must recognize and take action to address while remaining cognizant of current international trends in language assessment.

Technology for Assessment Implementation

Modern computer-based technologies are increasingly being used in language assessment. Among English language proficiency tests, TOEFL has already introduced a computer-based version, while the "International English Language Testing System" (IELTS) is also trialing an online format. In China, the New HSK can be partially done online, while SCT has been particularly innovative in that it is a fully automated oral exam. It seems clear that the trend toward the use of digital technology in language assessment will continue. Recognizing this trend is the first step toward harnessing technology to enhance language testing, whether it is for the HSK or to develop a new proficiency test of academic Chinese. Making better use of scientific and standardized technologies will also enable language assessment to benefit from the advances of modern science and technology.

Future Directions for CSL Assessment

The system of CSL assessment is a key part of the Chinese government's broader strategy of promoting the Chinese language around the world. There are still many aspects of Chinese language testing that need to be strengthened or improved. Ensuring that test designs are consistent with scientific evidence and provide an accurate measurement of language proficiency is one way that CSL assessment can make a greater contribution toward promoting not only the Chinese language but also Chinese culture. In light of the preceding analysis and discussion, we will finish by offering suggestions on future directions for CSL assessment, which address five aspects of its development.

Incorporating Language Theories

Language research and theoretical perspectives on language have played an important role in the development of language assessment across different periods. A changing understanding of language has driven a gradual shift in language assessment from an initial exploratory phase to its current stage, which is characterized by communicative language testing and based on scientific research. Additionally, we can see that language research and its theoretical perspectives influence language assessment in diverse and complex ways, affecting concepts, aims, testing methods, and scoring systems of language assessment among other aspects. At the same time, it is important to recognize that theories of language assessment from different eras are not necessarily incompatible. On the contrary, different theories often prove complementary. It is then imperative that language assessment integrates different perspectives and workable measurements of proficiency and also focuses on holistic thinking and synthesis in both theoretical and research works.

In the twenty-first century, language assessment has entered the Information Age; it is now based on cognitive linguistics, psychometrics, and information technology. In addition, there is a diverse array of theories informing contemporary language assessment research. Looking ahead, the developers of new proficiency tests must ensure that they follow the latest developments across different disciplines and seek to use such breakthroughs to strengthen the scientific and digitalized basis of modern language testing.

The fact that China has a range of CSL assessments that cater to different groups in society makes it evident that there are still a number of differences between current Chinese proficiency tests and other foreign language examinations administered overseas. Not least, further research must be conducted in China to investigate theories of assessment in greater depth, and to verify them experimentally, as Chinese proficiency tests are still in need of stronger theoretical foundations. Other key areas to focus on include assessment technologies and the latest concepts of test research and development.

Optimizing Examination Development

Exploring different ideas on test research and development is undoubtedly a major focus of the current CSL assessment system. Only by developing multiple, highly targeted tests for different groups and different purposes will modern language assessment be able to meet the immediate needs of all stakeholders (including tertiary institutions and candidates). For example, the internationally renowned IELTS has developed four types of English proficiency test based on the diverse needs of its stakeholders: IELTS Academic, IELTS General, IELTS For Migration, and IELTS Life Skills. China should study and seek to emulate this focus on the language assessment requirements of different groups.

Considering Chinese tertiary institutions' current entrance requirements for international undergraduate students, we would suggest a step-by-step strategy in which China gradually introduces a full set of "Academic Foundation Examinations for International Undergraduate Students" to guide the tertiary admission process. This strategy should be based on research and investigation into existing admission procedures for international students and linked to the current state of CSL assessment. By "academic foundation," we mean that the exam should assess the basic scholastic qualities and skills required for the completion of an undergraduate degree, as well as the specialist knowledge and abilities needed within a particular discipline. The international student admission system currently used in the USA offers an instructive model. See Tables 4 and 5 below:

Table 4 US international student admission system (undergraduate level) (Chai 2018a)

Entrance requirement	Examination		Test of
1. (one of)	TOEFL	Test of English as a Foreign Language	Language proficiency
	IELTS	International English Language Testing System	
2. (one of)	ACT	American College Test	Academic aptitude/ attainment
	SAT I	Scholastic Assessment Test (Reasoning Test)	
	SAT II	Scholastic Assessment Test (Subject Tests)	
	GAC	Global Assessment Certificate	

Table 5 US international student admission system (graduate level) (Chai 2018a)

Entrance requirement	Examination		Test of
1. (one of)	TOEFL	Test of English as a Foreign Language	Language proficiency
	IELTS	International English Language Testing System	
2. (one of)	LSAT	Law School Admission Test	Academic aptitude/ attainment
	GRE (General)	Graduate Record Examination	
	GRE (Subject)	Graduate Record Examination of Subject	
	GMAT	Graduate Management Admission Test	
	MCAT	Medical College Admission Test	

If China had a similar structure, then its educational institutions could use the HSK to evaluate international students' Chinese language proficiency and then make use of an "Academic Foundation Examination for International Undergraduate Students" to assess academic aptitude for disciplinary studies conducted in Chinese. This sort of framework not only would meet international students and tertiary institutions' current demand for improvements to the undergraduate admission system but would also help to establish a universal and standardized system of Chinese language academic aptitude testing in China.

Enhancing Use of Technology

A major development in the field of assessment has been the rise of computer-based testing (CBT) in recent years. The advantage of CBT is that it is not limited to a certain examination time or location. Since assessment items are stored on a server by the test provider, they can be transmitted anywhere. This allows candidates to complete the test whenever it suits them, as there is no need to go to a designated place at a designated time as they would for a traditional written exam.

CBT can be divided into four categories based on the order in which questions are presented, that is, whether the presentation sequence is linear, random, adaptive (i.e., CAT), or based on a simulated scenario. Due to issues with question selection strategies and the creation of question banks, large-scale standardized tests seldom use CAT; instead, most continue to rely on linear computer-based tests.

The development and implementation of an "Academic Foundation Examination for International Undergraduate Students" could begin with linear CBT. Once the question bank expands and question selection strategies are refined, we will see steady improvements to question presentation methods and more student-focused computerized examinations overall.

The New HSK began to introduce online tests in 2010. This resolved several issues in examination management. First, it removed the need to distribute paper tests and answer sheets, gave candidates a longer registration and preparation period, and significantly increased the security and integrity of the exam. Second, its modern, streamlined interface makes it quicker and easier for candidates to complete the test, removing the need to fill in an answer sheet by hand. Third, automated processing and machine scoring of exam papers allow examinees to receive their results more quickly. And finally, online exams reduce the distribution and supply costs for the examination provider.

Online Chinese language testing can be said to be already fairly well developed. However, the ongoing development of contemporary examination theories and technologies, as well as scientific and technological advances more generally, means that new demands in the language assessment market continue to emerge. For example, we have seen growing demand for automated examination systems based on artificial intelligence (AI). AI will shape the world's future. The field of educational and psychological testing is already calling for AI technologies to be incorporated into examinations. But in practice, the first step of any shift toward AI-based assessment must be the widespread implementation of CAT. This would not only enhance candidates' exam experience but also make CSL tests more effective assessment tools. Initial results from trials of machine scoring in oral examinations have been encouraging. The SCT, for instance, has trialed machine scoring for some specific oral assessment tasks, the results of which have shown it to be valid and reliable. While there is still much to be explored and studied in the machine scoring area, a growing trend toward using big data in recent years begets confidence that machine scoring will be widely introduced, and the efficiency of exam marking will significantly improve as a result.

Expanding Student Support Services

Students' experiences in accessing post-examination support services and resources have received greater attention in recent times. In many ways, the SAT and ACT exams in the USA exemplify best practice in examination support services. The ACT examination report includes information sections about applying to tertiary institutions and career planning in addition to exam scores. It not only generates a detailed breakdown of the student's marks and ranking for each subject but also compares their results against application requirements of different tertiary institutions, helps them identify knowledge gaps, and provides a summary of suitable majors and future career choices, which is based on exam results and information provided by the student in a questionnaire. Similarly, the SAT provides students with a detailed examination report through its website. The report includes a summary table of results; detailed description, interpretation, and comparison of the scores; and scanned images of the students' essays with an explanation of how the mark was derived. This information helps students analyze their exam performance and gain a better understanding of areas for improvement. Clearly, the ACT and SAT provide extensive support and examination resources. The thoughtful, analytical, and insightful information they make available is not only essential for examinees but also indicative of the exams' advanced development and high standards (Qiu 2014). CSL assessments in China would do well to study and learn from their support services.

Adopting a student-focused approach is also particularly important for high-pressure examinations that are used to determine a candidate's educational or career advancement. For example, the SAT – widely known to be a high-stakes exam – has been in the process of reforming its score-based selection model since 2009. Students are able to take any or all of their exams at a time that best suits their circumstances and are permitted to choose which set of results and subject marks are sent to admission offices to support their college application. This approach can help alleviate some of the pressure placed on students, create a more relaxed atmosphere during exams, and ensure students are able to realize their true potential and also means that universities do not overlook any exceptional students. These supportive measures are worth considering as ways to improve the student admission and selection outcomes of language proficiency tests in China.

Furthermore, it is worth noting that examination support services may also include diagnostic reports of student ability. Test providers in other countries are often obliged to provide candidates, other stakeholders, and the general public with detailed information and interpretation of exam scores. For example, the current CEFR publishes a detailed "can-do" statement for each of its three bands and six levels of language proficiency. Meanwhile, the C.TEST has developed a results report that focuses on constructive feedback and support; it provides a breakdown of the student's scores, a personalized diagnostic evaluation of their spoken Chinese proficiency, and tips on learning Chinese. These additions have been welcomed and commended by students and exam stakeholders alike.

Teaching and Testing: A Virtuous Cycle

Thanks to the washback effect, exams act as a testing ground and feedback mechanism for teaching and then influence teaching in turn. In the future, it is imperative that different kinds of CSL assessments are developed and implemented in a way that maximizes their potential to play this positive and constructive role in education. In particular, Chinese proficiency tests can be used to shape the teaching curriculum.

This can be achieved by revising course content in a way that makes it more consistent with evidence-based concepts of assessment. The fact that language tests have set purposes, content, and marking criteria also means that they are not just selection tools, but an integral part of the teaching curriculum and teaching itself. In other words, testing and teaching collaborate in a mutually beneficial way, and this ultimately benefits students and fosters their talent.

Conclusion

By delving into the historical background of Chinese language assessment, this chapter has analyzed and summarized the development and current state of Chinese language proficiency testing. It offers several suggestions for future development and enhancement of CSL assessment. In particular, the field of Chinese language testing is in need of a stronger theoretical foundation. Optimization of examination types is needed. Testing technologies and procedures can be further improved. Additionally, Chinese proficiency tests require a more student-focused approach that includes more examination resources and support. All of this will help promote a virtuous cycle of language teaching and language testing for Chinese.

References

An, Lanpeng, and Ruijun Han (安兰朋、韩瑞军). 2018. Hanyu guoji jiaoyu xueke fazhan xianzhuang ji jiaoxue celüe sikao 汉语国际教育学科发展现状及教学策略思考 [International Chinese language education: A reflection on the current state of the discipline and its teaching strategies]. *Hebei Jingmao Daxue Xuebao (Zonghe Ban)* 河北经贸大学学报(综合版) [*Journal of Hebei University of Economics and Business (Comprehensive Edition)*] 2:81–85. https://doi.org/10.14178/j.cnki.issn1673-1573.2018.02.013.

Chai, Xingsan (柴省三). 2015. Zhong Mei liuxuesheng jiaoyu zhaosheng kaoshi tixi duibi yanjiu 中美留学生教育招生考试体系对比研究 [International student admissions: A comparative study of the admissions systems in China and the US]. *Zhongguo Kaoshi* 中国考试 [*China Examinations*] 5:39–46. https://doi.org/10.19360/j.cnki.11-3303/g4.2015.05.007.

——— (柴省三). 2018a. Lai Hua yanjiusheng jiaoyu zhaosheng kaoshi tixi goujian yanjiu 来华研究生教育招生考试体系构建研究 [The admissions system for international graduate students in China]. *Xuewei yu Yanjiusheng Jiaoyu* 学位与研究生教育 [*Academic Degrees and Graduate Education*] 9:60–66. https://doi.org/10.16750/j.adge.2018.09.011.

——— (柴省三). 2018b. *Di er yuyan ceshi wenti yanjiu* 第二语言测试问题研究 [Research on second language testing]. Beijing: Duiwai Jingji Maoyi Daxue chubanshe 对外经济贸易大学出版社 [University of International Business and Economics Press].

Chen, Hong (陈宏). 1996. Di er yuyan nengli goujian yanjiu huigu 第二语言能力结构研究回顾 [Second language proficiency: A review of research]. *Shijie Hanyu Jiaoxue* 世界汉语教学 [*Chinese Teaching in the World*] 2:46–52.

——— (陈宏). 1997. Zai yuyan nengli ceyan zhong ruhe jianli jiegou xiaodu 在语言能力测验中如何建立结构效度 [Establishing construct validity in language proficiency tests]. *Yuyan Jiaoxue yu Yanjiu* 语言教学与研究 [*Language Teaching and Linguistic Studies*] 2:77–92.

College Board. 2020. AP Chinese Language and Culture 2020 Scoring Guidelines. https://apcentral.collegeboard.org/pdf/ap-chinese-language-and-culture-2020-scoring-guidelines.pdf. Accessed 20 July 2020.

HSK Testing Center, Beijing Language and Culture University (北京语言大学汉语水平考试中心 "HSK 改进工作项目组"). 2007. Hanyu Shuiping Kaoshi (HSK) gaijin fangan 汉语水平考试 (HSK) 改进方案 [Scheme for a revised version of the Chinese Proficiency Test (HSK)]. *Shijie Hanyu Jiaoxue* 世界汉语教学 [*Chinese Teaching in the World*] 2:126–135.

Li, Xiaoqi (李晓琪). 2014. Hanyu guoji jiaoyu shiye de fazhan yu zhanwang 汉语国际教育事业的发展与展望 [The development of international Chinese language education and future directions]. *Huanan Shifan Daxue Xuebao (Shehui Kexue Ban)* 华南师范大学学报(社会科学版) [*Journal of South China Normal University (Social Science Edition)*] 5:46–49.

Li, Xiaoqi, and Jinghua Li (李晓琪、李靖华). 2014. Hanyu Kouyu Kaoshi (SCT) de xiaodu fenxi 汉语口语考试 (SCT) 的效度分析 [Analyzing the validity of the Spoken Chinese Test (SCT)]. *Shijie Hanyu Jiaoxue* 世界汉语教学 [*Chinese Teaching in the World*] 1:103–112. https://doi.org/10.13724/j.cnki.ctiw.2014.01.016.

Liu, Yinglin, Shujun Guo, and Zhifang Wang (英林、郭树军、王志芳). 1988. Hanyu Shuiping Kaoshi (HSK) de xingzhi he tedian 汉语水平考试(HSK)的性质和特点 [Features of the Chinese Proficiency Test (HSK)]. *Shijie Hanyu Jiaoxue* 世界汉语教学 [*Chinese Teaching in the World*] 2:110–120.

Peng, Hengli (彭恒利). 2005. Zhongguo Xiaoshu Minzu Hanyu Shuiping Dengji Kaoshi 中国少数民族汉语水平等级考试 [An overview of the Chinese Language Proficiency Test for China's Ethnic Minorities]. *Zhongguo Kaoshi* 中国考试 [*China Examinations*] 10:57–59. https://doi.org/10.19360/j.cnki.11-3303/g4.2005.10.019.

Qiu, Jingyuan (邱静远). 2014. Meiguo ACT he SAT yuyan ceshi jiqi dui kaizhan Hanyu nengli ceshi de qishi 美国ACT和SAT语言测试及其对开展汉语能力测试的启示 [The ACT and SAT language tests in the US: Insights for Chinese proficiency tests]. *Zhongguo Kaoshi* 中国考试 [*China Examinations*] 5:39–45. https://doi.org/10.19360/j.cnki.11-3303/g4.2014.05.006.

Sun, Dejin (孙德金). 2009. Hanyu Shuiping Kaoshi fazhan wenti lüelun 汉语水平考试发展问题略论 [A brief discussion on the development of the Chinese Proficiency Test (HSK)]. *Zhongguo Kaoshi* 中国考试 [*China Examinations*] 2009.6: 18–22. https://doi.org/10.19360/j.cnki.11-3303/g4.2009.06.004.

Tang, Xueting (唐雪婷). 2016. Xin HSK yanjiu shuping 新 HSK 研究述评 [The New HSK: A review of research]. *Kaoshi Yanjiu* 考试研究 [*Examinations Research*] 4:90–96.

Wang, Hanwei (王汉卫). 2018. Huawen Shuiping Ceshi de sheji yu chubu yanzheng 华文水平测试的设计与初步验证 [The Overseas Chinese Proficiency Test: Design, development, and a preliminary experimental verification]. *Shijie Hanyu Jiaoxue* 世界汉语教学 [*Chinese Teaching in the World*] 4:534–545. https://doi.org/10.13724/j.cnki.ctiw.2018.04.010.

Wang, Jimin, Libing Huang, and Shujun Guo (王佶旻、黄理兵、郭树军). 2016. 来华留学预科教育"汉语综合统一考试"的总体设计与质量分析 [An analysis of the design and efficacy of the Chinese Language University Entrance Examination for International Students]. *Yuyan Jiaoxue yu Yanjiu* 语言教学与研究 [*Language Teaching and Linguistic Studies*]. 2:53–58.

Zhao, Qifeng (赵琪凤). 2020. Hanyu guoji jiaoyu kaoshi tixi fazhan yanjiu 汉语国际教育考试体系发展研究 [The development of an examination system for international Chinese language education]. *Yuyan Zhanlüe Yanjiu* 语言战略研究 [*Chinese Journal of Language Policy and Planning*] 2:71–79. https://doi.org/10.19689/j.cnki.cn10-1361/h.20200206.

Zheng, Jindan (郑锦丹). 2019. Huawen Shuiping Ceshi jianshao 华文水平测试简介 [A brief introduction to the Overseas Chinese Proficiency Test]. *Huawen Jiaoxue yu Yanjiu* 华文教学与研究 [*TCSOL Studies*] 3:94–95. https://doi.org/10.16131/j.cnki.cn44-2669/g4.2019.03.015.

Zhou, Xiaowei, and Shengyao Feng (周晓炜、冯生尧). 2010. Meiguo AP Zhongwen Kemu Kaoshi zhuguanti de tese yu qishi 美国AP中文科目考试主观题的特色与启示 [The US AP Chinese Language and Culture Exam: Features of and insights into question design]. *Jiaoyu Celiang yu Pingjie (Lilun Ban)* 教育测量与评价(理论版) [*Educational Measurement and Evaluation (Theoretical Edition)*] 12:51–54. https://doi.org/10.16518/j.cnki.emae.2010.12.020.

Technology in Chinese Language Teaching 29

Shijuan Liu and Jun Da

Contents

Introduction	702
Technology and Its Use in Chinese Language Teaching in the Past Century: 1900–1999	703
1900–1949	703
1950–1999	704
Technology and its Use in Chinese Language Teaching in the Twenty-First Century: 2000–2020	708
2000–2009	708
2010–2019	711
2020	715
Mastering the Use of Technologies in Chinese Language Teaching: Categorizations, Frameworks, Guidelines	717
Categorization of Technologies in Chinese Language Teaching	717
Frameworks for Discussing the Use of Technologies	722
Guidelines for Use of Technology in Chinese Language Teaching	727
Concluding Remarks	731
References	733

Abstract

This chapter provides a comprehensive overview of the use of technology in Chinese language teaching and learning in the past 120 years, tracing its history back to 1900 when phonology technology was first used to teach Chinese remotely. It presents the evolution of the use of technology in a chronological order: the past century (1900–1949, 1950–1999), the present century (2000–2009, 2010–2019, 2020). Technologies developed and used in Chinese language

S. Liu (✉)
Indiana University of Pennsylvania, Indiana, PA, USA
e-mail: sliu@iup.edu

J. Da
Middle Tennessee State University, Murfreesboro, TN, USA
e-mail: jun.da@mtsu.edu

teaching during each of the time periods are reviewed together with a summary of their applications in Chinese language teaching practice and research. Given the huge variety of technologies available today, the chapter suggests two ways of categorizing instructional technologies, shares major frameworks for discussing technologies, and provides several guidelines for applying technologies in Chinese language teaching and learning.

Keywords

Technology · Chinese as a foreign language · Teaching and learning · Guidelines

Introduction

Numerous research articles can be found on the application of technology in teaching Chinese as a foreign language (TCFL). The biannual *Journal of Technology and Chinese Language Teaching* (JTCLT http://www.tclt.us/journal/) alone has published more than 100 articles since its launch in 2010, addressing the application of a variety of technologies in Chinese language teaching, such as on Second Life (Cheng et al. 2010; Grant and Huang 2010), VoiceThread (Chen 2011; Zhang 2011), tablets (Lin and Lien 2012; Liu 2013), smartphones (Chen 2013), WeChat (Luo and Yang 2016), flipped learning (Chen 2014; Li and Jiang 2017), MOOCs (Lin and Zhang 2014), virtual reality (Ye et al. 2017), augment reality (Zhang 2018), eye-tracking technology (Shi 2018), and machine translation (Xu 2020; Tian 2018, 2020). There are also several articles that provide an overview on the use of technology in Chinese language teaching and learning, such as the journal articles by Liu (2007) and Xu (2015), and book chapters by Xie (1999), Xie and Yao (2009), Wu (2015), and Da and Zheng (2018).

In this chapter, we first review the history of using technology in TCFL chronologically from 1900. Although there are retrospective publications that have discussed the history of computer-assisted language learning, "the history of the use of technology for language learning in the early decades of the 20th century is sparsely documented" (Otto 2017, p. 33). The same holds true with TCFL. Nearly all available articles addressing the history of technology in Chinese language teaching and learning started from 1970s when the computer was first used for educational purposes. To provide a complete picture, this chapter tracks the history further back to the very beginning of using phonography technology in TCFL. In addition, it also summarizes the use of technology in TCFL at present and in the future including the impacts of COVID-19 and trends after 2020.

Because countless technologies have been developed and used in TCFL, the chapter then provides suggested categorization of the technologies for helping make connections among the various technologies and understand new emerging ones. Given that specific technologies can become outdated while new technologies are emerging, this chapter concludes with general guidelines and suggestions for using technologies in TCFL.

Technology and Its Use in Chinese Language Teaching in the Past Century: 1900–1999

This section reviews the use of technologies in Chinese language teaching in the past century. It is divided into two periods: the first half of the century (1900–1949) and the second half of the century (1950–1999).

1900–1949

In the first half of the past century, phonographs (invented in 1877) and related gramophone records were the major technology tool and resources that could be used for language teaching. There was very limited formal teaching of the Chinese language during that period worldwide. Nonetheless, historical documents show that pioneers in Chinese language teaching adopted technology from the very beginning and used phonographs and gramophone records just like instructors teaching languages that were commonly taught (e.g., French and Spanish).

Dr. John Enicott Gardner, who worked at the Chinese Bureau of San Francisco, was reported to have used the phonograph in teaching Chinese as early as 1900. He taught two Chinese language classes at the University of California and the University of Pennsylvania during the same semester. Instead of traveling between San Francisco and Philadelphia, he recorded his lecture by phonograph cylinders for students in Philadelphia to study. The report published in the 1900 May issue of *Phonoscope* mentioned that students at University of Pennsylvania did not have difficulty with his phonographic lectures. This anecdotal record of using phonograph cylinders can be considered as the earliest example of distance education in Chinese language teaching and learning.

In 1925, the Commercial Press of Shanghai published *A Phonoscope Course in the Chinese National Language* authored by Yuen Ren Chao (赵元任). There were gramophone records accompanying the book, which were "remarkably clear, especially in the tones" and immeasurable "to the usefulness of the work" in that students could study without a Chinese teacher at presence (Bruce 1926, p. 197). The Depression in the 1930s prevented a wider use of the phonograph (Schueler 1961) in language teaching.

During 1920–1930, hundreds of radio stations were established in the United States (US history 2008) and a number of articles were published to address the use of radio in language teaching and learning, such as Young (1932) on resources, Koon (1933) and Meiden (1937) on benefits, and Bolinger (1934) and Cabarga (1937) on their practice of making broadcasts of Spanish lessons and lectures over the radio stations of their institutions. However, there is no record documenting the use of radio broadcasting in TCFL in the 1930s. The Voice of America (VOA)'s Mandarin Service was established in 1942. Its audience was Chinese-speaking population abroad and its language program aimed to help Chinese speakers to learn English. Even though there is a possibility that a few learners and instructors occasionally used it for Chinese language learning and teaching, it would be very rare given the low number of Chinese language learners and instructors during that period.

There were also papers published on the use of movies in foreign language instruction in the United States in the 1930s, such as Ginsburgh (1935) and Hendrix (1939). Palomo (1940) summarized the advantages and limitations of using sound films over radio, such as presenting the target language "in organic context" but being made for entertainment purpose. Hence the instructor had to make adaption for classroom use. It was estimated that over 600 Chinese movies were made in the 1920s and 1930s in China (Cheng 1963). However, there was no documented record found of using Chinese films in TCFL during this period. This might be due to several reasons, such as the high cost of equipment for showing movies in schools and the small number of Chinese learners and instructors.

During the 1940s, Robert Tharp played an important role in the Chinese teaching team of Institute of Far Eastern Languages (IFEL) at Yale University (Ling 2018). As the chair of the Audio-Visual program and director of the Air Force course of IEEL during that time, Tharp acquired a variety of audio/visual equipment and used them in assisting his teaching. According to one of his trainees, Tharp was able to "fashion a method of instruction and sequence of classes, recitations and exercises, many making use of audio/visual equipment that made successful learning inevitable for any student who followed the plan that Tharp devised" (Ling 2018, p. 73).

1950–1999

The second half of the twentieth century saw more variety of tools and abundant resources used in the teaching and learning of (foreign) languages including the Chinese language. The People's Republic of China was established in 1949 and formal instruction of the Chinese language to learners from outside China began in the 1950s when Tsinghua University arranged a special class for exchange students from Eastern Europe to learn Chinese in Beijing. More resources in Chinese language, including those based on authentic radio broadcasting, television broadcasting, Chinese movies, and TV plays, have been developed in Mainland China since then.

Audio Playing and Recording, Radio Broadcasting

Tape recorders gradually replaced phonograph because of their low cost and convenience of recording voice. The popular audio recorders in the 1950s and 1960s were reel-to-reel recorders. Cassette tape recorders did not flourish in the market until the late 1960s and 1970s. The advent of Sony's Walkman, the affordable and convenient portable audio player, helps language learning take place at any time and at any place. It is estimated that two hundred million cassette players were sold between 1979 (the first appearance of Sony's Walkman) and 2010 when Sony retired its cassette tape Walkman (Franzen 2014).

Historic documents from teaching plans dated 1951 for the exchange students from East Europe who studied Chinese in China showed that each lesson was recorded for students to review (Cheng 2005). Several articles were published on the use of tape recorders in language teaching and learning, though not specifically on the Chinese language. For instance, Schueler (1961) analyzed the capacities and limitations of tape recorders, and commented that the recorder can "only reproduce, it cannot produce; it can repeat, but it cannot react; it can communicate *to* but not *with*, and what it does communicate is only what has been fed into it" (290). He further pointed out that the functions of tape recorders gave birth to the language laboratory and provided the "electronic buttress" of the audio-lingual method of language teaching.

Starting from the 1980s, more publishers made audio recording materials available for Chinese language teaching and learning. Many published language learning books were accompanied with audio recordings, such as the audio cassettes made for the textbooks *Practical Chinese Reader* and *Modern Chinese-Beginner's Course* in the early 1980s. In addition, instructors and students were also able to use audio cassettes to record and produce audio materials for their own teaching and learning, because of their low cost and convenience of use.

Furthermore, radio broadcasting programs including news for native speakers can be recorded and used as authentic materials for TCFL. China Radio International (CRI) started to include Chinese language learning programs in its English and Japanese channels in 1962 but they were discontinued in 1966 due to the Cultural Revolution (Zheng 2012). Its Japanese channel restored the Chinese language learning program in 1972. Research on the use of broadcasting in teaching and learning of the Chinese language started from the 1970s or even earlier. For instance, Kuo (1982) reported that University of Kansas started to introduce news segments into Chinese classrooms in 1975 and used them intensively in its advanced-level classes.

Film, Video Recording, Television Broadcasting

Buck (1974) shared how the Inter-University Program for Chinese Language Studies used news broadcasts and TV broadcasts in advanced conversational classes to help students improve their Chinese language proficiency. In addition to using authentic TV programs and film made by media professionals for general users, some instructors also made videos specifically for Chinese language classes. As Dr. Perry Link (2018) recalled, Professor Ch'en Ta-tuan (陈大端) at Princeton University recruited him and several others to play roles in making videotapes for the textbook *Chinese Primer* in the 1970s. Because Princeton University did not have any back then, they had to go to Princeton Theological Seminary which owned video equipment to shoot scenes there.

Additionally, professional television broadcasting service providers also produced programs specially for Chinese language learning. For example, NHK, Japan's public broadcaster, started to offer programs on Chinese language learning from the 1980s through both radio and television broadcasting. While the NHK radio broadcasting mainly focused on reading texts as well as some conversations, its television broadcasting used short videos to introduce Chinese daily lives and teach related vocabulary and sentences (Zheng 2012). In addition, China Central

Television (CCTV) started to produce a series of programs on Chinese language teaching and learning from as early as the 1990 (Zheng 2012).

Walker (1982) shared how to use video as texts for helping Chinese language learners in intermediate- and advanced-level courses to practice discourse in Chinese in the United States. Similarly, several publications in China studied the use of television programs/plays for helping students to learn Chinese, and some further discussed the teaching of a course titled *Watching, Listening, and Speaking*. Among them, Li (1991) suggested criteria in the selection of TV short plays (e.g., 20–30 min) for teaching intermediate-level courses, shared her lesson plans and reflections including the benefits of using plays to provide authentic contexts, raise student interest, and help students understand Chinese culture and customs.

Computer, Chinese Input Methods, Multimedia

While mainframe computers began to appear on American university campuses during the late 1950s, their computing power was accessed primarily via paper punch cards, which limited their applications in education. It was not until the late 1960s and early 1970s when computers began to support multiple terminals that allowed interaction with the computer via keyboard, and hence opened the door for their use in language teaching and learning (Otto 2017). The first use of computer in teaching of the Chinese language is credited to Dr. Cheng Chin-Chuan (郑锦全) who was involved with the PLATO project at the University of Illinois (Da and Zheng 2018; Xu 2015).

According to Cheng (1972, 1973), PLATO, which stands for the Programed Logic for Automated Teaching Operations, was developed by the computer-based Education Research Laboratory of the University of Illinois. Since its inception in 1960, PLATO has been used to teach courses in at least 20 fields including foreign languages such as Chinese. Cheng used the system to help students learn stroke orders of Chinese characters because the system could display the stroke orders of characters, allow students to imitate the order by pressing a designated key, and provide Pinyin and meaning of the character after the display of the final stroke. He also used the system to help students improve their reading comprehension – students could look up the meaning of any unknown character in the paragraph with the assistance of the system and be tested on their understanding of the paragraph and receive test results instantly. In addition, he used the system to help students "tune" their ears. Students typed Pinyin and tones for the sounds they heard, and the computer would evaluate and indicate errors. Cheng's pioneering work, as Wu (2015) commented, set the stage for the application of computers in Chinese language learning and instruction in the following years.

Because the QWERTY keyboard was designed for Latin-script alphabets, how to input thousands of Chinese characters into the computer was a big obstacle for Chinese information processing in those early days. This also renewed the debates on abolishing Chinese characters that began in the late 1910s. On the other hand, this challenge attracted people from all walks of life to tackle it and led to the explosion of various Chinese input methods in the 1970s and 1980s (Mullaney 2017). Early discussions on the use of computers in Chinese language teaching often addressed

Chinese input issues (Dew 1996). For instance, Wang (1966) discussed problems in using computers for Chinese language concordances. Sung (1986) provided a linguistic evaluation of Chinese computer input systems. Technical issues in inputting and displaying Chinese were, to a great extent, solved in 1993 when Microsoft released its Windows NT operating system based on the Unicode standard (Liu 2007).

Many multimedia instructional materials based on computers were developed for Chinese language teaching and learning from the late 1980s on and throughout the 1990s. Yao (1996) reviewed some computer-assisted language learning software for Chinese. On an annotated list of CD-ROM based courseware and materials provided by Bourgerie (2003), there were several developed before 2000, such as *Hanzi Assistant* and *Chinese Writing Tutor* (both in 1991) for Chinese character learning; *Chinese Pronunciation* and *Pinyin Master* (both in 1998) for learning Pinyin and pronunciation; *Chinese Newspaper Reading* (in 1993) and *Authentic Reading Materials for Advanced Students of Chinese* (in 1992) for reading; and *Colloquial Chinese* (in 1998) with 14 multimedia lessons for beginners.

The strengths and limitations of using computer technology in Chinese language teaching were well summarized by Zhang (1998), according to whom the major strengths include (1) interactivity, allowing learners to control their learning and receive instant feedback; (2) multimedia (e.g., digitalized audio, animated graphics for Chinese characters and speech production), revolutionarily improving presentation and making possible what conventional textbooks cannot do; (3) consistence and patience, supporting learners for unlimited repeated exposure; (4) rapid and random access, allowing instant retrieval of vocabulary and grammar explanation; and (5) ability to track and store data, allowing the monitoring of learners' learning process. The major limitation Zhang (1998) discussed was that interactivity between computers and users was less capable, and consequently the learners' creative and open-ended output could not be easily evaluated.

The Internet, the World Wide Web, and Early Adopters

Even though the Internet was invented in the 1960s, it did not reach the general public until the mid-1990s when it was popularized by the invention and adoption of the HTTP protocol. For the convenience of discussion henceforth, we use these three terms, the Internet, online, and the web, interchangeably. The wide adoption of the Internet supported by the HTTP and W3 (https://www.w3.org/) opened new and unlimited possibilities for human society beginning from the middle 1990s. As a tool that can be used for language learning and teaching, the Internet allows students to interact with their instructors and peers beyond the boundaries of the physical classroom and beyond class time. By using emails and other communication tools (e.g., instant messenger), students and instructors could interact and receive feedback from people outside the classroom and their community.

In addition, new resources have been continuously developed for Chinese language learning and teaching on the Internet, and many existing resources are digitized and made available online that instructors and students can use at any time and at any place with Internet access. Xie (1999) shared how to use emails, mailing lists, online reading and interactive exercises, and student webpages in Chinese language teaching. He set up a comprehensive website, *Learning Chinese Online* (http://www.learningchineseonline.net), providing a collection of resources for Chinese language teaching and learning. This website became one of the most recognized sites in the late 1990s and early 2000s (Da and Zheng 2018).

The First International Conference on Internet Chinese Education (ICICE) was held in Taipei in 1999 (http://edu.ocac.gov.tw/discuss/academy/netedu01/testdefault2.htm). The presentations given at the conference covered different categories. Some presentations discussed online Chinese resources (authentic materials) that can also be used by general users including native speakers. For instance, Luo and Zhang (1999) shared their experience in building a hypermedia (combination of hyperlinks and multimedia) database for the Chinese classic novel *Dream of the Red Chamber*. Some presentations discussed resources that are especially developed for Chinese language teaching and learning. For instance, Yao (1999) shared a website that collected sources for the *Integrated Chinese* textbook series. There were also presentations on using various tools. For instance, Fleming (1999) discussed the use of interactive TV (iTV) for distant teaching Chinese to students. Tools particularly targeted for learning and teaching of the Chinese language were also discussed at the conference. For instance, Gan and Cheng (1999) developed an open online tool on lexical co-occurrence in teaching and learning of the Chinese language.

Technology and its Use in Chinese Language Teaching in the Twenty-First Century: 2000–2020

While only 20 years have passed in the twenty-first century, what happened to modern technologies in these 20 years appear to have much surpassed that of the whole twentieth century. This section describes the use of technology in Chinese language teaching chronologically in three periods: 2000–2009, 2010–2019, and 2020. It is worth pointing out that as a happy coincidence in 2000, the first biennial International Conference and Workshop of Technology and Chinese Language Teaching (TCLT) was held, and in 2010 its journal, JTCLT, was launched.

2000–2009

The First 5 Years of 2000s: Multimedia, Internet, Online Resources, and Online Courses

During the first 5 years of the twenty-first century, audio cassettes (tape recorders), CDs/DVDs, and VHS tapes from the last century continued to be used in Chinese language teaching. At the same time, with the increasing popularity of personal computers and the Internet, more resources were digitalized and made available online. Authentic Chinese resources that were intended for native speakers could be more easily found and adopted. Berman (2001) discussed the use of authentic TV broadcast videos and images in elementary Chinese language classrooms, and Wang

(2002) explored the use of TV talk-show programs to construct pedagogical materials for advanced Chinese language learners.

In additional to authentic materials, more products were developed specially for the teaching and learning of Chinese as a foreign language. For example, Bai (2003) introduced several products which he and his colleagues developed, including a set of video materials titled *Real People on Real Topics*, and another set titled *Video Clips of Survival Chinese* (in both web and CD version); Chu (2004) introduced ChineseTA – an integrated software for Chinse language teachers; Xu and Jen (2005) developed *Penless* software for helping students to learn Chinese characters. When providing a comprehensive annotated bibliography of various multimedia and Internet resources, Bourgerie (2003) stated that with sheer quantity there was uneven quality among the resources. Liu (2002) noted that many of the existing resources were not widely used and information among different countries/regions needed to be further exchanged. Yao (2009) and Xie and Yao (Xie and Yao 2009) compiled a list of tools useful for Chinese language teaching and learning.

Xie (2000) analyzed the pros (e.g., convenience, flexibility, variety) and cons (e.g., cost of money, time, and energy) of using computers in Chinese language teaching. Similarly, Bai (2003) summarized the advantages (e.g., providing multi-stimulus environments and never getting tired of mechanical drills) and limitations (e.g., cost, not contributing much to productive skills) of using the multimedia technology in TCFL.

Liu (2004) applied the ten-level web integration continuum described by Bonk et al. (2000) in TCFL with real examples. Most activities and cases in TCFL fell into the range from level one (*marketing/syllabi on the web*) to level eight (*web as alternated delivery system for resident students*) on the continuum. Only few offered entirely online courses (level nine), such as the online Chinese courses offered by East China Normal University in China and La Trobe University in Australia in early 2000. It was even rarer to offer a series of courses and the entire program (level 10). eBLCU.net (网上北语), established by Beijing Language and Culture University in December 2000, was the first one that offered B.A. degrees online to learners around the globe, in addition to providing nondegree online courses.

Bai (2003) shared how two small liberal arts colleges used web-based distance-learning instruction for students to take courses across colleges. Chen (2006) conducted a critical review of distance learning courses via iTV. Liu (2002) mentioned companies that offered online Chinese courses, such as chineseon.net (which was already shut down) and their struggles in making profit (very few users were willing to pay). Among the discussions on Chinese distance education, Zheng (2001) and Li (2007) are two important pieces. Zheng (2001) analyzed the roles of human beings and technology situated in teaching Chinese as a foreign language, and discussed how to innovate classroom teaching with the assistance of the Internet technology and how to make the online class more effective by stressing the human aspect instead of only moving the content online. Li (2007) discussed characteristics of quality online courses based on surveys of experienced Chinese instructors and learners.

The Second 5 Years of 2000s: Web 2.0, Virtual Worlds, and Emergence of Mobile-Assisted Learning

Web 2.0 became a catch phrase in 2005 when a new set of web-based applications and services such as blogs, wikis, Facebook, YouTube, and Google Docs were developed and made available for social networking, sharing, and collaboration. While there is no unanimous consensus on the definition of Web 2.0, it can be characterized as "share," "participate," and "collaborate," which stands in contrast to Web 1.0 that focuses on "access" and "find" (O'Reilly 2005). By participating in various user-friendly platforms, users in the Web 2.0 era are not only consumers of information but also content creators at the same time (Kennedy et al. 2007; Liu 2008).

Honggang Jin (2009) explored the use of three participatory tools (discussion boards, blogs, Skype) in teaching Chinese to 21 undergraduate students. Li Jin (2009) researched the use of Facebook in Chinese culture learning. A comprehensive review of Web 2.0 empirical research was conducted by Wang and Vásquez (2012), which analyzed 43 empirical studies selected from peer-reviewed journals, dissertations, and two edited books.

Virtual worlds, popularized by Second Life (https://secondlife.com/), falls under the realm of Web 2.0 (Liu 2008, 2010; Wang and Vásquez 2012), whereas some (e.g., O'Driscoll 2007) called it 3D Internet and used it parallel to Web 2.0. Developed by Liden Lab in 2003, Second Life received wide attention from the public in 2006 and 2007. The inaugural issue of JTCLT included several case studies of the use of Second Life in Chinese teaching and learning, such as Chen (2010), Cheng et al. (2010), and Grant and Huang (2010). Liu (2010) provided a comprehensive review of using Second Life in teaching and learning of languages including Chinese, in which she shared resources and analyzed real cases of using it in Chinese teaching and learning; she also identified five advantages (e.g., making online teaching and learning more engaging and strengthening an online learning community), and five disadvantages (e.g., the high learning curves that demand time and energy, the monetary cost for high-end computers and purchase of the virtual space), and provided some suggestions (e.g., weighing the investment and outcomes in using the tool).

Another thread during this period was mobile-assisted language learning (MALL). Apple released its first generation of iPods in 2001, which functioned as a portable digital audio recorder/player. Chinnery (2006) summarized the benefits and challenges of mobile technologies based on his observation of the use of iPods, cell phones, and personal digital assistants (PDA). For instance, one great benefit of the mobile devices is their portability, and one challenge inherent in the portability is the reduced screen size. Limited audiovisual quality was another challenge he mentioned, but it is worth pointing out that his summary was made before the iPhone and the iPod touch were released in 2007. Zhang (2009) discussed values for incorporating MALL activities and shared her exploration of using video podcasts in Chinese learning and teaching. She stressed that "technology such as digital audio and video has made rendering of auditory and visual enhancement easy and effective. Yet, it is the mobile technology that allows these benefits to be fully utilized" (p. 55).

In summary, during the first 10 years of the twenty-first century, the use of computers and the Internet have gradually become normalized in education. While materials in the form of audio cassettes and CD/DVDs were still used in some places, web-based teaching and learning materials were becoming more common. In the early days of the World Wide Web, webpage authoring required special technical skills, which prevented mass participation in the creation of web content (Da and Zheng 2018). The maturing of Web 2.0 technologies in the mid-2000s empowered instructors and learners to create and share contents and helped them better connect and interact with each other around the globe.

2010–2019

Mobile Devices and MALL

Apple released a new version of iPhone each year after its initial release in 2007. In addition, Apple launched iPads in 2010. With its powerful multimedia functions and reasonable pricing, iPads became a hit in education and was regarded as "a solid education tool" in mass media (Bonnington 2012). Hundreds of K-16 schools purchased iPads for students to use. At the same time, smartphones and tablets (e.g., Samsung Galaxy, Amazon Kindle Fire) that run Google's Android and Microsoft Windows systems entered the market in 2010 and 2012, respectively. The share of Americans who say they own a smartphone increased from 35% in 2011 to 81% in 2019 (Pew Research Center 2019).

Numerous resources and publications addressing MALL are available. Some of them (e.g., Godwin-Jones 2011) reviewed apps that could be used for language learning and teaching. Reinders and Pegrum (2015) further provided an evaluative framework (including five categories and several sub-criteria) for mobile language learning resources. In addition, Stockwell and Hubbard (2013) suggested ten principles for MALL, such as *Limit multi-tasking and environmental distractions*, and *Push, but respect boundaries*. Another finding associated with MALL from the literature (e.g., Huhn et al. 2016) was that although learners used mobile devices frequently, they did not use them much for academic purposes, and hence instructors need to design corresponding instructional activities.

In the case of using mobile devices for Chinese language and learning, Chen (2013) shared an experiment she conducted in her introductory Chinese II course. She asked students to use Nokia Lumia smartphones (Windows system) provided by the university for practice activities. Benefits she found included the variety of apps enhancing student learning with element of play, and the convenience of using electronic dictionaries. Challenges included that there were much fewer language learning apps on Windows phones and some compatibility issues (such as playing audio files on the phone and connecting to the classroom projector). Lin and Lien (2012) shared a number of apps that can be used for Chinese teaching and learning on iPads. Liu (2013) provided a comprehensive overview of tablets in Chinese language teaching and learning, including the history, types, as well as advantages and limitations. Advantages include being multifunctional, and intuitively interactive (screen multi-touch with fingers), whereas limitations include heavy reliance on quality of the apps and Wi-Fi, students' limited access to tablets and different apps on different tablets, and the lack of pedagogical guidelines and support.

Based on a review of 22 popular apps for learning Chinese vocabulary, Lin et al. (2018) found that most of the apps lacked clear pedagogical goals, explanations, and feedback, and were insufficient for providing scaffolding for learning new words. Yang and Yin (2018) found the following from the activities they designed with WeChat (a robust social networking app widely used among Chinese communities worldwide) for students to learn Chinese vocabulary: Although sending vocabulary to students through WeChat helped increase incidental vocabulary learning, more variety of instructional activities should be used to engage learners to promote their long-term retention. Their findings expanded Luo and Yang's (2016) discussion on the benefits of using WeChat (e.g., expansion of time in learning, linguistic gains, promotion of cultural learning, enhancement of learning motivation, and establishment of a supportive Chinese language learning community). In addition to research on apps developed by companies, some researchers studied the apps they themselves designed. For example, Lu et al. (2014) designed an app based on iPod touch and used it to help primary school students in a bilingual school in Hong Kong to learn 200 fundamental Chinese characters. Similarly, Wei and Zhang (2018) reported the findings of using a Chinese pronunciation app designed and developed by Beijing Language and Culture University to help students practice pronunciation and receive instant feedback. Wong et al. (2010) shared a design-research study of using smartphones to help primary school students in Singapore to learn Chinese idioms.

Virtual Reality and Augmented Reality

In addition to computers, smartphones, and tablets mentioned, there are other emerging devices that can be used in language teaching and learning after 2010. Among them are products related to virtual reality (VR) and augmented reality (AR) technologies. Ye et al. (2017) reviewed the history and evolution of VR technologies and categorized head-mounted display (HMD) VR devices into four types. They also discussed the applications of VR in education and language teaching and learning, and further shared examples and resources for using VR in CLT. Liu (2018a) examined the relationships between VR and 3D virtual worlds (e.g., Second Life) and reviewed some sources and apps that can be used for teaching and learning of the Chinese language and culture; she also summarized the benefits of using VR headsets (immersive) and challenges (e.g., lack of quality apps for language learning, cost, and eye fatigue), and argued that VR was more advantageous for some other areas (e.g., biology, history, and tourism) in terms of effectiveness and efficiency. Xie et al. (2019) explored the use of interactive virtual reality tools (Google Cardboard and Expeditions) in an advanced Chinese language class and found that VR offered an authentic context for Chinese language learning, sparked interest in the virtually presented locales, and encouraged students to further explore the target culture. They also identified several challenges that Liu (2018a) mentioned such as technical difficulties and physical dizziness. Their study (including pairing up students, providing scaffolding) demonstrated the importance of the pedagogical aspects in applying technologies.

Zhang (2018) discussed relationship between AR and VR, explained AR technologies and their applications, and reviewed AR related empirical studies in language education. She further summarized reasons for the positive outcomes, such as increasing learners' active interaction with the learning materials, providing learners with a contextual and immersive learning experience, opportunities for dynamic negotiation and collaboration and using the language in a spontaneous and unplanned way. Some problems she identified included that learners need time to get familiar with the devices and be provided with scaffolding in both language learning and gameplay. Gao (2019) shared how to use such AR apps as IKEA Place and Amazon AR (https://www.amazon.com/adlp/arview) in teaching Chinese heritage learners. It is worth pointing out there is a wide range of AR and VR technologies, which can include smartphones and tablets that run VR and AR apps.

MOOCs

Another thread of using technologies in TCFL in 2010–2019 is MOOCs. The New York Times labeled 2012 as the "year of the MOOC" (Pappano 2012). Lin and Zhang (2014) analyzed what each component (Massive, Open, Online) means in MOOCs and discussed their characteristics regarding course materials, discussion forums, feedback, deadlines, and pedagogy. They also provided examples of MOOCs in Chinese language education and showed how to build a MOOC with Google Classroom. Wang and Bellassen (2017) reviewed existing Chinese MOOCs and identified some problems (e.g., monism-based teaching method, non-integration of cultural elements, and the lack of learner-learner interactions), and proposed some general design principles and applied them in the design of a 7-week Introductory Chinese MOOC for teaching beginning level Chinese to French-speaking learners. Liu (2018b) examined MOOCs and Open CourseWare (OCW) with the "iron triangle" (access, cost, quality) in higher education. She identified a wide variety of MOOCs courses available on different MOOCs platforms (e.g., Coursera, edX, XuetangX, Udemy) and noted dramatic differences among them, and further suggested ways of making use of MOOCs for formal Chinese language courses.

Continued Practice and Research on Topics from Previous Years

Apart from the emerging technologies and new topics, practices and research on other areas concerning the use of technologies in CLT mentioned from previous years continued in 2010–2019. For example, Yu (2012) reviewed existing pedagogical approaches in using feature films in foreign language classrooms and shared her experience of using the film *A Great Wall* together with the textbook *Integrated Chinese* in her intermediate-level course. Zhang (2013) grouped the use of short video clips in language classes into three categories based on their functions: "advance organizer," "core content," and "stimulus." She provided suggestions on selecting and editing clips and shared the design of instructional activities with a sampler video she created for advanced-level courses. Wu (2014) shared how to use comedy skits in lower-level Chinese classes.

Whereas the goal of using audio/video for language teaching and learning has remained the same, the quantity and quality of audio/video materials in the 2010s are unprecedented. In addition to online broadcasting programs produced by traditional professional broadcasting service providers (e.g., btime.com/btv/btvsy_index for Beijing TV Station), numerous new audio broadcasting services (e.g., the popular ximalaya.com/) have emerged in the Web 2.0 era when anyone can create audio/video podcasting and make them available online for the world to listen and watch. Furthermore, there are many individual channels on YouTube related to Chinese language and culture, and countless live videocasting platforms are active in China. Many quality sources are available and free to use including those developed by professionals, such as the series of programs produced by CCTV including *Growing Up with Chinese, Happy Chinese,* and *Travel in Chinese*. Liu (2012) shared a number of free online resources and tools that are especially useful for beginning levels.

During the 2010s, the prices of consumer cameras and camcorders have been dropping, and their functions have been greatly improved and become more user-friendly. Furthermore, the easy video recording functions on mobile phones and tablets allow anyone to produce videos at any time. In addition to viewing videos made by others, Chinese language students can produce videos to document and demonstrate their Chinese language learning outcomes. Liu (2014) summarized a number of benefits of incorporating video projects (including animation) in Chinese language classes, such as increasing student learning outcomes (most students practice many times outside of classes before their video shooting), expanding student learning beyond the physical classroom, encouraging collaboration with fellow students and other Chinese speakers, being more effective in giving feedback and assessing student learning and more efficient in managing class time, and encouraging holistic learning and student creativity. Limitations of the video projects include some students relying too much on scripts, not having much time to meet to practice, and spending too much time on the video editing process which could be better used for language learning.

There were continued practices and research on the use of Web 2.0-based social networking technologies during this period. For example, Zeng (2018) studied student learning of the Chinese language on Zhihu (知乎), a Chinese online question-answer community. Zhang (2019) studied the effectiveness of a wiki-enhanced task-based language teaching approach in the teaching of Chinese as a foreign language.

Similarly, more studies were conducted on online Chinese teaching and learning. For example, Wang (2012) reported a case study concerning how she designed and taught an online Chinese course at an Australian university. Stickler and Shi (2013) reported the practice and findings of an online Chinese course offered through the Open University in the UK. Sun et al. (2013) shared a case on how they designed and delivered online Chinese courses at University of Wisconsin. Li (2019) discussed issues concerning online program evaluation and quality assurance based on the online Chinese courses developed through the collaboration between Michigan State University and Michigan Virtual.

In addition to those that can be directly applied in classrooms, there are other tools and resources that can be used to support the practice and research on Chinese teaching and learning. For example, Shi (2018) introduced the eye-tracking technology and its applications in Chinese language teaching and learning research with examples and suggestions. Li and Guo (2016) provided an overview of the collocation analysis tools and corpora that can be used for Chinese language and discussed how to use these tools (e.g., AntConc) and design activities for teaching vocabulary with examples.

2020

Machine Translation and Artificial Intelligence

While research on machine translation related to Chinese language started from the 1950s and made significant progress in the late 1980s and 1990s, the application of machine translation did not receive much attention in the field of language teaching and learning until recently when the technology became more mature with advances in artificial intelligence (AI). Tian (2020) described machine translation as a double-edged sword: it is a valuable pedagogical tool but poses a challenge for traditional practices in language teaching. In addition to academic integrity issues when students use machine translation to assist their learning, the technology is also considered a threat to job security of language teachers. There have been a few studies (e.g., Tian 2018, 2020; Xu 2020) conducted on the applications of machine translation in TCFL, and more studies are needed on this topic with further development of the technological front.

COVID-19 and Its Impact

The COVID-19 pandemic that started in Spring 2020 affected nearly every nation and region. Almost all schools were forced to urgently switch their courses to the remote mode for preventing the spread of the virus. Hodges et al. (2020) used "emergency remote teaching" to refer to such a sudden switch due to the emergency and differentiate it from regular online teaching and learning when the courses are well planned. Nonetheless, nearly every Chinese language instructor and learner experienced online instruction to some degree. Many institutions in the United States decided to continue to offer Chinese language courses fully or partially online in the fall semester of 2020 due to the prolonged pandemic.

Several surveys have been conducted as of August 2020 on instructors and learners regarding their experiences and perspectives towards online teaching and learning. The results of these surveys consistently showed that overall students and instructors were satisfied with their online learning and teaching experiences and were willing to learn and teach Chinese online if they needed to. For example, nearly 70% of student respondents from different countries in Liu et al. (2020b) reported that they were willing (moderately willing, somewhat willing, or very willing) to take fully online courses and approximately 75% of the instructors from various countries in Liu et al. (2020a) indicated that they were willing (to different degrees) to teach fully online. Similarly, according to Su (2020), results from a survey conducted at Beijing Language and Culture University showed that more than 70% of the surveyed students who took the courses online during the pandemic chose to continue, and more than 75% of instructors accepted the online teaching mode. Zhu (2020) reported similar findings from a survey conducted at Beijing Normal University in that students highly rated online instruction and nearly 75% of the students chose to continue their Chinese learning.

The experience gained from the pandemic has helped make the online mode an option for future language teaching and learning and opened new opportunities for many programs. For instance, Harris (2020) mentioned that despite some challenges for the Middlebury summer immersion programs, the online mode opened a door and helped reach some students who might otherwise not attend the program. Similarly, Zhu (2020) shared that Beijing Normal University was planning to offer more online courses and training opportunities for Chinese language learners and instructors. Liu (2020) compared similarities and differences between onsite and online Chinese teaching and learning, and discussed opportunities and challenges of the online mode for K-16 instructors and learners in various courses and programs. Xing (2020) introduced the virtual summer camps organized by Oversea Chinese Association of China for heritage learners.

One positive impact of the online experience in 2020 on future teaching and learning is the increasing use of technology. Liu et al. (2020a, b) found that both the instructor and student groups rated "use more technology" as the first impact of their online experience on their future teaching and learning. It is expected that instructors and students become more familiar with technologies and increase their technology skills in the future.

Another positive impact of the pandemic is a significant increase of professional development opportunities for instructors through free webinars, workshops, and conferences. For instance, the National Chinese Language Conference made its 3-day conference in June 2020 free and open to anyone, and drew nearly 4000 participants, which is approximately three times the usual number for the onsite conference. The National Foreign Language Center at the University of Maryland offered a 3-day free virtual summit in July 2020 to language teachers worldwide. In addition, many institutions, organizations, and professional groups have organized numerous free talks and workshops through online platforms such as Zoom and YouTube. For example, the Chinese Language Association of Secondary-Elementary Schools (CLASS) organized more than 50 webinars synchronously and asynchronously from mid-March to August. Similarly, the DoIE Chinese Language & Exchange Programs at San Francisco State University organized a series of Zoom Presentations on Remote Chinese Teaching, reaching over 6,300 teachers from the United States and other countries from March to August. 2020.

Additionally, a variety of WeChat groups were formed among Chinese language teachers worldwide, such as the Special Interest Groups (SIG) associated with the Chinese Language Teachers Association (CLTA), which are open to any interested instructors. These groups also organized talks on a wide range of topics and members used WeChat to share resources and information and discuss issues related to Chinese language teaching. In addition to professional development, these WeChat groups facilitated information sharing, knowledge co-constructing, and collaborating in a more effective and efficient manner than ever, and hence benefit the field of Chinese language teaching in many aspects.

Mastering the Use of Technologies in Chinese Language Teaching: Categorizations, Frameworks, Guidelines

The above review of the use of technologies has revealed that numerous tools and resources have been used in Chinese language teaching in the past 120 years. While some tools and resources became outdated and phased out over time, many have continued and evolved in new formats. New technologies also keep emerging. Today's learners and instructors can be easily overwhelmed by the dazzling variety of technologies available to them. As more technologies become available, a teacher needs to be equipped with the knowledge and skills to decide which technology to use and how to use it. This section provides two ways of categorizing technologies and shares some frameworks that can be used in evaluating and discussing technology's potential use. General guidelines are then provided for applying technologies in teaching Chinese as a foreign language and conducting relevant research.

Categorization of Technologies in Chinese Language Teaching

As Bourgerie (2003) noted, it is difficult to categorize the variety of technologies because technologies are "increasingly diverse" in addition to the large quantity, and even the names used to refer to all kinds of technologies can "lead to confusion about their purposes" (18). This section suggests two ways of categorizing the technologies.

Devices Versus Materials

Technologies developed in the past 120 years can be grouped into two categories: devices versus materials. The categorization is visualized in the diagram below.

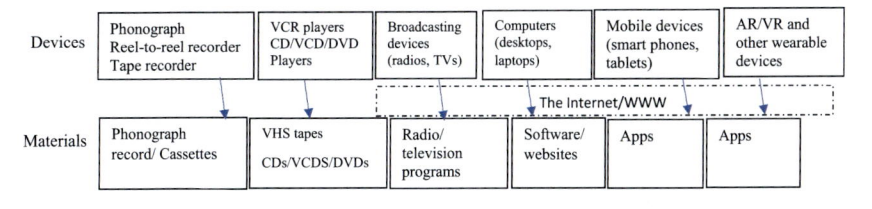

Devices	Phonograph Reel-to-reel recorder Tape recorder	VCR players CD/VCD/DVD Players	Broadcasting devices (radios, TVs)	Computers (desktops, laptops)	Mobile devices (smart phones, tablets)	AR/VR and other wearable devices
				The Internet/WWW		
Materials	Phonograph record/ Cassettes	VHS tapes CDs/VCDS/DVDs	Radio/television programs	Software/websites	Apps	Apps

As shown in the above diagram, devices include phonograph, recorders (reel-to-reel, tape recorders, and digital recorders), VCR players and CD/VCD/DVD players, broadcasting devices (radios, televisions), computers (desktops, laptops), mobile devices (smartphones, tablets), AR/VR, and other wearable devices. Materials, in contrast, refer to what devices support, including phonograph records, audio cassettes, VHS tapes, CDs/VCDs/DVDs, computer software, and apps on mobile devices and VR/AR devices. Both the Internet and the World Wide Web can also be grouped under software applications because websites and Internet-based services such as emails cannot stand alone but rely on hard devices to be hosted and accessed. The order listed in the diagram is roughly in alignment with the time periods when they were used in Chinese language teaching and learning as described in the previous sections.

This categorization has the following implications. First, materials can be considered as the "souls" for hardware devices. Without TV programs/channels, a television set is basically just a piece of hardware. Similarly, the power of computers can only be realized through the software running on them. Apart from the functions of built-in camera and other features, the usefulness of smartphones largely relies on the apps installed on them. In selection of hardware devices, one needs to consider what software and/or applications the devices support. Secondly, some software applications are tied with certain devices and not compatible across different devices and platforms. For example, Siri is only available on Apple's iOS products but not on Android and Windows phones. Apps developed by the same developer may have different functions and work differently on different devices and systems. It is advisable to choose cloud-based or the Internet-based software applications that can be used across different devices and platforms. For example, Kahoot (kahoot.com) is a good choice for using technologies in classrooms in that the instructor will not need to worry about the technical incompatibilities among different devices that students use.

Tools Versus Resources

The various technologies used in Chinese language teaching can also be categorized into tools versus resources depending on whether they provide content. Tools refer to technologies that do not come with content, whereas resources involve content.

Tools

Tools can be further categorized into three groups: (1) tools for general use, (2) tools for language teaching and learning, and (3) tools for Chinese language teaching and learning.

1. Tools for general use. General-purpose technology tools are not intentionally developed for language learning and teaching. They can be further divided into three subgroups:
    - Devices such as computers, smartphones, and TVs, which are invented for general use but not intentionally for language learning and teaching
    - Computer software (e.g., PPT), mobile applications (e.g., WeChat), and websites (e.g., Wiki, blog) that are not developed specifically for language learning

- Course (or Learning) Management Systems (CMS or LMS), such as Brightspace and Canvas, Moodle, Sakai
2. Language technology tools. Examples in this category include: (a) *tools related to speech technology*, such as Pratt (fon.hum.uva.nl/praat/), a tool developed by faculty from University of Amsterdam for the analysis of sounds; (b) *translation tools*, such as Google translate (translate.google.com/) and Baidu Fanyi (fanyi.baidu.com/); (c) *corpus analysis tools*, such as AntConc for concordancing and text analysis (laurenceanthony.net/software/antconc/) developed by faculty from Waseda University; and (d) other tools, such as WinCALIS developed by Duke University and SpeakEverywhere developed by Purdue University.
3. Chinese language tools. Various tools have been developed and/or used for Chinese language teaching and learning, such as websites used to help make conversions between simplified and traditional Chinese characters, and to help add Pinyin tone marks. There are also electronic toolkits and websites that include a variety of tools as well as related resources such as Chinese TA (a software package that helps teachers prepare learning materials), eStroke (eon.com.hk/estroke/), and Archchinese (archchinese.com).

Resources

Resources can also be further grouped into three subcategories: (1) general resources in Chinese language, (2) special resources for Chinese learning and teaching, and (3) Chinese corpus.

1. General Chinese language resources. This type of resources is often referred to as authentic materials that are created for native speakers and not specially developed for teaching and learning of the Chinese language. They include websites, TV shows, movies, YouTube videos, broadcast news, games, and others.
2. Special resources for Chinese teaching and learning. Resources in this category include CD courseware, websites, and applications that have developed for Chinese teaching and learning. Examples include CyberChinese, a multimedia aid for elementary Chinese language instruction associated with the textbook Practical Chinese Reader, Rutgers Multimedia Chinese Teaching System (http://chinese.rutgers.edu/index.htm), which includes four levels of online Chinese learning materials with free open access, and Chinese Reading World, which includes reading materials and quizzes for beginning, intermediate, and advanced levels, developed by the University of Iowa.
3. Chinese corpora. A number of Chinese corpora have been established in the past several decades, some of which have been be used for Chinese language teaching and learning. Li and Guo (2016) mentioned eight Chinese corpora such as *Chinese Learners Corpus* developed by the Advanced Center for the Study of Learning Science of National Taiwan Normal University, which collected written text samples from learners of Chinese at different levels; *HSK Dynamic Composition Corpus* developed by Beijing Language and Culture University, which collected more than 11,000 essays written by learners of Chinese from the HSK test; and *General Contemporary Chinese Corpus*, an online Chinese language

Fig. 1 *Categories of Technologies in Chinese Language Teaching: Tools vs. Resources*

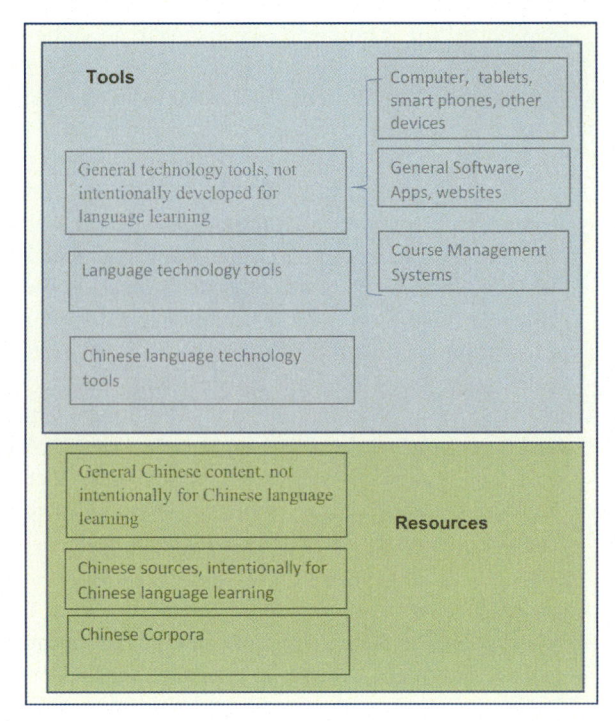

Tools:
- Computer, tablets, smart phones, other devices
- General technology tools, not intentionally developed for language learning
  - General Software, Apps, websites
  - Course Management Systems
- Language technology tools
- Chinese language technology tools

Resources:
- General Chinese content, not intentionally for Chinese language learning
- Chinese sources, intentionally for Chinese language learning
- Chinese Corpora

corpus sponsored by the State Language Affairs Committee of the Ministry of Education of China.

The division between tools and resources with six subcategories can be visualized in Fig. 1. In addition, the six subcategories can be regrouped into the following two categories: Chinese related versus not Chinese specific, as shown in Fig. 2. These categories and subcategories are based on the findings of Liu (2015), in which the author analyzed over 600 presentations related to technologies given at the annual meeting of Chinese Language Teachers Association (CLTA) and the biennial International Conference and Workshop of Technology and Chinese Language Teaching (TCLT) from 2000 to 2014.

Categorizing technologies into tools and resources with six subcategories help better analyze both technologies developed in the past and those newly emerging. For instance, before computers and the Internet became widely accessible, discussions on uses of technologies in Chinese language teaching in the last century mainly were on films, recorded audio/videos from radio/TV broadcasting programs or recorded by instructors for their own classroom teaching and/or published by publishers. Such technologies were content related and under the Resources category. The Internet and Web 2.0 technologies in the twenty-first century have made resources for Chinese language teaching and learning more accessible and sharable,

Fig. 2 *Categories of Technologies in Chinese Language Teaching: Chinese related vs. not*

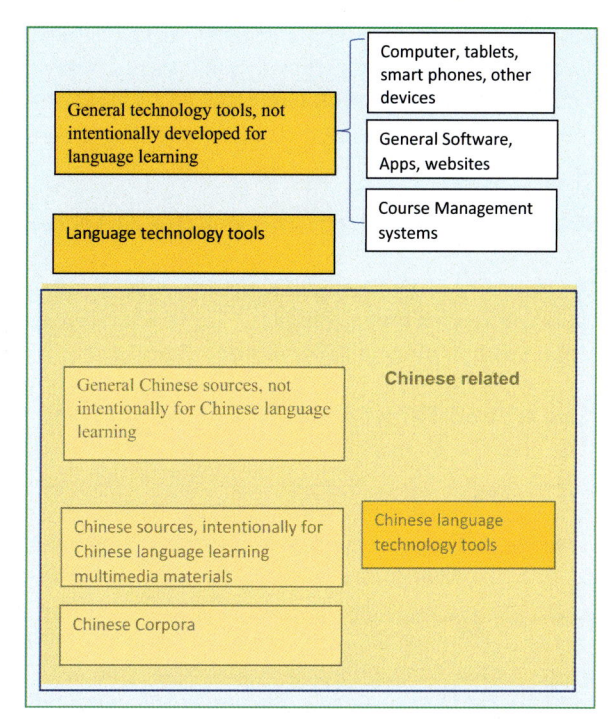

General technology tools, not intentionally developed for language learning	— Computer, tablets, smart phones, other devices
	— General Software, Apps, websites
Language technology tools	— Course Management systems

Chinese related
- General Chinese sources, not intentionally for Chinese language learning
- Chinese sources, intentionally for Chinese language learning multimedia materials
- Chinese language technology tools
- Chinese Corpora

and uses of resources such as films have become normalized and are even no longer considered as technologies by many instructors and researchers. The categorization suggested in this chapter helps clearly understand the relationships among various technologies involved. For instance, Cai and Chen (2020) shared their detailed pedagogical design (cooperative and collaborative method) of using two films (as content) in teaching Chinese language and cultures (instructional goals) to advanced Chinese language students with tools like Google Docs, Google Sheets, and Douban (https://www.douban.com/).

It is worth pointing out that that there are other ways to categorize technologies. For instance, Pitler et al. (2012) grouped technologies into nine categories: Word Processing Applications, Organizing and Brainstorming Software, Data Collection and Analysis Tools, Communication and Collaboration Software, Instructional Media (learner as consumer), Multimedia Creation (learner as producer), Instructional Interactives, Database and Reference Resources, and Kinesthetic Technology. Frank et al. (2008), based on their review of over 200 publications, identified four categories of technologies that are relevant for foreign language education: Classroom-based tools, such as course management systems, interactive white boards; Individual study tools, such as electronic dictionaries; Network-based social computing tools, such as blogs, wikis; and Mobile/portable, network-capable devices, such as tablet PCs and personal media players. These categorizations of technologies, though not specially addressing Chinese language teaching and learning, can still serve as useful references for Chinese language instructors and relevant practitioners.

Frameworks for Discussing the Use of Technologies

As Bransford (1979) has pointed out, having "some kind of organizational framework can keep us from getting lost" (p. 7). This section shares some frameworks that can be used in discussing the use of technologies.

Five C's and Three Modes of Communication

The Five C's (Communication, Cultures, Connections, Comparisons, and Communities) are the five goal areas identified in the *Standards for Foreign Language Learning: Preparing for the twenty-first Century,* by the American Council on the Teaching of Foreign Languages and several other associations of language teachers in 1996. It was adopted as the framework in the *World-Readiness Standards for Learning Languages* by the National Standards Collaborative Board in 2014.

Liu (2009) used the five C's framework to discuss how blogs, wikis, podcasting, video conferencing, online chat, and other Web 2.0 technologies could be used to enhance the 11 standards for foreign language learning. For instance, there are two standards under Communities: "Students use the language both within and beyond the school setting" and "Students show evidence of becoming life-long learners by using the language for personal enjoyment and enrichment." She provided sample activities for how to use the technologies based on the two standards.

Similarly, the three modes of communication (interpersonal, interpretative, presentational) are a new "Communicative Framework" that emphasizes the "content and purpose of the communication" (Cutshall 2012, p. 34). Wu et al. (2018) discussed how to use freely accessible technologies such as Edpuzzle and Google sites, to enhance the three modes of communication in Chinese language teaching and learning.

Four Essential Language Skills

Even though the practice and research on language learning have become more diverse, the four essential skills (listening, speaking, reading, and writing) are still the most familiar terms for language instructors and students. Technologies can be grouped based on how they are used to improve the four skills (Blake 2016). Liu and Xie (2010) shared a number of free synchronous and asynchronous tools for teaching and learning Chinese in the virtual world and categorized them based on instructional objectives. For example, tools such as blogs, wikis, and Google Docs are more effective for reading and writing related activities, whereas tools such as VoiceThread and online conference tools (such as Adobe Connect) are more suitable for listening and speaking practices.

Chinese characters are commonly acknowledged as a major challenge in learning to read and write Chinese. There have been many helpful websites and apps developed for teaching and learning Chinese characters. Liu (2017) grouped related tools and resources into the following five categories based on content and function: (1) input of Chinese characters, including Pinyin input, voice input, and handwriting, such as the built-in input methods and apps in smartphones; (2) dictionaries and translation and annotation tools, including online dictionary websites, such as Handian (zidc.net) and Apps such as Pleco, Hanping Chinese dictionary; (3) writing of Chinese characters, such as the Apps titled Word Tracer and Art Pen; (4) origins and stories of Chinese characters, such as Hanzi Graph and Chinese Etymology (hanziyuan.net/); and (5) instructional units and courses on Chinese characters, such as *A New Multimedia Course for Learning Chinese Characters* (http://chineseliteracy.net/content/contenttextbook.htm) and *Chinese Characters for Beginner* (https://www.coursera.org/learn/hanzi/).

Some preferred to consider culture as the fifth dimension (Damen 1987) or the fifth skill (Tomalin 2008) in addition to the four essential skills. While the importance of culture and related intercultural competence in language teaching and learning has been well acknowledged (Byram and Morgan 1994; Byram et al. 2001), it is still debatable whether to consider culture as the fifth skill (Thanasoulas 2001). As Kramsch (1993) articulates, "Culture in language learning is not an expendable fifth skill, tacked on, so to speak, to the teaching of speaking, listening, reading, and writing. It is always in the background, right from day one" (p. 1). Hence, it would appear more reasonable to categorize technologies related to culture under the framework on five "C" goal areas.

Fig. 3 NAEP model (http://carla.umn.edu/assessment/vac/Modes/p_2.html)

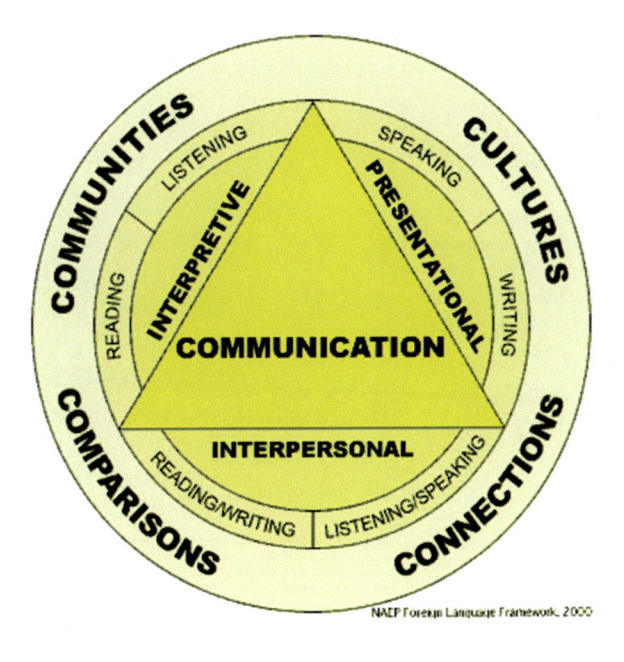

The Center for Applied Linguistics (2001) developed the NAEP (National Assessment of Educational Progress) model (see Fig. 3 below), which visualizes the three modes of communication, four essential language skills, and five "C" goal areas altogether. The model is a useful reference for grouping tools based on instructional goals and objectives.

Bloom's Taxonomy and Modified Digital Taxonomy

Bloom's taxonomy is another popular framework that can be used to evaluate and apply technologies based on instructional objectives. The original Bloom's Taxonomy (Bloom et al. 1956) includes six categories: Knowledge, Comprehension, Application, Analysis, Synthesis, and Evaluation. The revised taxonomy by Anderson and collaborators (Anderson and Krathwohl 2001) used action words (verbs) to describe the revised categories: Remember, Understand, Apply, Analysis, Evaluate, and Create. Churches (2008) advocated using digital Bloom's taxonomy for exploration of Bloom's taxonomy in digital context and added "collaboration" to reflect its importance in the twenty-first century. Ross et al. (2018) recommended using Bloom's Taxonomy as framework for Chinese language teachers in design and implementation of activities and assessments in their classes and discussed the Bloom's Digital Taxonomy.

Gagné's Nine Events of Instruction

Gagné's Nine Events of Instruction, which consist of the following nine instructional events, were identified by Robert Gagné based on conditions of learning (Gagné 1985): (1) Gaining attention, (2) Informing students of the objectives, (3) Stimulating recall of prior learning, (4) Presenting the content, (5) Providing learning guidance, (6) Eliciting performance (practice), (7) Providing feedback, (8) Assessing performance, and (9) Enhancing retention and transfer (Gagné et al. 1992). These nine

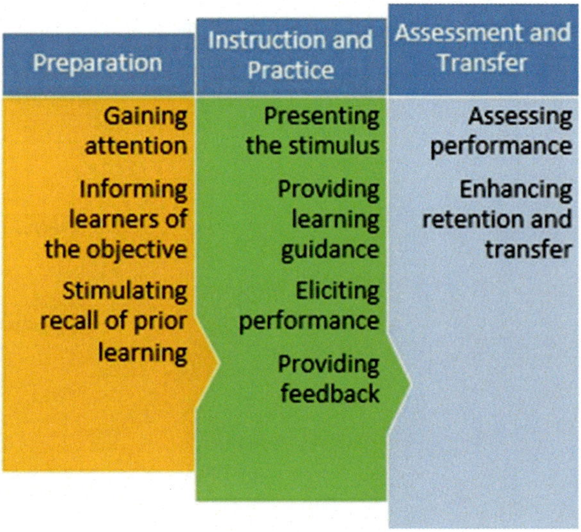

Fig. 4 Nine Events of Instruction (https://thepeakperformancecenter.com/business/learning/business-training/gagnes-nine-events-instruction/gagnes-nine-2/)

events, also known as nine steps, can be further divided into three segments/stages (see Fig. 4 below): Preparation (including the first three events), Instruction and Practice (including the middle four events), and Assessment and Transfer (including the last three events) (The Perk Performance Center 2020).

Gagné's Nine Events of Instruction, as an instructional model/framework, has been used widely in the design of instructional units (e.g., Krull et al. 2010), modules (e.g., Solanki 2014), and multimedia products (e.g., Al-Qassabi and Al-Samarraie 2013) in K-16 education and corporate training, and in both traditional and online modes (Binthabit 2019). It has also been applied in the language education field. For instance, Mei et al. (2015) applied it in the teaching of the Arabic language to non-native speakers. It is also used as a framework to organize technology tools in the course *Teach English Now! Technology Enriched Teaching* offered by Arizona State University through Coursera.org.

Second Language Acquisition Perspectives

Technology can also be discussed within the various frameworks based on second language acquisition theories. In the interactionist model, input and output are considered crucial in second language acquisition (Ellis 1997; Mackey and Gass 2015). Successful instructed language learning requires "extensive L2 input" and "opportunities for output" (Ellis 2005). Liu (2014) discussed how to integrate various tools and online sources in input (a variety of formats such as audio, video, and text with pictures), output (e.g., digital storytelling), and the process in a mixed-level Chinese language course. Da (2018) examined how to use speech technologies, such as text to speech technologies, machine translation, and automatic generated content technologies, to improve the quality and quantity of language input.

Frank et al. (2008) identified five functionalities of foreign language learning and teaching that technology has the potential to assist or enhance second language acquisition: Input, Output, and Interaction, Feedback, Collaboration, and Organization. Da (2017) selected the elements related to language acquisition (input, output, interaction, collaboration, and feedback) and used them as a framework to discuss technologies for Chinese language teaching and learning.

The SAMR Model

The SAMR model, proposed by Ruben Puentedura (2013), can be used as yet another framework to discuss the use of technologies. SAMR, which stands for Substitution, Augmentation, Modification, and Redefinition, refers to four different degrees of classroom technology integration. Among them, Substitution refers to the case where "Tech acts as direct tool substitute, with no functional change. Augmentation refers to the situation where "Tech acts as a direct tool substitute, with functional improvement." Modification refers to the case where "Tech allows for significant task redesign," and Redefinition refers to the situation where "Tech allows for the creation of new tasks, previously inconceivable." Substitution and Augmentation are considered as technology Enhancement, and Modification

Fig. 5 SAMR model. Puentedrua (2013) (http://www.hippasus.com/rrpweblog/archives/2013/05/29/SAMREnhancementToTransformation.pdf)

Redefinition
Tech allows for the creation of new tasks, previously inconceivable

Modification
Tech allows for significant task redesign

Augmentation
Tech acts as a direct tool substitute, with functional improvement

Substitution
Tech acts as a direct tool substitute, with no functional change

Transformation

Enhancement

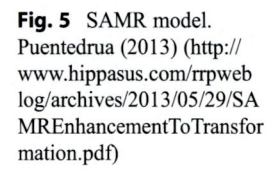

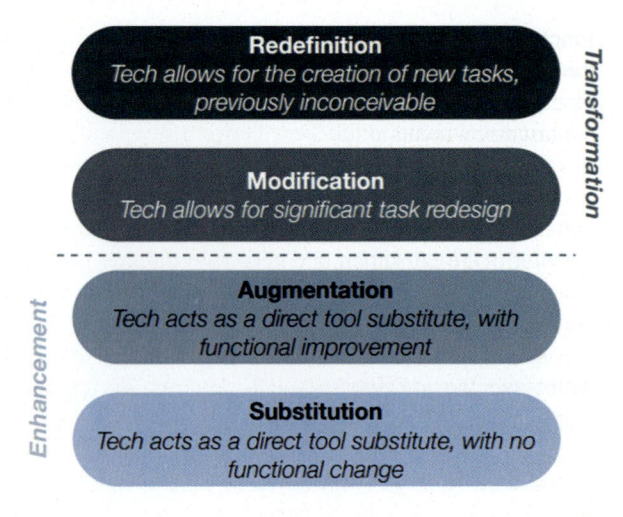

Redefinition are grouped as technology Transformation. The original diagram is shown below (Fig. 5).

For example, using a simple electronic version (such as scanned copy) of the same textbook is not much different from using a hard copy of the textbook. The electronic version is mainly just a Substitution of the hard copy. However, if each word in the vocabulary list is linked to an audio file and learners can hear its pronunciation and/or see the stroke order of each Chinese character by clicking on it, electronic textbooks will apparently be a strong Augmentation to the hard copy version. Using collaborative writing tools such as Google Docs and wikis will allow multiple students to work on the same document at the same time and will make their collaboration more effective. Using these technologies will allow for significant task redesign, which is Modification, a transformation from the paper and pencil-based writing process. With blogs and social media tools, learners will be able to provide and receive feedbacks from each other beyond the classroom and interact with native speakers worldwide. In the virtual word, such as Second Life, instructors can create new tasks such as driving a virtual car, getting in a virtual train, which otherwise they could not do in a traditional physical classroom. Such use of technologies will be Redefinition and allow for previously inconceivable new tasks. Jiang (2017) used SAMR as a framework in discussing the effective use of technologies in promoting active Chinese learning process.

It is worth pointing out that the four levels in the SAMR model are not clearly cut but a continuum. For example, one can argue that asking students to write blogs is augmentation while others can argue that it is modification. In addition, we need to keep in mind that it is not the technology itself belonging to which category but rather the tasks designed with such technologies. For example, one can use the chat function in Second Life, which is just a substitution of face-to-face conversation. In the design of instructional tasks and employment of technology tools, instructors are advised to ask themselves where the way of their use of technologies falls on the continuum, whether or not how they use the technology is just substitution, or more towards augmentation, modification, and redefinition.

Allan Carrington combined Bloom's taxonomy and SAMR together into a Pedagogical Wheel (https://educationtechnologysolutions.com/2016/06/padagogy-wheel/). In the Carrington's Wheel, the six levels described in Bloom's taxonomy are placed in the inner circle, and the four different degrees of technology integration of the SAMR model are placed in the outer circle. Ross et al. (2018) commented that the Wheel was designed to help instructors understand how to use mobile apps in their teaching and discussed the use of the Wheel for Chinese language teaching and learning.

Summary

This subsection provides several frameworks that can be used for understanding and organizing technologies for Chinese language teaching and learning from various perspectives. *The 5C', the three modes of communication*, and *the four essential language skills* are frameworks based on the goals of language teaching and learning; *Bloom's taxonomy* and *the modified digital taxonomy* also address the objectives/goals of the instruction though at a more general level. *Gagné's Nine Events of Instruction* attends to the instructional process while second language acquisition related framework focus on the language acquisition process. The SAMR model helps one deliberate on the selection and design of technology activities. All these frameworks highlight the fact that technology is not a goal in and of itself when being applied in language learning and teaching. Its application is dictated by pedagogical goals, objectives, and designs.

Guidelines for Use of Technology in Chinese Language Teaching

Egbert (2005) provided five guidelines for using educational technologies in language classrooms: (1) use technology to support the pedagogical goals of the class and Curriculum, (2) make technology accessible to all learners, (3) use technology as a tool, (4) use technology effectively, and (5) use technology efficiently. Frank et al. (2008) summarized three principles from second language acquisition research (i.e., provide rich target language input, engage learners in interactive tasks, and provide feedback) and four principles from cognitive psychology research (test frequently with minimal retrieval cues, distribute practice across small sessions, vary learning activities, and make learning challenging to require students process information deeply). They further provided five best practices for using technology use in the classroom: (1) maximize target language practice, (2) maximize the efficiency of classroom contact time with the teacher, (3) enable customization, shareability, reusability, and anytime/anywhere access to target language content and instructional materials, (4) archive and analyze interaction, learner output, to individualize input and feedback, and (5) motivate students and engage them in meaningful target language use. While these principles, guidelines, and practice for technology use do not specifically address the case of Chinese language teaching, they still serve as useful references for Chinese language instructors and other practitioners in the field.

Because instructors know their own specific situations best, this section provides some essential guidelines or principles, which can be also helpful for other practitioners and researchers. These guidelines are based on those suggested by Egbert (2005) and Frank et al. (2008) and the review of practice and research on using technologies in Chinese language teaching in the past 120 years and the frameworks presented in previous sections.

(1) *Technology Should Be Used to Serve the Goal of Language Teaching and Learning.*

As stated by the American Council on the Teaching of Foreign Languages (2017), "the use of technology is not a goal in and of itself." Rather, technology is one tool that supports language learning. This position has been well acknowledged in both research and practice (e.g., Chun et al. 2016; Egbert 2005; Golonka et al. 2014). Language instructors need to always put language learning first and technology after. In making plans for each lesson, they are advised to first focus on objectives and learning outcomes for the lesson and then decide whether and how technology can assist to reach the objectives and outcomes. Instructors should not feel obliged to use technology in each lesson. As Golonka et al. (2014) commented, using technology in delivering a lesson will not make bad pedagogy good, nor does "a lack of technological tools or applications prevent effective teaching" (p. 93). In addition to the objectives related to the textbook and specific courses, instructors are also advised to design lesson and units in alignment of goals, such as the Five C's, three modes of communication, and Bloom's taxonomy discussed in the previous section.

(2) *Instructors Should Stay Updated and Understand the Benefits and Limitations of New Technologies.*

Like practitioners in other professions, language instructors should stay current on new trends of technology in addition to pedagogy. As Xu (2009) states, "Technology will never replace teachers, but teachers who use technology will replace those who don't" (p. 103). Welch (2005) argues that an instructor should work as a learning broker, asking oneself "how can I broker for services and appropriate technologies for my student(s)?" Jung (2005) concurs that it is the instructor's job "to deploy all the media, old and new, to prepare the student for the onslaught of real-life communicative situations and the concomitant sanctions" (15). Similarly, Kessler (2016) points out that language teachers must become comfortable with "what are currently new, intelligent, and increasingly sophisticated resources as well as with those that will succeed them and make time to understand the great opportunities and obligations that will form the new landscape of world language teaching and learning" (p. 216).

Instructors should be aware of both the strengths and limitations of the technologies when using them to aid language teaching and learning (Zhang 1998; Bai 2003). They should maximize the strengths of each technology and avoid its limitations when designing instructional units. In addition to strengths and limitations of specific technologies, instructors are advised to refer to the SAMR model in the planning of integration of technologies into the courses, i.e., whether the integration belongs to substitution, augmentation, modification, or redefinition.

(3) *Instructors Should Consider Various Factors in the Selection of Tools and Resources.*

Bates (2015) recommends using the SECTIONS model (see Fig. 6 below) for instructors in the selection of technology and media, which includes eight factors: students, ease of use, cost/time, teaching media characteristics and instructional strategies, interaction, organizational issues, networking, and security and privacy.

Instructors need to know when to use what technology in which situation to achieve optimal outcomes. For example, although selected authentic video clips can provide students with rich inputs and benefit them in multiple ways, it will be questionable to use a large amount of class time to view video clips. Rather, class time could be used more effectively for students to practice, for example, their speaking skills and interact with each other and with the instructor.

Trade-offs should be expected in making decisions concerning the use of technology. For example, online courses and programs can provide an option for

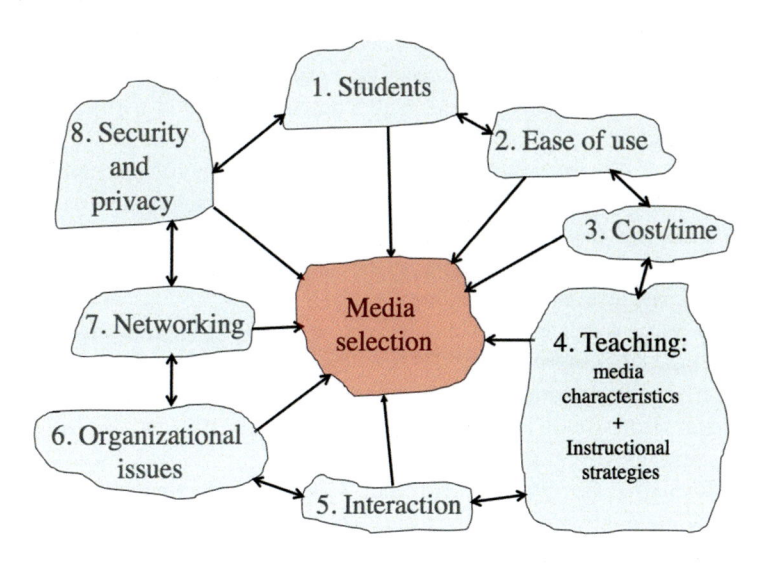

Fig. 6 The SECTIONS model (Bates 2015, 306)

students who cannot attend residential courses and programs. Learners who do not have other options can take free MOOCs courses to learn Chinese (Liu, 2018). Similarly, if students cannot afford to purchase apps such as Skritter to help them practice Chinese characters, instructors can recommend some free alternatives, which provide similar functions though they may not be as powerful as commercial ones.

Furthermore, as Bates (2015) suggests, instructors need to understand student preferences, previous experiences, access to technology, and other factors in selection of technologies to assist design and delivery of instructional units. For example, some apps are only available on iOS system. If using them for classroom activities, instructors need to make plans for how to get students who do not have iOS devices involved. If asking students to use the iOS app as part of their homework, instructors need to make alternative plans for those who do not have iOS devices or any devices.

(4) *Instructors Are Advised to Start with Simple, Low-Cost, and Highly Effective Technologies, and to Be Cautious with Unproven Technologies.*

Beatty (2003) cautions that "teachers need to be concerned about investing time and money in unproven technology" (72). This is particularly advisable for adopting new technologies associated with high learning curves and monetary cost such as the Second Life, VR, and AR gadgets. Unless for research and exploration, instructors are recommended to start with simple, low-cost, and highly effective tools in their daily teaching, especially considering the limited funds and human resources in most TCFL programs.

(5) *Instructors Should Adjust Their Teaching Based on Student Needs and in Accordance with the Evolution of Technologies.*

Students in the twenty-first century have access to numerous resources and tools in learning the Chinese language, and the instructor needs to provide guidance and to help students make the best use of these resources and tools. For instance, typing Chinese characters on computers and mobile devices can help students build connections between sound and the written text. While it is still under discussion how much handwriting practice should be emphasized, it is widely acknowledged that instructors would need to teach students how to input characters and design related activities, such as those suggested in He et al. (2007) and curriculum to respond to the evolution.

Similarly, instructors are advised to teach students how to take advantage of intelligent tools and services such as Siri for their Chinese learning. In addition, given the wide availability of the translation tools such as Google translate, it would be productive for instructors to revise or develop new activities to help students make use of these new technologies rather than simply forbidding or punishing students to use them.

(6) *One Should Be Specific in Describing What the Technology Is and How It Is Used When Discussing Its Effectiveness.*

Three reasons are identified for this suggested guideline. First, the same term is often used to cover or refer to rather different situations. For instance, online teaching is found to be used to cover fully online courses, partially online courses, and in-person courses that include online components (Liu 2020a). Similarly, while the hybrid mode usually refers to the situation where some sessions are conducted in a physical classroom with others online (usually asynchronously), the term is used by some institutions to refer to a mode in which some students attend the class in the physical classroom while other students join the class at the same time via online conferencing tools such as Zoom due to the pandemic.

Second, technology has been constantly improving. For instance, while the computer used in 2020 is called by the same name as the machine used in 1970s, its technology has become much more versatile and intelligent than it was 50 years ago. Third, the same technology can be used in different ways. Whether it is effective largely depends on how the technology is integrated and the instruction is designed. Therefore, it would be problematic to discuss whether a technology is effective without specifying what is referred to and how it is applied in teaching and learning.

Concluding Remarks

This chapter provides a comprehensive historical review of technology in Chinese language teaching, dating back to 1900 when the phonograph was first used in language classes and up to the current year of 2020. This review adopts a chronological order to describe the history instead of using any other labeled stages (e.g., Mark Warschauer's three phases of CALL, see Warschauer 1996; Warschauer and Healey 1998) based on the considerations that technologies are just technologies and can be used for different purposes depending on the instructional design.

This review of the use of technologies in Chinese language teaching reveals that technologies have evolved dramatically, though gradually, in the past 120 years. The only technology available for teaching in 1900 was to use the phonographs to record the instructor's voice for students to listen to. In contrast, in 2020, instructors and students have *too many* tools and resources to choose from. The challenges for instructors and students nowadays are how to select and use them effectively for their teaching and learning.

While earlier technologies were mainly used to provide students with resources, the development of computer technology in the late twentieth century allowed students to interact with a computer and receive feedback. However, due to the limitations with the technology, what the computer could do in the last century was mostly low-level mechanical drills which lacked communicative interactivity (Bai 2003; Zhang 1998). The computer is becoming more intelligent with the advancement of the artificial intelligence technology in the twenty-first century. The AI technology, big data, and other emerging technologies provide potential for Chinese language learning to be more interactive and increasingly personalized. Learners will be able to have real conversations and high-level interaction with AI, represented by Siri and other AI chatbots (Liu 2013; Cai 2019). Additionally, Learners' learning progress can be recorded and analyzed by these advanced technologies. Instructors can tailor their instruction to students' individual needs (Sung 2018; Zheng 2019), and learners will be provided with learning materials and instruction that match their proficiency level, learning style, personal interest, and individualized objectives.

The Internet has reached and impacted on nearly every aspect of human society in the past 30 years since the invention of the World Wide Web in the 1990s. In Chinese language education, Chinese language instructors and learners were able to find an abundance of resources for teaching and learning of the Chinese language. The Web 2.0 technologies of the 2010s further allow users to produce and share resources themselves. Students are now not only able to interact with their classmates and receive feedback from their instructors, but also able to connect, interact, and receive feedback from peers, instructors, and many others including native speakers from anywhere in the world. In other words, the advancement of technologies has not only improved learner interaction with machines and enhanced personalization of their learning process, but also widened and strengthened learners' connections and interactions with other human beings on their language learning journey.

In addition, Chinese language teaching and learning will become more open. The improvement of the Internet technology, such as the synchronous online conferencing tools, has revolutionized distance education. The various functions of the virtual classrooms allow instructors and students to hold classes online in a way very similar to a physical classroom, and bring more benefits compared to physical classrooms in some aspects. COVID-19 has helped make many institutions experience and realize the potential of online instruction and consider offering an online mode as an option for their students in the future. In addition to taking courses at their own institutions, interested students can also enroll and audit MOOCs offered by other institutions, especially those prestigious universities that have more available funds and resources.

Technology has also made language learning more convenient and fun. The smartphones and tablet technologies that emerged in the 2010s can be considered as microcomputers in hand. In addition to having nearly full functions of computers (except for its smaller screen), the convenience and robustness of built-in camera, video recorder/player, and audio recorder/player make the smart devices a most powerful, handy, and integrative technology. Their touch-screen feature, that allows learners to practice writing Chinese characters with fingers or electronic pens, helps solve the issues that typing on computers decrease the traditional handwriting practice on physical paper.

Along with the positive and exciting aspects of these new trends, challenges for Chinese language teaching and learning still remain. For instance, the close connection and convenience of communication among humans have shortened people's attention span and made them easily distracted, and their learning more fragmented (Lay 2015). Therefore, instructors need to keep up with the trend and adjust their teaching accordingly. They need to continue to think how to use the technologies, such as AI and machine learning, to assist their teaching rather than seeing them as a threat to their job. They need to continue to reinvent themselves and reimagine themselves as teachers to enhance students' learning experiences. The suggested categorization, frameworks, and general guidelines and principles are intended to help TCFL instructors and other practitioners meet with future challenges.

The practices of Chinese language instructors and related research in the past 120 years have well embodied what Last (1989) has commented, "language teachers as a body have been more ready than most to accept and explore the pedagogical potential of new technologies as they have emerged" (p. 15). Social networking tools, particularly WeChat, have strengthened the connections and collaborations among Chinese language instructors worldwide. Instructors are able to share information about resources and tools quickly (in minutes) and widely (a single WeChat group can include 500 members), discuss issues related to teaching with colleagues from different institutions in different countries, and reach and consult experts in their specialized areas more conveniently. The pandemic has prevented people from traveling to professional conferences and workshops. However, on the flip side, the online conferencing platforms have made most of the workshops and conferences which are usually expensive and inaccessible open and free to any interested in-service and pre-service teachers. It is estimated that in the TCFL field more than 500 free talks, workshops, and conference presentations have been given online between March and August 2020 alone and some talks attracted more than 1000 attendees. These contribute largely to the professional development of Chinese language teachers including pre-service teachers and can make positive impact on the overall quality of Chinese language instruction. To conclude, despite challenges, one can be optimistic that Chinese language teachers will be able to continue to make their teaching more effective, efficient, and enjoyable, and help more learners learn Chinese better, faster, and happier with the assistance from technologies.

References

Al-Qassabi, Hamed, and Hosam Al-Samarraie. 2013. Applying Gagne's nine events in the design of an interactive ebook to learn 3D animation. *Advances in Computing 2013* 3 (3): 60–72. https://doi.org/10.5923/j.ac.20130303.05.

Anderson, L.W., and D. Krathwohl, eds. 2001. *A taxonomy for learning, teaching and assessing: A revision of Bloom's taxonomy of educational objectives*. New York: Longman.

Bai, Jianhua. 2003. Making multimedia an integral part of curricular innovation. *Journal of the Chinese Language Teachers Association* 38 (2): 1–16.

Bates, A. 2015. *Teaching in a digital age: Guidelines for designing teaching and learning*. Available at https://www.tonybates.ca/teaching-in-a-digital-age/.

Beatty, Ken. 2003. *Teaching and researching computer-assisted language learning*. Essex: Pearson Education Limited.

Berman, P. 2001. The use of authentic TV broadcast videos and pictures in elementary Chinese language classroom. In *Presentation given at the annual meeting of Chinese language teachers association conference*. Washington, D. C. November 15–19, 2001.

Binthabit, N.M. 2019. Integration of language learning strategies and self-efficacy enhancing strategies for second language acquisition: A design and development study. Unpublished doctoral dissertation, Virginia Polytechnic Institute and State University. https://pdfs.semanticscholar.org/e72c/e6ae9b9185fb663d7f2577d3560387c0b88a.pdf.

Blake, Robert. 2016. Technology and four skills. *Language Learning & Technology.* 20 (2): 129–142.

Bloom, B.S., M.D. Engelhart, E.J. Furst, W.H. Hill, and D.R. Krathwohl. 1956. *Taxonomy of educational objectives, handbook I: The cognitive domain*. New York: David McKay.

Bolinger, Dwight. 1934. Spanish on the air in Wisconsin. *The Modern Language Journal*. 18 (4): 217–221.

Bonk, C.J., J.A. Cummings, N. Hara, R.B. Fischler, and S.M. Lee. 2000. A ten-level web integration continuum for higher education. In *Instructional and cognitive impacts of web-based education*, ed. B. Abbey, 56–77. IGI Global. https://doi.org/10.4018/978-1-878289-59-9.ch004.

Bonnington, C. 2012, January 23. iPad a solid education tool, study reports, https://www.cnn.com/2012/01/23/tech/innovation/ipad-solid-education-tool/index.html.

Bourgerie, Dana Scott. 2003. Computer assisted language learning for Chinese: A survey and annotated bibliography. *Journal of the Chinese Language Teachers Association* 38 (2): 17–48.

Bransford, J. 1979. *Human cognition: Learning understanding and remembering*. Belmont: Wadsworth.

Bruce, Joseph Percy. 1926. A phonograph course in the Chinese National Language. By Yuen Ren Chao, Ph.D. commercial press, Shanghai, China. *Bulletin of the School of Oriental and African Studies* 4 (1): 197–200. https://doi.org/10.1017/S0041977X00102836.

Buck, David. 1974. The use of broadcast television programs as a means of advanced conversational instruction in Chinese. *Journal of Chinese Language Teachers Association* 9 (2): 93–67.

Byram, Michael, and Carol Morgan. 1994. *Teaching-and-learning language and culture*. Clevedon: Multilingual Matters.

Byram, Michael, Adam Nichols, and David Stevens. 2001. *Developing intercultural competence in practice*. Clevedon: Multilingual Matters.

Cabarga, Demetrio A. 1937. Teaching Spanish by radio. *The Modern Language Journal* 22 (3): 189–200. https://doi.org/10.1111/j.1540-4781.1937.tb00585.x.

Cai, Q. 2019. The applications of digital games and AI chatbots in foreign language instruction. In *Paper presented in the ACTFL Annual Convention and World Languages Expo* in Washington, DC, November 20–22, 2019.

Cai, Jingjing, and Su-I. Chen. 2020. The application of films in advanced Chinese language courses: A cooperative-collaborative learning model. *Journal of Technology and Chinese Language Teaching* 11 (1): 84–113.

Chen, Y.-F. May, 2006. Is it worth it? A critical review of distance learning courses via interactive TV. TCLT4. University of South California, May 5–7.

Chen, Dongdong. 2010. Enhancing the learning of Chinese with second life. *Journal of Technology and Chinese Language Teaching* 1 (1): 14–30.

Chen, Joanne. 2011. Application of voice thread in Chinese teaching and learning: Some examples. *Journal of Technology and Chinese Language Teaching* 2 (1): 81–94.

Chen, Dongdong. 2013. What can a smartphone offer to learners of Chinese? *Journal of Technology and Chinese Language Teaching* 4 (2): 86–95.

Chen, Henny. 2014. Blend your lessons through flipped and seamless learning. *Journal of Technology and Chinese Language Teaching* 5 (1): 75–82.

Cheng, Jihua. 1963. *The development history of Chinese film*. Beijing: Chinese Film Publisher. [程季华. 1963.《中国电影发展史》中国电影出版社].

Cheng, C.-C. 1972, November. Computer-based Chinese teaching program at Illinois. In Paper presented at the 1972 Annual Meeting of the Chinese Language Teachers Association. Atlanta. Retrieved from http://files.eric.ed.gov/fulltext/ED071528.pdf.

Cheng, Chin-Chuan. 1973. Computer-based Chinese teaching program at Illinois. *Journal of Chinese Language Teachers Association* 8 (2): 75–79.

Cheng, Yuzhen. 2005. *The history of teaching Chinese as a foreign language in new China*. Beijing: Beijing University Publisher. [程裕祯. 2005年01月.《新中国对外汉语教学发展史》.北京大学出版社].

Cheng, Hsiu-Jen, Hong Zhan, and Andy Tsai. 2010. Integrating second life into a Chinese language teacher training program: A pilot study. *Journal of Technology and Chinese Language Teaching* 1 (1): 31–58.

Chinnery, G.M. 2006. Going to the MALL: Mobile assisted language learning. *Language Learning & Technology* 10 (1): 9–16.

Chu, C. 2004. Chinese TA-An Integrated Software for Chinese Language Teachers. In Given at the 3th international conference and workshops on technology and Chinese Language Teaching (TCLT3). Columbia University, New York, May 28–30.

Chun, D., R. Kern, and B. Smith. 2016. Technology in language use, language teaching, and language learning. *Modern Language Journal* 100 (s1): 64–80. Supplement issue for Celebrating 100 years of the Modern Language Journal. Retrieved from https://onlinelibrary.wiley.com/doi/epdf/10.1111/modl.12302.

Churches, A. 2008. *Bloom's digital taxonomy*. Available at http://www.ccconline.org/wp-content/uploads/2013/11/Churches_2008_DigitalBloomsTaxonomyGuide.pdf.

Cutshall, Sandy. 2012. More than a decade of standards: Integrating "communication" in your language instruction. *The Language Educator* 2012: 34–39. https://www.actfl.org/sites/default/files/publications/standards/Communication.pdf.

Da, Jun. 2017. Facilitating Chinese language acquisition with technology. In *Keynote speech given at the 3rd Online Chinese Teaching Forum and Workshop (OCTFW)*. East Lansing: Organized by the Confucius Institute of Michigan State University.

Da, Jun. 2018. Language processing technologies and the changing face of input. In *Invited keynote speech at the 4th Online Chinese Teaching Forum & Workshop*. East Lansing, MI: Michigan State University.

Da, Jun, and Yanqun Zheng. 2018. Technology and the teaching and learning of Chinese as a foreign language. In *The Routledge handbook of Chinese second language acquisition*, ed. Chuanren Ke, 432–447. New York: Routledge.

Damen, Louise. 1987. *Culture learning: The fifth dimension in the language classroom*. Reading: Addison-Wesley.

Dew, James. 1996. Advances in computerization of Chinese. *Journal of Chinese Language Teachers Association* 31 (3): 15–32.

Egbert, J. 2005. *CALL essentials: Principles and practices in CALL classrooms*. Alexandria: TESOL.

Ellis, Rod. 1997. *SLA research and language teaching*. Oxford: Oxford University Press.

———. 2005. Principles of instructed language learning. *System* 33 (2): 209–224. https://doi.org/10.1016/j.system.2004.12.006.

Fleming, S. 1999. A web-based resource for foreign language distance education teacher training. Presentation given at the First International Conference on Internet Chinese Education, Taipei, Taiwan. May 22–24, 1999.

Frank, Victor, Ewa Golonka, Anita Bowles, Emily Becker, Suzanne Freynik, and Dorna Richardson. 2008. *Optimal foreign language learning: The role of technology*. College Park: Center for Advanced Study of Language at the University of Maryland. Retrieved from http://www.casl.umd.edu/sites/default/files/Frank08_RoleofTechnology.pdf.

Franzen, C. Jul 1, 2014. The history of the walkman: 35 years of iconic music players. https://www.theverge.com/2014/7/1/5861062/sony-walkman-at-35.

Gagné, R.M. 1985. *The conditions of learning and theory of instruction*. 4th ed. New York: Holt, Rinehart & Winston.

Gagné, Robert M., Leslie J. Briggs, and Walter W. Wager. 1992. *Principles of instructional design*. 4th ed. Forth Worth: Harcourt Brace Jovanovich College Publishers.

Gan, K.W., and N.L. Cheng. 1999. *An online Chinese lexicon teaching system*. [顏國偉, 鄭雅麗 (1999). 一個網路的中文詞彙教學系統] Presentation given at the First International Conference on Internet Chinese Education, Taipei, Taiwan, May 22–24, 1999. http://edu.ocac.gov.tw/discuss/academy/netedu01/doc/%E8%AB%96%E6%96%87/%E4%B8%83%E8%AB%96%96E6%96%87%E7%99%BC%E8%A1%A8-19.htm.

Gao, Y. 2019. Using smartphone Apps to integrate authentic materials and cultural learning. Fall conference of Foreign Language Association of Virginia, October 3–5.

Ginsburg, Edward B. 1935. Foreign talking pictures in modern language instruction. *The Modern Language Journal* 19 (6): 433–438. https://doi.org/10.1111/j.1540-4781.1935.tb05848.x.

Godwin-Jones, Robert. 2011. Emerging technologies mobile apps for language learning. *Language Learning & Technology* 15 (2): 2–11. http://llt.msu.edu/issues/june2011/emerging.pdf June 2011.

Golonka, Ewa M., Anita R. Bowles, Victor M. Frank, Dorna L. Richardson, and Suzanne Freynik. 2014. Technologies for foreign language learning: A review of technology types and their effectiveness. *Computer Assisted Language Learning* 27 (1): 70–105. https://doi.org/10.1080/09588221.2012.700315.

Grant, Scott, and Hui Huang. 2010. The integration of an online 3D virtual learning environment into formal classroom-based undergraduate Chinese language and culture curriculum. *Journal of Technology and Chinese Language Teaching* 1 (1): 2–13.

Harris, M. 2020. Personal communication. July 1, 2020.

He, Wayne, Dela Jiao, Qiuxia Shao, and Christopher M. Livaccari. 2007. *Chinese for tomorrow: A new five-skilled approach*. Vol. One. Boston: Cheng & Tsui.

Hendix, William. 1939. Films in the learning of modern languages. *The Journal of Higher Education* 10 (6): 308–311. https://doi.org/10.2307/1973853.

Hodges, C., S. Moore, B. Lockee, T. Trust, and A. Bond. 2020, March 27. The difference between emergency remote teaching and online learning. *Educause Review*. https://er.educause.edu/articles/2020/3/the-difference-between-emergency-remote-teaching-and-online-learning.

Huhn, C., J. Dassier, and S. Liu. 2016. Realities of mobile learning technologies in foreign language classes. *The IALLT Journal for Language Learning Technologies* 46 (1): 36–62. Available at http://ialltjournal.org/index.php/ialltjournal/article/view/250.

Jiang, Z. November, 2017. *Effective use of technologies in promoting active Chinese learning process*. Workshop given at the 3rd Online Chinese Teaching Forum and Workshop. East Lansing: Michigan State University.

Jin, Honggang 2009. Participatory Learning in Internet Web Technology: A Study of Three Web Tools in the Context of CFL Learning. *Journal of Chinese Language Teachers Association* 44 (1): 25–50.

Jin, L. 2009. *Facebook-based Chinese culture learning*. Annual Conference of Chinese Language Teachers Association. San Diego. November 20–22.

Jung, Udo O.H. 2005. CALL: Past, present and future –A bibliometric approach. *ReCALL* 17 (1): 4–17. https://doi.org/10.1017/S0958344005000212.

Kennedy, G., B. Dalgarno, K. Gray, T. Judd, J. Waycott, S. Bennett, K. Maton, K.L. Krause, A. Bishop, R. Chang, and A. Churchward. 2007. The net generation are not big users of Web 2.0 technologies: Preliminary findings. In ICT: Providing choices for learners and learning. Proceedings ascilite Singapore 2007. http://www.ascilite.org.au/conferences/singapore07/procs/kennedy.pdf.

Kessler, Greg. 2016. Technology and the future of language teaching. *Foreign Language Annals*. 51: 205–218.

Koon, Cline M. 1933. Modern language instruction by radio. *The Modern Language Journal* 17 (7): 503–505. https://doi.org/10.1111/j.1540-4781.1933.tb05781.x.

Kramsch, Claire. 1993. *Context and culture in language teaching*. Oxford: Oxford University Press.

Krull, E., K. Oras, and E. Pikksaar. 2010. Promoting student teachers' lesson analysis and observation skills by using Gagne's model of an instructional unit. *Journal of Education for Teaching* 36 (2): 197–210.

Kuo, J. 1982. University of Kansas radio broadcast research. *Journal of Chinese Language Teachers Association* 17 (1): 67–76.

Last, Rex W. 1989. *Artificial intelligence techniques in language learning*. Chichester: Horwood.

Lay, Chua Chee. 2015. The global Positioning of Chinese Language Education in Digital Wave (数码狂潮中华语文教育的全球定位). Keynote speech given at the 9th International Conference on Internet Chinese Education (ICICE9), Massachusetts Institute Technology, Boston, MA, June 19–21, 2015.

Li, Jiahang. 2019. Online Chinese program evaluation and quality control. *International Chinese Language Education* 4 (3): 62–70.

Li, Jiahang, and Zilu Jiang. 2017. Students' perceptions about a flipped online Chinese language course. *Journal of Technology and Chinese Language Teaching* 8 (2): 25–38.

Li, Shouji, and Shulun Guo. 2016. Collocation analysis tools for Chinese collocation studies. *Journal of Technology and Chinese Language Teaching* 7 (1): 56–77.

Li, Shubin. 2007. The characteristics of online Chinese language teaching and learning in higher education: Perceptions of teachers and students. Unpublished doctoral dissertation of Alliant International University.

Li, Shuqing. 1991. 李淑清. TV drama lessons in teaching intermediate Chinese. 对外汉语教学中级阶段的"电视短剧"课. Chinese teaching in the world 世界汉语教学 1991 (2): 117–119.

Lin, Chin-Hsi, and Yujen Lien. 2012. Teaching and learning Chinese with an iPad. *Journal of Technology and Chinese Language Teaching* 3 (2): 47–63.

Lin, Chin-Hsi, and Yining Zhang. 2014. MOOCs and Chinese language education. *Journal of Technology and Chinese Language Teaching* 5 (2): 49–65.

Lin, C.H., K. Zhou, and S. Yang. 2018. A survey of mobile apps for learning Chinese vocabulary. *Journal of Technology and Chinese Language Teaching* 9 (2): 98–115.

Ling, Vivian. 2018. Institute of far eastern languages at Yale university. In *The field of Chinese language education in the U.S.: A retrospective of the 20th century*, ed. V. Ling, 67–80. Andover: Routledge.

Link, Perry. 2018. He took the road less travelled by historians: The story of T.T. Ch'en. In *The field of Chinese language education in the U.S.: A retrospective of the 20th century*, ed. V. Ling, 424–430. Andover: Routledge.

Liu, Peter. 2007. Computer technology and Chinese language teaching: Looking into the past and the future. *Journal of Chinese Language Teachers Association* 42 (3): 81–100.

Liu, Shijuan. 2002. Modern technologies in Chinese teaching and learning. In *Proceedings of the 2nd annual IST conference*. Bloomington: Indiana University. Available online: http://www.indiana.edu/~istb/conferences/2002/IST_Conf_2002_liu.pdf.

———. (2004). Ten levels of integrating the internet in teaching Chinese as a foreign language. In P. Zhang, T., Xie & J. Xu (Eds.). The studies on the theory and methodology of the digitalized Chinese teaching to foreigners (pp. 81–86). Beijing: Tsinghua University Press.

Liu, Shijuan. 2008, Web 2.0 and 5Cs: Opportunities modern technologies bring to foreign language and learning in the 21st century. Presentation given at the Second International Symposium on Chinese Regional Culture and Language, Bloomington, IN. August 8, 2008

———. 2009, Enhancing national standards for foreign language learning with web 2.0 technologies. Presentation given at the Conference on Language Technology at Pomona College, CA. April 3–5, 2009

———. 2010. Second life and its application in Chinese teaching and learning. *Journal of Technology and Chinese Language Teaching* 1 (1): 71–93.

———. 2012, *Free online tools and resources for beginning level Chinese courses*. Invited workshop given at the 7th International Conference and Workshops on Technology and Chinese Language Teaching, Manoa, HI. May 25–27, 2012

———. 2013. Tablets and Chinese language teaching and learning. *Journal of Technology and Chinese Language Teaching* 4 (1): 64–75.

———. 2014. *Content-based instruction with Technology in a Mixed Level Chinese Language Class. Presentation given at the inaugural conference of Chinese language teachers Association of Western Pennsylvania*. Pittsburgh: Carnegie Mellon University. May 4, 2014

———. 2015. Evolution of technology in Chinese language teaching in the United States in the 21st century. Invited presentation given at the 2015 International Conference on Internet Chinese Education (ICICE), Massachusetts Institute Technology, Boston, MA. June 19–21.

———. 2017, Teaching and learning Chinese characters (Hanzi) in the era of smart technologies: Tools, resources, and Pedagogy. Keynote speech given at the 3rd Online Chinese Teaching Forum and Workshop (OCTFW). Organized by the Confucius Institute of Michigan State University, East Lansing, MI. October 27–28, 2017

———. 2018a, Is virtual reality technology useful for language education? Presentation given at the 2018 Autumn Symposium of Chinese Language Teachers Association of Western Pennsylvania (CLTA-WPA) held at Carnegie Mellon University, October 14, 2018.

———. 2018b. Teaching and learning Chinese language online: What and why? *International Chinese Language Education* 3 (2): 11–26.

Liu, Shijuan. 2020. Similarities and differences between teaching and learning of the Chinese language online and onsite: Opportunities and challenges. Invited presentation given at the "A Series of Zoom Presentations on Remote Chinese Teaching", organized by DoIE Chinese Language & Exchange Programs, San Francisco State University, May 29, 2020.

Liu, Shijuan, and Tianwei Xie. 2010, Teaching and learning Chinese in the virtual world: Free asynchronous and synchronous tools. Invited workshop given at the 6th International Conference and Workshops on Technology and Chinese Language Teaching, Columbus, OH. June 12–14, 2010

Liu, Shijuan, Yanlin Wang, and Hong Zhan. 2020a. A survey of student perspectives on learning Chinese online: Preliminary results. Presentation given at The International Forum of Textbook Development and Virtual Conference on Teaching Chinese as an International Language, Organized by The Education University of Hong Kong, via Zoom, July 11, 2020.

———. 2020b. A survey of instructors on teaching Chinese online in 2020: Preliminary results. Presentation given at The International Forum of Textbook Development and Virtual Conference on Teaching Chinese as an International Language, Organized by The Education University of Hong Kong, via Zoom, July 11, 2020.

Lu, Jie, Sue Meng, and Vincent Tam. 2014. Learning Chinese characters via mobile technology in a primary school classroom. *Educational Media International*. 51 (3): 166–184. https://doi.org/10.1080/09523987.2014.968448.

Luo, Han., and Chunsheng Yang. 2016. Using WeChat in teaching L2 Chinese: An exploratory study. *Journal of Technology and Chinese Language Teaching* 7 (2)., Integrating Mobile Technologies, Social Media and Learning Design: 82–96.

Luo, Fengzhu, and Ruying Zhang. 1999. The red mansion on the Internet: Discussions on building online virtual museums. [羅鳳珠, 張如瑩《紅樓夢》文化藝術網上流芳—網路虛擬博物館建構機制探討.] Presentation given at the First International Conference on Internet Chinese Education, Taipei, Taiwan, May 22–24, 1999.

Mackey, Alison, and Susan M. Gass. 2015. *Second language research: Methodology and design*. London: Routledge.

Mei, F.S.Y., S.B. Ramli, and N.A. Alhirtani. 2015. Application of Gagne's nine approaches to teach Arabic language for non-native speakers: Experimental study at Sultan Idris Education University Malaysia (UPSI). *European Journal of Language and Literature* 3 (1): 32–37.

Meiden, Walter E. 1937. A technique of radio French instruction. *The Modern Language Journal* 22 (2): 115–125. https://doi.org/10.1111/j.1540-4781.1937.tb00568.x.

Mullaney, Thomas. 2017. *The Chinese typewriter: A history*. Boston: MIT Press.

O'Driscoll, Tony. 2007. Welcome to the era of the free range learner. Keynote speech at the World Conference on E-learning in Corporate, Government, Healthcare, & Higher Education, Quebec City, Canada. October 15–19, 2007

O'Reilly, Tim. 2005. What is web 2.0 design patterns and business models for the next generation of software. http://www.oreilly.com/pub/a/web2/archive/what-is-web-20.html.

Otto, Sue E.K. 2017. From past to present: A hundred years of technology for L2 learning. In *The handbook of technology and second language teaching and learning*, ed. C. Chapelle and S. Sauro, 33–49. Hoboken: Wiley-Blackwell.

Palomo, Jose R. 1940. A desired technique for the use of sound films in the teaching of foreign languages. *Modern Language Journal* 24 (4): 282–288. https://doi.org/10.1111/j.1540-4781.1940.tb02915.x.

Pappano, Laura (2012). The Year of the MOOC. Retrieved from https://www.nytimes.com/2012/11/04/education/edlife/massive-open-online-courses-are-multiplying-at-a-rapid-pace.html.

Pew Research Center. 2019. Mobile Technology and Home Broadband. Retrieved from https://www.pewresearch.org/internet/2019/06/13/mobile-technology-and-home-broadband-2019/.

Pitler, Howard, Elizabeth Hubbell, and Matt Kuhn. 2012. *Using technology with classroom instruction that works*. Alexandria: ASCD Publisher.

Puentedura, Ruben R. 2013. SAMR: Moving from enhancement to transformation. Retrieved from http://www.hippasus.com/rrpweblog/archives/2013/05/29/SAMREnhancementToTransformation.pdf.

Reinders, Hayo, and Mark Pegrum. 2015. Supporting language learning on the move. An evaluative framework for mobile language learning resources. In *Second language acquisition research and materials development for language learning*, ed. B. Tomlinson, 116–141. London: Taylor & Francis.

Ross, Adam, Jiahang Li, and Ann Marie Gunter. 2018. Learning Chinese in the digital age. In *CELIN briefs series*, ed. S.C. Wang and J.K. Peyton. New York: Asia Society. Available at https://asiasociety.org/sites/default/files/inline-files/learning-chinese-in-the-digital-age-celin-brief-edu-20180530-en.pdf.

Schueler, Herbert. 1961. Audio-lingual aids to language training-uses and limitations. *Quarterly Journal of Speech* 47 (3): 288–292.

Shi, Lijing. 2018. Eye-tracking technology and its application in Chinese teaching and learning research. *Journal of Technology and Chinese Language Teaching* 9 (1): 96–107.

Solanki, M.R. 2014. Developing instructional multimedia module incorporating Gagne's nine events of instruction. *The Journal of Education* 2 (1): 1–16.

Stickler, U., and L. Shi. 2013. Supporting Chinese speaking skills online. *System* 41 (1): 50–69. http://eprints.lse.ac.uk/49513/.

Stockwell, Glenn, and Philip Hubbard. 2013. *Some emerging principles for mobile-assisted language learning*. Monterey: The International Research Foundation for English Language Education. Retrieved from http://www.tirfonline.org/english-in-the-workforce/mobile-assisted-language-learning.

Su, Yingxia. 2020. Teaching Chinese online to non-degree seeking students. Presentation given via Zoom on July 18, 2020, organized by Beijing Language and Culture University Publishing House.

Sun, M., Y. Chen, and A. Olson. 2013. Developing and implementing an online Chinese program: A case study. In *Computer-assisted foreign language teaching and learning: Technological advances*, ed. B. Zou, M. Xing, Y. Wang, M. Sun, and C. Xiang, 160–187. Hershey: IGI Global. https://doi.org/10.4018/978-1-4666-2821-2.ch010.

Sung, Margaret. 1986. Linguistic evaluation of Chinese computer input systems. *Journal of the Chinese Language Teachers Association* 21 (2): 43–58.

Sung, Yao-Ting. 2018. *Applying advanced technologies for learning Chinese as a foreign language*. Keynote speech given at the *10th* International Conference and Workshops on Technology and Chinese Language Teaching (TCLT10), National Normal University, Taipei. June 1–3, 2018.

Thanasoulas, Dimitrios. 2001. The importance of teaching culture in the foreign language classroom. http://radicalpedagogy.icaap.org/content/issue3_3/7-thanasoulas.html.

The Perk Performance Center. 2020. Gagné's Nine Events of Instruction https://thepeakperformancecenter.com/business/learning/business-training/gagnes-nine-events-instruction/gagnes-nine-2/.

Tian, Ye. 2018. The challenge of machine translation to traditional translation homework in Chinese language learning. *Journal of Technology and Chinese Language Teaching* 9 (1): 78–95.

———. 2020. Error tolerance of machine translation: Findings from failed teaching design. *Journal of Technology and Chinese Language Teaching* 11 (1): 19–35.

Tomalin, Barry (2008). Culture the fifth language skill, British Council. https://www.teachingenglish.org.uk/article/culture-fifth-language-skill culture-fifth-language-skill.

Walker, Galal (1982). Videotext: A Course in intermediate to advanced Chinese. *Journal of the Chinese Language Teachers Association* 17 (2): 109–122.

Wang, Fred Fang Yu. 1966. Report on Chinese language concordances made by. *Computer* 1 (2): 73–76.

Wang, Jianqi. 2002. Using TV Talk-show Programs to Construct Pedagogical Materials for Advanced Chinese language Learners. Annual Meeting of Chinese Language Teachers Association. Salt Lake City, Utah. November 21–24.

Wang, Yuping. 2012. E-language learning and teaching in Australia: A case study. http://www98.griffith.edu.au/dspace/bitstream/handle/10072/48843/80825_1.pdf;jsessionid=3C9D4A47655AD2A416FBD375283C6ECB?sequence=1

Wang, Shenggao, and Camilla Vásquez. 2012. Web 2.0 and second language learning: What does the research tell us? *CALICO Journal* 29 (3): 412–430.

Wang-Szilas, J., and J. Bellassen. 2017. Dualism-based design of the Introductory Chinese MOOC 'Kit de contact en langue chinoise'. In *Beyond the language classroom: researching MOOCs and other innovations,* ed. Q. Kan and S. Bax, 43–57. Research-publishing.net. https://doi.org/10.14705/rpnet.2017.mooc2016.670.

Warschauer, M. 1996. Computer-assisted language learning: An introduction. In *Multimedia language teaching,* ed. S. Fotos, 3–20. Tokyo: Logos International.

Warschauer, M., and D. Healey. 1998. Computers and language learning: An overview. *Language Teaching* 31 (2): 57–71.

Wei, W., and J. Zhang. 2018. An intelligent Chinese pronunciation teaching app and the preliminary result of a teaching experiment. *Journal of Technology and Chinese Language Teaching* 9 (2): 83–97.

Welch, T. 2005. Uniting the world through technology innovations. Keynote presentation given at the 5th Conference of Foreign Language Education and Technology (FLEAT5), Provo, Utah.

Wong, L., Chee-Kuen Chin, Chee-Lay Tan, and May Liu. 2010. Students' personal and social meaning making in a Chinese idiom mobile learning environment. *Journal of Educational Technology & Society* 13: 15–26.

Wu, Yongan. 2014. Using sketch comedy skits to develop language in lower level Chinese classes. *Journal of Chinese Language Teachers Association* 49 (3): 65–85.

———. 2015. Technology in CFL education. In *Chinese language education in the United States*, ed. J. Ruan, J. Zhang, and C.B. Leung, 97–122. New York: Springer.

Wu, Ching Hsuan, Lily Childs, and I-Ling Hsu. 2018. Using educational technology to enhance the three modes of communication. *Journal of Technology and Chinese Language Teaching* 9 (1): 62–77. http://www.tclt.us/journal/2018v9n1/wuchildshsu.pdf.

Xie, Tianwei. 1999. Using computers in Chinese language teaching. In *Mapping the course of the Chinese language filed: Chinese teachers association monograph series volume III*, ed. M. Chun, 103–119. Kalamazoo: Chinese Language Teachers Association. Retrieved from https://web.csulb.edu/~txie/papers/Using_computers.htm.

———. 2000. The pros and cons of teaching Chinese with computers. In *Modern education technology and Chinese teaching to foreigners,* ed. Pu Zhang, 3–9. Guilin: Guang Xi Normal University.

Xie, Tianwei., and Tao-Chung Yao. 2009. Technology in Chinese language teaching and learning. In *Teaching Chinese as a foreign language: Theories and applications*, ed. M. Everson and Y. Xiao, 151–172. Boston: Cheng & Tsui Company.

Xie, Y., L. Ryder, and Y. Chen. 2019. Using interactive virtual reality tools in an advanced Chinese language class: A case study. *TechTrends* 63: 251–259. https://doi.org/10.1007/s11528-019-00389-z.

Xing, Bin. 2020. Post pandemic online teaching for K-12 learners. Annual National Chinese Language Conference, held online. June 24–26, 2020.

Xu, Debao. 2009. Incorporating DVD into powerpoint for language and film studies instruction. *Journal of the Chinese Language Teachers Association.* 44 (1): 103–122.

———. 2015. Issues in CALL studies. *Journal of Technology and Chinese Language Teaching* 6 (2): 1–16.

Xu, Jun. 2020. Machine translation for editing compositions in a Chinese language class: Task design and student beliefs. *Journal of Technology and Chinese Language Teaching* 11 (1): 1–18.

Xu, Ping, and Theresa Jen. 2005. "Penless" Chinese language learning: A computer-assisted approach. *Journal of the Chinese Language Teachers Association* 40 (2): 25–42.

Yang, J., and C. Yin. 2018. Learning Chinese colloquialisms through mobile technology. *Journal of Technology and Chinese Language Teaching* 9 (1): 35–47.

Yao, Tao-chung. 1996. A review of some computer-assisted language learning (CALL) software for Chinese. In *Chinese pedagogy: An emerging field. Chinese language teachers association monograph II*, ed. S. McGinnis, 255–284. Columbus: Foreign Language Publications.

———. 1999. Integrated Chinese Website: An Introduction. Presentation given at the First International Conference on Internet Chinese Education, Taipei, Taiwan. May 22–24, 1999.

———. 2009. The current status of Chinese CALL in the United States. *Journal of Chinese Language Teachers Association.* 44 (1): 1–23.

Ye, Weibing, Shijuan Liu, and Fei Song. 2017. History and current state of virtual reality technology and its application in language education. *Journal of Technology and Chinese Language Teaching* 8 (2): 70–100.

Young, Grace P. 1932. Bibliography of modern language methodology in America for 1931. *The Modern Language Journal.* 16 (8): 667–677.

Yu, Li. 2012. Integrating film narration into the lower intermediate level curriculum. *Journal of Chinese Language Teachers Association.* 47 (2): 33–63.

Zeng, Zhini. 2018. Situated learning in a Chinese question-and-answer online community: The case of Zhihu. Chinese as a second language. *The Journal of the Chinese Language Teachers Association* 53 (3): 222–256. https://doi.org/10.1075/csl.18002.zen.

Zhang, Zhengsheng. 1998. CALL for Chinese—Issues and practice. *Journal of Chinese Language Teachers Association* 33 (1): 51–82.

Zhang, Phyllis. 2009. Video podcasting: Perspectives and prospects for mobile Chinese learning. *Journal of the Chinese Language Teachers Association* 44 (1): 51–67.

———. 2011. Using voice thread to boost proficiency development: Performance-based activity design. *Journal of Technology and Chinese Language Teaching* 2 (1): 63–80.

———. 2013. Using video to promote the acquisition of advanced proficiency. *Journal of Technology and Chinese Language Teaching (JTCLT)* 4 (2): 69–85.

Zhang, Shenglan. 2018. Augmented reality in foreign language education: A review of empirical studies. *Journal of Technology and Chinese Language Teaching* 9 (2): 116–133.

———. 2019. "The effectiveness of a wiki-enhanced TBLT approach implemented at the syllabus level in the teaching of Chinese as a foreign language". *World Languages and Cultures Publications.* 214. https://lib.dr.iastate.edu/language_pubs/214

Zheng, Yanqun. 2001. Network in the classroom and classroom on the network. [郑艳群. 课堂上的网络和网络上的课堂—从现代教育技术看对外汉语教学的发展. 世界汉语教学. Chinese Teaching in the World. 2001(4):98–104.]

———. 2012. *Introduction to educational technologies for teaching Chinese as a foreign language.* Beijing: The Commercial Press. [郑艳群.《对外汉语教育技术概论》(2012)商务印书馆.]

———. 2019. 70 years of Chinese language teaching-the influence and role of educational technology. *Journal of International Chinese Teaching* 4: 69–76. [郑艳群 (2019) 汉语教学70年–教育技术的影响及作用,《国际汉语教学研究》第4期, 69-76.].

Zhu, Ruiping. 2020. Chinese language teaching in Beijing Normal University under Pandemic. Presentation given via Zoom on July 18, 2020, organized by Beijing Language and Culture University Publishing House.

Innovating the Design of Lower-Level Reading and Writing Curriculum for the Digital Age

Junqing Jia and Zhini Zeng

Contents

Introduction	744
Background: Research on Reading and Writing Instruction in the CFL Curriculum	745
Research on Character Instruction in the CFL Curriculum	747
Research on Reading Instruction in the CFL Curriculum	748
Studies on Writing Instruction in the CFL Curriculum	750
Embracing New Technology to Improve Reading and Writing Instruction	752
Pedagogical Goals of Lower-Level CFL Reading and Writing Classes	755
Differentiate Curricular Objectives for Various Skills Such as Character Recognition and Production	755
Help Learners Establish Successful Stories of Accomplishing Authentic Text-Based Communication Events	756
Incorporate Digital Technology into the Classroom	757
A New Lower-Level Reading and Writing Curriculum Design to Integrate Contextualized Digital Performance	758
Features of the Proposed Curriculum	758
Instructional Cycle	759
Assessments and Learning Outcomes	764
Student Evaluation of the Proposed Curriculum	766
Students' Attitudes Regarding Handwriting and Typing Chinese Characters	766
Digital Contextualized Reading and Writing Activities	767
The Current Approach and Learner Motivation	768
Conclusion	769
Appendix: Interview Scripts	770
References	771

J. Jia (✉)
Hamilton College, Clinton, NY, USA
e-mail: jjia@hamilton.edu

Z. Zeng
University of Mississippi, Oxford, MS, USA
e-mail: zzeng@olemiss.edu

Abstract

This chapter proposes an innovative curricular design that centers around the concept of digital performance to address three major shortcomings of lower-level reading and writing tasks as implemented by existing Chinese curriculums in the United States and beyond. First, the core emphasis on literacy development focuses students' attention on rote memorization of characters rather than the development of text-based communicative skills. Second, decontextualizing reading and writing events from their social contexts leads to a deep misunderstanding where handwriting grammatical sentences is equated with meaningful writing. Third, there is an underutilization of technology-assisted reading and typing tasks as learning activities. To fill the gap between our vision of what a successful Chinese language learner needs to accomplish in the target culture and the existing instructional methodologies, we propose a new pedagogical concept of "digital performance" and design an innovative lower-level reading and writing curriculum around this concept. By "digital performance," we mean using text in combination with other digital semiotic resources for computer-mediated forms of communication. The proposed curriculum consists of three features. First, it encourages students to read extensively with a goal of developing reading skills at a faster pace than handwriting skills. Second, it devotes valuable class time to conduct reading-based speaking activities and reading-based writing activities instead of focusing on memorization of Chinese characters. Third, it actively adopts contextualized performances in a digital environment such as texting, responding to emails, and posting on social media as class activities. The proposed approach has been applied in the first-year curriculum at a private college in North America. The assessment results show that most students were able to read and comprehend Chinese characters that they were not required to handwrite. Students also demonstrated strong ability and willingness to participate in a series of authentic text-based reading and typing tasks at the end of the academic year.

Keywords

Chinese as a foreign language · Reading and writing · Lower level · Digital performance

Introduction

This chapter proposes an innovative lower-level reading and writing curriculum that encourages beginner Chinese learners to perform authentic text-based tasks utilizing digital communication technologies. A review of the history of reading and writing instruction and related scholarship in the context of Chinese as a foreign language (CFL) reveals that literacy development has topped the agenda of most CFL educators at the beginner level. Under this pedagogical paradigm, comprehension is always the ultimate goal in any reading class, and understanding the cognitive processes involved in decoding the text has become the paramount goal in reading research. The unique script-speech relationship in Chinese writing requires great emphasis on CFL literacy development. However, this exclusive concern with the semantics of a text has caused CFL educators to neglect reading as a social activity. Compared to studies on reading, writing research is even scarcer as it has been considered to be the least utilized skill (Kubler 1997; Liu 2014). Similarly, in lower-level writing instruction, teachers have encouraged students to generate output of target grammatical constructions and lexical items. Although some scholars have recognized that the exclusive focus on form has led to a lack of authentic purpose which discourages students from writing (Liu 2014; Zhang 2016), CFL educators still debate about what genre type should be practiced in lower-level writing classes. In addition, despite society's increasing dependence on computer-mediated communication (CMC), it seems Chinese instructors have not fully integrated digital technology into their curriculum to promote learning or to include the markedly changed methods of textualization to enrich communication.

The proposed reading/writing curriculum is designed to bridge this gap by addressing the challenges encountered in lower-level literacy development. By lower-level, we mean the first two or three semesters of a North American university Chinese language program where classes meet 1 h a day, 3 to 5 h a week, between 12 and 15 weeks per semester. This curriculum which includes around 28 reading and writing classes (43 percent of total instruction hours) per semester has been put into practice at a private liberal arts college in which one of the authors teaches. We will highlight the pedagogical features of this curriculum and provide concrete examples of its instructional cycle. Suggestions based on feedback collected from a brief survey given to students taking this class and the reflections of the researchers and teacher of this curriculum are included at the end of this chapter to guide its future implementation in lower-level CFL classrooms at other institutions.

Background: Research on Reading and Writing Instruction in the CFL Curriculum

Reading and writing are central to our personal experiences, professional careers, and social relationships. In recent years, considerable attention has been placed on teaching these essential skills due to the increasingly critical roles they play in the development of Chinese language proficiency. However, a close review of the previous literature shows that many researchers who claim to have dealt with these skills have equated learning to read or write with simply learning to process or produce written symbols. Such a straightforward process might reflect a type of "native assumption" (Noda 2003a) as mastering written symbols is a major focus throughout the educational path of a native speaker, especially for those whose native writing system involves a unique script-speech relationship, like Chinese and Japanese. To be sure, learning to read and write includes developing an awareness of the basic unit of spoken language that is represented in the writing system. However, reading and writing should not be restricted to recognizing and producing orthography, even for introductory level language learners for whom the inventory of orthography has not yet been fully established.

In the 1970s and 1980s, scholars of reading and writing exclusively focused on understanding the Chinese writing system (Barnes 1970; N. Ching and E. Ching 1975; Mickel 1980). According to Everson (1993), "The research has focused so finely on 'bits' of reading that an overall grasp of the more holistic process of reading remains elusive" (p. 210). It seems that as long as educators master the secret of teaching this "heavenly script" (Moser 1991), the challenge of teaching reading and writing will be solved. It was not until the last three decades that researchers' collaborative efforts to debunk the "ideographic myth" of Chinese orthography began to redirect attention to other facets, such as learners' cognitive processes and strategies (Everson and Ke 1997; Lee-Thompson 2008), semantic knowledge (Lin 2000; Shen 2005), curriculum design (Walker 1989; Mou 2003; Zhang 2013), and technology mediated reading and writing activities (Mou 2003; Zhang 2009a). Meanwhile, as language educators' views of the nature of language began to evolve, some scholars gradually abandoned their long-held perception of reading/writing activities as solitary acts of processing information embedded in a written word or transforming one's thoughts into a static written text. Rather, they began to see both reading and writing as interactive social behaviors in which meaning is negotiated between readers and writers.

This move toward seeing reading and writing activities has led to a distinction between character instruction and reading/writing instruction. We need to clearly define the pedagogical goals of these three types of instruction at each level, especially at the lower level because email and texting have blurred the lines between oral and the written communication in the electronic age. In this chapter each of these three types of instruction will be reviewed separately in the context of the history of CFL in the classroom as well as in relevant research. Prior to a detailed review of the relevant scholarship, it is important to discuss the definitions of these seemingly simple concepts.

Character instruction refers to the type of instruction that focuses on developing a solid inventory of orthography and honing students' recognition and production skills that allow them to skillfully map between spoken and written forms of Chinese. Common pedagogical practices in this area usually involve acquiring knowledge of radicals and recognizing and producing characters at the word, sentence, or, sometimes, paragraph level.

Reading is a socially motivated activity (Christensen and Warnick 2006; Noda 2003b; Zhang 2013) in which the reader uses his or her personal experience to relate to the text. In this sense, the reader does not just passively process information in an attempt to comprehend it. Rather, as Warnick (1996) points out, the reader does more than simply achieve comprehension; he or she is also "contributing something original" (p. 26). Reading instruction in the context of CFL, therefore, is to develop learners' ability to participate in the social functions associated with the texts of the target-language community and interact with its members using the texts.

Writing is also a social activity through which a writer negotiates meaning with a reader through the medium of text (Liu 2014, p. 5). The goal of CFL writing instruction is for the students to learn how to write appropriately in specific genres, ranging from interpersonal correspondence to professional or academic documents in various fields.

Research on Character Instruction in the CFL Curriculum

As scholars continue to debunk the "ideographic myth" of Chinese orthography, linguists and CFL educators have also reached a consensus that written Chinese is based on its spoken form as a secondary tool of communication. Learning Chinese orthography is a major challenge for students whose first language uses an alphabetic writing system (e.g., Everson 1998; Xu and Padilla 2013). Therefore, many scholars have attempted to apply theoretical frameworks and experimental methods in order to find the most effective method for helping them to develop orthographic knowledge for reading.

In the last two decades, researchers have come to recognize the contribution of radical knowledge in the acquisition of Chinese characters including retention, recognition, and production (e.g., Taft and Zhu 1995, 1997; Taft and Chung 1999; Shen 2000, 2004; Shen and Ke 2007; Xu and Padilla 2013). In particular, Taft and Chung (1999) found that beginner learners were better able to memorize characters if radical knowledge was included when they were first presented with them. According to Shen (2000), beginner and intermediate learners with more in-depth radical knowledge performed significantly better on both character recognition and production tasks. To tackle the challenge of effectively teaching non-transparent phono-semantic compound characters, Xu and Padilla (2013) further suggested dividing them into *bujian* (chunks) to provide students with more reliable contextual clues to help them learn and retain Chinese characters. The introduction of the Meaningful Interpretation and Chunking (MIC) method provided the students with an economic way of memory storage that allowed them to learn Chinese characters in a more systematic way.

While these studies focused on searching for effective teaching techniques, attention has also been paid to understanding the developmental trajectory of learners' orthographic awareness. Based on Ke (1998)'s orthographic awareness model, Shen and Ke (2007)'s study yields valuable findings on adult non-native learners' development of radical awareness. Their experiments identified that in their first year, students were able to visually deconstruct Chinese characters. However, they reached a plateau after the first year in their abilities to apply radical knowledge in the learning process. The accuracy rates of applying radical knowledge at various levels, as shown in their studies, indicated that it would take approximately 3 years of intensive CFL study for students to complete the accumulation stage (Ke 1998), in which they acquire a sufficient number of characters to allow them to abstractly recognize their recurring components. Scholars who have utilized Ke's orthographic awareness model urge CFL teachers to fully take advantage of adult learners' cognitive maturity by introducing the radical knowledge of compound characters at the very beginning. More importantly, educators need to respect students' individual cognitive development patterns to avoid suppressing their motivation and enthusiasm, particularly at the beginning stages of study. As Ke (1998) proposed, differentiating curricular objectives for various skills by limiting the number of characters that the students must produce at the initial stages might be a solution (p. 98).

Although the studies discussed above shed light on development of orthographic knowledge, the one question that remains is: "What should instructors do in class to support literacy development?" Unfortunately, none of the class activities that have been identified as effective for facilitating the development of orthographic knowledge, such as copying characters or watching animation of character deconstruction, are interactive. As Walker (2010) suggests, literacy instruction should "eliminate all learner activities that do not support textually motivated spoken performance" (p. 72). This could not be truer in the digital age with all the technological advances in word processing systems, machine translation, and the presence of online dictionaries. It is time to rethink the antiquated teaching methods of brush and pen that take up valuable class time in the digital era and instead determine how to most effectively utilize the new technology in a digital age.

Research on Reading Instruction in the CFL Curriculum

Despite efforts to integrate all aspects of foreign language instruction, reading remains on the periphery of lower-level classes because decoding the orthography still receives the most attention in comprehension-oriented reading. Scholars such as Pressley (2000, 2002) tend to associate decoding and vocabulary with lower-level learning while prior knowledge of text structure, cognitive strategies, metacognition strategies, and affective factors have been traditionally part of higher-level reading instruction. At the peak of the communicative approach's popularity, Phillips (1978) proposed a method to better integrate reading into the foreign language classroom. In his article entitled "Reading is Communication, Too!" the somewhat defensive tone of the word "too" obviously points to the long-standing tendency to emphasize oral over written modes of communication. However, in this field, scholars of reading have exclusively focused on word recognition and other strategies and failed to view reading and writing as social and communicative activities.

Walker's two articles (1984, 1989) can be seen as the earliest attempt to initiate a discussion of how to adopt a social and performative perspective to expand the reading curriculum for CFL learners. Concerned about the exclusive focus on teaching students how to recognize characters, Walker suggested that Chinese reading instruction should go beyond this "simplest, most rudimentary stage of training readers" by incorporating more difficult tasks and highlighting the social aspects of reading. He encouraged Chinese educators to closely observe and record the ways native speakers use Chinese texts and to integrate these methods into the classroom. Specifically, he proposed that reading curriculum at lower level should concentrate on developing literacy skills by using pedagogical materials to rapidly equip students with an inventory of graphs so that they can retrieve and encode *koutouyu* (spoken Chinese) words that they have already learned into orthography. Walker stressed that such training should always include textually motivated communicative performances that are realistic in the target culture community. When moving to the higher level, the reading instruction will gradually proceed from enhancing literacy skills to developing both knowledge and strategies in dealing with *shumianyu* (written Chinese). The portion of authentic texts written for native speakers can be gradually increased until they completely replace pedagogical materials at the advanced level. Similarly, Walker stressed that after students have accomplished the action of reading, instructors should always engage them with appropriate performances in spoken Chinese to demonstrate their command of information retrieved from the written Chinese.

However, it was not until the twenty-first century that Walker's proposal would garner any serious consideration. Inspired by Walker, Zhang (2013) proposed that reading should be treated as a social activity. In her study, she proposed "a spiral, performance-based pedagogical approach" for lower-level CFL reading instruction. Moving beyond simply encoding and reading comprehension, Zhang's approach focuses on how to socially contextualize reading tasks and incorporate authentic reading performances into classroom learning. Specifically, she proposed a "CIP spiral model" comprised of three phases: comprehension, interpretation, and performance. The first phase consists of warm-up exercises to refresh students' prior knowledge including reading aloud, asking content questions, and having students ask and answer questions as well. In the second phase of interpretation, teachers design authentic text-based activities for students to demonstrate their understanding of what they have read. The last phase focuses on performance, consisting of a more complete text-based social activity, discussions, and text-based writing activities. Throughout these three phases, students are able to interact with the text in a nonlinear way, which helps them lower linguistic barriers by analyzing the text from multiple perspectives and enhance their memory through repetition and prolonged exposure. Moreover, these authentic reading tasks also enable students to see how the texts are relevant to their own lives, which increases their motivation to learn more about them.

Aside from the studies listed above, very little attention has been given to examining the essence of CFL learners' reading performance, especially from an empirical standpoint. The only two topics that have received some attention are the selection of reading materials (Huang 2018a) and CFL learners' reading strategies (Chang 2010; Huang 2018b). Huang (2018a)'s case study has revealed learners' positive perception of using authentic texts as supplementary reading materials to the textbook in a third-year Chinese class. When provided with effective scaffoldings from their instructors, intermediate-level learners believe that authentic materials can boost their motivation to read outside class and build their confidence in reading. Huang's preliminary findings echoed Walker (1989)'s proposal that CFL reading instruction should proceed from using pedagogical materials exclusively to build an inventory of orthography at lower level to utilizing authentic materials from the target culture to develop reading strategies.

Walker (1984, 1989) stresses the importance of developing students' reading strategy from the beginner level. Two studies have shed light on CFL learners' reading process. Chang (2010) investigated reading strategies used respectively by proficient and less proficient CFL learners at three different proficiency levels (after 1, 2, and 3 years of Chinese study). Her study identified a trend of moving from local to global processing strategies as learners' general linguistic level increases. The active engagement with global processing activities is also what distinguish proficient CFL readers from the less proficient ones. Although proficient CFL readers at lower level also adopt some global strategies, they are limited to compensation strategies to cope with learners' deficiency in lexicon knowledge. However, with enhanced linguistic ability, proficient CFL readers at higher level demonstrated more variety in the global strategies they use, including anticipating more incoming information, better assimilating text information into their existing knowledge, and better recognizing the text organization.

Huang's (2018b) study examined a group of third-year CFL learners' reading performances and found that successful CFL readers always use two or more strategies in pairs or clusters to facilitate comprehension. These effective strategy combinations include inferring words or phrases together with using context and decoding characters, constantly monitoring one's interpretation through context, text structure, translation, and paraphrasing, segmenting words with the help of mental lexicon, dictionary, and grammar analysis. Huang's study further pointed out the importance of explicitly introducing a wide range of reading strategies to CFL leaners and encouraging them to exploit these strategies from the lower level.

In addition to these existing studies on CFL reading, Ke (2012) has identified a number of new areas in CFL reading instruction that deserve more empirical researches including gathering data on CFL learners' experiences of using modified texts, investigating training reading fluency through intensive, extensive and graded reading, and examining how reading skills interacts with other skills, such as one's speaking ability or prosodic knowledge. However, these suggested research directions were rarely taken up within the recent decade after Ke's proposal. In the meanwhile, neither those existing studies on CFL reading nor these suggested future research directions have gone beyond seeing the achievement of reading comprehension, rather than engaging social functions, as the ultimate goal or reading instruction.

Studies on Writing Instruction in the CFL Curriculum

As early as two decades ago, writing was still considered to be "the least used of the four skills" for CFL students (Kubler 1997, p. 108). However, with enrollment growing beyond the introductory level and the development of online communication accelerating at a faster pace than ever before, writing has become increasingly integral to our daily lives. This has undoubtedly led to an increased interest in writing instruction with regard to today's CFL curriculum. However, according to Liu's (2014) thorough review of the research published in the *Journal of Chinese*

Language Teachers Association (renamed as *Journal of Chinese as a Second Language* since 2016) from 1966 to 2011, of the small number of articles that focus on writing, even fewer of them touch upon designing an improved CFL curriculum. According to our search, only 20 articles were published on the subject of writing from 1960 to 2019 in the 3 highly recognized journals in our field: *Journal of Chinese Language Teachers Association, Foreign Language Annals*, and *The Modern Language Journal*. Within these 20 identified articles, the term "writing" does not even necessarily refer to composition. Instead, it is used in a broad sense to refer to either Chinese character instruction or composition. In fact, there were more articles (i.e., 11) on writing Chinese characters, including one on the topic of calligraphy instruction. Among the other nine articles that focus on composition, most of the instructional activities fall into the category of "writing to learn" (Hedgcock and Lefkowitz 2011). Thus, in these studies, writing is primarily seen as a vehicle for practicing grammar points and eliciting learned vocabulary. The writing-to-learn approach proposed by some researchers emphasizes a certain aspect of students' linguistic abilities, such as using cohesive devices (Yang 2013; Li 2014), syntactic maturity (Jin 2007), and pragmatic errors (Hong 1998). It must be noted that this exclusive focus on linguistic output is always at the expense of context and communicative intentions. Presently, typical writing activities in foreign language classrooms include reading-based summary writing (Shen 2000; Zhang 2009b), writing picture-book stories (Zhang 2016), creating dialogue for cartoon strips (Mou 2003), essay writing based on artificial topics that align with the content of the textbook (Zhang 2009b), and freestyle journaling to encourage self-expression (Mou 2003). For the foreign language educators who design and propose such activities, it seems that as long as the writing tasks can generate original output of grammatical constructions and lexical items, instructional goals are fulfilled.

Among the nine articles reviewed above, only two touch upon beginner-level writing performance (Mou 2003; Zhang 2016). Mou (2003) is among the very first CFL studies to propose that writing instruction can begin as early as the introductory level and to recognize the benefit of utilizing word processors to lessen the burden of students' handwriting Chinese characters. Her integration of technology (NJStar) into the beginner-level classroom can be seen as an innovative step at that time. However, she was not able to detail how the writing activities aligned with the word processor NJStar despite her brief comment on students' being "overjoyed with the convenience of typing" (p. 117). The only writing activity in which students utilized the word processor mentioned in her article was when they made vocabulary lists for their oral presentations. In addition, other routine in-class and out-of-class writing activities listed in the article were not effective or even authentic based on her own report. For example, one writing assignment for students to summarize classmates' presentations did not achieve its goal because, as Mou recalled, "some students actually take notes in pinyin, while others use a mixture of pinyin and English." In addition, she designed a take-home writing assignment in which students would create dialogue for cartoon strips. Although such activities may give students immediate opportunities to practice transcribing what they are able to say, such as classroom expressions and simple greetings, activities such as filling in speech balloons for Calvin and Hobbes, the protagonists of the popular American comic strips, will never provide reliable and usable memories for their future interactions with native Chinese.

Zhang (2016) also recognized the problems with writing instruction at the beginning stage as focusing exclusively on form while lacking authentic purpose to motivate students to write. To encourage introductory and lower-intermediate level learners (i.e., first-year and second-year CFL learners) to do more meaningful writing, Zhang proposed a picture book-writing project and performed an empirical study to evaluate its feasibility. In addition to the writing project itself, Zhang also shared students' compositions with a group of beginner and lower-intermediate level learners. Based on the survey results and interviews with the student participants, Zhang concluded that not only were they able to successfully complete the story-writing assignment if given appropriate guidance and assistance, they also benefited from the project linguistically, culturally, and intellectually. Furthermore, both the student writers and readers perceived the assignment very positively and were motivated to continue their Chinese study. These results are encouraging to CFL educators who wish to adopt innovative approaches for enhancing reading/writing experiences. However, these results lack both the academic underpinnings and empirical data support other than participants' opinions. For example, scholars have pointed out that narrative writing requires more demanding linguistic processing at varying levels than descriptive writing tasks (Koda 1993). According to the ACTFL writing proficiency guidelines, narrative writing is an advanced level skill. In order to learn more about what students have truly gained from this project, both a quantitative and a qualitative analysis of their writing assignments are necessary. Moreover, many Chinese instructors are hesitant to have lower-level language learners read documents written by fellow students due to the potential negative influences from the unavoidable errors, which therefore requires more empirical evidence from the study.

Collectively, the limited number of studies on writing has highly aligned with Liu's (2014) findings: None of these published articles include a detailed description of an innovative writing curriculum. Unlike the other three skills which have relatively clear instructional objectives, the goal of a CFL writing curriculum seems vague and lacks consensus among CFL educators. Particularly, the two studies conducted on beginner-level writing make us wonder which genres would be both manageable and beneficial for students to focus on in class. Staying within the rudimentary stage of the development of a foreign language does not imply any lack of intellect on the part of the students. Therefore, Chinese educators should think beyond childish assignments such as writing dialogue for cartoon strips and creating picture books.

Embracing New Technology to Improve Reading and Writing Instruction

Thanks to the rapid development of digitalization, foreign language educators today have unprecedented opportunities to engage learners with authentic and enriched meaning making through new types of social encounters and connections with communities all over the world. This creates both exciting possibilities and unexpected challenges. Scholars and foreign language educators in recent decades have developed robust research agendas around computer-aided language learning (CALL) while calling for innovative approaches that embrace new technology to improve the skills of struggling readers and writers. Reflecting foreign language educators' attempts to incorporate students' digital competences into the classroom, new terms like "multimodal literacy" and "multimodal composition" have emerged to better describe how students use technology to expand their literacy skills in their daily life beyond the school curriculum. Smith (2014) discovered 6 recurring features of multimodal composition after reviewing 76 studies on this pioneering topic both in the classroom and in out-of-class settings, including (1) enhanced motivation, (2) high-quality collaboration, (3) specific benefits for struggling writers, (4) opportunities for explicit instruction, (5) scaffolding, and (6) similarities between the multimodal composition process and traditional forms of writing. Among these features, enhanced motivation can be attributed to the authenticity of the writing product and the personal significance of the text (Jiang and Luk 2016). Several studies on ESL writing indicated that the advantage of providing learners with an authentic communicative intention and connecting them with a real-world audience other than their teacher can go beyond motivation by encouraging the writing process and producing superior writing products (Lammers et al. 2014; Magnifico 2010).

In the field of CFL, the embrace of electronic writing mainly emerged from a pragmatic consideration to set realistic goals to ensure enrollment (Zheng 2010). Given the severely limited time and resources available, some CFL scholars came to recognize the needs to increase the role of typing in CFL curriculum and to downplay the importance of handing writing character (Wang 2010; Zheng 2010; Xie 2011; Zhang 2009b, 2018). However, despite their obvious values, all of these initial proposals are purely based on those scholars' intuitions, introspection of previous teaching experiences, and collecting learners' opinions. For example, based on an anecdote of his student's being scared away from his class by the requirement of handwriting Chinese characters, Zheng (2010) raised his hypothesis of "giving up handing writing" (*qixie* 弃写)to enhance the efficiency of teaching Chinese and to boost Chinese learners' confidence. Similarly, Wang (2010) conducted a survey among international students studying in Beijing Foreign Studies University, including both beginner and advanced Chinese learners, to verify learners' desire for incorporating typing into Chinese classes as an addition to the default four skills. Some scholars even took a step further to summarize a list of operational "tips" on how to manage the proportions of handwriting and typing in CFL curriculum. For example, Zheng (2010) proposed "recognition only and no handwriting; facilitate recognition through reading; replace handwriting with typing" (只认不写, 以读促写, 以打代写). Disagreeing with completely abandoning handwriting, Xie (2011) further revised Zheng's tips by proposing a new version "facilitate recognition through typing; facilitate handwriting through recognition" (以打促认, 以认促写). Xie (2011) also outlined a 4-year roadmap for implementing this plan – "first semester, handwriting only and no typing; second semester, handwriting and typing simultaneously; second year, typing only and no handwriting; third and fourth year, typing or handwriting as desired" (一上只写不打, 一下又写又打; 二上二下不写只打; 到了三四要写就写, 要打就打). Although these "tips" seem intuitively right, the lack of research underpinnings has prevented the field from reaching a consensus on what to do.

Studies on computer-mediated reading and writing performances in the field of CFL are extremely rare probably due to a general lack of interest among instructors in utilizing technology to learn and to teach Chinese (Ihde and Jian 2003). Ke (2012) also suspected that such technology might not be widely accessible to Chinese language learners and the software might not be as effective as it is in other languages. Lee's (2011) discussion of the design and implementation of a web-based reading class for advanced learners is the only study to be found on incorporating technology into CFL reading instruction. This examination of the feasibility of incorporating a task-based online reading activity has both quantitative and qualitative data that show the mixed approach's effectiveness; however, it is difficult to determine the extent of the hypertext's role. It seems that most of the positive comments from the students focused on the task-based approach rather than the utilization of technology.

Another shortcoming arising from these current studies is that all have narrowly viewed technology as a computer or certain software and hardware products, which change very rapidly. As Jenkins (2009) warned, foreign language educators cannot simply treat emerging technologies as an addition to our instruction. Instead, they require a whole paradigm shift in which the curriculum, teaching methods, and pedagogical materials may all need to be reshaped so as to help students become competent learners in the digital age. Since our job is to help students understand how linguistic and cultural conventions operate in the target society so they will be able to interact successfully, it is essential for FL educators to incorporate new ways in which language is used across various media and technology. That being said, in addition to discussions of specific tools, we must take a broad view of technology-mediated communication that has become an integral part of our daily lives. In sum, we must ask ourselves how technology-mediated text may change how we prepare students to become capable target-language readers and writers in the twenty-first century.

In the age of digital media, it is important to remember that our daily electronic communication is mainly based on reading and writing. Each tool we incorporate into our lessons "brings its own material properties, feel and techniques of use, affordances and limitations," defining "a particular relationship between writers (or readers) and texts" (Chun et al. 2016, p. 65). We must also realize that the connections among forms, contexts, meanings, and literacy practices in a variety of media is culturally encoded which requires critical attention and deliberate analysis as well.

As stated above, studies on reading instruction prior to this century most often dealt with introductory level instruction, and orthographic knowledge development became the only goal for beginners. This is partially due to the fact that authentic pedagogical materials available at that time were still mainly *shumianyu* or written Chinese. For example, Walker (1989) suggests that in the first year, instructors should focus on literacy and that textual materials should be pedagogical in order to purposefully "create representations of *koutouyu*" (p. 75) for their students. In other words, texts consisting of *koutouyu* are never read by native speakers, while authentic materials are written in *shumianyu*, which were deemed as inappropriate for introductory level students. This was indeed the case 30 years ago; however, things have greatly changed today. In fact, traditional distinctions between the forms and functions of *koutouyu* and *shumianyu* can be rendered problematic by an unprecedented degree of overlap in many online environments, making us rethink the dichotomy of spoken and written language. At the same time, while multimodal texts that incorporate sound, graphics, animation, or video clips create an intriguing reading experience, they might also require more symbolic sophistication and critical thinking skills than ever.

Pedagogical Goals of Lower-Level CFL Reading and Writing Classes

The literature review in the previous section has identified three major shortcomings in implementing lower-level Chinese reading and writing assignments into the curriculum. First, the exclusive focus on orthographic knowledge development and character production instruction at this stage has left no time for meaningful communicative activities to support text-based spoken performances. Second, the serious lack of scholarship on the overall design of a reading/writing curriculum has resulted in unclear pedagogical goals and a lack of consensus among CFL educators. Third, current pedagogical approaches and current studies have overlooked how technology-mediated communication has changed the way we interact and CFL instructors have failed to take advantage of the fruitful studies on multimodal literacy.

To bridge the gap in the literature as described above and to overcome these challenges, the following three pedagogical goals are proposed:

Differentiate Curricular Objectives for Various Skills Such as Character Recognition and Production

As discussed above, Ke's (1998, 2012) orthographic awareness model has made it clear that successful character recognition does not necessarily require having knowledge of all the character's details. Instead, the graphic provides a visual context in which students will identify and remember the character. This model also explains Ke's finding in 1998 that heritage learners during the first year of studying Chinese had no more significant advantage regarding character recognition and production performance than did non-heritage students. These findings will hopefully motivate CFL practitioners to develop a more realistic understanding of the timeframe required to reach the stage when students are able to simultaneously produce characters and perform other skills such as speaking and reading. It is important to keep in mind that realistically our students are only able to commit a certain amount of time to studying. Therefore, particularly in lower-level classes, spending a disproportionately large amount of time on learning to write characters is neither productive nor motivational.

In this chapter, we propose that CFL educators should make a clear distinction between the target characters students will produce and those they will be able to merely recognize, while the number of handwritten characters they are required to produce at the lower level is carefully limited. In fact, although the objectives of recognition and production are quite different, students will be able to produce all the characters that they can recognize in typing as long as they can pronounce them correctly. That being said, students' abilities to encode and decode are not much different when they are engaged in digital communication. Meanwhile, the time saved from reducing the workload on the production side should be invested in two interrelated activities: increasing the number of extensive reading assignments during which the students will become more familiar with the target characters and engaging in a variety of text-based speaking activities in which students will verbalize their comprehension through conversations. It must be noted that extended speaking assignments in reading classes will not only improve students' pronunciation which will further enhance their typing skills by increasing typing accuracy; they will also enhance the authenticity of the reading tasks.

Help Learners Establish Successful Stories of Accomplishing Authentic Text-Based Communication Events

Developing reading skills involves using the texts of a language community to engage in communication with its members. In this chapter, we advocate that both reading and writing should be seen as social activities that need to be properly placed within the context of the target culture so that the intentions of the readers and the writers can be easily inferred. Comprehension should not be the final stage of reading. Instead, it is critical to teach students how to demonstrate their comprehension through action, whether in spoken or written form. It is imperative for CFL educators to create authentic text-based communication events that will engage their students. Just as Christensen and Warnick (2006) argue, authenticity is essential "in the use of orthography, in the texts themselves, and in the tasks based on those texts" (p. 110). Instructors should make sure that post-reading activities that students would participate in are widely shared by native speakers of the target culture instead of being designed for language learners only. When students are fully absorbed in these activities, as Christensen and Noda (2002) propose, they will be able to compile thought-provoking stories that they can easily remember and retrieve in the future "at the right moment while participating in the target culture" (p. 13).

Incorporate Digital Technology into the Classroom

Technology has already changed and will continue to change the semiotic spaces in which learning takes place. Therefore, CFL educators must embrace the development of new ways of textualizing information. For example, Chinese pinyin keyboards and other input methods should be introduced during the first week of classes, and numerous exercises should be designed for students so that they may continue to hone their typing skills. Based on our observations and empirical studies of extensive surveys (Allen 2008; Liu 2014), we have found that "Skills of accurate pronunciation/spelling and character recognition are of the highest value" (Allen 2008, p. 239). Because students are already familiar with technology, they will be able to quickly master Chinese word-processing skills without much effort. However, introducing these new skills does not mean we are suggesting eliminating handwriting from the CFL classroom. Despite the diminishing practical need for it, we do recognize other benefits of practicing handwriting characters as many psychologists have verified that the writing of characters develops a motor memory which plays a significant role in graphic recall (e.g., Tseng 1989). Moreover, due to the long history of Chinese calligraphy, a high cultural value has been placed on the ability to produce a large number of characters from memory in beautiful handwriting which continues to serve as a symbol of social status and sophistication. Therefore, we not only encourage students to practice handwriting outside of class but also urge teachers to design authentic communication contexts, such as writing a thank-you note or filling out a registration form, in which handwriting is an optimal medium to use.

In the meantime, students should also be introduced to digital communication tools that are widely used among native Chinese people such as WeChat (the top workplace communication apps for the majority of Chinese professionals) and Line (widely used in Taiwan) and be encouraged to practice communication on these platforms either in-class or as an extracurricular activity. These digital platforms along with various types of multimodal discourse can motivate students to draw upon their rich prior experiences while interacting with people in this digital environment and will reduce their anxiety about composing in a foreign language. Of course, we must remember that while the multimodality will enrich methods of representation, the priority is still to focus on the linguistic mode as it continues to play a pivotal role, even in cyberspace.

Finally, new technology also requires students to negotiate various unfamiliar cultural conventions. For example, while having students compose emails, instructors should not only attend to the quality of text students produce but also make them aware of culture-dependent discourse as well as community-based norms and conventions, which might have to be renegotiated in the presence of the new media. For example, what is the appropriate amount of time to wait to answer an email from his/her Chinese professor or internship supervisor in China? Does the email require a reply if it does not contain a request or an inquiry? What constitutes an excessive delay? Our cross-cultural communication experiences have taught us that answers to these questions may vary significantly from one cultural context to another and sometimes from one work environment to another, which therefore requires clear explanations from the instructor and participatory observations from the students.

A New Lower-Level Reading and Writing Curriculum Design to Integrate Contextualized Digital Performance

Based on the discussion above, we have developed a lower-level reading and writing curriculum that encourages beginner learners of Chinese to perform authentic digital learning tasks, what we refer to as "digital performances." More specifically, students are taught to incorporate their knowledge of language with their personal experiences to accomplish contextualized social activities during their first year of study. The lower-level instruction referred to in this chapter corresponds to the Novice to Intermediate-Low on the ACTFL proficiency scale.

Features of the Proposed Curriculum

Unlike most prevailing beginner-level reading and writing classes in which character instruction is the major or even the only focus, in the proposed approach, learning Chinese characters will be considered a critical but initial step to achieving reading and writing proficiency. It is worth mentioning that the adapted curriculum, in the long run, will allow students to achieve a high level of literacy, while they develop related learning mechanisms such as self-regulating behaviors and knowledge of cultural appropriateness. Some pedagogical features that differentiate this curriculum from its predecessors are as follows: First, students will be taught to read "extensively" even though the number of Chinese characters they are required to write will be highly controlled (12 characters per week). In other words, students develop their character recognition skills more quickly than their character production skills. In addition, they will only read and write the Chinese characters that they have previously learned in speaking and listening class. Although the extensive reading method has always been embraced in advanced-level language instruction, when it is adapted for a lower-level CFL curriculum, students should be provided with materials that include characters that they have recently learned through speaking activities. The basic assumption of this approach is that students do not need to know how to write a character to be able to read it in a highly familiar context. Although recognizing a large number of characters should not be the only goal at this stage, the ability to recognize more characters at an early stage helps students to accomplish complex digital tasks and thus to achieve a strong sense of achievement.

Second, character recognition will take place outside of the classroom so that valuable class time can be devoted to oral and written quizzes, extensive reading activities, and contextualized performances. Unlike a traditional reading and writing class where students spend most of their time reading or producing texts silently and individually, they will be allotted the same amount of time that they receive in a speaking and listening class to devote to oral communication in the target language. In other words, reading and writing tasks will be presented as integrated social events that naturally include oral communication. To achieve this objective, useful genres and contexts will be designed using a set number of vocabulary words to elicit successful performances. Furthermore, students will be asked to direct their performances to a specific audience and have a clear purpose. For example, the task of practicing using recently learned vocabulary and characters will be contextualized by asking them to handwrite a postcard to a friend who has studied Chinese for 6 months in Beijing. The more specific and authentic the context, the easier it will be for them to integrate their personal experiences and produce meaningful texts. They will then exchange their postcards and use them as reading materials for extended oral discussion. Ideally, the process of producing texts should be a homework assignment.

Lastly, some of the contextualized tasks will be purposefully designed to incorporate digital performance such as texting, typing emails, posting on social media, and conducting text-based conversations on a Chinese online Q&A platform. Regarding task authenticity as one of the most fundamental rules, students will not be required to participate in these digital tasks while handwriting characters. Instead, they will use the Chinese typing system that they have been learning since the second week of class. It has been proposed that knowing how to type Chinese and using these skills to perform authentic tasks will motivate beginner language learners. It is also worth mentioning that incorporating typing activities into the curriculum will allow students to envision themselves engaging in practical daily tasks in the target culture. Taking the highly digital lifestyle in modern Asia into consideration, an added bonus to this approach is the sense of familiarity learners will gradually develop with the target community, rather than feeling disoriented when being in the target culture for the first time. Due to the fact that performance-based reading, writing, and typing tasks are the main objectives of the curriculum, students will be allowed to use their personal digital devices in class when needed. These activities will be done at the end of each class to give students the best opportunity to integrate their recently acquired linguistic and cultural knowledge.

Instructional Cycle

At the private institution where one of the authors teaches, first-year Chinese language students attend five classes every week. In the three spoken sessions, they converse and participate in drills and various types of communicative exercises. Two weekly sessions emphasize reading and writing. As the adopted learning materials suggested, the speaking and writing tracks are separate but integrated: "A day's spoken lesson is based on a conversation typically introducing one to three new grammar patterns and 15 to 20 new spoken words, while the corresponding written lesson introduces six new high-frequency characters used to write the basic conversation of the corresponding lesson" (Kubler 2017, p. 11). In other words, the learning materials used in the reading and writing classes are familiar to students because they have learned to speak and comprehend the target conversation in a previous listening and speaking class. Before each class, students are required to memorize six characters by utilizing a character practice sheet, reading the corresponding exercises in the textbook out loud, and reading the Chinese character version of the conversation. Students will not be required to memorize all the individual characters that appear in the conversation. Instead, they will be expected to recognize them in context.

A typical reading and writing class will consist of four segments inspired by Zhang's (2013) "CIP Spiral Model": quizzes (both written and oral), reading and discussing a conversation as scaffolding activity, reading performance, and contextualized digital performance. Each segment will be briefly introduced. Students will take quizzes on characters at the beginning of each class to assess their preparation for that day's session and remind them that they should memorize characters outside of class. The *written quizzes*, most often dictated, will focus on handwriting skills with the new and previously learned characters. Students will be asked to dictate two short sentences that include at least six characters that they have been required to handwrite. They will also be required to include the Roman spelling (*pinyin*) with tones. The purpose of the oral quiz, on the other hand, will be to check if students completed their oral reading assignments. It must be noted that it is important to correct each individual's pronunciation errors even in a reading and writing class because their ability to pronounce a character correctly directly contributes to their typing skills, as *pinyin* is their preferred method for inputting characters. In fact, for the purpose of assessment, each individual's performance should be clearly demonstrated and discussed after the oral quizzes so that instructors can gain a sense of each student's preparation and assign them a daily grade.

To expand the students' reading comprehension, they will be asked to participate in *scaffolding activities* which aim at building the knowledge and skills students need for later complicated learning tasks. Some examples of scaffolding activities in a reading and writing class include character recognition, vocabulary-based drills, sentence comprehension, and introduction of a certain genre. A typical scaffolding activity can be discussing a conversation that students are required to read in a Chinese character version before class. This version will also be shown on PowerPoint during class. We have observed that it is easier for students to read the text as a chunk since they have memorized and performed it recently. However, when attempting to answer detailed content questions and trying to locate specific aspects within the texts, they are usually not able to recognize some of the characters. Students should be reminded that this type of scaffolding activity is designed to help them review the previously learned dialogue and become more familiar with the characters in the context of a word or phrase, rather than to assess their ability to recognize each individual character. This session of the class may appear similar to an integrated lower-level language class in which students read the texts in Chinese character form and participate in a drilling exercise or discussion. However, the fundamental difference between the two is that the purpose of an integrated class is to help students develop speaking skills based on their prior knowledge of the Chinese characters. Instead, the approach proposed in this chapter will teach students to use their prior knowledge of spoken Chinese to gain a deeper understanding of the written texts.

To further the goal of helping students establish a relatively high level of reading proficiency, *reading performance* is an integral classroom activity. The materials used in this session will be very similar to those found in the textbooks students read prior to class. However, instructors sometimes edit these texts so that students can review the characters they have already learned and recognize them in various contexts. Characters in dialogues, which students are only required to read (but not to write), are also included in the reading performance. Both conversation-based readings and narratives will be introduced. When reading scripts, students are assigned to play various parts by reading the lines out loud. The instructor asks or answers questions to help students comprehend the text. When reading a paragraph of narrative, students are encouraged to focus on their reading speed and skimming skills. This exercise also helps instructors gain better control of how much time students spend on silent reading in class which is not a priority. Although it is inevitable that some authenticity of reading materials will be lost due to the fact that they have been adapted for lower-level learners, the reading performances can still be designed within an authentic context. For instance, some possible contexts for time-limited tasks include reading street signs while driving down the road, reading subtitles of a movie, and skimming a text message on a phone before class. Such a top-down reading model where a general hypothesis of the text meaning is first generated exposes lower-level learners to various genres and texts with various communication intentions (Goodman 1973). Some typical reading materials used during the first semester could include business cards, a work schedule, hand-written memos, train tickets, and event posters. Once students' grammar knowledge and vocabulary expand, they will be provided with more complex reading materials such as email exchanges, restaurant reviews, travel tips, and postcards. In the middle of the second semester of this curriculum, the instructor will use the email below for a reading performance in class (see Table 1) to encourage students to spend more time reading and writing and remind them how much they have already accomplished.

In this reading piece, there are altogether 75 different characters, 49 are those that students have learned to read and write, and 22 are additional ones that they have only learned to read in a familiar context but not required to write. Yet four others are characters that they haven't learned. Prior to the reading and writing class where this email was introduced, students recently learned how to introduce their experiences of learning Chinese in a speaking and listening class. During the previous scaffolding activities, students had also practiced most of the vocabulary items used in this email including those consisting of new characters. The email as a reading material was designed with three major purposes: first, improving students' reading ability by using characters they have learned but with different lexical meaning. For instance, students had learned the character *yao* (要) as "going to" to indicate that something is going to happen. However, in the email, *yao* was used in sentences such as "*ni yao zhidao* (你要知道)" and "*Qing nimen yao duo xie zhongguo zi* (请你们要多写中国字)" where *yao* means "need to." Based on their performance of reading the text and answering the text-based questions, it was also clear that learners do not need to know how to handwrite a character to be able to read it in context. Second, the email provides an opportunity for students to review some of the vocabulary they have learned in a recent speaking and listening class. After reading the email as a group, the instructor asks content questions, conducts vocabulary drills, and corrects students' pronunciation errors. If time allows, students are asked to reply the email by asking the instructor one or two questions that they have about learning Chinese. Last but not least, this email as a reading text in class is drafted to motivate students through clearly demonstrating which authentic tasks they have already learned to accomplish in the past several months. Instead of focusing on how many characters they are able to read and write, we suggest they use their language skills to accomplish authentic tasks with meaningful social purpose, such as reading an authentic email. Students have demonstrated a particularly high interest in reading emails and participating in extended text-based discussions.

In everyday life, reading and writing as social behaviors are often accompanied by oral communication. We propose that a complete instructional cycle of a reading, writing, and typing assignment should always include speaking activities. Whether handwritten or typed, the contents prepared by the instructors or produced by students should also be adopted as reference materials for speaking activities. With this understanding, the last segment of a class is designed to generate *contextualized digital reading and writing performance* in which students are expected to integrate what they have learned in order to successfully interact in a particular social context. Whether the task is completed by handwriting or typing, students should not be asked to do things that are unrealistic or inappropriate in the target culture. It is also critical for students to understand their social roles when completing such exercises. For example, when typing an email to their Chinese instructors, they will be expected to learn how to appropriately begin and end the email. There will be typically two types of activities in this segment: text-based speaking performance and text-based writing performance. In other words, both speaking and writing activities will be elicited by reading activities. The former is usually an extension of previously studied readings. For instance, after students read the email in Table 1, the instructor will facilitate discussion about how they should reply to it. A text-based writing performance, on the other hand, is structured around an essay on that day's reading or typing activity, such as commenting on a friend's social media post. For example, after searching for local restaurants and reading their ratings and menus, students will be asked to text one of their classmates to discuss which restaurant to go to and what to order. The instructor will walk around the classroom giving each group feedback on their writing or typing performances. Then, the instructor will ask each group to give oral reports about their texts.

Whether the activity is done in handwriting or by texting depends on the nature of the task. For example, it would not make sense for students to be asked to type a postcard or to handwrite an email. Since these exercises will take place near the end of class, instructors should make a special effort to coordinate the entire lesson plan so that the previous activities can add to the language and cultural knowledge needed for the last performance. One of the major reasons for emphasizing students' reading

Table 1 Example of *reading performance* in class

Email screenshot	Details
请多读、多写中国字 (Inbox) 同学们好， 你们学中文已经有一段时间了，中文难学吗？ 你要知道，这几个月里你已经学会了自我介绍、找人、坐火车、买东西、介绍自己的学校、介绍学过的语言、跟中国人聊天……你也已经会写几百个中国字了，认识的字更(geng4, even) 多。你们的中文更好了！ 可是，请你们要多写中国字，你可以在电脑上写，也可以在书上写。每天要读要写二十分钟的中国字，那你的中文会更好！ 加油 (jia1 you2) ！ 老师	Characters that students have learned how to read and write: 49 characters Characters that students have learned to read in a familiar context: 22 characters Characters that students haven't learned how to read or write: 4 characters (读, 己, 更, 油) Genre: Email Roles: Teacher-students Communication purpose: Instructor encouraging students to read and write more Chinese and reminding them of what they have achieved Possible post-reading activity: Students reply the email and ask curriculum-related questions

capacity is so that they can successfully engage in authentic typing tasks. In fact, typing emails and using smartphones to text are two of the most practiced skills taught in this last segment of class. There are both pedagogical and psychological rationales for including these types of exercises. Pedagogically, digital reading and typing activities expose lower-level students to practical genres and some formal written languages that they would not have the chance to learn in spoken Chinese. For example, when introducing students how to appropriately end an email in Chinese, they will learn to use the terms "祝好 (*zhùhǎo*)" or "秋安 (*qiūān*)" which are not covered in typical lower-level spoken Chinese classes. Without the introduction of typing into the lower-level curriculum, students would be able to perform a very limited number of authentic tasks in social contexts, such as handwriting a card. Instead, reading and typing activities in digital contexts will give them practice for using Chinese to successfully function in the modern world. Psychologically, being able to complete authentic tasks that incorporate one's own experiences is a motivational boost for beginner Chinese learners. Unlike most current lower-level CFL curricula which provide few opportunities for learners to gain a sense of accomplishment through reading and writing, the proposed approach will integrate rewarding experiences into each class. For the same reason, students will not be restricted from using auto-correct which could potentially increase their typing speed and accuracy. As discussed above, recognizing and memorizing characters is a procedural learning activity which should be done outside of class.

After each reading and writing class, students will be assigned homework to help them review the lessons of the day. It will include exercises such as transcribing audio files into Chinese characters, answering questions based on familiar texts, translating from English into Chinese by handwriting, and reading authentic materials. Students who cannot finish the contextualized writing and typing activity in class will be required to submit their work to the instructor after class. While the instructors may have already given group feedback in class, it is helpful if they can play a role in completing the digital performance and provide individual feedback. Feedback on a written homework assignment should encompass as many genuine written and oral comments as possible. For instance, when students respond to an email sent by their instructor, it is common for them to make typing or various grammatical mistakes in their reply. The instructor will evaluate the response by marking the errors and providing correct expressions. However, a more effective approach would be if the instructor replied to the email in a corrective but encouraging tone by saying "Did you mean..." to allow the student to clarify any misunderstandings. Students will be more motivated to respond to feedback and self-correct their mistakes if the instructor takes part in the performance.

Assessments and Learning Outcomes

When the proposed curriculum was tested at a private college in North America, speaking and listening skills were assessed through performance in each speaking class, overall achievement during mid-terms, and final oral interviews. Reading and writing capabilities were evaluated based on four areas.

Students' ability to handwrite Chinese characters. Although students are encouraged to develop typing skills at an early stage, the current approach suggests that being introduced to a limited number of Chinese characters is also essential for long-term motivation. As briefly discussed, if students memorize six characters before each reading and writing class, at the end of the first year of curriculum, they will have learned to handwrite a total of 288 characters and approximately 700 associated words and expressions. In order for the proposed approach to be successful, students must be prepared for each day's class. To achieve this goal, both written and oral quizzes are conducted at the beginning of each reading and writing class. The written quiz was in the form of dictation which tests students' preparation for the day's characters. Their ability to handwrite Chinese characters and compose phrases and sentences was also assessed in unit tests and the final exam. In these tests, students were expected to clearly write each stroke of the characters and produce their Roman spellings (*pinyin*) with tones. At the university where the proposed curriculum was tested at the end of the school year, students took a final written test including dictating 5 sentences made up of 88 characters (total points for dictation was 20). Each character or spelling mistake was -0.2, and each tone mistake was -0.1. Of the 25 students who took the test, 10 scored 92% or higher on the dictation section, 9 received 82–92%, 4 scored between 72% and 82%, and only 2 received 72% or lower. Students' scores on the dictation section corresponded closely to their overall grade on the written test. In addition, if students' tone errors were not counted, 14 of them scored 92% or higher in this section, 7 scored 82–92%, 3 scored between 72% and 82%, and only 1 received 72% or lower. Given that students do not need to know the tones of the characters to type them correctly, the above results indicate that students would have demonstrated an even better performance if the task were to ask them to type the sentences.

Students' reading ability. One key objective of the curriculum is to help beginner students establish a relatively large vocabulary by being able to recognize and comprehend more characters than they can handwrite. In every reading and writing class, students were asked to read texts out loud. Part of their daily grade was based on these performances. They were also asked to transcribe short dialogues written in characters into *Pinyin*. Students had experience writing some of these characters, but others only appeared in reading assignments. They were able to recognize characters that they were not required to learn how to write. For example, in Unit 8's test, the third reading and writing unit test during the second semester of the curriculum, students were asked to transcribe a 57-character dialogue. However, they had only learned how to handwrite 46 of them. One of the sentences was "请您再耐心等一会儿 (*Qǐng nín zài nàixīn děng yìhuier*)." Students at the time had not learned how to handwrite "再" or "耐心", but they had encountered this exact expression in one of their memorized dialogues and had read the Chinese character version of it. Among the 25 test participants, only 3 did not write the character "再", and 70% were able to transcribe "耐心" correctly. This is a strong indication that even without formally requiring students to master certain reading vocabulary, they are still able to acquire on their own some of the characters thanks to repeated exposure and typing tasks.

Ability to accomplish contextualized reading, writing, and typing tasks. Students were encouraged to participate in these tasks twice a week in class, which was part of the grade they received for the day. When assessing students' overall ability to perform in an authentic task, both their language knowledge and understanding of the context were evaluated. For instance, after reviewing Unit 8, a unit that focuses on the contents of giving directions, taking transportation, and traveling, students were asked to use their smartphone or computer to type an email in characters to their Chinese instructor. The prompt was presented on a PowerPoint slide in Chinese only: "你今年七月要去中国了，你想问老师什么问题? (Translated as: Since you are going to China this July, what would you like to ask your instructor?)." Students were assessed on how they began and ended their email. Among the 25 submitted emails, 23 included an appropriate beginning. Although there were basic errors in grammar, typing, and word choice, their writing was easily understood by native speakers of Chinese. Students were able to combine previously learned grammatical structures to formulate socially meaningful questions. Several students incorporated personal life experiences into the email when they asked about social media, regional cuisines, and pest control. More importantly, this assignment showed that students would be able to interact with native speakers of Chinese with ample linguistic and intercultural competence.

Ability to retrieve spoken Chinese digitally. Due to inadequate vocabulary and grammar, lower-level students were not required to create sentences on a random topic. Instead, they were responsible for retrieving phrases and expressions already familiar to them to produce conversational-style writing. The proposed curriculum includes an integrated term project in which 2 students were asked to write 60 lines of a script and act it out. Each line was a short conversation. The project was designed to encourage students to review all the dialogue they have learned in the spoken classes while demonstrating their ability to retrieve spoken language. They were required to type the script in both character and *pinyin* form, revise it based on feedback provided by instructors, and then video record their acting. Overall, students did well on the assignment. They were also able to organize the storyline so that the writing was coherent and highly comprehensible. Most errors students made were related to their oral production and included problems such as sentence structure and word choice. Given that most characters students used to write the script were included in the character version of their spoken material, many were able to fix the typos on their own. For this reason, we believe that instructors should not correct such typographical mistakes. Instead, students should be required to revise their own drafts before the instructor provides the final feedback before their filming. Based on analysis of the final video products, we also found that this assignment was successful in helping students correct some of the grammatical errors they repeatedly made in spoken Chinese, such as using 是 (shì) to connect a subject and an adjective as predicate.

Overall, although more types of instruments are needed to assess students' typing skills, current assessment results have demonstrated this approach to be effective in helping students develop the capacity to retrieve spoken Chinese and accomplish contextualized digital performance.

Student Evaluation of the Proposed Curriculum

At the end of the school year, students were asked to participate in a general course evaluation conducted by the college. Overall, they had positive opinions and favorable comments regarding the curriculum. Seven students were invited to participate in a face-to-face interview with the instructor to discuss details of the course and give feedback on the new approach (see Appendix for the interview questions). These students were selected based on two factors, their performance in reading and writing classes, and their previous background of learning Chinese. Four of the participants were recruited merely based on their course performance (two top students and two below average). Two heritage Chinese learners were recruited. Both of them also performed extremely well in this course. One of the subjects is an international student from Indonesia. This section includes an analysis of the preliminary qualitative data on learners' attitudes toward this lower-level reading curriculum that incorporates digital contextualized performance.

Students' Attitudes Regarding Handwriting and Typing Chinese Characters

The proposed approach is designed to develop both students' handwriting and typing skills. While the number of characters that students are required to handwrite is very limited, they are introduced to an extensive vocabulary so they can accomplish typing tasks in various contexts. Educators who support replacing handwriting with typing believe that students will appreciate and benefit from the more practical method. While the younger generation all over the world is embracing a highly digital lifestyle and handwriting Chinese characters does not seem to be a very useful skill anymore, we propose to reconsider the role of handwriting in the reading and writing curriculum. We suggest that all the skills that contribute to students' overall ability to accomplish authentic reading and writing tasks should be valued. Therefore, handwriting should be considered as one of the helpful techniques to further develop learners' reading and writing skills, rather than the ultimate objective.

Students' responses to the interview questions indicate that they value both their handwriting and typing skills. Six students agreed that being able to handwrite Chinese characters contributes to their understanding of the target culture. They believe that characters are closely tied to Chinese history, art, and culture, and knowing how to write them also provides a glimpse into the development of Chinese language and culture. Two students commented that handwriting characters helps them to remember them and read better. Correspondingly, only four believe that being able to type Chinese contributes to their understanding of the culture. They agreed that being able to type Chinese helps them to contextualize a word and participate in day-to-day activities in modern times. Two of them mentioned that being able to use social media is a big part of Chinese culture now, and it is important for them to have those skills.

When asked what types of writing and typing activities they would like to emphasize, five mentioned that they would like to write more emails. One recalled a particularly helpful contextualized performance in class in which they were required to write an email to an airline company to retrieve their luggage. Two mentioned that they would like to spend more time practicing handwriting skills in class or at home because it contributes to their long-term memory of the characters. It is clear that while these students generally embrace the idea of developing their typing skills so that they can engage in digital tasks, they also attach importance to their ability to handwrite characters.

Digital Contextualized Reading and Writing Activities

One major goal of the new curriculum is to incorporate technology-assisted authentic learning tasks into reading and writing class. All students agreed that these types of activities motivated them to explore how to use a Chinese keyboard, practice recognizing characters on Quizlet (a mobile and web-based study application), and practice typing often to accomplish day-to-day tasks. Two students mentioned that these real-world applications helped them to feel more confident about using Chinese and better prepared them to study abroad.

When asked to compare these tasks with other classroom activities in terms of difficulty level and usefulness, four students said that while these activities are at about the same difficulty level, they require a very different skill set than writing characters. Two students believe the tasks to be relatively easy because the students were using what they had already learned. According to one student, the tasks were more difficult as they required him to use his own knowledge to figure out what was required in the context to carry on a conversation. Six students agreed that these tasks were much more helpful than other classroom activities as they were able to apply what they learned to real-life tasks outside class. One of them commented that these activities helped students to review the characters.

Students' answers demonstrated a high level of acceptance of these digital contextualized tasks. It is also obvious that their understanding of the pedagogy affected their evaluation of the approach. Students who view the contextualized performance only as a technique for acquiring or practicing Chinese characters would be less likely to find the activities helpful. On the contrary, students who wish to use their language skills to engage with native speakers of Chinese and function in the target culture were more likely to appreciate this approach. We suggest that instructors make a special effort to introduce the learning cycle to the students and emphasize that recognizing and memorizing characters should take place outside the classroom.

The Current Approach and Learner Motivation

Student feedback also indicated that each individual could be motivated by different activities. One of the students reported that she did not enjoy handwriting postcards because she cannot see herself doing it in real life. However, another student claimed that being able to handwrite characters made him feel that he was actually learning the language rather than just participating in typing activities. Recent studies of language learner motivation have revealed the importance of helping students to establish a detailed vision of themselves using the target language (Dörnyei 2009; Kubanyiova and Crookes 2016; Jia 2019). Although how to efficiently use technology-assisted innovations to create motivational pathways remains controversial, the need for improving learning experiences in reading and writing classes is undeniable in the digital age. We suggest including learning tasks that are relevant to learner's personal life. For example, one of the most popular activities was providing students with a menu from a local Chinese restaurant and having them text one of their friends to discuss what to order. Followed by this reading and writing activity, they were asked to orally practise ordering food from the restaurant. Students were fond of this activity because it was based on familiar experiences that are useful to them at the personal level. Thus, we must incorporate learning materials and activities that will help students envision themselves using the language in real life, which will help them to remain motivated. In addition, it would be even more effective if we could personalize these activities by including students' individual interests, such as incorporating their favorite athletic teams.

In Thompson's (2017) investigation of two L1-English-speaking advanced language learners, one participant demonstrated an anti-ought-to attitude while learning

Chinese. A language leaner with an anti-ought-to vision tends to be motivated by proving stereotype wrong. In the context of learning Chinese as a foreign language, these learners are not willing to accept the typical impression Chinese people have about a foreign speaker of Chinese. For instance, the subject in Thompson (2017) stated, "The other related thing that drove me for many years was the absence of positive/ideal NNS role models who could speak Chinese in a way that drew praise from Chinese people" (p. 44). This attitude is particularly observed among beginner Chinese learners who are told that studying Chinese characters is the most challenging part of learning Chinese. Those ones with the anti-ought-to attitude then devote most their study time into memorizing characters. Instead, the proposed approach attempts to motivate students by focusing their attention on how many authentic tasks they could accomplish within a relatively short period of time instead of emphasizing on how many characters they have learned to read and write.

Conclusion

As reviewed at the beginning, some scholars of Chinese language pedagogy have emphasized the need to distinguishing character, reading, and writing instruction. However, few studies have addressed the respective pedagogical goals of these different skills in lower-level reading and writing curricula; even fewer have touched upon how to help learners develop these capacities in the context of the digital age. In addition, as some scholars pointed out (Walker 1989; Zhang 2016), most existing classroom activities that were identified to be effective in facilitating learners' orthographic and writing skills are not designed with a purpose of helping students conduct text-based communication. As a result, being able to generate grammatical construction by handwriting characters has long been the instructional goal of most existing lower-level curriculum.

In this chapter, we propose a curriculum that identifies various learning objectives for different skills such as character recognition and production. While we agree to control the number of characters students must be able to handwrite in lower-level instruction, the preliminary quantitative data revealed in the chapter suggests that students are able to recognize and comprehend more characters than they can write. One of the rationales to encourage students to develop their reading skills more quickly than their handwriting skills is so that they can participate in digital communicative interactions with native speakers of Chinese at an early stage, such as responding to email. Technology has changed how written language can be generated and consumed in social contexts. If we can agree on the importance for students to communicate in the target culture, then we need to incorporate digital reading and typing activities into the CFL curriculum. The proposed approach also suggests that task authenticity should be taken into serious consideration. To effectively achieve this goal, both reading and writing tasks should be designed as communicative social events where proper textual genre and task form needs to be adopted.

The qualitative interview data with seven students indicates that the current approach meets both their needs of learning handwriting characters and using their Chinese skills to accomplish real-life tasks. All students agreed that participating in technology-assisted authentic learning tasks motivated them to further explore more methods to practice their Chinese. Some stated that the current approach improves their confidence in using the language and better prepares them for study abroad. The interview results also shed some light on how students' different understanding of pedagogy affected their acceptance of the approach. We suggest to motivate students by focusing their attention on how many authentic tasks they could accomplish within a relatively short period of time rather than the amount of characters they are able to memorize.

The approach proposed in this chapter is an initial attempt to integrate authentic digital performance into lower-level CFL reading and writing curriculum. One limitation of this chapter is that the data of students' feedback came from only seven students and was constructed with one single method. Future study enrolling more participants and multiple research methods (questionnaire, classroom discourse analysis, and interview) is needed to investigate the effectiveness of the suggestions. Particularly, the current approach has only been applied in a small class learning environment (12–13 students); thus, the individual attention and feedback each student receives may contribute to the positive perception of the new curriculum. Moreover, additional thought needs to be put into designing digital task-based homework. Several students mentioned their frequent use of Quizlet to study Chinese characters. To encourage them to be prepared beyond character level, an interactive learning community where pre- and post-class efforts could be recognized and monitored needs to be established. Last but not least, although students' performance of authentic digital tasks is closely evaluated in class, a task-based online test needs to be developed to assess students' reading and typing skills.

Appendix: Interview Scripts

Statement: As mentioned in the consent form, this interview will focus on your Chinese learning experiences and behaviors. You can pause the interview and skip a question that you do not feel comfortable answering at any time. Our interview will be voice-recorded for the purpose of data review. The recorded file will be saved in researcher's work computer with restricted access. You could also pause or stop the recording during the interview by simply asking the researcher to do so. This interview will last about 15 min. Are you ready? Let's start.

1. Do technology-assisted learning activities (such as sending contextualized emails, WeChat messages, and using APP to search restaurants) encourage you to practice Chinese more?
2. In your reading and writing classes, you were asked to use your computers and smartphones to accomplish real-life reading and typing tasks, such as replying emails or ordering Chinese food with friends. Please evaluate these tasks:

- Are those tasks difficult compared with other classroom activities? (1, much easier than other activities; 3, the same difficult; 5, much more difficult than other activities)
- Are those tasks helpful? (1, much less helpful than other activities; 3, the same helpful; 5, much more helpful than other activities) In which sense?
3. What do you like and dislike most about the above reading and typing activities?
4. How would you rate the amount of handwriting practice in your class? (1, insufficient emphasis on handwriting; 3, appropriate; 5, too much emphasis on handwriting)
5. How would you rate the amount of handwriting practice in your homework? (1, insufficient emphasis on handwriting; 3, appropriate; 5, too much emphasis on handwriting)
6. How would you rate the amount of handwriting in test? (1, emphasis on handwriting; 3, appropriate; 5, too much emphasis on handwriting)
7. What types of writing/tying activity would you like to spend more time on? (Practice handwriting? Writing emails? Posting on social media? Writing essays)
8. What types of writing/tying activity would you like to spend less time on? (Practice handwriting? Writing emails? Posting on social media? Writing essays)
9. Do you think being able to type Chinese contributes to your understanding of Chinese culture? If so, how?
10. Do you think being able to handwrite Chinese contributes to your understanding of Chinese culture? If so, how?
11. Besides doing your homework, have you had a chance to type Chinese on your computer or smartphone in your daily life? In which contexts?
12. Do you use any learning APP or websites to study Chinese? If yes, please provide details (which learning tools and in which contexts do you use them?).

References

Allen, Joseph R. 2008. Why learning to write Chinese is a waste of time: A modest proposal. *Foreign Language Annals* 41: 237–251.

Barnes, Dayle. 1970. Writing in Chinese. *Journal of the Chinese Language Teachers Association* 4 (1): 8–14.

Chang, Cecilia. 2010. See how they read: An investigation into the cognitive and metacognitive strategies of nonnative readers of Chinese. In *Research among learners of Chinese as a foreign language*, Chinese Language Teachers Association monograph series, vol. 4, ed. M. E. Everson & H. H. Shen, 93–116. National Foreign Language Resource Center.

Ching, Nora, and Eugene Ching. 1975. Teaching the writing of Chinese characters. *Journal of the Chinese Language Teachers Association* 10 (1): 20–24.

Christensen, Matthew B., and Mari Noda. 2002. *A performance-based pedagogy for communicating in cultures: Training teachers for East Asian languages*. Columbus: National East Asian Languages Resource Center at The Ohio State University.

Christensen, Matthew B., and J. Paul Warnick. 2006. *Performed culture: An approach to East Asian language pedagogy.* Columbus: National East Asian Languages Resource Center at The Ohio State University.

Chun, Dorothy, Bryan Smith, and Richard Kern. 2016. Technology in language use, language teaching, and language learning. *The Modern Language Journal* 100: 64–80.

Dörnyei, Zoltán. 2009. The L2 motivational self system. In *Motivation, language identity and the L2 self*, ed. Zoltán Dörnyei and Ema Ushioda, 9–42. Bristol: Multilingual Matters.

Everson, Michael E. 1993. Research in the less commonly taught languages. In *Research in language learning*, ed. A. Omaggio Hadley, 198–228. Lincolnwood: National Textbook Company.

———. 1998. Word recognition among learners of Chinese as a foreign language: Investigating the relationship between naming and knowing. *Modern Language Journal* 82: 194–204.

Everson, Michael E., and Chuanren Ke. 1997. An inquiry into the reading strategies of intermediate and advanced learners of Chinese as a foreign language. *Journal of the Chinese Language Teachers Association* 32 (1): 1–20.

Goodman, Kenneth S. 1973. Comprehension-centered reading. In *Reading in education: A broad view*, ed. Malcolm P. Douglass, 251–260. Columbus: Charles E. Merrill Publishing.

Hedgcock, John, and Natalie Lefkowitz. 2011. Exploring the learning potential of writing development in heritage language education. In *Learning to write and writing to learn in an additional language*, ed. Rosa M. Manchón, 209–233. Amsterdam: John Benjamins.

Hong, Wei. 1998. An empirical study of Chinese business writing. *Journal of the Chinese Language Teachers Association* 33 (3): 1–12.

Huang, Sha. 2018a. Authentic texts as reading materials in a Chinese as a foreign language classroom: Learners' perceptions and pedagogical implications. *Journal of the National Council of Less Commonly Taught Languages* 23: 1–40.

———. 2018b. Effective strategy groups used by readers of Chinese as a foreign language. *Reading in a Foreign Language* 30 (1): 1–28.

Ihde, Thomas W., and Ming Jian. 2003. Language learning and the on-line environment: The views of learners of Chinese. *Journal of the Chinese Language Teachers Association* 38 (1): 25–50.

Jenkins, Henry. 2009. *Confronting the challenges of participatory culture: Media education for the 21st century.* Cambridge, MA: MIT Press.

Jia, Junqing. 2019. Chinese language learner motivation: Vision, socialization and progression. *Studies in Self-Access Learning Journal* 10 (1): 44–60.

Jiang, Lianjiang, and Jasmine Luk. 2016. Multimodal composing as a learning activity in English classrooms: Inquiring into the sources of its motivational capacity. *System* 59: 1–11.

Jin, Honggang. 2007. Syntactic maturity in second language writings: A case of Chinese as a foreign language (CFL). *Journal of the Chinese Language Teachers Association* 42 (1): 27–54.

Ke, Chuanren. 1998. Effects of language background on the learning of Chinese characters among foreign language students. *Foreign Language Annals* 31: 91–100.

———. 2012. Research in second language acquisition of Chinese: Where we are, where we are going. *Journal of the Chinese Language Teachers Association* 47 (3): 43–114.

Koda, Keiko. 1993. Task-induced variability in FL composition: Language-specific perspectives. *Foreign Language Annals* 26 (3): 332–346.

Kubanyiova, Magdalena, and Graham Crookes. 2016. Re-envisioning the roles, tasks, and contributions of language teachers in the multilingual era of language education research and practice. *Modern Language Journal* 100 (1): 117–132.

Kubler, Cornelius. 1997. *NFLC guide for basic Chinese language programs: Pathways to advanced skills.* Vol. III. Columbus: Foreign Language Publications.

———. 2017. *Basic mandarin Chinese: Speaking and listening*, revised edition. Clarendon: Tuttle Publishing.

Lammers, Jayne C., Alecia M. Magnifico, and Jen S. Curwood. 2014. Exploring tools, places, and ways of being: Audience matters for developing writers. In *Exploring technology for writing and writing instruction*, ed. Kristine E. Pytash and Richard E. Ferdig, 186–201. Hershey: IGI Global.

Lee, Siu-Lun. 2011. Online components for advanced Chinese reading classes: Design and implementation. *Journal of Technology and Chinese Language Teaching* 2 (1): 1–22.

Lee-Thompson, Li-Chun. 2008. An investigation of reading strategies applied by American learners of Chinese as a foreign language. *Foreign Language Annals* 41 (4): 702–721.

Li, Shouji. 2014. The gap in the use of lexical cohesive devices in writing between native Chinese speakers and second language users. *Journal of the Chinese Language Teachers Association* 49 (3): 25–48.

Lin, Yi. 2000. Vocabulary acquisition and learning Chinese as a foreign language. *Journal of the Chinese Language Teachers Association* 35 (1): 85–108.

Liu, Ying. 2014. Understanding readership: American students' perceptions of evidence in Chinese persuasive composition. Ph.D. Dissertation. Ohio State University.

Magnifico, Alecia M. 2010. Writing for whom? Cognition, motivation, and a writer's audience. *Educational Psychologist* 45: 167–184.

Mickel, Stanley L. 1980. Teaching the Chinese writing system. *Journal of the Chinese Language Teachers Association* 15 (1): 91–98.

Moser, David. 1991. Why Chinese is so damn hard? *Sino-Platonic papers* 27: 59–70. http://www.sino-platonic.org/complete/spp027_john_defrancis.pdf. Accessed 15 June 2019.

Mou, Sherry J. 2003. Integrating writing into elementary Chinese. *Journal of the Chinese Language Teachers Association* 38 (2): 109–136.

Noda, Mari. 2003a. Learning to read as a native speaker. In *Acts of Reading*, ed. Hiroshi Nara and Mari Noda, 9–23. Honolulu: University of Hawaii Press.

———. 2003b. Reading as a social activity. In *Acts of Reading*, ed. Hiroshi Nara and Mari Noda, 24–37. Honolulu: University of Hawaii Press.

Phillips, June K. 1978. Reading is communication, too! *Foreign Language Annals* 11: 281–287.

Pressley, Michael. 2000. What should comprehension instruction be the instruction of? In *Handbook of reading research*, ed. Michael L. Kamil, Peter B. Mosenthal, P. David Pearson, and Rebecca Barr, vol. 3, 545–561. Mahwah: Erlbaum.

———. 2002. Metacognition and self-regulated comprehension. In *What research has to say about reading instruction*, ed. Alan E. Farstrup and S. Jay Samuels, 3rd ed., 291–332. Newark: International Reading Association.

Shen, Helen H. 2000. The interconnections of reading text based writing and reading comprehension among college intermediate learners of Chinese as a foreign language. *Journal of the Chinese Language Teacher's Association* 35 (3): 29–48.

———. 2004. Level of cognitive processing: Effects on character learning among non-native learners of Chinese as a foreign language. *Language and Education* 18: 167–182.

———. 2005. Linguistic complexity and beginning-level L2 Chinese reading. *Journal of the Chinese Language Teachers Association* 40 (3): 1–28.

Shen, Helen H., and Chuanren Ke. 2007. Radical awareness and word acquisition among non-native learners of Chinese. *Modern Language Journal* 91 (1): 97–111.

Smith, Blaine E. 2014. Beyond words: A review of research on adolescents and multimodal composition. In *Exploring multimodal composition and digital writing*, ed. Richard E. Ferdig and Kristine E. Pytash, 1–19. Hershey: IGI Global.

Taft, Marcus, and Kevin Chung. 1999. Using radicals in teaching Chinese characters to second language learners. *Psychologia* 42: 243–251.

Taft, Marcus, and Xiaoping Zhu. 1995. The representation of bound morphemes in the lexicon: A Chinese study. In *Morphological aspects of language processing*, ed. Laurie B. Feldman, 293–316. Hillsdale: Erlbaum.

———. 1997. Submorphemic processing in reading Chinese. *Journal of Experimental Psychology: Learning, Memory, and Cognition* 23: 761–775.

Thompson, Amy S. 2017. Don't tell me what to do! The anti-ought-to self and language learning motivation. *System* 67: 38–49.

Tseng, Ovid J. 1989. Neuropsychological studies of spoken and written Chinese: A decade's effort. Keynote lecture in first northeast conference on Chinese linguistics, Columbus.

Walker, Galal. 1984. "Literacy" and "reading" in a Chinese language program. *Journal of the Chinese Language Teachers Association* 19 (1): 67–84.

———. 1989. Intensive Chinese curriculum: The EASLI model. *Journal of the Chinese Language Teachers Association* 24 (2): 43–83.

———. 2010. Designing an intensive Chinese curriculum. In *The pedagogy of performing another culture*, ed. Galal Walker, 51–95. Wuhan: Hubei Jiaoyu Chubanshe.

Wang, Zulei. (王祖嫘) 2010. "书写"还是"输入"?——留学生汉字输入学习刍议 ("Writing" or "inputting" – discussion on overseas students' learning of Chinese characters inputting). *Journal of Modernization of Chinese Language Education* 1(1): 71–77.

Warnick, Paul J. 1996. A phenomenology of reading performances: Reading Japanese as a foreign language. The Ohio State University Ph.D. ProQuest, UMI Dissertations Publishing.

Xie, Tianwei. (谢天蔚) 2011. "手写"还是"电写"——电脑输入中文引起的讨论 ("Hand-writing" vs. "electronic-writing": Which way we should go in teaching Chinese?). *Journal of Chinese Teaching and Research in the U.S.*: 98–102.

Xu, Xiaoqiu, and Amado M. Padilla. 2013. Using meaningful interpretation and chunking to enhance memory: The case of Chinese character learning. *Foreign Language Annals* 46 (3): 402–422.

Yang, Chunsheng. 2013. Textual conjunctives and topic-fronting devices in CFL learners' written summaries. *Journal of the Chinese Language Teachers Association* 48 (1): 71–90.

Zhang, De. 2009a. Essay writing in a mandarin Chinese WebCT discussion board. *Foreign Language Annals* 42 (4): 721–741.

Zhang, Zhengsheng. 2009b. Myth, reality and character instruction in the 21st century. *Journal of the Chinese Language Teachers Association* 44 (1): 69–90.

Zhang, Yongfang. 2013. A CIP spiral performance-based model for beginner level Chinese L2 reading instruction. *Journal of the Chinese Language Teachers Association* 48 (2): 91–118.

Zhang, Shenglan. 2016. Killing two birds with one stone? Turning CFL learners into book writers: An exploratory study. *Journal of the Chinese Language Teachers Association* 51 (2): 164–190.

Zhang, Zhengsheng. 2018. My never-ending education. In *The field of Chinese language education in the U.S.*, ed. Vivian Ling, 281–289. London, UK/New York: Routledge.

Zheng, Qiwu (郑启五). 2010. 对外汉语教学"弃写"的思索 (Thinking about "giving up handwriting" in teaching Chinese as a foreign language). Retrieved February 15, 2020, from http://www.chinalanguage.org/showinfo.asp?id=645

Developing Advanced CFL Learners' Academic Writing Skills: Theory and Practice

31

Yang Wang

Contents

Introduction	776
Current Chinese Writing Pedagogy at the Advanced Level: A Literature Review	777
The Role of Chinese Academic Writing Instruction in Advanced Chinese Curriculum	784
The Definition of Chinese Academic Writing	784
Importance of Academic Writing Instruction in Advanced Chinese Curriculum	785
Targeted Students and Preparation for Chinese Academic Writing Instruction	788
Instructional Design of Academic Writing Pedagogy in Advanced Chinese Courses	789
Overall Curricular Planning and Designing Scaffolded Writing Assignments	789
Connecting Critical Writing with Critical Reading	791
Designing Effective Academic Writing Tasks	792
Supporting Activities for CHINESE Academic Writing Instruction	795
Assessment and Feedback	800
Conclusion	803
Appendix	804
Appendix I: Writing Assignment for an Interview Report for Chapter 6 in the Textbook *China in Depth: An Integrated Course in Advanced Chinese*	804
Appendix II: Instructions for Final Independent Research Report	805
References	807

Abstract

For truly advanced learners of the Chinese language, training in Chinese academic writing has numerous benefits: It brings depth to the teaching content, improves students' overall ability to use the language, and more importantly, enhances their analytical and research skills as well as critical thinking skills. It is therefore worth exploring how to fully recognize the importance of teaching Chinese academic writing and integrate an academic writing component into an advanced Chinese language curriculum. The first part of this chapter reviews the current literature on Chinese writing pedagogy at the advanced level. The second

Y. Wang (✉)
East Asian Studies, Brown University, Providence, RI, USA
e-mail: yang_wang@brown.edu

part argues for the importance of including academic writing instruction in advanced-level Chinese courses. The third part discusses the instructional design of a writing curriculum and various pedagogical practices to support the development of students' academic writing skills.

Keywords

Writing pedagogy · Advanced Chinese · Academic writing, Curriculum design

Introduction

In 1871, a Yale University librarian named Addison Van Name taught a Linguistics course with "elements of Chinese." This marked the beginning of Chinese language instruction in the United States (Yao and Yao 2010). 170 years have since passed and the field of Chinese language instruction has undergone tremendous changes. According to the Modern Language Association, as of 2016, at least 794 institutions of higher learning in North America offered Chinese courses, with an enrollment of 53,069. With the end of the Cold War and China's rapid economic growth in the 1990s, the field of Chinese language pedagogy in the United States has experienced two decades of rapid growth. College enrollments for Chinese language courses peaked in 2013. Although enrollments have since been on a decline, it is clear that in the past 15 years, along with the establishment of Chinese AP courses and the steady development of study abroad programs, Chinese language instruction in North America has continuously grown in breadth and depth. This growth is reflected mainly by the steady increase in students enrolling in advanced Chinese courses and by significant improvements in their Chinese proficiency. According to MLA's most recent 2016 report on enrollments in foreign language courses in American higher institutions (2016, 32), although there was a 13.1% decrease in university course enrollments from 61,084 in 2013 to 53,069 in 2016, the ratio of introductory to advanced course enrollments in the fall semester actually changed from four to one in 2013 to three to one in 2016, indicating an increasing need for more advanced Chinese language courses.

In light of both the increase in the number of students taking advanced Chinese courses and the increase in the proficiency level of such students compared to a decade ago, the future direction of advanced-level Chinese instruction is one of the most vigorously discussed issues for instructors and scholars in the field. It is in this context that this chapter calls for a reconsideration of the role of Chinese writing instruction in advanced-level Chinese courses, investigates its pedagogical approaches, and proposes some tried and tested teaching practices. This chapter aims to fill a lacuna in the current critical conversation on advanced Chinese pedagogy and to contribute to the research and practice of advanced Chinese writing instruction in the future.

Current Chinese Writing Pedagogy at the Advanced Level: A Literature Review

Writing is widely acknowledged as the most difficult and time-consuming skill to acquire and perfect when learning a foreign language (Richard and Renandya 2002, 303). This is particularly true for Chinese language learners, especially those who have had no exposure to nonalphabetic writing systems. Learning how to write Chinese characters alone, the first step in gaining Chinese literacy skills, takes a tremendous amount of time.

More than three decades ago, Galal Walker (1989), one of the pioneers in the field of Chinese language pedagogy in the United States, already identified an oversight in the academic research on Chinese writing instruction. He pointed out that although learning to write Chinese characters has been viewed by several generations of Chinese learners as most representative of the language learning experience, in fact, "composing in *shumianyu*" is the least effective teaching practice. Twenty years later, scholars continue to express similar views (Liu 2014; Luo 2011; Piao 2007; Shi and Chen 2019; Song 2015; Tao 2007; Wang 2010; Wu 1999, 2012; Xu and Wu 2016; Zhang 2009). A search of papers on writing instruction published from January 2001 to March 2020 in *World Chinese Teaching* (*Shijie hanyu jiaoxue*) and the *Journal of Chinese Language Teachers Association*, two major journals on Chinese language pedagogy in mainland China and North America, yielded some interesting results. 849 papers were published in *World Chinese Teaching* (excluding newsletters and book reviews) during this period and only nine papers focused on Chinese writing instruction (excluding those on character instruction), whereas 337 papers (excluding newsletters and book reviews) were published in the *Journal of Chinese Language Teachers Association*, but only six papers focused on Chinese writing instruction (excluding those on character instruction). Papers related to the other three skill areas (reading, speaking, and listening) greatly outnumbered those focusing on the writing skill. Whether in China or abroad, it is evident that theoretical and practical research on Chinese writing instruction has generally been ignored and disregarded by the field. Among the few papers on writing, those focusing on Chinese writing instruction at the advanced level are even more rare.

Zhu (2007) interviewed eight instructors of Chinese writing courses and 33 international students from three Chinese universities in order to learn about the current pedagogical practices in advanced writing instruction. The survey found that there were considerable differences among the eight instructors in terms of their course objectives, teaching methods, and assessment criteria. With regard to course objectives, the biggest issue they disagreed on was whether a writing class should aim to improve the compositional technique of students or to improve their Chinese accuracy and fluency when expressing themselves in writing. In contrast, the 33 international students seemed to think that the reason for taking advanced Chinese writing courses was not much different from that of taking other Chinese courses, which was to improve their ability to express ideas in Chinese. However,

54.8% of the students revealed in their interviews that they had made little progress by the end of one semester. The lack of consensus among teachers on course objectives, course content, and teaching methodologies of advanced Chinese writing instruction reflected in Zhu's study probably captured the status quo of the field at the time and is likely to still accurate.

Regarding the learning objectives and content of Chinese writing courses, some early scholars believe that writing courses should focus on consolidating students' linguistic knowledge, such as vocabulary and grammar, regardless of their proficiency level. In a sense, writing is seen as a type of supplementary practice: The purpose of learning to write is to improve overall proficiency in Chinese (Zhao 1996; Zhu 1984; Nan 1994). Others believe that advanced writing courses should prioritize the study of Chinese rhetoric, compositional techniques, and different modes of written discourse – narration, description, exposition, and argumentation (Yang 1982). But this view has not actually departed from writing pedagogy that caters to native speakers of Chinese. Comparing Chinese native language writing instruction to Chinese foreign language writing instruction, Tao (2007) emphasizes the importance of discourse competence, knowledge of style, and essay structures of different genres in CFL writing instruction, reminding instructors to maintain a balance between language instruction and composition guidance. Luo (2002) points out that the fundamental goal of advanced writing instruction is to develop students' ability to express themselves in written discourse, which was invariably influenced by their overall language proficiency, genre knowledge, discourse awareness, and their preexisting writing skills in their native language. Luo's view has been widely accepted by other scholars in the field.

Luo (2002) was the first scholar to systematically articulate the importance of competence in written discourse (书面语篇能力). He argues that developing an awareness of discourse should be the focus of advanced writing instruction. He divides various discourse strategies into three categories (2002, 113): explicit strategies (such as the use of conjunctions or transitional words and phrases that demonstrate logical and temporal connections), semi-explicit strategies (such as omission and reference), and implicit strategies (which utilize the logical order of sentences, choice of sentence patterns and words, and coordination of the overall style of the text). Building on Luo's body of work, some scholars have proposed that instructors should teach discourse grammar during writing classes rather than teach vocabulary and grammar only at the sentence level (Peng 2004; Zhang 2003). Xin (2001) conducted a statistical analysis of the errors in eleven argumentative essays written by advanced-level Chinese students. She organized the errors found into three categories: simple structural and lexical errors, pragmatic errors, and errors in cohesion and coherence. Excluding 4.8% of unclassifiable errors, the first category accounted for 58.5%, the second for 21.9%, and the third for 14.6% of all the errors. Xin discovered that errors of the second and third categories could not simply be resolved at the sentence level, but required a broader understanding of the context of the paragraph or essay. She also emphasized that sometimes the language in students' writings was semantically disconnected and even incoherent, which was influenced by their logical thinking rather than their language ability. However, the

CFL field currently lacks systematic research on discourse grammar and textual coherence devices, which has directly affected the quality of advanced Chinese writing instruction.

The traditional methods of Chinese writing instruction align with the prevailing product approach in the West in the 1980s, where model essays were analyzed in the early stages of writing and extensive corrections would subsequently be given on the final piece of writing a student produced. The product approach places great emphasis on word choice and sentence construction. In the prewriting stage, students are taught technical writing knowledge, such as different textual styles and rhetorical strategies. They learn how to analyze the textual organization and linguistic features of the model essays and have ample opportunities to imitate them (Ma 1999; Wu 1999; Xin 2001; Wang 2009; Yang 2018). When evaluating students' written work, the instructors pay great attention to the accuracy of the final products, making sure that all the errors and mistakes are corrected (Piao 2007; Wu 1999; Chen 2003). The planning of the writing content, such as how to develop ideas for an essay, usually goes beyond the scope of a writing course (Xin 2001). Yang (2018) describes in detail how to analyze and highlight the textual features of model essays, provide comments on students' written work, and teach Chinese rhetoric in advanced writing courses.

The process approach to writing, which is influenced by the communicative language teaching approach, offers an alternative to the product approach. In recent years, this approach, which grew out of research on writing theory for English speakers and English learners, has had a profound impact on the teaching practices for Chinese writing instruction. Guided by this approach, the focus of writing instruction has shifted from the study of linguistic knowledge and model texts to the content and process of writing. The process approach divides writing into four basic stages: prewriting, drafting, revising, and editing. It emphasizes the exchange of ideas between teachers and students as well as among students. Luo (2002) believes that creating an equal, interactive, and cooperative writing environment to motivate students to write is the key to the success of the process approach. This requires teachers to improve their ability to facilitate classroom discussions in order to help students brainstorm ideas and provide feedback. Fu (2014) compared the differences between the product approach and the process approach in light of teaching content, assessment methods, student performance, and teachers' roles. She conducted a survey of 20 students who took elective writing courses and found that 84% of the students would prefer to move away from the traditional instruction model where "teachers talk, students write" and would instead like to have the opportunity to share their experience of the writing process and their own writing with their peers. Yang (2004) elaborates the theoretical foundation and the teaching principles of the process approach and uses concrete examples from an advanced writing class to demonstrate how to implement the guiding principles of the process approach when designing an overall writing course curriculum and devising both in-class and take-home writing activities. She also discusses the drawbacks of the process approach: Students are often at different proficiency levels or have different needs, some students have a tendency to fuss over minor issues during discussion, teachers may drone on, and the pace of the class may thus be hampered. Furthermore, the process approach overemphasizes fluency in the expression of content at the cost of accuracy and the correction of students' errors is not always prompt and clear. These are the issues that instructors must pay particular attention to when adopting the process approach.

The genre approach to writing emerged after the process approach. This approach assumes that writing is a social activity – a dialogue between authors and their readers, rather than a one-directional flow of thoughts from an individual author to their readers. Therefore, if students wish to successfully engage their target audience, they need to develop "readership awareness," producing texts that fulfill the expectations of their readers in terms of language, organization, and content. Liu (2015) tracked nine American students and their experiences with learning Chinese persuasive writing at an intensive study abroad summer program and found that the American students' perception of effective evidence differed greatly from those presented by their textbook and instructor in class. Liu argued that a reexamination of the role of readers in CFL writing instruction is crucial, since the purpose of persuasive writing is to convince and persuade readers in the target culture instead of reaffirming the writer's position in the base culture. She also called for more genre analysis that characterizes typical or conventional form-function correlations in different types of argumentative writing. The genre approach also emphasizes the relationship between the social purpose, language use, and schematic structure of the written texts. Wu (2008) advocated for combining the process approach with the genre approach, creating the process-genre approach. He argued that during the prewriting stage, the instructor needs to teach students how essays of different genres effectively communicate different purposes, building connections between discourse structures and certain linguistic forms, as well as analyzing its language style. Wu believed that advanced students should be required to learn different argumentative genres, beginning with short essays, followed by reading reflections, commentaries and reviews, and finally an introduction to academic writing. The genre approach is an emerging method in the field of Chinese writing instruction and is easier to adopt when teaching practical writing tasks, such as a cover letter for a job application. But at the advanced writing level, which genres of argumentative writing should be introduced to students? What are the schematic structures of such genres? These questions still need to be answered in further research.

In addition to the goals, content, and methods of writing instruction, students' language use in their writings is also an essential part of research on writing pedagogy. There is limited research on the linguistic development in L2 Chinese writing, especially at the advanced level, but recent research has yielded some interesting findings. Syntactic fluency, accuracy, complexity, lexical diversity, lexical complexity, frequency of lexical errors, and sometimes lexical density are usually used to assess linguistic development in students' writings. An 2015 randomly chose 90 advanced-level essays from the HSK composition database and divided them into three groups, based on the these essays' HSK test scores. He found that the higher the test scores were, the better their fluency in writing, and the higher their syntactic complexity. There was also some positive correlation between test scores and syntactic accuracy. However, there was no correlation between lexical accuracy and test scores – the incidence of lexical errors remained constant, despite higher test scores. Research conducted by both Wu (2016a) and Zhang (2019) found that students' lexical diversity and complexity in writing were directly proportional to students' writing scores. However, advanced students still mainly relied on class A vocabulary (甲级词汇) and some class B vocabulary (乙级词汇) from *Syllabus of Graded Words and Characters for Chinese Proficiency* (汉语水平词汇与汉字等级大纲) to express themselves, avoiding less commonly used and low frequency vocabulary. Wu's research shows that at the advanced level form errors in vocabulary gradually decrease, but substantial semantic errors appear. Zhang's study found that at the advanced level form errors and semantic errors in vocabulary decline, but pragmatic errors remain high, and more importantly, lexical errors are not correlated to writing quality in her study. Qi and Liao (2019) investigated the linguistic development of 34 advanced learners in their narrative and argumentative writing over a period of 9 weeks. The results show that by the end of 9 weeks students displayed no significant improvements in lexical accuracy in either narrative or argumentative writings. All of these studies demonstrate that lexical accuracy at the advanced level takes much longer time to develop than teachers have expected; therefore, teachers should adjust their expectations and not rush students in this respect. Wang (2017) studied the correlation between different lexical factors and the scores of 360 upper intermediate and advanced-level sample essays from the HSK Dynamic Corpus of Writing Papers. Her research found that lexical errors are the most prominent parameters predicting the writing test scores of students, which reveals that the instructors tend to ignore other markers of lexical richness such as lexical diversity, sophistication, and density when evaluating students' writing performance. T-units, or minimally terminable units, are widely used to evaluate syntactic development in English. Both An (2015) and Jiang's (2013) studies adopted the T-unit index to examine the development of syntactic maturity and found that at the advanced level there was a significant correlation between students' writing scores and both T-unit length and error-free T-unit length. These studies showed that, compared to lexical accuracy, students' syntactic accuracy had greatly improved at the advanced level. In his research, Wu (2016b) found that the number of topic chains, the number of topic chain clauses, and the number of empty categories were more effective indices for grammatical complexity across different levels of proficiency. However, Wu also stressed that syntactic complexity should not be used as an absolute index to evaluate language proficiency because advanced-level speakers (especially native speakers) are often able to produce sentences that are more concise and cohesive.

The type of writing assignment also has a significant effect on students' linguistic writing performance, which has pedagogical implications when designing advanced-level writing tasks. Two recent studies investigated the linguistic features of advanced L2 Chinese writing performances in narrative and argumentative writing. Qi and Liao's (2019) research shows that the lexical accuracy in narrative writing samples is higher than in argumentative ones, which may be due to the use of more difficult vocabulary in argumentative essays. However, the lexical density of argumentative essays is higher than that of narrative essays, which indicates that argumentative essays use a greater number of content words. In addition, the average length of T-units in argumentative essays is significantly longer than that of narrative essays, indicating that the syntax of argumentative writing is more complex than that of narrative writing. Shi and Chen's (2019) research revealed that there are significant differences in complexity and fluency at the lexical level, with students producing more words in narrative writing within a set time (90 min) and employing a greater number of advanced words in argumentative writing. These two studies affirm that, from a linguistic perspective, argumentative writing tasks are more challenging than narrative writing tasks.

What, after all, is so hard about writing? In her illuminating research, Wu (2011) interviewed students from beginner ($N=18$), intermediate ($N=18$), and advanced ($N=20$) levels in the same Chinese program and asked them to evaluate their own writing abilities and writing habits. The study found that students generally judged their Chinese writing skills to be inept, but it is worth noting that the percentage of students who thought their Chinese writing skills were not as good as their writing skills in their native language increased as they progressed through the course levels (55.6%, 61.1%, 75%). The more advanced they were, the more inept students felt about their ability to recall past experiences, thoughts, or opinions they could write about when given a writing assignment. Wu believes that this is because at the lower levels, students are generally given narrative writing tasks for which they can use either their own personal experiences or the experiences of people around them. As they enter the more advanced levels, the writing content becomes more and more abstract, requiring more rationalization skills, which results in students lacking ideas as to what to write. Wu surprisingly found that fewer students were clear on what they wanted to write about at the advanced levels when more prewriting plotting activities, which aimed to help students brainstorm key vocabulary and ideas, had been added to instruction. One explanation might simply be that having more ideas for what to write can actually be a burden, serving to muddle rather than clarify how to begin writing the paper, making it more confusing to determine how to organize the content of the paper as well as how to develop a compelling and focused thesis. Wu's research suggests that in addition to the linguistic factor, the other two factors that make writing more challenging at the advanced level are the content of the paper and the organization of ideas.

In recent years, with the advent of more and more advanced level Chinese learners, the curriculum design and pedagogy of advanced Chinese courses have become hot topics in the field. Against this backdrop, targeted research on "academic Chinese writing" has begun to receive more scholarly attention. Among existing studies, the research focus of academic Chinese writing has been concentrated on writing the degree thesis. This is largely influenced by the increase in international students pursuing bachelor or graduate degrees at Chinese universities because they have to complete a thesis in order to graduate. At Renmin University, for example, undergraduate students, including both Chinese and international students, are required to write a thesis of 8,000 words (Gao and Li 2018). Different scholars (Gao and Li 2018; Gao and Liu 2016; Guo 2016; Qi 2006; Yao 2005) have pointed out similar problems in the theses written by international students. In terms of thinking, some students are too "simplistic" and lack argumentative reasoning skills. Yao (2005) refers to this problem as "lacking thesis thinking": Papers reveal unclear arguments, illogical delivery of ideas, and insufficient substantiation. Furthermore, the inadequacy of research skills results in superficial content. In terms of language use, apart from grammatical and lexical errors, language styles are often inconsistent, mixing colloquial and formal expressions. Some expressions used are inappropriate for the genre and read more narratively. Many students still lack an understanding of the stylistic and writing conventions of Chinese academic papers. In order to help students improve their academic writing skills, scholars agree that it is necessary to design a specific academic thesis writing course. However, there are many different views on what the focus should be when teaching such a course. Qi (2006) believes that the focus of such a Chinese academic writing course is for instructors to primarily comment on students' work with numerous and incisive edits, and only secondarily teach students how to properly conduct research in a discipline, so students are able to produce articles with solid content. Yao (2005) posits that an academic writing course should not continue to emphasize linguistic training, but should instead combine the reading of critical essays with writing, using reading as a precursor to writing, therefore allowing writing to develop organically from the reading process. He argues that the benefit of this method is that students will learn content from the reading materials and thereby develop more ideas for their own papers. Through textual analysis of the reading materials, they will learn how to use the Chinese language in analysis and evaluation. He also emphasizes that the instructor should guide students in understanding the overall organizational logic, as well as the logic of expression in assigned readings. Gao and Li (2018) thinks that the focus of the academic writing course should be to nurture the writing ability of students, rather than to teach academic writing conventions. At the same time, they argue that cultivating students' academic research and critical thinking skills is not within the teaching scope of an academic writing course. Guo (2016, 47) holds a different view. She believes that academic writing instruction should "aim at enhancing students' research awareness and research thinking," and also that the cultivation of critical thinking should be prioritized throughout the entire writing process, which includes drafting, peer editing, and tutorial discussions with the advisor. She suggests that before guiding students on how to write their degree theses, the instructor should employ the principles of the task-based approach to segment the writing task – breaking it down into manageable sub-tasks, such as writing an introduction, crafting a paper outline, conducting surveys and statistical data analysis, and drawing conclusions.

Looking back on the past 20 years, advanced writing instruction has made significant progress. Empirical research and feedback from students have confirmed that argumentative writing is much more difficult than narrative writing, largely because the content and critical thinking required in argumentative writing is more challenging. Therefore, teachers must pay particular attention to the writing process when it comes to teaching argumentative writing. However, because language use, rhetorical style, and readers' expectations can be very unique in Chinese writing, it is necessary to focus on exemplifying model texts, studying Chinese writing conventions, and building reader awareness. This requires that the teaching process not rely on one approach only, but draw on the strengths of multiple approaches. Several empirical studies show that at the advanced level students display significant improvements in syntactic accuracy and fluency. However, the improvement of lexical accuracy is a slow process, so teachers need to adjust their expectations and realize that students will inevitably make a lot of lexical errors along the way. Furthermore, they should not overemphasize accuracy alone; the richness of lexicon is also an important dimension of language development.

Academic writing is one of the most challenging and advanced forms of argumentative writing. At present, the very limited research in this area still focuses on describing existing problems and investigating macro guidance strategies. In contrast, research on teaching methods and teaching content is relatively scattered, lacking specific, systematic, and operable discussions on classroom activities and instructional materials. In the end, the most pressing questions remain: Why do we want students to write? Why do they need to write academically, especially if they do not have a thesis requirement?

The Role of Chinese Academic Writing Instruction in Advanced Chinese Curriculum

The Definition of Chinese Academic Writing

Academic writing prioritizes an argumentative style while incorporating elements of the expository style. It does not necessarily take the form of an academic journal paper, a thesis, or a presentation for professional conferences, but encompasses a variety of written assignments in college courses across different disciplines. Some examples include short term papers, response papers, reading reflections, and research analysis reports. In the context of Chinese as a foreign language, Chinese academic writing generally refers to a type of writing in which students read primary materials from different sources on a topic in a certain academic discipline; conduct research on or investigate the subject; and then analyze, evaluate, and synthesize the materials, following which they then either express their own opinions or supplement, revise, or refute the opinions of predecessors.

The American Council on the Teaching of Foreign Languages (ACTFL)'s Proficiency Guidelines, first published in 1982, divides language proficiency into four levels: novice, intermediate, advanced, and superior. According to the revised guidelines for writing proficiency published in 2012 (ACTFL 2012), academic writing is a requirement for superior-level foreign language learners. Below is ACTFL's description of superior-level writing proficiency:

> Writers at the Superior level are able to produce most kinds of formal and informal correspondence, in-depth summaries, reports, and research papers on a variety of social,

academic, and professional topics. Their treatment of these issues moves beyond the concrete to the abstract.

Writers at the Superior level demonstrate the ability to explain complex matters, and to present and support opinions by developing cogent arguments and hypotheses. Their treatment of the topic is enhanced by the effective use of structure, lexicon, and writing protocols. They organize and prioritize ideas to convey to the reader what is significant. The relationship among ideas is consistently clear, due to organizational and developmental principles (e.g., cause and effect, comparison, chronology). These writers are capable of extended treatment of a topic which typically requires at least a series of paragraphs, but can extend to a number of pages.

Writers at the Superior level demonstrate a high degree of control of grammar and syntax, of both general and specialized/professional vocabulary, of spelling or symbol production, of cohesive devices, and of punctuation. Their vocabulary is precise and varied. Writers at this level direct their writing to their audiences; their writing fluency eases the reader's task.

Compared to the guidelines for advanced-level writers, the writing tasks for superior-level learners have shifted from being based in a narrative or descriptive mode to an argumentative mode. In-depth summaries, reports, and research papers all fall into the category of academic writing. When it comes to language use, lexical accuracy and richness are important benchmarks. When it comes to content, the ability to be logical and clear and to express complex and abstract ideas in an organized way is crucial for writers at this level. No matter how you approach the subject, the three core tenets – in-depth content, clear logic, and rich language – are absolutely integral to good academic writing.

Importance of Academic Writing Instruction in Advanced Chinese Curriculum

In reality, advanced-level students have always been slowest in developing their writing skills compared with other language skills. Why has writing always been the "Achilles' heel" for Chinese learners? The reasons are complex. To begin, writing is a language skill that requires comprehensive understanding of the multiple facets of language. It is not as easy as "My hand writes what my mouth says." Many students fear and resist writing. Secondly, the development of one's writing ability is a gradual process. It takes conscientious and long-term training to transform writing knowledge into practical writing skills. Different from the other three skills, it is not realistic to expect a learner to achieve immediate results from short-term intensive writing training. Meanwhile, instructors have limited contact hours with students, especially in advanced Chinese courses in the United States, which typically only have three classroom instruction hours per week. For students, writing does not seem to be the skill most closely aligned with their practical needs. Therefore, writing instruction tends to be largely "marginalized." Even in an advanced-level course, it is still often treated as an auxiliary teaching method to help students consolidate their knowledge of Chinese grammar and vocabulary and to improve their overall ability to use the language. Not to mention that throughout the history of the development of foreign language teaching approaches, from the Direct Method, to the Audio-lingual Method, to the currently popular Communicative Approach, emphasis has always been placed on developing the oral skills of students. As a result, writing instruction has always been secondary.

In the past decade, some profound changes have taken place in the field of Chinese language instruction in the United States. There are more and more high schools offering Chinese AP courses. In 2010, 6,388 students took the AP Chinese test in the United States, and this number doubled to 13,825 students in 2018. At the same time, short-term training programs for K–12 Chinese teachers, such as Startalk and graduate programs on Chinese language pedagogy (like the one offered by Middlebury College), have greatly improved the quality of K–12 Chinese instruction. The rapid developments in K–12 Chinese language programs have made it possible for more and more first-year college students to enroll in advanced Chinese courses. In addition, although the total number of college students studying abroad in mainland China and Taiwan has experienced a slight decline from 14,760 students during 2009–2010 to 12,470 students in 2017–2018 according to *Opendoors: Report on International Educational Exchange* (2019), intensive Chinese language programs and scholarships funded by the US government, such as the Chinese flagship programs and the Critical Language Scholarships, have nurtured quite a number of advanced Chinese learners through intensive training. All these factors have contributed to a steady rise in terms of proficiency among college students in advanced Chinese courses, while students enrolling in these courses are increasingly of a younger age. By their sophomore or junior year, it is not unusual now for some students to have already completed the full regular four-year Chinese language curriculum. Many of them plan to pursue careers in Chinese-speaking regions or graduate studies in China-related fields after college. For these students, honing their academic writing skills is of paramount importance – it will not only improve their overall language proficiency, but also bring depth to the content taught in the language courses. Therefore, it is important to recognize which benefits writing instruction can bring to advanced Chinese learners.

First, *shumianyu* (书面语), the formal written Chinese, is the key focus of advanced Chinese instruction. Zhang (2005) used statistical methods to observe the frequency of a select number of *shumianyu* vocabulary and structures in a variety of genres, such as screenplays, political editorials, popular science articles, and novels. The results showed that although the use of *shumianyu* is not entirely limited to writing, the frequency with which *shumianyu* occurs in oral contexts is significantly lower than in written contexts. Unlike the neutral style of syntax and vocabulary learned in beginning- and intermediate-level courses, the use of *shumianyu* is more demanding. Compared to common classroom oral practice, such as grammar drills and group discussions, writing practice not only gives students more time to think about how to use certain Chinese written vocabulary and structures, but also provides more organic and appropriate contexts to use them in.

Second, the general pedagogical trends for advanced-level language courses center on content-based instruction. Language becomes the medium by which a subject matter is learned; thus, what is taught in the language is as important as the language itself. The research element in academic writing makes it fundamentally different from the more general short essays students typically write in an upper intermediate or advanced-level course. Rich, varied, complex, or stylish language does not make a good academic paper. When it comes to the academic essay, content is more important than language. In order for their writing to be meaningfully grounded in evidence, logic, and argumentation, students' understanding of the topic must go beyond the superficial or anecdotal. Students must first read extensively, research and gather information online, and understand alternative viewpoints before writing. Such Chinese writing is more like using the Chinese language as a tool: In order to write academically, students must conduct in-depth research and learn how to express themselves clearly in their topics of interest and chosen disciplines. In this way, academic writing requires independent thinking and self-conscious reflection. This is fundamentally different from supplementary writing exercises, designed around grammar and vocabulary, and is more likely to stimulate students' interests and help them develop a deeper understanding of the various social and cultural issues of the target society.

Third, academic writing is a process that requires comprehensive use of all facets of language-related skills. When students use the target language to construct their train of thought in order to express themselves, they are simultaneously training their critical thinking skills. Here, a fundamental question related to language learning is whether or not the training of critical thinking skills should be one of the learning objectives of a language course. Before we can answer this question, we need to address two other related questions. First, what is critical thinking? Second, in the context of higher education, why should students learn a foreign language? According to the Foundation of Critical Thinking, critical thinking is "the intellectually disciplined process of actively and skillfully conceptualizing, applying, analyzing, synthesizing, and/or evaluating information gathered from, or generated by, observation, experience, reflection, reasoning, or communication, as a guide to belief and action" (Scriven and Paul 1987). Although philosophers, psychologists, and educators look at it from different angles, all theorists agree that skilled critical thinkers possess a healthy skepticism toward the information presented to them. As Oliver and Utermohlen (1995) point out, critical thinking skills are particularly crucial in the present moment because the amount of information available today is massive, due to the rapid development of technology. Students need to learn how to weed through the information and not just passively accept it as true. Whether they are studying in school or working in the workplace, the ability to avoid hasty conclusions and make well-justified decisions is essential for their success. Because of the importance of critical thinking, American higher education institutions have witnessed a wave of critical thinking movements since the 1970s. Regardless of discipline, all educators should consider the cultivation of critical thinking skills as one of their prime teaching objectives (Bean 2011, 19–21). Foreign language learning should be no exception.

This brings us to the second question: Why should a college student learn a foreign language in the first place? To learn another language is to learn how to think and behave in that culture. Learning a foreign language allows us to recognize the cultural basis for our own behaviors and ideologies, be willing to look past our own views of the world, and approach things from different perspectives. Because they have mastered the foreign language, students are able to absorb first-hand information without the original meanings or implications getting lost in translation; they are exposed to different perspectives in approaching a single issue and different interpretations of a set of facts. In a certain sense, learning a foreign language has unparalleled advantages over other disciplines in developing one's critical thinking skills. All foreign language teachers know many of their students will not be able to use the foreign language they have learned in their careers after they graduate from college. But the critical thinking skills that students develop in a foreign language class will benefit them for the rest of their lives. Therefore, to think that the goal of foreign language teaching is simply to teach pronunciation, vocabulary, and grammar or to deliver lessons on the achievement culture of the country, but not help students become critical thinkers would indeed be to miss the forest for the trees.

Much research has shown that writing academic essays, especially on the topics that were already discussed in class, greatly enhances students' critical thinking skills (Wade 1995; Condon and Kelly-Riley 2004; Benesch 2001; Chaffee 2014; Mehta and Al-Mahrooqi 2015). In order to produce academic essays, students read extensively, analyze and compare other people's opinions, decide on their own positions and opinions, and provide evidence and arguments. This very process is itself the best exercise for critical thinking.

Targeted Students and Preparation for Chinese Academic Writing Instruction

Before engaging students in academic writing activities, instructors should have an intimate understanding of their target audience in order to be able to teach students effectively at their aptitude level. It should be pointed out that the challenge of Chinese academic writing requires that the target students be advanced-level learners, namely, those who are at or above the advanced-mid-level on the ACTFL proficiency scale, or level 2 or above on the Interagency Language Roundtable (ILR) scale. These could be students who started as absolute beginners in high school or college but have had language-immersion experiences in a native Chinese environment or they may be advanced heritage students. Most Chinese learners still have to master the fundamentals of writing (e.g., writing for practical purposes, narrative essays, short argumentative essays). If they are given writing tasks that are overly demanding, it may have the converse effect of hindering their learning.

Schultz (1991a, b) pointed out that some foreign language students could write beautifully at the intermediate level, but the quality of their essays became disappointing when they moved up to the advanced literature class. Schultz believed that this was because at the intermediate level, writing tasks consisted of mainly narrative or descriptive style writing, which required relatively simple, linear thinking. However, argumentative writing, especially lengthy argumentative essays typically assigned in advanced-level courses, requires a more complex and abstract higher order thinking. This finding was also echoed in Qi and Liao (2019) and Shi and Chen's research (2019), which found that argumentative essays are more linguistically difficult and complex in terms of vocabulary and syntax than narrative essays. Therefore, when teaching the fundamentals of writing, the traditional approach, which starts with narrative writing before gradually moving to argumentative essays, is in line with the order in which CFL learners acquire writing skills.

The most common writing tasks related to argumentative writing that students are first exposed to require them to express their own opinions on a topic related to course reading materials. In the United States, this type of writing is typically assigned in a third-year Chinese college course. Unlike academic writing, these argumentative essays are shorter and do not include much of a "research" component. Students need to clearly represent their own point of view, but do not need to extensively address alternative viewpoints, nor do they need to recount, quote, summarize, or offer commentary. At this stage in the learning process, the key learning points are the basic textual organization of argumentative writing, argumentation methods, and the common vocabulary and grammatical structures characteristic of the argumentative texts, all of which lay a solid foundation for academic writing.

Compared to previous writing instruction, academic writing is considerably more challenging in terms of depth of content, complexity of higher order thinking, richness of language, and essay length. How should an advanced Chinese course incorporate academic writing components? The next section will offer an in-depth analysis of four pedagogical aspects of such an addition: overall curricular planning, design of writing tasks, supporting pedagogical activities, and assessment.

Instructional Design of Academic Writing Pedagogy in Advanced Chinese Courses

Overall Curricular Planning and Designing Scaffolded Writing Assignments

Due to the slow development of academic writing skills, instructors should assign shorter essays with easier content at the beginning of the semester, before gradually transitioning to broader and more complicated topics. Although assignments for different topics will be different in terms of content, they should remain as similar as possible in terms of style and essay organization. If it is an assignment with greater length and more complex content or research methodology, such as an independent research report (独立研究报告), the instructor should divide the writing task into several steps for students to complete over a period of time, making the task more manageable. Academic writing is most characterized by its critical writing and research aspects. Therefore, it is crucial to combine academic writing with other activities, such as interviews, extensive reading, online research, discussions, and oral presentations, so as to let students go beyond the textbook. Doing so also enables an organic integration and synthesis of writing skills with other language skills.

Each Chinese program designs advanced Chinese courses that are suited to and compatible with its unique institutional environment and the background of its students. In addition to the regular general Chinese courses, from first-year Chinese to fourth-year Chinese, there are usually also some thematic courses at the advanced level. Different types of courses have different ways of incorporating Chinese academic writing instruction. Many universities, for example, offer an advanced course called "Media Chinese," similar to courses from two decades ago, known as "Newspaper Reading," except the course materials are no longer limited to newspaper articles and include various videos or articles from traditional print media and new digital media. The content of such a course usually reflects the latest changes and trending topics in Chinese society, economy, popular culture, and values. A typical academic writing assignment for such a course would be an independent research report requiring the use of relevant online research and extensive reading, typically due at the end of the semester. For this assignment, students would need to choose a topic based on their interests and areas of academic specialization. They should conduct online research to find multiple relevant reports, commentaries, and even academic papers on the topic, which need to be of sufficient depth and length. After reading these various sources, students should analyze, summarize, and evaluate these viewpoints in a 2,000- to 3,000-character reading report. The instructor can adjust the number and length of articles to be read according to the students' proficiency levels. This kind of academic writing task is a big leap from the previous sorts of writing exercises that students have been exposed to in terms of length, content, and research requirement. Therefore, before assigning such writing tasks, the instructor may design some scaffolding exercises, such as interview reports, reading reflections, and response papers, so as to provide guidance along the way and to help them approach the final assignment. Although the interview reports, reading reports, and response papers are different types of writing assignments, they nonetheless share the same basic stylistic characteristics as the argumentative essay. They also require the same logical thinking skills needed to analyze, elaborate on, and refute alternative viewpoints as well as clarify one's own position. In addition, these three different types of reports employ basically the same organizational structure, which is that of "introduce (引), discuss (议), link (联), conclude (结)" (Li 2007, 35). To introduce is to describe the perspective of the interviewee or the article; to discuss is to take certain views as the starting point in order to elaborate on, counter, debate, and expound on one's own point of view; to link is to tie in relevant personal experiences or similar current affairs in the author's own country as additional or counterexamples and further develop and support one's own views; to conclude is to offer a closing statement, bringing together and summarizing all prior points. This organizational framework is similar to that of the traditional Chinese argumentative text, which uses the "introduction (起), elucidation (承), transition (转), summary (合)" framework. Both frameworks reflect the general conventions of Chinese argumentative writing. Thus, in writing their interview reports or reading responses, students are being trained in the textual structures and organizational thinking required in Chinese academic writing, which prepares them to approach the final independent research report.

If a student is writing a lengthy paper for the first time, the instructor should consider dividing the writing task into several smaller subtasks. Taking the independent research report discussed earlier as an example, the instructor should lay out the writing task at the beginning of the semester, so that students can start to think about their research topic. After 2 or 3 weeks, students would be required to submit a preliminary research proposal, explaining which topic or question they would like to explore, the significance of their research, the existing critical conversation on the topic, and possibly an outline of the essay. This is an important step that will help the instructor get an idea about where students are with language and content and to evaluate how well they understand their chosen topic. In addition, the instructor can also point the students to supplementary sources and give feedback on whether the topic is feasible or appropriate for the course. Experienced teachers all know that one of the most common mistakes students make is choosing a topic or a problem that is too broad for a term paper, such as "the current state of NGOs in China," or "the phenomenon of 'leftover women'." The narrower and more specific the focus of a term paper, the easier it will be to manage. After this step, students will have to find one or more related articles every 2–3 weeks or so, and write a reading report on the chosen materials. The content of this reading report can then be used as part of the final independent report. This way, by the end of the semester, they would have already completed two or three reading reports. They would only need to read one more article, combine their own analysis with the previously written reports, then complete the final independent research report. By breaking down and parceling out the overall assignment over the course of the semester, the instructor makes the assignment becomes more manageable for students.

Connecting Critical Writing with Critical Reading

Critical writing depends on critical reading because the academic essays that students produce inevitably involve their reflections on chosen readings. Yao (2005) argued that an effective method for developing critical thinking skills is to design a critical reading and writing course with argumentative and expository essays as primary course materials. Yet, good reading materials alone are not enough. At the advanced level, reading materials should not only be treated as a vehicle for teaching new vocabulary or grammar, nor should the instructor aim to merely ask students to comprehend the author's viewpoints. A critical thinker does not simply passively accept anything they read. As all authors have individual motives and agendas in their writing, it is crucial for the readers to find out what the intention behind a particular author's writing is and what effect they are hoping to achieve. Only by answering these questions can students truly understand and evaluate the text and put forward their own positions and views. Research by Mehta and Al-Mahrooqi (2015) shows that students were able to write more argumentatively with a clear thesis statement that focused on the topic when they were guided in identifying authorial intentions and tones in the reading materials provided for the writing prompts in addition to having had sufficient discussions about the focus of the writing task.

So how does one read critically? The fundamental principle of critical reading is not just to read what the article says, but more importantly to evaluate why and how the argument is being made. In tandem with the core tenets of foreign language teaching, the instructor can consider engaging students with the following five sets of questions that are modified based on Knott's five steps of critical reading (Knott 2005):

1. Determine the *central claims* or *purpose* of the text. How are the central claims developed or argued? Is the writer's purpose to inform, report, deliver an opinion, or criticize? How does the author achieve their purpose through their choice of content and language? What is the writer's tone (e.g., critical, sarcastic, objective, approving)?
2. Make some deductions about *context*. Who is the target audience? In what historical context was the text written? Who is the author and what is his/her background?
3. Distinguish the kinds of *logic* the text employs. Is it deductive or inductive? How is the text organized? What argumentation methods are used in the text? How has the author analyzed the material?
4. Examine the *evidence* the text employs. What counts as evidence used to support the argument? Is the evidence statistical, literary, or historical? What are the sources of the evidence used?
5. What is your *evaluation* of the arguments in the text? What are the strengths and weaknesses? Could the arguments be better or differently supported? Are there gaps, leaps, or inconsistencies in the arguments? Is the method of analysis problematic? Could the evidence be interpreted differently? Are the conclusions warranted by the evidence presented? What might a counterargument be? Can you detect any bias in the author's position or language?

Only when these questions are addressed can students form their own arguments with an open mind and write critically. This critical reading practice also provides a unique opportunity for students to develop a deeper understanding of the vibrancy of the emotional palette and cultural connotations conveyed by advanced vocabulary and formal structures, which are important for deciphering the author's positions.

Designing Effective Academic Writing Tasks

Traditional writing tasks do not include any specific writing prompts. They tend to be simple, involving straightforward titles, such as "My Hometown," "Generation Gap," or "The Relationship between Tradition and Modernization." Students often find it difficult to begin writing because they not only have to consider which topic to write on, but also how to write it, given their limited language ability. This is especially true for more abstract topics. Therefore, designing an effective writing prompt is particularly important when it comes to teaching foreign language writing. Many scholars have pointed out that if writing prompts are able to offer some tips and recommendations on the content, writing process, and relevant vocabulary for the paper, the difficulty of writing the paper will be appropriately reduced (Reid and Kroll 1995; Scott 1996; Way et al. 2000; Yuan 2010). Way, Joiner, and Seaman (2000) compared the effects of three writing prompts – the bare prompt, the vocabulary prompt, and the prose model prompt – on the performance of beginner- and intermediate-level French students. The bare prompt is a simple explanation of the writing task offering only the bare minimum, sometimes including only the title of the essay; the vocabulary prompt contains a list of key vocabulary words that students are likely to use in their essays; the prose model prompt provides a reading passage that not only contains key vocabulary and grammar structures, but also is meant to invoke a setting, emotions, or plot ideas. The results showed that when the students were given a prose model prompt, the overall score, essay length, sentence length, and language accuracy of their papers were better than the papers they wrote when given the other two types of prompts. Yuan (2010) conducted a similar study on writing prompts, but on intermediate- and advanced-level Chinese language students. The study found that if suitable grammar and vocabulary suggestions are provided with the writing prompt, the complexity of the students' language increased, but the fluency and accuracy of the papers were negatively affected. In contrast, if instructors provided guidance on the content of the paper prior to writing but did not limit or regulate the vocabulary or sentence structures used, the fluency and accuracy of the students' papers improved, but the complexity of the language used decreased. Both studies show that writing prompts have a direct impact on the writing produced by students.

Academic writing is challenging on all fronts. Which aspects should writing prompts focus on, then? Walker (1989) pointed out that at lower levels, the instructor should pay attention to students' grasp of specific linguistic knowledge. As their proficiency increases, the emphasis of teaching should shift from specific linguistic knowledge to the students' overall ability to use the language. The difference between academic writing and general nonacademic writing is that the primary goal of general, nonacademic writing is to master a language through writing practice, such that linguistic competence is the most important concern for the writer. However, the primary goal of academic writing is to present an author's informed thesis or argument, which is based on thorough and thoughtful research. Thus, writing prompts for academic writing should aim to pave the way for the generation of ideas in terms of content, overall structure of the paper, and research methods, instead of focusing solely on the use of certain specific sentence structures and vocabulary. The latter will not only restrict students' thinking, but it is also unrealistic because the instructor cannot accurately anticipate what students will write, making it impossible to decide which structures or vocabulary they should use.

Based on the principles of designing effective ESL academic writing tasks in Reid and Kroll study (1995), this chapter proposes the five categories of guidelines below to help instructors shape writing tasks for their courses.

Purpose	The purpose of the writing assignment is clear and adequately explained, reducing any ambiguousness as to why students are being asked to write. The writing assignment aligns with the teaching objectives of the course
Tasks	The topics should be organically combined with the course content and/or the individual research interests of students. They should be targeted and specific and not overly broad. Topics should also have a certain element of research involved. Topics should be engaging and inclusive; they should not make any student feel excluded
Prompt	The writing prompt should offer some form of clues, tips, or other guidance in terms of content and research methods in order to help students better grasp these facets of academic writing
Language	The language of the writing assignment should exemplify and reflect the characteristics of academic writing: It should be formal, clear, specific, and unambiguous. At the same time, the use of too many uncommon words should be avoided
Criteria	Evaluation criteria should be comprehensive: Students should pay equal attention to the quality of their written language, the depth of content, the use of critical thinking, essay organization, etc. The grading rubric should reflect these requirements. It is best for instructors to make the marking criteria and grading rubric available and clear to students before they begin to write, demystifying how their writing will be evaluated

Appendix I is a writing assignment for an interview report. This type of writing task often requires students to interview native speakers about the events, topics, or controversial viewpoints addressed in their course readings. In the report, students need to not only paraphrase, evaluate, and comment on the interviewee's position, but they also need to integrate their own opinions. This practice makes it an effective scaffolding exercise for academic writing. However, if students have no previous experience in conducting an in-depth interview in Chinese, their questions may lack depth or appear insufficiently challenging, improperly worded, even off-topic or overly subjective. It is best to provide a few sample interview questions in the writing prompts for students to reference. Many instructors must have encountered situations in which students started the interview report by launching immediately into an engagement with the interviewee's comments without first providing any context or introduction, leaving the readers completely disoriented: Who is the interviewee? Why is he or she being interviewed? Under what circumstances did he bring up such a view? To help students avoid such problems, writing prompts can also include certain content requirements for the interview report. For example, students should (1) introduce and offer background information on the interviewees in the report, (2) objectively and briefly introduce the background of the interview (e.g., why the interview is being conducted), (3) describe and outline the interviewees' opinions on the event or topic, (4) comment on and evaluate the interviewees' opinions based on students' personal experiences and opinions, and (5) summarize the entire essay and offer possible solutions for the problem or offer projections of potential implications and consequences of the situation. The sequence of these five requirements in fact reflects and suggests what the structure of the paper should be.

Appendix II is a writing assignment for a final independent research report. As it is significantly more challenging than the writing assignment in Appendix I, it is divided into several steps to be completed throughout the semester. The first step is to encourage students to think carefully about what topic they would like to choose. When deciding on the topic, in addition to being guided by their own interests, students should also consider the current understanding of the chosen issue and the significance of conducting research on the issue. Once they have decided on a topic, students need to think about what the most coherent and effective way of organizing their main argument (such as an outline) would be. This step is crucial for the successful completion of the project. In order to combine writing with reading and oral presentation, the writing prompt contains detailed instructions and deadlines for each step. This helps prevent students from procrastinating and then haphazardly trying to complete the assignment all at once. When it comes to longer writing tasks like this one, providing detailed and clear instructions for each step ensures that students will treat the process seriously instead of simply focusing on the end product. When students understand that the reading report they produced early on in the course will ultimately make up a portion of the final report, they will be more serious about each step of the process.

Supporting Activities for CHINESE Academic Writing Instruction

Writing competence is not the same as linguistic competence. Besides language use, other aspects of writing need to be addressed in the curriculum in order to help students write a high-quality academic paper. These include an understanding of appropriate academic writing styles and formatting rules, research skills (such as how to search for Chinese primary sources), and the ability to think independently and critically. However, practically speaking, without specialized writing courses, teachers only have very limited time in the classroom to spend on writing-specific instruction. Therefore, the acquisition and development of some of these skills need to be facilitated through the expansion of other activities in class. Following are four kinds of supporting activities to facilitate academic writing instruction: model essay guidance, individual tutorials, workshop on stylistic and formatting conventions, and workshop on online research methodology.

Model Essay Guidance

It is a common and effective practice that instructors teach vocabulary, grammar, and even speaking skills by demonstrating good examples for students to emulate. Writing is usually considered a creative act, so some instructors may feel hesitant to provide students with writing models. While there is definitely a certain amount of merit to this view, academic writing is *not* creative writing. It is only through the vigorous reading of many model essays that one is able to get a real feel for and understanding of the elements of good academic writing, such as overall organization, textual cohesion, and various argumentation methods.

When choosing model essays for Chinese academic writing, the instructor should keep several things in mind. First, the language used in the model essay should be easier than the primary reading materials of the course and of shorter length. The regular articles studied in class are used for intensive reading. One of the main purposes of doing so is to teach new language knowledge. These articles tend to include a certain amount of new grammatical structures and vocabulary. The purpose of reading model essays is to learn Chinese writing conventions and rhetoric; therefore, if the language used in the model essay is too difficult, students will spend a lot of time simply deciphering the meaning of the text instead of concentrating on the writing techniques demonstrated in the model essay. Model essays for academic writing often come from authentic materials, so naturally they might include a good amount of jargon or obscure vocabulary. In these cases, instructors can make some edits to simplify the language at their discretion, as long as the meaning of the original text remains unchanged. Second, the model essays should highlight the features of academic writing which include essay organization, common vocabulary and formulaic expressions used in academic writing, common argumentative methods, and academic writing conventions. Based on the learning needs of students, a model essay can be chosen if it allows students to focus on one or more of the above aspects of writing. The instructor can use different symbols or colors to highlight the key points of instruction in the model essay. It would be ideal if the content of the model essay is related to what the students are expected to write about. By reading the model essay, students could then acquire more background information on the topic and brainstorm ideas based on reading the opinions of others – all while acquiring relevant vocabulary.

If time permits, it would be ideal if the instructor could go over the model essays with students in class. More ideally, students would read the model essay before class to grasp the overall meaning of the essay. The instructor need not spend too much time helping students decipher the language in class; instead, they should draw students' attention toward other aspects of the text. It is also a good idea to ask students to discuss which aspects of the essay are valuable and worth modeling before the instructor explains and expounds on the essays.

Evaluation, analysis, and synthesis are key elements in academic writing (Liu 2014). Even if students have grasped these concepts in their native language, demonstrating them in Chinese is still a challenge. As a form of argumentative writing, academic writing cannot be separated from argumentation. There are many Chinese argumentation strategies, such as exemplification (例证法), quotation (引证法), compare and contrast (对比法), use of metaphor (喻证法), proof by contradiction (反证法), and the use of common sense and logic (道理论证法). Some of these argumentation strategies are presented somewhat differently than in English, whereas others – such as using quotes from famous figures as evidence – are not common practice in English argumentation (Liu 2015). Instructors should be sensitive to and perceptive of these writing techniques. Aside from complete essays, instructors may also excerpt paragraphs or sections from the students' course reading materials that model aspects of writing that students should understand and emulate. If instructors chance upon a section of the text worth modeling in students' regular course readings, they should highlight it in class and call students' attention to it.

Individual Tutorials

As discussed earlier, when students move from acquiring foundational writing skills to learning academic writing skills, the quality of content and the logic of expression become more and more important. Yet good ideas or arguments do not simply appear out of thin air. The author must undertake the process of "reading, thinking, and reflecting." Even so, it is probably only after doing so several times that they can come to a thoughtful conclusion. As American writing pedagogical expert Stephen North puts it, "nearly everyone who writes likes—and needs—to talk about his or her writing, preferably to someone who will really listen, who knows how to listen, and knows how to talk about writing too. Maybe in a perfect world, all writers would have their own ready auditor—a teacher, a classmate, a roommate, an editor—who would not only listen but draw them out, ask them questions they would not think to ask themselves" (North 1984, 439–440). Here, North perfectly encapsulates the importance of discussing and exchanging ideas with others to the writing process. Many scholars dedicated to writing pedagogy have since echoed him and argued that the more students communicate with teachers and their peers during the process of writing, the higher the quality of their writing will be (Sperling 1991; Williams 2002; Mehta and Al-Mahrooqi 2015; Chen 2016a).

Individual tutorials are an effective means for students to exchange ideas with and receive feedback from their instructor. They can be arranged after students conduct their preliminary reading and research but before they begin to write, or after they finish their first drafts. The purpose of individual tutoring is to help students organize their ideas, outline their arguments, and solve language problems. This will not only facilitate the writing process, but also combine speaking practice with writing practice. During the tutorials, the instructor's task is not to interfere with students' arguments by expressing support or disapproval. The instructor should instead adopt a neutral position while they listen to students as they freely express their thoughts about what they are going to write and how they are going to write it. Academic writing is fundamentally based on reading. During the tutorials, the instructor can first get students to summarize the main ideas of the essays in their preliminary readings before getting them to comment on the ideas or arguments in the preliminary readings. If they find that students have misread or misunderstood the articles, or their representation of the articles is not clear, instructors can correct students by asking them guiding questions. When reading, students may be particularly affected by certain details, yet they may still fail to grasp the article as a whole. This may cause them to be overly invested in these details and their representation of the article may then miss its larger governing logic. If this happens, the instructor should encourage students to finish presenting the materials they prepared, then, by prompting them, check if they have grasped the main claims of the text. The instructor can also debate students by playing devil's advocate and challenging their views in order to guide them to reflect upon any anomalies or inadequacies in their views and arguments. If time permits, at the end of the tutorial, the instructor can model another way of representing what the student intends to convey. Although people express themselves differently and have differing ways of thinking, and while indeed teachers do not necessarily have the perfect version of presenting the issue at hand, the purpose of doing so is to provide students with an example of how an educated native speaker would convey their thoughts on the topic.

Another issue to note during tutorial sessions is the fact that, in the context of cross-cultural communication, instructors and students may have different understandings of their individual roles and expectations. Chen (2016b) finds that in postacademic writing tutorial sessions, when encountering a problem, Chinese instructors tend to give implicit criticism or corrections to "save their students' face." For example, they often use words or phrases, such as "possibly 可能" or "I feel 我感觉," that have an implicitly subjective and equivocal tone to point out students' mistakes. However, students may not pick up on the instructor's actual intentions because from their own experience, Chinese teachers are always direct and strict when correcting errors in class. The relationship between teacher and student varies from person to person. On the one hand, the instructor should engage students on equal footing and demonstrate how Chinese people give "gentle critique," which is a useful skill for students to learn; on the other hand, they should make sure that when giving feedback, students truly understand what they mean and avoid "ambiguous" comments.

It is inevitable that students will make mistakes in language use during the tutorial session, especially in terms of vocabulary. Due to time constraints, it is impossible for the instructor to correct all language errors. More thorough corrections will have to be made when students turn in their drafts. If the instructor does not want to let error corrections interrupt the student's train of thought, one solution would be to note the major language errors on a piece of paper and hand the written notes to the student at the end of the tutorial, so that they can work on these errors in their drafts on their own time.

The key to the success of the individual tutorials is sufficient preparation on the part of the student prior to the meeting. Not only should they have read the articles thoroughly, they should also have prepared their summary of and reflection on the articles earnestly and thoughtfully. As Christensen (2013, 57) points out, individual tutorials (or individual instruction, the term used at his program at Brigham Young University) are "ideally suited to advanced-level learners who are typically highly motivated and mature enough to do a good deal of their study and preparation independently." He argued that this kind of intensive learning setting is "essential for continued advanced-level domain training in a purely Chinese academic or work environment" (2013, 58).

Instruction of Stylistic and Formatting Requirements

Academic writing is characterized by its formal style, so the development of academic writing skills cannot be separated from *shumianyu* instruction. Shi (2002) analyzed the samples of more than 100,000 characters of essays written by 89 fourth-year foreign students majoring in Chinese at undergraduate degree programs in China. These corpus samples came from the notes written by students on their in-class discussions on social issues as well as their at-home reading reflections and written reports. Her study found that although these foreign students were fluent in expressing themselves both verbally and in writing, they all demonstrated a tendency to avoid using formal vocabulary, especially idioms, new vocabulary, and function words. Advanced students do not necessarily *intentionally* avoid using formal vocabulary, but they may simply be unable to recall the words when they write. Thus, it is necessary for the instructor to regularly summarize and review the *shumianyu* that students have learned in order to establish the connections between the *shumianyu* vocabulary and their themes or their discourse functions.

There are specific requirements for the organizational structure and format of academic writing. Essay organization varies, depending on the writing task (i.e., a survey report is different from a response paper). There are also writing conventions, such as transitions and other formulaic expressions, conjunctions and linking words, transitional sentences, transitional paragraphs, serial numbers, or equivalent formulaic expressions, which ensure that the article is clearly and rigorously organized and ideas are logically and lucidly conveyed. In terms of format, proper citation and bibliographic referencing are very particular for academic writing. The overall organizational structures of different types of academic writing and various means for smooth transitions need to be taught with examples and models explicitly in class. For other writing conventions, such as citations or references, there are quite a few reference books and textbooks available on the market, such as *Thesis Writing Course for International Students* 留学生毕业论文写作教程 (Li and Deng 2012) or *Writing Essays in Chinese* 外国人汉语过程写作 (Yang 2006), which the instructor can refer students to directly so as not to spend too much classroom time on explaining citations and formatting.

Workshop on Online Research Methodology

Many Chinese students have been typing Chinese characters on computers, listening to Chinese podcasts, and reading Chinese news through translation software since they were young. They are tech savvy and virtually do not need any help from their teachers in this respect. However, although students are proficient in various technological tools, they do not necessarily know how to find relevant information on the Internet in Chinese. In today's modern society, conducting research is inextricable from searching for accurate and reliable information on the Internet. It is necessary for teachers to introduce this important skill to students. Workshops are a great way to introduce students to these online resources and online research techniques. Today's students belong to the "Internet generation" and are adept at picking up such skills. They only require a brief demonstration by the teacher in order to master these skills, so not a lot of class time need be allotted to this.

For academic writing, students need to use keywords to search for background information, relevant news reports and commentaries, and even scholarly papers online. Therefore, they need to know which credible Chinese news networks and video websites are available and how to utilize online tools, such as online dictionaries and character conversion tools. When students first begin their search with Chinese keywords, they may be overwhelmed by the staggering number of search results and become unsure of where to start. Here, instructors can help students narrow down the scope of their search. For example, before studying a new text on how the Internet has affected people's social lives, the teacher can select a few related keywords, such as "人肉搜索," "微公益," and "网红经济," and ask the students to search for these terms on some recommended Chinese encyclopedia websites, such as *baidu baike,* to gather information and explain these terms in class.

Zhihu (知乎), a Chinese online social question and answer platform similar to Quora in the United States, is another important resource students are likely to use when conducting online research. When first beginning to write academic essays in Chinese, American students understandably tend to rely on Western websites, such as the Chinese version of the New York Times, for their primary research on the topic. However, Western media inevitably uses Western perspectives or ideologies to report on and analyze issues related to China. In order to think cross-culturally, it is necessary to understand Chinese people's perspectives and opinions on the matter. Whether it is asking Chinese netizens for information, their opinions, or obtaining Chinese netizens' feedback on their (the American students') viewpoints, *Zhihu* can be very helpful in crafting their preliminary ideas as they begin to write. *Zhihu* can also be overwhelming for new users; it is vital for instructors to spend some dedicated time teaching students how to find posts, how to post questions, and how to exchange ideas with Chinese netizens. Zeng (2018) discusses in detail how to integrate this online Q & A community into an advanced Chinese language curriculum. In her pilot study of three types of class activities using *Zhihu*, she argued that participating in a "group 圈子," also referred to as a "community of practice" in the situated learning theory, is an effective channel for students to achieve autonomous learning. Students who participated in her study confirmed in their course feedback that learning how to use the *Zhihu* platform to obtain information, seek help, and establish relationships with native speakers was a crucial skill to acquire in order to conduct China-related research.

Assessment and Feedback

Fair evaluation of students' work is important because it is not only a reflection of the quality of students' present work, but also provides guidance and encouragement for future writing assignments. Evaluation criteria can be presented to students either holistically or analytically. Holistic grading gives students one score that reflects the instructor's overall impression, involving all categories of criteria at once, whereas analytic grading provides separate evaluations for each criterion. While some instructors object to analytic grading because they argue that writing cannot be separated into individual elements and then analyzed, others prefer holistic grading simply because it is easier and faster (Bean 2011, 270). However, Haswell (1991) noted that holistic grading tends to flatten a student's performance – to focus on weaknesses and conversely undervalue their accomplishments. Condon and Kelly-Riley (2004, 68) further argued that "a holistic score provides the basis for a rough ranking and nothing more," because one single grade conceals a considerable variety in a writer's strengths and weaknesses. Unlike lower-level writing assignments for which the criteria primarily focus on language use, many nonlanguage factors contribute to the quality of academic writing. Thus, analytical grading is more suited to evaluate students' strengths and weaknesses in each category. It also allows the instructor to weight some criteria more heavily than others, depending on what the instructor wishes the students to focus on.

One important criterion for academic writing that deserves special attention is critical thinking, as it is a key element in academic writing. Even though critical thinking is sometimes added to the grading rubrics, it is often treated as "a measurement of whether the logic is clear rather than whether the thinking is complicated, open-minded, or contextually appropriate" (McLaughlin and Moore 2012, 146). In their widely cited study that explored the relationship between academic writing and critical thinking in which 123 students participated, Condon and Kelly-Riley discovered a surprising inverse relationship between the scoring of students' work in their Writing Assessment Program (the entry-level Writing Placement Exam) as well as the junior-level timed writing portion of the Writing Portfolio and the evaluation of the same work using their school's *Guide to Rating Critical Thinking* rubric, even though the grading criteria for writing included elements of critical thinking, such as focus and support. The study found that the better the writing score, the lower the critical thinking score. The more problematic was the language use perceived by the writing instructors, the higher the critical thinking score. One factor that might have contributed to the lack of automatic connection between students' overall writing scores and their critical thinking scores might be that the writing instructors weighted aspects of language use more heavily than the use of critical thinking skills. McLaughlin and Moore (2012) found that many English instructors weighted their grading to focus more on the voice or the style presented in students' writings than on the demonstration of higher order thinking skills, such as focus and logic. One lesson derived from both studies is that although writing is believed to act as a vehicle for critical thinking, writing itself is not necessarily an act of critical thinking – if instructors do not explicitly call students' attention to doing so and train them to do so, students will not make deliberate efforts to think or write critically. Ultimately, if instructors truly value critical thinking, they should make a conscious effort to increase the weight of critical thinking skills in grading rubrics and alert students to this change.

Critical thinking is a very abstract higher order thinking ability. It is not easy to objectively and effectively evaluate its presence, depth, quality, and extent in a paper. Instructors may consider adopting the *Guide to Rating Critical Thinking* (Condon and Kelly-Riley 2004), the rubric developed by Washington State University's Critical Thinking Project. The guide identifies seven key elements of critical thinking: issues, personal perspective, alternative perspectives, assumptions, data or evidence, context of the issue, and conclusions or implications. The instructors can choose to adopt some or all of the elements to fit their disciplines or the needs of particular writing assignments.

Besides critical thinking, there are other aspects of writing that should also be evaluated. Below is a list of criteria that the instructor can consider when evaluating students' academic writing. It should be pointed out that the evaluation of critical thinking is placed in the content/ideas category because the quality of content and ideas is a reflection of the author's critical thinking ability. Teachers can also evaluate the two separately.

1. Language use: including accuracy, diversity, and complexity of both grammar and vocabulary
2. Content and ideas: whether the essay indicates synthesis of different views on the topic, in-depth analysis, and demonstrates original perspectives on the topic; whether the essay substantially develops ideas through the use of specific examples and rigorous analysis
3. Organization and cohesion: whether the essay includes a strong introduction, body paragraphs, and conclusion, with clear transitions between sentences and paragraphs; whether the structure of the essay adheres to Chinese readers' expectations
4. Research: whether the author selects appropriate and sufficient resource materials and correctly integrates information from external sources
5. Mechanics: including accuracy of written Chinese characters, use of punctuation, the division of headings, and the citation format of the bibliography

In addition to grading, the instructor has to comment on students' writing. When and how much feedback teachers should provide on form as well as content is something every frontier instructor needs to consider. Leki (1991) found that ESL college students tend to equate good writing in English with error-free writing and argued that students need to adjust their expectations about what constitutes improvement in writing in order to fully benefit from their teachers' feedback. It is crucial for the instructors to point out the weaknesses as well as the strengths of a student's essay and convey their own reflections and responses to the essay. The benefit of doing so is that students will feel that teachers are not merely an "error police," but also enthusiastic readers who wish to have intellectual exchanges with them. This very sense of a genuine author-reader relationship motivates students to write.

In actual fact, many instructors do provide feedback to students that covers such areas as content and organization, but their comments are often too simple or too general. As a result, such comments have less effect on students' rewrites compared to text- and location-specific feedback on the form of the essay (Fathman and Whalley 1990). Cohen (1987) studied ESL instructors' comments on students' academic writing assignments. Typical comments included short phrases written in the margins next to the paragraphs in question, such as "this is unclear," "transition is lacking," or "please explain further." But students reported that such comments were too vague or too general, resulting in further confusion: What does the teacher mean by "not clear"? How should I transition from one paragraph to another? Most of the time, students did not know how to revise and improve their essays, even when they knew what the problem was. CFL students face similar problems (Xu and Wu 2016). A better approach would be for the instructor to be as specific as possible in their comments and provide clear suggestions for revision. For example, if the meaning of a paragraph is not clear, besides pointing out the problem, the teacher should also let students know where they need to provide more explanation. Writing detailed comments is indeed time consuming. The instructor may choose to discuss students' writing assignments face-to-face if that is more efficient. The instructor should require a revised version of the paper to be submitted (at least for major writing assignments) so as to make sure that students understood the instructor's corrections and comments and incorporated them into their written work.

Time permitting, peer review should also be incorporated for academic writing assignments. Online course management platforms such as Canvas have made it easier for students to engage in the practice of peer review. Encouraging students to provide constructive critique to their classmates not only contributes to a collaborative learning environment and helps students learn to write for a wider audience beyond their instructors, but more importantly, it gives students another opportunity to think about the texts more critically, which will have the added effect of helping them to identify the strengths and weaknesses in their own writing.

Conclusion

For many years, language teaching professionals in higher education have faced the dilemma of the distinction between and division of content courses and language courses in their area studies departments. This compartmentalization is a result of perceiving content courses as offering "the real intellectual challenge" and language courses as merely "practical and technical" (Fandrych 2010). This division, in turn, has deepened the academy's biases in its perception of language learning. Recognizing the possibilities and importance of Chinese academic writing instruction in advanced-level Chinese courses, particularly its role in developing students' critical thinking abilities, will illuminate new paths for the expansion and growth of the discipline. Meanwhile, it will also allow language instructors and researchers to reflect on the importance of language learning to a liberal education.

This chapter explores both the theoretical and practical aspects of a series of issues related to Chinese academic writing instruction. How instructors choose to integrate Chinese academic writing into different types of courses will vary. The Chinese proficiency levels of students will also influence the instructor's choice of teaching methods. However, no matter what teaching methods instructors adopt or what teaching activities they design, one thing should always be crystal clear: Neither teachers nor students should expect quick results when it comes to producing good academic essays. They should not be impatient for success, but should focus on the process instead – the writing process is far more important than the final product.

Some liken academic writing to discussing a film with friends. If you want to have a vibrant, lively, and engaging conversation, you have to first watch the film carefully, then listen to the opinions of others, read film reviews, then add your own opinions to the mix, and finally listen to what others have to say about your thoughts. You may end up arriving at a new or deeper understanding of the film. It may be important to determine who has the correct or more novel point of view, but it is the process of banter and dialogue, debate and discussion which brings such pleasure. The same is true for academic writing. The final product of a long period of writing, the tangible fruit of students' labor, is certainly important, but it is the process of reading, thinking, discussing, drafting, revising, and rewriting, that will ultimately hone their ability to use Chinese to express themselves, their understanding of Chinese society and culture, and even their ability to think critically.

Appendix

Appendix I: Writing Assignment for an Interview Report for Chapter 6 in the Textbook *China in Depth: An Integrated Course in Advanced Chinese*

孙云晓在《寻找上海男孩》这篇文章中谈到当代中国城市里存在的有关男生教育的问题,并提出了"拯救男孩"的呼吁。请你采访一名身边的中国留学生,看看他对这个问题有什么看法。

以下是一些拟用的采访问题,供参考。

1. 你上中学的时候,老师有没有公平对待男生和女生?请举一个具体的例子。
2. 在你班上,男生受宠还是女生受宠?你认为是什么原因让他们受宠?
3. 现在很多家长、教育学家和社会学家认为中国的教育制度不利于男孩成长,孙云晓为此提出了"拯救男孩"的口号。你认为实际情况是这样吗?请谈谈你对这个问题的看法。
4. 在你看来,男生和女生在学业表现、心理特征、社会对其期待值等方面有哪些区别?为什么会有这样的不同?这些实际存在的区别是造成性别不平等的原因吗?
5. 很多中国的父母认为男孩子应该有"男子汉"的气质,不应该带"娘娘腔"。还有人认为"没有运动就没有男孩,更没有男子汉。不喜欢运动的男孩一定是问题男孩。"什么是"娘娘腔"?你是否同意这样的观点?
6. 我听说中国有些地方,一些有"贱养男孩"、"富养女孩"的看法。你能解释一下是什么意思吗?你觉得这样做有道理吗?

采访后,请根据采访内容写一篇采访报告。在这篇采访报告里:

1. 简单解释一下采访背景:"拯救男孩"这个口号是怎么来的?在你看来,深入探讨这个问题有什么意义?
2. 介绍被采访人的背景(年龄、性别、成长经历);

3. 被采访人对"拯救男孩"这个话题的看法(你不必把采访的全部内容记录下来, 挑选让你印象最深刻的一两点即可);
4. 结合你自己的经历、看法对被采访人的观点进行评论。譬如，他为什么会有这样的看法，你是否同意他的观点；在美国或你自己的国家有没有这样的问题；你认为不同文化是如何处理与解决这些问题的；通过这次采访以及对两种文化的比较与反思, 你有什么更深刻的思考?
5. 对这次采访做一个总结。

写作要求：

1. 格式:14号字体, 隔行写
2. 字数:800–1000字左右(包括标点)
3. 语言:多用课上学过的生词、语法, 语言准确和语言丰富同等重要，语句通顺, 语体正式。如果语言能做到生动活泼, 具有个人色彩, 会加分。
4. 内容:"言之有物, 言之有理", 观点明确, 论述有理有据, 表述具体充实, 条理清晰, 有逻辑性
5. 结构:完整清晰, 段与段之间过渡自然流畅

Appendix II: Instructions for Final Independent Research Report

题目：
　　根据你的个人兴趣、生活经历或专业研究方向选择一个研究题目。这个题目可以是一个跟当代中国社会(或你自己的国家)有关的热点话题, 也可以是一个文化现象，或者是我们上课所学内容的延伸。选题应有新意和研究意义，要小而具体, 不要过于宽泛, 否则容易泛泛而谈。选好题目后, 请先跟老师商量，得到批准后再开始动笔。
　　这个学期你一共需要查找并阅读六篇和选题相关的高质量的文章，内容可以是跟话题有关的深度调查报告、评论或学术文献。每过三个星期，你需要找两篇相关文章, 跟老师做一次个别谈话, 讨论你查找到的文章。个别谈话后需上交一篇1000字左右的阅读报告。学期中的阅读报告可以作为期末报告的部分内容。
　　期末应整合学期中阅读报告的内容，加上引言和结论，完成一篇3000-4000字左右的大型报告。
　　具体步骤：

第三周	个别谈话, 交独立研究开题报告 跟老师一起选一个合适的题目, 写一篇1000字左右(含标点)的开题报告。在报告中请解释你要研究什么问题, 为什么研究这个问题, 你对这个问题已有的了解, 及研究这个问题的意义。研究报告题目决定好以后, 开始考虑文章的结构(大纲)。
第六周	查找两篇文章, 个别谈话, 交阅读报告(一) 阅读两篇不少于2000字的相关文章。请务必在个别辅导的前三天把文章的链接发给老师。个别谈话后, 写一篇阅读报告。
第九周	查找两篇文章, 个别谈话, 交阅读报告(二) 要求同上

(continued)

第十二周	查找两篇文章，个别谈话， 准备下周的期末口头报告**(30分钟)** 再阅读两篇不少于2000字的文章，跟老师做完个别辅导后，结合前两篇阅读报告的内容，完成最终的独立研究报告。准备下个星期的口头报告，向全班汇报你的研究成果。
第十三周	交阅读报告 全文(第一稿)
第十五周	交阅读报告 全文(第二稿)

阶段性报告要求**:**

1. 格式:14号字体，隔行写，
2. 字数:1000字左右(包括标点)
3. 语言:多用课上学过的生词、语法，语言准确和语言丰富同等重要，语句通顺，语体正式
4. 内容:"言之有物，言之有理"，观点明确，论述有理有据，表述具体充实，条理清晰，有逻辑性
5. 结构:完整清晰，段与段之间过渡自然流畅

其它说明**:**

1. 研究题目选好以后，请先考虑一下你期末报告的大纲，文章需要几个部分，根据文章的大纲查找文献。下面有一些之前学生写过的研究报告大纲，供参考。
2. 做个别辅导以前，请仔细阅读文章，并提前至少三天把文章链接寄给老师。
3. 做个别辅导的时候，你你需要(1) 概括文章大意，转述作者的主要观点及立场; (2) 结合文章中的观点，阐述你自己的看法; (3) 和老师讨论这篇文章的优缺点，比如你觉得哪些地方写得好，哪些地方可以写得更好(如是否具有观点性，观点表达是否清楚，论述是否严谨); (4)如果阅读过程中遇到不懂的问题，请老师老师答疑。

前人独立研究阅读报告大纲范例

范例一: 中美两国民众在"环保"问题上的态度之异同

1. 引言
2. 中美两国现行的环保措施
3. 研究方法:问卷调查和采访
4. 调查结果及分析
5. 结论

范例二: 浅谈中国"蚁族"现象

1. 引言
2. "蚁族"的定义与生存现状

3. "蚁族"产生的社会背景与群体特点
4. "蚁族"的网络行为
5. "蚁族"对中国社会发展的影响

范例三: 中美大学校园流行语之比较

1. 引言
2. 调查方法
3. 调查对象
4. 调查结果
5. 调查结果分析
6. 研究结论及不足

References

ACTFL (American Council on the Teaching of Foreign Languages). 2012. ACTFL Proficiency Guidelines-writing. https://www.actfl.org/publications/guidelines-and-manuals/actfl-proficiency-guidelines-2012/english/writing. Accessed 8 Mar 2020.

An, Fuyong (安福勇). 2015. Butong shuiping CLS xuexizhe zuowen liuchangxing, jufa fuzadu he zhunquexing fenxi – yixiang jiyu T danwei celiangfa de yanjiu 不同水平 CLS 学习者作文流畅性、句法复杂度和准确性分析 — 一项基于T单位测量法的研究 [Analysis of fluency, grammatical complexity and accuracy of CSL writing: A study based on T-unit analysis]. *Yuyan Jiaoxue yu Yanjiu* 语言教学与研究 [Language Teaching and Linguistic Studies] 3:11–20.

Bean, John C. 2011. *Engaging ideas: The Professor's guide to integrating writing, critical thinking, and active learning in the classroom*. 2nd ed. San Francisco: Jossey-Bass Press.

Benesch, Sarah. 2001. *Critical English for academic purposes: Theory, politics, and practice*. London: Routledge.

Chaffee, John. 2014. *Critical thinking, thoughtful writing: A rhetoric with readings*. 6th ed. Stamford: Cengage Learning.

Chen, Xianchun (陈贤纯). 2003. Duiwai hanyu jiaoxue xiezuoke chutan 对外汉语教学写作课初探 [On teaching a Chinese as a second language writing course]. *Yuyan Jiaoxue yu Yanjiu* 语言教学与研究 [Language Teaching and Linguistic Studies] 5: 59–63.

Chen, Shiqi (陈诗琦). 2016a. Duiwai hanyu jiaoxue "xiezuo nan" xianzhuang zhi yuanyin fenxi yu duice 对外汉语教学"写作难"现状之原因分析与对策 [The reasons for writing difficulty in TCFL and the counter-measures]. *Yunnan Shifan Gaodeng Zhuanke Xuexiao Xuebao* 郧阳师范高等专科学校学报 [Journal of Yunyang Teachers College] 8: 136–138.

Chen, Yu (陈钰). 2016b. Liuxuesheng lunwen zhidao celüe de youxiaoxing yanjiu 留学生论文指导策略的有效性研究 [The effectiveness of instructional strategies in academic writing supervision for learners of Chinese as a foreign language]. *Yuyan Jiaoxue yu Yanjiu* 语言教学与研究 [Language Teaching and Linguistic Studies] 6: 19–27.

Christensen, Matthew B. 2013. Chinese for special purpose: An individualized approach. In *Individualized instruction in East Asian languages*, ed. Etsuyo Yuasa, 39–60. Columbus: The Ohio State University National East Asian Languages Resources Center.

Cohen, Andrew. 1987. Student processing of feedback on their compositions. In *Learner strategies in language learning*, ed. Anita Wenden and Joan Rubin, 57–70. Englewood Cliffs: Prentice-Hall International English Language Teaching.

Condon, William, and Diane Kelly-Riley. 2004. Assessing and teaching what we value: The relationship between college-level writing and critical thinking abilities. *Assessing Writing* 9: 56–75.

Fandrych, Christian. 2010. Language and subject matter reunited: A bilingual approach for teaching modern foreign languages at higher education institutions. *Forum Sprache* 3: 20–32.

Fathman, Ann K., and Elizabeth Whalley. 1990. Teacher response to student writing: Focus on form versus content. In *Second language writing: Research insights for the classroom*, ed. Barbara Kroll, 178–190. Cambridge: Cambridge University Press.

Fu, Ailing (付爱玲). 2014. Guocheng xiezuofa yingyong yu duiwai hanyu xiezuo jiaoxue de kexingxing fenxi 过程写作法应用于对外汉语写作教学的可行性分析 [The feasibility analysis of applying process approach to teaching Chinese as a foreign language]. *Jilinsheng Jiaoyu Xueyuan Xuebao* 吉林省教育学院学报 [Journal of Educational Institute of Jilin Province] 9: 54–55.

Gao, Zengxia (高增霞), and Shuo Li (栗硕). 2018. Xueshu hanyu xiezuo jiaocai jianshe chuyi 学术汉语写作教材建设刍议 [On the compilation of textbooks for Chinese academic writing]. *Yunnan Shifan Daxue Xuebao* 云南师范大学学报 [Journal of Yunnan Normal University] 6: 12–21.

Gao, Zengxiao (高增霞), and Fuying Liu (刘福英). 2016. Lun xueshu hanyu zai duiwai hanyu jiaoxuezhong de zhongyaoxing 论学术汉语在对外汉语教学中的重要性 [On the role of the CAP teaching in teaching Chinese to speakers of other languages]. *Yunnan Shifan Daxue Xuebao* 云南师范大学学报 [Journal of Yunnan Normal University] 3: 44–51.

Guo, Hanning (郭涵宁). 2016. Liuxuesheng benke biye lunwen xiezuoke jiaoxue moshi tantao 留学生本科毕业论文写作课教学模式探讨 [The teaching mode of the thesis writing course for international undergraduates majoring in Chinese]. *Guoji Hanyu Jiaoxue Yanjiu* 国际汉语教学研究 [Journal of International Chinese Teaching] 4: 44–52.

Guojia Duiwai Hanyu Jiaoxue Lingdao Xiaozu Bangongshi, Hanyu Shuiping Kaoshibu (国家对外汉语教学领导小组办公室、汉语水平考试部). 1996. *Hanyu Shuiping Dengji Biaozhun yu Yufa Dengji Dagang* 汉语水平等级标准与语法等级大纲 [Chinese proficiency guidelines and syllabus of graded grammar]. Beijing: Gaodeng jiaoyu chubanshe.

Guojia Hanyu Shuiping Kaoshi Weiyuanhui Bangongshi Kaoshi Zhongxin (国家汉语水平考试委员会办公室考试中心). 2001. Hanyu Shuiping Cihui yu Hanzi Dengji Dagang 汉语水平词汇与汉字等级大纲 [Syllabus of Graded Words and Characters for Chinese Proficiency]. Beijing: Jingji kexue chubanshe.

Haswell, Richard H. 1991. *Gaining ground in college writing: Tales of development and interpretation*. Dallas: Southern Methodist University Press.

IIE (Institute of International Education). 2019. *Open Doors®: Report on International Educational Exchange*. https://www.iie.org/Research-and-Insights/Open-Doors. Accessed on 15 Apr 2020.

Jiang, Wenying. 2013. Measurements of development in L2 written production: The case of L2 Chinese. *Applied Linguistics* 34 (1): 1–24.

Knott, Deborah. 2005. Critical reading towards critical writing. https://advice.writing.utoronto.ca/researching/critical-reading/. Accessed 15 Apr 2020.

Leki, Ilona. 1991. The preferences of ESL students for error correction in college-level writing classes. *Foreign Language Annals* 24 (3): 203–218.

Li, Ying (李英), and Deng, Shulan (邓淑兰). 2012. Liuxuesheng Biye Lunwen Xiezuo Jiaocheng. 留学生毕业论文写作教程 [Thesis Writing Course for International Students]. Beijing: Beijing daxue chubannshe.

Li, Zengji. (李增吉). 2007. *Hanyu Gaoji Xiezuo Jiaocheng* (xia ce) 汉语高级写作教程 (下册) [*A Course on Advanced Chinese Writing* (vol. 2)]. Beijing: Beijing daxue chubanshe.

Liu, Hong. (刘弘). 2014. Guoji hanyu jiaoxuezhong xiezuo fangshi duiyu xiezuo xiaoguo yingxiang de ge'an yanjiu 国际汉语教学中写作方式对于写作效果影响的个案研究 [A case study of the effect of writing methods in TCSL]. *Huawen Jiaoxue yu Yanjiu* 华文教学与研究 [TCSOL Studies] 2: 43–50.

Liu, Ying. 2015. Understanding readership: American students' perceptions of evidence in Chinese persuasive compositions. Ph.D. dissertation, The Ohio State University.

Luo, Qingsong (罗青松). 2002. *Duiwai Hanyu Xiezuo Jiaoxue Yanjiu* 对外汉语写作教学研究 [*On Teaching Chinese Writing*]. Beijing: Zhongguo shehui kexue chubanshe.

——— (罗青松). 2011. Duiwai hanyu xiezuo jiaoxue yanjiu shuping 对外汉语写作教学研究述评 [Review of studies on teaching writing in TCFL]. *Yuyan Jiaoxue yu Yanjiu* 语言教学与研究 [Language Teaching and Linguistic Studies] 3: 29–36.

Ma, Zhongke (马仲可). 1999. Guanyu ruhe peiyang gaoji hanyu xiezuo rencai de wojian 关于如何培养高级汉语写作人才的我见 [On How to train advanced Chinese writing talents]. *Diliujie Guoji Hanyu Jiaoxue Taolunhui Lunwenxuan* 第六届国际汉语教学讨论会论文选 [Conference Proceedings of 6th International Conference on Chinese Language Teaching and Research]. 120–125. Beijing: Beijing daxue chubanshe.

McLaughlin, Frost, and Miriam Moore. 2012. Integrating critical thinking into the assessment of college writing. *Teaching English in the Two-Year College* 40 (2): 145–162.

Mehta, Sandhya Rao, and Rahma Al-Mahrooqi. 2015. Can thinking be taught? Linking critical thinking and writing in an EFL context. *RELC Journal* 46 (1): 23–46.

MLA (Modern Language Association). 2016. Enrollments in languages other than English in United States Institutions of Higher Education 2016 final report. https://www.mla.org/content/download/110154/2406932/2016-Enrollments-Final-Report.pdf. Accessed 8 Mar 2020.

Nan, Yong (南勇). 1994. Liuxuesheng de hanyu xiezuo jiaoxue chuyi 留学生的汉语写作教学刍议 [On the teaching of Chinese writing to foreign students]. *Hanyu Xuexi* 汉语学习 [Chinese Language Learning] 6: 52–53.

North, Stephen M. 1984. The idea of a writing center. *College English* 46 (5): 432–446.

Oliver, Helen, and Robert Utermohlen. 1995. An innovative teaching strategy: Using critical thinking to give students a guide to the future. Eric Document Reproduction Services No. 389 702. https://eric.ed.gov/?id=ED389702. Accessed 05 May 2020.

Peng, Xiaochuan (彭小川). 2004. Guanyu duiwai hanyu yupian jiaoxue de xin sikao 关于对外汉语语篇教学的新思考 [A new consideration about TCSL text teaching]. *Hanyu Xuexi* 汉语学习 [Chinese Language Learning] 2: 49–54.

Piao, Dejun (朴德俊). 2007. Dui han xiezuo jiaoxue celüe 对韩写作教学策略 [Strategies for teaching writing to Korean students]. *Dibajie Guoji Hanyu Jiaoxue Taolunhui Lunwenxuan* 第八届国际汉语教学讨论会论文选 [Conference Proceedings of 8th International Conference on Chinese Language Teaching and Research] 126–136. Beijing: Gaodeng jiaoyu chubanshe.

Qi, Hua (亓华). 2006. Liuxuesheng biye lunwen de xiezuo tedian yu guifanhua zhidao 留学生毕业论文的写作特点与规范化指导 [The features of Graduation-thesis writing by foreign students and its standardized instruction]. *Yunnan Shifan Daxue Xuebao* 云南师范大学学报 [Journal of Yunnan Normal University] 1: 6–11.

Qi, Haifeng (亓海峰), and Jianling Liao (廖建玲). 2019. Jiyu jixuwen he yilunwen de hanyu eryu xiezuo fazhan yanjiu 基于记叙文和议论文的汉语二语写作发展研究 [An investigation into Chinese linguistic development in L2 narrative and argumentative writing]. *Shijie Hanyu Jiaoxue* 世界汉语教学 [Chinese Teaching in the World] 4: 563–576.

Reid, Joy, and Barbara Kroll. 1995. Designing and assessing effective classroom writing assignments for NES and ESL students. *Journal of Second Language Writing* 4 (1): 17–41.

Richards, Jack C., and Willy A. Renandya. 2002. *Methodology in language teaching: An anthology of current practice*. Cambridge: Cambridge University Press.

Schultz, Jean Marie. 1991a. Mapping and cognitive development in the teaching of foreign language writing. *The French Review* 64 (7): 978–988.

———. 1991b. Writing mode in the articulation of language and literature classes: Theory and practice. *The Modern Language Journal* 75 (1): 411–417.

Scott, Virginia Mitchell. 1996. *Rethinking foreign language writing*. Boston: Heinle & Heinle Publishers.

Scriven, Michael, and Richard W. Paul. 1987. *Defining critical thinking*, Draft statement written for the National Council for Excellence in Critical Thinking Instruction. http://www.criticalthinking.org/pages/defining-critical-thinking/766. Accessed 11 June 2020. http://www.criticalthinking.org/pages/defining-critical-thinking/766

Shi, Yanlan (史艳岚). 2002. Sinianji liuxuesheng cihui shiyong qingkuang diaocha 四年级留学生词汇使用情况调查 [An investigation of fourth-year foreign students' lexical abilities].

Zhongguo Duiwai Hanyu Jiaoxue Xuehui Diqici Xueshu Taolunhui Lunwenxuan 中国对外汉语教学学会第七次学术讨论会论文选 [The conference proceedings of the 7th Symposium of Chinese Language Teaching Association]. 75–85. Beijing: Renmin jiaoyu chubanshe.

Shi, Wen (师文) and Jing Chen (陈静). 2019. Hanyu eryu xiezuo yuyan tezheng de ticai chayi yanjiu 汉语二语写作语言特征的体裁差异研究 [The linguistic features of L2 Chinese writing performance in different genres]. *Hanyu Xuexi* 汉语学习 [Chinese Language Learning] 6: 76–85.

Song, Jingyao (宋璟瑶). 2015. Hanyu yilunwen pianzhang xide yanjiu 汉语议论文篇章习得研究 [A study on the acquisition of Chinese argumentative text]. *Huawen Jiaoxue yu Yanjiu* 华文教学与研究 [TCSOL Studies] 3: 18–28.

Sperling, Melanie. 1991. Dialogues of deliberation: Conversation in the teacher-student writing conference. *Written Communication* 8 (2): 131–162.

Tao, Jiawei (陶嘉炜). 2007. Renshi he chuli duiwai hanyu xiezuo jiaoxuezhong de san da wenti 认识和处理对外汉语写作教学中的三大问题 [Recognizing and solving three major problems in teaching Chinese writing]. *Dibajie Guoji Hanyu Jiaoxue Taolunhui Lunwenxuan* 第八届国际汉语教学讨论会论文选 [Conference Proceedings of 8th International Conference on Chinese Language Teaching and Research] 137–141. Beijing: Gaodeng jiaoyu chubanshe.

Wade, Carole. 1995. Using writing to develop and assess critical thinking. *Teaching of Psychology* 22 (1): 24–42.

Walker, Galal. 1989. Designing an intensive Chinese curriculum. In *Chinese pedagogy: An emerging field*, ed. Scott McGinnis, 181–228. Columbus: The Ohio State University Foreign Language Publications.

Wang, Yang. (汪洋). 2009. Zhongji hanyu xiezuo jiaoxue de teshu diwei ji xiezuo fanwen de daodu 中级汉语写作教学的特殊地位及写作范文的导读 [The special role of intermediate level writing instruction and the textual analysis of model essays]. *Dijiujie Shijie Huayuwen Jiaoxue Yantaohui Lunwenji* 第九届世界华语文教学研讨会论文 [Conference Proceedings of the 9th World Conference on Chinese Language Teaching] 4:241–253.

——— (汪洋). 2010. Guanyu zhongji hanyu jiaocai xiezuo lianxi bianxie linian yu fangfa de tantao 关于中级汉语教材写作练习编写理念与方法的探讨 [A discussion on the principles and methods of designing writing tasks in intermediate writing textbooks]. *Guoji Hanyu Chuanbo yu Guoji Hanyu Jiaoxue Yanjiu* 汉语国际传播与国际汉语教学研究 [Research on Chinese Teaching around the World], 303–316. Beijing: Zhongyang minzu daxue chubanshe.

Wang, Yixuan (王艺璇). 2017. Hanyu eryuzhe cihui fengfuxing yu xiezuo chengji de xiangguanxing 汉语二语者词汇丰富性与写作成绩的相关性 [The correlation between lexical richness and writing score of CSL learner – The multivariable linear regression model and equation of writing quality]. *Yuyan Wenzi Yingyong* 语言文字应用 [Applied Linguistics] 2: 93–101.

Way, Denise Paige, Elizabeth G. Joiner, and Michael A. Seaman. 2000. Writing in the secondary foreign language classroom: The effects of prompts and tasks on novice learners of French. *The Modern Language Journal* 84 (2): 171–183.

Williams, Jessica. 2002. Undergraduate second language writers in the writing center. *Journal of Basic Writing* 21 (2): 73–91.

Wu, Ping (吴平). 1999. Cong xuexi celüe dao duiwai hanyu xiezuo jiaoxue 从学习策略到对外汉语写作教学 [Students' learning strategies in CFL writing classes]. *Hanyu Xuexi* 汉语学习 [Chinese Language Learning] 3: 34–37.

Wu, Shuang (吴双). 2008. Lun guocheng ticai xiezuo lilun zai duiwai hanyu xiezuo jiaoxue zhong de yingyong 论过程体裁写作理论在对外汉语写作教学中的应用 [On the application of the process genre approach in CFL writing instruction]. *Xiandai Yuwen* 现代语文 [Modern Chinese] 3: 21–24.

——— (吴双). 2011. Liuxuesheng hanyu xieqian gousi huodong dui qi zuowen zhiliang de yingxiang 留学生汉语写前构思活动对其作文质量的影响 [On the influence of plotting in Chinese on the quality of compositions of Chinese learners]. *Shijie Hanyu Jiaoxue* 世界汉语教学 [Chinese Teaching in the World] 1: 99–109.

Wu, Jian (吴剑). 2012. Laihua yuke liuxuesheng hanyu xiezuo celüe tansuo 来华预科留学生汉语写作策略探索 [A research on Chinese writing strategies of foreign preparatory students in China]. *Huawen Jiaoxue yu Yanjiu* 华文教学与研究 [TCSOL Studies] 2: 47–55.

Wu, Jifeng (吴继峰). 2016a. Yingyu muyuzhe hanyu xiezuo Zhong de cihui fengfuxing fazhan yanjiu 英语母语者汉语写作中的词汇丰富性发展研究 [Research on lexical richness development in CSL writing by English native speakers]. *Shijie Hanyu Jiaoxue* 世界汉语教学 [Chinese Teaching in the World] 1: 129–142.

——— (吴继峰). 2016b. Yingyu muyuzhe hanyu shumianyu jufa fuzaxing yanjiu 英语母语者汉语书面语句法复杂性研究 [The grammatical complexity in English native speakers' Chinese writing]. *Yuyan Jiaoxue yu Yanjiu* 语言教学与研究 [Language Teaching and Linguistic Studies] 4: 27–35.

Xin, Ping (辛平). 2001. Dui 11 pian liuxuesheng hanyu zuowenzhong pianwu de tongji fenxi ji dui hanyu xiezuoke jiaoxue de sikao 对11篇留学生汉语作文中偏误的统计分析及对汉语写作课教学的思考 [A statistical analysis of errors in 11 compositions by international students and flections on Chinese writing pedagogy]. *Hanyu Xuexi* 汉语学习 [Chinese Language Learning] 4: 67–71.

Xu, Xiyang (许希阳), and Yongyi Wu (吴勇毅). 2016. "Chanchu daoxiang fa" lilun shijiao xia de duiwai hanyu xiezuo jiaoxue moshi zhi tansuo "产出导向法" 理论视角下的对外汉语写作教学模式之探索 [Exploring a new mode of teaching Chinese writing from the perspective of Production-Oriented Approach]. *Huawen Jiaoxue yu Yanjiu* 华文教学与研究 [TCSOL Studies] 4: 50–59.

Yang, Jianchang (杨建昌). 1982. Qiantan waiguo liuxuesheng hanyu zhuanye de xiezuoke jiaoxue 浅谈外国留学生汉语专业的写作课教学 [On the teaching of writing courses for foreign students majoring in Chinese]. *Yuyan Jiaoxue yu Yanjiu* 语言教学与研究 [Language Teaching and Linguistic Studies] 3: 110–113.

Yang, Li (杨俐). 2004. Guocheng xiezuo de shijian yu lilun 过程写作的实践与理论 [Process writing: Theory and practice]. *Shijie Hanyu Jiaoxue* 世界汉语教学 [Chinese Teaching in the World] 1: 90–99.

——— (杨俐). 2006. *Waiguoren Hanyu Guocheng Xiezuo* 外国人汉语过程写作 *Writing Essays in Chinese*. Beijing: Beijing University Press.

Yang, Lipei (杨理沛). 2018. Gaoji jieduan duiwai hanyu xiezuoke jiaoxue silu quetan 高级阶段对外汉语写作课教学思路榷谈 [A discussion on teaching methods in advanced CFL writing courses]. *Xiezuo* 写作 [Writing] 2: 78–80.

Yao, Shujun (幺书君). 2005. Hanguo liuxuesheng hanyu xueli jiaoyu gaonianji xiezuoke jiaoxue tansuo 韩国留学生汉语学历教育高年级写作课教学探索 [Designing advanced Chinese writing courses for Korean students majoring in Chinese]. *Haiwai Huawen Jiaoyu* 海外华文教育 [Overseas Chinese Education] 3: 25–29.

Yao, Tao-chung, and Kuang-tien Yao. (2010). Meiguo hanyu jiaoxue lishi huigu yu xianzhuang 美国汉语教学历史回顾与现状 [Chinese language instruction in the United States: A look at its history and current status]. In *Beimei Zhongguoxue: Yanjiu Gaishu yu Wenxian Ziyuan* 北美中国学:研究概述与文献资源 [*Chinese studies in North America: Research and Resources*], ed. Haihui Zhang, Zhaohui Xue and Shuyong Jiang, 773–784. Beijing: Zhonghua shuju.

Yuan, Fangyuan. 2010. Impacts of task conditions on learners' output in L2 Chinese narrative writing. *The Journal of Chinese as a Second Language* 45 (1): 67–88.

Zeng, Zhini. 2018. Situated learning in a Chinese question-and-answer online community: The case of Zhihu. *The Journal of Chinese as a Second Language* 53 (3): 222–261.

Zhang, Nian (张念). 2003. Duocheng fenxiang, xunhuan jianjin — xinxing xiezuo jiaoxue moshi tantao 多层分析,循环渐进— 新型写作教学模式探讨 [Multi-layered analysis, cyclical progress: An exploration of new writing instruction mode]. *Haiwai Huawen Jiaoyu* 海外华文教育 [Overseas Chinese Education] 3: 20–23.

Zhang, Zhengsheng (张正生). 2005. Shumianyu dingyi ji jiaoxue wenti chutan 书面语定义及教学问题初探 [On the definition of Chinese written language and the pedagogical issues in teaching Chinese written language]. In *Duiwai Hanyu Shumianyu Jiaoxue yu Yanjiu de Zuixin*

Fazhan 对外汉语书面语教学与研究的最新发展 [*The Latest Development of Teaching and Research of the Formal Written Chinese as a Foreign Language*], ed. Shengli Feng (冯胜利), & Wenze Hu (胡文泽), 323–338. Beijing: Beijing Language and Culture University Press.

Zhang, Baolin (张宝林). 2009. "Hanyu xiezuo rumen" jiaoxue moshi chuyi "汉语写作入门"教学模式刍议 [On teaching a "Chinese language writing primer" course]. *Yuyan Jiaoxue yu Yanjiu* 语言教学与研究 [Language Teaching and Linguistic Studies] 3: 54–59.

Zhang, Juanjuan (张娟娟). 2019. Dongnanya liuxuesheng jixuwen cihui fengfuxing fazhan yanjiu 东南亚留学生记叙文词汇丰富性发展研究 [On the developmental features of lexical richness in Chinese writing by students from Southeast Asian countries]. *Yunnan Shifan Daxue Xuebao* 云南师范大学学报 [Journal of Yunnan Normal University] 1: 33–44.

Zhao, Wen (赵文). 1996. Duiwai hanyu jiaoxue zhong de xiezuo jiaoxue 对外汉语教学中的写作教学 [The writing pedagogy of teaching Chinese as a foreign language]. *Waiyu yu Waiyu Jiaoxue* 外语与外语教学 [Foreign Languages and Their Teaching] 92: 74–76.

Zhu, Bingyao (祝秉耀). 1984. Qiantan xiezuoke jiaoxue 浅谈写作课教学 [On writing pedagogy]. *Yuyan Jiaoxue yu Yanjiu* 语言教学与研究 [Language Teaching and Linguistic Studies] 3: 96–105.

Zhu, Xiangyan (朱湘燕). 2007. Duiwai hanyu xiezuo jiaoxue diaocha ji yanjiu. 对外汉语写作教学调查及研究 [An investigation of the writing pedagogy of Chinese as a foreign language]. *Xiandai Yuwen* 现代语文 [Modern Chinese] 6: 101–103.

Calligraphy Education in Teaching Chinese as a Second Language

32

Yu Li

Contents

Introduction	814
Why Teach Calligraphy	816
Importance	817
Benefits	821
What to Teach: The Educational Goals of a Calligraphy Curriculum	823
Current Goals	823
Goals of a Liberal Education	827
How to Teach: The Pedagogical Approach	828
Leading with Cultural Perspectives	828
Conclusion	833
Appendix I	833
Appendix II	838
References	839

Abstract

This chapter focuses on Chinese calligraphy education offered by Chinese or (East) Asian language and culture programs in US universities at the undergraduate level. Addressing the why, what, and how of calligraphy teaching in this particular context, the chapter begins with a discussion on the broad significance of calligraphy instruction, including its artistic value, intellectual import, and cultural meaning as well as the contribution of calligraphy courses to student growth and overall curriculum development. Then, to address the question of what to teach, the chapter examines the current student learning outcomes and proposes to further the goals of calligraphy education in alignment with those of the broader liberal education. Finally, focusing on the pedagogical aspect, that is, the "how" of calligraphy instruction, the chapter discusses the methodology that

Y. Li (✉)
Department of Modern Languages and Literatures, Loyola Marymount University, Los Angeles, CA, USA
e-mail: yu.li@lmu.edu

© The Author(s), under exclusive licence to Springer Nature Singapore Pte Ltd. 2022
Z. Ye (ed.), *The Palgrave Handbook of Chinese Language Studies*,
https://doi.org/10.1007/978-981-16-0924-4_8

may be used to support such efforts. It argues that calligraphy education at the undergraduate level should be restructured to focus on cultivating student dispositions by teaching cultural perspectives first and foremost. The chapter proposes a cultural-perspective-led approach to calligraphy instruction and offers specific examples to illustrate this approach.

Keywords

Chinese calligraphy · Chinese as a second language · Interdisciplinary education · Student learning outcomes · Dispositions · Cultural perspectives · Curricular development

Introduction

Regarded the highest form of Chinese art, *shufa* (书法/書法), or brush calligraphy, holds a central place in the visual culture of China. For millennia, in premodern China, the instruction and learning of calligraphy had formed a distinguished tradition among the educated elite. Calligraphy practice typically started early in life along with literacy development and literary training, and being able to write a beautiful hand years later was often the prerequisite for passing civil service examinations and pursuing a successful bureaucratic career, the ultimate accomplishment of an imperial education. Many scholar-officials were master calligraphers, and their ability to produce brush writing of the highest caliber (品 *pin* 'quality, character') inevitably commanded admiration from their peers and the educated society. In modern China, although brush writing is no longer the main method of written communication, calligraphy learning and practice, thanks to the promotion of literacy and education, find a much broader array of participants: from elementary-school students to retirees, from calligraphy doctorates to self-taught enthusiasts. Calligraphy-related activities are supported and stimulated by a wide variety of school curricula, training courses, certificate programs, degree programs, competitions, semiprofessional associations, amateur interest groups, art exhibits, conferences, and, of course, constant publication and production of books, magazines, and multimedia materials. New subgenres of calligraphic art or venues of calligraphy creation have also emerged, from the Modernist to the avant-garde, from popular ground calligraphy (地书 *dishu* "ground writing," aka "water calligraphy") seen in public parks to robot-generated works using artificial intelligence. It may be fair to say that, since its inception some 2000 years ago (Knight 2012, p. 28; Li 2009, p. 72), the artistic practice of calligraphy has almost never stopped flourishing in China.

In the USA, opportunities to study Chinese calligraphy are available to many. University students may take a calligraphy course from the Chinese language, culture, and literature programs; Asian or East Asian studies programs; or visual art programs at their institutions. High school as well as college students may participate in calligraphy club activities on campus. Those who live in the vicinity of a university may take a calligraphy class from its continuing-education offerings. Children of Chinese heritage are often sent to weekend schools that teach Chinese language and calligraphy. Major art museums periodically put on Chinese calligraphy exhibitions (e.g., *Out of character: Decoding Chinese calligraphy*, the Metropolitan Museum of Art, 2014; *Calligraphy in Chinese and Japanese art*, Saint Louis Art Museum, 2014–2015). These museums as well as local art organizations may offer calligraphy-related lectures or workshops to the general public. Calligraphy writing is often seen in celebrating traditional Chinese holidays, especially the Spring Festival, among the Chinese communities, and it is also enjoyed by students learning Chinese language or culture during those occasions. Calligraphy supplies are increasingly available in Asian markets, bookstores, or art-supply shops as well as on major shopping websites (e.g., Amazon). Learning and practicing Chinese calligraphy is becoming a lively part of the multicultural life of many in America.

Calligraphy education in the USA has been bolstered, in recent decades, by the continued development and dedicated work of academic associations. The American Society of *Shufa* Calligraphy Education (ASSCE), most notably, was established in Maryland in 1992 as the Calligraphy Education Group (CEG) under the Chinese Language Teachers Association (CLTA). At the time, its membership was primarily Chinese-as-a-Second-Language (CSL) educators at the college level. It has now grown to include secondary-school teachers, graduate students, researchers, and artists. Since 1998, the association has organized in the USA as well as in East Asia 11 biennale international conferences on calligraphy education and has become an increasingly recognized platform for calligraphy educators, students, artists, and enthusiasts to share their works and ideas. Its most recent conferences, including the 2018 symposium hosted by Beijing Normal University and the 2014 conference by the Guangxi Art Institute, produced collections of select conference papers. The publication of conference proceedings, together with the association's work on establishing its own academic journal, will further increase its visibility in and impact on the CSL field.

In the broader context of calligraphy education and practice in China and the USA, it is particularly interesting and utterly necessary to consider calligraphy education in US universities. As these learning opportunities may be among the most rigorous available on this subject matter in the country, the conceptualization and implementation of calligraphy education in such settings may indeed determine the quality of Chinese-calligraphy-related intellectual and cultural life in the USA. This chapter focuses specifically on Chinese calligraphy education offered by Chinese (or (East) Asian) language and culture programs in US universities at the undergraduate level, addressing the why, what, and how of calligraphy teaching in this particular context. Although calligraphy courses rarely occupy a central position in Chinese language and culture curricula, they have enjoyed sustained popularity among students and have contributed substantially to the development and enrichment of these academic programs. In what follows, the chapter will begin with a discussion of the broad significance of calligraphy instruction, including its artistic value, intellectual import, and cultural meaning, as well as the contribution of calligraphy courses to student growth and overall curriculum development. This part of the discussion focuses on why calligraphy instruction is meaningful and important. Then, to address the question of what to teach, the chapter examines current student learning outcomes and proposes to further the goals of calligraphy education in alignment with those of the broader liberal education. Finally, focusing on the pedagogical aspect, that is, the "how" of calligraphy instruction, the chapter discusses the methodology that may be used to support such efforts. It argues that calligraphy education at the undergraduate level should be restructured to focus on cultivating student dispositions by teaching cultural perspectives first and foremost and proposes a cultural-perspective-led approach to calligraphy instruction.

The chapter draws information from course syllabi and survey data, and it may be useful to say a few words about them here. The syllabi are for introductory calligraphy courses offered by Chinese or (East) Asian studies programs of 14 US universities. The course syllabi provided basic information about the calligraphy courses, including course objectives, assessment methods, topics, and textbooks and other materials. Based on this information, a survey was conducted to further probe instructors' perception and practice in calligraphy instruction. In particular, it asked about student learning outcomes, instructors' preparation, course format and topics, relationship of course to Chinese language learning, teaching materials, contribution of the calligraphy course to the program, and challenges in teaching. (See survey questions and a summary of the responses in Appendix I.) Survey participants were current calligraphy instructors, i.e., professors who taught Chinese calligraphy over the past 10 years, from 2000 to 2019. Information was first collected as to which institutions offered calligraphy courses (see Appendix II for a list of these universities) and 31 universities and colleges were identified. Survey invitations were then sent to the calligraphy instructors or, in the case when instructor information was not available, to program directors. The survey received 17 responses.

Why Teach Calligraphy

Brush calligraphy is a quintessential Chinese art form. Over its long, evolving history, it has become intimately intertwined not only with how the Chinese communicate, how they express feelings and emotions, but also how they behave toward each other, and what they value as good and beautiful. It is the "ultimate art form" (Knight 2012, p. 17) practiced and admired by educated Chinese in almost all of China's recorded history. With its remarkable depth and nuances, it has also enchanted numerous non-Chinese scholars and dignitaries literate in the Chinese script, including those in Korea, Japan, and Vietnam, until this day. Yet there is no counterpart in the Western culture that one can rely on to directly "translate" the meaning of Chinese calligraphy. Even the word "calligraphy," as a common translation for *shufa*, is deemed inaccurate and insufficient, given that the Western calligraphic tradition and Chinese *shufa* are drastically different systems of aesthetic and implementary preferences. Thus, it is essential for students in the West to understand Chinese calligraphy in order to fully grasp the significance of the Chinese civilization. Indeed, calligraphy should be taught because, first and foremost, it is important.

Importance

Chinese calligraphy, in its various historical forms, has been the production means and method of recorded culture and history in China. Although calligraphy did not become a personal art form until the *Han* dynasty (Knight 2012, p. 28; Li 2009, p. 72), that is, after the third century BCE, it traces its origin to the earliest confirmed archeological evidence of writing in Chinese on oracle bones some 3,500 years ago. The kings of late *Shang* (c. 1600–c. 1046 BCE) and early *Zhou* (c. 1046–256 BCE) made use of Chinese text carved on ox bones and tortoise shells to consult with higher beings and their ancestors on a wide range of topics concerning the royal house. Often, their inquiries and the eventual outcomes were both recorded. Hundreds of thousands of pieces of oracle-bones have been discovered and excavated since the end of the twentieth century (Qiu 2000, p. 61; Keightley 1978, p. xiii). Once decoded, oracle-bone text offered precious yet concrete glimpses of remote *Shang-Zhou* history to modern researchers. Following the oracle bones, the form of writing that came to dominate historical records was Chinese inscriptions cast on ritual bronzeware known as the bronze script. Bronzeware inscriptions, like the oracle-bone text, recorded valuable historical information. One prime example is Duke Mao's Cauldron, one of the most prized items in the collection of the Taipei Palace Museum. Its 497-character text, beautifully arranged on the inside of the vessel, tells the circumstances and reason for the casting of the bronze cauldron, which in turn memorializes the occasion and the people involved. Compared to the oracle-bone script, the characters are much more sophisticated, and their forms are elegant and harmonious, demonstrating the aesthetics of writing that was no doubt already highly valued at this stage. Writings in seal script, which is yet chronologically closer, continued the calligraphic evolution. They mostly survived as stone inscriptions, including poetry written in the great seal script found on stone drums of the Spring and Autumn period and the small seal inscriptions on stone steles commemorating the accomplishments of the First Emperor. Driven by the increased need to record legal and administrative information in the united China in the *Qin* (221–206 BCE), the technology of writing quickly developed, and Chinese text came to be written most dominantly with soft brushes on surfaces such as bamboo and silk, and eventually paper by *Eastern Han* (25–220 CE). It was during the *Han* dynasty (206 BCE–220 CE), along with the rise of the educated elite (Knight 2012, p. 28), that clerical script, running script, cursive script, and the standard script all developed. These scripts, together with the seal script, formed the complete repertoire of the five canonical styles in the grand tradition of Chinese calligraphy. Writing came to be admired for its personal styles, and Chinese text written in brush and ink came to take on a double role of both pragmatic communication and elevated artistic creation. Chinese calligraphy began to flourish as an art form.

From a linguistic perspective, writing, and by extension calligraphy, has almost always been important in Chinese culture thanks to the fundamental characteristics of the Chinese writing system. The Chinese script, like all full writing systems in the world, represents speech sounds (DeFrancis 1989); yet unlike most of the other ones, it also encodes meaning in its symbols. To illustrate this point, it may help to compare the Chinese writing system with an alphabetical script, such as the familiar Roman alphabet. In the Roman alphabet, the basic units, or graphemes (Rogers 2005), of the script are the individual letters. Each letter represents a conceptually single sound, or a sound category that linguists call a phoneme. For example, the letter <t> represents the sound /t/, and the letter <m> represents the sound /m/. The speech-grapheme correspondence in an alphabetical writing system is at the phonemic level; or in other words, in an alphabet, each grapheme represents a phoneme. An alphabetic writing system is a phonemic script, though not necessarily a perfect one depending on the spelling rules of each language. The Chinese script, however, is what John DeFrancis calls a morphosyllabic writing system (DeFrancis 1984). The graphemes are individual characters, and each character, in the great majority of cases, represents a meaningful unit of speech sounds that linguists call a morpheme. For example, the character <天> represents /tian/ "sky, heaven," and the character <书> encodes /shu/ "book, writing." For the Chinese writing system, therefore, the speech-grapheme correspondence is at the morphemic level, and the character-based Chinese script is a morphemic writing system (Rogers 2005). A language generally has a stable and limited set of phonemes, but it has a constantly changing and unlimited (at least conceptually) number of morphemes. It takes hundreds of years for sound changes to materialize, yet the addition of new words may happen overnight. Because of this key linguistic difference, the Roman alphabet only requires 20 something (or 50 something if taking into consideration the difference between capital and lower-case letters) graphemes, while the Chinese writing system requires thousands to be fully functional. This large inventory of signs, though cumbersome to learn and to use, provides rich materials for the creative manipulation of calligraphers. Under the Chinese empire, children were taught to practice calligraphy by copying from model albums from the very beginning of their literacy training (Wilkinson 2015, p. 47). For example, the *Thousand Character Classic* (千字文), a children's primer popular since the early Tang (618–907 CE), was a rhymed poem consisting of exactly 1000 characters, each appearing only once (with one exception in some editions; Wilkinson 2015, p. 295). It introduced the reader to the meaning and usage of basic Chinese characters. In this capacity, it inevitably became the content of persistent calligraphy practice, and a good number of outstanding calligraphers left behind treasured works writing the *Thousand Character Classic* (Wilkinson 2015, p. 295). These, in turn, became calligraphy models for those that came after.

The art of calligraphy was historically ubiquitous in the visual cultures of East and Southeast Asia. Brush writing, along with Classical Chinese, the Chinese writing system, and the classical canon of Chinese philosophy, literature, and religion, spread to Korea, Japan, and Vietnam. When these cultures came into contact with China, they did not yet utilize writing. Koreans became aware of

Chinese writing before the second century BCE (Taylor and Taylor 2014, p. 172). Scholars from Korea were probably the first non-Chinese to study the Chinese language and text written in the character-based script, including Confucian classics and Buddhist scriptures. Centuries later, they spread their learning to the Japanese (Taylor and Taylor 2014, p. 172), and, for millennia after that and until the late nineteenth century, educated Koreans and Japanese read and wrote in Classical Chinese using the Chinese writing system for almost all formal and official purposes. Similarly, in Vietnam, writing was primarily done in Classical Chinese and in Chinese characters, especially during the 1000 years or so of Chinese rule (111 BCE–938 CE) (Li 2020, p. 101). The Chinese script was adapted to write the Korean, Japanese, and Vietnamese languages, despite the fact that none of them was related to Chinese. Chinese characters later served as the basis on which Koreans, Japanese, and Vietnamese people created their native scripts: *hangeul*, *kana*, and *chữ nôm*. As one can imagine, the impact of the Chinese literate culture on Korea, Japan, and Vietnam was profound. The art of calligraphy took roots as well in this broader area, especially in Korea and Japan. For example, the Silla Dynasty (668–935 CE) calligrapher Kim Saeng (金生 *fl. c.* 800 CE) was the earliest Korean calligraphic master and was greatly influenced by the Chinese calligraphers Wang Xizhi (303–361 CE) and Ouyang Xun (557–641 CE) (Chen 2011, p. 116). The Choson Dynasty (1392–1910 CE) calligrapher Kim Jeong-hee (金正喜 1786–1856 CE), a.k.a. Chusa (秋史), was a highly innovative grandmaster (Pratt and Rutt 1999, p. 209). In Japan, the three greatest masters of the *karayo* (Chinese) tradition were known as the *sampitsu*, the "three brushes": they were the monk Kukai (空海 774–835 CE), the Emperor Saga (嵯峨天皇 786–842 CE), and the courtier Tachibana no Hayanari (?–842 CE) (Shively and McCullough 1988, p. 416). After the ninth century, a Japanese style of calligraphy, the *wayo* tradition, emerged that made use of more gentle and delicate lines (Nakata 1973). Zen calligraphy, referred to as *bokuseki* (墨跡 "traces of ink") in Japanese, was another highly original calligraphic style developed in Japan. Calligraphy, along with swordsmanship and seated meditation, was one of the required disciplines taught to students at Zen monasteries from the fourteenth to the sixteenth centuries (Addiss 1989). Today, calligraphers in Korea, Japan, and Vietnam also write in *hangeul*, *kana*, or *quốc ngữ* scripts, and brush calligraphy has become a salient visual symbol of Asian culture that transcends Chinese writing.

In the social culture of China and a few other Asian nations, being able to write fine calligraphy has been held in high regard. The ability to write poetry and prose in a good calligraphic hand, as previously mentioned, would distinguish a candidate in the imperial examinations, and superior performance in the exams would lead one to bureaucratic appointments endowed with economic success and a high social status (Barrass 2002, p. 16). This strong association of calligraphic ability with socioeconomic positioning is still evident today, because, in essence, the direct link between a good education and a successful career never weakened in Chinese society. The modern high school- and college-entrance exams have been likened to the civil service examinations. Having attractive or at least neat and clear handwriting is still very much emphasized in the preparation for these important exams. It may even be tested in the exam: the very first question of the 2013 Dalian city high-school entrance exam required the students to copy a Chinese passage "clearly and neatly" in the standard script, accounting for 2% of the exam grade. The quality of one's handwriting is thought to have a major impact on one's employment prospect and success as well (Dalian Evening News 2013). In Japan, employers prefer job candidates who submit handwritten resumes instead of printed ones, and those with elegant handwriting are perceived more favorably (Crowley and Li 2016, p. 20). Proficient calligraphers in China are able to obtain extra income, if not making a good living entirely on calligraphy, by writing at others' request or selling their works on the art market. Those who are highly accomplished are socioeconomic winners as well as "cultural heroes" (Knight 2012, p. 18).

Yet the significance of calligraphy goes beyond a form of communication or art. "*Ren ru qi zi, zi ru qi ren* (人如其字,字如其人)" – in traditional China and East Asia, one's calligraphic ability was indicative of and integral to one's personhood, including one's dispositions and moral character (Yen 2005). This may have to do with the strict discipline that calligraphy students adhere to and the exceedingly large – and sometimes legendary – amount of time and effort they dedicate to perfecting their skills. For example, Wang Xizhi, the calligraphy sage, is said to have turned the pond in front of his residence black just by washing his brushes in it in his daily practice. Persistence through hard work is seen as required not just to grow one's calligraphic prowess, but also to develop one's personal qualities such as perseverance, endurance, and dedication, which are highly valued in the Chinese cultural tradition. Stories of ancient calligraphers continue to be told today to encourage children and young students to be persistent in their artistic, athletic, or academic pursuits. Nonetheless, "skill, even when refined to its ultimate, is not enough to make a great calligrapher" (Knight 2012, p. 23). Only by writing beautiful calligraphy, one may not yet be respected as a great calligrapher. "*Xin zheng bi zheng* (心正笔正)*"* – it is also required that the person be a model in his action and his moral character for one's calligraphy work to be truly great. In traditional China, esteemed calligraphers were usually scholar officials, generals, or artists known for their loyalty, uprightness, or generosity; in modern society, those calligraphers who care about and contribute to the greater good are the most respected (Li 2009b, p. 76). One's calligraphy is an extension of one's moral character, and the ultimate path to good calligraphy is to work on one's quality as a person. This understanding is reflected very early in Chinese art theory and criticism, the emphasis of which "is almost entirely on the artists and how their works reflect their personalities" (Knight 2012, pp. 24–25). The perception of calligraphy as an extension of one's moral person is also reflected in calligraphy students' choice of copying models. The works of the loyal officials Yan Zhenqing (颜真卿 709–785 CE) and Liu Gongquan (柳公权 778–865 CE) are among the most popular, while those of Cai Jing (蔡京 1047–1126) and Qin Hui (秦桧 1091–1155), who have been regarded as treacherous traitors, have few emulators even though they may be equal or more superior in artistic merit.

In sum, studying calligraphy is essential for understanding the intellectual culture of China and East Asia, the world view of its people, and their humanistic values and perspectives in both the traditional and the modern world. Calligraphy writing not only has a rich and distinguished legacy from the past but is also a vibrant practice in present-day China, other parts of Asia, and among diaspora communities. Serious students of China and Asia must at least know the basics about calligraphy, and many students in a Chinese or (East) Asian studies program are well-positioned to do so. Integrating calligraphy education in the curricular offerings of these programs is beneficial to the students as well as to the programs.

Benefits

Given the rich artistic and social traditions of calligraphy in China and (East) Asia, offering opportunities to engage in calligraphy learning and practice will no doubt benefit students interested in China or (East) Asian studies. Learning about calligraphy may serve as a link to a broad range of knowledge, perspectives, and experiences. Besides the challenge, fun, and excitement in handling brush and ink, students may learn about the linguistic foundation of the art form, including the Chinese language and writing system, the art history of calligraphy, the aesthetic preference and critical theory of traditional Chinese art, the philosophical thought and religious practice associated with calligraphy, and the art and literary works in the related genres of painting and poetry. They may also use calligraphy study as an entryway into inquiries in the broader social sciences, learning about the anthropological significance of handwriting, the socioeconomic impetus of calligraphy practice, the power relations encoded in calligraphy in historic and modern politics, and the development of calligraphy education. They may compare Chinese calligraphy with other artistic traditions including Western and Arabic calligraphy, modern dance, and abstract expressionist painting. Depending on the time and resources available, the scope may be further expanded to include more specialized topics such as the physical and psychological health benefits of calligraphy, or computer-generated calligraphy works. Hands-on experiences, in addition to brush writing, may also include seal carving, brush making, and paper making. Students may learn beyond the classroom by attending public talks or taking field trips to museums, galleries, calligraphers' studios, or local art and cultural festivals that involve calligraphy. In terms of topics and content, there is little chance that a calligraphy course or course module should run out of materials or ideas, and students in a Chinese or (East) Asian studies program will benefit intellectually and artistically from the rich variety of ways to learn about calligraphy.

The academic programs in which calligraphy teaching and learning are situated will also benefit. Offering calligraphy courses may be an effective way to attract relevant majors and minors as well as to boost the overall program enrollment. Calligraphy courses are generally popular among students. For example, in a survey conducted of Chinese or (East) Asia studies programs that offered Chinese calligraphy courses or learning modules, 11 out of the 17 respondents reported that enrollment was easy, two indicated that it was moderate, and only two responded that it was challenging for them (Appendix I, Q12). One private university currently offers two sections of a calligraphy course every semester and still often has more than 100 students on the waitlist (Yu 2019). Students love learning about and writing calligraphy, and calligraphy courses are often a big draw among students of diverse academic backgrounds (Li and Yu 2015). This is an important consideration in the reality of the current national decline in language enrollment. Based on the MLA's recent surveys, students taking Chinese in the USA decreased by 13% from 2013 (61,084) to 2016 (53,069) (MLA 2016). Chinese programs at institutions that do not have a language requirement may be especially challenged in this respect. Calligraphy learning has a broad appeal, and students taking calligraphy courses are not limited to those who study Chinese language and culture. Fourteen out of seventeen responded in the aforementioned survey that their calligraphy courses attracted non-Chinese, Asian studies, or East Asian studies majors or minors, or other non--China-related students (Appendix I, Q18). Some of these students may become majors or minors through taking the calligraphy course. Calligraphy teaching also creates opportunities to engage those students in the program who are not taking the course, and student engagement as such contributes to community building in and beyond the program, increases its visibility, and encourages its growth. Calligraphy-related events and activities, such as a calligraphy exhibit, contest, workshop, or demonstration, can be made part of a larger cultural festival and may involve all students interested in participating and trying it out.

Furthermore, offering a calligraphy course may be able to enhance the coherence of the overall curriculum (Li 2011). The calligraphy course, with its broad interdisciplinary potential, may connect to extant course offerings in Chinese or (East) Asian culture, literature, and linguistics. For instance, students taking a modern Chinese film course may be interested in attending a presentation given by a calligraphy student on Chinese calligraphy in martial art films. Students who have been exposed to such connections, when they are made explicit in the relevant courses (e.g., calligraphy, film), may be more inclined to take the other courses (e.g., film, calligraphy) to explore further. The exposure may come from a single class session or a series of classes that can be flexibly integrated into a course as appropriate. Among the possibilities in this regard are Chinese language courses. Calligraphy instruction offers occasions for students to learn new Chinese characters, words, and expressions, and also to systematically learn about the nature and characteristics of the Chinese writing system. The calligraphy instructor may collaborate with language course instructors in holding calligraphy-writing workshops for students enrolled in language courses with students in the calligraphy course participating or assisting, and the calligraphy students may take part in certain sessions of the Chinese language courses to learn about the basics of the Chinese script. Experience as such, while increasing students' interest in Chinese language and culture, will also strengthen the curricular coherence of the program.

The above discussion on the importance and benefits of calligraphy education clarifies the *why* of calligraphy teaching. The next section will focus on the *what*, that is, the student learning outcomes or the educational goals of calligraphy instruction in the context of a Chinese or (East) Asian studies program of a US university. It will address questions such as what students of calligraphy should know, understand, be able to do, and value by the end of their learning. These goals or outcomes, of course, are inevitably constrained by the time and resources available. As indicated in the survey (Appendix I, Q5), the great majority of instructors teach calligraphy as an independent course. The current discussion will thus be based on a typical full course of calligraphy that meets 3 hours a week for a total of 15 weeks, and readers may adjust based on their specific situations.

What to Teach: The Educational Goals of a Calligraphy Curriculum

The contents of a calligraphy course are directly determined by its student learning outcomes (SLOs). Formulating the SLOs is the first step in course design, and it lays the foundation for clarifying the assessment methods, identifying the topics, selecting or creating the materials, and deciding on the pedagogical approaches. This section first reviews the current SLOs or course objectives by looking at the syllabi of 14 calligraphy courses offered from 2002 to 2017, all in the context of a Chinese or (East) Asian studies program. It then proposes to expand these goals in alignment with the general goals of liberal education, in particular with an emphasis on the cultivation of student dispositions through restructuring calligraphy instruction. In the next section on the pedagogical approach, it will discuss further how to restructure calligraphy teaching.

Current Goals

The syllabi collected for this study are for various versions of more or less the same course, an "introduction to Chinese calligraphy" course at the undergraduate level. They are full, stand-alone courses with a hands-on brush-writing component. These syllabi reveal that calligraphy instruction focus on five major areas: brush-writing skills and practice, Chinese writing system and characters, history and development, aesthetics and appreciation, and cultural heritage. Table 1 shows the specific topics in these areas. Although no two courses are identical, they are highly similar – each course covers almost all of the main topics and more than half of the sub-topics.

Almost all (13 out of 14) of the syllabi include clear statements of course objectives or SLOs. An examination of these statements shows that these courses

Table 1 Topics of current calligraphy courses

Main topics	Sub-topics
Skills and practice	Posture, tools and materials, writing techniques
Chinese writing system and characters	Form, meaning, and pronunciation of Chinese characters, radicals and components, stroke orders, Xu Shen's *Shuowen jiezi*
History and development	Origin of calligraphy, Evolution in calligraphic styles, masters and masterpieces
Aesthetics and appreciation	Composition, aesthetic principles, connoisseurship, poetry, painting
Cultural heritage	Philosophy and religion, customs and festivals, architecture

share similar educational goals. The parts of the statements that contain key phrases (in bold) are provided below in Table 2.

To evaluate the stated course goals, we may use a framework developed for formulating student learning outcomes for second language programs (Norris 2006). Based on this framework, the educational goals of a course need to specify what students should know, understand, be able to do, and value by the time they complete the course. Or, simply put, student-learning outcomes need to consist of the following aspects: *knowledge*, *skills*, and *dispositions*:

> ...the outcomes of college FL programs promise (a) critical *knowledge* of cultures, our own and others, and of the tremendous richness of content entailed within an understanding of culture; (b) useful *language skills* – especially the ability to act as successful communicative agents in tasks ranging from the most mundane (if essential) survival needs to the most extraordinary literary accomplishments and everything in between, that is, the reasons we use language as human beings; and (c) thoughtful *dispositions* – toward an appreciation of pluriculturalism and plurilingualism, and a tolerance for ambiguity and diversity in language use (second language or first language), and away from critical, naive, monolithic, elitist, and even racist attitudes about the roles of individuals and societies as they communicate with each other in the world. (Norris 2006, p. 577)

It is important, first of all, to recognize that current calligraphy instruction more heavily focuses on the teaching of knowledge and skills. The course objective statements in Table 2 may be coded into the three categories of knowledge, skills, and dispositions, as shown in Table 3. As it is evident, significantly more courses have clear statements of knowledge and skills than on dispositions. This configuration may be a carryover of how calligraphy is usually taught in its native language and cultural environment. Calligraphy instruction in China, more often than not, is offered as art education. Calligraphy courses generally focus on improving students' brush writing skills, aesthetic sensitivity, and knowledge of the art history. Native Chinese students already speak the language, have intimate user knowledge of the writing system, and have ready access to rich resources for learning about the history and sociocultural significance of calligraphy practice. Even if they may not yet know or understand much, pre-high-school students may rely on their Chinese or history courses, and college students may be completely proficient in self-guided learning. Thus, a typical calligraphy course in China focuses on narrowly and traditionally defined disciplinary contents. Its instructor is often an accomplished calligrapher with rich personal experience to draw from.

To be sure, it is also common for calligraphy courses offered in the USA, whether intentionally or unintentionally, to follow the general practice of calligraphy instruction in China. This may have to do with the fact that most calligraphy instructors have a Chinese background and have been trained in their native cultural and academic context. As a result, however, their courses are heavily focused on mastering knowledge and skills. The course objectives regarding "skills," in particular, suggest much emphasis placed on training brush writing proficiency: for example, to "*master* the skills of writing Chinese characters," "to be able to write *decent* brush works," and to develop "*facility* with the Chinese characters in the regular style." It is important to recognize pedagogically that most of our students do

Table 2 Learning goals or outcomes stated in syllabi

1	"It is hoped that by the end of the session students will have developed **facility with the Chinese characters** in the regular style. In addition, the students are expected to acquire an **understanding and appreciation of the aesthetics** of Chinese calligraphy and its role in Chinese culture." (Hsieh 2017)
2	It focuses on teaching the student the basic writing techniques for standard script (*kaishu*). In addition, the course exposes students to some **basic knowledge of Chinese characters and elements of Chinese culture**. By the end of the semester students should learn the basic rules of doing Chinese calligraphy, obtain the skills of how to do brush work, and be able to create decent calligraphy pieces. They will also learn how to **appreciate fine works** of this unique Chinese art form. (Liu 2016)
3	This course is designed to introduce the importance of calligraphy in its **development integrated with Chinese art, history, and culture**, with emphasis on incorporating calligraphy into the more comprehensive objective of **achieving cultural proficiency**. Consequently this class provides students the opportunity not only to learn how to appreciate and to write Chinese calligraphy, but also to understand how the calligraphy **intertwining with various aspects of culture element** in China and its influence to the neighboring countries such as Japan and Korea. (Sheng 2016)
4	It will teach students **how to produce Chinese calligraphy** during hands-on practice in class. In addition to **the history, development, aesthetics, and appreciation** of Chinese calligraphy, it also includes many aspects of the culturally fascinating heritage of China. (Chen 2011)
5	"The object(ive) of this course is to encourage students to develop **a deeper interest and understanding of the Chinese language and its culture** through the **practice of Chinese calligraphy**." (Pittman 2010)
6	"It is hoped that by the end of the session the students will have **developed facility with the Chinese characters in the regular style**. In addition, the students are expected to acquire an **understanding and appreciation of the aesthetics of Chinese calligraphy and its role in Chinese culture**." (Wu 2009b)
7	"The primary task aims to enable students to **understand the fundamental rules in the structural formation of Chinese characters**, and on that basis, to **master the skills of writing Chinese characters**." (Lan 2009)
8	"... students should have a well-rounded and essential **knowledge of the history and theory of Chinese calligraphy** and will also **be able to write decent brush works**." (Wu 2009a; Hsieh 2009)
9	"The course aims to incorporate calligraphy into the more comprehensive objective of **achieving cultural proficiency**. Students will have the opportunity to **learn Chinese characters** not only as linguistic symbols but also **as culture emblems and art forms**." (Sheng 2009)
10	... focuses on the **history, development, aesthetics, and appreciation** of Chinese calligraphy, and teaches students **how to produce Chinese calligraphy** during hands-on practice sessions in class. ... This course does not only teach the simple skill of writing with a brush, but also attempts to introduce students to Chinese poetic artistry and to the interrelation of **poetry, calligraphy, and painting** in the Chinese tradition, with each element complementing the other two. Moreover, as writing is an expression and outpouring of ideas and emotions of the individual, it is expected that many aspects of China's **culturally fascinating heritage** will be introduced through the practice of calligraphy. (Huang 2006)
11	Chinese Calligraphy 101 is an introductory course that teaches the **fundamental techniques and aesthetic values** of Chinese Calligraphy...You will be introduced to the **history and art** of Chinese calligraphy through an exhibition of different calligraphy styles

(continued)

Table 2 (continued)

	and display of calligraphy supplies, then you can put your knowledge to use in a workshop with hands-on instruction. (He and Chew 2003)
12	"It is expected that through the **practice of calligraphy**, many **aspects of the culturally fascinating heritage of China** will be introduced." (Wu 2003)
13	"It will teach students the **basic training techniques** of writing Chinese characters in the "standard script" (kai-shu). First, students will be introduced to **materials** in calligraphy, and the proper **ways to manipulate the brush**. They will then learn the **concept** of "tracing the vermilion" (miao-hung), and ways to avoid the common errors in technique (ba-bing). Finally, students will be required to familiarize themselves, through practice, with the **proper sequence of strokes** of characters presented in this course…" (Yang 2002)

Table 3 Current calligraphy course goals in knowledge, skills, and dispositions. (The numbers are consistent with those in Table 2)

Knowledge	Skills	Dispositions
History of calligraphy art (#3, #4, #8, #10, #11) Knowledge of the Chinese writing system (#2, #7, #9) Knowledge of Chinese culture (#2, #10, #12) Basic knowledge of calligraphy practice (#13) (no statements): #1, #5, #6	Brush writing techniques through hands-on practice: #1, #2, #4, #5, #6, #7, #8, #10, #11, #12, #13 (no statements): #3, #9	Understanding and appreciation of the aesthetics (#1, #2, #4, #6, #10, #11) (vague statements: #3, #9 ("achieve cultural proficiency"), #5 "develop deeper interest in Chinese language and culture") (no statements): #3, #7, #8, #9, #12, #13

not aspire to become calligraphers. Given the time and energy demanded for being able to master calligraphy, and the limited time of practice during the course, we probably cannot expect the majority of the students to become very good at writing using brush and ink. Therefore, training students to become highly proficient brush writers of Chinese characters may not be the predominant goal to accomplish with such a course. The hands-on component of the course may be sufficiently characterized as *structured practice* of calligraphy. *Structured practice* means that, with careful planning, the students practice on a limited set of calligraphy assignments that progress from simpler strokes or characters to the more complex. In order to guide students in such practice, the instructor needs to be reasonably proficient in the specific tasks that the students will perform. With focused practice, instructors with moderate training and experience may become comfortable with demonstrating these tasks in class even though she or he may not be a highly skilled calligrapher. For performance-oriented demonstration, the instructor can effectively supplement the course with recorded video material and occasional visits by local calligraphers.

Calligraphy study provides excellent opportunities for intellectual inquiries across the related disciplines beyond the scope defined by traditional calligraphy training. Liberating the calligraphy course from the predominating goal of training good brush writers will help us pursue more important and meaningful objectives. We may reevaluate the SLOs or learning objectives of calligraphy courses by considering the overall goals of a liberal education.

Goals of a Liberal Education

Since the mid-twentieth century, liberal education philosophy and approach have been central to much of US higher education. Its guiding philosophy can be used to inspire the educational goals of calligraphy courses. According to the Association of American Colleges and Universities, liberal education is "an approach to learning that empowers individuals and prepares them to deal with complexity, diversity, and change.... A liberal education helps students develop a sense of social responsibility as well as strong and transferable intellectual and practical skills such as communication, analytical and problem-solving skills, and a demonstrated ability to apply knowledge and skills in real-world settings" (AACU 2019). Liberal education has also been defined as education for its own sake, with a strong emphasis on personal enrichment and the teaching of values (Harrison 2007, p. 191). A liberal education is thought to "produce[s] persons who are open-minded and free from provincialism, dogma, preconception, and ideology; conscious of their opinions and judgments; reflective of their actions; and aware of their place in the social and natural worlds" (Project on Liberal Education and the Sciences 1990).

The concept of *dispositions* as defined by Norris (2006), in the context of liberal education, coincides to a great extent with the set of *developmental imperatives* proposed by the pedagogue and advocate for liberal education Marshall Gregory (2009). Gregory argues that the end of a liberal education is in turning students into "different" people, and the learning of knowledge and the acquisition of skills are merely the means by which students approximate their developmental capacities as autonomous human beings. More specifically, he proposes that a meaningful liberal education should aim to address a set of "developmental imperatives" in areas including language, physicality, rationality, aesthetic responsiveness, imagination, tolerance for ambiguity, appreciation of diversity, moral and ethical deliberation, introspection, sociability, and capacity for shame. The basic idea is that the ultimate goal of a liberal education is not to accumulate knowledge about a certain subject or mastering the skills of particular tasks; rather, it is in developing higher-level perspectives, capacities, and capabilities in becoming compassionate, mature, and autonomous human beings. What should be the contents and goals of calligraphy education, then, in order to facilitate student growth in this direction?

The teaching of knowledge and training of skills in a calligraphy course are both important. However, they must not be taken as the only goals in the context of American higher education. This is because the end of a liberal education is not in learning knowledge or acquiring skills per se. In a dominantly Chinese cultural environment, the humanistic and developmental values of a calligraphy education are tacitly agreed upon and may be taken for granted, but in the USA, we cannot assume the same of our students. Therefore, we need to be more explicit in expressing such goals. If a student tries to recall what she has learned in college a few years ago, she actually retains a very small percentage of the knowledge and has possibly lost much of the skills that she has not been putting to practice. One year after taking the calligraphy course, the student may have difficulty remembering all the names for the calligraphic styles and may also find it difficult to pick up a brush and write a Chinese character well if she has not been practicing. This is not to say that a knowledge- and skill-focused calligraphy course may not implicitly cultivate student dispositions – it certainly does to some extent – but calligraphy instruction that *aims at* cultivating student dispositions and *explicitly* does so will be maximally meaningful for undergraduate students.

To translate the educational goals into pedagogical practice, it will be useful to return to the "three P's" framework in second language education (ACTFL 1999): cultural perspectives, cultural practices, and cultural products. Cultural perspectives are "culture's view of the world," its meanings, attitudes, values, and ideas. Cultural practices are "patterns of behavior accepted by a society and deals with aspects of culture such as rites of passage, the use of forms of discourse, the social 'pecking order,' and the use of space." Cultural products are both tangible artifacts and intangible information of a particular culture. This paper argues that cultivating student dispositions requires that calligraphy instruction restructure to focus more on cultural perspectives. The section below will discuss how this may be accomplished by using a few examples.

How to Teach: The Pedagogical Approach

Leading with Cultural Perspectives

To integrate the study of cultural perspectives into learning about cultural practices and products, this chapter proposes to develop calligraphy course modules, i.e., units of study, in which cultural perspectives play a leading role in both structure and content. Pedagogically speaking, "leading" here refers to at least two aspects: precedency in timing and primacy in importance. In other words, within a perspective-led course unit, student engagement in the study of cultural perspectives comes first, and it also receives the most attention and effort. The course module may begin with learning about an articulated cultural perspective through substantial reading and explicit discussion that directly addresses the cultural understanding and insights the course expects the students to acquire. It may then engage students in related cultural practices and in learning about the relevant cultural products. During this second stage, students should be guided and encouraged to reflect on the cultural perspective that they have carefully thought about and discussed in the first stage. Structured like this, a course module places primary focus on cultivating student dispositions, and the development of skills and the learning of knowledge only serve to enhance the learning toward the primary goal. Again, this is not to say that knowledge and skills are not important, useful, or beneficial; rather, their importance lies in serving the primary objective of nurturing students in becoming compassionate, mature, and autonomous persons.

What may such a course module look like? Below are two examples.

1. The Chinese personhood, writing *cangfeng*, and the calligraphy brush

    In writing the standard script, one basic technique is the so-called "hidden tip" (*cangfeng* 藏锋). In creating the hidden-tip effect at the onset of a stroke, the brush starts out pointing toward the opposite direction of the intended writing movement. As the tip of the brush makes contact with the writing surface, the brush is pressed down only slightly, making a pointed mark, and it is then brought to move in the intended direction, pressed further down at that time with full force and rigor. This movement makes a wider mark that covers the pointed start made by the tip of the brush and thus "hides" it from being visible. The result is a start of the stroke that is more or less rounded, heavy, and modest rather than sharp, light, and showy. The *cangfeng* technique may be used at the end of a horizontal or a vertical stroke as well. Although not all well-known calligraphers made extensive use of this technique (e.g., Ouyang Xun 欧阳询), many did. In modern day calligraphy classes for beginners, it is widely taught as a basic – and perhaps one of the most important – brush technique in writing the standard script.

    It has been argued that the *cangfeng* technique embodies some of the essential qualities of the Chinese personhood from a traditional cultural perspective. "The predominant preference of hidden brush tip (*cangfeng*) over exposed brush tip (*lufeng* 露锋) reflects the traditional Chinese ideal about the proper bearing of a gentleman (*junzi* 君子), a man with noble character – people with lofty character should not show off their talents" (Yen 2005). As Yen (2005) further elaborates in her book, according to the Confucian conceptualization of what constitutes a noble person, qualities such as depth, roundedness, modesty, and being controlled are highly preferred over shallowness, sharpness, empty sort of charm, and being unchecked. Such conceptualization of the Chinese personhood is fully reflected in calligraphic brush techniques, and the hidden tip technique is one such example. Before demonstrating to students how to use this technique and having them practice, the instructor may have them read about Confucian thoughts and ideals, in particular the concept of *junzi* and the preferred qualities of an ideal person in traditional Chinese culture. It may also be effective to have them compare such cultural expectations with those of the West and those of the modern China they know. Students should be able to come up with specific examples from their own life or from popular media, and their discussion should lead to a more in-depth understanding of what the Chinese culture is like in this regard and why.

    The cultural practice of this unit may focus on learning and applying the *cangfeng* technique in brush writing. For example, the instructor may demonstrate how to write a horizontal line in the standard script for which the technique is used at both ends. Students can observe and practice using their own brushes. It will be important that the instructor offer one-on-one guidance. It is also worth pointing out that, although the course is perspective-led, it does not mean that hands-on practice will become haphazard. Indeed, systematic practice of calligraphy writing is still important, and if the instructor carefully structures the calligraphy practice, it is also possible to accomplish. For example, if *cangfeng* is the very first unit in the course, then in the practice work, students may be asked to write simple characters that consist of the most basic strokes, the horizontal and the vertical. Characters such as 一, 二, 三, 十, 土, and 王 will be suitable. Then the next assignment in the same unit may introduce two more strokes, the down-left sweep and the down-right sweep, so characters like 八, 人, 大, 木, 本, 禾 may be good options. If *cangfeng* is scheduled toward the middle of the course (it probably should be introduced no later than midterm, as *cangfeng* is a basic technique used by many calligraphers), then the instructor may use more complex characters that also include bends, hooks, and dots that the students will have already practiced and for all of which the *cangfeng* technique is often employed.

The cultural product in this module may be the calligraphy brush, because the *cangfeng* technique has much to do with the construction of the brush. The instructor may lead the students in learning about this tool, offering, first of all, information related to the physical object. He/she may show various kinds of brushes, have students examine and handle the brushes, and talk about the materials brushes are made of. He/she may also teach the students how to choose a good brush, how to hold and use a brush, and how to maintain a brush. He/she may demonstrate the production of a rich variety of forms and the various degree of difficulty in handling the brush accordingly. Students may be encouraged to experiment with the brush to gain more in-depth understanding of its simplicity in structure and versatility in application. Furthermore, it may be an interesting exercise for them to compare and contrast the Chinese brush and the Western paint brush and talk about the differences in the art-making process and the artistic effect that the tools may create.

2. Beauty in nature, appreciating calligraphic forms, and calligraphy models

The aesthetic principles of calligraphy are rooted in nature. One basic criterion for evaluating the quality of a calligraphy work is to see if its brush strokes and character forms possess life energy, i.e., if it is *qiyun shengdong* (气韵生动). The most prized masterpieces are those whose strokes and characters have the same profound energy as found in nature. In fact, "nature is the inexhaustible resource of artistic inspiration. Through observation of nature, calligraphers discover the principles of every type of movement and rhythm and try to convey them using the calligraphic brush" (Yen 2005). One of the most often told stories is, again, that of Wang Xizhi. It is said that he devoted a large amount of time and attention to observing how geese walked, because he discovered that the neck of the geese, in their swaggy gait, was inexplicably graceful. He wanted his brush-written characters to have the same kind of elegance, an energy that is reserved yet remarkably refined. Through repeated observation and practice, he was eventually able to transfer that elegance and energy into his own calligraphy writing. Of course, looking at Wang's calligraphy, one will probably not see any goose-neck forms, because imitating the forms of living creatures or natural phenomena is not the point. Rather, the beauty of calligraphy is in its nature-inspired and nature-like vitality, dynamism, and balance. Observing nature is only the beginning, and being able to capture and instill the beauty of nature into one's brush writing requires a highly perceptive mind and great aesthetic sensitivity. The same principles apply to Chinese painting as well as poetry, which, together with calligraphy, are the "three perfections" that the Chinese literati aimed to accomplish. The shared aesthetic principles of these art forms underpin much of the humanistic creation of imperial China and have a profound impact on that of modern China.

Calligraphy criticism written by ancient Chinese calligraphers may serve as excellent teaching materials for this unit. Two of the best-known critics were Lady Wei (Wei Shuo 卫铄; 272–349 CE) of the Eastern Jin Dynasty and Sun Guoting (孙过庭; 646–691 CE) of the early Tang Dynasty, and they both wrote about how calligraphers may draw inspiration from the natural world. Below are two passages, one from each author, as examples. They are filled with metaphors that draw parallels between natural phenomena or living creatures and calligraphic strokes or character forms, and they may very vividly inspire students' imagination. The instructors may have students read such passages and discuss their understanding of the metaphors. After that, he/she may have them practice reading calligraphy models and coming up with their own nature-inspired metaphors. Of course, reading, in this case, is not about comprehending the textual meaning; rather, it is a practice to more fully appreciate the beauty of calligraphy by carefully observing, sensing, and feeling the brush strokes and the characters:

An elongated horizontal line should convey the openness of an array of clouds stretching for thousands of miles. A dot should contain the energy of a rock falling from a mountain peak. A *pie* (撇 sweeping-left stroke) ought to resemble an ivory tusk in its luminous smoothness and unrestrained curvature; a vertical line an ancient cane drooping from the tree in its stability and serenity. A *na* (捺 sweeping-right stroke) has to contain the orgiastic vigor of rolling waves, or crushing thunder and lightening. – Lady Wei, excerpt of *Chart of brush maneuvers* (*Bizheng tu* 笔阵图). (*Lidai* 历代 1979, p. 27 quoted from Yen 2005, p. 85)

Consider the difference between the "suspended needle (*xuanzhen* 悬针)" and "hanging dewdrop (*chuilu* 垂露)" brush strokes, and then consider the marvels of rolling thunder and toppling rocks, the postures of the wild geese in flight and beasts in fright, the attitudes of phoenixes dancing and snakes startled, the power of sheer cliffs and crumbling peaks, . . . sometimes heavy like threatening clouds and sometimes light like cicada wings; . . . – Sun Guoting, excerpt of *Treatise on Calligraphy* (*Shupu* 书谱). (translation adapted and modified from Chang and Frankel 1995; quoted from Yen 2005, p. 84)

The cultural product in this unit can be the calligraphy models that the students read. Where did these models come from? What purposes did they serve in the past and do they still serve similar purposes now? There may be a rich history that the instructor can talk about with the students. The fact that copying is a foundational practice may be likened to that of classical training in Western drawing and painting. The didactic relationship between imitating past masters and creating one's own style may apply to both cases. This discussion may serve as the larger context of the students' practice in class. It may help them understand why it is good to write the same stroke and the same character multiple times in their homework, and why each time is an opportunity to improve upon the previous attempt. The instructor may also talk about the creation of calligraphy models from rubbings of stone steles. Knowledge about what a stele is and how rubbings are made may seem trivial, but it is necessary for the students to understand in a more concrete manner the tradition in which they participate in every class meeting. As yet another example, the instructor may discuss the choosing of calligraphy models and how such choices, as discussed earlier in this chapter, are intimately linked to one's personhood and moral preferences. A fourth possibility in learning about calligraphy models is to look at the production and availability of model books in today's China. The fact that almost every other major publisher produces calligraphy model books shows the popularity of the art form and the profitability of related economic activities.

In addition to the two examples given above, thanks to its interdisciplinary nature, rich and long tradition, and vibrant modern practice, calligraphy affords almost inexhaustible possibilities for engaging students in such perspective-led studies. In selecting the specific cultural perspectives to incorporate in the calligraphy course, preference should be given to those that have relatively broad and profound implications and that may offer insights into the Chinese or (East) Asian people and society beyond the study of calligraphy. Here are a few more examples. First, the political prowess of calligraphy practice: for example, the perceived quality of one's calligraphy may be intertwined with one's political power (Hay 2005; Kraus 1991; Ledderose 1986), and, in contemporary China, calligraphy writing may be used as a means to exchange for economic gains by taking advantage of one's political position (Gaur 2000). Calligraphy practice and, along with it, the use of the Chinese writing system, were deeply gendered: in traditional East Asian cultures, Chinese characters were associated with the dominant male gender in contrast with later created alternative scripts predominantly used by the female (Hamlish 1999; Fan 1996). A third perspective may be the spirituality of calligraphy: how it is used in Daoist, Buddhist, and Confucian traditions (Kieschnick 2003; Li 2000; Stevens 1996). It may also be about pushing the boundaries in calligraphy writing in the modern context: how modern artists have used the vocabulary of traditional calligraphy in creating new artistic languages that speak to a more culturally diverse audience (Ingold 2007; Barrass 2002; Ericson 2001).

Ideally, a comprehensive textbook may be written to support using the perspective-led approach in teaching a full calligraphy course. Many of the current calligraphy textbooks represented in the survey (Wen 2014; Chen 2011; Li 2009; Ouyang and Fong 2008; Chin 2001; Billeter 1989; Kwo 1981; Chiang 1973) already incorporate cultural contents, but such contents are more heavily focused on knowledge than perspectives, and there is also a general lack of connection between cultural perspectives, practices, and products. In the proposed textbook, each chapter may start with discussing a particular cultural perspective and then introduce the cultural practices and products that may be understood and appreciated through that perspective. Then, a related aspect of calligraphy practice may be introduced and students may be encouraged to work on their brush-writing skills while reflecting on the cultural perspective of the chapter, linking the "what" and "how" to the "why" of the calligraphy practice. The challenge of compiling such a textbook may be to maintain the systematicity and comprehensiveness in the training of calligraphic skills. Indeed, it may require some compromise in this regard, but if the reward is to overall better serve the students in achieving their learning outcomes, then such compromise may be well justified.

Conclusion

Calligraphy instruction in the context of Chinese or (East) Asian studies programs, though still at its beginning stage, has been gaining popularity and importance. After reviewing the educational significance and benefits of teaching Chinese calligraphy from both the students' and the programs' perspectives, this chapter has proposed to further calligraphy education in order to cultivate student dispositions in alignment with the educational goals of a liberal education. Furthermore, it has proposed to teach calligraphy using a cultural-perspective-led approach, in which the practice of skills and learning of cultural knowledge serve the development of cultural understanding and insights. Overall, the chapter addresses the why, what, and how of calligraphy instruction with the aim to envision the near future of calligraphy education in US universities.

With the above vision in mind, we may begin to explore what it would require to prepare ourselves for teaching calligraphy in the perspective-led approach. The basic ingredients may in fact not be that different from the more traditional approach; only that they are in an inverted relationship: cultural understanding and insights now outweigh brush-writing skills or knowledge of cultural products. This is not to say that the latter is not important; they are, too, but the former should become our priority. Thus, as calligraphy educators, we will need to devote a lot more time and energy toward the development of calligraphy-related cultural perspectives. This should be a basic training for us. Ideally, there could already be comprehensive textbooks prepared for us to learn and to use. Before such books become available, however, our preparation may include broad-ranging reading and research using scholarly books and articles, films, primary-source materials, and so on. The selection of topics may be determined by the specific characteristics and goals of the student audience. Some of the cultural perspectives outlined in the previous section may be considered, but they are not mandatory. As we accumulate ideas through our reading and research, we will be able to come up with the topics most suited to our students. Ultimately, we will need to digest and distill the insights we have gained into teachable lessons complemented by the training of calligraphic skills.

Appendix I

Survey: Chinese Calligraphy Instruction in US Universities
    The survey was conducted in July, 2019. Below are the survey questions with a summary of the results. 17 responses were collected.

Q1. In what discipline did you receive your advanced degree (s) (PhD and/or Masters)?

Linguistics (3)
Applied linguistics (2)
Comparative literature (3)
Art history (1)
Chinese linguistics (1)
Chinese literature (1)
Chinese language and literature (1)
East Asian literature (1)
Education (1)
Japanese literature (1)
Language and literacy education (1)
Language, literacy, and culture (1)

Q2. In general, what kinds of courses do you teach at your institution? Choose all that apply.

Chinese culture (15)
Chinese language (13)
Chinese linguistics (4)
Chinese literature (3)
(None of the above) (1)

Q3. What is your skill level in calligraphy techniques?

Beginning (0)
Intermediate (5)
Advanced (12)
Mastery (0)

Q4. If your skill level is beginning or intermediate, did you collaborate with someone with more advanced skills in teaching? Based on your teaching experience, do you believe advanced calligraphic skills are required to effectively teach a calligraphy course to US undergraduate students in the context of Chinese or Asian studies? Why or why not?

1. Yes. (2); No (2); no response (12)
2. Advanced skills are: desirable but not necessary (7), required (4), depends (3), no or nonsensical response (3)

Q5. In what form (e.g., a course module, a full course) did you last teach calligraphy?

A full course (14), a one-credit-hour course (2), a course module (1)

Q6. If you taught a course module on calligraphy, what was the course within which this module was offered?

Chinese Culture and Civilization

Q7. If you taught a full course, what was the name of the academic program that offered this course?

Chinese (11); Chinese Language (1), Modern Languages (1), East Asian Languages and Cultures (1); cross-listed: Asian Studies (1), Art History (1), East Asian Studies (1), Anthropology (1); N/A (1); no or nonsensical response (2)

Q8. How long had you taught Chinese calligraphy before your last experience? What level was the course (e.g., second year)? How often was it offered?

1. 1 year (1), 2 years (1), 3 years (1), 4 years (1), 5 years (2), 6 years (1), 10+ years (2), 15 years (1); 18 years (1), almost 20 years (1), 20+ years (1); no or nonsensical response (4);
2. 1st year (2), 2nd year (2), 3rd year (2), 4th year (1), no or nonsensical response (10)
3. Every semester (5), once a year (4), depends on instructor availability (1), no or nonsensical response (7)

Q9. What was the language of instruction?

Chinese (0)
English (10)
A mixture of Chinese and English (7)

Q10. Was there a Chinese language learning component in the calligraphy course or course module you teach? If so, what was it like?

Yes (3): learning traditional characters, new words, Tang poems; listening and speaking skills of Chinese language; reading Chinese articles, discussing in Chinese, vocabulary quizzes, conversation practice, presentation in Chinese
No (14)

Q11. What kinds of students did you usually have in this course (module)? Choose all that apply.

Chinese majors or minors (14)
East Asian Studies or Asian Studies majors or minors (11)

Nonmajors or minors taking Chinese courses (8)
Other non-China-related students (8)

Q12. How was the enrollment?

Easy (11)
Moderate (4)
Challenging (2)

Q13. Below are topics from calligraphy course syllabi found online. Which of these topics did your course (module) cover? Choose all that apply.

Brush-writing knowledge and techniques (17)
Understanding of the Chinese writing system (17)
History of calligraphy including its masters and masterpieces (15)
Aesthetic principles and criticism (16)
Calligraphy and the Chinese personhood or worldview (10)
Connections with painting and poetry in traditional culture (14)
Calligraphy in Confucian, Daoist, or Buddhist practices (10)
Calligraphy and the civil service examination (9)
Calligraphy and modern politics (5)
Modernist calligraphy (5)
Calligraphy in avant-garde art (6)
Connections with modern dance, music, or graphic art (4)
Connections with martial art or human health (7)

Q14. If you did not teach one or more of the above topics, would you be interested in including them in your future teaching? If some of the topics you taught were not in the list above, please list them below.

Yes (1): health and martial arts;
No (3)
Maybe (2): (no elaboration)
No response or nonsensical response (11)

Q15. How important was each of these student learning outcomes (SLOs) to your teaching of calligraphy?

Write beautiful calligraphy	
Very important (6)	Moderately important (7)
Slightly important (4)	Not at all important (0)
Appreciate calligraphy works in artistic terms	
Very important (11)	Moderately important (4)
Slightly important (2)	Not at all important (0)

(continued)

Understand the social, cultural, and historical context of calligraphic practice	
Very important (10)	Moderately important (5)
Slightly important (1)	Not at all important (1)
Learn Chinese characters	
Very important (2)	Moderately important (7)
Slightly important (4)	Not at all important (4)
Read and understand Chinese text on calligraphy-related topics	
Very important (2)	Moderately important (5)
Slightly important (1)	Not at all important (9)
Discuss calligraphy-related topics in Chinese	
Very important (3)	Moderately important (3)
Slightly important (4)	Not at all important (7)

Q16. Did you use textbooks for your course? If so, what were they? If you did not use a main text, what did course readings consist of? What other teaching materials (e.g., online videos) did you use to supplement course readings?

Yes: Wendan Li 2009 (6); Yee Chiang 1973 (3); Ouyang Zhongshi and Wen Fong 2008 (1); Jin Cong, unpublished (1)
No textbook: self-created course packs (4); self-created handouts (1); individual readings (3); online videos (11); websites (2); self-created videos (2); DVDs (2); calligraphy models (2); no readings at all (3)
Nonsensical response (2)

Q17. Did you usually incorporate the following activities in your teaching? Choose all that apply.

Visits to art museums (6)
Guest calligraphers (5)
Guest speakers (4)
Student calligraphy show (14)
(None of the above) (2)

Q18. Do you think your calligraphy course (module) contributed to your program in one or more of the following ways? Choose all that apply.

Boosted program enrollment (14)
Attracted majors or minors (11)
Built connections between courses (11)
Increased visibility (13)
Created opportunities for student engagement (12)
(None of the above) (0)

Q19. What have been the greatest challenges for you in teaching calligraphy? What support will be the most helpful from either your institution or the field?

Better facilities including more convenient classrooms (3)
Wish to improve instructors' own calligraphy writing skills (2)
Demanding in instructor's energy (2)
Nothing (2)
Teaching approach (1)
Varied student needs (1)
Lack of class time (1)
Need for TAs (1)
Lack of opportunities for regular offering (1)
Need for an art museum nearby (1)

Appendix II

Universities and colleges offering Chinese calligraphy instruction

This list is compiled based on internet searches of departmental course listings or university course bulletins conducted in July 2019. The calligraphy courses or course components are offered by the Chinese language, culture, and literature programs or (East) Asian studies programs. Calligraphy teaching provided by art or other programs is not included.

26 National Universities:

University of Chicago
Massachusetts Institute of Technology
Yale University
Stanford University
Vanderbilt University
Washington University in St. Louis
Emory University
University of Southern California
Carnegie Mellon University
University of Michigan, Ann Arbor
University of North Caroline, Chapel Hill
University of California, Santa Barbara
University of Florida
Boston University
Villanova University
Rutgers University
Purdue University
University of Maryland, College Park
University of Pittsburg
Florida State University
University of Minnesota, Twin Cities
Texas Christian University
Indiana University

University of Delaware
University of Iowa
Cleveland State University

3 National Liberal Arts Colleges:

Oberlin College
Bard College
Kalamazoo College

2 Regional Universities:

Loyola Marymount University
California Polytechnic Pomona

References

AACU. 2019. What is liberal education? https://www.aacu.org/leap/what-is-liberal-education. Retrieved July 25, 2019.
ACTFL. 1999. *Standards for foreign language learning in the 21st century.* Lawrence: Allen Press.
Addiss, S. 1989. *The art of Zen: Paintings and calligraphy by Japanese monks.* New York: H. N. Abrams.
Barrass, G.S. 2002. *The art of calligraphy in modern China.* Berkeley: University of California Press.
Billeter, J.F. 1989. *The Chinese art of writing.* New York: Rozzoli.
Calligraphy in Chinese and Japanese Art. 2014–2015. September 12, 2014 – February 22, 2015. Saint Louis: Saint Louis Art Museum.
Chang, C.-H., and H. Frankel. 1995. *Two Chinese treaties on calligraphy.* New Haven: Yale University Press.
Chen, T. 2011a. *Chinese calligraphy.* Cambridge, UK: Cambridge University Press.
Chen, Y. 2011b. Chinese calligraphy. University of Wisconsin – Madison course syllabus. Retrieved July 1, 2019 from https://www4.uwm.edu/schedule/syllabi/110047036.docx
Chiang, Y. 1973. *Chinese calligraphy.* Cambridge, MA: Harvard University Press.
Chin, T. 2001. *Eight hundred characters in the Standard Script for beginning Chinese brush writing.* Hong Kong: China Culture Development Foundation Fund Publication Company.
Chin, T., and W. Li. 2004. *East Asian calligraphy education.* Bethesda: University Press of Maryland.
Crowley, C., and Y. Li. 2016. Calligraphy in East Asia: Art, communication, and symbology. *Education about Asia* 21 (3): 19–21.
Dalian Evening News. 2013. Handwriting affects performance in high school and college entrance exams, employment, and even one's life. [写字关乎中高考、就业甚至影响人一生]. September 24, 2013. Retrieved July 15, 2019 from http://ln.sina.com.cn/edu/school/2013-09-24/094714615.html?from=ln_xgbd
DeFrancis, J. 1984. *The Chinese language: Fact and fantasy.* Honolulu: University of Hawaii Press.
———. 1989. *Visible speech: The diverse oneness of writing systems.* Honolulu: University of Hawaii Press.
Ericson, Britta. 2001. *Words without meaning, meaning without words: The art of Xu Bing.* Washington, DC: Sackler Gallery.

Fan, C.C. 1996. Language, gender, and Chinese culture. *International Journal of Politics, Culture, and Society* 10 (1): 95–114. https://doi.org/10.1007/BF02765570.

Gaur, Albertine. 2000. *Literacy & the politics of writing*. Bristol: Intellect.

Gregory, Marshall. 2009. Notes from the 13th Marshall Gregory pedagogy workshop. Emory University. May 2009.

Hamlish, Tamara. 1999. Calligraphy, gender and Chinese nationalism. In *Gender ironies of nationalism: Sexing the nation*, ed. Tamar Mayer. New York: Routledge.

Harrison, J.F.C. 2007. *A history of the working men's college (1854–1954)*. London: Routledge.

Hay, Jonathan. 2005. The Kangxi Emperor's brush-traces: Calligraphy, writing, and the art of imperial authority. In *Body and face in Chinese visual culture*, ed. Wu and Tsiang, 311–334.

He, C., and Chew, S.M. 2003. Chinese calligraphy 101. University of Chicago course syllabus. Retrieved October 1, 2010 from http://calligraphy.uchicago.edu/meetings.html

Hsieh, Becky. 2009. Chinese calligraphy. Course syllabus prepared for the Johns Hopkins University. Retrieved from www.ltc.jhu.edu/syllabi/fall09/373_303.pdf

Hsieh, Y. 2017. Chinese calligraphy. Rutgers University course syllabus. Retried July 1, 2019 from https://asianstudies.rutgers.edu/images/Documents/111-Spring-2017-syllabus-section-01.pdf

Huang, N. 2006. Language, art, and culture: Chinese calligraphy. University of Southern California course syllabus. Retrieved July 1, 2019 from https://web-app-test.usc.edu/ws/soc_archive/soc/syllabus/20073/25401.doc

Ingold, Tim. 2007. *Lines: A brief history*. New York: Routledge.

Keightley, D.N. 1978. *Sources of Shang history: The oracle-bone inscriptions of Bronze Age China*. Berkeley: University of California Press.

Kieschnick, John. 2003. *The impact of Buddhism on Chinese material culture*. Princeton: Princeton University Press.

Knight, M. 2012. Introduction: Decoding Chinese calligraphy. In *Out of character: Decoding Chinese calligraphy*, ed. M. Knight and J.Z. Chang, 17–52. San Francisco: Asian Art Museum.

Kraus, R.C. 1991. *Brushes with power: Modern politics and the Chinese art of calligraphy*. Berkeley: University of California P.

Kwo, D.-W. 1981. *Chinese brushwork in calligraphy and painting: Its history, aesthetics, and techniques*. Newburyport: Dover Publications.

Lan, F. 2009. *Chinese calligraphy and poetry*. Course syllabus prepared for Florida State University. Retrieved from https://mailer.fsu.edu/~flan/chinese/documents/syllabus_callig.pdf

Ledderose, L. 1986. Chinese calligraphy: Its aesthetic dimension and social function. *Orientations* 17 (10): 35–50.

Li, C. 2000. *The sage and the second sex: Confucianism, ethics, and gender*. Chicago: Open Court.

Li, W. 2009a. *Chinese writing and calligraphy*. Honolulu: University of Hawaii Press.

———. 2009b. 书法课对汉语教学的辅助作用. [The faciliatory effect of calligraphy courses on Chinese language teaching]. *Journal of Chinese Language Teachers Association* 44 (2): 63–79.

Li, Y. 2011. A broader interdisciplinary approach to calligraphy-related course development: Towards an integrated Chinese curriculum. *Journal of the Chinese Language Teachers' Association* 46 (3): 49–60.

———. 2020. *The Chinese writing system in Asia: An interdisciplinary perspective*. Abingdon/New York: Routledge.

Li, H., and Y. Yu. 2015. Assessing needs for a Chinese calligraphy course in the university context. *Teaching Chinese in International Contexts* 1: 3–26.

Lidai: *abbreviation for Lidai shufa lunzen xuan [An anthology of early texts on the art of calligraphy]*. 1979. Shanghai: Shanghai Shuhua Chubanshe [Shanghai Calligraphy and Painting Publishing House].

Liu, X. 2016. Chinese calligraphy. Vanderbilt University course syllabus. Retried July 1, 2019 from https://www.coursehero.com/file/20142801/syllabus-CHIN-1231-02/

MLA. 2016. Language enrollment database, 1958–2016. https://apps.mla.org/flsurvey_search

Nakata, Y. 1973. *The art of Japanese calligraphy*. New York: Weatherhill.

Norris, J.M. 2006. The why (and how) of assessing student learning outcomes in college foreign language programs. *The Modern Language Journal* 90 (4): 576–583. https://doi.org/10.1111/j.1540-4781.2006.00466_2.x.

Out of Character: Decoding Chinese Calligraphy. 2014. April 29 – August 17, 2014. The Metropolitan Museum of Art, New York City.

Ouyang, Z., and W.C. Fong. 2008. *Chinese calligraphy*. New Haven: Yale University Press.

Pittman, Melissa. 2010. Introduction to Chinese calligraphy. Course syllabus developed for Weber State University. Retrieved from https://faculty.weber.edu/melissapittman/Classes/2010/Spring/CHNS%201000/chinese_1000.htm

Pratt, K., and R. Rutt, eds. 1999. *Korea: A historical and cultural dictionary*. Richmond: Curzon Press.

Project on Liberal Education and the Sciences. 1990. *The liberal art of science: Agenda for action*. Washington, DC: American Association for the Advancement of Science.

Qiu, X. 2000. *Wenzixue gaiyao* [Chinese writing]. Transl. by Gilbert L. Mattos and Jerry Norman. Berkeley: Society for the Study of Early China.

Rogers, H. 2005. *Writing systems: A linguistic approach*. Malden: Blackwell.

Sheng, Ruth. 2009. Calligraphy. Course syllabus prepared for University of Florida. Retrieved October 1, 2010 from www.languages.ufl.edu/syllabi/calli_syl_2009.doc

Sheng, R. 2016. Chinese calligraphy. University of Florida course syllabus. Retried July 1, 2019 from http://languages.ufl.edu/files/CHI3403-calligraphy-sheng.pdf

Shively, D.H., and McCullough, W. (eds.). 1988. Heian Japan. In *The Cambridge history of Japan*, Vol. 2, ed. J.W. Hall, et al. Cambridge, UK/New York: Cambridge University Press.

Stevens, John. 1996. *Sacred calligraphy of the east*. Boston: Shambhala.

Taylor, I., and M.M. Taylor. 2014. *Writing and literacy in Chinese, Korean and Japanese*. Amsterdam/Philadelphia: John Benjamins.

Wen, X. 2014. *Hiding the tip: Gateway to Chinese calligraphy*. Portland: Merwin Asia.

Wilkinson, E.P. 2015. *Chinese history: A new manual*. 4th ed. Cambridge, MA: Harvard University Asia Center.

Wu, Z. 2003. *Language, art and culture: Chinese calligraphy*. Course syllabus prepared for the University of Southern California. Retrieved from https://www.usc.edu/dept/ealc/chinese/newweb/course_page.htm

Wu, S. 2009a. Chinese calligraphy. Course syllabus. University of Toronto. Retrieved from http://www.bsaminoritystudyabroad.com/documents/Chinese_Calligraphy%2D%2DBSA_Summer_2009.doc

Wu, Y. 2009b. Chinese calligraphy. Course syllabus prepared for Rutgers University. Retrieved from https://asianlanguages.rutgers.edu/courses/spring2009/165111.pdf

Yang, M. 2002. Calligraphy. Course syllabus prepared for Grinnell College. Retrieved October 1, 2010 from https://web.grinnell.edu/courses/chi/S02/CHI211-01/syllabus_b.html

Yen, Y. 2005. *Calligraphy and power in contemporary Chinese society*. London/New York: Routledge Curzon.

Yu, Y. 2019. Calligraphy course enrollment. Private email correspondence.

The Performed Culture Approach

Li Yu

Contents

Introduction	844
Key Players and Intellectual Underpinning of the Performed Culture Approach	845
Core Concepts of the Performed Culture Approach	847
Pedagogical Practices of the Performed Culture Approach	853
Curriculum Design	853
Material Preparation	859
Assessment	861
Teacher Training	862
A Comparison of PCA to Other Foreign Language Approaches and Methods	865
Major Contributions to the Foreign Language Field	869
Conclusion	871
References	872

Abstract

This chapter provides a comprehensive overview of the theory and practice of the Performed Culture Approach (PCA). PCA is a paradigm-shifting pedagogical framework developed in the field of East Asian language pedagogy. It has its origins in the millennia-old Confucian philosophy and draws inspiration from the knowledge fields of cognitive studies, linguistic anthropology, cultural psychology, and performance studies. While the foreign language education field has been dominated by methods and approaches developed for teaching European languages, PCA offers a unique perspective on how a learner can be trained to acquire intercultural communicative skills in a language and culture that is drastically different from their own. Through meticulously executed curricular design as well as "performances" that put learners in authentic linguistic and cultural contexts, this pedagogical approach successfully integrates language instruction and culture instruction. It aims to enable language learners to not

---

L. Yu (✉)
Department of Asian Studies, Williams College, Williamstown, MA, USA
e-mail: Li.Yu@williams.edu

© The Author(s), under exclusive licence to Springer Nature Singapore Pte Ltd. 2022
Z. Ye (ed.), *The Palgrave Handbook of Chinese Language Studies*,
https://doi.org/10.1007/978-981-16-0924-4_7

only gain advanced linguistic skills, but also develop intercultural communicative competence, successfully navigating in the target culture and effectively communicating with native speakers in culturally appropriate ways. The main body of this chapter consists of five sections. The first section traces the key players and intellectual underpinnings of PCA. The second section explains the core theoretical concepts of PCA. The third section introduces the pedagogical practices of PCA, including curriculum design, material preparation, assessment, and teacher training. The fourth section compares PCA with some other major language teaching approaches and methods, including the Audio-lingual Method, the Direct Method, the Silent Way, and the Communicative Approach. The fifth section discusses the major contributions of PCA to the field of foreign language education.

Keywords

Chinese language teaching · Intercultural communication · Performance · Second-language acquisition · Second-culture learning

Introduction

The Performed Culture Approach (PCA) is a philosophy of language learning and teaching. It is a paradigm-shifting pedagogical framework in foreign language education that aims to help learners achieve both linguistic proficiency and intercultural communicative competence when studying a second language. This approach is theoretically grounded in the millennia-old Confucian philosophical tradition and bases its theories on the knowledge fields of cognitive studies, linguistic anthropology, cultural psychology, and performance studies. It was developed by a group of Chinese and Japanese language educators in the United States. Equipped with a complete set of field-tested teaching methods and instructional techniques, the Performed Culture Approach is positioned to succeed the Communicative Approach and initiate a new phase in foreign language education. This phase is marked by a paradigm shift to prioritize the central importance of culture, especially behavioral culture, in language learning and to adopt the concept of performance to achieve the goal of truly integrating second-language acquisition and second-culture learning.

This chapter provides a comprehensive overview of the theory and practice of the Performed Culture Approach. The key players and theoretical underpinnings of PCA are first introduced, followed by detailed explanations of its core concepts and pedagogical practices, especially with regard to how they are applied in the context of Chinese language teaching. A comparative analysis of this approach with other major foreign language teaching approaches and methods is then offered. The chapter concludes with a discussion of the major contributions of PCA to the field of foreign language education.

Key Players and Intellectual Underpinning of the Performed Culture Approach

The key players of the Performed Culture Approach are associated with two institutions that are prominent in the field of East Asian language pedagogy in the United States – Cornell University and The Ohio State University. Cornell was one of the few selected universities that began to develop the then so-called Far Eastern Studies in the early twentieth century (Cameron 1948; Latourette 1955). In 1972 the Division of Modern Languages and Linguistics at Cornell founded a unique Full-Year Asian Language Concentration (FALCON) designed to bring beginning learners of Chinese and Japanese to advanced working proficiency within a short period of time. FALCON was the only full-year, full-time, intensive East Asian language program offered at an American university (McCoy 1979; Jorden et al. 1991). Likewise, The Ohio State University was a pioneering institution during the 1980s Individualized Instruction Movement and still offers a vibrant Individualized Instruction program in various languages. Ohio State houses a full-fledged Department of East Asian Languages and Literatures with a long intellectual tradition in East Asian language pedagogy. In the late 1980s the department was the institutional home for the Chinese Language Teachers Association. The first, and so far the only, Ph.D. program in East Asian language pedagogy in the United States is housed in the department. The department offers a one-of-a-kind summer intensive teacher training program (SPEAC) that focuses on both theory and practice in East Asian language teaching. It is not coincidental that Cornell and Ohio State became the breeding grounds of the Performed Culture Approach.

The largest intellectual debt of the Performed Culture Approach is owed to Galal Walker and Mari Noda, the chief architects of the theoretical framework and pedagogical practices of the approach. Both Walker and Noda received their doctorates from Cornell and have worked at The Ohio State University since the 1980s. Through their work in material preparation and teacher education, the duo are dedicated to the cause of improving the national ability to provide high-quality instruction in less commonly taught languages, particularly Chinese and Japanese. Both Walker and Noda have been instrumental in initiating explorations into the use of multimedia for Chinese and Japanese curriculum design and pedagogical material preparation. Originally trained in literary studies, Walker brought an interdisciplinary perspective to the field of language pedagogy, drawing on research from a wide range of disciplines such as cognitive studies, linguistic anthropology, cultural psychology, and performance studies. He is known for spearheading efforts to bring innovative instructional models to the Chinese language field, such as Individualized Instruction, overseas internship programs, and advanced language and culture programs. He is also the master designer of a series of PCA-informed learning materials including *Chinese: Communicating in the Culture* and *Chinese Out of the Box*. Mari Noda specializes in curriculum design, material preparation, and language assessment. Trained as a linguist, she has a keen eye for structure and context in language learning and teaching. She has co-authored a series of Japanese learning materials that are influential in the Japanese field (Jorden and Noda 1987, 1988, 1990; Noda et al. 2020). A veteran classroom teacher of the Japanese language, she stresses the importance of meticulous design of lesson plans and the execution of fine-tuned teaching techniques to ensure the highest instructional quality and the most effective use of classroom time.

Two Cornell-associated linguists, Eleanor H. Jorden (1920–2009) and A. Ronald Walton (1943–1996), influenced much of Walker and Noda's earlier work. A legendary Japanese linguist and language educator trained at Yale and a major principal at the US State Department's Foreign Language Institute, Jorden was instrumental in shaping the Japanese language field (Jorden et al. 1991). She brought the work of anthropologist Edward T. Hall into language pedagogy which led to the integration of cross-cultural communication with the study of the linguistic code (Hall 1977). As the leading author of several widely successful sets of Japanese learning materials that dominated the Japanese language field for decades, Jorden developed a practice-based pedagogy that focused on both structural analysis and social context of the language (Jorden and Chaplin 1962; Jorden and Noda 1987, 1988, 1990). In 1988 Jorden founded the Japanese Teacher Training Institute under Exchange: Japan, an organization designed to increase the number of professionally trained instructors in American classrooms by providing universities an economical means to establish or enhance Asian language programs. Through Exchange: Japan, she trained hundreds of Japanese language teachers for the field. The curricular design as well as the training methods that Jorden developed influenced the teacher preparation model used later in the Performed Culture Approach. She had considerable intellectual influence on a cohort of graduate students at Cornell who later became pillars in the emerging Chinese and Japanese pedagogy fields in the 1990s.

A member of that cohort, A. Ronald Walton, later collaborated with Jorden. Walton is highly regarded as a visionary figure in the Chinese language field who promoted the teaching of foreign languages, especially the less commonly taught ones, at the national level (Ramsey 1996; Walton 1989; Brecht and Walton 1994). Walton was one of the main forces behind the creation of the National Foreign Language Center in Washington DC and served as its deputy director from its inception in 1987 until his untimely death. Walton's work in the intersections between the field of Chinese language pedagogy and that of national language policy and language planning inspired the practitioners of the Performed Culture Approach to take a panoramic and holistic view when building the infrastructure of the emerging Chinese language field.

Working together, Jorden and Walton proposed the theoretical concept of "truly foreign language" – non-Indo-European languages that are spoken within societies that are markedly different from the English-speaking world – and advocated for a set of pedagogical practices to address the linguistic and cultural challenges faced by learners of these languages. In a seminal article, Jorden and Walton argue that those languages that are linguistically unrelated to English pose an enormous challenge to Western learners, not only because of the difficult linguistic codes and unique writing systems but also, more importantly, each with a complex cultural code "which pervades the linguistic code at every level—phonological, morphological, syntactic, and discourse" (Jorden and Walton 1987, 111). They suggest that culture, or more precisely what they refer to as "socioverbal behavior," "must be an integral part of a good teaching program from the first hour of instruction" (ibid.). They support a clear distinction between oral and written languages to expedite the language learning process. To ensure high instructional quality, they recommend a fact-act dichotomy when assembling a teaching team that consists of both base-native and target-native instructors. They advise that in language teaching materials vocabulary should not be regarded as the primary unit of instruction and that all vocabulary and structural patterns "must be introduced within contexts." Therefore a basic dialogue "should be the introductory pedagogical unit, subsequently analyzed, drilled, manipulated, adapted, and applied" (Jorden and Walton 1987, 123). The groundbreaking concept of "truly foreign language" that Jorden and Walton propose as well as the pedagogical methods and teaching techniques developed to address the instructional challenges of teaching these languages paved the way for the formulation of the core concepts and pedagogical practices of the Performed Culture Approach.

Core Concepts of the Performed Culture Approach

One of the two fundamental concepts of the Performed Culture Approach is "culture." The term has a myriad of definitions in different disciplines, but for the purpose of language pedagogy, it is defined as "what people do" (Walker and Noda 2000). Culture is the framework for how meaning is negotiated in a given linguistic community. Under this definition, culture is linked to language with this formula: culture creates contexts – contexts provide meanings – meanings produce intentions – intentions define individuals – language provides a medium to express and explain intentions (Walker 2000). In other words, intention precedes the linguistic code, and the interpretation of the latter presupposes a functional understanding of the cultural code. Contexts provide the key to meaning negotiations through the linguistic code. To learn a language means to learn to do things in a new culture. For second-language learners to be effective when communicating with native speakers, they must acquire the cultural code along with the linguistic code. But what is the cultural code? What is it composed of?

The Performed Culture Approach dissects culture into learnable components that can be gradually acquired in a language curriculum. It categorizes culture in two ways. The first method of categorization is based on the importance of the components of the target culture to second-language learners, following the conceptualization of Hector Hammerly, a forward-looking language teacher and pedagogue (Hammerly 1986). For language learners, various components of the target culture usually fall into three ways of talking about culture: behavioral culture, informational culture, and achievement culture. A language learner must first and foremost acquire behavioral culture in order to function successfully in the target culture. Behavioral culture refers to "the common daily practices and beliefs that define an individual and dictate behavior in a specific society" (Christensen and Warnick 2006, 13). Behavioral culture is often embedded in the linguistic code and determines how speech acts (e.g., greetings, requests, complaints, commands, apologies) are codified and ritualized in a specific language. The emphasis in second-language programs, especially at the beginning and intermediate levels, should be on behavioral culture (Hammerly 1986; Walker 2000). A learner must then also identify elements of informational culture and achievement culture in the target society. Informational culture refers to the "kinds of information that a society values, as well as historical and other facts and figures about the society" (Christensen and Warnick 2006, 12; Hammerly 1986). Examples of informational culture include history and geography of the target society or the annual occurrence of college entrance examinations in East Asian countries. Achievement culture represents "the great achievements of a society" such as literature, philosophy, or religious tradition (Christensen and Warnick 2006, 12; Hammerly 1986). While informational culture and achievement culture are important, they should only be added into a language program incrementally as the learner's linguistic skills progresses (Walker 2010). All too often language teachers, programs, and pedagogical materials in the current Chinese field equate "culture" with "achievement culture" and place too much emphasis on it while ignoring the importance of behavioral culture (Yu 2009). The most prominent elements of behavioral culture play a role in communication that often differs from learners' base cultures. This is a situation that the Performed Culture Approach sets out to rectify and has done so successfully in the classroom (Qin 2013, 2017a, b).

The second method of categorization of culture in the Performed Culture Approach is based on the native speaker's attitude toward or perception of their own culture when interacting with a non-native. Analyzed under this lens, culture can be viewed as consisting of revealed culture, ignored culture, and suppressed culture. Revealed culture is the type of cultural knowledge that a native is generally eager to communicate to a non-native (Walker 2000). Ignored culture is the kind of cultural knowledge that a native is generally unaware of until the surprising odd behaviors or expressions of a non-native bring it to light. This is what Edward T. Hall has called "hidden or covert culture" (Hall 1977; Walker 2000). Suppressed culture is knowledge about a culture that a native is generally unwilling to communicate to a non-native (Walker 2000). Research has shown that ignored culture is the most useful type of cultural knowledge for beginning-level language learners, whereas advanced-level learners can be trained to have more dealings with suppressed culture (Christensen and Warnick 2006). The fact that ignored culture tends to be hidden from native speakers and most native-born language teachers makes it a fertile field for further research in intercultural communication and second-language acquisition, making foreign language teachers the expected experts in this area. In the current Chinese field, too much emphasis is put on revealed culture in pedagogical materials which creates unfortunate lacunae in the communicative abilities of foreign speakers of these languages.

Hammerly's three discourses of culture – achievement, informational, and behavioral – share intersections and overlaps with Walker's three categories of culture, revealed, ignored, or suppressed. By nature, most elements of achievement culture (e.g., philosophy, canonical literature, classic films or music, gourmet food, festivals) and many aspects of informational culture (history as written in standard textbooks, geographical knowledge) are good examples of revealed culture since native speakers tend to eagerly share these aspects of their culture with non-natives. However, native speakers may ignore some aspects of achievement culture and informational culture. For instance, people from other countries may remind Americans how great their plumbing systems or their national park system are. For another instance, Americans tend to not view Richard Nixon as a great president due to his scandalous resignation and may thus be puzzled by Chinese people's reverence for him. Likewise, Chinese people did not realize how great the inventions of papermaking, the compass, gunpowder, and printing were until Western writers and scholars pointed out their importance and attributed them to China in the nineteenth century. Contemporary Chinese may still be surprised to learn that the millennial-old Chinese multiplication table that even young preschoolers in China can effortlessly recite amazes today's Western math educators. Examples of suppressed achievement or informational culture include the not-so-glorious part of history that natives tend to not reveal to non-natives, award-winning films that make natives feel that they have lost face on the international stage, or blatant racism and ethnic tensions that natives feel uncomfortable talking about, especially in front of non-natives.

The analytical lens of revealed, ignored, and suppressed culture proves to be especially useful in further probing of the intricacies of the often-elusive behavioral culture. Based on many years of experience helping American learners navigate the Chinese workplace, Xiaobin Jian and Eric Shepherd (2010) summarize three sets of behavioral cultural rules in the Chinese context: those that Chinese people are happy to discuss (*jinjin ledao de guize*), those that are present but invisible (*shi er bu jian de guize*), and the "dirty laundry" that natives are not eager to air in front of cultural outsiders (*"jiachou buke wai yang" de guize*). *Jinjin ledao de guize* basically comprise revealed behavioral cultural rules. For example, Chinese people are usually quick to use the terms *mianzi* (face), *guanxi* (connections, interpersonal relationships), *hanxu* (nonexplicit, nonconfrontational), or *huhui* (reciprocity) to explain Chinese social behaviors to foreigners. Chinese do not usually realize the importance of such rules as *ke sui zhu bian* (a guest conforms to the host's wishes) or the dominant use of third-party mediation in Chinese culture. When greeting or engaging in small talk, Chinese locals tend to state the obvious (e.g., "*shangke qu a?* Going to classes?" "*Kanshu na?* Are you reading?" "*Shanghuo le.* You have a cold sore.") without realizing they are doing so. Such rules constitute *shi er bujian de guize*, hidden or ignored behavioral culture. A good example of suppressed behavioral culture that Jian and Shepherd have discovered is Chinese people's attitude and treatment of Chinese Americans, who are usually viewed as neither part of the Chinese in-group nor as "real" Americans. As a result some Asian American students experience a form of racism when navigating Chinese society.

In order to help learners acquire the cultural knowledge of the target society, especially the usually unspoken rules of behavioral culture and ignored culture, the Performed Culture Approach relies on its second fundamental concept –

performance. Simply put, to learn a foreign language is to perform a foreign culture. A learner has to know what is expected of him or her in a given sociocultural situation and how to respond in linguistically accurate and culturally appropriate ways. Basically, learning the culture of a foreign language is learning how to predict what to say and how to behave in social situations that are novel to the speaker. Here culture is the end, while performance is the means. Like "culture," the term "performance" has been overly used and hotly contested in many academic fields. Within the Performed Culture Approach, "performance" takes on multiple layers of meanings. At the theatrical level, performance is an act of staging an event, which consists of five elements – specified time, specified place, specified roles, appropriate script, and accepting or acceptable audience – conveniently acronymized as PARTS by followers of the Performed Culture Approach. Translated into classroom practice, the idea of performance is implemented through contextualization of all language practice. No vocabulary or grammatical items are practiced, and no tasks are completed without providing a social-verbal context consisting of the five elements of a performance. One classroom teaching method commonly used in PCA is to have learners act out a scene carefully set up by the instructor. A speech act (e.g., greeting, leave taking, apologizing) or a communicative event (e.g., completing a transaction in a store, purchasing a ticket at a train station, doing a presentation on China's environmental policies, or writing a film review) is always contextualized in a setting that simulates a familiar situation in the target world.

At a deeper level from a sociological and anthropological perspective, an individual's behavior in society reflects the role each person plays, either consciously or unconsciously, while carrying out the duties of a social role. Everyone's actions and behaviors are influenced by their interrelationships with other people (Goffman 1959). Conversely, these relationships are also affected by each individual's actions and behaviors. In this sense, language is a social performance between human beings on the stage of a community, a society, or group. When second-language learners communicate with members of the target society, they are performing the social roles expected of them in the target culture. In this sense, second-language learning is a process of socialization for adult learners and of building C2 memory and constructing a C2 worldview (Duff and Doherty 2018). In other words, to learn a language is to visualize oneself in a possible future and "remember the future" (Walker and Noda 2000). Performance is a key pedagogical step in this learning process.

Performance is used not only as a teaching methodology in the Performed Culture Approach but also as a hallmark PCA assessment tool. Linguistically and philosophically speaking, PCA practitioners believe that to know something is to do it, that is, linguistic performance is the common goal for both language learners and language pedagogues. Here the millennia-old Confucian philosophy has nourished PCA with its epistemological injunction on the nature of knowledge (Walker 2010). According to Wang Yangming (1472–1529), the quintessential Neo-Confucian thinker and educator, to know is to carry out one's knowledge through one's action, i.e., *zhi xing he yi* ("knowledge and action is but one"). This insistence on the integration of knowledge and action dates farther back in time to the idea of *xi* (to try out, to put into practice) in the Confucian *Analects* and is applicable in the context of second-language acquisition. Linguistic knowledge or competence is inseparable from one's ability to actually use the language at the right time, i.e., proficiency. Proficiency, however, is not the main goal of the Performed Culture Approach. Separating itself from the Proficiency Movement, PCA recognized the shortcomings of the proficiency-driven paradigm early on. Through field observations, the practitioners of the Performed Culture Approach have discovered that a proficiency-driven curriculum tends to produce learners who are "+linguistic code, −cultural code," that is, fluent speakers who have a high level of linguistic proficiency but lack in intercultural communicative skills, i.e., "fluent fools." Such learners often offend native speakers without intending to do so by saying/doing the wrong thing in accurately delivered language. They are not aware of these blunders and thus cause permanent damage to their interpersonal relationships with locals (Christensen and Warnick 2006; Jian and Shepherd 2010). It is critical to use performance, such as contextualized dialogue performance, scripted communicative drills and exercises, prochievement interviews, and portfolios, in assessing a learner's progress and to predict their future success in the target culture (McAloon 2015).

A related concept to the five elements of performance is game. In the Performed Culture Approach, games are defined as "performances with a defined scoring system" (Walker and Noda 2000, 202). A game is used both as a metaphor of culture and as a motivational idea in the pedagogical design of learning activities. Metaphorically speaking, operating in different cultures is like playing different games. Navigating American culture is like playing the game of basketball, whereas doing things in the Chinese culture is like playing the game of volleyball. Each game has its own set of rules and objects (e.g., nets, balls). One cannot use basketball (American cultural) rules when playing volleyball (in the Chinese culture) and vice versa. At the same time, the idea of game adds a hint of playfulness and teamwork in language practice, which helps relieve some adult learners' uneasiness when acquiring a second culture. Learning a second language and acquiring its value system does not mean students have to let go of the core values forged by their base culture. As the sociological and anthropological sense of performance implies, acquiring a new culture is similar to putting on a new piece of social skin (or assuming a new persona) without the need to change one's core personality and identity. The idea of game also helps teachers design pedagogical activities that are interesting and motivating to learners (Kapp 2012; Bell 2017). If the concept of performance fully utilizes human beings' basic learning mechanism of mimesis, then the idea of game capitalizes on humans' desire to earn recognition and merit. Games have proven to be a valuable motivational and pedagogical instrument for Chinese language learners (Jia 2017; Li 2018).

This focus on the psychological and cognitive aspects of human learning brings up yet another important core concept in the Performed Culture Approach, namely, the learner. A learner is differentiated from a student in that the former is an autonomous individual who has acquired the good habits of learning and engaging in lifelong learning, whereas the latter denotes an administrative role, someone who studies at an educational institution with a number attached. In the eyes of PCA practitioners, learning needs to be a lifelong process. Therefore acquiring cultural and linguistic knowledge is as important for the learner as acquiring effective learning habits and strategies. The Performed Culture Approach postulates that throughout the second-language learning process, a learner is basically constructing his or her second-culture worldview through a series of memory-constructing activities (Walker and Noda 2000). First, a learner brings his or her own persona into the learning process and acquires cultural and linguistic knowledge. Through a series of performances and games, he or she comes up with some stories about using the language in contexts in the target culture. Here story is defined in PCA as a basic cognitive unit of personal memory. For instance, if the learner has practiced asking for directions on a Beijing street, he or she would have a personal memory (story) of doing just that. The PCA practitioners believe that the more appropriate C2 stories learners can participate in, the more intelligent they will appear to be to members of the target society. As such memories/stories accumulate in the learning process, the learner consciously or experientially compiles them into larger units of memory – cases, sagas, and themes – that can overlap or remain discrete. Cases are a series of stories about doing something in the target culture (e.g., asking for directions, requesting information, describing things). Sagas are a series of stories about a set of people (e.g., classmates saga, supervisor saga, doctor saga) or at a specific location (e.g., the Starbuck saga, the restaurant saga, the office saga, the classroom saga, the hospital saga, the street saga, the Beijing saga). Themes, however, are acts in the target culture and target language that convey cultural values (e.g., politeness, hierarchy). As a learner's inventories of memories about the target culture increase, his or her second-culture worldview is also constructed and expanded while acquiring a new layer of social skin for his or her persona in the target culture. This learning process constitutes an ever-growing and potentially endless cycle of "C2 Cultural Compilation" (Walker and Noda 2000; Noda 2020).

A well-trained teacher plays a critical role in this spiraling cycle of learning and memory construction. According to PCA, the teacher is not someone who merely imparts knowledge to the learner. PCA advises teachers to not view their job as "teaching a language," but rather a manager of learners and a facilitator of learning (Chai 2018). Essentially, the performances of the learners constitute the goals of the teacher. In order to assume this, a teacher plays multiple roles in the PCA classroom – she is a playwright, director, prop master, stage manager, actor, interlocutor, and an evaluator all at once. When designing the lesson plan, the teacher must come up with scripts that fit appropriately in various scenes and contexts (Qin 2013; Chai 2020; Meng 2020). She needs to contextualize the script and set up the scene with aptly designed visual aids and props so as to elicit performances from the students and enhance their memories of the events. At times, she needs to play a role in a scene to provide a model for the students, while at other times she needs to move out of that role in order to give timely feedback as an audience and to assess the students as an evaluator. Beyond the classroom, PCA encourages teachers to see themselves not only as teachers but also as program builders. Thus they are not focused only on their day-to-day teaching but adopt a way of thinking as a team and program builder who knows how to design a holistic, conducive language-learning environment that supports learning both within and outside the classroom.

The ultimate reason teachers have to do all this is time. Life is short, and effective and efficient instruction can drastically shorten the time that a learner needs to progress from beginning to advanced or even superior levels in Chinese. The United States does not have a national foreign language policy, and most federally funded foreign language study (e.g., Critical Language Programs, Flagship Programs, government language institutes) has generally been restricted to languages useful for defense and intelligence purposes. Most children in the United States do not have the opportunity to learn Chinese until they enter college. Research in a government full-time language institute has found that it takes an average of 2,200 hours of full-time, intensive training for a learner to progress from zero to "professional working proficiency" (or a score of "Speaking-3/Reading-3" on the Interagency Language Roundtable scale, not considering writing abilities) of Mandarin Chinese (Foreign Service Institute 2019). This amount of time is not available for most college students, who are busy with other coursework while studying a foreign language. The best undergraduate Chinese programs in the United States can only provide between 150 and 200 instructional hours at each level per school year. Therefore helping American students achieve truly advanced language skills within a short period of time requires an effective model of curricular design, usable content, and a practiced classroom teaching methodology. To this end, PCA has developed a comprehensive stock of teaching methods and instructional techniques (Yu 2008; Yu et al. 2020b). The theories of the Performed Culture Approach are fully implemented in its pedagogical practices, which include the four major aspects of curriculum design, material preparation, assessment, and teacher training.

Pedagogical Practices of the Performed Culture Approach

Curriculum Design

The PCA curricular design is guided by one central principle, namely, a curriculum does not stand alone but belongs to a program. A program, whether a computer program, a concert program, or a language program, is a series of events leading to a planned outcome or result. By definition, a language program must be goal-oriented with clearly spelled-out learning objectives to achieve the goal over a reasonable amount of time. All instructors within the same program should make concerted efforts toward the shared goal. All too often many language programs lack a clearly defined, consistent goal that states the mission of the program and pulls all courses, instructors, and resources together. The ultimate goal for a PCA-based curriculum is to train learners to become effective communicators in identifiable parts of the target culture.

In order to build an effective language program and a supportive language-learning environment, PCA recommends the combination of two instructional models at the macro-level – Learning Model Instruction (LMI) and Acquisition

Model Instruction (AMI). Also referred to as item-based instruction, LMI is used primarily at the beginning level. Its proportion in a language program decreases as students' proficiency level increases. At the lower level, the linguistic and cultural knowledge that a learner accumulates is presented item by item in a discrete manner. He/she learns the sound system, the vocabulary, the grammatical patterns, the speech acts, and communicative events of a new language one by one, item by item. Instructional instruments such as the textbook, audio materials, the vocabulary list and exercises in the textbook, the dictionary, the glossary, instructional performances, and games are all present in this learning environment to support the learning of the linguistic items. However, as the learner makes progress in his or her linguistic proficiency, it is also very important for the teacher to incorporate Acquisitional Model Instruction, or strategy-based instruction, in this learning environment. Instead of focusing on individual items, AMI aims to train learners to use strategies and tactics in order to deal with the unexpected in the target language and target culture. To borrow an old Chinese saying, if LMI is to give fish to the learner, then AMI is to train the learner how to fish on their own. For instance, AMI trains learners not to become anxious or panicky when they are overwhelmed by a string of unfamiliar sounds that they cannot comprehend. A teacher using AMI would deliberately train the learners to listen for keywords and use contextual clues to deal with an unfamiliar situation or to use language they have already learned in order to seek further clarification. Another example of AMI is to teach students early on how to deal with ambiguity and how to catch keywords or look up new words in dictionaries (e.g., online dictionary, concordance software, smartphone apps).

The most important objective of AMI is to train learners to become autonomous learners who are equipped with the ability to learn a language effectively on their own. Autonomous learners recognize the importance of audio materials and auditory learning in acquiring a new language. They know how to manage their time for learning and how to maximize deliberate practice in order to develop muscle memory for automaticity in using a language. To this end, PCA advises all teachers to prepare a learning schedule that spells out in detail what the learner needs to do in order to prepare for each class. This way, while following the curriculum, students also acquire good habits to assure their success in the long journey of language learning or any type of learning for that matter. In this sense, PCA has preceded the so-called flipped classroom by decades. A PCA classroom is always and by definition a flipped one: the learner spends 2–3 hours studying before every hour of class, and in class the teacher maximizes the use of the contact hour to have the students practice and provide immediate feedback, helping the learner improve and improvise their performance in contextualized, culturally authentic settings (Chai and Chen 2017; Chai 2020).

At the micro level, PCA adopts an instructional model that combines ACT classes with FACT classes. The forerunner of the ACT/FACT dichotomy was devised as a stopgap measure during World War II when the United States needed to train a large number of military personnel for war efforts under extreme time pressure. Since the country lacked language specialists at the time, government language schools and some elite universities hired target-native speakers who had not received any formal linguistic or pedagogical training to conduct drill sessions with students. Base-native linguists and sometimes area specialists untrained in linguistics or pedagogy were responsible for the course design and for explaining grammatical rules in lecture sessions (Light 1987, 1993). Having served for years in US government language institutions, Eleanor Jorden adopted and revised this instructional structure to meet the challenges of teaching East Asian languages. According to Jorden and Walton (1987, 121), "The fact component, which includes thorough, detailed, objective linguistic and cultural analysis in terms that are meaningful to a foreign language learner, is best handled in the base language, guaranteeing that the students understands and is able to ask questions whenever necessary. The act component requires drill and practice in sufficient amounts to enable students to internalize the fact component—to use it accurately and meaningfully, not simply to talk about it." In comparison to the practices of the wartime US government language schools, Jorden and Walton believe that both ACT and FACT sessions "require highly skilled instruction offered by professionally trained pedagogues (ibid.)." They recommend that a teaching team be composed of both base-natives who usually handle FACT classes and target-natives who usually teach ACT classes. Both types of teachers should be well trained in conducting either ACT or FACT classes.

The ACT/FACT dichotomy is supported by psychological studies on learning. Language learning, like many other types of skill learning, needs two types of knowledge – procedural and declarative. Simply reading a car manual and studying all the components of a car does not teach a person how to drive. It is the procedural knowledge, i.e., actually doing it, that helps a person acquire the necessary skill set for driving. Language learning is no exception. Recent research in the field of second-language acquisition has confirmed the importance of both implicit and explicit learning (Jin 2018). Merely learning the vocabulary and grammatical rules on paper and translating from one linguistic code to another does not guarantee skillful use of the language when interacting with native speakers. The learner needs to spend a large amount of time on deliberate practice with the language in contextualized settings in order to acquire the skill of communicating successfully in the target society. For instance, terms of address are a complex system in the Chinese social context. At the beginner level, students usually learn that terms such as *xiansheng* (Mr.), *xiaojie* (Miss), *nüshi* (Ms.), or *taitai* (Mrs.) are placed after a surname and used in more formal settings, whereas the prefix of *xiao* (Little) or *lao* (Old) are placed before a surname and used between people who are on familiar terms with each other. To acquire the skills of properly addressing a person in a given Chinese social context would take hours of practice, if not more. PCA suggests that an ideal ratio for ACT/FACT is at least four ACT sessions per FACT session (Christensen and Warnick 2006). This ratio ensures that students get ample practice in the language before moving on to metacognition to understand the whys behind the linguistic form and socioverbal behaviors. A FACT class is offered only after a series of ACT classes because learners can then bring questions to the class based on their learning experiences in the ACT classes. They can also experience the language first and then discover and formulate grammatical rules through inductive reasoning. In FACT sessions, students discuss not only the grammatical structures but also the social and cultural implications of the language. FACT classes offer support to the ACT classes and may include short written quizzes to check the learners' progress (Christensen and Warnick 2006). As the learner progresses to higher levels, FACT classes can also be conducted in the target language. When that happens, the differences between ACT and FACT classes decrease. At the advanced level, the need to conduct an entire FACT class in the base language also diminishes.

A typical ACT class in PCA is composed of two parts: dialogue performance and dialogue manipulation. Dialogue is a basic learning unit in PCA. Often set in a realistic scene, the dialogue (or sometimes called core conversation or basic conversation) provides the necessary context for how communication is achieved and meanings are crossed in a given situation in the target society. Students come to an ACT class ready to perform the dialogue, and the teacher elicits the performance of the dialogue from the students through a well-contextualized scene. After the core dialogue is elicited, the teacher organizes a series of communicative activities in order for students to have more practice (Meng 2020; Chai 2020). Such communicative activities often include dialogue substitution, expansion, and variation, referred to as a whole as dialogue manipulation. As students' skill level increases, content questions are added to check their comprehension of the dialogue and to train narrative skills. For example, in a basic conversation revolving around visiting a sick classmate, students learn to use such target expressions as *tingshuo ni bingle, xianzai hao dianr le ma?* (I heard you were sick. Are you better now?), *ni zuijin mang shenme ne?* (What have you been busy with the last few days?), *hai bu shi mangzhe qimo kaoshi* (I've been busy with final exams), or *dou kuai ba wo kaoyunle* (Soon I'm going to get dizzy from all this testing). In class activities, the teacher can set up a scene in which a classmate who was absent the day before shows up and either a teacher or a fellow classmate enquires about their welfare. She may use a similar context but with changed roles (e.g., a sick teacher and a student or a worried mother and their child in college). In these different scenes, students need to apply the target phrases (e.g., *jintian hao dianr le ma, ni zuijin mang shenme ne, mangzhe...., doukuai....le*) but use different vocabulary and conversational or socioverbal strategies when playing different roles. The practice of various grammatical structures (e.g., adjective+le, *mangzhe* something, *doukuai...le*) are always conducted in appropriate contexts.

The use of the ACT/FACT instructional model is not possible in all language programs, especially in programs where there is a small number of instructors, an imbalanced ratio of target-native and base-native instructors, or a limited number of instructional hours. Under such circumstances, it is unrealistic to offer separate FACT classes. PCA recommends the periodic insertion of a FACT segment at the end of an ACT class. A FACT segment typically occupies 10% of an ACT class hour (i.e., 5 min out of a 50-min class). This practice has proven to be extremely effective in smaller programs in which there are limited teaching resources.

In terms of modality or skill mix, PCA adopts a curricular model that has been time tested in government language institutions and many successful elite university and college programs in the United States. At the beginning level, the most effective method for teaching English-speaking natives is to have separate tracks for teaching speaking/listening and reading/writing skills (Kubler 2006). PCA recommends a time lag between the introduction of the spoken language and the introduction of the written language so that learners can have a solid foundation in listening and speaking before they begin working with the challenging aspects of the Chinese language, namely, literacy skills. This time lag can be as short as a few days or as long as a few weeks. Reading/writing skills can catch up to listening/speaking skills at the late beginning or early intermediate level.

The idea of Performed Culture is applied not only to listening and speaking classes but also in reading and writing instruction as well. In the Performed Culture Approach, reading and writing are viewed as social activities. To use reading as an example, reading is differentiated from decoding, a process of translating from one linguistic code to another without any understanding of the sociocultural context of the text. Successful reading requires five types of knowledge – cultural knowledge, orthographic knowledge, linguistic knowledge, domain knowledge, and strategy knowledge (e.g., how and when to scan or skim, monitoring what words one understands and does not understand). In other words, reading activities occur in social contexts and serve social functions. Learners who are equipped only with linguistic knowledge but not the other types of knowledge are not going to be successful readers.

PCA classifies all reading activities into two major categories according to the purpose of reading: informational reading (reading for information or perspective) and affective reading (reading for aesthetic fulfillment) (Noda 2002a). Teachers should incorporate both genres of reading into their instructional design. This distinction corresponds with Augustine's conceptions of "ascetic reading" and "aesthetic reading" as well as with the modern distinction between "labor reading" and "ludic reading" (reading for pleasure) (Stock 1996; Nell 1988). When designing reading activities, the teacher needs to bear in mind the social aspect of reading and ensures the task authenticity surrounding a written text (Zhang 2012, 2013). A text, whether it is an email message or a novel, is a performance with its five key elements. For instance, an email message has a sender, a recipient or recipients, and sometimes an audience. The relationship between these roles would determine the tone and content of the message. The sender has a reason or purpose to send the message. Likewise, the recipient(s) may respond in different ways or decide not to do anything, depending on the tone and content of the message. Alerting the teacher to the social and performative aspects of reading and writing activities is a key instructional feature in the Performed Culture Approach.

At the advanced to superior levels, PCA believes that the learner needs to develop expertise in a specific domain that is of personal, academic, or professional interest to him or her (Yu 2020). Moreover, the advanced learner should be able to carve out a "third space" in the target culture for him or herself (Kramsch 2006, 2009; Zeng 2015, 2018a; Yu 2008). The field of foreign language education has long recognized the fact that reaching the proficiency level of an educated native speaker is no longer the goal for second-language learner (Kramsch 2014; Kramsch and Whiteside 2008) because it is neither a practical nor useful goal, not to mention the fact that there is no single definition of what "an educated native speaker" actually means in a target society. However, to reach working proficiency and expertise in a specific domain and to be perceived as an expert in that domain by the native speakers is a more realistic and attainable goal (McAloon 2008; Zeng 2015, 2018b, Jia 2017). Cognitive studies on learning, expertise, and motivation have shown that experts excel mainly in their own domain (Glaser and Chi 1988) and that learners push themselves harder and learn most effectively if the topic can bring them intrinsic pleasure or reward (Zull 2002; Jia 2017). PCA advises teachers to be ethnographers of the learner and observe the advanced learner in action (i.e., when they carry out professional duties in the target society) in order to better adjust the curricular design, teaching methods, and pedagogical material preparation at the advanced/superior levels.

The PCA deems study abroad, or in-country language learning, as a critical step toward advanced proficiency and C2 worldview construction. The field of foreign language education has long acknowledged that quality time spent in a well-designed study-abroad program can help learners gain linguistic proficiency and cultural understanding, bring about improvement in their overall communicative and learning skills, and positively affect their learning attitude (Brecht 2014; Kinginger et al. 2018). In addition to these benefits, for PCA, study abroad also provides a unique opportunity for learners to be socialized in the target culture and acquire essential skills to establish and maintain positive interpersonal relationships in a professional context (He and Qin 2017). In fact, PCA practitioners were among the first to operate study abroad programs in China. As early as the 1990s they established and managed the first in-country internship program (*Zhongmei niudai*, or US-ChinaLinks in Shandong), which ultimately served as a prototype for many current study abroad internship programs. PCA postulates that successful study-abroad programs must carefully design curricular components of experiential learning in the target society and integrate them with structured learning in the classroom.

The learner-centered tasks of Field Performance and Performance Watch have been developed as effective pedagogical methods to achieve such integration. Field Performance tasks are designed for learners to engage in live interactions with native speakers in various social situations and help them develop a network of relationships in the target community (Noda et al. 2017; Chai et al. 2017). In order for deep learning to be achieved, learners should not only complete those real-life tasks but also reflect on their experiences performing those tasks and report on their findings. A Performance Watch is a goal-oriented task to observe and analyze the unfolding of a communicative event in the target culture in terms of how contextual factors such as time, place, and the social roles of speakers and audience contribute to meaning-making in verbal and nonverbal communication. Both Field Performance tasks and Performance Watch tasks train learners to become ethnographers of socioverbal behaviors in the target culture and to grow more sensitive to contextual clues that facilitate meaning negotiation and relationship building. Preliminary evidence has indicated that through such exercises, learners' awareness about the connections between form and context is heightened, which may enhance their ability to recognize speakers' intentions (Cornelius 2015).

Material Preparation

In the Performed Culture Approach, using both linguistically and culturally authentic materials for teaching and learning a second language is of paramount importance. However, PCA defines "authenticity" in a different way than the prevalent understanding of this concept. The conventional wisdom stipulates that a pedagogical material can only be called "authentic" if it is "originally produced by and intended for native speakers of the target language" (Frye and Garza 1992, 225). This is a dominant view held by most foreign language teachers. Mari Noda (2002b) points out several problems with this narrow view of "authentic materials." One key problem is that this definition excludes all materials (and language) produced and intended for non-native speakers (e.g., language textbooks). Rejecting the conventional understanding of authenticity, Noda (2002b, 202) defines it as "the use of language that would create a story that is contextually and culturally feasible." Here "story," as one of the aforementioned key concepts of PCA, refers to the basic memory unit about how to do certain things in the target culture. It is closely related to the idea of cultural script, which means a knowledge structure "that describes an appropriate sequence of events in a particular context" (Schank and Abelson 1977, 41; Schank 1990). This new definition of "authenticity" has broadened the scope of authentic materials. They can include not only linguistic artifacts used by native speakers in the target culture (e.g., novels, films, newspaper articles, television shows, the "authentic materials" as understood by conventional wisdom) but also those prepared by language teachers (regardless if they are native or non-native speakers of the target language) for non-native speakers (e.g., language learners), as long as these materials can create a feasible "story." In other words, a dialogue taken from a feature film from the target culture as well as a dialogue written by a language teacher for second-language learners can be equally "authentic" as long as they can create a story that is contextually and culturally feasible.

In fact, PCA encourages teachers to choose either route when preparing learning materials for students. They can either attempt to locate materials from existing cultural artifacts (i.e., linguistic and textual samples in the target society intended primarily for the use of native speakers) or write their own materials based on acute observation and analysis of real language use. The key is that the learning materials should provide good models for learners to emulate (Noda 2020; Meng 2020). The former route is extremely challenging when creating materials for the beginning and intermediate levels. Successful attempts at using this method have been made in *Chinese: Communicating in the Culture* and *Chinese Out of the Box*, two sets of learning materials that have found usable and learnable dialogues for beginning-level learners from Chinese films, television shows, and cartoons (Walker and Lang 2004; Walker and Zhao 2013). When authoring their own materials, PCA practitioners are strongly encouraged to conduct "performance watch" – an exercise of ethnographic observation of speech acts and communicative events in the real target world – in order to come up with the most authentic script for language learners to study. Excellent examples using the second route can be found in *Basic Spoken Chinese* and *Intermediate Spoken Chinese* (now renamed *Basic Mandarin Chinese:*

Speaking and Listening and *Intermediate Mandarin Chinese: Speaking and Listening*) (Kubler 2011, 2013). Basic conversations in this series of materials are based on the author's years of performance watch and sophisticated linguistic analysis of the spoken language.

In PCA teacher-authored scripts are called "Oral Performance Script" (OPS) and "Textual Performance Script" (TPS), the former providing a good model for the learner to study spoken language and the latter written language. A performance script is defined as "a course of action that learners can imitate, learn, and use in various contexts, including those that are suggested in the materials" (Noda 2003). Examples of OPS include a dialogue or the script for a contextualized communicative drill or exercise. Examples of TPS include menus, email messages, letters, and short narratives. High-quality OPS and TPS have to be authentic, useful (i.e., fit the purpose of language instruction), and feasible (i.e., can be learned and rehearsed by the learner at this level).

To check authenticity of a performance script, PCA has a list of criteria to evaluate its quality. The first dimension is story or contextual authenticity. Does the script (either spoken or written) match the time, place, roles, and audience of the story presented in the target culture? At what time and in what place will the learner encounter or produce a similar dialogue or text in the target culture in the future? Is the course of action suggested in the material imitable? The second dimension is linguistic authenticity. Is the usage of grammar and vocabulary accurate and does it reflect real usage in the target society? For OPS, another dimension is pronunciation authenticity. Does the audio presented in the material sound natural? For TPS, in addition to story/contextual authenticity and linguistic authenticity, task authenticity ("Do readers and writers in the target culture tend to do similar things with the text?") and textual authenticity are two other important considerations. Textual authenticity can be further divided into three subcategories: orthographic authenticity ("Is the orthography used authentic?" "Are the symbols and punctuation marks authentic?"), generic authenticity ("Is the proper genre and its conventions including layout being adopted for the text?"), and discursive authenticity ("Do sentences link together naturally and cohesively?" "Does the textual flow follow the convention of this genre?").

As a pedagogical approach that is informed by and deeply rooted in actual teaching practices, PCA has inspired the development of a series of innovative Chinese language learning materials. In addition to the aforementioned *Chinese: Communicating in the Culture* and *Chinese Out of the Box* which are targeted at beginning-level adult and K-5 learners, other materials have been designed to address learners' needs at different levels and for different learning purposes. *Perform Suzhou*, a course for intermediate and early advanced levels, follows the saga of a group of American students who study and learn to adjust to life in Suzhou. It is the first installment of the "Perform China" series, which aims to use "community-specific and contextually defined performances" to train learners to become expert in doing things in a specific locale in the Chinese-speaking world (Jian et al. 2016). *Perform Chun Cao: A Multimedia Advanced Chinese Course* adopts the Chinese novel *Spring Grass* together with its eponymous television drama as the main learning materials and designs various pedagogical activities to enhance advanced learners' oral and written communication skills (Zeng 2019). *Tell It Like It Is: Natural Chinese for Advanced Learners* uses episodes from a popular Chinese television show and trains learners to acquire advanced listening skills to understand authentic and unrehearsed speech uttered by a broad variety of individuals representing different genders, ages, accents, and registers (Wang 2005). For study-abroad programs, *Action! China: A Field Guide to Using Chinese in the Community* provides an inventory of meticulously designed authentic tasks for intermediate to advanced students to develop meaningful connections with the locals and optimize their immersive experience in the target culture (Chai et al. 2017).

Assessment

As mentioned earlier, in PCA, performance is used not only as an instructional tool for learners to acquire the target culture but also as an assessment tool for the teacher to evaluate the learner. Performance-based assessment in the Performed Culture Approach is carried out at three levels for different purposes: hourly performance evaluation, periodic prochievement evaluation, and portfolio evaluation. The term prochievement refers to a type of achievement test that takes the form of a proficiency test. Hourly performance evaluation, also known as daily grading in some programs, is a commonly used evaluative method in PCA. For every contact hour, learners are evaluated based on their performance in class, using a rubric with clearly defined criteria. Hourly performance evaluation is a type of formative assessment and considered an alternative to frequent testing without the typical drawbacks of the latter (high cost, high stress levels for students). It provides constant diagnosis of students' learning and prompts the teacher to give concrete feedback on a regular basis. This assessment has higher reliability than a single summative test (Noda 2002c). It holds learners accountable on a daily basis and encourages them to fully prepare for every day's class, making the contact hour more effective (Christensen and Warnick 2006).

Periodic prochievement evaluation is usually conducted in the middle or at the end of a semester. To assess students' listening and speaking skills, PCA strongly recommends the use of a Prochievement Oral Interview (Chai and Chen 2017; Yu et al. 2020a). The testing contents are based on materials that have been covered in the curriculum, including both items and strategies, but the testing format is similar to a proficiency oral interview. The test is a one-on-one conversation between the student and the instructor. The instructor/interviewer sets up the context(s) for this conversation in one scene or multiple scenes and plays the role of a natural interlocutor to interact with the student. The student needs to respond to the prompts and visual aids presented by the interviewer. In terms of format, this conversation is very much like a communicative drill or exercise in a regular class, but with two major differences: First, it is much longer than the latter. While a communicative drill or exercise in a regular class period might happen in one context and lasts about 2–5 min, the conversation in a prochievement oral interview usually lasts between 5 and 20 min and cover more materials. Second, while the teacher gives corrective feedback in a regular class period, in the prochievement oral interview, the teacher/interviewer has to stay in the actor mode without making any corrections to errors. Like the hourly performance assessment, the prochievement oral interview is also graded with a rubric of criteria that covers various aspects of the performance, including listening comprehension, pronunciation, delivery, structural accuracy, cultural appropriateness, and content. To assess reading and writing performance, PCA also strives for authenticity, including linguistic, contextual, textual, and task authenticity, when designing test items. Students should be given a good reason (other than the fact that they have to take this test) why they are doing a reading or writing activity. The text presented should meet all the standards for a high-quality Textual Performance Script. When possible, the reading and writing task should be as authentic as possible.

The portfolio has proven to be yet another excellent assessment tool that the PCA advocates, especially at the advanced and superior levels. Traditionally used in the field of visual arts (e.g., drawing, architecture, design, photography), the portfolio has been increasingly used in a variety of professional settings for a wide range of purposes, including the educational field to assess the effectiveness of teaching and learning (Barnhardt et al. 1998; Linström 2005). Since the Performed Culture Approach sets the goal of language learning as being able to do things in the target culture, it is only natural that a working portfolio, defined as a collection of a learner's speaking and writing samples over a period of time that the learner assembles during the learning process, can be an effective alternative way of measuring an advanced learner's domain expertise in a given target language and culture. Used as a form of formative assessment, the portfolio can also be a motivating factor for learners and make the learning process more engaging and more personalized. Studies have shown that this form of assessment is reliable in measuring domain expertise of advanced- and superior-level Chinese-language learners (McAloon 2008; Zeng 2015).

Teacher Training

The teacher-training model used in the Performed Culture Approach is also performance based (Christensen and Noda 2002; Yu 2006). The distinction between declarative knowledge and procedural knowledge, as well as the process of expertise development mentioned earlier, also applies to teacher development (Walker and McGinnis 1995). Knowledge about linguistics, second-language acquisition, or any subject area in the field of China studies (e.g., literature, art, history, political science) does not automatically translate into active and effective teaching skills in the foreign language classroom. While such knowledge bases, especially in the domain of applied linguistics and second-language acquisition, are necessary in preparing the future language teacher, they are not sufficient. An effective teacher also needs to equip herself with knowledge in psychology, philosophy, sociology, anthropology, cognitive studies, and so forth. Additionally, classroom teachers need hands-on preservice and in-service training to acquire teaching methods and instructional techniques in order to conduct effective instruction and manage students in the real classroom. Therefore the ideal teacher-training model in PCA consists of two interdependent components, a language program paired with a teacher-training program. The language program is there to provide a teaching lab for teacher trainees so that they can not only observe master teachers and real learners in action but also, more importantly, practice what they have learned in pedagogical theory and receive immediate feedback from trainers.

Christensen and Noda (2002) suggest seven required courses for a teacher-training institute. Three of these courses are core courses. The first course focuses on the learner and the theory of language learning; the second examines the presentation of languages and cultures, including material selection and adaptation, as well as classroom management; and the third, which is the most time-consuming course, is a teaching practicum. In the teaching practicum, trainees would regularly observe classes taught by master teachers and write observation reports. They design lesson plans for their own classes and implement them in the real classroom. Follow-up discussions on their teaching are held. In addition to these three core courses, Christensen and Noda also propose that trainees take courses in pedagogical syntax, material development, Second Language Acquisition research, and research methodology.

Christensen and Noda's teacher-training model has been successfully implemented in two programs: the SPEAC Teacher Training Program at The Ohio State University and the ALLEX Teacher Training Institute currently based at Washington University in St. Louis. These are the only two summer programs in the United States that have seamlessly integrated an intensive teacher-training program with an intensive language program. SPEAC was first established as a summer intensive language program. In 1997 a Japanese teacher training program was integrated into the program, and in 1999 a Chinese teacher-training component was added. The target trainees for SPEAC are graduate students in America who aspire to become college-level East Asian language teachers but otherwise do not have the opportunity to receive hands-on training in their home institutions. In SPEAC, trainees take 3 intensive courses totaling 15 credits in 7 weeks: "Learning East Asian Languages in Cross-Cultural Contexts" (four credits), "Presentation of East Asian Languages and Cultures" (four credits), and "Practicum in East Asian Languages and Cultures" (seven credits). The first two courses meet 1 h a day and feature lectures and discussions. The third and the most time-consuming course meets 2.5 h every day, of which 1 h is devoted to language class observation in the morning and 1.5 h are used for post observation discussions, teaching rehearsals, and demo teaching. There are also lectures on lesson plan design and demonstrations of teaching technique, such as how to elicit a performance, how to contextualize a communicative drill, how to ask content questions, how to give feedback, and how to conduct hourly performance evaluations.

ALLEX Teacher Training Institute is a successor organization to the Teacher Training Institute under Exchange: Japan, a unique program founded in 1988 by Eleanor Jorden and Akio Terumasa, a Japanese philanthropist. Exchange: Japan aimed to improve the national capacity of Japanese teaching resources by recruiting and training high-quality Japanese-language teachers for American universities. In 2006 ALLEX (Alliance for Language Learning and Educational Exchange) was established to continue the important work started by Exchange: Japan and expanded its scope to include the training of novice Chinese and Korean language teachers. ALLEX recruited bright young people from Chinese-speaking regions, Japan, and Korea, trained them, and placed them in educational institutions throughout the United States. In exchange for their service to teach language courses and in some cases build or enhance language programs, the ALLEX trainees received waived tuition and free room and board from their host universities. Although most of the ALLEX trainees had very limited or no experience in language teaching prior to participating in the Summer Teacher Training Institute, many of them were able to make rapid progress in acquiring the theory and teaching techniques of the Performed Culture Approach after 7 weeks of intensive training. Since the organization's inception, over 200 universities have collaborated with ALLEX to start new language programs or to enhance established language offerings. More than 900 instructors have been trained and placed at North American universities, and a large percentage have completed master's degrees while teaching (Mason 2018). Some ALLEX alums have gone on to earn Ph.D. degrees in East Asian studies and posts as professors and lecturers at leading US institutions.

The ALLEX Teacher Training Institute offers three courses for novice Chinese-language teachers. "Teaching Chinese as a Foreign Language" offers a critical introduction to the principles of the Performed Culture Approach; "Pedagogical Grammar" discusses the pedagogical grammar of the Chinese language; and "Teaching Preparation and Practice" prepares the trainees to become independent classroom teachers in foreign language programs in American institutions of higher education. Through discussions, model class observation, lesson plan development, rehearsal teaching, as well as demo teaching of real classes in a parallel language course, trainees receive hands-on training in all aspects of a language program. They also have the opportunity to learn a new language (Japanese or Korean) for a few weeks in order to gain the learner's perspective and reflect on the Performed Culture Approach as a beginner language learner.

In both the SPEAC teacher training program and ALLEX Teacher Training Institute, trainees conduct demo teaching in incremental stages: 5 min, 10 min, 25 min, and possibly an entire 50-min class. While SPEAC trains some of the trainees on how to conduct a FACT class and ALLEX trains some novice teachers on how to conduct a FACT segment, one main objective of both programs is to ensure that novice teachers will have acquired the skills to conduct ACT classes by the end of the training period. Each stage of the demo teaching has a subset of objectives to meet in terms of the teaching methods and instructional techniques that trainees need to acquire. For instance, the 5-min demo teaching focuses on developing trainees' skills in executing some basic instructional techniques, such as keeping their voice loud and clear, maintaining good positioning in the classroom, modeling for repetition, random calling, and giving positive and corrective feedback. In addition to these general techniques, the 10-min demo teaching requires trainees

33 The Performed Culture Approach

to pay more attention to the cohesive arrangement and sequence of classroom activity as well as the skill of eliciting target cultural performances. For the 20-min demo teaching, trainees not only take note of the teaching techniques but also apply the Performed Culture Approach theories to their lesson plan design. Those trainees who excel in passing these three stages of demo teaching are given the opportunity to teach an entire ACT class. At every stage of the demo teaching, the trainees conduct a practice run 2 days before their actual teaching.

During these performance rehearsal sessions, trainees receive concrete advice and suggestions from their trainers and fellow trainees. After they teach in the real classroom in the parallel language program, they watch a video of their teaching and write a self-reflection report before receiving a detailed comment sheet together with grading of their teaching performance from their trainer. The work for both the trainees and trainers is intensive. It is no exaggeration to say that SPEAC and ALLEX run the most rigorous training programs for novice teachers who aspire to teach East Asian languages in the United States. Just like language learners, trainees in these programs understand well that the journey of acquiring expertise in the art of teaching does not end at the conclusion of the summer program. They need to keep practicing and reflect on their teaching in order to further improve and perfect their teaching skills.

A Comparison of PCA to Other Foreign Language Approaches and Methods

When an observer comes to a PCA ACT class, he or she often sees that the first 10–15 min of the class are devoted to dialogue performance. The students, usually two of them at a time, take turns performing a dialogue from memory, either sitting in their seats or preferably on a "stage" area of the classroom. Because of the similarity between this dialogue performance procedure and dialogue memorization in the Audio-lingual Method (ALM), the Performed Culture Approach is often mistaken as ALM or a "glorified" version of the latter. Granted, dialogue performance and dialogue memorization may seem similar to untrained eyes. Moreover, both PCA and ALM prioritize listening and speaking skills over reading and writing skills. Both believe in the adage "language is habit" and adopt many similar drilling techniques. However, the rationale and principles behind the seemingly similar teaching techniques as well as the implementation of these techniques in the real classroom are drastically different. As a language learning and teaching approach, PCA has adopted and revised some effective teaching techniques used in many second-language teaching methods. This section focuses on the similarities and the major differences between PCA and some of the more influential methods, namely, the Audio-lingual Method, Direct Method, Silent Method, and Communicative Approach.

The biggest difference between PCA and the Audio-lingual Method lies in the fact that while dialogue memorization serves as the end point in an ALM class, dialogue performance is merely a starting point in a PCA ACT class. Put another way, if memorization of grammatical patterns in their linguistic context is the goal in ALM, then in PCA performing and not merely memorizing the dialogue is used as a means to the end of helping learners acquire culturally appropriate communicative skills to respond to various situations and contexts in the target culture. In an ALM class, the teacher usually first presents a dialogue, asks the students to repeat it after her line by line, and then uses repetition drills, substitution drills, and transformative drills throughout the class period in order for students to learn and memorize the entire dialogue (Larsen-Freeman and Anderson 2015). The focus is on the acquisition of grammatical patterns and new vocabulary items rather than on how to communicate effectively in a given situation in the target culture. The ALM teacher may not take cultural elements or contextualization into consideration when presenting the dialogue and conducting the drills. They may not even care whether the target dialogue or the drills resemble actual conversations in the target culture. What they do care about is the accuracy of the grammatical patterns as presented in the dialogue and the drills. As long as the students can remember the dialogue at the end of the day and apply those grammatical rules correctly in the future, the teacher's job is done. In PCA, however, the teacher and the pedagogical materials always contextualize the main dialogue as well as all the other supporting drills, mostly communicative drills and exercises. This kind of contextualization ensures that learners are not relying on rote memorization but processing the information at a deeper, meaningful level. Classroom research has affirmed the effectiveness of using meaningful dialogue recitation techniques to help learners acquire formulaic expressions in Chinese (Yang 2016; Zhang 2016). Formulaic expressions, including fixed phrases such as "*ziwo jieshao yixia* (Let me introduce myself)," structural patterns such as "...... *zuo shenme gongzuo?* (What does someone do for a living?)," "*xing X, mingzi jiao Y* (surname is X, called Y)," or Chinese idioms and common sayings, are high-context linguistic forms that require deliberate, contextualized practices.

In PCA attention is paid not only to structural accuracy but also to cultural authenticity, including the authenticity of the script that the learners have to memorize. When students come to a PCA ACT class, they have already memorized the dialogue. The teacher sets up a context or scene for the students to perform this dialogue. The rest of the class activities are then built on the foundation of this first performance. The teacher presents other scenes and other contexts to elicit more improvised performances from the students. Both linguistic points, including new vocabulary items and grammatical patterns, and cultural points such as pragmatic patterns and intercultural communicative strategies are always drilled in culturally appropriate contexts. The end result is an improved target culture worldview in the learners' mind that will enable them to successfully complete similar communicative tasks in the target culture in the future.

In addition to this major difference, PCA and ALM are very different in many other aspects. Chen (2017, 2020) identifies 17 areas in which the two methods are different. For instance, while ALM considers structure and form to be paramount, PCA emphasizes culturally appropriate language behaviors. While ALM encourages rote memory, PCA promotes memory construction, especially memory construction of the target culture, through meaningful performances and interactions. Although

PCA adopts some of the mechanical drilling techniques in ALM, it uses more communicative drilling and real communicative activities. ALM forbids the use of the student's native language or translation as a teaching technique, whereas PCA allows the use of students' base language in FACT classes. When translation facilitates the students' understanding of the differences between the target and base languages, it can also be used in PCA to draw contrasts between the two languages and cultures.

This last point brings in the major difference between PCA and the Direct Method and many other immersive teaching methods. At first glance, a PCA ACT class is very similar to a Direct Method or an immersive language class. No base language is allowed and the teacher uses gestures, body language, and visual aids (including pictures, props, and realia) to make her meanings come across. Challenging the dominant Grammar-Translation Method, practitioners of the Direct Method believe in the primacy of the spoken language. One early advocate claimed, "Speaking must be an essential and important part of all modern language instruction as the best means of learning the foreign language" (Krause 1916, 26). Teachers using the Direct Method believe that meaning can be conveyed in the target language through the use of demonstration and visual aids without the need to use the students' native language (Larsen-Freeman and Anderson 2015). This focus on meaning making and emphasis on the power of human communication without the aid of one's base language is very much shared by PCA practitioners. After all, studies on human communication have shown that nonverbal messages transmit 65% of meaning, and contextual clues play a key role in meaning negotiation (Applebaum et al. 1974; Kramsch 1993). Direct Method's insistence on good pronunciation, oral work, and genuine reading based on realia from the target society, or "authentic reading materials" in today's term, are also key features of PCA as discussed earlier. However, unlike the Direct Method and many other similar immersive teaching methods, PCA believes in the value of using FACT classes and allows the use of base language to support learners, especially adult learners, in their learning of structural and intercultural items. Unlike children who may be more readily receptive to a completely inductive approach to teaching grammar as advocated by the Direct Method, many adult learners benefit from a mixture of inductive and deductive learning. Some structural patterns and cultural issues can be better explained in depth in the student's base language, especially at the beginning and intermediate levels. Therefore PCA does not exclude the use of the learner's base language in pedagogical materials and classroom instruction, especially in FACT classes.

Although PCA does not exclude deductive learning of linguistic and cultural rules in its FACT classes and pedagogical materials, in ACT classes it encourages students to think on their feet to respond to various situations. An important teaching technique in PCA is contextualized elicitation of target expressions and target behaviors. When moving into a new drill or exercise about a new linguistic or cultural pattern, the teacher uses visual aids or oral instructions to set a scene and the context. Students are expected to come up with the correct target expression and/or target behaviors to act out the scene. To the uninitiated, this contextualized elicitation seems unnecessary and leads to moments of silence in class when students are thinking hard to come up with the correct answer, while the teacher simply waits for the students' response. Why not present the new patterns to the students directly and then have them practice the new patterns? The principle behind this important elicitation technique in PCA is very similar to the major principle adopted by the Silent Way, that "teaching should be subordinated to learning" (Gattegno 1972, Larsen-Freeman and Anderson 2015). This is exactly PCA's teaching philosophy. Students should not be "given" items to learn in class. They need to figure out how to respond to a situation with what they have prepared before class. In this way, both PCA and the Silent Way aim to foster autonomous learning habits. Although there are no elaborate color-coded Fidel Charts or the use of rods to elicit students' performance in PCA, it is similar to the Silent Way in terms of giving autonomy to the learner, maximizing students' speaking time in class, and encouraging self- and peer correction (Larsen-Freeman and Anderson 2015).

Since the 1970s the dominant approach in the foreign language field has been the Communicative Approach, which emphasizes learning a language for real communication. "Communication" is defined as "a continuous process of expression, interpretation, and negotiation" (Savignon 1983, 8). With this definition, communication has to have a purpose, and context refers to the setting and the participants (VanPatten 2017). PCA shares this emphasis on meaning negotiation and purposeful, contextualized interaction. Communication is indeed one of the major premises of PCA. However, PCA does not stop at that level. For PCA, communication in the target culture is more important. Merely using a second language to communicate in one's base culture or an idiosyncratic "classroom culture" is not sufficient for PCA, even though those settings constitute contexts as well. All too often in classrooms that claim to have adopted the Communicative Approach, students learn Chinese to speak English, that is, they use the Chinese linguistic code to do things the American way. Either the acquisition of the linguistic code is completely separated from the acquisition of the target cultural code, or the latter is completely ignored.

This language/culture separation is manifested in the 5C logo of the ACTFL Proficiency Guidelines (ACTFL 2012). In the widely publicized original ACTFL 5C logo, the circle that stands for "cultures" is linked with two of the other four Cs, communication and connections, while the other two, communities and comparisons, are disconnected from it. In the newly updated 5C logo, although the 5Cs are joined by the common center, the element of cultures is still perceived as something external to communication, connections, communities, and comparisons. If PCA had its own logo it would be composed of four concentric circles, with the largest one being "culture," the second one being "community," the third one "context," and the smallest one in the center "communication." This view of the role of culture in language learning and in communication is the key difference between PCA and the Communicative Approach. This is also the main feature that distinguishes PCA from many other contemporary language-teaching methods that are derived from the principles of the Communicative Approach and form the subcategories of

Communicative Language Teaching (CLT), such as proficiency-based language teaching, TPR (Total Physical Response), TPRS (Teaching of Proficiency through Reading and Storytelling), and Task-Based Language Teaching (TBLT).

Yet another major difference between the Communicative Approach and PCA is in the area of pedagogical practices. Scholars have observed that there is a certain "fuzziness" about Communicative Language Teaching in that it lacks prescribed classroom techniques (Klapper 2003). Bill VanPatten, one of the leading figures promoting CLT, comments that "for whatever reason, communicative language teaching became a buzzword everyone thought they understood but maybe really didn't. The outcome: communicative language teaching became whatever people wanted it to be" (VanPatten 2017, 46). In contrast, the Performed Culture Approach is based on real classroom practices and over the years has formed a set of pedagogical practices and classroom techniques that go hand in hand with the theoretical principles. As discussed earlier, the theories of PCA are fully implemented in its pedagogical practices, ranging from curriculum design, pedagogical materials, and assessment tools to teacher training. With PCA novice teachers will not have to rely on some nebular theories, but will be able to use concrete classroom techniques and practices to guide and reflect on their teaching.

Major Contributions to the Foreign Language Field

It is worth noting that all the major foreign language teaching methods and approaches that have been dominant in the United States as well as the rest of the world were developed from the teaching of European languages, especially English as a foreign or second language. The Performed Culture Approach is the first, and so far the only, foreign language teaching approach to be developed in the field of East Asian language pedagogy. In recent years its theories have also been applied to the teaching of English as a foreign language in China. The historically one-way street, theories, and methods coming from European language fields to the less commonly taught language fields is slowly being changed into a two-way exchange. The real-world wisdom and philosophy of the Confucian world is being recognized for its educational values that are starting to influence Western civilization. This is no small change in the history of foreign language education both in the United States and elsewhere in the world.

Putting aside the issue of Eurocentric arrogance and bias, the Performed Culture Approach was developed with learners' actual learning needs in mind when their base language and base culture require communicative behaviors that drastically differ from the target language and target culture. The Performed Culture Approach was born out of a dissatisfaction with two influential intellectual movements in the twenty-first century, one within the field of foreign language education and the other well beyond it, namely, the Proficiency Movement and the Cognitive Revolution.

The Proficiency Movement gained momentum in the 1970s when foreign language educators turned their attention to linguistic performance in their teaching and assessment and attempted to introduce proficiency assessment instruments developed by government language institutions into academia. These assessment tools, such as the Foreign Service Institute (FSI) Oral Proficiency Rating Scale and the Interagency Language Roundtable (ILR) Scale, later became the basis for the ACTFL Proficiency Guidelines (Thompson 1991). The Cognitive Revolution grew out of linguist Noam Chomsky's rejection of the behaviorist approach and the rise of cognitive science since the 1950s. Since then, linguists, psychologists, anthropologists, and computer scientists have turned their attention to the mind in order to explore how it works and how human beings learn. Language learning is no longer viewed as simply the result of repetition and habit formation, but rather of a complex cognitive process (Larsen-Freeman and Anderson 2015). The ACTFL Proficiency Guidelines had a huge impact on classroom practices, and many teachers turned to standards-based or performance-based instruction. The Cognitive Revolution gave rise to the Cognitive Code Approach which prompted educators to redefine the learner and learning.

However, in the midst of the initial excitement over these two intellectual movements, culture somehow managed to get lost. The enthusiasm over the cognitive processes and computational models of the mind during the early years of the Cognitive Revolution resulted in a lack of attention to how culture works on the mind. This lack can be best illustrated by psychologist Jerome Bruner's critique of Noam Chomsky's language acquisition hypothesis. According to Chomsky, children are born with a Linguistic Acquisition Device (LAD) which enables them to learn their first language without any reliance on any factors in the learning environment. Bruner proposes that "the infant's Language Acquisition Device could not function without the aid given by an adult who enters with him into a transactional format. That format, initially under the control of the adult, provides a Language Acquisition Support System, LASS" (Bruner 1985, 2009). Based on his work on developmental psychology, Bruner further criticizes the Cognitive Revolution's overzealousness regarding information processing while overlooking the role that culture plays in the human meaning-making process (Bruner 1990). These findings are also supported by anthropological studies on how culture works and what culture means. According to cultural psychologist Bradd Shore, "It is the business of culture to replace personal symbols with conventional ones so that we might talk together. And it is the business of the mind to struggle to overcome the arbitrariness of the forms on which this collective life depends, investing conventional forms with personal meaning" (Shore 1998, 54). Linguistic anthropologists have long held the view that the key to communication is understanding the context and culture of conversation (Agar 1993). Recent breakthroughs in studies of the origins of human language have also pointed to the indispensable role that culture has played in the invention of language as a communicative tool in human society (Everett 2012, 2017).

In the field of foreign language education, while the Proficiency Movement and the Communicative Approach have drawn language teachers' attention to performance and communication, the Performed Culture Approach has added culture to the equation in proficiency development and assessment. PCA's maxim is that language teachers' ultimate goal is to help the learner effectively communicate in the target culture through the means of performance. In recent years globalization and migratory movements in the world have prompted researchers in European language teaching fields to critically examine the old framework of foreign language teaching. Many researchers have noted the Communicative Approach's lack of attention to culture and thus revived the push for intercultural communicative competence in language teaching and learning (Byram 1997; López-Rocha and Arévalo-Guerrero 2014). In 2012 the US government Interagency Language Roundtable (ILR) added "competence in intercultural communication" as a new skill area for assessment in the ILR scale, a set of descriptions of abilities to communicate in a language (Interagency Language Roundtable 2012). The positive washback effect of the ILR scale is sure to influence the field of foreign language education in the years to come. This change reflects a full circle around from what Eleanor Jorden first proposed during her time at Foreign Service Institute back in the 1950s. The Performed Culture Approach is the first comprehensive language-teaching approach that aims to simultaneously develop both linguistic competence and intercultural communicative competence. It has successfully done so not only with a set of well-developed pedagogical theories that combines the wisdom of the East (the Confucian philosophy of social interaction) and the West (cognitive theories) but also with a gamut of clearly prescribed and acquirable pedagogical practices. It is in this sense that it is a paradigm-shifting language-teaching approach.

Conclusion

The Performed Culture Approach is a foreign language pedagogical framework developed by a team of language specialists under the leadership of Professors Galal Walker and Mari Noda based on decades of theoretical and practical work in the classroom. It was born out of a frustration of the old mindset regarding language learning and teaching and aspires to help learners truly master a second language/ culture in order to navigate the native workplaces and engage in social interactions in the target culture (Walker 2020). Teachers and programs that have adopted this approach successfully integrate instruction in both language and culture through meticulously designed performances that place learners in authentic linguistic and cultural contexts. Although the approach has been developed in the field of East Asian language pedagogy in the United States over the past three decades, its influence has begun to reach the field of teaching English as a foreign language in China. This approach draws on the ancient Confucian recognition that "knowledge and action are but one" and builds on the language philosophy of Western cognitive studies, linguistic anthropology, cultural psychology, and performance studies. More importantly, the Confucian educational philosophy of combining learning with pleasure is profoundly embodied in the practice of the Performed Culture Approach, making it a fundamentally humanistic pedagogical approach to language acquisition. It is opening a new era in the field of teaching Chinese as a foreign language in the twenty-first century.

References

ACTFL (American Council on the Teaching of Foreign Languages). 2012. ACTFL proficiency guidelines 2012. https://www.actfl.org/publications/guidelines-and-manuals/actfl-proficiency-guidelines-2012. Accessed July 7, 2019.

Agar, Michael H. 1993. *Language shock: Understanding the culture of conversation*. New York: William Morrow and Company, Inc.

Applebaum, Ronald L., Edward M. Bodaken, Kenneth K. Sereno, and Karl W.E. Anatol. 1974. *The process of group communication*. Chicago: Science Research Associates.

Barnhardt, Sarah, Jennifer Kevorkian, and Jennifer Delett. 1998. *Portfolio assessment in the foreign language classroom*. Washington, DC: National Capital Language Resource Center.

Bell, Kevin. 2017. *Game on: Gamification, gameful design, and the rise of the gamer educator*. Baltimore: Johns Hopkins University Press.

Brecht, Richard D. 2014. A framework for discussing the uniqueness of in-country language learning. 探讨在目标语国家语言学习特征的整体框架. In *Chinese as a foreign/second language in the study abroad context* 留学生在华汉语教育初探, ed. Kunshan Li, 1–8. Beijing: Peking University Press.

Brecht, Richard D., and A. Ronald Walton. 1994. National strategic planning in the less commonly taught languages. *The Annals of the American Academy of Political and Social Science* 532: 190–212.

Bruner, Jerome. 1985. *Child's talk: Learning to use language*. New York: W.W. Norton & Company.

———. 1990. *Acts of meaning*. Cambridge, MA: Harvard University Press.

———. 2009. *Actual minds, possible worlds*. Cambridge, MA: Harvard University Press.

Byram, Michael. 1997. *Teaching and assessing intercultural communicative competence*. Clevedon: Multilingual Matters.

Cameron, Meribeth E. 1948. Far Eastern studies in the United States. *The Far Eastern Quarterly* 7 (2): 115–135.

Chai, Donglin. 2018. Knowledge transmission within the analects of Confucius and its implications for the training of CFL teachers. *Journal of Chinese Teaching and Research in the U.S* (May 2018): 85–110.

———. 2020. Eliciting cultural performances at the intermediate level. *American Journal of Chinese Studies* 27 (October, 2020): 155–160.

Chai, Donglin, and Hanning Chen. 2017. Performed Culture in the classroom (video project). Columbus: National East Asian Languages Resource Center at the Ohio State University. https://www.youtube.com/playlist?list=PLxrMNt4zd0aVwVZcsMa6CE9qa8Q1Zdn6O. Accessed August 8, 2019.

Chai, Donglin, Crista Cornelius, and Bing Mu. 2017. *Action! China: A field guide to using Chinese in the community*. 体演日志:中文实地应用指南. London/New York: Routledge.

Chen, Guangyan. 2017. Performed culture: An approach to US collegiate Chinese language education. *Quarterly Journal of Chinese Studies* 5 (1): 84–102.

———. 2020. *Performed culture in Chinese language education: A culture-based approach for U. S. collegiate contexts*. Lanham: Lexington Books.

Christensen, Matthew B, and Mari Noda. 2002. A performance-based pedagogy for communicating in cultures: Training teachers for East Asian languages. Columbus, OH: National East Asian Languages Resource Center at The Ohio State University.

Christensen, Matthew B., and Paul Warnick. 2006. Performed culture: An Approach to East Asian Language Pedagogy. Columbus: The Ohio State University East Asian Language Resource Center.

Cornelius, Crista Lynn. 2015. *Language socialization through Performance Watch in a Chinese study abroad context*. MA thesis. The Ohio State University. https://etd.ohiolink.edu/pg_10?::NO:10:P10_ETD_SUBID:105494. Accessed Nov 9, 2019.

Duff, Patricia A., and Liam Doherty. 2018. Chinese second language socialization. In *The Routledge handbook of Chinese second language acquisition*, ed. Chuanren Ke, 82–99. Abingdon: Routledge.

Everett, Daniel L. 2012. *Language: The cultural tool*. New York: Pantheon Book.
———. 2017. *How language began: The story of humanity's greatest invention*. New York/London: Liveright Publishing.
Foreign Service Institute. 2019. Foreign language training. https://www.state.gov/foreign-language-training/. Accessed August 8, 2019.
Frye, Robert, and Thomas J. Garza. 1992. Authentic contact with native speech at home and abroad. In *Teaching languages in college: Curriculum and content*, ed. Wilga Rivers, 225–243. Lincolnwood: National Textbook.
Gattegno C 1972. Teaching foreign languages in schools: The silent way. 2nd edition. New York: Educational Solutions.
Glaser, Robert, and Michelen T.H. Chi. 1988. Overview. In *The nature of expertise*, ed. Michelene T.H. Chi, Robert Glaser, and Marshall J. Farr, xv–xxviii. Hillsdale: Erlbaum.
Goffman, Erving. 1959. *The presentation of self in everyday life*. New York: Anchor Books, Doubleday.
Hall, Edward T. 1977. *Beyond culture*. New York: Anchor Books.
Hammerly, Hector. 1986. *Synthesis in language teaching: An introduction to linguistics*. Blaine: Second Language Publications.
He, Yunjuan, and Xizhen Qin. 2017. Students' perceptions of an internship experience in China: A pilot study. *Foreign Language Annals* 50 (1): 57–70.
Interagency Language Roundtable. 2012. Interagency Language Roundtable skill level descriptions for competence in intercultural communication. https://www.govtilr.org/Skills/Competence.htm. Accessed June 12, 2020.
Jia, Junqing. 2017. *Motivating experiences in an extended Chinese as a foreign language learning career: Identifying what sustains learners to advanced-skill levels*. Ph.D. dissertation, The Ohio State University.
Jian, Xiaobin, and Eric Shepherd. 2010. Playing the game of interpersonal communication in Chinese culture: The "rules" and the moves. In *The pedagogy of performing another culture*, ed. Galal Walker, 96–143. Columbus: National East Asian Languages Resource Center at the Ohio State University.
Jian, Xiaobin, Jianfen Wang, Junqing Jia, and Chenghua Feng. 2016. Perform Suzhou. 体演苏州: 中高级中文听说教程. Suzhou, Jiangsu: Soochow University Press.
Jin, Hong Gang. 2018. Implicit and explicit learning, knowledge, and instruction in CFL studies. In *The Routledge handbook of Chinese second language acquisition*, ed. Chuanren Ke, 393–414. London/New York: Routledge.
Jorden, Eleanor H., and Hamako I. Chaplin. 1962. *Beginning Japanese: Part 1*. New Haven: Yale University Press.
Jorden, Eleanor H., and Mari Noda. 1987. *Japanese: The spoken language: Part 1*. New Haven: Yale University Press.
———. 1988. *Japanese: The spoken language: Part 2*. New Haven: Yale University Press.
———. 1990. *Japanese: The spoken language: Part 3*. New Haven: Yale University Press.
Jorden, Eleanor H., and A. Ronald Walton. 1987. Truly foreign languages: Instructional challenges. *The Annals of the American Academy of Political and Social Science* vol. 490, *Foreign Language Instruction: A National Agenda* (Mar. 1987): 110–124.
Jorden, Eleanor H., Richard D. Lambert, and Jonathan H. Wolff. 1991. *Japanese language instruction in the United States: Resources, practice, and investment strategy*. Washington, DC: National Foreign Language Center, Johns Hopkins University.
Kapp, Karl M. 2012. *The gamification of learning and instruction: Game-based methods and strategies for training and education*. San Francisco: Pfeiffer.
Kinginger, Celeste, Qian Wu, and Sheng-Hsun Lee. 2018. Chinese language acquisition in study abroad contexts. In *The Routledge handbook of Chinese second language acquisition*, ed. Chuanren Ke, 301–317. London/New York: Routledge.
Klapper, John. 2003. Taking communication to task: A critical review of recent trends in language teaching. *The Language Learning Journal* 27: 33–42.
Kramsch, Claire. 1993. *Context and culture in language teaching*. Oxford: Oxford University Press.

———. 2006. From communicative competence to symbolic competence. *The Modern Language Journal* 90: 249–252.

———. 2009. Third culture and language education. In *Contemporary applied linguistics: vol.1, Language teaching and learning*, ed. L. Wei and V. Cook, 233–254. London: Continuum.

———. 2014. Teaching foreign languages in an era of globalization: Introduction. *The Modern Language Journal* 98: 296–311.

Kramsch, Claire, and A. Whiteside. 2008. Language ecology in multilingual settings: Towards a theory of symbolic competence. *Applied Linguistics* 29: 645–671.

Krause, Carl. 1916. The Direct method in Modern languages. New York: Charles Scribner's Sons.

Kubler, Cornelius. 2006. *NFLC guide for basic Chinese language programs*. 2nd ed. Columbus: National East Asian Languages Resource Center at the Ohio State University.

———. 2011. *Basic spoken Chinese: A practical approach to fluency in spoken Mandarin*. Rutland: Tuttle Publishing.

———. 2013. *Intermediate spoken Chinese: A practical approach to fluency in spoken Mandarin*. Rutland: Tuttle Publishing.

Larsen-Freeman, Diane, and Marti Anderson. 2015. Techniques & principles in language teaching. 3rd edition. Oxford: Oxford University Press.

Latourette, Kenneth Scott. 1955. Far Eastern studies in the United States: Retrospect and prospect. *The Far Eastern Quarterly* 15 (1): 3–11.

Li, Cong. 2018. *Gamification in foreign language education: Fundamentals for a gamified design of institutional programs for Chinese as a foreign language*. Ph.D. dissertation, The Ohio State University.

Light, Timothy. 1987. *Xiandai waiyu jiaoxuefa: Lilun yu shijian* 现代外语教学法:理论与实践 [Modern language pedagogy: theory and practice]. Beijing: Beijing yuyan xueyuan chubanshe.

———. 1993. Chinese language training for new Sinologists. In *American studies of Contemporary China*, ed. David Shambaugh, 239–263. Washington DC/Armonk: Woodrow Wilson Center Press/M. E. Sharpe.

Linström, Lars. 2005. The multiple uses of portfolio assessment. *Studies in Educational Policy and Educational Philosophy* 2005 (1): 1–15.

López-Rocha, S., and E. Arévalo-Guerrero. 2014. Intercultural communication discourse. In *The Routledge handbook of Hispanic applied linguistics*, ed. M. Lacorte, 531–550. New York: Routledge.

Mason, Thomas. 2018. Annual report of the ALLEX Foundation. Personal communication.

McAloon, Patrick. 2008. *Chinese at work: Evaluating advanced language use in China-related careers*. Ph.D. dissertation, The Ohio State University.

———. 2015. From proficiency to expertise: Using HR evaluation methods to assess advanced foreign language and culture ability. In *To advanced proficiency and beyond: Theory and methods for developing superior second language ability*, ed. T. Brown and J. Brown, 153–170. Washington, DC: Georgetown University Press.

McCoy, John. 1979. The full-year Asian Language Concentration Falcon program in Chinese language at Cornell University. ERIC Number: ED192619.

Meng, Nan. 2020. Creating an immersive cultural environment at the beginning level. *American Journal of Chinese Studies* 27 (October 2020): 151–155.

Nell, Victor. 1988. *Lost in a book: The psychology of reading for pleasure*. New Haven: Yale University Press.

Noda, Mari. 2002a. Reading as a social activity. In *Acts of reading: Exploring connections in pedagogy of Japanese*, ed. Hiroshi Nara and Mari Noda, 24–37. Honolulu: University of Hawaii Press.

———. 2002b. Evaluation in reading. In *Acts of reading: Exploring connections in pedagogy of Japanese*, ed. Hiroshi Nara and Mari Noda, 197–222. Honolulu: University of Hawaii Press.

———. 2002c. Selection and development of learning materials. In *Acts of reading: Exploring connections in pedagogy of Japanese*, ed. Hiroshi Nara and Mari Noda, 223–244. Honolulu: University of Hawaii Press.

———. 2003. On authenticity and scaffolding. Lecture at SPEAC Summer Teacher Training Program, The Ohio State University.

———. 2020. The role of pedagogical material in the Performed Culture Approach. *American Journal of Chinese Studies* 27 (October 2020): 146–150.

Noda, Mari, Yui Iimori Ramdeen, Stephen D. Luft, and Thomas Mason Jr. 2017. *Action! Japan: A field guide to using Japanese in the community*. London/New York: Routledge.

Noda, Mari, Patricia J. Wetzel, Ginger Marcus, Stephen D. Luft, Shinsuke Tsuchiya, and Masayuki Itomitsu. 2020. *NihonGO NOW!: Performing Japanese culture-Level 1 Volume 1 textbook and activity book*. London/New York: Routledge.

Qin, Xizhen. 2013. Xingwei wenhua de ketang jiaoxue shijian yu fankui [Teaching Behavioral Culture in Chinese Classroom]. *Journal of the Chinese Language Teachers Association* 48, 3: 1–24.

———. 2017a. Improving Chinese language learners' intercultural communicative competence through performing culture. *International Chinese Language Education* 2 (2): 25–30.

———. 2017b. *Understanding intercultural misunderstandings between Chinese and American cultures: Applying the Performed Culture Approach* 中美跨文化交际误解分析与体演文化教学法. Beijing: Foreign Language Teaching and Research Press.

Ramsey, S. Robert. 1996. Obituary: A. Ronald Walton (1943–1996). *The Journal of Asian Studies* 55 (4): 1114–1115.

Savignon, Sandra J. 1983. *Communicative competence: Theory and classroom practice*. Reading: Addiso-Wesley Publishing Company.

Schank, Roger C. 1990. *Tell me a story: A new look at real and artificial memory*. New York: Macmillan Publishing Company.

Schank, Roger C., and R. Abelson. 1977. *Scripts, plans, goals and understanding: An inquiry into human knowledge structures*. Hillsdale: Erlbaum.

Shore, Bradd. 1998. *What culture means, how culture means*. Heinz Werner lecture series XXII. Worcester: Clark University Press.

Stock, Brian. 1996. *Augustine the reader– Meditation, self-knowledge and the ethics of interpretation*. Cambridge, MA: The Belknap Press of Harvard University.

Thompson, Irene. 1991. The proficiency movement: Where do we go from here? *The Slavic and East European Journal* 35 (3): 375–389.

VanPatten, Bill. 2017. *While we're on the topic: BVP on language, acquisition, and classroom practice*. Alexandria: American Council on the Teaching of Foreign Languages. Digital edition.

Walker, Galal. 2000. Performed culture: Learning to participate in another culture. In *Language policy and pedagogy: Essays in honor of A. Ronald Walton*, ed. Richard D. Lambert and Elana Shohamy, 221–236. Amsterdam: J. Benjamins.

———. 2010. *The pedagogy of performing another culture*. Columbus: National East Asian Languages Resource Center at the Ohio State University.

———. 2020. Mindreading and the performed culture approach. *American Journal of Chinese Studies* 27 (October 2020): 142–146.

Walker, Galal, and Yong Lang. 2004. *Chinese: Communicating in the culture*. Columbus: National East Asian Languages Resource Center at the Ohio State University.

Walker, Galal, and Scott McGinnis. 1995. *Learning less commonly taught languages: An agreement on the bases for the training of teachers*. Columbus: Foreign Language Publications.

Walker, Galal, and Mari Noda. 2000. Remembering the future: Compiling knowledge of another culture. In *Reflecting on the past to shape the future*, ed. Diane Birckbichler, 187–212. Lincolnwood: National Textbook Company.

Walker, Galal, and Huanzhen Zhao. 2013. *Chinese out of the box: For teachers of K-5 Mandarin Chinese* 非常中文.Columbus: Foreign Language Publications at the Ohio State University.

Walton, A. Ronald. 1989. Chinese language instruction in the United States: Some reflections on the state of the art. *Journal of the Chinese Language Teachers Association* 24 (2): 1–42.

Wang, Jianqi. 2005. *Tell it like it is: Natural Chinese for advanced learners*. New Haven: Yale University Press.

Yang, Jia. 2016. Dialog recitation and CFL learners' production of formulaic expressions. *Journal of the National Council of Less Commonly Taught Languages* 19: 115–145.

Yu, Li. 2006. *Zhongwen jiaoshi peixun xiangmu SPEAC moshi shuping* 中文教师培训项目 SPEAC 模式述评 [A description and critique of a Chinese teacher training program: The SPEAC model]. *Guoji hanyu jiaoxue dongtai yu yanjiu* [International Chinese Education and Research] 4: 3–11.

———. 2008. Exploring learners' intercultural behaviors: Applying the performed culture approach in Chinese language pedagogy. *The Journal of the Chinese Language Teachers Association* 43 (3): 121–146.

———. 2009. Where is culture: Culture instruction and the foreign language textbook. *The Journal of the Chinese Language Teachers Association* 44 (3): 73–108.

———. 2020. Designing specialized domain courses at the advanced level. *American Journal of Chinese Studies* 27 (October 2020): 160–163.

Yu, Li, Donglin Chai, Nan Meng. 2020a. *Zhenshi ketang zhong de tiyan wenhua: Ceshi pinggu* 真实课堂中的体演文化: 测试评估 [PCA assessment at the beginning and intermediate levels]. Columbus: National East Asian Languages Resource Center at the Ohio State University. https://www.youtube.com/playlist?list=PLxrMNt4zd0aVBF24LLLPYku8f0YHuCjSd. Accessed August 20, 2020.

———. 2020b. *Zhenshi ketang zhong de tiyan wenhua: Chuji zhongji ketang jiaoxue jiqiao* 真实课堂中的体演文化:初级、中级课堂教学技巧 [PCA basic teaching techniques at the beginning and intermediate levels]. Columbus: National East Asian Languages Resource Center at the Ohio State University. https://www.youtube.com/playlist?list=PLxrMNt4zd0aU9XPsJOgMG8Z2b78DlbOp2. Accessed 20 Aug 2020.

Zeng, Zhini. 2015. Demonstrate and evaluate expertise in communicating in Chinese as a foreign language. Ph.D. dissertation, The Ohio State University.

———. 2018a. Striving for the third space: An American professional's unconventional use in Chinese workplace. *Foreign Language Annals* 51: 658–683.

———. 2018b. Beyond advanced proficiency: Using portfolio assessment to evaluate expertise in communicating in Chinese as a foreign language. *Journal of the National Council of Less Commonly Taught Languages* 24: 1–42.

———. 2019. *Perform Chun Cao: A multimedia advanced Chinese course* 体演春草: 高级汉语视听说教程. Beijing/Columbus: Foreign Language Teaching and Research Press/Foreign Language Publications at The Ohio State University.

Zhang, Yongfang. 2012. Integrating authentic written texts into beginning Chinese reading instruction. *Journal of the Chinese Language Teachers Association* 47 (1): 43–60.

———. 2013. A CIP spiral performance-based model for beginner level Chinese L2 reading instruction. *Journal of the Chinese Language Teachers Association* 48 (2): 91–118.

Zhang, Xin. 2016. *Four-character idioms in advanced spoken Chinese: Perception and reaction of native speakers and a pedagogy of C2 expectations*. Ph.D. dissertation. The Ohio State University.

Zull, James E. 2002. *The art of changing the brain: Enriching the practice of teaching by exploring the biology of learning*. Sterling: Stylus Publishing.

Part V

Transference from Chinese to English

Transference of Chinese to English: Introduction

34

Jock Onn Wong and Zhengdao Ye

Contents

Introduction	879
Chapter Summaries	880
Conclusion	881
References	881

Abstract

This introductory chapter provides an overview of the chapters included in the section on transference of Chinese to English. These chapters deal with Singapore English, Malaysian English, Academic English, Chinese loanwords in English, and Business English.

Keywords

Chinese English · Singapore English · Malaysian English · Academic English · Chinese loanwords · Business English

Introduction

The five chapters in this section of *The Palgrave Handbook of Chinese Languages Studies* explore the dynamic interactions between varieties of Sinitic (Chinese) languages and English(es). They approach the topic from three angles. The first two

---

J. O. Wong (✉)
Centre for English Language Communication, National University of Singapore, Singapore, Singapore

Z. Ye
School of Literature, Languages and Linguistics, The Australian National University, Canberra, Australia
e-mail: zhengdao.ye@anu.edu.au

© The Author(s), under exclusive licence to Springer Nature Singapore Pte Ltd. 2022
Z. Ye (ed.), *The Palgrave Handbook of Chinese Language Studies*,
https://doi.org/10.1007/978-981-16-0924-4_50

chapters (by Jock Onn Wong and Siew Imm Tan) come at it from the point of view of interactions between "Global Chineses" and "Global Englishes" in two Southeast Asian countries, where Chinese and English languages are both used on a daily basis.

Chapter Summaries

Jock Onn Wong's chapter looks specifically at the influences of Cantonese and Hokkien on Singlish, a collective term that refers to informal varieties of English used in Singapore, at the phonological, lexical, semantic, and cultural levels. An underlying theme of his chapter is that culture is inseparable from language. Singlish is a prime example of what Wierzbicka describes as the answer to "a felt cultural need" (Wierzbicka 2003, p. 340). It appears that some of the languages spoken locally in Singapore have played a major role in shaping Singlish; the culturally salient meanings and values embodied in the locally spoken languages, such as Cantonese and Hokkien, enter Singlish because Singaporeans have every need to express them in their daily lives. To give the reader a better appreciation of the intertwined nature of language and culture, Wong examines in detail the historical context in which Singlish developed. He also touches upon the future of Singlish in the age of social media.

Siew Imm Tan's chapter focuses on borrowing, a particular type of contact-induced change, as a case study of the Chinese language presence in Malaysian English (MYE). Using the Malaysian component of the corpus of Global Web-based English (GloWbE MY) and adopting a corpus-based approach, Tan shows the influences on MYE from multiple Sinitic languages, most noticeable from Hokkien, Cantonese, and Mandarin, three of the most dominant Sinitic languages spoken in Malaysia. These influences, as Tan illustrates, are also domain specific, reflecting particular activities with which the speakers of these languages are engaged in the Malaysian society. The chapter shows, on the one hand, the collective influences of the three languages on MYE and, on the other hand, the different social statuses these three languages enjoy and how they compete to enter the lexicon of MYE.

While Wong's and Tan's chapters depict the dynamic language contact situations between Sinitic languages and Englishes in multilingual societies, the next two chapters examine specific sites of contact – academic writing and business communication. In these sites, English functions as a lingua franca or, in the sociologist Abram de Swaan's words, "a hyper-central language" because it sits at the apex of the pyramid of the global language system due to its greater communicative reach compared with all other languages (de Swaan 2001). This means that more and more people use English as their additional language across various domains, of which academic English writing and international business are two noted areas where there have been growing numbers of L1 Chinese speakers.

Focusing on the domain of academic English writing, Feng Cao's chapter offers an overview of findings in academic discourse research that identify the ethnolinguistic features characteristically produced by L1 Chinese academic writers. The many features revealed at various levels – lexical, syntactical, pragmatic, and information organizational – raise the interesting question of how to reconceptualize domain-specific features distinctive to academic English discourse when academic writers are increasingly non-L1 English users.

Per Dust-Andersen and Xia Zhang's chapter gives a rare glimpse into the influences of Chinese as a mother tongue on its speakers' production and perception of a range of communicative acts commonly encountered in the international business settings. These acts relate to problem-solving strategies, conflict resolution, and apology. Based on data drawn from the Global English Business Communicating Project headed by Per Durst-Andersen, the chapter not only presents different communicative strategies and resources used by L1 Mandarin speakers, L1 British English, and L1 Russian speakers but also introduces a theoretical framework to account for these differences.

The final chapter by Will Peyton approaches the topic on the transference of Chinese to English by surveying Chinese loanwords in Anglo English. It focuses on both well-established terms of Chinese origin included in the authoritative *Oxford English Dictionary*, such as *kowtow* and *gung-ho*, and recent Mandarin loanwords used in major English newspapers. While exploring reasons for the dearth of Chinese loanwords in English, Peyton investigates how they were initially used in media in order to be accepted into the English language and traces the trends of linguistic borrowing from past to the present. By examining Chinese terms codified in the English language, the chapter shows how major English language media project and interpret China to Anglophone audiences. One can only expect that there will be increased interest in the social, cultural, and political life of Chinese people from English language media as China plays an increasingly important role in global affairs.

Conclusion

Today, the influence of English is felt in many corners of the world, and much of the discussion on language contact involving English tends to be concerned with how English influences or shapes other languages (e.g., Rapatahana and Bunce 2012). The chapters in this section provide a fresh perspective by examining reverse influences, and collectively they show the deep interconnectedness between the Chinese- and English-speaking worlds.

References

de Swaan, A. 2001. *Words of the world: The global language system.* Cambridge: Polity.
Rapatahana, V., and P. Bunce, eds. 2012. *English language as hydra: Its impacts on non-English language cultures.* Bristol: Multilingual Matters.
Wierzbicka, A. 2003. Singapore English: A semantic and cultural perspective. *Multilingua* 22: 327–366.

The "Chineseness" of Singlish

Jock Onn Wong

Contents

Introduction: Singapore English as a Cultural Entity	884
Linguistic Evidence	885
Tones in Singlish	886
The Interdental Fricatives	886
The Dark /l/	887
The Voiceless Plosives	888
Color Terms	888
Particles and Other Cultural Keywords of Chinese Origin	889
Semantic and Cultural Evidence: *Kiasu* and *Chope*	889
Historical Context: The Rise of English in Singapore	892
Further Historical Context: The Eradication of Non-Mandarin Chinese Languages	897
The Future of Singlish	899
English Language Teaching	899
Emerging Trends	901
Conclusion	903
References	903

Abstract

The linguistic scene in multicultural Singapore is, like in many multicultural societies, complex. Even though for decades the Singapore government has actively promoted the use of Standard English among the ethnically diverse Singaporeans, the dominant language used for everyday purposes, including speaking and writing in messengers, continues to be Singlish. To appreciate why Singlish is thriving despite government efforts to eradicate it, one needs to understand that a language is not merely a set of forms that can exist in a cultural vacuum; it is inevitably linked to the culture in which it is used. To examine the cultural setting of Singlish, this chapter surveys studies on Singaporean English

J. O. Wong (✉)
Centre for English Language Communication, National University of Singapore, Singapore, Singapore

© The Author(s), under exclusive licence to Springer Nature Singapore Pte Ltd. 2022
Z. Ye (ed.), *The Palgrave Handbook of Chinese Language Studies*,
https://doi.org/10.1007/978-981-16-0924-4_52

by scholars of various persuasions and from a variety of perspectives, including historical, political, educational, sociological, formal, and cultural. It also reviews linguistic evidence, synthesizes findings, and discusses the key linguistic and cultural forces that shape Singlish, which have their roots in Southern Chinese languages, primarily Cantonese of the Yue (粵) family and Hokkien of the Southern Min (閩南) family, and their cultures.

Keywords

Chinese English · Singapore English · Singlish · Cantonese · Hokkien · Language and culture · *kiasu* · *chope* · Bilingual education · Language policy

Introduction: Singapore English as a Cultural Entity

For the purposes of this chapter, unless otherwise stated, "Singapore English" refers to the non-standard, informal, colloquial varieties used in multicultural Singapore, where the majority of the population are ethnically Chinese, Malay, and Indian. These colloquial varieties of Singapore English are collectively called "Singlish." Singlish has been extensively studied, and many of its phonological, morphological, syntactic, and sociolinguistic features have been well-described. Further, a number of studies have also attested that Singlish has now developed ethnic varieties (Lim 2000, 2001; Yeo and Deterding 2003) at least among "slightly less well-educated" people (Deterding and Poedjosoedarmo 2000, p. 6), showing that Singlish is very firmly grounded in Singapore culture. For example, one study suggests that "certain patterns of grammatical usage are slightly different for Malay and Chinese secondary pupils in Singaporean schools" (Yeo and Deterding 2003, p. 83). Another study notes that an "alternation" is observed in the Singlish of Tamil Singaporeans, "where /v/ is sometimes realised either as the voiced labial-velar approximant or the voiced labiodental approximant," but not in the Chinese and Malay Singaporeans – e.g., the /v/ in *very* is realized as a /w/ among some native Tamil speakers (Lim 2004, p. 32). It should be noted that while the word "Chinese" is often used to refer to members of the largest ethnic group in Singapore, most of these people are children, grandchildren, or great grandchildren of those who migrated there from South China, especially from Guangdong and Fujian province, whose home language is a Southern Chinese language (e.g., Cantonese, of the Yue (粵) family; Hokkien, a Southern Min (*mǐnnán*, 閩南) Chinese language).

Among the numerous studies on Singlish, however, not many identify it as a culturally distinct language variety in its own right. Some sources, such as Schneider (2007), use terms akin to "English variety" to describe the kind of English used in Singapore, even though such terms do not mark Singlish as a culturally distinct language variety. Admittedly, Schneider (2007), like Kachru (1992), goes a step further and uses the term "Englishes," of which "Singapore English" is one, in recognition of the kinds of "English" spoken around the world. However, the term

"Englishes" does not mark the different kinds of "English" as culturally distinct varieties either. In fact, all such terms ("English varieties," "Englishes") can imply that the entities that they refer to are culturally similar, when in fact what binds them together is little more than their colonial past. These entities have since gone culturally separate ways and are without a doubt becoming increasingly divergent and distinct.

Interestingly, the first author who gives Singlish a cultural interpretation is not a Singaporean, but the Polish-Australian linguist Anna Wierzbicka. In a paper entitled "Singapore English: a semantic and cultural perspective," Wierzbicka presents the idea that "the shared language of Singapore society," which is Singlish, reflects "Singapore culture," one that is "over and above the Chinese, Malay and Indian cultures" (2003, p. 328). Singlish is particularly interesting because it has evolved and is still evolving in a multicultural setting in which the various cultures "interpenetrate" (Wierzbicka 2003, p. 330). It is culturally shaped by a number of languages used in Singapore: Southern Chinese languages (in particular Cantonese and Hokkien), Singapore Mandarin (which is itself shaped by the local varieties of Cantonese and Hokkien), and, perhaps to a lesser extent, Malay and Tamil. Thus, one finds in Singlish features and cultural components of the Chinese languages, Malay and Tamil (Pakir 1992).

This chapter focuses on Chinese transference in Singlish. As we shall see, many of the so-called grammatical errors associated with the colloquial form Singlish, far from being "grammar gaffes" as the Singapore government, through the "Speak Good English Movement," would have people believe, are in fact examples of transference from Cantonese and Hokkien. Many words used in Singlish are either borrowed from them or originally English words given a Cantonese or Hokkien interpretation.

The rest of the paper presents and explains Chinese transference in Singlish. It first presents linguistic evidence (formal, lexical, semantic, and cultural) to show that Singlish is to a significant extent a product of Chinese transference, particularly that of Hokkien and Cantonese, and to a lesser extent Mandarin, by highlighting several Singlish features of Chinese origin. In particular, the paper also examines the meaning of a Singlish keyword, *chope* (very roughly, "reserve"), that reflects an originally Singapore Chinese value. It then presents a historical context to explain this phenomenon of Chinese transference. The paper then concludes by discussing implications of the "Chineseness" of Singlish for language teaching and emerging trends.

Linguistic Evidence

This section presents several representative phonological, lexical, and syntactic Singlish features of Chinese origins, in particular Cantonese and Hokkien, and to a lesser extent Mandarin, to showcase Chinese transference in Singlish.

Tones in Singlish

Chinese languages are tonal languages, and it is noted that a number of Singlish words, especially direct loans, are tonal. An example is the set of particles, such as *lah* and *lor*. In the words of Lim, "tone is an integral feature of the particles" (Lim 2007, p. 446). The tones can help us identify homophonous particles from one another (Wong 2014); studies which do not consider tones, such as Gupta (1992), tend to underestimate the number of particles there are in Singlish. Other examples of tonal Singlish words include *kiasu* (see below), *Ah Beng*, *kiasi* (literally "afraid of dying," "cowardly"), *bojio* (roughly "didn't invite (me)"), *limpeh* ("your father") (Liew 2015), and the reduplication of Chinese names (Wong 2014). These words are pronounced in the same way that they are pronounced in the language of origin. For example, Singlish speakers pronounce *kiasu* the way it is in Hokkien, with a falling pitch on the first syllable and a high pitch on the second. For further discussion, see section on cultural keywords below.

When Chinese words enter English, they tend to lose their tones (e.g., *ketchup*, *kowtow*, *Canton*, *Peking*, *Shanghai*, *Taiwan*). Generally, when they enter Singlish directly (i.e., not via another language), they tend to retain the tones. As a result, in Singlish, place names *Canton*, *Shanghai*, and *Taiwan* are not tonal because they came from English, whereas *Guǎngzhōu*, which is "Canton" (Wee 2014), and *Guìlín*, for example, are tonal, because they entered Singlish directly from Mandarin. Thus, an interesting observation is that Singaporeans can use the atonal *Canton* or the tonal *Guǎngzhōu* to refer to the same place. The tones of some Singlish words constantly remind us of their Chinese origin. Such words are evidence of Chinese transference in Singlish.

The Interdental Fricatives

The dental or interdental fricatives are of course common sounds in English (e.g., *the*, *this/that*, *they/them/their*). However, according to Low and Brown (2003, p. 73), they are "notorious" in "English teaching circles," including in Singapore. It thus comes as no surprise that many researchers report the avoidance of dental fricatives as a distinctive feature of Singlish (Lim 2004). Low and Brown (2003) note that, in practice, many Singaporeans substitute the interdental fricatives with alveolar stops (/t, d/), or what some authors call "TH-stopping" (Phoon et al. 2013, p. 14), in the syllable-initial position and with the labiodental fricatives (/f, v/), or what some might call "TH-fronting" (Phoon et al. 2013, p. 18), in the syllable-final position. They write:

> In Singapore, many speakers substitute the *th* sounds /θ, ð/ with /t, d/ in the initial position (thus *thought* and *though* are pronounced like *taught* and *dough*) and with /f, v/ in the final position (thus *death* and *clothe* sound like *deaf* and *clove*). (Low and Brown 2003, p. 74)

Below are further examples of how the interdentals are realized in Singlish:

Thief /t/ (TH-stopping)
Those /d/ (TH-stopping)
Teeth /f/ (TH-fronting)
Breathe /f/ or /v/ (TH-fronting)

As a result, in Singlish, the words *thief* and *teeth* ([tɪf]) are often homophones, as are *those* and *dose* ([dos]) and *breathe* and *brief* ([brɪf]). In sum, the interdental fricatives /Ɵ, ð/ are usually replaced by /t, d/ in syllable-initial position and by /f, v/ in syllable-final position (Brown and Deterding 2005), although voiced consonants at the final position often become voiceless. In this respect, it is interesting to note that The Coxford Singlish Dictionary spells the word *grandfather* as "grandfudder" (Goh and Woo 2009, p. 89). The interdentals are not a feature of the Chinese languages. Chinese Singaporeans who have not gone through much formal education, and even many of those with tertiary education, are not familiar with the interdentals. Thus, it can be said that this replacement of interdentals with other consonants is an example of Chinese transference.

The Dark /l/

Another feature associated with Singlish is the omission of the dark or velarized /l/, which is found at the syllable-end position. According to Low and Brown (2003, p. 76), "many Singaporeans lose the tongue tip contact for (what elsewhere is a) dark /l/, leaving in effect a high back vowel of the /ʊ/ type." In Singlish, the phoneme /l/ is thus "vocalized" (Brown and Deterding 2005, p. 12; Lim 2004, p. 31), and some call the phenomenon "/l/ vocalization" (Tan 2005, p. 43) or "vocalization of /l/" (Phoon et al. 2013, p. 10). As a result, in Singlish, the word *bell*, for example, is pronounced "[beʊ]" and *little*, "[lɪtə]" (Low and Brown 2003, p. 76). The word *also* is pronounced "[oso]" (Lim 2004, p. 31). Evidence comes from spelling too; the word *uncle* is sometimes spelt "unker" in Singlish (Nair 2019).

The omission or vocalization of the dark /l/ is a Chinese phenomenon and is also found in Malaysian English (presumably the variety spoken by ethnically Chinese people) and "Chinese Englishes," such as China English, Taiwan English, and Hong Kong English (Phoon et al. 2013, p. 10). In a study on the ethnic varieties of Malaysian English, the authors note that "omission of final dark /l/ occurred inconsistently for the Chinese speakers" but "none of the Malay or the Indian speakers omitted final dark /l/" (Phoon et al. 2013, p. 19).

In a book chapter entitled "Vocalisation of /l/ in Singapore English," the author presents his investigation of the claim that Singlish speakers "frequently vocalise their dark /l/" (Tan 2005, p. 43). Perhaps unremarkably, the study suggests that the dark /l/ is not a feature of Singlish. The author, Tan, writes:

> Evidence from the auditory survey confirms that Singaporeans indeed do vocalise their dark /l/, Furthermore, some speakers use vocalized /l/ widely, while others generally avoid [the] use of this sound.

Interestingly, Tan draws a conclusion about how Singaporeans speak when "only ethnically Chinese subjects" were employed (Tan 2005, p. 44), unwittingly associating the Singlish feature with Chinese Singaporeans. This is not intended as a criticism of Tan's work, but it is evidence to suggest that the omission of the dark /l/ is an instance of Chinese transference.

The Voiceless Plosives

There is a variable rule in Singlish that is rarely discussed in literature. While in Standard English voiceless plosives are almost always aspirated, this is not the case in Singlish. In Singlish, a plosive may or may not be aspirated. The word *pin*, for example, may be pronounced with an initial aspirated [p^h] or unaspirated [p]. Where this variable rule comes from seems obvious. As stated in a study:

> The first thing that can be pointed out is that Chinese languages have a phonemic distinction between aspirated and unaspirated plosives. Since English does not have this phonemic distinction, it is understandable that in [colloquial Singapore English] these two sounds may then be used in free variation with impunity (...). However, one should also note that in Malay, voiceless plosives are not aspirated in any position (...). (Lim 2004, p. 37)

What the author says in the quote suggests that while the use of the aspirated and unaspirated voiceless plosives in free variation is considered a Singlish feature, it is in fact an instance of transference from Chinese.

Color Terms

It is not often documented, but as Brown (1999) points out, Singlish speakers have a tendency to combine a color term with the word *color*. He gives an example: *I want to buy a red colour dress* (Brown 1999, p. 53). Singapore's "Speak Good English Movement" website presents another example, "I like blue colour bags," which they consider a "blunder."

While the sources do not say anything about the possible origin of this combination, there is evidence to suggest that it is a Chinese feature. The combination is unremarkable in the Chinese languages; Chinese allows the use of a color term on its own or in combination with the word for *color*, which is "色" (*sè* in Mandarin, *sik1* in Cantonese). To illustrate using Mandarin (although one finds the same combinations in other Chinese languages), *red* could be *hóng* ("red") or *hóngsè* ("red color"). For example, one could say *hóng méiguì* ("红玫瑰," literally "red rose") or *hóngsè de méiguì* ("红色的玫瑰," literally, "rose that is red-color"). The Singlish combination (a color term + the word "color") seems to be another instance of transference from Chinese.

Particles and Other Cultural Keywords of Chinese Origin

Cultural keywords are words that embody "culture-specific meanings" that reflect ways of living and thinking that are characteristic of a society (Wierzbicka 1997, p. 5). There is abundant evidence to suggest that many Singlish cultural keywords are in fact Chinese in origin. A quote may be in order: "A main feature of colloquial [Singapore English] is its stock of substrate-derived lexical items. Chinese languages have provided the strongest input (...)" (Leimgruber 2013, p. 67). The section surveys a few Singlish words of Chinese origin.

One of the more prominent aspects of the Singlish lexicon is the set of pragmatic particles. Their use is said to be "one of the most distinctive features" of Singlish (Gupta 1992, p. 31). These particles have been discussed by numerous authors, including but not limited to Gupta (1992), Smakman and Wagenaar (2013), Wee (2004), Wierzbicka (2003), and the present author (Wong 2014). Several authors recognize that these particles come from Southern Chinese languages (Gupta 1992; Lim 2007; Wong 2014); Gupta (1992, p. 31), for example, calls them "loans from Southern Chinese varieties." Additionally, as mentioned above, they are tonal (Lim 2007; Wong 2014), which betrays their Chinese origin. In fact, in a study entitled "Discourse particles in Colloquial Singapore English," the authors refer to them as "Chinese particles" (Smakman and Wagenaar 2013, p. 317). The same authors point to "a Chinese tendency to use particles that have a Chinese origin" in Singlish (Smakman and Wagenaar 2013, p. 320).

Wierzbicka's study on Singlish is telling, too. Her paper is arguably the first in-depth study of Singlish words from a semantic and cultural perspective. Wierzbicka studies the following words: *kiasu*, two particles, *filial*, *ang moh*, *cheena*, and *Ah Beng*. Other than the particles, most of the remaining words seem to be of Chinese in origin. *Kiasu* is a tonal word of Hokkien, a Southern Chinese language, origin (Cheng and Hong 2017; Ho et al. 1998), literally meaning "fear of losing." Although the Singlish word described is *filial*, it originates in the phrase *filial piety*, which embodies a "Confucian ideal" (Wierzbicka 2003, p. 348) and which is frequently used in public discourse. The phrase *filial piety* is the English translation of the Chinese word *xiào* ("孝"), which embodies a top-ranking traditional Chinese value and which refers to one's indebtedness and obligations to one's parents and, by extension, parent-like figures (e.g., grandparents). The tonal noun phrase *ang moh* is an originally Hokkien phrase that literally means "red hair" but which refers to Caucasians. *Ah Beng* is a tonal Hokkien name and a Singlish social category (Wong 2014).

Semantic and Cultural Evidence: *Kiasu* and *Chope*

This section examines semantic and cultural evidence of Chinese transference in Singlish. It briefly discusses the Singlish cultural keyword *kiasu* and the value it embodies, followed by a study of another cultural keyword *chope*, which relates to the value of *kiasu*. It can be shown that the *kiasu* mindset has a Chinese origin.

Presumably, anyone with some basic knowledge of Singapore culture would know that Singaporeans are associated with *kiasu* behavior. Singaporeans themselves are all too familiar with this value. As some authors put it, *kiasu*, more than a "personal value," is in fact a "salient," "shared cultural norm" for Singaporeans (Cheng and Hong 2017, pp. 876, 886). It is said to be "a prominent part of the national culture of Singapore" (Kirby and Ross 2007, p. 109). Moreover, a 2018 survey (Devadas 2018) conducted on the how Singaporeans view their society and workplace and the ones conducted in 2012 (Tay 2012) and 2015 (Tan 2015) show that *kiasu* has consistently been "among the top three societal values" (Wong 2018). In other words, Singaporeans see themselves as, first and foremost, *kiasu*. *Kiasu* might thus be seen as a de facto national value, and its cultural significance cannot be overstated. A number of academic writers have discussed this word, the value it embodies, and its associated behavior and implications (Cheng and Hong 2017; Ho et al. 1998; Hwang 2003; Kirby and Ross 2007; Wierzbicka 2003), and so not much more will be said here. However, there is strong evidence to suggest that the Singaporean *kiasu* is very much a Southern Chinese value; the tonal word *kiasu* is a transliteration of the Hokkien phrase "惊输" (literally "afraid to lose"). Its meaning has been described by Wierzbicka (2003, pp. 335–336), who proposes that it may be stated in this way (with some modifications for readability):

Kiasu =
Some people often think like this:
    'Many people can do something now.
    If someone does it, something good will happen to this person.
    If I do it, this the good thing will happen to me.
    If I don't do it, this the good thing will not happen to me.
    If it doesn't happen to me, I will have to think:
    "This good thing happened to other people, it didn't happen to me."
    I don't want to have to think like this.
    Because of this, I have to do something now.
    I want to do it.'
Because these people think like this, they do many things.
I think: 'It is bad if people think like this.'

Interestingly, this originally Chinese value is now rather deeply rooted in Singapore culture and can also be observed in another Singlish word, *chope*. The word *chope* is undoubtedly a Singlish keyword. Its meaning is usually said to be "reserve." For example, a website by the name "Chope" that helps people make reservations in restaurants in Singapore and other places in the region states that *chope*, a "Singapore slang," is a "transitive verb," the meaning of which is "to reserve, such as a seat in a fast food restaurant, by placing a tissue pack or paper on it." However, it is obvious that the given definition is at best an approximation. In Anglo English-speaking cultures, the act of reservation does not include using an object. A better way of stating the meaning of *chope* might be to say that it is "to place one's belonging (usually something small and not valuable, like a packet of tissues or an umbrella) on a seat or a table (in a place where people sit to do things, not uncommonly an eating place, and where reservation is not accepted) to show other people that it is taken." As a result, when people see an empty seat with someone's belonging on it, they will know that it is taken and look for a seat elsewhere. This behavior is not accepted in Anglo culture, which requires a person to be seated at a table to indicate that the seats there are taken.

When a Singaporean *chopes* a seat, it usually happens in an eating place where reservation is not accepted and where ordering does not take place at the table, such as in a fast food joint or a food court. In such a place during busy hours, a single person may not be able to find a seat after ordering food. If they find an empty seat and want to eat there, the only way is to *chope* the seat, then go and order the food, and later come back to the seat. Of course, one could *chope* a seat in a library or a place where free seating is the norm. It is believed that when Ho et al. are talking about "reserving library seats" (Ho et al. 1998, pp. 365, 367), which they associate with *kiasu* behavior, they are referring to the act of *choping* seats. Wherever the place may be, the act is a Singapore's practical solution that may not be accepted elsewhere. Interesting, some Singaporeans actually find the *choping* of seats an anti-social behavior. They started an "anti-chope movement" which was apparently endorsed in 2019 by the Minister for Community Development, Youth and Sports at that time, Grace Fu.

It is not difficult to see how *chope* reflects a *kiasu* attitude. The essence of the meaning of *kiasu* seems to be this (taken from Wierzbicka's definition presented above):

If I do it, this the good thing will happen to me.
If I don't do it, this the good thing will not happen to me.
If it doesn't happen to me, I will have to think:
    'This good thing happened to other people, it didn't happen to me.'
I don't want to have to think like this.
Because of this, I have to do something now.
I want to do it.

If the person does not *chope* an empty seat during busy hours, it will go to someone else. It is not a thought that the person wants to entertain. As a result, they *chope* the seat. The meaning of the word is thus proposed.

John *choped* a seat (in the library, food court, McDonald's) =
John was in a big place of one kind at that time.
At one moment, he did something there like people often do in places of this kind.
They do it because they think like this about the place where they are:
    'It is like this here now:
        There are many people here now, I am one of these people.
        All these people want to be here for some time
            because they want to do some things here.
        They all want to sit somewhere when they are doing these things.
        There are many places here where someone can sit,
            I want to sit in one of these places.
        I can sit in one of these places now, it is good for me.

At the same time, it is like this:
> Before I sit in this place, I want to do something somewhere else for a short time.
> I don't want anyone else to sit in this place during this time.
> I want all people here to know that they can't sit in this place
>   when I am not here.
> Because of this, I want to do something in this place now.
> After I have done this, people can see something here. This thing belongs to me.
> I want people to see it here because I want them to think about it like this when I am not here:
>> "This thing is there because someone else wants to say:
>> I will sit here soon, I don't want anyone else to sit here.
>> I can't sit here because of this."
> This will be good for me.
> If I don't do it, someone else can sit in this place when I am not here.
> This will be bad for me. I will feel something bad because of this.'

John did something to one of the places in the big place where he was
    because he thought like this.

"Seeing others in a better position or obtaining an advantage which too could have been attained by oneself presents substantial mental distress" to a *kiasu* person (Ho et al. 1998, pp. 363–364). The attitude is manifested in the act of *choping* a seat. The person *choping* a seat does not like the idea of the empty seat they come across in a crowded setting being taken by another party when they are away (e.g., buying food). The act is motivated by the originally Chinese value of *kiasu*. Given the cultural importance of the value of *kiasu* in Singapore, it might be said that the extent of Chinese transference in Singlish cannot be overestimated.

Historical Context: The Rise of English in Singapore

This section explains why, against all odds, a form of English, once a "foreign" language in this region, has eventually become the dominant language used in Singapore, where the important languages among the majority Chinese population in the 1960s and 1970s were Southern Chinese languages, especially Cantonese and Hokkien, and where the national language is Malay. In doing so, it surveys aspects of Singaporean history, paying attention to aspects of the history of Singapore before its independence in 1965, its historical ties with Malaysia, the nation-building years after independence, and the rise of English as the medium of instruction in the education system. It also highlights the tensions between official language policies and the non-Mandarin Chinese languages that emerged in the process of promoting a form of English as the dominant language to be used in Singapore. The historical context discussed in this section will help to explain the phenomenon of Chinese transference (particularly Cantonese and Hokkien) in Singlish to be discussed in the next section.

Historically, Singapore was often associated with peninsular Malaysia (previously called "Malaya"), at least geographically. According to one source:

Singapore was often included when people referred to "Malaya" as a place or a state (generally speaking, and not specifically the "Federation of Malaya") and to things "Malayan". Even when, after the Second World War, Singapore was reconstituted as a Crown Colony by itself, it was still considered inseparable from Malaya, and was expected to be merged with Malaya one day. The experience of separate independent statehood was completely new for Singaporeans when it happened, abruptly, on that fateful 9 August 1965. (Lee 2008, pp. 21–22)

In fact, for a brief period from 16 September 1963 to 8 August 1965, Singapore was officially a part of Malaysia (Abshire 2011; Lee 2008), which is of course largely a Malay nation, where the language of the Parliament was and still is Malay. Singapore has thus enjoyed strong historical, cultural, social, and familial ties with Malaysia and its people. The indigenous people in Singapore are the Malays. After it became an independent country in 1965, Singapore's national language has remained Malay. While the language of the parliament is English, the national anthem is sung in Malay, and military commands are given in the same language. Prior to the 1980s, there were Malay-medium schools. Thus, one could say that Singapore and Malay cultures have intimate historical ties. Partly because of these ties, in the 1960s and 1970s, many Chinese and Indian Singaporeans could speak at least basic Malay for intercultural communicative purposes in public spaces, such as a "wet" market.

Despite Singapore's Malay origins, however, the first prime minister of Singapore Lee Kuan Yew envisioned a society that was not Malay (nor Chinese, nor Indian); Lee's "abiding nation-building imperative" was to treat every citizen equally, "regardless of race, religion, or creed" (Lee 2008, p. 239). As a new prime minster then, Lee envisioned a multicultural society. In fact, even in 1960, before Singapore gained independence, the primary school leaving examination was conducted in all the four official languages, English, Chinese [Mandarin], Malay, and Tamil (Ng 2008). Nevertheless, the newly established Singapore government soon recognized the importance of English. When it left Malaysia and became independent in 1965, Singapore lost a hinterland but saw itself gaining "the whole world" instead as its hinterland (Hong 2016, p. 15). Obviously, to connect with the world, Singapore needed English, the language of the colonial master. English was seen as "a necessary tool in Singapore's effort to make the world its marketplace" (Goh and Gopinathan 2008, p. 14). By 1966, the study of two languages, English and another official language, was made compulsory in all schools (Ang 2008; Hong 2016; Ng 2008). In most cases, one language was studied at a higher level, and it was called the "first language," while the other, the so-called second language, was studied at a lower level. In the Chinese, Malay, and Tamil schools, English was taught as a "second language." In English schools, it was taught as a "first language," and the language that represented the student's ethnicity (Mandarin Chinese, Malay, or Tamil, which the Ministry of Education now calls the "mother tongues") was somewhat ironically taught as a "second language" (Ang 2008, p. 75).

The compulsory bilingual education was (and still is) a necessity and "a key component in Singapore's education system," as English was needed for economic purposes and the local languages were needed for cultural purposes (Goh and Gopinathan 2008, p. 14). However, there was apparently a caveat. According to one source, in 1969, the government decided that "second language papers would be set and marked at a level two years below that of the student's first language" (Ng 2008, p. 46). Apparently, even that was not enough as there were Chinese Singaporean students who still could not meet the lower standards (Wu 2017); as we shall see, the Chinese Singaporean students struggled with the study of Mandarin even as a "second language" (which, as mentioned, was studied at a lower level than the so-called "first language" English). Subsequently, in 2001, the Ministry of Education further introduced a Chinese syllabus which "imposed a lower level of content" to help students who found learning Chinese difficult (Wu 2017, p. 48). All this meant that Chinese Singaporean students in an English-medium school studied English language at a much higher level than Mandarin Chinese.

As mentioned, even though the Singaporean government tried to treat each official language equally after Singapore's independence, there was an obvious bias towards English. When the British ruled Singapore before 1965, they only provided "education in English" (Ang 2008, p. 69). The British schools were set up in "good buildings" (Goh and Gopinathan 2008, p. 15), unlike many other, vernacular schools offering classes in Hokkien and Cantonese (Leimgruber 2013), which were "not well resourced" (Goh and Gopinathan 2008, p. 16). This, coupled with the transformation of the Singapore economy that began in 1959, gave parents strong motivations to enroll their children in English-medium schools (Goh and Gopinathan 2008). Furthermore, especially in the decade following independence, parents increasingly saw a future in giving their children an education in English. As one source puts it, "This dramatic drift was brought about by the free choice of pragmatic parents in response to the nation's drive toward high-value-added industrialization and to an economy where the language of business is English" (Goh and Gopinathan 2008, p. 15).

As some authors point out, at that time, it was obvious that an English-medium education would yield greater financial returns than vernacular-medium education (Goh and Gopinathan 2008). In fact, in 1979, shortly before all schools in Singapore became English-medium, English-medium schools accounted for 91 percent of all primary one students (Goh and Gopinathan 2008). The importance of an English education could not be overestimated. In fact, Lee, Singapore's first Prime Minister, went through his university education in English; he studied law at Cambridge, England (Lee 2008).

Further, the new Singapore government had foresight in other ways. Being pragmatic, "the Singapore government has always exhibited a bias toward science and engineering education because of the economic policies and industrialization needs," and by 1968, shortly after Singapore gained independence, science and arithmetic were taught in English in most non-English-medium primary schools (Ng 2008; Gwee et al. 1969). This meant that many Singapore school students learned logical thinking and reasoning skills in English.

Recognition of the importance of English became even more widespread by the mid-1980s, to the extent that all schools became English-medium (Goh and

Gopinathan 2008; Ang 2008). Among other things, this meant that (i) English was the main medium of instruction in all schools (Goh and Gopinathan 2008), (ii) it was universally studied as "the first language," and (iii) the language representative of the student's culture, correctly or incorrectly called the "mother tongue," was studied as "the second language" (Ang 2008). It also meant that *all students* studied English language at a higher level than the so-called mother tongues (unless the student opted to study the "mother tongue" at the first language level, which was uncommon).

Studying the so-called mother tongue as a second language was of course not problematic if the language was indeed the mother tongue, like Malay was for Malay students or Tamil for some Indian students. However, this was not the case for Chinese Singaporean students. In the words of Ang:

> Among the Chinese population there was another problem. Most were more conversant in dialects rather than Mandarin, which was designated as the official mother tongue in Chinese. For non-Mandarin-speaking Chinese students, Mandarin was a new spoken language, even though the written form was the same. (Ang 2008, p. 75)

This point is not often discussed in literature. Even though the Ministry of Education called Mandarin Chinese the "mother tongue" of Chinese students, Mandarin was not in fact their mother tongue. The home languages of the Chinese students were Southern Chinese languages, the so-called dialects, such as Cantonese and Hokkien. Many of these students could perhaps speak some Mandarin, but only as a second language, and it certainly was not their dominant language. The government considered Mandarin Chinese as the language that represented the Chinese population, but that was (and still is) only because it was (and still is) the national language of China and thus had (and still has) official status.

The task of learning Mandarin for Chinese students during those years was challenging. The early Chinese Singaporeans did not go through much formal education and, as mentioned, spoke Southern Chinese languages, of which Mandarin is not one. Mandarin, while now spoken throughout China, was originally a northern Chinese language. In the early years of Singapore, until at least the mid-1980s, Hokkien was the common language among Chinese Singaporeans (Leimgruber 2013). Mandarin, for many Chinese Singaporeans, was a second or third language. Most parents in those days (like this author's) did not speak Mandarin to their children. Thus, many students were confronted with learning what in significant ways seemed to be a new language in school. Mandarin is mutually unintelligible with Cantonese and Hokkien. It has a very different phonology and tonology (de Sousa, this volume). For example, the voiced velar nasal /ŋ/ is found in the word-initial position in Cantonese but not in that position in Mandarin. On the other hand, the voiceless "apical post-alveolar" affricates and fricative found in Standard Mandarin (Lee and Zee 2003, p. 110), represented in *hànyǔ pìnyīn* (the modern Mandarin pronunciation system) as *zh*, *ch*, and *sh*, are not Cantonese sounds and prove particularly challenging for Cantonese speakers (including this author).

Mandarin has four lexical tones, while Cantonese has six "contrastive" or "distinctive" ones (Francis et al. 2008, p. 271; Jia et al. 2015, p. 351).

Writing presented a challenge too in the 1970s, even after the Singapore government, following Mainland China, replaced traditional Chinese characters with the "simplified" ones; the simplified characters have been used in Mainland China since the 1950s, but the traditional ones are still used in Hong Kong, Taiwan, and Macau today (Yang and Wang 2018). Even then, Chinese characters are not easy to learn and constitute the most challenging aspect of learning Chinese (Cui et al. 2018, p. 111). Firstly, although all Chinese languages largely share the same writing system, there are lexical and semantic differences, and the meanings of some Mandarin characters are unknown to Singaporean Chinese. For example, the Mandarin word for "walk" is "走" (zǒu), but the same character in Cantonese (/tsəʊ/) means "run," retaining the meaning in classical Chinese. The various meanings of the word "青" (qīng, a color and visual term) are not known to many Chinese Singaporeans (Tao and Wong 2019). As a result, Chinese Singaporean students had to learn a number of new words because words for a number of everyday concepts, like *see*, *eat*, *drink*, *table*, *chair*, *stand*, *walk*, *run*, and *he/she*, are different in Cantonese and in Mandarin. There are differences in pragmatics too. For example, in Cantonese, there are two phrases used, depending on the situation, for thanking people, only one of which is available in Mandarin (Wong and Liu 2019). Southern Chinese Singaporean students thus faced major linguistic challenges in learning a formal variety of Mandarin in class then.

Further, while the Malays and Indians in practice only needed to learn one new language (English) to connect with Singaporeans of other ethnicities, the Chinese had to learn two new languages, Mandarin for communicating with other Chinese Singaporeans and English for non-Chinese Singaporeans, which, as we shall soon see, is very much related to the crux of the problem. For the pragmatic Chinese, it seemed unnecessary. After all, they could easily and more efficiently use English to speak with *both* Chinese and non-Chinese Singaporeans. Why master English and Mandarin for communicative purposes when English alone would do? Moreover, as discussed, many Chinese Singaporeans were more proficient in English because they studied it at a higher level in school than Mandarin. Presumably as a result, the pragmatic Chinese adopted a form of English as their home and dominant language rather quickly and eventually gave up the non-Mandarin Chinese language that their forefathers spoke. It has been noted that many young ethnically Chinese Singaporeans do not speak Mandarin well, do not like the language, and speak mainly English (Ng 2014).

At any time, it seems that a high proportion of the Chinese Singaporean majority speak English. In fact, data from the Singapore Department of Statistics (Singapore Department of Statistics 2000, 2005, 2010, 2015), presented in Table 1, show that for over a decade, proportionally more Chinese Singaporeans started speaking a form of English as a home language than the next dominant ethnic group, the Malays. In Tan's words, the Chinese have "the fastest shift into speaking English as a dominant family language" (Tan 2016, p. 11).

Table 1 Proportion (percent) of Singaporeans speaking English at home over two decades by ethnicity

Ethnicity	Proportion of Singaporeans speaking English at home by ethnicity (%)				
	1990	2000	2005	2010	2015
Chinese	19.3	23.9	28.7	32.6	37.4
Malays	6.1	7.9	13.0	17.0	21.5
Indians	32.3	35.6	39.0	41.6	44.3

Rubdy sums it up well when she says, "The rise of English is most pronounced in Chinese and Indian homes, with many Singaporeans describing themselves as literate only in English. Chinese homes citing English as the home language rose from 10.2% in 1980 to 23.9% in 2000" (Rubdy 2012, p. 226). The drastic change took place in a mere 20 years! Thus, we begin to see why a form of English has become the dominant language in Singapore. However, as anyone who has some knowledge of Singapore presumably knows, this form of English is semantically and culturally very different from Anglo varieties of English (Wierzbicka 2003; Wong 2008). The next section explains why.

Further Historical Context: The Eradication of Non-Mandarin Chinese Languages

The next question to address is specifically why Cantonese and Hokkien have played a major role in shaping Singlish. On the surface, it would seem that the compulsory bilingual education is a good move and, in many ways, it is. The advantages of bilingualism are obvious. The bilingual individual has access to two cultural worlds and the ability to navigate them (Besemeres and Wierzbicka 2007). There is also evidence to suggest that bilinguals outperform monolinguals in certain linguistic tasks (Bartolotti and Marian 2012; Escudero et al. 2016; Kaushanskaya 2012; Kaushanskaya and Marian 2009). It thus makes sense that the Singapore government proactively promoted the use of English and Mandarin among Chinese Singaporeans. It *could have meant* that a Chinese Singaporean would eventually speak three languages – English, Mandarin, and a non-Mandarin Chinese language.

However, it was not meant to be. To promote the use of Mandarin among Chinese Singaporeans, the government orchestrated the Speak Mandarin Campaign, launched in 1979 by the then Prime Minister Lee. While it was clear that the campaign was organized to promote the use of Mandarin, there was what might be seen as a sinister aspect to it – the eradication of non-Mandarin Chinese languages, which have been incorrectly called "dialects," in Singapore (Wong 2020). In the opening ceremony of the "Promote the use of Mandarin" Campaign, as it was called then, in 1979, PM Lee began his speech by saying:

> Chinese Singaporeans face a dilemma. The Chinese we speak is divided up among more than twelve dialects. Children at home speak dialect; in school they learn English and

Mandarin. After 20 years of bilingual schooling, we know that very few children can cope with two languages plus one dialect, certainly not much more than the 12% that make it to junior colleges. (Lee 1979)

In a later part of the speech, Lee stated, "Once it is clear to the government that parents want their children to learn and to use Mandarin, not dialects. The government will take administrative action to support their decision." This was exactly what they did. Measures were taken, the "most drastic" of which in Rubdy's (2012, p. 230) words was "the banning of the dialects from the media." This means that the "use of dialects on television and radio programmes was officially forbidden" and non-Mandarin Chinese programs "were dubbed in Mandarin" (Rubdy 2012, p. 230).

Interestingly, while the government was rather successful in eradicating the non-Mandarin Chinese languages, their efforts in promoting the use of Mandarin have met with only limited success, as seen above. Many Chinese Singaporeans only have functional competence in the language and do not use it as their dominant language. According to a study (Chinese Language Curriculum and Pedagogy Review Committee 2004) cited by Ng (2014), around 77% of primary six Chinese Singaporean students from English-speaking background found learning Mandarin difficult. Even the prestigious "Special Assistance Plan" schools, whose objectives "were to preserve the ethos of the Chinese-medium schools and to promote the learning of [Chinese language] and culture," did not meet their objectives because students disliked Mandarin and spoke mainly in English (Ng 2014, p. 367). Chinese Singaporean students continue to speak mostly English with their friends and only occasionally use Mandarin (Tan 2016).

The reason is not difficult to understand. The crux of the matter is that the Singapore government did not (and still do not) appear to recognize that language and culture are intrinsically linked, which, in this author's opinion, can at least partly explain why the government and its language policies and other efforts have not fully attained the goal of producing new generations of Chinese Singaporeans who are proficient in both Standard Mandarin and Standard English. The Singapore government does not seem aware that every natural language has evolved in a particular cultural setting and is thus culture-specific; English is not culturally neutral but embodies Anglo culture (Wierzbicka 2013). Furthermore, they have also incorrectly assumed that Mandarin, a northern Chinese language in China, can universally express all Chinese meanings and values, including Southern Chinese ones. This is of course not true; to adequately express Southern Chinese cultures, speakers need the languages that have evolved in those cultures, such as Cantonese and Hokkien. When one is deprived of one's home language, one is not able to express one's culture-specific meanings and values. This is presumably what happened to many traditional Southern Chinese Singaporeans, many of whom were deprived of an important tool of communication, their home language, and faced communication problems both at home and outside home. For example, as noted by Rubdy, one of the consequences "of the replacement of the dialects with Mandarin is the intergenerational communication gap that has made it difficult for grandchildren to communicate with their grandparents" (Rubdy 2012, p. 231).

The fact seems to be that, even when their mother tongue was taken away from them, the Chinese Singaporeans, as cultural beings, did not lose the need to express Southern Chinese meanings and values, many of which could not be expressed in Standard Mandarin or Standard English. For example, many of the meanings associated with Cantonese and Hokkien particles were not expressible in Mandarin or English. What followed was inevitable. The Chinese Singaporeans found ways to express Cantonese and Hokkien meanings and values through Mandarin and especially English, their dominant language. While learning Mandarin and English in school, they interpreted them from a Southern Chinese perspective. The results are a form of Singapore Mandarin and Singapore English (or Singlish) which are shaped in significant ways –phonologically, grammatically, semantically, pragmatically, and culturally – by Cantonese and Hokkien. In summary, Southern Chinese Singaporeans had their home languages (e.g., Cantonese and Hokkien) taken away from them, replaced by a form of English. However, they needed to express their Southern Chinese cultural values that are not expressible in any Anglo English and could only do it through the form of English they learned in school. Singlish was thus born.

The Future of Singlish

This section looks at some implications of the "Chineseness" of Singlish in English language teaching and other areas. In particular, it presents implications for the eradication of Singlish (if it is ever possible) and clears its name by exposing the reason that many Singaporeans are not mastering Standard English. The section then presents implications for language policy in Singapore before briefly discussing the future of Singlish.

English Language Teaching

Wierzbicka argues that there is a Singapore culture "over and above the Chinese, Malay, and Indian cultures" that is reflected in "Singapore English" (Wierzbicka 2003, p. 328). To an extent, she is right. However, on closer examination, one finds that many aspects of Singlish can be traced back to Southern Chinese languages (mainly Cantonese and Hokkien).

For a long time, purists, including the Singapore government, have demonized Singlish and given it a bad name; Singlish "has faced relentless attacks by an ongoing government campaign intent upon its eradication, been banned from local media, and been branded a major obstacle to Singapore's global competitiveness" (Cavallaro et al. 2014, p. 378). It is "viewed as an obstacle to the development of students' literacy skills in standard English" (Rubdy 2007, p. 308). Some Singaporeans have even developed "hostile feelings" towards Singlish (Cavallaro et al. 2014, p. 393). The idea is that speaking Singlish can stop the speaker from effectively learning so-called Standard English, which is apparently so crucial to the development of the country and the individual.

The view that Singlish is an obstacle to the learning of a standard variety of English is of course an ill-conceived cliché. One source recognizes it as a myth. According to Lu (2016), the idea that "Singlish is a hindrance to learning Standard English" is "a myth that has been repeated ad nauseam without any actual linguistic evidence." After all, Standard English was introduced to Singapore by its colonial master the British, and it was used in Singapore long before the emergence of Singlish. It had the upper hand. From a chronological perspective, it is difficult to imagine how Singlish could have become an obstacle to the learning of Standard English.

In fact, Singlish is evidence that many Singaporeans were not effectively learning the form and culture of Standard English. If it can be accepted that Singlish is to a significant extent shaped by Southern Chinese linguistic and cultural components, it would not be difficult to see where the obstacle or interference comes from. Ironically, the Chinese cultural roots that the government wants Chinese Singaporeans to remain in touch with constitute the "obstacle" (so to speak) to the learning of Standard English. Standard English does not meet the collective cultural needs of Chinese Singaporeans, and the only way to meet the needs is to express Chinese meanings and values using a form of English now known as Singlish. The logic seems quite straightforward. As long as Chinese Singaporeans remain culturally Chinese and use a form of English as their dominant language, Singlish will stay. Singlish thus very clearly showcases the strong link between language and culture.

Instead of demonizing Singlish and trying to eradicate it, it might be more helpful for language planners and educators to find ways to address the challenges Singaporeans face when learning Standard English. Instead of saying that Singlish is bad English while Standard English is good English, it might be more helpful to draw Singaporeans' attention to the two sets of rules, one associated with Singlish and the other Standard English. The teacher could point out differences. For example, Singlish speakers tend to exaggerate (Wong 2014), while speakers of Standard English, which is a variety of Anglo English, tend to understate (Fox 2004; Wierzbicka 1991). While Singlish speakers have a tendency to express uncertainties as certainties (Wong 2014), Anglo English speakers are predisposed to present certainties as uncertainties (Fox 2004; Wierzbicka 2006). Singlish speakers try to minimize the personal space and distance between people using a number of interactive devices, such as the Singlish particles. By contrast, Anglo English speakers prefer to maintain a measure of social distance (Wierzbicka 1991). Wierzbicka notes the relevance of Singlish to the Singapore cultural context:

> What matters here is the *interactive* character of Singapore English particles, engaging the addressee and including him or her in the illocutionary structure of the speaker's own utterance. When one watches, for example, the interaction of Singaporeans in the squeeze and the bustle of a hawker centre (with people's elbows touching, and loud conversations criss-crossing the dense network of tables and food stalls), one can well imagine the

functionality of interactive Singaporean particles (as well as expressive interjections); and the contemporary Anglo notions of 'privacy', 'personal space', 'autonomy', and 'non-imposing on other people' seem scarcely germane. In this kind of setting, an outsider would perhaps be less surprised to hear strangers addressing an unknown person as 'auntie' or 'uncle' than they would be in a cafeteria in Britain or in the United States. (Wierzbicka 2003, p. 346)

It is also important for the authorities to realize that there are three "faces" of language: form, meaning, and culture (Wong 2010). When it comes to English, the Singapore education system pays much more attention to form, and some aspects of form at that, than to meaning and virtually none to culture. As a result, Singaporeans might match an Anglo English form or a modified version with a Singaporean meaning. For example, Singaporeans who do not understand the meaning of the English question tags may modify their forms to the invariant *is it* and use it to express a Chinese meaning (Wong 2008). However, given that Chinese Singaporeans want to express Chinese meanings and that Standard English was not evolved in a Chinese cultural context, it might be unreasonable to expect Singaporeans to use Standard English as the dominant English variety in Singapore. The dominant use of Singlish simply tells us, quite unremarkably, that Chinese Singaporeans are merely being culturally Chinese.

Emerging Trends

In April, 2019, *The Straits Times*, the dominant English newspaper in Singapore, published a series of articles on the theme of millennials. In one of the articles, entitled "Millennial speak: A brief lexicon of the generation's slang," the author, Olivia Ho (2019), a millennial herself, presents ten words that are associated with the generation. There is also a video version, in which she asks an older Singaporean, assistant sports editor Rohit Brijnath, the meaning of the ten words. He only gets two right. The ten words are:

1. *Shook* ("Having a very intense, emotional reaction of shock, fear, excitement and more")
2. *On fleek* ("Perfectly executed")
3. *Bae* ("Your significant other")
4. *Basic* ("a person so mainstream that they are beyond uninteresting")
5. *Lit* ("Really amazing")
6. *Thirsty* ("Desperate, especially sexually")
7. *Salty* ("Upset, bitter and, therefore, rude")
8. *FOMO* ("Fear of missing out" and "can be used as a verb")
9. *tl;dr* ("Too long; didn't read," commonly used in Reddit and Tumblr)
10. *Woke* ("Aware about issues of social injustice")

There are other words associated with millennials too, as presented in other articles in the series, such as *adulting* ("things adult have to do on a regular basis") (Parker 2019); *memes* ("the format of text superimposed on photos, to communicate an idea or make fun of a situation") (Abdulla 2019); *snowflake, strawberry generation,* and *slacktivism* ("the act of signalling support for a cause while doing very little for it concretely") (Zhuo 2019); and *hashtag, yolo, avocado toast,* and *burnout generation* (Ho 2019; Vaughn-Hall 2018).

Two things seem to stand out. Firstly, the overwhelming majority of these words or phrases are not Chinese or even Asian in origin. For example, the phrase *avocado toast* apparently originated in Australian millionaire and property mogul Tim Gurner's admonition to millennials in a 2017 episode of the Australian current affairs program "60 Minutes" for spending too much on "smashed avocado" and coffee (Levin 2017). The acronym FOMO possibly originated in the description of an emerging consumer motivation by Herman (2000), which is "ambition to exhaust all possibilities and the fear of missing out on something" (p. 365) revealed by data accumulated in "Israel and elsewhere" (p. 330), and is now typically referred to "in a negative light, especially in relation to social media and smart phone usage" (Hodkinson 2019, p. 65). The association of FOMO behavior with social medial seems clear; Przybylski et al. (2013, p. 1841) say it is likely that social media is "a particular boon for those who grapple with fear of missing out." Further, it is said in Ho's article referred to above that *tl;dr* is commonly used in Reddit and Tumblr to "to call out somebody for being long-winded, or to signal in advance that a post will be lengthy so people do not complain later." The word *woke* (as in *stay woke*) was originally associated with the Black Lives Matter movement, and "the concept of 'stay woke' infers that blacks, as an oppressed people in U.S. society, must always pay attention to inequalities in society" (Edmondson et al. 2019). Secondly and relatedly, nowhere in *The Straits Times* articles do the authors refer to these words and phrases as Singlish, which suggests that the Singapore millennial speakers do not see them as Singapore culture-specific. This could be a sign that Singlish is becoming less Chinese or at least that there is another dominant cultural force that is shaping it.

As the phrase "millennial speak" implies, the parents of millennials and older generations are not familiar with many or most of the words presented above, as this author can testify; what they know, they learned it from the millennials but will probably not use it routinely, if at all. The form of English that the Singapore millennials use may thus not be considered totally Singlish, simply because it comprises many words that are commonly used by young English users in the Internet world, in Singapore, and elsewhere and because it is not used by older Singaporeans.

The future of Singlish is thus to some extent uncertain. As discussed, Singlish has been shaped to a significant extent by Cantonese and Hokkien, and many of the values embodied in these languages are not compatible with contemporary values that are propagated by social media. Social media is a leveler; social media tools in principle allow anyone who has access to the Internet to speak their mind, regardless of race, gender, religion, sexual orientation, and so on, and is in this sense egalitarian. This value of egalitarianism is obviously incompatible with many traditional Chinese values embodied in Singlish that emphasize a social hierarchy that is based on age, generation, and status and may thus be a cultural force counteracting the Chineseness of Singlish. Millennials, at least the Internet-savvy ones, are familiar with and possibly supportive of a number of so-called "Western" values related to, for example, human rights (e.g., the Black Lives Matter movement, gay rights), which many older Singlish-speaking Singaporeans may not care about. The use of "millennial speak" among young Singaporeans thus suggests that Chinese transference may play a lesser role in shaping their Singlish. Then again, it is also possible that the millennials are exhibiting bilingualism and biculturalism; they use Singlish, with all its Chineseness, to communicate with older Singaporeans and "millennial speak" to communicate with fellow millennials in Singapore and in other parts of the world.

Conclusion

It is somewhat ironic that the Singapore government wants Chinese Singaporeans to be in touch with their cultural roots but at the same time does their best to eradicate the dominant language (i.e., Singlish) that allows Singaporeans to express Chinese values. However, the government's wish may just come true in the distant future, if Singapore's millennials decide to pass on only contemporary values to their children, and not traditional Chinese values. It is possible that in a few generations' time, much of contemporary Singlish, with all its Chineseness, will become history.

References

Abdulla, Z. 2019, April 20. *Striking a chord through humour: What's in a meme?* Retrieved from The Straits Time, Lifestyle: https://www.straitstimes.com/lifestyle/arts/whats-in-a-meme

Abshire, J.E. 2011. *The history of Singapore.* Santa Barbara: Greenwood.

Ang, W.H. 2008. Singapore's textbook experience 1965–97: Meeting the needs of curriculum change. In *Toward a better future: Education and training for economic development in Singapore since 1965*, ed. S.K. Lee, C.B. Goh, B. Fredriksen, and J.P. Tan, 69–95. Washington, DC: The World Bank and National Institute of Education, Singapore.

Bartolotti, J., and V. Marian. 2012. Language learning and control in monolinguals and bilinguals. *Cognitive Science* 36 (6): 1129–1147.

Besemeres, M., and A. Wierzbicka, eds. 2007. *Translating lives: Living with two languages and cultures.* St Lucia: University of Queensland Press.

Brown, A. 1999. *Singapore English in a nutshell: An alphabetical description of its features.* Singapore: Federal Publications.

Brown, A., and D. Deterding. 2005. A checklist of Singapore English pronunciation features. In *English in Singapore: Phonetic research on a corpus*, 7–13. Singapore: McGraw-Hill Education.

Cavallaro, F., B.C. Ng, and M.F. Seilhamer. 2014. Singapore Colloquial English: Issues of prestige and identity. *World Englishes* 33 (3): 378–397.

Cheng, C.Y., and Y.-Y. Hong. 2017. Kiasu and creativity in Singapore: An empirical test of the situated dynamics framework. *Management and Organization Review* 13 (4): 871–894. Retrieved from https://ink.library.smu.edu.sg/soss_research/2420.

Chinese Language Curriculum and Pedagogy Review Committee. 2004. *Report of the Chinese language curriculum and pedagogy review committee*. Singapore: Chinese Language Curriculum and Pedagogy Review Committee.

Cui, J., H.H. Goh, C. Zhao, and K.C. Soh. 2018. Effective ways in teaching Chinese characters without phonetic clues. In *Teaching Chinese language in Singapore: Efforts and possibilities*, ed. K.C. Soh, 111–120. Singapore: Springer.

Deterding, D., and G. Poedjosoedarmo. 2000. To what extent can the ethnic group of young Singaporeans be identified from their speech? In *The English language in Singapore: Research on pronunciation*, ed. A. Brown, D. Deterding, and E.L. Low, 1–9. Singapore: Singapore Association for Applied Linguistics.

Devadas, D. 2018. *IPS-aAdvantage roundtable on the national values assessment for Singapore (2018)*. Retrieved from Institute of Policy Studies: https://lkyspp.nus.edu.sg/docs/default-source/ips/enews_ips-aadvantage-roundtable-2018-report_130818.pdf

Edmondson, V.C., B.S. Edmondson, and T.B. Perry. 2019. Stay woke: The Black Lives Matter movement as a practical tool to develop critical voice. *Communication Teacher* 33: 1–6. https://doi.org/10.1080/17404622.2019.1575433.

Escudero, P., K.E. Mulak, C.S. Fu, and L. Singh. 2016, August 15. More limitations to monolingualism: Bilinguals outperform monolinguals in implicit word learning. *Frontiers in Psychology* 7. https://doi.org/10.3389/fpsyg.2016.01218.

Fox, K. 2004. *Watching the English: The hidden rules of English behaviour*. London: Hodder and Stoughton.

Francis, A.L., V. Ciocca, L. Ma, and K. Fenn. 2008. Perceptual learning of Cantonese lexical tones by tone and non-tone language speakers. *Journal of Phonetics* 36 (2): 268–294.

Goh, C.B., and S. Gopinathan. 2008. The development of education in Singapore since 1965. In *Toward a better future: Education and training for economic development in Singapore since 1965*, ed. S.K. Lee, C.B. Goh, B. Fredriksen, and J.P. Tan, 12–38. Washington, DC: The World Bank and National Institute of Education, Singapore.

Goh, C., and Y.Y. Woo. 2009. *The Coxford Singlish dictionary*. Singapore: Angsana Books.

Gupta, A.F. 1992. The pragmatic particles of Singapore colloquial English. *Journal of Pragmatics* 18 (1): 31–57.

Gwee, Y.H., J. Doray, K.M. Waldhauser, Z. Ahmad, and T.R. Doraisamy. 1969. *150 years of education in Singapore*. Singapore: Teachers' Training College Publications Board.

Herman, D. 2000. Introducing short-term brands: A new branding tool for a new consumer reality. *The Journal of Brand Management* 5: 330–340.

Ho, O. 2019, April 13. *Going beyond millennial stereotypes of avocado toast and #adulting*. Retrieved from The Straits Times, Lifestyle: https://www.straitstimes.com/lifestyle/arts/going-beyond-millennial-stereotypes-of-avocado-toast-and-adulting

Ho, J.T., C.E. Ang, J. Loh, and I. Ng. 1998. A preliminary study of kiasu behaviour - is it unique to Singapore? *Journal of Managerial Psychology* 13 (5/6): 359–370.

Hodkinson, C. 2019. 'Fear of missing out' (FOMO) marketing appeals: A conceptual model. *Journal of Marketing Communications* 25 (1): 65–88. https://doi.org/10.1080/13527266.2016.1234504.

Hong, M.T. 2016. *The rise of Singapore: The reasons for Singapore's success*. Vol. 1. Singapore: World Scientific Publishing.

Hwang, A. 2003. Adventure learning: Competitive (Kiasu) attitudes and teamwork. *Journal of Management Development* 22 (7): 562–578.

Jia, S., Y.-K. Tsang, J. Huang, and H.-C. Chen. 2015. Processing Cantonese lexical tones: Evidence from oddball paradigms. *Neuroscience* 305: 351–360.

Kachru, B. 1992. World Englishes: Approaches, issues and resources. *Language Teaching* 25 (1): 1–14. https://doi.org/10.1017/S0261444800006583.

Kaushanskaya, M. 2012. Cognitive mechanisms of word learning in bilingual and monolingual adults: The role of phonological memory. *Bilingualism: Language and Cognition* 15 (3): 470–489.

Kaushanskaya, M., and V. Marian. 2009. The bilingual advantage in novel word learning. *Psychonomic Bulletin & Review* 16 (4): 705–710.

Kirby, E.G., and J.K. Ross. 2007. Kiasu tendency and tactics: A study of their impact on task performance. *Journal of Behavioral and Applied Management* 8 (2): 108–121.

Lee, K. Y. 1979, September 7. *Press Room.* Retrieved from Promote Mandarin Council: https://www.languagecouncils.sg/mandarin/en/-/media/smc/documents/goh-pm-lee-kwan-yew_smc-launch-speech_070979.pdf

Lee, E. 2008. *Singapore: The unexpected nation.* Singapore: Institute of Southeast Asian Studies.

Lee, W.-S., and E. Zee. 2003. Standard Chinese (Beijing). *Journal of the International Phonetic Association* 22 (1): 109–112.

Leimgruber, J.R. 2013. *Singapore English: Structure, variation, and usage.* Cambridge: Cambridge University Press.

Levin, S. 2017, May 15. *Millionaire tells millennials: if you want a house, stop buying avocado toast.* Retrieved from The Guardian: https://www.theguardian.com/lifeandstyle/2017/may/15/australian-millionaire-millennials-avocado-toast-house

Liew, K.K. 2015. 'I am limpeh (your father)!' Parodying hegemony, anti-nostalgic cultural insurgency and the visual amplification of Lee Kuan Yew in late authoritarian Singapore. *Journal of Creative Communications* 10 (1): 21–38.

Lim, L. 2000. Ethnic group differences aligned? Intonation patterns of Chinese, Indian and Malay Singaporean English. In *The English language in Singapore: Research on pronunciation*, ed. A. Brown, D. Deterding, and E.L. Low, 10–22. Singapore: Singapore Association for Applied Linguistics.

———. 2001. Ethnic group varieties of Singapore English: Melody or harmony? In *Evolving identities: The English language in Singapore and Malaysia*, ed. V.B. Ooi, 53–68. Singapore: Times Academic Press.

———. 2004. Sounding Singaporean. In *Singapore English: A grammatical description*, ed. L. Lim, 19–56. Amsterdam: John Benjamins.

———. 2007. Mergers and acquisitions: On the ages and origins of Singapore English particles. *World Englishes* 26 (4): 446–473.

Low, E.L., and A. Brown. 2003. *An introduction to Singapore English.* Singapore: McGraw-Hill Education.

Lu, L. 2016, May 30. *The 4 myths of Singlish.* Retrieved from National Institute of Education: https://www.nie.edu.sg/about-us/news-events/news/news-detail/the-4-myths-of-singlish

Nair, A. 2019, March 14. *Singlish isn't dying. It's going to replace English.* Retrieved from Rice: https://www.ricemedia.co/culture-life-singlish-isnt-dying-its-going-to-replace-english/

Ng, D.F. 2008. Strategic management of education development in Singapore (1965–2005). In *Toward a better future: Education and training for economic development in Singapore since 1965*, ed. S.K. Lee, C.B. Goh, B. Fredriksen, and J.P. Tan, 39–68. Washington, DC: The World Bank and National Institute of Education, Singapore.

Ng, C.L. 2014. Mother tongue education in Singapore: Concerns, issues and controversies. *Current Issues in Language Planning* 15 (4): 361–375.

Olivia, H. 2019, April 13. *Millennial speak: A brief lexicon of the generation's slang.* Retrieved from The Straits Times, Lifestyle: https://www.straitstimes.com/lifestyle/millennial-speak-a-brief-lexicon-of-the-generations-slang

Pakir, A., ed. 1992. *Words in a cultural context: Proceedings of the lexicography workshop.* Singapore: UniPress.

Parker, P. 2019, April 14. *Adulting through the ages.* Retrieved from The Straits Times, Lifestyle: https://www.straitstimes.com/lifestyle/adulting-through-the-ages

Phoon, H.S., A.C. Abdullah, and M. Maclagan. 2013. The consonant realizations of Malay-, Chinese- and Indian-influenced Malaysian English. *Australian Journal of Linguistics* 1: 3–30.

Przybylski, A.K., K. Murayama, C.R. DeHaan, and V. Gladwell. 2013. Motivational, emotional, and behavioral correlates of fear of missing out. *Computers in Human Behavior* 29 (4): 1841–1848.

Rubdy, R. 2007. Singlish in the School: An impediment or a resource? *Journal of Multilingual and Multicultural Development* 28 (4): 308–324. https://doi.org/10.2167/jmmd459.0.

———. 2012. English and mandarin in Singapore: Partners in crime? In *English language as Hydra: It's impacts on Non-English language cultures*, ed. V. Rapatahana and P. Bunce, 221–243. Bristol: Multilingual Matters.

Schneider, E.W. 2007. *Postcolonial English: Varieties around the world*. New York: Cambridge University Press.

Singapore Department of Statistics. 2000. Chapter 4: Language and literacy. In *Census of population 2000, Advance data release*. Singapore: Singapore Department of Statistics. Retrieved May 15, 2020, from https://www.singstat.gov.sg/publications/cop2000/cop2000adr.

———. 2005. Chapter 2: Education and language. In *General household survey 2005, statistical release 1: Socio-demographic and economic characteristics*. Singapore: Singapore Department of Statistics. Retrieved May 15, 2020, from https://www.singstat.gov.sg/publications/ghs/ghrs1.

———. 2010. Chapter 2: Education, literacy and home language. In *Census of population 2010, statistical release 1: Demographic characteristics, education, language and religion*. Singapore: Singapore Department of Statistics. Retrieved May 15, 2020, from https://www.singstat.gov.sg/-/media/files/publications/cop2010/census_2010_release1/cop2010sr1.pdf.

———. 2015. Chapter 3: Literacy and home language. In *General household survey 2015*. Singapore: Singapore Department of Statistics. Retrieved May 15, 2020, from https://www.singstat.gov.sg/-/media/files/publications/ghs/ghs2015/findings.pdf.

Smakman, D., and S. Wagenaar. 2013. Discourse particles in colloquial Singapore English. *World Englishes* 32 (3): 308–324.

Tan, K.L. 2005. Vocalisation of /l/ in Singapore English. In *English in Singapore: Phonetic research on a Corpus*, ed. D. Deterding, A. Brown, and L. Ee Ling, 43–53. Singapore: McGraw-Hill Education.

Tan, M.-W. (2015). *Closed-Door discussion on the 2015 national values assessment for Singapore*. Retrieved from Institute of Policy Studies: https://lkyspp.nus.edu.sg/docs/default-source/ips/2_national-values-assesment-cdd-report_5_170815.pdf

Tan, C. 2016. The present: An overview of teaching Chinese language in Singapore. In *Teaching Chinese language in Singapore: Retrospect and challenges*, ed. K. Soh, 11–26. Singapore: Springer Science and Business Media.

Tao, J., and J. Wong. 2019. The confounding Mandarin colour term 'qing': Green, blue, black or all of the above and more? In *Studies in ethnopragmatics, cultural semantics and intercultural communication: Minimal English (and beyond)*, ed. L. Sadow, B. Peeters, and K. Mullan, 95–116. Singapore: Springer.

Tay, E.K. 2012, August 23. *IPS-aAdvantage roundtable on the national values assessment (Singapore 2012)*. Retrieved from Institute of Policy Studies: https://lkyspp.nus.edu.sg/docs/default-source/ips/aadvantage-rt_national-values-assessment_230812_report.pdf

Vaughn-Hall, J. 2018, July 26. *These avocado toast instagram captions are all you avo wanted*. Retrieved from Elite Daily: https://www.elitedaily.com/p/these-avocado-toast-instagram-captions-are-all-you-avo-wanted-9879019

Wee, L. 2004. Reduplication and discourse particles. In *Singapore English: A grammatical description*, ed. L. Lim, 105–126. Amsterdam: John Benjamins.

Wee, D. 2014, February 21. *Changing Guangzhou's name to Canton will raise city's international profile, delegate suggests*. Retrieved from South China Morning Post: https://www.scmp.com/news/china/article/1432508/changing-guangzhous-name-canton-will-raise-citys-international-profile

Wierzbicka, A. 1991. *Cross-cultural pragmatics: The semantics of human interaction*. Berlin: Mouton de Gruyter.

———. 1997. *Understanding cultures through their key words*. New York: Cambridge University Press.

———. 2003. Singapore English: A semantic and cultural perspective. *Multilingua* 22: 327–366.

———. 2006. *English: Meaning and culture*. New York: Oxford University Press.

———. 2013. *Imprisoned in English: The hazards of English as a default language*. New York: Oxford University Press.

Wong, J. 2008. Anglo English and Singapore English tags their meanings and cultural significance. *Pragmatics & Cognition* 16 (1): 88–117.

———. 2010. The "triple articulation" of language. *Journal of Pragmatics* 42: 2932–2944.

Wong, J.O. 2014. *The culture of Singapore English*. Cambridge: Cambridge University Press.

Wong, C. 2018, July 30. *Top values of Singapore society include 'kiasu', 'complaining', 'self-centred': survey*. Retrieved from Yahoo News Singapore: https://sg.news.yahoo.com/top-values-singapore-society-include-kiasu-complaining-self-centred-survey-093021416.html

Wong, J. 2020. Semantic challenges in understanding Global English: Hypothesis, theory, and proof in Singapore English. In *Studies in ethnopragmatics, cultural semantic, and intercultural communication: Minimal English (and beyond)*, ed. L. Sadow, B. Peeters, and K. Mullan, 117–142. Singapore: Springer. https://doi.org/10.1007/978-981-32-9979-5.

Wong, J., and C. Liu. 2019. Two ways of saying 'thank you' in Hong Kong Cantonese: m-goi vs. do-ze. In *Further advances in pragmatics and philosophy: Part 2 theories and applications*, ed. A. Capone, M. Carapezza, and F. Lo Piparo, 435–447. Cham: Springer.

Wu, Y. 2017. *Teaching Chinese as an international language: A Singapore perspective*. Cambridge: Cambridge University Press.

Yang, R., and W.S. Wang. 2018. Categorical perception of Chinese characters by simplified and traditional Chinese readers. *Reading and Writing* 31: 1133–1154.

Yeo, N.P., and D. Deterding. 2003. Influences of Chinese and Malay on the written English of secondary students in Singapore. In *English in Singapore: Research on grammar*, 77–84. Singapore: McGraw Hill.

Zhuo, T. 2019, April 13. *Not just' snowflakes': Millennials are changing the face of civil society*. Retrieved from The Straits Times, Lifestyle. https://www.straitstimes.com/lifestyle/not-just-snowflakes-millennials-are-changing-the-face-of-civil-society

Chinese Languages and Malaysian English: Contact and Competition

36

Siew Imm Tan

Contents

Introduction	910
The Status of English in Malaysia	910
Language Contact	912
Method	914
Results	920
Borrowings from Hokkien, Cantonese, and Mandarin	920
Integrative Potential of Chinese Borrowings in Malaysian English	924
Discussion and Conclusion	928
References	931

Abstract

The transference of features from Chinese languages is one of the more recognizable aspects of Malaysian English (MYE). This case study explores the Malaysian component of the Corpus of Global Web-Based English (GloWbE MY) for quantitative and qualitative evidence of the influence of Hokkien, Cantonese, and Mandarin on MYE. Framed against a discussion of the status of English in Malaysia and an overview of relevant language contact theories, the chapter focuses on a particular type of contact-induced change – borrowings. The data extracted from GloWbE MY demonstrates that Hokkien, Cantonese, and Mandarin have been key in shaping the lexicon of MYE. Analyses of frequencies, concordances, and collocates, however, revealed salient differences in the influence of these languages. Furthermore, there are significant differences in the integrative potential of the borrowings that each contribute. These differences reflect, firstly, the dynamics of the contact between English and Chinese in Malaysia and, secondly, the competition among the three Chinese languages in this intensely multilingual contact situation. It is postulated that any study on the

S. I. Tan (✉)
Faculty of Education, University of Canberra, Canberra, ACT, Australia
e-mail: Siewimm.Tan@canberra.edu.au

© The Author(s), under exclusive licence to Springer Nature Singapore Pte Ltd. 2022
Z. Ye (ed.), *The Palgrave Handbook of Chinese Language Studies*,
https://doi.org/10.1007/978-981-16-0924-4_53

influence of Chinese on MYE must consider the multiplicity of Chineseness, as well as the evolving statuses, roles, and functions of the Chinese languages spoken in the country.

Keywords

Malaysian English · Language contact · Corpus linguistics · Hokkien · Cantonese · Mandarin

Introduction

Malaysia is a multilingual and multicultural country of 31.7 million people (Current Population Estimates 2014–2016) speaking an estimated 138 living languages (Lewis, Simons and Fennig 2015, p. 6). Although Malay is the national language, English and various southern Chinese and southern Indian languages are widely spoken. It is this multilingual contact situation that has given rise to Malaysian English (MYE), a variety characterized by influence from the other languages spoken in Malaysia.

This chapter presents a corpus-based case study of MYE, focusing on the influence of Hokkien, Cantonese, and Mandarin, three of the most dominant Chinese languages in Malaysia. Framed against a discussion of the status of English in this country and the language contact situation within which this variety has evolved, the chapter offers wide-ranging examples of authentic language use extracted from the Malaysian component of the Corpus of Global Web-Based English (GloWbE MY) to illustrate the collective and comparative influence of these languages on the lexicon of MYE. The patterns of variation and change reflect the dynamics of the contact between English and Chinese in Malaysia, as well as the competition among Hokkien, Cantonese, and Mandarin in this intensely multilingual contact situation.

The Status of English in Malaysia

A significant proportion of Malaysia's population speak English. A rather dated estimate by David Crystal in 2003 put the number of L1 speakers of English in the country at 380,000 and L2 speakers at a further seven million (p. 63). The status of English is however a thorny issue, and many education and language policies have been implemented to curtail it. English is regarded by many Malay nationalists as the language of the former colonial masters, a threat to their ethnic and religious identities (Rajadurai 2011, pp. 28–36), as well as an impediment to the development of Malay, the national language. This has led to predictions that the number of English speakers would decrease, a concern that appears to be largely unfounded. As observed by Edgar Schneider (2007), "English is still widespread and deeply rooted in the country, most notably in urban environments" (p. 149).

L1 speakers of English form a small minority in Malaysia. They include people of mixed European and Asian parentage (Eurasians), Malaysians of Portuguese descent who have shifted from Kristang (a Portuguese creole) to English (Baxter 2012), people whose parents do not share a common ancestral language, and a small group of Chinese and Indian urbanites "for whom English has become the first language and by whom the original ancestral language has been discarded" (David 2000, p. 65). There is however a much larger group of English speakers – those who are bilingual, maintaining English and at least one other language. Typically middle-class urbanites, they learned English in childhood and continue to use it in the home, work, and social domains (Pillai 2006). In addition to English, they may also speak their ancestral language(s), as well as Malay and/or Mandarin, depending on their ethnicity, educational background, place of residence, and socioeconomic status. Transcending ethnicity, these people make language choices depending on interlocutor, context, and communicative purpose, frequently alternating between languages or varieties of a language.

A recent survey by Stephanie Pillai and Lok Tik Ong (2018) suggests that there is a revival in the status of English in Malaysia, particularly in the domain of education. The development of an English-medium private education sector in Malaysia since the 1990s (Tan and Santhiram 2017) has produced a growing group of privileged English-speaking Malaysians who view proficiency in English as the pathway to better employment opportunities in an increasingly globalized marketplace. In government schools, English has remained a compulsory subject, defying forces that have sought to challenge its roles in the education system. A recent example of the challenge that English has had to face is the opposition to the 2003 policy of using English as the medium of instruction for the teaching of mathematics and the sciences in primary and secondary schools (David 2004, p. 8). This resulted in the policy being overturned in 2009 and Malay being reinstated as the language of instruction for these subjects. The discourse of English and globalization has nevertheless prevailed in the Malay-medium public education sector. In 2016, the Ministry of Education rolled out a series of reforms aimed at "uphold(ing) Bahasa Melayu and strengthen(ing) the English language" (Malaysia Education Blueprint 2013–2025, cited in Pillai and Ong 2018, p. 148).

Despite the ebb and flow of the status of English in Malaysia, the language has continued to thrive, functioning as an important second language in the domain of education, a language of the corporate workplace (Nair-Venugopal 2003; and Rahmat et al. 2015) and online communication (Hassan and Hashim 2009), and a marker of prestige, identity, and solidarity (Tan 2013). As the sociolinguistic landscape of Malaysia is so pervasively multilingual, both at the societal and individual levels, it is inevitable that the variety of English spoken in Malaysia exhibits many features from the other languages of the society. Some of the diverse and typologically distinct languages that MYE has come into contact with include dialects of Malay; numerous southern, and more recently northern, Chinese languages; varieties of south Indian languages; and languages of the indigenous groups of Peninsular Malaysia and Borneo. This contact situation has produced systematic and stable changes in the vocabulary, pronunciation, grammar, and discourse structure of the variety of English spoken in Malaysia (Abdul Rahim and Haroon 2003; Hashim and Tan 2009; Hashim and Leitner 2011; Tan 2013; Yamaguchi and Deterding 2016).

The Chinese communities form a little less than a quarter of the population of Malaysia. The Current Population Estimates, 2014–2016, classify the country's population according to three "ethnic groups," Bumiputera (68.6%), Chinese (23.4%), and Indian (7%), and others (1%). These categories are at best reductionist; they do not capture the ethnic diversity and linguistic heterogeneity within each category. The Chinese communities alone have ancestral links to at least ten languages: Hokkien, Cantonese, Hakka, Teochew, Hainanese, Kwongsai, Foochow, Henghua, Hokchia, and "Other" (General Report of the Population and Housing Census of Malaysia 2000).

This chapter explores the influence of Hokkien, Cantonese, and Mandarin, focusing on a particular type of contact-induced change called borrowings. Based on data extracted from GloWbE MY, it is demonstrated that these languages have been key in shaping the lexicon of MYE. Analyses of frequencies, concordances, and collocates, however, reveal that these languages do not influence MYE in the same way, nor to the same degree. Furthermore, there are significant differences in the integrative potential of the borrowings that each language contributes to MYE. These differences reflect, firstly, the dynamics of the contact between English and Chinese in Malaysia and, secondly, the competition among the three Chinese languages in this contact situation. It is postulated that any study on the influence of Chinese on MYE must consider the multiplicity of Chineseness, as well as the evolving statuses, roles, and functions of the Chinese languages spoken in the country.

Language Contact

The foundation of this study is based on theories of language contact. The field of contact linguistics is concerned not just with describing contact-induced change but also with the extensive social parameters that can impact the outcomes of language contact. Peter Siemund's (2008) summary below captures the importance of studying language contact in context:

> It has been observed that the number of speakers in the respective language groups, the relative social status of the groups involved as well as the relative prestige of the languages to a great extent determine the linguistic outcome of language contact. In addition, it matters a lot how long two communities with different languages stay in contact and, above all, how intense the social and linguistic contact between the groups is. One of the best predictors of contact-induced language change is probably the degree of bilingualism found across the communities in contact and, ... whether one community is gradually shifting to the language of the other community. (Siemund 2008, p. 4)

Borrowings are just one of the many ways in which a language in contact with another can change. In contact linguistics, borrowings are typically defined as the "reproduction in one language of patterns previously found in another" (Haugen

1950, p. 212). The phenomenon, lexical borrowing, is commonly associated with language maintenance – the maintenance of a minority language by its speakers in the face of competition from a majority language (Thomason and Kaufman 1988, p. 37; Winford 2003, p. 11). In other words, the minority language speakers retain the use of their language in various domains but borrow some features from the majority language with which it is in contact.

The concepts of majority language and minority language are, however, difficult to apply in the context of MYE. While English, the recipient language of this study, is not numerically dominant in Malaysia, it is certainly socially dominant in that it is spoken by an educated middle class and is associated with upward mobility, employment prospects, and modernization. The relative statuses of Hokkien, Cantonese, and Mandarin, the source languages, are equally difficult to define using these concepts. Hokkien and Cantonese have the status associated with being two of the most dominant ancestral languages among the Chinese in Malaysia (Wang 2016), while Mandarin is the basis of standard Chinese, the official language of the People's Republic of China. What we do know is that MYE does face competition from the other languages of Malaysia. In spite of this, it has been retained across many domains of language use. The continuing relevance of MYE – in other words, the maintenance of MYE – requires it to be receptive to the incorporation of features from the other local languages. Lexical borrowing is thus conceptualized as a deliberate incorporation of lexical

Table 1 The composition of the Corpus of Global Web-Based English (GloWbE)

Country	Code	General (words)	Blogs (words)	Total (words)
United States	US	253,536,242	133,061,093	386,809,355
Canada	CA	90,846,732	43,814,827	134,765,381
Great Britain	GB	255,672,390	131,671,002	387,615,074
Ireland	IE	80,530,794	20,410,027	101,029,231
Australia	AU	104,716,366	43,390,501	148,208,169
New Zealand	NZ	58,698,828	22,625,584	81,390,476
India	IN	68,032,551	28,310,511	96,430,888
Sri Lanka	LK	33,793,772	12,760,726	46,583,115
Pakistan	PK	38,005,985	13,332,245	51,367,152
Bangladesh	BD	28,700,158	10,922,869	39,658,255
Singapore	SG	29,229,186	13,711,412	42,974,705
Malaysia	MY	29,026,896	13,357,745	42,420,168
Philippines	PH	29,758,446	13,457,087	43,250,093
Hong Kong	HK	27,906,879	12,508,796	40,450,291
South Africa	ZA	31,683,286	13,645,623	45,364,498
Nigeria	NG	30,622,738	11,996,583	42,646,098
Ghana	GH	27,644,721	11,088,160	38,768,231
Kenya	KE	28,552,920	12,480,777	41,069,085
Tanzania	TZ	24,883,840	10,253,840	35,169,042
Jamaica	JM	28,505,416	11,124,273	39,663,666
Total		**1,300,348,146**	**583,923,681**	**1,885,632,973**

Source: Corpus of Global Web-Based English (GloWbE)

features from a source language to enhance the communicative and expressive powers of the recipient language so that the latter can function fully in the contact situation.

Haugen's (1950) landmark study identifies three categories of borrowings: loanwords, loanblends, and loanshifts. These are distinguished on the basis of the degree to which the features are morphemically, phonemically, and semantically adapted in the process of being borrowed (a summary is available in Tan 2013, pp. 54–59). There are considerable differences of opinion as to how distinct these categories really are. Hilts (2003), for instance, suggests that what appears as loanblends may well be attempts at integrating loanwords into the linguistic system of the recipient language. In his study of lexical borrowing, Winford (2003) devotes an entire section to the phonological and morphological integration of borrowed features in diverse languages, arguing that "borrowed items are manipulated so that they conform to the structural and semantic rules of the recipient language" (p. 50). Following Hilts (2003) and Winford (2003), this chapter does not make the fine-grained distinction among categories of borrowings proposed by Haugen. Instead it discusses all borrowed features under the broad umbrella of borrowings, paying special attention to how these features are distributed across diverse semantic fields and the degree to which they are morphosyntactically integrated in the linguistic system of the English language.

Method

Corpus-based approaches have been shown to be very effective in promoting our understanding of how lexical, morphological, syntactic, and discourse features are used (see Kennedy 1998, pp. 88–203 for a comprehensive review of corpus-based studies of the English language since the 1960s). A major advantage of working with corpora is that it allows aspects of the language to be represented quantitatively as well as qualitatively. For the present case study, corpus data facilitated the development of functional interpretations regarding why particular features are borrowed and how they are used in Malaysia.

This chapter presents data extracted from GloWbE MY, a component of the 1.9-billion-word GloWbE corpus which represents English from 20 countries across the world. Table 1 shows the composition of GloWbE: the countries represented, the size of each subcorpus, the size of the general component (comprising newspapers, magazines, and company websites), and the size of the informal component (predominantly informal blogs). The corpus texts were all sourced from the web in December 2012. The Malaysian component consists of 42,420,168 words extracted from 45,601 webpages.

The GloWbE MY corpus offered two additional advantages specific to case studies. Compared to other smaller corpora, it yields many more lexical features, and many more instances of each lexical feature, making it a robust source of data for examining culture-specific lexical and morphosyntactic variation and change (Davies 2017). Furthermore, its informal component, the blogs, is a viable alternative to unguarded, spoken vernacular – the segment of language most likely to yield creative, non-standard lexical features.

Table 2 Chinese borrowings in MYE (n = 108)

FOOD (n=56)					
Hokkien (n=20)	**Freq.**	**Cantonese (n=23)**	**Freq.**	**Mandarin (n=13)**	**Freq.**
angku "oval glutinous rice cake with a sweet filling"	28	bok choy "Chinese white cabbage"	12	chow mien "fried noodles"	2
bak chang "glutinous rice in bamboo leaves"	7	char siew "barbequed pork"	122	guotie "kind of pan-fried dumpling"	3
bak chor "minced meat"	1	chee cheong fun "rice noodle snack"	27	jiaozi "kind of dumpling"	4
bak kut teh "medicinal pork bone soup"	90	chow mein "fried noodles"	4	la mien "hand-pulled wheat noodles"	20
bak kwa "kind of jerky"	4	choy sum "flowering cabbage"	12	liu sha bao "kind of custard bun"	1
chai poh "pickled radish"	5	dim sum "snack of bite-sized dumplings and pastries"	228	nian gao "glutinous rice cake"	1
chap chai "stir-fried mixed vegetables"	3	fu chuk "dried sheet of beancurd"	1	pan mien "flat wheat noodles"	4
chye sim "flowering cabbage"	2	ham choy "picked mustard greens"	2	shuijiao "kind of boiled dumpling"	4
hay bee "dried shrimp"	2	har gao "kind of prawn dumpling"	4	tang yuen "glutinous rice balls"	1
kiam chye "pickled mustard greens"	1	hor fun "kind of broad rice noodle"	37	you tiao "kind of cruller"	7
kua chi "sunflower seed"	2	kailan "chinese broccoli"	16	yusheng "raw fish salad"	1
koay teow "kind of broad rice noodle"	224	mai fun "thin rice noodle"	2	xiao long bao "kind of steamed bun"	22
mee "yellow wheat noodle"	755	mui choy "dried mustard greens"	6	zha cai "pickled stem of mustard greens"	2
mee hoon "thin rice noodle"	67	sang har mein "prawn noodle"	9		
popiah "spring roll"	76	siew mai "kind of pork dumpling"	20		
tau huay "sweet silken beancurd"	7	sui gou "kind of boiled dumpling"	1		
tau sar "mung bean paste"	7	tau fu fa "sweet silken beancurd"	26		
tau yu bak "braised pork"	2	tong sui "dessert of sweet soup"	9		
taugeh "beansprout"	11	tong yuen "glutinous rice ball"	1		
tauhu "beancurd"	6	tung mein fun "wheat starch"	4		
		wantan "kind of dumpling"	54		
		yau char kwai "kind of cruller"	7		
		yee sang "raw fish salad"	6		

(continued)

The GloWbE corpus is, however, not without its critics. In her work comparing it with the International Corpus of English (ICE), Loureiro-Porto (2017) cautions that the fact that GloWbE is sampled entirely from the Internet may affect its representativeness. She nevertheless acknowledges that its size and reliability resulting from the careful selection of webpages are clear strengths.

The present study utilized existing glossaries of MYE words (Tan 2013), Singapore English words (Lee 2004), and Chinese borrowings (Moody 1996) as a starting point in the search for Hokkien, Cantonese, and Mandarin borrowings in GloWbE MY. Not all features in these glossaries are relevant to MYE. Moody's glossary, for example, includes many Chinese borrowings that have entered English dictionaries,

Table 2 (continued)

PEOPLE AND TRAITS (n=25)					
Hokkien (n=16)	**Freq.**	**Cantonese (n=6)**	**Freq.**	**Mandarin (n=3)**	**Freq.**
ah beng "uncouth Chinese man"	41	amah "Chinese female domestic worker"	8	lao wai "foreigner"	2
ah kow "typical Chinese male name"	4	chap chong "mixed breed"	3	majie "Chinese female domestic worker"	5
ah gua "male transvestite"	4	chee sin "crazy"	2	shifu "expert"	4
ah lian "uncouth Chinese woman"	4	gwailo "foreigner" (derogative, literally 'devil person')"	8		
ah long "loan shark"	22	hamsap "lecherous"	17		
ah pek "middle-aged Chinese man"	20	sifu "expert"	95		
ah soh "middle-aged Chinese woman"	6				
angmoh "Caucasian (literally 'red hair')"	26				
kancheong "anxious, nervous"	2				
kay poh "nosy"	18				
kiasi "afraid of taking risks"	6				
kiasu "afraid of losing out"	118				
paiseh "embarrassed, ashamed"	28				
sinseh "traditional Chinese doctor"	11				
suay "unlucky, jinxed"	3				
towkay "owner of a business"	19				
FESTIVALS AND FOLK RELIGION (n=9)					
Hokkien (n=5)	**Freq.**	**Cantonese (n= 2)**	**Freq.**	**Mandarin (n=2)**	**Freq.**
ang pow "gift of money in a red envelope"	85	gong hei fatt choi "may you attain greater wealth (lunar new year greeting)"	6	gong xi fa cai "may you attain greater wealth (lunar new year greeting)"	3
chap goh meh "fifteenth day of the lunar new year"	13	lai see "gift of money ion a red envelope"	6	hong bao "gift of money ion a red envelope"	4
chingay "parade held during the lunar new year"	5				
kau ong yeh "nine deities of the Taoist tradition"	1				
tua pek kong "a Chinese deity"	24				

(continued)

Table 2 (continued)

MARTIAL ARTS AND GEOMANCY (n=8)					
Hokkien (n=0)	Freq.	Cantonese (n=2)	Freq.	Mandarin (n=6)	Freq.
		kungfu "kind of martial art"	111	feng shui "Chinese geomancy"	517
		tai chi "kind of martial art"	82	gongfu "kind of martial art"	3
				qi "vital energy"	62
				qigong "kind of meditative exercise"	115
				tai ji "kind of martial art"	3
				wushu "kind of martial art"	10

MISCELLANEOUS (n=10)					
Hokkien (n=1)	Freq.	Cantonese (n=6)	Freq.	Mandarin (n=3)	Freq.
kangtau "contacts, connections"	4	cheongsam "traditional Chinese dress"	27	guan xi "networks and connections"	1
	31	gao tim "completed"	5	pinyin "romanization system for standard Chinese"	26
		kowtow "to show obsequious deference"	31	qipao "traditional Chinese dress"	4
		tapau "takeaway"	45		
		wok hei "charred aroma of food cooked over very high heat"	21		
		yum seng "drink to success"	10		

but have never actually gained currency in MYE (e.g., *loquat, chopsuey,* and *taipan*). Yet others are words that were originally Chinese but have lost their etymological significance in Malaysia. The word *sampan* "small, wooden boat" originated in Cantonese 三板 *san ban* (literally, "three planks") and was subsequently borrowed into Malay. Today, it is recognized as a Malay word and is not widely used by Cantonese speakers in Malaysia. Some words that were identified by Lee (2004) were excluded because their orthographic representation does not signal their precise origin. The word *tai tai* "lady of leisure," for instance, reflects the pronunciations of 太太 "madam" in all three Chinese languages (but their tonal differences are not captured in the orthographic representation). As the precise source language was important for the broader aims of this study, words of indeterminate origin, such as *tai tai*, were excluded. My own 2013 list was also subject to culling. In my 2013 volume, I identified the word *kongsi* as a Hokkien borrowing because the orthographic representation matches the Hokkien pronunciation of 公司 "clan association" and because I thought the concept was strongly associated with the Hokkien community in Malaysia (e.g., Tan and Ooi 2014). Since then, I have learned that this word and its associated concept are also relevant to Cantonese speakers in Malaysia. As the precise source language of this borrowing is unknown, it was not included in this study. Altogether, 108 borrowings were identified. Table 2 shows these features, grouped under four semantic fields and three source languages. Each borrowing is accompanied by a brief definition (in cases where multiple definitions exist, only the most common meaning is listed) and its frequency in GloWbE MY.

Table 3 Concordance lines of ah so*

1	so the fault lies with them and no one else. #	Ah so	its the KingofBullShit again? Seriously they
2	# Car for rich chics lah, the amoi, lenglui,	ah so	, anak datok who got money but dad says "
3	escu said the same thing as Najib has said.	Ah so	what happen to Nicolae # Guys &; Gals, The
4	goose should be good for the gander as well,	ah so	? # Like it or not, in less than half a year's
5	me the feeling like she's a big sister or an'	Ah So	'. # Jamie: I think you're like the sunshine
6	on the ground with the 1.0 electorate -- the	Ah Soh	hawkers and the Ah Chek petty traders
7	I have personally seen and enquired were	Ah Soh	aunties who barely can see over the car
8	is very telling of the chinese psyche. The	ah soh	learnt all about Nasi Padang recipes from
9	gain pounds # Ed: Yah some maggi, some	ah soh	chap fan, some kway teow # Ront: All are
10	In loud pidgin Cantonese or Hokkien, the	Ah Sohs	would liberally throw comments about the

Once words of ambiguous etymologies were discarded, the process of identifying the source language of each of the 108 borrowings was fairly straightforward. Hokkien, Cantonese, and Mandarin, often considered dialects of Chinese, are linguistically quite different and are generally mutually unintelligible. These differences are often significant enough for us to identify the precise source language of a MYE borrowing. We know, for instance, that *dim sum* "snack of bite-sized dumplings and pastries" is borrowed from Cantonese 点心 because the orthographic representation *dim sum* replicates the Cantonese pronunciation of the term. If it had been borrowed from Mandarin, it would have been spelled *dian xin* in MYE, and if it had been from Hokkien, the spelling would more likely have been *diam sim*. The Mandarin variants, *dian xin* and *dianxin*, do occur in GloWbE, but there is an overwhelming preference for the Cantonese variants, *dim sum* or *dimsum*, across all 20 varieties of English represented in the corpus. Neither of the Mandarin variants occurs in GloWbE MY.

In principle, it is possible for all the borrowings identified in this study to have variants from the other two source languages. For instance, the Cantonese borrowing *cheongsam* "traditional Chinese dress" (n = 27) has a Mandarin equivalent *qipao* (n = 4), the Hokkien borrowing *ang pow* "gift of money in a red envelope" (n = 85) has a Cantonese variant *lai see* (n = 6) as well as a Mandarin one *hong bao* (n = 4), and so on. In practice, this is rare, and even when it occurs, there is always a strongly preferred variant.

In order to establish the frequency of each borrowing, all orthographic variants of the borrowing had to be identified, and all concordance lines of each variant had to be analyzed to ensure that only actual instances of the feature were counted. For instance, in the case of the borrowing *ah soh* "middle-aged Chinese woman," a wildcard search of *ah so** was performed to capture all possible orthographic variants of this feature and their morphosyntactic adaptations. Table 3 shows the five concordance lines of *ah so*, four of *ah soh*, and one of *ah sohs* (*ah soh* with an -*s* plural marker). Of these, only line 2 and lines 5–10 are legitimate occurrences of this borrowing. The others are instances of the exclamation *ah* followed by the conjunction *so* and were therefore not counted. Because *ah soh* is the preferred variant in MYE, it is the one listed in Table 2. The frequency (n = 6) does not

Table 4 Immediate left and immediate right collocates of popiah/pohpiah

No.	L1 collocate	Freq.	R1 collocate	Freq.
1	the	18	skin	13
2	making	5	,	7
3	this	3	skins	6
4	white	3	and	6
5	of	3	(	4
6	mini	3	stall	3
7	,	3	is	3
8	love	2	"	2
9	likes	2	in	2
10	"	2	party	2
11	,	2	or	2
12	one	2	.	2
13	wrap	1	with	2
14	with	1	type	1
15	was	1	turned	1
16	simple	1	to	1
17	selling	1	there	1
18	roll	1	that	1
19	our	1	style	1
20	original	1	roll	1

include *ah sohs*, the plural form of the word. This and other instances of morphosyntactic adaptations were analyzed separately as evidence of the integrative potential of a borrowing.

Besides the spread, frequency, and use of Hokkien, Cantonese, and Mandarin borrowings, this study also examined their integrative potential, that is, the degree to which these borrowings are morphosyntactically adapted to accommodate the grammar of the English language. In most cases, wildcard searches were sufficient to identify the application of inflectional and derivational morphemes to Chinese stems (e.g., see *Ah Sohs* in line 10 of Table 3). Less straightforward was the identification of creative compounds using Chinese borrowings, an important lexicalization strategy in MYE. This study used the collocate function of GloWbE to identify collocations based on Chinese borrowings. Table 4 shows the first 20 immediate left and immediate right collocates of the Hokkien borrowing *popiah* (variant *pohpiah*) "spring roll." The vast majority of these collocates are not lexicalized and cannot be considered creative compounds. Extracts 1 and 2 show such collocations in context:

1. She told me that she was *making popiah* skin and it did not turn out so well.
2. Unlike the usual roll *popiah style*, which results in the sauces coming out of every orifice, this method holds everything in.

Collocates that do contribute to compounds based on *popiah* are *skin(s)* and *stall* as illustrated below:

3. Finally managed to take a photo of uncle making his fresh *popiah skin* and 'kong th'ng' (peanut cookie).
4. Which was all nice and dandy until we drove by this *pohpiah stall* and had to just cry in agony cos we were too full to even have a bite.

Results

The results are presented in two parts. The first part examines the spread, frequency, and use of words that have been borrowed from Hokkien, Cantonese, and Mandarin into MYE, and the second part explores the integrative potential of these borrowings. These outcomes of the English-Chinese contact in Malaysia will be discussed alongside the extra-linguistic factors that shaped them.

Borrowings from Hokkien, Cantonese, and Mandarin

The comparative spread, frequency, and use of Hokkien, Cantonese, and Mandarin borrowings are useful indicators of the social ecology of the contact situation that gave rise to the emergence and continuing evolution of MYE. In this section, how the varied outcomes of contact summarized in Table 2 can be related to the history and social dynamics of the English-Chinese contact in Malaysia will be discussed.

Ninety-eight of the 108 borrowings identified in this study can be subsumed under the semantic fields of food (n = 56), people and traits (n = 25), festivals and folk religion (n = 9), and martial arts and geomancy (n = 8). The ten remaining borrowings are not readily classifiable and are grouped under "miscellaneous." The vast majority of these features – those categorized under the first three semantic fields, in particular – are words referring to local objects and cultural constructs for which there are no pre-existing English words. Schneider (2018) considers them to be "the most straightforward and directly evident manifestations of cultures . . . typical objects and artefacts; food and clothes; social roles, values, and the like" (p. 103).

The best examples of such manifestations of cultural contact in MYE are borrowings under the semantic category of food and festivals and folk religion. The borrowing of words related to Chinese festivals (*ang pow, chap goh meh,* and *chingay* from Hokkien, *lai see* from Cantonese, and *hong bao* from Mandarin) and quasi-religious concepts (*kau ong yeh* and *tua pek kong* from Hokkien) are clearly manifestations of the continuing relevance of Chinese festivals, not only in Malaysia but in many other parts of the world. The borrowing of food terms is also motivated by cultural maintenance. This includes various types of noodles (*koay teow, mee,* and *mee hoon* from Hokkien, *hor fun* and *mai fun* from Cantonese, and *la mien* and *pan mien* from Mandarin), kinds of dumplings (*har gao, siew mai, sui gou,* and *wantan* from Cantonese and *guotie, jiaozi,* and *shuijiao* from Mandarin), kinds of vegetables (*chye sim* and *taugeh* from Hokkien and *bok choy, choy sum,* and *kailan* from Cantonese), kinds of pickles (*chai poh* and *kiam chye* from Hokkien, *ham choy* and *mui choy* from Cantonese, and *zha cai* from Mandarin), ceremonial food (*angku* and *bak chang* from Hokkien, *tong yuen* and *yee sang* from Cantonese, and *nian gao* and *tang yuen* from Mandarin), food dishes (*bak kut teh*, *chap chai*, and *tau yu bak* from Hokkien, *sang har mein* from Cantonese, and *chow mien* from Mandarin), and soy products (*tau huay* and *tauhu* from Hokkien and *fu chuk* and *tau fu fa* from Cantonese).

These borrowings are a sign of the retention of diverse Chinese culinary cultures in the Malaysian gastronomic scene. There is however another social dimension that has shaped the outcome of the English-Chinese contact in this space: the overwhelming preference for Hokkien and Cantonese borrowings over Mandarin ones. In his sociological analysis of Chinese food in Malaysia, Chee Beng Tan (2001) emphasizes that in this country, Chinese food is tightly connected to the ancestral identities of the diverse "speech groups" (p. 127). For most Chinese Malaysians, whose ancestral ties are with the southern provinces of present-day China, it is Cantonese, Hokkien, Hakka, Teochew, and other southern Chinese cuisines that are the most relevant (see also Yoshino 2010). Mandarin food terms are however slowly entering the vocabulary of MYE, often in reference to food from other regions of China. Tan (2011) explains it thus:

> With the end of the Cold War and the global movement of people from and to China, and the emergence of China as an economic power house, we find that since the 1980s, more and more restaurants in Southeast Asia have employed cooks from China. *Lanzhou lamian* (Lanzhou hand-made noodles), for example, is now easily available in Singapore whereas as recent as in the 1980s this was not a common dish in Southeast Asia. Chinese restaurants serve to perpetuate "standard" Chinese food as well as globalize Chinese food that originated from different regions of China. (Tan 2011, pp. 29–30)

In MYE, this dimension of English-Mandarin contact – one resulting from the recent migration of people from China – has produced borrowings like *guotie* "kind of pan-fried dumpling," *jiaozi* "kind of dumpling," *la mien* "hand-pulled wheat noodles," *pan mien* "flat wheat noodles," *xiao long bao* "kind of steamed bub," and *zha cai* "pickled stem of mustard greens." These refer to newly imported foods for which there are no existing Hokkien or Cantonese words. Within the semantic field of food, Mandarin borrowings remain less frequent than Hokkien and Cantonese borrowings.

The dominance of Hokkien and Cantonese borrowings in the food sphere can also be explained by the ceremonial and symbolic functions of food for the Chinese and the Peranakans (Straits Chinese descended from the seventeenth- and eighteenth-century southern Chinese immigrants and their local wives) in Malaysia. For example, *angku* "oval glutinous rice cake with a sweet filling" is not just a cake; it also symbolizes "bliss and abundancy" and is presented to family and friends to announce the arrival of a new baby (Ng and Karim 2016, p. 98). The word *angku* is borrowed from Hokkien 红龟, literally "red tortoise." Red represents prosperity, while the tortoise represents longevity. Such symbolism is an important aspect of many diasporic Chinese communities, and Malaysia is no exception (Tan 1998). It is a distinctive aspect of MYE, as can be seen from the extract about "full moon" (an event marking a baby's first month of life) from GloWbE MY below:

5. In bigger cities such as Kuala Lumpur, you can now find catering companies, trendy cafes and even bakeries that provide full moon packages. They have a menu according to your budget, but will always include traditional *ang ku* and red eggs.

Other borrowings with similar significance include *bak chang* from Hokkien, *tong yuen* and *yee sang* from Cantonese, and *nian gao* and *tang yuen* from Mandarin.

The semantic field of people and traits is also represented by borrowings for which there are no convenient English equivalents. These are words referring to traditional occupations and activities (*ah long*, *sinseh*, and *towkay* from Hokkien, *amah* and *sifu* from Cantonese, and *majie* and *shifu* from Mandarin) and a range of affective borrowings that express attitudes ranging from mild disapproval to outright derision. Examples of affective borrowings are words referring to local subcultures (*ah beng* and *ah lian* from Hokkien), kinship terms that have acquired derogatory connotations (*ah pek* and *ah soh* from Hokkien), racial categories that can be pejorative (*angmoh* from Hokkien, *gwailo* from Cantonese, and *lao wai* from Mandarin), and derisive words (*kancheong*, *kay poh*, *kiasi*, and *kiasu* from Hokkien and *chap chong*, *chee sin*, and *hamsap* from Cantonese). As with the semantic field of food, the semantic field of people and traits is also dominated by Hokkien and Cantonese borrowings.

The affective borrowings extracted from GloWbE MY are interesting because they offer a glimpse into the various subcultures of Malaysia (and Singapore) and societal attitudes toward them. The borrowing *ah beng* is a case in point. Originally a common Hokkien male name *ah beng* acquired a novel referential function in Singapore English in the 1980s during the rise of its consumption-driven culture (Chua 2003). Chua (2003) postulates that *ah beng* and his female counterpart, *ah lian*, "represent, for the English-speaking middle-class, caricatures of youth who are working class or otherwise failures in the competitive education system and market economy. (They) stand as the 'Other' of the self-appointed sophisticated English-speaking cosmopolitans" (p. 10). In contemporary MYE, *ah beng* is more commonly used in reference to a subculture characterized by working-class membership, a distinct social dress, conspicuous consumption, and a tendency to orient oneself "towards the Chinese-speaking world" (Chan 2013, p. 93). The two extracts from GloWbE MY below illustrate the use of *ah beng* and *ah lian* in MYE:

6. the GT gives people the feeling that it is a sport sedan without trying too hard to be one. I dont think the look is '*Ah Beng*'.. the *ah beng* factor is a different story, i think it is safe to say that '*ah beng*' cars are more for those who pimped their rides (I am sure you have seen an *Ah Beng* BMW or Merc too).
7. Don't give excuses – nyonyas not lembah lembut, half-dead wimpy cry-baby *Ah Lian* type one, can tumbuk belacan by hand anytime

Another example is the Hokkien borrowing *kiasu*, which means "afraid of losing out." *Kiasu* is another import from Singapore English; it was initially borrowed from Hokkien into Singapore English to describe a stereotypical Singaporean attitude. The following is a good description of *kiasuism* by Lim (1989):

Kiasuism may be defined as an attitude by which a person undergoes, on the one hand, extreme disquiet if he discovers that he has not got full value for his expenditure of money, time and effort, and on the other, a distinct sense of exhilaration if he discovers that he has got much more than the full value for that expenditure. The ultimate distress is when he has got nothing for something, and the ultimate joy when he got something for nothing. (cited in Brown 1999, p. 123)

In Singapore English and MYE, *kiasuism* is viewed with a mixture of disdain and admiration; "the *kiasu* person is selfish," but he is also more likely to do well in life because he is quick to "seize opportunities" as they arise (Leo 1995, pp. 18–19). The extract below shows the word in context:

8. Competition fuels progress, collaboration guarantees excellence, *kiasuism* is one practice to be discarded – you keep one thing from 10 colleagues, they deprive you of 10 things, who at the losing end now? Say for example, when building a CV for a training position – why fight for a research project, do a solo presentation in a professional conference? Why not teaming up with 2 other buddies, do 3 inter-related studies, get all 3 published and presented with each other's help, attend an international conference as a group.

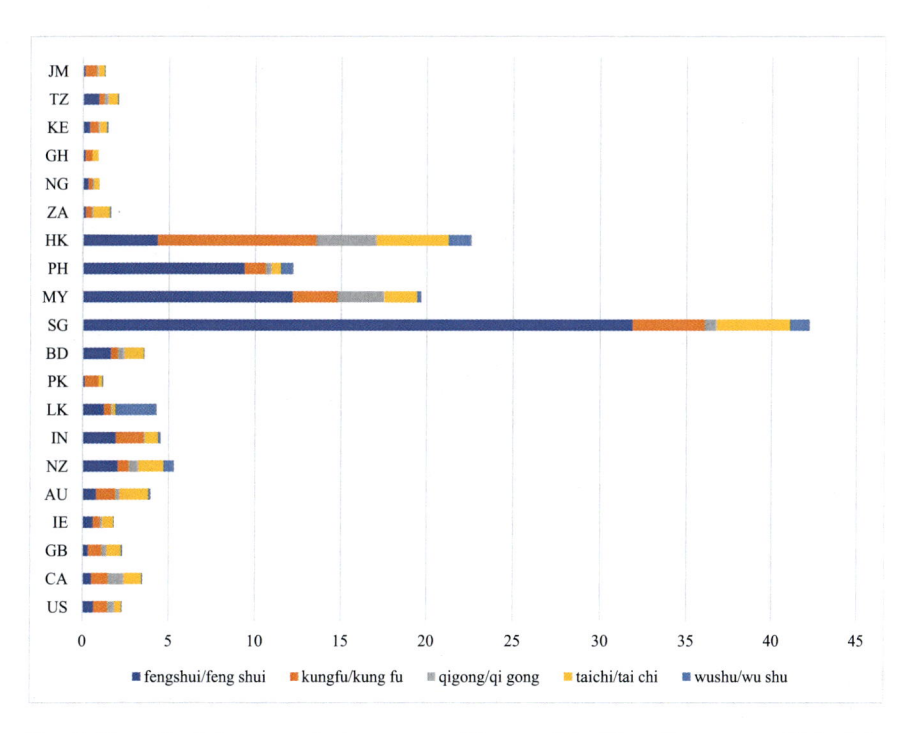

Fig. 1 Normalized frequency (instances per million words) of key Cantonese and Mandarin borrowings in the GloWbE corpus

Martial arts and geomancy is the only semantic field where the influence of Mandarin surpasses that of Hokkien and Cantonese. These are words referring to "Chinese 'high' culture" (Moody 1996, p. 405) that evolved out of other English-Chinese contact situations (e.g., in colonial Hong Kong and Shanghai) and were dispersed all over the world in different historical periods. Many of these borrowings have become so widespread that they are often regarded as standard international English words. A reliable indication of this is the admission of (variants of) words like *fengshui*, *kungfu*, *qigong*, *taichi*, *wushu*, and many others into the Oxford English Dictionary (see Cannon 1988, and Yang 2009, for other Chinese words that have penetrated the lexicon of international English). Another indication is their distribution across the 20 varieties of English represented in GloWbE (see Fig. 1). As can be seen, these words occur across the entirety of the GloWbE corpus, with the highest frequencies seen in Singapore English, Hong Kong English, MYE, and Philippine English, varieties spoken in countries with relatively large Chinese communities. Although constituting an aspect of MYE and how it has been influenced by Chinese, this semantic field is perhaps the least illuminating in this study.

Integrative Potential of Chinese Borrowings in Malaysian English

In theory, Chinese borrowings have the potential to be fully integrated into the linguistic structure of the English language. In other words, a borrowed noun can, and often does, function like other conventional English nouns, a borrowed verb like other verbs, and a borrowed adjective likewise. Extract 9 below demonstrates the use of the noun *ang pow* as a direct object, while extract 10 shows the use of the noun *Ah Pek* as the head of the noun phrase [The elderly *Ah Pek*] which functions as the subject of the sentence:

9. Businessman Lee Aik Chooi, 37, was among those at the temple who offered prayers and distributed *ang pow* to the poor.
10. The elderly *Ah Pek* wept in full view of all present at the developer's office!

In extract 11, the verb *tapau* is inflected for tense, while in extract 12, *kiasu* and *kiasi* function as adjectives and are used to modify the noun *mentality*:

11. ... they *tapau-ed* 5 packs of Sarawak laksa for us to take back to KL!
12. Its a chaotic situation there and the whole place is jammed full of Malaysians and Singaporeans who have kept the '*Kiasu* and *Kiasi*' mentality so vibrantly alive!

In addition to integration at clause-level grammar, the borrowings identified in this study also exhibit evidence of morphosyntactic integration. An analysis of the context and usage of the 108 borrowings identified in this study revealed four main types of morphosyntactic adaptation: (i) the use of the *-s* inflectional morpheme to mark nominal plurality; (ii) the use of the *-ed* and *-ing* inflectional morphemes to

Table 5 Borrowings that take the plural -s inflectional morpheme and frequency

Hokkien	Freq.	Cantonese	Freq.	Mandarin	Freq.
angku	2	dim sum	6	xiao long bao	6
chai poh	1	kailan	1		
mee	2	wantan	2		
popiah	1	amah	2		
ah beng	7	chap chong	2		
ah kow	1	gwailo	1		
ah lian	2	sifu	14		
ah long	25	cheongsam	5		
ah pek	1	yum seng	1		
ah soh	1				
angmoh	14				
kay poh	2				
kiasi	1				
kiasu	7				
sinseh	1				
towkay	21				
ang pow	44				

mark verbal tense and aspect; (iii) the use of derivational suffixes to create new word classes; and (iv) the combination of these borrowings with English morphemes to form creative compounds.

Of the 108 borrowings listed in Table 2, only 27 combine with the plural -*s* inflectional morpheme (see Table 5), and even then, only irregularly. Although unexpected (given the fact that the vast majority of the 108 borrowings are nouns), this pattern is not inexplicable. In MYE, even when the referent intended is clearly plural, the noun does not have to take the plural -*s* morpheme, as can be seen in extract 13 below (compare extract 14):

13. Everyone in town knows that this is the spot for delectable *popiah* (spring rolls) ...
14. Girlfriend decided to make this miso soup and some really yummy *popiahs* ...

This can be attributed to the high proportion of informal language in GloWbE MY, but it may also be because Hokkien, Cantonese, and Mandarin are classifier languages that do not make a grammatical distinction between count and non-count nouns. Chinese common nouns do not typically have plural morphology, and plurality is often marked using a combination of cardinal numbers and classifiers. Some scholars have suggested that nouns in Chinese languages are all non-count nouns that have their "plurality already built in" (e.g., Chierchia 1998, p. 53). It seems likely that lexical borrowing from Chinese into MYE is accompanied by the incorporation of some Chinese morphology.

Verbs are rarely borrowed, and so not much can be said about the use of verbal inflectional morphemes in MYE. Verbal borrowings that are marked for tense and/or aspect are as follows: *kowtow* (n = 31), *kowtows* (n = 3), *kowtowed* (n = 2), and *kowtowing* (n = 6); *tapau* (n = 45), *tapaued* (n = 12), *tapauing* (n = 2); and *yum seng* (n = 10), *yam senging* (n = 1). Extracts 15–17 offer us a glimpse into how the inflectional suffixes combine with these verbs in MYE:

> 15. Tun made a mistake for putting anwar there, but Tun was misled by anwar into believing he could deliver. Anwar made himself king, because politicians only *kowtowed* the guy who had the money.
> 16. We *tapau-ed* Mcd to go eat at my place.
> 17. If the gluttonous Chinese are not out eating and *yam senging*, something must be very wrong.

In addition to these, inflectional suffixes are occasionally used in creative ways on nouns: *feng shui* (n = 517), *feng shui-ed* (n = 1), and *tai chi* (n = 82), *taichi-ing* (n = 1). In extract 18, the businessman is hoping to enhance the auspiciousness of his house; he wanted his house "feng shui-ed" using a "water dragon":

> 18. The owner, a businessman wanted his house to be *feng shui-ed* by water dragon

In extract 19, the gerund *taichi-ing* describes the sparring between "Zaid Ibrahim and the press editors":

> 19. Except for a small hiccup by a member of the floor and the ongoing *taichi-ing* between Zaid Ibrahim and the press editors on the floor...

The use of derivational suffixes to create new word classes is equally uncommon. Of the 108 Chinese borrowings, only 3 are used to derive new words: *ah beng* (n = 41), *ah bengish* (n = 1); *feng shui* (n = 517), *feng shui-ish* (n = 1); and *kiasu* (n = 118), *kiasuism* (n = 12) and *kiasuness* (n = 12). Extracts 20–22 illustrate these derived forms in context:

> 20. and now, being global as u though, New Proton using R3 such as Toyota-TRD.. using their good name to sell more car without upgrading anything but eye candy!! This gotta be "Symptom of *Ah Bengish*" disease.
> 21. the Feng Shui Master made some interesting points that made me feel all... *feng shui-ish*.
> 22. Birthed, into wealth raised in the comfort of a foreign land, suckled and grown on a life of ease and fattened by the spoils of wanton *kiasuism* and all things illegal, the new generation of pigs stood on the threshold of their small kingdom, surveying the vast expanse beyond and dreaming of vast riches and easy pickings.... but not through a suicidal frontal assault this time.

By far the most productive process of integration of Chinese borrowings in MYE is compounding. This involves the creation of new lexical items by combining a Chinese borrowing with an English morpheme. In MYE, this typically results in hybrid nouns: lexical items comprising a Chinese and an English base. Table 6 lists some of the hybrid compounds extracted from GloWbE MY.

As might be expected, within the semantic fields of food, people and traits, and festivals and folk religion, hybrid compounds based on Hokkien and Cantonese borrowings dominate, while in the semantic field of martial arts and geomancy,

Table 6 Hybrid compounds based on Chinese borrowings

Food					
Hokkien	**Freq.**	**Cantonese**	**Freq.**	**Mandarin**	**Freq.**
bak kut teh soup	6	baby *bok choy*	1	8-flavor *xiao long bao*	2
dry *bak kut teh*	3	*char siew* filling	7		
claypot *bak kut teh*	1	*dim sum* breakfast	7	*shuijiao* skin	1
chai poh omelette	1	*dim sum* lunch	6		
fried *koay teow*	10	*dim sum* buffet	2		
koay teow soup	3	fried *hor fun*	2		
prawn *mee*	21	stir-fried *kailan*	6		
fried *mee*	13	peanut *tong sui*	2		
fried *mee hoon*	8	*wantan* noodles	5		
curry *mee hoon*	1	*wantan* skin	2		
mini *popiah*	3	*wantan* soup	2		
popiah skin(s)	19	*wantan* wrappers	1		
popiah stall	3	fried *wantan*	3		
tauhu brain	1				

Peopleand traits					
Hokkien	**Freq.**	**Cantonese**	**Freq.**	**Mandarin**	**Freq.**
Chinese *sinseh*	3	*hamsap*ress	1		
		*hamsap*eror	1		

Festivals and folk religion					
Hokkien	**Freq.**	**Cantonese**	**Freq.**	**Mandarin**	**Freq.**
ang pow packet(s)	5				
ang pow money	2				
chap goh meh festival	1				
tua pek kong temple	17				

Martial arts and geomancy					
Hokkien	**Freq.**	**Cantonese**	**Freq.**	**Mandarin**	**Freq.**
		kungfu master	8	*feng shui* master(s)	31
		kungfu grandmaster	2	*feng shui* analysis	17
				feng shui consultation	12
		kungfu movie	4	*feng shui* expert	10
		tai chi lessons	3	*feng shui* consultant(s)	10
		tai chi master	1	*feng shui* practitioner	6
		tai chi exercise	1	*feng shui* arrangement	5
				feng shui app	6
				feng shui audit(s)	9
				feng shui book	4
				qi energy	4
				qigong exercise(s)	22
				qigong master(s)	7
				qigong practitioner(s)	4
				qigong class	2
				qigong group	2

Miscellaneous					
Hokkien	**Freq.**	**Cantonese**	**Freq.**	**Mandarin**	**Freq.**
				pinyin class	3
				Chinese *pinyin*	4

Mandarin ones do. These compounds appear to be motivated by the need to enrich the vocabulary of MYE, in particular with regard to socioculturally relevant concepts for which there are no English equivalents. For instance, "wrapped" food (the various dumplings and spring rolls) is an important part of Chinese culinary traditions. These food items typically have two components: the outer dough, also known as the *skin* or *wrapper*, and the filling. Consequently, MYE speakers talk about *popiah skin, shuijiao skin, wantan wrapper*, and *char siew filling* (see Table 6 and extracts below) to distinguish the different types of dough and the various kinds of filling:

23. It should be tacky so that the *popiah skin* will be as thin as possible and it has to be elastic too, otherwise it will tear when wrapped with all the fillings.
24. I was so sweaty and tired by then, so I got my hubby to run down to the supermarket to get the *Sui Jiao Skin* to wrap the fillings.
25. Trim *wanton wrappers* into round shape and wrap some filling in it.
26. Here's my new found express way to make delicious *char siew filling* for my husband's favourite buns.

Within the semantic field of martial arts and geomancy, experts of different branches of martial arts are distinguished using compounds such as *kungfu master* and *qigong master*:

27. But the moment the Londoner set eyes on Quek, any hope he had of a calm elderly *kungfu master* was immediately dispelled.
28. By chance or as fate would have it, one day I accompanied a friend to a Qigong class conducted by Mak Chung Man, a Chinese herbalist and *Qigong master* who is regarded as one of the foremost authorities on both these traditional practices.

The relevance of Chinese geomancy, or *feng shui*, to MYE speakers is evidenced by the wide range of compounds based on this Mandarin borrowing (see Table 6). Extracts 29–31 below demonstrate the use of *feng shui arrangement* "layout based on the principles of Chinese geomancy," *feng shui app* "an app that assesses the layout of a living space," and *feng shui audit* "the assessment of the layout of a living space conducted by a *feng shui* professional" in MYE:

29. The correct *feng shui arrangement* of the kitchen brings wonderful money luck coupled with health.
30. I met Benny Khoo in last mid August 2011, he is so excited when got to know he going to develop a *Feng Shui app* for me.
31. Master Lee came to my house for *Feng Shui audit* a few days ago. I must say that he is very knowledgeable in Feng Shui and very professional.

Discussion and Conclusion

As can be observed from the results of this case study, Hokkien, Cantonese, and Mandarin have contributed many borrowings to MYE. For the most part, lexical borrowing from these languages appears to be motivated by the need to enhance the communicative and expressive powers of the English language, as evidenced by the fact that these features usually refer to concepts for which there are not adequate English equivalents. There are however critical differences in the way that these Chinese languages have respectively influenced MYE, differences that can be linked to the dynamics of the English-Chinese contact situation in Malaysia, as well as the competition among these source languages in this intensely multilingual contact situation.

From the distribution of the borrowings identified in this study, it is clear that Hokkien and Cantonese borrowings are more prevalent in the semantic fields of food, people and traits, as well as festivals and folk religion, while Mandarin borrowings are more prevalent in the semantic field of martial arts and geomancy. To a certain extent, these findings reflect Yang's (2009) observation, based on his analysis of English dictionaries, that Cantonese and Hokkien ("Amoy") tend to contribute food names to English, while Mandarin tends to contribute concepts relating to high culture. Yang predicts that the influence of Mandarin on English will continue to grow. He cites two reasons: (i) the growing dominance of Mandarin, the official language of China, which is spoken by 660 million people in the country as their native language (p. 97), and (ii) the "increasing contact between English and Mandarin speakers" brought about by the rapid expansion of the Chinese economy and mounting interest all over the world in learning Mandarin as a second or foreign language (pp. 100–101). This mirrors De la Cruz-Cabanillas' (2008) prediction that Mandarin will be increasingly important as a source language of Chinese borrowings in English.

The rising influence of the People's Republic of China and consequently Mandarin, its official language, is obviously important forces that will continue to shape the outcomes of English-Chinese contact, but the data of the present study suggest that for English varieties spoken in countries with significant Chinese diasporic communities, such as Malaysia (and Southeast Asian nations, in general), the influence of Hokkien and Cantonese cannot be underestimated. For diasporic Chinese communities in Southeast Asia, "the use of ... native-place categories has been important in the social organization of the ... community" (Crissman 1967, cited in Strauch 1981, p. 239). Judith Strauch's (1981) elaboration below is worthy of note:

> In the contemporary context, the most inclusive meaningful units are national units (Malaysian Chinese, Thai Chinese, Indonesian Chinese, etc.). Within each of these national Chinese communities, the first degree of internal segmentation follows linguistic lines (Cantonese, Hokkien, Hakka, etc.) which largely coincide with territorially bounded native place divisions .... Further segmentation is likely to occur within any given host locality according to smaller geographical units of origin, which also display some cultural diversity in custom and dialect usage. (Strauch 1981, p. 239)

The heterogeneity of Chineseness in Malaysia, at both the linguistic and cultural levels, is clearly manifested in MYE. This reflects the significance of ancestral ties to many Chinese Malaysians, who consider their ancestral language to be a marker of their "dialectal identity" and clan membership, both of which can open doors to jobs and other economic opportunities (Ting and Puah 2015, p. 134).

The dominance of Hokkien and Cantonese, in particular, is well-attested. These languages are spoken by descendants of the large number of Chinese immigrants who have been moving to the Malay Peninsula for centuries and, particularly, from the second half of the nineteenth century to work in tin mines and plantations and to establish businesses. The Hokkiens and the Cantonese formed migrant majorities in many urban centers, and this contributed to their success at maintaining their respective ancestral language. The size of their communities and their participation in important sectors of the economy granted their languages some status, which induced other Chinese communities to acquire these two languages. John Platt's (1977) observation of the distribution of these languages in 1970s Malaysia is informative:

> The Cantonese are predominant in the tin mining areas and, in fact, generally in the central part of West Malaysia from Seremban in the state of Negri Sembilan north to around Ipoh in the state of Perak and including the capital, Kuala Lumpur. They are also dominant in Cameron Highlands, Pahang, and over to the east coast, e.g. Kuantan in Pahang. South of this area, the dominant dialect is generally Hokkien, e.g. in Malacca and through the state of Johore. In the north-eastern states of Kelantan and Trengganu, Hokkien is the dominant dialect among the Chinese and in the north-western region, from around Taiping, Perak northward through the states of Kedah and Perlis and including Penang, a sub-variety of Hokkien is dominant. (Platt 1977, p. 365)

Irene Wong and Henry Thambyrajah's (1991) description of the association between these urban centers and a socially dominant Chinese language is also illuminating:

> ... in each geographical location, there is usually a dominant Chinese dialect spoken not only by its native speakers but also other Chinese living there. A Chinese living in an area with a dominant dialect that is not his native one is usually forced by circumstances to acquire this dominant dialect at least sufficiently for basic communication. (Wong and Thambyrajah 1991, p. 4)

More recently, there have been suggestions of some weakening of this association between language and region. Wang (2016, pp. 206–207) attributes this to the breaking down of "barriers between dialect groups," movement of people away from their traditional places of residence for purposes of employment, intermarriage, and language shift to Mandarin.

Although increasingly dominant, Mandarin only arrived in the sociolinguistic landscape of Malaysia in the 1920s. It was first introduced as a medium of instruction of Chinese schools, replacing various ancestral languages in response to the promotion of Mandarin as the official language of China (Wang 2016) despite the fact that no major group in Malaysia can claim an ancestral connection to Mandarin. Today, Mandarin is a marker of a supra-Chinese identity (Hsiao and Lim 2007; and Ting and Chang 2008) and a language of "prestige," especially for those educated in Chinese schools (Wong and Thambyrajah 1991, p. 4). With the rising influence of the People's Republic of China, knowledge of Mandarin is seen as a pathway to career opportunities (Cheng 2003). Whether Mandarin constitutes a threat to the long-term survival of Hokkien and Cantonese in Malaysia is difficult to say. The ancestral languages of the Chinese communities in Malaysia are important, not just as tools of communication but also as transmitters of cultural identity, knowledge, and values. They have also remained central to the domains of family, friendship, and local transactions, such as those performed in markets and neighborhood shops. Mandarin does not have an official status in Malaysia. Its capacity to function as an inter-ethnic lingua franca is curtailed by the fact that it is not usually acquired by people from other ethnic groups. Even among the Chinese, there are many who do not speak Mandarin. People who went through English-medium and Malay-medium education (those most likely to use MYE) tend to have, at most, a superficial knowledge of Mandarin.

In contemporary MYE, Hokkien, and Cantonese borrowings are more widespread and better integrated than Mandarin ones. This reflects the long and intense contact between English, on the one hand, and Hokkien and Cantonese, on the other. Whether the influence of Mandarin will eventually surpass that of the ancestral languages of Chinese Malaysians is difficult to predict.

Future research tracking the *comparative* influence of these languages on MYE is crucial if we are to have a better understanding of how *competition* among source languages affects the outcomes of contact. It is clear that any investigation into the influence of Chinese must consider the multiplicity of the notion of Chineseness and the competition among the diverse Chinese languages.

References

Abdul Rahim, Hajar, and Harshita Aini Haroon. 2003. The use of native lexical items in English texts as a codeswitching strategy. In *Extending the scope of corpus-based research: New applications, new challenges*, ed. Sylviane Granger and Stephanie Petch-Tyson, 159–175. Amsterdam: Editions Rodopi B.V.

Baxter, Alan. 2012. The creole Portuguese language of Malacca: A delicate ecology. In *Portuguese and Luso-Asian legacies in Southeast Asia, 1511–2011: Vol. 2. Culture and identity in the Luso-Asian world – Tenacities and plasticities*, ed. L. Jarnagin, 115–142. Singapore: Institute of Southeast Asian Studies.

Brown, Adam. 1999. *Singapore English in a nutshell: An alphabetical description of its features*. Singapore: Federal Publications.

Cannon, Garland. 1988. Chinese borrowings in English. *American Speech* 63 (1): 3–33.

Chan, Rachel Suet Kay. 2013. The Ah Beng subculture: The influence of consumption on identity formation among the Malaysian Chinese. In *Trending now: New developments in fashion studies*, ed. Laura Petican, Mariam Esseghaier, Angela Nurse, and Damayanthie Eluwawalage, 93–101. Oxford: Inter-Disciplinary Press.

Cheng, Karen Kow Yip. 2003. Language shift and language maintenance in mixed marriages: A case study of a Malaysian-Chinese family. *International Journal of the Sociology of Language* 161: 81–90.

Chierchia, Gennaro. 1998. Plurality of mass nouns and the notion of 'semantic parameter. In *Events and grammar*, ed. Susan Rothstein, 53–103. Dordrecht: Kluwer.

Chua, Beng Huat. 2003. *Life is not complete without shopping: Consumption culture in Singapore*. Singapore: Singapore University Press.

Corpus of Global Web-Based English (GloWbE). Accessed 9 Mar 2021 from https://www.english-corpora.org/glowbe/.

Crissman, Lawrence. 1967. The segmentary structure of urban overseas Chinese communities. *Man* 2: 185–204.

Crystal, David. 2003. *English as a global language*. 2nd ed. Cambridge: Cambridge University Press.

Current Population Estimates, Malaysia, 2014–2016. Putrajaya: Department of Statistics Malaysia. Accessed 9 Mar 2021 from https://www.dosm.gov.my/v1/index.php?r=column/pdfPrev&id=OWlxdEVoYlJCS0hUZzJyRUcvZEYxZz09.

David, Maya Khemlani. 2000. The language of Malaysian youth: An exploratory study. In *English is an Asian language: The Malaysian context*, ed. Halimah Mohd Said and Keat Siew, 64–72. Kuala Lumpur/Sydney: Persatuan Bahasa Moden Malaysia/Macquarie Library.

———. 2004. Language policies in a multilingual nation: Focus on Malaysia. In *Teaching of English in second and foreign language settings: Focus on Malaysia*, ed. Maya Khemlani David, 1–15. Frankfurt: Peter Lang.

Davies, Mark. 2017. Using large online corpora to examine lexical, semantic, and cultural variation in different dialects and time periods. In *Studies in corpus-based sociolinguistics*, ed. Eric Friginal, 19–82. New York: Routledge.

De la Cruz-Cabanillas, Isabel. 2008. Chinese loanwords in the OED. *Studia Anglica Posnaniensia* 44: 253–274.

General Report of the Population and Housing Census of Malaysia 2000. Putrajaya: Department of Statistics Malaysia.

Hashim, Azirah, and Gerhard Leitner. 2011. Contact expressions in contemporary Malaysian English. *World Englishes* 30 (4): 551–568.

Hashim, Azirah, and Rachel Tan. 2009. Malaysian English. In *English in Southeast Asia*, ed. Ee-Ling Low and Azirah Hashim, 55–74. Amsterdam/Philadelphia: John Benjamins.

Hassan, Norizah, and Azirah Hashim. 2009. Electronic English in Malaysia: Features and language in use. *English Today* 25 (4): 39–46.

Haugen, Einar. 1950. The analysis of linguistic borrowing. *Language* 26: 210–231.

Hilts, Craig. 2003. From taxonomy to typology: The features of lexical contact phenomena in Atepec Zapotec-Spanish linguistic contact. *Ohio State University Working Papers in Linguistics* 57: 58–99.

Hsiao, Hsin-Huang Michael, and Khay Thiong Lim. 2007. The formation and limitation of Hakka identity in Southeast Asia. *Taiwan Journal of Southeast Asian Studies* 4 (1): 3–28.

Kennedy, Graeme. 1998. *An introduction to corpus linguistics*. London/New York: Longman.

Lee, Jack Tsen-Ta. (2004). *A dictionary of Singlish and Singapore English*. Accessed 27 Feb 2019 from http://www.singlishdictionary.com

Leo, David. 1995. *Kiasu kiasi: You think what?* Singapore: Times Books.

Lewis, M. Paul, Gary F. Simons, and Charles D. Fennig (eds.). 2015. *Ethnologue: Languages of the World, Eighteenth edition*. Dallas, Texas: SIL International. Online: http://www.ethnologue.com

Lim, Catherine. 1989. *O Singapore! Stories in celebration*. Singapore: Times Books International.

Loureiro-Porto, Lucía. 2017. ICE vs GloWbE: Big data and corpus compilation. *World Englishes* 36 (3): 448–470.

Moody, Andrew. 1996. Transmission languages and source languages of Chinese borrowings in English. *American Speech* 71 (4): 405–420.

Nair-Venugopal, Shanta. 2003. Malaysian English, normativity and workplace interactions. *World Englishes* 22 (1): 15–29.

Ng, Chien Y., and Shahrim Ab Karim. 2016. Historical and contemporary perspectives of Nyonya food culture in Malaysia. *Journal of Ethnic Foods* 3: 93–106.

Pillai, Stephanie. 2006. Malaysian English as a first language. In *Language choices and discourse of Malaysian families: Case studies of families in Kuala Lumpur, Malaysia*, ed. Maya Khemlani David, 61–75. Petaling Jaya: Strategic Information and Research Development Centre.

Pillai, Stephanie, and Lok Tik Ong. 2018. English(es) in Malaysia. *Asian Englishes* 20 (2): 147–157.

Platt, John. 1977. A model for polyglossia and multilingualism (with special reference to Singapore and Malaysia). *Language in Society* 6: 361–378.

Rahmat, Noor Hanim, Normah Ismail, and Azizah Daut. 2015. Use of English at the workplace: How far is this true in Malaysia? *Indonesian Journal of English Language Teaching* 10 (1): 72–82.

Rajadurai, Joanne. 2011. Crossing borders: The linguistic practices of aspiring bilinguals in the Malay Community. *Australian Review of Applied Linguistics* 34 (1): 24–39.
Schneider, Edgar W. 2007. *Postcolonial English: Varieties around the World*. Cambridge: Cambridge University Press.
———. 2018. The interface between cultures and corpora: Tracing reflections and manifestations. *ICAME Journal* 42: 97–132.
Siemund, Peter. 2008. Language contact: Constraints and common paths of contact-induced language change. In *Language contact and contact languages*, ed. Peter Siemund and Noemi Kintana, 3–11. Amsterdam: John Benjamins.
Strauch, Judith. 1981. Multiple ethnicities in Malaysia: The shifting relevance of alternative Chinese categories. *Modern Asian Studies* 15 (2): 235–260.
Tan, Chee Beng. (1998). Chinese Peranakan food and symbolism in Malaysia. Proceedings of the 5th Symposium on Chinese Dietary Culture, Foundation of Chinese Dietary Culture, Taipei (pp. 185–210). Accessed 31 Oct 2019 from http://www.airitilibrary.com/Publication/alDetailedMesh?docid=P20160628001-199806-201609010016-201609010016-185-210
———. 2001. Food and ethnicity with reference to the Chinese in Malaysia. In *Changing Chinese foodways in Asia*, ed. Y.H. Wu David and Chee Beng Tan, 125–160. Hong Kong: Chinese University Press.
———. 2011. Cultural reproduction, local invention and globalization of Southeast Asian Chinese food. In *Chinese food and foodways in Southeast Asia and Beyond*, ed. Chee Beng Tan, 23–46. Singapore: NUS Press.
Tan, Siew Imm. 2013. *Malaysian English: Language contact and change*. Frankfurt: Peter Lang.
Tan, Kim Hong, and Bok Kim Ooi. 2014. *The story of Hokkien Kongsi, Penang*. Penang: Hokkien Kongsi.
Tan, Yao Sua, and R. Santhiram. 2017. Globalization, educational language policy and nation-building in Malaysia. In *Globalization in Southeast Asia: Issues and challenges*, ed. Hock Guan Lee, 36–58. ISEAS: Singapore.
Thomason, Sarah G., and Terrence Kaufman. 1988. *Language contact, creolization, and genetic linguistics*. Berkeley: University of California Press.
Ting, Su-Hie, and Chang, Yee Shee. (2008). Communication in a close-knit extended Hakka family in Kuching, Sarawak: Mandarin or Hakka? *Proceedings of 9th Biennial Borneo Research Council (BRC) Conference*, Kota Kinabalu, Sabah.
Ting, Su-Hie, and Yann-Yann Puah. 2015. Sociocultural traits and language attitudes of Chinese Foochow and Hokkien in Malaysia. *Journal of Asian Pacific Communication* 25 (1): 117–140.
Wang, Xiaomei. 2016. The Chinese language in the Asian diaspora: A Malaysian experience. In *Communicating with Asia: The future of English as a global language*, ed. Gerhard Leitner, Azirah Hashim, and Hans-Georg Wolf, 205–215. Cambridge: Cambridge University Press.
Winford, Donald. 2003. *An introduction to contact linguistics*. Malden: Blackwell.
Wong, Irene F.H., and Henry Thambyrajah. 1991. The Malaysian sociolinguistic situation: An overview. In *Child language development in Singapore and Malaysia*, ed. Anna Kwan-Terry, 3–11. Singapore: Singapore University Press.
Yamaguchi, Toshiko, and David Deterding. 2016. English in Malaysia: Background, status and use. In *English in Malaysia: Current use and status*, ed. Toshiko Yamaguchi and David Deterding, 3–22. Leiden: Brill.
Yang, Jian. 2009. Chinese borrowings in English. *World Englishes* 28: 90–106.
Yoshino, Kosaku. 2010. Malaysian cuisine: A case of neglected culinary globalization. In *Globalization, food, and social identities in the Pacific region*, ed. James Farrer, 1–15. Tokyo: Sophia University Institute of Comparative Culture.

Chinese as a Mother Tongue in the Context of Global English Business Communication

37

Per Durst-Andersen and Xia Zhang

Contents

Preliminary Remarks	936
Introducing the GEBCom Speech Production Corpus	937
Scenarios Involving Permission	939
Scenarios Involving Cancellation of an Obligation	942
Introducing the Theoretical Framework	945
Chinese and English Belong to Different Communicative Supertypes	945
Chinese as a Speaker-Oriented Language	948
Russian as a Reality-Oriented Language	950
British English as a Hearer-Oriented Language	951
Final Discussion of Data	952
Problem-Solving: The Use of Directive Utterances	952
Conflict Resolution and Conflict Avoidance	958
The Main Results of the Reception Test	966
Conclusions	968
References	970

Abstract

This chapter introduces the Global English Business Communication Project and presents its data and analysis concerning Mandarin Chinese, British English, and Chinese English. It shows how aspects of ways of speaking, thinking, and hearing in the Chinese linguaculture are transferred to English as a lingua franca. The data presented include scenarios of solving problems, resolving conflicts, and expressing apology and thus give access to the participants' view on communication, rhetoric, politeness, face, and societal logic. The chapter shows that the theory of communicative supertypes developed by Durst-Andersen (2011), which

P. Durst-Andersen (✉) · X. Zhang
Copenhagen Business School, Copenhagen, Denmark
e-mail: pd.msc@cbs.dk; xz.msc@cbs.dk

draws on semiotic theory (Durst-Andersen 2009; Durst-Andersen and Bentsen 2021), could be a useful framework to account for the reasons why the various transferences take place in the way they do. The results of the project question the usefulness of English as a lingua franca, because what is called Global English is not a neutral language without cultural load, but different linguacultural varieties of English, where the semiotic direction from one's mother tongue plays a crucial role when both producing and perceiving English speech.

Keywords

Lingua franca · Global English · Global English Business Communication · Chinese English · Pragmatics · Directive speech acts · Face · Societal logic · Apologetic expressions · Thanking expressions

Preliminary Remarks

With the globalization of world economy, English has become the lingua franca of global business (Neeley 2012; Ku and Zussman 2010). The following questions naturally arise: Is English an effective tool for communication in the age of globalization? Or is it a potential source of intercultural clashes, because it results in miscommunication? Some studies show that the use of English as business lingua franca leads to various problems (Charles and Marschan-Piekkari 2002; Marschan-Piekkari et al. 1999). Other studies find that English just works fine, because the authors put emphasis on the more practical side of communication, i.e., a kind of "mutual intelligibility" (Crystal 2003) and "getting the things done" (Bjørnman 2009).

This chapter introduces the Global English Business Communication Project (2012–2019) (henceforth the GEBCom Project), which aims to answer the question posed above and understand the nature of Global English spoken by non-native speakers of English. The project tests three levels of communication, i.e., speech production (Ibsen 2016; Zhang 2019), speech reception (Bentsen 2018), and understanding of emotional words in English (Mosekjær 2016) by different non-native English speakers with a British English group as the benchmark comparison group. The overarching research objective for the whole GEBCom Project was to investigate whether or not the culture-specific mental universe of the mother tongue affects the speech production and the understanding of English by non-native English speakers and, if so, to identify the actual kinds of influence.

The GEBCom Project is based on Durst-Andersen's (2011) theory of communicative supertypes which says that although all languages speak about reality, some do it through the speaker's direct experience of a situation in reality (called speaker-oriented languages such as Chinese), others do it through the speaker's and the hearer's shared understanding of the situation (called reality-oriented languages such as Russian), and still others do it through the hearer's memory of the situation in reality (called hearer-oriented languages such as British English). Orientation is equivalent to semiotic direction, and there are three possible directions corresponding to the three obligatory participants in the communication situation: (1) the speaker; (2) the situation in reality; and (3) the hearer. The semiotic direction can be revealed by the grammatical categories of a language and their prominence in that language. The main question is: Is the category of mood, aspect, or tense the dominating category when speaking of its verbal system? Is the category of classifiers, case, or articles the dominating category when dealing with its nominal categories?

If, without any doubt, the answer is mood and classifiers, as it is in the case of Chinese, then the semiotic direction points toward the speaker, and the language is probably a speaker-oriented language. If, still without any doubt, the answer is aspect and case, as it is in the case of Russian, then the semiotic direction points toward the situation in reality, and the language is probably a reality-oriented language. And if the answer is tense and (definite and indefinite) articles, as it is in the case of British English, then the semiotic direction points toward the hearer, and the language is probably a hearer-oriented language. However, a thorough analysis of the entire language is needed in order to be confident with one's preliminary classification. It is, however, important to emphasize that a semiotic direction yields a certain way of communicating and a certain way of thinking. In this specific sense, grammar and communication as well as grammar and culture go hand in hand.

Members of a speech community have to make a choice among the three semiotic directions in order to avoid miscommunication. The theory of communicative supertypes builds on the underlying assumption that "our mother tongue goes into our body and blood in a way that a foreign language is not able to, creating a bond between language, mind and body as well as anchors a given language in its culture" (Bentsen 2018, p. 1). A section is devoted to the presentation of the theory that places Chinese and English in different supertypes. It turns out that this theory is the only one that can describe and explain the data of the GEBCom Project that show that the speech production and speech perception of Chinese speakers of English were influenced by both their mother tongue and their mother culture.

The data were collected in both organizational settings and at universities in the UK and four other countries, including China. One strand of data collection was carried out in collaboration with the Carlsberg Group and their employees in China, Russia, England, and Denmark, respectively. Another strand of data collection was carried out with university students in Japan, China, Russia and England, respectively. The total number of participants sums up to 275 (for a detailed description of the entire project and its various tests and results, see Ibsen 2016; Mosekjær 2016; Bentsen 2018; Zhang 2019).

Introducing the GEBCom Speech Production Corpus

We chose to employ closed role play with cartoon prompts as our primary data collection instrument in the GEBCom speech production test (Ibsen 2016; Zhang 2019), because it has the advantage of eliciting semi-natural oral data (Golato 2003) and controlling for social variables in the scenario design. The participants were business employees in the Danish headquarters and British, Chinese, and Russian subsidiaries of the Danish-based multinational company, the Carlsberg Group. In each country, a group of 25 people participated in the closed role play in their mother tongue, and another group of 25 people participated in the closed role play in English. Zhang (2019), who focused on business people's verbal performance in business scenarios, redefined the method as "closed role enactment," because business professionals took on the roles in some real-life business scenarios they are familiar with and the oral data are thus "semi-experimental and semi-ethnographic data" (Zhang 2019, p. 25).

We designed each scenario in the production test according to one out of four deontic modalities (permission, prohibition, obligation, or cancellation of an obligation) within the conceptual framework of imperative (directive) frames developed by Durst-Andersen (1995) on the basis of von Wright's distinction between alethic and deontic logic (von Wright 1951). Alethic modality deals with knowledge of the laws of nature (i.e., possibility, impossibility, necessity, and non-necessity), whereas deontic logic is concerned with knowledge of the laws of society. It goes without saying that culture is more attached to deontic modality than to alethic modality, and this was the reason why the main focus was on deontic modality. The idea was that the imperative frames involving permission, prohibition, obligation, and cancellation of an obligation could be used to "identify cross-cultural differences in societal rules" (Zhang 2019, p. 52). Seventeen scenarios were designed involving either permission, prohibition, obligation, or cancellation of an obligation, but the different scenarios involving, for instance, permission differed from one another with respect to power, social distance (cf. Brown and Levinson 1987; Beebe and Cummings 1996; Blum-Kulla et al. 1989; Wood and Kroger 1991; Wolfson 1988), and degree of formality of the social context (cf. Van Dijk 1977). In this way, we ended up having four different scenarios of each imperative frame all based on deontic modality and one single scenario based on alethic modality (possibility).

In the present chapter, we will mainly use Chinese mother tongue data and Chinese English data from the GEBCom speech production corpus with emphasis on the scenarios involving the imperative frame of permission, i.e., the copy-machine scenario, the seat scenario, and the trolley scenario, and the imperative frame of cancellation of an obligation, i.e., the moving scenario and the meeting scenario. The reason for looking upon these five scenarios is that they show systematic differences and, moreover, they were felt to be more or less natural by all the participants irrespective of country of origin. Other scenario descriptions were found to be unnatural and culturally inappropriate by some of the participants. This concerns, for instance, the train scenario, which implied a Danish way of buying tickets that was in contrast to the Chinese way. Apart from the data mentioned above, data from British English and Russian will also be presented to show contrast, comparison, and importance.

Scenarios Involving Permission

Description of Scenarios

It is worth mentioning that all scenarios named "permission" were not predicted to give an output which contained *I allow you* ... or *You may* .... We anticipated a wide variety of direct speech acts (involving the imperative mood) or indirect speech acts (involving the declarative or interrogative mood). However, all of them should be traced back to some sort of an offer, i.e., a directive speech act, where the speaker acts in the interest of the hearer (the speaker presumes that the hearer wants what the speaker offers her/him) – not in the interest of the speaker her–/himself, which would have been a request, where the speaker cannot presume that the hearer wants what the speaker wants her/him to do.

Table 1 shows an overview of the parameters involved in the three scenarios involving permission. Table 2 illustrates the scenario descriptions themselves.

Table 1 Parameters involved in the four situations based on *permission*

Scenario	Modality	Social context	Hearer	Power	Social distance
Copy-machine	Permission	Informal	Colleague	None	None
Trolley	Permission	Formal	Boss	Hearer	Some/great
Seat	Permission	Public	Stranger	Speaker	Some/great

Table 2 Descriptions of the three scenarios involving *permission*

Scenario	Description
Copy-machine	**Instructor:** Imagine that you are at work. You are printing out a lot of copies, when your colleague, John, comes into the copy room with just one piece of paper in his hand. There is only one printer in the room, and you can see that John is in a hurry to use the machine, but does not like to ask you. You want to tell him that it is ok, so you say:_____ **Instructor's response:** _____
Trolley	**Instructor**: Imagine that you are at the airport. You are on a business trip with your boss, Mr. John. The plane has just landed, and you have collected your luggage and put it on a trolley. You see your boss walking with a very heavy suitcase. You want to let Mr. John know that there is space on your trolley, so you say:____ **Instructor's response:** _____
Seat	**Instructor:** Imagine that you are at the airport. You are sitting in front of the departure gate and waiting for your delayed flight. All seats are taken by impatiently waiting passengers. You notice a young tired-looking mother with a baby wandering around looking for a place to sit. She stops in front of you, and you can see her frustration. You feel sorry for her and want to let her know that it's ok for her to use your seat, so you get up and say: _____ **Instructor's response:** _____

Preliminary Discussion of Data

The data analysis of these three scenarios shows some interesting results in terms of the choice of sentence forms, viz., the imperative sentence form, the declarative sentence form, and the interrogative sentence form, among the group of British English speakers and among the group of Chinese English speakers. The three scenario descriptions all involve the speakers offering the hearer something that will solve her/his problem in the present situation. In the copy-machine scenario, the hearer is offered access to the photocopier; in the seat scenario, the hearer is offered the chair that the speaker is sitting on; and in the trolley scenario, the hearer is offered an empty space for his heavy suitcase on the speaker's trolley. If we look at Table 3, it is evident that the distribution of the three possible sentence forms between the two groups is very different.

From the British English data, it appears that all three sentence forms are used and, surprisingly enough, almost equal. Since the three sentence forms in the three scenarios are employed without any noticeable difference in degree of politeness, the speaker's specific choice among the three forms does not seem to be based on politeness alone. We checked for other parameters that could explain the distribution, but it turned out that neither age, gender, nor profession could explain the distribution.

Ever since Brown and Levinson (1978, 1987) introduced their influential model of politeness, the bulk of research on speech acts in pragmatics has been grounded in the social variables of distance, power, and degree of imposition of the act as the most important factors influencing speakers' linguistic choices. The authors claim that there is a positive correlation between these variables and the degree of indirectness (where the interrogative sentence form is the most indirect one of the three forms employed by the speaker. According to Brown and Levinson (1978, 1987), in cases when the hearer's status, power, and social distance are bigger than the speaker's, the greater degree of indirectness will be employed by the latter. However, our test does not show this. The only remaining factor that seemed to be able to explain why people choose different sentence forms is the personal factor, i.e., how the speakers are as persons and how they view themselves in relation to the hearers in the specific scenario description.

Table 3 Comparing British English and Chinese English offers

	Imperative		Declarative		Interrogative	
	British English (n = 24)	Chinese English (n = 25)	British English	Chinese English	British English	Chinese English
Copy-machine scenario	7 (29%)	5 (20%)	10 (42%)	16 (64%)	7 (29%)	4 (16%)
Seat scenario	13 (54%)	9 (36%)	2 (8%)	13 (52%)	9 (38%)	3 (12%)
Trolley scenario	12 (50%)	9 (36%)	4 (17%)	11 (44%)	8 (33%)	5 (20%)

We argue that British English speakers and hearers follow a specific decision tree for directives (to be examined later). Since choice of sentence forms in connection with directive speech acts is traditionally considered to be determined by degree of politeness alone, we lack tools to explain our data. Our data do not demonstrate that interrogative forms are more polite than declarative forms and that imperative forms are more impolite than declarative forms. In the scenarios mentioned above, the various forms used by the British English participants are all felt to be polite. In short, we could not find an explanation for these differences in the existing literature and had to develop our own model, i.e., a decision tree for directives, which could explain the reasons why some chose one form instead of the other two alternatives.

The interesting data from the Chinese English speakers clearly demonstrate that the declarative sentence form is their unmarked choice (see examples 1a, 1b, 1c and 1d). This is a significant difference between the two groups. Because nobody so far has been interested in the function of each of the three sentence forms in connection with directive speech acts, we have no access to any possible explanation for this difference. We may, of course, conclude that Chinese people do not seem to be aware of the decision tree for directives, which is assumed to be possessed by all British English speakers. This may be due to the fact that the three sentence forms are not part of the pragmatic grammar of Chinese. If this is true, then there is bound to be a considerable amount of transfer from Chinese to English. This is furthermore confirmed by data from the Chinese mother tongue test, where we only found declarative sentence form:

(1) a. If it is urgent, I can interrupt here and let you print first
   b. I will log out first and then you can use it
   c. John! Are you very urgent with the copy? I am OK. You can first
   d. I have many copies to print, but if your copy is urgent, I will let you use the machine

It is worth noting that we also found a kind of argumentation in the Chinese English data that are not present in the Chinese mother tongue data, at least not on the surface (cf. 1a, 1b, 1c and 1d). The question is twofold. Why do the Chinese participants use the declarative sentence form when speaking Chinese? And why do the Chinese participants prefer the declarative sentence form when speaking English? The declarative sentence form is first person oriented, i.e., speaker-oriented, while the interrogative sentence form is second person oriented, i.e., hearer-oriented. Could these two things have something to do with Chinese belonging to the speaker-oriented supertype, i.e., another communicative supertype than English being hearer-oriented? We believe that and will later try to explain it in more detail, when we have looked more carefully into the Chinese mother tongue data from the same production test, where we only find declarative sentence forms. Here we frequently meet the Chinese sentence-final particle *ba* that, of course, cannot be transferred directly into English, since English does not use particles. Nevertheless, *ba* seems to appear in disguise on the surface. There's more about this later on.

Scenarios Involving Cancellation of an Obligation

Description of Scenarios
Zhang (2019) defined the data in the scenarios involving cancellation of an obligation as a complex speech act, which presupposes a previous conversation composed of a request and a promise (or a positive compliance to the request). From the point of view of the literature on speech acts, it belongs to directive speech acts or simply directives. However, it is a new and complex type of a directive, which has not been reported in the speech act literature or communication literature. Since it forms a natural part of deontic modality being the negation of obligation, it is impossible to ignore the speech act that reflects this type of modality. As a matter of fact, it turned out that this type of speech act gave us access to something that has never been observed, described, or explained before.

Tables 4 and 5 show the parameters and the scenario descriptions in the two scenarios involving cancellation of an obligation that we chose to analyze in detail.

Preliminary Discussion of Data
The data analysis of these two scenarios shows interesting use of expressive components, such as apologetic expressions and thanking expressions, and their internal distribution in the discourse.

Table 4 Parameters involved in the two scenarios involving cancellation of an obligation

Scenario	Modality	Social context	Hearer	Power	Social distance
Moving	Cancellation of an obligation	Informal	Friend	None	None
Meeting	Cancellation of an obligation	Formal	Boss	Hearer	Some/great

Table 5 Descriptions of the two scenarios involving *cancellation of an obligation*

Scenario	Description
Moving	**Instructor:** Imagine that you are at home. You have asked your good friend, John, to help you move into a new apartment tomorrow. He has happily agreed to do so and even taken the day off from work. But then your family surprises you by arriving to help you move. This now makes John's help unnecessary, so you have to call him and say: _____ Instructor's response: _____
Meeting	**Instructor**: Imagine that you are at work. The manager of the Carlsberg IT department in England, Mr. Johnson, has arrived at your work this morning. You would like his opinion on a project and ask him if he could possibly attend your meeting this afternoon. The time does not suit him too well, but he agrees to come anyway. Unexpectedly, the meeting is cancelled. You have to call Mr. Johnson on his mobile to inform him about the situation, so you say: _____ Instructor's response: _____

The Meeting Scenario

In the meeting scenario (cf. Tables 4 and 5), we are dealing with cancellation of an obligation, which is part of what is termed *deontic modality* concerned with laws of society. The scenario presupposes the following three situations:

- A subordinate had asked a foreign superior to attend an important meeting.
- The superior agreed to do so, although he had other plans.
- Due to unforeseen circumstances, the subordinate has to call the superior on the phone in order to cancel the meeting.

The results themselves did not reveal any immediate significant differences between the various groups, if we disregard the Danish participants: 96% of the Chinese speakers, 59% of the Danish speakers, and 83% of the British English speakers used apologetic expressions in the mother tongue test, while 72% of the Chinese speakers and 48% of the Danish speakers used apologetic expressions in their English test. The interesting figures appear, when the percentage of the use of apologetic expressions is compared to the percentage of the use of thanking expressions. Here we find that 60% of the Chinese English speakers, 48% of the Danish English speakers, and 29% of the British English speakers used thanking expressions. We deliberately focused on the data from the English tests, because here we can make a sharp distinction between an apologetic expression and a thanking expression. This is not possible in the same way in Chinese. It turns out that Chinese participants have a tendency to apply a twofold strategy: an apologetic expression in the beginning and a thanking expression at the end. This tendency is not observed in any other group. Moreover, the answers clearly reveal that the scenario has been perceived very differently by the different groups. It seems that cancellation of an obligation is regarded as more face-threatening by the Chinese and British groups than by the Danish groups. However, there is no doubt that the Chinese speakers, in the mother tongue group as well as in the Chinese English group, are united in the sense of treating cancellation of an obligation with big caution. This could be explained by the fact that a subordinate acted upon a superior in the meeting scenario, but the same was true in the case of the moving scenario that involves equals (see below). Again, we could not find answers for that in the existing literature, because cancellation of an obligation has not been introduced as a subject at all. Therefore, we had to work it out by ourselves. We let us guide by Chinese being a speaker-oriented language, which we believed could explain, among other things, the specific Chinese notion of *face*.

The Moving Scenario

In the moving scenario, the British respondents did not seem to have any problem with cancelling their mutual agreement. Some seemed to interpret it positively from the point of view of the hearer (2a), and others seemed to be more or less indifferent to it (see 2b).

(2) a. John
Good neeews
I don't need you tomorrow
my whole family have just turned up
so there'll probably be more people than we need
so you can either go to work or you can have a day to yourself
b. John
Thank you so much for offering to help me move today
and I know you took the time off from work
but the thing is my family is here now
They can give me a hand
So you now have a day off to do something else
but thanks so much

In contrast to this, most Chinese respondents had trouble in cancelling their mutual agreement. They seemed to interpret the cancellation as something negative, which could have serious consequences for the speaker's and the hearer's friendship (see 3).

(3) 不好意思 bù hǎo yì sī
I am sorry
我家里有人来帮忙, 不用你来了 wǒ jiā lǐ yǒu rén lái bāng máng, bù yòng nǐ lái le
My family has arrived to help, so you don't need to come
谢谢你, 麻烦你了!xiè xiè nǐ, má fǎn nǐ le!
Thank you, and sorry for having troubled you.

In the mother tongue data, 48% of the Chinese mother tongue speakers used apologetic expressions, whereas none of the Danish native speakers used apologetic expressions. In the English data sets, 92% of the Chinese English speakers, 35% of the Danish English speakers, and 21% of the British English speakers used thanking expressions. While the English and the Danish participants viewed cancellation of an obligation as yielding a possibility to do something else (and possibility is good), the Chinese participants conceived cancellation of an obligation as breaking a law, which might have serious consequences for the personal relationship between the speaker and the hearer. These differences between the British English, on the one hand, and Chinese mother tongue data and Chinese English data, on the other hand, are so interesting that they need an explanation. Again, we could not find an explanation in the existing literature, because the subject has not been treated at all.

If we want a unified explanation for the peculiarities found in the Chinese English data compared to all other data sets, we cannot find it in existing theories. In this respect, all theories lack *explanatory adequacy*, but not only that – they also lack *descriptive adequacy*, because they have not the necessary means to describe the various phenomena found, and, in some places, they, moreover, lack *observational adequacy*, because they have not observed this type of data at all. This is a big problem and also the reason why we have to look a little bit closer at the theory of communicative supertypes that has at its disposal principles and concepts being able to describe and explain the fundamental differences between English and Chinese as well as the peculiarities in the Chinese English data. We believe that this theory will live up to the following basic requirements to any theory, namely, that it should be *economical* (use as few principles and rules as possible), *exhaustive* (be able to observe, describe, and explain all data), and *consistent* (there should be no discrepancies and no casuistic explanations).

Introducing the Theoretical Framework

If one compares what people with different mother tongues say in exactly the same situation, it is striking that they seem to verbalize the situation in different ways, cf. (1) where a Chinese (a), Russian (b), and British (c) person came late for a job interview and said the following to the CEO of the company:

(4)   a.   对不起，我来晚了，因为坐出租车时交通堵塞。
            duìbuqǐ, wǒ lái wǎn le. Yīnwéi zuò chūzūchē shí jiāotōng dǔsè.
            Sorry, I am late. It is because I was sitting in a cab during a traffic jam
       b.   Izvinite za opozdanie. Ja zastrjal v probke v taksi.
            Sorry for being late. I got stuck in a traffic jam in a cab.
       c.   Sorry for the delay. I was stuck in a cab.

One might think that the differences between the three examples are quite accidental, but they are much more fundamental. The Chinese, Russian, and British speakers simply communicate in different ways. In (4c) the English speaker gives new information to the hearer without telling what everyone knows, namely, that traffic jams cause delays. In (4b) the Russian speaker gives a description of the situation as such with clear indication of the source of her/his being stuck and with a clear indication that she/he got involved in that situation without having any idea of it. In (4a) the Chinese speaker gives her/his experience of the situation that caused her/his being late as well as her/his experience of her/his involvement in it.

Chinese and English Belong to Different Communicative Supertypes

Reality Exists in Three Modalities in the Human Mind
Normally, people believe that human beings communicate directly about external reality, but, in fact, the speaker and the hearer communicate directly with one another using the same communication channel without even touching external reality. There are three communication channels (see Fig. 1).

CHANNEL ONE: SPEAKER-ORIENTED LANGUAGES LIKE CHINESE CHINESE SPEAKERS SPEAK ABOUT REALITY THROUGH THE SPEAKER'S EXPERIENCE OF IT
CHANNEL TWO: REALITY-ORIENTED LANGUAGES LIKE RUSSIAN RUSSIAN SPEAKERS SPEAK ABOUT REALITY THROUGH THE SPEAKER'S AND THE HEARER'S SHARED UNDERSTANDING OF IT
CHANNEL THREE: HEARER-ORIENTED LANGUAGES LIKE BRITISH ENGLISH ENGLISH SPEAKERS SPEAK ABOUT REALITY THROUGH THE HEARER'S MEMORY OF IT

Fig. 1 Communicative supertypes as different communication channels

STEPS	MODALITY	TYPE OF MEMORY	PRODUCT
FIRST STEP	EXPERIENCE	SENSORY MEMORY	INPUT
SECOND STEP	UNDERSTANDING	WORKING MEMORY	INTAKE
THIRD STEP	MEMORY	LONG TERM MEMORY	OUTCOME

Fig. 2 Reality's existence in three modalities in the human mind

These three communication channels reflect the fact that reality exists in three different modalities in the human mind (cf. Durst-Andersen 2012), because human beings process visual stimuli from situations in external reality in three steps, which leave three different products (see Fig. 2).

Let us assume that you see a house you never saw before at a place, which is familiar to you. You see a stable picture on your perceptual screen, and this stable picture has "house" as the figure and the surroundings as the ground. This is your visual experience. This is the input produced by your sensory memory (first step, cf. Figure 2). However, your understanding of it is far more complex than that. Although you have not seen more than a new house, you understand that some people have been engaged in the activity of building it and this activity caused the creation of the house, i.e., the effect of the building activity. In other words, your intake, the product of your working memory, is very different from your input, although we are actually dealing with the same situation in reality (second step, cf. Figure 2). You have your visual experience and your understanding of this situation, but, besides that, you also have your total memory of the experience and the understanding, i.e., the outcome of the two other processes, which is located in what is traditionally called long-term memory (third step, cf. Figure 2).

Members of a Speech Community Must Choose a Common Voice
What is called external reality thus exists in three modalities in the human mind: (1) as your experience of a situation; (2) as your understanding of the experience of that situation; and (3) as your memory of the understanding of the experience of that situation. When members of a speech community want to communicate with one another, they have to choose a common voice among the three different ways in which reality exists in the human mind. All members must agree in order to make sure that they talk about the same thing. If this is not the case, they will not be able to understand one another. Members of the Chinese speech community have agreed to speak about reality through the speaker's experience of it. Members of the Russian, French, or German speech community have agreed to speak about reality through the speaker's and the hearer's shared understanding of it. And members of the British English and Scandinavian speech communities have agreed to speak about reality through the hearer's memory of it (for more about this, see Durst-Andersen and Cobley 2018). Members of different speech communities have agreed to use different communication channels (cf. Figure 1).

It is important to emphasize that a language shows its channel to its users by having specific categories that all point in the same semiotic direction, i.e., toward the speaker, which indicates that channel one is used; toward reality, which indicates that channel two is used; and toward the hearer, which indicates that channel three is used. The semiotic direction in each of the three communicative supertypes plays a crucial role. The semiotic direction is something all speakers of a speech community have in common, and this affects many areas of culture and society, for instance, the conceptions of face, societal logic, negotiation, humor, commercials, advertising in general, film making, and many other things such as ordinary problem-solving and conflict resolution in the society. Due to this, in the following we shall also employ the terms speaker-oriented, reality-oriented, and hearer-oriented linguacultures.

Reality Exists in Three Modalities in the Communicative Setting
If one turns to the communicative setting itself, which requires three participants, i.e., speaker, hearer, and reality, one witnesses the same: reality exists in three modalities. We have (1) the speaker's experience of it; (2) the speaker's and the hearer's shared understanding of the situation they experience; and (3) the hearer's experience of it. The fact that the structure of our mind and the structure of the communicative setting itself fit nicely together must be the reason why communicative supertypes exist and why they exist as exactly three different supertypes and not as two or as four supertypes. Speaker-oriented languages have the voice of the speaker as their common voice; reality-oriented languages have reality's voice as

their common voice; and hearer-oriented languages have the hearer's voice as their common voice.

We are accustomed to think that all people irrespective of language use the same universal communication channel and that this channel is used to convey information. However, this universalistic understanding does not enable us to explain why people with different mother tongues say different things in exactly the same situation in external reality (cf. 4), why the grammars of languages differ fundamentally from one another, and why certain grammatical categories are present in some languages and absent in others. As a matter of fact, only hearer-oriented languages such as British English convey information, while Chinese conveys the speaker's experiences and Russian conveys the situations referred to.

This is the reason why it is almost impossible to find overlapping categories among these three languages. In short, they do not share grammatical categories, which should be taken as a clear indication of the fact that Chinese, Russian, and British English are grounded in different principles.

Chinese as a Speaker-Oriented Language

The Importance of Having Serial-Verb Constructions

Chinese belongs to the supertype of speaker-oriented languages that talk about external reality through the speaker's experience of situations in reality. This speaker orientation seems to yield a certain way of structuring a sentence. One of the clearest indications of the Chinese orientation toward the speaker's own experiences is the existence of so-called serial-verb constructions. The result is that all complex situations, such as "give" or "go to," but not simple situations, i.e., states such as "stand" or activities such as "walk," are referred to by means of more than one verb. For instance, "I went to France on vacation," which needs one verb in English, must have at least three verbs in Chinese: *wǒ zuò fēijī qù fǎguó lǚyóu*. One says: "I sat down in an aeroplane, flew to France in order to travel around', because that is what you experience." You experience that you perform two different actions in a certain order, and when they have been carried out, you can do what you find is the purpose of your visit, namely, to travel around in France, which is an activity. It is as if the Chinese speaker has taken three pictures, each of them reflecting the different phases of the entire journey.

Lexicon and Grammar

The focus on experience, which applies to all human senses, is visible at several places in Chinese, for instance, at word level. While English uses one verb, e.g., *understand*, Chinese uses two verbs, e.g., *tīng dǒng* "listen + understand" and *kàn dǒng* "see + understand," to indicate the senses involved. Moreover, Chinese makes use of so-called classifiers before nouns to show that the thing named by the noun has been or can be experienced by the sense of vision, e.g., *zhāng* (used for things with a flat top such as pictures), *běn* (used for objects with many pages such as books), *jiā* (used for container-like objects such as restaurants), *zhī* (used for long, thin inflexible objects such as sticks), *gēn* (used for not thin objects such as flag poles), and the more general, *ge* (used in connection with pure quantification, e.g., *wǒ yǒu yī gè péng yǒu* "I have a friend"). There are almost 80 different classifiers in all. If one looks closely into their semantics, it seems that their content is based on looking at specific pictures of objects which have subsequently been transformed into images or percepts, that is, into a form that can capture all possible pictorial manifestations (for a similar view on classifiers in general, see Aikhenvald 2000; Allan 1977, p. 308f). Once again, we observe a specific Chinese interest in experience, i.e., a kind of image-based thinking which is opposed to the purely idea-based thinking characteristic of English (see Durst-Andersen and Barratt 2019).

Image-Based Thinking

The distinction between idea-based and image-based thinking becomes evident if we compare English *be* and *have* with the Chinese equivalents. It appears that *be* covers both existence (*She is in* Paris) and quality (*She is* beautiful) while *have* covers both possession (*I have a* car) and experience (*I have a need/pain/dream,* etc.). The overall idea that unites these two meanings is difficult to see, but it seems to be the case that the intransitive uses are non-controllable, while the two transitive uses are controllable, i.e., one can do something if one has a car or has a need. In Chinese, *zài* and *yǒu* are far more concrete and directly linked to visual perception which is the most important of all senses. The former is used when the speaker has a figure on a ground, and the latter is used when the speaker has a ground with a figure or figures on her/his perceptual screen. That is why *yǒu* is ambiguous from the point of view of the English language – it means both "have" and "be." The Chinese categorization seems to be based on the basis of images, not on the basis of ideas. This must be the reason why Chinese lacks a character/word for the subordinating conjunction *that* of English as well as for relative pronouns such as English *which* and *who*. They cannot be experienced; only the nouns they point to can be experienced and, as already mentioned, are indeed classified according to the type of pictures they leave on the perceptual screen.

Sentence-Final Particles

Chinese has at its disposal a set of sentence-final particles (*a, ya, ba, ma, le,* and *ne*) that occur at the end of a sentence in order to show the exact place from where it is verbalized in the speaker's universe of discourse, e.g., *ne* gives or demands information from the hearer and thus shows something about the speaker's universe of knowledge; *le* tells the hearer that the speaker's world has changed and thus shows something about her/his universe of experiences; *a* indicates that the speaker tells someone about her/his direct impression and thus shows the hearer her/his universe of opinions; and *ya* tells the hearer that the speaker is in a state of wondering and thus shows the hearer something about her/his universe of beliefs.

Other Things that Are Relevant

Another important feature of Chinese is its lack of the category of tense. This lack only makes sense if one views Chinese in the light of a speaker-oriented language, which focuses on experience: any visual, auditory, gustatory, olfactory, or somatosensory experience will always be concrete parts of the speaker's actual world. It is impossible to experience something in the past. When a concrete visual experience has become past, it is not a pure experience anymore. In this case, it is part of long-term memory where visual experiences are mixed up with the speaker's understanding and with her/his beliefs and opinions. Chinese is not at all interested in the stored experiences – only in the direct experiences, the pure ones that are processed online. In short, Chinese categories ensure that the speaker can speak of her/his experiences at all possible levels.

This leads us to conclude that, from the point of view of semiotics, the Chinese utterance is in every way a *symptom* of the speaker's experiences of a situation. A symptom is an index that points backward in time. This means, for instance, that "smoke" is a symptom of "fire," i.e., "smoke" points back to "fire" – in this case "fire" can be said to have caused "smoke." Likewise, "experience" is a symptom of "situation," i.e., "experience" points back to "situation," which means that a certain situation caused a certain experience. In this way, people know that there was a situation, but they speak about it indirectly through their experience of it.

This means that the semiotic direction of Chinese utterances goes toward the speaker. From this point of view, it seems predictable that when producing or receiving English utterances, Chinese speakers will use the communication process acquired through her/his Chinese mother tongue and, in doing that, they will produce what we shall call *semiotic transfer*. Note that semiotic transfer takes place all the time, but if L1 and L2 belong to the same communicative supertype, such as Danish and English do, the semiotic transfer will not be noticed in the same way – one will just think that a Dane speaks more or less perfectly well English, because the way in which British people and Danes communicate in their mother tongue is quite identical (this does not exclude the possibility of grammatical mistakes and lexical errors).

Russian as a Reality-Oriented Language

Russian Complex Verbs
In the supertype model, Russian, German, and French are classified as reality-oriented languages that speak about external reality through the situation being shared by the speaker as well as the hearer. This means that there is a big difference between simplex verbs such as state verbs (*sit, stand, lie,* and *hang*) and activity verbs (*walk, run, swim* and *climb*), on the one hand, and complex verbs, on the other (*sit down, stand up, walk to sth,* and *run to sth*) – the former name one single situation, be it stable or unstable, and the latter name two situations, i.e., an activity related to a state, a complex situation. The focus is on complex verbs, which appears from the category of aspect – the most important verbal category in all reality-oriented languages. In Russian this focus is particularly clear, because all complex verbs are represented by two verbs, for instance, "give sth to sb" is both *dat'* (perfective aspect) and *davat'* (imperfective aspect). The perfective form *dat'*

presents an action as an event, i.e., a state caused by an activity. The imperfective form *davat'* names the same action but presents it as an ongoing process, i.e., an activity intended to cause a change of state. It is impossible to find an action as such in external reality; it is only possible to find its manifestations, i.e., events and processes. In English, there is a verb for the action as such, i.e., the idea itself, whereas in Russian one cannot give a name to something that does not exist in external reality. The situation rendered by the English infinitive "*give sth to sb*" is something that you can imagine in your mind, but it will always instantiate an event (Eng. *She gave him a kiss*) or a process (*Eng. She is giving him a kiss*) when the verb occurs in a finite form. This is a huge and significant difference between Russian and Chinese, on the one hand, and between Russian and English, on the other.

Other Important Features

Russian has a specific mood system that distinguishes the real word (the indicative mood) from the imagined world (the subjunctive mood) and a very elaborate case system, where the nominative and the accusative cases show that the entity named by the noun is present in the situation referred to (*Mama* (nom.) *doma* "Mummy is home") while the genitive case shows that the entity named by the noun is not present in the situation referred to (*Mamy* (gen.) *doma* net "Mummy is not home"). Furthermore, Russian has a range of different syntactic constructions with and without a subject to be able to name the many varieties of the same prototypical situation. By choosing one of them, the speaker is able to precisely refer to the situation as it appeared in the real world.

This means that from a semiotic point of view, the Russian utterance is in every way a *model* of the situation referred to in external reality, i.e., any Russian utterance will always point to a specific situation in reality, be it in the real world or in an imagined world. A *model* is an index that is not dependent on time, because it applies at any point in time. In this way, it points neither forward in the future nor backward in the past, but simply points ahead, to what is in front of you.

British English as a Hearer-Oriented Language

The Importance of Tense and Articles

British English and all Scandinavian languages are hearer-oriented languages that speak about external reality through the hearer's mind. This means that the speaker must compare what she/he has in her/his memory to what the hearer has in her/his memory. The result of this comparison is either old information or new information – old information, if the speaker and the hearer share the same knowledge, and new information, if they do not share the same knowledge. The orientation toward the hearer yields specific grammatical categories which are completely absent in Chinese and Russian. This concerns, for instance, the crucial distinction within the verbal tense system between the simple past (*bought*, old information) and the present perfect (*has bought*, new information), the corresponding distinction within the nominal system between the indefinite article (*a car*, new information) and the definite article (*the car*, old information), and the equivalent distinction within the syntactic system between *there*-sentences (new information) and *it*-sentences (old information). New information has higher rank than old information, which means that new information, i.e., what counts as the most important information to the hearer, is often foregrounded, e.g., *He failed the exam, because he did not prepare.* The order of clauses will often contradict the actual order of the situations referred to, viz., "He did not prepare (cause) and failed the exam (effect)." Moreover, British English has at its disposal two present tense forms, i.e., the present and present perfect, as well as two past tense forms, i.e., the simple past and pluperfect, in order to be able to show to the hearer which kind of information is before or after another kind of information. That is why the complex category of tense can be said to be the most important category in hearer-oriented languages.

A Distinction Between New and Old Information at All Levels

New and old information is, however, not only a matter of grammar – it is a way of thinking that is present in any utterance and therefore in any kind of communication. For instance, one will often find *I met the director of the hospital* the other day... *The hospital's director told me that* ... Thus the distinction between *of-* and *'s-* genitive is also a matter of new and old information. Moreover, the extensive use of pronouns of all kinds, for instance, personal pronouns and relative pronouns, instead of repeating the noun, as one does in Chinese, is also an indication of the desire to downgrade old information. Everything mentioned above ensures that English is capable of making a sharp distinction between new and old information to the hearer at any sentence level.

From a semiotic point of view, this is tantamount to saying that the English utterance always points to the hearer and in every way is a *signal* to the hearer. A signal is an index that points forward into the future. This means that the hearer must do something in order to find the old information in her/his long-term memory and recreate it as a symptom in her/his own mind or to create a symptom in her/his mind which matches the new information given by the speaker in her/his utterance.

Final Discussion of Data

Problem-Solving: The Use of Directive Utterances

What Is Problem-Solving?

Inspired by the ideas found in Leech (1983, p. 35), the use of directives will be linked to problem-solving. If a directive utterance is uttered in a context where the speaker has a problem that can be solved by the hearer, the directive is called a request, e.g., *Could you pass me the salt, please?* or *Give me a glass of water, please?* in English; if the directive utterance is used in a context where the hearer has a problem that can be solved by the speaker, the directive is named an offer, e.g., *Have a seat!* or *Why don't you sit down?* in English. A problem will always involve a clash between the actual state and the state desired by the speaker alone or via the hearer. In this way, the actual state is experienced by the speaker as a defective state, i.e., a state that is opposite to the desired state. This means that there exists an imbalance between the real world and the imagined world. This imbalance is eliminated when the speaker creates either an equilibrium inside her–/himself if a request is carried out by the hearer or an equilibrium inside the hearer if the speaker carries out an offer. Whenever a speech act is successfully carried out, the result will be re-established consensus between the speaker and the hearer, which feeds into the equilibrium of society (cf. Finlayson 2005, pp. 143–144; Habermas 1998). In short, directive speech acts play an important role, not only among individual members of a speech community but also in society as a whole.

How Problems Are Solved in the Chinese Linguaculture

Discussing and Defining Various Terminological Issues

When dealing with requests and offers, Chinese differs from English and Russian. The Chinese variant of the imperative, e.g., *chī*! "Eat quickly!," seems only to be used at home, and the interrogative sentence form cannot be used for directives – *ma* will always need an answer in direct contrast to English, where a non-verbal reaction is the right answer to a question used as a directive, e.g., *Could you pass me the salt, please*? or *Why don't you sit down*? This leaves the declarative sentence form as the only sentence form in Chinese to be employed when the Chinese speaker requests the hearer to solve her/his problem or when the speaker offers her/his help to solve the hearer's problem.

Earlier Chinese studies confirm that the declarative sentence form is the only form to be used when dealing with directive speech acts, but Chinese scholars use a different terminology from their Western colleagues. They use the term *direct request*, which might give the impression that Chinese speakers use the imperative form, since a direct speech act in connection with a directive will always be an imperative, while declarative and interrogative forms count as indirect speech acts. When Chinese scholars (for instance, Song 1994; Gao 1999) argue that Chinese native speakers tend to be more direct than native English speakers when making a request, they do not mean that Chinese speakers use the imperative mood form. For them a direct request will be a request where you use a verb in the first person (they are performatives or performative-like), which involves some sort of request-meaning, for instance, *qǐng zuò* "I beg you to sit down," which counts as a very polite utterance (It is, unfortunately, translated by "Sit down, please," i.e., by an imperative form). All variants of this kind (I want you, I tell you, I ask you, etc.) are called direct requests by Chinese scholars (cf. above), but, from the point of view of sentence form, they all belong to the declarative sentence form. They all make good sense, since Chinese is a speaker-oriented language that focuses on the speaker's experience or mental state with a certain effect on the hearer. Note, however, that utterances of this kind have another function in English. They are normally used to discuss why the hearer did not do what he was requested to do, for instance, *I ask you to be quiet, but you just continue to shout*.

Problem-Solving in the Chinese Linguaculture

We shall argue that the declarative sentence form used as a directive is the speaker's *fixed proposal for solving a problem*, but this proposal has to be negotiated with the hearer. In this connection, the sentence-final particle *ba* plays a crucial role. As it should appear from (5), *ba* can be said to have three distinct functions:

(5) a. 他是美国人吧 tā shì méigóu rén ba. "He seems to be an American."
 b. 好吧 hǎo ba! "It is OK with me!"
 c. 我们坐吧 wǒmen zǒu ba! "Let us sit down!"

These three functions are intimately linked to first, second, and third person, respectively. In (5a) the speaker has given her/his best bid for a description of a situation in reality based on her/his impression of a person. This function is third person oriented. In (5b) the speaker has been presented with a proposal that she/he accepts. This function is first person oriented. In (5c) the speaker tells the hearer what she/he recommends both of them do together. This function is second person oriented. The point is that these functions do not exclude one another. They rather presuppose one another and enter into a certain order. The function and meaning of (5c) presuppose (5b); the function and meaning of (5b) presuppose (5a). This means that when a speaker gives a recommendation to a hearer, the recommendation to sit down presupposes that the speaker accepts the hearer and, moreover, that she/he finds that "they sit down" is the best proposal at hand. The direction is only backward, i.e., second person meaning presupposes first person meaning that presupposes third person meaning. Let us illustrate this by (6) which was used by 15 out of 25 Chinese speakers in the copy-machine scenario (cf. Tables 1 and 2). Here the speaker is making a lot of copies, when the hearer enters the printing room with a single piece of paper and she/he seems to be in a hurry:

(6) 你先复印吧 nǐ xiān fuyìn ba. "You copy first BA."

In saying so, the Chinese speakers all presented their best bid for solving the problem (third person-oriented meaning), they all accepted the sudden change in the situation (first person-oriented meaning), and against this background, they all gave their recommendation to the hearer to copy first (second person-oriented meaning). Since we did not find two persons from the UK, Russia, Denmark, or Japan who said exactly the same in the same scenario, and since 20 out of 25 Chinese persons all used ba in another scenario, the lunch scenario, it seemed to be interesting to work out the reason "why?"

The consistent results from the Chinese speakers participating in the mother tongue test seem to suggest that they have learned a rule that *if a person is more in a hurry than the speaker, then that person should do what s/he wants to do first*. This general rule lies hidden in the person's mind and is only activated when "the hearer is more in a hurry than the speaker is," and the result will be *You go first*. In short, what we heard in all utterances ending with the particle ba was the result of a deductive inference. Neither the rule (If you are more in a hurry than I am, then you go first) nor the Case (You are more in a hurry than I am) was uttered directly – only the result of the deductive inference (You go first) was uttered directly. Moreover, it was quite interesting to observe that not a single Chinese considered the photocopier to be her/his machine and not a single Chinese speaker specified how to execute the offer.

This result is in sharp contrast to that concerning participants from Denmark and the UK. The Danish and British English speakers gave their permission to the hearer to use the photocopy machine by using *may* or *can* and thus showed that they considered themselves to be the owners of the photocopier, because they were using it when the hearer entered the room. In addition, some Danish and English speakers told the hearer to execute the offer in a rapid manner, e.g., Eng. *It's OK. You can take a copy, but do it quickly!* Da. *Du kan da* (acceptance) *bare* (permission) *tage en hurtig* (condition) *kopi* "You may just take a quick print."

Problem-Solving in the Russian Linguaculture

The Russian data are only mentioned to put the Chinese and the British English data into perspective. When speaking of requests and offers, Russians use the imperative mood without any flavor of impoliteness. A military order is issued by using the infinitive form – the imperative sentence form can never count as an order. Russians use the imperative in the private sphere of life as well as in the public sphere of life. Russian speakers tend to use a direct speech act instead of one of the indirect alternatives when issuing directives (Ibsen 2016). The preference for the imperative mood appears very clearly from all extensive studies (cf. Larina 2009; Bolden 2017) (for more about this, see Durst-Andersen 2019.) This means that since Russian is a reality-oriented language, Russians prefer the imperative sentence form, but may also use the declarative sentence form. Only in specific circumstances, i.e., when the hearer has no idea of what the speaker might want and when the speaker is forced to make the hearer get something for him, can the speaker use the interrogative sentence form when issuing a directive (e.g., *Would you be so kind as to give me the book over there?*). Note, however, that it will also need an answer before the hearer executes the speaker's request.

Problem-Solving in the British English Linguaculture

If we turn to the British English speech community, it appears that its members use all three sentence forms. Traditionally, it has been claimed (cf. Searle 1969, 1983; Brown and Levinson 1987; Haugh 2015) that the interrogative sentence form is the most polite one, since it does not present a threat to the hearer or does not involve any pressure from the point of view of the speaker. As pointed out earlier, the British English data involving all three sentence forms (cf. Table 3) did not reveal any difference in politeness degree. Therefore, we had to analyze the UK data for offers in a new way, and the result of our analysis can be seen in Fig. 3, which forms a decision tree for directives that is meant to account for the logic of the use of directives in British English based on our production test.

The decision model for directives includes three levels formed as a hierarchy, which places the imperative sentence form at the upper level, the interrogative

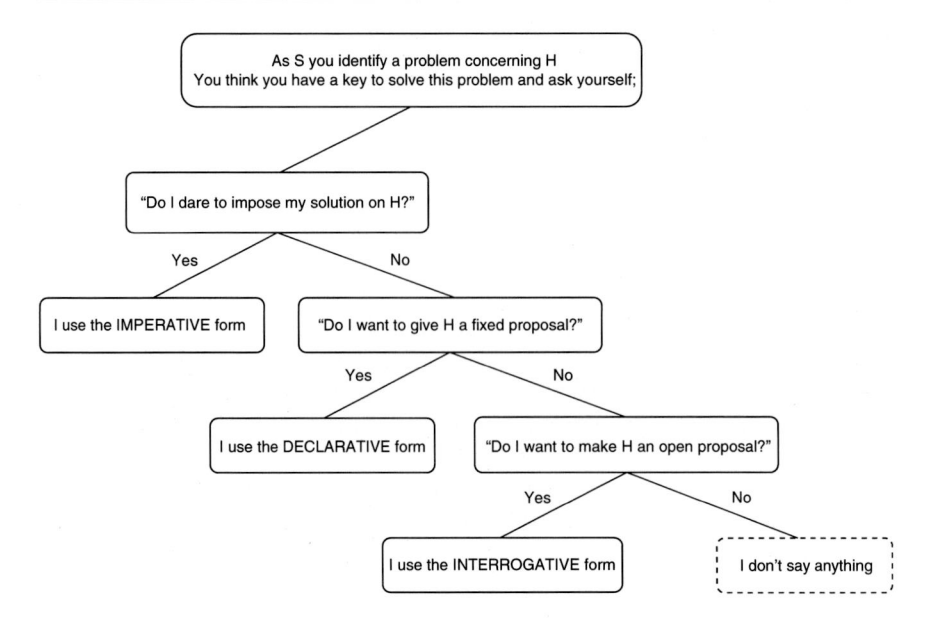

```
                 As S you identify a problem concerning H
                 You think you have a key to solve this problem and ask yourself;

                      "Do I dare to impose my solution on H?"
                         Yes                No
    I use the IMPERATIVE form    "Do I want to give H a fixed proposal?"
                                     Yes                No
                         I use the DECLARATIVE form    "Do I want to make H an open proposal?"
                                                         Yes                No
                                    I use the INTERROGATIVE form    I don't say anything
```

Fig. 3 The decision model for directives in British English

sentence form at the lower level, and the declarative sentence form in between. The hierarchy reflects two things. First, the imperative-declarative-interrogative hierarchy can be said to indicate that the strength of illocutionary force is viewed as a continuum from strong to weak. Secondly, the imperative-declarative-interrogative hierarchy can also be regarded as a continuum ranging from directness to indirectness.

Faced with the hearer's problem, the speaker has to consider the question: *Where does the right solution to the problem lie?* Due to the fact that we are dealing with a communicative situation with three obligatory participants, i.e., the speaker, the hearer, and a situation in reality, the question can be answered in three distinct ways:

- The solution lies in the situation itself – if the speaker thinks that the right solution is found in the situation itself (in our case, the photocopy machine), then she/he will use the imperative form: *Just make your copy quickly!*
- The solution lies with the speaker – if the speaker thinks that she/he her–/himself has the right solution to the problem, she/he will use the declarative form: *You can go first if you want to.*
- The solution lies with the hearer – if the speaker thinks that the hearer her–/himself has the right solution to the problem, she/he will use the interrogative form: *Would you like to use the machine?*

This is how we explain why the British English speaker uses an imperative form, a declarative form, or an interrogative form in connection with directives. The degree of politeness or impoliteness caused by the specific choice can be explained by the fact that when using the imperative, the speaker gives a solution that is non-negotiable; when using a declarative, the speaker gives a proposal that has to be negotiated with the hearer; and when using the interrogative, the speaker gives the hearer an open proposal, i.e., the hearer may write the text of the contract just as she/he wants to. In other words, the speaker is not guided by the concept of politeness or impoliteness when choosing a specific sentence form. The hearer may feel that a sentence form is more or less polite, but it will be a concomitant effect of using a specific form.

The Chinese English Data Compared to the Data from the UK

We now return to Table 3 (see Table 3), which shows that British English speakers employ all three sentence forms. However, the data from the Chinese English speakers demonstrate very clearly that the declarative is their unmarked choice – the declarative sentence form is first person oriented just like the Chinese language as a whole (see examples below repeated from above):

(1) a. If it is urgent, I can interrupt here and let you print first
 b. I will log out first and then you can use it
 c. John! Are you very urgent with the copy? I am OK. You can first
 d. I have many copies to print, but if your copy is urgent, I will let you use the machine first

The Chinese speakers that employ the declarative sentence form demonstrate a kind of *transfer of negotiation strategy* from Chinese to English. Note that in British English problem-solving, the declarative mood represents a fixed proposal, while the interrogative mood constitutes an open proposal. The Chinese English use of the declarative sentence form seems to be a combination of these two varieties, since the proposal is fixed, but, nevertheless, open for negotiation and for refusal. This combined with the fact that the utterances consist of two clauses makes the strategy even more interesting. When analyzing (6) with ba, we argued that the ba utterance contained the conclusion (the result) based on a deductive inference, which is represented as follows:

(6') Rule: If you are more in a hurry than I am, then you go/copy first
 Case: You are more in a hurry than I am
 Result: You go/copy first

If we were dealing with direct transfer from Chinese, participants would have said *You go/copy first*, but only a few said that. The majority verbalize the entire rule (see 1 compared to 6') and thus directly show the hearer the rule that forms the basis for the decision. This is inductive reasoning. In this way, the hearer can be confident if she/he decides to step in and use the photocopy machine for a couple of seconds. This kind of assuring the hearer and negotiating directly (cf. transfer of negotiation strategy) with the hearer is completely foreign to the British English speakers. We call it *rhetorical transfer*, since it involves logical argumentation in normal discourse, where one normally sees the result, q (effect), or the case, p (cause), but, seldom, the entire rule (if p, then q). Our findings are in line with Kirkpatrick and Xu (2002), Zhang (1995), and Kong (1998), who all emphasize the inductive way of reasoning in various varieties of Chinese discourse. From a more general point of view, Gumperz (1982) and Odlin (1989) have mentioned transfer of rhetorical norms and prototypical discourse structures from L1 to L2.

Conflict Resolution and Conflict Avoidance

Politeness and Face

Politeness has been a hotly debated topic in the literature of pragmatics and communication studies after Brown and Levinson's (1987) seminal work which was based on Goffman (1967). Since then it became a tradition to employ "face" as the main universal concept, including the distinction between positive face and negative face. In the 1990s Eastern scholars challenged the universality of politeness and tried to highlight the content of face in an Asian context. This concerns, for instance, Lim and Bowers (1991, p. 142) who distinguish three types of face, i.e., fellowship face, competence face, and autonomy face. Gu (1990) and Mao (1994) both point out that pragmaticists in Western cultures overemphasize the negative face, and they attempt to draw our attention to the meaning of Chinese face by referring to the distinction made by the Chinese anthropologist Hu (1944) between moral face, i.e., *liǎn*, and social face, i.e., *miànzi* (corresponding more or less to Bourdieu's notions of internal and external habitus, cf. Bourdieu 1994). In the discursive approach to face and politeness, the focus has shifted to areas such as identity, relational work, and social interaction with dialogues as the main empirical data (see, for instance, Watts 2003; Spencer-Oatey 2007; Haugh 2007). Despite the disputes and the use of different approaches, there has been a general consensus that apologetic expressions and thanking expressions are devices meant to disarm potential aggression (Brown and Levinson 1987, p. 1), to resolve potential interpersonal conflicts, and to maintain rapport or interpersonal harmony (cf. Spencer-Oatey 2008; Goffman 1967; Aijmer 1996).

Three Types of Face

It is an unfortunate fact that the original Chinese concepts of face (Hu 1944) have been distorted at the expense of the distinction between negative and positive face. The result has been that the Western concept of politeness and the Eastern concept of face have been mixed into something universal. We cannot turn back this development, but we can argue that it is crucial to differentiate three distinct understandings of face: first person's face, second person's face, and third person's face. All types of face are, of course, present in all types of societies and cultures, but one of them will be prioritized in a particular linguaculture.

Face in the British English Linguaculture

We shall argue that the British English linguaculture is oriented toward the face of the second person, the hearer's face. This is a characteristic of all hearer-oriented languages. This also appears from the fact that in the literature on the subject, the concepts of positive and negative face relate to the hearer, not to the speaker. When people from a hearer-oriented linguaculture think about face, the only kind of face that comes to their mind is the hearer's face. It is not a coincidence that the slogan "The customer is always right" was coined by a British English-speaking person from London. The focus is on the other person, be it the hearer in communication or the customer in business. The mere possibility of using the interrogative mood in connection with directives to give the hearer an open proposal points to the same interest in the other part, i.e., hearer. This possibility is excluded in the two other communicative supertypes. Note that if a person receives a letter/e-mail from a sender working at an official institution or in a private company, and if this sender uses "I," then the receiver will perceive this as being very polite (this was also the intended effect). This is because the sender in this way invites the receiver into the private sphere of the sender. In other linguacultures, the same private way of writing would be considered impolite and an undesirable way for the sender to promote her−/himself. The primary focus on the hearer has the implication that the speaker is afraid of "touching" the hearer in the sense of imposing something on the hearer. The speaker so to say does not want to step on the hearer's toe. This is the shortest way to explain how Western politeness functions.

Face in the Russian Linguaculture

We shall further argue that the Russian linguaculture is oriented toward the face of the third person, the situation itself, or society's face. The speaker, the hearer, and the situation in reality make up a microsociety within the big society. This focus on the (neutral) third person is typical of reality-oriented languages. As already mentioned, Russians prefer the imperative mood, because they are focused on the solution of a problem. The problem constitutes a problem shared by the speaker and the hearer – a problem that has to be solved immediately in order to remove imbalances or obstacles in society. To employ the imperative form as a completely neutral form presupposes that the speaker and the hearer are together in solving the problem and that they prefer balance to imbalance and harmony to disharmony. Thus the imperative form itself can be argued to create contact between two people and to have a binding effect in Russian society, as odd as it may sound in the ears of people from other linguacultures, i.e., speaker-oriented and hearer-oriented linguacultures.

Face in the Chinese Linguaculture

We shall argue that the Chinese linguaculture is oriented toward the face of the first person, the speaker's face, which is a characteristic feature of all speaker-oriented languages. In the Chinese society, one cannot use the imperative outside one's home, one cannot use the interrogative as a request, and instead the speaker uses the declarative form with focus on her−/himself. This is a clear sign of the fact that the Chinese linguaculture prioritizes the speaker's face, either the speaker's moral or inner face (liǎn) or the speaker's social or outer face (miànzi). According to Hu (1944), a child is born with a full moral face, but with an empty social face. This means that from the very birth, a Chinese person tries to keep the inner face and to gain some outer face. To lose one's moral face has far more serious consequences than to lose one's social face. If one loses the moral face, then one might lose one's place in a group or in society as a whole. If one loses the social face, one loses influence, but one will still have a place within the group or within society.

Discussing the Results of Data from the Meeting Scenario
In this section, we will compare how native speakers of English and how Chinese persons speaking in their mother tongue or in English behave linguistically, when they are faced with a potential or a real conflict. Again all scholars agree that disarming is involved, but the question is how it is executed and achieved.

As already pointed out, the Chinese participants have a tendency to apply a twofold strategy: an apologetic expression in the beginning and a thanking expression in the end. Moreover, the answers clearly reveal that cancellation of an obligation is considered a face-threatening act by the Chinese group. However, before going into a detailed discussion of the data, we will try to draw attention to an important difference between Chinese and British English that has been overlooked so far.

Chinese as Image-Based and British English as Idea-Based
If we compare Ch. *bùhǎoyìsī* and Eng. *I am sorry*, nothing suggests that they should differ much from one another. That they should be similar is also confirmed by the fact that the Chinese apologetic expression is translated into the English apologetic expression and vice versa. But if we stipulate that any expression unit has an idea content which derives from the brain as well as an image content which derives from the body (see Durst-Andersen and Bentsen 2021; Durst-Andersen and Barratt 2019), we at the same time open the door to a possible difference. Although one could question the existence of idea (thoughts) and image (feelings) in connection with the ordinary lexicon, it should be obvious that apologetic and thanking expressions must involve both thoughts and feelings. The question is where the focus is.

The Meaning of bùhǎoyìsī
If we take a close look at *bùhǎoyìsī*, it literally means "it does not make good sense." However, the place where it does not make sense is not the brain, but the body. A Chinese speaker using this expression will say that she/he is not at all in balance or in harmony with her–/himself; she/he is in a state of feeling ashamed. It is a serious condition for a Chinese person. The Chinese speaker tells the hearer that "I am crying inside my body" and "It hurts inside my body." If we compare this to the meaning involved in *I am sorry*, it should be obvious that they differ fundamentally from one another.

The Meaning and Function of "I Am Sorry"
The English apologetic expression does not necessarily point to the body. It could point to the body, but in order to make it do so, one has to say *I am really, really sorry* or something like that. If one says *I am sorry*, it points to the brain of the speaker. The meaning is not "I am crying inside my body" or "It hurts inside my body," but "I hereby commit myself to the thought *I am sorry*." In a hearer-oriented

speech community, this is enough to disarm the hearer and to save the hearer's face. In Western cultures, an apology emphasizes personal responsibility for the fault or mistake (Goddard and Wierzbicka 1997; Villadsen 2014), so you need not say anything more. No British English speaker will think about her/his own face when saying *I am* sorry. The question would never pop up, if they did not do the act on purpose.

However, the speaker will be afraid that she/he hurt the hearer. The hearer's answer will often be *That is alright*, thus indicating that the hearer accepts the apology and fully understands the situation. After that, the potential conflict has been resolved. There's no harm done.

Facework in a Conflict Situation

In a Chinese setting, the speaker knows that her/his cancellation of the meeting will offend the hearer, that is to say, she/he will be hurt, become annoyed. This is also partly the Chinese speaker's concern, but the speaker's main concern is to save her/his own face. This is done by saying *bùhǎoyìsī* (see 7).

(7) 不好意思, 约翰逊先生 bù hǎo yì sī, yuē hàn xùn xiān sheng
I am sorry, Mr. Johnson.
我们因为某种原因, 今天这个会不开了 wǒ mén yīn wéi mǒu zhǒng yuán yīn, jīn tiān zhè gè huì bù kāi le
We, because of a temporary reason, have to cancel the meeting today.
麻烦你了! má fán nǐ le!
Sorry for having troubled you!

Imagine a child who comes crying to you and says "I am sorry" after having done something wrong – Would you hit them? Would you show your anger toward them? No, you would not, because when somebody literally lies down, i.e., breaks down and confesses, you are disarmed and become paralyzed. This is the effect on the hearer when the speaker says *bùhǎoyìsī*. The speaker saves her/his face by "laying down" or surrendering and by paralyzing the hearer and making her/him feel sorry for the speaker. In many cases, this will not be sufficient, because the hearer may still feel that she/he has been bothered, although she/he is paralyzed and feels sorry for the speaker. In order to remove the hearer's feeling, the speaker may say *máfán nǐ le*, literally "I troubled you" (see 7). That is to say, by using this expression, the speaker matches the hearer's feeling and succeeds in removing the hearer's negative feeling. The words make the feeling dissolve, i.e., disappear into the air.

Traditionally, *máfán nǐ le* is treated as a thanking expression, but this is questionable. To thank somebody is to put some extra weight on the hearer who has lost "weight," i.e., the hearer lacks appraisal for something she/he did. The expression *máfán nǐ le* is an apologetic expression. It means "Sorry!" By saying so, one removes some extra weight from the hearer who feels that she/he has been bothered. That is the reason why the Chinese expression has nothing to do with thanking. By using two apologetic expressions, one (*bùhǎoyìsī*) verbally oriented toward the speaker, but with a paralyzing effect on the hearer, and another (*máfán nǐ le*) verbally oriented from the speaker toward the hearer and with an effect on the hearer, the harmony inside the speaker and inside the hearer is re-established and thus also the harmony between them. No one lost face. No conflict arose. A potential conflict was avoided (see 7).

The message sequence of (7) seems to be quite in line with Scollon and Scollon's (1994, p. 135) description of Chinese conversational patterns composed of "facework + topic+ facework." Gu's (1990) maxim of self-denigration seems also to apply to our mother tongue data: first, denigrating oneself by using an apologetic expression oriented toward oneself (cf. I am sorry) and, then, elevating another person by using another apologetic expression (cf. Sorry!). Spencer-Oatey (2007) seems also to be in line with our view, because she regards emotion (feelings) as the implicit thread that links faces. In (7) we witness that emotions are the starting point and the end point for mutual face saving; the Chinese speaker verbalized her/his own feelings and the hearer's feeling of being troubled.

Chinese English Data

In the Chinese English data, we observe the influence of Chinese face culture, especially the influence of emotion-based politeness strategies deriving directly from the Chinese mother tongue. We witness *facework transfer* where the Chinese English speakers use apologetic as well as thanking-like strategies (note that *má fán nǐ le*! is always translated as "Thanks," cf. above):

(8) (a) Oh sorry about that
(be)cause my meeting is cancelled
and thank you for your support as well
(b) I am very, very sorry, Mr. Johnson
I am sorry for disturbing you for the meeting
But the meeting has been cancelled
Thanking for your coming
(c) Hello Johnson
I am sorry that the meeting is cancelled for some reason
I am terribly sorry that I booked your time, but cannot make it
Sorry and thank you again
(d) Oh Mr. Johnson, I do apologize
Looks like the meeting we got scheduled this afternoon has been cancelled
I am very sorry
You can join another one
I will let you know when next time is rescheduled

Moreover, it is clear that the apologetic expressions in Chinese English are intensified – presumably because Chinese people know that English differs from Chinese by being focused on thoughts, not emotions as in Chinese. We interpret this as an overall strategy to show the non-Chinese hearer that the speaker is emotionally affected. If this is true, then we are dealing with *semiotic transfer* where the Chinese speaker tries to compensate for things in her/his English speech that in Chinse are felt to be necessary to express in order to obtain harmony between the hearer and the speaker. It might also be the case that this intensification leaves the hearer with the impression that the speaker takes full responsibility for the cancellation, although the cause is to be found elsewhere and is outside the speaker's control. As Zhang (2019, p. 170) argues, "the use of apology intensifiers is closely related to the emotion-laden social-psychological dimension of the notion of face." What we witness is transfer of expressiveness.

Societal Logic and Types of Culture

Introducing the Problem

In the moving scenario (see Tables 4 and 5), a person had asked his friend to help him move to a new flat, because his family could not help him on that day. The friend agreed and would take one day off to help him. Unexpectedly, his family turns up. On the day before his actual moving, the speaker is supposed to call his friend to tell him that his help is no longer needed. The British respondents did not seem to have any problem with cancelling their mutual agreement. Some seemed to interpret it positively from the point of view of the hearer (9a); others seemed to be more or less indifferent to it (see 9b).

(9) a. John
Good news
I don't need you tomorrow
my whole family have just turned up
so there'll probably be more people than we need
so you can either go to work or you can have a day to yourself
b. John
Thank you so much for offering to help me move today
and I know you took the time off from work
but the thing is my family is here now
They can give me a hand
So you now have a day off to do something else
but thanks so much

In contrast to this, most Chinese respondents had trouble in cancelling their mutual agreement. They seemed to interpret the cancellation as something negative, which could have serious consequences for the speaker's and the hearer's friendship (see 10).

(10) 不好意思bù hǎo yì sī
I am sorry
我家里有人来帮忙, 不用你来了wǒ jiā lǐ yǒu rén lái bāng máng, bù yòng nǐ lái le
My family has arrived to help, so you don't need to come
谢谢你, 麻烦你了!xiè xiè nǐ, má fán nǐ le!

Thank you, and sorry for having troubled you.

These differences between the British English and Chinese data are so interesting that they need an explanation. We shall try to give one just below.

Modality in British English

In the English language, we find a sharp distinction between epistemic modality dealing with beliefs and non-epistemic modality concerned with knowledge. Non-epistemic modality consists of alethic and deontic modality. *Alethic modality* deals with knowledge of the laws of nature, while *deontic modality* is concerned with the laws of society. British English does not distinguish between alethic and deontic modality, which appears from the fact that utterances involving the modal verb *can* are ambiguous. A third person-oriented utterance such as *He can come tomorrow* is unambiguously alethic and means that *it is possible for the person to come tomorrow*. However, if the utterance is made second person oriented, one gets a deontic reading: *You can come tomorrow*. Now it involves a permission, i.e., *You have my permission to come tomorrow*. British English and all Scandinavian languages have at their disposal a lot of epistemic means (in the shape of particles, adverbials, and word order), but they do not distinguish between alethic and deontic modality as, for instance, Russian does (cf. Durst-Andersen 2019; for a general description, see Palmer 2014). Since epistemic modality concerns beliefs, whereas alethic and deontic modality concerns knowledge, we shall argue that English distinguishes between epistemic modality and non-epistemic modality, i.e., between beliefs and knowledge, but with no distinction between knowledge of laws of nature and knowledge of laws of society.

British English Societal Logic Is Based on Alethic Possibility

It is a fact that Western logicians derive all kinds of modalities from the alethic notion of possibility (represented by a diamond, $\Diamond$), i.e., possibility is $\Diamond$ (it is possible to do so), impossibility is $\neg\Diamond$ (it is not possible to do so), necessity is $\neg\Diamond\neg$ (it is not possible not to do so), and non-necessity is $\Diamond\neg$ (it is possible not to do so). They thus consider alethic possibility to be something fundamental from which one may derive other modalities. Western societies are claimed to represent individualist cultures in which the focus is on the individual person (Triandis 2018; Hofstede 1991; Trompenaars and Woolliams 2003; House et al. 2004). An individualist culture is grounded in the notion of possibility, because what is possible for someone may be impossible for someone else. This means that Western societies seem to be built on the alethic notion of possibility. In other words, the logic that applies to the laws of nature has been transferred to the logic of Western societies and has become their societal logic.

The focus on nature is evident in Western countries. People do not want to destroy nature, and they want to visit nature without losing the illusion of being inside nature itself. At the same time, Western architecture stresses the importance of building houses or buildings that nicely fit with the surroundings – the ideal being that the boundaries between nature and buildings are not visible.

If people's knowledge of the laws of nature has been used to build the rules and laws of society, it is completely understandable why people in their language do not distinguish between alethic and deontic modality. One might argue that the mixture of nature and society is reflected in the English language. The focus on the possibilities of the individual in the society manifests itself in the crucial role epistemic modality, i.e., subjective beliefs, plays in the English language. In short, we shall argue that British English societal logic derives from alethic logic based on the notion of possibility from which all other modalities are derived, i.e., impossibility, necessity, and non-necessity. If one cancels something obligatory, the result will be that it is possible not to do something ($\neg\,(\neg\Diamond\neg) => \Diamond\neg$) and a possibility is always good.

Chinese Societal Logic

The Chinese linguaculture seems to be grounded in quite another distinction. Data from the GEBCom Project involving cancellation of an obligation suggest that Chinese societal logic is built on obligation, i.e., a deontic type of modality (cf. von Wright 1951). From obligation (represented by $\Box$) all other modalities are derived, i.e., cancellation of obligation is $\neg\Box$ (it is not obligatory to do so), permission is $\neg\Box\neg$ (it is not obligatory not to do so), and prohibition is $\Box\neg$ (it is obligatory not to do so). It is important to emphasize that obligation plays a crucial role in Confucianism. Not only does this emphasis on obligation explain why the Chinese respondents use apologetic expressions and in general express that they feel uncomfortable in making something that is obligatory non-obligatory. It also explains why the Chinese society is claimed to belong to collectivist cultures (Triandis 2018; Hofstede 1991; Trompenaars and Woolliams 2003; House et al. 2004): obligation is for everyone without exception – in sharp contrast to the alethic notion of possibility which is linked to each single case and each single individual. It would also explain why permission to do something is understood as a signal to do it in Western countries, but completely lacks this appellative element in the Chinese society: *It is not obligatory not to do so* makes room for contemplation rather than immediate action. Moreover, it would explain why sights in nature are often transformed into microsocieties. The Chinese societal logic that builds on deontic modality seems to be transferred into nature, i.e., the opposite direction from that taken by Western countries, where nature is transferred into society.

This difference between British English and Chinese societal logic has many implications in our study and probably also in other areas that we haven't discussed. In the given case, we argue that the way in which Chinese English speakers perform the entire communicative act involving cancellation of an obligation can be explained by the societal logic acquired in their Chinese mother tongue and in their Chinese mother culture. This is transfer of Chinese societal logic grounded in the deontic notion of obligation which is directly opposed to the alethic notion of possibility, which makes up the fundamentals of the logic of the British society.

Russian Societal Logic

Just to complete the picture of different kinds of societal logic, we shall briefly describe the Russian variant, i.e., the third possible variant. The Russian language sharply distinguishes between alethic modality and deontic modality by having perfective and imperfective infinitives and imperatives. Outside finite forms in the present and past tense, the category of aspect signals modality. In Russian both the infinitive and imperative are employed to issue directives, and the choice between the perfective aspect and imperfective aspect is thus identical to a choice between alethic and deontic modality. In short, non-epistemic modality, i.e., knowledge, is more important than epistemic modality, i.e., beliefs. The sharp split between knowledge of laws of nature and knowledge of laws of society indicates that the Russian societal logic derives neither from the alethic notion of possibility nor from the deontic notion of obligation. The Russian society and culture seem to belong to a third variety where society itself is regarded as mixture of pure nature and pure society, i.e., a mixture of what is possible and impossible because of nature and what is permitted and prohibited by laws. We do not dare to give a name to this kind of society, but we shall argue that the notion of togetherness plays a big part in it. It presupposes the notion of individualism and the notion of collectivism and implies the notion of contact between two or more people.

The Main Results of the Reception Test

It should be mentioned that the GEBCom reception test shows exactly the same as the production test, namely, that the semiotic orientation from the Chinese participants' mother tongue is transferred to British English. This means that the communication process acquired in their Chinese mother tongue is followed when acting as a speaker and when acting as a hearer (see Table 6).

While the Chinese speaker's role is to use the Chinese grammar to produce an utterance that involves her/his experiences (output) caused by certain situation in the real world (speaker's input), the hearer's role is to use the grammar to find the information value of the output utterance (hearer's intake). In other words, the hearer is directly shown the speaker's symptom that contains her/his experiences, but she/he has to work out the signal itself that will give her/him access to the information value of the utterance. Speaker-oriented languages are characterized by having this type of communication process, viz., from a situation in reality (speaker's input) via speaker's experience (output) to information (hearer's intake). The visible/audible output that connects the speaker and the hearer defines the semiotic direction and thus the communicative supertype, i.e., Chinese is a

Table 6 The Chinese communication process

Step one	Step two	Step three
Speaker's input	**Output**	Hearer's intake
Situation	**Experience**	Information
Model	**Symptom**	Signal

Table 7 The British English communication process

Step one	Step two	Step three
Speaker's input	**Output**	Hearer's intake
Experience	**Information**	Situation
Symptom	**Signal**	Model

speaker-oriented language that communicates the speaker's experience of a situation to the hearer.

If we look at the corresponding process in British English, which is a hearer-oriented language, we find another type of process (see Table 7).

Here the British English speaker's role is to use English grammar to produce an utterance that contains new and old information to the hearer (output), while the hearer's role is to find the situation referred to (hearer's intake). Thus the hearer is given a signal (information) and has to work out the model, i.e., the situation that produced a symptom in the speaker, but was shown to the hearer through a signal. Hearer-oriented languages have this specific kind of communication process: from speaker's experience (speaker's input) via information to the hearer (output) to situation in reality (hearer's intake). The output, i.e., information to the hearer, defines the supertype, and this is the reason why English is a hearer-oriented language.

The input, however, also plays an important role, because it is the speaker, and not the grammar, who chooses what to say. The grammar frames what the speaker has named into a code that is comprehensible to the hearer. By choosing specific words, the British English speaker names her/his experience of a situation, for instance, *I feel stuck in this cab* (symptom). This input is then transformed into an output by the English grammar that is hearer-oriented, i.e., *I was stuck in a cab* (signal), which is interpreted by the hearer as a situation in which "the speaker felt that he was stuck in his cab that was caught in a traffic jam" (model) (see 4c). This means that in British English the way in which the speaker names things has the function of pointing her/his feelings, thoughts, knowledge, etc. caused by a certain situation. That is to say, naming reflects the speaker's mental state in a past or present world.

The Chinese respondents were asked in multiple ways to describe the function and meaning of various indirect and direct speech acts in English where a university professor sends e-mails to students who have sent their papers to him. All utterances involved the speaker's acceptance of the paper's general quality, but the speaker had also some reservations with respect to its degree of details:

- It's OK, but why don't you include more details?
- It's OK, but couldn't you include more details?
- It's OK, but it needs to include more details.
- It's OK, but I would probably include more details.
- It's OK, but perhaps include more details!

The Chinese respondents had no difficulty in understanding the signal itself, namely, that they were requested to include some details in their paper, although all three sentence forms were used and all three persons functioned syntactically as subjects (cf. above). Their answers to questions about their willingness to change the paper, to the politeness degree of the utterances, and to the specific speech act involved (order, suggestion, opinion, request, piece of advice, urge, experience, obligation) demonstrated that they had no clear idea of the symptom (the speaker's input) behind the signal (the output). The Chinese respondents did not seem to be aware of the fact that the many indirect ways of requesting somebody to do something in English have the function of pointing back to the speaker's mental state, i.e., the symptom behind the signal, which could be the speaker's universe of knowledge, opinions, beliefs, or various kinds of experience, such as emotions or taste. This means that in this context *perhaps* and *probably* should neither be understood in their literal sense nor be interpreted as vagueness markers or mitigating devices, which seemed to be the case for many of the non-native speakers of English. They simply have the function of pointing back to the speaker's universe of beliefs.

Thus our study shows that *It needs to include more details* points back to the speaker's universe of knowledge. By naming her/his request in this way, the speaker shows the hearer that he acts as a university professor who perfectly knows how one writes an assignment at universities. The utterance *I would probably include more details* points back to the speaker's universe of beliefs and her/his universe of experiences. When the speaker was reading the paper, he experienced a lack of details, and against this background she/he put her–/himself in the hearer's place in her/his imagined world and concluded that she/he would include more details in her/his paper. All these symptoms were perceived by the native speakers of British English, but neither the Russian speakers nor the Chinese speakers seemed to fully understand them. In the case of *I would probably include more details*, some Chinese participants regarded this utterance as belonging to the speaker's personal opinion and would not make change in the paper in contradistinction to the British English participants who all wanted to make changes.

The only natural explanation is that the Chinese respondents applied the communication process learned in Chinese where they focus on the signalling value of any utterance, where *perhaps* is interpreted as a politeness marker and *probably* is interpreted as "it is not obligatory to change the paper." They did not apply the communication process in English, where the British English hearer knows that behind any signal is an important symptom, which is indicated by the very way in which the speaker names something that is framed as a signal, in this case as a request to the hearer to include more details. In short, also here we notice *semiotic transfer* from the mother tongue to the foreign language (for a detailed analysis of all this, see Bentsen 2018).

Conclusions

As has become clear from the previous paragraphs, many of the differences between British English speakers, on the one hand, and Chinese mother tongue speakers and Chinese English speakers, on the other, can be explained by their belonging to different communicative supertypes, which yield different linguacultures. British English is a language that in all respects focuses on the hearer. Chinese is a language that in all respects focuses on the speaker. The differences occur in all areas, i.e., when persons act as speakers and when they act as hearers. Because Chinese is a speaker-oriented language and British English is a hearer-oriented language, we have witnessed transparent semiotic transfer in speech production and non-transparent transfer in speech perception.

However, semiotic transfer is a broad concept that can be specified according to the two areas being examined in this chapter, viz., problem-solving and conflict resolution – areas that are intimately connected to the culture-specific mental universe of members of a speech community.

With respect to how Chinese people solve problems when speaking English, we identified transfer of rhetorical strategies from Chinese. It was argued that the Chinese sentence-final particle ba always incorporates a hidden rule, and this rule surfaced in their English speech – presumably in order to be able to persuade the British English hearer to execute the Chinese speaker's offer. We also identified a kind of *transfer of negotiation strategy* from Chinese to English. The Chinese English speakers preferred the declarative mood and thus showed the British English hearer that the offer must be accepted by the hearer – the contract is negotiable, not imposed on the hearer as non-negotiable. Note that in British English problem-solving, the declarative mood represents a fixed proposal, while the interrogative mood constitutes an open proposal. The Chinese English use of the declarative sentence form seems to be a combination of these two varieties, since the proposal is fixed, but, nevertheless, open for negotiation and for refusal.

As to how Chinese people resolve conflicts when speaking English, we identified *transfer of facework strategy*. The Chinese English speakers tried to obtain harmony, but also tried to avoid a potential conflict. This was achieved by employing a twofold strategy in which they do not only apologize to save their own face but also thank the hearer with the purpose of establishing internal harmony in her/him. We also identified what could be called *transfer of expressiveness* from Chinese to English. Because apologetic expressions in Chinese are much more emotionally laden than their English counterparts, Chinese speakers seemed to intensify their English speech, presumably to compensate for what would be lost, if they hadn't made use of intensification. Last, but not least, we identified *transfer of societal logic*, where Chinese speakers transfer their reluctance to cancel an obligation from the Chinese societal rules to their English speech. Whether or not this transfer can be traced back to the semiotic direction of Chinese or is a mere reflection of Chinese culture remains an open question. If it turns out that it is not possible to find a speaker-oriented linguaculture that is based on alethic modality and all appear to be grounded in obligation, then it will be possible to answer this question.

All the other kinds of transfer can be traced back to Chinese being a speaker-oriented linguaculture, where the semiotic direction goes toward the speaker. This does affect not only ordinary communication where the speaker transmits a message to the hearer but certainly also problem-solving and conflict resolution, where the focus on the speaker's face and consensus between the speaker and the hearer are evident.

When a Chinese person produces semiotic transfer when speaking English, in many cases the utterances and the discourse will be something that the English hearer has never heard before, and she/he will not know how to decode them, although the utterances contain completely normal English words and have no obvious grammatical mistakes. The same will be true of all persons having a mother tongue that do not belong to hearer-oriented languages. If one ignores this fact, Global English communication will necessarily lead to miscommunication, because the information derived from the context will not be enough to avoid it, especially if one considers that communication in the globalized business world often takes the form of a short e-mail or a short text message, i.e., a written utterance produced by a person in complete isolation from the hearer and without any sign of her/his intentions or feelings. This kind of communication produces miscommunication rather than prevents it.

The results of the GEBCom Project thus question the usefulness of English as a lingua franca, because what is called Global English is not a neutral language without cultural load. Global English consists of different linguacultural varieties of English, where the semiotic direction from one's mother tongue plays a crucial role when both producing and perceiving English speech. Non-native speakers of English transfer the communication process of their mother tongue to their English speech with the result that Chinese speakers of English communicate their experiences and Russian speakers of English their situation descriptions as they would do it in their mother tongue, but nobody is aware of it, because they use English words and English grammar. Native speakers of English are convinced that non-native speakers communicate new and old pieces of information. It is often the case that one cannot find a lexical error or a grammatical mistake in their speech. The only thing that is felt is a certain pragmatic accent (an expression that is meant to cover all the various kinds of transfer mentioned just above). This pragmatic accent may even be difficult for non-linguists to distinguish from the speaker's phonetic accent. If this confusion takes place, then the focus will be on the expression side and not on the content itself. In this way, the hearers will try to comprehend what is actually said, but not try to understand what is meant by what is said. This, of course, constitutes a big problem for communication in Global English.

References

Aijmer, K. 1996. *Conversational routines in English: Convention and creativity.* London: Longman.

Aikhenvald, A.Y. 2000. *Classifiers: A typology of noun categorization devices.* Oxford: Oxford University Press.

Allan, K. 1977. Classifiers. *Language* 53: 285–311.

Beebe, L.M., and M.C. Cummings. 1996. Natural speech act data versus written questionnaire data: How data collection method affects speech act performance. In *Speech acts across cultures: Challenges to communication in a second language*, ed. S.M. Gass and J. Neu. Berlin: Mouton de Gryter.

Bentsen, S.E. 2018. The comprehension of English texts by native speakers of English and Japanese, *Chinese and Russian speakers of English as a Lingua Franca. An empirical study.* Copenhagen: Copenhagen Business School (PhD dissertation).

Bjørnman, B. 2009. From code to discourse in spoken ELF. In *English as a lingua franca: Studies and findings*, ed. A. Mauranen and E. Ranta, 225–251. Newcastle upon Tyne: Cambridge Scholars Publishing.

Blum-Kulla, S., J. House, and G. Kasper, eds. 1989. *Cross-cultural pragmatics: Requests and apologies.* Norwood: Ablex.

Bolden, G. 2017. Requests for here-and-now actions in Russian conversation. In *Imperative turns at talk: The design of directives in actions*, ed. M.-L. Sorjonen, L. Raevaara, and E. Couper-Kuhlen, 175–211. Amsterdam: John Benjamins.

Bourdieu, P. 1994. *Raisons pratiques. Sur la théorie de l'action.* Paris: Editions du Seuil.

Brown, P., and S. Levinson. 1978. Universals in language usage: Politeness phenomena. In *Questions and politeness: Strategies in social interaction*, ed. E.N. Goody, 56-289. Cambridge, UK: Cambridge University Press.

Brown, P., and S. Levinson. 1987. *Politeness: Some universals in language usage.* Cambridge: Cambridge University Press.

Charles, M.L., and R. Marschan-Piekkari. 2002. Language training for enhanced horizontal communication: A challenge for MNCs. *Business Communication Quarterly* 65 (2): 9–29.

Crystal, D. 2003. *English as a global language.* 2nd ed. Cambridge: Cambridge University Press.

Durst-Andersen, P. 1995. Imperative frames and modality. Direct vs. indirect speech acts in English, Danish and Russian. *Linguistics and Philosophy* 18: 611–653.

———. 2009. The grammar of linguistic semiotics. Reading Peirce in a modern linguistic light. *Cybernetics & Human Knowing* 16 (3/4): 38–79.

———. 2011. *Linguistic Supertypes. A cognitive-semiotic theory of human communication.* Berlin/New York: De Gruyter Mouton.

———. 2012. What languages tell us about the structure of the human mind. *Journal of Cognitive Computation, 4* [Pointing at boundaries: Integration cognition and computation on biological grounds] (1), 82–97.

———. 2019. The Russian imperative as a mirror of societal logic. *Human Being: Image and Essence. Humanitarian Aspects* 37 (2) [The Semantics of Russianness]: 73–97.

Durst-Andersen, P., and S.E. Bentsen. 2021. The word revisited: Introducing the CogSens Model to integrate semiotic, linguistic and psychological perspectives. *Semiotica* 238: 1–35.

Durst-Andersen, P., and D. Barratt. 2019. Idea-based and image-based linguacultures: Evidence from American English and Mandarin Chinese. *International Journal of Language and Culture* 6 (2): 351–387.

Durst-Andersen, P., and P. Cobley. 2018. The semiotic wheel: Symptom, signal and model in multimodal communication. *Semiotica* 225: 77–102.

Finlayson, J.G. 2005. *Habermas: A very short introduction.* Oxford: Oxford University Press.

Gao, H. 1999. Features of request strategies in Chinese. In *Working papers*, vol. 47. Department of Linguistics: Lund University.

Goddard, C., and A. Wierzbicka. 1997. Discourse and culture. In *Discourse as social interaction*, ed. T.A. van Dijk, 231–257. London: Sage.

Goffman, E. 1967. *Interaction ritual: Essays in face-to-face behaviour.* New York: Pantheon Books.

Golato, A. 2003. Studying compliment responses: A comparison of DCTs and recordings of naturally occurring talk. *Applied Linguistics* 24 (1): 90–121.

Gu, Y. 1990. Politeness phenomena in modern Chinese. *Journal of Pragmatics* 14: 237–257.

Gumperz, J. 1982. *Discourse strategies.* Cambridge: Cambridge University Press.

Habermas, J. 1998. *On the pragmatics of communication.* Cambridge: Policy Press.

Haugh, M. 2007. The discursive challenge to politeness theory: An interactional alternative. *Journal of Politeness Research* 3 (2): 295–317.

———. 2015. *Im/politeness implicatures.* Berlin/Munich/Boston: De Gruyter Mouton.

Hofstede, G. 1991. *Cultures and organizations: Software of the mind*. London, UK: McGraw-Hill.
House, R.J., P.J. Hanges, J.M. Javidan, P.W. Dorfman, and V. Gupta, eds. 2004. *Culture, leadership, and organizations. The GLOBE study of 62 societies*. Thousand Oaks, CA: Sage.
Hu, H.C. 1944. The Chinese concepts of "face". *American Anthropologist* 46: 45–64.
Ibsen, O. R. 2016. *An empirical cross-linguistic study of directives. A semiotic approach to the sentence forms chosen by British, Danish and Russian speakers in native and ELF contexts*. Copenhagen: Copenhagen Business School (PhD dissertation).
Kirkpatrick, A., and Z. Xu. 2002. Chinese pragmatic norms and 'China English'. *World Englishes* 21 (2): 269–279.
Kong, K.C. 1998. Are simple business request letters really simple? A comparison of Chinese and English business request letters. *Text-Interdisciplinary Journal for the Study of Discourse* 18 (1): 103–141.
Ku, H., and A. Zussman. 2010. Lingua franca: The role of English in international trade. *Journal of Economic Behaviour & Organization* 75: 250–260.
Larina, T.V. 2009. *Kategorija vezhlivosti i stil' kommunikacii. Sopostavlenie anglijskix i russkix lingvokul'turnyx tradicij*. Moskva: Rukopisnye pamjatniki Drevnej Rusi.
Leech, G. 1983. *The principles of pragmatics*. Harlow: Longman.
Lim, T., and J.W. Bowers. 1991. Facework, solidarity, approbation and tact. *Human Communication Research* 17: 415–450.
Mao, L.R. 1994. Beyond politeness theory: "Face" revisited and renewed. *Journal of Pragmatics* 21 (5): 451–486.
Marschan-Piekkari, R., D. Welch, and L. Welch. 1999. In the shadow: The impact of language on structure, power and communication in the multinational. *International Business Review* 8 (4): 421–440.
Mosekjær, S. 2016. *The understanding of English emotion words by Chinese and Japanese speakers of English as a lingua Franca. An empirical study*. Copenhagen: Copenhagen Business School (PhD dissertation).
Neeley, T. 2012. Global business speaks English. *Harvard Business Review* 90 (5): 116–124.
Odlin, T. 1989. *Language transfer: Cross-linguistic influence in language learning*. Cambridge: Cambridge University Press.
Palmer, F.R. 2014. *Modality and the English modals*. New York/London: Routledge.
Scollon, R., and S.W. Scollon. 1994. Face parameters in east-west discourse. In *The challenge of Facework: Cross-cultural and interpersonal issues*, ed. S. Ting-Toomey, 133–157. Albany: State University of New York Press.
Searle, J.R. 1969. *Speech acts: An essay in the philosophy of language*. Cambridge: Cambridge University Press.
———. 1983. *Intentionality. An essay in the philosophy of mind*. Cambridge: Cambridge University Press.
Song, M.L.-W. 1994. Imperatives in requests: Direct or impolite – Observations from Chinese. *Pragmatics* 4 (4): 491–515.
Spencer-Oatey, H. 2007. Theories of identity and the analysis of face. *Journal of Pragmatics* 39 (4): 639–656.
———. 2008. Rapport management: A framework for analysis. In *Culturally speaking: Culture, communication and politeness theory*, ed. H. Spencer-Oatey, 11–47. London: Continuum.
Triandis, H.C. 2018. *Individualism and collectivism*. New York/London: Routledge.
Trompenaars, F., and P. Woolliams. 2003. *Business across cultures*. West Sussex: Capstone Publishing.
Van Dijk, T. 1977. *Text and context: Explorations in the semantics and pragmatics of discourse*. London: Longman.
Villadsen, L.S. 2014. Apologi og undskyldningsretorik. In M. Lund & H. Roer (Eds.), *Retorikkens aktualitet: Grundbog i retorisk kritik* (3.udgave). København: Hans Reitzels Forlag.
von Wright, G.H. 1951. *An essay in modal logic*. Amsterdam: North-Holland.

Watts, R. 2003. *Politeness. Key topics in sociolinguistics*. Cambridge: Cambridge University Press.

Wolfson, N. 1988. The bulge: A theory of speech behaviour and social distance. In *Second language discourse: A textbook of current research*, ed. J. Fine, 21–38. Norwood: N.J.: Ablex.

Wood, L., and R. Kroger. 1991. Politeness and forms of address. *Journal of Language and Social Psychology* 10 (3): 145–168.

Zhang, Y. 1995. Strategies in Chinese requesting. In G. Kasper (ed.), *Pragmatics of Chinese as native and target language* (pp. 23–68). Second Language Teaching and Curriculum Center: University of Hawaii.

Zhang, X. 2019. *Obligation, face and facework: An empirical study of the communicative act cancellation of an obligation by Chinese, Danish and British professionals in both L1 and ELF contexts*. Copenhagen: Copenhagen Business School (PhD dissertation).

Chinese Ethnolinguistic Influences on Academic English as a Lingua Franca

38

Feng Cao

Contents

Introduction	976
Chinese English	977
Approaches to Chinese English	979
The Chinese Ethnolinguistic Influences on English Writing in Academic Settings	981
Chinese Influences on Lexico-grammatical Features of L2 English Writing	981
Lexical Bundles	982
Linking Adverbials	984
Chinese Influences on L2 English Syntax	987
Chinese Influences on Discourse Organization of L2 English Writing	988
Hedging and Boosting in L2 English Writing	993
Conclusion	997
References	998

Abstract

As the number of Chinese-speaking academics and students who use English as an additional language for academic communication keeps growing, there has been much research on the cross-linguistic/cross-cultural influences between Chinese and English in written academic discourse, particularly on how Chinese academics and students differ from their Anglophone counterparts in English academic writing. This chapter focuses on Chinese ethnolinguistic influences on academic English produced by writers from various Chinese speech communities and presents a summary of the Chinese characteristics of English L2 academic writing. The chapter begins with a brief overview of the concept of Chinese English and the approaches to studying Chinese English. Next, it discusses Chinese influences on academic L2 English in various aspects, including lexico-grammar, syntax, discourse organization, and pragmatics. In view of

---

F. Cao (✉)
Centre for English Language Communication, National University of Singapore, Singapore, Singapore
e-mail: elccf@nus.edu.sg

English as a lingua franca, the chapter ends by calling for a reconceptualization of the divergences from English L1 norms as potential features of a culture-specific English variety in academic writing.

Keywords

Academic lingua franca · Academic writing · Chinese · Chinese English · English as a lingua franca (ELF) · Ethnolinguistic · Language contact · Hybrid academic discourse

Introduction

This chapter focuses on Chinese ethnolinguistic influences on English L2 (second language) written communication for academic purposes. The term "ethnolinguistics" refers to "the relationship between language and culture, communicative practices, and cognitive models of language and thought" (Riley 2007, p. 11). The aim of ethnolinguistics is to examine such relationships between language, society, and culture. Chinese ethnolinguistic influences thus describe how Chinese social and culture norms have shaped the use and communicative practices of another language.

While much has been written about how English language has impacted on Chinese language and culture (Hu 2005; Pan and Seargeant 2012), relatively little has been discussed on how Chinese language and culture have influenced English language. Most existing scholarly work in second language acquisition focused on the language transfer (for the use of first language in a second language context, see Gass and Selinker 2008) from Chinese L1 to English L2 (Mohan and Lo 1985) or examined the differences between English L1 and English L2 through the lens of contrastive rhetoric (Connor 1996). From these perspectives, the impact of Chinese L1 on English L2 is often seen as a deficit on the part of the Chinese learners of English. According to Gass and Selinker (2008), language transfer makes a distinction between positive and negative transfer. While positive transfer refers to the facilitation of L1 in L2 learning, negative transfer refers to the interference of L1 in L2 learning. In these terms, the cross-linguistic influences from Chinese to English are largely a result of negative transfer or interference from Chinese.

Recent research on "China English" or "Chinese English," however, has begun to shift the focus away from this deficit view and shed new light on the development of the variant of English used by Chinese speakers (Xu et al. 2017b). In this chapter, such ethnolinguistic influences from Chinese language and culture on L2 English are explored through English as a lingua franca (ELF) perspective. More specifically, the Chinese characteristics in the English used by Chinese speakers are perceived as potential features of an ELF as a result of the rising status of English as a common language of intercultural communication.

Mauranen (2018, p. 8) defines ELF as "a contact language between speakers or speaker groups when at least one of them use it as a second language." As a vehicular language, ELF is often used by speakers of different L1s for different purposes, such as business communication or academic exchange. Horner (2018) introduces the concept of "written academic English as a lingua franca" (WAELF) and further argues that WAELF is a more dynamic model than the model of so-called standard written English (SWE) and shows more tolerance toward the deviations from the SWE due to the influence of writers' L1 culture and literacy. While SWE mainly dominates Anglophone academia, Chinese English in academic settings can be conceived as an alternative form of written English which is emerging from the literacy practices of Chinese-speaking academics who use English as a lingua franca for knowledge production and exchange. Such a perspective will inform the discussion of Chinese ethnolinguistic influences on L2 English academic writing in this chapter.

The chapter begins with a brief overview of the term of "Chinese English" and the differences between Chinese English and so-called Chinglish. It then discusses potential influences from Chinese language on various aspects of L2 written English, including lexical and syntactic features, discourse organization, and pragmatics. It should be noted that this chapter uses the term of "Chinese language" to refer to written Chinese rather than spoken Mandarin; and the scope of "Chinese language" includes both the simplified Chinese as used in mainland China and the traditional Chinese as used in other places such as Hong Kong and Taiwan. The chapter concludes with the implications for pedagogy and future research.

Chinese English

According to Xu, Deterding, and He (2017a), Chinese English, against the backdrop of research on World Englishes, is currently a "developing variety of English" in the sense that "features of Chinese English are yet to be systematically codified, and people's perceptions of it and their attitudes towards it are markedly divided" (p. 7). It is "a variety that has emerged as a result of language and cultural contact between 'native' varieties of English and Chinese language" and thus shows various degrees of "Chinese-specific linguistic features and cultural conceptualizations" (Ma and Xu 2017, p. 191).

The development of Chinese English, as pointed out by a number of researchers (Ma and Xu 2017; Wei and Fei 2003; Xu 2017), has started as a result of the language contact between Chinese and English and evolved through several stages toward a localized variety of English in mainland China. According to Bolton (2002, 2003), English first arrived in China in 1637 in a British trade expedition and was subsequently used between the English-speaking traders and the local Chinese in coastal areas in South China such as Guang Zhou and Macau (Bolton 2002). This early variety of English was later referred to as "Chinese Pidgin English," showing "a limited vocabulary, a reduced grammar, and a simplified phonology" in comparison with native English (Wei and Fei 2003, p. 42).

According to Eaves (2011, p. 64), an "English boom" (promotion of English learning from the late 1970s to 1990s) began in China since the late 1970s together with the development of a market economy. Radtke (2012, p. 147) also notes the existence of an "English fever" in mainland China, which is a nation-wide enthusiasm for English teaching and training after 1990. The growing importance of the role of English has "spurred the development of localized varieties, as well as the so-called 'Chinglish' phenomenon" (Eaves 2011, p. 64).

The term "Chinglish" is often used pejoratively as it is perceived as an interlanguage or learner English with interference from Chinese (Eaves 2011; Wei and Fei 2003; Xu 2017). Interlanguage is a psycholinguistic term referring to the development process in acquiring a target language (Selinker 1972). It is often a result of the negative transfer from learner's L1 or learner's overgeneralization of L2 rules. Chinglish is usually produced by poor translation from Chinese or, as Eaves (2011) claimed, is "a product of errors made by learners as they advance in fluency level" (p. 66). Eaves (2011) cited a typical example of Chinglish from a deficient translation where the Chinese expression *zhili wanju* (智力玩具) was translated as "mental toy," which is nonsensical to native English speakers (more appropriate translation would be "intellectual toy" or "educational toy"). In addition, Wei and Fei (2003) offered some typical Chinglish expressions used by Chinese learners, such as *good good study, day day up* ("好好学习，天天向上" or "study hard: do have a progressive spirit everyday") and *people mountain people sea* ("人山人海" or "very crowded").

In a summary of the existing work on the relationship between "Chinglish" and "Chinese English," Xu (2017) noted that most scholars have drawn a distinction between the two. Although both Chinglish and Chinese English show varying degrees of transfer from Chinese, Chinglish is an interlanguage and hinders communication, while Chinese English is a legitimate local English variety. In the current literature, Chinese English is viewed as an emerging variety of English with explicit Chinese characteristics at different linguistic levels (e.g., phonology, lexis, syntax, and discourse). As Xu et al. (2017a) argued, Chinese English, similar to other varieties of English such as Indian English or Singapore English, is a developing variety of English with strong influences from the Chinese language and culture.

While "Chinese English" has been used to refer to the recent development of language contact between Chinese and English, whether this developing variety should be named as "Chinese English" or "China English" has caused considerable debate (Eaves 2011; Li 2019; Xu 2017). For some, "Chinese English" is an interlanguage similar to "Chinglish," whereas "China English" is a "maturing cultural variety of English" (Eaves 2011, p. 66). Others, however, argue that there is no clear conceptual difference between "Chinese English" and "China English" as both refer to the variety of English with Chinese characteristics, but "China English" should replace "Chinese English" in naming the variety because the latter was perceived negatively by native English speakers due to its historical association with "Chinglish" (Wei and Fei 2003; see Xu 2017, for a summary of the debate). More recently, it is argued that "Chinese English" should be restored as a more appropriate term for the Chinese variety of English to be on par with other varieties of World Englishes (Li 2019; Xu 2017). Following Xu (2010) and Xu et al. (2017a), this chapter will use the term "Chinese English" for this emerging Chinese variety of English.

In terms of application, Xu (2017) points out that Chinese English is mainly used in news broadcast, Chinese-English translation, intercultural communication, as well as for constructing Chinese cultural identity. However, an under-researched area is Chinese academic English which is widely used in scientific publication or academic writing. As a large number of Chinese students and scholars need to conduct research and write for publication by employing English as a medium of communication, how their L1 language (Chinese) and culture affect their L2 academic English remains an interesting topic for Chinese English research. The next section therefore presents a focused review of the main research strands and key findings on the Chinese ethnolinguistic influences on L2 academic English produced by Chinese students and faculty in academic contexts.

Approaches to Chinese English

Chinese English has been studied from various perspectives, such as sociolinguistics, intercultural communication, and translation studies. Two main approaches, namely, World Englishes and Intercultural Rhetoric, are typically used for identifying characteristics of Chinese English. In particular, research in World Englishes is concerned with whether Chinese English has gained the status as a variety of English similar to other varieties spoken in places such as India and Singapore and how its speakers use Chinese English to construct their identities. In comparison, intercultural rhetoric approaches Chinese English with more emphasis on linguistic differences, such as how Chinese English diverges from native varieties of English in terms of rhetorical features. While both approaches are equally valid, this chapter will put more emphasis on research in intercultural rhetoric as this approach is more relevant to Chinese academic English in written communication. Below, these two approaches are introduced one by one.

World Englishes provide a general framework to understand different varieties of English around the world as a result of language contact between English and local languages. One of the leading researchers of the World Englishes, Kachru (1985), for example, developed a well-known model of classification of the varieties of English by conceptualizing them in different nation states as belonging to one of three concentric circles: "inner," "outer," and "expanding." The majority of people from the inner circle countries, such as the UK and the USA, speak English as a native language, whereas those from the outer circle countries, such as India and Singapore, speak English as a second language. In comparison, in the expanding circle countries like China and Japan, English is spoken mainly as a foreign language. From the perspective of World Englishes, Chinese English currently belongs to the expanding circle, but as Xu (2010) pointed out, it has the potential to develop into a "nativized" variety (varieties of English which have been influenced by local languages and cultures) in the outer circle. As an emerging variety, Chinese English has shown some distinctive linguistic features transferred from Chinese language and cultural norms. The presence of these Chinese features, according to Kirkpatrick and Xu (2002, p. 278), should not be seen as "deviations" from L1 English norms but as "an inevitable result" of the development of a variety of English "with Chinese characteristics."

Another approach that has looked into Chinese English is contrastive rhetoric or intercultural rhetoric in its more recent development (Connor 1996; Connor et al. 2008). Based on the assumption that language and rhetorical conventions are culture-specific, contrastive rhetoric provides "systematic analysis of the influence that one's first language (L1) and culture has on the structure of L2 writing" (McIntosh et al. 2017, p. 13). The origin of this approach can be traced back to Kaplan's (1966) pioneering work which showed differences in the organizational patterns between the English essays written by native and non-native students. He proposed that the rhetorical conventions of students' first language interfered with their second language writing in English. In particular, he argued that students from Chinese culture, as well as those from other Asian backgrounds, adopt an "indirect" or circular style of writing whereas students from Anglophone cultures write in a linear and direct style. For example, an "indirect" essay may unfold by presenting many digressive discussions around the subject but delaying the subject matter or thesis. In Kaplan's words, topics are developed "in terms of what they are not, rather than in terms of what they are" (Kaplan 1966, p. 10). This hypothesis of "indirectness" in Chinese writing was subsequently supported by other scholars (Cai 1993; Matalene 1985; Scollon 1991) and inspired extensive empirical research on Chinese influences on L2 English writing.

Contrastive rhetoric, however, has also been criticized for essentializing language and cultural patterns but ignoring variations within individual language and cultural communities (Kubota 1999; Spack 1997). To address such criticisms, Connor and colleagues (Connor 2002; Connor et al. 2008) expanded the approach and replaced the term "contrastive rhetoric" with "intercultural rhetoric" to reflect the development of this field of research. According to Connor et al. (2008), the new model is more sensitive to the social context of writing and presents a more complex and dynamic view of culture with an emphasis on discourse negotiation and accommodation. Building on the development of genre analysis (Swales 1990), intercultural rhetoric has expanded its focus from essay writing to many other academic and professional genres. This approach has shed much light on how Chinese students and academics transfer the rhetorical conventions specific to the Chinese language and culture into their English L2 writing, which will be detailed in the next section.

In terms of specific methodology, corpus linguistics has offered useful techniques for inquires in World Englishes and intercultural rhetoric. For example, it is noted that World Englishes research has made extensive use of corpus data in investigating linguistic features of varieties of English and addressing sociolinguistic issues such as language evolution (De Costa et al. 2018). Similarly, intercultural rhetoric literature is replete with analyses of corpora constructed for specific research purposes because corpus-based methods "enable multiple means of analysing huge amounts of authentic language use data, making it possible for intercultural rhetoric studies to be far more evidence-based than ever before" (Belcher and Nelson 2013, p. 2). In cross-cultural studies of written discourse, Moreno (2008) emphasized the importance of designing comparable corpora to identify similarities and/or differences in discourse structure and rhetorical features across writing cultures. The principle of comparable corpora is also applicable to researching characteristics of Chinese English in academic writing. For example, many studies reviewed in this chapter designed and built comparable corpora to investigate L1 Chinese-related influences on aspects of L2 academic English (e.g., Bychkovska and Lee 2017; Chen and Zhang 2017; Hu and Cao 2011; Lu and Deng 2019; Wu and Rubin 2000; Yeung 2019). It should be noted that where appropriate in the following sections, short excerpts of corpus data from published work will be used to illustrate the characteristic features of Chinese English. In addition, the current chapter also uses examples from NUS Learner Corpus of Academic Writing (hereafter referred to as NUS learner corpus), which is an ongoing corpus project on English L2 academic writing by advanced learners in a Singaporean university. As over 90% of the corpus data was contributed by L1 Chinese learners, a part of the project aims to determine the degree to which L2 English diverges from comparable L1 English in academic writing.

The Chinese Ethnolinguistic Influences on English Writing in Academic Settings

Over the past few decades, research in intercultural rhetoric and second language writing has described Chinese cultural impacts on English L2 writing produced by Chinese writers, particularly in academic settings. It is found that the Chinese English academic writing diverges in some aspects from English L1 academic writing in the same genre or discipline (Connor 2011). It has been further argued that such divergences are largely a result of the ethnolinguistic differences between Chinese and English (Kaplan 1966; Matalene 1985; Scollon 1991). In this section, some well-researched Chinese ethnolinguistic influences on English L2 academic writing are discussed at different linguistic levels. The following section begins with Chinese influences on the lexico-grammatical features of English L2 writing with an emphasis on lexical bundles and linking adverbials. Next, the discussion will focus on the modifying-modified sentence structure and the discourse organization of English L2 writing at the syntactic and the discoursal level, respectively. Finally, at the level of pragmatics, the section discusses the pragmatic strategies of hedging and boosting in English L2 writing.

Chinese Influences on Lexico-grammatical Features of L2 English Writing

Research in Chinese English has described various Chinese lexical features transferred into English. Lexical borrowing from Chinese, for example, has become a prominent feature of Chinese English (e.g., Yang 2009; Zhong 2019). In a corpus-based study, Yang (2009) compared Chinese borrowings in English language dictionaries published in the 1980s with those published 20 years later. He found that there was an overall increase of Chinese borrowings by 54%, and he further noted that the newly added borrowings are mostly from Mandarin rather than other varieties of Chinese (e.g., Cantonese, Amoy), are *Pinyin*-based, and are semantically from "high" cultural domains such as politics and history. Yang's (2009) findings about *Pinyin*-based loanwords and the borrowings across semantic fields are consistent with Zhong's (2019), who analyzed the top 100 Chinese loanwords in a survey of news media and among English speakers from inner circle countries. By consulting lexicographical data, Zhong found that nearly half of those Chinese borrowings are included in the entries or sub-entries of *Oxford English Dictionary Online* (*OED Online*), indicating their general acceptance in English. Moreover, Zhong showed that these Chinese loanwords are distributed in a wide range of semantic fields from daily expressions like *nihao* (你好 "hello") to terms associated with cultural, social, and political discourses such as *shaolin* (少林), *gaotie* (高铁 "high-speed rail"), and *zhongguomeng* (中国梦 "Chinese dream"). While it is observed that only 15 out of the top 100 Chinese borrowings are included in the *OED Online* with their *Pinyin* forms, Zhong (2019, p. 14) predicted that "the Pinyin spelling may be more popular and more recognizable to English users" in future.

Although the Chinese borrowings reflect growing Chinese ethnolinguistic influences on English language in non-academic contexts, relatively little attention has been paid to the transfer of Chinese lexical features in L2 academic English. A few recent cross-linguistic/cross-cultural studies, however, have uncovered some potential Chinese influences on L2 academic English in terms of lexico-grammatical features. In particular, this body of research focuses on the specific features such as lexical bundles and linking adverbials in English L2 writing produced by Chinese students and academics, which will be discussed in the following sections.

Lexical Bundles

Lexical bundles have drawn considerable interest from research in corpus linguistics, discourse analysis, psycholinguistics, second language acquisition, and others (Wray 2013). Lexical bundles are recurrent multiword sequences regardless of their idiomaticity and structural status (e.g., *on the other hand*, *the degree to which*) (Biber et al. 1999). Studies on the use of lexical bundles between L1 and L2 English writers have revealed some interesting findings regarding the transfer from L1 Chinese to L2 English.

In a corpus-based study comparing L1 Chinese and L1 English undergraduate students' use of lexical bundles in their English argumentative essays, Bychkovska and Lee (2017) found that Chinese students tend to overuse a few bundles containing vague expressions in making general reference to people and quantity (e.g., *people who do not*, *people do not have*, *more and more people*, *nowadays more and more*, *a lot of people*, *a lot of time*). They also found that the word *people* ranked ninth on the top word list in their Chinese data and believed that since Chinese culture highly values collectivist thinking, Chinese students may prefer to use more bundles with *people* in their writing. According to Paquot (2013), if a lexical bundle is more frequent in a learner's L1, it is more likely to be borrowed into the learner's L2. Bundles such as *more and more* and *a lot of* may be directly translated from the equivalent Chinese phrases *yue lai yue duo* (越来越多, "more and more") and *hen duo* (很多, "much more"), respectively, as both are commonly used expressions in Chinese. The authors therefore argued that "the abundance of such bundles in L1-Chinese students' texts could be attributed to the influence of Chinese language and culture" (p. 46).

A similar effect of transfer from L1 Chinese to L2 English is also evident in Lu and Deng (2019)'s analysis of lexical bundles in the thesis abstracts written by L1 Chinese doctoral students from Tsinghua University and their L1 English counterparts from the Massachusetts Institute of Technology. They found that Chinese students used a few bundles which rarely occurred in the American students' abstracts (e.g., *with the development of*, *with the increase of*). For example, *with the development of* was used more than five times by Chinese students than by American students. Similarly, *with the increase of* occurred only once in the American students' abstracts but was used 325 times by the Chinese students. While *with the rapid development* was used 152 times by the Chinese students, this bundle was absent from the American students' abstracts altogether. The authors also found qualitative difference between Chinese and American students in the use of the same bundle, *with the development of*. As illustrated by Example 1 below, Lu and Deng (2019) argued that whereas "development" was used by the American students to refer to "the creation of new product or procedure," the same word was used by the Chinese students to refer to "the growth or advancement of something" (p. 28):

Example 1

> THz spectroscopy has emerged as an important probe for a wide variety of systems *with the development of* pulsed THz radiation sources and time-domain detection methods. (MIT Corpus)
>
> *With the development of* computer technology, complex numerical simulations of different flow become possible. (Tsinghua Corpus) (Lu and Deng 2019, p. 28)

According to Lu and Deng (2019), the higher frequency of the bundles like *with the development of* in the Chinese students' abstracts may be a result of language transfer from the corresponding Chinese concept of "development" (*fazhan* 发展), which refers to social, economic, and technological progress in the Chinese discourse.

These findings from lexical bundle studies between Chinese English and native English showed clear influences of the Chinese language on students' academic writing in L2 English. Similar influences have been found in the use of another lexico-grammatical feature, linking adverbials, in Chinese academic English, which is the topic of the next section.

Linking Adverbials

Linking adverbials (e.g., *however, furthermore*) are important cohesive devices for connecting different units of a discourse (Biber et al. 1999). Rather than adding information to a discourse unit, the primary function of linking adverbials is to explicitly signal relationship between two discourse units. In written discourse, they can be used to mark a variety of different relationships, such as addition, contrast, and result/inference. Contrastive language analysis of linking adverbials in academic discourse has shown that these devices are likely to be influenced by the writer's first language (e.g., Appel and Szeib 2018; Granger and Tyson 1996). More specifically, previous research has shown how Chinese academic writers used linking adverbials in constructing their English texts and the extent to which their use of these devices differed from native English writers or writers from other language backgrounds. This line of research finds that Chinese writers share a few general tendencies in using linking adverbials in L2 English writing, including the preference for informal linking adverbials (e.g., *besides, what's more*) and overuse of particular linking adverbials (e.g., *on the other hand*). These general patterns of usage will be presented below.

Informal Linking Adverbials A general finding in the L2 English writing of Chinese students is that many Chinese student writers tend to choose linking adverbials from informal registers, thus rendering their writing speech-like (Field and Yip 1992; Leedham and Cai 2013). Following an analysis of linking adverbials in the written assignments of Chinese and British undergraduates in four UK universities, Leedham and Cai (2013) identified three overused items (*what's more, besides, last but not least*) by Chinese students to be "incongruous with the expected formality of academic writing" (p. 378). The students' use of these informal linking adverbials is illustrated below in Examples 2 and 3. As shown in Example 2, *what's more* is used here to indicate an additional point to the previous statement. However, with a contracted verb form, *what's more* is considered to be more common in speech rather than in writing and can be replaced by *furthermore* or *in addition*. The linking adverbial *besides* appears to be another informal item which is particularly favored by the Chinese student writers (Leedham and Cai 2013; Yeung 2009). As Example 2 illustrates, *besides* is often used by Chinese undergraduates as an additive connector in their academic English writing. This usage, according to Yeung (2009), is stylistically inappropriate because there is evidence that *besides* is used more frequently in informal registers in native English data, such as in dialogues and monologues of narratives. More formal substitutes for *besides* should be used in academic texts similar to Example 2, where *moreover* might be more appropriate. As Wong (2018) points out, the logical connector *moreover* is not only stylistically more formal than *besides* but also primarily used in arguments to support a conclusion.

Example 2

> In my opinion, this essay is well-organized because the author successfully provides three sub-points to support his main argument. He clearly states that the flawed methods,

ignorance of diversity and inequality caused by internationalization make rankings' disadvantages dominate. *What's more*, with the appropriate citations, the author offers the effective explanations which are taken from source texts, thus incorporating sources and his ideas properly. *Besides*, the evidence he provided is closely correlated to each supporting view in each paragraph. (NUS learner corpus)

Another linking adverbial, *last but not least*, is also rarely used by L1 English writers in formal academic writing. In Example 3, while *last but not least* is used to signal the conclusion part of an argument, the more formal expression such as *finally* or *in conclusion* might be used instead. Notably, the use of *actually* and the sentence-initial *but* has added to the degree of informality:

Example 3

Last but not least, many people argue that when universities focus on rankings, they may forget how important teaching is. *But actually*, it may push both teaching and research at the same time. (NUS learner corpus)

The use of informal linking adverbials, as demonstrated above in the L2 English essays of Chinese undergraduates, is also found in the English writing of more advanced Chinese learners. For example, Chen (2006) identified the inappropriate use of *besides* in the final papers submitted by Taiwanese master's students, and Lei (2012) reported similar findings in the doctoral theses written by mainland Chinese PhD students. The use of such informal linking adverbials, as Chen (2006) pointed out, "gives an unintended colloquial tone to the academic paper" (p. 124). The speech-like writing style in the Chinese students' English writing is possibly a result of the influences of teacher language, textbooks, and classroom instruction in Chinese contexts (Chen 2006; Field and Yip 1992; Liang et al. 2019). When one student participant in Liang et al. (2019)'s study reflected on the use of *last but not least* in her writing, she reported that she picked up the phrase from her English language class in China where the teacher emphasized the importance of cohesion and provided focused instruction on conjunctions. In her words, "[I]f I do not use conjunctions in my writing, I will lose marks" (p. 44).

Overuse of Linking Adverbials Apart from their informality in register, all three linking adverbials (*what's more, besides, last but not least*) in the above examples were found to be overused in the Chinese students' writing (Leedham and Cai 2013), which appears to be another characteristic of the linking adverbials in Chinese English. Past research showed that in L2 English academic texts, Chinese student writers tend to overuse specific linking adverbials as compared with their L1 English counterparts (Appel and Szeib 2018; Chen 2006; Gardner and Han 2018; Lei 2012; Leedham and Cai 2013). For example, it is well-documented that *on the other hand* is overused by Chinese writers (Field and Yip 1992; Gardner and Han 2018; Leedham and Cai 2013). Below are several typical uses of this linking adverbial in L2 English writing of Chinese students:

Example 4

From those figures we can see on one hand more experiment data results in a more accurate profit kernel density, indicating a better estimation of the underlying choice model. *On the other hand*, the performance difference between the two figures indicates that our approach is able to identify the true underlying consumer's choice model and then achieve the true optimal profit. (NUS learner corpus)

Example 5

Hence, students must compete.... And, *on the other hand* parents will put pressure on their children.... (Field and Yip 1992, p. 25)

Example 6

Firstly, as dividends and tax liabilities are cash transactions, there are risks that IHG would be incapable to pay the proposed dividends to shareholders. *On the other hand*, it also implies that there would be financial problems for IHG to repay the amounts owning in the short term to their suppliers. (Gardner and Han 2018, p. 874)

As shown in Examples 4, 5, and 6, while in each case a linking adverbial may be used to signal additional relationship (e.g., *moreover*), in each case *on the other hand*, a marker of contrastive relationship, was used instead. In Example 4, *on the other hand* was used together with *on one hand* to form a pair of adverbials equivalent to the literal translation of the Chinese transitional phrases "一方面" (*yi fang mian*, "for one thing") and "另一方面" (*ling yi fang mian*, "for another thing") (Field and Yip 1992; Gardner and Han 2018; Leedham and Cai 2013). According to Li (2015), the pair of Chinese transitions are commonly used to express similarity rather than contrastive relations between linguistic units. Thus, the overuse of *on the other hand* among Chinese students is likely influenced by the students' reading and writing practices in Chinese language. Examples 5 and 6 show that *on the other hand* often occurred in tandem with other signal words like *and* and *also* in context, which highlights additional argument or discourse unit. This usage, as Gardner and Han (2018) argued, is "not only superfluous as the relationship between the sentences is already indicated by '*also*' and '*and*', but it is rather misleading, as readers are looking for a contrast" (p. 874).

In a similar fashion, Chinese academics are also found to overuse particular linking adverbials in their published English research articles when compared with their native English-speaking colleagues (Gao 2016). For example, it was found that *meanwhile* was overused by Chinese writers in clause-initial positions to mark additive relationship, whereas their English L1 counterparts used it to show simultaneity of two events. Examples 7 and 8 below illustrate the typical uses among the Chinese scholarly writers:

Example 7

Employing the Mulliken analysis, a C-Cu(Pd) bound was found between CO and metal atom, suggesting the chemisorption of CO on the surface. *Meanwhile,* we notice that upon absorption, the bond distance of CO is slightly extended compared to that in gaseous state. (Gao 2016, p. 22)

Example 8

> The results indicate that tilting direction tends not to influence user performances. This is similar to the study findings with the Tilt Menu. *Meanwhile*, our results show that tilting actions cause incidental co-variations in the panning and pen tip movement. (Gao 2016, p. 22)

As the above examples show, the Chinese academics used *meanwhile* to report additional findings rather than what happens at the same time. Instead, *furthermore* or *moreover* might have been used to indicate such additive relationship. Similar to *on the other hand*, the overuse of *meanwhile* by the Chinese academics can be seen as another example of the transfer from Chinese language (Gao 2016). The Chinese literal translation of *meanwhile*, i.e., "同时"(*tongshi*), can be used to as either an adverb or a linking adverbial in a Chinese context. While it signals simultaneity (at the same time) as an adverb, the same expression marks additive relationship (moreover) as a linking adverbial. It is likely that the linking adverbial use of *meanwhile* in Chinese has been carried over into English L2 writing. Such influences from the Chinese language, together with a lack of awareness of the semantic meaning of *meanwhile* in English, probably have resulted in the overuse of this linking adverbial (Gao 2016).

Chinese Influences on L2 English Syntax

One basic principle of Chinese rhetorical organization operating at sentence level, according to Kirkpatrick and Xu (2012), is the modifying-modified sequence. In Chinese *pianzheng fuju* (偏正复句), namely, complex sentences, the modifying clauses or subordinate clauses are followed by the modified clauses or the main clauses. Drawing from the theories of Chinese linguistics, Kirkpatrick and Xu (2012) argued that the modifying-modified sequence or subordinate-main clause is the unmarked, natural order in Chinese, whereas the marked order (modified-modifying or main-subordinate clause) is used for giving prominence or emphasis in style.

To verify whether the modifying-modified sequence of Chinese is transferred into L2 English, Jiang (2017) investigated four types of adverbial clauses, temporal, conditional, concessive, and causal (beginning with subordinators *when*, *if*, *although*, and *because*, respectively), in 188 English essays written by Chinese graduate students. Although the adverbial clauses can occur at the initial, medial, or final position in complex sentences, the study found that 61.3% adverbial clauses with subordinators of *although*, *if*, and *when* were in the initial positions whereas 38.7% were in the final positions. The sentences in Example 9 illustrate, respectively, how Chinese student writers placed concessive, conditional, and temporal adverbial clauses in the sentence initial positions.

Example 9

1. *Although* many evidences show that it might be a mistake and cruelty giving high pressure to the kids when they are very young, I think it is true for elementary school students to have English courses.
2. Especially in the intense competition of modern society, *if* you have accepted good education, you will have more chances to get a job than others.
3. *When* I was a child, I had the closest relationship with my family. (Jiang 2017, pp. 98–99, with errors unedited)

Jiang (2017) attributed the tendency for Chinese students to place adverbial clauses in sentence-initial position to the influence of Chinese syntactic features. This finding is supported by Wang's (2002) analysis of adverbial clauses in spoken and written Chinese, where the temporal, conditional, and concessive clauses typically precede the main clauses. For example, Wang reported that in her written data, more than 95% of the cases of temporal, conditional, and concessive clauses occur before their main clauses. As Example 10 illustrates, the initial conditional clause is used to establish an optional situation frame for the following assertion:

Example 10

Daxue qiuxue qijian, kongpa shi yige ren yisheng zhong dui rensheng yiyi de tanqiu, dui jingshen shenghuo de xiangwang, zui qiangsheng de shiqi. Ruguo zhe shihou you xing du dao yiben hao shu, zhe ben shu keneng hui yingxiang yi ge ren de xin lu licheng.

("College life would be the strongest phase for a person to explore the meaning of life and long for a spiritual lifestyle. If he is lucky enough to read a good book at this time, the book may influence his mental processes for life.") (Wang 2002, p. 155)

As Huang (2010, p. 155) noted, the preference for sentence-initial adverbial clauses in Chinese discourse is different from English discourse conventions because "Chinese does not provide the main statement at the beginning, but instead prefers to delay it until after background information has been given." Jiang (2017, p. 105) argued that such Chinese-specific features "may serve as evidence of nativization of English in the Chinese context at the syntactic level."

Chinese Influences on Discourse Organization of L2 English Writing

According to Kirkpatrick and Xu (2012), Chinese language and culture have strong influences on the discourse organization of the L2 English writing produced by Chinese writers. A most significant aspect of Chinese influences, as previous research shows (Kaplan 1966; Cai 1993; Yang and Cahill 2008; Wu and Rubin 2000), is the discourse organization of argumentative/persuasive writing (*yilunwen* 议论文). The genre of argumentative writing appears to be culturally sensitive, and Chinese and English languages differ in what counts as a persuasive argument and in how to formulate the argument effectively (Liu and Du 2018). It is generally found in previous cross-cultural research on argumentative writing that Chinese discourse tends to be more "indirect" or "circular" than English discourse and such indirectness has been transferred into the English texts produced by Chinese students (Kaplan 1966; Yang and Cahill 2008; Wu and Rubin 2000).

In one early contrastive rhetoric study, Kaplan (1966) posited that "Oriental" writing, specifically those of Chinese and Korean, are marked by indirectness in rhetorical structures, whereas L1 English composition follows a linear development. In terms of paragraph organization, according to Kaplan, a writer "usually begins with a topic statement, each supported by example and illustrations, proceeds to develop that central idea and relate that idea to all the other ideas in the whole essay" (1966, pp. 4–5). By contrast, writers of L2 English often compose in a circular way by "turning and turning in a widening gyre" and "the circles or gyres turn around the subject and show it from a variety of tangential views, but the subject is never looked at directly" (Kaplan 1966, p. 10). Kaplan's indirectness hypothesis offered a useful starting point for comparing different rhetorical conventions across cultures and was supported by other scholars such as Matalene (1985), Scollon (1991), and Cai (1993) who similarly found distinctive rhetorical patterns in L2 English writing of Chinese students, such as delayed thesis, digressive discussion, and frequent appeals to history and authority. For example, Matalene (1985) analyzed the structure of an argumentative essay written by one of her Chinese students and found that the author began by rhetorical questions and narrative examples and continued with historical background before arriving at the thesis in the third paragraph. To Matalene, holding off the thesis suggests indirectness in essay development, and she claimed that "to be indirect in both spoken and written discourse, to expect the audience to infer meanings rather than to have them spelled out is a defining characteristic of Chinese rhetoric" (1985, p. 801). Likewise, Cai (1993) argued that Chinese written academic discourse has always been conditioned by the dominant social and political ideologies in Chinese culture. Moreover, he showed evidence that Chinese students are likely to apply rhetorical strategies typical of Chinese academic discourse in their L2 English writing, such as the avoidance of personal views and emotions and the preference for an "indirect" approach in topic development.

The line of discussion has inspired widespread interest in comparing rhetorical conventions between Chinese and English academic writing. Much empirical work (e.g., Chien 2011; Hsiung 2014; Ji 2011; Jia and Cheng 2002; Wu and Rubin 2000; Yang and Cahill 2008 to name just a few) has been done to determine whether Chinese is more "indirect" than English in terms of discourse organization, as well as whether the rhetorical style of Chinese L1 writing has been transferred to English L2 writing. It is noted by some that Chinese rhetoric prefers an "indirect" rhetorical structure, particularly concerning the position of the thesis statement, which has influenced English L2 writing in academic contexts (Hsiung 2014; Ji 2011; Jia and Cheng 2002). In more specific terms, Chinese writers are more likely to postpone the thesis statement in their writing, which appears indirect and incoherent to L1 English readers. For example, various degrees of "indirectness" have been observed in the published research articles of Chinese academics in a cross-linguistic comparative study (Hsiung 2014). By analyzing the placement of thesis statements in

600 paragraphs extracted from 60 Chinese and 60 English journal articles in the field of psychology, Hsiung (2014) found that while English articles overwhelmingly placed the thesis statement in the first sentence of a paragraph, Chinese articles displayed greater variation in the placement of thesis statements. The analysis showed that whereas about 97% of the English paragraphs place the thesis statement in the first sentence, only about 3% place the thesis statement in the middle or toward the end of the paragraph. In comparison, although 56% of the Chinese paragraphs placed the thesis statement in the first sentence, the remaining Chinese paragraphs placed the thesis statement between the second and the last sentences. Hsiung thus concluded that the position of thesis statement in Chinese paragraphs appeared to have more flexibility than in English paragraphs and a large proportion of Chinese paragraphs followed an "indirect" approach in organization.

The "indirect" feature of Chinese rhetoric seems to have also transferred to English L2 writing of Chinese students. For example, in an analysis of 46 expository English essays written by prize-wining Chinese university students who participated in a college writing competition, Jia and Cheng (2002) found the majority of the writers (31 out of 46 essays) adopted an "indirect" approach by organizing the discourse inductively and delaying the thesis statement to the end. The authors claimed that the "indirectness" in Chinese English is by no means accidental but a result of influence from Chinese cultural patterns. This finding is supported by Ji's (2011) analysis of 26 English texts written by Chinese undergraduate students from Tsinghua University, where it was found that a third of the student English essays displayed rhetorical features of "indirectness," such as delayed thesis statement and digressions from main argument. Ji noted that an "indirect" essay typically "starts from something peripheral, moves from surface to core, and only gradually gets to the point" (2011, p. 81).

Interestingly, the pattern of "indirectness" in Chinese writing and Chinese English writing has found limited support in other studies. Wu and Rubin (2000), for example, compared the essays written by 40 Taiwanese with 40 US college students and found although the Chinese writers tended to be more "indirect" than the native English writers in discourse organization, half of them followed a direct model as their American counterparts. More importantly, no significant differences were found between L1 Chinese and L2 English in the placement of the thesis statement, suggesting that there is no cross-language transfer concerning indirectness. In other words, Chinese L1 writing seems to have little effect on the English L2 writing of the Chinese writers in discourse organization. Similarly, in a comparative analysis of 80 Chinese L1 essays, 103 English L2 essays, and 72 English L1 essays in the university setting, Yang and Cahill (2008) reported that both Chinese L1 and English L1 writers preferred a direct way of organization despite showing variability in degrees. In terms of the influence from Chinese on English, they found little variation in the placement of thesis statement between Chinese L1 writing and English L2 writing, suggesting no transfer between Chinese and English. The absence of influence from Chinese on English in relation to position of thesis statement is also reflected in high school students' essays. Following an analysis of two sets of essays written in Chinese and English classes in a Taiwanese high school, Chien (2011) showed that the majority of students adopted a direct approach in both their Chinese and English essays. In other words, most students put the thesis statement in the beginning rather than the middle or conclusion of the essay. Although a few remaining students used an indirect approach in their Chinese essays, they shifted to a direct approach in their English essays, which suggests that their English writing was not influenced by their Chinese writing.

While the findings on "indirectness" in L2 English writing appear inconclusive, scholars often attributed such characteristic in Chinese rhetoric and Chinese English to the influences from classical Chinese rhetorical styles. Following an analysis of four English essays written by Chinese university students, Kaplan (1972) ascribed the indirectness of the Chinese English writing to the impact of the traditional Chinese writing style called *baguwen* (八股文) or "the eight-legged essay." The eight-legged essay was a regulated style of writing in Chinese civil service exams during the Ming (1368–1644) and Qing (1644–1911) dynasties (Krikpatrick and Xu 2012). Test candidates must follow strict rules of the composition by structuring their texts in accordance with different components or "legs." The *baguwen* is characterized by rhetorical features such as "elaborate parallelism and harmonious contrast," and it advances the argument by giving indirect answers to the rhetorical questions (Wu and Rubin 2000, p. 153).

Although scholars like Kaplan (1972) and Scollon (1991) argue that the *baguwen* has exerted influences on L2 English writing of Chinese university students, Kirkpatrick (1997) disagreed because (i) it is a difficult and complex model for contemporary Chinese students to master; (ii) it has been rejected by contemporary Chinese scholars due to the imperial associations; and (iii) it has come to an end after the Qing dynasty and was replaced by a new vernacular style of writing initiated during the May Fourth Movement in 1919. Likewise, You (2005) maintained that traditional Chinese rhetoric in fact shared similar rhetorical values, such as originality and directness of discourse, as Western rhetoric. He argued that the introduction of Western rhetoric into China has enriched traditional Chinese rhetoric, and together they have influenced modern Chinese writing. Clearly, apart from traditional Chinese rhetoric, other factors may also have played a part in shaping the discourse organization of the English writing of Chinese students. For example, according to Ji, the "indirectness" can be attributed to multiple sources, such as the influence of Chinese textbooks in secondary schools, the requirement of *gaokao* (高考), or the Chinese national college entrance exam, and the Chinese sociocultural norms.

Besides the placement of thesis statement, another feature of the "indirectness" of Chinese English seems to be related to the style of reasoning. Liu (2005), for instance, compared online instructional materials of argumentative writing in Chinese and English and found some fundamental differences between the emphases of the Chinese and the English materials. While the English instructional materials emphasized the need to address the opposition in argumentative writing, the Chinese materials focused on analogy (to make use of a parallel case in argument) and dialectical argument (to consider issues from multiple perspectives and reconcile seemingly contradictions) (Liu 2005). As a result, it can be expected that English argumentative writing tends to be more "direct" in anticipating the opposite positions, whereas Chinese argumentative writing tends to be "less direct" by relying on dialectical argument. In other words, Chinese argumentative writing stresses on considering both sides of an issue and attempts to reach a conclusion by resolving the conflict between the opposites.

If such differences exist between Chinese and English writing instructions, it is natural to ask whether Chinese rhetoric follows a different logic in composing argumentative essays. Liu (2009) examined the notion of "logic" (逻辑 *luoji*) in contemporary Chinese scholarly work on rhetoric and pedagogical materials from a comparative perspective. It was found that both inductive logic (from specific cases to a general conclusion) (归纳, *guina*) and deductive logic (from general premises to a specific inference) (演绎, *yanyi*), in addition to dialectical logic (a balance between opposites) (辩证, *bianzheng*), co-exist in scholarly discussion of rhetoric, but a strong emphasis has been put on dialectical logic in the contemporary Chinese language textbooks for secondary students. For example, in a relevant section of the textbooks published by People's Education Press (an authoritative publisher for primary and secondary textbooks in mainland China), students are taught to look at issues from different perspectives and to analyze things "in an all-around manner and avoid one-sided, extreme views" (Liu 2009, p. 103). The instructions make explicit reference to dialectical thinking, which attempts at a reconciliation of contrasting views and seeks a "middle" way. Example 11 is taken from the first paragraph of an argumentative essay written by a Chinese student, which exemplifies how dialectical argument is used.

Example 11

> It is a very common phenomenon that more and more people, especially children, play video games in their spare time as mentioned in text A. Thus, the writer holds the view that video games have bad effects. But the author of text B shows an opposite view that video games are good for our body and mind. The debate between text A and text B is whether video gaming is harmful or beneficial. Personally, I believe that video games are like coins which have two sides. The main point is not in itself, but in the way how to use it
> ...
>
> In summary, video games have drawbacks as well as merits. From my point of view, I'm not entirely convinced of one-sided argument. (NUS learner corpus)

Although the prompt requires the student writer in Example 11 to adopt an either-or stance toward the topic, namely, playing video games is either beneficial or harmful, the student appears to transcend a one-sided view and instead has taken a dialectical stance in positioning herself. Following the opening paragraph, she continues to discuss both advantages and disadvantages of playing video games in the body parts of the essay and reiterates her stand at the end of the essay by arriving at a compromised conclusion.

As illustrated above, the use of dialectical logic is not uncommon among Chinese students' English writing. As noted by Liu (2009, p. 104), "dialectical logic holds a special place, often unrecognized by Westerners, in theory and practice in contemporary Chinese rhetoric and composition." Thus, the deep-rooted influences of dialectical logic on contemporary Chinese literacy practices may have predisposed Chinese students to the preference for dialectical argument.

Given such a strong emphasis on dialectical argument in Chinese literacy, it is expected that the style of dialectical reasoning would be carried over to L2 English writing of Chinese students. To verify this hypothesis, Yeung (2019) examined the use of dialectical model by comparing 30 Chinese L1 academic articles published by scientists from mainland China and 30 English L1 academic articles published by native English-speaking scientists on the same topic (human cloning research), as well as 30 English L2 argumentative essays written by Chinese university students in Hong Kong. The study showed that while L1 English articles consistently developed the argument in a linear way and adopted a polemical model of argument, a significant number of L1 Chinese articles used "a dialectical style which examines opposing positions without taking sides and yet rising above them to resolve conflicting issues" (p. 29). Similarly, a significant proportion of students used dialectical argument with dualistic conclusions in their L2 English essays, although less frequent than experts. Yeung (2019) pointed out that this dialectical style of argument demonstrated in the students' L2 English essays is significant, because these L2 English learners in Hong Kong "adopted such a style without any formal dialectical teaching" (p. 52), and despite the fact that they were taught the English linear style in their curriculum and exam training. On the whole, these findings indicate that the dialectical logic in contemporary Chinese rhetoric appears to have influenced the discourse of L2 English writing of student writers.

Hedging and Boosting in L2 English Writing

At the level of pragmatics, a notable characteristic of Chinese English in academic discourse is related to "hedging" and "boosting." According to Hyland (1998), hedging and boosting are pragmatic strategies for reducing or increasing the propositional strength. In academic discourse, they not only help the writer to express an appropriate degree of doubt and certainty toward the content but also signal the writer's interpersonal relationship with the reader.

Hedges, such as *possibly*, *may*, and *likely*, convey the writer's tentativeness in making an explicit claim and indicate the writer's humility and deference to colleagues, whereas boosters, such as *clearly* and *undoubtedly*, express the writer's confidence and conviction toward the statements. Together, hedges and boosters constitute two important rhetorical resources in constructing knowledge and establishing writer stance in academic discourse. As writing is a culturally bound activity, considerable research has been done on how hedging and boosting may differ across different languages and cultures (e.g., Fløttum et al. 2006; Loi and Lim 2013; Vassileva 2001; Vold 2006). Comparative studies between Chinese and English scholarly writing have shown that Chinese L1 writers appear to display more certainty in tone than English L1 writers, whereas the latter sounds more cautious and tentative. For example, in a corpus-based study of hedges and boosters in 649 article abstracts from 8 prestigious L1 Chinese- and L1 English-language journals in applied linguistics, Hu and Cao (2011) discovered that L1 Chinese writers used significantly fewer hedges (on average 0.84 hedge per abstract) than their L1 English counterparts (on average 2.04 hedges per abstract). Similarly, in a cross-cultural study of metadiscourse in 40 research articles published in L1 Chinese and L1 English, Mu et al. (2015) reported that L1 Chinese writers used more boosters than expected (observed count 288 > expected count 272, $p < 0.001$), whereas the Anglo-American writers used more hedges in comparison (observed count 1618 > expected count 1376, $p < 0.001$).

Examples 12 and 13 from Hu and Cao's (2011) corpus data illustrate the typical pattern of hedges and boosters in L1 English and L1 Chinese abstracts (hedges are boldfaced and boosters are underlined). Example 12 is authored by a L1 English researcher in applied linguistics, whereas Example 13 is authored by a L1 Chinese academic in the same discipline. As can be seen, the author of Example 12 used both *suggest* and *may* as hedges to offer a cautious interpretation of research results. By contrast, the author in Example 13 not only avoided hedges but used boosters such as 表明 ("show") and 指出 ("point out") to strengthen the force of the conclusion, thus leaving little space for negotiation.

Example 12

> The Mainland Chinese IPA-trained participants performed better than both the Hong Kong participants and the Mainland Chinese non-IPA-trained participants in initial phoneme deletion. However, both Mainland Chinese groups outperformed the Hong Kong group on a phoneme-grapheme nonword matching task. This pattern of results **suggests** that phonemic awareness in Chinese L1 readers of English is not simply an effect of orthography, but rather, **may** be interpreted in terms of access to explicit demonstration of phonemes. Further, tests carried out in L2 which are intended to assess metalinguistic awareness **may** be susceptible to artefacts introduced by the participants' L1 spoken language. (Hu and Cao 2011, p. 2803)

Example 13

> 本文报告了一项实证研究--中国英语学习者的二语预制词块识别能力和二语水平之间的关联性。研究结果表明:预制词块识别能力和二语水平之间具有明显的关联性。同时,材料难度也是制约学习者词块识别能力的重要因素,尤其对二语水平较高的学习者的词块识别能力影响较大。文章最后指出:二语预制词块识别能力不能自动生成,即使是二语水平较高的学习者,其预制词块识别能力也必须有意识地加以系统训练才能获得.
> (The paper reports an empirical study on the correlation between Chinese L2 learners' ability to identify L2 prefabricated chunks and their L2 proficiency. The findings show that there is an obvious correlation between the two. Meanwhile, the material's degree of difficulty is also an important factor inhibiting learners' ability to identify L2 prefabricated chunks, especially that of comparatively advanced learners. Finally, the paper points out that L2 learners cannot develop the ability to identify prefabricated chunks by themselves, and even for comparatively advanced learners only systematic training can enable them to acquire the ability.) (Hu and Cao 2011, unpublished corpus data)

The variation in the hedging and boosting strategies between Chinese and English academic discourse can be attributed to the ethnolinguistic influences on the literacy practices between different cultures (Hu and Cao 2011; Hu and Wang 2014). Chinese literacy practices view language as a vehicle of transmitting knowledge rather than a means for constructing knowledge. Therefore, knowledge and truth are believed to be self-evident, and "verbal debate and argumentation are not meaningful tools for understanding truth and reality" (Peng and Nisbett 1999, p. 747). This belief about the reduced role of language in knowledge-making can be reflected in Confucius' words, such as "I transmit but do not innovate; I am truthful in what I say and devoted to antiquity" (述而不作, 信而好古; p. 57), and his belief that "it is enough that the language one uses gets the point across" (辞达而已矣; p. 159) (author Confucius 1983). These well-established cultural beliefs and views of knowledge-making practices have cohered well with the modern influences from scientism since it was introduced to China during the early twentieth century and embraced enthusiastically by Chinese intellectuals (Kwok 1965). According to Ouyang (2003), scientism is "the idealist belief in the power of science to modify society using the methods, values, and ideas underlying science" (p. 117). In other words, "science" was worshiped and regarded as the ultimate solution of all problems to Chinese society (Hu and Cao 2011). The combined beliefs about Chinese literacy and language, as well as the influences from scientism, have shaped the rhetorical and discursive practices in modern Chinese academic discourse in general and in knowledge-making practices in particular. As a result, such beliefs and practices may call for a higher degree of conviction and confidence in representing knowledge as certain and uncontested in Chinese academic texts.

In contrast to the traditional Chinese beliefs, the awareness of the role of language in knowledge construction has been well entrenched among the Anglo-American intellectual communities (Hu and Wang 2014). For instance, Fairclough (1992) argues that discourse has "the constructive effects" on "social identities, social relations and systems of knowledge and belief" (p. 12). Wierzbicka (2006, p. 247) also points out that the frequent occurrence of epistemic features in English language can be linked with "the post-Lockean emphasis on the limitations of human knowledge, on the need to distinguish knowledge from judgment." In addition, research on sociology of science has increasingly recognized the role of language and discourse in constructing scientific knowledge (Latour and Woolgar 1986). Even though scientism has its origins in western intellectual traditions, there has been a growing rejection of such epistemology in contemporary Anglophone academic cultures (Hu and Wang 2014). These developments have underpinned the literacy practices of contemporary Anglo-American societies, which may account for a higher degree of tentativeness and circumspection in negotiating and constructing knowledge in English academic discourse.

Research in intercultural rhetoric also shows that the culture-specific features of hedging in Chinese academic discourse have transferred into L2 English by Chinese writers. For instance, Hu and Cao (2011) found that L2 English abstracts used significantly fewer hedges than L1 English abstracts but showed no difference from L1 Chinese abstracts from the same Chinese-language journals. These findings indicate a transfer of a pragmatic feature from L1 Chinese into L2 English. Likewise, Chen and Zhang (2017) compared hedging strategies in the conclusion section of the published applied linguistics papers written in L1 and L2 English between

Anglophone and Chinese writers, reporting significantly more frequent hedges employed by L1 English writers than by L1 Chinese writers, especially in the subcategory of "*I/we* +non-factive verb" such as *I/we suggest* and *I/we argue*, which are illustrated in Example 14 below:

Example 14

1. ...we **would suggest** that further research investigate its effect on other linguistic error categories.
2. In this article, I have **argued** that tails are a surprisingly consistent and durable feature of spoken English... (Chen and Zhang 2017, p. 12)

Chen and Zhang (2017) argued that the Chinese writers' avoidance of such type of hedges in English writing is influenced by collectivism in Chinese culture. Based on interviews with their Chinese informants, Chen and Zhang (2017, p. 16) claimed that "Chinese academic writers, due to the influence of collectivism in Confucian philosophy, give prominence to authors' group attribute and weaken their individual characteristics."

Chinese influences on hedging and boosting are also evident in student writing in L2 English. For example, in a corpus-based study comparing hedges and boosters in the English academic essays written by L1 Chinese students in Hong Kong high schools and their English counterparts in Britain, Hyland and Milton (1997) found L2 English students used 60% more boosters than their L1 English counterparts where the latter used 73% more hedges than the former. Thus the writing of L2 English students shows characteristics of "firmer assertions, more authoritative tone, and stronger writer commitments" as compared with the discourse of L1 English students. These findings are generally consistent with other studies on hedging and boosting in university students' academic English writing. Lee and Deakin (2016), for example, in a corpus-based investigation of interactional metadiscourse in successful and less successful English argumentative essays by L1 Chinese and the highly rated essays by L1 English university students, reported that both L1 English essays and successful Chinese students' essays contained more hedges than the less successful Chinese students' essays. In another study on metadiscourse in undergraduate students' writing, Li and Wharton (2012) compared the written work by two groups of Chinese students in different contexts, one from a Chinese university and another from a British university. They found that the students from the Chinese university were more inclined to boost their claims, whereas the students from the British university were more likely to hedge their statements. As a result, hedging and boosting in learner discourse present a complex picture, and factors such as English proficiency level and local context may also play important roles. However, despite the interaction of different factors, the academic discourse of Chinese English may be generally characterized by less doubt and more certainty across different studies, indicating measured pragmatic influences from Chinese rhetoric and culture.

Conclusion

The review of Chinese influences on L2 academic English has offered clear evidence of how Chinese-speaking students and academics transferred L1 rhetorical and literacy preferences to their English writing processes and products. Such ethnolinguistic influences are often perceived as interference of L1 in research of intercultural rhetoric or as interlanguage/learner English in World Englishes inquires. Rather than seeing such language- and culture-specific features in L2 English as departures from the standard, native English, scholars like Jenkins (2009) proposed to explore these divergences as potential features of English as a lingua franca (ELF). Through the lens of ELF, all native and non-native English varieties, instead of being dependent on some "standard" English norms, are equally acceptable, and consequently, L2 Englishes are just seen as different rather than deficient (Jenkins et al. 2011). Moreover, from an ELF perspective, the rhetorical differences between L1 English and L2 English writing can be conceptualized as a tension between the well-established Anglophone academic conventions and the culture-specific rhetorical practices transferred from the writer's L1 to English (Mauranen et al. 2010). Such a tension between L1 and L2 Englishes, according to Mauranen et al. (2010), is likely to result in the hybridization of academic discourse. In this case, the "Chineseness" in L2 English writing can be conceptualized as features of a hybrid academic discourse, which may have the potential to develop into an academic lingua franca.

In an ELF context, the differences between L1 English and L2 English with L1 mother tongue influences have important implications for the current practices in academic writing and the pedagogy for English for academic/specific purposes (EAP/ESP). While academic writing has been conventionally tied with L1 English norms, the rise of English as a lingua franca in academic settings has shown the presence of different linguistic and rhetorical varieties which may not necessarily conform to native English standards (McIntosh et al. 2017). Interestingly, growing evidence of non-native grammar and lexical expressions have been found in the high-stakes genre of research articles published in renowned international journals in science and engineering (Martinez 2018; Rozycki and Johnson 2013). A case in point is the increasing use of *besides* as a sentence-initial transition marker. As previously illustrated in Example 2, such usage of *besides* may be seen as deviating from L1 English norm; however, Martinez (2018) found that between 2000 and 2015, an increasing number of multilingual academics in food science have used *besides* as a logical connector in the sentence-initial position in their published English journal articles. Although Flowerdew (2015) has cautioned against the recognition of ELF status only on the basis of a few non-native lexico-grammatical features, these shifting norms of academic writing may call for a rethinking of the usefulness of the L1 English standards in an academic community where most practitioners are working in a L2 English context. Pedagogically, linguistic and cultural differences between L1 English and L2 English may also find their way into EAP/ESP classrooms, where novice writers can be guided to making informed choices in L2 English academic writing.

To conclude, the above review of studies in World Englishes and intercultural rhetoric shows that Chinese language and culture have cast strong ethnolinguistic influences on the variety of English used for academic purposes. Chinese learners and academics have transferred a range of L1 linguistic and rhetorical features at various levels into their English L2 written discourse. Such variations from L1 English norms mark the hybridity of L2 English in Chinese context and can be seen as an integral part of an academic ELF.

Although the research so far has laid some ground work about the characteristics of Chinese academic English, more empirical work is necessary to further describe the linguistic and rhetorical characteristics of this variety which may become a culture-specific variety of ELF (Mauranen et al. 2010). One important direction for future research is the systematic codification of distinctive features of Chinese academic English through corpus-based analysis. By using well-designed comparable corpora (Moreno 2008), it is possible for researchers of Chinese English to perform multi-dimensional analysis of L2 English academic writing at lexico-syntactic, discoursal, and pragmatic levels, which can provide further evidence of how Chinese L2 writers adapt and negotiate L1 English norms in using English as an academic lingua franca.

References

Appel, Randy, and Andrzej Szeib. 2018. Linking adverbials in L2 English academic writing: L1-related differences. *System* 78: 115–129.
Belcher, Diane, and Gayle Nelson. 2013. Why intercultural rhetoric needs critical and corpus-based approaches: An introduction. In *Critical and corpus-based approaches to Intercultural Rhetoric*, ed. Diane Belcher and Gayle Nelson, 1–6. Ann Arbor: The University of Michigan Press.
Biber, Douglas, S. Johansson, G. Leech, S. Conrad, and E. Finegan. 1999. *Longman grammar of spoken and written English*. London: Longman.
Bolton, Kingsley. 2002. Chinese Englishes: From Canton jargon to global English. *World Englishes* 21 (2): 181–199.
———. 2003. *Chinese Englishes: A sociolinguistic history*. New York: Cambridge University Press.
Bychkovska, Tetyana, and Joseph J. Lee. 2017. At the same time: Lexical bundles in L1 and L2 university student argumentative writing. *Journal of English for Academic Purposes* 30: 38–52.
Cai, Guanjun. 1993. *Beyond "bad writing": Teaching English composition to Chinese ESL students*. Paper presented at the annual meeting of the conference on College Composition and Communication, San Diego.
Chen, Cheryl Wei-yu. 2006. The use of conjunctive adverbials in the academic papers of advanced Taiwanese EFL learners. *International Journal of Corpus Linguistics* 11 (1): 113–130.
Chen, Chenghui, and Lawrence Jun Zhang. 2017. An intercultural analysis of the use of hedging by Chinese and Anglophone academic English writers. *Applied Linguistics Review* 8 (1): 1–34.
Chien, Shih-Chieh. 2011. Discourse organization in high school students' writing and their teachers' writing instruction: The case of Taiwan. *Foreign Language Annals* 44 (2): 417–435.
Confucius. 1983. *The analects* (D.C. Lau, Trans.). Hong Kong: The Chinese University Press.
Connor, Ulla. 1996. *Contrastive rhetoric: Cross-cultural aspects of second language writing*, Cambridge applied linguistics. Cambridge, MA: Cambridge University Press.
———. 2002. New directions in contrastive rhetoric. *TESOL Quarterly* 36 (4): 493–510.
———. 2011. *Intercultural rhetoric in the writing classroom*. Ann Arbor: University of Michigan Press.

Connor, Ulla, Ed Nagelhout, and William Rozycki. 2008. *Contrastive rhetoric: Reaching to intercultural rhetoric*. Amsterdam: John Benjamins.

De Costa, Peter, Jeffrey Maloney, and Dustin Crowther. 2018. In *The Palgrave handbook of applied linguistics research methodology*, ed. Aek Phakiti, Peter De Costa, Luke Plonsky, and Sue Starfield, 719–739. London: Palgrave Macmillan.

Eaves, Megan. 2011. English, Chinglish or China English?: Analysing Chinglish, Chinese English and China English. *English Today* 27 (4): 64–70.

Fairclough, Norman. 1992. *Discourse and social change*. Cambridge, MA: Polity Press.

Field, Yvette, and Lee Mee Oi Yip. 1992. A comparison of internal conjunctive cohesion in the English essay writing of Cantonese speakers and native speakers of English. *RELC Journal* 23 (1): 15–28.

Fløttum, K., T. Dahl, and T. Kinn. 2006. *Academic voices: Across languages and disciplines*. Amsterdam: John Benjamins.

Flowerdew, John. 2015. Some thoughts on English for research publication purposes (ERPP) and related issues. *Language Teaching* 48: 250–262.

Gao, Xia. 2016. A cross-disciplinary corpus-based study on English and Chinese native speakers' use of linking adverbials in academic writing. *Journal of English for Academic Purposes* 24: 14–28.

Gardner, Sheena, and Chao Han. 2018. Transitions of contrast in Chinese and English university student writing. *Educational Sciences: Theory & Practice* 18 (4): 861–882.

Gass, Susan, and Larry Selinker. 2008. *Second language acquisition: An introductory course*. 3rd ed. New York/London: Routledge.

Granger, Sylviane, and Stephanie Tyson. 1996. Connector usage in the English essay writing of native and non-native EFL speakers of English. *World Englishes* 15 (1): 17–27.

Horner, B. 2018. Written academic English as a lingua franca. In *The Routledge handbook of English as a lingua franca*, ed. Jennifer Jenkins, Will Baker, and Martin Dewey, 413–426. London/New York: Routledge.

Hsiung, Hui-yu. 2014. *Contrastive rhetoric study: The placement of paragraph thesis statements in English and Chinese articles*. Unpublished doctoral thesis, Texas A&M University.

Hu, Guangwei. 2005. English Language Education in China: Policies, Progress, and Problems. *Language Policy* 4 (1): 5–24.

Hu, Guangwei, and Feng Cao. 2011. Hedging and boosting in abstracts of applied linguistics articles: A comparative study of English- and Chinese-medium journals. *Journal of Pragmatics* 43 (11): 2795–2809.

Hu, Guangwei, and Guihua Wang. 2014. Disciplinary and ethnolinguistic influences on citation in research articles. *Journal of English for Academic Purposes* 14: 14–28.

Huang, Li-shih. 2010. The potential influence of L1 (Chinese) on L2 (English) communication. *ELT Journal* 64 (2): 155–164.

Hyland, Ken. 1998. Boosting, hedging and the negotiation of academic knowledge. *Text* 18 (3): 349–382.

Hyland, Ken, and John Milton. 1997. Qualification and certainty in L1 and L2 students' writing. *Journal of Second Language Writing* 6 (2): 183–205.

Jenkins, Jennifer. 2009. English as a Lingua Franca: Interpretations and Attitudes. *World Englishes* 28 (2): 200–207.

Jenkins, Jennifer, Alessia Cogo, and Martin Dewey. 2011. Review of developments in research into English as a lingua franca. *Language Teaching* 44 (3): 281–315.

Ji, Kangli. 2011. The influence of Chinese rhetorical patterns on EFL writing: Learner attitudes towards this influence. *Chinese Journal of Applied Linguistics* 34 (1): 77–92.

Jia, Yuxin, and Cheng Cheng. 2002. Indirectness in Chinese English writing. *Asian Englishes* 5 (1): 64–74.

Jiang, Wendong. 2017. A study on modified-modifying sequence in the compositions by Chinese advanced users of English. In *Researching Chinese English: The state of the art*, ed. Zhichang Xu, David Deterding, and Deyuan He, 93–107. Cham: Springer.

Kachru, B.B. 1985. Standards, codification and sociolinguistic realism: The English language in the outer circle. In *English in the world: Teaching and learning the language and literatures*, ed. R. Quirk and H.G. Widdowson, 11–36. Cambridge: Cambridge University Press.

Kaplan, Robert B. 1966. Cultural thought patterns in intercultural education. *Language Learning* 16 (1): 1–20.

———. 1972. *The anatomy of rhetoric: Prolegomena to a functional theory of rhetoric: Essays for teachers*. Philadelphia: Centre for Curriculum Development.

Kirkpatrick, Andy. 1997. Traditional Chinese text structures and their Influence on the writing in Chinese and English of contemporary mainland Chinese students. *Journal of Second Language Writing* 6 (3): 223–244.

Kirkpatrick, Andy, and Zhichang Xu. 2002. Chinese pragmatic norms and 'China English'. *World Englishes* 21 (2): 269–279.

———. 2012. *Chinese Rhetoric and writing: An introduction for language teachers*. Anderson, S. C; Fort Collins, Colo: Parlor Press.

Kubota, Ryuko. 1999. Japanese culture constructed by discourses: Implications for applied linguistics research and ELT. *TESOL Quarterly* 33 (1): 9–35.

Kwok, Danny Wynn Ye. 1965. *Scientism in Chinese thought 1900–1950*. New Haven: Yale University Press.

Latour, Bruno, and Steve Woolgar. 1986. *Laboratory life: The construction of scientific facts*. Princeton: Princeton University Press.

Lee, Joseph J., and Lydia Deakin. 2016. Interactions in L1 and L2 undergraduate student writing: Interactional metadiscourse in successful and less-successful argumentative essays. *Journal of Second Language Writing* 33: 21–34.

Leedham, Maria, and Guozhi Cai. 2013. "Besides…on the other hand": Using a corpus approach to explore the influence of teaching materials on Chinese students' use of linking adverbials. *Journal of Second Language Writing* 22 (4): 374–389.

Lei, Lei. 2012. Linking adverbials in academic writing on applied linguistics by Chinese doctoral students. *Journal of English for Academic Purposes* 11 (3): 267–275.

Li, Jinxia. 2015. *Xiangsi fuju guanxi ciyu duibi yanjiu ("Contrastive analysis of conjunctions in similarity complex sentences")*. Beijing: China Social Science Press.

Li, Yiyang. 2019. China English or Chinese English: Reviewing the China English movement through the Kachruvian lens. *English Today* 35 (2): 3–12.

Li, Ting, and Sue Wharton. 2012. Metadiscourse repertoire of L1 Mandarin undergraduates writing in English: A cross-contextual, cross-disciplinary study. *Journal of English for Academic Purposes* 11 (4): 345–356.

Liang, Li, Margaret Franken, and Shaoqun Wu. 2019. Chinese postgraduates' explanation of the sources of sentence initial bundles in their thesis writing. *RELC Journal* 50 (1): 37–52.

Liu, Lu. 2005. Rhetorical education through writing instruction across cultures: A comparative analysis of select online instructional materials on argumentative writing. *Journal of Second Language Writing* 14 (1): 1–18.

———. 2009. Luoji (Logic) in Contemporary Chinese Rhetoric and Composition: A contextualized glimpse. *College Composition and Communication* 60 (4): W98.

Liu, Ying, and Qian Du. 2018. Intercultural rhetoric through a learner lens: American students' perceptions of evidence use in Chinese Yìlùnwén writing. *Journal of Second Language Writing* 40: 1–11.

Loi, Chek Kim, and Jason Min-Hwa Lim. 2013. Metadiscourse in English and Chinese research article introductions. *Discourse Studies* 15: 129–146.

Lu, Xiaofei, and Jinlei Deng. 2019. With the rapid development: A contrastive analysis of lexical bundles in dissertation abstracts by Chinese and L1 English doctoral students. *Journal of English for Academic Purposes* 39: 21–36.

Ma, Qing, and Zhichang Xu. 2017. The nativization of English in China. In *Researching Chinese English: The state of the art*, ed. Zhichang Xu, David Deterding, and Deyuan He, 189–201. Cham: Springer.

Martinez, Ron. 2018. Specially in the last years…': Evidence of ELF and non-native English forms in international journals. *Journal of English for Academic Purposes* 33: 40–52.

Matalene, Carolyn. 1985. Contrastive rhetoric: An American writing teacher in China. *College English* 47 (8): 789–808.

Mauranen, Anna. 2018. Conceptualising ELF. In *The Routledge handbook of English as a lingua franca*, ed. Jennifer Jenkins, Will Baker, and Martin Dewey, 7–24. London/New York: Routledge.

Mauranen, Anna, Carmen Pérez-Llantada, and John M. Swales. 2010. Academic Englishes: A standardized knowledge? In *The Routledge handbook of world Englishes*, ed. Andy Kirkpatrick, 634–652. Abingdon, Oxon: Routledge.

McIntosh, Kyle, Ulla Connor, and Esen Gokpinar-Shelton. 2017. What intercultural rhetoric can bring to EAP/ESP writing studies in an English as a Lingua Franca world. *Journal of English for Academic Purposes* 29: 12–20.

Mohan, Bernard A., and Au-Yeung Lo. 1985. Academic writing and Chinese students: Transfer and developmental factors. *TESOL Quarterly* 19 (3): 515–534.

Moreno, Ana I. 2008. The importance of comparable corpora in cross-cultural studies. In *Contrastive Rhetoric: Reaching to Intercultural Rhetoric. Eds Ulla Connor, Edward Nagelhout and William Rozycki*, 25–45. Amsterdam: John Benjamins.

Mu, Congjun, Lawrence Jun Zhang, John Ehrich, and Huaqing Hong. 2015. The use of metadiscourse for knowledge construction in Chinese and English research articles. *Journal of English for Academic Purposes* 20: 135–148.

Ouyang, Guangwei. 2003. Scientism, technocracy, and morality in China. *Journal of Chinese Philosophy* 30 (2): 177–193.

Pan, Lin, and Philip Seargeant. 2012. Is English a threat to Chinese language and culture?: The 'Threat' of English in China might be balanced by the promotion of Chinese language and culture. *English Today* 28 (3): 60–66.

Paquot, M. 2013. Lexical bundles and L1 transfer effects. *International Journal of Corpus Linguistics* 18 (3): 391–417.

Peng, Kaiping, and Richard E. Nisbett. 1999. Culture, dialectics, and reasoning about contradiction. *American Psychologist* 54 (9): 741–754.

Radtke, Oliver. 2012. More than errors and embarrassment: New approaches to Chinglish. In *Chinese under globalization: Emerging trends in language use in China*, ed. Jin Liu and Hongyin Tao, 145–170. Hackensack: World Scientific.

Riley, Philip. 2007. *Language, culture and identity: An ethnolinguistic perspective*. London: Continuum.

Rozycki, William, and Neil H. Johnson. 2013. Non-canonical grammar in best paper award winners in engineering. *English for Specific Purposes* 32: 157–169.

Scollon, Ron. 1991. Eight legs and one elbow: Stance and structure in Chinese English compositions. In *International reading association, second North American Conference on Adult and adolescent literacy*, Banff.

Selinker, Larry. 1972. Interlanguage. *International Review of Applied Linguistics* 10: 209–231.

Spack, Ruth. 1997. The rhetorical construction of multilingual students. *TESOL Quarterly* 31 (4): 765–774.

Swales, John. 1990. *Genre analysis: English in academic and research settings*. Cambridge: Cambridge University Press.

Vassileva, I. 2001. Commitment and detachment in English and Bulgarian academic writing. *English for Specific Purposes* 20: 83–102.

Vold, E.T. 2006. Epistemic modality markers in research articles: A cross-linguistic and cross-disciplinary study. *International Journal of Applied Linguistics* 16: 61–87.

Wang, Yu-fang. 2002. The preferred information sequences of adverbial linking in Mandarin Chinese discourse. *Text* 22 (1): 141–172.

Wei, Yun, and Jia Fei. 2003. Using English in China. *English Today* 19 (4): 42–47.

Wierzbicka, Anna. 2006. *English: Meaning and culture*. New York/Oxford: Oxford University Press.

Wong, Jock Onn. 2018. The semantics of logical connectors: Therefore, moreover and in fact. *Russian Journal of Linguistics* 22 (3): 581–604.

Wray, Alison. 2013. Formulaic language. *Language Teaching* 46 (3): 316–334.

Wu, Su-Yueh, and Donald L. Rubin. 2000. Evaluating the impact of collectivism and individualism on argumentative writing by Chinese and north American college students. *Research in the Teaching of English* 35 (2): 148–178.

Xu, Zhichang. 2010. *Chinese English: Features and implications*. Hong Kong: Open University of Hong Kong Press.

———. 2017. Researching Chinese English: A meta-analysis of Chinese scholarship on Chinese English research. In *Researching Chinese English: The state of the art*, ed. Zhichang Xu, David Deterding, and Deyuan He, 235–266. Cham: Springer.

Xu, Zhichang, David Deterding, and Deyuan He. 2017a. What we know about Chinese English: Status, issues and trends. In *Researching Chinese English: The state of the art*, ed. Zhichang Xu, David Deterding, and Deyuan He, 1–14. Cham: Springer.

Xu, Zhichang, Deyuan He, and David Deterding. 2017b. *Researching Chinese English: The state of the art*. Cham: Springer.

Yang, Jian. 2009. Chinese borrowings in English. *World Englishes* 28 (1): 90–106.

Yang, Ling, and David Cahill. 2008. The rhetorical organization of Chinese and American students' expository essays: A contrastive rhetoric study. *International Journal of English Studies* 8 (2): 113–132.

Yeung, Lorrita. 2009. Use and misuse of 'besides': A corpus study comparing native speakers' and learners' English. *System* 37 (2): 330–342.

———. 2019. Dialectics versus polemics in Chinese rhetoric: A study of indirection in Chinese and Chinese ESL argumentative writing as compared with English argumentative writing. *Chinese as a Second Language Research* 8 (1): 29–55.

You, Xiaoye. 2005. Conflation of rhetorical traditions: The formation of modern Chinese writing instruction. *Rhetoric Review* 24 (2): 150–169.

Zhong, Ai. 2019. The top 100 Chinese loanwords in English today: Can one recognise the Chinese words used in English? *English Today* 35 (3): 8–15.

A Sketch of Mandarin Loanword Use from *kowtow* to *gaokao*

39

Will Peyton

Contents

Introduction 1004
Kowtow to *Gaokao* 1008
Conclusion 1018
References 1021

Abstract

This chapter is a literary survey of the adoption of Mandarin loanwords in English media since the introduction of the term *kowtow*. The study of linguistic borrowing is limited to loanwords themselves, though some discussion of other forms is given. Moreover, the Chinese loanwords that are reviewed are specifically Mandarin terms, rather than those of other Sinophone dialects, as prior broader studies have included. The principal criterion for selecting such terms is their inclusion in the Oxford English Dictionary, though others that could potentially be included in the future are considered. Since recent linguistic borrowing from Chinese to English has largely been in the form of Mandarin loanwords, and scholars generally agree that this shall be the future trend of such borrowing, this survey aims to give a sketch of what the general principles underlying loanword adoption appear to be. The chapter provides a literary study of the use of terms following *kowtow*, including *gung-ho*, as well as *tuhao*, *gaokao*, and *jiayou*, among others. It should show that while not much has changed since prior studies of Chinese borrowings, we are likely to see a higher use of Mandarin pinyin loanwords in the future. It encourages further studies of borrowings as increasing bilingual contact, particularly through online media, increases linguistic transference between the Chinese- and English-speaking worlds.

W. Peyton (✉)
The Australian National University, Canberra, ACT, Australia
e-mail: william.peyton@anu.edu.au

Keywords

Chinese loanwords · Borrowings · English · Reception · Semantics

Introduction

Having a long history of direct contact with a vast number of languages, English has incorporated a variety of calques and concepts. When considering this fact, what is peculiar about the history of English borrowings is the relative lack of terms of Chinese origin. Indeed, some attention has been paid to this question, particularly the 1988 study by Garland Cannon and another from 2009 by Yang Jian. In a more recent discussion on the use of Chinese terms in English, Xu Mingwu and Tian Chuanmao noted how "it was rumoured around the end of 2013 that Chinese words, such as *tuhao*, *dama*, and *hukou*...would enter the Oxford English Dictionary, which was still not the case" in 2015 (and is still not so in 2020). As they write, these three terms are examples of "China English" (or rather Chinglish), the use of "English with Chinese characteristics," that is, "the use of Chinese words that have been translated literally into English." The reason for this arises out of the fact that "English is extensively used in China in many areas, including education, foreign publicity, tourism and intercultural communication. The rapid development of the country has given rise to many new social, economic and cultural phenomena" and so we now find many "Chinese neologisms and buzzwords" (Xu and Tian 2017: 2).

Many have commented on the increasing use of English within Chinese language, that is, the reverse of the form described above. In Chinese language use, particularly on the Internet and on social media over the last decade, the increased incorporation of English terms has resulted in expressions such as "you can you up" (which advises one not to criticize the work of others). But for Chinese terms used within English, as Xing Hongbing has noted, terms like *dama* are there to "fill a gap" by signifying foreign social phenomena both briefly and expediently (China Daily 2014). This reflects the fact that the English-speaking world is looking increasingly at Chinese society and culture and, in doing so, requires a larger vocabulary to express its less translatable aspects. Some linguists say that this phenomenon suggests that "Chinglish" is now being accepted by the rest of the world and has been integrated into daily life. This, however, is perhaps an overstatement. Indeed, as a 2013 language column of *The Economist* asks on the relative dearth of Chinese words in English: "'Twenty years from now, how many Chinese words will be common parlance in English?' I replied that we've already had 35 years since Deng Xiaoping began opening China's economic, resulting in its stratospheric rise – but almost no recent Chinese borrowings in English" (The Economist 2013).

As this chapter shall discuss, the use of such terminology remains limited and tends to be the result of a contest between awkwardness in English translation and awkwardness in romanized transliteration. The purpose of this chapter is to isolate one phenomenon: Mandarin loanwords. The reason for focusing on the main Chinese dialect and this single form of linguistic borrowing is that it is ultimately what scholars see as Chinese language's potential influence on English going into the future. Thus, the aim is to trace this trend from the past to the present. Prior linguistic studies, because they have not isolated this important point, have lost sight of the forest for the trees, overlooking an important aspect of the importation of Chinese words into English, namely, the influence of the central dialect of one country on the most widely spoken language globally. In this sense, the chapter should be understood less as a theoretical linguistic analysis but rather as a chronological literary survey of one case of cross-lingual transference.

So what types of terms have made their way into English thus far? As Yang Jian (2009) explains, "during the past 50 years, Mandarin's dominant position has been further consolidated, with its standard form, Putonghua, serving as China's official language and the language of instruction." Because of its cultural stature as the language of the educated, *borrowings* from Mandarin Chinese tend to be high culture words, and the language "accounts for only two food and drink loanwords..." (Yang 2009: 104). Garland Cannon, in his study entitled "Chinese borrowings in English," defined 19 semantic areas of borrowing, which include "190 food and drink including utensils, 175 biota, 110 geography, 100 arts, 49 religion and philosophy, 48 government and politics, 34 ethnography, 34 status and occupation" among other lesser categories (Cannon 1988: 15).

A major issue with these prior studies is that they have considered borrowings in the broadest sense, from calques, foreign words, and loanwords. As was established by Einar Haugen in the 1950s, *borrowing* encompasses a number of types, which he divided into two fundamental categories, substitution and importation, the former being based on the importation of a morpheme and the latter the phoneme. Foreign words and loanwords, which borrow the phonetics directly from the original language, are principal examples of importation. Substitution is rather using the target language's words to substitute for the original languages, calques being the principal example. Of these different means of borrowing, Haugen said that "all linguistic features can be borrowed, but they are distributed along a scale of adoptability which somehow is correlated to the structural organization" of the target language (Haugen 1950: 224).

For the purposes of this chapter, I shall focus on **loanwords**, namely, those words that incorporate the semantic meaning of foreign words into the receiving language while molding the target language's sound patterns (e.g., the English *music* from the French *musique*). Here, some might argue that pinyin terms adopted in English are simultaneously "foreign words" and "loanwords." This is a consideration that falls outside the scope of this study. For the purposes of this chapter, they shall be considered loanwords insofar as the pinyin system is still originally a foreign phonetic system, derived by a group of Chinese linguists, and diverges to some degree from the English pronunciation system for roman characters. The chapter does not look at types of *substitution*, such as *calques*, primarily because this is a separate area of inquiry. A shortcoming of Cannon's earlier study is its quest for breadth, where he considered all types of linguistic borrowing, failing to separate these different borrowing categories (naturally because he was aiming to consider the broadest scope of borrowing between Chinese and English). Loanwords, considered on their own, are as accurate a measure as any of the influence and limitations of cross-lingual transference between Mandarin Chinese and English. Other forms of borrowing, like *calques*, might be left to other scholars. On this note, "Mandarin" and "Mandarin Chinese," for the sake of brevity, shall be used interchangeably with "Chinese" in this chapter. (Pinyin refers to the phonetic transcription system implemented by the Chinese government in 1956 which superseded the various phonetic systems that had been used prior, which includes Wade-Giles. Pinyin has been the standard romanization system used internationally from the 1990s onwards. It is also used in Singapore.)

Fundamentally, the chapter asks why, when we encounter Chinese concepts in English prose, do we tend to find transliterated romanizations (i.e. *kowtow*), while we do not see translations of such phrases as "prostration." The emergence of terms like *hukou* and *dama* in English are, for Xu and Tian, important because the adoption of romanizations of Chinese terms in English is historically inconsistent. Yet a number of such terms, though still very few, have begun to appear over the last several decades. On the romanization of Chinese characters (*hanzi*), they write that "this method, which uses the phonetic alphabet to transcribe Chinese characters, may be called the Chinese pinyin method...Transliterations of Chinese words and expressions are frequently used in China English [sic]" (Xu and Tian 2017: 4). Other examples of neologisms (though they may not be eligible for our study here, for lack of wider usage in English) are *hongbao* 红包, *shanzhai* 山寨, and *geili* 给力, meaning "red envelope," "fake," and "awesome" or "cool," respectively.

Generally speaking, publishers of literary material are reluctant to adopt foreignized terms. However, in journalistic media, this sort of language usage is readily adopted for the sake of expedience, particularly in foreign correspondence where Chinese ideas and concepts need to be explained with brevity through the use of a single term or phrase. The task will therefore be to see how this process occurs with English and Mandarin Chinese. The criterion for choosing terms is their listing in the *Oxford English Dictionary* (OED herein), as this measure is often taken as a standard for entry into the English lexicon. Thus, in looking at Chinese terms' functional usage in media, the emphasis shall be on *how* they were initially used in such media in order to be accepted into the English language. This chapter shall also consider those romanized **pinyin** terms that have wider usage in English writing than those that appear solely in the OED, though its terms will also be included. Thus, English language media, particularly news and reportage, will be the primary material of inquiry. The intention behind this survey is to gain some perspective on the semantic and pragmatic changes which occur with Chinese terms in English as Chinese terms undergo the process of romanization. As mentioned, this does not include terms which do not transliterate through romanization, for example, "red envelope" which fits the model of polysemy calquing (I shall return to this point at the end of the chapter).

Linguistic contact between Chinese and English was historically indirect. Perhaps the oldest term *silk*, which originates from Old English, came through Latin and, beforehand, Greek in which it is called *Serikos*. As the OED explains, the "Chinese *si* [丝] "**silk**," Manchurian *sirghe* and Mongolian *sirkek,* have been compared to this, and the original name in Greek might be a rendering via Mongolian of the Chinese word for "silk," but it states that this derivation is uncertain (OED). Names for things, particularly commodities, would increasingly enter English via other languages from the early modern period onward. **Japanese**, French, and Dutch were later languages of transmission. Through Japanese, Chinese terms, such as *tofu*, *rickshaw*, and *soy*, were able to enter English usage, but were all borrowings from Chinese. *The Economist* points out this peculiar and comparatively larger number of Japanese derivations in English since the Second World War: "kamikaze, futon, haiku, kabuki, origami, karaoke, tycoon, tsunami, jiu-jitsu, zen and honcho are all common English words that nowadays can be used without any reference to Japan." Other terms are "karate, judo, sumo, bonsai, manga, pachinko, samurai, shogun, noh and kimon, say, not to mention foods from the bland (tofu) to the potentially fatal (fugu)" (The Economist 2013).

Cannon has noted that "**French** lags far behind Japanese as a direct transmitter, accounting [only] for *kuchean, souchong, kaolin*, and the latter's five derivations (*-ic, -ite, -itic, -ization*)" (Cannon 1988: 12). Indeed, the term *Kaolinite* (*gaolingtu* 靠领土), a term for a type of clay that derives from 1712 through its mention in French texts (Online Etymology Dictionary 2001) demonstrates an example of Chinese loanword transmission through French. This phenomenon aside, Yang Jian has pointed out that "most of the recent Chinese borrowings have been imported directly" with the increase in bilingual speakers of Mandarin and English. (Yang 2009: 93). Thus, direct borrowings between Chinese and English are the consideration of this survey.

It was during *kowtow*'s introduction into the English language during the 1800s that the greatest number of Chinese borrowings was adopted. As Cannon explains, "in the second half of the nineteenth-century and the first half of the twentieth century...[Chinese] was an appreciable word supplier to English" (Cannon 1988: 25). This is true of many **culinary terms**. The first, and most notable, is *tea*, which was introduced from **Hokkien** dialect to English, its meanings changing from "'eaves' in 1598, 'drink' by 1601, 'the plant' by 1663, and 'meal' in 1738" (Cannon 1988: 14). The term ketchup 茄汁 came into English "in 1711, said to be from Malay (Austronesian) *kichap*, but probably not original to Malay. It might have come from Chinese *koechiap* 'brine of fish,' which, if authentic, perhaps is from the Chinese community in northern Vietnam...." Much later, "Tomato ketchup emerged c. 1800 in U.S. and predominated from early 20c" (Online Etymology Dictionary 2001). Another, *Oolong* tea 乌龙茶, came into English in the mid-1800s, being written in trade registers (OED). Presumably it was less awkward than writing "black dragon tea." *Ginseng*, a variety of the *panax* from the term 人蔘, comes from the Hokkien phonetics *jin-sim* and appeared in a 1666 Royal Society transaction and has remained consistent since, with minor variations. Other food terms obviously come from Cantonese, such as *chop-suey* 杂碎 and *dim sum* 点心. The former appeared in English from 1898 and the latter from 1945 (OED).

But as noted by Yang Jian, higher register **cultural words** generally come from Mandarin, rather than the local dialects. Take, for example, *Yin-yang* 阴阳, collocating the Daoist symbols for dark and light. These terms appear separately in the

Atlas Chinensis from 1671. However, the compound term derives from the western missionary journal *Chinese Repository* in 1850. *Feng-shui* 风水, referring to the traditional concept of geomancy in Chinese, appears in the Encyclopedia Britannica in 1797 as *fong-choui* before the more familiar "feng-shui" form is adopted elsewhere in 1883. *Kung-fu*, from the Chinese martial art *gongfu* 功夫, appeared in 1966 in *Punch* magazine, *The Clarendonian*, and later K. Platt's *Pushbutton Butterfly*. In each of these uses, the context of *gongfu* is assumed as it is referred to as a "martial art," implying that its meaning was already well-established in English speech by then.

These loanwords, appearing before and after *kowtow*, were either cultural terms or physical commodities, deriving from a combination of Sinophone dialects. But what occurs with the end of the Second World War and the founding of the People's Republic of China in 1949 is the standardization of Mandarin Chinese (it had previously been referred to as "national language" or *guoyu* 国语 but changed to "common/standard speech" or *putonghua* 普通话) and the standardization of the romanized pinyin system used today (it took up until the 1990s for pinyin to become the standardized romanization for foreigners dealing with Chinese language). Mandarin loanwords have since become the prevailing source for importing Chinese words – hence this chapter's specific emphasis on these words, unlike Cannon's original study.

This question of cross-linguistic transference bears upon **areal semantics**, an emerging area of linguistic inquiry that deals with "the diffusion of semantic features across language boundaries in a geographical area" and "is a potentially vast field, spanning the convergence of individual lexemes" or, for our purposes, individual words (Koptjevskaja-Tamm and Liljegren 2017: 204). As Koptjevskaja-Tamm and Liljegren note, this research field has "great potential for historical and areal linguistics, but is still awaiting systematic research. This is partly related to the relatively limited cross-linguistic research on lexical issues in general" (Koptjevskaja-Tamm and Liljegren 2017: 227). Overall, this chapter's approach emphasizes chronology first, going from the term *kowtow* (from the early nineteenth century) up until the most recent introductions of terms into English usage like *gaokao* (from the twenty-first century). This angle aims to give a sense of the change in their usage, as well as the relationship between the Chinese and English languages themselves.

Kowtow to Gaokao

Kowtow itself derives from the Mandarin Chinese term *ketou* 叩头, which entered the English language through various accounts of the Lord Macartney's famous 1793 embassy to China, in which he was made to prostrate before the Qianlong Emperor. (Prior to the pinyin (1949–) and the Wade-Giles systems (1859–), there was no uniform romanization system for Chinese terms, though many had been devised by missionaries from the seventeen century onward.) This symbolic act was naturally a controversial one and became, in many regards, a symbol of the political difficulties between China and the West since. As the historian James Hevia has discussed, "the word *koutou* [was] a name for the notorious act that so outraged several generations of 19th-century European and American diplomats and...'kowtow' continues in common usage up to the present as a scornful term for shamefully subservient behaviour." Hevia, in his analysis, pays "particular attention to the refiguring of...the *koutou* over time, paying attention to what the term "meant in Europe and America." Generally, Hevia says that the *kowtow* was "omnipresent in 19th-century English language sources on China. In fact, it would be quite an accomplishment...to find among these writings one that does not mention the *koutou*. Indeed, by the 1840s it had become so thoroughly fetishized by western observers that former American president John Quincy Adams claimed that it, rather than opium, was the real cause of the first Anglo-Chinese war" (Hevia 1993: 57–58).

The first mention of *kowtow*, listed in the OED, is attributed to John Barrow who published his personal account of his travels to China alongside Macartney. As he recounts of one portion of the journey, "whether at home or in the palace, the Chinese were determined they should be kept in the constant practice of the koo-too, or ceremony of genuflexion and prostration" (Barrow 1804). This definition of the initial 1804 use of the term, as "koo-too," is subsequently listed as "the Chinese custom of touching the ground with the forehead in the act of prostrating oneself, as an expression of extreme respect, submission, or worship" (OED). This spelling of the term is maintained in other definitions, *Fraser's Magazine* similarly mentioning the "koo-too" in 1834. However, in other contemporary and subsequent definitions, the romanization of the term changes to "ko-tou," as in a subsequent 1817 mention of Macartney's journey. The British journal *Athenæum* later mentions that "he felt some reluctance when called upon to perform the ko-tow." There is also an 1864 spelling by the historian Thomas Carlyle in which it is rendered as "kowtoos" (OED). It is close to the turn of the century that the more familiar spelling of "kow-tow" appears, with an 1898 mention of the same Macartney voyage. Into the twentieth century, the OED lists usages from 1905, 1920, and 1966 following the same form. It is at last in 1972, in an excerpt from the *Hong Kong Supplement* of *The Times*, that one finds the final form of kowtow, without the hyphenation or alternate vowel spellings, where it describes "President Nixon's dignified *kowtow* and the belated entry of the people's republic into the United Nations" (OED).

This reflects historical changes, as noted by Hevia, where, in diplomatic protocol at the beginning of the twentieth century, the *kowtow* began to disappear as a part of political practice and as "a living political issue between European powers and the Qing empire. In turn, representations of the *koutou* also changed. Certainly, people still saw it as a humiliation for westerners, but some began to argue that it had other meanings for the Chinese." In fact, "the Empress Dowager's portrait painter Catherine Carl argued in 1909 that the *koutou* did not 'imply any slavelike inferiority'...but was rather a 'time honored' way of expressing thanks to the sovereign." Despite these practical changes, "the more negative representation of the *koutou* did not disappear," and many writers have since "continued to treat the *koutou* as a distasteful and scandalous act typical of China's pre-modern sense of universal superiority." The historian John King Fairbank "himself led the way here…eventually characterizing the *koutou* as one of those 'rituals of abject servitude' common in traditional China. This line of representation dovetailed in part with the English word 'kowtow,' which remains a term of derision and ridicule, and gives it a history outside of its incidence as a Chinese act" (Hevia 1993: 63–65).

The grammatical uses of the term have remained consistently varied over this time, whether as a noun or a verb. As a noun, *kowtow* has appeared often as the object of a verb, "to perform the ko-tow," or just generally as a subject with the definite article, "the ko-tou," "the kow-tow," and so on. Where the distinction in usages matters is in the conceptual domain, insofar as the usage of kowtow from its earliest appearances in English had both literal and figurative functions. In the earliest accounts, the usage of *kowtow* referred explicitly to the event of the famous *kowtow*-ing by Macartney. However, by the 1830s, the term had clearly become used in other contexts as literary examples provided suggest, in which the term is employed without any reference to China, Lord Macartney, or even the physical act of prostration. Rather, the term comes to describe the very act of deference in abstract terms or, in the OED's words, "an act of obsequious respect." Carlyle's 1864 usage, for example, describes "Voltaire from of old had faithfully done his kowtoos to this King of the Sciences." A 1905 use mentions that "professors and students all kow-towed and sounded the hew-gag before him." Lastly, President Nixon's *kowtow* mentioned in *The Times* also demonstrates that separation of the term from its particular temporal and historical circumstances to more broad usage (OED).

Along with this abstract distinction of the *kowtow* as an act that can be performed not solely physically but through any symbolic gesture is the use of the term as an intransitive verb. Rather than appear as a noun, in which one "performs the kowtow" or as a transitive verb in which one "kowtows to someone," the phrase simply "to kowtow" begins to emerge. Thus, the term eventually becomes a figurative verb, that is, to act obsequiously in *any* circumstance and in *any* form. Thus prior to this semantic evolution, *kowtow* referred to a specific Chinese form of prostration, but came to mean, in English, the very act of being servile, whether to another person or just as a general attitude of being. This is exemplified in two figurative definitions given by the OED: "the Marquess *kotooed* like a first-rate mandarin, and vowed 'that her will was his conduct'," and the "doctor kowtowed to him." Given that these two examples come from 1826 and 1883, it is interesting to note just how early the term for Chinese prostration became separated from the particular cultural context from which it was drawn, toward a wider and more generalizable abstraction. The OED even lists something called "kow-towism" for the very practice of *kowtow*-ing. This would eventuate to a nominalized form – that is, "one who kowtows," as shown in an example in the 1961, where "the Russians [are described] as…kow-towers to the West." There are, lastly, adjectival forms, akin to verbal states, in which the *kowtow*-ing is attributed to other states, an early example being of "an almost Koo-too-ing kindness" (OED).

The next example of a common loanword is also an historical one. During the Second World War, the term **gung-ho** was "was introduced into American English early in 1942 by lieutenant Colonel Evans Fordyce Carlson, United States Marine Corps Reserve" (Moe 1967). The term was originally "a contraction in Chinese [and] means...*Indusco*, the acronym for *Industrial Cooperatives*. The two Chinese characters are translated individually as 'work' and 'together,' but they do not join together to form a phrase. In Chinese, this is neither a slogan nor a battle cry; it is only a name for an organization" (Moe 1967: 30). Rather *gung-ho*, or *gonghe* 共合 in proper pinyin, simply stands for the "Chinese industrial Cooperatives Movement [which] was organized in Hankow [Hangzhou] during July and was formally established on August 5, 1938, as the Chinese Industrial Cooperatives Association. [The romanization gung-ho is an approximation of the Wade-Giles form, kung ho.] The mission of the Chinese Industrial Cooperatives Society [zhongguo gongye hezuo she] was to assist in economic reconstruction by the production of daily necessities for both military and non-military uses and to establish a sound cooperative basis for small industries to be scattered throughout the country." Importantly, the miscomprehension of the two characters was simply made "by giving the dictionary translation of each character and by linking the translation together to form a phrase. *Gung* ... is a noun and is translated as: 'labor,' 'laborer,' 'worker,' 'job,' and 'time occupied in doing a piece of work.' *Ho*...may be used as a verb, a noun, or an adjective, and has several translations: 'to shut,' 'to enclose,' ''to join,' 'to pair,' 'to agree,' 'to total,' 'in agreement with,' 'side by side,' 'joined,' and 'the whole'" (Moe 1967: 27–28).

During the war, "Marines other than those in the Second Raider Battalion used *gung ho* as a term of disparagement to describe anyone whose conduct or behaviour was obnoxious or offensive." Because of their unruly behavior, "the *Gung Ho* Battalion and its personnel soon were referred to as the '*gung ho* bastards'." The term would eventually be defined by the Marine Corps itself as "GUNG-HO 1. Aggressive spirit de corps; 2. Sometimes sardonically employed to characterised cocky indiscipline or contempt toward orthodox procedures and regulations" (Moe 1967: 24). Following its usage in the Pacific Theatre, the term "survives today as a Marine Corps contribution to American speech. It currently seems to have settled down to usage as an adjective with the meanings 'enthusiastic,' 'zealous,' 'eager,' 'officious,' 'spirited,' and 'ambitious'; as a noun with the meanings 'enthusiasm,' '*espirit de corps*,' 'zealousness,' 'officiousness,' 'busybody,' 'eager beaver,' and 'ambition'; and as a verb with the meanings 'to move aggressively,' 'to bulldoze one's way,' 'to plunge headlong or recklessly,' and 'to act energetically'" (Moe 1967: 30). In becoming a commonly used Americanism, "its connotations have been numerous, ranging from derogatory to complimentary. Its various linguistic uses, as they have developed in the United states, have been peculiar to American speech" (Moe 1967: 19).

What helped its transition from the military to popular culture was the eponymous 1944 film *Gung-ho!* which, upon its release, "contributed greatly toward making the term popular in general American speech." Moe lists the earliest dictionary entries from this time, the first being "work together (A U.S. Marine slogan of World War II)" and "work tougher – a Chinese slogan used as a battle cry." Generally, most other dictionaries used the phrase "work together" as the translation. These dictionaries were the *American College Dictionary, A Comprehensive English-Chinese Dictionary, Funk & Wagnalls New Practical Standard Dictionary, Webster's New*

World Dictionary of the American Language, and *The World Book Encyclopedia Dictionary* (Moe 1967: 22). The OED has likewise listed the pragmatic evolution of the term, showing its earliest usage in November 1942. The term changes into an adjective for enthusiasm, particularly "for" something else. Examples from 1967 to 1970 include "he was very gung ho for National Socialism," "one of the most 'gung-ho' (exceptionally keen to be personally involved in combat) characters I ever met," and "the enthusiastic 'Gung-ho' units of a few years back" (OED).

In view of these changes, Moe argues that "the various meanings that have been ascribed to *gung ho* as an Americanism are fully acceptable. As we have seen, it may be used as a term of commendation, disparagement, or condemnation...." Drawing upon examples from sports, politics, and other cultural domains, Moe writes: "This sampling seems to indicate that *gung ho* is employed primarily as an adjective, but that it has also become both a noun and a verb. The term is widely used, and...it is particularly favoured by sports writers" (Moe 1967: 25–26). By the time Carlson was aware of the error in his translation of *gung-ho*, Moe notes the "term as an Americanism was so firmly entrenched that any effort on his part to correct that error probably would have been futile." In fact, he notes, *hezuo* 合作 would have functioned as a more appropriate translation for the concept since it "not only meant 'cooperative' as both a noun and an adjective but also 'to cooperate' as a verb" (Moe 1967: 29).

As China underwent an extended period of isolationism and political turmoil through the Great Leap Forward (1958–61) and the Cultural Revolution (1966–1976), there was little reporting in Western media on China, since foreigners had limited access to the country. Thus, Chinese loanwords would only continue to be introduced into English from late 1970s with the country's economic opening. The principal literature on China during the 1950s to the 1970s came out of the academy (often referred to as "China watching" literature), which focused on explaining political goings-on inside the PRC. Example of loanwords being introduced in this period are the terms for the Chinese currency, *renminbi* 人民币 and *yuan* 元, the former being for the currency itself, the latter being for the unit of currency, which were both implemented in 1948. The earliest usage one can find is from the eminent scholar G. William Skinner, who writes in 1951 that "the new authorities wanted to replace the silver dollars then in use, the first stable currency that Chengtu [Chengdu] had had in years, with paper money, the so-called Jen Min Pi." He also adopts the acronym JMP in the same way that RMB is used today (Skinner 1951: 69). An earliest example from Britannica puts it as *jenminpi* before it is listed in 1971 in Whitaker's Almanac as **renminbi**.

The next term, **hukou**, comes in the 1980s, the earliest English usages also being academic. This term can refer either to China's internal system of residence registration or the actual internal residence permits used by individuals in China. Rather than adopt such translated names as "local or regional residency permit," *hukou* was employed in transliterated pinyin, in English so as to highlight the foreignness of the concept as well as its specificity. Like place-name terms, *hukou* was also introduced into English prose alongside explanations of the term. The earliest piece from the *New York Times* to use the term states: "Legal residence, defined by a 'hukou,' or household registration, brings with it most importantly coupons to buy grain at subsidized prices, an urban ration system. For China's floating population, no urban hukou means no rice coupons" (Gargan 1987).

It is difficult to collect further early examples of the term except from academic literature in the 1990s. In one piece from *The China Quarterly*, the necessity for using the term *hukou* is made clear where it describes how, in the 1950s, "China implemented a code of laws, regulations and programmes whose effect was formally to differentiate residential groups as a means to control population movement and mobility and to shape state developmental priorities." It then calls this "the *hukou* system, which emerged in the course of a decade [and] was integral to the collective transformation of the countryside, to a demographic strategy that restricted urbanization, and to the redefinition of city-countryside and state-society relations" (Cheng and Selden 1994: 645). The necessity or the use of *hukou* is to indicate that any other equivalent term in English would be lacking to describe this particular type of massive state-regulated system through which China's population movement is maintained. Perhaps reflecting this thinking, the collocation "hukou system" was subsequently used in another article (Chan and Zhang 1999).

The eventual common usage of "hukou," without collocation or qualification, can be seen by the end of the 1990s in one article from the *Los Angeles Times*, which describes the difficulties had by many Chinese people in gaining residency in Shanghai: "the city can be considered among China's most exclusive clubs: One must be born here to belong. Yet as desirable a place to live as Shanghai is, without a coveted residence permit, or hukou, outsiders like Huang can't settle permanently, marry or send their children to the city's free public schools." The term is then used in the possessive form, as in Chinese: "In January, along with the one $20,000 apartment he bought on the outskirts of Shanghai, Huang got the key to the city: his hukou." There is also the phrase: "Huang's new hukou." Here, the *hukou* is not describing a system, but rather an article, a document, which an individual can own – it is a personal possession. But as shown in the following sentence, it can carry both connotations simultaneously: "With more and more people leaving their homes in search of better jobs the hukou came to mean split families, lost job opportunities and missed marriages" (Farley 1999). Despite its increased usage, it has yet to enter the OED, as had earlier been expected.

There are, however, more recent Chinese romanizations which have entered the OED, the principal one being **guanxi**. This term, which designates the implicit rules governing social relationships between people in Chinese culture, began to appear in western journalistic writing during the 1980s. This arose out of the necessity for explaining the nuances of political and business relationships as well as government corruption, as foreigners had increasing involvement and insight into Chinese society from the late 1970s. In a piece in *The Globe and Mail* in 1982, Stanley Oziewicz writes that "the French say copinage and sous la table. Italians refer to it as aggancio and raccomandazione. The Chinese, too, have their own terms: guanxi and hou men. . . . Whatever the language, the words refer to the time-honored practice of using connections for personal gain, and circumventing the system through the back door. . . . Guanxi and hou men have been perfected in Communist China." It is important to note there that *guanxi*, rather than *houmen*, is the appropriate term to discuss as the latter was never widely adopted in English speech. It was necessary to introduce outside readers to the term since it was a fundamental aspect of Chinese culture. Oziewicz writes that "the nature of the [Chinese] society and its centralized economy, with inefficiency, delays, bottlenecks, class structure and strangling bureaucracy, have served to refine guanxi and hou men to a fine art." He continues that "it is publicly acknowledged that guanxi and hou men are rampant, pervading even the upper reaches of the Communist Party" (Oziewicz 1982a).

Interestingly, it was never typical for *guanxi* to be mentioned alongside *houmen*, as *guanxi* is usually deemed sufficient for explaining connections and favors, implying such things as personal favors or nepotism. This can be observed in the OED's examples of the usage of the term. *Guanxi*, defined here as "in Chinese contexts: a network of personal connections and social relationships one can use for professional or other advantage." The earliest example given is from a December 1979 Washington post piece describing how "a new marriage becomes part of the village web of relationships, what the Chinese call *guanxi*" (OED). Note here how, as with *hukou*, where the term is used in earlier English transliterations, it is done provided with an explanation. This can be seen as well in its usage throughout the 1980s. As one letter in *The New York Times* explains, "the key word in the Chinese language in commercial, and in all kinds of relationships is summarized in the Confucian concept of 'guanxi' (rapport)" (Leung 1987).

However, at a certain point, the term begins to be employed in a manner that implies assumed knowledge. One article, for example, describes a visit to a skating rink in Beijing with some locals: "The admission fee for two hours of ice time is the equivalent of nine cents. 'Never mind that,' an acquaintance says proudly. 'I've got guanxi here. We needn't pay'" (Abel 1984). Yet the author does not provide an explanation for it. This change is reflected in the OED's later examples of the term. A 1989 example states that: "for doctors without guanxi, the price is five thousand yuan." From 1996: "You never know the extent of an individual's *guanxi*, so you must cultivate a temperate approach in your dealings with all people." And from 2014: "My father was ushered in through the back door, thanks to a good guanxi there [sic]" (OED). The fundamental difference in these non-contextualized usages is that *guanxi* is phrased more naturally within the flow of the prose while also be couched within the reader's supposed assumption that it is a sort of social commodity.

The same type of pattern, moving from *explaining* context to *assuming* context, can be seen in the earliest writing on the Chinese rice wine **baijiu**. In one piece from 1993, by an American poet on a Fulbright scholarship at Renmin University, the term is used. He writes about his relationships with his Beijing friends: "Wang [a local poet] began to invite me to his parties, and Jin and I began to meet in restaurants, in my apartment, occasionally in his, to work on far more translations than we finally chose to keep." He writes that "the common denominator in this work is, finally, a bowl of rice, a plate of Jin Hong's special noodles, green tea, Wang's Beijing Beer and gin-like baijiu..." (Haven 1993). The distinction is that where the term is not given an explanation – here, "gin-like" – given in italics to denote foreignness. Another case where the term appears without explanation, in order to capture the atmosphere of a scene, is a piece on state banquets (the drink's reputation being most notorious here): "Li Peng's 1995 decree replacing baijiu, Chinese grain spirit, with wine at banquets just led to officials drinking shots of wine 'down the hatch,' like baijiu. Other obstacles include lack of refrigerated storage" (Napack 1999).

However, most uses tend to qualify *baijiu* with an explanation of the substance, where the foreignness of the term is likewise highlighted. Most of the articles are to do with the growth of the baijiu industry in China for news publications. One example goes: "Beijing actively supported beer industry growth in the 1980s through such efforts as helping to convert a large portion of Chinese wine (*baijiu*) consumption to beer" (Steinman 1999). Another writer states that "the Chinese government recently has actively encouraged consumers to shift away from highly alcoholic wines like baijiu to protect their health" (Kissel 2005). Another example is seen in one piece on drinking habits within Chinese business culture mentions that "the liquor of choice is always baijiu, a fiery spirit that is up to 60 per cent alcohol" (Ridder 2001). Finally, a satirical commentary of sanitization campaigns in Beijing, particularly the way locals feel about nanny-state laws, points out: "In this 'people's war' spitting and gum chewing are out; hand-washing and assiduous disinfecting is in. After doubling the penalty for spitting in public to 100 yuan...hygiene police are out in force watching everyone." But the "citizens in this plague-stricken capital are circulating their own counter-propaganda.....Drink as much baijiu – the local grain spirit brew – as possible and keep smoking. 'Of course it works, and it's cheaper than medicine too,' said...a housewife who drinks a beaker a night" (Becker 2003).

As we now approach the twenty-first century, it is important to emphasize here how observations of local social phenomena became more and more common in journalistic coverage, as opposed to the prior emphasis on business and regime politics. As China became wealthier and more prominent an economic power, one can assume that an interest in the local society became of deeper interest to foreign readers. This would naturally accompany the increasing attention to China with the lead up to the 2008 Beijing Olympics and after, as well as the rising immigration of westerners into China during the Jiang Zemin and Hu Jintao years. It was in this period that the aforementioned terms **dama** and **tuhao** began appearing in reportage. The first of these, *dama*, "literally means auntie in Mandarin, which is a suitable word to describe the women who appear very much like the average housewife" (Chen and Liu 2017). The term "has no authoritative definition, with various fields of research offering their own explanations. Traditionally in Chinese, Dama is defined as 1. Auntie; 2. A term of respect for elderly women." In general, "calling someone Dama can be both affectionate and also indicate old age and a lack of understanding of contemporary fashion and culture" (Li 2017: 798).

The term, however, would not have been adopted in English if not for a peculiar story. As Li Qin explains, the term "was coined in 2013, when it gained great traction on mainstream social media. The term can be traced back to the Chinese bestseller *Currency War*, in which the author Song Hongbing mentions the 'gold rush' of

Chinese Dama; however, it was on Jan. 23, 2013, when *The Wall Street Journal* published an article" about a large number of middle-aged Chinese women purchasing gold online. When this story appeared, "the transliteration of Dama appeared in the article and was used to refer to middle aged Chinese women who 'rush to purchase gold,' in this case, three hundred tonnes worth" (Li 2017: 798). Since many *dama* are "suffering from empty-nest syndrome, or simply have too much time on their hands... they have become known as the 'hall monitors' of China." While they have so far been notorious for their financial choices, "it is only very recently that they have been associated directly with patrolling" (Chen and Liu 2017). For these reasons, the term has gained very negative connotations with "the wide coverage of a Beijing Dama blackmailing a young foreign man on the street in 2013 and a Wuhan Dama beating a young lady on the subway in 2015 [which] are two typical cases of the media's distortion of the Dama image." This negative press also reflects wealth disparity in China today, as the "emergence of Chinese Dama is a reflection of the disparity between the rich and the poor and the uneven distribution of wealth in today's China" (Li 2017: 798).

Humor nevertheless surrounds the term. One story reports of a Terry Crossman who "became an online celebrity after a video of him applying to become a 'Xicheng Dama' went viral." Chinese media interviewed the long-time expat, a "62-year old American man who describes himself as a 'real Beijing gentleman.'" The man was clearly peculiar to locals as "'Xicheng Dama' refers to a group of volunteers who roam Xicheng District in central Beijing acting as public security volunteers. Some 70 per cent of them are females aged 58 to 65" (Liu 2017). Likewise humor, irony, and sarcasm surround the other common Chinese social categorizations that are used in English media, the principal example being *tuhao*. An example of a *tuhao* is one Mr. Xu who said he "would give away up to 1000 pairs of shoes to anyone who showed up to his marriage proposal, the internet "denouncing M. Xu as a tuhao, a neologism used to describe Chinese nouveaux riches who have lots of cash but not the class to go with it" (Century 2013). As with *dama*, its humor encompasses negative judgment as to a certain state of being.

It is "a popular expression used by netizens to refer to people with great wealth who spend money freely" and "has been included in a newly revised [Chinese] dictionary, triggering controversy among language experts [because] the word referred originally to landlords living in rural areas who had a lot of money, land and power and who often bullied others" However, "in the modern context, it also refers to people who have a great deal of money, but who are lacking in education and correct values, according to the third edition of the Standard Dictionary of Modern Chinese" (Zhao 2014). Xu and Tian have noted that "the spelling of transliterations is not consistent in some texts. For example, *tuhao* is sometimes spelt as *Tuhao* and sometimes as *tuhao*, even in the same essay. It is assumed that there is no need to capitalize *tuhao* because it is not a proper noun" (Xu and Tian 2017: 7). Generally, the lower-case version has assumed the standard form.

This attention to social categorizations is perhaps reflected also in the increased usage of the term *gaokao,* the most recent Chinese loanword addition to English, meaning China's high school entrance exam. Prior to the adoption of the pinyin term, the concept was simply in English translated form, as shown from a 1979 example: "A group of about 120 students who received pass marks in the recent university entrance examinations but who have not been given places marched from the municipal headquarters to the Zhongnanhai compound with a petition..." (Reuters 1979). The pinyin transliteration *gaokao*, however, would begin to be used during the 2000s as increasing reportage on China's exam system and the subsequent pressure it put on students. This reflects the increasing attention to contemporary Chinese social phenomena in this period.

Grammatically, what is interesting about the introduction of the term into English discourse is the usage (or lack thereof) of the definite article. In the first examples, the term appeared as such: "Ruizhe is preparing for **gaokao**, the college entrance exam, the annual chance for Chinese children to improve their fates. The entrance exam will open doors – or close them, depending on your score – to the best colleges and the most popular majors there" (Spak 2006: 11). Note here the use of explanation with the pinyin transliteration. In this usage, where the *gaokao* is assumed to be unknown to the foreign reader, it appears without an article, solely as "gaokao," in order to denote a concept. This is also reflected in this *China Daily* piece: "There are so many things about you that gaokao cannot test... We can freely admit we hate gaokao, but do we have something better to replace it?" [sic] (Zhou 2006). The term is thus used in the same manner as *guanxi* when introduced into English.

However, it appears with the definite article increasingly during the first decade of the 2000s, both with and without the definite article. In later writing, it appears with the definite article as well as explanations of what *the* gaokao exam factually is. In one piece, it is first mentioned as "the two-day test that ended Friday," then as "the college entrance exam," and then finally as: "the gaokao [that] determines everything" (Chang 2007). Likewise, another piece writes that "the gaokao is like millions of warriors rushing to cross a one-plank bridge" and also contains a translated quote from a Shandong student: "The gaokao is the de facto embodiment of knowledge. It is like a do-or-die moment" (South China Morning Post 2005). Importantly, the term rarely appears as the collocation "gaokao exam" but rather as a singular noun, reflecting the fact that its meaning is assumed within the discourse. Much like in English, one does not supplement the noun "exam" over the names for them (e.g., the SATs in the USA, the A-Levels in the UK, or the HSC in Australia).

There is one last Chinese term worth discussing that can be used in English: *jiayou*, often used as an equivalent to "good luck." Several examples can be found in the coverage of sports events. Regarding an international volleyball match held in China in 1988, an early piece states: "with another 10,000 fans wedged into the Wutaishan Gymnasium shouting, 'Jiayou!' (which, literally translated, means 'pour oil on the fire'), the Chinese did just that, pounding the Americans in three games" (Brown 1988). However, the term only began to appear frequently in media with the 2008 Beijing Olympics, an example being: "'Jiayou' or 'add oil,' the four-step routine is designed to help spectators cheer in a 'smooth and civilized manner' at the August 8-24 Games. The chant will be promoted by television programs, video presentations and squads of cheering volunteers..." (Edmonton Journal 2008).

As the term increased in appearance in media, the *China Daily* would report that "netizens were thrilled to learn that the popular Chinese phrase 'add oil' has been entered into the Oxford English Dictionary." Stating that it is the translation of *jiayou* "which is commonly used when people want to show encouragement, support or excitement [and] acts like a verbal pat on the back," the term has an equivalent meaning to "'cheer up,' 'come on' or 'go for it,' depending on the situation" (China Daily 2019). What is significant about the entry into the OED is that the English translation "add oil," rather than the pinyin transliteration *jiayou*, was chosen. This decision likely results from the fact that the English translated form is used in Hong Kong and Singapore already. Indeed, the OED lists it, not as a Chinese or Mandarin term, but principally as an expression from "Hong Kong English. add oil!: expressing encouragement, excitement, or support, excitement, or support: go on! go for it! [After Chinese (Cantonese) gā yáu, with reference to petrol being injected into an engine...corresponding Mandarin jiāyóu.]." The entry includes examples from 1964 all the way up to 2016. (OED).

Jiayou is thus an example of where the Chinese loanwords in English diverge from pinyin transliterations to become actual translated terms. This can be seen with other Chinese expressions, as noted by Koptjevskaja-Tamm and Liljegrenm who mention two Chinese calques in Singaporean English (Singlish): "eat salt" or *chiyan* and "give face" or *geimian* (Koptjevskaja-Tamm and Liljegren 2017: 207). Another example is "bare branches," from the Chinese *guanggun* 光棍, a term for single men who are the product of male overpopulation (and thus female underpopulation), lacking social prospects and opportunities for marriage. Rather than use the romanized pinyin term in English, as has happened with the other words discussed, "bare branch" is used (The Economist 2015). Likewise, the term "leftover women" is used in English writing rather than the Chinese *shengnü* 剩女. Ultimately, this is a much wider question to consider which extends across standard English usage as well as dialectical forms of English. One might suppose that the reason for translated forms of Chinese words being adopted rather than pinyin loanword forms is simply that the former, like "bare branches" and "leftover women," do not read or sound awkward, rendering pinyin transliteration unnecessary.

Conclusion

What has been the focus of this chapter are those Chinese terms, concrete nouns, and verbs that have snuck into English in their romanized forms and which have appeared in English media. *Kowtow* and *Gung-ho* came to stand for very foreign concepts initially, and it is unsurprising that its early users went straight to the Chinese transliterations, or their renditions of them, to express the already evidently foreign concept. The same point can be made for the adoption of *guanxi* in English much later. Simply put, there is no way of expressing this particular conception of relationships in English. Thus, employing the Chinese form for "relationships" –

guanxi – self-evidently shows that the term means that particular Chinese conception of social relationships. The same can be said for *baijiu*, which is more expedient than Chinese "white wine," "white alcohol," "white spirit," or "rice wine." The nouns *hukou*, *dama*, *tuhao*, and *gaokao* all identify ideas more complex than a single word in English can convey. Thus, the pinyin form, when employed in English, can stand in for a wider concept that might only be expressible as a convoluted or imprecise phrase, thus condensing more information into spoken or written discourse.

This point is perhaps not adequately appreciated in Tian and Xu's discussion about such terms, where they conceive of a "CEV" or Chinese English Vocabulary. "As a major source for the production of CEV," they explain, "the *China Daily* website has created a large number of neologisms via transliteration, literal translation, free translation and other translating methods." Such words "carry a strong Chinese flavor and represent writers' growing cultural awareness" (Xu and Tian 2017: 8). However, the issue is that CEV words and those that appear in the *China Daily* are a much narrower category of words. They are limited to translated terminology used among Chinese people and within the Chinese culture sphere. It is not yet clear whether those concepts are able to be universalized so as to enter the English lexicon. As can be seen with *kowtow* and *gung-ho* – the terms have to embody usage and concepts *outside* of the Chinese culture context in order to become useful, it appears at least. In short, Chinese loanwords become of great value and interest when they are adopted by outsiders to Chinese culture, as this chapter has explored.

It may be difficult for terms like *tuhao* and *dama* to enter the English lexicon sufficiently enough to be accepted into the OED because they are terms generally used only by native Chinese speakers or foreigners who have already acquired Chinese language, rarely if ever being employed outside of these circles. *Gung-ho* and *kowtow* were able to comfortably enter English usage precisely because they assumed meanings that had nothing to do with the Chinese cultural context. It is hard to imagine how *dama* would be used when it refers to a specific type of Chinese woman and her social circle's activities. The closest equivalent for *tuhao* is the Australian/New Zealand slang for an uncouth wealthy person – "cashed-up bogan" – such that there is little space for the term already.

It seems that retaining "**foreignness**" is not the criteria for entry into English but rather a by-product of a loanword's entry, since the authors are not always trying to play up the foreignness of a word in their prose. It is rather that by adopting singular pinyin terms, the web of social, cultural, political, and historical context in which these terms exist can be condensed merely by the romanization of the foreign phonetics. None of these aforementioned terms can be accurately expressed in less than a phrase. Thus, in using the foreign word, that foreignness can stand in for a wider set of ideas than simply the closest English cultural equivalent. *Kowtow*, rather than "Chinese ceremonial prostration," more adroitly represents the larger set of assumptions of that idea. This reason appears consistent for each of the loanwords that have appeared since.

Ultimately, the two earliest loanwords, *gung-ho* and *kowtow*, are the most interesting because they eventually came to designate universalizable ideas which had no strict dependence on their source language. More precisely, English speakers would never have had the notion that these words have a Chinese etymology (and usually never know this fact). That this has not been repeated perhaps indicates the closer contact and understanding between China and the West in recent decades, where there are many more bilingual speakers of both English and Mandarin, thus allowing for conceptual overlap through the usage of the two tongues. As *The Economist*'s erudite summation from 2013 asked, "whether future Chinese borrowings will be new edibles, cultural items or even philosophical terms will depend on China's development and how the West responds. In other words, we should hope Chinese terms we will adopt will be more of the guanxi than of the flak variety." What terms like *kowtow* and *gung-ho* show is that abstract Chinese concepts that are eventually adopted by English speakers, even if they are greatly distorted in their being loaned, have to be wrestled away from Chinese cultural specificity. Is it plausible that the notions of *guanxi* or *jiayou* could eventually be wrestled away from their singular cultural context? Only time will tell.

To an extent, we are still left with the same conclusion that Garland had thirty years ago: "While many of the Chinese words are well established and probably form a part of general international English, the evidence suggests that Chinese is not likely to rival [then more dominant] Japanese and become a major word supplier in the near future" (Cannon 1988: 25). But as Yang Jian has since put forth, "Chinese borrowings will occur more frequently, more borrowings will come from Mandarin, the majority of the new loanwords will be Pinyin-based, and both the new loanwords and loan translations will occur in various semantic fields." He also predicts that loan translations will be high in number (Yang 2009: 104–105). What has changed during the period between the writing of these two articles is the introduction of the Internet and the increase of China's presence in global cultural discourse. Perhaps these changes will represent the tipping point wherein pinyin terms enter English speech. Online media of the last decade, as shown, might be the earliest indication of this.

Overall, the need to adopt pinyin terms will match the need to explain the particulars of Chinese culture and society among English speakers, whether Chinese or foreigners. But unlike the original term *kowtow*, it is hard to predict just exactly what concepts can be internalized in English to have meaning outside of the Chinese context. Romanizing Chinese terms is ultimately a way of avoiding the awkwardness of rendering translated collocations or concepts that have a clumsy ring to them when sounded out in English. But the introduction of Mandarin loanwords is a mostly embryonic issue at this stage. As for other forms of linguistic borrowing, such as calques, as well as their relevance to areal linguistics, this is more of a question for scholars of Singlish and Hong Kong English, of which there has been much continuous work. In general, Chinese loanword transference will certainly be a question for future language scholars to explore. *Jiayou*!

References

Abel, Allen. 1984. China: Nine-cent skating amid the city's high-rise hegemony. *The Globe and Mail*, February 11.

Barrow, John. 1804. *Travels in China, containing descriptions, observations, and comparisons, made and collected in the course of a short residence at the imperial palace of Yuen-min-yuen, and on a subsequent journey through the country from Pekin to Canton*. London: Cadell and W. Davies.

Becker, Jasper. 2003. Beijing launches a people's war against the plague. *The Independent*, May 6, p. 11.

Brown, Dennis. 1988. One eye on sights, the other on Seoul Olympics in mind, U.S. Women's Volleyball Team Tours Asia. *Los Angeles Times*, June 11.

Cannon, Garland. 1988. Chinese borrowings in English. *American Speech* 63: 3–33.

Century, Adam. 2013. Popping the question, Tuhao-Style. *New York Times*, November 25.

Chan, Kam Wing, and Li Zhang. 1999. The Hukou system and rural-urban migration in China: Processes and changes. *The China Quarterly* 160: 818–855.

Chang, Anita. 2007. College-entry test brings China to standstill. *Orlando Sentinel*, June 9.

Chen, Chia-lun, and Kuan-lin Liu. 2017. 'Chinese damas': China's new secret weapon. *Focus Taiwan – CAN English News*, September 13.

Cheng, Tiejun, and Mark Selden. 1994. The origins and social consequences of China's Hukou system. *The China Quarterly* 139: 644–668.

China Daily. 2019. Add oil! Popular Chinese expression added to Oxford English Dictionary. *China Daily*. http://www.chinadaily.com.cn/newsrepublic/2018-10/24/content_37132382.htm. Accessed September 1.

China Daily USA. 2014. Chinglish gains popularity overseas. *China Daily USA*, May 12.

Edmonton Journal. 2008. China brings order to Olympic chants. *Edmonton Journal*, June 6.

Farley, Maggie. 1999. Chinese cities opening up to country's Elite; Asia: New rules allow a select few to receive coveted residence permits. *Los Angeles Times*, March 15.

Gargan, Edward A. 1987. China's 'wave of the future': Millions moving to the big city. *New York Times*, May 4.

Haugen, Einar. 1950. The analysis of linguistic borrowing. *Language* 26: 210–231.

Haven, Stephen. 1993. Beijing Journal. *The American Poetry Review* 22: 39.

Hevia, James L. 1993. The Macartney embassy in the history of Sino-Western relations. In *Ritual & diplomacy: The Macartney Mission 1792–1794*, ed. Robert Bickers, 57–79. London: The British Association for Chinese Studies.

Kissel, Mary. 2005. China's winemakers grab spotlight. *Wall Street Journal*, April 5.

Koptjevskaja-Tamm, Maria, and Henrik Liljegren. 2017. Semantics patterns from an areal perspective. In *The Cambridge handbook of areal linguistics*, ed. Raymond Hickey, 204–236. Cambridge: Cambridge University Press.

Leung, Frankie. 1987. Chinese business is westernizing. *New York Times*, May 18.

Li, Qin. 2017. Characteristics and social impact of the use of social media by Chinese Dama. *Telematics and Informatics* 34: 797–810.

Liu, Ning. 2017. Passionate about China, 62-year-old American "Xicheng Dama" becomes online hit. CGTN. https://news.cgtn.com/news/3d3d514e3551544f7a457a6333566d54/share_p.html. Accessed 2 Sept 2019.

Moe, Albert F. 1967. Gung Ho. *American Speech* 42: 19–30.

Napack, Jonathon. 1999. China toasts fledgling wineries. *New York Times*, October 15.

Online Etymology Dictionary. 2001. Online etymology Dictionary. https://www.etymonline.com/. Viewed 4 Sept 2019.

Oxford English Dictionary Online. 2000. OED. https://www-oed-com.virtual.anu.edu.au/. Viewed 1 Sept 2019.

Oziewicz, Stanley. 1982a. Even in China, you say? The Globe and Mail, February 15.
Reuters. 1979. Peking protest ends in scuffle. *The Guardian*, September 31.
Ridder, Knight. 2001. Mainland state sector players booze or they lose. *South China Morning Post*, November 25.
Skinner, G.William. 1951. Aftermath of communist liberation in the Chengtu plain. *Pacific Affairs* 24: 61–76.
South China Morning Post. 2005. National exam like a do-or-die moment; changing faces. *South China Morning Post*, July 2005.
Spak, Kara. 2006. Pressure fierce on China's children. *Daily Herald*, April 26.
Steinman, Glen. 1999. Trying times amid spectacular growth. *The China Business Review* 26: 26–31.
The Economist. 2013. Why so little Chinese in English? *The Economist*, June 6.
———. 2015. Why China and India face a marriage crisis. *The Economist*, April 23.
Xu, Mingwu, and Chuanmao Tian. 2017. So many *tuhao* and *dama* in China today. *English Today* 130: 2–8.
Yang, Jian. 2009. Chinese borrowings in English. *World Englishes* 28: 90–106.
Zhao, Xinying. 2014. Tuhao joins a wealth of modern words. *China Daily*, August 28.
Zhou, Raymond. 2006. Don't let Gaokao seal your fate. *China Daily*, June 10.

Index

A
Academic lingua franca, 997, 998
Academic writing, 977, 979, 981, 983, 985, 989, 997, 998
Accent, 33
Achievement culture, 848
Acoustic phonetic study, 22–25
Acquisition Model Instruction (AMI), 854
ACT, 854
Agrammatism, 71, 82
Alethic modality, 964
Alexia with agraphia, 70
Altaic languages, 127, 300, 339–341, 352, 360
American Society of *Shufa* Calligraphy Education (ASSCE), 815
Amsterdam model, 124
Analogy, 176
Analogy innovation, 170
Anglo English, 881
Anomic aphasia, 80
Anyang dialect, 175
AP Chinese Language and Culture Course, 660
Aphasia, 68–71, 73–74, 76, 80, 82, 86, 90, 92
Aphasia Battery of Chinese, 70
Apologetic expressions, 942–944, 958, 960–962, 965, 969
Approaches to writing
  genre approach, 780
  process approach, 779–780
  product approach, 779
Archaic Chinese, 195–198
Areal linguistics, 306, 308, 318, 322 451
Areal semantics, 1008
Argumentative writing, 780, 783, 788, 790, 796
Artificial intelligence (AI), 68, 715
Asian Festival of Children's Content (AFCC), 576
Aspect markers, 166, 201–202
Aspectual particle, 160
Asserting an event, 171
Asserting a state, 171
Asymmetric contributions, 402
Audio-visual narratives, 622
Audiovisual translation (AVT), 462
  academic and social impacts, 618
  ad hoc participatory translation, 630
  advantages and disadvantages, 618
  Chinese translation and language studies, 619
  constraints, 624–625
  danmaku-driven participatory subtitling, 619
  3D film with subtitles, 625
  dominant form, 620
  Fansubbing communities, 621
  interlingual mediation, 619
  multimodal approach, 622
  research and pedagogy, 619
  subtitling, 618
  subtitling challenges, 621
  Western universities' courses and programs, 630
Auditory cortex, 95
Augmented reality (AR), 712, 713
Austroasiatic language, 127
Austronesian language, 126, 127
Authenticity, 756, 859
  communication context, 757
  contextual authenticity, 860
  linguistic authenticity, 860
  personal experiences, 759
  post-reading activities, 756
  pronunciation authenticity, 860
  reading and writing accompanied by oral communication, 762
  task authenticity, 860

Authenticity (cont.)
  textual authenticity, 860
    discursive authenticity, 860
    generic authenticity, 860
    orthographic authenticity, 860
Autism, 77, 94
Autonomous learners, 854
Autosegmental-metrical (AM) model, 25
Auxiliaries
  grammatical phenomenon, 359, 360
  post-positioned quotation verbs, 357
  sentence-final *yǒu*, 358
  sentence-final *zhe*, 358
  zhe, 356, 357

B
Baijiu, 1014, 1015, 1019
Behavioral culture, 847
Beihai Cantonese, 415
Beijing dialect, 160, 172
Beijing Language and Culture University Corpus Center (BCC) corpora of Modern Chinese, 472
Biaohua, 312–314
Bilateral cingulate cortex, 89
Bilingual education, 893, 897
Bilingual speakers, 95, 306, 1007
Bilingualism, 95, 97, 254, 897, 912
Binyang dialect, 170
Blending, 53–55
Borrowings, 222, 224, 381, 385, 390, 400, 920–928, 1005, 1007, 1020
Boston Diagnostic Aphasia Evaluation, 70
Brain lesions, 68
British Centre for Literary Translation (BCLT), 585
British English, 951–952, 955–959, 964–965, 967
Broca's aphasia, 70, 80, 83, 90
Brodmann area 44 (BA44), 86
Buddhism
  Biddhist translation, 216
  spread in China, 214–216
Buddhist Chinese
  characteristics of, 219–220
  four-character phrases, 221–222
  Indic Source Languages, 222–224
  influence of Buddhist Concepts, 224–225
  reasons behind formation of characteristics of, 220–225
Buddhist scriptures, 342–351

Buyang, 312
Bunu, 319
Burmese, 326

C
Calligraphy Education Group (CEG), 815
Cantonese, 126, 301, 410, 411, 413, 414, 884, 885, 888, 892, 894–899, 902, 910, 912–913, 915–927, 929–931
  Beihai, 415
  dialect, 162, 166
  Guangzhou, 415, 428
  Hong Kong, 307, 414, 415, 427, 428
  jurisdictions, 426–430
  Kuala Lumpur, 429
  Lannang, 390
  Macau, 416
  Nanning, 416–425
  Written, 430–433
  See also Yue
Case, 852
Causation, 166
Cebu Lánnang-uè, 397
Chabacano, 394
Chadong, 315
Changdu dialect, 168
Chen Bochui International Children's Literature Awards, 579
China Central Television (CCTV), 706
China Cognitive Linguistics Association (CCLA), 44
China Fiction Book Club, 585
China Foreign Languages Publishing Administration, 565
China International Forum on Cognitive Linguistics (CIFCL), 44
China Radio International (CRI), 705
China Shanghai International Children's Book Fair (CCBF), 579
China Writers Association, 569
Chinese speech acts, 949, 958
  apologetic expression, 960
  borrowings into English, 1004, 1007
  Buddhist Chinese, 220
  bùhǎoyìsī, meaning of, 960
  characters 11
  Chinese English data, 962–963
  Classical Chinese, 220
  communication process, 966
  corpora, 472–473, 719
  dictionaries, 472–474
  early Modern Chinese, 188, 222
  face, in Chinese linguaculture, 959–960

Index

four-character set phrases, 89, 221–222
history, 307
image-based thinking, 949
lexicon and grammar, 948–949
Middle Chinese, 11, 188, 310, 349–351
Modern Chinese, 139, 188, 336
Netspeak, 277–278
Northern dialects, 307
Old Chinese, 188, 306, 336–342 (*see also* Archaic Chinese)
problem solving, in Chinese linguaculture, 953–955
Proto-Sinitic 188
sentence-final particles, 949
serial-verb constructions, 948
societal logic, 965
Southern dialects, 307
two-part allegorical sayings, 88
Chinese academic writing instruction
assessment and feedback, 800–803
curriculum design, 789–791
definition of Chinese academic writing, 784
individual tutorials, 797–798
model essay guidance, 795–797
online research methodology, workshop on, 799–800
stylistic and formatting requirements, instruction of, 799
Chinese as a foreign language (CFL), 648, 672, 673, 744
AP workshops, 660–662
Backward Design approach, 650, 652
CARLA, 659
character instruction, 746
Chinese cinema, 668
Chinese orthography, 747
classical Chinese poetry, 668
dream as a literary mode, 669
essential competencies, 662–664
learners' orthographic awareness, 747
Middlebury Chinese School, 649–656
multimodal literacy, 753
pedagogical grammar, 667
phonetics, 665
reading, 746
the role of typing, 753
social and performative perspective, 748
sociolinguistics, 665, 667
STARTALK, 657–659
teaching Chinese vocabulary, 667
teaching methodology, 666
writing, 747
writing-to-learn approach, 751
Chinese as a second language (CSL), 33–34, 815, 824, 828
contemporary assessments in China, 682
historical development, 676–680
Chinese Books for Young Readers, 588
Chinese borrowings, 1004, 1007
Chinese calligraphy
calligraphy student on, 822
courses/learning modules, 821
learning and practicing, 815
meaning of, 816
tradition of, 817
vs. Western and Arabic calligraphy, 821
Chinese character processing, 71
Chinese children's literature, in English translation, 565–567
academic conferences, 577
AFCC, 576
Bologna Children's Book Fair, 577
CCBF, 579
China Fiction Book Club, 585
China in Britain: Myths and Realities Project, 586
Chinese Books for Young Readers, 588
Chinese folk and supernatural tales, twentieth century, 558–565
GLLI, 588–589
IBBY, 575–576
IYL, 575
Leeds Centre for New Chinese Writing, 586–588
Literary Translation Summer Schools, 585–586
LTC, 585
missionaries, late nineteenth century, 554–559
Moganshan Literary Translation Residency, 584
Paper Republic, 583–584
picture books, 570, 571, 573
translator-led initiatives, 580–583
Chinese cinema, 604, 668
Chinese cultural keywords, 495, 497, 498, 506, 508
Chinese diaspora, 301
Chinese English, 977–979
contrastive rhetoric, 980
intercultural rhetoric, 980
Chinese/English legal translation
legal terms, 545–549
sources of difficulty, 541–544
Chinese English Vocabulary, 1019
Chinese ethnic minorities, 462

Chinese ethnolinguistic influences, academic
ELF, 981
  discourse organization, of L2 English
    writing, 988–993
  hedging and boosting, in L2 English
    writing, 993–996
  informal linking adverbials, 984–985
  L2 English syntax, 987–988
  lexical bundles, 982–983
  linking adverbials, 984
  overuse of linking adverbials, 985–987
*Chinese Fairy Tales and Folk
    Tales*, 561
Chinese folk tale, 555
Chinese grammar
  auxiliaries, 356
  case markers, 354
  causal postposition, 354, 355
  cognition-based study on, 43
  comparative analysis of Sanskrit-Chinese
    versions, 343
  evolution of Old Chinese and Modern
    Chinese word order, 336
  functional words, 346–348
  genesis, 344
  language contact, 352
  linguistic type, 361
  location words, postpositions, 355
  metaphrastic documents, 352
  modern, 47
  Modern Chinese, SOV language, 336–341
  Mongolian-Chinese vernacular
    inscriptions, 335
  plural markers, 343
  postpositional conjunction, 350
  postposition hypothetical conjunction, 356
  post-preposition phrase, 349
  pronouns, 353, 354
  structural linguists to analyse, 45
  traditional, 45
  translation of Buddhist scriptures, 335
  vocative order, 349
  written history, 334
  yún hé, 345
Chinese historical syntax
  aspect markers, 201–202
  Construction Grammar in, 193–194
  future research, 207
  grammaticalization, 188–193
  publications on, 206
  verb categorization advances in, 198–200

Chinese Internet Language (CIL)
  2001–2005, 280–281
  2006–2010, 282–284
  2011–2019, 284–289
  description, 276
  pre-2000, 279
Chinese kinship, 461
  terms, 515, 516, 520–524
Chinese Language Association of Secondary-
    Elementary Schools (CLASS), 716
Chinese language planning, 261, 265–267
Chinese Language Teachers Association
    (CLTA), 720
Chinese law, 545, 549
Chinese legal language, 461, 541, 542, 544
Chinese loanwords, 881, 1012, 1016,
    1018–1020
Chinese meaning, 460
Chinese neurolinguistics
  aphasia research, 70–71
  development of, 76–79
  establishment of, 75–76
  future research, 97–99
  language acquisition, development and
    decline, 91–92
  language processing in atypical populations,
    93–95
  language processing research, 71–73
  mental lexicon, 79–83
  phonology, 90–91
  pragmatics, 88–89
  second language learning, 95–97
  semantics, 86–88
  syntax, 83–86
  theories, methods and case studies, 73–74
Chinese pedagogical grammar, 654
Chinese picture books, 573, 583
Chinese proficiency tests
  for ethnic minority groups, 688
  for medical profession, 692
  for overseas Chinese, 688–689
  teaching curriculum, 697
Chinese social interaction, 504, 507
Chinese Spanish Pidgin, 394
Chinese Tagalog Pidgin, 395, 403
Chinese Text Project database, 472
Chinglish, 978
Chope (Singlish), 885, 889–891
Cinematic exhibitions, 620
Cíyuán 辭源, 474
Classical Chinese poetry, 668

Classifier(s), 82–83, 422, 424
Classifier-noun constructions, 167
Cleft palate 28
Clinical neuropsychological assessments, 69
Cochlear implant, 29, 95
Code-switching, 372, 373, 387, 393–400, 610
Cognition, 158
Cognitive linguistics, 4, 518, 519, 534, 535
    achievement in Chinese, 48–59
    category view of, 49
    concepts of, 43, 51
    development of, 44
    features of Chinese, 59
    paradigm in, 47
    perspective of, 46
    philosophical basis of, 43
    principles and methods of, 43
Cognitive prototype, 119
Colloquial Taiwanese, 174
Comitative prepositions, 160
Common voice, 947
Communication channels, 945–948
Complex verbs, Russian, 950–951
Compound words, 81, 90
Conduction aphasia, 80, 90
Conjunctives, 357
Constructional copying, 317–320
Construction Grammar, 55–57, 193–194
Contact-induced change
    areal linguistics and grammaticalization area, 322–324
    from Chinese to minority languages of Southern China, 311–320
    constructional copying, 317–320
    contact-induced grammaticalization, 310
    genitive construction, 312
    locative phrase, 314
    from minority languages of Southern China to Chinese, 321–322
    narrowing, 315–317
    relative clause, 313
    reordering, 312–315
    verb-complement constructions, 314–315
Content questions, 856
Content words, 156, 158
Contextualized elicitation, 867
Contrastive rhetoric, 980
Cooperative principle, 175
Corpus of Contemporary American (COCA), 473
Corpus of Global Web-Based English (GloWbE), 915–917, 919, 923–924
Covert translation, 626
Creativity, 479
Critical reading, 792
Critical thinking, 783, 787, 788, 801–802
Critical writing, 791
Cross-linguistic comparison, 120, 127, 129, 134
Culinary terms, 1007
Cultural diversity, 534
Cultural keywords, 460
    definition, 492
    defy conventional translation, 494
    English expression, 492
    mind, 495
    subject of, 493
Cultural perspectives, 816, 828–833
Cultural Revolution, 16, 493, 566
Cultural words, 1007
Culture, 543, 544, 546, 547, 848
Curricular development, 821, 822

D

Dai language, 312–315, 608
Dama, 1004, 1006, 1015, 1016, 1019
Danmaku function, 630
Daohua, 127, 308
Data glossing principles, 119
Deafness, 93–95
Deep alexia, 80
Deontic modality, 964
Description de la Chine, 555
Determinative-sentence, 348
Developmental dyslexia, 93
Diachronic semantic maps, 143–147
Diachronic specialization, 159
Dialogue manipulation, 856
Dialogue performance, 856
Dictionaries, 474, 480, 481, 514, 516, 521, 524, 526, 527, 529–532
Diffusion tensor imaging, 77
Digital performance
    assessments and learning outcomes, 764–766
    contextualized tasks, 759
    current approach and learner motivation, 768–769
    digital contextualized reading and writing activities, 767–768

Digital performance (*cont.*)
    new ways of textualizing information, 757
    practical genres, 763
    reading performance, 761
    unfamiliar cultural conventions, 757
    writing performance, 762
Diminutive noun, 167
Directional complement, 166
Directional marker, 159
Directional verbs, 164
Directive speech acts, 939, 941, 942, 953
Discourse markers, 175
Disposal construction, 156, 204, 348
Dispositions, 816, 820, 823, 824, 826–828, 833
Domestication (translation methods), 546, 547
Dongguan Yue, 417
Dorsal medial frontal cortex, 89
3D subtitlers' creative input, 626
Duiwai hanyu 对外汉语, 640
Dysgraphia, 71
Dyslexia, 93
Dysphemia, 76

E
Early Manila Hokkien, 381, 402
Early Modern Chinese period, 188
Elliptical constructions, 85
Emerging Translators Network (ETN), 586
English, 1004–1008, 1010, 1012–1020
    borrowings, 1004
    language teaching, 899–901
English as a lingua franca (ELF)
    Chinese ethnolinguistic influences (*see* Chinese ethnolinguistic influences, academic ELF)
    definition, 976
English L2 academic writing, Chinese ethnolinguistic influences, *see* Chinese ethnolinguistic influences, academic ELF
Ergative verbs, 199
Ethnocentrism, 518
Ethnolinguistics, 460
    cognitive, 474
    cultural semantics and, 485
    and worldviews, 469–471
Event-related potential (ERP), 27
Eye-tracking, 19, 702–715

F
Face, 958
    in British English linguaculture, 958–959
    in Chinese linguaculture, 959–960
    in conflict situation, 961–962
    in Russian linguaculture, 959

FACT segment, 854, 856
Fansubbers, 625
Fansubs, 629
Field performance, 858
Filipinization, 372, 400
Filipino, 373, 382
Film dialogue, 626
Films
    audiovisual translation, 605
    Chinese subtitles, 610
    code-switching, 614
    directors, 607
    dubbing, 605
    ethnic minorities, 604, 605
    filmmakers, 606, 609
    Hajab's Gift, 610
    interview, 614
    linguistic diversity, 614
    multilingual country, 606
    multilingual films to English, translating, 610–612
    subtitling, 605
    translation, 605
First International Conference on Internet Chinese Education, 708
First Opium War, 413
Flourishing, 217–219
Foreignization (translation method), 546, 547
Foreign language strategy, 257, 259
Foreignness, 1012, 1015, 1019
Forensic phonetics, 30–31
Formal typology, 119
Form–sound–meaning, 80
French, 1005, 1007, 1013
Frontal-temporal language network, 94
fùcì, 347
Full-Year Asian Language Concentration (FALCON), 845
Functional magnetic resonance imagining, 76
Function words, 156, 158, 224

G
Game, 851
Gan and Min, 448–449
Gan language, 168
Gǎnqíng 感情 ('attachment, feelings'), 496
Gaokao, 1008, 1016, 1017, 1019
Genitive construction, 312
German legal texts, 543
Geshiza rGyalrong, 317
Giving verbs, 164
Global aphasia, 80
Global English, 936, 970

Global English Business Communication Project (GEBCom Project), 936, 937, 965, 970
   British English as hearer-oriented language (*see* British English)
   cancellation of an obligation, scenarios, 942–945
   Chinese as speaker-oriented language, 948–950
   permission, scenarios, 939–941
   reception test, 966–968
   Russian as reality-oriented language, 950–951
Global Literature in Libraries Initiative (GLLI), 588–589
Global Web-based English, 880
Glottalization, 159
Gōngmín 公民 ('citizen'), 472, 475–478, 480–482, 484–487
Gōngmín liángxīn 公民良心, 475
Grammatical change, 307, 324
Grammatical deficits, 82
Grammaticalization, 5, 188, 324, 425
   bleaching, 157, 159
   Boom period (2005 to present), 160, 161
   Chinese dialects, 158
   concurrent phenomena, 161
   constructions, 170
   content words, 156
   cycle, 168, 176
   function words, 156
   historical documents, 162, 163
   language contact, 172, 173
   linguistic typology, 171, 172
   mechanisms, 157, 176
   monographs and conferences on, 192–193
   notions, 157
   patterns, 176
   phases, 160
   pragmatics and discourse, 175, 176
   research development, 177
   research from typological perspective, 191–192
   subjectivity and subjectivisation, 174, 175
   synchronic materials, 163
   theoretical innovations based on case studies of, 189–191
Grapheme, 90, 93, 818, 994
Guangshan dialect, 167
Guangzhou Cantonese, 415, 428
Guangzhou dialect, 174
Guānxì 关系, 496, 1013, 1014, 1017, 1018, 1020
Guiqiong, 317
Guiyang dialect, 157, 159, 309
Gung-ho, 1010, 1012, 1018–1020
Guómín 国民, 477, 480

H
Hakka, 166, 173, 446–448
Han-er Yanyu" 汉儿言语, 335, 352
Hans Christian Andersen Awards, 576
Hanyu Shuiping Kaoshi (HSK), 676
Haplology, 53–55
Head-mounted display (HMD), 712
High-functioning autism (HFA), 94
Historical development, CSL assessment, 676–680
Hmong-Mien languages, 308, 311, 445
Hokkien, 381, 403, 884–886, 889, 890, 892, 894, 895, 897–899, 902, 910, 912–913, 915–927, 929–931, 1007
Hong Kong Cantonese, 170, 307, 414, 415, 427, 428, 434
Hsin-Yi Foundation, 570
Huangxiao dialect, 176
Hui'an dialect, 168
Huihuihua, 308, 319–320
Hukou, 1004, 1006, 1012–1014, 1019
Hybrid academic discourse, 997

I
Identifiability-precedence principle, 124
Ignored culture, 848
Iloilo Chinese English, 386
Iloilo Lánnang-uè, 398
Imperative marker, 358–359
Indigenized varieties, 405
Indo-European languages, 12, 50, 80, 127
Inferior Frontal Gyrus, 82
Inflection, 223–224
Informational culture, 848
Instructional technology, 642
Instrumental-comitative prepositions, 142
Integrative potential, of Chinese borrowings, 924
Inter-clausal code-switching, 399
Intercultural communication, 848, 871
Intercultural communicative competence, 644
Intercultural competence, 723, 871
Intercultural mediation, 619, 626
Intercultural rhetoric, 979–981, 995, 997, 998
Interdental fricatives, 886–887
Interdisciplinary education, 822, 832
Interlanguage, 978
Internalized noun-verb equilibrium, 125

International Board for Books for Young People (IBBY), 575–576
International Corpus of English (ICE), 917
International Translation Day, 585
International Youth Library (IYL), 575, 577
Interrogative construction, 307
Interrogative pronouns, 196–197, 345, 354, 939–940
Interrogative sentence, 358, 940, 941, 953, 955–957, 969
Intonation 27, 34
Inverted double object construction, 307
Irrealis clause marker, 163

J
Japanese, 1007, 1020
Jiangxian dialect, 168
Jiayou, 1017, 1018, 1020
Jin dialect, 166
Jinghua, 320
Jintan dialect, 306
Journey to the West, 564, 566
Judicial language, 541
Jurchen language, 353

K
Kadiwéu language, 316
Keqiao dialect, 167
Keywords and keyword approach, 467–469
Khams Tibetan, 317, 325
Kiasu (Singlish), 886, 889–892
Kinship database
　large-scale, 516
　reliable, 524
Kinship relations, 514, 516, 526, 528, 530
Kinship system
　Chinese, 516, 520–522, 530
　classical Chinese, 525
　classificatory, 516, 522, 523, 525, 534
　Cocopa's, 522
　documenting/studying, 531
　macro-taxonomy of, 517
　multilayered, 529
　Murdock's, 516
　typology, 519
　world, 516
Khitan language, 353
Kowtow, 1006–1010, 1018–1020
Kuala Lumpur Cantonese, 429
Kunming dialect, 159
Kymograph 13

L
L2 learning, 95–96
Lajia, 315
Language and culture, 898, 900
Language as resources, 261, 266
Language contact, 158, 172, 173, 300–302, 410, 416, 425–427, 429, 434, 449, 912–914, 977–979
　definition, 372
　grammatical patterns, 310–320
　history of research in China, 304–309
　internal and external factors, 304
　Sino-Philippine, 380
Language control, 97
Language creation, 400, 401
Language decline, 77, 92
Language deficits, 76–77, 90, 98
Language description, 120, 405
Language diffusion, 171, 305–307
Language disorders, 69–70
Language ecology, 268, 371, 387
　Philippine, 388, 404
Language economy, 261–263
Language impairment, 68, 70–71, 79
Language industry, 262–265
Language life, 258, 260, 263, 267
Language management theory, 266–268
Language policies, 892, 898, 899
Language policy and planning (LPP)
　language economy, 261–263
　language service in, 263–265
　language strategy as Chinese innovations in, 265–269
　language strategy studies with, 256–259
　Li and Li, 255
Language resource, 261, 262, 268
Language services, 263–265
Languages of southern China, 309–310
Language strategy, 256–259
Languoids, 117, 125–127
Lannang Cantonese, 390
Lannang Hiligaynon, 386
Lannang Mandarin, 392
Lannang Tagalog, 384
Lannang Taishanese, 390
Lao qida 老乞大, 352, 359, 360
Learner, 851
Learning Model Instruction (LMI), 853
Leeds Centre for New Chinese Writing, 587–588
Legal concepts, 543–545, 547
Legalese, 430
Legal language, 541–544, 550

Legal terminology, 541, 545
Legal translation, Chinese and English
  cultural differences, 543
  judicial language, 541
  legal concepts, legal norms and application of laws, 543
  legal documents, 544
  legal scholarly texts, 541
  legal terms, 544–549
  legal texts, 543
  legislative language, 541
  modern Chinese legal language, 542
  nature of law and legal language, 542
  terminological challenges, 544
Legal translation, 461
Legal translator, 543, 549
Legislative language, 541
Lánnang-uè, 396, 403
Less Commonly Taught Language (LCTL), 644
Lexical access, 80
Lexical deficits, 70
Lexical-functional grammar, 83
Lexicalization, 5
Lexical tone, 90, 92, 94
Lexico-grammatical features, of L2 English writing, 981–987
Lexicography, 515, 524–532, 535
Lianjiang dialect, 169
Liberal education
  calligraphy education, 643
  humanistic value of language learning, 644
Libo dialect, 309
Liheci 离合词, 81
Lingua franca, 936, 970
Linguistic area
  Mainland Southeast Asia, 324
Linguistic development in advanced CFL writing, 780–782
Linguistic ecology, 301
Linguistic interactions, 380, 381, 404
Linguistic inventory typology (LIT), 128
Linguistic normativity, 514
Linguistic typology
  data-based methodology, 120
  data glossing principles, 119
  generalizations, 119
  grammaticalization, 171, 172
  LIT, 127, 128
  markedness, 119
  meta-knowledges, 119
  reviews and textbook compilations, 117, 118
  semantic maps and grammaticalization, 120
  Sinitic languages, 120–127
  stages, 118
  theoretical development, 127, 128
  translation, 116, 117
Linhai dialect, 167
Linwu dialect, 306
Lisu language, 318, 609
Liyang dialect, 306
Literary Translation Centre (LTC), 585
Literary Translation Summer Schools, 585–586
Loanwords, 1005, 1008, 1010, 1012, 1018–1020
Locative phrase, 314, 355
London Book Fair, 585
Loudi dialect, 167
Lower-level Reading and Writing, 744
  extensive reading method, 758
  instructional cycle, 759
  pedagogical goals, 755
  proposed curriculum, 758
  scaffolding activity, 760
  student evaluation, 766
  students' attitudes, 766

M

Macau Cantonese, 416
Mainstream Tagalog, 383
Malaysia, status of English in, 910–912
Malaysian English (MYE), 880, 930
  Chinese borrowings, 915, 924–928
  collocations, 919–920
  Hokkien, Cantonese and Mandarin borrowings, 920–923
  language contact, 912–914
Manchu, 352, 359
Manchu-Tungusic languages, 353
Mandarin, 29, 125, 127, 139, 337–339, 353, 378, 391, 404, 910–913, 915–925, 927–931
  Fusui, 324
  Jianghua, 306
  Lannang, 390
  loanwords, 881, 1004, 1008, 1020
  Northern, 353
  Phonemes, 306
  Standard, 127, 149, 174
  Southwest (SWM), 306, 308–310
  tone, 32
  word order, 337
Mandarin Excellence Programme (MEP), 587
Manila Chinese English, 385, 392, 405
Manila Lánnang-uè, 397

Manila Mainland Hokkien, 388
Mannan dialect, 169
Maonan, 313, 318
Markedness, 119
Markers, 166
Mashi wentong 马氏文通, 156, 195
Measure words, 167, 168
Mebzang nDrapa, 317
Mental lexicon, 72, 76, 79–83
Metaphor 52–53, 159
Metonymy 52–53
Miànzi 面子 ('face'), 496, 497
Miao-Yao languages, 125, 609
Middle Chinese period, 188
Middle frontal gyrus, 70
Middle Mongolian, 354, 355
Migrating meanings, 466–467
Miluo dialect, 175
Min
    branches of, 444
    classification of, 443–444
    Eastern, 170, 323
    Far-Western, 323
    and Gan, 448–449
    and Hakka, 446, 447
    homebase of, 442
    origin, 442
    vs. Sinitic languages, 444
    Southern, 165, 323, 1303(*see also* Hokkien)
    Wu and, 445
    Xianghua and, 448
Minhe Gangou dialect, 173
Minyag language, 316
Mismatch negativity (MMN), 27
Missionaries, 552–559
Mixed language, 127, 301, 308, 323, 394, 396
Mobile assisted language learning (MALL), 710, 711
Modal particles, 163, 166
Modern Chinese period, 188
Modern Chinese Scientific Terminology database (MCST), 472
Modern phonetics, 15–19
Moganshan Literary Translation Residency, 584
Mongolian, 352, 606
Mongolic languages 123
MOOCs, 713
Morpheme, 142
Morphosyntax, 434, 924
Multidimensional scaling (MDS), 134, 147
Multi-functionality, 134, 167
Multi-functional usage, 167

Multi-functional words, 167
Multilingual cinema, 462
Multilingualism, 462

N

Nanjing Hsin-yi Children Cultural Development Company, 570
Nanning Cantonese, 172, 416–425, 434
Narrowing, 315–317
National Assessment of Educational Progress (NAEP) model, 723, 724
National Security Language Initiative (NSLI), 257
Natural semantic metalanguage (NSM), 460, 494, 497–499, 503, 508, 517, 520, 534, 535
nDrapa language, 325
Negator, 421
Negotiating commensurability, 479
Netspeak
    future research, 289–291
    grammar of, 286–288
    in media, 285–286
    neologisms, 285
    origins and development, 277–289
    regulation, 288–289
    words and constructions, 285
Neural mechanisms, 68–69, 71, 73–78, 82, 88, 91
Neural representations, 76
Neurobiology, 68, 99
Neuroimaging technology, 76
Neurolinguistics, 4
New trends of learner needs, 641
NFLC Guide for Basic Chinese Language Programs, 663
Ningbo dialect, 162
Non-checked tones, 159
Neologism 291
Non-perfective negator, 421
Nouns, 163, 168–170
Noun-verb distinction, 124
Null object constructions (NOC), 85

O

Old Book of Tang, 555
Old Chinese period, 188
Open CourseWare (OCW), 713
Oral performance script (OPS), 860

P

Paper Republic, 583–584
Parallel Encoding and Target Approximation (PENTA) model, 24
Paratactic grammar, 85
Particles, 163, 166
Parts of speech, 40, 81–83, 86, 125
  definition, 49–50
  modal particles, 166
  multi-functional words, 167
  nouns, 51, 163, 168–170
  particles and markers, 166
  prepositions, 165
  substantive words, 167–170
  verbs, 49, 163–165
Passive construction, 156, 205, 347
Pedagogical approach
  Communicative Approach, 643
  Performed Culture Approach, 644
  Presentation, Practice and Production (PPP), 643
  Task-Based instruction (TBI), 643
  TPRS, 643
Pedagogical grammar, 667
Pekinese Rhymes, 564
Perfective negator, 421
Performed culture approach (PCA)
  categorization of culture, 848
  Comparison of PCA to Other Foreign Language Approaches and Methods, 865
    Audio-lingual Method, 865
    Communicative Approach, 865
    Direct Method, 865
    Silent Method, 865
  fundamental concepts, 847
  key players, 845
  saga, 852
  story, 860
  theme, 852
  theoretical underpinnings, 844
  time, 853
Philippine-based languages, 301
Philippine Hokkien, 382
Philippine Hybrid Hokkien, 396
Philippines, 371
Phonetic(s), 665
  forensic, 30
  lenition, 159
  reduction, 159
Phonetic studies, in China
  acoustic phonetic study, 22–25
  experimental phonetics, 12–15
  forensic phonetics, 30–31
  modern phonetics, formation and development of, 15–19
  speech acquisition and cognitive development, in infants and young children, 31–33
  speech learning of Chinese as a second language, 33–34
  speech pathology, 28–30
  speech perception, 26–27
  speech production mechanism, 20–22
Phonographs, 703
Phonological alexia (form–sound), 72, 80
Phonology
  acquisition, 34
  generative, 11
  historical, 10, 12
  Mandarin Chinese, 12, 24
  Min, 451
Pidgins, 294, 394, 403
Pinghua, 309, 310, 325–327
Pinyin, 1006, 1008, 1011, 1012, 1017–1020
Plural suffix, 353
Polyfunctional model, 311
Posterior cingulate cortex (PCC), 95
Post-focus compression (PFC), 25
Postposition, 335–337
Pragmatics, 88–89, 158, 940, 958, 970
Prepositions, 163, 165
Problem solving, 952–953
  in British English linguaculture, 955–957
  in Chinese linguaculture, 954–955
  in Russian linguaculture, 955
Prochievement oral interview, 861
Proficiency-based testing, 653
Proficiency level, 97
Programed Logic for Automated Teaching Operations (PLATO), 706
Pronouns, 163, 167, 168, 354
Pronunciation, teaching of, 34
Prosodic boundaries, 86, 91
Prosody, 33, 34
Proto-Sinitic perios, 188
Psychological reality, 518, 520, 532, 534, 535
Pure agraphia, 70
Pure alexia, 70
Putonghua, 13, 18, 26, 28–34

Q

Qiang, 125, 126, 308, 317, 325
Qieyun 切韵 10, 309
Qin Dynasty, 411

Qinglan Project, 78
Qingwen zhiyao 清文指要, 335, 352
Qixia dialect, 166
Quantifiers, 167
Quanzhou dialect, 162, 163

R
Reading deficits, 76
Reading impairments, 70
Reduplication, 169
Reference grammar, 125–127
Relational grammar, 83
Relative clause, 85, 313
Rènào 热闹, 506
Renminbi, 1012
Rénqíng 人情 ('favor'), 496, 497
Resting-state functional connectivity (RSFC), 96
Revealed culture, 848
Russian, 950–951, 955, 959, 966

S
Sabde Minyag, 316
Sangleys, 376, 381, 402
Sanskrit, 216, 222–224, 230, 236, 240, 248–250
Second-culture learning, 844
Second-generation semantic maps, 147
Second language acquisition (SLA), 640, 666, 844, 848, 851, 855, 862
Second language learning, 95–97
SECTIONS model, 729
Semantic(s), 1008
    analysis, 141, 516, 518, 519, 534
    bleaching, 159
    category, 83
    change, 189
    explications, 520, 534
    fields, 80
    integration, 81
    kinship terms, 515
    molecules, 534
    primitives, 532
    typology, 135, 138, 149, 150
Semantic map model (SMM), 120, 135, 147, 149, 158, 171
    diachronic semantic maps, 143–147
    origin of, 134
    research domains, 136–138
    research methods, 139–143
    second-generation semantic maps, 147

Sentence-final particle (SFP), 160
Separable words, 81
Shanggu Hanyu, 188
Shanghai dialect, 163
Shanyin dialect, 175
Shaowu dialect, 166
She, 319
shengse 生色
    development of, 237–239
    origin of, 235–236
Shenmu dialect, 175
Shici 实词, 156
Sign languages, 94
Silk, 1006
Singlish, 880, 885
    colour terms, 888
    dark /l/, 887–888
    emerging trends, 901–903
    English language teaching, 899–901
    interdental fricatives, 886–887
    kiasu and *chope*, 892
    particles and cultural keywords of Chinese origin, 889
    tones in, 886
    voiceless plosives, 888
Sinicization, 372, 400
    in Buddhist studies, 227–228
    modes in Buddhist Language, 228–246
    research on Buddhist Language, 225–227
Sino-British Joint Declaration, 540, 541
Sino-Philippine contact languages, 394
Sino-Philippine language creation, 372
Sino-Philippine language ecology, 406
Sino-Philippine linguistic varieties, 372, 391, 401, 406
Sinophone, 401
Sino-Tibetan language family, 125, 127
Societal logic
    British English, 964–965
    Chinese, 965
    Russian, 966
Sociolinguistics, 665, 667
Sound change, 189
Southern dialects, 160
Southern Min dialect, 162
Southern Qiang, 308
Spanish, 381
Speech communities, 542
Speech community planning, 268, 269
Speech pathology, 28–30
Speech perception, 26–27
Speech production mechanism, 20–22
Speech verbs, 163

Index   1035

Standard Cantonese, 389, 410–412, 414, 417–425, 429, 434
Standard written English (SWE), 977
Stative complement, 159
Structuralism, 43, 49, 53, 127, 517
Student learning outcomes, 816, 822, 824, 836
Stuttering, 93, 94
Subcortical aphasia, 75
Subjectivization, 158, 174, 175
Subjectivity, 174, 175
Substantive words
   measure words, 168
   nouns, 168–170
   pronouns, 167, 168
Substitution, Augmentation, Modification, and Redefinition (SAMR), 725–727
Subtitlers, 629
Subtitling, 460, 462, 626
Subtitling software, 624
Sui, 309
Surface alexia (form–meaning), 72
Susong dialect, 166
Sustainable multilingualism strategy, 261
Synchronic balance, 159
Synchronic materials, 163
Syntactic ambiguity, 81
Syntactic context, 156
Syntactic distribution, 189
Syntactic violations, 85

T
Tagalog, 295, 383
*Taīdiōkâ*s, 378
Tai-Kadai languages, 309, 326
Tai languages, 421
Taishanese, 389
Taiwan Southern Min, 165
Tang Dynasty, 411, 412
Tangwanghua, 127
Tangut language, 317
Teacher training, 862
   ALLEX Teacher Training Institute, 863
   SPEAC Teacher Training Program, 863
Teacher training, CFL
   AP workshops, 660–662
   essential competencies, 662–664
   in-service teacher training, 649–656
   summer professional development programs, 657–660
   systematic and integrated approach, 664–672

Teaching Chinese as a Foreign Language (TCFL)
   academic writing skills, 643
   advanced-level curricula, 643
   assessment, 642
   beginner-level literacy curriculum, 643
   character instruction, 642
   curricular design, 642
   emergence of, 640
   interdisciplinarity, 644
   primacy of speech, 643
   teacher training, 641
   trends in teaching, 641
Technology, in Chinese language teaching
   audio playing and recording, radio broadcasting, 704–705
   Bloom's taxonomy and modified digital taxonomy, 724
   Chinese corpora, 719
   Chinese language tools, 719
   computer, Chinese input methods, multimedia, 706–707
   COVID-19 and impact, 715–717
   devices *vs.* materials, 717–718
   film, video recording, television broadcasting, 705–706
   five C's and three modes of communication, 722
   four essential language skills, 722–723
   Gagné's Nine Events of Instruction, 724–725
   general Chinese language resources, 719
   general-purpose technology tools, 718
   guidelines, 727–731
   Internet, world wide web and early adopters, 707–708
   language technology tools, 719
   machine translation and artificial intelligence, 715
   MALL, 710, 711
   mobile devices, 711–712
   MOOCs, 713
   multi-media, internet, online resources and online courses, 708–709
   phonographs, 703
   SAMR model, 725–727
   second language acquisition perspectives, 725
   special resources, 719
   virtual worlds, 710
   Web 2.0, 710
Terms of address, 527, 529, 532
   See also Vocative terms

Textual performance script (TPS), 860
Thanking expressions, 942–944, 958, 960, 961
The Horse Thief, 608
Theory-of-mind, 89
Tiantai Wu dialect, 159
Tibetan, 308, 316, 317, 325, 326, 606
Tibeto-Burman languages, 123, 318
Tone
    acquisition, 33
    distinctions, 31, 306
    experimental studies, 12
    inference scales, 13
    lexical, 90, 92, 94
    merger, 416
    perception, 31–33
    qusheng, 26
    sandhi, 27
    shangsheng, 26
    shusheng, 159(*see also* Non-checked tones)
    Singlish words, 886
    yangping, 26, 29, 305
    yinping, 26, 29
Tōngzhì tiáogé 通制条格, 352
Topicalization constructions, 84
Trace theory, 83
Transcortical motor aphasia, 80
Transcortical sensory aphasia, 80
Transfer, 400
    discourse particle system, 403
Transference verb, 166
Transformational grammar, 75
Transitivity, 198–199
Translatability, kinship vocabulary, 515–517, 530, 533
Translator-led initiatives, 580–583
Transnational filmmaking, 623
Tuhao, 1004, 1015, 1016, 1019
Tujia language, 308
Tungusic languages, 353
Two-part allegorical Chinese sayings, 88

U

Unaccusative verbs, 199
Unergative verbs, 199
Universal grammar, 84
Universality, 534
US foreign language strategy, 259

V

Vascular cognitive impairment, 92
Velopharyngeal insufficiency (VPI), 29
Verb(s), 163
    of arrival, 165
    of cognition, 165
    of holding, 165
    of placing, 165
    phrase, 163, 424
Verb-belonging-to-noun paradigm, 125
Verb categorization
    based on temporality, 200
    based on transitivity, 198–199
    relations between nouns and verbs, 200
    unaccusative verbs *vs*. unergative verbs, 199–200
Verb-complement constructions, 314–315
Verb-complement structure, 202
Vernacular, 354
    Cantonese, 335
    Chinese, 335
    Hokkien, 380
    literature, 220
    Malaysian English, 914
    medium of education, 894
    Ming and Qing, 190
    Mongolian-Chinese, 335
    style of writing, 991
    Swatow, 556
Virtual reality (VR), 712, 713
Virtual worlds, 710
Visual cortex, 95
Vocabulary acquisition, 92
Vocative terms, 515, 523
    See also Terms of Address
Voice onset time (VOT), 23
VP-ellipsis, 85

W

Waxiang dialect, 308, 448
Web 1.0, 710
Web 2.0, 710
Wenshui dialect, 175
Wernicke's aphasia, 80
Wernicke-Lichtheim model of aphasia, 73
Western laws, 543, 545
Western tradition, 119
Word class typology, 124, 125
Word order, 224, 312, 336–342, 349–351
Word order typology, 121–124, 194–198
Word storage, 80
Working memory, 84
World Englishes, 977–980, 997, 998
Writing system, 818
Written academic English as a lingua franca (WAELF), 977
Written Cantonese, 430–433
Wu dialect, 13, 126, 158, 162, 166, 172, 445–446
Wujiang Tongli dialect, 175
Wutunhua, 127, 308

Index

X

Xiàndài hànyǔ cídiǎn 现代汉语词典, 474, 480
Xiandai Hanyu, 188
Xianghua, 448
Xiang dialects, 167
xianzai 现在
    origin and its use in Medieval Buddhist texts, 239–241
    prior to the Mid-Ming Period, 241–243
    reappearance as Time Word, 243–245
    usage in Medieval domestic literature, 241
Xiào 孝 'filial piety, 497
Xīn 心 ('heart/mind'), 496
Xīnhuá cídiǎn 新华词典, 474, 480, 481
Xinshao (Cunshi) dialect, 175
Xuci 虚词, 156

Y

Yongxin dialect, 306
Younuo, 318
Yuan dianzhang 元典章, 335
Yuan Dynasty, 412
Yuangu Hanyu, 188
Yue, 434
    Baise, 310
    earlier history, 411–412
    Guangxi, 310
    Lianjiang, 173
    Nanning, 170, 310
    Zhongshan, 417
    See also Cantonese

Z

Zhengding dialect, 166
Zhonggu Hanyu, 188
Zhongshan Yue, 417
Zhuang, 310, 318–319, 325–326
Zhuang-Dong language, 173
ziji 自己
    development of the function and usage of, 232–234
    origin of, 229–232
Zizhong dialect, 167
Zongyang dialect, 176